THE LIBRARY
GUILDFORD COLLEGE
of Further and Higher Education

AS AND A Level

Advanced DESIGN AND TECHNOLOGY

THIRD EDITION

WITHDRAWN

AS AND A Level

Advanced DESIGN AND TECHNOLOGY

THIRD EDITION

EDDIE NORMAN JAY CUBITT
SYD URRY MIKE WHITTAKER

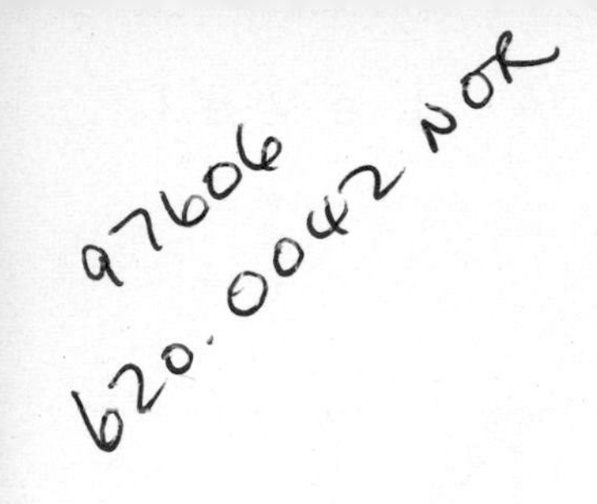
97606
620.0042 NOR

Pearson Education Limited
Edinburgh Gate
Harlow
Essex CM20 2JE
England
and Associated Companies throughout the world

© Pearson Education Limited 2000

All rights reserved; no part of this publication may be reproduced, stored in a retrieval system, or transmitted in any form or by any means, electronic, mechanical, photocopying. recording or otherwise without either the prior written permission of the Publishers or a licence permitting restricted copying in the United Kingdom issued by the Copyright Licensing Agency Ltd, 90 Tottenham Court Road, London, W1P 9HE.

First published 1990
Second Edition 1995
Third Edition 2000
ISBN 0 582 32831 4

The publisher's policy is to use paper manufactured from sustainable forests.

Set in 10 point Times
Printed in Italy by G. Canale & C.S.p.A Borgano T.se-Turin

Contents

Acknowledgements

A book of this kind must owe a great deal to the writings of others and the present authors have derived much benefit from the works of those who are cited in the references and bibliographies. Our thanks are also due to many organisations and individuals who responded to our requests for help.

We received unstinting help from colleagues at Loughborough University of Technology and NEMEC (the National Electronics and Microtechnology Education Centre). Professor Ken Brittan and Brian Bullock gave considerable assistance in the preparation of the Materials section; Steve Garner read the Design chapters and made several valuable suggestions; and Clive Mockford supplied Figs. 2.49–2.53. Chapter 17 owes much to Graham Bevis and the chapters on Electronic systems and Control systems have benefitted from the ideas and examples provided by Patricia Etheridge, David S C Thompson and Steven Hodson. We are indebted to the following students in the University's Department of Design and Technology who prepared diagrams and photographs and whose project work is featured in some of the illustrations:

Albert Au, Nick Bowden, Christopher Choy, Peter Forse, Anthony Grace, Ian Gravestock, John Greenhalgh, David Hallam, Nigel Harper, Neil Harris, Steven Harvey, Charles Hill, Mark Hinnells, Dave Jones, Jamie McCartney, Graham Roberton, Andrew Rogers, Paul Saville, John Smith, John Sutton, Helen Taylor, Sue Thompson, Phil de Vries, Harvey Woodall, and Richard Wragg.

Neil Harris of Loughborough Technical College assisted with Chapter 18 and staff of the College prepared the automotive mechanisms for photographs in Chapter 11.

The Design Council prizewinners – Michael Butterick, Lee Davies, Simon Evans, Rachel Forster, Marc Jackson and Richard Tyler – kindly allowed their student projects to be used in the case studies. Our gratitude goes as well to Darren Fenton, Rhian Chapman, Arthur Parke, Neil McGaughey, Peter Killops, Matt Levell, James Edgehill, Simon Todd and Joanne Smith, and their respective schools, for permitting us to make extracts from the course work they submitted as part of their A-level examinations.

Our thanks also go to Margaret Norman for typing and Geoff Cubitt for photography.

We are grateful to them all.

EWLN
JLC
SAU
MW

We are indebted to the following Examination Groups for permission to reproduce questions that have appeared in their examination papers. Whilst permission has been granted to reproduce these questions, the answers, or hints on answers, are solely the responsibility of the Authors and have not been provided or approved by a Group: The Associated Examinations Board (AEB); University of Cambridge Local Examinations Syndicate (UCLES); University of London Examinations and Assessment Council (ULEAC); Northern Examinations and Assessment Board (NEAB) (Formerly the Joint Matriculation Board); University of Oxford Delegacy of Local Examinations (UODLE) and Welsh Joint Education Committee (WJEC).

We are grateful to the following for permission to reproduce photographs: Argos Distributors, page 30; Bridgeman Art Library, page 176 *right* (Private Collection); Camera Press, page 371; J Allan Cash, page 372 *below*; Trevor Clifford Photography, pages 2, 369, 464, 522, 588; Collections, pages 396 and 398 (photos Brian and Sal Shuel); Compair Broomwade, page 714 (2); Design Council, London, xvii–xviii; Dyson UK, page 34; C.M. Dixon, page 177 (2) (British Museum); Department of the Environment Graphic Design Service, Northern Ireland, pages 349 and 350; Hoover plc, page 34 (2); Mallet and Sons Antiques, page 176 *left*; Norgren Martonair Ltd, pages 710, 715, 716 (2), 717 and 731; Oxford University Press, page 5 *above*; Rover Group, pages 268, 337 *left and below right*; Trustees of the Science Museum, London, page 34 (5); Science Photo Library, pages 5 *below*, 6, 99 and 338; Schrader Bellows, pages 718, 724, 736 and 738; TQ International, page 339; Wood Visual Communications, pages 335 *left and right* and 337 *above right*.

Cover photo by Arcaid.

Additional picture research by Marilyn Rawlings and Louise Edgeworth.

We gratefully acknowledge use of the following:

Fig. 1.7 Zanussi
Fig. 1.24 from 'Introduction to Design', by Walker and Cross. Published by Open University Educational Enterprises, 1983.
Fig 2.1 from Product and Technological Innovation, A Course Reader'. Published by Open University Press, 1986.
Fig. 2.29 from 'Design Processes and Products Course', by N Cross *et al.* Published by Open University Educational Enterprises, 1983.
Figs 1.26, 2.71 from 'Guide to Design for Production', by Cooke, Corbett, Pugh, Weightman. Published by the Institution of Production Engineers, 1984.

Extracts from British Standard publications are reproduced with the permission of BSi Sales, Linford Wood, Milton Keynes, MK146LE; telex 825777 BSIMK G: telefax 0908 320856:
Fig. 2.23 PP 7317: 1987
Fig. 2.24 PP 7317: 1987
Fig. 2.26 PP 7310: 1984
Fig. 2.58 PP 6470: 1981
Fig. 2.59 PP 6470: 1981
Fig. 2.60 PP 6470: 1981
Fig. 2.63 PP 6470: 1981
Fig. 2.74 BSi
Fig. 2.75 BS 4863: 1973
Fig. 2.76 BS 4863: 1973
Fig. 2.77 BS 4863: 1973
Fig. 2.54 Bite Design
Fig. 5.1 Adapted from D R Askeland's 'The Science and Engineering of Materials' and *Fig. 5.9* Published by PWS – Kent Publishing Company, a Division of Wadsworth Inc., © 1984. Figures 1.1 page 3 and 18.17 page 615.
Fig. 8.16 BR European Passenger Services.
Figs 10.11, 10.15, 10.16, 10.17, 10.18, 10.30, 10.41, 10.44 from 'Schools Council Modular Courses in Technology: Mechanisms', by Bailey. Published by Oliver and Boyd/Longman.
Fig. 11.6 BP Statistical Review of World Energy, June 1993.
Figs 17.3, 17.4, 17.5, 17.7, 17.9 from Schools Council Modular Courses in Technology: Pneumatics, Teachers Guide. Published by Oliver and Boydl Longman.

Figs 18.23, 18.50, 18.67 from Schools Council Modular Course in Technology: Pneumatics, Pupils Book. Published by Oliver and Boyd/Longman.
Table 1.1 World Bank UNEP, The Economist.
Table 1.2 UK Ecolabelling Board.
Table 1.3 New Scientist 24-5-73.
Table 1.8 IT futures, NEDO, 1985.
Table 2.2 PP 7310:1984
Table 2.3 H. Dreyfuss 1978 The Measure of the human factors in design. Published by Whitney.
Table 2.4 PP 7317: 1987
Table 2.5 BS 5940: 1980
Table 2.6 PP 7317: 1987
Table 2.7 from Design evaluation by A M Brichta in *Engineering Designer*.
Tables 4.4, 4.6, 4.7 from 'Creative Design and Technology', by Jordan. Published by Longman.
Table 4.8 Engineering Metallurgy Part 1 by R A Higgins. Published by English University Press, 1950.
Tables 4.9, 4.10 from 'The Structure of Wood', by F W Jane. Published by A & C Black (Publishers) Ltd., 1970.
Tables 5.1, 5.2, 5.3, 5.5–5.9, 5.11 from 'Engineering Tables and Data, by A M Howatson, P G Lund and J. D Todd. Published by Chapman and Hall, 1972. The source of Table 5.1 was originally 'Structures and Properties of Engineering Materials' by R. Brick, A. Pense and R. Gordon. Published by McGraw-Hill, 1977.
Tables 5.10, 5.14, 5.15 from 'Properties of Engineering materials', by R A Higgins. Published by Hodder and Stoughton, 1977.
Tables 5.4, 5.12 Data from 'materials' from Mitchell's Building Series pages 52 and 54, by A Everett. Published by B T Batsford, 1986.
Tables 5.13, 5.3, 5.4 adapted from D R Askeland's, 'The Science and Engineering of Materials'. Published by PSW – Kent Publishing, a division of Wadsworth Inc., © 1984.
Table 6.1 from 'Manufacturing Technology', by Timings and Savage. Published by Heinemann Professional Publishing (Oxford).
Table 6.2 from 'Joining and Materials'. Published by The Welding Institute, Cambridge, July 1988.
Tables 6.6, 6.7 Adapted from material supplied by Boxford Ltd.
Table 7.1 from 'Engineering Product Design/Design Processes and Products Course, by N. Cross *et al.* Published by Open University Educational Enterprises, 1983.
Table 7.4 from L E Wingfield 'Essential Information for Product Design', Published by RCA 1979.
Table 11.1 'Reproduced by permission of the Watt Committee on Energy'.
Tables 18.3–18.5, 18.7, & 18.9 from 'Pneumatics' Teachers Guide. Published by Oliver and Boyd/Longman.

We are grateful to the following for permission to reproduce copyright material:

The Design Council for the Figure 'The Design Spectrum' from *Industrial Design Education in the United Kingdom* (1977).

Preface

This book is intended for students who are preparing for A- and AS-level examinations in Design and Technology, or related subjects such as Engineering, Engineering Science and Elements of Engineering Design. It should also prove useful to first year undergraduated in engineering and industrial design, and college students following BTEC courses.

The requirements of most examining boards include course work projects, a design examination and a written paper on materials, mechanics, thermodynamics, control and electronics. Our aim has been to cover all these elements in a single volume. We have assumed little previous knowledge so that the book is generally self-contained but we have added references and bibliographies at the ends of chapters for those readers who wish to explore topics in more detail.

Throughout the book we have drawn attention to the human, social, environmental and safety aspects of the subject and we have added some historical notes to show how design and technology have evolved in past centuries.

We have included a large number of examination questions and design briefs that have been set in recent years and their sources are indicated as follows:

AEB	The Associated Examining Board
Cambridge	University of Cambridge Local Examinations Syndicate
London	University of London Examinations and Assessment Council
NEAB	Northern Examinations and Assessment Board (Formerly the Joint Matriculation Board)
Oxford	University of Oxford Delegacy of Local Examinations
Welsh JEC	Welsh Joint Education Committee

We are indebted to the respective authorities for permitting us to reproduce this material but the responsibility for the answers is entirely ours.

We have used SI units throughout but we have provided conversion factors for a few long-established units that are still widely used. Appendices 1 and 2 list the symbols and abbreviations used in this book but the reader should be prepared for alternative forms in examination papers and other texts. The differences are explained where the quantities first appear.

Our thanks are due to the Publishers for their encouragement during the writing of the book and their care in its production.

Despite careful checking it is too much to hope that all errors have been removed and we shall gratefully acknowledge any comment or correction.

EWLN
JLC
SAU
MW

September 1990

Preface to the Third Edition

Students reading this book could be taking AS or A2-level courses based on syllabuses used in England, Wales and Northern Ireland, Scottish Higher and Higher Still examinations, the International Baccalaureate, GNVQs and undergraduate programmes in industrial, product or engineering design in the UK and other countries.

It might be thought that this is an impossibly wide group of readers to serve, but designing and making is a more generic human activity than such a view would imply. Design products share many of the same characteristics in many countries and cultures, although it has to be remembered that technology and values will change.

This book is about design and technology; the processes involved in reaching decisions about how to create our material culture and the technology involved in implementing such decisions. It is also about 'learning by doing'. The learning of a particular technology is often best approached through carrying out a project which uses it, although there are some matters which are better learnt before the project starts.

We hope this book will support you in whichever learning style you are adopting, but particularly if you are learning through completing a design project. We know that our colleague, Professor Syd Urry, believed strongly in this mode of learning and we hope that this third edition remains true to this goal. Syd Urry edited the first two editions, but died before work was able to start in earnest on the third. We have made significant changes, but we hope these are those he would have wanted and supported. The third edition is dedicated to his memory.

EWLN
JLC
MW

September 2000

The authors

Eddie Norman is a senior lecturer in the Department of Design and Technology at Loughborough University. A former teacher, he is particularly interested in the application of science to design.

Jay Cubitt (formerly Riley) has extensive school and university teaching experience, and is now a consultant writing flexible learning materials and lecturing on Design and Technology.

Syd Urry was Emeritus Professor and former Head of Building Technology at Brunel University. He was also a Chair of Examiners for AS and A level Design and Technology with a major examining board.

Mike Whittaker was a senior lecturer for the National Electronics and Microtechnology Education Centre (NEMEC) at the University of Southampton and is now employed by a leading manufacturer of science and technology teaching equipment.

Introduction

The book is arranged in five sections:

- design,
- materials,
- systems at rest,
- mechanical systems in motion,
- control systems.

The design section has been written to help you understand the range of issues that you must deal with when undertaking a course in design and technology and to help you undertake your design projects. The later sections contain considerable information concerning technologies. These may be studied either before you start a project or whilst completing one. They have been arranged in groups which generally match current syllabuses and a detailed index has also been provided to help you find the section you need. However, before reading any of these sections you might find it helpful to learn from the experiences of other students.

0.1 Some examples of design projects

Fig. 0.1 shows six projects that have won the Design Council schools design prize competition. Together they illustrate something of the possible range of project work in design and technology and the technologies that can be involved.

Fig. 0.1(a) shows slalom poles for artificial ski slopes, which were designed by Richard Tyler at St Andrew's Church of England School in Croydon, Surrey. The project was based on materials and manufacturing technology – the production of aluminium castings of appropriate form to retain them beneath the matting and in which a steel spring could be secured.

Figure 0.1 (a)

Fig. 0.1(b) and (c) show two structures – a hiking tent by Lee Davies of Shrewsbury Sixth Form College and a baby carrier day pack by Rachel Foster of The Lakes School in Windermere, Cumbria. These are different kinds of structures in that the hiking tent has a supporting framework, but they both use textiles in an innovative way. In both cases the weight of the structure is a crucial issue, and hence knowledge is required of the strengths and densities of the materials used.

Fig. 0.1(d) shows a project involving the design of a mechanical system – a lorry wheel alignment aid by Michael Butterick of Boston Spa Comprehensive School in

Figure 0.1 (b)

Figure 0.1 (c)

Figure 0.1 (d)

Figure 0.1 (e)

Figure 0.1 (f)

West Yorkshire. The system requires the design of a mechanism in that it has to locate the wheel at the correct height, but it is also a structure (system at rest) in that it must support the stresses resulting from the lorry wheel's weight.

Fig. 0.1(e) shows an electrically assisted bicycle by Simon Evans of Clowne Comprehensive School in Derbyshire. This is an example of the design of both a structure and an electromechanical system. Apart from supporting the loads it was necessary to consider the issues of power, weight, the range provided by the batteries and appropriate gearing.

Fig. 0.1(f) shows a multi-use photographic lightbox by Marc Jackson of Burleigh Community College in Loughborough, Leicestershire. It is an electrical project based on the use of an electroluminescent panel. This had to be evenly illuminated in order to enable good quality prints to be produced.

The technological content of these projects was not necessarily their most significant feature – for example, human factors were clearly vital in the design of the baby carrier day pack – but technologies were some of the resources used in the completion of the projects. Many of the technologies that you might need for your project are described in the last four sections of this book. However, designing comes first!

0.2 Getting your design project underway

The first problem that students face is identifying a need which their project can address. (This is discussed further in section 3.2.) Three of the projects shown in section 0.1 resulted from a particular interest of the student involved – skiing (the slalom poles for artificial ski slopes), outdoor pursuits (the hiking tent) and photography (the multi-use photographic light box), but even so there was still much work to be done to clarify a project area. The other three projects had different origins – browsing through *Design* magazine (the electrically assisted bicycle), observing parents and young children (the baby carrier day pack) and watching *Tomorrow's World* (the lorry wheel alignment aid). As an indication of the kind of things you might have to do to get started, the early stages of one of the projects relating to the development of a personal interest and the other three are discussed briefly below.

0.2.1 The multi-use photographic light box – Marc Jackson

Identifying the need

Marc considered a number of possible topics for his project, all connected with his major pastime, photography. He recognised that the bulky and expensive contact printers currently on the market were unsuited to the home environment, and finally set his brief as a multi-use photographic light box with the primary function being a contact printer.

Getting started

To substantiate the need for a light box which could also function as a contact printer, Marc wrote to a number of photographic darkroom suppliers and other relevant companies. He also sent questionnaires to local press photographers and the Loughborough Photographic Club to define the specific problems involved in making contact prints. Replies to the questionnaires showed that the average size of photographic paper used for printing contact sheets was 25 × 30 cm, so Marc decided that the dimensions of his screen should be 27.5 × 32.5 cm, allowing for alignment of the paper.

Marc received considerable help from two experts at Thorn EMI, Mr Baker and Mr Wharnby. Jon Baker gave particular assistance in suggesting alternative solutions to the lighting problem and with the design of associated circuits. A major problem was the expense of specialist materials, such as lights and Perspex for the screen. This difficulty was largely resolved when Thorn EMI provided Marc with £100 worth of lights for experimentation, including an electroluminescent panel – a flat panel that just plugs in and needs no controls. The various light sources had to be carefully tested, as an even distribution is essential for contact sheets to be printed successfully. Marc tested his lights by placing a grid over the screen and taking light readings across its surface for different lighting set-ups. Graphs were then plotted from the grid readings and superimposed onto each other to compare the variations in light intensity.

0.2.2 The electrically assisted bicycle – Simon Evans

Identifying the need

Like many students beginning their project work, Simon had considerable difficulty in choosing a topic for his major project. He looked at his school and home environments, but nothing really occurred to him. As a last resort he decided to browse through the magazine *Design*, and he discovered that a group of companies had come together to build an electrically assisted bicycle. This was to be launched when the new legislation, allowing teenagers over 14 to drive electrically powered vehicles on the public highway without a driving licence or insurance, was introduced in 1984. Simon thought the design was rather complex, and felt he could find a simpler approach; he was also aware of the failure of the Sinclair C5 project.

So Simon set himself the brief of designing ‘an electrically assisted bicycle for the 14–16-year-old age group, that was of simple design and could be built in the school workshops’. It was intended for use on the road as transport and also off the road purely for enjoyment.

Getting started

Simon started by identifying the four main areas that would need research – the frame, battery, motor and electronics. All these were investigated at the same time and Simon also began sketching some preliminary ideas. All the companies he approached for advice and help were very generous in the assistance they gave.

CBS Batteries of Liverpool gave him information on the 12-volt batteries suitable for electric bicycles. They also provided two GKP28 batteries and an empty one as a safety precaution for use during the development of the design. The GKP28 battery is incorporated into the final design – it is particularly suitable as it provides the maximum discharge for as long as possible and can be recharged overnight. It may also be recharged for over 300 life cycles.

For advice on forming the frame of the bicycle, Simon approached Raleigh of Nottingham. He went on a factory tour so that he could see the welding techniques carried our by robots in the manufacture of BMX machines. Raleigh subsequently supplied all the metal tubing needed for the electric bicycle.

Simon also sought help from EMD Halstead of Essex, who provided him with a 12-volt motor for the bicycle, and Eddison Cycles of Clowne who gave him advice on the construction. They made a special dish-shaped rear wheel, designed to work alongside the widened crank that the size of the battery had made necessary.

Consider Simon's comments on the effect of his research on his thinking:

> As the research progressed different ideas had to be ruled out because of limited components available on the market. After the research had finished, I was well into the process of finding a final design, because of the limitations on things like the shape of motors and the size of batteries ...

0.23 The baby carrier day pack – Rachel Forster

Identifying the need

Rachel chose her project after seeing the problems faced by parents of young children visiting the Lake District encumbered by pushchairs and bags. She had noticed that looking after a baby on a day out requires a lot of equipment to keep the child clean and fed, and she set out to design a product which would hold all the equipment and incorporate a baby carrier to make life easier for the parent and baby.

Rachel had wanted to use fabrics in her project work, and to design something for use with babies. These interests, together with the importance of tourism in her local area, led to her choice of project.

Getting started

Rachel undertook a very thorough research programme. This ranged from measuring babies and photographing the ways in which people coped with their babies' needs while out for the day, to investigating relevant products already on the market and searching out professional advice about the way a baby should be supported. She also looked into the suitability of different materials and paid particular attention to joining and attachment methods. The project was undertaken during the second year of her course and the research was carried out in the period up to Christmas. During this period she also used a sketchpad – recording initial ideas as well as her research findings.

0.2.4 The lorry wheel alignment aid – Michael Butterick

Identifying the need

An item on BBC television's *Tomorrow's World* caught Michael Butterick's attention. It showed a device to aid the alignment of a car wheel when it was being replaced after a puncture. However, this device broke.

Michael's father works for a haulage company, Smith and Robinson, in Rothwell, West Yorkshire, and Michael wondered how tyre fitters manage to align the bulkier and heavier wheels on a lorry. He found that the most common answer was brute force and a crowbar.

Getting started

Michael began by writing to many companies to obtain details of standard lorry wheel sizes. Even though his father was in the haulage business, he found it difficult to obtain all the information he needed to gain access to lorries. Michael analysed the strength needed for an alignment aid and decided to use steel tubing for his prototype. He also broke the problem down into two sections as he felt it was too difficult to tackle as a whole. These were:

i) how the wheel was to be supported;
ii) the lifting mechanism.

0.2.5

It is not possible to state exactly what you will have to do in order to get your project underway – it depends on the nature of your project – but reading these four examples should have given you some ideas. This kind of activity is part of the 'preparation' phase of designing (see section 2.1). Thorough preparation is essential for effective design. Section 2.2 identifies three aspects of preparation – writing a design brief, information retrieval and preparing a specification. All of these are continued attempts to understand the problem and begin the process of modelling a solution, initially through the use of language. Chapter 2 goes on to describe the processes of generating ideas, synthesis, prototype completion and manufacture, evaluation and communication, and these are also evident in the case studies discussed in Chapter 3.

However, it is worth giving the last words in this section to two of the prizewinning students. Consider first Michael Butterick's comments on his project – particularly concerning the real nature of his research activity:

> Research was an ongoing process and a great deal of complex and detailed information was required. Different things are needed at different stages of the design process.
>
> It is important to identify the requirements. The device or 'final' solution has to do what it is intended to do.
>
> Many skills were learnt or developed by doing this project – information retrieval, talking to people, looking at things from a design perspective, problem solving and how to work hard!

Now read Rachel Forster's comments on how to get the project done!

> Project goals were important to keep to as it helped you to keep up with your work. It was sometimes difficult to keep to the deadlines but looking back you feel you should have been stricter on yourself maybe, so one area of the project hadn't been so rushed in completion. It was difficult to decide on the best final design. It was easy to continue churning out initial ideas without actually committing yourself to the 'best' idea. Perhaps a lot more work with mock up models would bring confidence into choosing the best idea.

0.3 Some general comments

Courses in design and technology generally require students to undertake project work and they often allow a wide choice concerning the type of product or system to be designed. The end point is generally a working model or prototype, but if architectural design studies or similar tasks are undertaken, the result may be graphic images or visual models and feasibility assessments. The main factors which influence the choice of project are:

- the capabilities and interests of the students and their supervisors,
- the requirements of the examination boards,
- the availability of specialist outside assistance.

Commercial product design and development is often carried out by teams whose members specialise in particular aspects, such as graphics and product modelling, or electronic circuit design; other specialists can be brought into the team if the project requires the use of new skills. Your projects, on the other hand, need to be selected carefully because you will normally deal with all aspects yourself. You can, of course, seek expert help – from teachers or from outside your school or college.

Facing up to a major project for the first time often produces a feeling of not knowing where to start. There is so much to do that you end up doing nothing – paralysed by fear of the unknown! The experienced designer has confidence gained from many successful past projects. When you are starting you need landmarks, and Chapter 2, 'Designing', was written with this in mind. There is some measure of broad agreement on the way you should structure your early design projects, but industrial designers and engineering designers often disagree about professional practices. As you grow in confidence you should read some of the books and watch some of the videos suggested in the bibliographies. These will increase your awareness of techniques that have proved successful for designers with a wide range of backgrounds, who work on different types of products and systems.

Clearly, different projects require different technological information, but a knowledge of technology is essential if you are to make effective decisions when designing. It is worth telling the story of Henry Dreyfuss, who was one of ten artists offered one thousand dollars in 1930 by the Bell Telephone Company for ideas on the form of future telephones. Dreyfuss was convinced that speculative designs of external form alone were irrelevant and he asked to work closely with Bell's engineers. At first this request was refused because the company felt that his artistic scope would be limited. However, after the other nine sets of design ideas were submitted the company changed its mind and hired Dreyfuss to work in the way he wanted. His successful association with the company lasted for the next two decades. The inclusion of a 'core' knowledge of technology in design and technology syllabuses reflects what Dreyfuss knew many years ago, and you should be able to exploit technological knowledge within your own designing.

Technological knowledge is clearly essential for the detailed decisions that need to be made concerning the selection of materials, manufacturing processes, component dimensions, energy sources and electrical and mechanical systems, and you will need to use your knowledge of technology to make these decisions effectively. But even in the conceptual design stage there are likely to be many technical, economic and moral issues associated with the use of different technologies. The selection of materials for vehicle bodies is discussed in detail in Chapter 7, but, as an example of some of the issues that must be resolved, consider the following questions.

- Which materials could be used?
- Which materials are most suited to mass manufacturing techniques?
- Which materials and manufacturing techniques would use the greatest quantity of energy?
- Is the total energy consumed in producing the raw materials and during manufacture more or less important than the level of employment?

These are not easy questions and the answers selected have very wide implications, not only for the individuals involved in the manufacture and use of the product, but also for society, in terms of both the national economy and the ecosystem. Designers cannot anticipate everything, but they are morally bound to take into account all the factors they can. They must make the best decisions they are able to in the light of their knowledge, using the Earth's resources wisely in the pursuit of a good quality of life for this and succeeding generations. Understanding technological alternatives and their implications is vital.

0.4 The design spectrum

Fig. 0.2 shows the 'design spectrum'. This was developed by a group working on a Design Council report on 'Industrial Design Education in the United Kingdom', which was published in 1977. On the horizontal axis it shows the range of product types which might be thought of as representing the design field. On the vertical axis it shows the proportions of industrial and engineering design activities which might be included in different types of product.

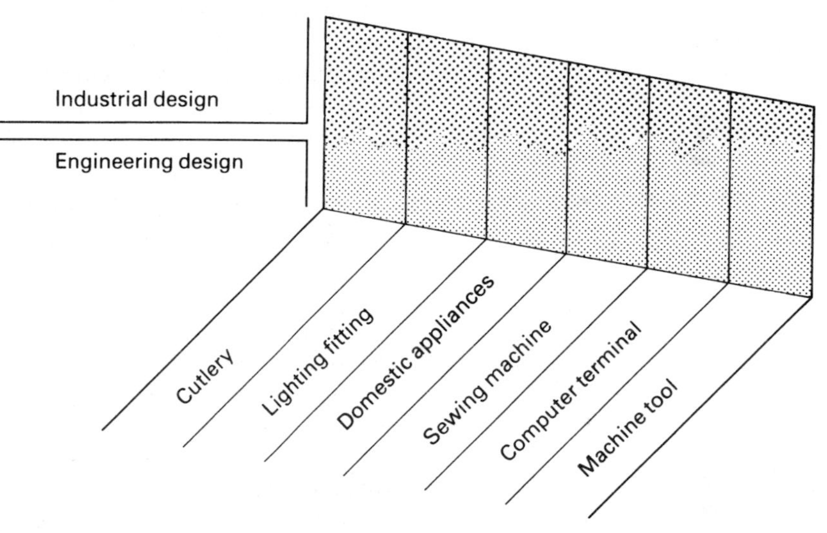

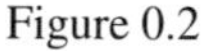

Figure 0.2

The distinction between 'industrial design' and 'engineering design' is rarely significant in design and technology courses at 16+, but it is useful in focusing the mind when selecting a project. Industrial design is often thought essentially to concern issues relating to aesthetics and human factors, and engineering design essentially to concern technological matters. However, neither category is really that exclusive and the design of any product requires both areas to be considered to some extent. What kind of balance is involved in the project you are considering? Does this fit well with the requirements of the Examination Board? Can you cope with all the necessary technology? If not, can you find someone to help you? These kinds of questions need to be answered early in the project selection process. If you go on to study more advanced courses in the design field, these will often have titles like 'industrial design' and 'engineering design'. Before you decide where to undertake further study, it is very important to find out whether or not the course has the balance that you expect. 'Product design' is an even more ambiguous term and you must be particularly careful about finding out what courses using this title have to offer.

SECTION 1 Design

The three chapters in this section have different aims and objectives.

- ***Chapter 1 Design and technology in society*** discusses some of the general issues which will influence your work and on which your work may have an influence, for example 'green' issues concerning sustainable development.
- ***Chapter 2 Designing*** has been written to help you get started on your design projects. It describes designing as a 'systematic activity' but also gives enough examples to demonstrate that this is not always true. You must be sure to leave scope for creativity and imagination.
- ***Chapter 3 Design project work*** has been written to provide you with case studies describing current 'best practice' for students aged around 18. Clearly, it would be possible to include case studies from undergraduates and professional designers, but this could result in students setting themselves unrealistic goals.

It is important to recognise that there is no natural limit to what might be written in Chapters 1 and 2. Designing concerns the creation of the material culture and so most issues impinge on it. Our understanding of designing as an activity is continually advancing, so Chapter 2 could never be complete. As your career in design and technology develops you must not only be prepared to keep up to date with changing technology, but also with the key issues relating to society and design practice itself.

Chapter 3 has been written by Mike Hopkinson who has been the Principal Moderator for A-level Design and Technology projects with the Oxford and Cambridge Examination Board (OCR) for many years. He is very aware of the kinds of problems that students can experience in completing their projects and has chosen case studies to help you avoid difficulties. We hope that you will be able to do a better project as a result of reading this chapter.

The figure below shows shows the very first Royal Enfield motorbike, which was made in the UK. You can compare it with Simon Evan's electrically assisted bicycle shown in the introduction to this book. Both designers faced the same issues, but there were very different outcomes. In the more modern design the motor no longer used petrol, the structure was a steel alloy and the lights no longer used acetylene gas. However, the issues remained the same. What would be the appropriate power source? From what material should the frame be made? How could lighting be provided? It is not only time that can change the answers to such questions, but also culture. Would the answers be the same in Europe and Africa? Reading these chapters should help you get started towards answering these questions, but be prepared to do more. Examples of excellent design can be seen by visiting a Design Museum, like the one in London, and websites such as those given below. The Design Council has commissioned special, web-based resources to enable you to learn more from their Millennium Products collection. The International Design Resource Awards is a competition run from the USA in order to promote design for sustainable development. Their website shows numerous examples of the ingenuity of both professional and student designers in using recycled and reclaimed materials.

The first Royal Enfield motorcycle.

- **Design Council** on *http://www.design-council.org.uk*
- **International Design Resource Awards** on *http://DesignResource.org*

1 Design and technology in society

Man is distinguished from other animals by his imaginative gifts. He makes plans, inventions, new discoveries, by putting different talents together; and his discoveries become more subtle and penetrating, as he learns to combine his talents in more complex and intimate ways.
(J Bronowski 1973 *The Ascent of Man*)

1.1 Human needs and achievements

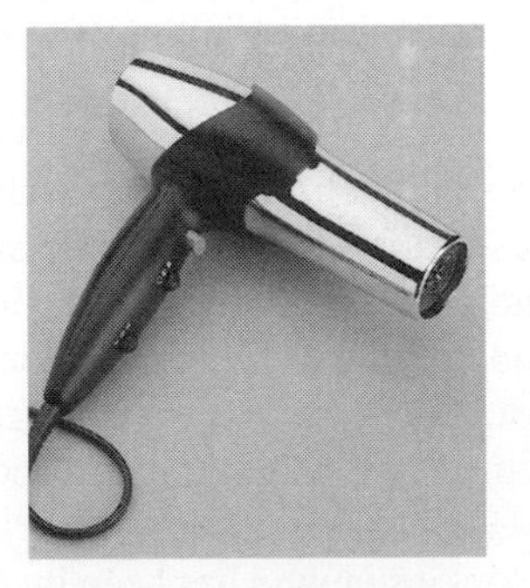

If you make a list of items you consider necessary for a full and satisfying life it is likely to run to many pages. Which of the products shown in Fig. 1.1 do you consider necessary for your life? The following five-level hierarchy of needs was drawn up by Maslow and it recognises that we seek much more from life than the fundamentals of existence. In descending order of priority he identified the following types of need:

- physiological needs
- safety needs
- belonging and love needs
- esteem needs
- need for self-actualisation.

According to Maslow, the earlier needs in this list must be satisfied before much energy can be devoted to meeting the later ones. Physiological needs such as hunger and thirst are the most important. Safety needs require us to feel free from attack and sheltered from the environment. If these fundamental needs are being met then the need for affection, friendship and belonging to a group will influence the way we behave. The esteem needs are for self-respect, self-confidence or prestige; self-actualisation refers to fulfilment, creativity and self-expression. When the earlier needs are met, these become the driving forces behind human behaviour.

Consider Maslow's analysis in relation to your own life and that of people around you. What proportion of your time do you normally spend meeting these various needs? How different are they for single-handed sailors, astronauts and refugees from wars and famines?

Products or systems can be intended to meet any of the needs in Maslow's hierarchy. The concept of fashion relates to the higher levels, belonging to a group and self-esteem, and as such may be considered wasteful. These needs are not

Figure 1.1 Common modern products.

fundamental and hence the resources used to satisfy them could have been saved. The realisation that people's esteem needs will lead them to change their clothes and their cars before they are actually worn out provides the justification for built-in obsolescence. What is the point of designing a car to last twenty years if no one will want it after seven? It is clear that there is a relationship between people's needs and the culture in which they live. The products or systems they use must bring them esteem in the eyes of those they live near. Short journeys to school or work are often undertaken on high-powered motorcycles or lightweight racing bicycles. Whilst these do meet the basic transportation requirement their real importance lies in other people seeing them being used. Ownership of sophisticated hi-fi equipment and computer systems often reflects the same kind of need, although the owner will never admit it, of course.

Design as an activity clearly offers tremendous potential for self-actualisation. Designers have more opportunity than most professionals to express themselves and act creatively. The normally recognised characteristics of self-actualising people – openness to experience, non-defensiveness, independence of mind, little fear of the unknown, childlike perception and maturity – are those that a designer should strive for, but Maslow's analysis should be remembered. Physiological, safety, belonging and esteem needs must come first. The right environment, in every sense, is vital for good design.

As a species we have been remarkably successful in harnessing natural resources to meet our needs. Our ability to make use of our intellectual skills, and to adapt the materials we have found around us, have given us the potential for a healthier, more comfortable and fulfilling life if we use them wisely. Table 4.1 on pages 174–175 shows how different materials have been exploited throughout history and other chapters describe the structures, machines and control systems that have been developed through human ingenuity.

A fundamental result of mankind's success has been the vast increase in the number of people on the planet. In the last ten thousand years the world's population has grown from around six millions (6×10^6) to six billions (6×10^9), a thousandfold increase. The growth has not been uniform and there has been a doubling in the last 50 years alone. Fig. 1.2 illustrates the UK census results since records were started during the industrial revolution.

The general improvement in health care (despite setbacks in many countries) has led to new problems. More people means that some natural resources are being used faster than they can be renewed by nature. There is also evidence that the large-scale

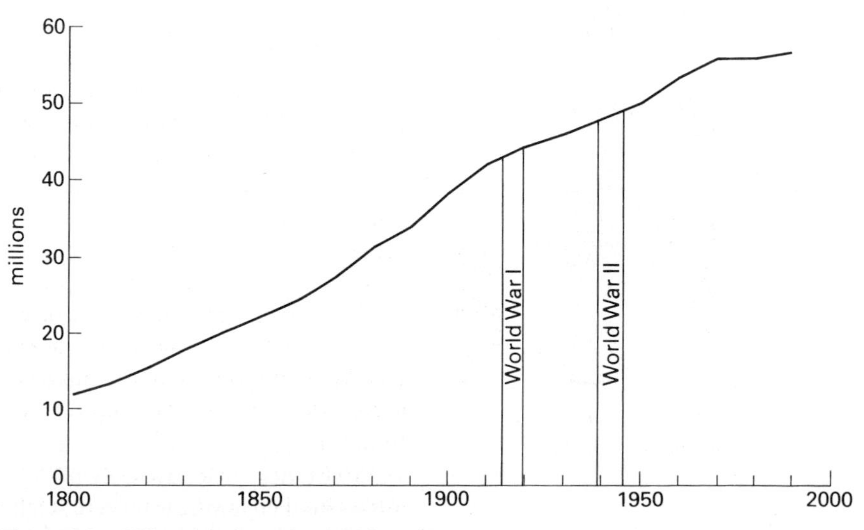

Figure 1.2 UK population growth 1801–1991.

manufacture and use of products such as cars, fertilisers and refrigerators can cause detrimental effects to the air we breathe and the soil on which our food is grown.

The dilemma has been stressed by David Layton in his book[1] *Technology's Challenge to Science Education:*

> Technology frequently seems to lead a double life: whilst it can fulfill the most exorbitant intentions of its practitioners, it also, and simultaneously, produces unintended outcomes. Medical technologies have reduced the death rate in developing countries, but have also contributed to uncontrollable population growth. Food production has become more efficient through genetic and chemical technologies, but at the cost of damage to related ecosystems. No one planned or wanted the effects that DDT has had on bird populations, for example, and no one intended to create a hole in the ozone layer. In an attempt to exercise greater control of the effects of technological developments, a new branch of study called Technology Assessment has come into being.

There are no simple answers to many problems arising from technological developments and you should be prepared to look at all sides of the argument when you consider such issues. A commuter travelling from home to the office may enjoy a more comfortable journey by car than by public transport but the unrestricted use of cars in a city may cause pollution that diminishes the quality of life for the whole community. The impact of the car in our lives is examined in greater detail in section 1.3.4. In an examination you should be prepared to discuss such issues and reinforce your answers with examples from your own reading and experience.

1.2 Technology and designing

Technology is a concept with a broad range of meanings. It can refer to the way in which a society produces the goods and services it requires, e.g. the use of the terms 'high tech' and 'low tech'. It is this general meaning which relates to terms like 'alternative' and 'intermediate' technology. If you get the chance you should try to visit centres such as the Centre for Alternative Technology in Machynlleth, Wales, in order to understand the different point of view that they represent. If you cannot get to Wales, then you can probably find a way of visiting their website (http://www.cat.org.uk). Alternatively, Intermediate Technology is an international organisation whose mission is to build the technical skills of poor people in developing countries, enabling them to improve the quality of their lives and that of future generations. They have also launched a new website (http://www.stepin.org). This is an interactive site aimed at younger students (KS3), which has been developed by the Sustainable Technology Education Project (STEP). The way we currently do things is not in some way 'inevitable', but results from the collective decision-making of this and previous generations. It is this kind of decision-making that lies at the heart of designing and it is therefore not surprising that many design and technology syllabuses at 16+ require students to study such issues. This chapter introduces many of the issues, but you should read and investigate as widely as possible in order to begin to grasp the complexity of these matters.

Some products can challenge our conceptions of the way things should be. For example, it is now possible to recycle coffee cups into pencils and biros, and sell them commercially. Have a look at one such manufacturer's website at: www.remarkable.com/html/frame/html. The very existence of these products makes us think about the issues associated with polymer recycling and appropriate power sources. Recycling polymers is discussed further in section 4.1.

Technology can also be thought of as meaning 'applied science'. Figs. 1.3 and 1.4 show advances that have only been possible as a result of scientific developments. Although there are cases where scientific ideas underpin designing, there are many instances where this is not so: there are notable examples in the history of technology

Figure 1.3 CD-ROM.

of successful products being developed before the corresponding scientific theory was fully understood. The steam engine and powered flight are just two. On the other hand, science may predict results that can lead to profound changes in our lives many years later. Einstein's work seemed to have little relevance to everyday life when it was first published, but numerous products and systems have been designed as a result of his theories. We have all been affected by the military and civil applications of nuclear energy.

Figure 1.4 Repair of the Hubble space telescope, December 1993.

Sometimes a new scientific idea will have consequences in many fields. Chaos theory is a case in point. We can predict the times of tides with great accuracy for many years ahead but forecasting our weather is fraught with uncertainty. Yet the motions of air and water are subject to the same basic principles of fluid mechanics (Chapter 13). It has been found that this kind of randomness can arise from very simple equations, and can be demonstrated on a pocket calculator, but it has possible applications that include the output of electronic circuits, the behaviour of forced vibrations, economic changes and movements in the solar system. Chaos theory was put forward in the early 1970s to explain some unexpected mathematical results, and the first claim of a commercial application – in connection with the tangling of clothes in a tumbler drier! – was made in February 1994. An intriguing spin-off of chaos theory is the generation of striking computer graphics (Fig. 1.5).

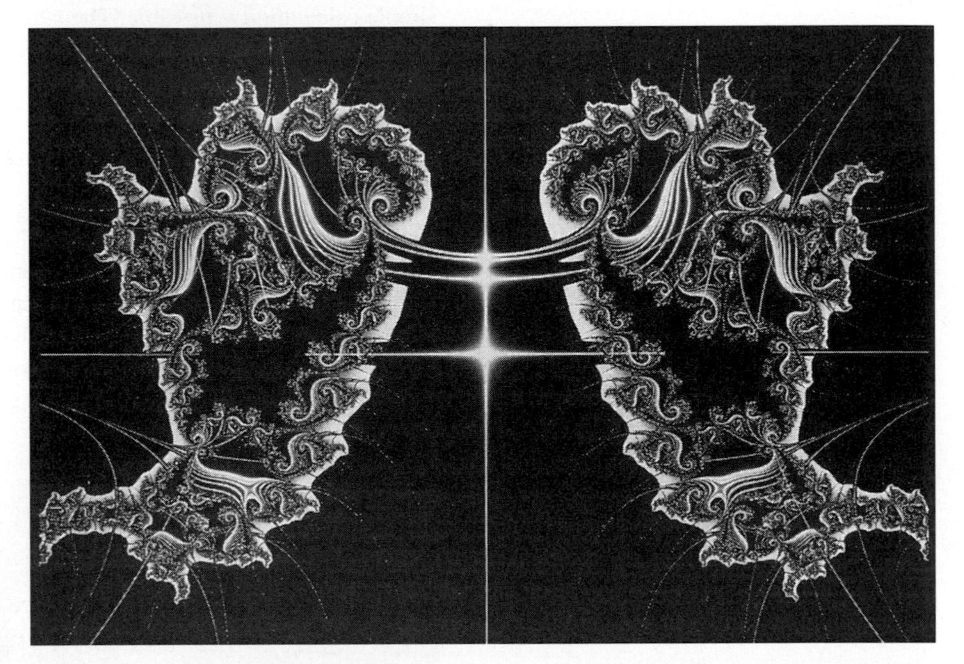

Figure 1.5 Chaos theory graphics.

Figure 1.6 A new design for a polymer acoustic guitar.

So how can designers proceed without science? Where there are no scientific principles to build on, the designer must make decisions based on judgement and experience. As a recent example, consider the polymer acoustic guitar shown in Fig.1.6.

Owain Pedgley designed this guitar at Loughborough University between 1995 and 1999. Before starting, engineers and scientists were consulted and it was found there was no possibility of predicting the vibration behaviour of polymers in such a complex structure. The mathematical modelling was too demanding and too little was known about the vibration properties of polymers. Rob Armstrong, an internationally renowned luthier (guitar maker), was also consulted. With his guidance a successful instrument was developed and the new technology patented in 1999. The vital breakthrough was the use of foamed polycarbonate sheet for the soundboard. In this context however, the crucial point is that the design was based on Rob Armstrong's knowledge and experience gained from building hundreds of guitars, and not on science. It would have been similar, practically-based understanding that facilitated the development of the steam engine and powered flight. Hence, it is possible to interpret technology for designing more broadly than the application of science.

Wherever possible the designers of structures, machines and vehicles rely on scientific theory, experiments and calculations, but there are problems for which neither theory nor experiments exist. Then they have to make decisions based on judgement and experience. This intuitive approach is sometimes called 'feel' or designing 'by eye' and every designer has to resort to it on occasion. In your own work you should seek the advice of designers who have had opportunities to tackle similar problems before. The engineer's adage that 'if a design looks right, it is right' may be too drastic but the converse is almost certainly true, and you should develop a sense of what is reasonable when making design decisions. If a calculation to find the diameter of a car transmission shaft leads to an answer of 1mm or 1m it is almost certainly wrong! This book contains a great deal of technological information derived from science, but it may not always provide the answer you need. The judgement and knowledge of experienced designers may sometimes be the only way forward. However, this should not be read as an excuse to take a wild guess where a simple calculation is possible. The later chapters of this book should help you to find the appropriate way forward for your project.

1.3 Environmental issues

There is now concern at international level that the growth of the world's population, coupled with the expectations for higher standards of living, is depleting natural resources and damaging the environment. Many people and organisations are working towards the concept of 'sustainable development' to ensure a better future for all. In 1992 the United Nations called a Conference on Environment and Development[2] in Rio de Janeiro known as Earth Summit '92 at which the participating nations drew up a declaration pledging support for measures to protect the environment. Among other things the declaration calls for countries to enact legislation on environmental standards. The UK government has responded with a series of reports of its own.[3]

Table 1.1 shows some consumption and waste levels for various countries based on data compiled by the World Wildlife Fund.[4] When using numerical data of this kind it is important to take account of the units, and whether the figures are quoted

	CO_2 Emissions[a]	*Energy*[b]	*Water*[c]	*Paper*[d]	*Food*[e]	*Chemicals*[f]	*Waste*[g]
US	19.68	295	2162	287	3642	99	1 071 000
Germany	10.48	156	668	176	3800	384	80 319
Netherlands	8.43	195	1023	171	3258	642	11 952
Japan	8.46	118	923	169	2858	418	354 196
UK	9.89	147	507	134	3257	350	70 568
Italy	6.82	111	983	93	3494	151	50 000
Switzerland	5.94	107	502	174	3432	426	2620
Malaysia	2.82	41	765	32	2723	157	n/a
China	2.16	23	462	n/a	2628	262	n/a
India	0.77	9	612	2	2204	69	n/a
Ecuador	1.47	19	561	n/a	2058	34	n/a
Nigeria	0.77	5	44	2	2114	12	n/a

n/a = not available
All columns represent per capita consumption levels, with the exception of CO_2 emissions.
[a] industrial emissions metric tons/per capita/per annum (1989)
[b] measured in gigajoules per annum 1989
[c] cubic metres per annum 1990
[d] paper consumed per annum 1984
[e] calorie intake per day
[f] tons of fertilisers consumed per year
[g] thousands of tons of waste generated per annum, including household and municipal, industrial and hazardous waste, with the exception of Switzerland (minus industrial) and Italy (minus hazardous waste).

Table 1.1 Consumption and waste levels per country

for the whole country or are averages for each person ('per capita' as it is usually called), and whether they are annual or daily values.

You must also allow for the time lag in the collection and publication of results. Most of the items in the table relate to 1989 or 1990 and there is evidence that values have changed in recent years.

The standard and style of living are roughly represented by the figures for food and waste. There are some striking differences between the countries when their populations are taken into account. The United States represents 5% of the world's population, uses 25% of its energy, and is responsible for 22% of all CO_2 emissions. In contrast the figures for India are 16%, 3% and 3% respectively.

1.3.1 Energy issues

Many activities at work and in the home have energy implications. Factory production and transport of all kinds depend on machines that convert energy from one form to another. By the principle of conservation of energy (section 11.2.2) it does not disappear in these processes but it changes to a form that is no longer useful. The energy obtained from burning petrol in a car is not lost in the scientific sense because it is converted to heat but it is said to be *degraded*. In temperate climates we depend on artificial heating in buildings and there is an energy cost to pay for standards of personal comfort. Note that artificial cooling, like heating, also leads to an energy cost (see section 12.4).

For many purposes we depend on fossil fuels such as oil, coal and natural gas. These account for 70–80% of our energy usage and strenuous efforts are made in the exploration for new sources. There is a second environmental problem. The energy in fossil fuels is released during the burning of hydrocarbons and, in this process, the carbon combines with oxygen to form carbon dioxide (CO_2) or carbon monoxide (CO). Table 1.1 includes two related measures – energy consumption based on all forms of generation and CO_2 emissions from the burning of fossil fuels.

Carbon dioxide plays a key role in determining the Earth's climate. It affects the incoming solar radiation and the energy radiated by the Earth differently, and the net effect is a warming of the Earth's surface. This is known as the 'greenhouse effect' and CO_2 levels have increased substantially since the industrial revolution in the nineteenth century. It is predicted that there will be a steady warming of the planet with mean temperature rises of 5°C or more in the next 50 years if these increases are not checked. Such changes would seriously affect the ice caps and disrupt rainfall patterns.

Coal-burning also leads to the discharge of sulphur dioxide (SO_2), a gaseous product of combustion which can dissolve in rain and form a corrosive acid. This acid rain can destroy trees and make lakes lifeless far away from the source of the pollution, even in a different country.

1.3.2 Air and soil pollution

In addition to the emissions of CO_2 and SO_2 the natural composition of the atmosphere is changed in other ways by human activities.

CFCs and the ozone layer

We are protected against the Sun's radiation by a thin layer of ozone (O_3) in the upper stratosphere. This shield is not uniform and it can vary naturally from place to place but there is evidence that it is weakened by chlorofluoromethanes (CFCs) still used in some aerosols, refrigerators and air-conditioning plant. The US has taken the lead in reducing the use of these substances and suitable alternatives are now available.

Lead emissions from cars

The development of car engines in the last 50 years has caused another form of chemical pollution. The thermal efficiency of a petrol engine can be increased by an increase in the compression ratio but this can cause 'knocking' or 'detonation' (see section 12.3). It was discovered that the effect can be reduced by the addition of tetraethyl lead to the fuel but the presence of lead in the atmosphere can damage the health of people (particularly children) in areas of high traffic density. An indication of the size of the problem is that roughly 500 000 tonnes of lead is released into the air annually of which half comes from vehicle exhaust and only 3500 tonnes from natural sources.

Chemical pollution of the soil

There is a wide range of substances used in factories and agriculture that can be health hazards. Some of them such as fertilisers and pesticides have the short-term aim of increasing food production but they may depend on the use of chemicals that can accumulate in the soil. The dumping of industrial waste in landfill sites can also result in the presence of toxic materials such as heavy metals including mercury, lead, cadmium, chromium and nickel. These can enter the food chain through microscopic creatures in water and pass through plants and animals into our food.

1.3.3 Ecology and the green movement

The relationship between plants and animals, with one another, and with the environment, is studied under the name 'ecology'. As the previous sections show, there are many aspects to this relationship and fierce debates take place about the size of the problems and the actions that should be taken. Many people have formed groups to study particular issues and political parties have been set up in some countries to argue for changes in policies through legislation.

There are fundamental conflicts in all discussions of this kind on two grounds:

- the differences between individual needs and those of society,
- present and future needs.

Most of us can think of possessions that would improve the quality of our own lives but many of them depend on scarce resources or would otherwise be to the detriment of others in society. We may also accept that selfish actions now will reduce the quality of life in the future but it is difficult to translate this into individual decision-making.

As with many contentious decisions financial costs have a considerable influence. Are we as individuals and as a nation prepared to pay more for goods and services in order to reduce the adverse environmental effects they may cause? To what extent can decisions on pollution and the environment be left to individuals and companies, and how much has to be enforced by legislation? Raising environmental standards costs money, at least initially; does this make companies or even whole countries uncompetitive?

The effect of environmental legislation on product design has been considered in a report by Kate Langley[5] in which she describes the proposal for a European 'eco-label'. The EC directive was published in December 1991 with the aims of encouraging products with the lowest environmental impact and supplying the consumer with environmental information on all products. The label will be awarded to those products that achieve satisfactory ratings on a 'cradle to grave' assessment matrix (Table 1.2).

Environment fields	*Product life cycle*				
	Preproduction	*Production*	*Distribution*	*Utilisation*	*Disposal*
Waste relevance					
Soil pollution and degradation					
Water contamination					
Air contamination					
Noise					
Consumption of energy					
Consumption of natural resources					
Effects of ecosystems					

Table 1.2 The Eco-label assessment matrix

Figure 1.7.

Specific criteria are being drawn up for particular product types such as dishwashers, washing machines, light bulbs and hair sprays. In the study of washing machines it was concluded that the three most damaging features were energy consumption, water consumption and detergent consumption. Furthermore the investigators suggested other criteria for consideration, such as the user-friendliness of instructions, reliability and durability, manufacturing policy, packaging, recyclability and the re-use of materials.

Some manufacturers recognise the concern of many consumers for environmental issues and reflect this already in their advertising (Fig. 1.7).

1.3.4 Transport issues

The importance of transport in modern society can hardly be overstated. Many of us make journeys to school or work, to visit friends or the shops, and to go on holiday, that are far longer than we could manage on foot. In addition, much of our food and the other goods on which we depend travels long distances before reaching us from the producers. Directly or indirectly we spend a great deal of our money on transport. Furthermore we all contribute to national transport developments through our taxes.

The total cost of the Channel Tunnel and its associated rail links was more than £10 billion and in the UK, despite cutbacks, the government's roads programme cost £23 billion during the 1990s. What are the average costs of these projects for each person in the country?

In addition to the huge capital expenditures, the building of roads, airports and other transport projects uses large areas of land. It can also seriously disrupt the lives of local people before, during and after the construction phase. Each major scheme is subject to a public enquiry and there can be years of uncertainty before a decision is reached. During the period of this planning 'blight' residents may not be able to sell their homes and when construction begins there may be interruptions and long traffic delays. It has also been found that motorways consistently attract more traffic than was predicted by the surveys used at the planning stage.

Recent changes in modes of travel

Since 1950 there have been drastic changes in the extent and methods of transportation in the UK. At that time the rail network retained almost all the lines that existed at the beginning of the twentieth century with duplicate routes running between many major towns and freight trains delivering to almost every station. By the 1940s trams and trolley buses were being replaced by diesel buses due to their greater operational flexibility and the low price of oil at the time. Few towns had by-passes and there were no motorways so that average speeds of 50km/h (30mph) were commonplace. Indeed, this figure was the legal maximum for some lorries.

Fig. 1.8 shows the changes in UK passenger and freight journeys by road and rail for the 40-year period up to 1992. The growing dominance of the car for personal travel throughout the world is also illustrated by car ownership figures. It has been estimated that there are now more than 400 million cars in the world with nearly 20 million being added each year.

In Britain the trends were established in the 1950s. Higher living standards led to increasing car ownership. Street parking was still possible on many city-centre roads but the first parking meters were installed in London in 1958. The first stretch of UK motorway was opened in the same year and the network has expanded steadily since that time. The completion of the M25 in 1986 was regarded as a notable achievement yet it quickly proved inadequate for the volume of traffic using it, and within seven years extra lanes were being added to some sections.

There have been equally significant changes in international transport. In the early post-war years most people took their holidays within the UK and currency restrictions limited overseas journeys. Travel between continents relied mainly on ships and it was not until the mid-1950s that the number of people crossing the Atlantic by air exceeded those travelling by sea. There were also scheduled passenger services by sea from the UK to South America, West and South Africa, Australasia, India and the Far East. All of these disappeared in the face of competition from the airlines.

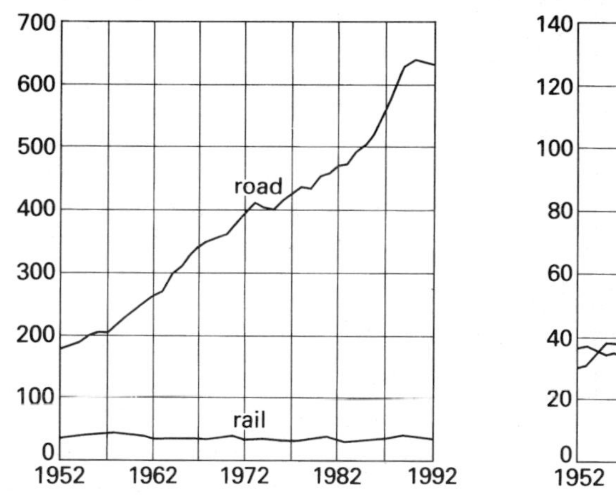

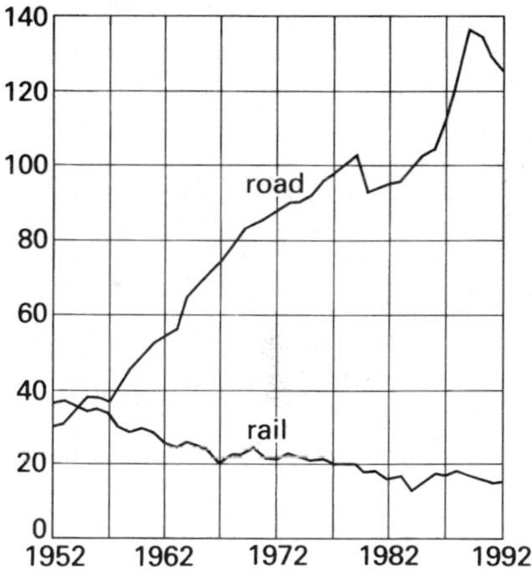

Figure 1.8 Transport growth.

The growth of personal transport

The most significant change in our travelling habits has been brought about by the car. Its ability to make door-to-door journeys without changes, its protection against adverse weather, its availability at all times and the opportunity to carry large amounts of luggage have made it indispensable in the eyes of most people. The central place of the car in our lives is indicated by the statistics on car ownership.[6] In the UK in 1994 the total number of cars was 22.9 million. Almost one household in four had two or more cars, with 20% having two and 4% having three or more. Only 31% had no vehicle.

The car has fundamentally altered the pattern of life for many people. Commuters can make daily journeys that would be impossible by other means, the motorway network enables fans to support sports events in distant towns, holiday-makers can take a range of luggage and equipment with them and, by towing a caravan, can be independent of hotels. The opportunities for motorists to travel in Europe have been greatly extended by the development of RoRo (roll-on, roll-off) ferries and the opening of the Channel Tunnel in 1994. Every improvement of this kind so far has been matched by an increase in the number of cars and a demand for more roads.

Other factors are at work. There is fierce competition between rival car manufacturers, with aggressive advertising to promote one model against another and to increase the total market for new vehicles. Much of this advertising aims to meet Maslow's 'esteem needs'. Design features such as improvements in safety, security, reliability and economy are accompanied by fashion changes that reflect current preferences in shape and colour, or that stress luxury and high performance.

The effects of car production and use are summarised in Table 1.3. It was based on American experience in 1973 and you should consider to what extent it applies to other countries today. In the UK, for example, car use has doubled during the intervening years.

Selected impacts of the automobile (1895 to present)

Values

Geographic mobility
Expansion of personal freedom
Prestige and material status derived from automobile ownership
Over-evaluation of automobile as an extension of the self – an identity machine
Privacy – insulates from both environment and human contact
Consideration of automobile ownership as an essential part of normal living (household goods)
Development of automobile cultists (group identification symbolised by type of automobile owned)

Environment

Noise pollution
Automobile junkyards
Roadside litter

Social

Changes in patterns of courtship, socialisation and training of children, work habits, use of leisure time, and family patterns
Created broad American middle class, and reduced class differences
Created new class of semi-skilled industrial workers
Substitution of automobile for mass transit
Ready conversion of the heavy industrial capability of automobile factories during World War II to make weapons
Many impacts on crime
Increased tourism
Changes in education through busing (consolidated school versus 'one room country schoolhouse')
Medical care and other emergency services more rapidly available
Traffic congestion
Annual loss of life from automobile accidents about 60 000
Increased incidence of respiratory ailments, heart disease and cancer
Older, poorer, neighbourhood displacement through urban freeway construction

Institutional

Automotive labour union activity set many precedents
Decentralised, multi-divisional structure of the modern industrial corporation evident throughout the auto industry
Modern management techniques
Consumer instalment credit
Unparalleled standard of living
Emergence of US as foremost commercial and military power in world
Expansion of field of insurance
Rise of entrepreneurship
Basis for an oligopolistic model for other sectors of the economy
Land usage for highways – takes away from recreation, housing, etc.
Land erosion from highway construction
Water pollution (oil in streams from road run-off)
Unsightly billboards
Air pollution – lead, asbestos, HC, CO, NO_x, SO_x

Demography

Population movements to suburbs
Shifts in geographic sites of principal US manufacturers
Displacement of agricultural workers from rural to urban areas
Movement of business and industry to suburbs
Increased geographic mobility

Economic

Mainstay and prime mover of American economy in the 20th century
Large number of jobs directly related to automobile industry (one out of every six)
Automobile industry the lifeblood of many other major industries
Rise of small businesses such as service stations and tourist accommodations
Suburban real estate boom
Drastic decline of horse, carriage and wagon business
Depletion of fuel reserves
Stimulus to exploration for drilling of new oil fields and development of new refining techniques, resulting in cheaper and more sophisticated methods
Increased expenditure for road expansion and improvement
Increased federal, state and local revenues through automobile and gasoline sales taxes
Decline of railroads (both passengers and freight)
Federal regulation of interstate highways and commerce as a pattern for other fields
Highway lobby – its powerful influence

Table 1.3 The effects of car production (from *New Scientist*, 24 May 1973)

Car-making is an international industry and many vehicles are produced in some countries for sale in others. Manufacturers may locate production in particular countries to take advantage of low wage rates or avoid import tariffs. Some manufacturers collaborate with others to achieve cost savings through greater production volumes. Several countries have launched car industries by buying successful but older designs from established companies. The importance of the car industry to a nation's economy has been recognised by governments through incentives to overseas firms who establish new plants in their countries.

The dominance of the car for personal transport has been accompanied by a rapid growth in lorry traffic which has been able to benefit from the same road improvements. Freight movement between the UK and Europe has grown in the same way as private motoring, and door-to-door deliveries between different countries can reduce transport costs by eliminating handling at the ports. Altogether, transport accounts for 60% of petroleum consumption and it therefore contributes to many of the environmental problems described earlier (sections 1.3.1 and 1.3.2).

1.3.5 Conservation of resources

Increasing populations and rising standards of living lead to more and more demands on limited resources, and many initiatives have been taken by individuals, local and national authorities, and the international community to try and achieve 'sustainable development'. The aim is to meet human needs now but still preserve the resources that will be needed by future generations. This objective can only be met by a combination of several strategies:

- more efficient use of existing materials and other resources,
- avoidance of materials and manufacturing methods that damage the environment,
- substitution of materials and processes by more effective ones,
- recycling of materials.

Energy conservation

The problems associated with the wasteful use of energy can be tackled in many ways. We can reduce our total energy requirements by changing to energy-efficient products, particularly in the areas of transport and buildings. Fuel economy in cars has improved through developments such as the 'lean burn' engine and the adoption of multi-speed gearboxes. Diesel engines with their greater thermal efficiency are now widely available in cars of all sizes.

Much greater reductions in energy are obtained if individual vehicles are replaced by 'mass transit' systems. A bus full of passengers uses far less fuel than separate cars carrying the same number of people, and a few cities in the UK are now installing trams and light railway systems to encourage their use. There is a further energy benefit from changing from road to fixed track networks, arising from reductions in friction. At the moment, however, the flexibility and convenience of the private car is overriding in the eyes of many people.

Heating houses accounts for large quantities of energy (see section 11.4 and Table 11.1) and substantial savings can be made by increasing insulation standards in the walls and roof, and controlling the ventilation rate. Human beings generate about 100W each, and in a specially designed school at Wallasey it was found possible to maintain comfortable internal temperatures with the sun, the occupants and the electric lighting as the only heat sources. Solar panels can be fitted to existing houses but installation costs have to be set against savings in energy costs.

In power stations and other industrial plants some of the heat carried away in water cooling installations can be utilised in waste heat recovery schemes. The principle has been used on a limited scale in the UK to heat flats located near power stations. District heating schemes in Scandinavia and some other European countries enable individual householders to receive heat generated centrally as a 'utility' comparable to gas, water and electricity. In some cities the heat is generated in plants that burn household rubbish.

Substantial energy savings can also be made by using fluorescent tubes for lighting. These tubes also have a much longer life than the filament bulbs used in most domestic installations. The savings in both energy and replacement costs have led to their being widely adopted in large buildings and for street lighting. However, some people consider that fluorescent lighting does not render colours as well as filament lamps.

The harmful effects of meeting our energy needs from the burning of fossil fuels (see section 1.3.1) have drawn attention to alternative sources (see section 11.3). Although they avoid the problems associated with the gases of combustion, some of them have other environmental disadvantages. The construction of dams for generating hydroelectric power may cause adverse ecological changes, for example, and accidents in nuclear power plants such as that at Chernobyl can cause radioactive fallout over a wide area.

Materials recycling and substitution

Our excessive use of raw materials gives rise to concerns similar to those associated with the use of energy. However, the consumption of steel, cement, paper and some other raw materials has been growing less rapidly than our use of energy in recent years, at least in industrialised countries. The demand for timber and wood products remains high and the clearing of forests can lead to soil erosion, flooding and the destruction of wildlife habitats. The extraction of minerals for the manufacturing industry together with sand, gravel and stone used in construction also leads to environmental damage.

The initiatives to counteract shortages and reduce the impact of these activities on the environment fall into three groups:

- recycling,
- materials substitution,
- development of renewable sources.

The collection, sorting and re-use of household and trade waste is widely practised. Many local authorities provide collection points for cans, glass, plastic, paper and fabrics, and techniques have been developed for segregating individual materials from general household rubbish. In some areas the residual waste is used as a fuel for power generation. Metals reclamation is widespread in industry and substantial quantities of steel are produced from salvage, notably from scrapped cars and lorries. To encourage this it has been suggested that a salvage tax be paid on each new car to cover the cost of scrapping it at the end of its useful life. Recently the recycling of plastics in motor cars has been aided by good design, clear marking of the type of plastic on each component, and a commitment from the automotive and waste industries to increase substantially the proportion of materials recycled.

Efforts are being made to find alternatives to materials that are dangerous or in short supply. Mercury and cadmium are being eliminated from dry batteries, for example, and new gases are being introduced to replace CFCs in refrigerators and freezers to help reduce the greenhouse effect.

The decline of the world's forests is a major problem but wood is a renewable material. Schemes to plant a new tree for each one felled help to maintain stocks and have the advantage that new plantings are more efficient than mature forests in improving the quality of the atmosphere. In many countries wood is a primary fuel so that large areas of forest are denuded for energy purposes. In other places trees are

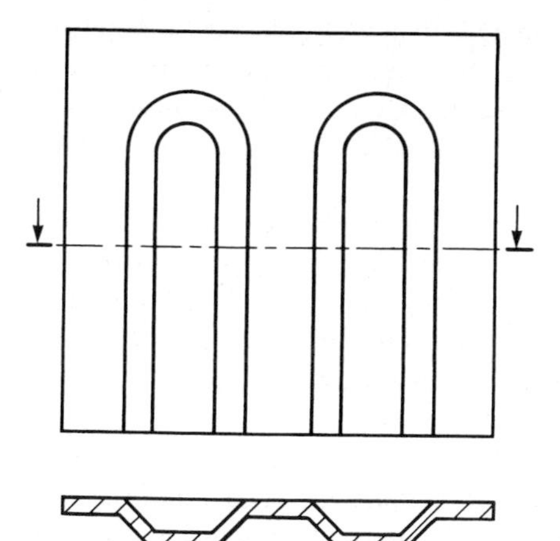

Figure 1.9 Stiffening ribs.

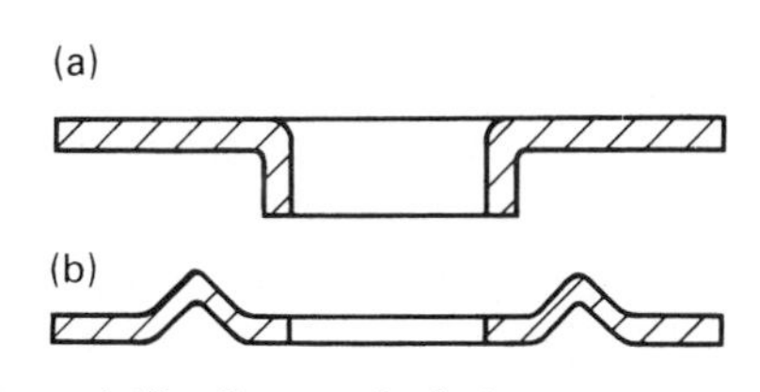

Figure 1.10 Support for holes
(a) flanged hole
(b) local reinforcement.

cut down to make room for more profitable activities such as ranching. Faster-growing softwoods have ousted other species in many countries and hardwoods are now too expensive for some purposes. As a result furniture and built-in kitchen units now rely to a large extent on chipboard and veneer.

1.4 Economic issues

There are many manufacturing and other activities in which the total cost is made up of two parts known as the fixed and variable costs (see section 7.2). The fixed costs are those that remain the same however many items are produced or consumed. In factory production the fixed costs include the capital cost of equipment and outgoings such as rent and rates, often referred to as overheads. The variable costs include the material, labour and energy costs that arise as a direct result of the production. In many cases these are directly proportional to the number of items made.

If the fixed cost is F and the variable cost per item is V, then the total cost C for a quantity q is:

$$C = F + qV$$

The amount V can be regarded as the extra cost to be added for each additional item, often called the *marginal cost*. A graph of the total cost C plotted against the quantity q is a straight line, Fig. 1.11(a).

Many household bills follow this pattern. Gas, electricity and telephone accounts are made up in two parts – a fixed (or standing) charge together with the variable cost based on the number of units used.

Motoring costs can be analysed on the same basis. The fixed charge can be thought of as the cost of standing still and the variable charge arises from the running cost for each mile or km travelled. Under which headings would you place the following costs – petrol, oil, tax, insurance, servicing, depreciation, tyres and the replacement of batteries?

Consider, for example, an issue often faced by designers – that of whether or not to stiffen metal or plastic panels. If sheet panels are used flat, then they may have to be up to twice as thick as when they are stiffened by using bends, ribs and flanges. Such stiffening requires extra processing operations, and their cost must be balanced against the cost of the thicker sheet. The greater weight may cause significant disadvantages in the product's use, for example, increased transportation or fuel costs but, equally, there may be advantages like the greater thickness to corrode away giving the product longer life. Fig. 1.9 shows the kind of ribs which are often used.

If it is decided to use thin sheet and stiffen it, then it is also important that the panels should be locally stiffened in the areas surrounding functional holes. This can be done by forming flanges or by using local reinforcement, as shown in Fig. 1.10.

Where the average cost per item, A, is the total cost, C, divided by the quantity, Q:

$$A = \frac{C}{q} = \frac{F + qV}{q}$$

Table 1.4 and Fig. 1.11(a) and (b) show the total and average costs for various production quantities in the stiffened panel example where the fixed cost is £1000 and the variable cost is £2 per unit.

Production quantity	*100*	*200*	*300*	*400*	*500*	*600*
Fixed cost (£)	1000	1000	1000	1000	1000	1000
Variable cost (£)	200	400	600	800	1000	1200
Total cost (£)	1200	1400	1600	1800	2000	2200
Average cost per item (£)	12.00	7.00	5.33	4.50	4.00	3.67

Table 1.4 Fixed, variable and average costs

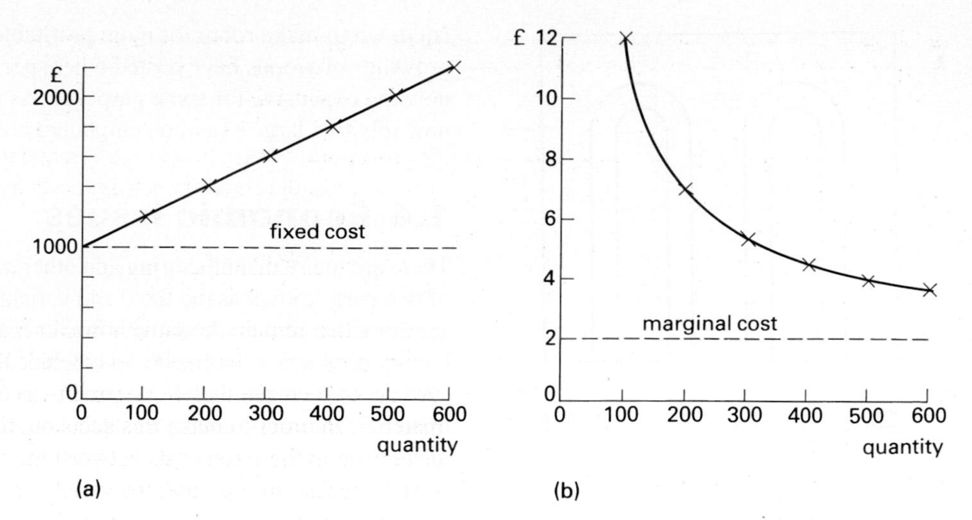

Figure 1.11 (a) Total cost and (b) average cost.

The results show that the average cost decreases as the production quantity increases, an effect known as the *economy of scale*. It is also clear that this effect becomes smaller as production is expanded. It can never fall below the marginal cost of £2 with the present values.

In relation to producing a panel of a given stiffness from flat sheet or by using stiffening ribs, these costs might be listed as shown in Table 1.5.

Such a list clarifies some of the issues, but does not greatly assist the decision-making. In order to make further progress it is necessary actually to put a cost figure on each of these factors. Clearly, with an item like a bandsaw it might be used on a number of products making it difficult to apportion the cost, but assume that the numbers shown in Table 1.5 provide a good estimate.

	Flat panels	*Stiffened panels*
fixed costs	storage bandsaw	storage bandsaw press press tooling
	£500	£1000
variable costs	material used labour for cutting operation blade replacement transportation	material used labour for cutting and pressing operation blade replacement transportation
	£3 per unit	£2 per unit

Table 1.5 Cost factors for stiffened panels

If the fixed costs are F_1 and F_2 for the two strategies and the variable costs are V_1 and V_2, then the cost of producing a given production quantity, q, can be written as:

the cost of strategy 1, $C_1 = F_1 + qV_1$
and the cost of strategy 2, $C_2 = F_2 + qV_2$

At the break-even quantity, q_b, these two costs will be the same, that is:

$$C_1 = C_2$$

and, therefore,

$$F_1 + q_bV_1 = F_2 + q_bV_2$$

from which

$$q_b = \frac{F_2 - F_1}{V_1 - V_2}$$

Expressed in words, the break-even quantity can be calculated by dividing the difference in the fixed (or set-up) costs by the difference in the variable costs. For the example given in Table 1.5,

$$q_b = \frac{1000 - 500}{3 - 2} = 500$$

The cost variations are shown graphically in Fig. 1.12.

For production quantities less than 500 it is cheaper to use flat sheet. If more than 500 are to be made then it pays to invest in the tooling for stiffening the thinner sheet material. In order to make this decision, the designer need only consider the difference in the fixed costs between the two strategies and the difference in the variable costs. In this case, the variable costs are more difficult to assess because there will be a saving in the material usage and the transportation cost, but an increased labour cost resulting from the pressing operation.

There are many products for which the variable cost is a small proportion of the total. The mould for a plastic model aeroplane kit may cost thousands of pounds but the material for each unit produced may be worth only a few pence. Again, the work put into writing the software for a computer game or business application can run to millions of pounds but the disk on which it is stored can be made for less than one pound. As a result, the price of such products may be many times the value of the raw materials of which they are made.

Manufacturers go to great lengths to prevent others copying their products and software, and most countries offer them protection through their patent and copyright laws.

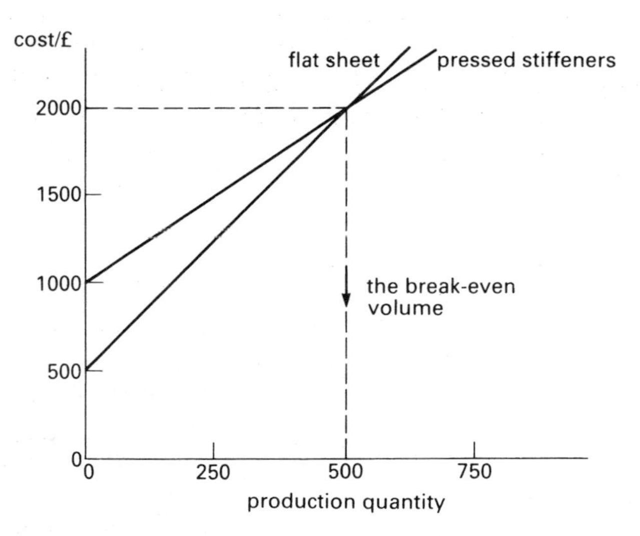

Figure 1.12 The break-even volume.

1.4.1 One-off, batch and mass production

The results of the calculations shown in Table 1.4 show that the average cost of a product decreases as more and more are made. It is therefore beneficial to manufacturers and consumers to use standardised designs for many items and our high standard of living in developed countries has been achieved by producing large numbers of identical products.

As a result manufacturers compete fiercely with one another to increase their sales and benefit from the economies of scale. The history of the motor car in Britain is a striking example of this trend. In the first few decades of the twentieth century scores of car manufacturers were established in the UK but many survived for only a few years. For example, only three thousand cars of the famous Bentley marque were made throughout the life of the company before it was taken over by Rolls-Royce. This represents only a few days' output from a modern production plant making a popular model for the mass market.

The design, the materials, the production method and the resulting cost of a product can be greatly affected by the number to be produced. A large building such as a hospital, a shopping complex or a power station is a 'one-off' product that is not repeated. In contrast, millions of cars may be built to the same design. In both cases the design stage may take several years but the car can be developed and tested as a prototype before it goes into production, a benefit not enjoyed in the construction of major buildings. The concept of 'batch' production falls between the extremes of one-off and mass production and shares some of the advantages and disadvantages of each. The chief characteristics of the three approaches are:

One-off

- The product can be designed and made to meet the needs of one customer or client.
- Often it is sold and paid for as soon as it is made.
- Design and development costs have to be covered by the income from a single sale.

Mass production

- The design, tooling and setting-up costs can be spread over many items.
- The selling price can be less, perhaps only a small fraction of that for a comparable one-off product.
- The low selling price will lead to increased sales.
- If production and sales rates match exactly there are no costs arising from stockholding.

Batch production

- Some of the advantages of mass production can be obtained by manufacturing a larger number of items than can be sold immediately.
- An assembly line can be switched from one product to another.
- Storage costs will be incurred while the batch of products is used up or sold.

The choice of one-off, batch or mass production often decides which materials and fabrication methods are used for the product and cost is often the criterion. There may be a break-even quantity (Fig. 1.12) at which the advantage changes from one fabrication method (or one material) to another.

If items are produced in large batches the storage costs may be prohibitive. The output from a car manufacturing plant may fill several hectares of parking space each working day and an efficient dispersal system is needed.

At the other extreme very small batches lead to frequent changeovers and re-tooling which are reflected in the total cost of the product. Suppose that it costs £1200 to set up each batch of an item and that there is a steady demand of 100 units a week. If the storage cost is £1 a week for each unit in stock, how big should the batches be and how frequently are they required?

Fig. 1.13(a) shows the variation in the stock level for this item. Starting with a new batch, the level falls steadily day by day until all the items have been sold. A new batch is then made and the process is repeated. If we assume that each batch can be made very rapidly the graph follows a 'saw-tooth' pattern and the average stock level equals half the batch size as shown.

Batch size	100	200	300	400	500	600	700	800	900
Interval (weeks)	1	2	3	4	5	6	7	8	9
Stock level (mean)	50	100	150	200	250	300	350	400	450
Weekly costs (£):									
Storage	50	100	150	200	250	300	350	400	450
Changeover	1200	600	400	300	240	200	171	150	133
Total	1250	700	550	500	490	500	521	550	583

Table 1.6 Weekly storage and changeover costs.

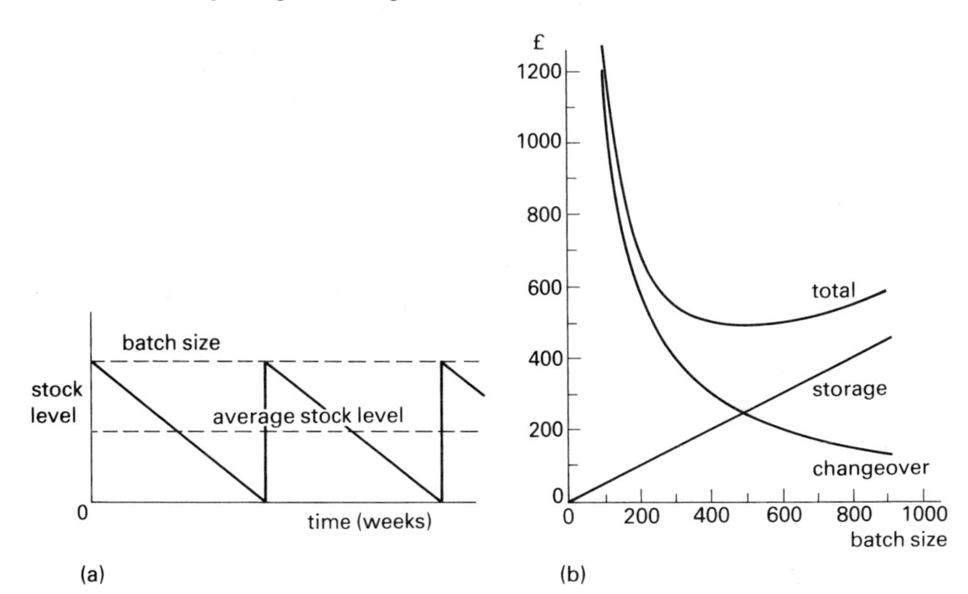

Figure 1.13 (a) Batch production and (b) weekly costs.

The average weekly costs are shown in Table 1.6 for a range of batch sizes and are illustrated by Fig. 1.13(b). With frequent small batches the changeover cost is high (on a weekly basis) but the storage cost is low. With large batches the changeover cost on a weekly basis becomes small but the storage cost is large. In comparing the costs for different batch sizes it is important to work on a weekly basis using the average stock level for calculating storage costs. For a batch of 300, for example, there would be an average of 150 items in stock and at £1 per item per week the cost would be £150.

As the table shows, the sum of the storage and changeover costs is smallest with a batch of approximately 500 every five weeks. (It can be shown by calculus that 490 is slightly more accurate but there is very little variation in the total cost for batches between 500 and 600.)

In the calculations shown in Table 1.6 it is assumed that the demand for this item is steady, the production of the batch is effectively instantaneous and that the item does not deteriorate during storage. In practice the number required may vary from day to day, some items may be sold while the rest of the batch is being made and products such as foodstuffs may have a limited shelf life.

Nevertheless there are many production examples in which the total cost is made up of two parts that change in opposite directions as the scale of the operation varies. The analysis is very similar for stock bought from a supplier in batches.

1.4.2 Changes in manufacturing systems

Batch and mass production became possible through a number of related developments dating from the time of the industrial revolution – roughly 1760–1860 in Britain. These included:

1 The organisation of production in factories that led to concentrations of the population close to the sources of the raw materials used in textiles, iron and ceramics.

2 The availability of steam power that made possible fabrication processes that were too heavy or slow to be carried out by hand.

3 The invention of machine tools, notably Whitworth's screw cutting lathe, that led to manufacturing precision, the accurate assembly of components and the smooth running of machinery.

4 The use of moulds, jigs and templates that made batch produced components interchangeable.

In order to speed up production the moving assembly line was introduced in the first part of the twentieth century, notably in the car industry, and it has been applied to many products such as radio and television sets. Each operator added one component or carried out one process as the product moved forward, emerging in its complete state at the end of the line.

It was argued that this division of labour, applied to large numbers of identical products, would lead to low costs. Henry Ford wrote in his book *My Life and Work* (1922) that the aim in mass production was to reduce the operative's task to doing 'as nearly as possible one thing with only one movement'.

This development reduced the human contribution to simple and repetitive tasks and it was satirised in books and films such as Charlie Chaplin's *Modern Times* (1936). There is evidence that the lack of job satisfaction in some mass-production plants cannot be compensated by high wages. It is also argued that the breakdown of tasks in assembly-line working removes the sense of responsibility among operators and leads to poor quality in manufacturing.

This has lead to some employers in the car industry adopting 'team working' in which groups of employees operate in 'cells' that take responsibility for the complete assembly of vehicles. At the same time a 'just-in-time' system of stock control has been adopted in which components are only produced or delivered when they are required.

The introduction of robots is bringing changes in production as far-reaching as the assembly-line system of earlier years. Robots can undertake repetitive tasks accurately and operate in areas such as paint shops that may be unsafe for humans. It is noteworthy that writers of fiction saw the possibilities of these machines at an early stage. The name 'robot' was coined by the Czech author Karel Capek for his play *R.U.R.* (Rossum's Universal Robots) in 1920.

Each of the developments outlined here has spread to many countries. The industrial revolution started in England in the eighteenth century and initially gave this country many economic advantages. Furthermore, its colonial empire provided a market for many of the goods produced here. Britain's economic successes in the early part of the nineteenth century stimulated industrialisation in other countries, and the achievements celebrated in the Great Exhibition of 1851 faced growing competition in Europe. 'The Continent will not suffer England to be the workshop of the world,' Disraeli prophesied in 1838.

In America emigrants brought with them the technological achievements begun in Europe and applied their organisational skills to develop large enterprises. As a result the US became a leading industrial country by the beginning of the twentieth century. The use of mass-production and assembly-line techniques pioneered by entrepreneurs such as Henry Ford were quickly copied in other countries.

The tendency towards larger but fewer manufacturers has been boosted by the growth in international trade. Shipping costs between countries can be offset by the cost savings gained from increased production runs. Cars, television sets, computers, cameras and household goods are designed so that they can be sold in different countries. Owners' manuals often carry instructions in several languages. In this respect Britain and the US have benefited from the widespread use of English as a world language.

A feature of the world economy in the second half of the century has been the growth of manufacturing industry in countries of the Far East. Japan has rapidly expanded its production of cars, electronics, ships and cameras to become a leading manufacturer in these fields and its success has spread to other Far Eastern countries – Korea, Malaysia, Taiwan and Singapore. Japan itself has extended its influence by establishing factories in Europe and America, from where the industrialisation process began over a hundred years earlier. Many observers believe that China, with a population greater than Western Europe and North America combined, is at the beginning of a transformation at least as great as any so far.

1.4.3 Recent developments in manufacturing systems

The drive towards ever more efficient manufacturing systems has led to the development of new stategies which can help organisations reduce their costs. The three major new strategies are described here:

- total quality management (TQM)
- just-in-time manufacturing (JIT)
- best practice benchmarking (BPB).

As in every other area, manufacturing management strategies are continually developing and this discussion provides no more than an introduction. The key matters for a designer to appreciate are that such strategies are complementary and that they have implications for the designer.

Total quality management

The origins of TQM are generally traced to Edwards Deming in the 1950s. He was an American, but his work was largely ignored in the US. In 1950 he was invited to teach quality control to engineers, plant managers and research workers by the Union of Japanese Scientists and Engineers. This is often seen as the beginning of the 'quality revolution' that lies behind the success of many Japanese companies. The cost of quality is made up of three key elements: the cost of 'inspection' i.e. checking that the quality is right; the cost of rectifying any mistakes found; and the cost of preventing mistakes occurring in the first place. The central concern of TQM is to eliminate the costs of inspection and rectifying mistakes and to get things 'right first time'. The implementation of TQM rests on changing the culture of the whole organisation and this is usually done by getting everyone to see themselves as part of a *quality chain*. Whatever their job within an organisation, everyone has both customers and suppliers as illustrated by Fig.1.14.

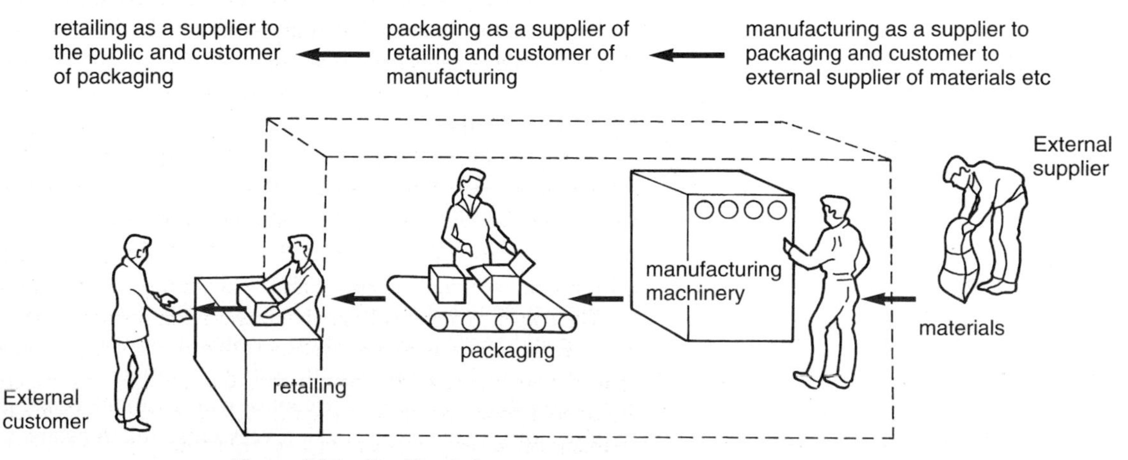

Figure 1.14 Quality chains.

Individuals are able to measure the quality of their work by how satisfied their customers are. If a dialogue between 'suppliers' and 'customers' is promoted throughout an organisation, then mistakes and their causes can be systematically eliminated. Such a strategy rests on individuals taking responsibility for their own work and hence the need for cultural change within the whole organisation. The traditional question for addressing quality issues was 'have we done the job correctly?', but this leads to a culture where errors are caught by someone else and large amounts of time are spent putting things right. The more appropriate modern question is 'are we capable of doing the job right first time?'. It is for this reason that employers invest in the training of all employees, and employees are encouraged to improve their education and performance. It is important to realise that TQM does not only refer to direct manufacturing operations. Other parts of the organisation are involved e.g. market research, design, sales and customer support. It is just as essential for people in these areas of the company to be involved in its TQM strategy, so designers must ask how they too can get the design right first time.

Just-in-time manufacturing

JIT focuses on the elimination of waste from the direct manufacturing operations. Observation of factories has shown that up to 95% of production time can be associated with non-value added operations, e.g. stock being held in 'buffers' and moving components from one workstation to the next. It is surprising, but many efforts to improve production efficiency have tended to focus on the 5% of active processing rather than the less obvious inefficiencies. JIT is a development of the 'Toyota manufacturing system' and has the central intention of eliminating all waste from the manufacturing process. 'Waste' is given the broadest possible definition, e.g. the elimination of anything other than the minimum amount of resources (equipment, materials, space, labour etc.) which are absolutely essential to add the required value to the product.

A number of techniques are employed to implement just-in-time manufacturing, but the most common approach is the reduction of stock levels. Problems will show up in the manufacturing operation as the stock 'buffers' are removed and these need to be tackled as they arise. The aspiration is to eventually hold no stock in the system so that manufacturing is 'pulled' by the customer. The customer places an order and the request travels through the manufacturing system from supplier to supplier, from machine to machine to meet the order. For example, parts are only ordered from a car component manufacturer when the customer places the order in the car showroom. 'Kanban' is the Japanese word for card and this approach is often referred to by this name, because it was cards that once carried the requests for parts through the manufacturing systems from customers to suppliers. Now, of course, it is computers that carry the information.

Another key factor concerns the reduction of the movement of parts. Considerable time can be wasted as parts travel from one machine to another. Group technology is the name given to the strategy of locating machines so as to minimise the necessary movement. Savings made as a result of group technology are sometimes referred to as *economies of scope* as opposed to *economies of scale* (see the introduction to section 1.4). It is widely recognised that the ideas of mass production and economies of scale have implications for the product designer and it is becoming better understood that 'designing for group technology' also has implications for the product designer. A company needs to produce a range of products using the same technology so that it can respond easily to changes in the market.

It should be clear that a Kanban system requires components to be right first time. Any error will cause the manufacturing system to stop. So TQM and JIT are complementary strategies; JIT cannot work without TQM. It is also obvious that JIT requires manufacturers to develop very close relationships with their component suppliers and to involve all their employees if it is to be successful.

Best practice benchmarking

Best practice benchmarking requires an organisation to compare its performance with that of others that perform similar roles. It may be that two factories or retail outlets belonging to the same company compare their performance, but companies also find it useful to compare their performance with that of their competitors. So whereas TQM tends to look at the company's culture and JIT seeks to eliminate waste from manufacturing, BPB is seeking improvements in performance by looking outside the organisation rather than inside it. All three are therefore complementary strategies.

Best practice benchmarking is normally described as having four key stages.

1. Deciding what is to be benchmarked (e.g. delivery times, product consistency)
2. Deciding which organisation to benchmark against (e.g. a store in the same chain)
3. Analysing the information obtained
4. Using the information to improve performance.

1.5 Consumer issues

In making a decision about buying a product the customer is likely to consider several factors. In some cases the choice can be made quickly on the basis of experience or a general feeling about its suitability, in others it is the result of a long deliberation. Among the questions likely to be considered are:

- Is it suitable for my purpose?
- Can I afford it?
- What will the running and maintenance costs be?
- Is it safe?
- Is it reliable?
- How long will it last?
- Is it attractive and fashionable?

In a few instances there may be only one product giving acceptable answers to these questions but in the majority of cases there are several, often many, competing models. For items such as cars, mobile phones, hi-fi, home computers and cameras the choice can be bewildering. Manufacturers' advertisements and catalogues may give answers to some of the questions but are likely to stress the particular strengths of a product and minimise its shortcomings.

Consumers are protected by law from some dangerous or substandard products. There is a general legal requirement of shopkeepers and manufacturers that the goods they offer for sale are safe and fit for their intended purpose but there are fewer safeguards for buyers of secondhand items purchased privately. In addition there are hundreds of legal requirements for the manufacture and use of individual products. For example, inflammable materials are banned in the manufacture of clothes or furniture, and the use and disposal of toxic substances and corrosive materials are subject to various restrictions. The consumer also has responsibilities where the public are at risk, as in the case of defective steering, brakes and tyres in older vehicles.

The British Standards Institution (BSI) publishes national standards for industrial and consumer products covering most sectors of industry and trade. Some of these standards are voluntary but others are covered by legislation. The Institution undertakes the certification of products and manufacturers, and grants the 'Kitemark' and 'Safety Mark' to those that comply with its requirements (see section 2.6.2 and Fig. 2.74). The Institution is financed by government grants, voluntary subscriptions, certification fees and the sale of its publications. Bodies such as the ASA (American

Standards Association) and DIN (Deutsche Industrie Norm) carry out similar work in other countries and the ISO (International Standards Organisation) exists to coordinate these activities.

Some manufacturers and suppliers have formed associations to regulate their own industries and to establish standards. They issue codes of practice that are expected of their members. They are financed by levies on their members and they recompense consumers who suffer losses as a result of breaches of their codes.

A number of publications publish the results of tests between competing products and these help individual consumers select items to meet their needs. The Consumers' Association is financed entirely by the subscriptions of its members. It tests many domestic products and preserves its independence by accepting no advertising in its magazines. Its interests include motoring and services such as holidays and health.

There are commercial journals such as *What Car?* and *What HiFi?* that carry extensive advertising and publish performance tables. There are also TV and radio magazine programmes that review new products on a regular basis.

1.5.1 Price factors

For many products price is an important consideration; in some cases it is overriding. In a competitive market there is a close link between the cost of an item and the number sold. If the price of the article is raised the demand is likely to decrease and, from the manufacturer or supplier's point of view, the decision on selling price is crucial. Conversely, a reduction in price can lead to greater sales, new uses or even completely new markets. An outstanding example is the growth in the sales of computers in recent years. The way in which the increase in sales is related to the reduction in price is known as the elasticity of demand.

Setting the price for a product is a crucial decision for a manufacturer. Suppose that the company making the stiffened panels of section 1.4 finds from market research that it can sell 100 units at £8 each and that the sales increase by 100 for each reduction of £1 in the price. If the firm sets the price at £8 the total income is £800 but with a price of £7 and sales of 200 the income is £1400. Table 1.7 shows the results for sales in the range 100–600. It also gives the total cost for each number of units, taken from Table 1.4, and the corresponding profit. The results are illustrated by the graphs of Fig. 1.15.

The table of results and the graphs show that the total income starts to go up as the sales increase. After a time the increase in sales is outweighed by the reduction in price and the total income is reduced. By calculation, or an estimate from the graph, the maximum income is £2025 with a price of £4.50 and sales of 450. The operation shows a profit for sales in the range 200–500 but outside this band the total cost is greater than the income from sales. What is the maximum possible profit, and what are the corresponding values of the price and sales? Why is it that the maximum income does not lead to the maximum profit?

Sometimes products are made in very small numbers deliberately to keep prices high, or manufactured for a short period of time. Swatch watches and sports clothes from leading brands come into this category, where 'rarity' allows higher prices to

Selling price (£)	8	7	6	5	4	3
Sales	100	200	300	400	500	600
Total income (£)	800	1400	1800	2000	2000	1800
Total cost (£) (Table 1.4)	1200	1400	1600	1800	2000	2200
Profit (£)	–400	0	200	200	0	–400

Table 1.7 Income, cost and profit at different sales levels

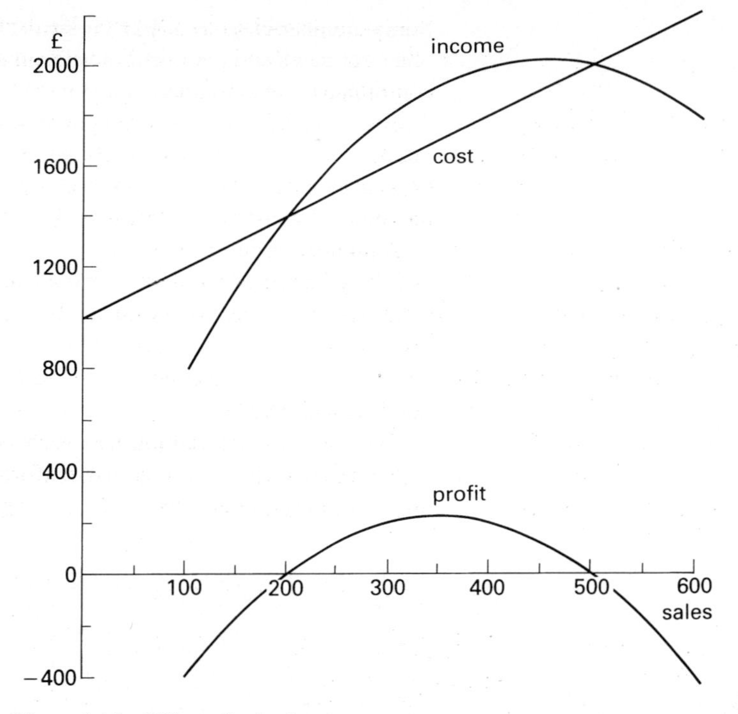

Figure 1.15 Effect of sales level.

be charged. Some customers are prepared to pay a great deal of money for exclusive products such as *haute couture* clothes or limited editions of works of art.

1.5.2 Depreciation and replacement decisions

Much of the equipment we use at school, work and in the home reduces in value year by year and we cannot expect to sell an item at the price we paid for it when it was new. Its value has depreciated. The process of allowing for this reduction in the value of our assets is generally known as 'writing down'. There are two methods that are widely used in calculating the amounts involved and they are illustrated in Fig. 1.16.

The *straight-line method* is based on the assumption that the reduction in value is the same each year. It makes the arithmetic very simple but it is unrealistic in many cases. A new £10 000 car may reduce in value to £8000 in its first year, but we would not expect it to fall from £2000 to zero in the fifth year.

In practice, the reduction in value is likely to be smaller year by year and a common assumption is to take it as a fixed percentage of the value at the beginning of the year. This approach is known as the *declining balance* or *reducing balance method.* As the diagram shows, the straight-line method leads to a zero value after a

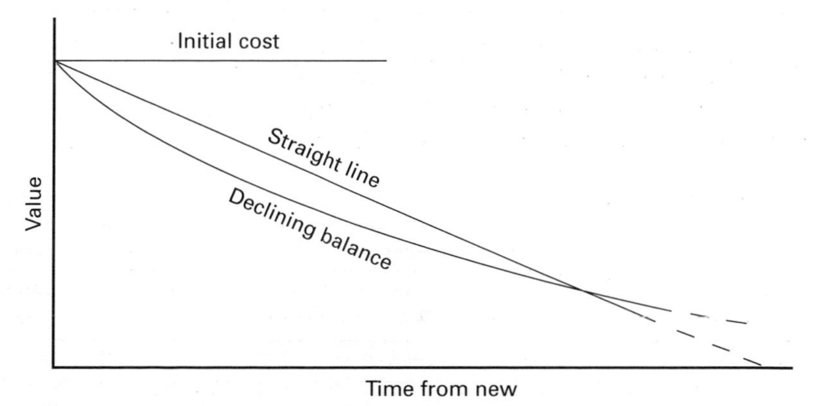

Figure 1.16 Depreciation.

finite number of years – and a negative one if taken further! With the declining balance method some positive value always remains.

Suppose there is a permanent need for an item of equipment such as a vehicle or machine tool that wears out over a period of several years. In the first year or two, depreciation is likely to be high but maintenance costs should be low. Towards the end of its useful life the depreciation may be small but heavy repair bills can arise. Between these extremes there is likely to be an economic lifespan at which the item should be replaced.

Additionally, the cost of replacing the item rises year by year, meaning the gap between the depreciated value and the replacement cost might become significantly larger.

In contrast, some components such as light bulbs fail suddenly. At home we usually wait until failure occurs before replacing them. In a hospital or factory, however, it may prove cheaper to replace them on a routine basis before they reach the age at which failures are expected. In these circumstances decisions will be made on the basis of probability.

1.5.3 Marketing and advertising

A new product can only be successful if consumers are aware of its existence and convinced that it meets one or more needs in Maslow's hierarchy (section 1.1). Manufacturers therefore spend much time and money on informing and persuading the public about their goods. It has been estimated that as much as 5% of the price of a new car is spent in advertising it.

Market research is carried out at an early stage in the design of a new model to identify features that will appeal to customers and ensure large sales. Some manufacturers do this for themselves but there are specialist organisations such as MORI that undertake this for individual companies or trade associations. The general approach is to seek the opinions of consumers through questionnaires and personal or telephone interviews. The technique was originally developed by the Gallup organisation to test political opinions in the 1940s. It was found that the accuracy of such surveys depended not on the size of the sample but how closely it represented the make-up of the population in terms of age, gender, family, social class, education and income. In political polls a sample of one or two thousand may be sufficient to predict the outcome of national elections with good accuracy but it has been found that the time lag between sampling and actual voting can lead to differences between the forecasts and the actual outcome.

In applying market research to commercial products there are other considerations. If several manufacturers carry out similar surveys on half a dozen possible designs for a car they may all come to the same conclusion as to which receives the most votes. Those who choose this design may have to share the market between them but a manufacturer of a niche product can become very successful. A notable example was the VW Beetle which was regarded as old-fashioned and utilitarian by many drivers (and other car-makers) but attracted loyal support and remained in production for 50 years with sales of 21 million units. Volkswagen have now introduced a new Beetle model to exploit the niche in the market it created.

Market research may also be misleading by reflecting existing tastes rather than future trends. In making choices about cars, houses, furniture and many other products we are likely to make comparison with examples that are familiar to us and we may start with a prejudice against the unfamiliar. Nevertheless, the designer or manufacturer who anticipates changes and helps to bring them about will benefit.

The growth of advertising

Each development in communications has led to new forms of advertising. Most newspapers and magazines depend on the income from advertisements to survive and they make strenuous efforts to increase their circulation through competitions, subscriptions and other incentives. Journals themselves are in competition with other advertising methods including:

- commercial terrestrial TV,
- satellite and digital TV,
- commercial radio,
- mailshots,
- salespeople calling on retailers,
- door-to-door salespeople,
- telephone and fax advertising,
- posters out of doors and in buildings,
- sponsorship of sports and the arts.

There are agencies that act for manufacturers and suppliers in each of these fields. They identify groups of potential customers and recommend effective ways of reaching them. Lists of customers with special professional interests or hobbies can be bought to enable firms to reach potential customers for their goods. These are often based on the membership lists of societies or the records of previous purchases.

Consumers receive some protection from misleading advertisements by the Trades Description Act and the code of the Advertising Standards Authority.

Inducements to customers

Various devices are employed by manufacturers to encourage the purchase of individual products. In former times many manufacturers had a policy of fixing the prices at which shops could sell their products but this was made illegal for many items and they now quote only a 'recommended' price. This is sometimes set deliberately at a high value to enable shopkeepers to offer large 'savings' to customers.

Retailers, individually or in collaboration with manufacturers, may offer reduced prices for a limited period in a 'sale' but there is a legal requirement that the goods must have been offered at the previous price for a certain time.

An alternative to a cash reduction is so-called 'free' extras. These can take many forms such as vouchers to be exchanged for goods from a catalogue, additional equipment, extended warranty, or travel and holiday vouchers. Generally the manufacturer or supplier can purchase these well below the price asked of the consumer. Prize draws are often linked with certain products but, for legal reasons, they carry the disclaimer 'no purchase necessary'.

Suppliers and retailers of products that are needed frequently often make special efforts to foster loyalty in their customers. Consider the following offer made by one supermarket. It announced that, at the end of each month, it would publish the date on which the store takings had been least and would then give a full refund to customers who presented checkout bills bearing that date.

Do you feel that the scheme would benefit the store? Does it have any advantages or disadvantages for the customer?

1.6 Information technology

In the *Glossary of Computing Terms* drawn up by the British Computer Society Schools Committee information technology has been defined[7] as:

> The application of appropriate technologies to information processing. The current interest centres on computing, telecommunications and digital electronics.

The information may be of various kinds – spoken, pictorial, sound, textual and numerical. Computing and telecommunications were previously developed separately and both now rely on digital signal processing (Chapter 16). From the 1940s computers progressed from single-purpose machines used for scientific purposes to general-purpose commercial computers that could be applied to many tasks using programming languages.

At the same time, telecommunications expanded from the existing applications in radio and the telephone to include long-distance dialling, television, tape recording, microwave links, optical fibre cables and satellites. These developments are summarised in Table 1.8.

Both technologies benefited from the introduction of transistors, microchips, printed circuits and integrated circuits. Since the 1970s computers and telecommunications have become closely intertwined, with progress in one field stimulating developments in the other. In the 1990s there has been a rapid growth in optical technologies as alternatives to microelectronics.

The outstanding feature of the changes has been an increase in power of the processing and storage devices but a spectacular reduction in their cost. This has been made possible through the improvements in the fabrication techniques for microelectronic components. Single transistors have been superseded by integrated circuits that are equivalent to thousands of transistors and components. At the same time, the adoption of digital techniques has made it possible for a single channel to carry thousands of signals.

The processing of information depends on the recognition and combining of signals using electronic signals of two kinds – analogue and digital. Information in the form of words and figures uses digital signals and it is shown in Chapter 16 that these are binary, each signal having only two possibilities, 1 and 0 (see section 16.1). The name 'bit' was coined to represent a single binary number. Two such bits lead to four possibilities, three give eight possible values and so on. For many purposes it is convenient to use eight-bit numbers and these are called bytes. Each byte therefore offers 256 possibilities (2^8) in our normal decimal (sometimes called 'denary') system and this is ample to represent all the characters of the alphabet in upper and lower case, together with the decimal digits 0 to 9 and special symbols such as exclamation marks, brackets and asterisks. A code for these characters known as ASCII (American Standard Code for Information Interchange) has been adopted to enable different computers to communicate with each other.

The storage and processing capacity of computers is measured by bytes or kilobytes (kb). The prefix k is often used to mean 2^{10} (=1024) but the difference with 1000 is small enough to be ignored in many practical cases. It is a measure of the progress in computer power that the BBC Micro launched in 1981 possessed 32k Ram (Random Access Memory) whereas 4Mb (megabytes) has become (in 1994) the standard for PCs (Personal Computers) and these now cost little more in real terms than the earlier machines. (It should be pointed out, however, that the BBC machine offered some control facilities that are not available in standard PCs.)

During this period the number of bits that can be stored on a chip has quadrupled every three years and the cost per bit of storage has halved every three years.

The two types of computer memory – RAM and ROM – are explained in section 16.5.1. In general the contents of the RAM memory are lost when the operating power is switched off and various devices have been used to store the data. The BBC Micro and other home computers were able to exchange programs and data with a cassette recorder but this technique was soon overtaken by the faster and more reliable 'floppy disk' drives. The present generation of PCs (in 1994) uses drives with disks that can store 1.44Mb. It is a measure of their capacity that the text of this book amounts to about 2Mb (though the illustrations would add considerably to this figure). Computers now have built-in hard disks with storage capacities measured in hundreds of megabytes. However, floppy disks are widely used for making back-up copies and exchanging files with other users.

Date	*Computer technology*	*Information technology*		*Telecommunications technology*
1940	single-function computers general-purpose computers	(military)		mobile radio for air-borne and land-based vehicles
1950	commercial computers programming languages transistors			tape recording cable TV microwave telephone links subscriber trunk dialling (STD) video tape recording
1960	integrated circuits microcomputers structural programming			communication satellites digital communications electronic switching
1970	database management systems large-scale integration applications generators microprocessors relational database management systems wordprocessors spreadsheets	remote sensing computer-aided diagnostics CAM/CAD electronic mail and teleconferencing materials planning stock control	on line enquiry management infor- mation systems integrated text and data processing transaction clearing systems	facsimile transmission (fax) mobile phones teletext/videotext optical fibres videodisks teleconferencing
1980	portable computers optical disk storage transputers video recognition dataflow processors wafer-scale integration gallium arsenide chips compact disc storage (CD–ROM)			local area networks cellular radio wide area networks private satellites digital networks mobile phones
1990	parallel processing learning capability natural language recognition optical chips biochips ultra-intelligent machines	multimedia processing authoring systems	world wide web and the internet	switched wideband services personal mobile communication (via satellite) video on demand videophones
2000				

Table 1.8 Developments in information technology (adapted from *IT Futures*, NEDO, 1985), (dates are approximate)

The latest development in storage is the CD-ROM (Fig. 1.4(a)). This uses a laser drive similar to that for audio compact discs. CD writers are available, but for most users it is a 'read only' system at present. One CD-ROM can store up to 660Mb (megabytes) of information, the equivalent of nearly 470 floppy disks, and it can carry speech, photographs, music, graphics and film clips as well as text and this makes it especially valuable for encyclopedias. There are other benefits. The whole text can be scanned in a few seconds and the direct cost of pressing each copy is little more than that of a floppy disk. This makes it economic to issue new editions at frequent intervals.

1.7 Design issues

The word 'design' is both a noun and a verb and it has many meanings. Most dictionaries list several definitions and you will find a range of synonyms if you look it up in a computer or printed thesaurus. An extensive discussion of the term 'design' is given by Morrison and Twyford in their book *Design Capability and Awareness*. It is generally agreed that design is a creative process involving imagination and ingenuity, and it is usually expressed through visual images. Design is associated with many different fields including:

- product design,
- architecture,
- interior design,
- landscape architecture,
- clothes and fashion,
- textile design,
- jewellery,
- graphics.

In all these aspects design is concerned with shape and form, materials, texture and colour and it is also related to the 'fine' arts of painting and sculpture.

The driving force for the development of a new product or system normally derives from one of two sources:

Demand pull – the requirements of the user(s) become so clear and urgent that the search for solutions is initiated;

Technology push – some change in the state of human knowledge or awareness provokes new approaches.

Figure 1.17 Alarm clocks.

Designers have a major role in both cases, but the demands on them and the techniques that need to be employed differ. You will be operating on different time scales and with different constraints from those which face professional designers, but there are a great many similarities between the ways in which you and they must behave in order to be successful.

The designer is constantly faced with decisions. In some cases the choice is limited but in others there will be a wide range of possibilities. The external shape and size of an airliner, for example, will be decided on the basis of lengthy calculations. In contrast, an alarm clock can take many external forms and be made of various materials; one showroom catalogue shows models with shapes as diverse as a motorbike and a rooster (Fig. 1.17). Sometimes the external appearance is intended to conceal the product's purpose as in the case of TV sets or hi-fi units hidden in cabinets made in the style of eighteenth-century furniture.

As well as technical considerations designers must allow for human factors – ergonomics and anthropometrics (section 2.4.1). However large we make our aeroplanes, ships and land vehicles they have to be controlled by human beings and it is essential that the control systems are within our physical capabilities. There is a corresponding problem with small products. The electronics components of portable computers can now be made so tiny that the overall size is governed by the keyboard.

Designs are not permanent. The response to a human need will change from time to time in the light of the new materials, new fabrication methods, developments in science and technology and changing fashions. The designer must also take into account local resources and local requirements. The family car may well be the same in different countries but the family house is likely to differ widely depending on the climate and local building materials.

The function of a water tower is to ensure the pressure of the supply to consumers by maintaining a water level at a high point in the system, and Fig. 1.18 shows four examples. The mock brick castle at (a) with its castellated turret diverts attention from this purpose; it dates from 1900. At (b) the principal material is again brick but the absence of windows (except at the top) and the lack of ornamentation have led to a simpler and much 'cleaner' design. The examples at (c) and (d) take advantage of

(a)

(b)

(c)

(d)

Figure 1.18 Water towers (a) New Milton, Hants; (b) Mariehamn, Åland Islands; (c) Lahti, Finland; (d) Örebro, Sweden.

modern materials and construction techniques to reflect the fact that the full width of the structure is only required at the top. The 'mushroom' at (d) involved a novel fabrication technique; the head was constructed at ground level and then jacked up in stages for the stem to be added a section at a time.

1.7.1 The evolution of the guitar

The evolution of the guitar as a product, and more recently as a 'system', provides an excellent example of the way in which human need resulted in a significant demand pull, and the invention of the electromagnetic pickup gave a major technology push. The first instrument with all of the general characteristics of the guitar appeared in a Hittite stone carving of around 1300 BC and during the following centuries many instrument makers experimented with different shapes and woods in an attempt to influence the tone and the projection of the sound. Instruments were made with varying numbers of strings, sometimes paired, with five becoming standard in the sixteenth century and a sixth eventually being added. The guitar's rise to popularity in the sixteenth century is often associated with the poet Vincente Espinel, and many similarities can be observed between the events then and the explosion of interest in the electric guitar in the period after the World War II. Consider the following quotation:

> Soon (the Spanish guitar) seemed to be everywhere, rapidly overshadowing the fragile and more complicated lute. Alarmed purists and lute lovers condemned the guitar, trying to give it a bad name by associating it with undignified frolicking in the streets, unrestrained body movements and a general spirit of joyful, sensual abandon. The more the guitar was identified with such pursuits, the more it dominated folk music in Europe. Romantics loved to serenade with it, ladies loved to hear it, painters loved to paint it. Those were grim days for lute makers.
>
> (Tom Wheeler 1981 *The Guitar Book*)

The Spanish guitar had achieved a cult following, and the associated demand led to a degree of freezing in the design development. To be associated with the 'folk culture' the guitar had to be of a certain form – soft feminine curves at the waist, a flat top and a long fretted neck. The first guitar to have six single strings tuned to the present arrangement appeared in 1810 – the Carulli guitar – and from here on advances in design have been a matter of refinement. The guitar's width was increased, the fretboard narrowed and the ornate decoration dropped in favour of the functional appearance now associated with the guitar aesthetic. Performers required greater volumes for larger audiences and in order to achieve this ever thinner sound boards (the front face) were used. Without support the front face would simply collapse under the pressure produced by the tension in the strings and consequently different forms of 'strutting' have been introduced. A modern concert guitar is a very sophisticated and critical structure. Great luthiers (guitar makers) must understand the technology – how the sound is produced and how the forces are carried – in order to make their instruments. It is essential to know that the sound of the treble and base strings will not be balanced unless the sound board is thinner on the treble side, that the sound board vibrates less near the neck, that the sounds produced by the front and back faces must be in phase and that close-grained timbers produce greater volume. There is a great deal of knowledge required in addition to the considerable craft skills.

The events concerning the evolution of the electric guitar since the 1930s echo these earlier happenings to a surprisingly large degree. Guitarists in big bands had always found that the guitar was swamped by other instruments. Steel strings had been used to increase the volume – with consequential reinforcement to the neck (with a steel bolt) and the body – but it was only when the first electromagnetic pickup was developed in 1931 that this situation changed. The 'technology push' led to a revitalising of guitar design, this time as a system, incorporating both the sound production and its

Figure 1.19 The culturally accepted forms of Spanish and electric guitars.

amplification. A period of rapid and varied development occurred between 1930 and the early 1950s primarily associated with two men – Leo Fender and Les Paul.

Although many radical designs were produced, notably by Leo Fender himself, and despite the fact that the guitar body shape has little influence on the tone of the electric guitar, two designs – Leo Fender's 'Stratocaster' and the Les Paul series produced by Gibson – came to dominate the market. Like the Spanish guitar these designs had achieved a cult following after their adoption by popular folk heroes such as rock musicians Jimi Hendrix, Eric Clapton and Jimmy Page. With television to boost the impact, these two designs have come to dominate the perception of the electric guitar concept and it is difficult for designers to break away from these images. Fig. 1.19 indicates the forms of the Fender Stratocaster and the Gibson Les Paul in comparison to the Spanish guitar.

In recent years, many manufacturers have tried different materials, and radical shapes and colours, but guitarists wish to imitate their heroes and this cultural influence has led to a major difficulty in getting the new forms accepted. When competitors wished to enter the market they did so by copying Gibson and Fender designs. Cynics might suggest that this was because they had no ideas of their own but, because of the cultural influence on the guitar market, it was the only realistic course to take. Now that the patent has expired on the Fender Stratocaster it is produced all over the world to a design very like the original, despite recent refinements in the detailed design of the pickup and bridge assemblies for example. They sell in thousands because they look right. Even with the advent of the new synthesiser technology these two forms are still dominating guitar design, despite their complete irrelevance to the sound produced.

1.8 Product analysis and product life cycles

All designers need to learn to see new projects in relation to their predecessors and near competitors. This is not just to avoid reinventing the wheel, but to learn from the work of others who have tried to meet the same or similar human needs. Such 'product analysis' will lead to a deeper understanding of the requirements the product must meet.

1.8.1 Looking back

Fig. 1.20 shows a series of vacuum cleaners which illustrate the way in which the demands on the product have evolved. They have become ever more sophisticated, incorporating more and more functions as well as the changes which have resulted from new materials and technology. The earliest models produced a vacuum by using bellows, initially operated by the foot and then later hand-held. A patent was granted to a French inventor in 1900 for a system where the bellows were worn by the maid as 'shoes'. The Griffith vacuum cleaner of 1905, Fig. 1.20(a), which was operated by two people, and the hand-held Star of 1911, Fig. 1.20(b), followed.

It became clear, however, that the human body was not a very suitable power source.[8] The first powered suction cleaner, made in 1901 by a British engineer, Hubert Booth, was so large that it was carried on a horsedrawn van and required a petrol-driven engine and long suction hoses running through the house windows. It caught the imagination of the public by being used to clean the carpet for Edward VII's coronation in Westminster Abbey.

The first upright electric machine was launched in 1907 by two Americans, J. Murray Spangler and William Hoover, but very few homes had electricity at that time. An early competitor was the Swedish company Electrolux that started production in 1912, specialising in cylinder machines. Hoovers arrived in the UK just after World War I and cost as much as a maid's annual wage. The growing concern with

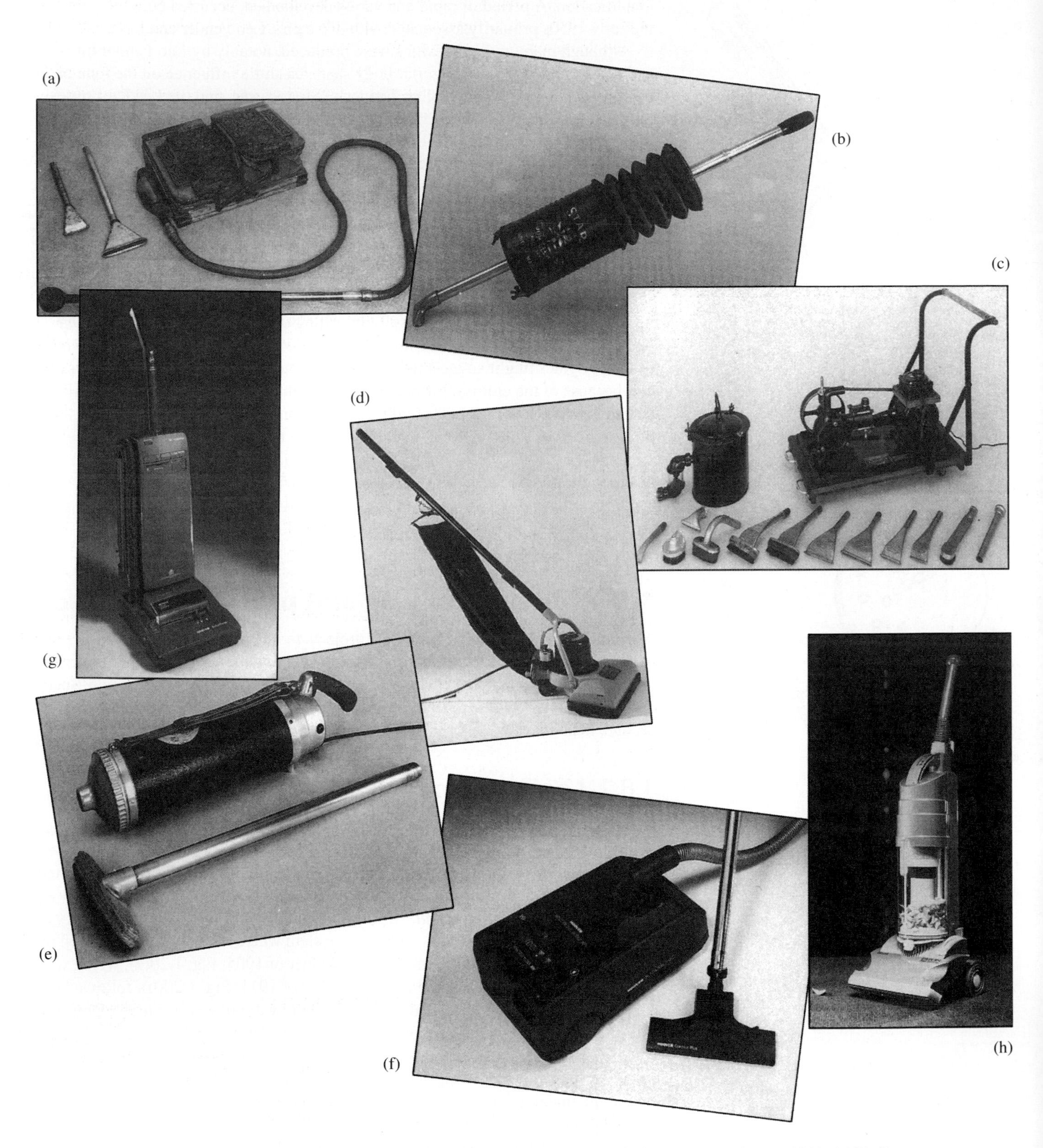

Figure 1.20 The evolution of vacuum cleaners. (a) The Griffith vacuum cleaner, 1905; (b) Star vacuum cleaner, 1911; (c) Trolley vacuum cleaner, 1906; (d) Early upright Hoover, 1920s; (e) Early Electrolux, horizontal cylinder type; (f) Modern cylinder vacuum cleaner; (g) Modern upright vacuum cleaner; (h) Dyson Dual Cyclone Vac.

hygiene and cleanliness (spurred on by the advertisements) and the wider availability of mains electricity stimulated sales of both cylinder and upright models – as did the trend towards fitted carpets after World War II.

From the 1950s the use of plastics and bright colours has led to style changes aimed at minimising the image of cleaning as a household chore. More recent developments have included wet-and-dry cleaners and the increasing popularity of cylinder models.

A notable newcomer is the Dyson Dual Cyclone Vac for which several advantages are claimed. It has no bag and the dust is collected in a bin after passing through a five-stage filtration process. As a result the suction is not reduced as more dust is accumulated. It is also convertible from an upright to a cylinder model. The Cyclone Vac shown in Fig. 1.20 show that the story is still not over.

A detailed study of the history of a product like a vacuum cleaner is relevant to all designers, not only those who are working on new designs for this product. The early designs using humans as a power source might well incorporate insights into designs for human-powered machinery in a more general sense. Physical exercise is now a significant cultural influence in the Western world and there could well be a market resurgence for products like pedal-powered lawnmowers or wood lathes if they were presented in this context. Equally, the change in styling from functional to streamlined, is a lesson for any product designer and can be compared to similar sequences for calculators and other products.

1.8.2 Looking around

Comparisons of similar products can be approached in a number of ways. Table 1.9 shows the basic characteristics (or attributes) of a number of barrows and carts currently on the market. This information was selected from a *Which?* survey conducted in 1984 and a *Sunday Times* survey conducted in 1987.

The order imposed on the information by tabulation makes comparisons easier but much of the subtlety of the differences in the designs is lost. Fig. 1.21 shows a second way in which the information can be organised – a graph of the major characteristic, capacity, against the cost. This is a good approach to considering questions concerning the basic feasibility of a proposal; for example, can a barrow of capacity 120 litres be manufactured for a cost of £20?

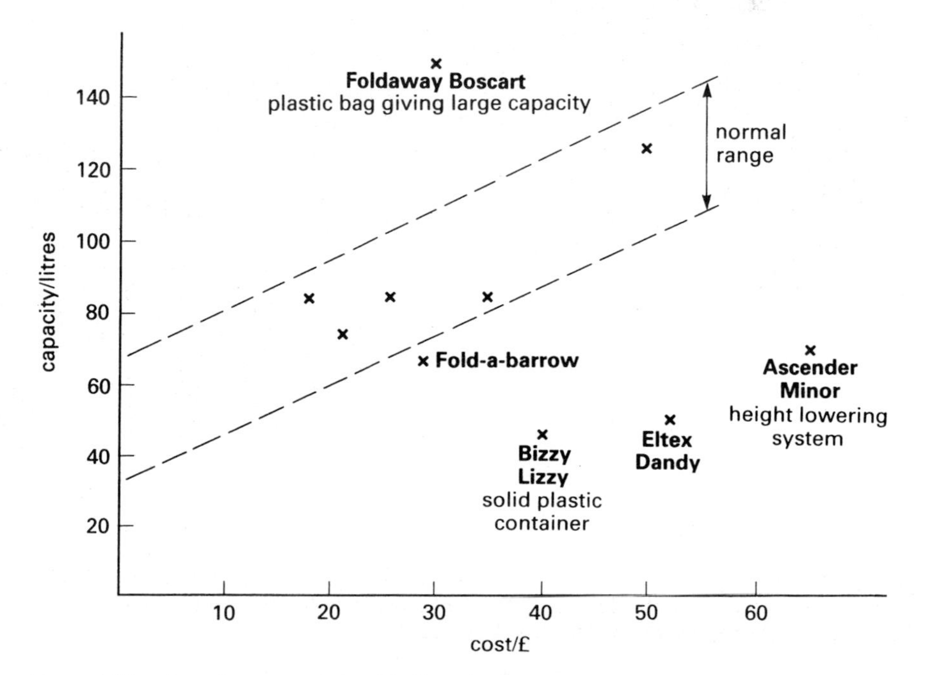

Figure 1.21 Capacity-cost relationship for wheelbarrows.

Make	*Weight (kg)*	*Capacity (l)*	*Cost (£) (approx.)*	*Frame material*	*Container/ material*	*Notes*
Ascender Minor	18	70	65	22 mm dia tubular steel	22 swg galvanised steel	Uses a 70 × 310 mm dia solid rubber tyre. Lowers for sweeping in leaves etc, and for heavy loads
Bizzy Lizzy	8	46	40	20 mm dia tubular steel	plastic	Two wheels 194 × 40 mm dia. Back rear castor
Corrie Easywheeler	8	84	26	tubular steel	galvanised steel	Uses two solid rubber tyres. Single handle to aid manoeuvrability
Corrie Giant	18	126	50	32 mm dia tubular steel	22 swg galvanised steel	Uses a 102 × 405 mm dia pneumatic tyre. Can be fitted with an extension
Corrie Syon	9	74	21	22 mm dia tubular steel	24 swg galvanised steel	55 × 310 mm dia wheel with solid rubber tyre
Eltex Dandy	6.5	50	52	tubular steel	galvanised steel	Pram style handle and uses two solid rubber wheels
Fold-a-barrow	9	67	29	tubular steel	galvanised steel	Pram style handle and folds flat for storage. Retractable stand
Foldaway Boscart	3.5	150	30	15 mm dia tubular steel	polythene sheet	Two wheels, 30 × 140 mm dia. Foldaway design using plastic sheet
Heath Eccles Builders' barrow	20	84	18	32 mm dia tubular	steel steel	Uses a 100 × 400 mm dia pneumatic tyre Traditional builders' barrow
Husbander Ballbarrow	8	84	35	22 mm dia tubular steel	plastic	Large plastic ball instead of wheel. Capacity can be trebled with extension

Table 1.9 Fundamental characteristics of some current barrow designs.

The graph indicates the expected cost range, and provides a good method of establishing realistic targets for the designer to aim at. Much greater costs can only be justified by special design features, such as the lowering mechanism of the Ascender Minor or the injection-moulded body of the Bizzy Lizzy.

When establishing detailed design criteria, it is useful to look at the alternative barrow designs in much more detail. Calculating the weight and the position of the centre of gravity when loaded with, for example, sand would allow the difficulty of tipping to be assessed and the ease of overturning to be compared and quantified (see section 2.4.2 on mathematical modelling). Measuring dimensions such as the container height, to allow the ease of loading to be compared, or the length and height of the handles to consider the lifting position and walking restrictions on the user would provide interesting ergonomic information and a good comparison for analysis based on anthropometric data (see section 2.4.1). Looking at the methods of manufacture to assess the easiest routes, the materials to assess the likely product life and the wheel- or ball-mountings to find simple, effective approaches would also yield useful information. The products might also be assessed in use – looking for safety hazards, sharp edges, handles which slip off or dangerous instability. You might even ask people with very different physiques to use the barrows and comment on them, if some barrows are available for testing. Good designs are normally produced by people who have got thoroughly involved with a product and are sufficiently inquisitive to investigate every possible factor they can imagine.

1.8.3 Seeing the context

A further way of thinking about product analysis is to consider the concept of product life cycles. Often the primary need does not change, but it is met in different ways. The changes in approach can be thought of as a series of overlapping life cycles. Fig. 1.22 shows the four stages normally identified in the life cycle of a product and the associated expected profitability. These stages – introduction, growth, maturity and decline – represent the general pattern which can be expected to occur, but the shape of the curve and the scales will depend very much on the specific product.

Introduction

Whether a product is satisfying an old need in a fresh way or meeting a newly identified requirement, there will be a period during which knowledge of the new product will be spreading. Potential users or customers will begin to evaluate the product's potential in relation to the alternatives they know, and competing manufacturers will begin to determine their responses. In the case of about two-thirds of new products, sales will eventually begin to increase rapidly and the growth phase will be entered. The other third will not be accepted by the market.

In the 1980s Sir Clive Sinclair was associated with products meeting both fates – the ZX computer series was a phenomenal success with very rapid acceptance in the market place, but the C5 electric vehicle failed dramatically. In the 1990s there is renewed interest in electric vehicles and in an experiment in Bordeaux a network of charging points has been established to enable battery-powered cars to be evaluated.

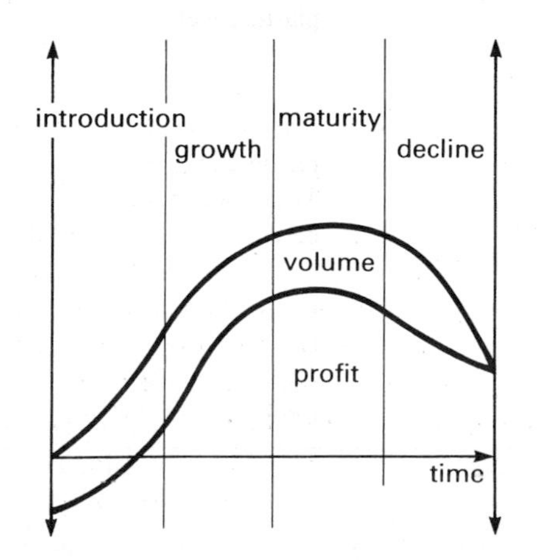

Figure 1.22 Product life cycles (Ward 1984).

Growth

As the sales increase rapidly, the product's life cycle enters its most dynamic phase. Opportunities for new entrants to the market, perhaps offering new features or combinations of features, or just direct imitations with little product differentiation, abound. During this period there is scope for experimentation with the product form and characteristics, because demand is high and there is a correspondingly large margin of error. The recent history of the home computer market demonstrated that machines with very different technical specifications can perform equally well in the market place, and it is only as the product category reaches maturity that some kind of a shakeout begins.

Maturity

Sooner or later, the period of rapid expansion will end. There are a number of possible reasons for this – for example, there are no further new customers to be found, other competing products have been introduced taking part of the market, or the product has become dated and is no longer able to attract new users. How long this mature phase lasts is very much dependent on the extent to which the product is influenced by fashion – skateboards and miniskirts were very short-lived although miniskirts enjoyed renewed popularity some years later but sports cycles and wedding gowns have had a much longer life.

Decline

The decline of a product category is normally a result of its being superseded by other products meeting the same or similar requirements. The decline of the fountain pen has been associated with the rise to maturity of the market for ball points, the growth in the market for felt tips and the introduction of erasable pens. Similarly, the decline in record sales can be viewed against the maturity of the market for audio tapes, the growth in the market for video tapes and the introduction of compact discs (Fig. 1.23). The product category and primary need for all of these are concerned with home entertainment, and expenditure on newer forms results in a decline for more traditional products.

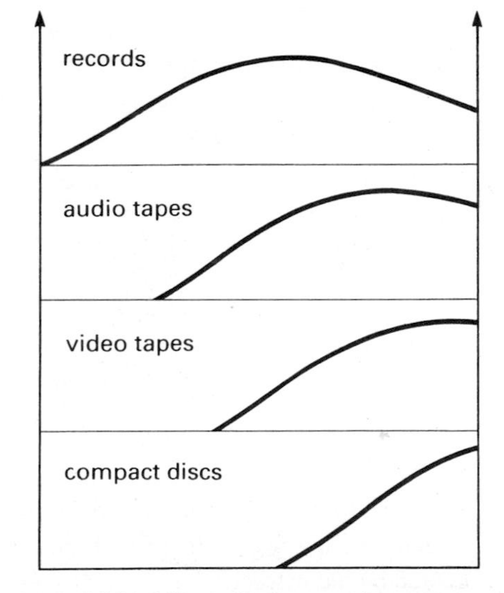

Figure 1.23 Life cycles for a product category.

It is also worth considering the financial returns which a company can expect to receive. Clearly during the development phase the company will be investing money which must appear as a 'loss', and this loss generally continues as a result of tooling-up, advertising and promotion costs until the market begins to grow. If a company is to have a long-term future it must make sufficient profit in the good years of product growth and maturity to cover the development and introduction costs of the product's successor when decline occurs. As product life cycles are generally now complete within a few years a company must continually innovate and develop new designs in order to survive.

The issues associated with advertising are also best discussed in relation to product life cycles. During the introduction phase, investment in advertising is essential to spread information about the product. As the market grows and matures, the strategy must change. In a mature market, the only way to gain extra sales is to take them from someone else, so advertising must reflect the inevitable competition and comparison of products. When decline sets in, some companies will drop out; this period can be quite a profitable one for the survivors, but large marketing and advertising budgets become superfluous.

1.9 The role of the designer

There are four parties to the product development process – the client, the designer, the maker and the user. The client sets the objectives for the project, or provides the brief. The designer must work to transform the brief from a product idea to a sufficiently detailed specific product proposal to allow the maker to carry out the next stage and deliver the product to the user. The possible ways in which the roles may be divided between one or more individuals or groups of people are illustrated in Fig. 1.24.

The four cases are presented in the order in which they have evolved historically, but all of them are still relevant.

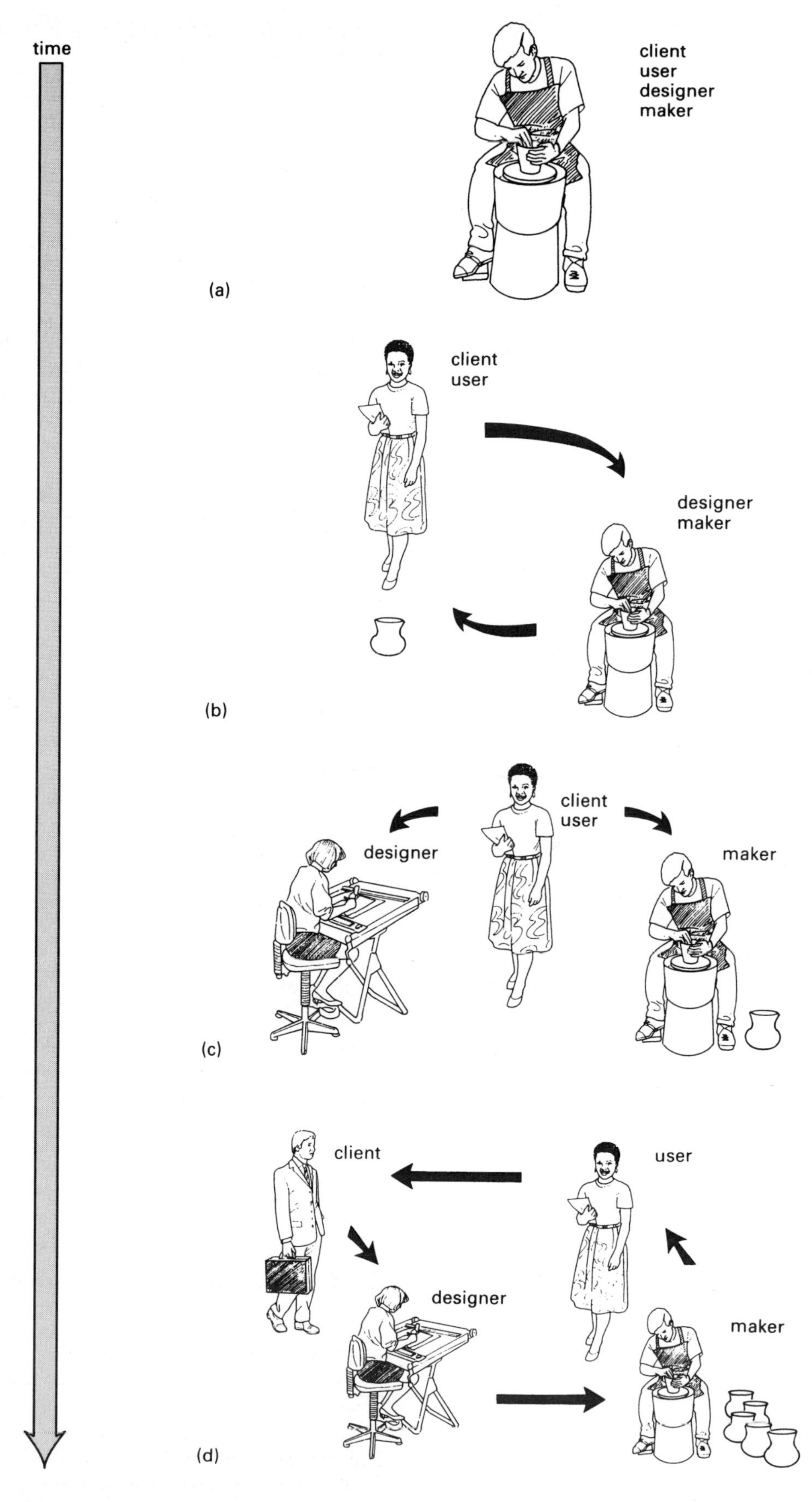

Figure 1.24 The changing design process.

1.9.1 Self-sufficiency (Fig. 1.24(a))

Here all four roles are combined in a single individual. Originally this applied to people in pre-industrial society designing and making products for their own use. Modern examples include:

- research workers who are involved in the design and development of specialist prototype equipment,
- sportsmen and outdoor pursuits enthusiasts evaluating and improving the performance of their hardware as it approaches its existing limits,
- DIY enthusiasts (although their work is often directed towards their relatives and friends).

With self-sufficiency, a single individual is providing the brief and the necessary resources, evolving a product specification and designing to meet it, manufacturing the product and evaluating the outcome. As the only person involved in the process, they do not need to tell anyone else what they want or what decisions they have made – the only records needed are for their own use, to back up their memory.

1.9.2 Designer-makers (Fig. 1.24(b))

In this case the roles of client and user are combined in one individual, and those of designer and maker in a second. It originally applied to rich patrons commissioning products from designer craftsmen. Modern 'patrons', who come from many sectors of society, are looking for uniqueness and individualism, or for solutions to problems which are only partially met by mass-manufactured goods. They include:

- professional musicians and cyclists who have their instruments and cycles specially constructed,
- people buying motor caravans with custom-built interiors laid out to meet their own specific family circumstances.

There are often good reasons for the roles of the designer and maker to be combined in these circumstances. Good design may be dependent on a detailed knowledge of specialist materials or user techniques, and perhaps legal constraints. The designer-maker may need to be something of an enthusiast.

1.9.3 Specialist designers (Fig. 1.24(c))

The separation between the roles of the designer and maker has been a natural development within an industrial community founded on the division of labour. It allows individuals to acquire specific knowledge and skills in complex areas such as:

- negotiation with the client to evolve a design specification,
- the relevant legislation and regulations in home and export markets,
- the market trends in relation to specific products,
- the capability of specialist manufacturing equipment,
- computer-controlled manufacture and flexible manufacturing systems.

The dangers in separating the roles are that designers may propose designs that cannot be economically produced, or production engineers may impose excessive limits on the product form to simplify the use of the machinery. Success in product development now depends on good liaison and communication between the participating specialists.

This model of the design process applies to:

- the ordering of equipment by major companies and government agencies,
- architects employed in the construction and restoration of buildings,
- the work of other design consultants.

1.9.4 Mass production (Fig. 1.24(d))

The division of the roles of the client and user relates to the organisation of a society based on mass manufacture. Economies of scale are made possible by large production quantities, aimed at a large population of users. Companies are organised to produce goods and distribute them to the users. Some individuals within the company act as clients, providing the design brief as a result of formal market research or an insight believed to represent market trends. The designers may be on the company staff or working for a design consultancy; similarly the manufacturing might be undertaken by the company itself or subcontracted to another company. The efficient operation of the total system depends on accurate communication of information between those performing the four roles. The greater specialisation provides greater opportunity for the handling of complexity, but equally there is more likelihood of the operation faltering. Communication problems now make it possible that:

- the client may become out of touch with the user's requirements,
- the designer may fail to interpret the brief accurately and in line with the market developments,
- the designer may remain unaware of current manufacturing technology,
- the manufacturers may be unable to meet the product specifications and deadlines.

It follows that communications and information technology lie at the heart of modern industrial society.

These models of the possible relationships between the client, designer, maker and user show that the designer's role is to interpret the client's brief and provide a definition of the product specification to be passed to the maker, but their activities and influence extend beyond these boundaries. If clients are very sure of their requirements, the designer's task is known. When clients have little more than a general awareness of an area of difficulty the designer must start by conducting a search, through negotiations with the client, to develop a more precise statement of the need; this will require good communication and analytical skills.

Similar skills will be needed to communicate with the maker, to explain the final design and discuss the manufacturing route to be taken. The design function is a link role between the project management and the manufacturers. Unless they work for themselves, designers must be good at communicating through a variety of media: informal and formal drawings, written and spoken language, project presentations and displays.

The Design Council has produced a series of videos and handbooks describing the nature of design and the work of particular designers. These are an excellent way to gain a perception of the designer's role in a variety of fields and some of those available are listed following the Bibliography.

1.10 Price and non-price factors and 'total' design

The commercial success of a product depends partly on 'price' factors such as its production cost and selling price, and partly on 'non-price' factors such as its quality and the number of unique features it offers. The view that products will be successful if their cost to the customers can be brought below that of their competitors' is only partly true. Really successful products cannot be developed simply by always opting for the cheapest alternative; the problem facing the designer is more complicated than that. It can be illustrated in two different ways, which look at the same idea from different angles.

The first approach is a comparison of the significance of price and non-price factors. Fig. 1.25 shows the result of an analysis of US–West German trade carried out in 1971 by two economists, Kravis and Lipsey.

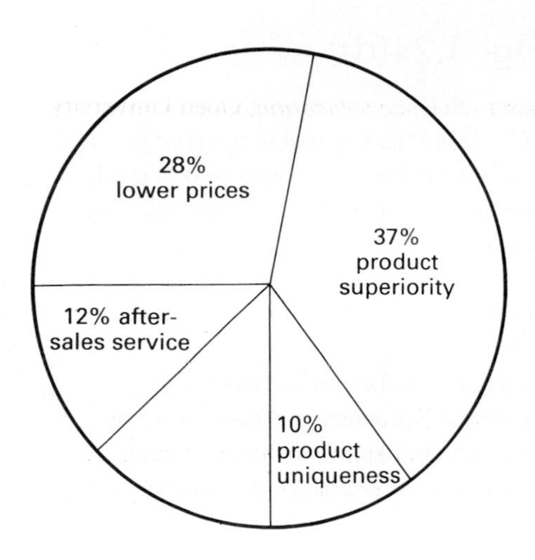

Figure 1.25 Factors affecting US–West German trade (Rothwell *et al* 1983).

It can be seen that non-price factors such as product superiority and uniqueness account for around 50% of a product's success – a figure which has been confirmed by several studies discussed in a report prepared for the Design Council (Rothwell, 1983). Being cheapest is no guarantee of customer approval because the product may not meet the user's requirements sufficiently well.

The second illustration considers the demands of a 'total design' approach. A designer must, of course, be cost-conscious, but the decisions which must be made are more subtle in nature than merely looking for the cheapest alternative – it is a matter of compromise. Fig. 1.26 shows 'Pugh's plates' by which Professor Stuart Pugh tried to express the designer's problem.

If you imagine a circus performer trying to keep all the plates spinning on poles and each one representing a factor the designer must take into account, then the designer must give them all the necessary attention to keep them aloft. If one plate crashes then the designer has failed. Clearly the designer needs to be aware of the priority which must be given to any one factor (or plate) at a given moment, and this will vary at different stages of the design process and for different kinds of product. The importance of such a 'total design' philosophy is that it does not allow undue attention to be given to cost, aesthetics, ergonomics, performance or any other factor.

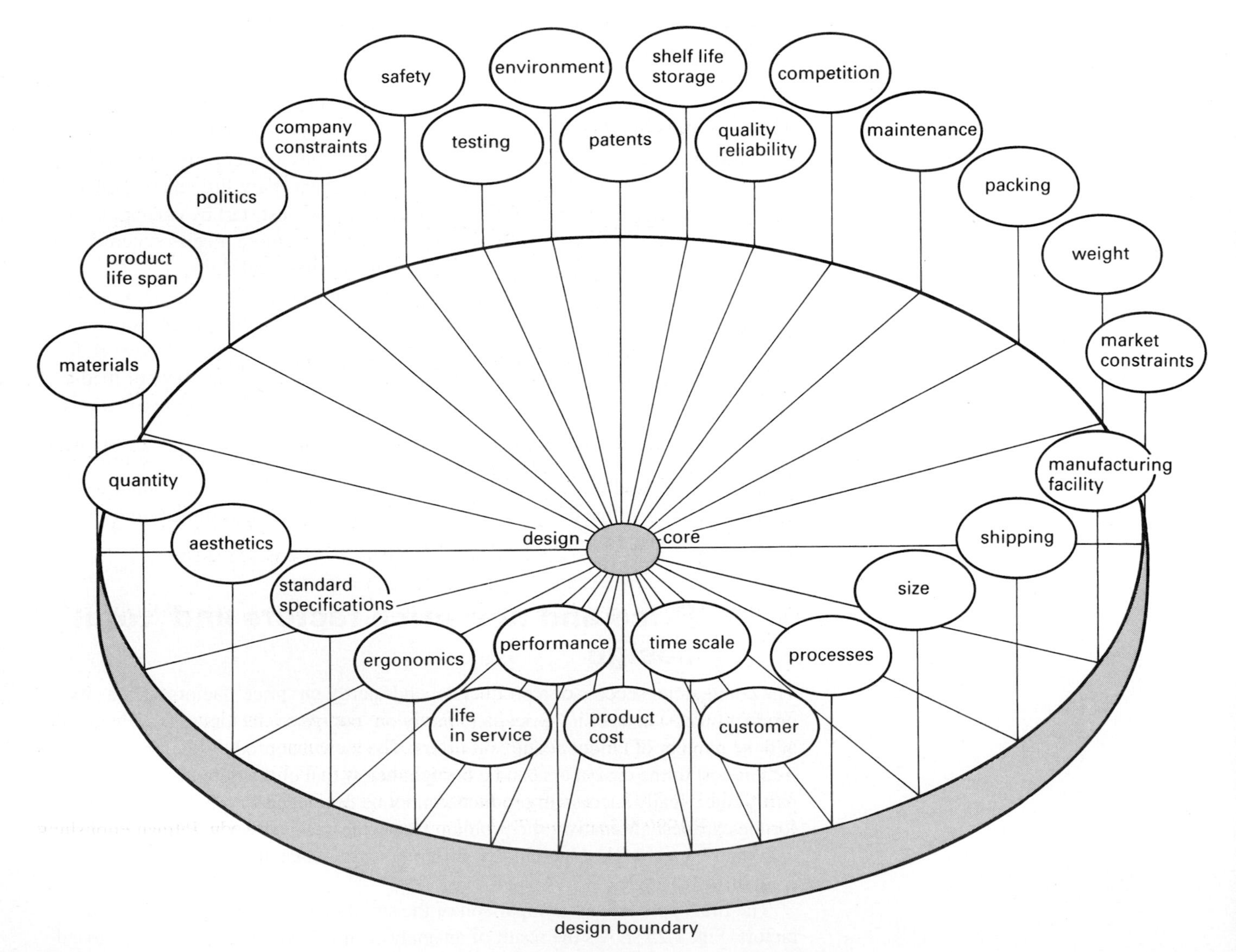

Figure 1.26 Pugh's plates – the elements of a design specification (a guide to design for production).

Notes

1. Layton D 1993 *Technology's Challenge to Science Education*. Open University Press
2. Quarrie J, ed. *Earth Summit 1992* The Regency Press
3. White papers on post-Rio developments HMSO 1994.
   Sustainable forestry
   Bio-diversity
   Climate change
   Sustainable development
4. United Nations NGLS E & D File Vol. ii, No. 15, December 1993
5. Kate Langley *An Assessment of the Impact of Environmental Legislation on Product Design, with Special Reference to British Product Design*. BA in Industrial Design and Technology dissertation, Loughborough University of Technology 1994
6. MORI Lex report on motoring 1994
7. *The British Computer Society Schools Committee Glossary of Computing Terms* 6th edn Cambridge University Press
8. Fay Sweet 'Clean-up operation' *Designweek* 15 April 1994

Bibliography

Bronowski J 1973 *The Ascent of Man*. British Broadcasting Corporation
Heskett J 1980 *Industrial Design*. Thames and Hudson
Mackenzie D 1991 *Green Design: Design for the Environment*. Laurence King
Morrison J and Twyford J 1994 *Design Capability and Awareness*. Longman
Myers N, ed. 1985 *The Gaia Atlas of Planet Management*. Pan Books
Packard V 1960 *The Waste Makers*. Penguin (Pelican)
Packard V 1978 *The People Shapers*. Futura Publications
Papanek V 1984 *Design for the Real World*. Thames and Hudson
Papanek V 1995 *The Green Imperative: Ecology and Ethics in Design and Architecture*. Thames and Hudson
Potter N 1989 *What is a Designer: Things, Places, Messages* (3rd edition). Hyphen
Pye D 1983 *The Nature and Aesthetics of Design*. Herbert Press
Rothwell R *et al* 1983 *Design and the Economy*. The Design Council
Roy R and Wield D 1986 *Product Design and Technological Innovation – A Reader*. Open University Press (Open University T362 course reader)
Royal Designers on Design, 1986. The Design Council
Sparke P 1986 *An Introduction to Design and Culture in the Twentieth Century*. Allen and Unwin
Toffler A 1970 *Future Shock*. Bodley Head
UN Earth Summit '92. Regency Press
Walker D and Cross N 1983 *An Introduction to Design*. Open University Press
Ward J 1984 *Profitable Product Management*. Heinemann
Wheeler T 1981 *The Guitar Book*. Macdonald Futura Publishers
Whitfield P R 1975 *Creativity in Industry*. Penguin (Pelican)
Zorkoczy P 1990 *Information Technology, an Introduction* 3rd edn. Pitman Publishing

Design videos, slidepacks and books

The former Design Council publications are now available from Michael Benn and Associates, PO Box 10, Wetherby, West Yorks, LS23 7EL.

Video multipacks with supporting booklets

Living by Design 1
Grundy and Northedge, Graphic designers
Christina Shannon, Textile designer
Nick Butler, Product designer
Paul Atkinson, Furniture designer

Living by Design 2
Katharine Wykes, Ergonomist
Luck Rowbotham, Optical engineer
Simon Holyfield and Mike Hodson, Engineering designers
Materials Matter
Brilliant or What? Making Design Work
Quiche, Keys and Roller Skates Contexts for design
Design is ...
Drawing for Engineers
Designers Talking

Slidepacks

Techniques for Sketching
Three-Dimensional Design for Schools
New Pillar Box for the Post Office
Product Analysis and Evaluation
Understanding Taste
Geraldine Clark Knitwear Designer

Books

Designing exhibitions
Lighting Design, Carl Gardner and Barry Hannaford
Computer Graphics, John Vince
Engineering Design: A Systematic Approach, G Pahl and W Beitz
Issues in Design series:
Style in Product Design
Green Design
Colour in Industrial Design
Packaging Design
Critial Paths

Assignments

1 Discuss the responsibilities of a designer of consumer products to society as a whole.

2 The dictionary defines technology as 'the total knowledge and skills available to any human society for industry, art, science, etc.' Within this prescription, develop your own ideas and thoughts on the subject of technology and its particular impact on your life and future well-being.

3 Developments in technology have had an inevitable impact on product design. Describe one historical and one current example where technological developments have had a major impact on product design.

4 Most major car manufacturers are currently carrying out research into electric cars. Discuss the potential benefits of an electric car to
(i) the manufacturer,
(ii) the customer,
(iii) the environment.

5 Review the present use of energy at a national level for the United Kingdom and detail how our performance in the use of energy as a nation might be improved.

6 Discuss the ways in which designers can ensure that energy is used efficiently in the following fields:
(a) private motor transport;
(b) school buildings.

7 Clearly indicate what you understand by the term green issues and explain why you consider this has become more important to us at the present time.

8 (a) Discuss the ways in which environmental problems may be caused by:
either, (i) motor vehicles;
or, (ii) heavy industry.
(b) (i) In the area you have chosen in (a), suggest ways in which designers can help to limit environmental damage.
(ii) Explain how legislation may assist designers in this task.

9 The use of coal-fired power stations for the production of electricity is a matter of increasing environmental concern.
(a) Why is this?
(b) What measures are currently taken to minimise energy losses or reduce consumption:
(i) at the point of generation;
(ii) during transmission;
(iii) in domestic appliances?

10 'Green' issues are becoming increasingly important in contributing to the commercial success of many consumer products.
To what extent do you think a concern for conservation is a passing trend or a fundamental criterion for a designer to consider for all products?
In your answer you should refer in detail to two dissimilar products.

11 Re-use of products can sometimes be a more effective way of conserving resources than recycling materials. Show how the design of products can increase the likelihood of their re-use.

12 (a) What do you understand by renewable sources of energy? Describe a project known to you in which a renewable energy source or sources has been made an economic reality.
(b) Apart from preserving natural resources what other reasons are there for recycling materials? Use examples to illustrate your answer.

13 With reference to your own experience of designing and making, explain how and why materials and production methods might need to be changed if your designs were to be batch or mass produced, rather than produced as one-off items.

14 (a) What is meant by 'batch production'?
(b) Describe examples of the appropriate use of batch production and explain why it is appropriate to the situations described.

15 The retail price of a product is, to a considerable degree, determined by costs incurred at various stages.
Discuss, with reference to specific examples, the ways in which aspects of each of the following activities may influence the retail price of a product.
(a) Designing.
(b) Research and development.
(c) Selection and procurement of materials.
(d) Manufacturing.
(e) Distribution and selling.

16 The costs involved in producing an article are many and varied. These costs fall into two categories:
(i) fixed costs;
(ii) variable costs.
(a) Explain, in general terms, the distinction between fixed costs and variable costs.
(b) With the help of suitable graphic techniques, show how the relevant cost-elements will vary in the production of:
(i) a volume-produced torch;
(ii) a piece of hand-made ceramic ware.

17 Discuss four specific needs of an elderly and infirm person living alone and give examples of how these might be met through the application of design and technology.

18 Discuss the case for designing some products to have a limited life. What are the possible consequences of planned obsolescence?

19 Advertising is a major business involving large expenditure to convey messages about products in order to increase the sales. For each of the forms of media set out below, describe using the examples the advantages and disadvantages of each form.
(a) Television
(b) Newspapers
(c) Leaflets distributed through direct mailing
(d) Radio
(e) Perimeter advertising at sporting events.

20 Whilst it is often said that a designer's role is to satisfy needs, others have argued that it is to promote dissatisfaction and hence create the desire for new products.
Discuss these two views of the role of a designer.

2 Designing

> Things have never come in a flash: they come only as a result of months, even years, of very heavy work.
>
> (Barnes Wallis)

2.1 Design as a systematic activity

The expectation of being able to leap from your bath one day crying 'Eureka! I've got it!' is quite realistic if you have put in weeks, months or even years of patient endeavour; quite unrealistic if the necessary groundwork has not been done. 'Chance favours only the prepared minds' was Louis Pasteur's view of the problem of scientific research, and perhaps something of the same feeling lies behind one of the definitions of design offered by J. Christopher Jones: 'The performing of a very complicated act of faith.' In order to clarify what is meant by design, consider the following passage written by Sir Monty Finniston for a conference address in 1987.

> In my definition *design is the conceptual process by which some functional requirement of people individually or en masse is satisfied through the use of a product or of a system which derives from the physical translation of the concept.* As examples of individual products which satisfy a public or a market need there is the motor car, the television set and the radio, the fridge and the dishwasher, shoes and socks and baby nappies but also the painting, the sculpture, the musical score and the other manifold realised expressions of the artist etc.: and as to systems there is the telephone and the railway, the motorway and the supermarket, the orchestra, the provision of utilities (gas, water and electricity) and so on.[1]

This quotation emphasises that designing concerns the conceptual processes which bring products or systems into being, an idea which is put more succinctly in another definition from J. Christopher Jones: 'The effect of designing is to initiate change in man-made things.' In your course of study you are required to go further than just initiating change – you must go on to bring it about, that is to realise your design, and then evaluate the outcome. Graphical models of designing are often used to try to clarify the stages and activities involved. Two models are shown here – firstly, one of the creative process in design and secondly, one of the total design activity.

In his book *How Designers Think* Brian Lawson summarises several descriptions and theories of the creative process and identifies a set of five stages which are common to all of them. These stages are shown in Fig. 2.1.

The process begins with a conscious recognition that a problem exists and the growth of the determination to tackle it (stage 1). A period of preparation (stage 2) must now follow, during which the designer must strive to understand the problem. As a means of deepening their understanding and continuing the process of immersing themselves in the problem designers often sketch tentative solutions, but the insight that will result in the final solution normally requires a period of incubation (stage 3). The insight (stage 4) comes from the subconscious mind. Exactly how this happens is not well understood, but it seems that time to absorb a problem is important. Consider the following remarks made by Gordon L. Glegg in *The Design of the Designer.*

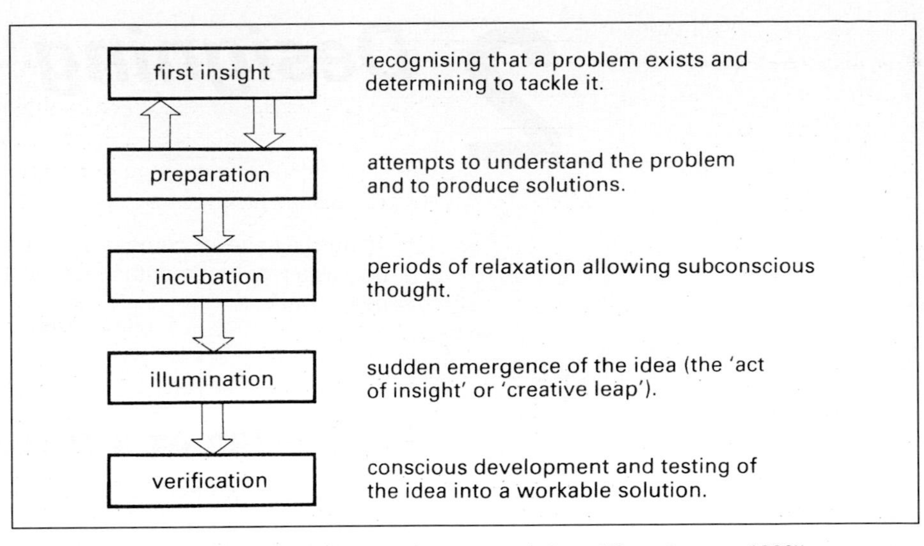

Figure 2.1 Five-stage model of the creative process (adapted from Lawson 1980).

> It is also important to realise that our subconscious minds will hand up their suggestions in the form of symbols or pictures. The subconscious has no vocabulary. To encourage communication between the conscious and the subconscious, we should practise their only common language, which is three-dimensional pictures. That is why all engineers should learn to do three-dimensional sketches.
>
> (Glegg 1969)

During these early stages the importance of recording ideas and communicating them as three-dimensional sketches should not be underestimated. This was, of course, the practice of Leonardo da Vinci as long ago as the fifteenth century. In his notebooks, Leonardo recorded information about machines and the human body as sketches. He also often switched from the written word to sketches and back again in explaining his ideas. Some things are better drawn.

Once the insight has occurred, a period of conscious development and testing of the idea (stage 5) follows until a workable solution has been found. This process of conscious development requires the modelling of the solution, using mathematical, graphical, physical and /or computer models as necessary or appropriate. The modelling process eventually results in the synthesis of all the streams of thought which have arisen into the best available solution. This stage is often seen as the end of the creative design process, but it is important to realise that design activities still continue, albeit at a lower level. As the chosen solution is worked out in detail, design decisions – for example, whether to facilitate manufacture by employing a special process or to use standard components – will be made. Even after manufacturing begins it may be necessary to make small alterations, perhaps in response to particular material characteristics: craftsmen in wood are, of course, not only able to anticipate and exploit the properties of the material they are using, but are very used to responding to the material as they work it.

Fig. 2.2 shows a model of the total design and development activity. There is a central *design core* consisting of the key phases of investigation, generating ideas, synthesis, manufacture and evaluation. These activities are surrounded by the *design boundary*, which encompasses all the factors which have to be taken into account – illustrated by Pugh's plates (section 1.10). These are also represented by the *product design specification* (PDS), formulated during the investigation phase, which details the characteristics of a solution (see section 2.2.4). Aspects of the PDS may change or evolve as designing progresses and more is learned about the problem, but it represents a starting point. It should be realised that if the PDS changes, then the design boundary has moved and a review of all design decisions taken on previous assumptions should be undertaken. Such reviews are, of course, only possible if adequate records of the design evolution have been kept on design sheets or in a logbook. This model shows the process of meeting the user need completed by the manufacture and evaluation of a working model. It is this total activity which represents the task you face.

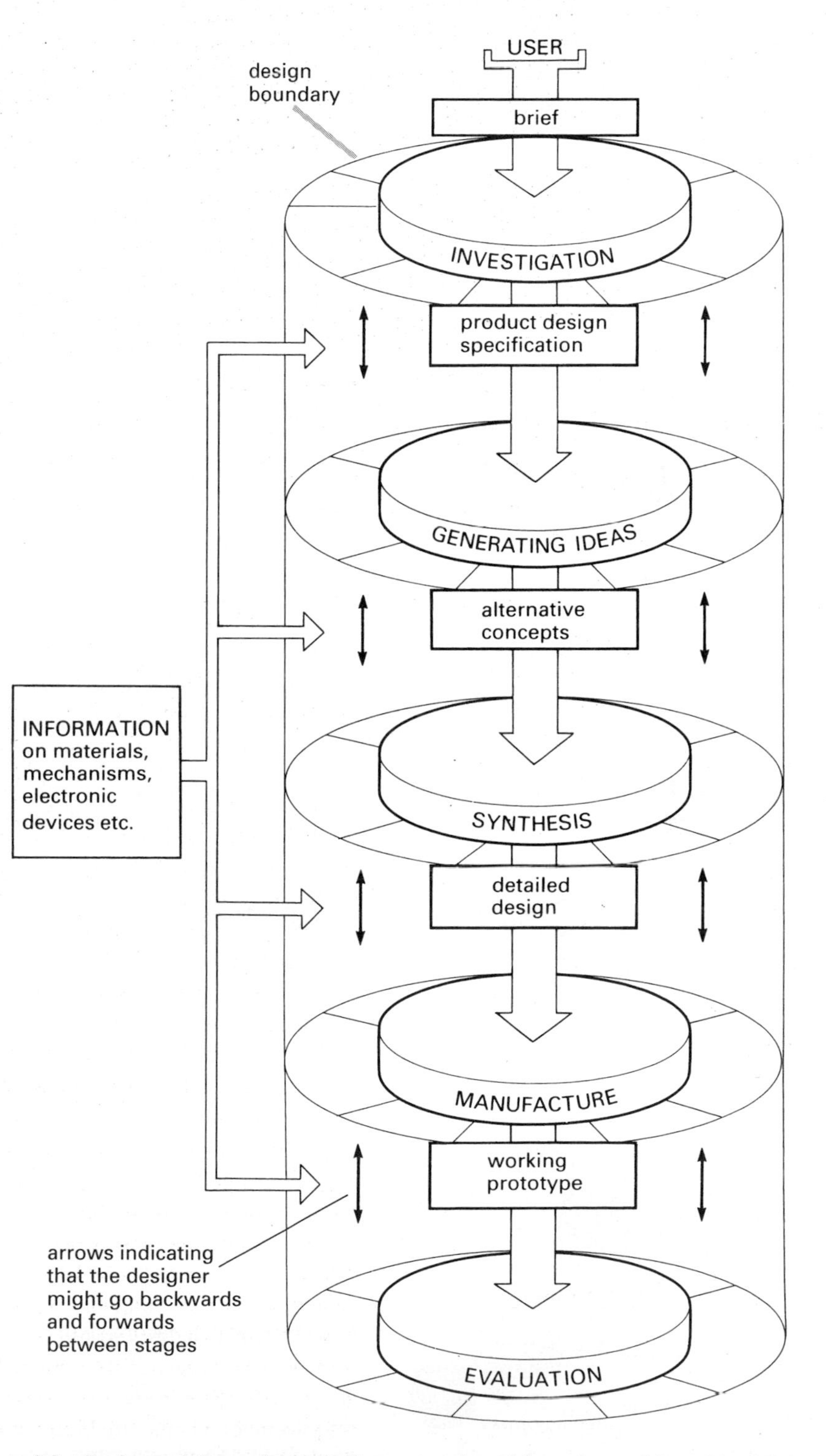

Figure 2.2 Design activity model (overall concept adapted from the model used by SEED in their Curriculum for Design publications).

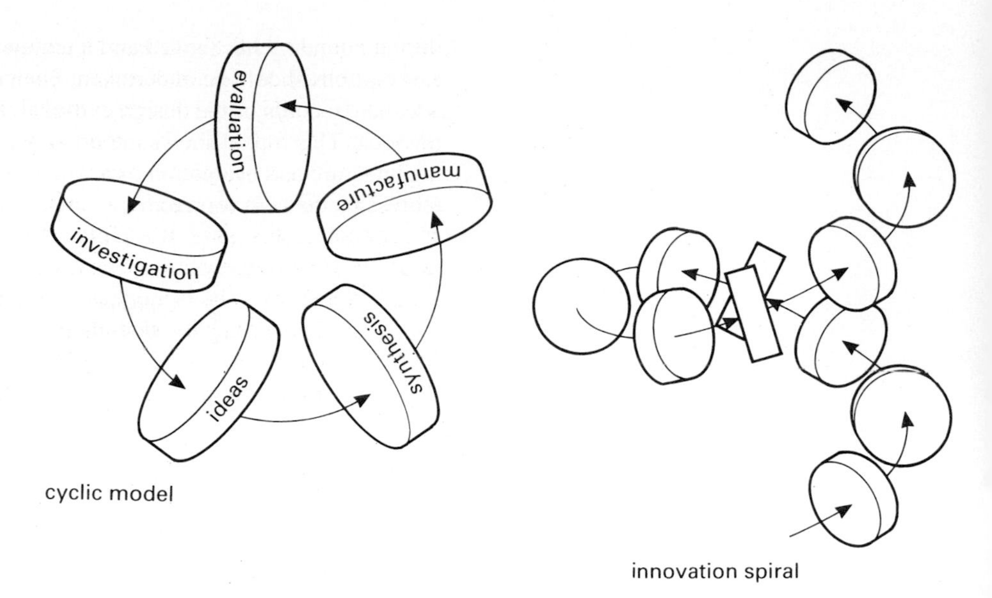

Figure 2.3 Other forms of design activity model.

Models like that shown in Fig. 2.2 are often dismissed as being linear, with a start and an end, whereas the product design process has feedback and is continuous. What is learned in the design and development process, and the final evaluation, might well lead to a new specification, just as an analysis of the deficiencies of an existing product might. For this reason models of the design and development process are often shown as circles or spirals, as illustrated in Fig. 2.3.

The model shown in Fig. 2.2 is perhaps best imagined as the circle 'straightened out', or one segment of an innovation spiral. Of course, the designer might go backwards and forwards between stages. For a particular designer working on a specific project, the design process can be seen as beginning with the design brief and ending with the evaluation of the proposed solution.

The convergent nature of the design process is shown in Fig. 2.4. The first nebulous idea evolves in the client's mind into an ill-defined problem. This is presented to the designer as a brief and the design process produces a well-defined solution, which eventually results in a specific object. The development or existence of this object might well spark off more nebulous ideas, but that is another story.

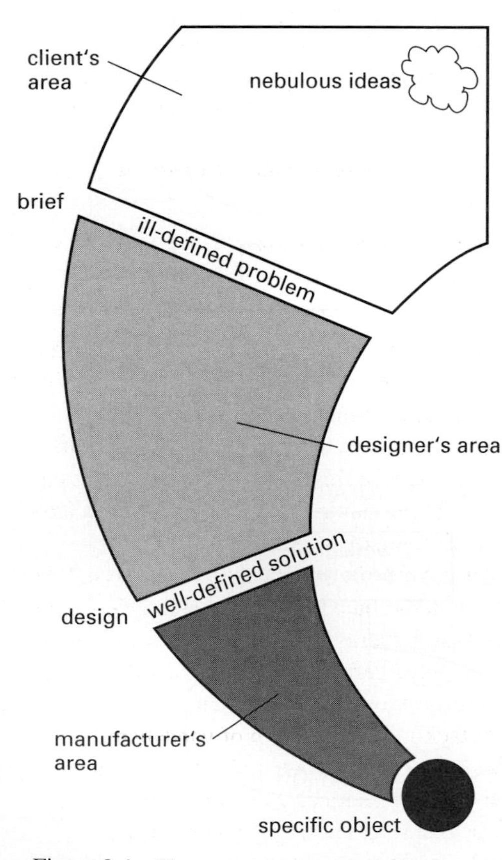

Figure 2.4 The convergent nature of the design process (Cross 1983).

2.2 Investigation

The designer's first task is to explore the design brief provided by the client. In some cases it may give a very clear picture of the requirements, but in others it will be necessary to help clients develop a full understanding of their needs. Briefs should not be regarded as tablets of stone; they may alter as designers begin their research and communicate their findings. A concept which many designers find helpful in analysing a design brief is that of primary and secondary functions. It is sometimes possible to draw up a list of functions that the product must fulfil, and others that would enhance its value but are not essential and could be sacrificed. When the PDS is written this analysis would be reflected in the weighting or priority given to a particular aspect when designing. A possible list of the primary and secondary functions is given in relation to the second design brief discussed below.

Information is the life blood of design activity, and never more so than at the beginning. The types and levels of information which are required change as the process progresses, beginning with broad concepts such as form, energy and control and moving on to the detailed implications of, for example, the use of particular manufacturing processes and precise anthropometric data. As it is essential at the start of designing, information retrieval is discussed as part of the investigation phase, but it should be understood to be ongoing.

The investigation phase ends when the designer and client have agreed on the PDS. This sets out all the characteristics of the desired solution as they are understood at this point. It is a place to start from, but as the design progresses and understanding deepens it may be necessary to alter the PDS. Producing a solution to a PDS you know to be inadequate might be justified if a deadline means that a poor solution is better than no solution, but other circumstances that could justify this are hard to imagine. Starting with a PDS that you are likely to alter may seem a major difficulty, but it is more of a problem in theory than in practice.

2.2.1 Design briefs

Design briefs must give sufficient information to put the problem in context and to indicate the requirements, but they must not impose unnecessary constraints on the solutions which might be proposed. Consider the following brief.

Bottle disposal and recycling

A large supermarket chain, which is very careful to preserve its image of social responsibility, currently operates a glass bottle disposal and recycling system based on the use of three large bins of different colours. Users are intended to place green, brown and colourless bottles in the appropriate bin. The company does not wish to abandon the service, but it is currently proving uneconomic for a number of reasons.

1. *Users occasionally put bottles into the wrong bin, which necessitates screening of the bottles before smashing.*
2. *The bins fill very rapidly with the bulky bottles which necessitates frequent exchange.*
3. *Vandals are attracted if there is any delay in emptying the bins and they have a ready supply of missiles.*

The directors feel that the system needs to be reviewed to simplify sorting, reduce storage problems and reduce abuse.

This brief gives a context to the problem and explains the requirements in broad terms. There is no implied solution to the difficulties of bottle disposal and recycling, merely an indication of the total system's deficiencies. The problem of sorting could be tackled at two different places - the point where the users dispose of the bottles, or in the factory. It could also be tackled in different ways: a small coin might be given as a reward for a number of bottles in the correct bin (providing the colour can be detected and recognised), or the bottles could be sorted, manually or mechanically, on the factory's conveyor belts. The storage problems would be significantly reduced (and the sorting problems would be multiplied) if the bottles could be smashed before they enter the bin, but could this be done safely? Could 'user energy' be employed or would it be necessary to provide an independent power supply? Perhaps the bottle smashing could be turned into an entertainment and act as a reward for using the system.

There are also, of course, many ways of tackling the problem of vandalism. Increasing security is always an option, but so is changing people's attitudes and involvement with the system. Product abuse can be as much a symptom of poor design as it is of disturbed minds. The bottle disposal and recycling brief is an invitation to explore, with the clients, a total system in order to improve its performance. A successful system would help to conserve the Earth's energy resources, but if it is to succeed the system must be acceptable to the users. It is particularly important that the users should not be inconvenienced in any sense; they should preferably be given an incentive at the point of use, in view of the voluntary nature of the system. It is also worth observing that a system which is economically successful, but consumes more energy than it saves, is unlikely to benefit the community. This is a difficult equation because of the cost and energy consumption inherent in waste disposal.

Now consider a second brief, which again looks to improve the performance of a system:

Portable drawing equipment

It has been common practice in schools to equip specific rooms as graphic studios. This leads to considerable inflexibility in the ways in which drawing experiences can be integrated with other aspects of Design and Technology and also to timetabling difficulties. The Head of the Design and Technology department has decided to tackle this problem by constructing a portable unit to house 20 of the new-style Rotring A3 boards and draughting heads. The unit should be secure, house all the necessary accessories and be easily transportable within the department. As it is to form part of a Design and Technology department's equipment it is essential that the design sets a good example in all respects.

Here there is a more specific requirement for a particular artefact. Clearly there will be a need for the design of specific items as a result of the review of the bottle disposal and recycling system, but there is a difference in emphasis. It is likely that one of the first questions to enter your mind when you read the brief for the portable drawing equipment was 'Why Rotring?'. This kind of reaction to an apparently unnecessary constraint is exactly the right kind of response. Designers must not just accept such statements, but set out to establish and search the boundaries of a brief. This brief is explored further in section 2.2.3, but an indication of what might be included in the list of primary and secondary functions is given below.

Primary functions: housing 20 Rotring boards,
housing all necessary sundry equipment,
providing a secure system of storage,
being easily transportable.

Secondary functions: improving the efficiency in using drawing boards,
facilitating an integrated approach,
improving the department's visual environment,
serving as a design teaching aid.

2.2.2 Information retrieval

Information is the raw material of designing. Information retrieval does not just mean visiting the libraries in your school, college or town, although this can be part of the exercise. Much of the information you require will not be recorded, so it will only be possible to obtain it by observation and experiment, or through discussions or questionnaires.

Recorded information

Libraries are storehouses of fascinating information and this can be a hindrance as well as a help. It is all too easy to spend half an hour looking at a book or article with an interesting title or attractive front cover, but of very little significance to your project. It is important to be clear about the purpose of your search and to exercise self-discipline.

The procedures for searching and locating information will depend on the particular library, and you should make good use of the library catalogues and any other referencing systems. There is often a database providing a *keyword search* facility.

The best starting point for your search is likely to be one of the following three sources:

(1) The BSI *Compendium of Essential Design and Technology Standards for Schools and Colleges* which links a number of standards with the type of products commonly manufactured as part of Design and Technology courses.
(2) Publications such as *Design* magazine and *Designing*. These contain discussions of up-to-date designs from both professional designers and school and college students.
(3) The magazine *Which?*, well-known for its reports on products currently available.

Nonrecorded information

Section 1.8 refers to the kind of product analysis it will be necessary to carry out in order to see the product in relation to its predecessors and competitors. Much of the information will not be published and obtaining it depends on the goodwill of the individuals who possess it. Apart from a general lack of courtesy, the easiest way to lose someone's sympathy is to waste their time, so it is important to prepare questions carefully and record answers accurately. Although you might find it easier to communicate by letter, telephone conversations are often more productive when seeking nonrecorded information. Visits take up more time, and you should remember this when seeking a meeting.

Useful sources of nonrecorded information include users in the general public and industry, manufacturers and suppliers, and experts in particular fields. Where domestic products are concerned, members of the general public are particularly useful. Questions to be put to them in questionnaires and surveys must be carefully constructed to yield the required information. You should also get the questionnaire checked by a supervisor, colleague or friend. If you are seeking information from industry you will find that most people are only too willing to help, but remember that everyone's time is valuable, so make appointments, be prompt and do not overstay your welcome. Useful literature can often be obtained from manufacturers and suppliers, and trade exhibitions provide particularly good opportunities to see and discuss products. You will also find that experts in the field in which you are working are normally willing to give their time, but the response you get will depend very much on the effort, interest and enterprise that you show.

As an example of the vital role of information in getting a project under way, consider the experience of one student – Andrew Rogers of Maidstone Grammar School. Andrew wished to undertake an architectural project concerned with a proposed office and shop development in the town centre. He conducted a survey among people using the town centre to discover what they felt was needed and, from their responses, he derived the outline requirements for a sports and leisure centre. The office and shop development was seen as unnecessary because there were already empty offices and shops in the town centre. Having established the need, Andrew went on to discuss the best approach with experts in the Architectural Department at Canterbury College of Art. They suggested drawing a *bubble diagram* and adding the necessary connections between spaces as shown in Fig. 2.5.

Having talked to local people about the need, and to experts about the right approach, Andrew's project was off to an excellent start and his design for the Vinter's Centre went on to win a Design Council prize. Some of the problems encountered during the development of the design are discussed in section 2.4, 'Synthesis'.

Designers also need to be prepared to make their own observations and conduct their own experiments. Ergonomics (see section 2.4.1) concerns the understanding of the way in which people use products and systems, and much information may

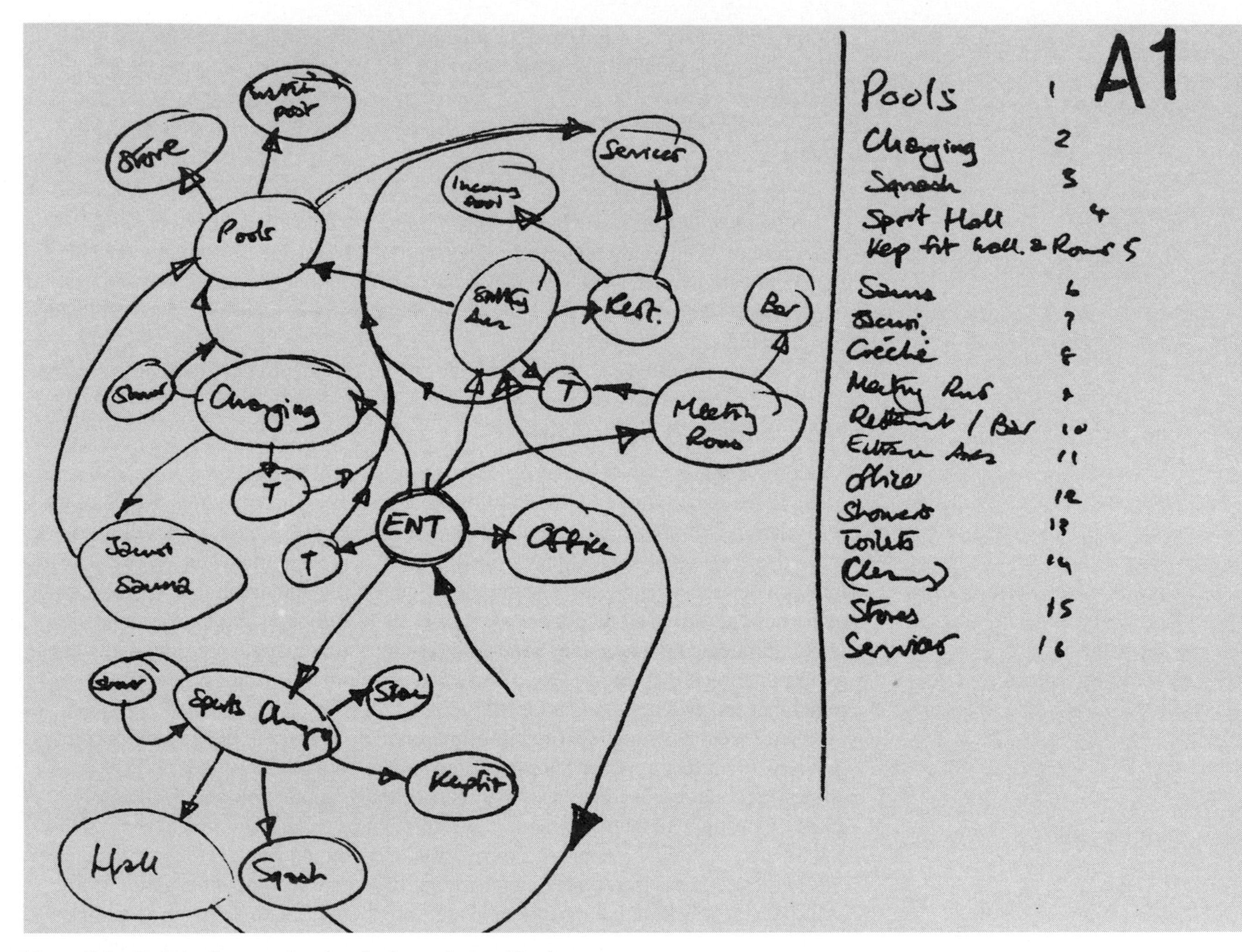

Figure 2.5 Bubble diagram showing the interrelationships between spaces.

already be available from these studies. However, conducting your own tests on similar existing products or mock-ups should always be considered. This will also give the designer an extra insight into the way people will use the product. Again, the technical information you need is not always readily available. Finding a value for a coefficient of friction from a textbook can take a very long time and, even when you find one, the test conditions may not be reported. For steel on steel the value can be 0.7 when the steel has been degreased, 0.3 for the steel 'as received' and less than 0.1 when it has been lubricated. How do you interpret a number like 0.4 for steel on aluminium without knowing the condition of the materials? Conducting your own tests on the materials you intend to use, and in the conditions you intend to use them, could probably be organised in less than half an hour (see section 8.4.2).

2.2.3 Searching on the Internet

For students in the new millennium the Internet has become a major source of both recorded and non-recorded information. It is often now the first option that students choose. There is a wealth of information available and it is also possible to seek others' help through email. This form of interaction often suits people well because they can control when they respond and how much time they spend. There are a number of references to websites in this book and companies often provide websites as a first point of contact. It is possible to find information about materials and components and their suppliers, as well as products and issues (e.g. the environment). If you have access to the Internet, then you should make every effort to exploit its potential.

2.2.4 Specifications

It is hard to overemphasise the importance of being completely aware of what you are trying to achieve. Writing a specification is the best way of clarifying your ideas, and updating it regularly will ensure that you remain on the required route as the project progresses. The specification is a statement of the characteristics that a design must possess in order to be a solution to the human need, and is more properly known as a product design specification (PDS). By itself the word 'specification' can have several meanings; consider the following passage from PD 6112 *Guide to the Preparation of Specifications* produced by the British Standards Institution in 1967.

> A specification is essentially a means of communicating the needs or intentions of one party to another. It may be a user's description, to a designer, of his requirements for purpose or duty; or it may be a designer's description, to a manufacturer, of an embodiment of these requirements; or it may be a manufacturer's detailed description, to his operator, of the components, materials, methods, etc., necessary to achieve that embodiment; or it may be a statement, by a seller, describing suitability for a purpose to satisfy a need or even a potential need, of a user or a possible user. It may, of course, be some or all of these in one.

In order to prevent confusion different words have been adopted to refer to these kinds of specification. A user's description to a designer is called a *brief*, a designer's description is the *product design specification* and the manufacturer's technical description to a purchaser is the *product specification*. The word 'specification' by itself needs to be interpreted with reference to its context.

Fig. 2.6 shows two designs resulting from the same brief which illustrate the significance of the specification phase to a designer. The brief was for a portable drawing office as given in section 2.2.1. The differences which exist can be largely

Figure 2.6 Drawing board storage units.

attributed to the detailed contents of the product design specifications that the students developed.

The specification for the lower of the two units emphasised aesthetics, equipment security and the avoidance of overcrowding when the pupils or students were getting and returning boards, compasses etc., and, at the same time, trying to meet the needs for a compact form, mobility, low cost, stability and robustness. The taller design emphasised the overall weight and width in order to facilitate the movement of the unit up and down staircases, and through narrow doorways, whilst trying to meet the other requirements. In the lower unit the detachable trays for the storage of extra items like compasses and stencils are a direct result of the desire to spread them out and avoid overcrowding. In the taller unit the trays are placed very low down to give greater stability. The castors chosen for the low unit and the larger wheels selected for the taller one clearly indicate the differences in their intended use. The selection of MDF (medium density fibreboard) for the low unit and plastic sheet panels on the taller one show the relative significance attached to security and weight. Such decisions are a result of differences in the content of various aspects of the PDS and the different emphases made by the two designers.

Not only does PD 6112 make it clear that the word 'specification' can have several meanings, but it goes on to state that the form of a specification, and its content, depend very much on who has developed it and for what purpose. A designer's specification to a manufacturer is most likely to take the form of detailed engineering drawings. The drawings would show allowable tolerances, the necessary surface roughness and any required manufacturing processes or other constraints as well as the form and dimensions of each component. It is often stressed that design briefs should not include unnecessary constraints, and this is also true of an engineering drawing. If a dimension is noncritical then it should only be covered by a general tolerance. If the manufacturing process does not matter then only the form, finish and accuracy required should be given. However if, for instance, a forging rather than a casting is necessary to give sufficient strength to a particular component, this must be specified. The appropriate form for engineering drawings is covered in section 2.7.1. The notes which follow concern the appropriate form of a product design specification – that which designers write to clarify their own ideas and for communication with their client – but before reading them consider the following passage, again from the foreword to PD 6112, and ask yourself how much it reflects your attitude to the writing of a specification.

> There is often a tendency to think of a specification as something lying in the applied scientific world, or more particularly in the field of engineering, but fortunately this view is being rapidly changed with the emergence of consumer bodies with widely spread interests and concerning themselves with common domestic articles. Nevertheless, to explain the persistence of such a view, it can be said that virtually any attempt to rationalise thought about requirements and to describe each need or each part of a need in unambiguous terms, involves mastering processes of accurate description which, in the end result, have the appearance of science or engineering. One can therefore refer to the 'performance' of a sweeping brush or of an aircraft engine.

When this was written in 1967 there was clearly an optimism about the speed of change, which has not, perhaps, lived up to expectations.

Writing a product design specification

The concept of Pugh's plates was introduced in section 1.10 when considering the total design approach, and the structure of a specification should represent a continuation of this kind of thinking. It is helpful if the document begins by setting the scene and the first eight sections mentioned in PD 6112 are suggestions of what might be included, They are summarised below.

1 Title The title of the specification needs to be well chosen to be unambiguous and informative.

2 List of contents This provides an easy way into the document for the reader.

3 Foreword A description of any relevant background information, the history of the project and how the brief originated.

4 Scope of the specification An overview of the material to be covered. The extent and limitations of the subject matter and a brief description of the product or system.

5 Role of the product or system If this has not been included in describing the scope of the specification, then it should be clearly stated.

6 Definitions Any unusual terminology, symbols, abbreviations or measuring units need to be defined.

7 Relevant authorities to be consulted Any authorities or bodies (for example, safety officers, health authorities, factory inspectors etc.) who must be consulted in relation to the product's design and use.

8 Related documents and references Any other documents, for example, statutory regulations, British Standards, codes of practice etc. which may be applicable.

The later part of PD 6112 illustrates the way in which some of the aspects of the PDS, as represented by Pugh's plate (see Fig. 1.26), can be tackled.

A bicycle sidecar

The importance of investigation and the writing of the resulting PDS can be well illustrated by a project set to Loughborough University undergraduates by Intermediate Technology (IT) Transport Ltd in 1999. IT Transport is an independent, international consultancy specialising in transport for rural and urban development. They have worked in many countries around the world and have gathered considerable expertise concerning transport requirements in different countries and cultures. Fig. 2.7 shows a bicycle trailer designed to be made and used in less-developed regions.

Figure 2.7 Bicycle trailer.

Bicycle trailers are also manufactured to be used in the developed world, but these, of course, use very different technology. All designing requires feedback and designing for other countries and cultures is only likely to be successful if an organisation like IT Transport can support the project through their expertise. However, the bicycle sidecar was different. IT Transport needed to generate funds in order to support their work abroad and Ron Dennis – one of their engineers – had thought that a bicycle sidecar made for, and sold in, the UK market might prove profitable. Four Loughborough undergraduates decided to take on this idea for their final year project.

They began to investigate and ask advice. Initially both the students and their tutors were concerned about the feasibility of the idea. The increased width of the vehicle was a cause of concern. Would the bicycle sidecar be less safe on the roads? Would it fit within cycle lanes? How would the steering be affected? However it was recognised that having the child next to the adult rider offered considerable potential advantages; the possibility of eye contact being comforting to both the adult and the child. They also discovered that bicycle sidecars had been made before in the UK (Fig. 2.8) and that there was a modern design for sale in the USA. When chatting on the phone Syd Urry told the story of some friends of his who, in the 1930s, cycled from London to Dorset to visit him and his wife. They rode a tandem bicycle with a sidecar attached!

Figure 2.8 Bicycle sidecar.

So the doubts concerning feasibility rapidly evaporated. In order to write the PDS, the students needed to find out more. What sort of bicycle sidecar would be suitable for the UK market in the new millennium? They found out some answers by conducting a survey with potential users in rural and urban areas. They built a 'lash-up' model to find out the effect on the bicycle's steering (Fig 2.9). The vintage

Figure 2.9 Lash-up model.

Figure 2.10 Vintage sidecar attachment.

UK design had a simple attachment that allowed the bicycle to rotate independently of the sidecar as shown in Fig. 2.10. The modern US design had a patented 'parallelogram' attachment so that the bicycle and sidecar rotated together. Is this an essential feature?

A PDS resulting from these investigations is shown below.

A bicycle sidecar for the UK market

Contents

1 Background information
2 Scope of the specification
3 Relevant authorities to be consulted
4 Performance
5 Environment
6 Competition
7 Customer needs
8 Aesthetics, appearance and finish
9 Ergonomics
10 Target product cost
11 Quantity
12 Materials
13 Manufacturing facility
14 Product life span
15 Life in service
16 Packaging and shipping
17 Maintenance
18 Standards and specifications
19 Safety and product liability
20 Intellectual property rights
21 Testing
22 Documentation
23 Disposal

1 Background information
Sidecars for bicycles can be purchased in the USA and were manufactured and marketed in the UK in the 1930s. This project is being conducted to explore the feasibility of developing a modern bicycle sidecar that can be profitably manufactured.

2 Scope of the specification
This specification covers the requirements for the design, manufacture and eventual disposal of the bicycle sidecar.

3 Relevant authorities to be consulted
Legal requirements for the use of bicycle sidecars on public roads are not currently known, but the police or another legal authority would need to be consulted before any trials of a finished prototype could take place.

4 Performance
The sidecar must be capable of travelling as fast and reliably as the bicycle so the wheel(s) and bearings must be similar in performance to those of the bicycle. The sidecar should not significantly increase the stopping distance of the bicycle; an increase of 10% would be acceptable. The sidecar should not adversely affect either the cornering of the bicycle at speed or the low speed manoeuvring. The sidecar needs some means of enabling the bicycle to lean both ways to facilitate cornering. The sidecar is unlikely to be used continuously and so it must be easily attached and detached from the bicycle. The attachment system should fit as wide a range of existing bicycles as possible. The maximum likely loading is two toddlers plus some small luggage i.e. approximately 400 N. The frame of the trailer should not deflect more than 20 mm under this loading.

5 Environment
The sidecar must protect the occupants from wind and rain. All components must be resistant to corrosion resulting from effects of the weather. The materials used should also be resistant to road salt and fluids associated with cycling, for example, oil and detergents. Vibrations will be caused by the road surface, and therefore all nuts, bolts, screws and other fastenings should be designed to cope with this. The passengers will also need to be comfortable when travelling over such surfaces. The sidecar is likely to be stored in a garage where space might be at a premium.

6 Competition
There are no bicycle sidecars currently on the UK market. There are, however, a number of cycle trailers, designed for both children and luggage. These are well-established products, selling typically for £400–£500, and are the main source of competition.

7 Customer needs
Potential customers felt that although they were unlikely to use the sidecar regularly on the road, they would like to use it as a leisure accessory. The main requirement of the potential market is that the cyclist is not affected by the presence of the sidecar. The bike must be able to lean, brake, etc. normally.

8 Aesthetics, appearance and finish
The sidecar needs to look well-designed, attractive, appealing, socially acceptable, fun, and easy to use. There are many different styles of bicycle that the sidecar might attach to, so the appearance (colour, graphics, wheel(s) etc.) will either need to be fairly neutral or a range of finishes will need to be manufacturerd.

9 Ergonomics
Fixings to the bicycle need to be easy for the cyclist to use. This must take into account hand sizes and acceptable magnitudes for any necessary forces (for example to obtain the appropriate tightness). The weight of the sidecar needs to be

such that the cyclist can cycle normally with it attached and fully loaded. The layout and dimensions inside the sidecar need to be appropriate to the size of passengers and take at least the 5%le to 9%le range of users into account.

10 Target product cost

The target selling price for the sidecar is between £160 and £200. This fits in with the products already sold by IT Transport and would provide a competitive advantage over trailers. This is a low price for a product of this nature and this could therefore have significant implications for the design of the sidecar.

11 Quantity

ITT currently manufactures products in low volumes, using mostly manual methods. The design should be compatible with this manufacturing mode, but take into account possible economies from larger scale manufacturing.

12 Materials

There are no pre-determined materials. Those chosen should have good strength and stiffness-to-weight ratios, good fatigue resistance, and be resistant to UV light and the temperature range expected in a UK outdoor environment. They should be durable and resist abrasion, water, common detergents and oils.

13 Manufacturing facility

The sidecar is most likely to be made and assembled at IT Transport's Ardington workshop where facilities are limited. However, IT Transport are willing to subcontract work if it is financially viable. There is the possibility that it could be sold in a flat pack and the customer could assemble it. This would save assembly costs. The main manufacturing processes would be cutting, bending, drilling and assembly of some component parts. Some manufacturing work, such as the making of panels, bodywork or anything that requires complex equipment, could be subcontracted.

14 Product life span

The product could be expected to remain in production for up to 15 years.

15 Life in service

The product would be expected to have a service life of approximately 20 years.

16 Packaging and shipping

The product is likely to be transported in small quantities and only within the UK, therefore by land. It is possible that the sidecar could be supplied to the consumer in kit form, e.g. packaged in a cardboard box with internal protective materials.

17 Maintenance

There should be few maintenance requirements. The wheel bearings and other moving parts may require lubrication. The wheel(s) should be easy to remove in order to enable the user to repair punctures. One type of fastener should be used throughout to minimise the number of tools needed. Cyclists are generally used to performing minor maintenance tasks on their bicycles so, assuming that the operations required are similar, this should not present a barrier to purchase.

18 Standards and specifications

There are no British Standards or legal requirements relating to sidecars or trailers. BS6102 relates to bicycle legislation but does not cover sidecars. However, it could provide useful tips on matters such as lights, brakes, reflectors and other safety related accessories.

19 Safety and product liability
The main safety issues relate to the young passengers, for example, sharp edges, small and detachable objects, safety belts and the attachment system. However the bicycle performance, such as braking distances, is also important for the product safety.

20 Intellectual property rights
The Burley Bicycle Trailer Company (USA) has patented a certain type of bracket for fixing to the back axle of a bicycle, but no other patents have yet been found.

21 Testing
The product will need to be tested in relation to all matters covered by this specification. No standard procedures exist and so tests will need to be devised.

22 Documentation
The product is likely to need instructions for the user concerning attachment to the bicycle, maintenance, use and limitations. If the product is supplied in flat pack, then it will need detailed instructions about assembly.

23 Disposal
The product is likely to have a long life in service, but it should be designed to be disassembled, so that parts can be recycled or reused when it comes to the end of its useful life. Polymer materials used should be clearly labelled to identify their type.

Figure 2.11 Completed sidecar prototype.

Such a PDS provides a secure basis for a project. It was decided that this project was too large to be completed by one student in the time available and so two pairs of students collaborated on the final designs and prototype construction. The prototype completed by Ross Braithwaite and Tim King for the 1999 Loughborough University Degree Show is shown in Fig. 2.11.

2.3 Generating ideas

Exploring a design brief and gathering information will, it is hoped, leave your mind full of possibilities and avenues which you might pursue. It is particularly valuable during these early stages to keep a logbook which records ideas and the project's progress. It will help you in the sifting process. If you have written your product design specification carefully, and become aware of all the requirements, you will probably see that many of the possibilities lead nowhere. Even so, you are likely to be left with several potentially viable ideas. If you have no ideas to pursue, it is clearly necessary to generate some, and the techniques described in this section have been developed to help you. They might be worth pursuing anyway, to ensure that the problem is explored exhaustively. Which techniques might help cannot be predicted in advance; it is just a question of trying them until one succeeds.

One of the hopes of all designers is that they will have a sudden vision of the solution to their problem. Eureka! It has already been emphasised that such insight results from total immersion in the problem. Analysing the brief and writing a PDS is part of that process. Morphological analysis, searching for analogies, group interaction and lateral thinking can all deepen the immersion process and make such insight more likely. As an example consider the problem of a signalling device intended for Third World lobster fishermen to locate their pots when they wished to collect them but not to show their position to poachers. A string can be attached to a submerged float (plastic bottle), but how do you make it come to the surface at the right time? Electronics or a viscous fluid could perhaps be used, but it took weeks of thought, research and discussion for a student to come up with the idea of using the expansion of rice in sea water to trigger a mechanical latch. Absolutely appropriate technology which deserved its appearance on the BBC's *Tomorrow's World* programme.

2.3.1 Morphological analysis

Morphological analysis is simply a mechanistic way of generating ideas, but is a very straightforward technique to apply and will follow quite naturally from the detailed product analysis of currently available products and their historical origins. Its mechanistic nature can really be regarded as an advantage because it avoids personal blind spots which can result from habit, familiarity or prejudice.

To see how the technique is applied, consider again the wheelbarrow information given in Table 1.9. The table shows some data about the barrows and also indicates some of their characteristics. In order to apply morphological analysis we need a list of the characteristics which describe a particular design, or the design attributes. These attributes are shown at the heads of the columns in Table 2.1.

Frame type	*Frame material*	*Movement aid*	*Container material*	*Number of wheels/balls*	*Number/type of handles*
fixed	tube	ball	mild steel	1	1/straight
lowerable	steel	wheel	galvanised steel	2	2/straight
foldaway			thin plastic sheet	3	1/pram
			moulded rigid		
			plastic moulding		

Table 2.1 Attributes of barrows currently on the market

There are six columns, covering some of the attributes which are evident from our market survey. In each column the characteristics for existing designs are shown, and it may come as a surprise to discover that there are already $3 \times 1 \times 2 \times 4 \times 3 \times 3$ or 216 possible combinations.

The market analysis was only based on ten designs, but simply listing their attributes in matrix format has generated many more possibilities. Some of these are strange or impossible, but others are quite thought-provoking,

- a folding, steel-framed barrow with one handle, one ball and a container made of thin plastic sheet?
- a lowerable steel-framed barrow with three wheels and a rigid plastic container?

Some, of course, represent existing designs, for example:

- a lowerable steel-framed barrow with one wheel and a galvanised steel container – the Ascender Minor.

Having listed the attributes, it is usually not too hard to expand the options. What about adding to the frame material possibilities – aluminium tubing and plastic tubing are obvious candidates? Equally, the container might be made from other materials, perhaps GRP (glass reinforced plastic) or aluminium sheet, and perhaps the mobility could come from a roller or caterpillar tracks, or even an air cushion! A 'hoverbarrow' seems rather a silly concept, but perhaps you could make an attachment for a lawnmower? If we just included two new frame materials and two new container materials, the number of possible combinations would increase to $3 \times 3 \times 2 \times 6 \times 3 \times 3$ or 972.

Thinking about the attributes of products leads very quickly to new possibilities, but in a sense they are all derivative ideas and we have not attempted to make any conceptual leap in our thinking. This approach to generating ideas is really just being very thorough – a highly commendable thing to be at any time.

2.3.2 Analogy and information transfer

Trying to find suitable *analogues* to the problem you are considering is another powerful means of coming up with new approaches. It is less mechanistic than morphological analysis, and depends on the ability to identify useful lines of thought. It is important that the conceptual area from which the analogy derives does not overlap too closely with the original problem; for example, an analogy to a chemistry problem might be better looked for in cookery rather than physics. If the two conceptual areas are too close, the language and conventions are likely to be too similar to achieve a completely new viewpoint.

As an example of the way in which thinking about an analogous problem can provide a breakthrough, consider how the problem of launching the Sea Harrier was solved. The decks of the aircraft carriers were too short for the planes to take off with a full load, unless the carrier steamed at full speed into the wind. This was obviously a serious operational restriction and many people knew of the problem, and were searching for a solution. It was an analogy with water-ski jumping that provided the breakthrough illustrated in Fig. 2.12.

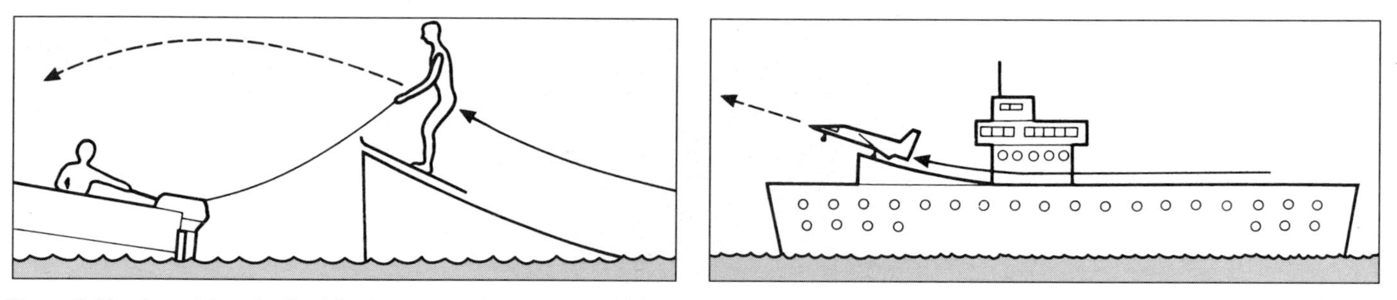

Figure 2.12 Launching the Sea Harrier – an analogy to waterski-jumping.

After this achievement, much work had to be carried out to find the precise angle required and to design the structure, but the right conceptual approach had been found.

For designers, the richest source of analogies is nature. Many people have observed that nature seems already to have solved almost every problem we come across. It is just a question of knowing where to look, and being able to transfer the information from one context to another. The natural world has been a continual source of inspiration for forms, patterns and decoration, but there has also been a transference of ideas of a more technical nature – most notably in relation to streamlining. Evolution has resulted in forms for the dolphin and porpoise which minimise the water's frictional drag. It was by seeing this relationship that designers of modern submarines came to decide on the forms they chose. Clearly the analogy between submarines and dolphins is conceptually quite close, but the concept of streamlining has been transferred to products which do not have any functional relationship to drag forces or fluid frictional resistance. The American designer Norman Bel Geddes popularised this approach in the 1930s and 1940s and it was applied to many products to symbolise efficiency and modernity. The Hotchkiss stapler of 1936, shown in Fig. 2.13, is perhaps a classical example where streamlining is functionally irrelevant, but results in a powerful image. Clean lines and continuity in profiles are often sought by designers to create an impression of polished efficiency or precision.

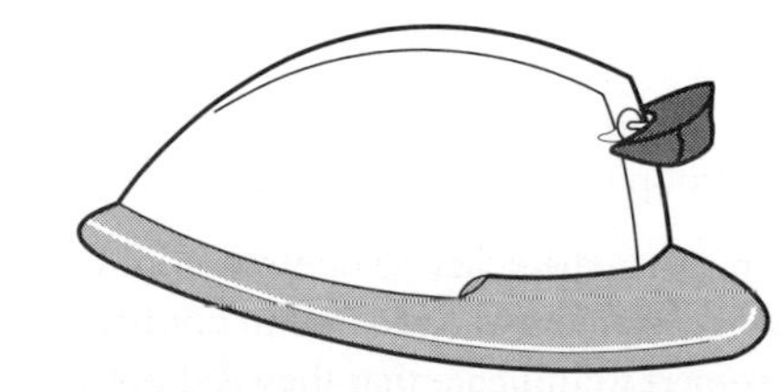

Figure 2.13 Hotchkiss stapler.

As an example of how powerful a method of thinking analogy can be, consider the approach taken by Mark Hinnells – a student in the Design and Technology department at Loughborough University – in designing a new garden barrow. He had used the Aquaroll for transporting large quantities of water on camp sites for several years and knew how effective a method it was. He therefore set out to use the same principle in the design of a new garden barrow as shown in Fig. 2.14.

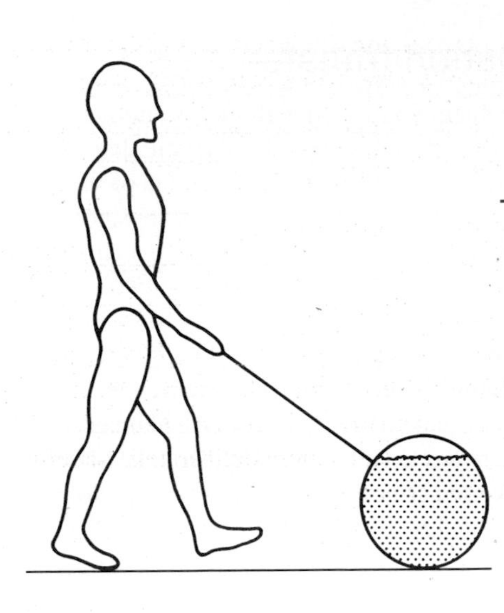

Figure 2.14 Analogous thinking – Aquaroll to wheelbarrow.

As with the Sea Harrier launching ramp there was still a tremendous amount of detail design to be done, but the fundamental concept of large-diameter wheels, a low-slung container and a counterbalanced handle had been formulated.

2.3.3 Group interaction and brainstorming

Brainstorming in groups is often seen as a solution when matters are becoming desperate, but group interaction should really be a more routine matter than this. Others may see aspects of a problem which the designer has overlooked. The more varied the backgrounds of the individuals the more likely it is that they will bring a different perspective and this reality underlies the popularity of multidisciplinary teams. Open plan work areas which facilitate free discussions with colleagues are the ideal environment for good design.

Brainstorming – where a group is formally asked to focus their attention on a specific issue or problem in order to generate ideas rapidly – originated in the advertising industry where Alex Osborn formulated some simple rules for promoting spontaneous expression. His requirements were:

- no criticism of any idea was allowed – judgement being withheld until later,
- all ideas were welcomed, however bizarre or frivolous they might appear,
- the emphasis was to be on producing a large number of ideas,
- building on someone else's idea to create a group chain reaction was encouraged.

Insisting on these rules led to a relaxed environment where people were not inhibited from thinking freely and adventurously. Some experiments conducted to confirm the synergistic effect of brainstorming sessions seemed to indicate that they did not generate more ideas than people working individually and then pooling their thoughts, and much clearly depends on how the session is conducted. In addition to applying the above rules firmly, it helps if the following conditions can be met:

- advance notice should be given of the problem so that people can tune in to it,
- a balanced team of about a dozen should be formed with a firm, experienced leader,
- a reporter should be appointed to record all the ideas,
- too wide a difference in the status of group members should be avoided as it tends to be inhibiting.

Brainstorming is one of the most widely used techniques for generating ideas, but clearly there will be a lot of analysis to complete after the session to assess the potential of each concept. When the number of alternatives is very restricted, brainstorming is unlikely to produce much progress and a more deliberately forceful technique like *lateral thinking* may provide the answer.

2.3.4 Lateral thinking

Lateral thinking should not be thought of as a magic formula, but as a more creative way of using the mind. All students and designers would benefit from studying the techniques developed by Edward de Bono, and the thinking skills learned would become part of their normal approaches to generating ideas. As with group interaction, lateral thinking should not be resorted to just in desperation.

The term 'lateral thinking' was chosen to distinguish it from what is perhaps the traditional or conventional form – vertical thinking. Vertical thinking represents the kind of logical approach where each step follows from a previous one. Any conclusion can be explained by looking at the path taken to reach it. Such vertical approaches have a significant place in design, but there is also a major role for lateral thinking as is indicated in the following quotation from de Bono – '(Design) is not so much a matter of linking up a clearly defined objective with a clearly defined starting position (as in problem solving) but more a matter of starting out from a general position in the direction of a general objective' (de Bono 1970 p. 282). The further you go in the design process the more closely you will have defined your objective and starting point, and the more like problem solving it becomes. 'Lateral thinking is for generating ideas, logical thinking is for developing, selecting and using them' is the way de Bono puts it and this contains the essence of the truth, although it should be recognised that logic has its place in generating ideas as do conceptual leaps in their development.

The principles of lateral thinking can be thought of in four broad areas:

- the recognition of dominant polarising ideas,
- the search for different ways of looking at things,
- a relaxation of the rigid logical control associated with vertical thinking,
- the use of chance.

It is not possible, or appropriate, to reproduce de Bono's work here, but some techniques will be mentioned to give the flavour of his methods.

Dominant ideas

In thinking about a situation or an issue it is quite usual for your thinking to be dominated by particular ideas. The more complicated the problem is the more likely this is to be the case; for example, chess players will tend to think of good moves in the opening stages as those which control the centre or free the major pieces. Moves which do not fit these general criteria look wrong – even if they are actually good moves. To liberate your mind from rigid patterns of thinking you must recognise what ideas dominate *your* thoughts, and these are not necessarily the same as those which dominate other people's thinking.

Looking at things differently

The discovery by Edward Jenner of the vaccination process and the protection it gave against smallpox followed a change in the way he looked at the issues. His attention shifted from asking why people got smallpox to asking why dairy maids apparently did not. He eventually concluded that exposure to the less virulent cowpox was somehow producing immunity. Such a change of emphasis might by itself provide a new perspective on your thinking, but designers also have the option of using visual images instead of language as de Bono commends. Line drawings, diagrams, graphs, charts and colour can all be used to explore relationships between aspects of a design brief, and this is, of course, exactly what designers try to do when exploring their ideas on *design sheets.*

Escaping from vertical logic

Vertical thinking depends on each step along a path of reasoning being right, but it is only necessary to be correct at the end – a good solution or idea remains a good solution or idea however it was arrived at. Marconi's first radio transmission across the Atlantic Ocean occurred because he kept on trying despite the quite logical objection that the radio waves would not follow the Earth's curvature and would travel in straight lines out into space. Neither the objectors nor Marconi knew of the ionosphere which would reflect the waves back towards the receiver. Aeroplanes would never have been built if the logical objection, that machines heavier than air could not fly, had not been challenged. (Patent offices once refused applications for flying machines in much the same way as they now refuse applications for perpetual motion machines.) The interested student must read the work of de Bono, but perhaps the essence can be described as the suspension of judgement on ideas in favour of their exploration to search for new patterns. The introduction of random words to force connections and find unusual analogies is one technique through which the search can be pursued.

Using chance

You cannot make things happen by chance, but you can allow chance to play its part. Chance events have played a major part in scientific discovery, for example, the discovery by Daguerre of the use of silver salts to make paper light-sensitive for photography and the discovery of X-rays by Roentgen. Designers can allow chance to play its part by wandering through exhibitions or shops and absorbing, indiscriminately, all the stimulating aspects of the environment. Consider the analogy which de Bono gives:

> The difference between the emergence of a new idea through chance interactions and the careful construction of a theory by logical means may be illustrated by an analogy with paper clips. A chain may be made out of paper clips by carefully and deliberately attaching one clip to another. A chain can also be formed by opening out the clips a bit and then tossing them in a pan. If they are tossed long enough and vigorously enough a chain-like structure may be lifted out at the end. A chain has been formed by the chance interaction and intertwining of the clips from themselves into a pattern which is always unexpected and usually original. Once the pattern is formed it can, of course, be trimmed and modified.
>
> (de Bono 1967 *The Use of Lateral Thinking*)

The random arrangement may not always be as strong as a deliberately constructed chain, but it will be new. To be efficient there needs to be a large number of clips in the pan and they need to be opened out a bit. The interaction of new information and stimulants to give new patterns works much the same way – they need to be numerous and not contained in tight and rigid groups.

2.4 Synthesis

In describing the phases which occur within different design stages J. Christopher Jones identified three elements: divergence, transformation and convergence. There is a time to concentrate on generating ideas and different conceptual approaches, and a time to begin the process of evaluating what has been developed, bringing together the key features and resolving conflicts. The following five sections deal with various aspects of this process of synthesis, initially looking at human factors and then modelling techniques and computer-aided design (CAD).

To give an example of synthesis consider the task which Andrew Rogers faced at this stage of the design of the Vinters Sports and Social Centre. The bubble diagram showing the necessary spaces and their connection has already been given (see section 2.2.2). The left-hand side of Fig. 2.15 shows how this functional requirement developed.

The photographs on the right-hand side of Fig. 2.15 show the aesthetic theme developed from an exploration of cylindrical form. When these two lines of thought had reached sufficient maturity there was a positive attempt at their marriage. Twenty-three design sheets were produced, including the overlay shown in Fig. 2.15, before a satisfactory synthesis was achieved. The final design is shown as a presentation drawing here, but a three-dimensional model was also produced.

In general there will be many more than two lines of thought, and graphical and three-dimensional visual models are unlikely to be the only kinds necessary. In fact the product design specification is really a verbal-numerical model of the desired product. It represents the beginning of the process of synthesis – the recognition of all the factors that must be taken into account. Mathematical models and analysis might help as might different kinds of two-dimensional and physical modelling. Computer modelling can contribute as well if the facility is available. With computers becoming more common and more powerful, designers must be prepared to use them in the process of synthesis as they would any other tool. It is a question of selecting the most appropriate modelling technique for your purpose.

2.4.1 Human factors – ergonomics and anthropometrics

All design is carried out to meet a human need and all products or systems must interact in some way with people. The reason that such a great deal of effort is expended in dealing with the interface between products and systems is that, whatever else can be standardised in the design process, the dimensions, capabilities and responses of humans certainly cannot. Probably the first scale drawings of humans intended for designers were made by Henry Dreyfuss. The 'average American man – Joe' and the 'average American woman – Josephine' were drawn in a number of situations, for example standing, sitting and crawling, and the set of drawings was published in 1955. The dimensions of small and large men and women were also marked on the drawings, but it was still found that most of the designs were being done for the 'Dreyfuss average man' or the 'Dreyfuss average woman'. Your task as a designer is to fit your design to the maximum number of people: just because it fits the average person is no guarantee of success. Placing a shelf at the right height for an average person is no guarantee that it will be suitable for the very tall or short, the elderly, the very young and the disabled. In fact it will probably not be suitable for the majority of the population and fit most badly those who are least able to adapt.

The average human is a dangerous concept for designers to employ and as part of the synthesis process it is really essential to consider all of the target user groups, the limits on their sizes or capabilities, and choose the design dimensions or build in sufficient adjustment so that the majority of all of them can use the product or system effectively. If this cannot be achieved then it will be necessary to produce a range of products – as when cycle manufacturers offer a series of different frame sizes or

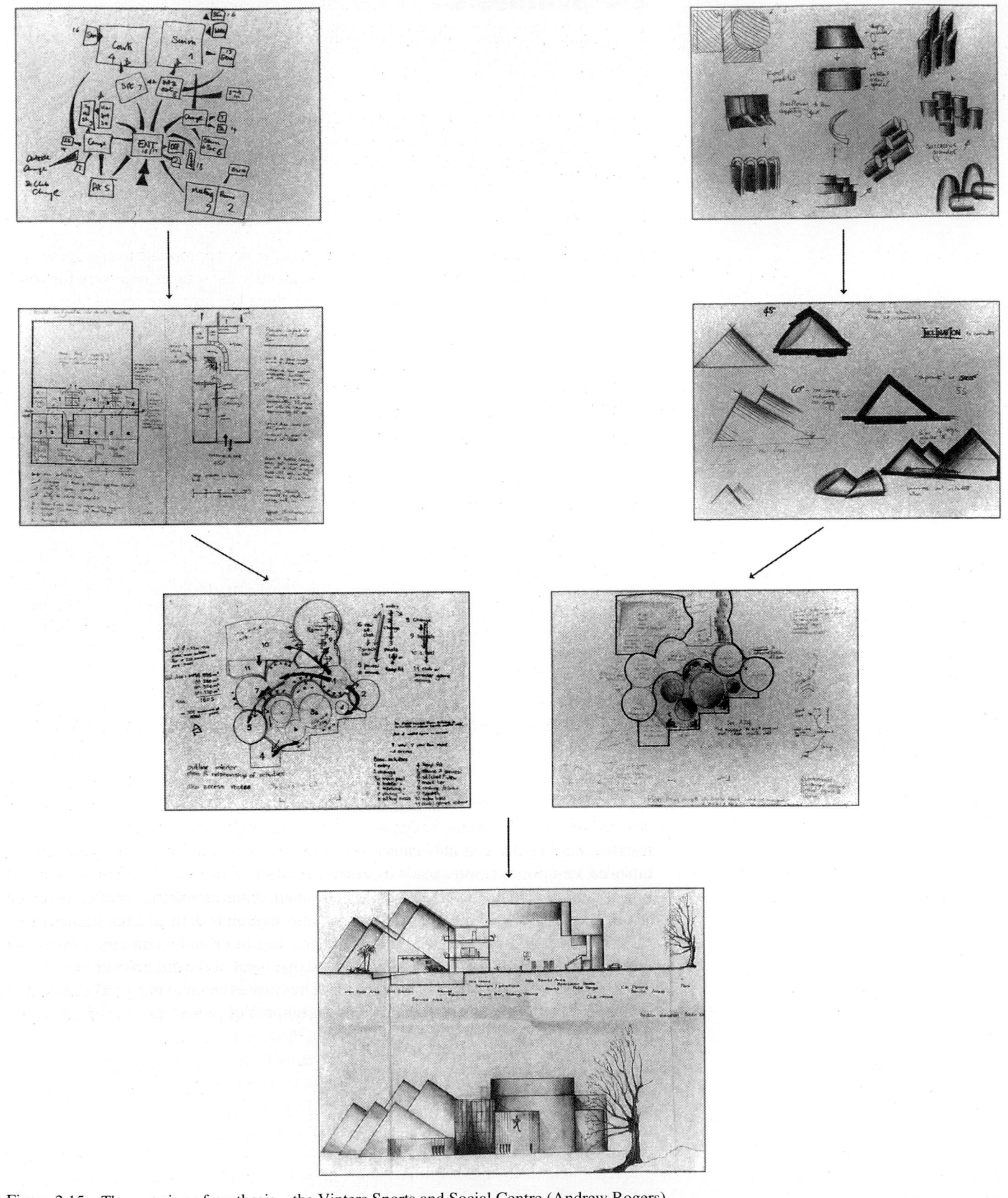

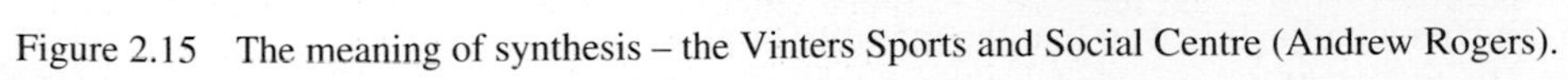

Figure 2.15 The meaning of synthesis – the Vinters Sports and Social Centre (Andrew Rogers).

clothing manufacturers small, medium and large or number designations (...10, 12, 14, 16...). Clearly there can be major additional costs in producing a range of sizes. The manufacturers incur the extra costs associated with increased tooling, set-up and distribution, and the retailers face the charges arising from holding and controlling larger stocks. If a range of sizes is to be produced, it must be kept as small as possible whilst still covering the market adequately.

There is now a great deal of information available concerning humans and their capabilities and a number of the most significant sources are given in the bibliography. The BSI *Compendium of Essential Design and Technology Standards for Schools and Colleges* is an important source, as are the DTI Publications *Childata* and *Adultdata* and the ergonomics books listed in the bibliography. The titles of these books are a good indication of their content. It is neither possible nor appropriate to attempt to reproduce all of this information and you must identify the sources necessary for your particular task. In order to facilitate this process, three key areas are covered here:

- definitions and discussions of the terminology used for human factors,
- the use of statistics in aiding decision-making,
- case studies which indicate the kind of information necessary in different design situations.

Terminology for human factors

Problems associated with human factors are essentially interdisciplinary and there are four particular disciplines on which designers will commonly need to draw: physiology, psychology, anthropometrics and ergonomics.

Physiology is concerned with the study of the systems within the human body, their responses, limitations and capabilities. Reaction times, responses to temperature changes, visual acuity and colour perception, strength, fatigue and muscle control and hearing thresholds are just some of the areas with which a designer may be concerned.

Psychology is concerned with the study of the human mind. Human senses are continually being stimulated and sending corresponding signals to the brain. There they are processed. Some signals are ignored and some provoke a rapid response. Some result in an accurate interpretation and some are misunderstood. Understanding some of the processes involved in this interpretation activity can be vital for developing good designs.

Anthropmetrics concerns the measurement of the physical characteristics of humans and in particular the determination of their physical dimensions, although other data on, for example, how far people can reach, how much space they need and how much force they can exert is also determined. Anthropometrics is often regarded as being military in origin, and much of the information is derived from studies of service personnel. If you are faced with the problem of providing uniforms and boots for millions of people, then clearly you have a desperate need for guidelines. Equally, because the group you are catering for is so large there is a good chance of the statistical approaches adopted being successful. You cannot be so confident if your target population is small.

Ergonomics is concerned with the relationships between people and the equipment and environments they use. Its origins were in the 1930s as part of the drive towards increased efficiency in manufacturing, and the term was coined by Professor Hywel Murrell in the 1940s. (The term 'human factors' is now used more or less interchangeably with 'ergonomics', particularly in the USA.) Prior to this time most industrial equipment had been designed without reference to the requirements of the operator who, it was assumed, would adapt. Humans are, of course, extremely capable of adapting, but it was realised that the limits of this approach had been reached and that further progress depended on improving the design of the equipment so that operators could use it more efficiently. It is evident that ergonomics will be dependent on much of the information gained from physiological, psychological and anthropometric studies, but its uniqueness lies in the application of this knowledge to equipment design. There is now a wealth of information concerning the relationship between people and products which has grown up as a result of ergonomic studies.

Safety in the use of products and systems is now primarily thought of in the context of ergonomics. Consider the following extracts from an article by Duncan Hopwood concerning ergonomics and safety issues.

> People cut hedges with rotary lawnmowers. They stand on chairs. They use fridge doors as stepladders to climb up to change lightbulbs. Believe it or not that case reached the US law courts. Also in the States, several people died from putting petrol in washing machines to clean dirty overalls. They died horribly when the machines burst into flames... At Three Mile Island, banks of instruments on each side of the control room were mirror images of each other, a natty design concept but hardly usable. Staff distinguished two vital levers which had entirely different functions by placing different brands of beer cans over each lever. The slowness in shutting down the reactor has been blamed on control room confusion. Ergonomics testing of the control equipment at Three Mile Island could have averted a nuclear disaster.[2]

There is an element of predictability in the way humans misuse products and systems. It was not just one person who put petrol in a washing machine. There are also remedies like the asymmetrical grouping, sizing and colouring of controls which are now well known to be effective. Ergonomists must face this problem of prediction and have a thorough knowledge of potential remedies.

Statistics and design decision-making

Dreyfuss' first publication concerning anthropometric data for designers recognised the need to define small, average and large for men and women and the statistical approach now adopted represents a formalisation of this concept. Fig. 2.16 shows a *normal* (or Gaussian) distribution curve.

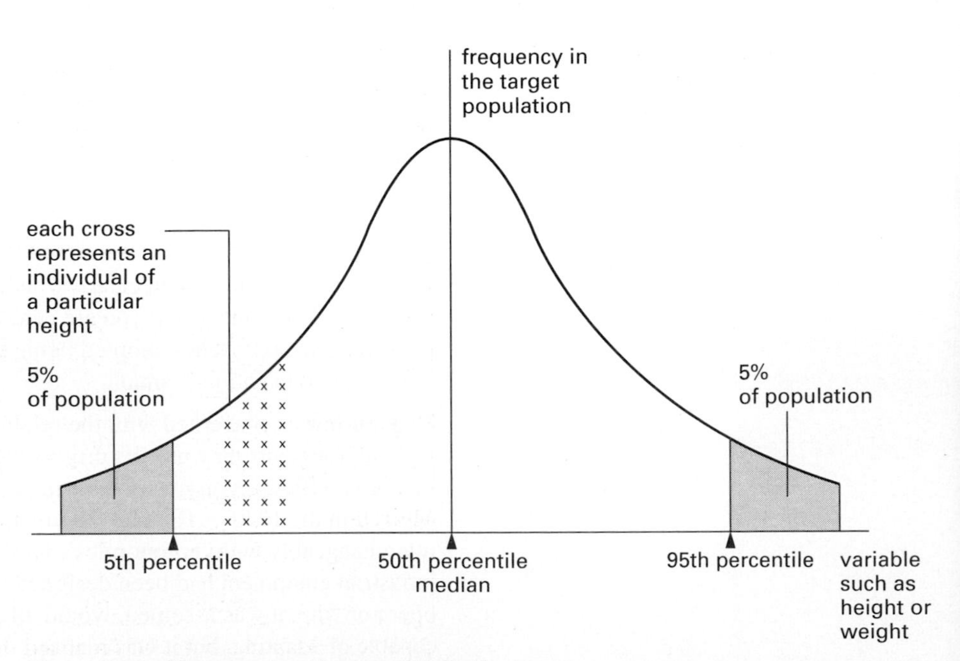

Figure 2.16 The normal or Gaussian distribution curve.

This distribution has been found to be the one which most commonly applies and is sometimes named after the mathematician Gauss. The horizontal axis represents the range of the variable you are considering, perhaps people's heights, with the smallest height value on the extreme left and the largest value on the extreme right. You can imagine generating the curve shape by making a cross for each individual of your target population with a particular height as is illustrated for part of the curve. If human heights follow a normal distribution then the most common height, known as the mode or the modal height, will be in the middle. The value where there is 50% of the population with a greater value and 50% with a lesser value is also known as the *median*. For a normal distribution the mode and the median can be seen to have the same value. As the curve is symmetrical this will also be the same as the arithmetical average or *mean*.

Although the most likely distribution of a population is the normal or Gaussian one, some populations are *skewed* as shown in Fig. 2.17.

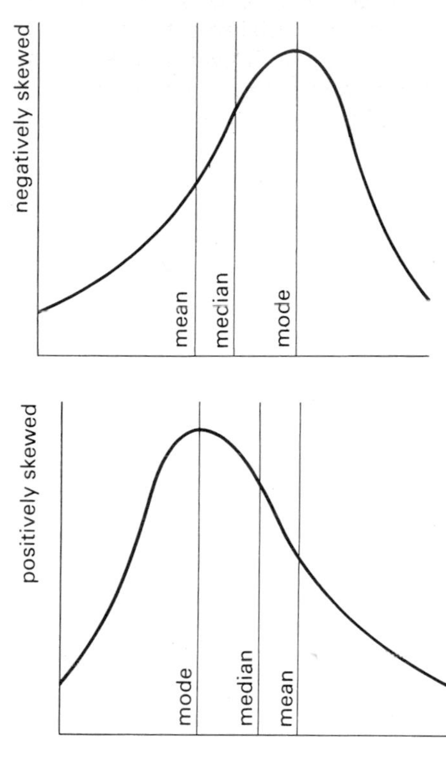

Figure 2.17 Skewed distributions.

This would be the result of a disproportionately large number of short (positively skewed) or tall (negatively skewed) people in the population. The most common value, the mode, will not now be the same as the median or the mean. Anthropometric data are usually based on a normal distribution and gaps in the information are often filled in on this basis. The information given usually refers to percentiles. The 1st percentile, for instance, is the value reached by all but 1% of the population and the 5th percentile is the value exceeded by 95% of the population. Fig. 2.18 shows two-dimensional manikin models of 5th percentile, 50th percentile and 95th percentile individuals which were made up to test a scale model of an exercise bicycle. Although such an investigation can often be carried out mathematically or graphically a physical model can make a thorough investigation of the total movement easier.

Table 2.2 shows anthropometric data on hand measurements which is used in the first case study.

Also shown in this table is the *standard deviation (s.d.)* which indicates how peaky or flat the normal distribution is. Fig. 2.19 shows the effect of moving away from the median in steps of one standard deviation.

The first step in either direction will cover about 34.1% of the population, the second step 13.6% and the third 2.15%. It follows that only 0.3% of the population (three in a thousand) is not covered by a width of six standard deviations – three each side of the median. The designer is therefore able to base a decision either on the low and high percentiles or on the median and the standard deviation. It should be emphasised, however, that the assumption of a normal distribution lies behind all of this information, and if there is any reason to believe the population is in fact skewed then it should be treated with considerable caution.

Percentile		*1st*	*5th*	*50th*	*95th*	*99th*	*s.d.*
males, 19–65	length	165	175	190	205	215	10
	breadth	75	80	90	95	100	5
females, 19–65	length	150	155	170	185	190	9
	breadth	65	70	75	85	85	5
male students	length	170	180	195	210	215	9
	breadth	80	85	90	100	100	5
female students	length	155	160	175	190	195	9
	breadth	65	70	80	85	90	5
males, 65 +	length	155	160	180	195	205	11
	breadth	70	75	85	95	95	6
females, 65 +	length	135	145	160	175	185	10
	breadth	60	60	70	80	85	5

Table 2.2 Anthropometric data for a hand exerciser (from PP 7310) (all dimensions in millimetres)

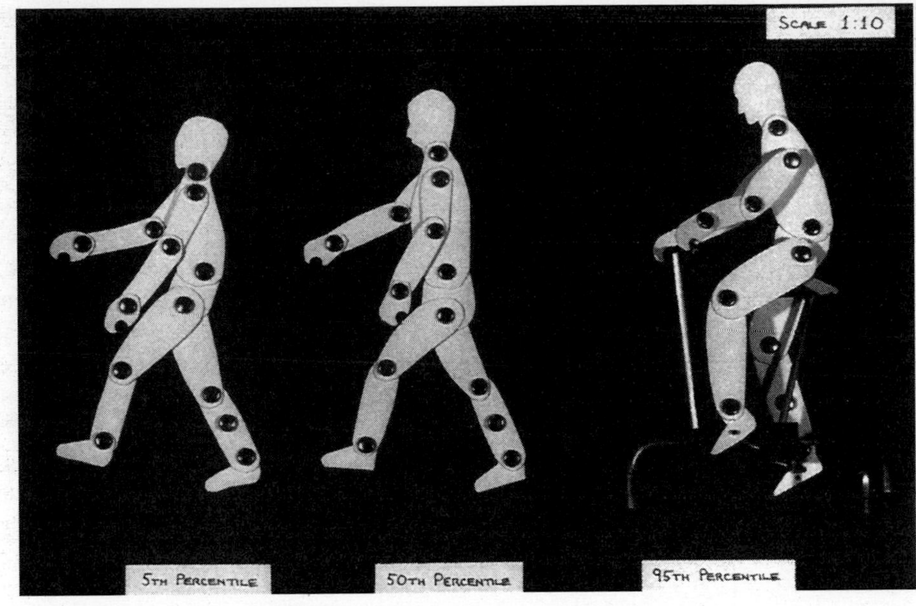

Figure 2.18 Manikin models for a cycle exerciser.

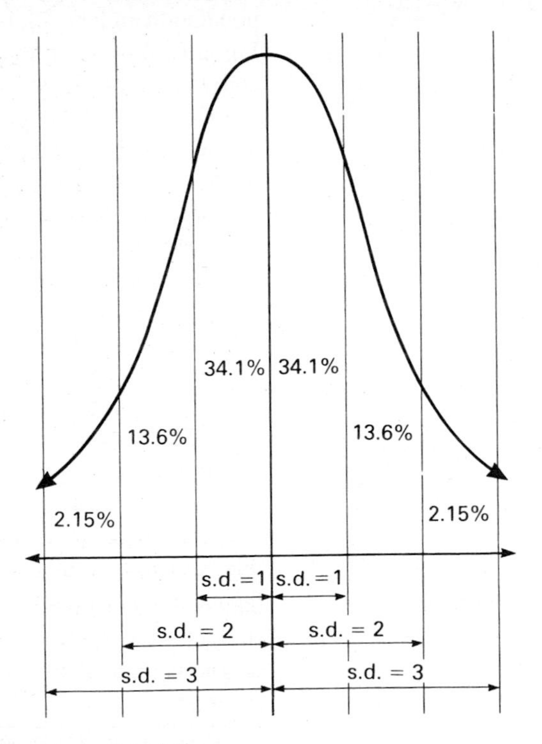

Figure 2.19 The standard deviation.

Case studies

Hand exerciser

Fig. 2.20 shows a prototype hand exerciser developed for use by accident victims whose hand tendons and ligaments of hands have been severed or damaged.

Following surgery it is essential that patients exercise their hands as early as possible if sensitivity and a full range of movement are to be recovered. This particular device was designed to keep the fingers flat as the hand is opened and closed which ensures that the damaged tendons are used through the maximum range of movement. The movable weight allows the load on the tendons to be increased as the hand recovers its strength.

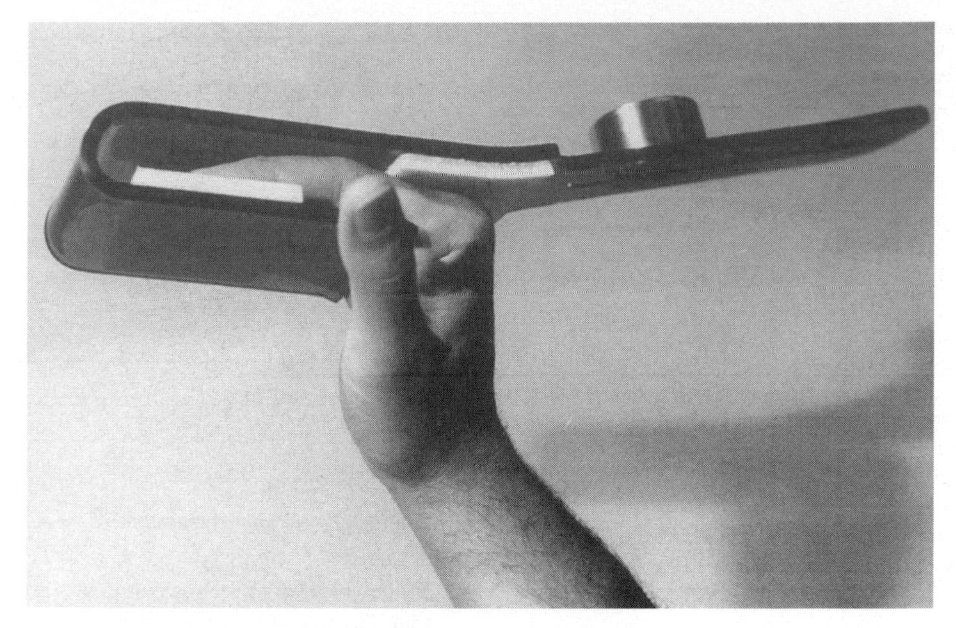

Figure 2.20 Prototype hand exerciser.

Anthropometric data are available in PP 7310 for men and women aged 19–65, male and female students, elderly men and women aged 65 years and over and for boys and girls in five different age bands. This prototype was to be developed for adults and hence the data for boys and girls were excluded. It was also decided that this prototype should fit at least 90% of this target population and hence for each measurement the 5th and 95th percentile dimensions need to be considered. Studying Table 2.2 will indicate that the limits might be set by the dimensions for male students and females over 65, that is:

Maximum length will be 210 mm (95th percentile male student)
Minimum length will be 145 mm (5th percentile female over 65)
Maximum breadth will be 100 mm (95th percentile male student)
Minimum breadth will be 60 mm (5th percentile female over 65)

However, this information only fixes one dimension of the device, the width, which should be 100 mm, If the breadth were less than 100 mm, the fingers of more than 5% of the male students would overlap the edges and hence could bend. A little less than 100 mm might well be acceptable though, because not all the width of the outside fingers would overlap the edges, and they would probably still be held straight. The other two critical dimensions, X and Y, indicated in Fig. 2.21, depend on knowing the depth of the fingers and their length rather than the hand length.

Information concerning the ratio of finger lengths to hand lengths can be found in, for example, Dreyfuss 1978 *The Measure of Man* from which the information in Table 2.3 was extracted.

It can be seen that the expected ratio of finger length to total hand length is approximately 0.6 and hence dimension X could be 0.6×210 mm or 126 mm if the

Percentile		*2.5th*	*50th*	*97.5th*
adult male	total hand length	173	193	208
	finger length (fraction)	102(0.59)	117(0.61)	127(0.61)
adult female	total hand length	–	175	
	finger length (fraction)	–	102(0.58)	

Table 2.3 Hand measurements in mm (Dreyfuss 1978)

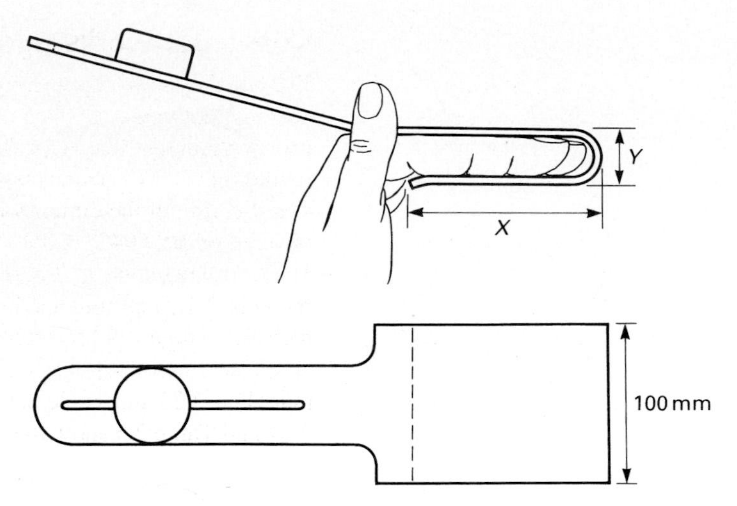

Figure 2.21 Critical dimensions of the hand exerciser.

fingers of 95% of the male students are to be fully enclosed. Again, a depth of less than this would probably be sufficient providing the first joint in the fingers was still held flat.

Information on finger thickness is very difficult to find, although PP 7317: 1987 gives the hand thickness at the palm. In any case, it should be remembered that the patients' fingers might well be swollen or bandaged following surgery, and so normal population data would be of little assistance. The aim, for the whole target population, is to grip the fingers as firmly as possible – an effect the foam included in the prototype was intended to achieve. However, to achieve sufficient adjustment, it might be necessary to use an inflatable bag or air cushion which could surround and support damaged and swollen fingers. The size eventually decided on for dimension *Y* was based on limited trials with the more extreme members of the student's immediate colleagues, family and friends.

Fig. 2.22 shows the average male and female hand dimensions and the 5th percentile female dimension for the over 65s drawn on these minimum guidelines.

Although the design would function with the 5th percentile female over 65 it is really swamping the small hand size and it would feel large. A universal size would have to be more like 90 × 115 mm in order to suit everyone, but it is already clear that there ought to be more than one size. Perhaps small (for children), medium and large would be more appropriate. Testing a prototype would yield more information which should be taken into account in reaching a final decision; for example, how tight does the fit on the fingers need to be? Or does it matter if the device appears unnecessarily large?

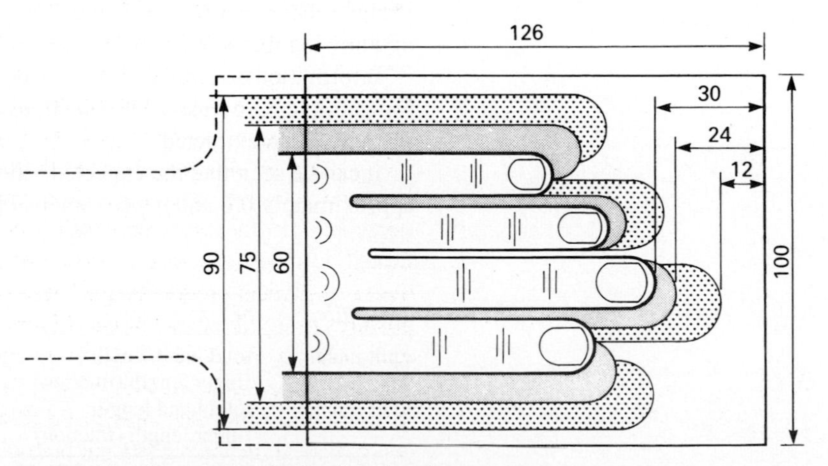

Figure 2.22 Considering the fit of the hand exerciser.

Computer workstation

The consideration of the design of a computer workstation demonstrates the use a designer may make of ergonomic information as well as anthropometric data. The primary problem lies in the determination of the precise nature of the users' requirements. The major market for computer workstations is in offices and the specification of the dimensions of office workstations, desks, tables and chairs is covered by BS 5940: Part 1: 1980, an extract of which can be found in PP 7302: 1987, *Compendium of British Standards for Design and Technology in Schools*. The recommendations contained in BS 5940 are based on anthropometric data for adult men and women and selected to fit 90% of the population. The limits on each dimension are therefore the 5th percentile adult woman and the 95th percentile adult man. Fig. 2.23 and Table 2.4 show the recommendations for office chairs and Fig. 2.24 and Table 2.5 show the recommendations for desks, worktops and foot rests.

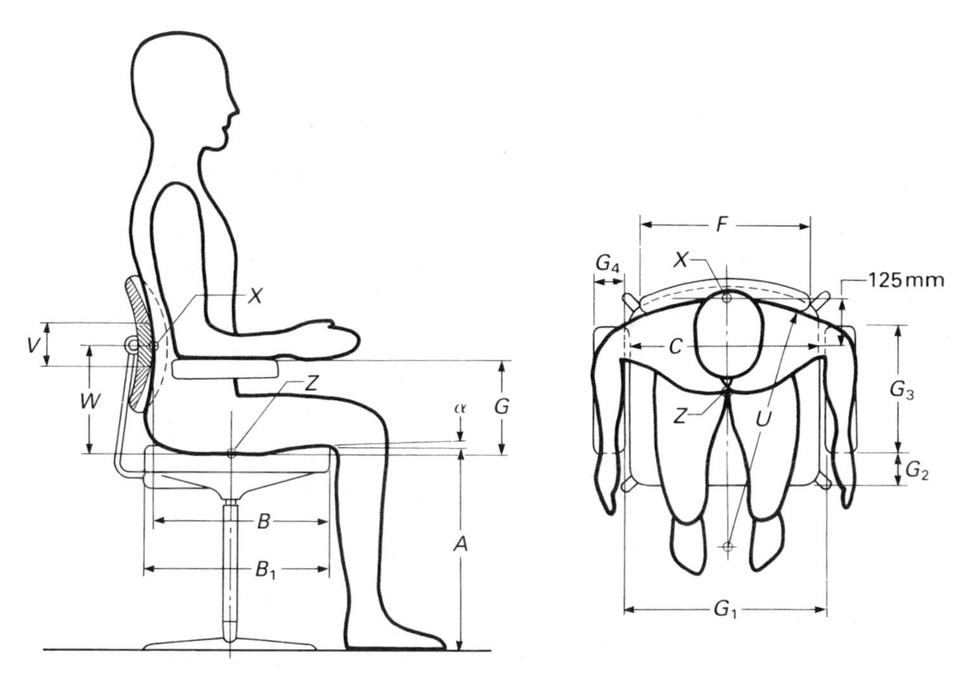

Figure 2.23 Office chairs (from BS 5940: part 1: 1980).

This information represents a great deal of accumulated experience in relation to office furniture, but it does not reflect the change from the paper to the electronic office. This change may not be incorporated in British Standards for a few years because revision takes time. The most significant changes are the length of time an operator spends at the workstation before moving, and the magnitude of the forces necessary to operate the equipment. The nature of these changes was revealed in a recent ergonomic study undertaken in Switzerland by Grandjean.[3] In the paper office people would move quite often to retrieve or place documents in filing cabinets, get information from colleagues, use desktop calculators etc. In the electronic office people do not need to move as frequently because a single machine can provide document storage, word processing, communications and calculating facilities. Equally, modern electronic keyboards require much smaller operating forces than their mechanical predecessors. These two factors result in operators setting body postures more like those of car drivers, when they are given fully adjustable workstations. Fig. 2.25(a) shows the mean and range of observed trunk postures for fifty-nine operators and Fig. 2.25(b) the consequential recommendation. Although the upright posture indicated in BS 5940 would still be appropriate for traditional equipment requiring significant operating forces, it is not suitable for operating computers over long periods.

Seat	
A seat height:	
chair with fixed height seat	440
chair with adjustable height seat, minimum range of adjustment	420 to 500 (see note 1)
B effective seat depth:	
chair with fixed back	380 min. 430 max.
chair with adjustable back	380 min. 470 max.
minimum range of adjustment	380 to 420 (see note 1)
B_1 seat pad depth	380 min.
C seat width	400 min.
** slope of the seat in relation to the horizontal	0° to 5° (see note 2)
Back	
W vertical height of X above Z:	
chair with fixed backrest	210 ± 15
chair with adjustable backrest	170 min. 250 max.
minimum range of adjustment	170 to 230 (see note 1)
V vertical height of area of essential lumbar support (having X at its centre)	100 min.
F width of essential lumbar support:	
general purpose chair	360 min.
machine operator's chair	360 min. 400 max.
U horizontal curvature of lumbar support, radius	400min.
Arm rests (if fitted)	
G height of arm rests above point Z of the seat	200 min. 250 max. (see note 3)
G_1 inside distance between arm rests	460 min. 525 max.
G_2 set back of arm rests in relation to the front of the seat	100 min.
G_3 length of the arm rests	200 min.
G_4 width of the arm rests	40 min.

Notes
1. The range of adjustment provided shall include at least the specified minimum range and may be larger.
2. Preferred angle 3 ± 1°.
3. Preferred height 215 min. 230 max.

From BS 5940: Part 1: 1980.

Table 2.4 Dimensions of chairs in mm (from PP 7317: 1987)

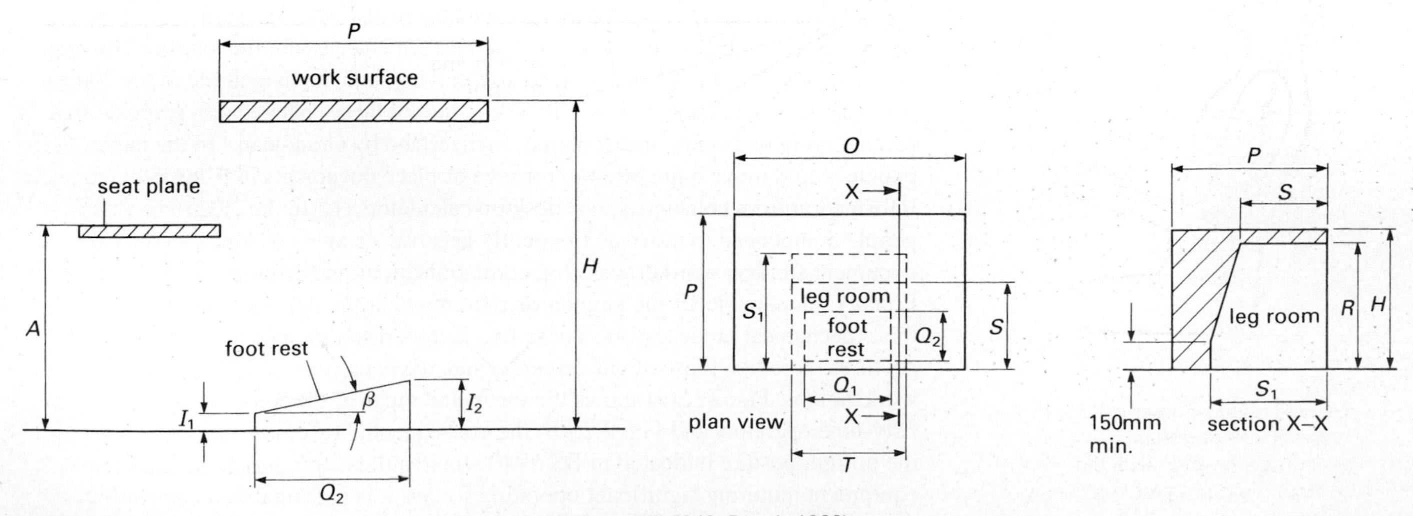

Figure 2.24 Desk, table, worktop, foot rest and leg room (from BS 5940: Part 1: 1980)

Desks, tables and worktops	*General purpose*	*Machine operators*
H height of top surface:		
fixed height top	720 ± 10	670 ± 10
adjustable height top recommended adjustment range (see note 1)	670 to 770	610 to 720
Leg room		
R clearance below desk top:		
fixed height top	650 min.	620 min.
adjustable height top	620 min.	580 min.
T clearance across the kneehole	580 min.	580 min.
S leg room, front to back	450 min.	450 min.
S_l leg room, front to back	600 min.	600 min.

Drawers (see note 2)	*Horizontal storage*	*Suspended filing*	*Other types*
Internal dimensions of usable space			
length, front to back	420 min.	420 min.	Not defined
width, nominal	330	330	330
height	120 min. (see note 3)	270 min. 290 max.	Not defined

Notes

1. If adjustment is provided in fixed steps, the steps should not exceed 30 mm.
2. Dimensioned to suit size in accordance with BS 1467 and BS 4264.
3. For horizontal storage of A4 files and vertical storage of A6 cards.

From BS 5940: Part 1: 1980.

Table 2.5 Dimensions of desks, tables, worktops and drawers

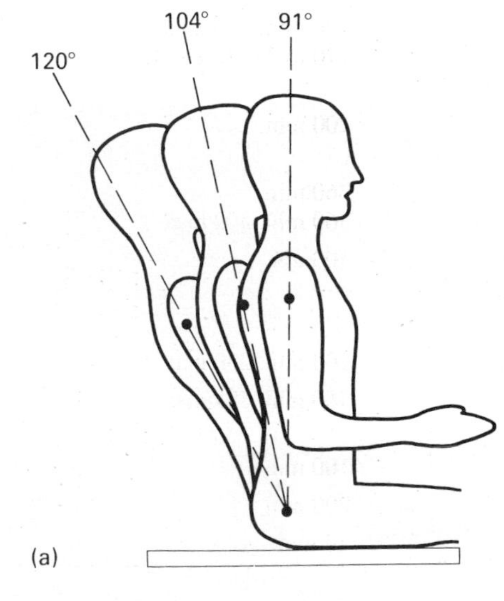

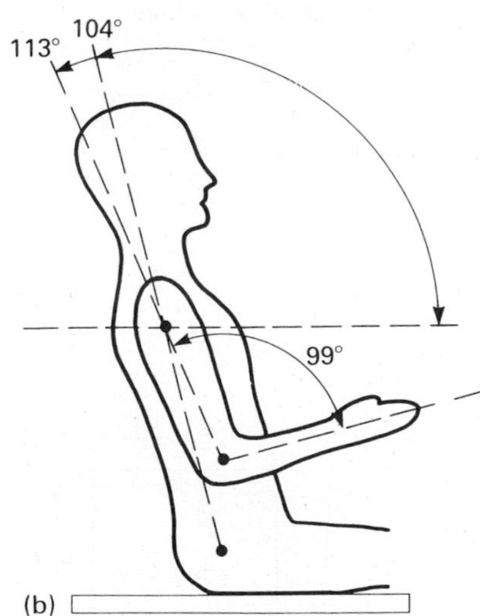

Figure 2.25
(a) Mean and range of observed trunk postures for 59 operators
(b) The 'average posture' with the workstation settings preferred by the operations.

Having resolved the question of posture, it remains to place the visual display unit appropriately. Fig. 2.26 shows the preferred and acceptable locations of visual displays as given in PP 7310: 1984. This information would need to be combined with the minimum and maximum dimensions for the eye position, in order to determine the possible location of the display relative to the keyboard. A graphical method for achieving this can be found in PP 7310.

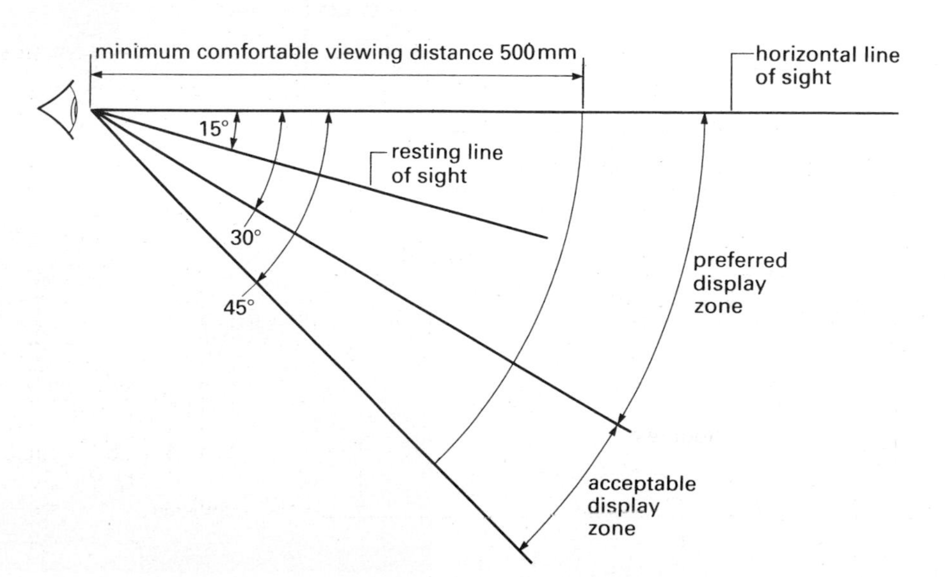

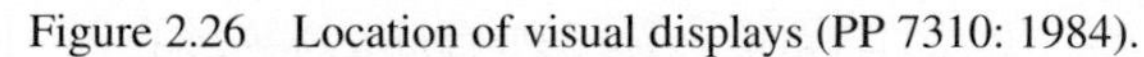

Figure 2.26 Location of visual displays (PP 7310: 1984).

The determination of the shape, remaining dimensions and layout of the workstation depends on a number of further factors, for example:

- the arm reach forwards and sideways,
- the size of the keyboard, terminal and other equipment, e.g. the telephone or printers,
- documents or manuals to be used,
- right- and left-handedness.

It should be remembered, however, that altering the body posture periodically helps to reduce fatigue, and the promotion of such movement should be part of the designer's thinking. Having everything within easy reach might not be the optimum arrangement.

Cab design study

In order to explore some of the broader ergonomic issues, consider the cab design study illustrated by the model shown in Fig. 2.27.

Figure 2.27 Concept model of a cab for a large digger.

The seat design problems are similar initially to those encountered with the computer workstation, but they are complicated by the need to provide reaction forces to those produced when operating the foot pedals and the necessity of dampening the constant jarring which would otherwise cause damage to the base of the spine. The latter problem is related to the characteristics of the vehicle suspension system, and the two must be considered together. In addition to the seat design issues, the environment needs to be controlled to provide suitable temperatures and noise levels for the operator, and the instrumentation and controls need to be selected and located to promote efficient and safe use. In this type of situation the human operator is continually receiving information, making decisions and operating controls in response to the perceived need. The design problems are in these respects similar whether the human is operating a power station, driving a vehicle or flying an aircraft, although each situation will also have its unique features.

Receiving information Information will come to an operator either directly or through an instrumentation system. Several of the human senses may be involved in receiving crucial information and it must be remembered that sight is not the only one being used. In the present example, changes in the sound of the engine or associated with the digger cutting action might be providing useful control information and warning signals. The driver might well be using the jolting or vibration of the cab to monitor performance, and perhaps even his sense of smell. In any vehicle good vision must be ensured, but in this case it must be in all directions, because the driver will be continually reversing and turning. It must also be possible to see the part of the vehicle where the digging action occurs. The polyhedral shape of the cab ensures good visibility.

Instrumentation systems provide information in one of two forms – as an analogue or digital display as illustrated in Fig. 2.28.

Both of these have virtues which are a direct result of their form. Digital displays normally take up the whole of the display area, and when you look at them you see their value immediately. They therefore promote rapid response and do not require prolonged periods of observation. Analogue displays normally use a pointer and a scale, and reading them takes longer. Several tenths of a second are required to absorb the image. Their virtue is that the information is shown in context, for example, seeing that the petrol tank is half full may well be more useful than knowing it contains 10 litres. With a digital display it is necessary to recall a reference value (for example that the total tank capacity is 20 litres) in order to appreciate the significance of the information. Either type of display can flicker or oscillate, but this problem can be more acute with the faster response of the digital system. For vehicles a mixture of analogue and digital instrumentation is likely to be most suitable – digital, where fast information is essential, analogue, where seeing the trends is what matters.

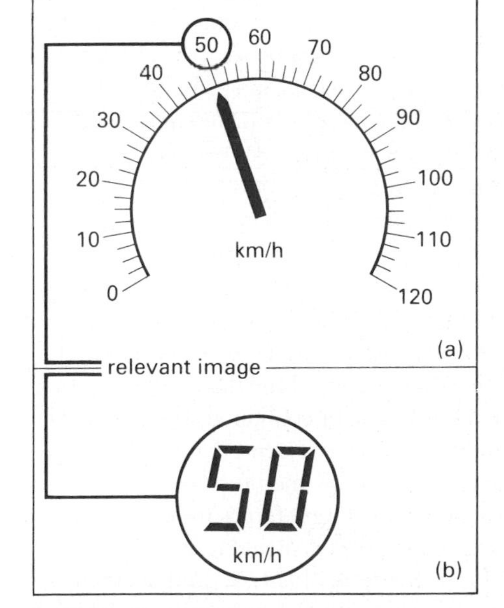

Figure 2.28
(a) Analogue display
(b) Digital display.

As with the computer screen, the instrumentation must be placed at the appropriate distance and viewing angle in the cab. Nearly all sources of ergonomic information give detailed data concerning the design of digital displays and analogue scales, for example, appropriate sizes for markings and lettering, scale lengths for particular reading distances and conventional pointer designs, but PP 7317: 1987 gives a particularly good account. One crucial general issue is that of stereotypes. People come to expect certain things to happen – sometimes as a result of long learning curves. If you turn a fluid control valve clockwise, for example a water tap or gas cooker control, you expect it to turn the fluid supply off. If you turn an electrical control, clockwise you expect the appliance to come on. Analogue displays are expected to indicate an increase in value when the needle goes clockwise, upwards or to the right as illustrated in Fig. 2.29.

In an emergency, or under stress, people tend to see what they expect to see, and respond as they are used to, and the design of instrumentation and control systems must take this into account.

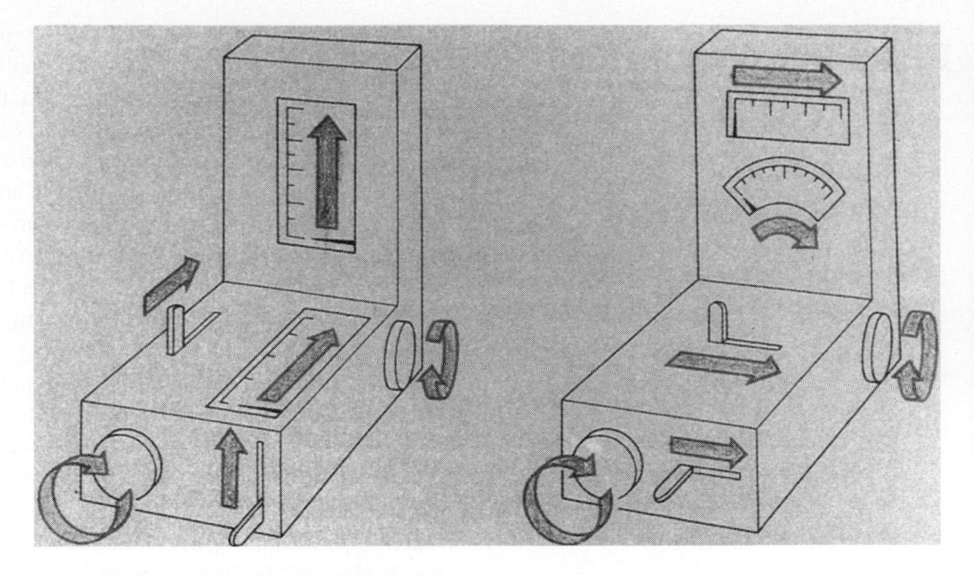

Figure 2.29 Stereotypes, illustrating the expected result following the movement of a control
(a) The pointer of the vertical or flat reading instrument would be 'expected' to move in the direction of the arrow following the movements shown on the levers or knobs.
(b) With horizontal reading instruments the appropriate control movements are shown.

Making decisions Good decision-making depends on the efficient operation of the human brain. Stress, perhaps as a result of concern for personal safety, will reduce the brain's effectiveness. Equally people's performance is very sensitive to temperature changes as indicated in Table 2.6.

Safety and efficiency in the digger operation is going to be as dependent on the design of the environmental control system as it is on any other aspect of the design. Any feeling of insecurity or concern in the driver's mind is likely to result in unsafe or inefficient use of the digger.

Controls The driver will respond to changing situations by operating controls. Apart from conforming to the human stereotypes, the controls must be selected to give the right mechanical performance and to have appropriate visual characteristics. They may be operated by either the hands or feet, but these have very different strengths and capabilities. Hands and arms are capable of very delicate and sensitive movements, but are prone to muscular strain, particularly if controls are not carefully positioned. The human foot is incapable of performing difficult manoeuvres and correct positioning of pedals is vital.

There is, of course, a wide variety of controls: cranks, handwheels, knobs, levers, pedals, push buttons, rotary switches, joysticks etc. They differ in the speed and accuracy with which they can be used and the magnitudes of the forces necessary to operate them. The best choice depends on the kind of control function required; for example, simple on–off control, a multi-position switch or continuous operation. The sensitivity will be determined by the required size of a movement; that is, larger diameter knobs allow bigger movements at the edge of the knob for fine adjustments. Equally the greater diameter would allow larger forces to be exerted with an increased risk of the operator causing damage. You should refer to ergonomic texts, for example, Murrell (1979) or Shackell (1974), to find detailed research findings and to manufacturers' catalogues to establish currently available products. For the digger a combination of twistgrips and foot pedals was selected as shown in Fig. 2.30.

°F	°C	
110	43	just tolerable for brief periods
90	32	upper limits of reasonable tolerance
80	26	extremely fatiguing to work in. Performance deteriorates badly and people complain a lot
78	25	optimal for bathing, showering. Sleep is disturbed
75	24	people feel warm, lethargic and sleepy. Optimal for unclothed people
72	22	most comfortable year-round indoor temperature for sedentary people
70	21	optimum for performance of mental work
64	18	physically inactive people begin to shiver. Active people are comfortable
60	16	manual dexterity impaired (stiffness and numbness of fingers)
50	10	lower limit of reasonable tolerance
32	0	risk of frostbite to exposed flesh

Notes
These figures are a rough guide only and should not be used for design purposes. Effective temperature (ET*) is equal to air temperature when relative humidity is 50%, air flow is small and radiant heat is negligible.

Table 2.6 Typical responses of effective temperature (ET*) (from PP 7317: 1987)

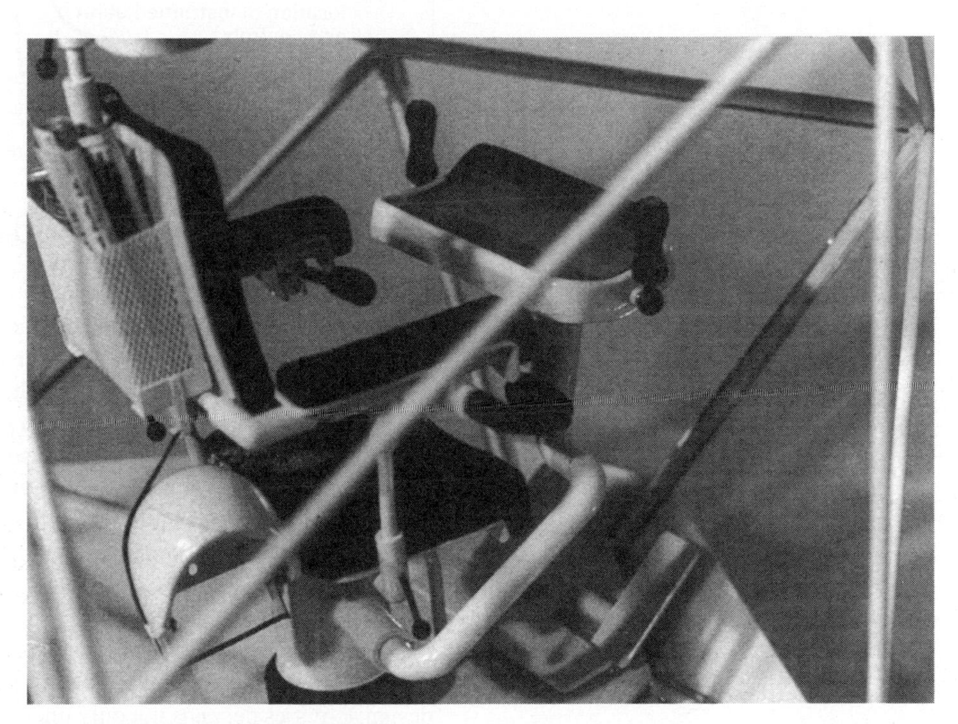

Figure 2.30 Interior of the cab model.

The layout of instrumentation and control panels is also a matter of crucial concern. The controls must be grouped by function and laid out in relation to the operating sequence. All zeros should be at a similar point to aid fast checking. Some switches might be flush with the surface and others recessed to prevent accidental operation. Colour might also be used to distinguish specific groups. Fig. 2.31 shows a summary of the ergonomic considerations relating to the instrumentation and controls for the cab design study.

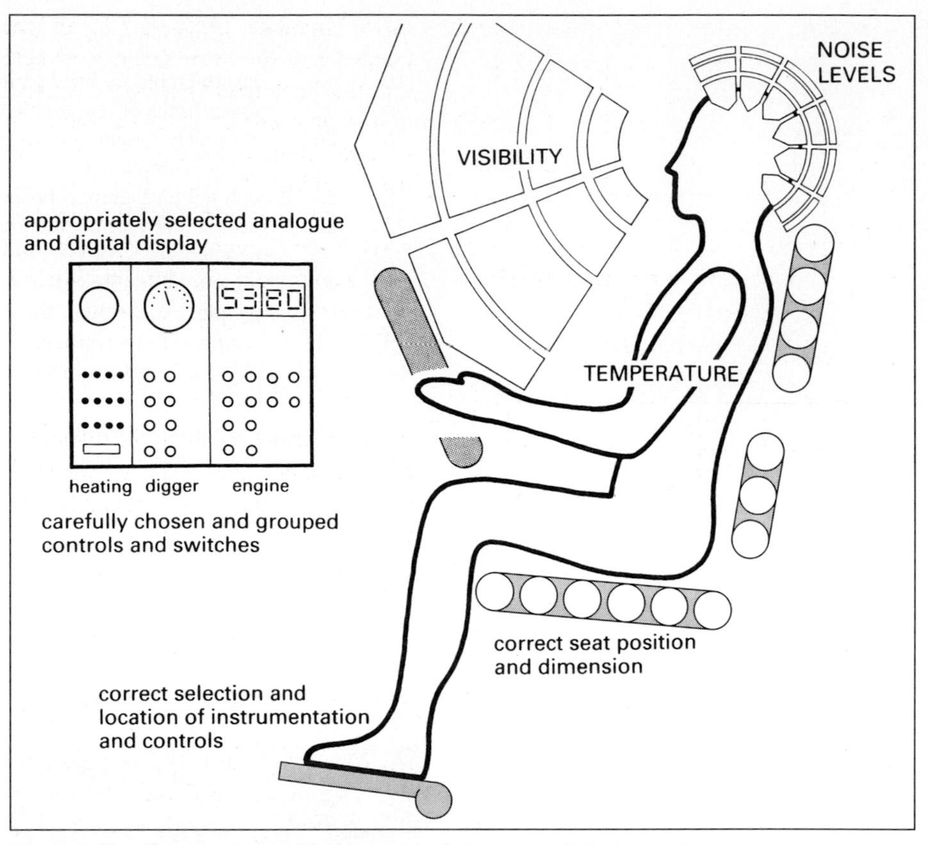

Figure 2.31 Ergonomic considerations in relation to a cab design study.

2.4.2 Mathematical models for performance and cost

Modelling concerns the construction and investigation of a limited representation of a product or system in order to discover something about the characteristics of a fully developed version. The necessity and the ability to use modelling is what distinguishes a modern designer from a traditional craftworker. Change in products was once brought about by a series of small evolutionary steps introduced by succeeding generations; now new products or systems come in to being as a result of the use of models which allow them to be tailored in advance to the perceived user need. The use of 2-D sketching as a modelling tool is well understood, but the potential of 3-D models of different types and mathematical modelling is much less often exploited. Much of the later sections of this book concern the mathematical modelling of various physical systems and it is not appropriate to dwell on this aspect here. However, the use you will be able to make of this information in your design activities depends not only on how well you understand the technology, but also on your conceptual understanding of the use of modelling in design.

The concept of a mathematical model

Anthropometrics concerns the mathematical modelling of humans. They are represented by statistical tables and graphs in order to aid the design process. Designers often find this an inadequate form in which to present the models and consequently turn to the use of 2-D sketches in scaled-down proportion or 2-D card models. These alternative forms of presenting anthropometric data do not however alter the reality that they are based on a statistical (mathematical) model of humanity. The statistical model has in-built assumptions, particularly that of a normal

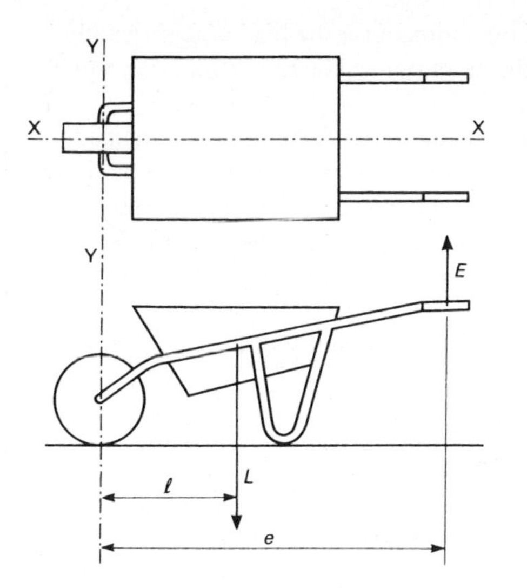

Figure 2.32 Critical axes for a wheelbarrow.

distribution (see section 2.4.1), and these must not be overlooked. You cannot use as a defence against your failure to meet the user's requirements that the design fitted the statistical model perfectly. Another way of putting this might be that you cannot blame the user for not fitting the model – where would fashion houses get if they tried to defend to the customer poorly fitting garments on the grounds that they fitted the manikin?

A mathematical model is a tool for assisting the design process. If the design of a barrow were being considered then taking moments about the two axes shown in Fig. 2.32 would yield vital information to the designer.

In the starting position shown, the equation established by taking moments about YY,

$$L \times l = E \times e$$

yields information about the difficulty of lifting a given load. It is important to emphasise however that even with such an apparently simple mathematical model the designer still has considerable influence over the results which will be obtained. It depends on whether the barrow is imagined to be full of earth or sand or concrete, or filled level with the top or piled high, what the magnitude of the force calculated to lift the barrow will be. The mathematical model does not yield a single unique result – rather a range of results dependent on the assumptions made.

If moments about YY are considered in later positions it might be superficially concluded that there will be no change in the effort required as the barrow is lifted. Fig. 2.33 shows how an excessively simplistic mathematical abstraction of the real situation can indicate this result – the lever arms for the forces appearing to remain proportionally related as the barrow rotates.

The model indicated in Fig. 2.33(a) ignored several key features of the situation and therefore only indicates a partial truth. What happens to the direction of the effort as the barrow rotates? The load will continue to act vertically, but the effort will not. It will in fact rotate to be perpendicular to the handle, thereby increasing the effective lever arm. The position of the point of rotation is also displaced vertically from the line joining the points of application of the load and the effort. Fig. 2.33(b) is an exaggeration of

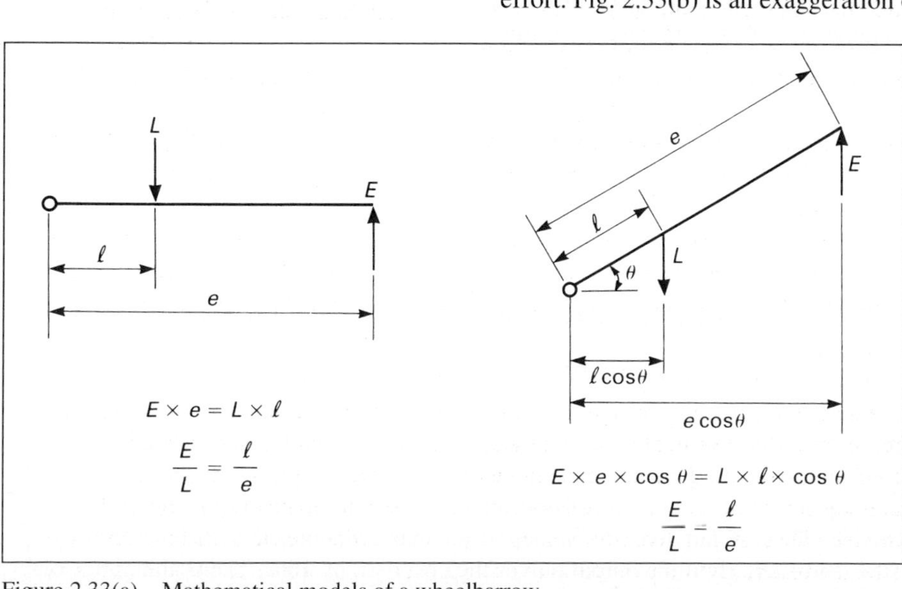

Figure 2.33(a) Mathematical models of a wheelbarrow.

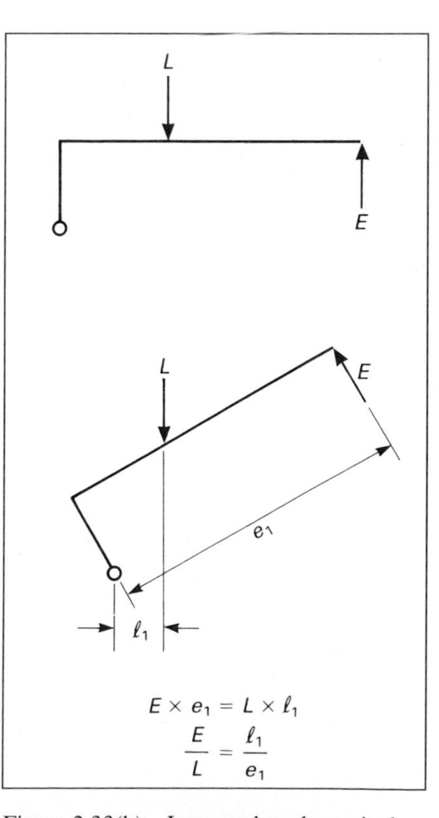

Figure 2.33(b) Improved mathematical modelling of the tipping action.

the real situation, but shows how the turning moment for the load is significantly reduced by the vertical displacement of the pivot point. The ratio l_1/e_1 is seen to be very much less than the original ratio l/e.

A designer who concluded that the force required to lift the barrow remained unaltered as it was tipped would produce a very inadequate design and would have shown no understanding of the fitness of a mathematical model for its purpose. Considering moments about the other axis in Fig 2.32, XX, yields information about the difficulty the user would have in keeping the barrow upright if it were disturbed from its vertical position.

The above discussion illustrates the key warnings which should be heeded when you use mathematical modelling:

- be wary of statistics,
- do not oversimplify,
- be critical of assumptions,
- do not expect a *right* answer, just an answer for the particular input information.

In Chapters 4 to 20 you will find a wealth of information in the form of mathematical models. If you choose models which fit your design situation sufficiently well, you can use them to aid your decision-making, confident that you will reach a position closer to the endpoint more quickly than by using empirical methods. As a final word of caution, however, it is worth noting that many ergonomists do not believe in the use of anthropometric data based on static measurements of humans at all, and argue that all design should be based on user trials. Clearly there is a risk of continually reinventing the wheel, but this view emphasises that the relationship between the user and the product or system is a dynamic one which may not be adequately represented by the anthropometric model.

Mathematical models for predicting costs

Common mathematical models for cost prediction depend on estimating the direct material costs and then relating these to the final product cost through some ratio dependent on the product type or industrial sector. One of the simplest rules of this type was proposed by Herbert Rondeau in the USA[4] – the *1-3-9 rule*. The numbers 1, 3, and 9 are the ratios of the material cost, manufacturing cost and retail price respectively. This rule was based on his many years' experience of estimating costs in the chemical industry, primarily for pressure vessels. The calculated direct material cost, C, is estimated as follows:

$$C = WMF_{w}F_{t}$$

where C is the component material cost
W is the weight of the component
M is the cost per unit weight
F_w is the wastage factor
F_t is the tooling factor

Rondeau suggests that F_w and F_t should both be taken as 1.1 if no other information is available.

If the direct material cost calculated in this way is multiplied by 3 it gives an estimate of the manufacturing cost to the factory. If it is multiplied by 9 it represents the cost to the customer or the retail price. The factor of three assumes that no profit is being shown at the manufacturing stage, and components bought in later are likely to show a ratio of around four. Rondeau also notes an interesting relationship between the cost and quantity, namely if the output doubles then the cost falls by 10%, and similarly if the output halves the cost rises by 10%. The usable range of this rule was found to be about six to eight doublings (or halvings).

The effectiveness of using the weight of a component as a predictor of its manufacturing cost has been the subject of much investigation. Graphs showing such relationships can be found in several British Standards, for example, BS 4360: 1972 for mild steel fabrications, BS 1490: 1970 for pressure die castings and BS 1452: 1981 for grey iron castings. Perhaps inevitably, this approach has been demonstrated to be effective only when the volume of material used is the major contributor to the component cost. It should be observed that, with castings and injection mouldings, the complexity of the product form exerts a significant influence on the product cost, and simple mathematical models for this type of product have not proved very useful.

A similar approach to the establishment of a mathematical model for cost prediction has been developed in Germany.[5] This relates the total direct material cost to the final product cost, but the ratio varies with the sector of industry. Table 2.7 shows some of the ratios which have been found to apply.

The total direct cost considered here includes that for materials and the cost of bought-in components, and it represents a very high proportion of the total cost of some consumer products; 75–80% for vacuum cleaners and cars. However, for the majority of products the ratio was found to be in the range of 30–60%. It may be possible for you to use these ratios directly, but it is equally important to note the approach. If you are making a particular kind of product it may be possible to estimate the direct costs and establish your own ratio, for example, for DIY tools or bicycles etc. The ratios found in Germany seemed to be good predictors for the particular industrial sectors, but they were, of course, based on large sample sizes.

Product	%
vacuum cleaners	77
motor cars	77
sewing machines	57
desk telephones	54
amplifiers	52
electrical controls	42
wall clocks	42
precision measuring instruments	38
telex typewriters	35
electronic measuring instruments	30
drawing instruments	20

Table 2.7 Material cost as a percentage of production cost for various products[6]

A recent research project carried out at Hull University looked again at the problem of estimating the component cost. The results were included in a book that was published in 1997 by K. G. Swift and J. D. Booker. Their investigation into product costs took into account not only the quantity of material used, but also the manufacturing process. The mathematical model they developed takes the form:

$$C_T = VM_v + RP$$

Where C_T is the component cost.
V is the volume of material used.
M_v is the cost of the material per unit volume.
R is the relative cost coefficient.
P is the basic processing cost for an ideal design by a specific process.

The research project indicated that the basic processing cost varied with the production quantity as shown in Fig. 2.34(a).

It was found that the relative cost coefficient assigned to a component design needed to take account of the complexity of the shape, the suitability of the material for the manufacturing process chosen, the section dimensions, the tolerances required and the specified surface finish. These factors are very similar to those discussed in Section 2.5.1. The resulting cost predictions were found to be very close to the actual component costs if all these factors are taken into account as shown in Fig. 2.34(b). Details of how to make such predictions can be found in their book.

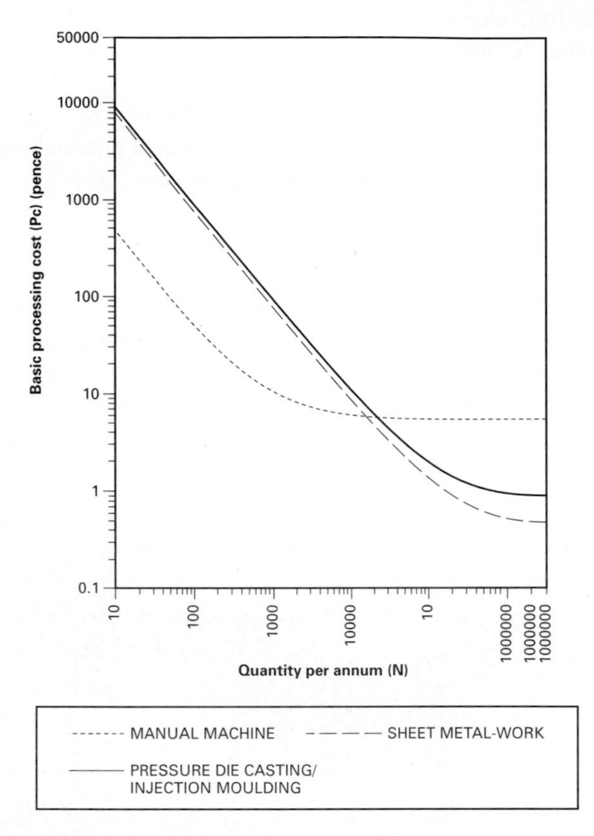

Figure 2.34(a)

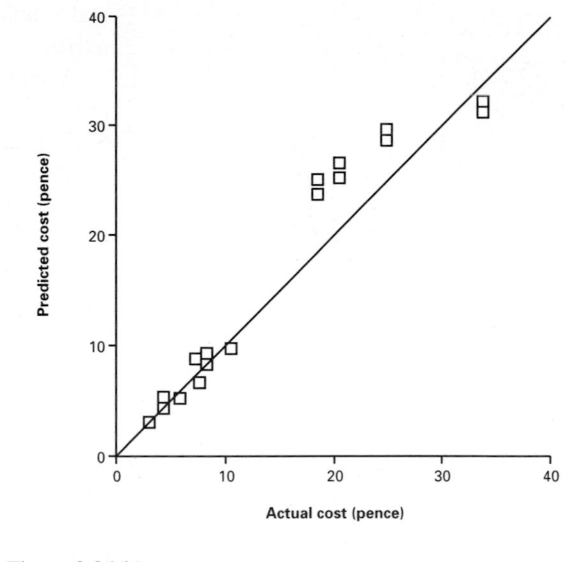

Figure 2.34(b)

2.4.3 Graphical modelling

As their thinking progresses, designers move between one kind of model and another – between verbal, numerical and graphical representations. Graphical techniques are used in the early stages to record information and ideas that are perceived to be relevant. During the divergent and speculative phases drawings are used to help concepts develop by giving them a concrete form which provides an image for the mind to work on. It should be recognised, however, that the act of putting a form to a concept is part of the process of convergence and should not be undertaken too early. Fig. 2.35 shows an early sketch indicating the idea of using a toggle mechanism for an embossing unit. It is almost deliberately vague in order to prevent a firm image becoming established. Fig. 2.36 shows the first appearance of the idea of combining the toggle mechanism with a pelican form, and this drawing still retains some ambiguity. Fig. 2.37 shows a drawing and a model of the pelican embosser in its final form, and here the lines are much firmer and the outline more precise.

Fig. 2.38 shows part of the early evolution of an easel design. It includes the specification and initial ideas, some joint details, and anthropometric and ergonomic considerations. The last drawing shows the easel folded to illustrate the problems of packaging and transportation. Design sheets like these form an effective record for the designer and for the communication of their thinking to others. When the process as well as the product is being assessed it is essential that all the key stages are adequately recorded.

Drawing diagrams like that of the folded easel, which show the product in a changed position, can be the quickest way of exploring movements, although the virtues of quick sketch models should not be overlooked (see section 2.4.4).

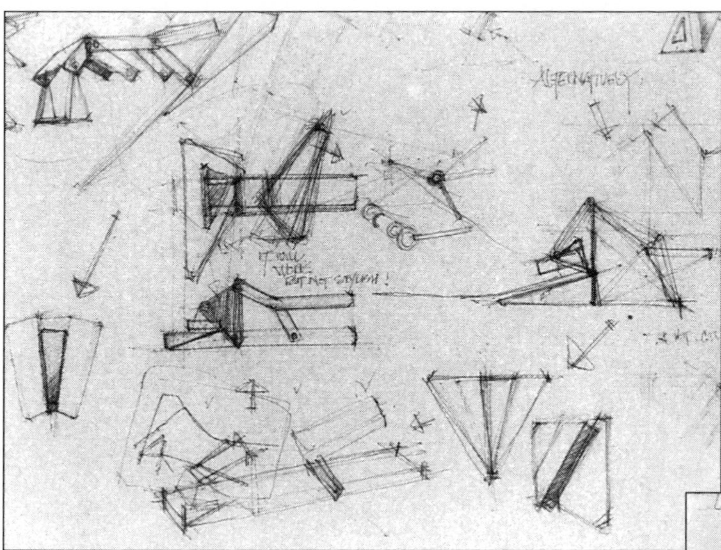

Figure 2.35 An early sketch of the idea of using a toggle mechanism.

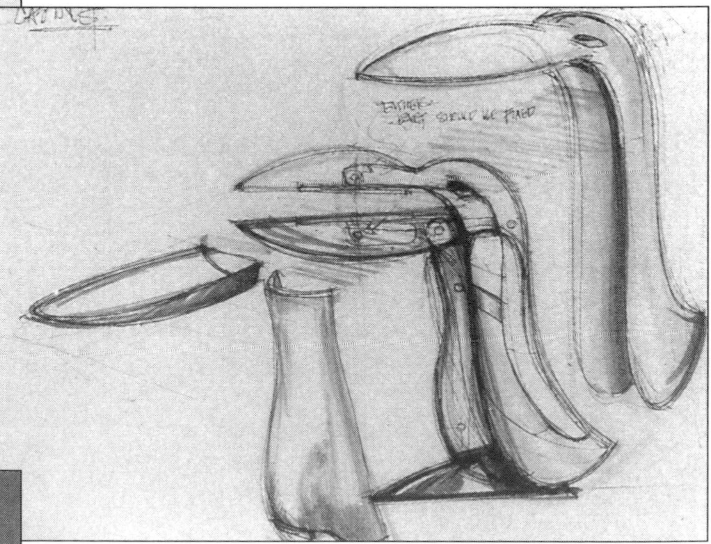

Figure 2.36 The conceptual synthesis of the pelican form and the toggle mechanism.

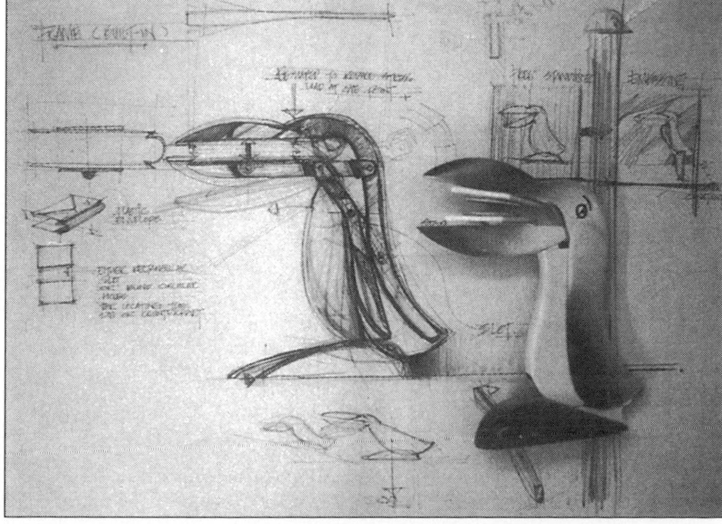

Figure 2.37 The pelican embosser in the final form.

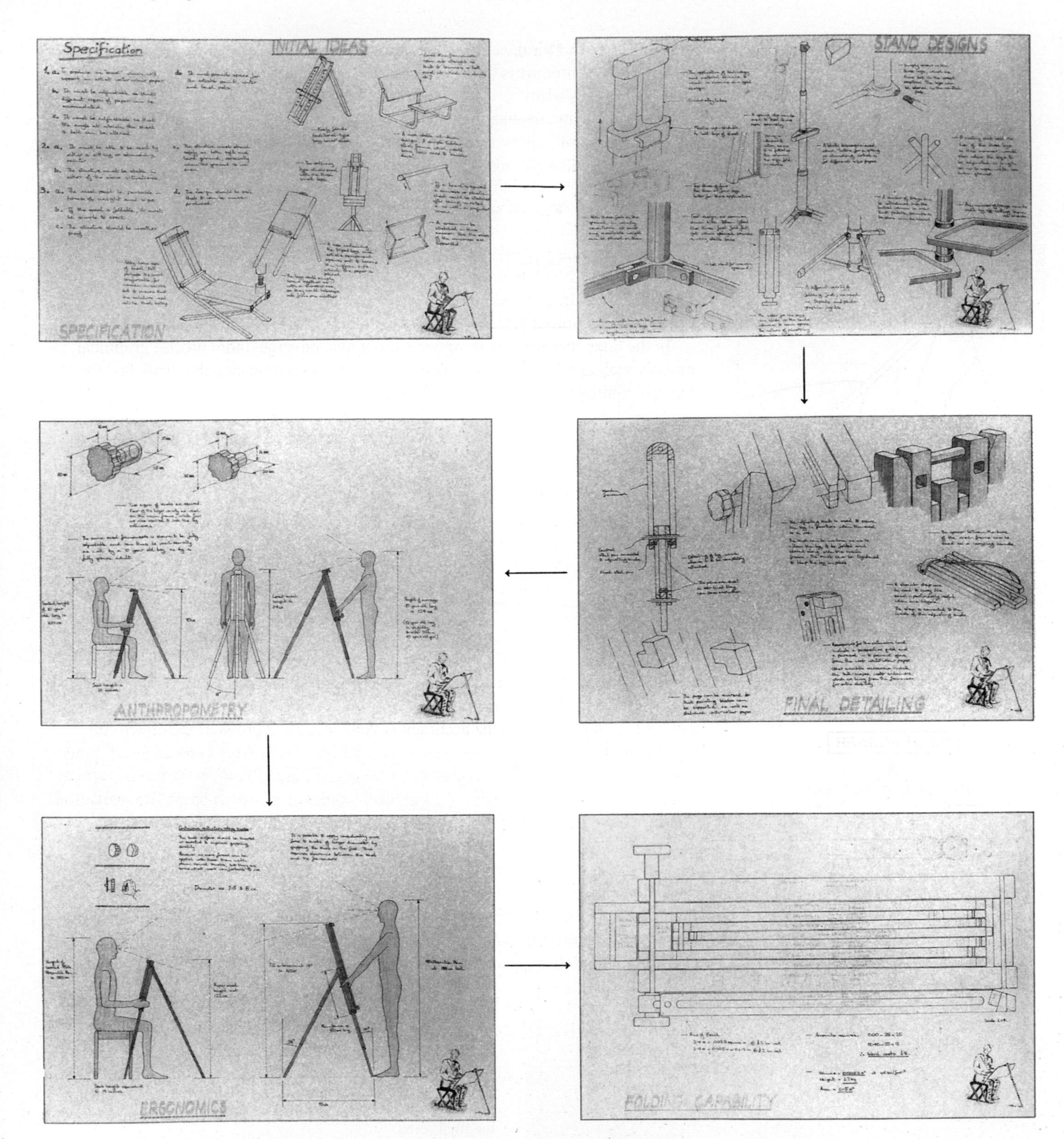

Figure 2.38 Early evolution of an easel design.

Similar examples occur when investigating the movement of mechanisms, by drawing the locus of a critical moving point, and the movement of people and their interaction with a product. Fig. 2.39 shows the approximate position of a 95th percentile man when pushing the excellently designed ballbarrow. The slight risk of inconvenience can be seen, and there might have been some advantage in increasing the lever arms. This has however probably been correctly resolved in relation to other factors like stability and the difficulties of emptying the barrow.

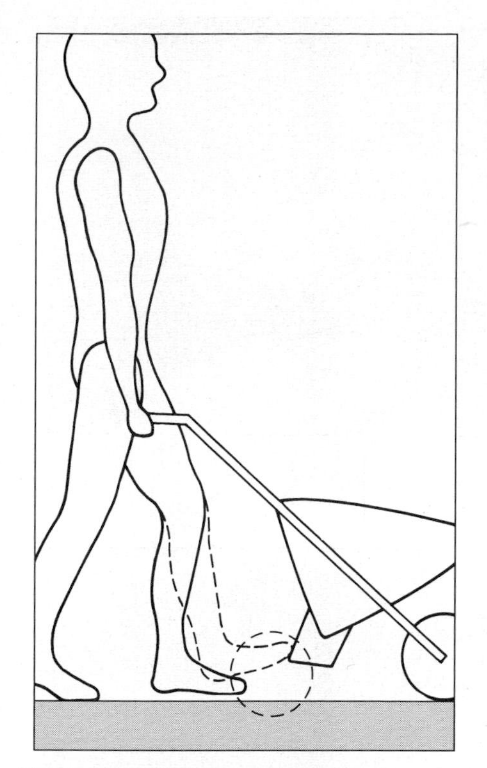

Figure 2.39 95th percentile man pushing a barrow.

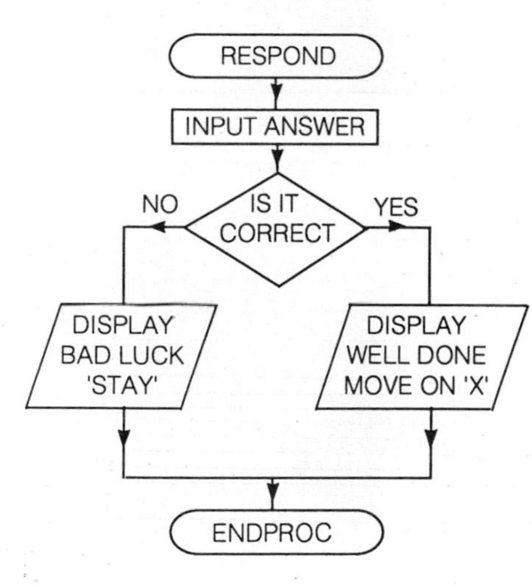

Figure 2.40 Example of a flow chart.

Graphical modelling is also an essential tool in the development of systems. In producing software or designing control systems it is common to start with a flow diagram as shown in Fig. 2.40.

This shows the operating sequence for software to be incorporated in an instructional game. Such diagrams represent the imagined system logic and allow this to be examined and analysed.

Circuit diagrams for electrical and pneumatic systems are used in much the same way in system design as product designers use design sheets. Initial circuits are sketched freehand and slowly refined through mathematical and physical modelling, that is, *breadboarding* and the use of modular component kits. Sophisticated computer modelling allows complex systems in pneumatics and electronics to be modelled quickly and easily. The general issue of physical modelling is the subject of the next section (2.4.4).

In the later stages as the designer's thoughts converge, more precise graphical models such as engineering drawings are used to communicate their conclusions. Design communication is covered in section 2.7.

Drawing to scale

For many products it is usual for accurate scale drawings to be used quite early in the design process. For example, in the car industry a full-size drawing, now mainly on CAD, is sometimes produced from the clay model of the concept car. This drawing then acts as a reference for all the designers who are working on subassemblies. Such drawings are usually known as layout drawings and are a very effective group management tool. They can be equally helpful to designers working on their own in providing a common starting point for all detail design.

In general the level of precision implied by accurate scale drawing is not necessary during the development of concepts, but there are occasions when it is useful or even essential. If the development of the functional requirement for the sports and social centre shown in Fig. 2.15 is studied, it can be seen that the accuracy of the scaling is increasing. In this context the accuracy is vital because it provides the effective constraint on the possible forms. As the design progresses from concept development to detailed work so the importance and relevance of scale drawings becomes greater. Much of the detail design concerns the accurate fitting of parts and the determination of appropriate dimensions. Such work is usually best done through scale drawings.

There is also often a requirement for an accurately scaled drawing in order to assess visual characteristics. Such drawings might be in orthographic, isometric or perspective form but, whichever is chosen, accurate scaling will result in a better impression of the final appearance. Giving the product some kind of context, as is done for the chair in Fig. 2.43 by the reading light and the wall, also helps in assessing the likely product image.

2.4.4 Physical modelling

Physical modelling concerns the making of limited three-dimensional representations of a product or system in order to reveal particular information. It is perhaps first thought of in the context of fluid flow, and associated with finding efficient forms for aircraft, cars, ships and racing cycles. The theory behind this type of model testing in wind tunnels and ship tanks is covered in section 13.9.3. Physical modelling is a key step in the development of most products and systems. Fig. 2.41 shows physical models of electrical and pneumatic circuits and a mechanical system, which have been assembled to get a better idea of how they will function.

Fig. 2.42 (overleaf) shows two very different kinds of model of the artist's easel. One of these was put together very quickly to see how the easel would fold. This type of model is sometimes known as a sketch model. It is intended to fulfil a similar

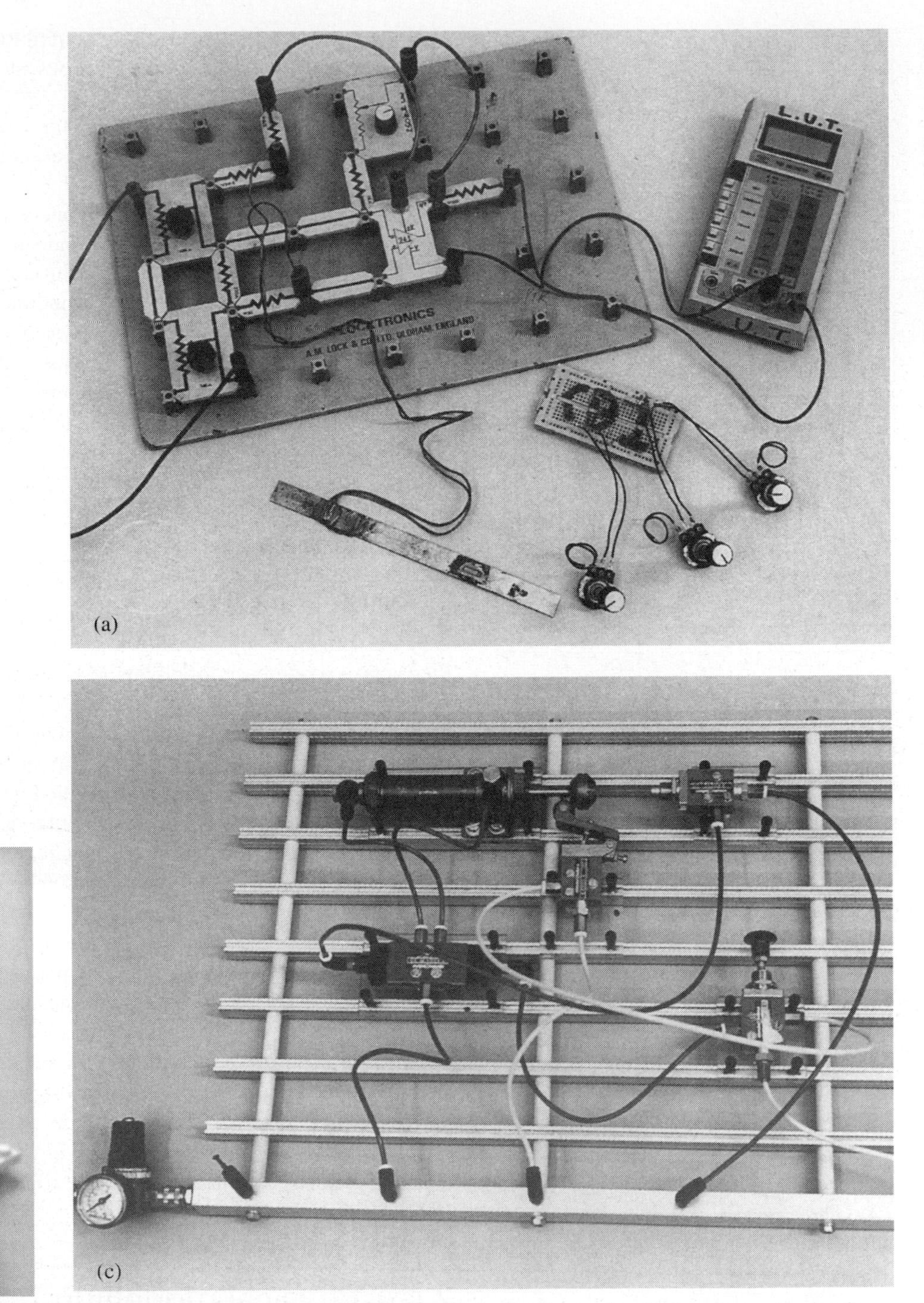

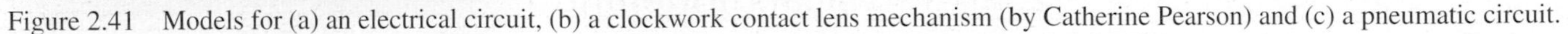

(b)

Figure 2.41 Models for (a) an electrical circuit, (b) a clockwork contact lens mechanism (by Catherine Pearson) and (c) a pneumatic circuit.

purpose to the diagram shown in Fig. 2.38, but it leaves less room for error in visualising the folding action. In this form it would also be very quick to put together, whereas a more accurate model would obviously take much longer to make. Graphical modelling is probably a better method of accurately checking dimensions. The other model of the easel was put together with much more care in order to give a good impression of the way it would look. It was made at a much later stage in the design process when the form had been considerably investigated on paper. The important thing is that the right amount of time and care was put into each of these models in order that they should be able to fulfil their purpose. Fig. 2.43 shows a model of a chair made from plastic sheet and tissue paper together

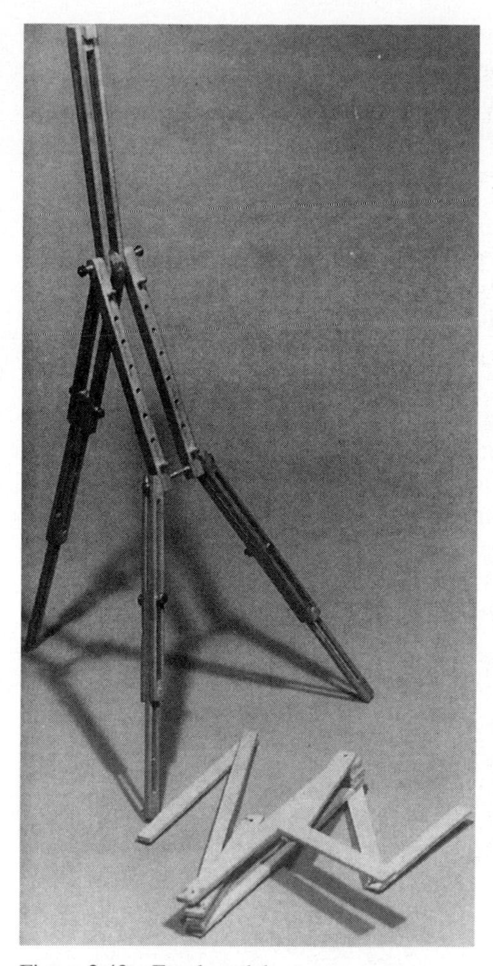

Figure 2.42 Easel models.

Figure 2.43 Chair models.

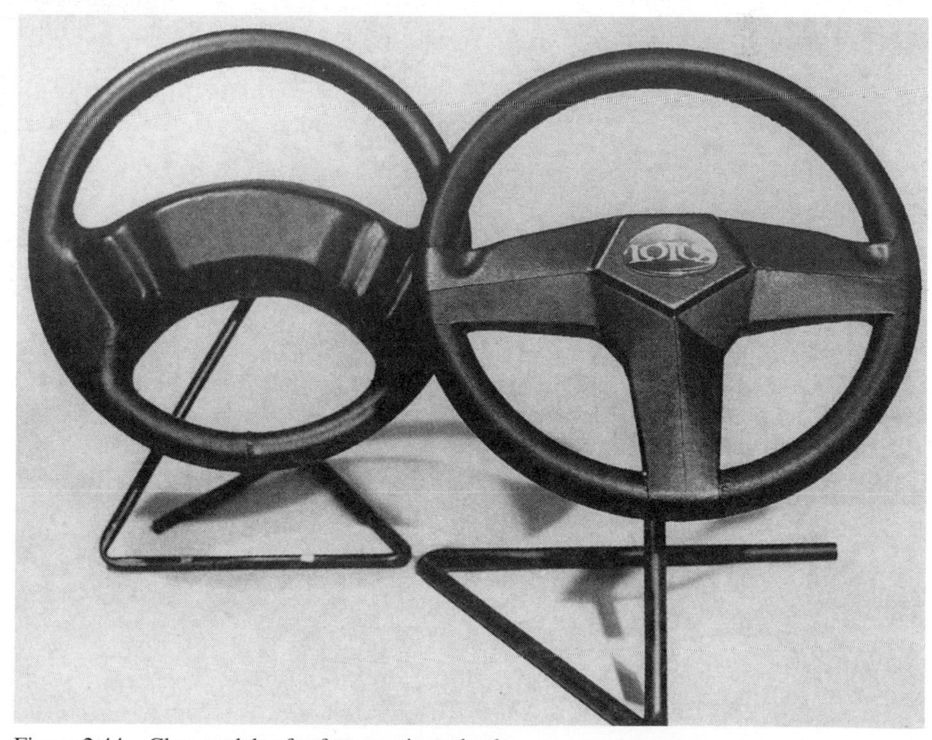

Figure 2.44 Clay models of safety steering wheels.

with a two-dimensional graphic image of the same design. Fig. 2.44 shows original clay models of safety steering wheels designed for the Lotus Esprit, for which the initial development work was carried out using a computer solid modeller (see section 2.4.5). Although made in different materials these later models all indicate how the final design will appear.

Physical modelling has also been long associated with the design of mechanisms and structures. Initial thoughts might again be of Meccano Fischertechnik or Legotechnic, but the use of prefabricated parts is not the only, nor necessarily the best, way of building such models. Fig. 2.45 shows a model of a flipover step for a caravan and the completed product. Although it would have been possible to plot the locus of the mechanism's movement, the cardboard model allows its function to be checked much more easily.

Fig. 2.46 shows four models of designs for trailer chassis. These were made from tinplate with soft soldered joints representing the welded joints of the real chassis. The thickness of the tinplate was chosen to give a good scale factor in relation to the thickness of the steel sections which calculations had indicated were appropriate. They were tested as shown in Fig. 2.47 using weights and a dial gauge in order to compare their stiffness-to-weight ratios. This also allows the deflection of the real structure to be predicted.

(a)

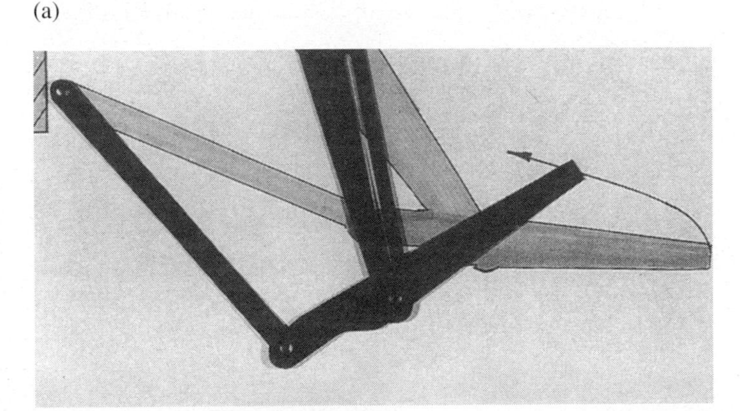

(b)

Figure 2.45 Flipover step for a caravan (a) card model (b) finished product.

Figure 2.46 Four models of designs for trailer chassis.

Figure 2.47 Testing the model chassis.

Relating the real and model deflections

It is shown in section 13.9.8 that the deflection of a structure under a particular load can be estimated from the deflection measured on a model using the following relationship:

$$d_r = d_m \frac{F_r L_m E_m}{F_m L_r E_r}$$

where

suffix m indicates the model,
suffix r indicates the real structure,
d is the deflection (mm),
F is the force (N),
L is the length (mm),
E is the modulus of elasticity (N/mm^2).

The construction of the real trailer was interrupted in order to check the predicted deflection as indicated in Fig. 2.48.

Figure 2.48 Checking the deflection during construction.

The results obtained were reasonably accurate, although the measured full-scale deflections were a little lower than predicted. This was attributed to the stiffening effect of the sides which had been built on when the real tests were conducted and were not present on the models. The correlation was much better than those found with mathematical predictions based on simple beam members, because these mathematical models did not take into account the complex interaction of the members and the consequential stiffening effects.

Physical modelling can be seen to be relevant to aiding the synthesis process for all types of products and systems. It is as relevant in the early stages as in the later ones and should be used in conjunction with graphical and mathematical techniques.

2.4.5 Computer modelling

Computers are now widely used in designing in industry and, with computer power becoming cheaper, they are beginning to play an increasing role in schools and colleges. As their use has become more widespread so the design activities where they can be applied most effectively have become clearer. The easy storage and retrieval of information through large databases is a major benefit as in many other fields. In addition the enormous analytical power of computers enables designers to tackle calculations that previously would have been too time-consuming. Such applications in CAA (*computer-aided analysis*) are widely associated with the idea of finite elements where the equilibrium and continuity equations are applied to small sections of a structure, mechanism or system but there are many other forms. Until recently such programs had to be left to run for hours – sometimes overnight – on large mainframe computers, but the rapid advances in the performance of PCs (*personal computers*) is making such programs widely accessible. Furthermore, the software has become more reliable and robust so that the need for expert interpretation of the results from the computer is less critical. CAD (*computer-aided design*) software is now the most commonly used form of computer modelling and numerous CAD packages are available to schools and colleges. The combined field of CAD and CAA is known as CAE (*computer-aided engineering*).

As well as its obvious advantage in assisting the design process, CAE can be linked to CAM (*computer-aided manufacture*). The development of CAM has altered the nature of some manufacturing operations and these changes are explored at the end of Chapter 6 'Materials processing'. CAA, CAD and the relative newcomer CAID (*computer-aided industrial design*) are discussed below. The direct link from CAD to CAM is important since dimensional information can be entered into the system during design and there is no further need to input data. This can result in major savings in time and improvement in quality through the reduction of errors. Some industries now have direct links between the design offices or studios and the manufacturing facilities. In fabrication shops, for example, shapes can be cut directly from steel plate by oxy-acetylene or plasma cutters from the data entered in the CAD system. Some companies now have fully integrated systems which link CAD, CAA and CAM, and which also allow several designers to work on different aspects or parts of the total design.

CAD refers to systems for both 2-D draughting and 3-D modelling. 3-D models can be based on data concerning the surface of the object (surface models) or the whole volume (solid models). 2-D draughting packages normally entail a long learning curve but, once they are mastered, drawings can be entered quickly. This remains the standard way of entering dimensions into the system, but it is possible to start with a 3-D model. 2-D engineering drawings can be extracted from the completed solid model if required.

Fig. 2.49 shows the fundamental method for entering a solid model. The object is made up of some basic shapes – often called primitives. Here there are four spheres, a cube and a cylinder placed in appropriate positions. These shapes can then be added or *unionised*, as has been done with the cube and the spheres, and subtracted or *differenced*, as in the case of the cylinder and the cube. The resulting shape is shown in Fig. 2.50. Once the basic form of the object has been produced the hidden lines can be removed as shown in Fig. 2.51 or it can be shaded as in Fig. 2.52. Fig. 2.53 shows a solid model of one of the steering wheels designed for the Lotus Esprit (compare this with the clay model shown in Fig. 2.44).

Creating a solid model in this way may be quite straightforward for some forms, but other shapes not related to the available primitives can require long-winded procedures. Once the image is completed, however, it can be rotated or colour-shaded and the lighting position(s) altered. Until a few years ago the colour-shading

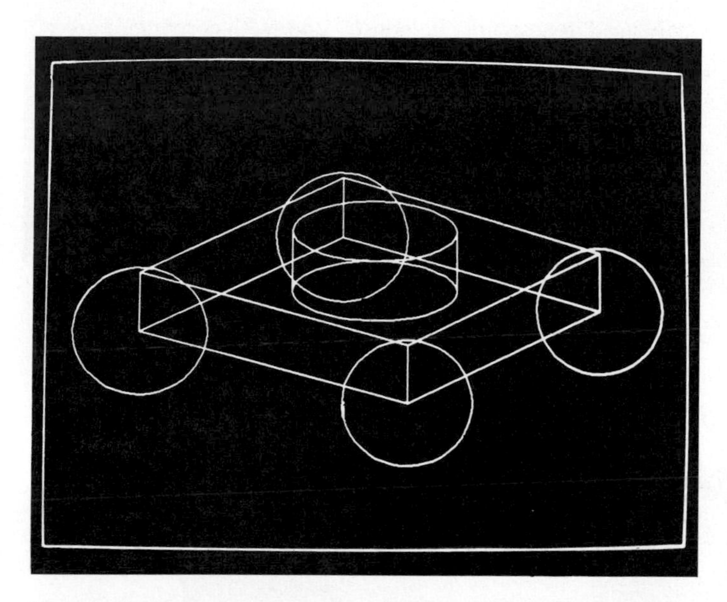

Figure 2.49 Primitive shapes for building up 3-D computer models.

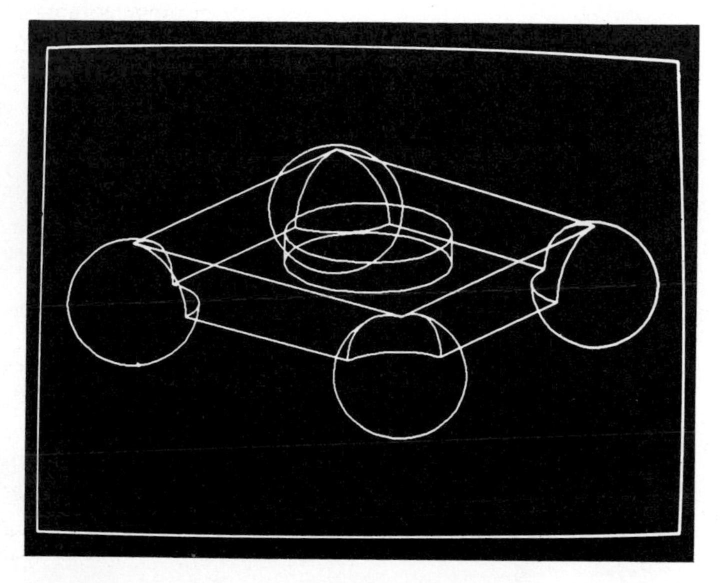

Figure 2.50 Primitive shapes added and subtracted.

Figure 2.51 The object with the hidden lines removed.

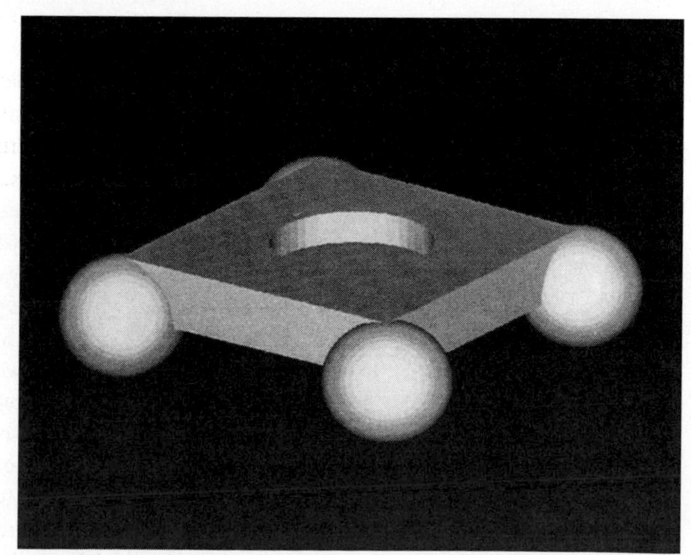

Figure 2.52 The rendered object.

Figure 2.53 A computer model of a Lotus steering wheel.

(or rendering) of solid models took hours of computing time but with modern systems it can be almost instantaneous. It is this aspect of CAD that has led to the development of CAID. In the early stages of the design process designers tend to sketch their ideas very rapidly and, although 'sketch pad' inputting software is being developed, CAD systems are not generally regarded as significant for this purpose. Towards the end of the design process, however, the graphics demands can be laborious and time-consuming as, for example, in producing a range of quality renderings with minor variations in form and colour. It is now possible to generate computer images that are photo-realistic such as the hair dryer shown in Fig. 2.54.

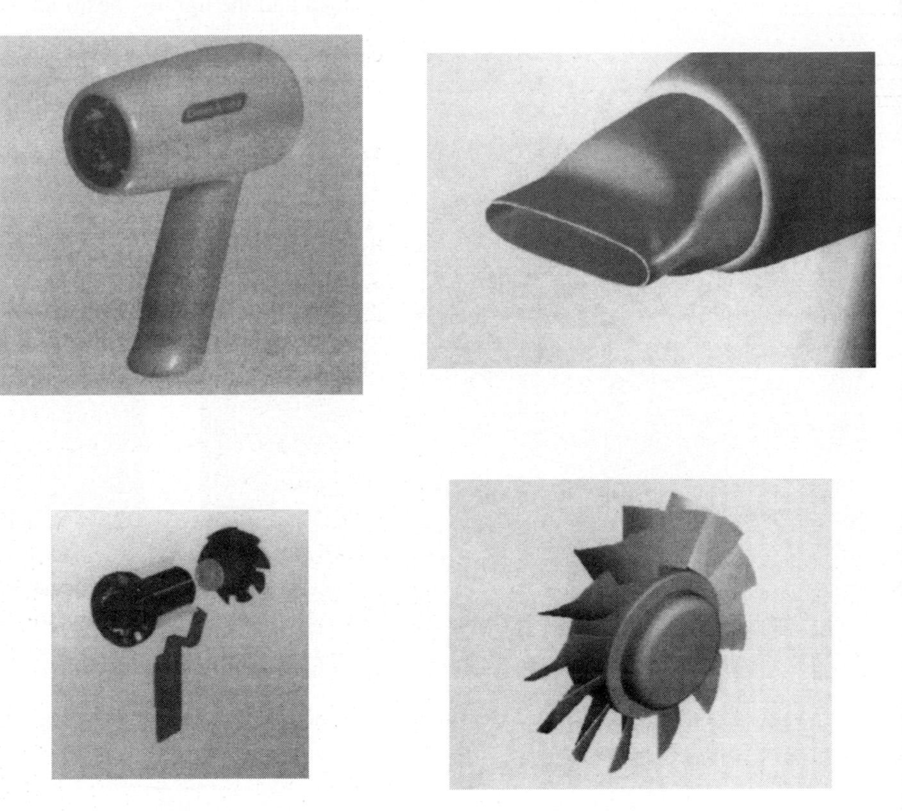

Figure 2.54 Computer-generated images of a hair dryer and some of its components.

The ability to alter the lighting position has led to the use of CAID in the motor industry to investigate 'highlights'. This is a vital aspect of car body design where it is essential to avoid forms that appear to show bumps or dents under some lighting conditions.

There are a number of current developments in the CAID field that enable the production of *rapid prototypes*. These are physical models that are generated directly from the CAD system. They can be used, for example, to produce working prototypes of the product or the tooling for castings (see section 6.31 – the full-mould process). Stereolithography – in which liquid polymer is 'set' at the point of intersection of two laser beams – is probably the most common approach, but by no means the only one. Paper models are produced by stacking large numbers of paper templates (cross-sections) to generate the required form. The current rapid prototyping techniques tend to result in 'stepped' models that require final sanding but this is likely to be overcome in the near future.

Rapid prototyping is becoming ever more important for designers in industry. Consider the following passage written by Katsundo Hitami in 1992 concerning trends and issues in Japanese manufacturing and management.

> Since consumers want products that are of high quality, specialised, fashionable, and different from others, product life cycles are rapidly decreasing. For example, Ford's T-model cars were produced for 20 years, during which time (1908–27) almost 15 million cars were made. In contrast, a recent hit product in Japan – Minolta's automatic focus camera (α-7000) – lasted only three years, supplying 3 million. The average life cycle of an air conditioner is one year, that of Japanese word-processors only three months. Hence manufacturing industries aggressively try to manufacture and supply a huge variety of products. Therefore, 'multiple-product, small-batch production' is now common. Almost 85% of Japanese products and 75% of US manufactured products fit this category.

With product life cycles and, consequently, product development times becoming ever shorter, it is essential for designing to accelerate. Rapid prototyping is one way that this can be achieved. Fig. 2.55 shows an example of rapid prototyping applied to cutlery design.

With CAD/CAM software and hardware becoming increasing available in schools and colleges, parallel developments have been taking place. Figure 2.56 shows examples of rapid prototyping as applied to project work.

The development of *concurrent engineering* (or *simultaneous engineering*) has been the result of the same pressures to reduce product lead times. The traditional approach to product design and development has often been referred to as the 'over the wall' approach. Designing was carried out without any reference to manufacturing and when it was complete it was passed to manufacturing engineers 'over the wall'. Concurrent engineering refers to the overlapping of some of the concept design and design for manufacture activities in order to reduce product development times. The introduction of concurrent engineering has a number of significant implications for the designer.

- There is a need for information exchanges between designers, manufacturing engineers and supply companies. The engineers and suppliers must have early access to the required information and this effectively requires designers to use CAD systems. It is information technology that makes the sharing of information both easy and possible and the introduction of CAID is largely a response to this requirement.
- The designer actually spends longer working on a product. Although being a member of a multi-disciplinary team is different to the designer's traditional pattern of working, they do have longer 'contact' with the product. They might be

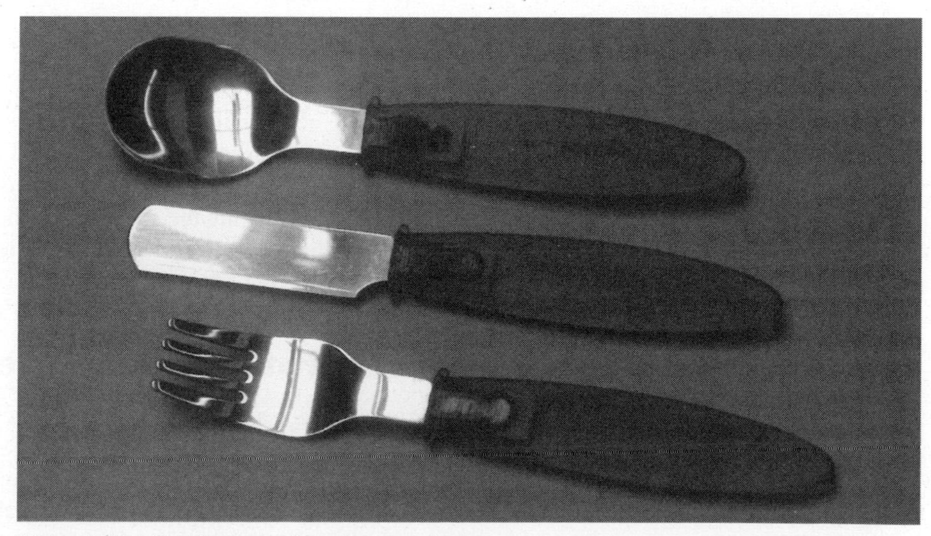

Figure 2.55 Rapid prototyping of cutlery design.

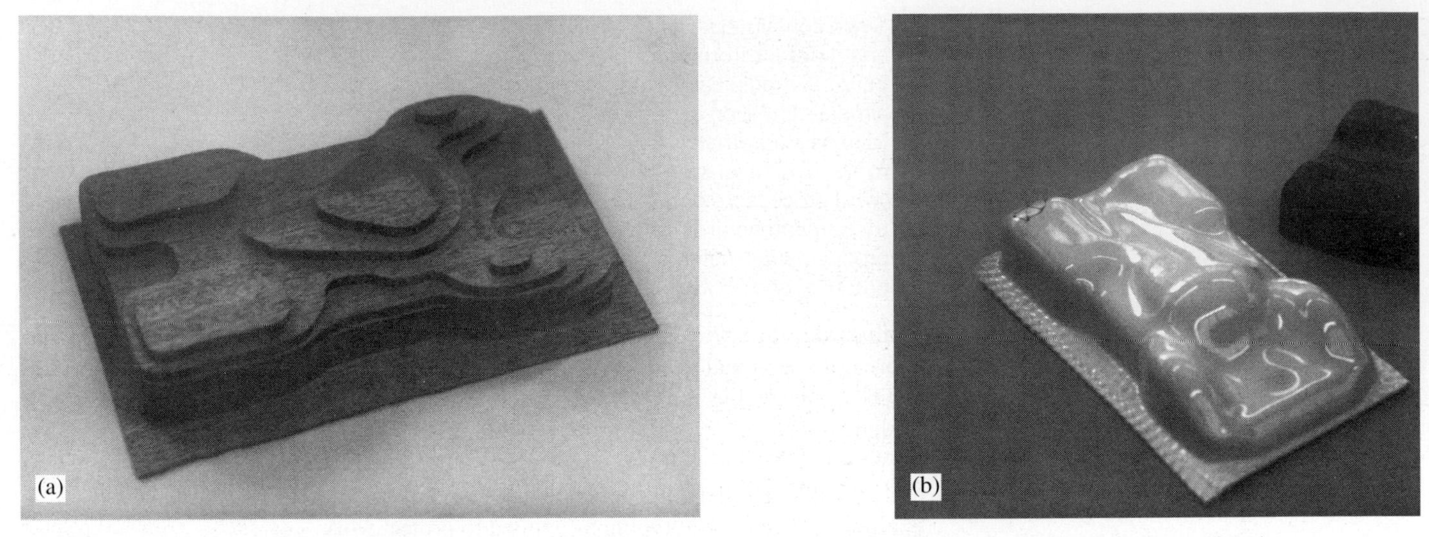

Figure 2.56 Rapid prototyping applied to project work (a) approximate shape for mould CNC routed (b) finished vacuum forming.

involved in early market research activities and still be part of the team making decisions about manufacturing installations and assembly lines. However, in order to participate fully in this wider range of activities, the designer needs a correspondingly broad education.

- The design needs to be right first time. With shorter product life cycles the market does not allow second chances and the penalty facing a company bringing a product to the market late are very severe.
- Short product life cycles bring the issues of recycling and reuse to the forefront. As well as getting the design right from a marketing perspective, the designer must also take an increasingly 'cradle-to-grave' responsibility for their product. If shorter product life cycles, and the concurrent engineering practices that allow them, are not to result in an increasingly 'throw-away' society, then designing for the eventual disposal of the product must become part of getting the design right first time.

CAA was first used in calculating the stresses due to applied loads or the temperatures due to heat flow using finite element programs, but there are many other kinds. CAA can be used to find the forces in frameworks (see Chapter 8) and the shearing forces and bending movements in beams (see Chapter 9). Programs are also available for kinematics – analysing the movement of mechanisms. Other programs can analyse 'fit and function' – the checking of the assembly of components. It is now possible to analyse the fluid flow around objects (see Chapter 13) using computational fluid dynamics, although such techniques are only possible at present using parallel processing and fast super-computers. Fig. 2.57 shows the analysis of fluid flow around a vehicle design displayed on a computer screen. One reason for producing the solid model shown in Fig. 2.53 would be its usefulness in CAA for finding the resonant frequencies of the wheel. These are important in avoiding steering-wheel vibrations at driving speeds.

2.5 Prototype completion and manufacture

A prototype is a fully working version of the product or system made to resemble as closely as possible the design intended for manufacture. The ease or difficulty found in the completion of a prototype will depend on two factors: firstly, the consideration of the manufacturing route which has taken place during the design process, and secondly, the planning of the design and manufacturing activity. The cost of the

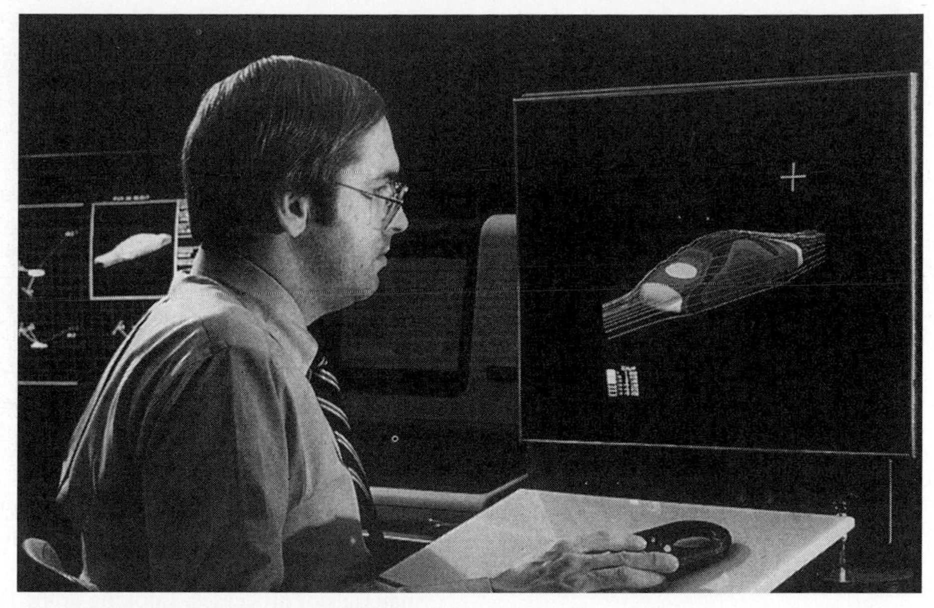

Figure 2.57 Computer-aided analysis of fluid flow around a vehicle.

prototype can also be radically altered by decisions concerning the materials selected and the manufacturing processes to be used. Designers need to acquire information concerning the manufacturing implications of their decisions and to ensure that their final design represents the most cost-effective approach which meets their objectives, Equally they must ensure that the entire design and manufacturing process is efficiently managed.

Making cost-effective decisions about the manufacturing route depends on a knowledge of the materials and processes to be employed. The detailed design must reflect the materials selected and the manufacturing method and hence it is important for these to be chosen as early as possible in the design process. The detailed design must also make effective use of standard components. Fittings for furniture, gears, electric motors etc. are all readily available, but the range of sizes, although wide, is not limitless.

Good planning of the design and manufacturing activity can make the difference between success and failure in a project. For example, if all materials and components are ordered at the right time, special journeys to collect small but critical items can be avoided. Equally, the completion of the project will not be held up for want of a scarce machine or other resource because it can be used when it is available. The most comprehensive technique developed to aid project planning (PERT) is described in section 2.5.2.

2.5.1 Cost-effective manufacturing decisions

In order to make cost-effective decisions concerning the manufacturing route, designers need to consider four key areas:

- the selection of materials and components,
- the designed shape,
- the standard of dimensional accuracy,
- the surface finish.

Much of the detailed information relating to these four areas is to be found in Chapter 6 'Materials processing' and Chapter 7 'Material and process selection', but the key issues relating to design are outlined here. One general issue, that of standardisation, has an impact in all four of these areas and is therefore considered first.

Standardisation

Standardisation has always been accepted with reluctance by British designers. Consider the following statement taken from the introduction to the BSI publication *The Management of Design for Economic Manufacture*.

> J. Williams Dunford, an architect, wrote in 1895 to *The Times* quoting the case of the frustrated contractor who complained that his order for iron girders had been passed from one British supplier to another and not one had been able to meet his specification. The order was eventually supplied from Belgium.
>
> A London iron merchant, John Skelton, replied:
> 'Rolled steel girders are imported into Britain... from Belgium and Germany, because there is too much individualism in this country... where collective action would be economically advantageous. As a result, architects and engineers generally specify such unnecessarily diverse types of sectional material for given work that anything like economical or continuous manufacture becomes impossible.'
>
> (PD 6470:1981)

As with any other factor in the design process standardisation must not be allowed to dominate the development of a solution at the expense of more important considerations, but that is not an excuse for pointless proliferation of sizes, parts, materials or processes. Looking at the British Standards Institution's own definition of standards and standardisation shows what important concepts they are:

> Standard... A technical specification or other document available to the public. drawn up with the cooperation and consensus or general approval of all interests affected by it, based on the consolidated results of science, technology and experience, aimed at the promotion of optimum community benefits and approved by a body recognised at the national, regional or international level.
>
> Standardization... An activity giving solutions for repetitive application, to problems in the spheres of science, technology and economics, aimed at the achievement of the optimum degree of order in a given context. Generally, the activity consists of the process of formulating, issuing and implementing standards.

It is a foolish designer who ignores all the effort and energy expended during the standardisation process without very good cause.

Selecting materials and components

The cost of materials and components has already been seen to be up to 75–80% of the total product cost for some consumer goods (see section 2.4.2) and making cost-effective decisions is therefore crucial. The material selected must ultimately be determined by the service requirements, but the influence this decision will have on the product cost is considerable. PD 6470: 1981 gives information not only on the relative costs of materials (for example, the cost comparisons for sheet materials shown in Fig. 2.58) but also on secondary consequences like the increased

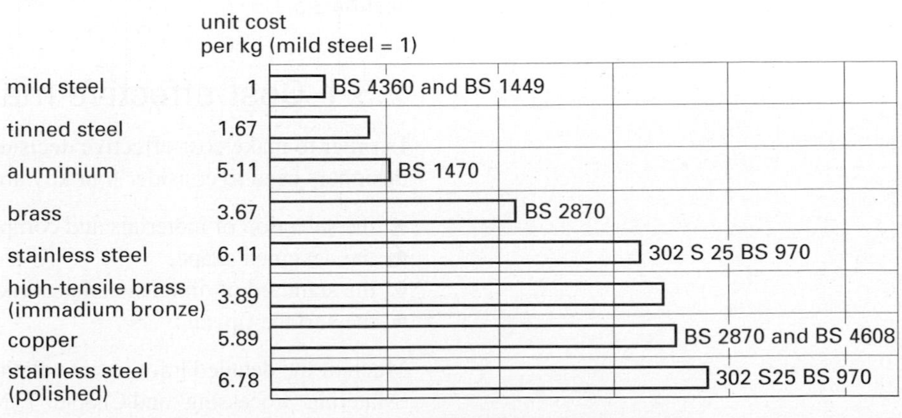

Figure 2.58 The relative cost of sheet materials (PD 6470:1981).

machining costs illustrated in Fig. 2.59. These increased costs result from the longer machining times for more difficult materials. As can be seen, making vehicle bodies from stainless steel rather than mild steel sheet involves nearly a sevenfold increase in the basic material cost, but if a turned component were being considered there would be a further increase of 30–40%. For students making the component themselves, this would be reflected in a longer machining time.

Clearly the selection of standard components is essential if the product cost is to be competitive, but it is important to note that the selected component quality can also have a major influence. Fig. 2.60 shows the relative costs of different grades of gear and there can be seen to be a factor of five between the highest and lowest indicated qualities. The use of standard components will also significantly reduce the time required to produce a prototype.

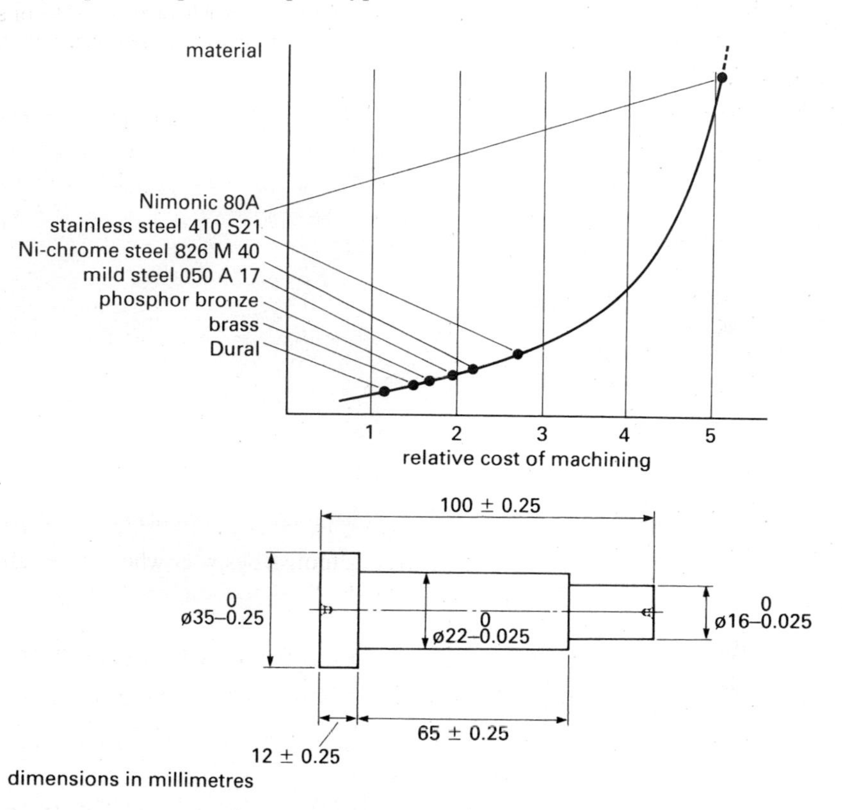

Figure 2.59 The influence of material choice on the machining cost.

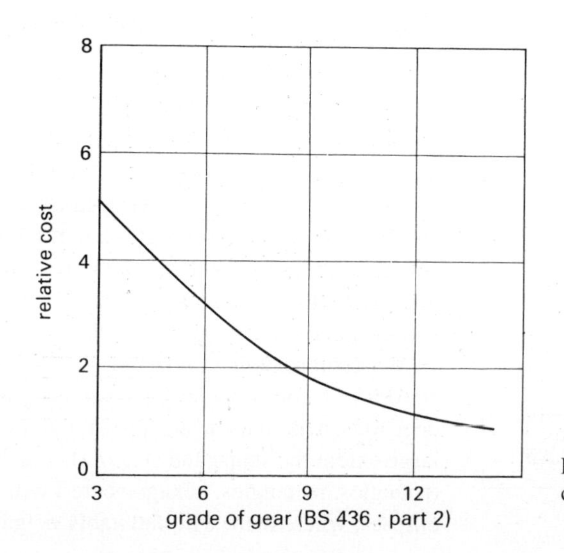

Figure 2.60 The influence of component quality on cost (PD 6470: 1981).

The designed shape

The detailed design of the product must reflect the material selected and the manufacturing process which is to be employed, and these relationships are explored in Chapter 5. One example is shown in Fig. 2.61, of chain-rivet extractors.

The example in the centre of Fig. 2.61 is the most common design – an assembly of machined steel components. A design which has been developed for casting in brass is shown on the right. The design on the left is also for manufacture by machining, this time from rectangular rather than hexagonal bar, but here there is no handle and a standard bolt has been modified to provide the force on the rivet. A comparison is made of these two designs as an example of value analysis in section 2.6.1. Clearly, many factors have influenced the designers' thinking in arriving at the final forms, but the results of changing the manufacturing process and material selected are evident.

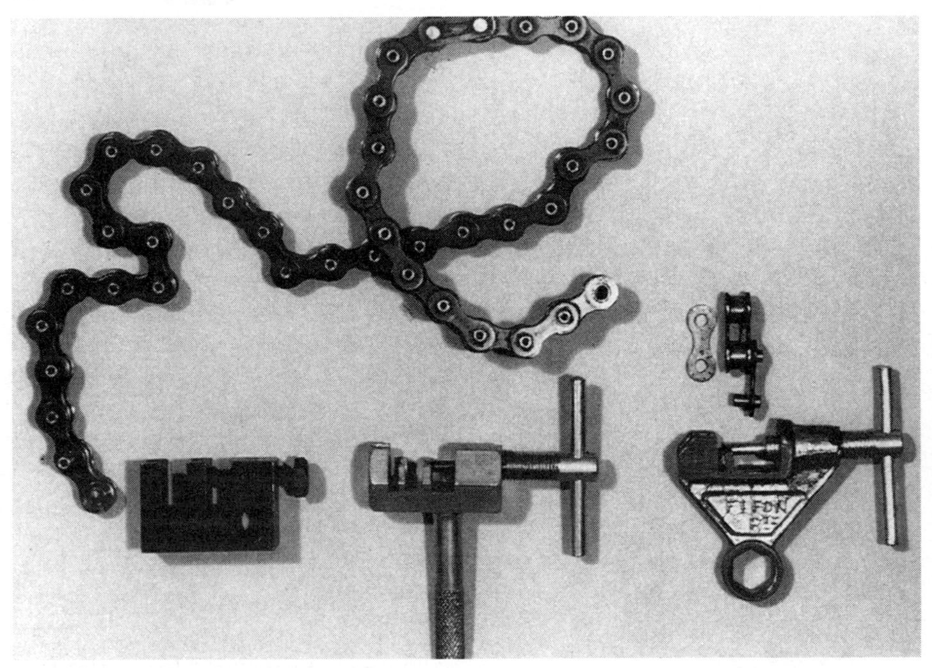

Figure 2.61 Three chain-rivet extractors.

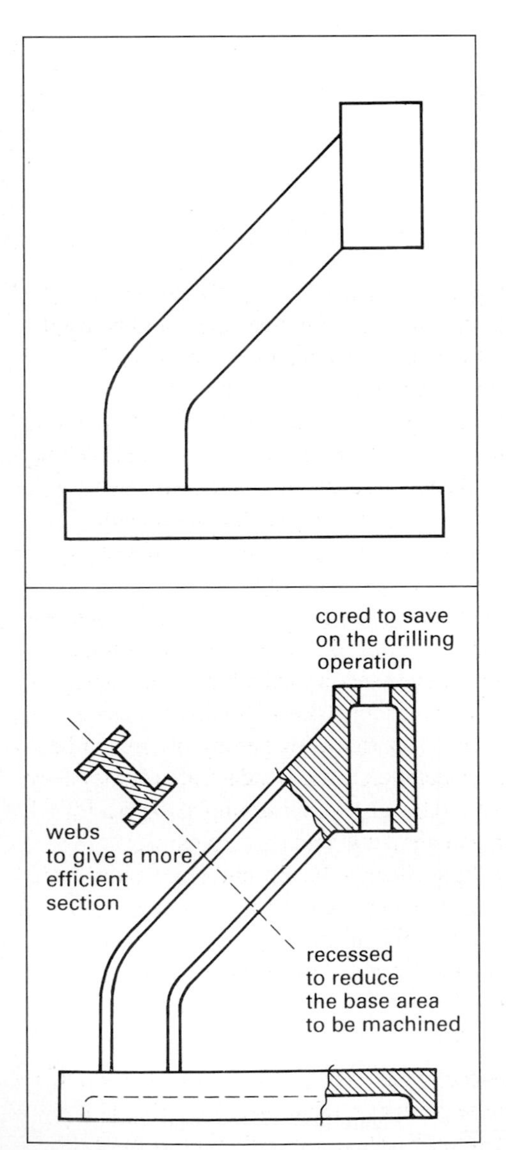

Figure 2.62 Savings in castings.

The material selected and the manufacturing process are not, however, the only influences on the designed shape, and other factors, such as transportation and packaging costs, can be equally significant. For example the weight of a casting can be reduced by using sand cores and a more complex mould, as shown in Fig. 2.62.

The mould will take more time to produce and each box will take longer to pack with consequential cost increases. These increases may possibly be recovered because of the material saved and the reduced machining time, but the potentially lower transportation costs must also be part of the equation. It is worth noting that the cost of time spent in making sure components are not overdesigned may be similarly repaid by the reduction in the quantity of materials used and the consequential savings. It should not be forgotten that such savings also contribute directly to the conservation of the Earth's resources, both in terms of raw materials and the energy used to process them.

Wasteful packaging is one of the easiest targets when attacking the irresponsible actions of a consumer society, but designers can add to the problem through their decisions. The shape of the product and whether it 'knocks down' determines to a large extent the shape and size of the packaging. Clearly shapes that tessellate (triangles, rectangles, hexagons etc.) will stack more economically than circles or ellipses, where there will inevitably be gaps between the items.

The standard of dimensional accuracy

The requirements put by the designer on the dimensional accuracy for a component can determine not only the feasibility of particular manufacturing routes, but also the complexity or number of the required operations. Fig. 2.63 shows the tolerances which can be achieved with various casting processes and indicates, for example, that if the accuracy required is 0.1 mm per metre, this can only be achieved by investment casting.

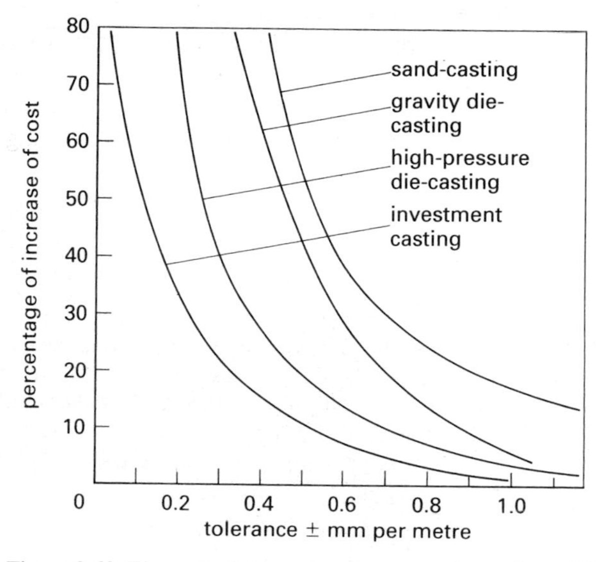

Figure 2.63 The cost of dimensional accuracy in castings (PD 6470: 1981).

Equally if the accuracy required is 0.8 mm per metre then any of investment casting, high-pressure die-casting, gravity die-casting or sand-casting could be used. It also shows the effect within a process of increasing the required accuracy, for example, if the tolerance on a sand-casting is decreased from 1.0 mm per metre to 0.6 mm per metre then there is a consequential increase in cost of around 20%. These increased costs result from the refinements necessary to the process, for example, the number of runners and risers used to control shrinkage and the greater accuracy required of the mould maker. For a student making a prototype these costs will primarily result in an increase in the time spent at all stages, particularly if a change from sand to a more accurate casting method becomes necessary.

It is not only in casting processes that increased costs can result from the designer calling for a higher standard of dimensional accuracy. Too close a tolerance on the diameter of a rod can mean that material cannot be used 'as delivered', as the designer intended, but must be purchased 'oversize' and machined down. Similarly a close length tolerance may result in a facing-off operation following parting-off on a lathe. It is essential that designers give careful consideration to the standard of accuracy they specify and do not put a general tolerance on a drawing of, for example ±0.05 mm, just because they had seen this number on another drawing. When you are manufacturing your own prototype you will see whether the machining tolerances you have indicated are reasonable and, if not, you can alter them in the workshop. Remember that this is not the normal way of working in industry where the design and manufacturing functions are separated and carried out by different personnel.

The surface finish

There is generally a need to control the surface finish on a product for both aesthetic and technical reasons. The appearance of the product is obviously determined by the visible surfaces and these may need to be smooth or textured, reflective or matt. Equally, the surface finish must be controlled to give the right frictional properties and to limit the initial wear. Sometimes it is also important to reduce the number of crevices from which fatigue cracks can initiate and in which corrosion can begin. The significance of the decision can be seen from Table 2.8 which shows that the relative cost increases by a factor of forty as the specified finish goes from very rough to super fine.

Surface texture		R_a *(μm)*	*Roughness grade number*	*Relative cost*
machined	very rough (very coarse feed)	50	N12	1
	rough	25	N11	3
	semirough	12.5	N10	6
	medium	6.3	N9	9
	semifine	3.2	N8	13
	fine	1.6	N7	18
ground	coarse	0.8	N6	20
	medium	0.4	N5	30
	fine	0.2	N4	35
lapped	super fine	0.1	N3	40

Table 2.8 Relative costs of surface textures (from PD 6470: 1981)

Specifying an increased fineness initially requires the machining of surfaces which otherwise might have been left ‘as cast’ or ‘as received’ and eventually the use of grinding or lapping processes. When you put N-numbers next to the roughness symbol on a drawing your decision will influence both the time expended on producing a component and its cost.

A case study – a computer keyboard converter

The QWERTY keyboard is difficult for young children to use and a project had therefore been undertaken to develop a keyboard converter for use in primary schools. The system had both software and associated hardware to allow children to learn about music, money and colours and to play educational games.

Fig. 2.64 shows the original equipment which had been produced to test the software during development. There was clearly a great deal of flexibility in the

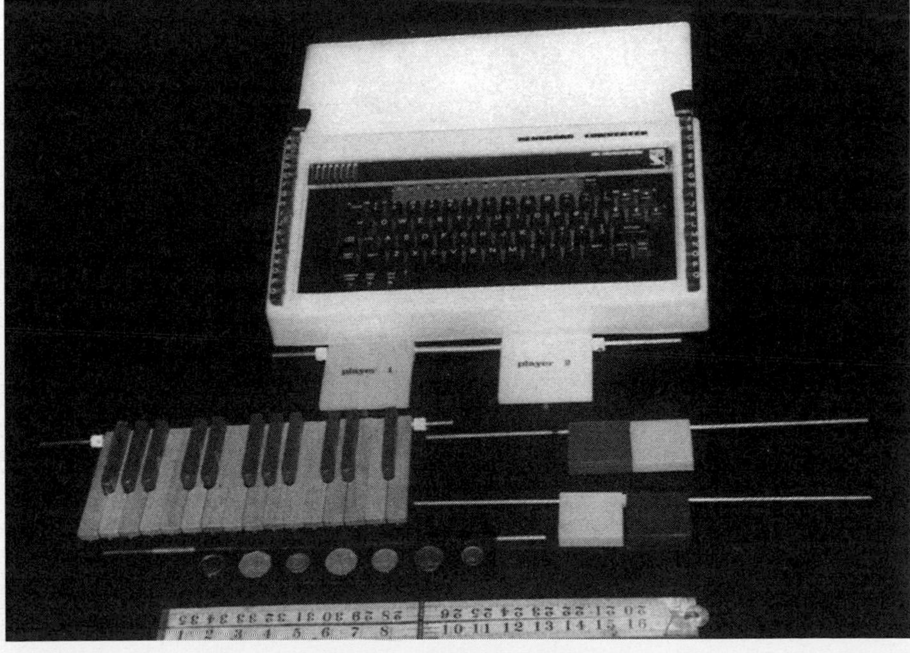

Figure 2.64 Prototype hardware for the keybord conversion.

possible arrangements of the components, but this was gained at the cost of considerable complexity and a lack of unity in the product. All the components had also been produced by one-off production methods in a small workshop.

Fig. 2.65 shows the prototype design proposed for batch manufacture which can be seen to be based on the vacuum forming and injection moulding processes.

Fig. 2.66 shows the basic vacuum formings – a front panel and different tops for each of the games. The design of the buttons was standardised for all but the music keyboard and based on the injection moulding shown in Fig. 2.67.

The buttons were made distinguishable by changing the colour of the plastic used, and by sticking different shapes, for example, plastic imitation coins, to the top surface. Fig. 2.67 also shows the two components considered for fixing the buttons to the tops. Both the white injection moulding and the metal ring are standard items. The ring was eventually chosen on the grounds of cost.

The important features of this design from a manufacturing point of view are that the major components are produced by the cheapest available process, vacuum forming. This process requires only comparatively cheap wooden tooling for the

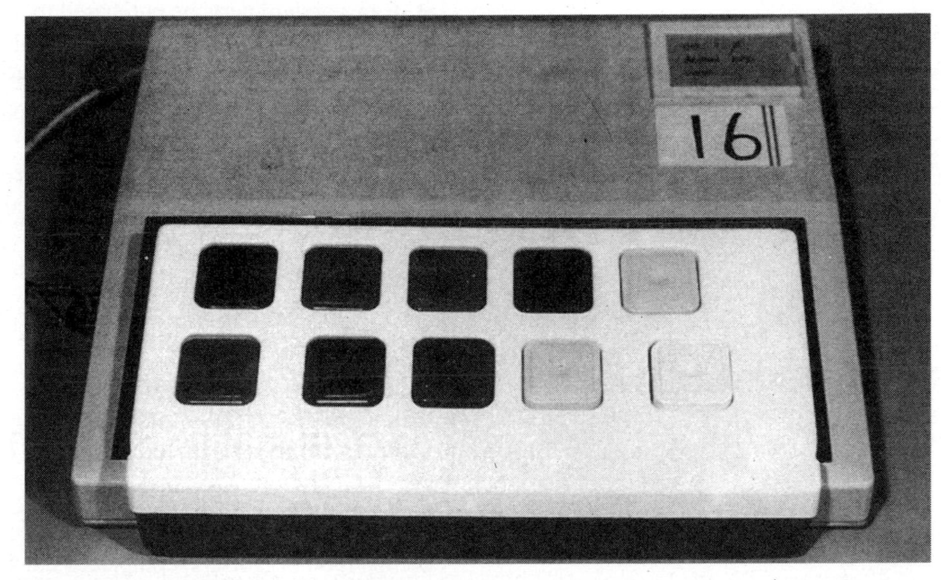

Figure 2.65 The redesigned keyboard converter.

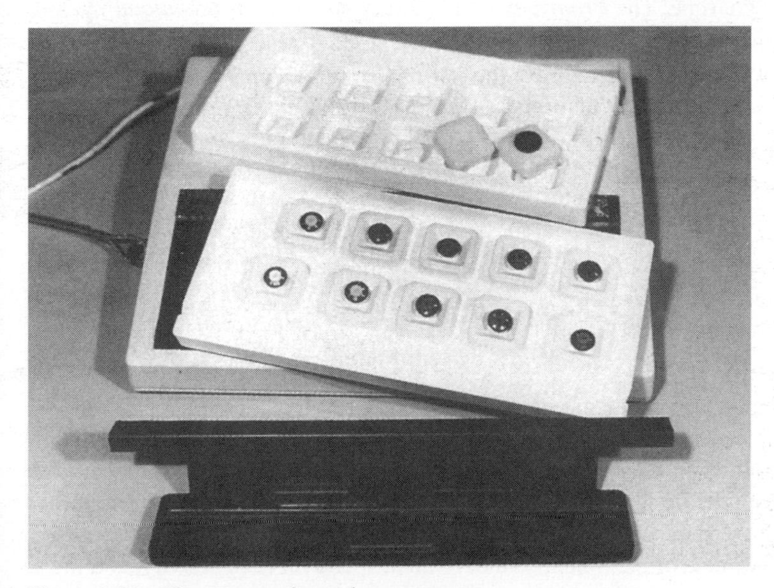

Figure 2.66 The vacuum-formed components.

Figure 2.67 The injection-moulded buttons.

batch quantities and accuracy needed here, and hence represents a considerable saving over injection moulding. There is, however, the need to trim the uneven edges and remove flashing which would not be required on injection moulded components. If the batch size were great enough it would eventually become worth investing in the more expensive, two-piece metal tooling required for the injection-moulding process and saving on these labour costs (see section 1.4 for a discussion of break-even quantities). Only the buttons for this design are produced using injection moulding and the tooling costs kept to a minimum by standardising the design. If plastic is to be used for the buttons then there is no realistic alternative, although they could, perhaps, have been made from wood using CNC machining as were the keys for the keyboard music game.

As an aid to reviewing your own designs you might like to consider the questions in the following checklist.

Detail design checklist

- Can the parts be simplified by altering the design?
- Can standard parts be purchased to replace any of those specified?
- Can the number of different materials required be reduced?
- Have more material sizes than necessary been used?
- Can the number of parts needed be reduced?
- Is the product easy to assemble?
- Are special tools needed to assemble or disassemble the components?
- Are special fixtures or tooling needed during manufacture?
- Can a cheaper material be used?
- Is all the machining indicated essential?
- Is the tolerance on dimensions tighter than it need be?
- Would a rougher surface finish be acceptable?
- Is there a cheaper alternative finishing method?

2.5.2 Planning and PERT analysis

Being able to complete a project by the set deadline is important both to students and industry. Students must comply with the dates set by examination boards and competition organisers: companies must meet the requirements of their customers. Planning is important if the actions necessary to achieve these objectives are going to be taken at the right time. The *Programme Evaluation and Review Technique* (PERT) was developed in America in the late 1950s to help with the planning of the Polaris programme. In its simplest form the technique is also known as *critical path analysis* or *network analysis*. It consists of three distinct stages:

- Listing the activities which need to be completed.
- Drawing a network showing the activities in order.
- Producing a chart showing the planned sequence.

The first two stages are equally appropriate for students and companies, but the final stage will be interpreted differently.

Listing the activities

The first and most obvious step in effective planning is to draw up a list of all the activities which need to be carried out. Associated with each of these activities the time taken to complete it must be estimated. Table 2.9 shows such a list drawn up for students beginning the design of the safety steering wheel for the Lotus Esprit.

Some notes have been added to explain unfamiliar items, and two times have been given – one an estimate of the time the student must spend on each activity, and the other an estimate of how much project time will have elapsed.

No.	*Activity*	*Time needed* *Student days*	*Project days*	*Explanatory notes*
1	research regulations	5	25	steering wheels must be designed to meet EU or UN regulations + time for replies.
2	analyse competitors' products	5	25	
3	analyse brief	1	6	visit Lotus + time to organise it.
4	develop solution	20	20	graphical, computer and mathematical modelling. Initial ergonomic assessment and make clay model.
5	learn DOGS-2D	5	5	CAD system for producing engineering drawings. (Drawing Office Graphics System – DOGS)
6	learn Boxer	5	5	computer solid modeller.
7	write PDS	2	2	to clarify ideas and confirm with the client.
8	learn clay modelling	1	1	essential for 3-D visualisation.
9	Lotus feedback	1	6	
10	Sheller-Clifford feedback	1	6	Sheller-Clifford make nearly all the steering wheels for the UK.
11	complete stress analysis	2	2	
12	complete vibration analysis	2	2	
13	make physical model	5	5	for destructive 'pendulum' impact tests.
14	organise tests	2	7	arranging to use a test rig + delays
15	ergonomic assessment	5	5	examining the final proposal in a Lotus Esprit mock-up
16	Lotus evaluation	1	11	industrial concerns must fit in helping students around profitable work.
17	Sheller-Clifford evaluation	1	11	an assessment of the manufacturing problems.
18	complete engineering drawings	5	5	
19	organise prototype manufacture	1	11	prototype must be professionally produced for car tests.
20	final preparations for course assessment	2	2	presentation graphics and display organisation.
21	organise Lotus feedback	1	1	
22	organise Sheller-Clifford feedback	1	1	

Table 2.9 List of activities for the design of a safety steering wheel for the Lotus Esprit

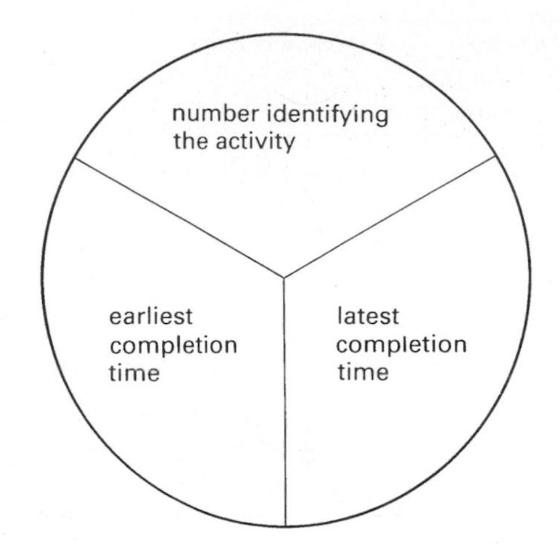

Figure 2.68 Symbol for use in networks.

It must be emphasised that there is no point in waiting until all the times are known exactly, because by then the project will be almost over. It is necessary to make estimates so that the feasibility of completing the project on time can be considered. If a project is to be abandoned because it is likely to take too long to complete, then this decision must be taken very early in the project's history. Clearly neither the design nor the manufacturing route will be fully established and an element of guesstimation will be necessary. Consulting those who have already done similar tasks will be the best way of getting reliable estimates – teachers, experienced designers, manufacturers, librarians etc.

Drawing a network

Before a decision concerning the total project time can be taken the order in which the activities can be carried out must be determined. It may be possible to carry out some activities during the same time period, that is in parallel. In other cases it may only be possible for an activity to start when a previous one has been fully completed. Each activity in the project is represented by a circular symbol in the network as shown in Fig. 2.68. The number in the central sector identifies the activity, the sector on the left will contain the earliest time by which the activity may be completed and the sector on the right the latest time.

Activities are joined by solid lines with arrows in the order in which they must be completed. If one or more parallel activities must be completed before another can begin, then they are joined by a dotted line with an arrow, sometimes known as a *dummy activity*.

Fig. 2.69 shows a network for the safety steering-wheel design project based on the project elapsed times. Several examples of dummy activities can be seen and the critical path is shown by a thickened line. This path represents the shortest time in which the project can be completed and can be easily identified because the earliest and latest completion times will be the same. On this route there is no slack time and any delays in a particular activity will delay the whole project. In this case the minimum total project time can be seen to be 96 days or about 19 weeks if a 5-day working week is assumed. Clearly if you are doing a project by yourself it is not possible to undertake activities strictly in parallel. You cannot do two things at once, but if you are aware of when delays outside of your control will occur, extra activities can be fitted in to these periods. If you add up the total number of days the student must spend in Table 2.9 it will be found to be 72 or just over 14 weeks. Hence, if everything were to fit in conveniently then 96 days is a realistic target. Clearly if the total number of student days had been greater than 96 then this could never be an achievable target if you were working by yourself. This could easily have been the case if there had been a large number of parallel activities.

Activity sequence chart

Fig. 2.70 shows an activity sequence chart which has been drawn up from the network analysis diagram.

This can be seen to consist of blocks of time – the shaded part of each block representing the time the student actually spends. Where an activity could take place in parallel, the block of time will appear together with an arrow which indicates the latest possible finishing time. The first set of parallel activities, 2, 6, 8 and 5, can be seen to fit satisfactorily in the slack time for activity 1. Time can also be found for activities 22 and 13. There is however no slack time available for activities 21, 11 and 12. In industry, extra personnel could be brought in at these times so that it would still be possible to finish the project in 96 days if required. When working on your own there would be an additional 5 days required – the total time necessary to complete activities 21, 11 and 12.

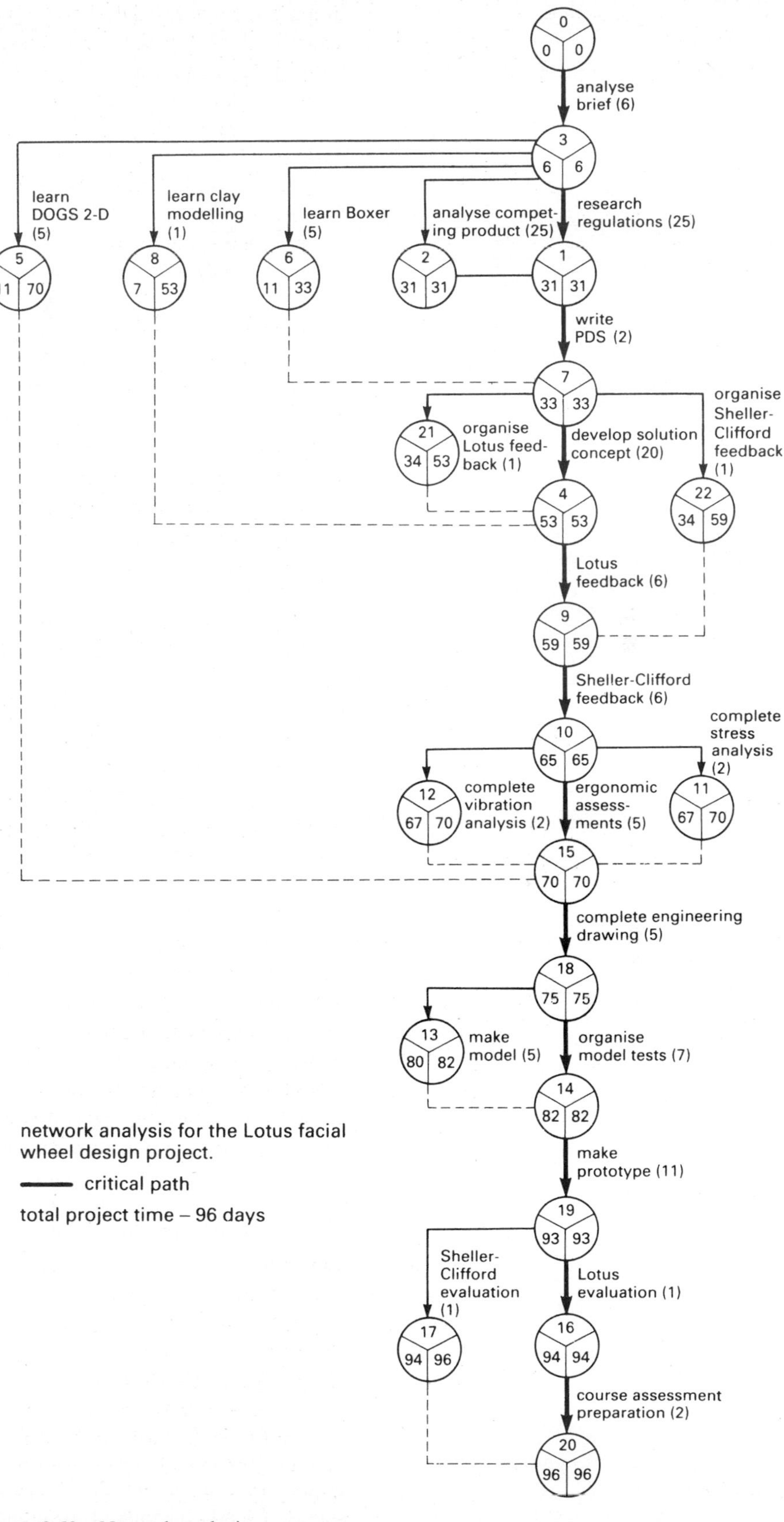

Figure 2.69 Network analysis.

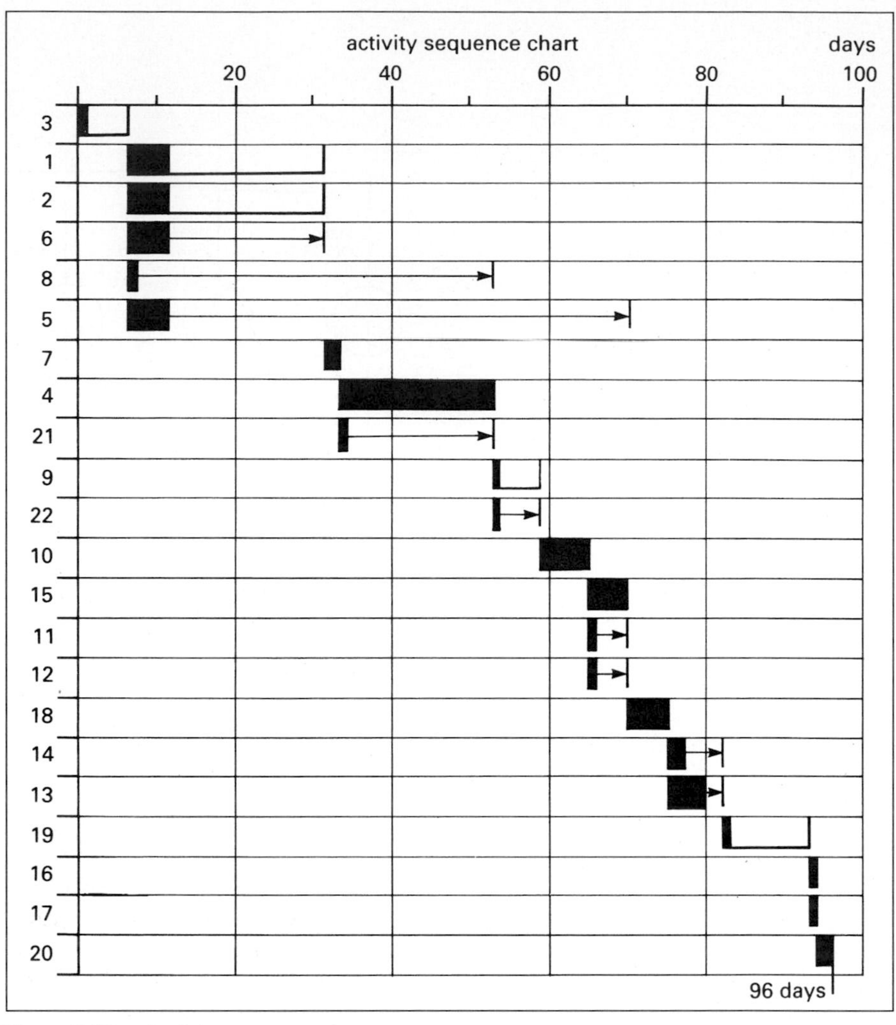

Figure 2.70 Activity sequence chart.

Students have even greater difficulties in planning their time than this analysis indicates, because they are working in more than one subject area and sometimes on more than one project. The important steps are listing all the activities to be completed and drawing the network. At this point the order in which the activities must be carried out is known and the critical path has been identified. An awareness of these two factors for all your projects will enable you to use your time effectively. If you go on to consider the activity sequences for each subject or project and note the slack times your chances of meeting all your deadlines will be significantly increased.

2.6 Evaluation

The vital importance of constant evaluation during the design process is best realised by reflecting on Fig. 2.71. This shows that in this company about 80% of the total product costs are committed during the design process, and most of it in the early stages.[8] Clearly the bulk of the money is actually spent much later during development and when the tooling up for manufacture occurs, but there is not a great deal that can be done at this stage to reduce the product costs significantly.

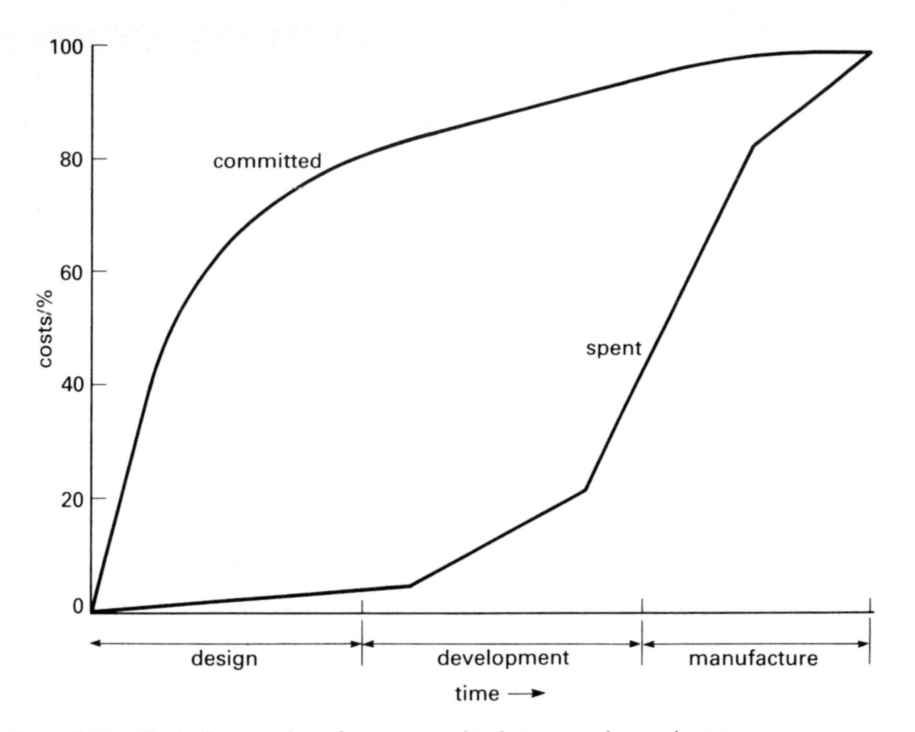

Figure 2.71 Typical proportion of costs committed at successive project stages.

As a student, you are in much the same position when you undertake your projects. You have to make a lot of crucial decisions in the early stages and there will be little you can do to alter the consequences later. For you the cost is not so much financial, but it will be reflected in the quality of the product and its contribution to your exam success. The importance of thinking very carefully about what needs to be done, and the alternative conceptual approaches and their implications, during the early stages of the design process cannot be overemphasised. Equally, the design must be constantly reviewed as it progresses and carefully evaluated when completed.

The process of constantly reviewing a design whilst it is in progress is known as *value engineering*. If the same kind of review is held of an existing product, then it is known as *value analysis*. Students must engage in both value engineering as their design progresses, and value analysis when it is completed. The following section describes these processes in relation to industrial practice, but the principles apply just as well to student projects, as the examples given afterwards illustrate.

It should also be emphasised that it is during the evaluation of a design as it progresses, and the eventual product, that some of the major benefits of a clearly written product design specification (PDS) are found. This allows each aspect of the design to be considered carefully and an assessment made of how closely that particular design objective has been met. The PDS itself is also subject to review, of course, particularly in the light of changes in either the market or the available technology that often occur while the project is in progress.

When a prototype is constructed it is possible for the product to be evaluated in use, or to apply tests that simulate the expected use of the product. Clearly this provides the most reliable form of feedback to the designer and is a key stage before a final decision to manufacture can be confidently implemented.

2.6.1 Value engineering and analysis

Value engineering and *value analysis* are concerned with finding ways in which a particular function or objective can be achieved at reduced cost. This is not the same as looking at ways for achieving cost reductions for particular components or assemblies. For example if the function required is 'generating a line sufficiently black for reproduction and of uniform thickness', cost reduction might look at cheaper manufacturing methods for pens; value analysis would look at this, but also alternatives like pencils using plastic leads, lasers and special paper.

The fundamental questions for either value analysis or value engineering are:

- What is it?
- What does it do?
- What does it cost?
- What else will do the job?
- What does that cost?

The process is generally carried out by teams and in a number of stages as described below. Teams of students might be formed to carry out value analysis on commercial products or to review each other's projects and help with the value engineering which everyone should undertake.

Team formation

A typical industrial team could consist of representatives from production, sales, purchasing, costing and design. As in brainstorming sessions (see section 2.3.3), multidisciplinary teams are likely to be much more effective than those composed of people from a single specialism.

Selection of a product to study

A suitable product may be indicated by high production costs, a competitor with a cheaper line, a difficult manufacturing method, high reject rates or the excessive production of scrap. A lot of effort is to be expended completing the value analysis and there must be a reasonable prospect of a good return. It should be noted, however, that early value analysis companies and practitioners in the USA prospered on a percentage of the savings they made for the companies who employed them.

Information gathering

The information required for the value analysis of a product will probably include most of the following items: manufacturing specifications and drawings, production rates, estimated production life, manufacturing costs, material suppliers and a set of parts of the article, together with subassemblies and completed assemblies.

Speculation stage

In a very similar sequence to that conventionally followed by designers, when all the necessary information has been gathered the team must consider it carefully and think creatively. The questions which must be asked are also much the same as those designers need to have constantly in mind:

- Can something else be used?
- Can dimensions be reduced?
- Can waste in manufacture be reduced?
- Are correct limits or finishes specified?
- Can standard components be used instead of manufactured parts?
- Can cheaper components be substituted?

As when organising a brainstorming session with a group, or trying to promote lateral thinking, it is vital to establish the kind of forum where all ideas are welcomed, and judgement on them is temporarily suspended.

Investigation stage

All the ideas which have appeared need to be looked at and those that look promising need to be thoroughly investigated. Having reached this stage there may be a feeling of obligation to change something, but there must be a very detailed analysis of both the benefits and the costs of any new proposal.

Recommendation stage

The team must present its findings to the management of the organisation for approval. The team members by this point will all feel committed to the proposal being put forward and an assessment by an independent group is essential to ensure that some key aspect has not been overlooked.

Implementation

If, following the independent assessment, the proposal is still considered a likely success it can be implemented and the organisation can, it will be hoped, reap the expected benefits.

Value analysis or value engineering is now written into many contracts particularly in America. It might be thought that value engineering would always be preferred to value analysis, but the issue is not that straightforward. In favour of value engineering it can be said that modifications are made on paper, so that no stock obsolescence, or costly retooling, is required and it prevents the existence of Mk I and Mk II of a product, with consequent spares problems. Against it is the fact that it introduces another delay to the start of the production with the consequential increased risks from competitors, changes in the market requirements and new technology.

Examples for value engineering and analysis

Fig. 2.72 shows the final design of a clamping system for drilling sheet materials. There are basically eight parts: a wooden base, a replaceable wooden block, two machined steel bars acting as side slideways, two steel bars acting as sheet restraints and slideways for the toggle clamps and two standard toggle clamp assemblies. The components are assembled using machine screws. As some parts are duplicates and because of the simplicity of their forms, this prototype was comparatively simple to manufacture.

What is much more difficult to convey is the months of agonising and discussion which went into the development of this design. The earliest versions were conceptually closely related to machine vices with one or two screw threads to provide the clamping force and stepped holes or teeth to provide adjustment. The refinement of the final prototype is the result of constant value engineering during the design process.

Fig. 2.61 showed three designs for chain-rivet extractors. The version which has been machined from hexagonal bar is the most complex and comparison with the other two immediately provokes the kind of questions which would be part of a value analysis exercise. Fig. 2.73 shows the detailed design of this extractor. The additional vertical support is to enable the extractor to be used to loosen tight rivets.

Figure 2.72 Clamping system for drilling sheet materials.

Figure 2.73 Details of the rivet punch screw thread.

Some questions	*Some responses*
Is the handle necessary?	This design is intended for cyclists, the rectangular bar version is intended for motorcyclists who are more likely to be using a vice.
Is the facility to loosen tight rivets needed?	Without it, this is not an easy operation, although purchasers may not appreciate this.
Is casting a realistic alternative?	The product form can be redesigned for casting, but aluminium and brass are weaker materials and the vertical supports might shear. Without the additional, vulnerable vertical support needed to loosen tight rivets this might be feasible.
Is the moving tip needed?	The moving tip reduces friction and wear at the contact point with the rivet, but considerably complicates the manufacture. For the number of occasions it is used, a fixed tip might be good enough.
Is the tommy bar needed?	Relying on the right sized spanner being available by using a bolt might be overoptimistic, but perhaps a sliding bar as on the cast version is better.
Why use hexagonal bar?	It is distinctive and there is some attractive symmetry. However, it does not particularly help the drilling operations.

If you were a member of a value analysis team reviewing the chain rivet extractor as a product you would have to grapple with all of the questions and these and similar responses. Ultimately it is success in the market place which counts, and all unnecessary manufacturing complexities must be removed. The difficulties lie in knowing which manufacturing operations are a result of necessary product characteristics, and which are merely a result of the influences on the designer at the time the product was designed.

2.6.2 Testing products and prototypes

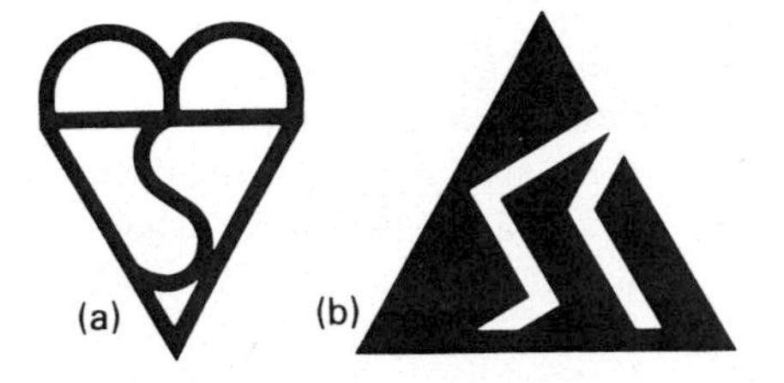

Figure 2.74 The British Standard
(a) Kite mark and
(b) Safety mark

Testing procedures for products and prototypes are among the most common topics covered in British Standards. The British Standards Institution (BSI) runs one of the largest independent test centres in Europe where tests are carried out on a wide range of products such as pharmaceutical containers, car windscreens and motorcycle safety helmets. If a product has been manufactured under a safety system drawn up by BSI and agreed with the manufacturer or supplier, it can carry either the Kitemark or the Safety Mark shown in Fig. 2.74.

Samples of Kitemark or Safety Mark products are tested regularly at the BSI testing centre to ensure that the level of quality demanded by the standard is maintained. Where a product just carries the BS number the manufacturer is claiming compliance with the requirements of that standard, although the product has not been approved by the BSI. Other organisations also mark products to show compliance with their test requirements, for example, the British Electrotechnical Approvals Board (BEAB) which tests most electrically powered consumer products.

Whatever project you undertake one of the final stages will be the testing and evaluation of a prototype. If the product or system is covered by a British Standard then appropriate procedures already exist. If there is no current standard then test procedures will have to be devised. It will almost certainly be possible for you to get a good idea of what is required by looking at the tests recommended for comparable products and systems. Two publications produced for schools and colleges are particularly helpful: PP 7302: 1987 *Compendium of British Standards for Design and Technology in Schools* and PP 888: 1982 *Safety and Performance of Domestic Electrical Appliances*.

The introduction to PP 7302 gives details of the testing procedures and acceptance limits for hacksaw blades. The safety section gives details of the tests required for toys, outdoor play equipment and playpens. The Toys (Safety) Regulations indicate, for example, suitable construction materials, details for finishing edges and joints, appropriate hinges, mechanisms and springs, tests for toys to be put into the mouth and those which the child can enter, or are intended to bear the child's weight. There are also details for the sizes of 'accessibility probes' for different age ranges and a 'sharp point tester'. As an example consider the regulations indicated for a folding mechanism:

> 3.1.6 *Folding mechanisms* Folding mechanisms on collapsible toys, e.g. toy ironing boards and dolls' pushchairs, shall have a safety stop or locking device to prevent unexpected or sudden movement or collapse of the article, or shall have clearance to give protection for the fingers, hands or toes against crushing or laceration in the event of sudden collapse of the article; such a clearance shall admit a 12 mm diameter rod. When a load of 50 kg is placed on the appropriate surface of the toy, e.g. the ironing board itself or the seat of a doll's pushchair, the safety stop or locking device shall not fail.
>
> NOTE: Collapsible toys should be designed so as to prevent the risk of injury during the folding or unfolding action.
>
> (BS 5665: Part 1: 1987)

Such regulations give general guidance and also specific design constraints, for example, the clearance must be at least 12 mm and the toy must be able to take a load of at least 50 kg without collapsing.

BS 5696: 1986 on outdoor play equipment concerns such matters as static test loads for swings, seesaws, trapezes, ladders, platforms and guardrails, appropriate clearances for swing seats and the avoidance of entrapment, recommended dimensions for climbing equipment and slides and tests to ensure that the potential damage caused by the effect of the impact of a seat on a child is minimised. For this last test the acceleration of the simulated child's head (a tenpin bowling ball) is measured and there must be no peak value greater than $50g$ (g is 9.81 m/s^2). This can be compared with the $80g$ deceleration which has been allowed for the tests on safety steering wheels described later in this section. Although the acceptance limit is much stricter for swing seats, the impacting speed is also much lower.

BS 4863: 1973 on safety requirements for playpens covers construction materials and general requirements, for example, '3.3 All bars or slats shall be vertical and no vertical open space shall be more than 100 mm or less than 75 mm wide.' They also cover tests for stability and strength which are indicated in Figs. 2.75, 2.76 and 2.77.

The playpen passes the stability test if it returns to its normal position after being lifted to the angle shown in Fig. 2.75. In order to pass the pull-strength tests the playpen must withstand a force of 150N on any of the bars and rails and in any of the directions shown in Fig. 2.76. The impact test requires the 15 kg sandbag to fall from the position shown in Fig. 2.77 ten times against each side of the playpen without failure. Again, what makes these tests easy to apply is that they are defined in very specific terms. It is particularly important that any evaluation criteria included in a PDS are written with equal precision.

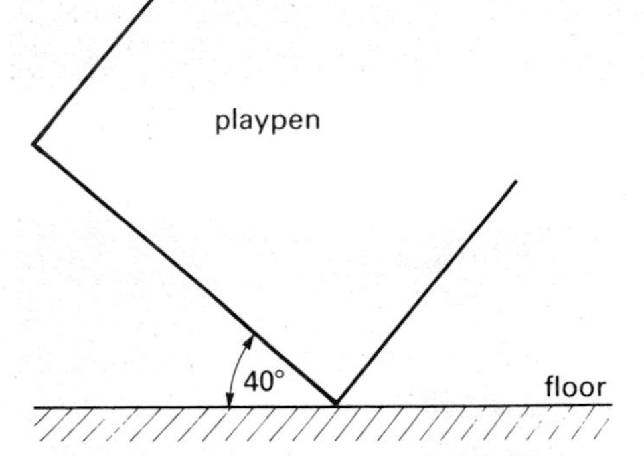

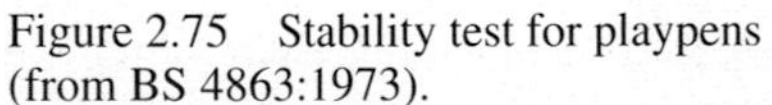

Figure 2.75 Stability test for playpens (from BS 4863:1973).

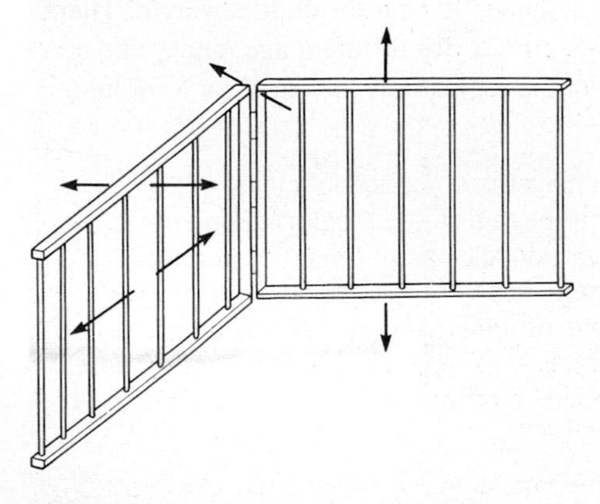

Figure 2.76 Directions of pull-strength test (from BS 4863:1973).

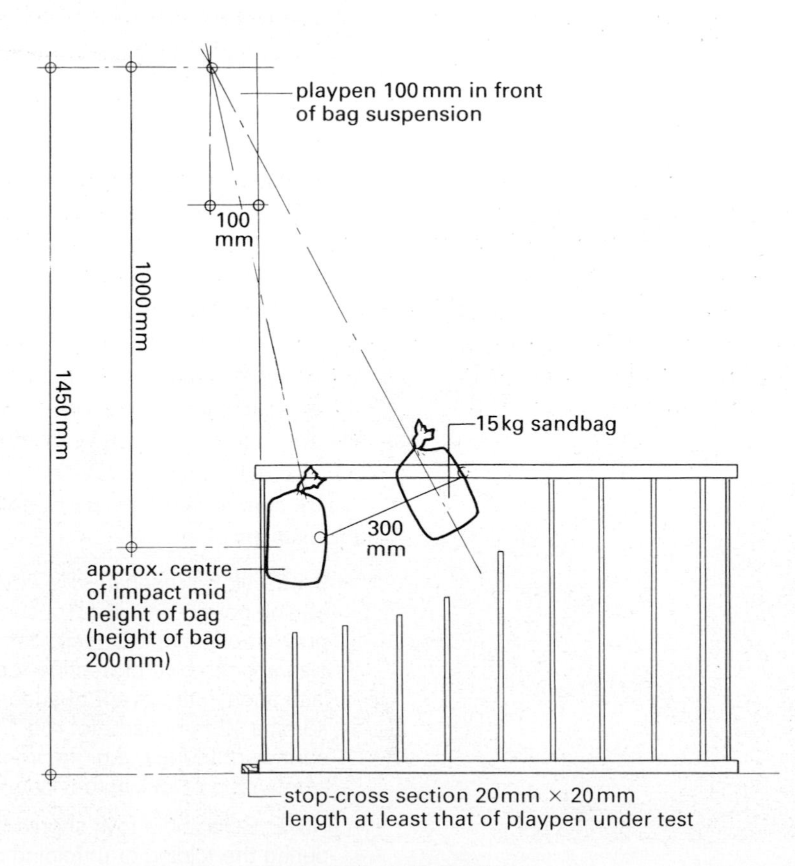

Figure 2.77 Impact test (from BS 4863:1973).

A summary of the recommendations in PP 888 appears in PD 7302 covering such areas as protection against electric shock, the starting, overheating and overloading of electrical appliances, and the effects of moisture and mechanical loads. These are covered in greater depth in the full document and clear examples of the tests for products like fan heaters, electric blankets and blenders are given. Fig. 2.78 shows a 'jointed finger' being used to check the accessibility of live parts.

Fig. 2.79 shows the kinds of test which are carried out on cooking hobs to ensure that neither spillages nor dropped pans will result in safety hazards. Fig. 2.80 shows an impact test on a microwave oven. After the test the door must not allow any microwaves to leak.

The latter part of PP 888: 1982 deals with tests concerning the performance of appliances. Means of establishing the effectiveness of cookers, toasters, washing machines, microwave ovens, spin dryers, food mixers, kettles and vacuum cleaners are all described. In order to test these products standardised procedures must be used; for example, a standard cake mix and the use of shade charts to evaluate cake browning or standard loads for washing machines and spin dryers. Many of the

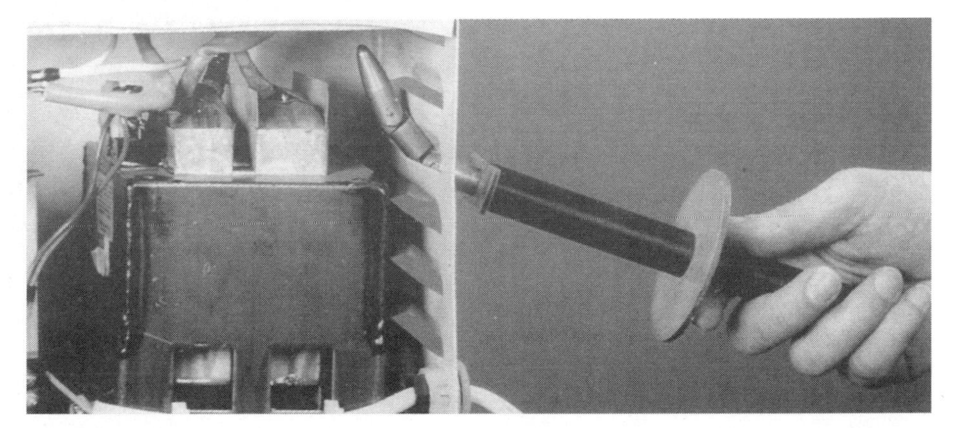

Figure 2.78 A 'jointed finger' testing the accessibility of live parts.

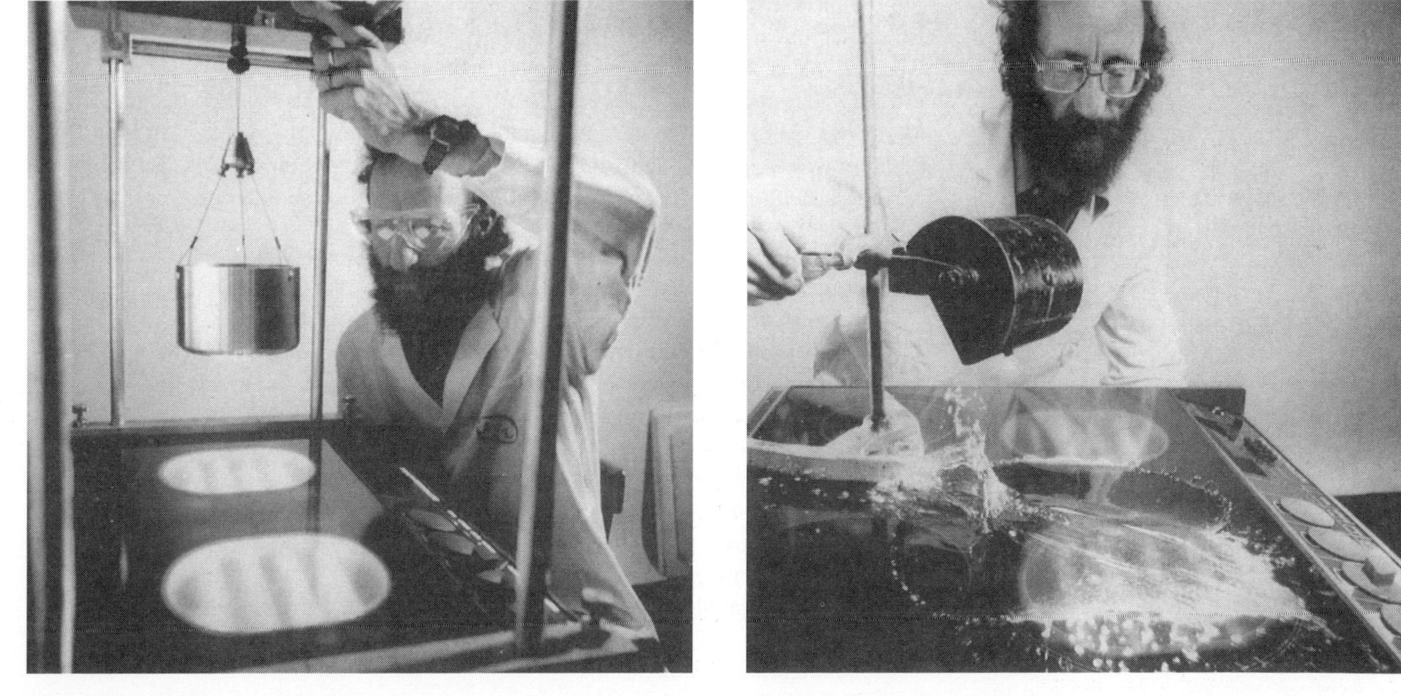

Figure 2.79 Testing a cooking hob.

Figure 2.80 Testing a microwave oven.

products are covered by internationally agreed standards, and properly valid tests require the use of their experimental specimens. For example, International Electrotechnical Commission (IEC) Specification no. 456 defines the standard washing machine load and IEC 312 standard dust for testing vacuum cleaners. The test for vacuum cleaners requires a measured quantity of the standard dust (or sand as suggested in PP 888) to be spread and rolled into a carpet area of 1.4 m long × 0.055 m wide. Using a standardised technique the dust is removed and the results recorded. PP 888 also indicates the kind of records which should be kept:

Results

Record:

1 Type of vacuum cleaner used
2 Tools used or the carpet adjustment position
3 Airflow control position
4 Percentage of sand collected in the bag and left on the carpet
5 Carpet material
6 Variables not controlled or measured
? Standard carpet
? Standard dust
? Preconditioning
? Initial dust application
? Voltage
....
....

Figure 2.81 Aluminium honeycomb used to simulate facial impacts.

Safety steering wheels

Steering wheels used to be tested by impacting a dummy on to the wheel to simulate the expected chest impact. Since the introduction of legislation to compel drivers to wear seat belts chest impacts of this kind have been very uncommon. In a crash the driver now usually rotates about the lap belt and it is the face which hits the steering wheel. Consequently the steering wheel is now tested on a pendulum impact rig and the face is simulated by the aluminium honeycomb material shown in Fig. 2.81.

The need to reduce facial injuries has led to the development of air-bags fitted to steering wheels in new cars. Safety-conscious manufacturers may also fit passenger and door-mounted air-bags. The design shown in Fig. 2.53 is of a sports steering wheel, without an air-bag but still engineered for safety. The performance of the wheel is defined by the depth of crushing of the honeycomb when it is impacted on the steering wheel at a specific speed. There is also a limit of 80*g* placed on the allowable deceleration of the simulated head which is included in order to prevent brain injuries.

These new requirements were the driving force behind the designs for the Lotus steering wheels shown in Figs. 2.44 and 2.53. The designers were attempting to increase the impact area (thus reducing the depth to which the honeycomb crushes) without obstructing the view of the instruments and maintaining suitable aesthetic qualities. The limit on the allowable deceleration means that the depth of the wheel must also be increased, which influences the driving position. Engineering and presentation drawings of the final design proposals are shown in section 2.7.

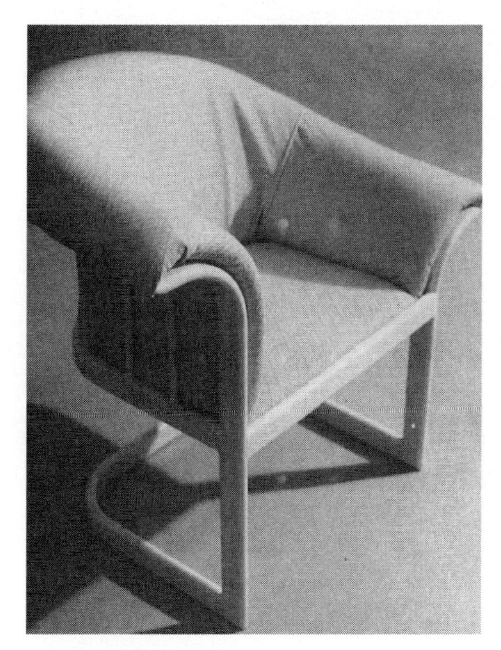

Figure 2.82 Prototype chair design.

A chair design

The preceding paragraphs have been concerned with the technical aspects of testing products and prototypes, but the visual aspects are equally important. Fig. 2.82 shows a prototype of the chair design shown earlier in Fig. 2.43.

The graphical and physical models, and the prototype, make an interesting visual comparison of the designer's intentions and the production reality.

The visual significance of the fabric selected is apparent as is the impact of the dimensions of the steel used in the frame. Decisions about such matters can have a profound influence on the aesthetics of the product, although other criteria like the available steel section sizes, the product weight and the required strength might tend to dominate.

2.7 Communication

The principles of good design communication are the same whether the means being employed are words or drawings:

- here must be a very clear understanding of the purpose, and the drawing or document must be carefully prepared and structured with this purpose in mind,
- the expectations and capabilities of the intended readers or viewers must be accommodated,
- the detailed presentation, linework or text, must not distract the user,
- personal style must be used to create interest but not to obscure the intended message.

The following sections contain a variety of examples of different forms of design communication and they are described with reference to these principles.

2.7.1 Engineering drawings

Engineering drawings are one of the most commonly recognised endpoints for a design project. Professional engineering drawings are produced as a record of a design which has been completed and released for production.

Amendments might still be made, but once the engineering drawings are completed these changes are formally recorded. Inherent in this requirement to act as a record of a design is a need for precision. The conventions summarised in PP 7308: 1986 have been developed to promote this accuracy in recording and communicating the design.

Conventions concerning the layout of views and their labelling, the representation of details, the line weights, types and sizes of lettering and methods of dimensioning help to ensure that the designer's intentions are accurately interpreted. It is not easy to define a three-dimensional form fully and there is considerable potential for confusion. It is essential that designers adhere to the British Standards conventions and do not try to invent their own if they wish to minimise the likelihood of being misunderstood.

Fig. 2.83 shows an example of an engineering drawing of a steering wheel design for a Lotus.

Several further examples can be found in PP 7308 along with the conventions most needed in schools and colleges. You should consult this document in order to produce professional engineering drawings.

The standard conventions of engineering drawing have been adopted in the development of CAD systems. This is mainly the result of the need to use conventions familiar to those who must read the drawings, but it should also be recognised that the conventions represent a highly evolved system. The approach detailed in PP 7308 is a very logical method of detailing three-dimensional forms for manufacture.

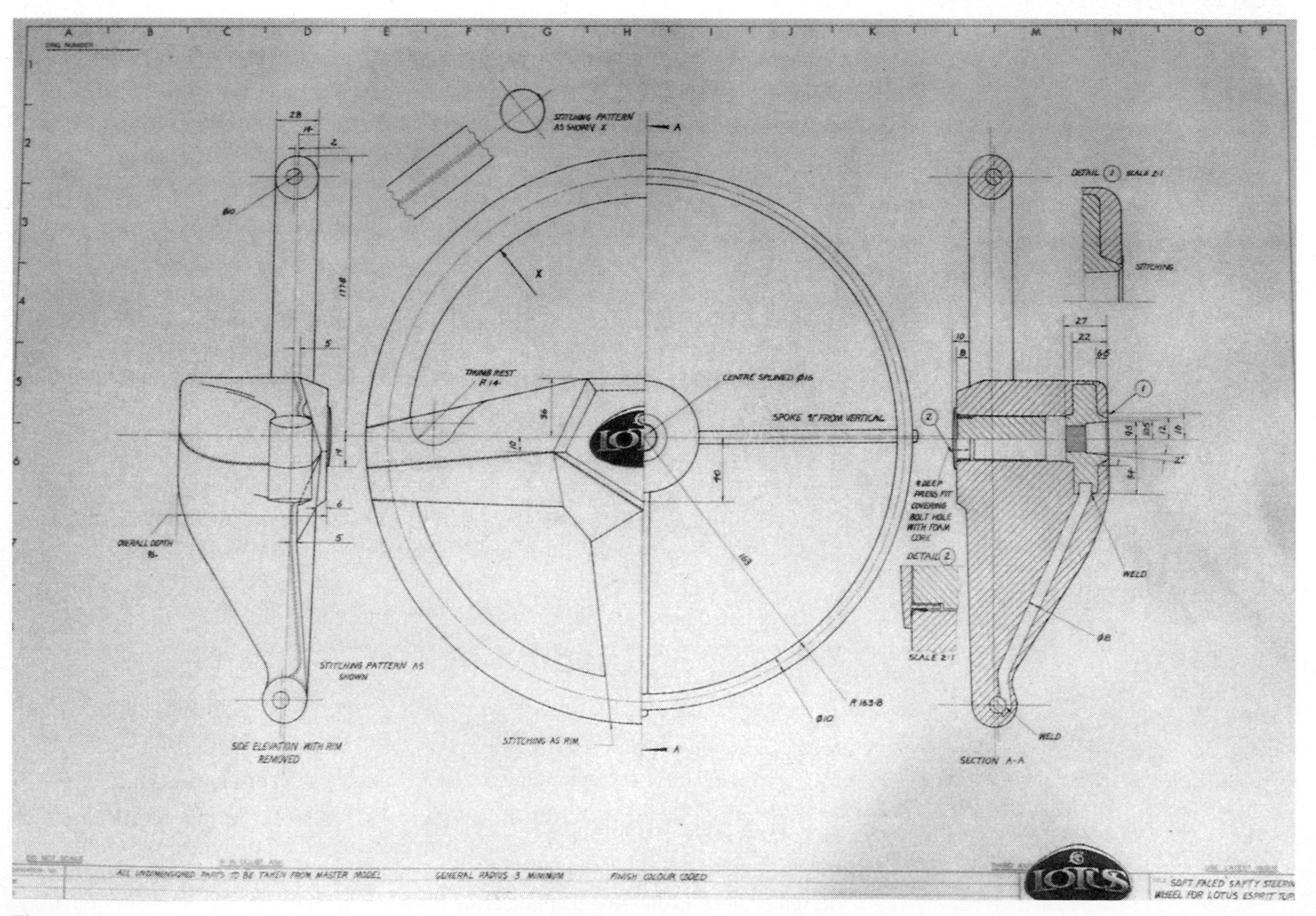

Figure 2.83 Engineering drawings of a prototype steering wheel for a Lotus.

Fig. 2.84 is included here as a reminder that a design project is just as likely to end with a circuit diagram giving a system statement, as with an engineering drawing defining a three-dimensional product form.

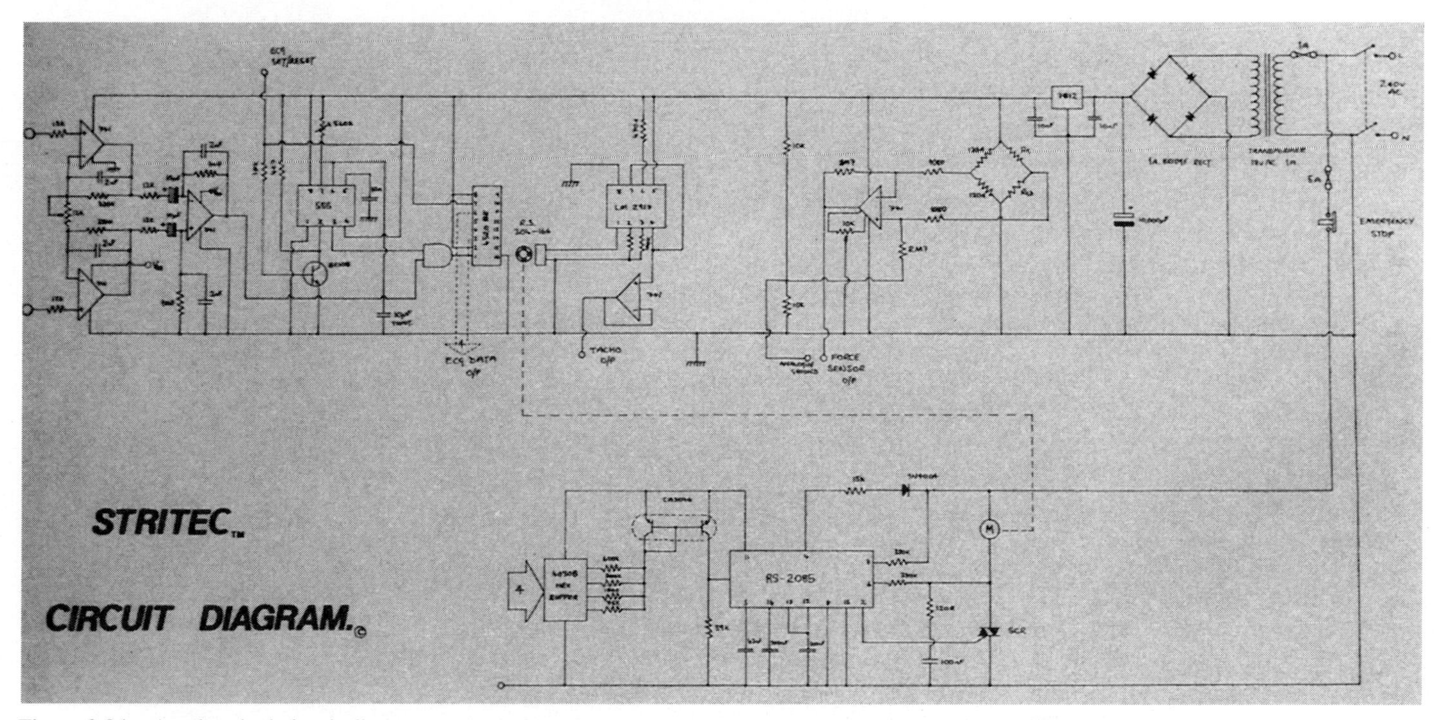

Figure 2.84 An electrical circuit diagram.

2.7.2 Project presentation and display

As a designer you will need to present your project through a display of sketches, drawings and models, and sometimes a prototype. In industry the display will be intended for your client, and in school or college its primary purpose is to show your supervisors and examiners what you have done.

The kinds of projects students might need to present will vary from minor projects, and product analysis exercises, to complete designs. Fig. 2.85 shows an investigation of natural form leading to a vacuum forming. The initial sketches are shown at the top of the sheet, and then development sketches leading to the vacuum forming which is attached to the bottom left-hand corner.

Figs 2.86 and 2.87 show product analysis exercises for plastic razors and clothes pegs. Each includes a dominating image which conveys the central theme and a variety of surrounding sketches showing the product's construction and use.

All these single sheet presentations seek to convey something of their purpose at first glance and to offer detailed information for those who wish to look more closely. This same idea is important when putting together displays of complete design projects. It must be immediately apparent what it is that has been designed and how it is intended to be used. Figs 2.88 and 2.89 show displays for the artist's easel design which was featured in section 2.4.3 and a shop display system.

These displays achieve their impact in different ways but they have aspects in common. The shop display system features a full-size prototype, whereas the artist's easel is shown as a full-size engineering drawing and a scale model. It is these features which initially attract the viewer's attention. Both displays then seek to draw in their audience by showing presentation drawings of the product in context. This is

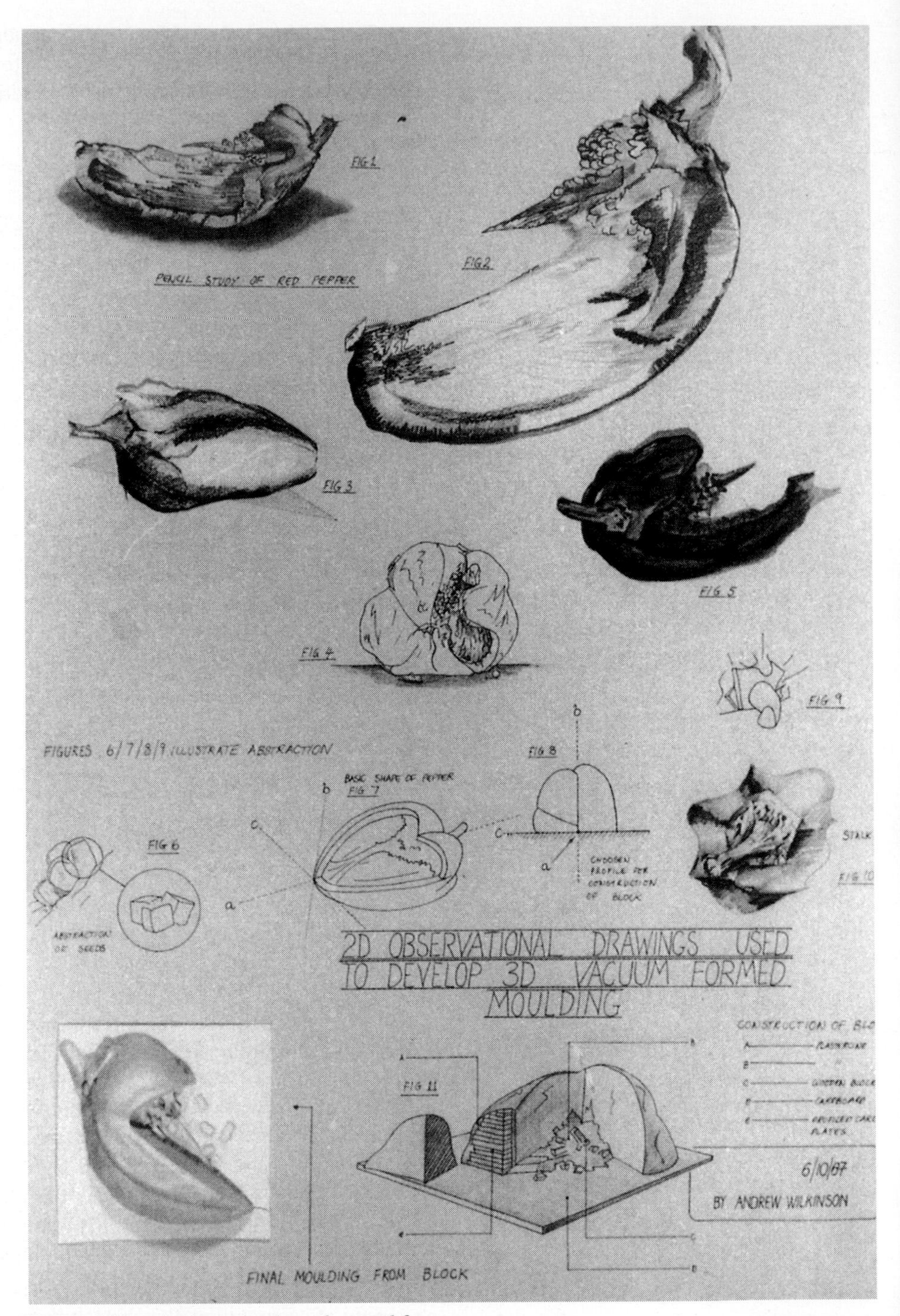

Figure 2.85 An investigation of natural form.

much more important for the shop display system, where the manner in which the product is intended to be used is not so immediately apparent. Figs 2.90 shows a further example – the Lotus steering wheel in its intended context.

It is very important to include such drawings in displays if casual viewers are not to misunderstand or dismiss new designs. This is particularly the case where an unusual or unexpected approach has been adopted. Important visitors to exhibitions and displays do not have time to read detailed explanations, nor should they be expected to do so, but they might well wish to encourage and explore innovative designs.

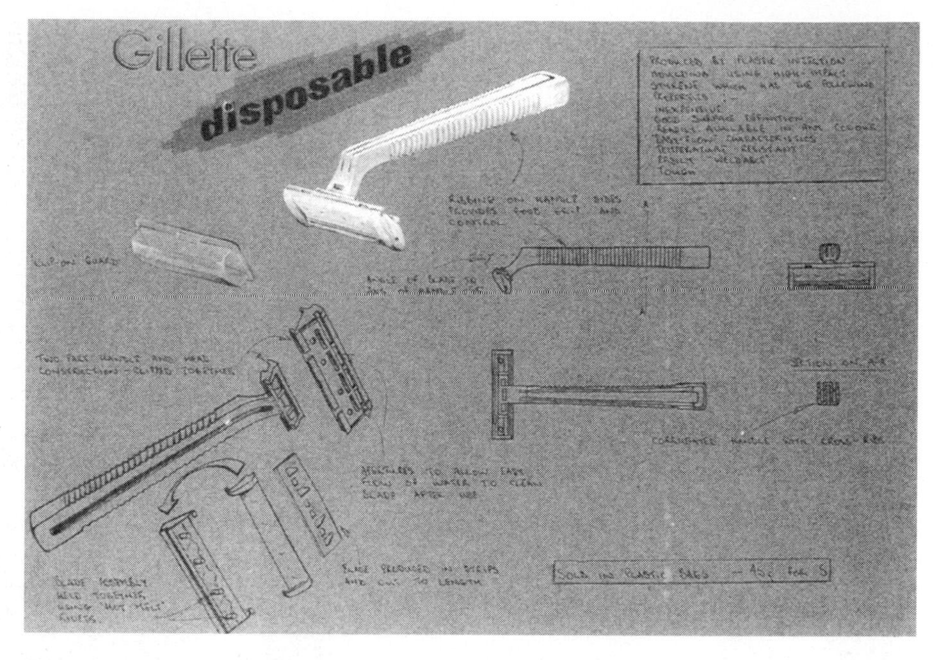

Figure 2.86 Product analysis of a plastic razor.

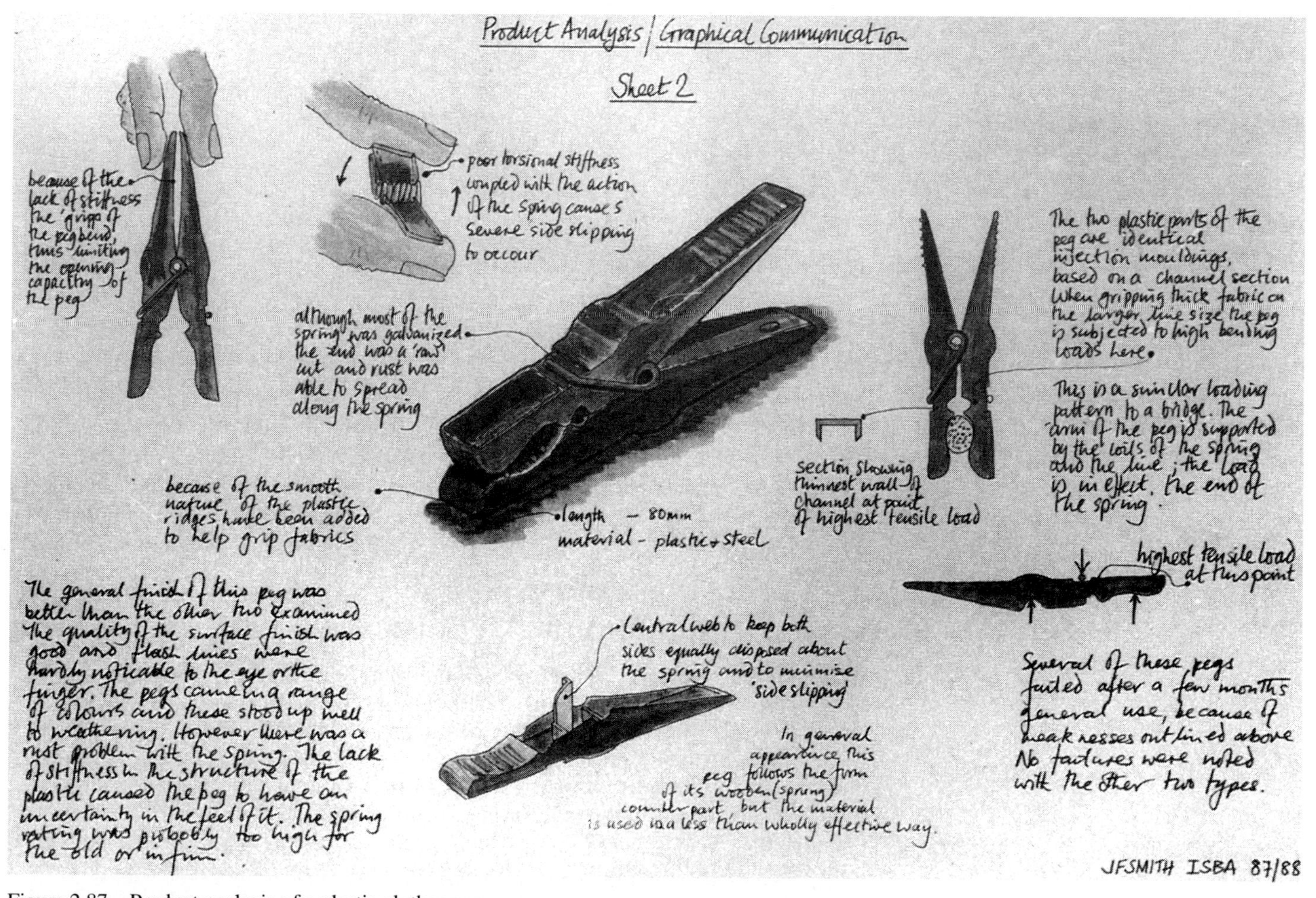

Figure 2.87 Product analysis of a plastic clothes peg.

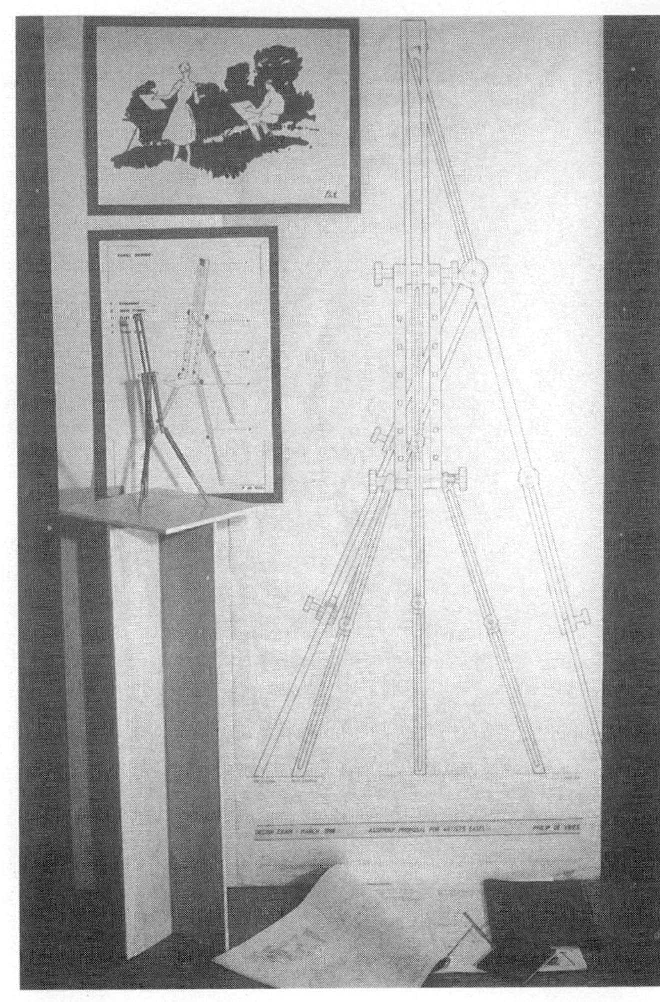

Figure 2.88 A design for an artist's easel.

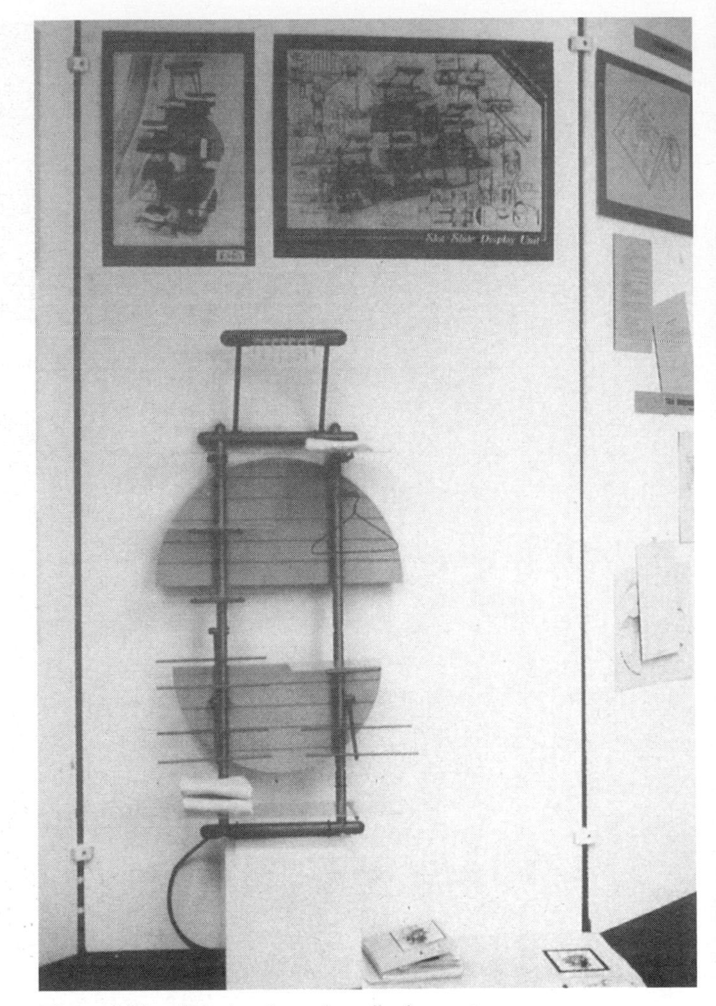

Figure 2.89 A design for a shop display system.

The full project displays incorporate the folders of initial research material and design sheets, and the designers' logbooks for inspection. Examiners have to assess the process as well as the final product and this documentation must be available to them. Other people who visit the display may wish to explore a designer's background thinking and seek to understand the influences which led to a particular design proposal. Both of the examples shown also contain written reports; these are the subject of the next section.

2.7.3 Written and oral reports

Written reports need to be accessible at a number of levels in the same way as a project display. The document must begin with a brief synopsis of a few hundred words to give an overview of the project and its achievements. This might begin with background information as at the beginning of the PDS (see section 2.2.4), indicating the origins and aims of the project. Key decisions and stages in the project should be highlighted and the final position stated. The synopsis should contain enough pointers to indicate to the reader which aspects of the report might be of further interest.

The remainder of the report is best presented as largely independent sections which are carefully referenced. The decimal numbering system used in this book for

Figure 2.90 Graphical presentation of a steering-wheel design for a Lotus.

sections and subsections is very common and is one that allows an index to be constructed easily. After reading the synopsis the reader should be able to turn to an index and locate a section of particular interest. It is not realistic to expect everyone who picks up the report to read it all and it must be structured so that readers can quickly find the items important to them.

PP 7317: 1987 contains an interesting section on the factors which influence the ease with which text can be understood. This section emphasises that the clarity of a passage depends primarily upon the orderliness of its structure. Some detailed advice is also given, for example that short sentences are easier to understand than long ones and that positive statements are easier to understand than negative ones.

Oral presentations of designs are normally backed by the kind of full project display indicated earlier. They are a way of introducing the design and provide an opportunity of attracting the audience to its strongest features. Again the structure of the presentation is crucial because listeners will often miss odd words or sentences. This can happen through simple lapses of concentration, as well as obvious causes such as inaudibility or confusion of accents. With an oral presentation there is no opportunity to look back and little chance to seek clarification. The general rules can be summarised by the following:

- say what you are going to say,
- then say it,
- then say what you have said.

With this type of presentation the listener is given a clear overview at the beginning when they are most likely to be attentive. If later they miss isolated details, or there is something that is difficult to understand, the presentation can still succeed because the general direction and intention is known. No more detail than is necessary should be given in an oral presentation because it can very easily distract the listener from the main points being made. It is important to end with a summary because the presentation should conclude with a general overview rather than matters of detail.

Notes

1. Address to the DES Conference at Loughborough University, 1987
2. Hopwood D 1987 Ergonomics – science for a safe future *Engineers' Digest* May.
3. Grandjean E 1984 Postures and the design of VDT workstations *Behaviour and Information Technology* Vol. 3 No. 4
4. Rondeau H R 1975 The 1-3-9 rule for product cost estimation *Machine Design* August
5. Tallenwerk, Technisch-Wirtschaftliches Konstruieren/VDI Richtlinien 2725/1&2 (1964)/VDI Fachgruppe Konstrucktion
6. Brichta A M 1976 Design evaluation *Engineering Designer*
7. Hitami K 1992 'Present trends and issues in Japanese manufacturers and management', Technovation, Vol. 12 No. 3 p. 179.
8. Cooke P, Corbett J, Pugh S and Weightman D 1984 *A Guide to Design for Production*. Institution of Production Engineers

BSI publications

The British Standards Institution (BSI) publishes a number of useful documents for schools and colleges. Details can be found at their website:

www.bsi.org.uk/education/

In particular the BSI publishes a *Compendium of Essential Design and Technology Standards for Schools and Colleges*. This publication is designed to assist the teaching of the National Curriculum programme of study for Design and Technology from Key Stage 3 onwards, although GCSE, A/AS level and Vocational A level students and teachers will find it particular useful.

The *Compendium* links a number of standards with the type of products commonly manufactured as part of Design and Technology courses. Chapter headings include; Quality management, Risk assessment, Anthropometrics, Textiles, Food products, Furniture, Toys, Graphical symbols, Packaging.

The *Compendium* includes sections on food, testing, toy safety, labelling, health and safety, and the ISO 9000 series of standards concerning quality management systems. The content of the *Compendium* has been gathered from British, European and International Standards, which have been precised and rewritten. Hence, although they are an excellent guide, they should not be viewed as substitutes for the full standards.

Bibliography

Baxter M 1995 *Product Design: Practical Methods for the Systematic Development of New Products*. Chapman and Hall

Beasley D 1979 *Design Illustration, Sketching and Shading Techniques*. Heinemann Educational Books

de Bono E 1967 *The Use of Lateral Thinking*. Penguin (Pelican)

de Bono E 1970 *Lateral Thinking – A Textbook of Creativity*. Ward Lock Educational

Breckon C J, Jones L J and Moorhouse C E 1987 *Visual Messages: An Introduction to Graphics*. David and Charles

Croney J 1980 *Anthropometrics for Designers (2nd edition)*. Batsford

Cross N 1984 *Developments in Design Methodology*. Wiley

Cross N, Elliot D and Roy R 1975 *Man-made Futures: Design and Technology Course Units*. Open University Press

Cross N *et al* 1983 *Design: Processes and Products Course Units*. Open University Press (particularly 'An introduction to design' and 'Everyday objects – ergonomics and evaluation')

Diffrient N, Tilley A R and Harman D 1981 *Humanscale* 4/5/6. MIT Press

Dreyfuss H 1978 *The Measure of Man – Human Factors in Design*. Whitney

Dull J and Weerdmeester B 1993 *Ergonomics for Beginners*. Taylor and Francis

Glegg G L 1983 *A Design Engineer's Pocket Book*. Macmillan Education

Haigh M J 1985 *An Introduction to Computer-aided Design and Manufacture*. Blackwell Scientific Publications

Hawkes B and Abinett R 1984 *The Engineering Design Process*. Pitman

Jones J C 1980 *Design Methods – Seeds of Human Futures*. John Wiley and Sons

Kraemer K and Grandjean E 1997 *Fitting the Task: the Human*. Taylor and Francis

Lawson B 1990 *How Designers Think (2nd edition)*. Butterworth

McCormick E J and Saunders M S 1986 *Human Factors in Engineering and Design*. McGraw-Hill

Miles L D 1961 *Techniques of Value Analysis and Engineering*. McGraw-Hill

Morrison J and Twyford J 1994 *Design Capability and Awareness*. Longman

Mucci P E R 1986 *Handbook for Engineering Design Using Standard Materials and Components*. Mucci

Murrell K F H 1979 *Ergonomics: Man in his Working Environment*. Chapman and Hall

Norman D A 1998 *The Design of Everyday Things*. MIT Press

Norris B and Wilson J R 1998 *Childata: The Handbook of Child Measurements and Capabilities: Data for Design Safety*. DTI

Panero J and Zelnik M 1979 *Human Dimension and Interior space*. Architectural Press

Parker M (ed) 1984 *Manual of British Standards in Engineering Drawing and Design*. Hutchinson

Peebles L and Norris B 1998 *Adultdata: The Handbook of Adult Anthropometric and Strength Measurement: Data for Design Safety*. (More information is available at www.openerg.com.)

Pheasant S T 1990 *Anthropometrics: an Introduction*. BSI

Pheasant S 1996 *Bodyspace: Anthropometry, Ergonomics and the Design of Work (2nd edition)*. Taylor and Francis

Powell D 1990 *Presentation Techniques*. Macdonald

Pipes A 1990 *Drawing the Three-dimensional design*. Thames and Hudson

Ray M S 1985 *Elements of Engineering Design – An Integrated Approach.* Prentice-Hall International

Sanders M S and McCormick E J 1993 *Human Factors in Engineering and Design (7th edition).* McGraw-Hill

Starkey C V 1988 *Basic Engineering Design.* Edward Arnold

Starkey C V 1993 *Engineering Design Decisions.* Edward Arnold

Swift K G and Booker J D 1997 *Process Selection – From Design to Manufacture.* Edward Arnold

Assignments

1 'It is eminently true of design that if you are not prepared to make mistakes you will never make anything at all.' (David Pye)
With reference to your own experience of designing explain what you think Pye meant by this statement. Indicate to what extent you believe Pye's statement is valid.

2 With reference to a personal design project you have completed, show how the process of information gathering was achieved and, with the benefit of hindsight, indicate how you might have improved your performance at this task.

3 The natural forces of our environment have a strong influence upon shape and form in nature. Discuss the ways in which a designer of jewellery and a civil engineer can both benefit from a close study of natural form.

4 Draw up a list of criteria for the purchase of a personal stereo with radio and tape facilities, which could be used by a teenager.

5 Dicuss the criteria a designer would use in developing each of the following consumer products.
(a) A relative cheap can opener for use by elderly people.
(b) A cheap pencil sharpener for use by children that retains the shavings.
(c) A pen to be given as a present by a pharmaceutical company to doctors.
(d) A table lamp to be used on a bedside cabinet in a teenager's room.
(e) A decorative breadboard for use in a country cottage.

6 The design of any product or system is influenced by a wide range of factors. Many of these are identified in the '*Product Design Specification*'.
The following factors appear in many Product Design Specifications and are important to designers.
(i) Reliability and Maintenance.
(ii) Safety.
(iii) Environmental Issues.
Discuss each of the above factors illustrating the point you make by referring, in each case, to two examples.

7 In product design, discuss the role of:
(a) adjustment
(b) a range of sizes.
Indicate the limitations of each idea, with particular reference to the variation in human sizes. Illustrate your answer by referring to specific examples.

8 (a) Discuss the significance of anthropometrics and ergonomics to the designer.
(b) Using a bicycle as an example show how designers have utilised data relating to anthropometrics and ergonomics to derive a final design.

9 (a) Designing is essentially related to people, consequently, a knowledge of 'Man as an Individual' is important.
Explain each of the following:
(i) anthropometrics;
(ii) 'other' physiological factors;
(iii) psychological factors.
(b) Choose *one* of the following fields:
– domestic appliances;
– domestic lighting;
– private housing.
Select *one* item and explain how the designer has applied ergonomic principles in its design. Make detailed references to specific examples to illustrate your answer.
(c) Choose an example of a product, from any field, in which you feel that the ergonomic factors have *not* been fully taken into account.
Explain your reasons for holding this view.

10 'Mathematics not only allows us to understand certain aspects of nature, but can help to give something of nature's efficiency to man-made objects.' (Rowland) Discuss this statement, referring to specific aspects of nature and man-made objects in the broadest sense and not solely in terms of mathematical calculations.

11 With reference to specific examples discuss the advantages and disadvantages of each of the following as means of communication.
(a) Physical models.
(b) Computer simulations.
(c) Orthographic drawings.
(d) Written reports.

12 The use of models, mock-ups, prototypes and test-rigs can be very useful, if not essential, in the development of new products. Explain, in detail, using wherever possible your own experience, where and why they would be used.

13 Explain the importance of the use of 'standard' components in manufacturing industry.

14 (a) Compile a breakdown of contributing factors that influence the final market price of a product.
(b) Under each category in (a) suggest strategies that could be adopted to minimise the costs.
(c) Using as an example a product or component you have analysed, show how the choice of material and manufacturing process have been influenced by production costs considerations.

15 'The computer can act as an extension of the hand and the brain, but it cannot replace either.' (Toft)
With reference to this statement discuss the advantages and disadvantages of CAD/CAM in the development of two dissimilar products.

16 (a) Explain four reasons why computers are said to be an improvement on previous methods of producing drawings and designs.
(b) Describe in detail an Industrial Computer Aided Design application that you have studied or researched.
(c) Give the following details of a piece of CAD software with which you are familiar.
(i) State its name and give a brief description of it.
(ii) Give full details of the hardware on which it runs.
(iii) Using notes and sketches give an example of the type of work that you have done, as an example of its best use.
(iv) Explain why the use of CAD was the most appropriate for this drawing.

17 The use of the computer to produce graphics is widespread in schools and industry.
(a) What benefits are there in this form of communicating information?
(b) What hardware is needed for computer-aided graphics for inputting information and what are the advantages and disadvantages?
(c) How can the information generated be used to produce programs for CNC machines?

18 (a) When planning the production of work in a school workshop it is important that resources are used efficiently. Identify and discuss the factors that must be considered in connection with each of the following:
(i) time; (ii) facilities; (iii) materials.
(b) Compare the factors identified in (a) with the equivalent planning procedures operated in industry.

19 A project is to be completed in a college by a number of students. One of the students breaks down the project into a number of activities, some of which can only start when others have been completed. The duration of each activity is given in the table below.

Activity	Time in days
A	3
B	8
C	8
D	15
E	6
F	11
G	10
H	7
J	6
K	9
L	10

Activities A, B and E may start at the beginning of the project.
When B is complete, D and C may start.
When A is complete, F may start.
H can only start when F and D are complete.
H is a final activity.
G is also a final activity
G can only start when C is complete.
K can start when E is finished.
L can only start when C and K are complete.
J can only start when L is complete and J is a final activity.
The project ends when G, H and J are complete.

Show, in diagrammatic form, the activities and determine the shortest possible time the project will take. Indicate the activities which are on the critical path.

20 (a) The ability to communicate is essential to a designer.
(i) Explain why the range of people with whom the designer communicates requires the designer to use a variety of communication techniques.
(ii) Describe the range of graphic techniques available to the designer and identify the situations where each could be used to greatest advantage.
(b) Discuss how the use of computers can improve the effectiveness of communication.

Design briefs

1 The August 1982 edition of 'Design' magazine carried an article entitled 'Ironing Board – A Pressing Problem. The makers of these infernal domestic devices can't do much ironing'. Design an ironing board which is free standing and easily stored. Special attention should be given to the safe use of the iron and its positioning during pauses in the ironing process.

2 Design and Technology is a fast growing subject in the Primary School. Design a work bench which can be used in the Primary School for Design and Technology activities. The work bench must show clearly housings for all the necessary tools and equipment as well as clamping devices for holding materials. Methods of transporting the work bench to different classrooms as well as varying the height for different children in the seven to eleven age range should be considered. In the analysis part of your answer the tools and equipment necessary for Design and Technology in primary schools must be clearly identified.

3 For many people suffering from arthritis or other disability in the hands, the ordinary domestic door lock can provide some difficult problems. Design or devise a system which will enable these people to overcome these difficulties in handling keys and locks but at the same time ensuring that security is not adversely affected.

4 The passenger in a car who acts as a navigator at night often finds difficulty when there is no provision for map reading. Design a pocket navigation aid which should incorporate illumination (this should not cause disturbance to the driver's vision), magnification, a compass and a simple mileage calculator. Materials and production methods are left to your discretion.

5 Hiking and camping are two of the most popular leisure activities. Design a framed rucksack capable of carrying the requirements for an individual adult hiker and camper. Part of the frame should form part of the tent construction. In your analysis the contents of the hiker's/camper's rucksack should be clearly identified.

6 Between the age at which children can first sit up and when they can sit in a normal chair, they are often placed in a 'high chair'. Design such a chair which is very stable, provides a secure safe seat, and has a surface upon which food and toys can be placed. The chair must be capable of being folded so as not to take up too much space when not in use. The chair should also incorporate a simple amusement aid which should keep the child happy when not being fed and which the child cannot detach.

Figure 2.91

7 For reasons of cost or lack of storage space, most householders would probably not wish to equip themselves with elaborate capital equipment but may often need to engage in DIY work above normal reach height.

Design an appropriate product, at least one function of which would be, to provide a *working platform* at a height of up to one metre above ground or floor level.

8 Each year even small gardens produce quite large volumes of soft and semi-hard fibrous waste material from weeds, old plants and hedge clippings etc.

Design a tool, suitable for garden use, capable of shredding all types of semi-hard fibrous waste, up to 25 mm in diameter, to produce material suitable for composting. Your design must be suitable for production at the rate of 20 000 per annum.

9 The quick and accurate weighing of ingredients is an important part of cooking today.

Design an electronic, battery-powered weighing device which will fit unobtrusively into a modern kitchen. The device should be ergonomically designed, safe and easy to clean. It will normally be used on a flat work surface, with a wall mounted option. The device must have a digital display with the option of switching between imperial and metric units. It should be accurate to 1/8 oz/2 g and take loads of up to 5 lb/2 kg. The weighing device must adjust automatically to the weight of the bowl being used. There must be a visible warning to indicate when the batteries are low in power as well as an automatic switching off when not in use. The retail price of the product should be £30. The design must include circuit diagrams; 'black boxes' may be used where appropriate, but at least as much emphasis should be given to the casing as to the electronics.

10 For a family taking a camping holiday the provision of hot water is a significant problem. This can be time consuming and moderately expensive if even the smallest quantity has to be heated from cold on a camping stove.

Design a device, suitable for camping purposes, capable of raising the temperature of up to 10 litres of stored water during the day. Your design must be appropriate for manufacture in batches of 2000.
You are to assume that permanent electricity and gas supplies are not available.

11 If, for some reason, the whole family has to be away from home for a number of days during a severe winter, the garden pond may become iced over for a long enough period to threaten the life of any fish present.

Design a product which would ensure that there was a safely maintained tea-tray sized area of the pond surface always free from ice.

Figure 2.92

12 Snow clearing can be a very heavy task and frequently needs to be done first thing in the morning before leaving for work.
Prepare an outline design of a powered snow-clearing machine suitable for manufacture for the domestic market.

13 Simple tasks like switching on room lights, opening and closing doors, drawing curtains, etc. can present considerable difficulty to people in wheel chairs.

Design a control system, complete with details of TWO of your specified actuating units, suitable for installation in an average home as an aid for such people.

Your solution must be suitably packaged, so as to fit unobtrusively into a variety of decor styles.

14 The growth in the use of XY plotters in schools and colleges has been very considerable in recent years.

Design a mobile unit which will transport the XY plotter, computer and associated equipment, as well as storing paper for use on the plotter and other peripheral equipment. The unit must provide an opportunity for the student to sit comfortably when using the keyboard or other input devices (the actual seat is not expected to be part of the unit). The unit will be stored in a room whose door is only 650 mm wide.

15 Learning to tell the time is a difficult task and, with both analogue and digital systems in use, the task is made no easier for young children.

Design a game or toy which will provide interest and enjoyment for young children while at the same time helping them to tell the time. The solution must help children understand both analogue and digital systems. The game needs to show only twelve hours and must divide the hours into blocks of 15 minutes or less.

16 Hose pipes are used for a wide range of purposes by householders, from cleaning cars and pavings, to watering gardens. A manufacturer wishes to produce a more sophisticated hose nozzle which has the capacity to create a more powerful pressure cleaner.

Design a hose nozzle which connects to a normal water supply through a standard click-fit hose fitting. The hose nozzle must have the capacity to generate high pressures for cleaning under cars, a range of different sprays for the garden and a built-in liquid soap provision which can easily be switched on or off as appropriate.

17 Driving lessons are becoming very expensive and, to begin with, taking a car out onto the road may not be the most cost-effective way of getting used to the vehicle controls.

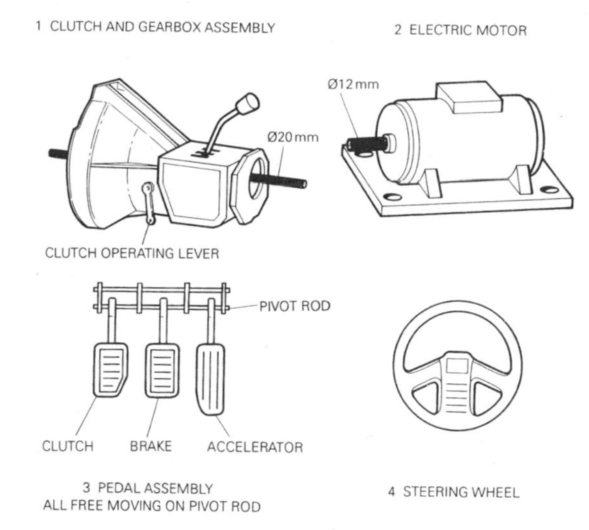

Figure 2.93

Use the electric motor shown in Fig. 2.93 to drive the gearbox and clutch assembly, also shown in Fig. 2.93, to form the basis of a simple gear-changing simulator. You are also to use the pedal assembly and steering wheel in your design along with any other items you think are necessary.

18 Design a tree pruning device which would give an operator a vertical reach of 4 metres but which would store easily in a cupboard when not in use (Fig. 2.94).

Figure 2.94

19 An amateur interested in lapidary work wishes to make a device for shaping semi-precious gemstones for rings, brooches and similar jewellery. From a rough-sawn slab of stone up to 5 mm in thickness, the amateur initially wishes to grind accurate outlines of a circle, ellipse, square and rectangle.

Design an attachment for a grinding machine to produce these outlines on stones within dimensions between 10 mm and 40 mm. Provision for doming of the stones after grinding the outlines is not required with the attachment you design.

Note: You may assume that the slab of stone may be dopped (fastened at right angles with sealing wax) to a circular metal rod 8 mm in diameter and of a length you think appropriate. The grindstone is mounted with its axis vertical on a bench top, surrounded by a water trough and safety guard as shown in Fig. 2.95. It is anticipated that your designed attachment will be screwed to the bench top at the side of the grindstone, the face of the grindstone (not the edge) being used for the shaping operation.

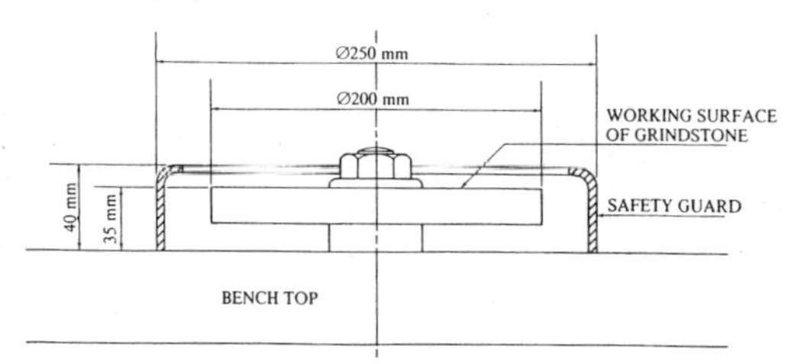

Use your discretion where dimensions are not given.

Figure 2.95

20 Stolen cars and thefts of articles from cars are major areas of crime. Increasingly insurance companies are requiring owners to have some form of security device to protect the car and its contents. Solutions include visual warning labels, mechanical devices, infra red sensors and audible alarms.

Design a security system for a motor car that will deter a criminal from stealing it and will help to protect the contents in some way. The system must not take more than 2 minutes to set up or remove as it may discourage drivers from using the system. In addition, it is important that potential criminals know the car has an alarm system, thus as part of the answer you should design a clear label to explain the existence of an alarm system. In your answer you should should provide a leaflet to explain how to set up and disarm the alarm, as well as illustrating your visual warning design.

3 Design case studies

Compiled by Mike Hopkinson, Principal Project Moderator OCR DT (Technology)

3.1 Design projects

This chapter is written to help candidates and centres that have not entered project work at an Advanced level before or who wish to refine their approach. By following advice it should be possible to improve project management skills and significantly raise your performance against required criteria.

The following areas will be covered:

- an introduction to design processes common to most assessment criteria suggesting possible context areas for project selection
- a sample of exemplar projects using different themes and approaches
- selecting from these projects good examples to meet assessment criteria in different sections of a specification mark scheme
- a case study of an exemplar project showing work from every section
- a 'moderator's tips' section – follow this advice and it could be possible to improve a project mark by up to a grade!

Design processes – mirrored in the assessment criteria

All Advanced level Design and Technology courses now involve candidates in a practical design project. The approach to these projects differs depending largely on candidates' backgrounds, the facilities offered by a school or centre and the influence of the opportunities offered in the locality. One requirement that they all have in common is to demonstrate an understanding of and an involvement in design processes, detailed in Chapter 2, which lead to the construction of a quality end product. There are many models of designing but most follow general stages, which can be approached in a linear or cyclic way:

Common Stages in Models of Designing	Typical Exam Specification (OCR A LEVEL 2002)	
• Identify Use(s) • Write a Brief • Investigation • Write a Product Design Specification	20	Recognition and Investigation of Design Opportunities
• Generating Ideas • Analysing Alternative Concepts	25	Generating Design Ideas
• Synthesis	20	Synthesis
• Detailed Design	25	Development/modelling
• Manufacture • Construction of a Working Prototype	70	Planning and Making
• Evaluation	20	Testing /evaluation
	180	*Total Marks*

Table 3.1

A design process model is likely to provide the underpinning ethos which will link the approach to all projects. Most examination specifications will mirror design processes when stating the assessment criteria for the project section.

In the new design specifications marks awarded for different sections may be split between different modules and marks may not be percentages. Candidates need to reflect on the relative award of marks when involved in initial time planning.

3.2 Possible context areas for project selection

Extract from OCR guidance notes – Advanced level 2002

The first problem many candidates face when starting a project is to find a need or context from which to identify a possible area for project development. The following list of context areas is not exhaustive, but it was drawn up by over 20 teachers with experience ranging from textiles, resistant materials, graphics products and control systems. Only a few ideas are listed under each heading, selected mainly from possibilities for projects involving control systems. Some examination specifications will contain fuller support material. Individual circumstances and locally-available resources may well influence the choice of project area:

- Leisure and sporting activities
  Sports – starting, timing and judging systems, photographic triggering, scoring, control and monitoring of exercise, safety clothing/equipment, camping equipment, mood lighting, theatre and stage design/lighting systems, control, visitor monitoring, pet care, gardening, environmental control, board games.
- Public open spaces
  Gardens and parks, children's play areas, litter, vandalism, seating, shelters, information signs, and visual display systems, safety equipment.
- The food industry
  Simulation-automated food production, date stamping, printing, shape pressing, conveyor belts, temperature monitoring and control.
- Fashion and personal adornment
  Seasonal extreme weather clothing, electronic jewellery, stamping/pressing shapes for personal adornment.
- Travel and transport systems
  Personal transport systems, car park systems control, automatic door/lifts, intelligent transport movement, sensing and control, monitoring/security.
- The environment, agri/horticulture, conservation
  Farming machinery/buildings, monitoring and automated feeding of livestock, planting, harvesting and sorting of food, drainage, irrigation, cultivation, energy conservation/alternative energy, monitoring/control of horticultural/biological environments.
- Historical contexts
  Preservation or protection of valuable/fragile artefacts.
- Industrial context/health and safety
  Conveyor/sorting systems, food production, pick and place robotics, sensing, locating guards, clamping and drilling, automated car wash systems, garage environments.
- Consumer products
  Automated consumer testing-electrical/mechanical/pneumatic.
- Personal/domestic environments
  Monitoring and control of environments for personal safety, young/aged-security, comfort convenience, heating, doors, windows, lighting, locks and alarms, energy efficient environments, remote control, designing for those with special needs, the young and elderly.

- Community
  Transport systems, caring for the young and elderly, shopping centres, health centres, hospitals, ambulance, police, fire service, libraries, nursery provision, service station, garage, church, temple, meeting place, community centre, display stands and exhibitions, shop window displays.
- Appropriate/intermediate technology
  Design for the developing world using appropriate materials/equipment, sustainable resources, recyclable resources, design sensitive to cultural diversity, energy efficient design, values and ethical considerations – cost of batteries – alternatives, e.g. clockwork.

When choosing areas for project development candidates should be cautious of raising client expectations. Often the relationship develops best when a client helps the candidate with development, with no other expectation than evaluating a working prototype. This is often more productive than promising a fully working usable end product.

3.2.1 Traffic control torch *(Patent pending application: GB 9920338.2)*

Darren Fenton, Grosvenor Grammar School, Belfast, Northern Ireland.
Taught by: Mr Ricky Cowan.

Living in a community in Northern Ireland where the stopping of cars at night had become commonplace, Darren was aware of the dangers for members of the police force involved in this activity. Wider research involving over 50 police forces internationally revealed a much larger problem, emphasised by this comment from Northamptonshire police:

> The problem of stopping vehicles in a roadside check is one that is inherently dangerous and is always a concern particularly since the death of an officer in 1991.

Working closely with the Royal Ulster Constabulary (RUC) Traffic Branch, and assisted by Shorts Bombardier, Aerospace Manufacturers, Darren researched and developed an effective answer to the problem. This resulted in an end product which has received praise from many sources and led to success as winner of the regional Northern Ireland and runner up in the National Young Engineer for Britain Competition.

Figure 3.1 Darren and the completed torch with the RUC Chief Constable, Sir Ronnie Flannagan.

The completed project, which is featured later in this chapter, consists of a fluorescent red tube for a 'stop' signal, a green light for 'go', a flashing blue light for hazard warning and a white light for inspection. A watertight polycarbonate tube and box protect the lights and associated control circuitry.

The Police Scientific Development Branch and the Police Research Group are now evaluating Darren's torch. He has applied for a patent and, following many requests, has written to all 55 forces and torch manufacturers contacted during the research to see if they would be interested in buying or producing the torch. The highlight of Darren's project experience was working with the RUC and Shorts Aerospace. He concludes

> I feel the experience gained in conducting my A Level Project will help me at university as it has given me a holistic view of the design process. When I finish university I want to become a product designer.

Darren is currently studying Technology and Design at the University of Ulster.

3.3 Examples of design projects

Section 3.3.1 shows four examples of projects linked to a manufacturer. Four other projects are shown in section 3.3.2.

3.3.1 Projects involving an industrial link

There are many advantages to working with local industry. An excellent relationship between Grosvenor Grammar School, Belfast and a local firm, RFD Limited, is summarised here by the comments of one candidate and a letter from RFD Production Engineer David Curry. The letter stands as a testament to the value of the partnership and also provides a valuable element of a third party evaluation. Further design sheets from all of these projects will be shown later.

> In order to build my jacket I had to enter the world of industry. I worked in a local company, RFD. The resources to which I was exposed were amazing and it was a totally invaluable experience. I was able to see the chain of production, the factory processes, and learn about all aspects of factory life.

RFD Limited Kingsway Dunmurry Belfast BT17 9AF Northern Ireland
Tel: +44 (0) 1232 301531 Facsimile : +44 (0) 1232 621765 Telex: 747264

For the second consecutive year R. Coburn (Plant Director) has been pleased, on behalf of RFD Dunmurry, to be able to offer company facilities in assisting pupils from Grosvenor Grammar School in pursuit of their 'A' level technology project.
The company, only too aware of the challenges faced by experienced engineers in converting concept designs into successful products, considers participation in such schemes to be essential for the long term future of the industry. As experience within the company's industry sector is not extensive an added appreciation of the challenges facing students designing such products further reinforces company commitment to assist these young pupils. We hope that the experience gained will encompass areas of behavioural, social, managerial and interactive aspects of industrial life that may not have been envisaged as a part of an engineer's activity. In reality however, an understanding of the human factors indicated above can have a significant benifit for any aspiring engineer. It is our hope that the assistance provided will be as relevant to this years student designs as it was in helping in the success of the last years projects.

THE LIBRARY
GUILDFORD COLLEGE
of Further and Higher Education

On behalf of the engineering and other RFD personnel with whom Peter and Neil interacted I take this early opportunity to pass on our admiration for their enthusiasm and hard work in preparing and implementing their innovative concept designs. The boys had obviously carried out considerable research into the requirement for and feasibility of their relevant concepts as was evident from their professional constructed and visually appealing presentations. Peter and Neil quickly adapted any preconceived constructional features and techniques with practical design solutions, based on the specific properties and characteristics of our materials and the capabilities of our manufacturing processes. We regret that in relation to some of the required inflation equipment necessary for their specific designs, namely gas cylinders and operating heads, we were unable to provide for all their needs. Unfortunately none of our existing equipment was fit for purpose in this instance, but their basic designs and prototype construction would function satisfactorily if the said equipment was obtained. Indeed I am sure the required equipment could be procured from a number of suppliers if the designs were to be progressed further.
In conclusion, it was most interesting and enjoyable to have assisted Peter and Neill in completing their projects, their intelligent and enthusiastic approach combined with a willingness to both listen and learn was encouraging.

Peter and Neil displayed considerable patience, as the busy schedule of engineering and other manufacturing activity within RFD meant many interruptions and frustrating periods waiting for support from temporarily unavailable personnel. Grosvenor Grammar School can be proud of the quality and calibre of the 'A' level pupils it is generating as the performance of this and the previous years students have proved.

David S. Curry (Production Engineer) on behalf of Clifford Wylie (Production Engineer), Gavan Gillespie (Project Engineer) and all other participating personnel at RFD.

Registered in England at Grove Mill Earby Colne Lancashire BB8 6UT

Figure 3.2 Report from engineering firm RFD Limited on cooperation with students from Grosvenor Grammar School.

Horse riding protective safety vest

Rhian Chapman, Grosvenor Grammar School, Belfast, Northern Ireland.

Rhian identified the reluctance of many young fashion-conscious horse riders to wear the correct protective clothing as an area for product development. His project provides riders with a lightweight waistcoat which converts into a robust, resilient protective vest in the time taken for a falling rider to hit the ground. The innovative design relies on a pin which is fixed to the saddle, activating a pressurised cylinder which inflates the vest in 0.4 secs.

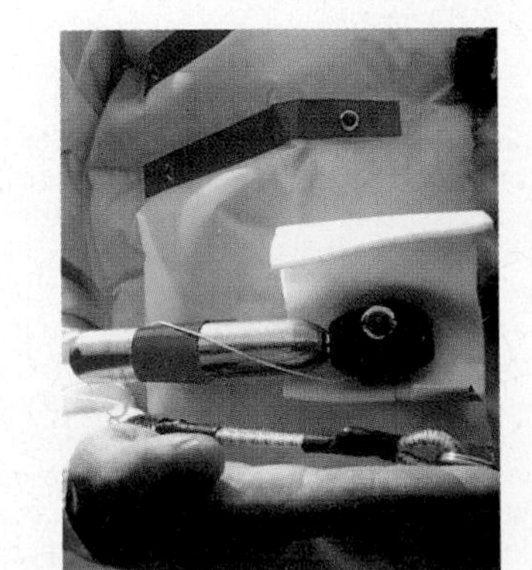

Figure 3.3 Rhian Chapman's horse riding vest.

If my project can help the disabled to have greater enjoyment of horse riding or help even one person to avoid a potentially dangerous accident, I will be more than happy with my work.

Following a sixth form course in Biology, Economics and DT (Technology), Rhian is now studying Business Studies at Bristol University, with aspirations of owning a PR or marketing business.

Home/nursery fire escape

Arthur Parke, Grosvenor Grammar School, Belfast, Northern Ireland.

Figure 3.4 Arthur Parke's inflatable fire escape.

There are many fire escapes on the market but probably only one which inflates in seconds from a window box on the first floor of a house. Arthur's innovative design uses the same technology as Rhian's but is applied in a totally different way. Careful consideration has been given to the unfolding of the device to provide optimum performance. Arthur was awarded second prize in The Young Innovator of the Year Competition.

Arthur studied English, Geography and DT (Technology). He is now studying Architecture at Queen's University, Belfast with aspirations to qualify as a chartered architect within the next seven years.

Arthur would like to continue with his design and is interested in contacting appropriate manufacturers.

Emergency burns stretcher

Neil McGaughey, Grosvenor Grammar School, Belfast, Northern Ireland.

The fire service requirement to remove a casualty quickly from a dangerous environment conflicts with the immediate care needs of burn victims. Neil's project relies on sound research of both areas to combine the rigid requirements of a conventional stretcher with the effective means of accommodating a burns victim. Neil was awarded first prize in the Advanced Level section of The Young Innovator of the Year Competition.

I am now studying Technology and Design at the University of Ulster. A large part of the course is project based and there is a placement year in industry.

Figure 3.5 Demonstrating the emergency burns stretcher.

Rapid rescue raft for frozen lakes

Peter Killops, Grosvenor Grammar School, Belfast, Northern Ireland.

The traditional method of rescue using a ladder or boat is much improved here with Peter's inflatable raft which keeps the rescuers safe and secures the victim in a harness before pulling them onto a platform for safe evacuation.

Figure 3.6 Rapid rescue raft for frozen lakes.

> The fire service have encouraged me to develop my innovation, as they believe it is a very marketable product.

Peter is studying Technology and Design at the University of Ulster.

3.3.2 Other examples of projects

Mountain bike braking system

Matt Levell, Brockenhurst College.
Taught by: Mr. Norman Andrew. (Milling assistance by Redmayne Engineering)

Matt noticed the reduced efficiency of his mountain bike brakes in wet and muddy conditions. He set himself the tight design brief of designing a system 'which would slow him down from 30 mph quicker than his current system'.

Although the final result is a most professional end product, which is beautifully engineered, life was not without its tribulations. The next page tells the tale, ending with a personal evaluation in Matt's own words. Further design sheets will be shown later in the chapter.

Matt is now working out his gap year at Southampton Ocean Technology Centre 'helping to redesign a small research submarine'! He will then take up his place at Swansea University on a Mechanical Engineering course. This will be his second step along the road to fulfilling his lifelong ambition to become a design engineer. The first step was completing the DT course:

> Looking back, having finished the course, I feel a sense of satisfaction and achievement. I learnt to use professional tools and to design and analyse my solutions – this project has helped me in my interviews!

Figure 3.7

Figure 3.8 When I was testing at 20 mph, I started to brake and then something happened, and I stopped really quickly. I looked down and saw a massive bent mess!!'

Matt had decided to test his braking system by initially fixing it to the bike with a jubilee clip. He had calculated the large forces involved and had even used aerospace grade aluminium in the disc. The weakest link in the chain was the temporary clip. The photograph (Figure 3.8) shows the results of the high forces acting on the system.

Figure 3.9 The final project after very necessary modifications.

A famous climber once said, 'High failure overleaps the bounds of low success': this is certainly true here. A meticulously researched project is evaluated with complete honesty and has provided a most valuable learning experience. Matt concluded his project with his own personal evaluation:

> When I started the major project I wasn't looking forward to it. As the project developed, I felt myself getting into the design problems, the designs, the production plans, and producing the system. The whole major project has been an

insight into the world of design and I'm really enjoying it. I am considering a career in mechanical engineering, and I feel that the project has played a part in my decision.

I have learnt so much while doing this project. I have taught myself to use a professional drawing package to a limited degree: AutoCAD LT. Also I have taught myself to use 'finite element analysis' to analyse my designs strengthwise. I have learnt to design to a high degree of specification, and how to alter my designs to suit production facilities. I can now confidently solve complex design problems and come up with a workable solution. Also in the workshop I have learnt to use machines and tools to my advantage – a lathe, mill, and a variety of tools to produce my disc brake. These skills will be useful in later life, especially if I become an engineer.

I would recommend this course to anyone!

Kayak launcher

James Edghill, St. Paul's School, London.
Taught by: Mr. K. M. Campbell, Mr. A. E. Goodridge, Mr. G. E. Nava and Mr. G. Collard.

Launching a kayak may not seem to be too complex a problem, but when you live in Barbados, on top of a cliff, 5 metres above the sea, and the nearest launch point is 4 kilometres away, a design brief can take on a very different complexion. In James's own words:

> The project provided several challenges, each giving the opportunity to learn new skills such as welding and brazing. Another considerable factor during design was the limitation placed on size and weight.

You could be forgiven for missing the importance of the last point but as the staff at the school point out, James's high motivation for the solution had to take into account that all of the separate elements had to be capable of being flown out to Barbados for final installation. The photographs show the success of this undertaking. A design sheet showing a test on one of the sections will be given later.

Figure 3.10 James Edghill's kayak launcher.

James studied Mathematics, Further Mathematics, Physics and DT (Technology). It may come as no surprise that James has aspirations to become a civil engineer. He is now studying Civil Engineering at Bristol and will be well prepared for any unusual problems which come his way.

Ski Speedometer

Simon Todd, Bolton School (Boys div.)
Taught by: Mr. Mike Whitmarsh and Mr. Chris Walker.

Simon's project is introduced by the fact that skiing is thought to have originated in Grecce around 2000 BC, where early records mention the use of elongated shoes or sliders to move across the snow. There is then a quantum jump to the requirements of the sophisticated skiers of today, who require to know how far and how fast they are travelling. His unique design involves a ski mounted speed sensor which communicates to a unit displaying information on the skier's wrist. Phillip's technical staff assisted Simon in developing the microwave module.

Figure 3.11 The device was tested and evaluated at Rossendale ski slope. No snow in Scotland!

Simon's project involved a system design with five discrete sections:

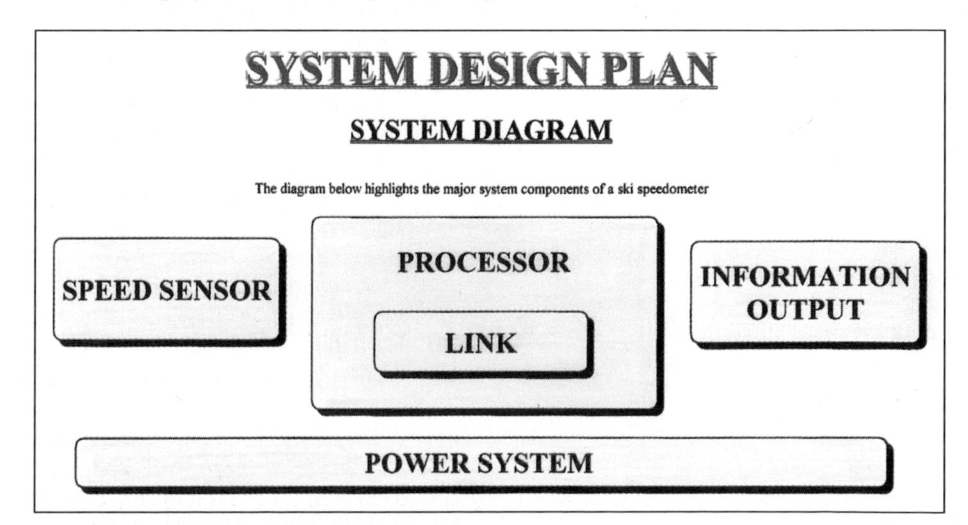

Figure 3.12 Simon Todd's system design.

Each section involved research and the selection of possibilities for development.

- The speed sensor is the means by which motion is detected – trundle wheels, altimeters and gyroscopes were looked at before a final choice was made between ultra-sonics and microwave Doppler shift.
- The information output looked at ways of displaying the information required. 'Head up' display on the ski and on top of the ski pole were rejected in favour of a cycle computer mounted on the wrist.
- Linking the system together considered wire and infra-red before a radio link was selected.
- The complete systems design is shown in Figure 3.13.
- The unit is powered from batteries and includes a regulator and a 'power/battery low' indicator in the wrist unit.

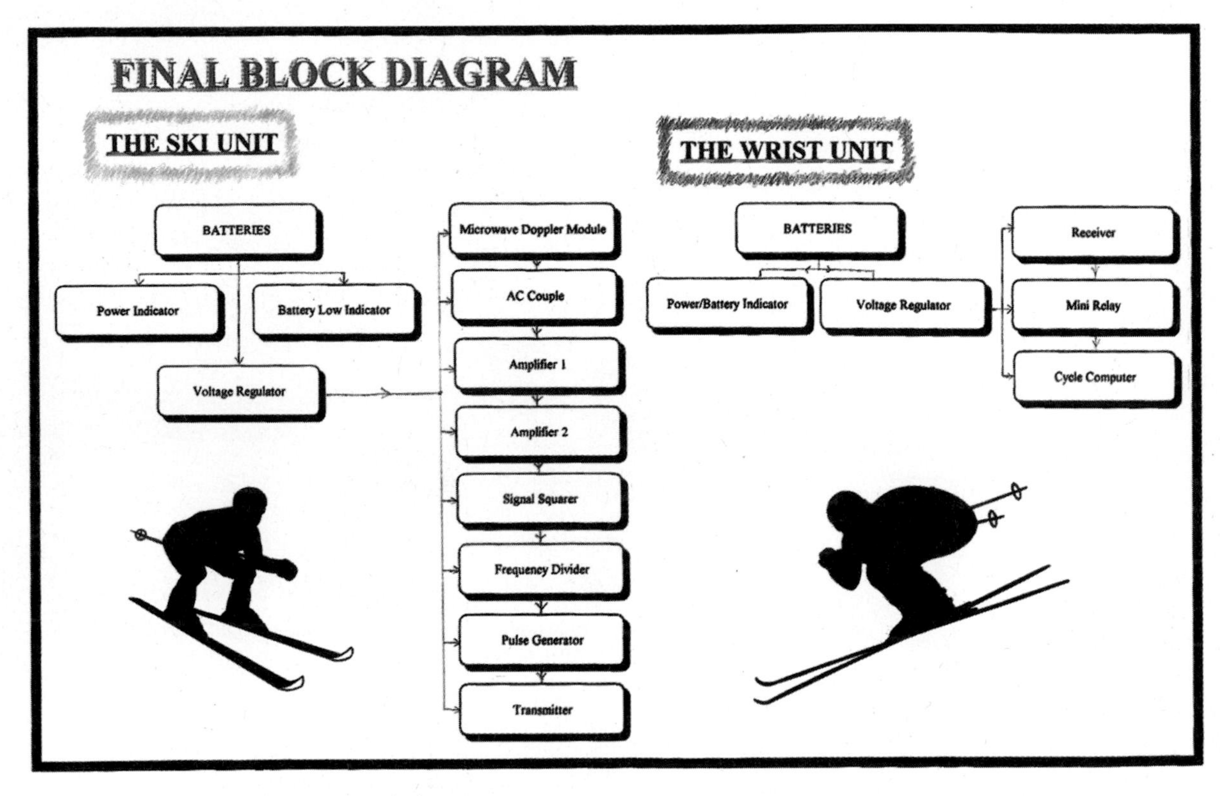

Figure 3.13 The complete systems design.

Simon concludes his project with his own personal evaluation:

> I have learnt a surprising amount from this project. For example, I have learnt a great deal about the techniques of research, in particular the need to organise, plan and keep track of research (an embarrassing situation occurred when I phoned the same company twice asking about the details of a cycle computer). I have also discovered many more sources of information which I never knew existed, in particular the existence of market research information such as MINTEL, reports and EIU surveys. I have developed a good understanding of the properties of microwaves and the phenomenon of Doppler shift. I have also learned much about the nature of aerials, as several weeks were spent attempting to increase the communication range between the two units by modifying the aerials. With my area of interest lying in the field of electronics, it is easy to forget the importance of other fields of engineering, such as the construction and engineering of materials to produce cases. The project gave me some much-needed practice in this field. I felt that I overlooked the difficulty and time involved

> to develop the housing for the electronic system, and I received quite a shock when it came to designing the casing and attachments.

Simon studied Maths, Physics and DT (Technology) and is now at Newcastle University studying Physics. The school staff label him 'a prolific inventor', being a past winner of the Duracell Competition and Finalist in the Young Engineer for Britain Competition. Simon concludes:

> The experience of project management has given me a tremendous start at university enabling me to make the most of the opportunities.

Children's art table

Joanne Smith, Bolton School (Girls div.)
Taught by: Rachel Langley-West and Mr. Mike Whitmarsh.

Whilst on a community service placement at St. Peters Primary School, Joanne was involved in helping out with art lessons. It was here she identified the need for a children's art table which was dual purpose and removed the need to clear away all specialist equipment at the end of each session.

The innovation here is that in the closed position the table is exactly the same height and dimensions as the other tables the children work on in the school. In the open position all of the coloured paint pots 'pop up' from a hidden recess.

The strength of this project is the extent to which Joanne interacted with the potential clients – both the teacher, Mrs. Pickary, and the children in her class – in coming to an excellent conclusion.

Figure 3.14 'The design process worked well and I was pleased with my final product and am confident that it could be used in a classroom to solve my initial project brief.'

Joanne's own personal reflections are given overleaf, together with the table shown in the 'closed' position.

Figure 3.15 Joanne Smith's art table in the closed position.

My initial interview with Mrs. Pickary was a worthwhile one. It gave me the information necessary for my questionnaire, along with my observation of the children doing work. My questionnaire told me that there was a need and assessed the most important problems. It would have been better to have given my questionnaire out to more teachers but is was hard to find a lot of primary school teachers to fill them in. I could have done more market research by writing to companies but the research I did was adequate to assess a need, for inspiration and to compare prices. The cost of my final design was a lot cheaper than those on the market and, if manufactured, could be a lot cheaper as I used the prices of screws, washers, rubber ends, etc. from the high street.

I was happy with my chosen design but I could have had more initial ideas as my other ideas did not fit my specification very well. I could also have extended my initial ideas a little more into solutions. My research into British Standards was sufficient but anthropometric research was not really useful. My working drawings were adequate for me to realise my design, as was my research into materials, manufacturing techniques and the sizes of each section. The realisation of the design went well and I did not encounter many problems.

The only problems encountered were small ones like filling chips in the wood and sticking down loose bits of veneer. The filing and filling of the frame joints took a long time but was a worthwhile job. It was hard to find rubber feet, but after a long search found some chair feet which were too big so I wrapped the ends round in tape to make them fit.

Problems with the design were minor, but if I made the design again I would alter the design to solve my suggested modifications. The design process worked well and I was pleased with my final product and am confident that it could be used in a classroom to solve my initial project brief.

Joanne studied Mathematics, Physics and DT (Technology) in the sixth form; a combination, which the staff say, prepared her well for all the calculations necessary in designing the mechanisms. Joanne is now studying Psychology at university.

Figure 3.16 Problem identification for some projects in section 3.3.

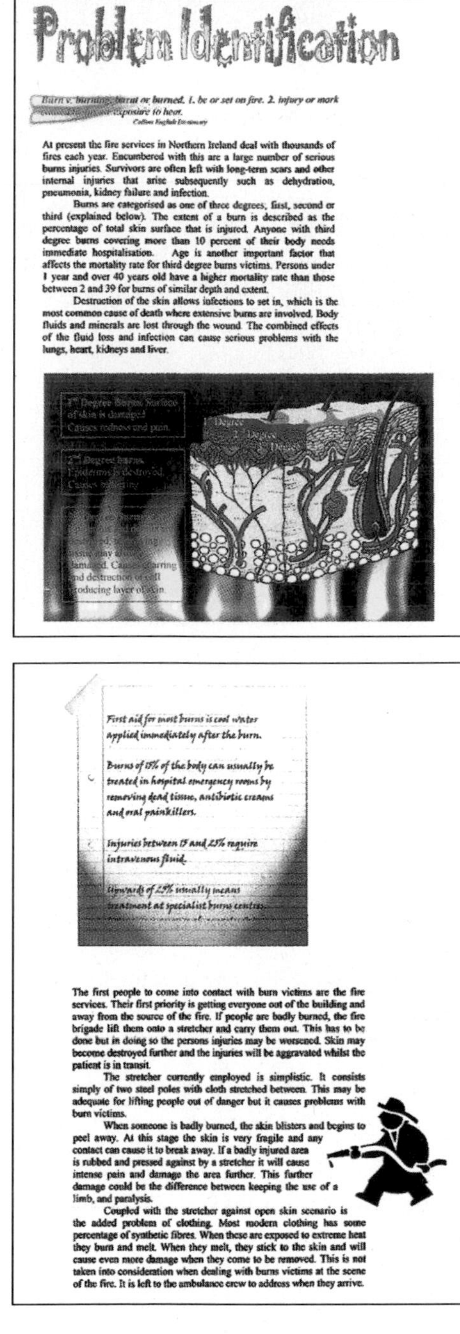

Problem Identification

Burn v. burning, burnt or burned. 1. be or set on fire. 2. injury or mark caused by ... exposure to heat.
Collins English Dictionary

At present the fire services in Northern Ireland deal with thousands of fires each year. Encumbered with this are a large number of serious burns injuries. Survivors are often left with long-term scars and other internal injuries that arise subsequently such as dehydration, pneumonia, kidney failure and infection.

Burns are categorised as one of three degrees, first, second or third (explained below). The extent of a burn is described as the percentage of total skin surface that is injured. Anyone with third degree burns covering more than 10 percent of their body needs immediate hospitalisation. Age is another important factor that affects the mortality rate for third degree burns victims. Persons under 1 year and over 40 years old have a higher mortality rate than those between 2 and 39 for burns of similar depth and extent.

Destruction of the skin allows infections to set in, which is the most common cause of death where extensive burns are involved. Body fluids and minerals are lost through the wound. The combined effects of the fluid loss and infection can cause serious problems with the lungs, heart, kidneys and liver.

First aid for most burns is cool water applied immediately after the burn.

Burns of 15% of the body can usually be treated in hospital emergency rooms by removing dead tissue, antibiotic creams and oral painkillers.

Injuries between 15 and 25% require intravenous fluid.

Upwards of 25% usually means treatment at specialist burns centres.

The first people to come into contact with burn victims are the fire services. Their first priority is getting everyone out of the building and away from the source of the fire. If people are badly burned, the fire brigade lift them onto a stretcher and carry them out. This has to be done but in doing so the persons injuries may be worsened. Skin may become destroyed further and the injuries will be aggravated whilst the patient is in transit.

The stretcher currently employed is simplistic. It consists simply of two steel poles with cloth stretched between. This may be adequate for lifting people out of danger but it causes problems with burn victims.

When someone is badly burned, the skin blisters and begins to peel away. At this stage the skin is very fragile and any contact can cause it to break away. If a badly injured area is rubbed and pressed against by a stretcher it will cause intense pain and damage the area further. This further damage could be the difference between keeping the use of a limb, and paralysis.

Coupled with the stretcher against open skin scenario is the added problem of clothing. Most modern clothing has some percentage of synthetic fibres. When these are exposed to extreme heat they burn and melt. When they melt, they stick to the skin and will cause even more damage when they come to be removed. This is not taken into consideration when dealing with burns victims at the scene of the fire. It is left to the ambulance crew to address when they arrive.

These design sheets are clear in identifying an appropriate problem area and justify this by asking if there is a problem or need. A project brief and initial specification have then been completed.

At this stage a detailed and realistic time plan for project completion should be drawn up using computer software or detailed diary targets.

3.4 Candidates' approaches to coursework projects

3.4.1 User, brief, investigation: recognition and investigation of design opportunities

The selection and justification of an appropriate problem or opportunity which offers scope for development at an Advanced level

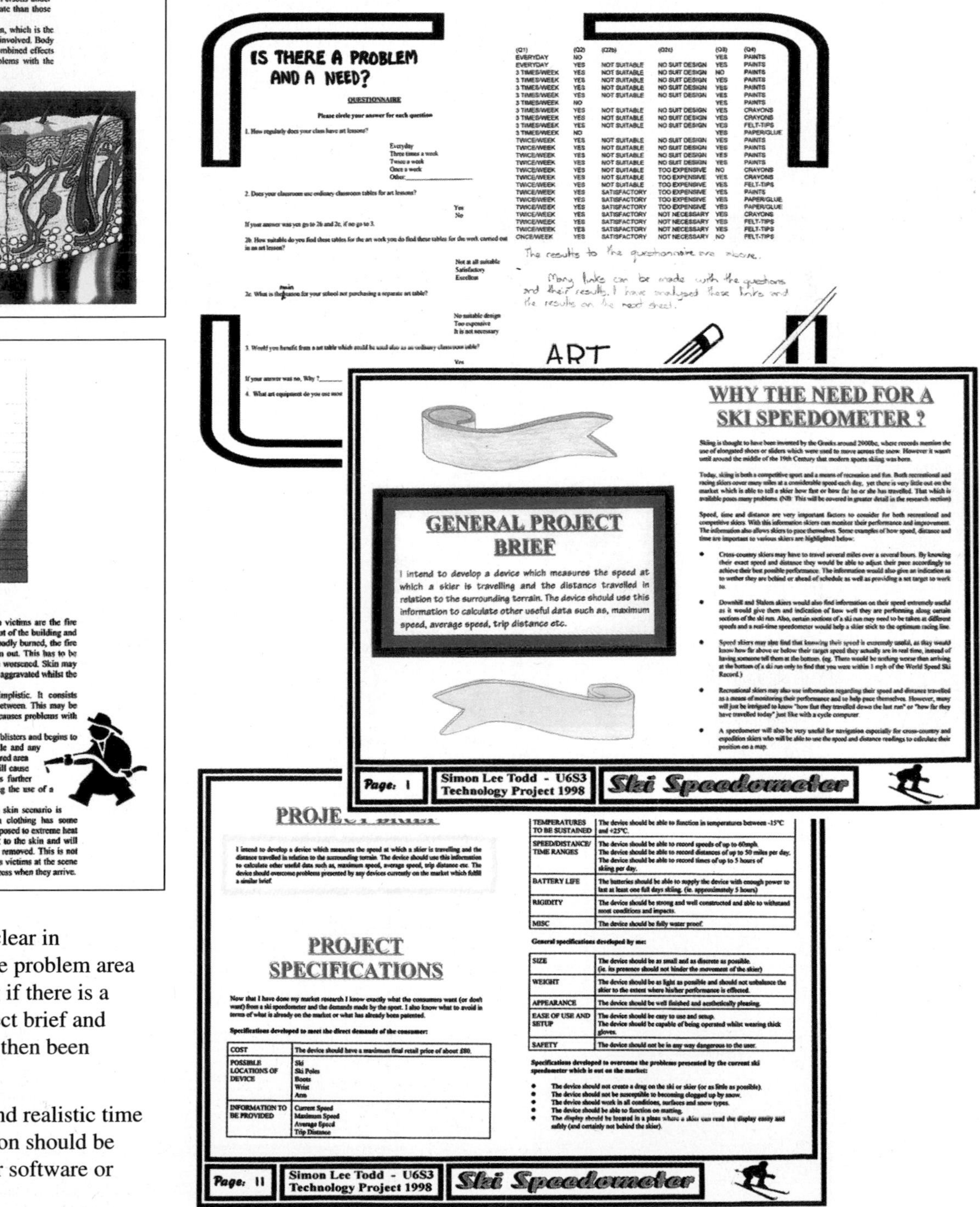

IS THERE A PROBLEM AND A NEED?

QUESTIONNAIRE

Please circle your answer for each question

1. How regularly does your class have art lessons?

Everyday
Three times a week
Twice a week
Once a week
Other:________

2. Does your classroom use ordinary classroom tables for art lessons?

Yes
No

If your answer was yes go to 2b and 2c, if no go to 3.

2b. How suitable do you find these tables for the art work you do find these tables for the work carried out in an art lesson?

Not at all suitable
Satisfactory
Excellent

2c. What is the main reason for your school not purchasing a separate art table?

No suitable design
Too expensive
It is not necessary

3. Would you benefit from a art table which could be used also as an ordinary classroom table?

Yes

If your answer was no, Why ?______

4. What art equipment do you use most

(Q1)	(Q2)	(Q2b)	(Q2c)	(Q3)	(Q4)
EVERYDAY	NO			YES	PAINTS
EVERYDAY	YES	NOT SUITABLE	NO SUIT DESIGN	YES	PAINTS
3 TIMES/WEEK	YES	NOT SUITABLE	NO SUIT DESIGN	NO	PAINTS
3 TIMES/WEEK	YES	NOT SUITABLE	NO SUIT DESIGN	YES	PAINTS
3 TIMES/WEEK	YES	NOT SUITABLE	NO SUIT DESIGN	YES	PAINTS
3 TIMES/WEEK	YES	NOT SUITABLE	NO SUIT DESIGN	YES	PAINTS
3 TIMES/WEEK	NO			YES	PAINTS
3 TIMES/WEEK	YES	NOT SUITABLE	NO SUIT DESIGN	YES	CRAYONS
3 TIMES/WEEK	YES	NOT SUITABLE	NO SUIT DESIGN	YES	CRAYONS
3 TIMES/WEEK	YES	NOT SUITABLE	NO SUIT DESIGN	YES	FELT-TIPS
3 TIMES/WEEK	NO			YES	PAPER/GLUE
TWICE/WEEK	YES	NOT SUITABLE	NO SUIT DESIGN	YES	PAINTS
TWICE/WEEK	YES	NOT SUITABLE	NO SUIT DESIGN	YES	PAINTS
TWICE/WEEK	YES	NOT SUITABLE	NO SUIT DESIGN	YES	PAINTS
TWICE/WEEK	YES	NOT SUITABLE	NO SUIT DESIGN	YES	PAINTS
TWICE/WEEK	YES	NOT SUITABLE	TOO EXPENSIVE	NO	CRAYONS
TWICE/WEEK	YES	NOT SUITABLE	TOO EXPENSIVE	YES	CRAYONS
TWICE/WEEK	YES	NOT SUITABLE	TOO EXPENSIVE	YES	FELT-TIPS
TWICE/WEEK	YES	SATISFACTORY	TOO EXPENSIVE	YES	PAINTS
TWICE/WEEK	YES	SATISFACTORY	TOO EXPENSIVE	YES	PAPER/GLUE
TWICE/WEEK	YES	SATISFACTORY	TOO EXPENSIVE	YES	PAPER/GLUE
TWICE/WEEK	YES	SATISFACTORY	NOT NECESSARY	YES	CRAYONS
TWICE/WEEK	YES	SATISFACTORY	NOT NECESSARY	YES	FELT-TIPS
TWICE/WEEK	YES	SATISFACTORY	NOT NECESSARY	YES	FELT-TIPS
ONCE/WEEK	YES	SATISFACTORY	NOT NECESSARY	NO	FELT-TIPS

The results to the questionnaire are above.

Many links can be made with the questions and their results. I have analysed these links and the results on the next sheet.

ART

GENERAL PROJECT BRIEF

I intend to develop a device which measures the speed at which a skier is travelling and the distance travelled in relation to the surrounding terrain. The device should use this information to calculate other useful data such as, maximum speed, average speed, trip distance etc.

WHY THE NEED FOR A SKI SPEEDOMETER ?

Skiing is thought to have been invented by the Greeks around 2000bc, where records mention the use of elongated shoes or sliders which were used to move across the snow. However it wasn't until around the middle of the 19th Century that modern sports skiing was born.

Today, skiing is both a competitive sport and a means of recreation and fun. Both recreational and racing skiers cover many miles at a considerable speed each day, yet there is very little out on the market which is able to tell a skier how fast or how far he or she has travelled. That which is available poses many problems. (NB: This will be covered in greater detail in the research section)

Speed, time and distance are very important factors to consider for both recreational and competitive skiers. With this information skiers can monitor their performance and improvement. The information also allows skiers to pace themselves. Some examples of how speed, distance and time are important to various skiers are highlighted below.

- Cross-country skiers may have to travel several miles over a several hours. By knowing their exact speed and distance they would be able to adjust their pace accordingly to achieve their best possible performance. The information would also give an indication as to wether they are behind or ahead of schedule as well as providing a set target to work to.
- Downhill and Slalom skiers would also find information on their speed extremely useful as it would give them and indication of how well they are performing along certain sections of the ski run. Also, certain sections of a ski run may need to be taken at different speeds and a real-time speedometer would help a skier stick to the optimum racing line.
- Speed skiers may also find that knowing their speed is extremely useful, as they would know how far above or below their target speed they actually are in real time, instead of having someone tell them at the bottom. (eg. There would be nothing worse than arriving at the bottom of a ski run only to find that you were within 1 mph of the World Speed Ski Record.)
- Recreational skiers may also use information regarding their speed and distance travelled as a means of monitoring their performance and to help pace themselves. However, many will just be intrigued to know "how fast they travelled down the last run" or "how far they have travelled today" just like with a cycle computer.
- A speedometer will also be very useful for navigation especially for cross-country and expedition skiers who will be able to use the speed and distance readings to calculate their position on a map.

Page: 1 | Simon Lee Todd - U6S3 Technology Project 1998 | Ski Speedometer

PROJECT BRIEF

I intend to develop a device which measures the speed at which a skier is travelling and the distance travelled in relation to the surrounding terrain. The device should use this information to calculate other useful data such as, maximum speed, average speed, trip distance etc. The device should overcome problems presented by any devices currently on the market which fulfill a similar brief.

PROJECT SPECIFICATIONS

Now that I have done my market research I know exactly what the consumers want (or don't want) from a ski speedometer and the demands made by the sport. I also know what to avoid in terms of what is already on the market or what has already been patented.

Specifications developed to meet the direct demands of the consumer:

COST	The device should have a maximum final retail price of about £80.
POSSIBLE LOCATIONS OF DEVICE	Ski Ski Poles Boots Wrist Arm
INFORMATION TO BE PROVIDED	Current Speed Maximum Speed Average Speed Trip Distance
TEMPERATURES TO BE SUSTAINED	The device should be able to function in temperatures between -15°C and +25°C.
SPEED/DISTANCE/ TIME RANGES	The device should be able to record speeds of up to 60mph. The device should be able to record distances of up to 50 miles per day. The device should be able to record times of up to 5 hours of skiing per day.
BATTERY LIFE	The batteries should be able to supply the device with enough power to last at least one full days skiing. (ie. approximately 5 hours)
RIGIDITY	The device should be strong and well constructed and able to withstand most conditions and impacts.
MISC	The device should be fully water proof.

General specifications developed by me:

SIZE	The device should be as small and as discrete as possible. (ie. its presence should not hinder the movement of the skier)
WEIGHT	The device should be as light as possible and should not unbalance the skier to the extent where his/her performance is effected.
APPEARANCE	The device should be well finished and aesthetically pleasing.
EASE OF USE AND SETUP	The device should be easy to use and setup. The device should be capable of being operated whilst wearing thick gloves.
SAFETY	The device should not be in any way dangerous to the user.

Specifications developed to overcome the problems presented by the current ski speedometer which is out on the market:

- The device should not create a drag on the ski or skier (or as little as possible).
- The device should not be susceptible to becoming clogged up by snow.
- The device should work in all conditions, surfaces and snow types.
- The device should be able to function on matting.
- The display should be located in a place where a skier can read the display easily and safely (and certainly not behind the skier).

Page: 11 | Simon Lee Todd - U6S3 Technology Project 1998 | Ski Speedometer

Figure 3.17 These design sheets show a variety of approaches to investigation.

The collection, editing and recording of relevant data, including user needs and existing product details

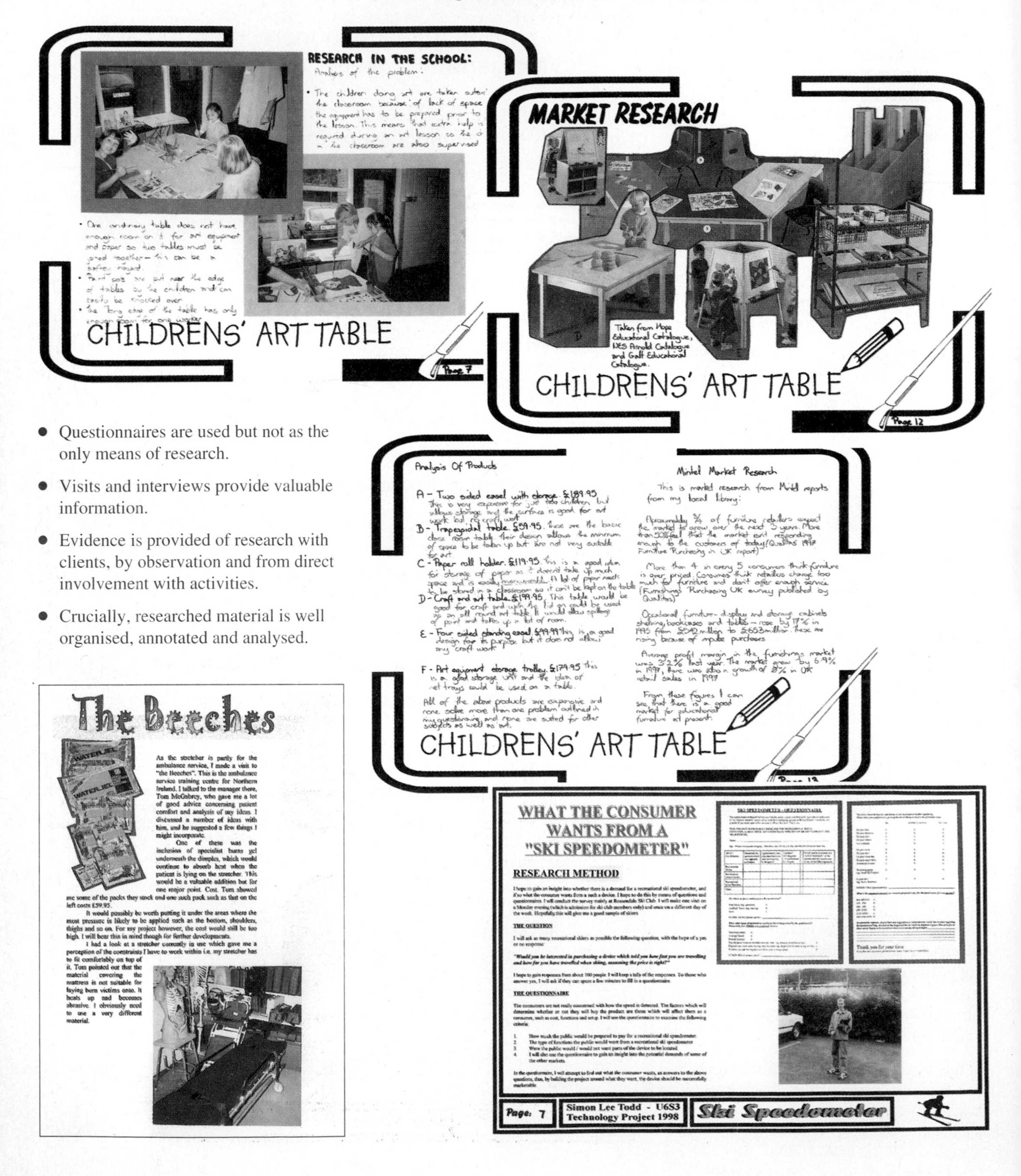

- Questionnaires are used but not as the only means of research.
- Visits and interviews provide valuable information.
- Evidence is provided of research with clients, by observation and from direct involvement with activities.
- Crucially, researched material is well organised, annotated and analysed.

The identification of the strengths and weaknesses in existing products, through analysis of edited research material, to provide information for later use

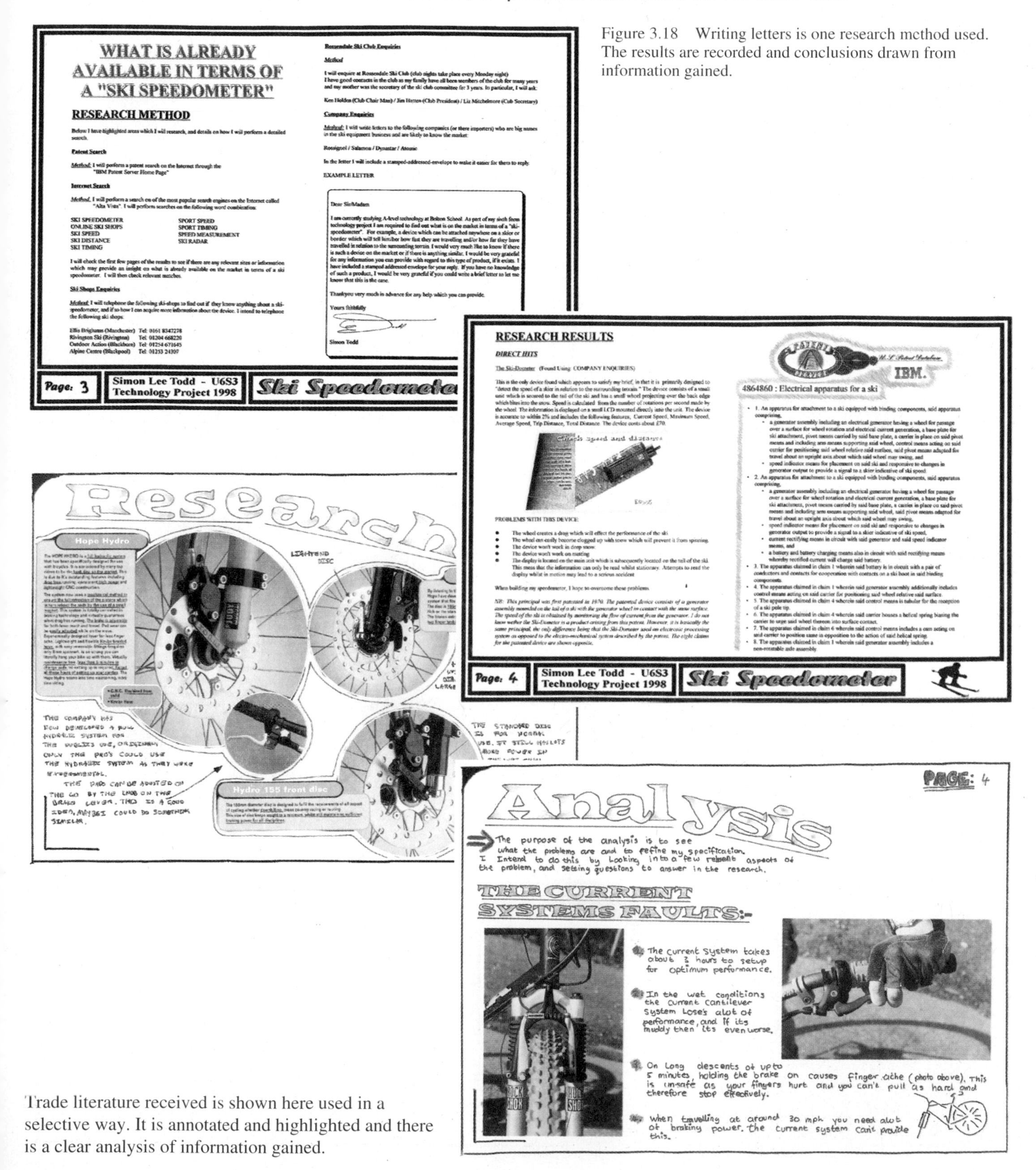

WHAT IS ALREADY AVAILABLE IN TERMS OF A "SKI SPEEDOMETER"

RESEARCH METHOD

Below I have highlighted areas which I will research, and details on how I will perform a detailed search.

Patent Search

Method: I will perform a patent search on the Internet through the "IBM Patent Server Home Page"

Internet Search

Method: I will perform a search on of the most popular search engines on the Internet called "Alta Vista". I will perform searches on the following word combination:

SKI SPEEDOMETER
ONLINE SKI SHOPS
SKI SPEED
SKI DISTANCE
SKI TIMING
SPORT SPEED
SPORT TIMING
SPEED MEASUREMENT
SKI RADAR

I will check the first few pages of the results to see if there are any relevant sites or information which may provide an insight on what is already available on the market in terms of a ski speedometer. I will then check relevant matches.

Ski Shops Enquiries

Method: I will telephone the following ski-shops to find out if they know anything about a ski-speedometer, and if so how I can acquire more information about the device. I intend to telephone the following ski shops:

Ellis Brigham (Manchester) Tel: 0161 8347278
Rivington Ski (Rivington) Tel: 01204 668220
Outdoor Action (Blackburn) Tel: 01254 671645
Alpine Centre (Blackpool) Tel: 01253 24397

Rossendale Ski Club Enquiries

Method

I will enquire at Rossendale Ski Club (club nights take place every Monday night). I have good contacts in the club as my family have all been members of the club for many years and my mother was the secretary of the ski club committee for 3 years. In particular, I will ask:

Ken Holden (Club Chair Man) / Jim Hatten (Club President) / Liz Mitchelmore (Club Secretary)

Company Enquiries

Method: I will write letters to the following companies (or there importers) who are big names in the ski equipment business and are likely to know the market:

Rossignol / Salamon / Dynastar / Atomic

In the letter I will include a stamped-addressed-envelope to make it easier for them to reply.

EXAMPLE LETTER

Dear Sir/Madam

I am currently studying A-level technology at Belton School. As part of my sixth form technology project I am required to find out what is on the market in terms of a "ski-speedometer". For example, a device which can be attached anywhere on a skier or border which will tell him/her how fast they are travelling and/or how far they have travelled in relation to the surrounding terrain. I would very much like to know if there is such a device on the market or if there is anything similar. I would be very grateful for any information you can provide with regard to this type of product, if it exists. I have included a stamped addressed envelope for your reply. If you have no knowledge of such a product, I would be very grateful if you could write a brief letter to let me know that this is the case.

Thankyou very much in advance for any help which you can provide.

Yours faithfully

Simon Todd

Page: 3 | Simon Lee Todd - U6S3 Technology Project 1998 | Ski Speedometer

Figure 3.18 Writing letters is one research method used. The results are recorded and conclusions drawn from information gained.

RESEARCH RESULTS

DIRECT HITS

The Ski-Dometer (Found Using: COMPANY ENQUIRIES)

This is the only device found which appears to satisfy my brief, in that it is primarily designed to "detect the speed of a skier in relation to the surrounding terrain." The device consists of a small unit which is secured to the tail of the ski and has a small wheel projecting over the back edge which bites into the snow. Speed is calculated from the number of rotations per second made by the wheel. The information is displayed on a small LCD mounted directly into the unit. The device is accurate to within 2% and includes the following features, Current Speed, Maximum Speed, Average Speed, Trip Distance, Total Distance. The device costs about £70.

PROBLEMS WITH THIS DEVICE:

- The wheel creates a drag which will effect the performance of the ski
- The wheel can easily become clogged up with snow which will prevent it from spinning
- The device won't work in deep snow
- The device won't work on matting
- The display is located on the main unit which is subsequently located on the tail of the ski. This mean that the information can only be read whilst stationary. Attempts to read the display whilst in motion may lead to a serious accident

When building my speedometer, I hope to overcome these problems.

NB: This principal was first patented in 1970. The patented device consists of a generator assembly mounded on the tail of a ski with the generator wheel in contact with the snow surface. The speed of the ski is obtained by monitoring the flow of current from the generator. I do not know wether the Ski-Dometer is a product arising from this patent. However, it is basically the same principal, the only difference being that the Ski-Dometer used an electronic processing system as opposed to the electro-mechanical system described by the patent. The eight claims for the patented device are shown opposite.

IBM.

4864860 : Electrical apparatus for a ski

- 1. An apparatus for attachment to a ski equipped with binding components, said apparatus comprising,
  - a generator assembly including an electrical generator having a wheel for passage over a surface for wheel rotation and electrical current generation, a base plate for ski attachment, pivot means carried by said base plate, a carrier in place on said pivot means and including arm means supporting said wheel, control means acting on said carrier for positioning said wheel relative said surface, said pivot means adapted for travel about an upright axis about which said wheel may swing, and
  - speed indicator means for placement on said ski and responsive to changes in generator output to provide a signal to a skier indicative of ski speed.
- 2. An apparatus for attachment to a ski equipped with binding components, said apparatus comprising,
  - a generator assembly including an electrical generator having a wheel for passage over a surface for wheel rotation and electrical current generation, a base plate for ski attachment, pivot means carried by said base plate, a carrier in place on said pivot means and including arm means supporting said wheel, said pivot means adapted for travel about an upright axis about which said wheel may swing,
  - speed indicator means for placement on said ski and responsive to changes in generator output to provide a signal to a skier indicative of ski speed,
  - current rectifying means in circuit with said generator and said speed indicator means, and
  - a battery and battery charging means also in circuit with said rectifying means whereby rectified current will charge said battery.
- 3. The apparatus claimed in claim 1 wherein said battery is in circuit with a pair of conductors and contacts for cooperation with contacts on a ski boot in said binding components.
- 4. The apparatus claimed in claim 1 wherein said generator assembly additionally includes control means acting on said carrier for positioning said wheel relative said surface.
- 5. The apparatus claimed in claim 4 wherein said control means is tubular for the reception of a ski pole tip.
- 6. The apparatus claimed in claim 4 wherein said carrier houses a helical spring biasing the carrier to urge said wheel thereon into surface contact.
- 7. The apparatus claimed in claim 6 wherein said control means includes a cam acting on said carrier to position same in opposition to the action of said helical spring.
- 8. The apparatus claimed in claim 1 wherein said generator assembly includes a non-rotatable axle assembly.

Page: 4 | Simon Lee Todd - U6S3 Technology Project 1998 | Ski Speedometer

PAGE: 4

Analysis

The purpose of the analysis is to see what the problems are and to refine my specification. I intend to do this by looking into a few relevant aspects of the problem, and setting questions to answer in the research.

THE CURRENT SYSTEMS FAULTS:-

- The current system takes about 3 hours to setup for optimum performance.
- In the wet conditions the current cantilever system loses alot of performance, and if its muddy then its even worse.
- On long descents of upto 5 minutes, holding the brake on causes finger ache (photo above). This is un-safe as your fingers hurt and you can't pull as hard and therefore stop effectively.
- When travelling at around 30 mph you need alot of braking power. The current system can't provide this.

Trade literature received is shown here used in a selective way. It is annotated and highlighted and there is a clear analysis of information gained.

3.4.2 Generating ideas

Using annotated sketching to generate a wide range of design ideas

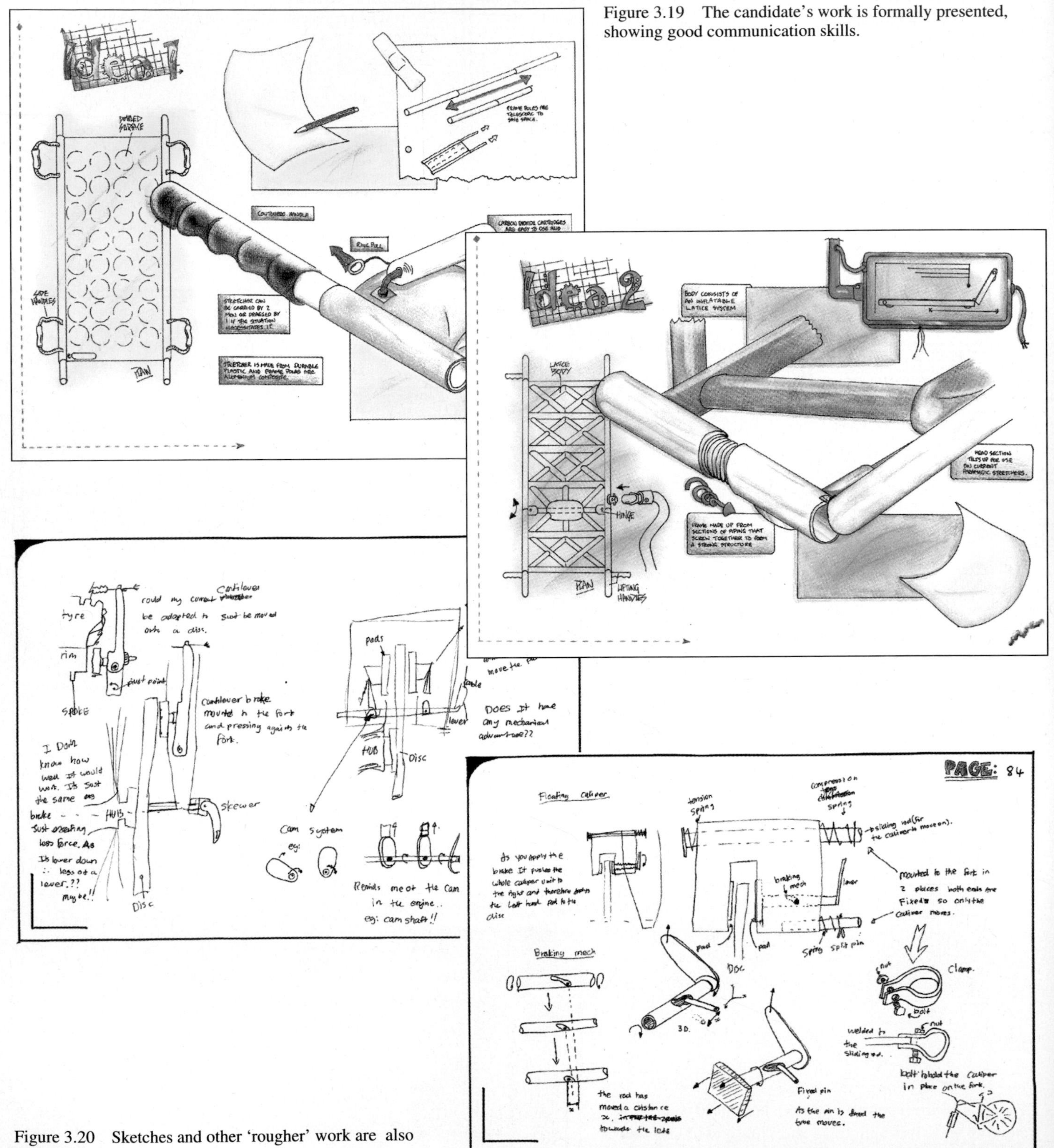

Figure 3.19 The candidate's work is formally presented, showing good communication skills.

Figure 3.20 Sketches and other 'rougher' work are also welcome!

Figure 3.21 This candidate gives a clear analysis of the merits of each idea, with reasons for selection and rejection

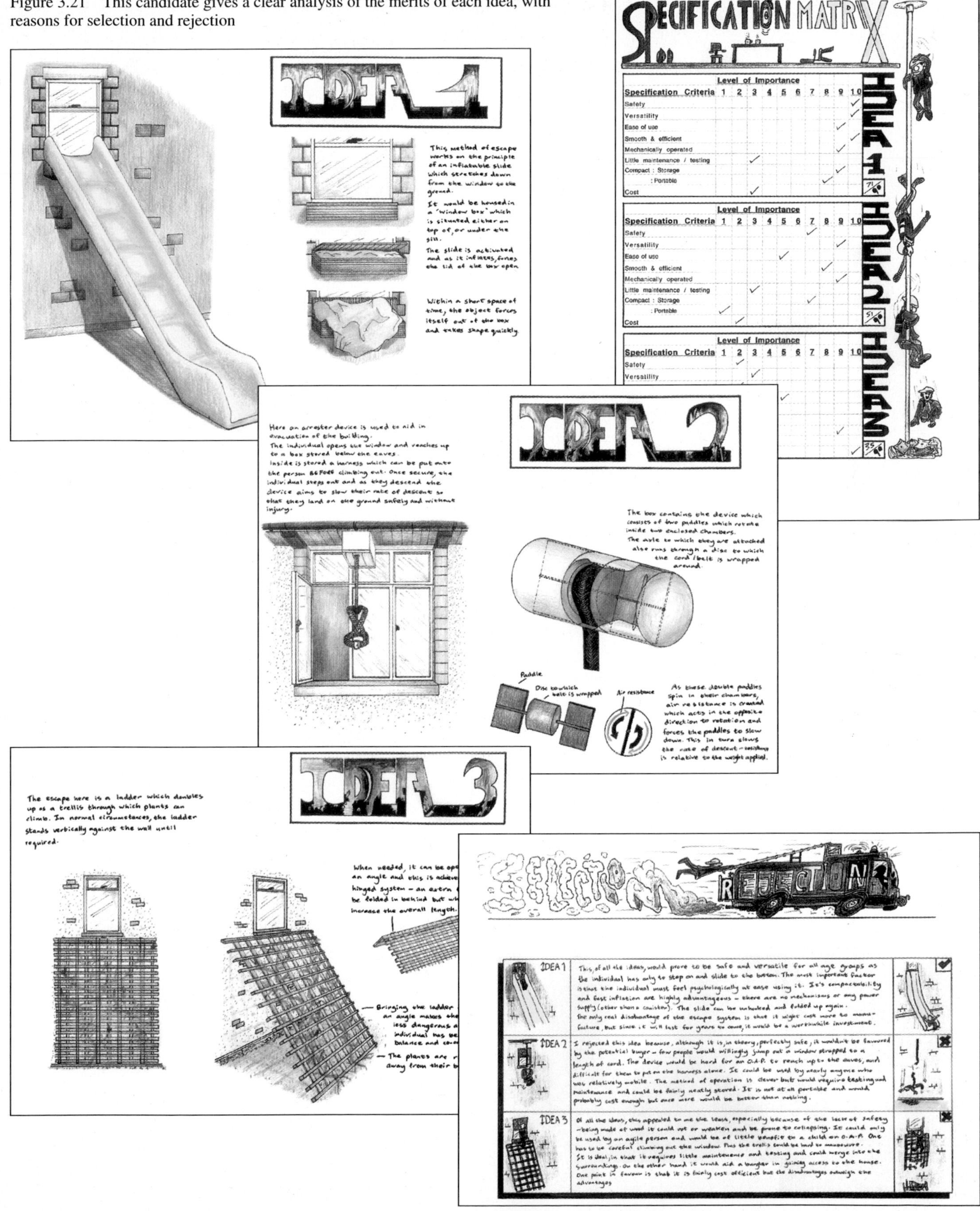

3.4.3 Synthesis – detailed design-development modelling

The presentation and analysis of information using a combination of text, graphical techniques and ICT as appropriate

Figure 3.22 A variety of modelling techniques clearly recorded.

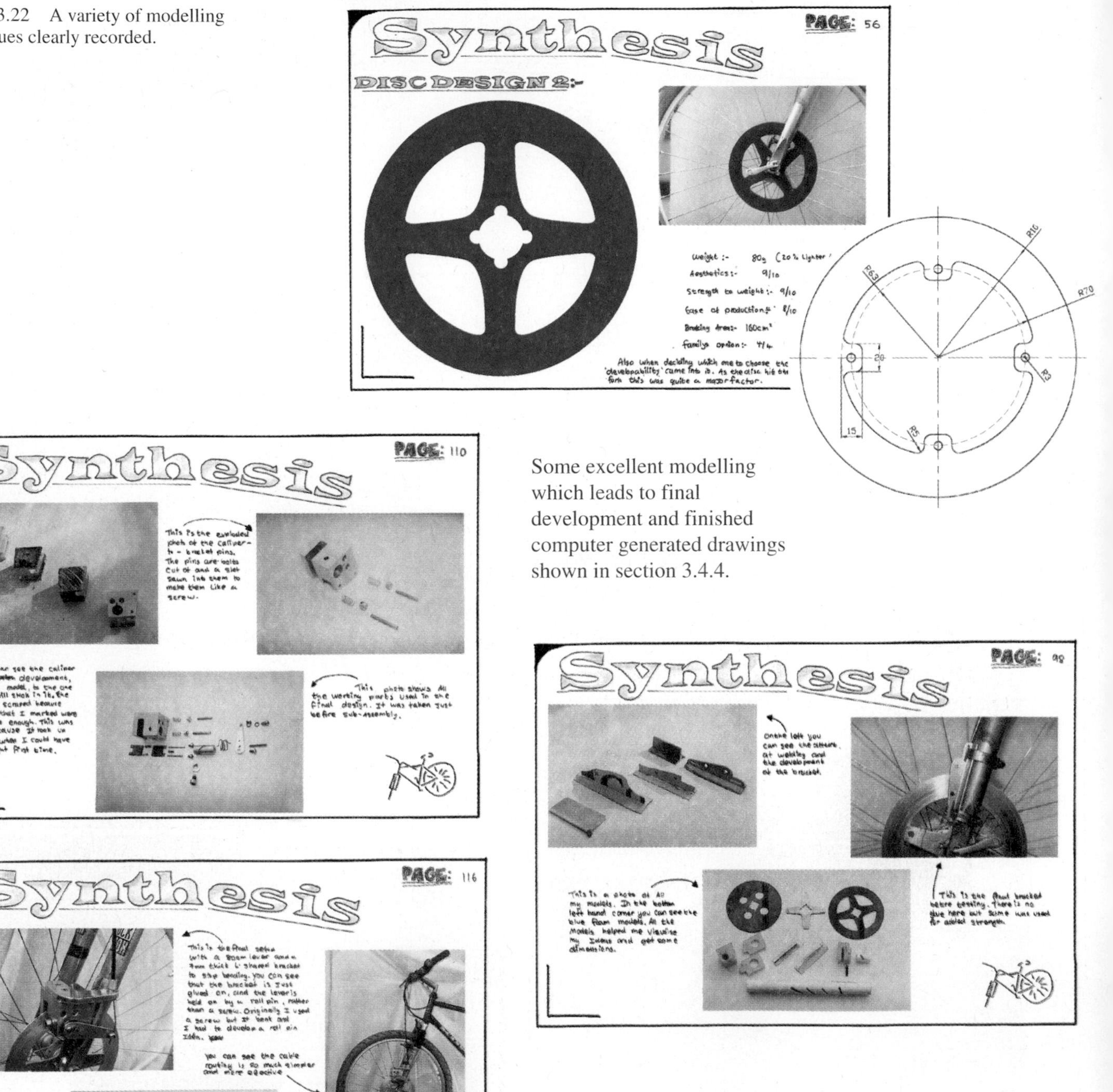

Some excellent modelling which leads to final development and finished computer generated drawings shown in section 3.4.4.

Figures 3.23 A variety of modelling techniques is shown. This stage of project development is vital in all cases.

The production of first generation 2D and 3D models suitable for establishing the validity of initial thinking and the chosen solution

Prototyping electronics circuits by software packages supported by prototype breadboard construction complements 3D modelling.

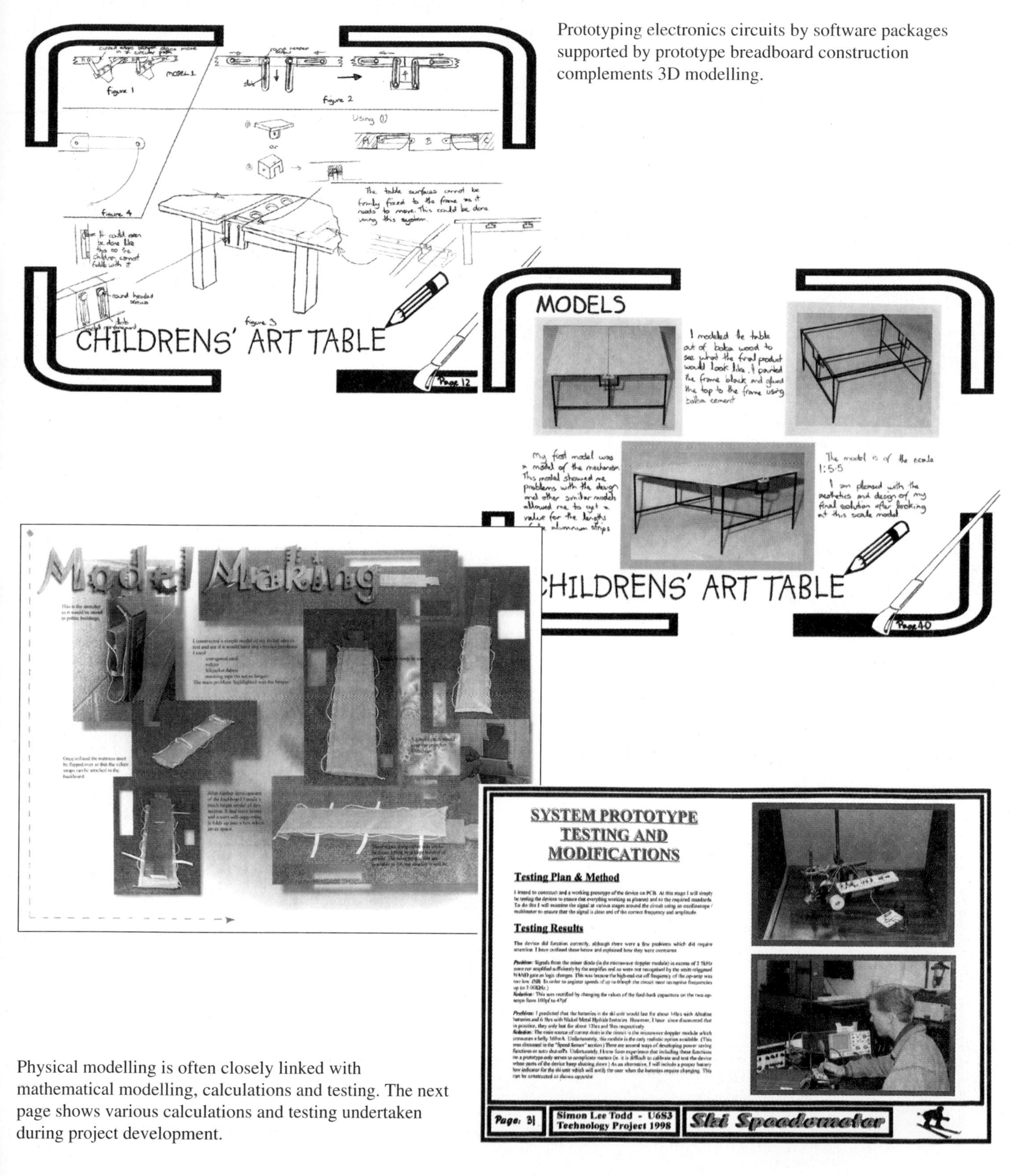

Physical modelling is often closely linked with mathematical modelling, calculations and testing. The next page shows various calculations and testing undertaken during project development.

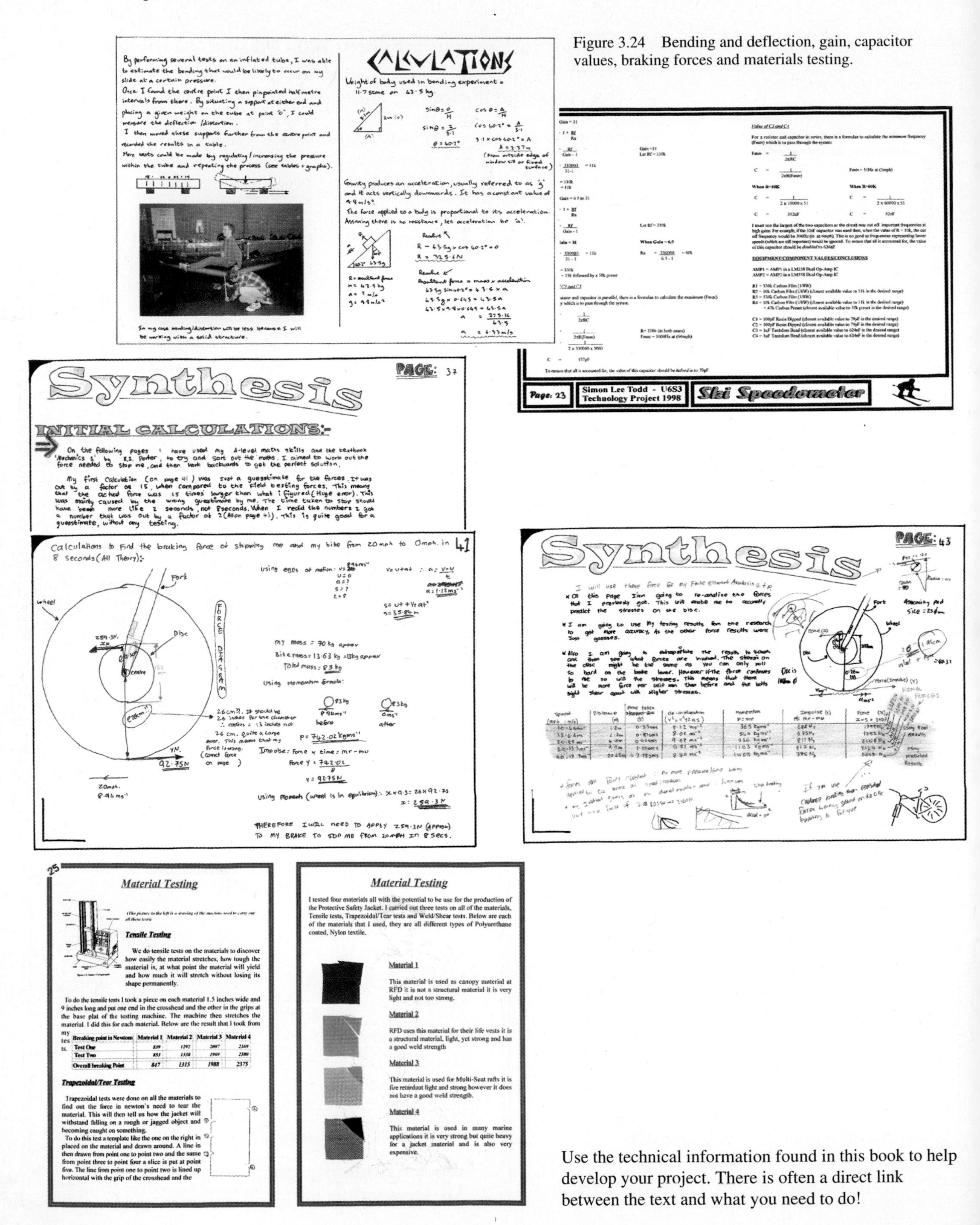

By performing several tests on an inflated tube, I was able to estimate the bending that would be likely to occur on my slide at a certain pressure.
Once I found the centre point I then pinpointed half metre intervals from there. By situating a support at either end and placing a given weight on the tube at point 'O', I could measure the deflection /distortion.
I then moved these supports further from the centre point and recorded the results in a table.
More tests could be made by regulating / increasing the pressure within the tube and repeating the process (see tables + graphs).

In my case bending/distortion will be less because I will be working with a solid structure.

CALCULATIONS

Weight of body used in bending experiment = 11·7 stone or 63·5 kg.

Gravity produces an acceleration, usually referred to as 'g' and it acts vertically downwards. It has a constant value of 9·8 m/s².
The force applied to a body is proportional to its acceleration. Assuming there is no resistance, let acceleration be 'a'.

Value of C1 and C4

Equipment/component values/conclusions

Page: 23 | Simon Lee Todd - U6S3 Technology Project 1998 | Ski Speedometer

Synthesis PAGE: 37

INITIAL CALCULATIONS:-

On the following pages I have used my A-level maths skills and the textbook 'Mechanics 1' by R.I. Porter, to try and sort out the maths. I aimed to work out the force needed to stop me, and then work backwards to get the perfect solution.

My first calculation (on page 41) was just a guesstimate for the forces. It was out by a factor of 15, when compared to the field testing forces. This means that the actual force was 15 times larger than what I figured (Huge error). This was mainly caused by the wrong guestimate by me. The time taken to stop should have been more like 2 seconds, not 8 seconds. When I redid the numbers I got a number that was out by a factor of 2 (All on page 43). This is quite good for a guesstimate, without any testing.

41

Calculations to find the braking force of stopping me and my bike from 20 mph to 0 mph in 8 seconds (All Theory):-

THEREFORE I WILL NEED TO APPLY 259·3N (APPROX) TO MY BRAKE TO STOP ME FROM 20MPH IN 8 SECS.

Synthesis PAGE: 43

25

Material Testing

(The picture to the left is a drawing of the machine used to carry out all these tests)

Tensile Testing

We do tensile tests on the materials to discover how easily the material stretches, how tough the material is, at what point the material will yield and how much it will stretch without losing its shape permanently.

To do the tensile tests I took a piece on each material 1.5 inches wide and 9 inches long and put one end in the crosshead and the other in the grips at the base plat of the testing machine. The machine then stretches the material. I did this for each material. Below are the result that I took from my tests.

Breaking point in Newtons	Material 1	Material 2	Material 3	Material 4
Test One	839	1292	2007	2369
Test Two	853	1338	1969	2380
Overall breaking Point	847	1315	1988	2375

Trapezoidal/Tear Testing

Trapezoidal tests were done on all the materials to find out the force in newton's need to tear the material. This will then tell us how the jacket will withstand falling on a rough or jagged object and becoming caught on something.
To do this test a template like the one on the right in placed on the material and drawn around. A line in then drawn from point one to point two and the same from point three to point four a slice is put at point five. The line from point one to point two is lined up horizontal with the grip of the crosshead and the

Material Testing

I tested four materials all with the potential to be use for the production of the Protective Safety Jacket. I carried out three tests on all of the materials, Tensile tests, Trapezoidal/Tear tests and Weld/Shear tests. Below are each of the materials that I used, they are all different types of Polyurethane coated, Nylon textile.

Material 1

This material is used as canopy material at RFD it is not a structural material it is very light and not too strong.

Material 2

RFD uses this material for their life vests it is a structural material, light, yet strong and has a good weld strength

Material 3

This material is used for Multi-Seat rafts it is fire retardant light and strong however it does not have a good weld strength.

Material 4

This material is used in many marine applications it is very strong but quite heavy for a jacket material and is also very expensive.

Figure 3.24 Bending and deflection, gain, capacitor values, braking forces and materials testing.

Use the technical information found in this book to help develop your project. There is often a direct link between the text and what you need to do!

Figure 3.25 Working drawings.

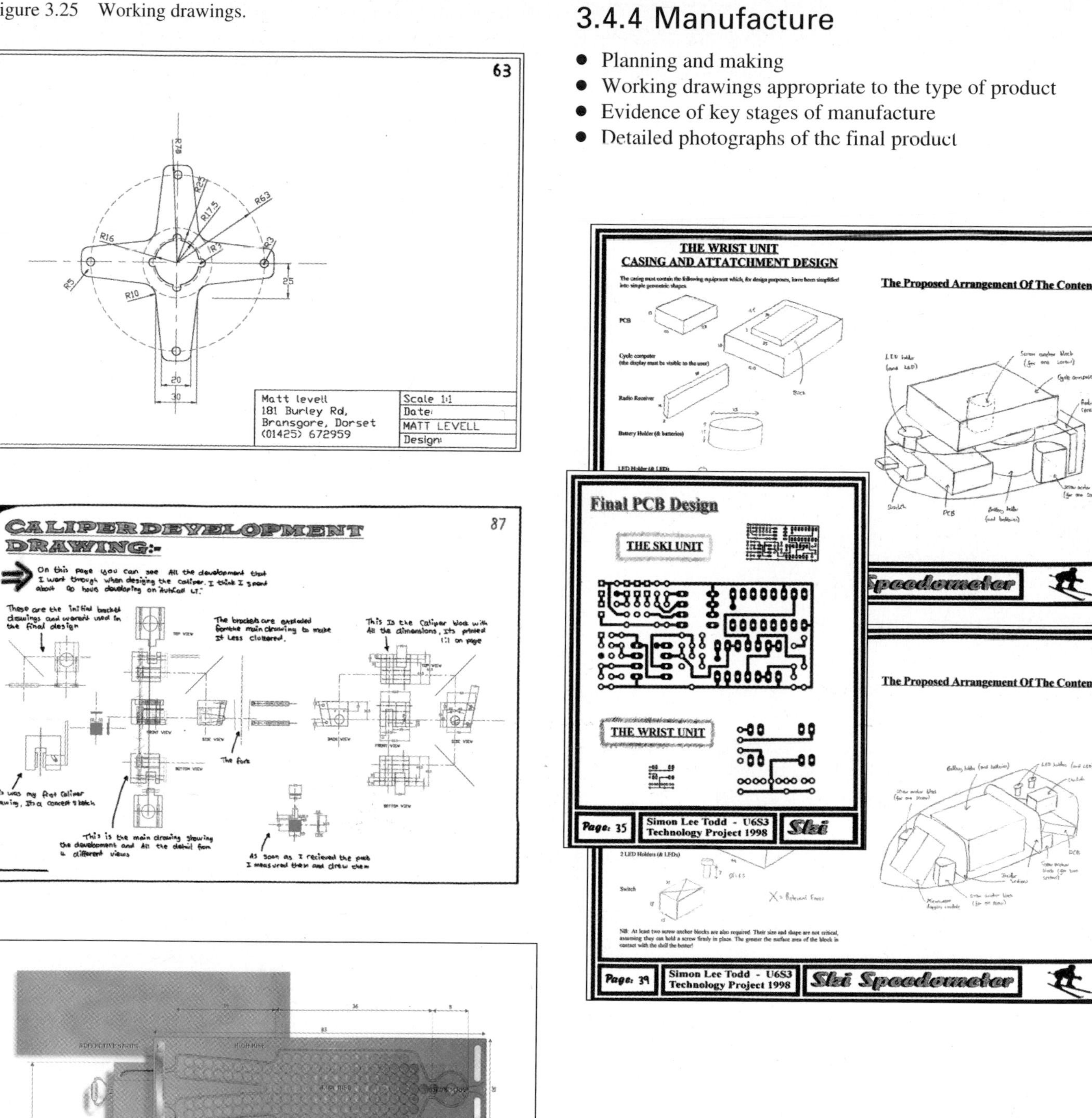

3.4.4 Manufacture

- Planning and making
- Working drawings appropriate to the type of product
- Evidence of key stages of manufacture
- Detailed photographs of the final product

Photographs of the final projects can be seen in section 3.3.

Figure 3.26 Final detailed drawings and a clear photograph.

Clear production stages are shown specific to the project.

The final design idea is shown with a photograph which clearly indicates to the moderator both the function of the product and the standard of manufacture.

Figure 3.27 Testing.

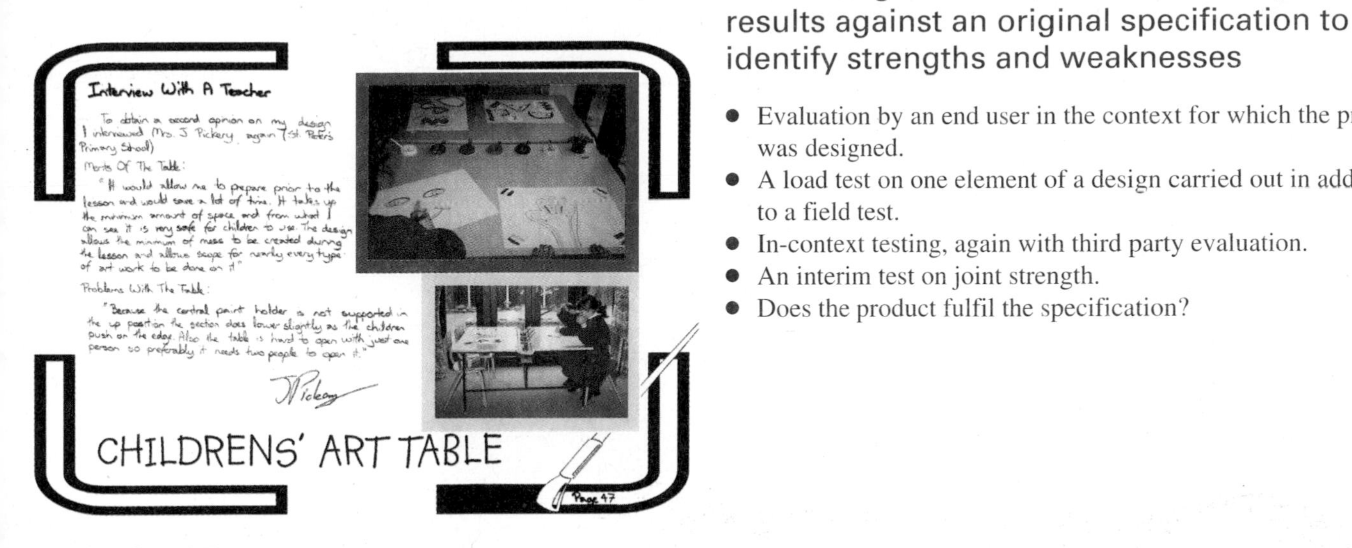
Interview With A Teacher

To obtain a second opinion on my design I interviewed Mrs. J. Pickery again (St. Peter's Primary School)

Merits Of The Table:

"It would allow me to prepare prior to the lesson and would save a lot of time. It takes up the minimum amount of space and from what I can see it is very safe for children to use. The design allows the minimum of mess to be created during the lesson and allows scope for nearly every type of art work to be done on it."

Problems With The Table:

"Because the central paint holder is not supported in the up position the section does lower slightly as the children push on the edge. Also the table is hard to open with just one person so preferably it needs two people to open it."

CHILDRENS' ART TABLE

Page 47

3.4.5 Evaluation

The testing of the solutions and analysis of results against an original specification to identify strengths and weaknesses

- Evaluation by an end user in the context for which the product was designed.
- A load test on one element of a design carried out in addition to a field test.
- In-context testing, again with third party evaluation.
- An interim test on joint strength.
- Does the product fulfil the specification?

Results of Load Test

The results of the load test were very positive and showed that the brackets could easily carry the required load (see Figs. 83.1 & 83.2). The device was loaded to 240 kg (10 x 20 kg weights and 4 x 10 kg weights were used). The bracket appeared to hold up well to the load with no cracking sounds during testing. The brackets were also visually investigated after the test with absolutely no signs wear or damage to the bracket. I can thus safely boast that the brackets have a factor of safety greater than 4 (see Page), which in my opinion is more than sufficient.

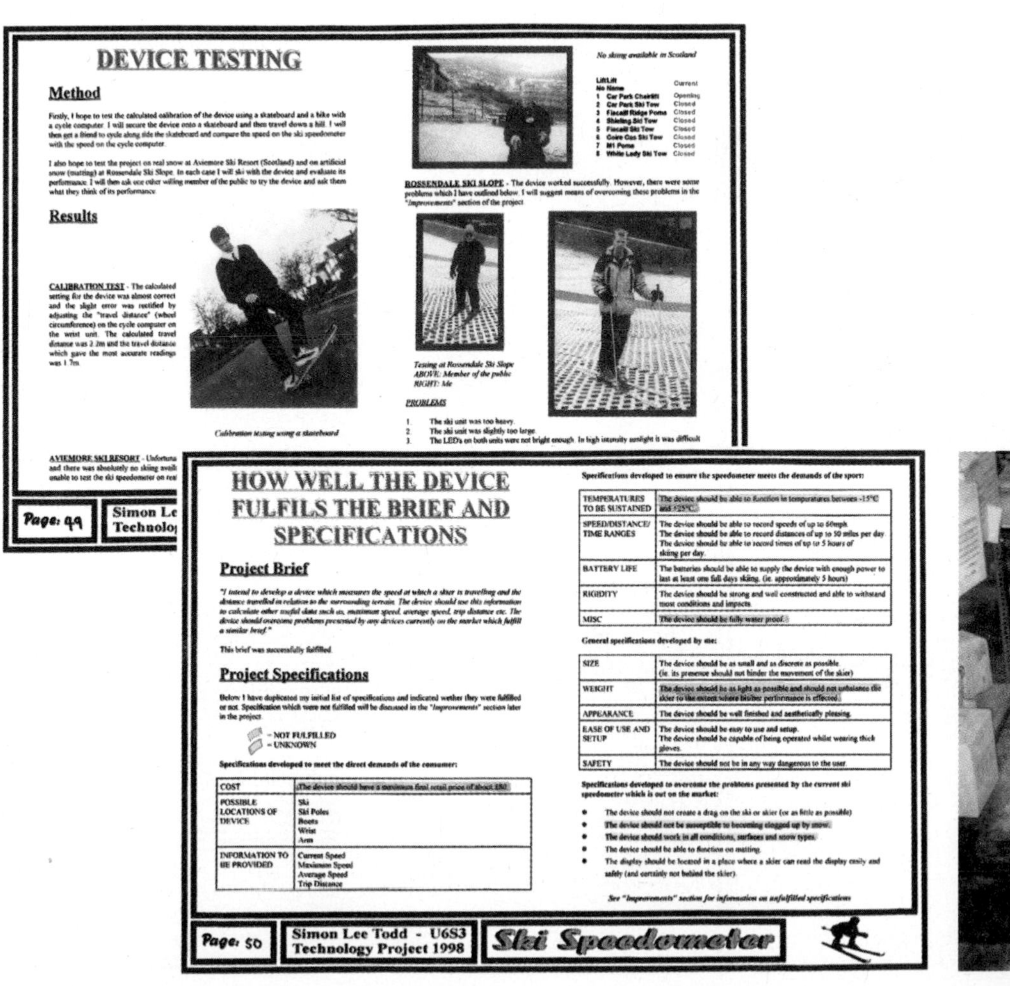
DEVICE TESTING

Method

Firstly, I hope to test the calculated calibration of the device using a skateboard and a bike with a cycle computer. I will secure the device onto a skateboard and then travel down a hill. I will then get a friend to cycle along side the skateboard and compare the speed on the ski speedometer with the speed on the cycle computer.

I also hope to test the project on real snow at Aviemore Ski Resort (Scotland) and on artificial snow (matting) at Rossendale Ski Slope. In each case I will ski with the device and evaluate its performance. I will then ask one other willing member of the public to try the device and ask them what they think of its performance.

Results

CALIBRATION TEST - The calculated setting for the device was almost correct and the slight error was rectified by adjusting the "travel distance" (wheel circumference) on the cycle computer on the wrist unit. The calculated travel distance was 2.2m and the travel distance which gave the most accurate readings was 1.7m.

Calibration testing using a skateboard

AVIEMORE SKI RESORT - Unfortuna... and there was absolutely no skiing avail... unable to test the ski speedometer on real...

No skiing available in Scotland

Lift No	Lift Name	Current
1	Car Park Chairlift	Opening
2	Car Park Ski Tow	Closed
3	Fiacaill Ridge Poma	Closed
4	Shieling Ski Tow	Closed
5	Fiacaill Ski Tow	Closed
6	Coire Cas Ski Tow	Closed
7	M1 Poma	Closed
8	White Lady Ski Tow	Closed

ROSSENDALE SKI SLOPE - The device worked successfully. However, there were some problems which I have outlined below. I will suggest means of overcoming these problems in the "Improvements" section of the project.

Testing at Rossendale Ski Slope
ABOVE: Member of the public
RIGHT: Me

PROBLEMS

1. The ski unit was too heavy.
2. The ski unit was slightly too large.
3. The LED's on both units were not bright enough. In high intensity sunlight it was difficult

Page: 49 | Simon Le... Technolo...

HOW WELL THE DEVICE FULFILS THE BRIEF AND SPECIFICATIONS

Project Brief

"I intend to develop a device which measures the speed at which a skier is travelling and the distance travelled in relation to the surrounding terrain. The device should use this information to calculate other useful data such as, maximum speed, average speed, trip distance etc. The device should overcome problems presented by any devices currently on the market which fulfill a similar brief."

This brief was successfully fulfilled.

Project Specifications

Below I have duplicated my initial list of specifications and indicated wether they were fulfilled or not. Specification which were not fulfilled will be discussed in the "Improvements" section later in the project.

- NOT FULFILLED
- UNKNOWN

Specifications developed to meet the direct demands of the consumer:

COST	The device should have a maximum final retail price of about £80.
POSSIBLE LOCATIONS OF DEVICE	Ski Ski Poles Boots Wrist Arm
INFORMATION TO BE PROVIDED	Current Speed Maximum Speed Average Speed Trip Distance

Specifications developed to ensure the speedometer meets the demands of the sport:

TEMPERATURES TO BE SUSTAINED	The device should be able to function in temperatures between -15°C and +25°C.
SPEED/DISTANCE/ TIME RANGES	The device should be able to record speeds of up to 60mph. The device should be able to record distances of up to 50 miles per day. The device should be able to record times of up to 5 hours of skiing per day.
BATTERY LIFE	The batteries should be able to supply the device with enough power to last at least one full days skiing. (ie. approximately 5 hours)
RIGIDITY	The device should be strong and well constructed and able to withstand most conditions and impacts.
MISC	The device should be fully water proof.

General specifications developed by me:

SIZE	The device should be as small and as discrete as possible. (ie. its presence should not hinder the movement of the skier)
WEIGHT	The device should be as light as possible and should not unbalance the skier to the extent where his/her performance is effected.
APPEARANCE	The device should be well finished and aesthetically pleasing.
EASE OF USE AND SETUP	The device should be easy to use and setup. The device should be capable of being operated whilst wearing thick gloves.
SAFETY	The device should not be in any way dangerous to the user.

Specifications developed to overcome the problems presented by the current ski speedometer which is out on the market:

- The device should not create a drag on the ski or skier (or as little as possible)
- The device should not be susceptible to becoming clogged up by snow.
- The device should work in all conditions, surfaces and snow types.
- The device should be able to function on matting.
- The display should be located in a place where a skier can read the display easily and safely (and certainly not behind the skier).

See "Improvements" section for information on unfulfilled specifications

Page: 50 | Simon Lee Todd - U6S3 Technology Project 1998 | Ski Speedometer

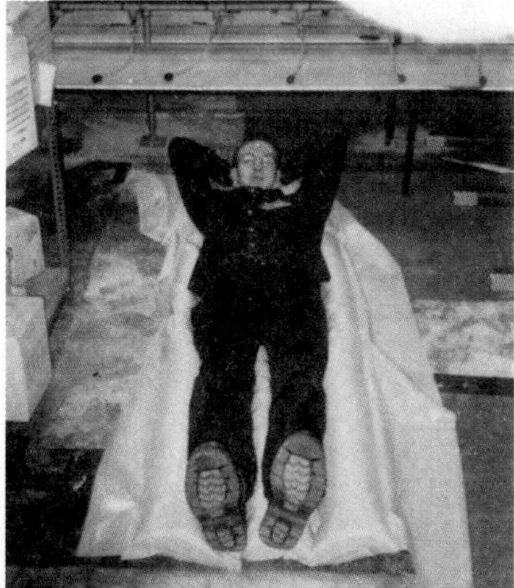

Evaluation

PAGE: 127

COSTING:-

The Table below shows the comparison with the estimated production cost and my cost.

ITEM	MY COST	MY TIME	ESTIMATED PROFESSIONAL COST	ESTIMATED PROFESSIONAL TIME
Disc and spider	£0·00	40 hours	£10·00	3 hours
Caliper	£0·00	80 hours,	£15·00	12 hours
Hub+wheel	£89.99	10 minutes on phone	£89·99	10 minutes on phone
Springs	£3·68	½ hour driving	£3·68	NIL
Stainless steel bolts	£[Approx]5·00	½ hour driving	£[APPROX]5·00	NIL
Other Misc bolts	£2·50	½ hour driving	£2·50	NIL
Other disposablels eg: sand paper, polish.	£7·00	½ hour driving	£7·00	NIL
	TOTAL: £107·18	Total £=122hrs 10minutes	Total = £132·18	Total = 15 hrs 10mins

You can see that It It cost me very little money to produce my caliper and parts but a massive amount of time. This is because, I made It by myself by Hand, not on a CNC mill as a professional would would do. The total cost to me was Half that of a new Hyd solution from Hope.

THE FINISHED PRODUCT

DEVICE COSTING

The overall cost of the system materials was... £101.57
The overall cost of the casing and attachment materials was... £7.10
Total cost of all materials: £108.67

This is fairly high. However, it must be noted that the device produced in this project is a prototype. There is little point in costing this prototype for manufacture, as many expensive modules have been used which would simply not be necessary in a final version. For example, a version intended for manufacture would not use a £20 cycle computer on the wrist unit, it would use its own on-board components to perform the necessary calculations and to control the display. Also, the transmitter and receiver would not be bought as modules, but would instead be built into the PCB. This would allow the manufacturer to reduce the cost by removing unnecessary functions provided by the pre-built modules. These would vastly cut dow the cost of the device.

Below is an approximate guide to the cost of materials for a manufactured version of the device, assuming the device is produced in large quantities (ie. 1000+).

The ski unit would consist of:

Microwave Doppler Module ≈ £10.00
Surface Mount PCB ≈ £5.00
Switch ≈ £0.10
2 LED's ≈ £0.10
Areal ≈ £0.10
Injection mould casing and attachment ≈ £2.00

The ski unit would consist of:

Surface mount PCB ≈ £5.00
Switch ≈ £0.10
LED ≈ £0.05
LCD ≈ £2.00
Areal ≈ £0.10
Casing & Attachment ≈ £1.00

This would give a total materials cost of about £25.00.
The cost of manufacturer would double this cost to about £50.

This would lead to a final retail price of about £120!

Page: 48 | Simon Lee Todd - U6S3 Technology Project 1998 | Ski Speedometer

Figure 3.28 An understanding of the potential of the final solution for industrial production in the context of manufacturing and commercial constraints.

Figure 3.29 Evidence of responding positively to external evaluation.

Weaknesses

The design only allows paint pots of diametre 60mm and depth of a maximum 70mm. If I were to design the table again I would research the dimensions of paint pots to find the most common and alter the lengths of the strips according to these dimensions

When the surface B is in the top position it does not stay level with the other two surfaces and whilst children are working it is pushed a little lower by the children reaching for paint. The surface could be fixed in the up position with the design below

rotates about screw

The only thing that my design does not fulfil in my specification is the desirable criteria that the table could hold paper. If I had more time I would encorporate a paper holder into the design

The table needs two people to change the position of the paint pot rack which is an inconvenience. When the surfaces are moved pressure is put on the plastic tube which is not very strong as it is only glued down with araldite. If I made the table again I would make this out of metal for strength

CHILDRENS' ART TABLE

Page 50

POSSIBLE IMPROVEMENTS TO THE DEVICE

Although the project fulfilled the brief, several of the specifications were not fulfilled. Testing also highlighted some problems which would have to be overcome before the device could be marketed. The areas which need improvements are highlighted below with suggestions on how to go about making the improvements.

Problem: The cost of the device is too high. An estimated retail price £120 has been predicted which is considerably higher than the £80 proposed in the specifications.
Solution: The device produced in this project is a prototype. The retail price has been developed from cost estimations for a final manufacturable version of the project (see "Project Costing" section). Consequently, this cost estimation may contain a degree of error. Much more research would be required in order to obtain a realistic idea of production costs. Despite this, the cost of the device would fall the greater the number of devices are produced. For example, the first digital cycle computers originally retailed at about £40 and were only purchased by dedicated cycle enthusiasts. Several years later, almost identical cycle computers are retailing at about £15 and are considered to be a "general cycle accessory"!

Problem: The ski unit is too heavy.
Solution: Approximately 80% of the weight of the ski unit is due to the 6 AA batteries used to power the unit. This large mass of batteries are required to meet the power demands of the microwave doppler module which consumes 160mA out of the 180mA used by the unit. This can be overcome by using a lower powered doppler module. Philips do make a module which consumes only 30mA but they are almost three times the size of the module used. Philips do say that they can custom design a doppler module to almost any specification assuming a large enough order will be placed. This will of course be the case if the device goes into production. If this modification is made then the 6 AA batteries can be replaced by a single 9V PP3 battery which is approximately one third as heavy as the 6 AAs.

Problem: The ski unit is slightly too large.
Solution: The 6AA batteries occupy about 60% of the ski unit. The modification described above will allow a single PP3 battery to replace the 6 AA batteries. A PP3 will require only about one third of the space required by the 6 AA batteries. Also, if the device goes into production, then surface mount components can be used. This would reduce the size of the circuitry allowing the unit to be made even smaller.

Problem: The LED's on both units are not bright enough.
Solution: The value of the resistors which governs the current flowing through the LEDs can be decreased. In both cases about 3mA are allowed to flow through the LEDs. By increasing this to 8mA the LED's would become much brighter. However, it must be noted that the battery life will be slightly reduced in both units.

R = V/I V = 5V
R = 5/0.008 I = 0.08A
R = 625Ω

Problem: The display is not updated frequently enough.
Solution: The cycle computer will only update the display every time the relay contacts are pulsed closed. This only occurs when a pulse is sent from the ski unit to the wrist unit (ie. when the unit travels the set "travel distance", which in this case is 1.7m). This problem can be reduced by setting the travel distance to its minimum value (ie. 1.3m) and adjusting the frequency dividing mechanisms in the processor to compensate. However, in order to eliminate this problem, the cycle computer unit and relay must be replaced completely by custom designed processing system and LCD. This would of course be the case if the device went into production. This would eliminate the limitations of the cycle computer and allow the "travel distance" to be reduced to a few centimetres. This would allow more speed readings to be calculated over a set distance. There may even be no need for the frequency divider, as the processor may be developed to calculate the speed from the frequency of the signal directly from the signal squarer (ie. 51Hz for every 1mph).

Problem: The device is unable to operate across the required temperature range (-15°C to +23°C)
Solution: The limiting factor is the cycle computer unit which can only function in temperatures down to 0°C. In order to eliminate this problem the cycle computer unit must be replaced by a custom made system which will be able to cope with the necessary temperatures. Again, this would be the case if the device went into production.

Problem: The device is not fully waterproof.
Solution: This can be overcome by adding waterproof seals to all joins or orifices in the casing. Unfortunately, I was not able to include such features on the device due to the limited time and resources available. I successfully managed to include a rubber seal around the base of the ski unit and a water resistant switch cover which made the ski unit fairly waterproof. Again, the seals would be included if the device went into production.

Page: 51 | Simon Lee Todd - U6S3 Technology Project 1998 | Ski Speedometer

Figure 3.30 Evaluation of end product, including details of any modifications necessary to overcome weaknesses.

3.5 Individual case study

Traffic control torch, Darren Fenton, Grosvenor Grammar School

This section follows the progress of Darren's project from the first identification of the problem through to the final costing. The titles are Darren's own. They follow the design process and mirror this in addressing the required sections of the examination specification.

Figure 3.31 Identification of the problem.

Figure 3.32 Examples of questionnaires, responses from various police forces and a summary of results.

- I had an interview with the Superintendent of the RUC's Traffic Branch to discuss problem
- I wrote to all the 55 police forces in the UK and even some from around the world enclosing a questionnaire on the methods they used
- I gave out questionnaires to a selected number of officers in Traffic Branch and to officers on patrol
- Formulated all the results in the form of graphs and used them to solve the problem
- Wrote to various companies like the
  -Airports- for information on what they use to marshall planes
  -Opticians- for information on colour blindness
  -Epilepsy Societies- for information on epilepsy because of using flashing lights
  -Police Research Groups- to see if any research has been carried out into this area
  -DOE Road Service- to see how they control traffic when working on the roads
  -European Commission- to see if there is an legislation on stopping cars
  -British Standards- to see if there are any standards the torch must comply with
- Researched different bulbs etc. to see which give of the brightest light
- Researched different batteries and packs to see which last the longest
- Researched different materials to see which are the most durable and strongest
- Researched different switches to see which would be best since the officer will have gloves on
- Researched other products and torches on the market
- Went to see different companies for help and advice

Methods

The following is the methods used by each of the police forces that I wrote to. The forces that use the same method as the RUC are in bold writing.

UK POLICE FORCES

Avon & Somerset Constabulary	Reflective clothing and police vehicle displays text message to drivers
Bedfordshire Police	Reflective clothing and a white light
Cambridgeshire Constabulary	Reflective clothing and f police vehicles
Central Scotland Police	Reflective clothing and a white light
Cheshire Constabulary	Reflective clothing and a white light
City of London Police	Reflective clothing and s
Cleveland Constabulary	Reflective clothing and a white light called a "Di
Cumbria Constabulary	Reflective clothing and a white light
Derbyshire Constabulary	Reflective clothing and a white light called a "B
Devon & Cornwall Constabulary	Reflective clothing and f police vehicles
Dorset Police	Reflective clothing and a white light to illuminat
Dumfries Constabulary	Reflective clothing and a white light to illuminat
Durham Constabulary	Reflective clothing and f police vehicles
Dyfed-Powys Police	Reflective clothing and f

WORLD POLICE FORCES

Australian Federal Police	**Reflective vest and a torch with a red cone**
Garda Siochana	Reflective clothing and a torch which emits a white light
Metropolitan Toronto Police	Reflective clothing and red and blue flashing lights on police vehicles
Netherlands Police	Only on highways, use a police car or take registration number and send a ticket
New York Police Dept.	Use of siren, flashing lights and directed by police officer
Ontario Provincial Police	Only on highways, take a reading and a police car pulls the car over

Note. In total I wrote to fifteen police forces around the world, the following nine police forces didn't write back to me:- **Denmark Police, Finland Police, French Police, Los Angeles Police, Japanese Police, San Francisco Police, South African Police, Spanish Police and Swedish Police.**

NI Results

The following shows the results of the questionnaires completed by the police in Northern Ireland.

1- Do you think the present method i.e. red torch is safe?	Yes-82%	No-18%
2- Have you ever been put in danger when using this method?	Yes-32%	No-68%
3- Do you know anyone that has been put in danger?	Yes-48%	No-52%
4- Would you like to see an improved method?	Yes-55%	No-45%
5- Do you think it could be improved?	Yes-52%	No-48%
6- Do you think it is easily seen at night against car brake lights?	Yes-66%	No-34%
7- Do you have problems with this method?	Yes-20%	No-80%

8- Indicate the most important factors in the design of the

PUBLIC QUESTIONNAIRE

I am currently an upper sixth student at Gros Belfast. For my A-Level technology project, I and design and make a solution for it. I have ch problem and I would like the opinion of the pu agreed to help me with my project. The police red cone attachment over the top of it (as show given in this questionnaire will be treated confi appreciate it if you would take the time to fill i
Thank-you Darren Fenton

1- Have you ever been stopped by the police using this torch, e.g. at a checkpoint, to check tyres, for speeding?
2- If the answer to number 1 was yes, was the signal clear?
3- If the answer to number 1 was yes, did you have sufficient time to stop?
4- Do you know the signal for 'stop', is the rotation of the torch with the red cone attached?
5- Do you know the signal for 'go', is the back and forth movement of the white light?
6- When you have seen the police carrying out vehicle or speed checks, were they wearing their yellow jackets?
7- Do you think the red light is easily seen at night against the car backlights, since they would blend in together?
8- Do you think a green light would be a good idea to indicate go?
9- What do you think would be best to indicate stop?
flashing red ☐ flashing red and blue alter
other ☐
If you have any comments please write them below

Thank-you

POLICE QUESTIONNAIRE

I am currently a lower sixth student at Grosvenor Grammar School in Belfast. For my A-Level technology project, I have to identify a problem and design and make a solution for it. I have chosen the police torch as a problem and I would like the opinion of the officers that would use the torch. Sergeant Spence of the RUC's Traffic Branch has agreed to help me with my project. I would appreciate it if you would take the time to fill in this questionnaire. Thank-you Darren Fenton

	Yes	No
1- Do you think the present method i.e. red torch is safe to use?	☐	☐
2- Have you ever been put into danger when using this method?	☐	☐
3- Do you know anyone that has been put in danger?	☐	☐
4- Would you like to see an improved method?	☐	☐
5- Do you think it could be improved?	☐	☐
6- Do you think it is easily seen at night against car back lights, since they would blend in together?	☐	☐
7- Do you have problems with this method? If yes please give details	☐	☐

8- Indicate using a scale from 1 to 8 the most important factors which you think would be the most important in the design of this project (1= most important and 8= least important) USE DIFFERENT NUMBER FOR EACH BOX
Reliability ☐ Size ☐
Weight ☐ Portability ☐
Visibility ☐ Appearance ☐
Material ☐ Safety ☐

9- What would you like the new method to be? (Tick one)
Glove ☐ Jacket ☐
Torch ☐ Hat ☐
Sign ☐ Other ☐

10- What would you like the new method to have on it? (Tick one)
Light ☐ Green light ☐
Red light ☐ Red light (flashing) ☐
Other ☐

If you can think of any solutions please draw them on the back of this page.
Thank-you
Thank-you

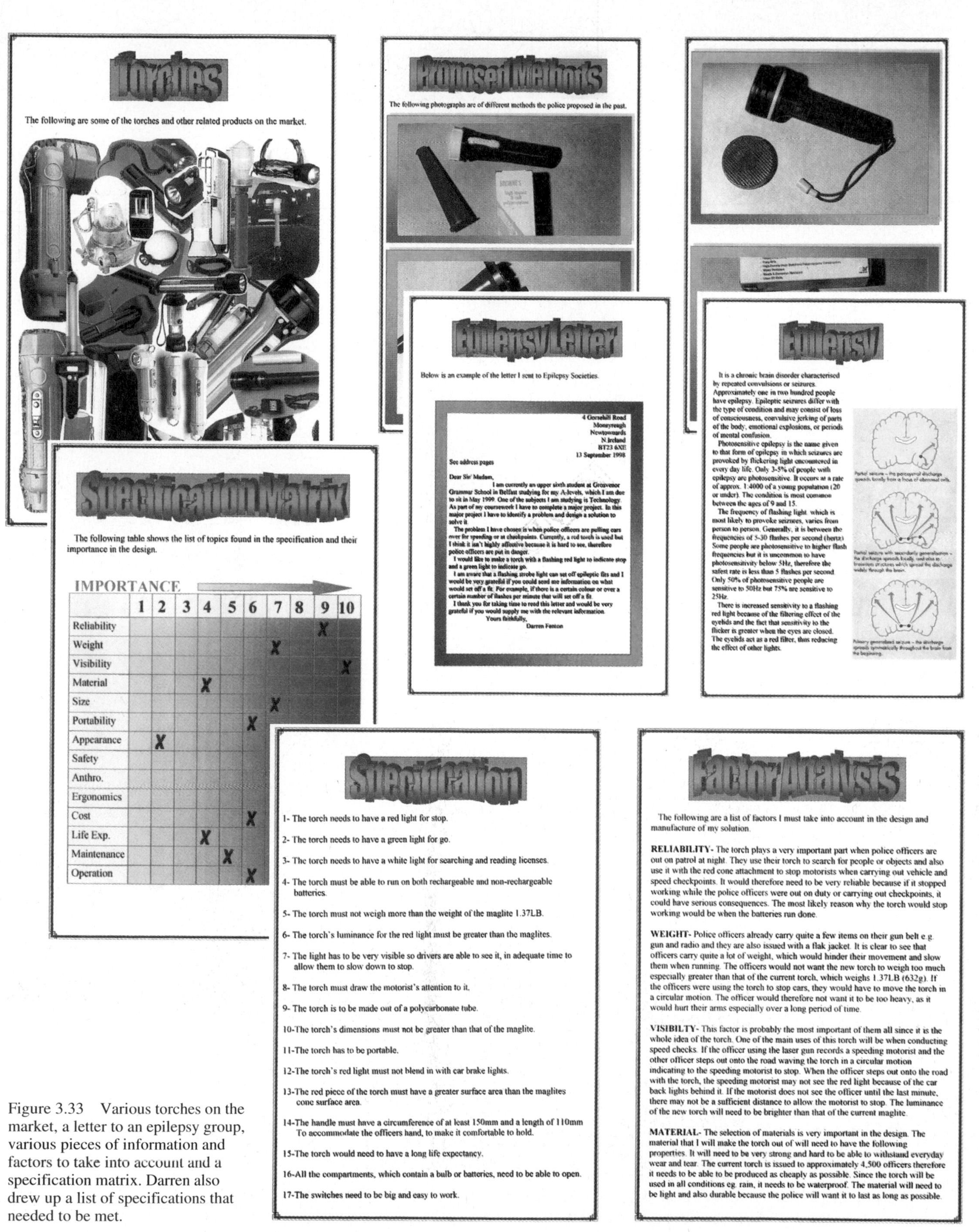
Torches

The following are some of the torches and other related products on the market.

Proposed Methods

The following photographs are of different methods the police proposed in the past.

Epilepsy Letter

Below is an example of the letter I sent to Epilepsy Societies.

4 Gorsehill Road
Moneyreagh
Newtownards
N.Ireland
BT23 6XE
13 September 1998

See address pages

Dear Sir/ Madam,

I am currently an upper sixth student at Grosvenor Grammar School in Belfast studying for my A-levels, which I am due to sit in May 1999. One of the subjects I am studying is Technology. As part of my coursework I have to complete a major project. In this major project I have to identify a problem and design a solution to solve it.

The problem I have chosen is when police officers are pulling cars over for speeding or at checkpoints. Currently, a red torch is used but I think it isn't highly affective because it is hard to see, therefore police officers are put in danger.

I would like to make a torch with a flashing red light to indicate stop and a green light to indicate go.

I am aware that a flashing strobe light can set off epileptic fits and I would be very grateful if you could send me information on what would set off a fit. For example, if there is a certain colour or over a certain number of flashes per minute that will set off a fit.

I thank you for taking time to read this letter and would be very grateful if you would supply me with the relevant information.

Yours faithfully,

Darren Fenton

Epilepsy

It is a chronic brain disorder characterised by repeated convulsions or seizures. Approximately one in two hundred people have epilepsy. Epileptic seizures differ with the type of condition and may consist of loss of consciousness, convulsive jerking of parts of the body, emotional explosions, or periods of mental confusion.

Photosensitive epilepsy is the name given to that form of epilepsy in which seizures are provoked by flickering light encountered in every day life. Only 3-5% of people with epilepsy are photosensitive. It occurs at a rate of approx. 1:4000 of a young population (20 or under). The condition is most common between the ages of 9 and 15.

The frequency of flashing light, which is most likely to provoke seizures, varies from person to person. Generally, it is between the frequencies of 5-30 flashes per second (hertz). Some people are photosensitive to higher flash frequencies but it is uncommon to have photosensitivity below 5Hz, therefore the safest rate is less than 5 flashes per second. Only 50% of photosensitive people are sensitive to 50Hz but 75% are sensitive to 25Hz.

There is increased sensitivity to a flashing red light because of the filtering effect of the eyelids and the fact that sensitivity to the flicker is greater when the eyes are closed. The eyelids act as a red filter, thus reducing the effect of other lights.

Specification Matrix

The following table shows the list of topics found in the specification and their importance in the design.

IMPORTANCE

	1	2	3	4	5	6	7	8	9	10
Reliability									X	
Weight							X			
Visibility										X
Material				X						
Size							X			
Portability						X				
Appearance		X								
Safety										
Anthro.										
Ergonomics										
Cost						X				
Life Exp.				X						
Maintenance					X					
Operation						X				

Specification

1- The torch needs to have a red light for stop.

2- The torch needs to have a green light for go.

3- The torch needs to have a white light for searching and reading licenses.

4- The torch must be able to run on both rechargeable and non-rechargeable batteries.

5- The torch must not weigh more than the weight of the maglite 1.37LB.

6- The torch's luminance for the red light must be greater than the maglites.

7- The light has to be very visible so drivers are able to see it, in adequate time to allow them to slow down to stop.

8- The torch must draw the motorist's attention to it.

9- The torch is to be made out of a polycarbonate tube.

10-The torch's dimensions must not be greater than that of the maglite.

11-The torch has to be portable.

12-The torch's red light must not blend in with car brake lights.

13-The red piece of the torch must have a greater surface area than the maglites cone surface area.

14-The handle must have a circumference of at least 150mm and a length of 110mm To accommodate the officers hand, to make it comfortable to hold.

15-The torch would need to have a long life expectancy.

16-All the compartments, which contain a bulb or batteries, need to be able to open.

17-The switches need to be big and easy to work.

Factor Analysis

The following are a list of factors I must take into account in the design and manufacture of my solution.

RELIABILITY- The torch plays a very important part when police officers are out on patrol at night. They use their torch to search for people or objects and also use it with the red cone attachment to stop motorists when carrying out vehicle and speed checkpoints. It would therefore need to be very reliable because if it stopped working while the police officers were out on duty or carrying out checkpoints, it could have serious consequences. The most likely reason why the torch would stop working would be when the batteries run done.

WEIGHT- Police officers already carry quite a few items on their gun belt e.g. gun and radio and they are also issued with a flak jacket. It is clear to see that officers carry quite a lot of weight, which would hinder their movement and slow them when running. The officers would not want the new torch to weigh too much especially greater than that of the current torch, which weighs 1.37LB (632g). If the officers were using the torch to stop cars, they would have to move the torch in a circular motion. The officer would therefore not want it to be too heavy, as it would hurt their arms especially over a long period of time.

VISIBILTY- This factor is probably the most important of them all since it is the whole idea of the torch. One of the main uses of this torch will be when conducting speed checks. If the officer using the laser gun records a speeding motorist and the other officer steps out onto the road waving the torch in a circular motion indicating to the speeding motorist to stop. When the officer steps out onto the road with the torch, the speeding motorist may not see the red light because of the car back lights behind it. If the motorist does not see the officer until the last minute, there may not be a sufficient distance to allow the motorist to stop. The luminance of the new torch will need to be brighter than that of the current maglite.

MATERIAL- The selection of materials is very important in the design. The material that I will make the torch out of will need to have the following properties. It will need to be very strong and hard to be able to withstand everyday wear and tear. The current torch is issued to approximately 4,500 officers therefore it needs to be able to be produced as cheaply as possible. Since the torch will be used in all conditions eg. rain, it needs to be waterproof. The material will need to be light and also durable because the police will want it to last as long as possible.

Figure 3.33 Various torches on the market, a letter to an epilepsy group, various pieces of information and factors to take into account and a specification matrix. Darren also drew up a list of specifications that needed to be met.

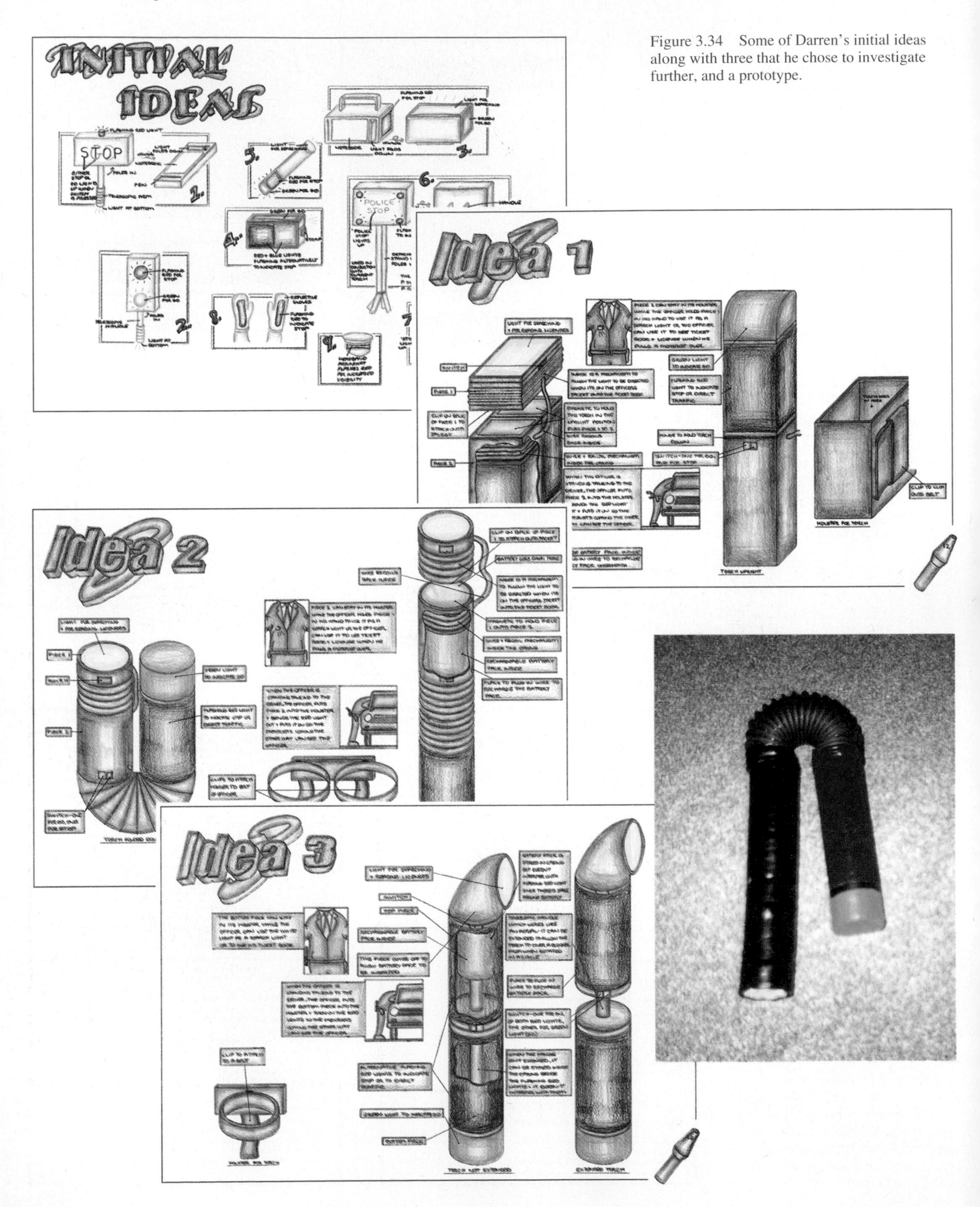

Figure 3.34 Some of Darren's initial ideas along with three that he chose to investigate further, and a prototype.

Figure 3.35 Other ideas that were developed further, and the final selection.

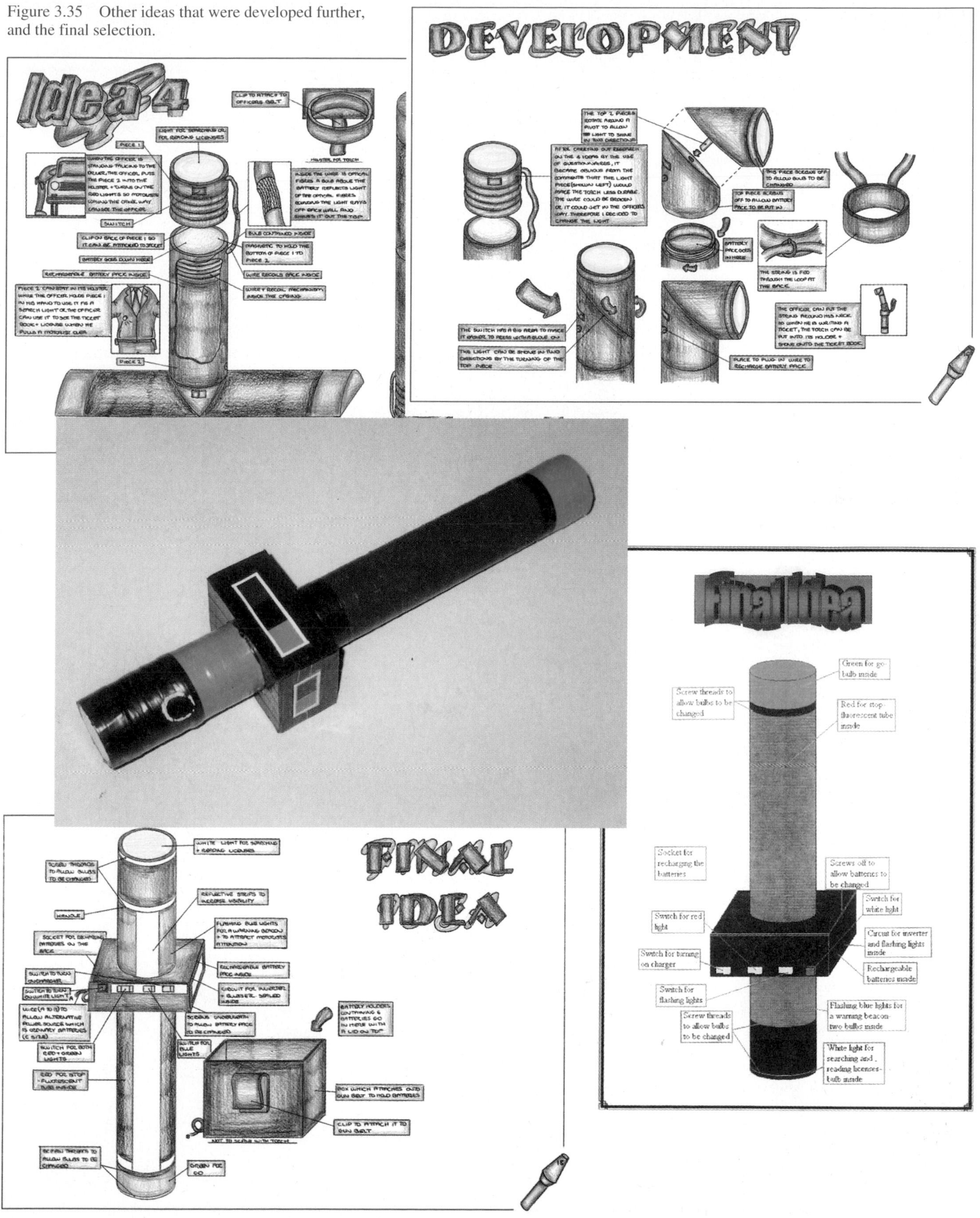

Figure 3.36 Further research work on the four shortlisted ideas, plus results of the questionnaire sent out asking for feedback on the four ideas; analysis of the four ideas.

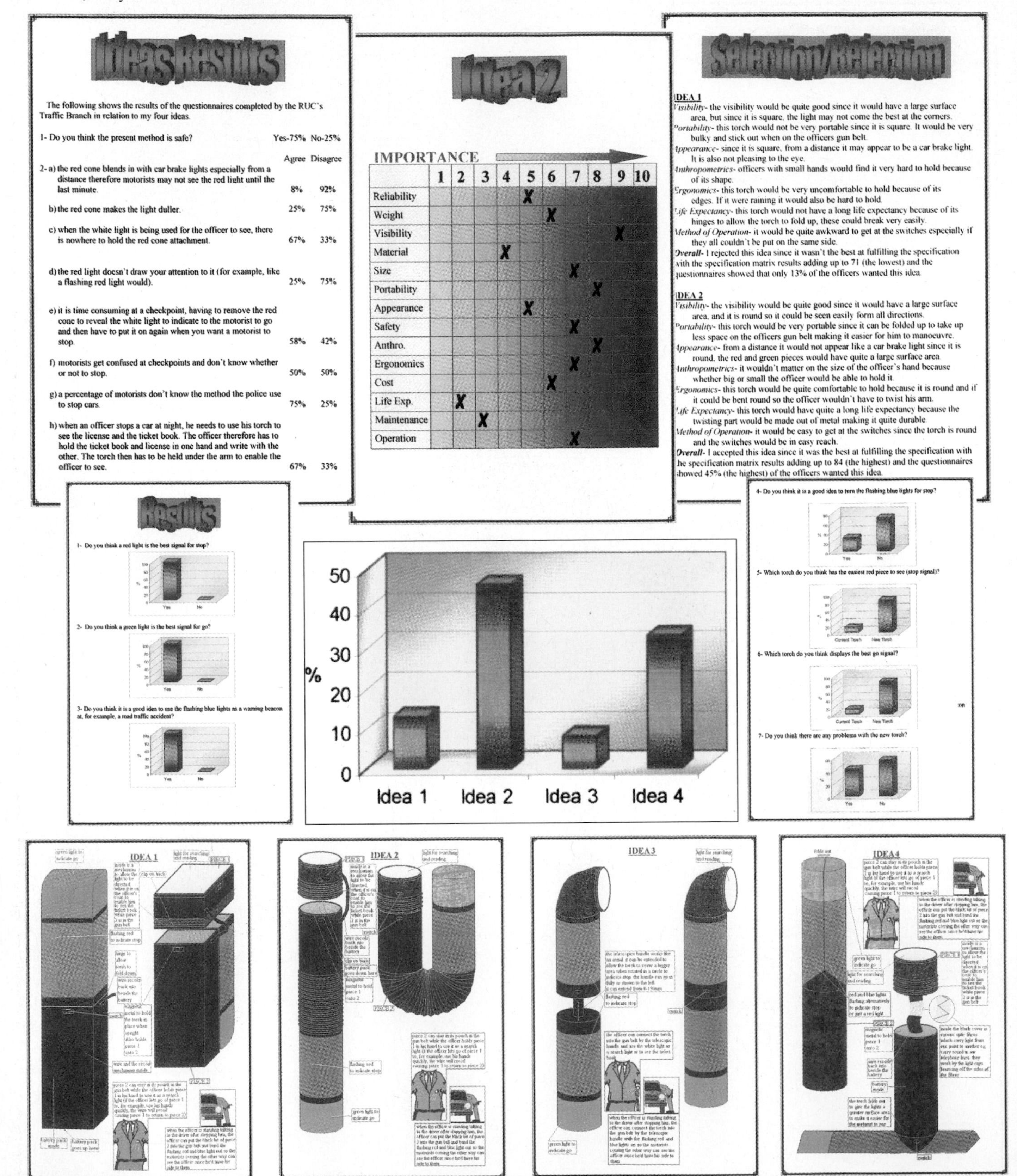

Ideas Results

The following shows the results of the questionnaires completed by the RUC's Traffic Branch in relation to my four ideas.

1- Do you think the present method is safe? Yes-75% No-25%

	Agree	Disagree
2- a) the red cone blends in with car brake lights especially from a distance therefore motorists may not see the red light until the last minute.	8%	92%
b) the red cone makes the light duller.	25%	75%
c) when the white light is being used for the officer to see, there is nowhere to hold the red cone attachment.	67%	33%
d) the red light doesn't draw your attention to it (for example, like a flashing red light would).	25%	75%
e) it is time consuming at a checkpoint, having to remove the red cone to reveal the white light to indicate to the motorist to go and then have to put it on again when you want a motorist to stop.	58%	42%
f) motorists get confused at checkpoints and don't know whether or not to stop.	50%	50%
g) a percentage of motorists don't know the method the police use to stop cars.	75%	25%
h) when an officer stops a car at night, he needs to use his torch to see the license and the ticket book. The officer therefore has to hold the ticket book and license in one hand and write with the other. The torch then has to be held under the arm to enable the officer to see.	67%	33%

Idea 2

IMPORTANCE

	1	2	3	4	5	6	7	8	9	10
Reliability					X					
Weight						X				
Visibility									X	
Material				X						
Size							X			
Portability								X		
Appearance					X					
Safety							X			
Anthro.								X		
Ergonomics							X			
Cost						X				
Life Exp.		X								
Maintenance			X							
Operation							X			

Selection/Rejection

IDEA 1

Visibility- the visibility would be quite good since it would have a large surface area, but since it is square, the light may not come the best at the corners.

Portability- this torch would not be very portable since it is square. It would be very bulky and stick out when on the officers gun belt.

Appearance- since it is square, from a distance it may appear to be a car brake light. It is also not pleasing to the eye.

Anthropometrics- officers with small hands would find it very hard to hold because of its shape.

Ergonomics- this torch would be very uncomfortable to hold because of its edges. If it were raining it would also be hard to hold.

Life Expectancy- this torch would not have a long life expectancy because of its hinges to allow the torch to fold up, these could break very easily.

Method of Operation- it would be quite awkward to get at the switches especially if they all couldn't be put on the same side.

Overall- I rejected this idea since it wasn't the best at fulfilling the specification with the specification matrix results adding up to 71 (the lowest) and the questionnaires showed that only 13% of the officers wanted this idea.

IDEA 2

Visibility- the visibility would be quite good since it would have a large surface area, and it is round so it could be seen easily form all directions.

Portability- this torch would be very portable since it can be folded up to take up less space on the officers gun belt making it easier for him to manoeuvre.

Appearance- from a distance it would not appear like a car brake light since it is round, the red and green pieces would have quite a large surface area.

Anthropometrics- it wouldn't matter on the size of the officer's hand because whether big or small the officer would be able to hold it.

Ergonomics- this torch would be quite comfortable to hold because it is round and if it could be bent round so the officer wouldn't have to twist his arm.

Life Expectancy- this torch would have quite a long life expectancy because the twisting part would be made out of metal making it quite durable.

Method of Operation- it would be easy to get at the switches since the torch is round and the switches would be in easy reach.

Overall- I accepted this idea since it was the best at fulfilling the specification with the specification matrix results adding up to 84 (the highest) and the questionnaires showed 45% (the highest) of the officers wanted this idea.

Figure 3.37 Work undertaken on the materials sections and testing; material data; calculations undertaken.

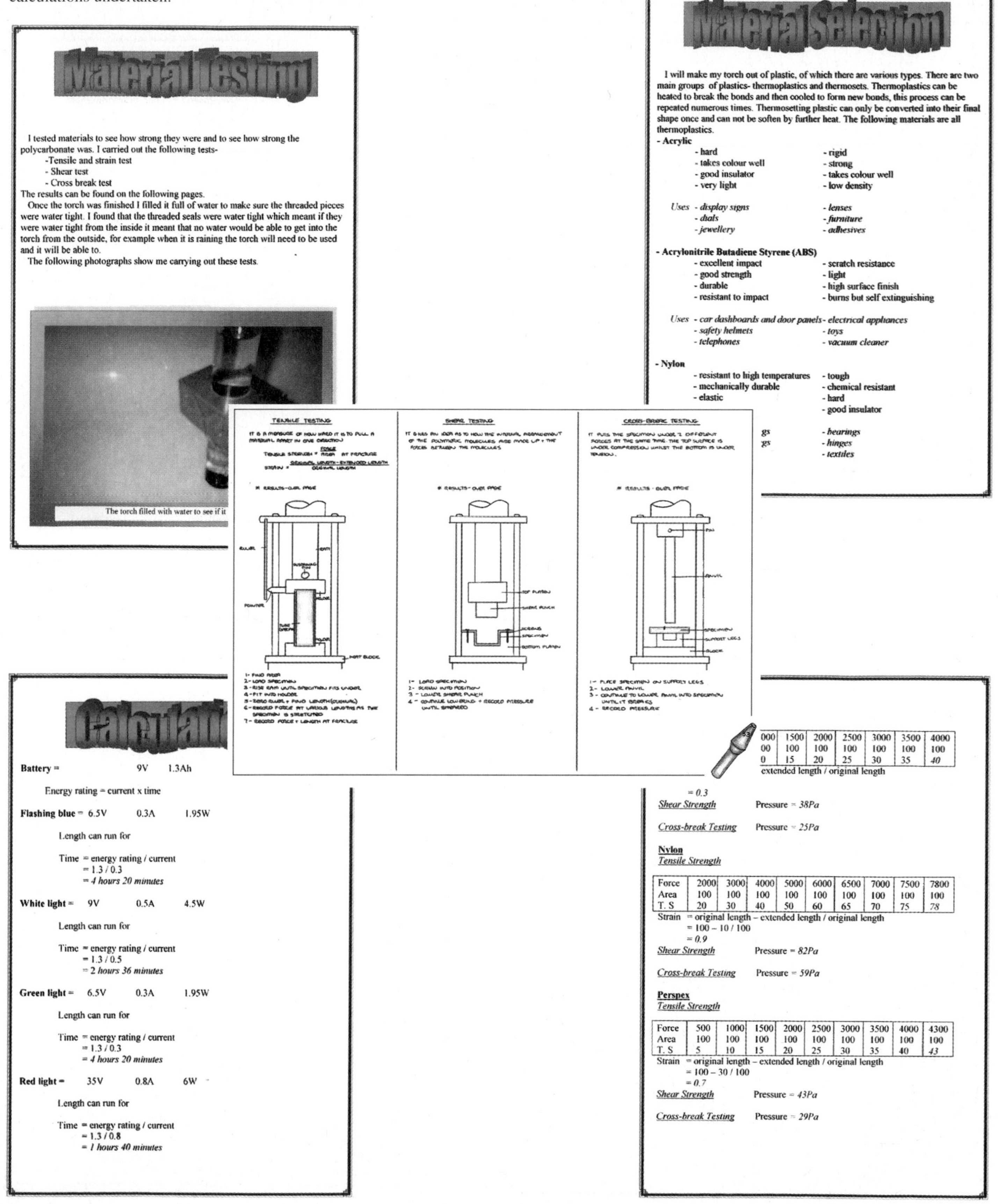

Material Testing

I tested materials to see how strong they were and to see how strong the polycarbonate was. I carried out the following tests-
- Tensile and strain test
- Shear test
- Cross break test

The results can be found on the following pages.

Once the torch was finished I filled it full of water to make sure the threaded pieces were water tight. I found that the threaded seals were water tight which meant if they were water tight from the inside it meant that no water would be able to get into the torch from the outside, for example when it is raining the torch will need to be used and it will be able to.

The following photographs show me carrying out these tests.

The torch filled with water to see if it

Material Selection

I will make my torch out of plastic, of which there are various types. There are two main groups of plastics- thermoplastics and thermosets. Thermoplastics can be heated to break the bonds and then cooled to form new bonds, this process can be repeated numerous times. Thermosetting plastic can only be converted into their final shape once and can not be soften by further heat. The following materials are all thermoplastics.

- Acrylic

- hard - rigid
- takes colour well - strong
- good insulator - takes colour well
- very light - low density

Uses - *display signs* - *lenses*
- *dials* - *furniture*
- *jewellery* - *adhesives*

- Acrylonitrile Butadiene Styrene (ABS)

- excellent impact - scratch resistance
- good strength - light
- durable - high surface finish
- resistant to impact - burns but self extinguishing

Uses - *car dashboards and door panels* - *electrical appliances*
- *safety helmets* - *toys*
- *telephones* - *vacuum cleaner*

- Nylon

- resistant to high temperatures - tough
- mechanically durable - chemical resistant
- elastic - hard
- good insulator

gs - *bearings*
gs - *hinges*
- *textiles*

TENSILE TESTING

IT IS A MEASURE OF HOW HARD IT IS TO PULL A MATERIAL APART IN ONE DIRECTION

TENSILE STRENGTH = FORCE / AREA AT FRACTURE

STRAIN = ORIGINAL LENGTH - EXTENDED LENGTH / ORIGINAL LENGTH

* RESULTS-OVER PAGE

1- FIND AREA
2- LOAD SPECIMEN
3- RISE RAM UNTIL SPECIMEN FITS UNDER
4- FIT INTO HOLDER
5- ZERO RULER + FIND LENGTH(ORIGINAL)
6- RECORD FORCE AT VARIOUS LENGTHS AS THE SPECIMEN IS STRETCHED
7- RECORD FORCE + LENGTH AT FRACTURE

SHEAR TESTING

IT GIVES AN IDEA AS TO HOW THE INTERNAL ARRANGEMENT OF THE POLYMERIC MOLECULES ARE MADE UP + THE FORCES BETWEEN THE MOLECULES

* RESULTS-OVER PAGE

1- LOAD SPECIMEN
2- SCREW INTO POSITION
3- LOWER SHEAR PUNCH
4- CONTINUE LOWERING + RECORD PRESSURE UNTIL SHEARED

CROSS-BREAK TESTING

IT PUTS THE SPECIMEN UNDER 2 DIFFERENT FORCES AT THE SAME TIME. THE TOP SURFACE IS UNDER COMPRESSION WHILST THE BOTTOM IS UNDER TENSION.

* RESULTS-OVER PAGE

1- PLACE SPECIMEN ON SUPPORT LEGS
2- LOWER ANVIL
3- CONTINUE TO LOWER ANVIL INTO SPECIMEN UNTIL IT BREAKS
4- RECORD PRESSURE

Calculati

Battery = 9V 1.3Ah

Energy rating = current x time

Flashing blue = 6.5V 0.3A 1.95W

Length can run for

Time = energy rating / current
= 1.3 / 0.3
= *4 hours 20 minutes*

White light = 9V 0.5A 4.5W

Length can run for

Time = energy rating / current
= 1.3 / 0.5
= *2 hours 36 minutes*

Green light = 6.5V 0.3A 1.95W

Length can run for

Time = energy rating / current
= 1.3 / 0.3
= *4 hours 20 minutes*

Red light = 35V 0.8A 6W

Length can run for

Time = energy rating / current
= 1.3 / 0.8
= *1 hours 40 minutes*

000	1500	2000	2500	3000	3500	4000
00	100	100	100	100	100	100
0	15	20	25	30	35	*40*

extended length / original length

= *0.3*

Shear Strength Pressure = *38Pa*

Cross-break Testing Pressure = *25Pa*

Nylon

Tensile Strength

Force	2000	3000	4000	5000	6000	6500	7000	7500	7800
Area	100	100	100	100	100	100	100	100	100
T. S	20	30	40	50	60	65	70	75	*78*

Strain = original length – extended length / original length
= 100 – 10 / 100
= *0.9*

Shear Strength Pressure = *82Pa*

Cross-break Testing Pressure = *59Pa*

Perspex

Tensile Strength

Force	500	1000	1500	2000	2500	3000	3500	4000	4300
Area	100	100	100	100	100	100	100	100	100
T. S	5	10	15	20	25	30	35	40	*43*

Strain = original length – extended length / original length
= 100 – 30 / 100
= *0.7*

Shear Strength Pressure = *43Pa*

Cross-break Testing Pressure = *29Pa*

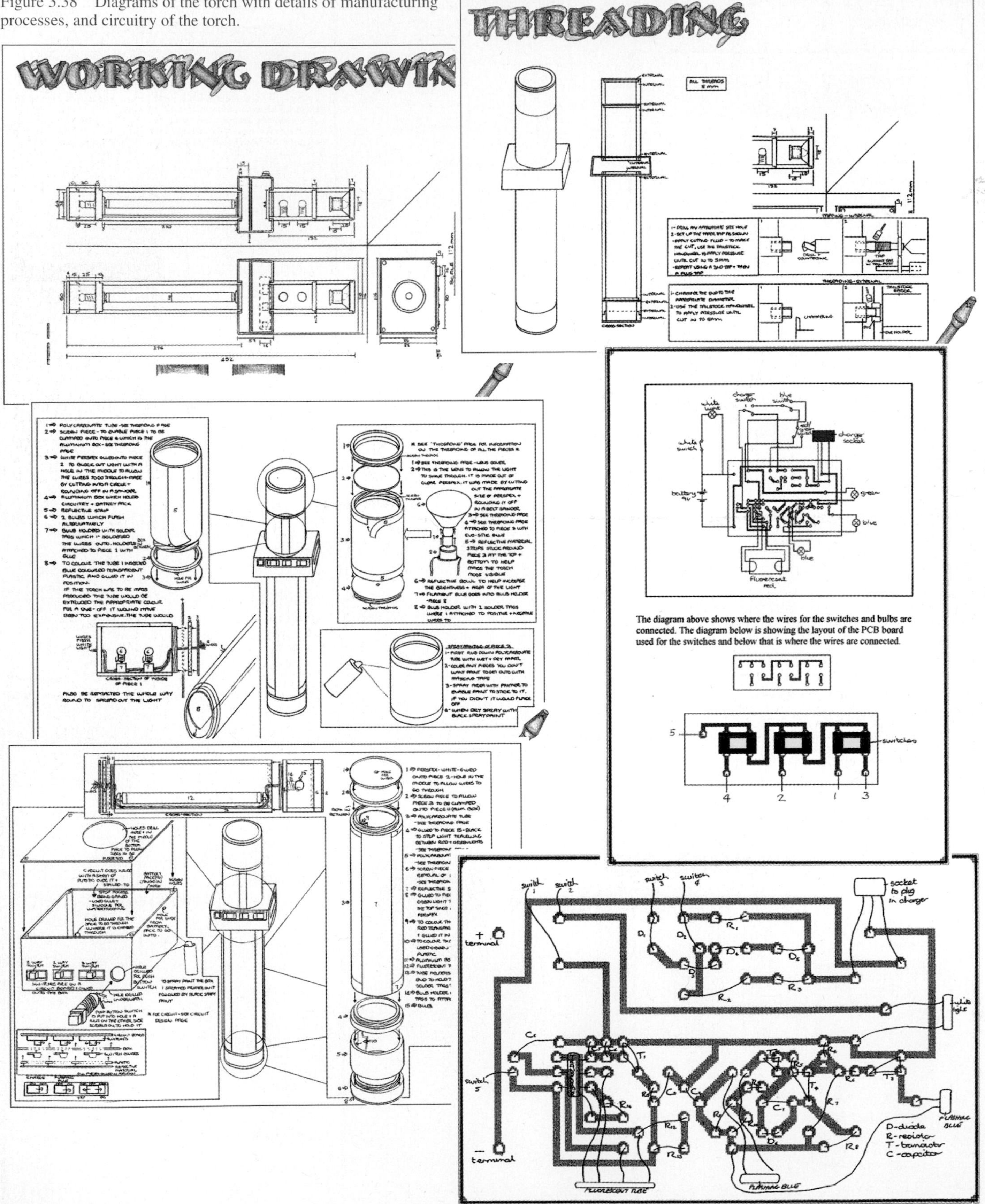

Figure 3.38 Diagrams of the torch with details of manufacturing processes, and circuitry of the torch.

Darren critically evaluated his final design; what its weaknesses were, how he could have done it better and what further development he would do.

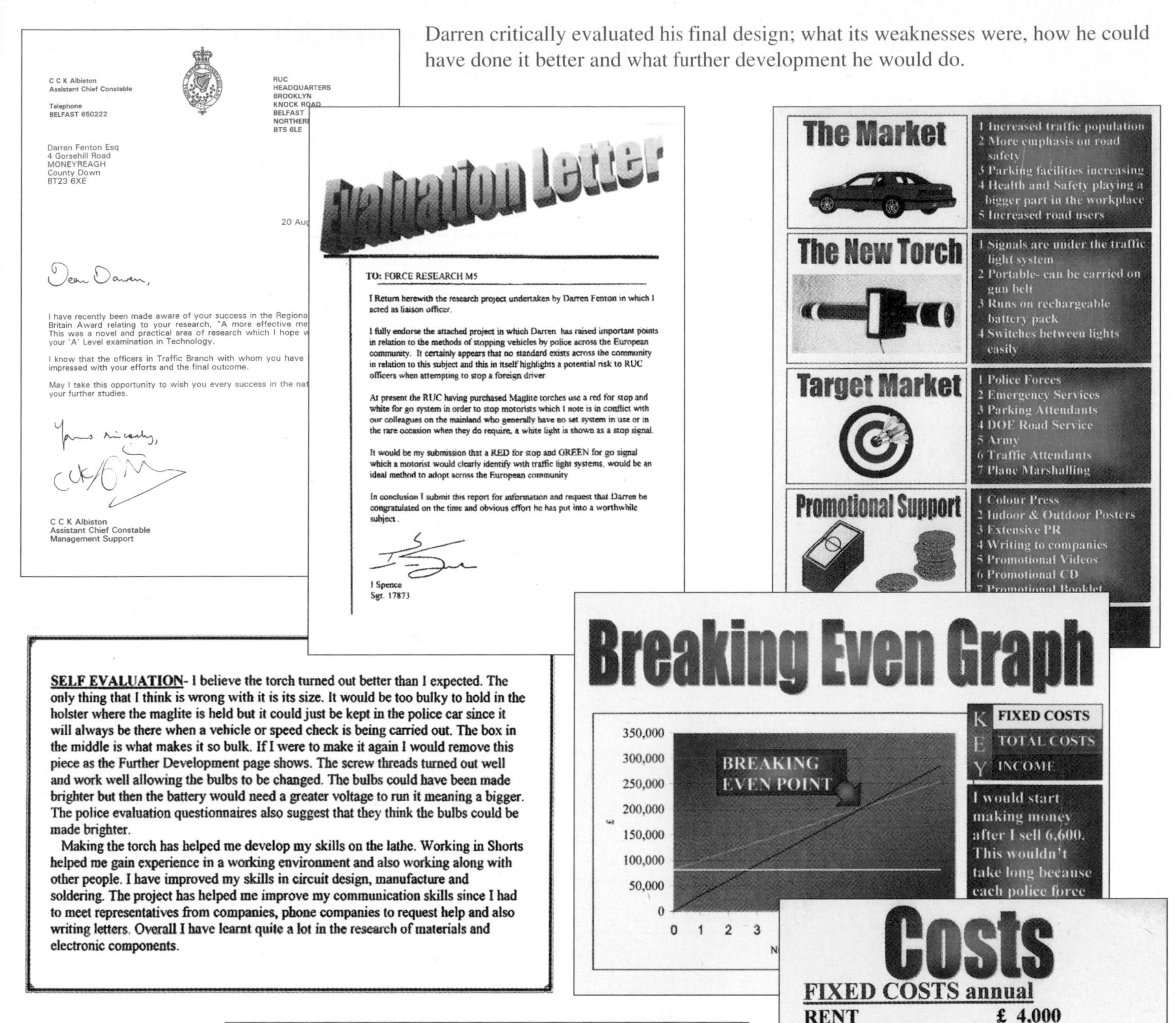

C C K Albiston
Assistant Chief Constable

Telephone
BELFAST 650222

RUC
HEADQUARTERS
BROOKLYN
KNOCK ROAD
BELFAST
NORTHER
BT5 6LE

Darren Fenton Esq
4 Gorsehill Road
MONEYREAGH
County Down
BT23 6XE

20 Au

Dear Darren,

I have recently been made aware of your success in the Regiona Britain Award relating to your research, "A more effective me This was a novel and practical area of research which I hope w your 'A' Level examination in Technology.

I know that the officers in Traffic Branch with whom you have impressed with your efforts and the final outcome.

May I take this opportunity to wish you every success in the nat your further studies.

Yours sincerely,

C C K Albiston
Assistant Chief Constable
Management Support

Evaluation Letter

TO: FORCE RESEARCH M5

I Return herewith the research project undertaken by Darren Fenton in which I acted as liaison officer.

I fully endorse the attached project in which Darren has raised important points in relation to the methods of stopping vehicles by police across the European community. It certainly appears that no standard exists across the community in relation to this subject and this in itself highlights a potential risk to RUC officers when attempting to stop a foreign driver

At present the RUC having purchased Maglite torches use a red for stop and white for go system in order to stop motorists which I note is in conflict with our colleagues on the mainland who generally have no set system in use or in the rare occasion when they do require, a white light is shown as a stop signal.

It would be my submission that a RED for stop and GREEN for go signal which a motorist would clearly identify with traffic light systems, would be an ideal method to adopt across the European community

In conclusion I submit this report for information and request that Darren be congratulated on the time and obvious effort he has put into a worthwhile subject.

I Spence
Sgt. 17873

SELF EVALUATION- I believe the torch turned out better than I expected. The only thing that I think is wrong with it is its size. It would be too bulky to hold in the holster where the maglite is held but it could just be kept in the police car since it will always be there when a vehicle or speed check is being carried out. The box in the middle is what makes it so bulk. If I were to make it again I would remove this piece as the Further Development page shows. The screw threads turned out well and work well allowing the bulbs to be changed. The bulbs could have been made brighter but then the battery would need a greater voltage to run it meaning a bigger. The police evaluation questionnaires also suggest that they think the bulbs could be made brighter.

Making the torch has helped me develop my skills on the lathe. Working in Shorts helped me gain experience in a working environment and also working along with other people. I have improved my skills in circuit design, manufacture and soldering. The project has helped me improve my communication skills since I had to meet representatives from companies, phone companies to request help and also writing letters. Overall I have learnt quite a lot in the research of materials and electronic components.

Costs

FIXED COSTS annual	
RENT	£ 4,000
ELECTRICITY	£ 3,000
INSURANCE	£ 3,000
WAGES (5 PEOPLE)	£65,000
TELEPHONE	£ 2,000
ADVERTISING	£ 3,000
TOTAL	£80,000

VARIABLE COSTS	
MATERIAL	£10.00
CIRCUITRY	£ 3.00
COMPONENTS	£ 5.00
TOTAL	£18.00

SELL FOR £29.95

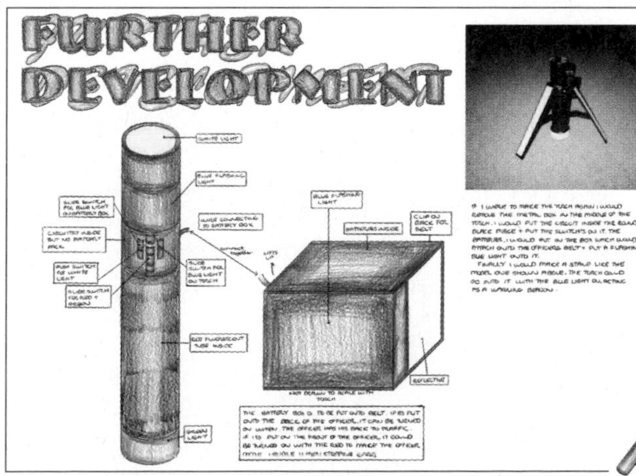

Figure 3.39 The final illustrations show factors to be considered if the torch were to go into production.

3.6 Tips for improving project grades

A message from the Principal Project Moderator OCR, DT (Technology)

It is important to realise that this report can only look back on the existing syllabus entries. The new examination 'specifications' for examination from 2002 have just been written and have not, at the time of writing, been entered by any candidates. A new specification for two 'project modules', which mirror a design process, are introduced in this chapter, in section 3.1. The text relating to candidates' projects used these titles where possible, but candidates were following the 1998/99 syllabus and the overlap was only approximate.

It is vital therefore that candidates use this guidance as an accumulation of past experience and reference it where relevant to the exact requirements of the new specifications and, in particular, the specific information individual Examination Boards often provide as guidance notes. Comments will be made under headings which combine a design process with an examination specification. Candidates work will be referenced.

3.6.1 User, brief, investigation, specification – recognition and investigation of design opportunities

(Candidates' work shown used these headings: introduction description, specification and analysis)

Good projects will have:

- Context-led introduction with supporting photographs.
  - An excellent example of this is the children's art table project, which has clear photographs of children involved with artistic activity.
- A clear description of the project – this can be re-written after completion and inserted into this section to give a clear idea to the moderator of project intentions. (Moderators report that they are often five to six pages into a project before it becomes clear what it is the candidate is engaged upon.)
- A specification which can be cross-referenced during evaluation.
  - The ski speedometer shows this dual purpose well.
- The discriminating factor for higher grades is the extent to which the specification points are analysed in depth.
  - The police torch shows a specification with part factor analysis.
- More than one method of research/investigation used – context visits, library, interviews combined with writing letters, completing questionnaires and surfing the internet. (This should be used with caution, in a highly selective way and only when directly relevant to the topic.)
  - The stretcher project involves a visit to specialist.
- Research material from a wide variety of sources used, which is annotated, and cross-referenced.
  - The bike brake project selects and annotates trade literature.
- The discriminator for the higher grades is to show that information received has actually been **used** and **analyse**d. A conclusion, which feeds the next section, containing a summary and individual reflections, is also to be encouraged.

Features of poor projects:

- These usually contain a very underdeveloped brief with poor specification.
- Letters and or questionnaires are the only research method used, usually with a percentage return rate graph and absolutely no attempt to analyse content.
- Internet material is reproduced in whole sections with no attempt to analyse it.
- Scanned text is used without acknowledgement – a very dangerous practice and also illegal!

3.6.2 Generating ideas, synthesis – generating design ideas, synthesis/development modelling

There is a tendency to include only the very best of design ideas – at its most refined this reduces to 'three ideas from which one is selected'. It is still possible to gain good marks in this section if the ideas are good, there is a progression of ideas and reasons for selection and rejection are given. However, a progression of styles is to be encouraged, i.e. do put in the rough sketches!

Good projects will have:

- Alternative design ideas with details for selection and rejection.
  - A good example of the 'formal approach' is given in the fire escape project where reasons for selection and rejection are given. When generating ideas a more realistic approach in the early stages is given under 3.4.2 where good 'bike brake' sketches are shown.
- Modelling – testing analysis development of prototypes with detailed photographic evidence.
  - Excellent examples of this are shown in the children's art table, and police torch. Outstanding use of a progression of modelling techniques is illustrated under section 3.4.3 in the bike brake project.
- Where electronics is involved in the solution to a problem, prototyping and modelling also form a vital part of project development as do detailed circuit diagrams and details of circuit fixing within a design environment.
  - The ski-speedometer project addresses all of these issues.

Features of poor projects – the lowest marks in this section of work are nearly all awarded to projects involving electronics:

- Single solutions are proposed, in some cases a single standard circuit.
- Alternative designs for 'electronics' are confined to boxes which contain circuits.

3.6.3 Manufacture – planning and making (quality of work submitted)

As the photographs in this chapter show, many candidates produce work of outstanding quality and are a credit to themselves, their centres and the subject. The requirements for excellence in design planning and construction is generally well understood.

Good projects will have:

- An assembled product using good or outstanding making skills.
- A good or outstanding product involving electronics will usually have loomed wires, fixing pillars, grommets, heat shrink sleeving, cable ties.
- Electronics elements incorporated within a product design context, with care and attention given to circuit fixing within a designed environment.
- Photographs of finished products, which are in focus and in detail.

Features of poor projects:

- Mechanical projects are usually unfinished, with unrelated separate parts.
- Involving electronics includes those housed in bought plastic boxes – dry joints, damaged tracks, bent leads, unshielded connections and unloomed wires are also a feature. Mains work which is not isolated and does not have the correct health and safety scrutiny cannot be marked in the higher categories. (Avoid mains if possible.)

3.6.4 Evaluation – testing evaluation

Good evaluations:

- Include detailed testing carried out in context – art table, ski-speedo, torch, stretcher.
- Specifications used to cross reference expectations and conclusions provided on the success of meeting objectives – the ski-speedometer shows this well.
- Third party evaluations provided to support personal evaluation – torch, table, industry.
- Suggested modifications by both designer and end user detailed – torch, art table.
- Costing provided with implications for volume production – the torch, bike brake.
- Personal evaluation provided as a summary and conclusion – emergency burns stretcher, ski speedometer.

Poor evaluations:

- Typically only contain generalities, running out of time, apologies.

The concluding comment for this whole chapter is the most important.

Time planning should reflect the relative apportionment of specification marks, (e.g. planning making = 70; evaluation =20). Spend 10 hours evaluating for every 35 hours making. Book this time by giving clients and critics a copy of your initial time plan.

Do this even if you have not quite finished the practical work. It could gain you a grade!

SECTION 2 Materials

It is significant that we divide the early history of the human race into eras that are named after the materials that were predominantly being used. The Stone Age, Copper Age, Bronze Age and Iron Age all reveal how important materials are to us. Table 4.1 shows the relationship of these eras and their timespans. It also shows that the discovery and development of materials has accelerated as time has progressed. We now live in the age of 'synthetic materials', in which materials are designed and developed to perform particular functions, for example polymer/fibre composites are replacing metal alloys in aeronautical applications.

Chapter 4 introduces the evolution and classification of materials in a conventional way, but there may soon come a time when this way of thinking about materials becomes obsolete and old-fashioned. We are moving towards the age of 'smart' materials. In a recent book discussing the development of smart materials and structures M. V. Gandhi and B. S. Thompson comment as follows:

> When this technology has been perfected, the materials scientist will be able to synthesise, design and create three-dimensional atomic arrangements which will render obsolete the categorisation of materials into such groups as insulators, metals, polymeric materials and biomaterials, for example.[1]

They also say:

> The quantum jump in materials technology seems certain to revolutionise the future in ways far more dramatic than the way the electronic chip has catalysed the evolution of our life-styles.[2]

If a fourth edition of this book appears, it may be necessary to reconsider the entire approach taken to materials technology. However, we have not got there yet.

- Chapter 4 provides an introduction to the evolution of materials and the categories currently used. It also explains the ways in which materials can be examined and something about their structures. A section on smart materials has been included in order to indicate the scope of this coming technology.
- Chapter 5 introduces some physical and mechanical properties of materials and describes ways in which these can be measured by materials testing. It also introduces materials selection charts, which is an approach developed at the Cambridge University Engineering Design Centre. This section was written by Hugh Shercliff and Andrew Lovatt of Cambridge University and they have also contributed to the improved treatment of 'toughness and brittleness'.
- Chapter 6 gives an overview of the way that materials can be processed by forming, casting, wasting and fabricating, as well as discussing corrosion and finishing techniques. The final section discusses computer-aided manufacture (CAM).
- Chapter 7 looks at the broad issues of material and process selection and, in doing so, demonstrates how all of the matters covered in this section are brought together when designers select materials and processes to meet the required design performance requirements. Hugh Shercliff and Andrew Lovatt have made a substantial contribution to this chapter, in particular by writing the sections concerning material selection charts.

Whatever else you learn concerning materials and processes from reading this book you must realise the need to keep up to date. The articles on page 172 show two recent innovations, from an issue of the *Sunday Times*.

Tents could be alternative to bricks and mortar

FABRICS

TENTS could provide a feasible alternative to conventional housing thanks to a new fabric, *writes Roger Dobson.*

The Teflon-coated coloured fabric is soundproof and can last indefinitely, according to its makers. The specially tensioned fabric is being used in new buildings around the world, including a Malaysian railway station, a Chilean shopping centre and an American theatre.

A specially designed cloth has also been developed for the Skyscape concert centre on the Millennium Dome site. The dome tent will also have a quilt lining designed to absorb the sound from pop concerts and audiences of more than 5,000 that will use the building after its opening at the end of the year.

Fabric has traditionally been associated with tents, awnings and temporary shelters, but rarely with permanent structures. However, the new fabric, designed by Landrell Fabric Engineering in South Wales, could now be used instead of bricks and mortar for permanent structures.

Bill Brown, managing director, says: "The new tents are also extremely solid. Unlike traditional tented structures these new buildings are semi-rigid with tensions of about four to six tons per square metre."

Fabric buildings can also be made from semi-transparent material, allowing natural light to shine in. Unlike traditional buildings, they can be dismantled easily.

Brown says: "Using fabric also means you can get a wide variety of shapes in a building that you could not achieve with brick and steel and traditional materials. Architects and designers love to use it because they can express themselves quite differently and quite dramatically."

Blade redesign will make helicopters safer

AEROSPACE

HELICOPTER blades that can change their shape have the potential to make the aircraft faster and safer, *writes Mark Prigg.*

Helicopters have a complex rotor assembly – to which the blades are attached – that pilots control to tilt the blades during flight. This alters the airflow over them so that the aircraft can manoeuvre.

However, researchers at the Massachusetts Institute of Technology (MIT) in Boston have created a prototype smart rotor with blades that are built from a shape-changing ceramic material.

The new rotor/blade assembly is far lighter because it has no moving parts and also far safer, its inventors claim.

Yet-Ming Chiang, MIT's Kyocera professor of ceramic engineering, says that the prototype blades would be grown in a laboratory rather than being manufactured.

"The principle behind the rotors is fairly simple. The rotor itself is lined with piezo-electric fibres, which expand when an electric current is applied," he says.

"We build the rotor and place these fibres at an angle of 45 degrees. If we then apply a current to the fibres, they expand and effectively twist the rotor and control the pitch of the helicopter."

Thousands of the electric fibres would be used and placed into a fibre-based material such as fibreglass. Each is attached to an electrode that can trigger the change, with the electrodes controlled by the helicopter's onboard computer system.

During manufacture each fibre is grown at a rate of one inch per hour. It can change its shape by up to 1%, which Chiang claims is more than enough to alter the airflow over the blades.

"This doesn't sound like much, but in a spinning rotor with thousands of these fibres twisting the blade, 1% can make a lot of difference," Chiang says.

Funding for the blades has come from the Defence Advanced Research Projects Agency (Darpa), which hopes to use the material in fighter helicopters.

Chiang says: "We are still at the early stages, so we are not entirely sure about the performance enhancements. However, we have already shown our rotor to be far lighter, which obviously increases the speed a helicopter can travel and makes it more efficient to run. The lack of complicated control mechanisms also makes this far safer if the helicopter were to be hit by gunfire, as there is less that can be damaged with our rotor."

His team works with MIT's aeronautical department, which is now building a working prototype of the smart rotor/blade assembly.

The American military wants to be one of the first users of the blades and Chiang hopes the new fibres will be used in thousands of other products.

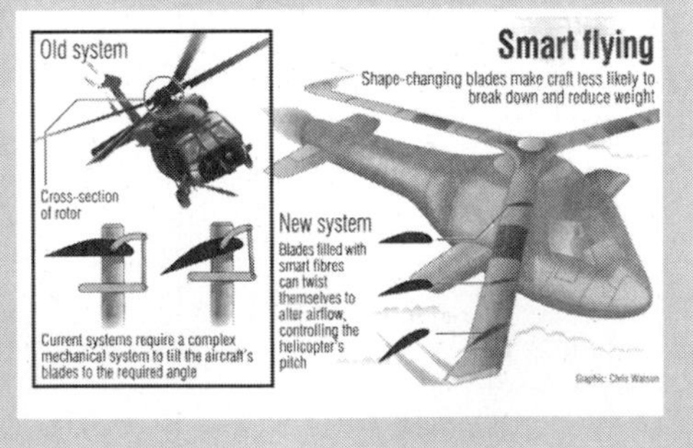

Technological innovations in the newspapers.

Source: *The Sunday Times*, 15 August 1999.

4 Materials

Poems are made by fools like me
But only God can make a tree
from Trees *by A J Kilmer (1886–1918)*

and parodied by Professor Charles Gurney as

Plastics are made by fools like me
But only God can make a tree

It has been necessary to include references to different material properties such as hardness, toughness and ductility in this chapter. If these terms are unfamiliar you will find explanations in Chapter 5.

4.1 The evolution of materials

The evolution of materials is discussed in three sections:

- a historical perspective concerning the period up to the late twentieth century
- a discussion of smart materials
- other materials developments.

4.1.1 A historical perspective

In the Old Stone Age, Fig. 4.1, humans could only use the materials they found at hand, such as stone, reeds, wood, clay, animal hide, hair and bones. To counteract their lack of speed and strength they fashioned tools and developed skills to enable

Figure 4.1 The importance of organic materials in the Stone Age.

Table 4.1 History of material development

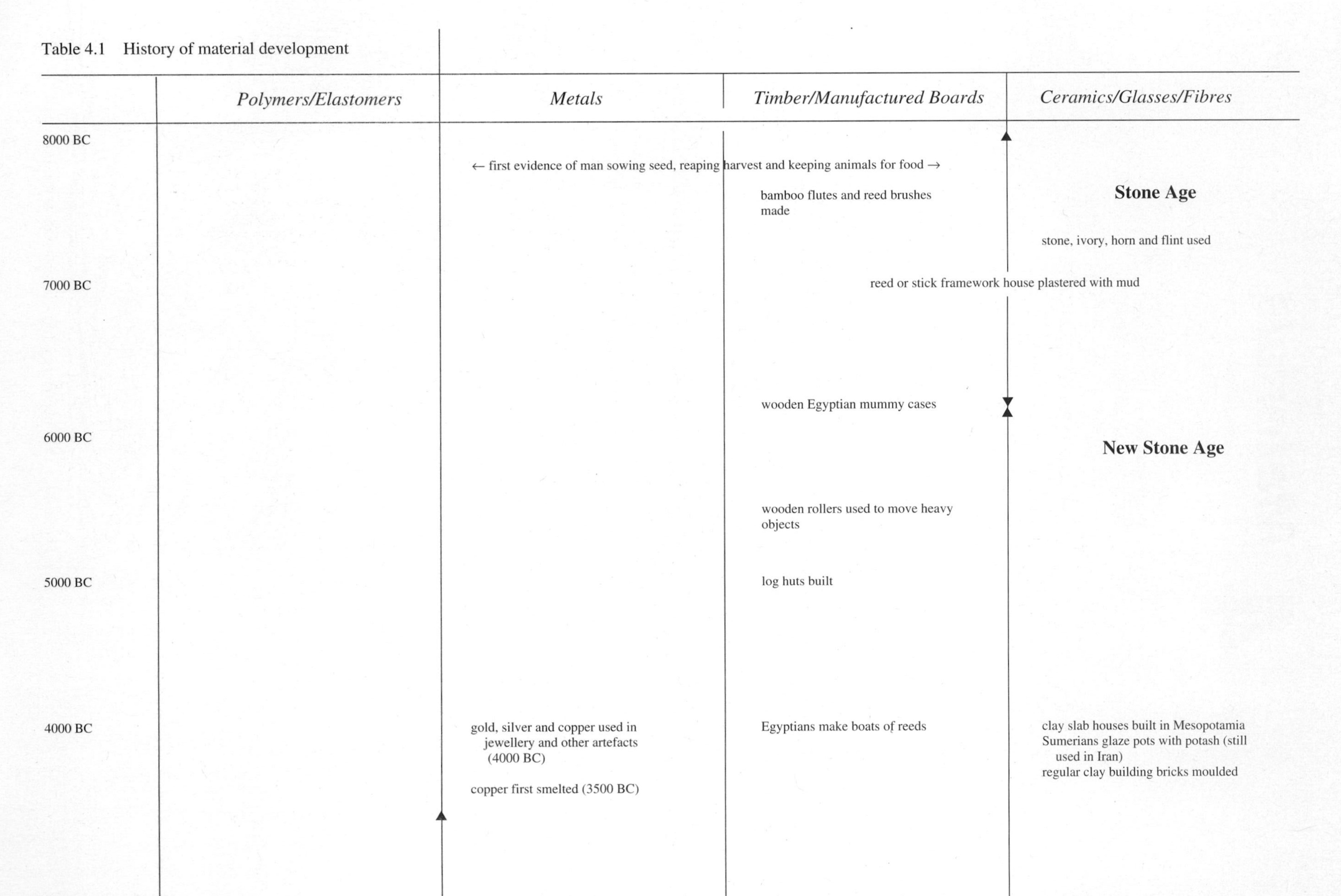

	Polymers/Elastomers	Metals	Timber/Manufactured Boards	Ceramics/Glasses/Fibres
8000 BC		← first evidence of man sowing seed, reaping harvest and keeping animals for food →		
			bamboo flutes and reed brushes made	**Stone Age**
				stone, ivory, horn and flint used
7000 BC			reed or stick framework house plastered with mud	
			wooden Egyptian mummy cases	
6000 BC				**New Stone Age**
			wooden rollers used to move heavy objects	
5000 BC			log huts built	
4000 BC		gold, silver and copper used in jewellery and other artefacts (4000 BC) copper first smelted (3500 BC)	Egyptians make boats of reeds	clay slab houses built in Mesopotamia Sumerians glaze pots with potash (still used in Iran) regular clay building bricks moulded

3000 BC		**Copper Age**	wooden boats built in Egypt wooden wheels and axles made	kiln firing of pots in Mesopotamia
			veneer, marquetry and inlay in Egypt	
2000 BC	bitumen used in Mesopotamia as house asphalt, mortar, drain lining		mortise and tenon joints, dowel and glue in Egypt	Stonehenge started
		Bronze Age		Stonehenge finished
		Iron obtained from haematite ore (1400 BC)	metal and wooden wheel rotating smoothly on fixed axis	
1000 BC		steel daggers, swords and jewellery forged in Mesopotamia		
	shellac resin used as dye and varnish		first recorded wooden watermill	Building bricks glazed – decorative and waterproof
			Chinese furniture first produced lathe turned chair and table legs in Roman wickerwork chairs	
0		**Iron Age**		
1000 AD			wooden furniture more common in Europe wood carving important in Europe (1400) European wood turning developed and English marquetry (1650) veneered articles introduced in Europe (1750) laminated woods used in structures (1840) plywood and hardboard (1909–35) laminates and composites MDF and new fibreboards (1979)	
2000 AD	first natural-based plastic developed (1862) first synthetic plastic (1907) then polystyrene, acrylic, PVC, nylon, polyethylene space age plastics e.g. polyamides used in place of metals	iron castings made (1500) large-scale iron and zinc production aluminium extracted (1845) stainless steel perfected (1918) high-strength steels (1930s) space age alloys developed		nonshatter glass (1905) prestressed concrete (1922) ceramic magnets, cookware and cutting tools (1946–70) carbon fibre developed (1963) glass-reinforced cement (1971)

them to survive. The skills required to work each material with limited tools and resources, particularly energy, became more and more refined and many of the techniques used today can be traced back to ancient times. Four thousand years ago the early craftsmen could forge, drill, saw, engrave, stamp, solder and cast in gold.

During the earliest periods, organic materials were the most important. Fur, leather and wool were used for clothing, and wood and reeds were used for frameworks, tools, and hunting weapons. Musical notes could be generated from bamboo, and paintings could be created using natural pigments and brushes made from reeds. Wood has continued to be an important material to this day. It has historically been a readily available material and, with long-term planning, can be a continuing resource. People discovered that they could use wood to build homes for their comfort, boats and carts for transport, and windmills and watermills to provide power.

In the Middle Ages wooden furniture, often with ornate carving, became more common. By 1650, inlay and marquetry could be combined with carving and turning to produce the elegant furniture so typical of the seventeenth and eighteenth centuries. Fig. 4.2 shows a mid-eighteenth-century turned chair and Fig. 4.3 shows a French lacquered commode of the same period with intricate inlay and a marble top. Craft workers have continued to extend their skills with wood and are now designing with laminated woods and fibreboards as well as the traditional timbers.

Human power was greatly increased by the discovery that mixing and heating materials changed their nature. Ceramics, the earliest inorganic materials to be worked, showed clearly how the properties of materials depend on their internal structure. A pliable clay could be moulded into the required shape and heated to create a solid, stable earthenware article because the microstructure of the clay had been altered in the heating process.

Metals have been the most important materials so far in the advancement of technology. Their strength and ease of working have been of vital importance. The earliest use of metal followed soon after clay firing became fairly widespread. The forming of ceramics and metals both began in the Middle East in the regions we now

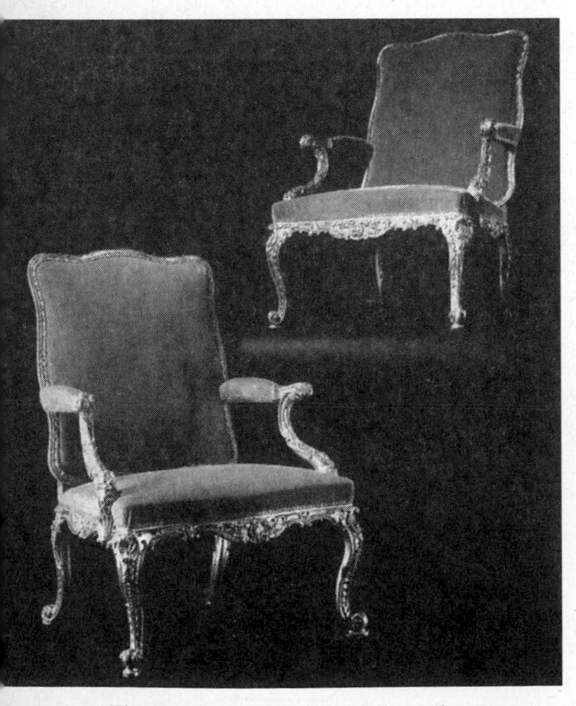

Figure 4.2 English mid-eighteenth-century turned-walnut chairs.

Figure 4.3 A mid-eighteenth-century French commode with intricate inlay and a marble top.

Figure 4.4 A gold pendant from the Aegina treasure, circa 600 BC.

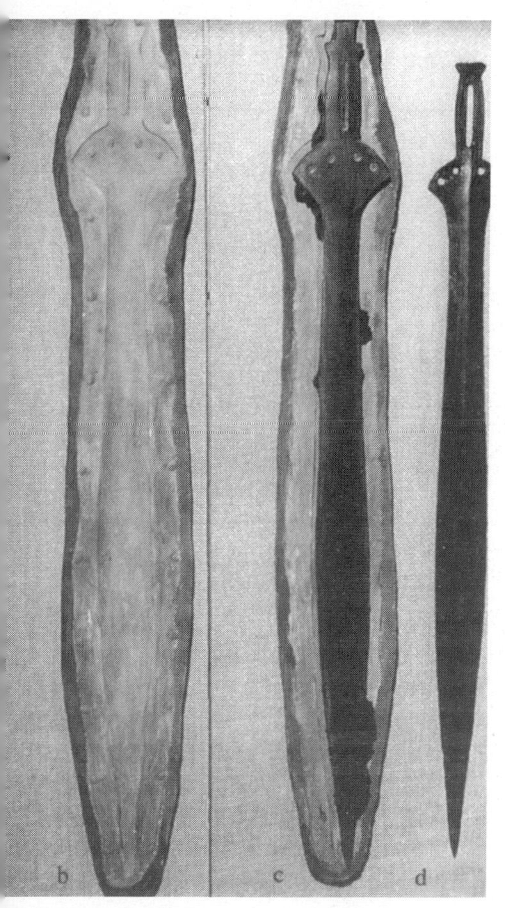

Figure 4.5 The stages in casting a bronze sword
(a) wooden prototype made
(b) clay mould made from prototype
(c) bronze casting
(d) finished sword.

know as Iran, Iraq and Turkey. The people there had hammered and cut local copper and gold to make jewellery and decorations since about 8000 BC. Being soft, these metals could be shaped easily, but they occurred in such small quantities that they were reserved for small, decorative items of high value.

By 4000 BC, the people of the Middle East were able to smelt copper from ore, and it became a plentiful and important material. By smelting ores that contained impurities the first copper alloys were created. Later it was found that arsenic and antimony additives created a harder metal than pure copper, hence better castings could be formed. At this time the techniques and artistry involved in making gold and silver artefacts were very well developed and the same methods continued to be used for many years. Fig. 4.4 shows a gold pendant made in about 600 BC on the Greek island of Aegina.

About 2000 BC, and in the same part of the world, it was found that another soft metal, tin, when added to copper created an alloy of an attractive colour. The alloy was easy to form and, since it was harder than its component metals, more practical hardware could be produced. This was the birth of the Bronze Age. The molten metals could be mixed in appropriate proportions to produce the correct composition of alloy for the different purposes for which it was to be used, from weapons to vessels and from statues to jewellery. Fig. 4.5 shows the stages in casting a bronze sword from the fashioned wooden prototype to the finished product.

The extraction of iron from haematite in about 1200 BC signalled the start of a new age in materials which, mainly for economic reasons, has dominated for 3000 years. Iron in its pure form was inferior to bronze in every way, but the Middle Eastern people found that by heating iron weapons in charcoal and hammering them a tougher metal was produced – carbon had been absorbed to form steel. They also found that plunging the metal into cold water produced a hard material superior to other metals. Today these processes are referred to as *hardening and quenching* the metal. It was also found that reheating and cooling the metal slowly, now known as *tempering*, produced a less hard, but tougher and less brittle material. These processes of quenching and tempering are still important techniques. Thus, by 1200 BC, all the basic metallurgical techniques had been discovered.

The last three centuries have seen tremendous innovations in the discovery and development of materials. From 1700 onwards, and especially in the last two hundred years, the complexity and variety of materials developed have grown enormously. Table 4.2 expands this later stage to show this period of development more clearly.

Many new commercial processes for extracting and producing metals and alloys were introduced at the beginning of this period. Although metals have been known and worked for over 5000 years, it is only in the last hundred years that a systematic examination of their microstructure has revealed a detailed picture of their structure-dependent properties and how these can be modified. As a result, new alloys have been produced with different properties and exciting applications.

Technical progress is dependent on finding suitable materials to translate innovative ideas into reality. In 1832, Charles Babbage designed the first automatic calculator with a stored program, the forerunner of the modern computer. However, the available technology was limited to mechanical systems and this proved far too clumsy for implementing his elaborate ideas. Despite spending his own fortune and a substantial government grant he was unable to complete the 'calculating engine' and he died a disappointed man. It was not until a hundred years later, with the advent of electronic valves, that his ideas could become a reality.

Similarly, Sir Frank Whittle, in producing his design for the jet engine in 1930, found that the materials necessary to withstand the intense heat and strain were not yet available. It is noteworthy that he first presented his ideas as a student project. His design was considered impractical and he received no support from government or industry but, within seven years, suitable alloys of nickel and chromium, able to

Table 4.2 Development of materials since AD 1700

	Polymers/Elastomers	*Metals*	*Timber/Manufactured Boards*	*Ceramics/Glass/Fibres*
1710		large-scale iron production – first coke fuelled blast furnace (1709)		
		zinc smelted commercially (1740)		
1750		nickel first isolated (1750)	wood predominant in manufacture (1750) veneered wooden furniture in Europe (1750s) rococo furniture at its height	
		cast iron used to build Ironbridge (1777) wrought iron puddling process patented (1784)		
1800		industrial revolution (1798–1832) wood and metal machines introduced to facilitate production		
		first hot blast furnace (1828)		first spectacles made from naturally occurring crystalline quartz
		first metal telescope mirrors of speculum, Cu_4Sn	laminated structure for viaduct (1837)	
	polystyrene discovered (1839)	aluminium extracted from bauxite (1845)	Kirkpatrick's wooden bicycle (1839)	

1850	Parkesine (celluloid) developed (1862) further casein and cellulose acetate plastics cellulose nitrate (1868)	Bessemer process for steel invented (1856)		
		Aluminium produced commercially (1886)		triplet lens, original photographic lens (1894)
1900	elastic rubber replaces whalebone in corsets (1905) first synthetic plastic Bakelite (1907) plastic adhesives developed urea formaldehyde and first amino plastic (1924) corrosion-preventing antioxidant addition to rubber (1924) PVC and first acrylic polymer, polymethylacrylate (1927) latex foam; neophrene, butyl and styrene synthetic rubbers (1930–33) acrylics, Perspex, polystyrene, PFTE, polyurethane, polythene, nylon developed (1930s) Melamine formaldehyde utensils introduced (1939) foam glass insulation blocks (1940) polyesters, polyethylene, silicones and fluorcarbons (1942/3) plasticised PVC for electrical insulation (1943) Teflon produced commercially (1948) first acrylic fibre garments made (1950)	first metal alloy unaffected by heat, Invar – 66% steel (1904) high strength aluminium alloy, Duralumin (1909) stainless steel perfected and used for knives (1913) Stellite, cobalt alloy, used for cutting tools (1915) stainless steel 18:8 Staybrite (1930) nickel-chromium super alloys developed used in high temperature, stress situations (1931–40) Lurex, aluminium-based yarn (1946) new strong, shock resistant cast iron (1947) pure titanium produced commercially (1948)	wooden aircraft – Wright flyer (1900) early luxury cars with timber bodywork laminates and plywood developed (1909) paper manufactured from hardwood (1922) fibreboard ceiling panels produced (c 1925) hardboard manufactured urea formaldehyde glues for timber such as Aerolite 306 (1942) chipboard manufactured development of wood composites	optical glasses made by Schott (1900) nonshatter glass with celluloid, later triplex (1905) wire-reinforced glass (1905) Pyrex glass for cookware and chemical plants (1915) bridge built from prestressed concrete (1922) Perspex commercially produced (1934) first glass fibre (1938) fused quartz-silica for engineering applications (1938) laminated safety glass with polyvinyl butyral interlayer (1938/9) development of silicon chips (1940s) silicones manufactured commercially (1943) ceramic magnets (1946)
1950	Araldite epoxy resin adhesives acetal resin automotive parts (1956) polypropylene industrial mouldings (1957) polycarbonate resin (1957) Lycra, artificial elastic (1958) superpolymer heat-resistant plastic (1961) Marlex plastic used as artificial human tissue (1965) Neoprene, synthetic rubber (1965) new range of 'melded' domestic and engineering fabrics (1973) thermoplastic natural rubber (1978)	friction welding of dissimilar metals (1954) maraging steels, high tensile strength, developed (1961) use of oxygen in steelmaking (early 1960s) beryllium used in aircraft part manufacture (1968) space age metal alloys and surface coatings plastic aluminium developed (1976) expanded aluminium foil (1976) superplastic metal alloys especially titanium alloys for aircraft (1980s) superconducting materials	blockboard manufactured Melamine-faced chipboard work surfaces (c 1955) plywood car made – Marcos (1960) Scandinavian influence in pine Sundealer board for pinboards MDF, medium density fibre board (1979) Medite fibre board for indoor and outdoor use (1987) Caberwood fibre board (1987)	silicon transistors and solar cells developed (1954) aircraft laminated windows incorporating polymeric materials (1955) optical fibres invented (1955) Pyroceram cookware, new crystalline ceramic (1957) carbon fibre developed (1963) ceramic suitable for cooker hob developed (1966) Sialon, high-speed ceramic for cutting metals, developed (1970) glass-reinforced cement invented (1971) strong cheap building ceramic produced from power station waste (1973) gallium arsenide solar cells (1977)
2000				

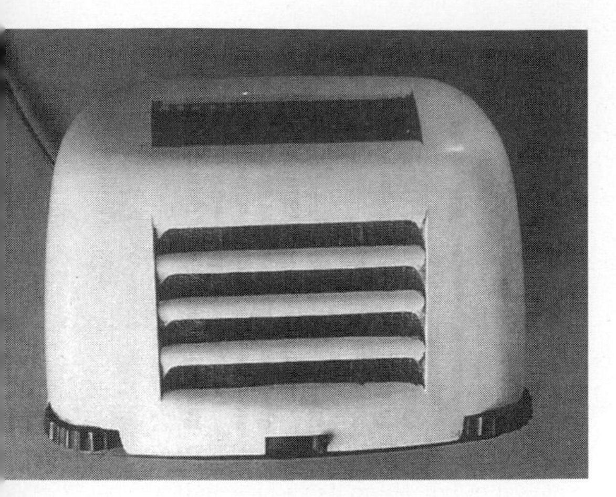

Figure 4.6 An early Bakelite radio.

resist creep and fatigue caused by the centrifugal forces at high temperatures, became available and made the jet engine a practical proposition.

Again, the first patents on tape recording were taken out as early as 1906 but, because of a lack of suitable materials, it was nearly half a century before magnetic tape recorders were available commercially.

Looking at the development of plastics, we find that naturally occurring plastics have been used for many years. Bitumen, a natural mineral pitch, has been used since 2000 BC for housing, irrigation projects and roads. Shellac, a crimson resin produced by the lac insect puncturing certain trees, was used as a dye and varnish in early Egyptian and Roman cultures and latterly for early gramophone records, wood varnish and electrical work. Rosin from pine trees was used as a resin, and a synthetic rosin is now used to give grip to the fibres of stringed instrument bows. Casein is obtained from milk and is still used to make buttons.

In 1862 the first manufactured plastic, initially called Parkesine and later Celluloid, formed from natural materials was developed from cellulose. Other casein and cellulose acetate plastics followed in the next few years. In 1906 the chemist Baekeland began a research programme that led to the first synthetic polymer, Bakelite. It proved superior to natural materials in many electrical applications and, in the 1930s, its advantages in mass production were widely exploited by the radio industry. Fig. 4.6 shows an early Bakelite radio. In the period between the wars new synthetic plastics with different colours and properties were developed. The most important of these were polystyrene, acrylic, Perspex, polyvinyl chloride (PVC), nylon, polypropylene and polyethylene.

Since 1940 there has been an influx of plastics into all aspects of our lives. Polyester, nylon and Terylene are commonly used in clothing and fabrics and are entirely plastic. Other fabrics use the polymers to create drip-dry or crease-resistant material based on cotton. One-size socks and gloves, nonscuff or patent shoes all use various plastics to advantage. Polypropylene, developed in 1935, proved to be an extremely versatile material and is used in such diverse products as buckets, chairs, blankets, clothing, rope and transparent film.

Today the impact of synthetic materials is enormous and some plastics are beginning to replace metals in applications where weight, corrosion resistance, colour and ease of forming are important design criteria. Polycarbonate is a hard, rigid plastic with good impact strength and can be used as a substitute for light metals such as aluminium. Also, polyethylene and PVC are now used to a great extent instead of lead, zinc, cast iron and steel for plumbing pipes and tubing. The medical world has been at the forefront of developments in plastic, and polytetrafluorethene, developed for medical use, has similar properties to human bone. It is compatible with human tissue and so the natural bone growth forms round a plastic implant, making it an ideal new material for hip joint replacements. Plastic resins have also revolutionised the manufacture of adhesives which, in turn, have enabled plywood and laminates to be produced.

Our understanding of materials is constantly growing, so that traditional materials are being improved and new materials are being synthesised. We now have glass with the strength of steel, ceramics that stay tough and rigid when white-hot, and light-conducting fibres that are strong yet flexible. Many of the new composite materials no longer fall into the traditional classifications. For instance, is glass reinforced plastic (GRP) a polymer or ceramic? The ceramic industry is using powder technology, in which metallic powders are packed into a mould and then pressure and heat are applied. Fig. 4.7 shows two powder metallurgy filters. The powder metallurgy process produces a cohesive solid without melting the metal, but is this new material, with all the properties of a metal, actually a metal or is it a ceramic?

Figure 4.7 Powder metallurgy filters.

How can we continue to develop materials for our future benefit? Bearing in mind our depleting natural resources of certain materials and our growing ability to create polymers to meet design criteria, our future would appear to be in sophisticated new reinforced polymers or fibres, and the transfer of this expertise to enable us to improve and extend traditional materials. The exploration of space has led to the development of new materials such as Teflon. We are now beginning to create materials in space itself and, with no gravitational forces, perhaps we shall see a new range of fantastic, if expensive, materials emerging. We should also remember our traditional materials and methods, taking care to ensure the future availability of resources such as timber that we have taken for granted for so long.

4.1.2 Smart materials

In order to be truly regarded as 'smart', a material needs to have one or more of the following features;

- *Sensors* which can be embedded in the material or bonded to its surface and which can detect a change in a key parameter, e.g. strain or temperature.
- *Actuators* that, again, can be embedded in the material or bonded to its surface and which enable the material to respond appropriately.
- *Control capabilities* that determine the appropriate response of the material, e.g. microprocessors.

It is difficult initially to imagine how such materials could exist and so some recent materials developments are described below in order to provide the necessary background.

Shape-memory alloys and plastics

Shape-memory alloys (SMAs) regain their shape if they have been deformed at low temperature and are subsequently reheated. The most common alloy which behaves in this way is the nickel-titanium alloy, Nitinol, named after the Naval Ordnance Laboratory (NOL) in the US where it was developed, i.e. Ni-Ti-NOL. Plastic strains of 6–8% can be recovered by these alloys. However other alloys, e.g. Cu-Zn, Au-Cd, Cu-Sn, Fe-Pt, etc., also exhibit the effect. These materials have been employed in a variety of applications, including simple fasteners, pipe connections, fuses, water sprinklers, space and medical components. Some examples are shown in Fig. 4.8.

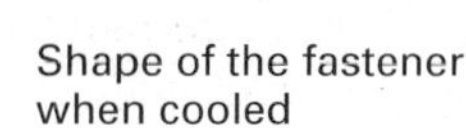

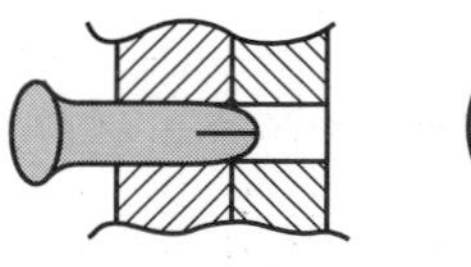

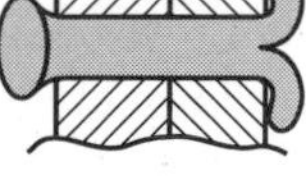

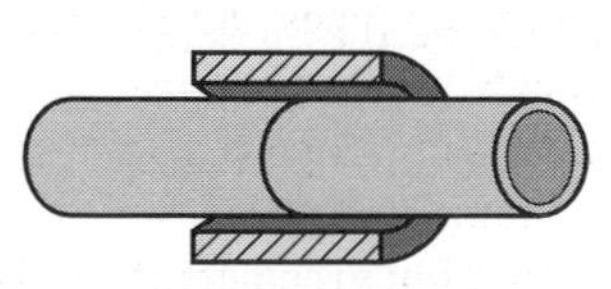

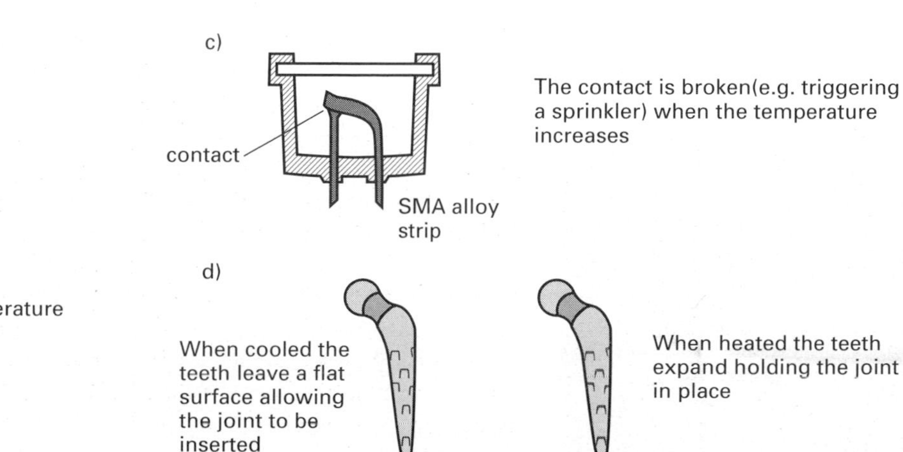

Figure 4.8 Examples of the use of SAMs a) simple fasteners b) couplings c) temperature fuse d) hip joints.

Shape-memory plastics (SMPs) have also been developed, e.g. Zeon Shable (a trade name). This plastic is available in five different grades which feature five different shape recovery temperatures ranging from 40°C to 80°C. The material can be processed by injection moulding, extrusion and blow moulding, for example.

Because they change their shape when heated SMAs and SMPs could be used as actuators in smart materials. SMAs could be heated electrically in order to respond to a detected change in the condition of the smart material. They could also be used as sensors of temperature changes because their shape alters when they are heated. However, SMAs and SMPs take time to heat and this will limit their suitability in some applications because of the time delay. There is also a significant energy requirement and they change their shape at a specific temperature that may not be suitable.

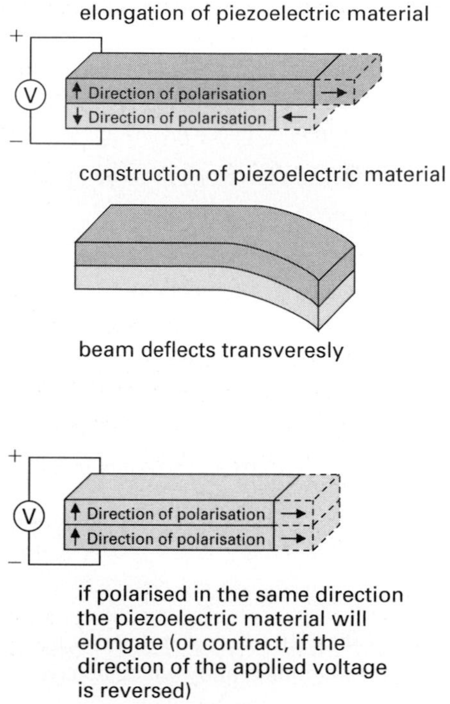

Piezoelectric materials

Piezoelectric materials could also be used as both sensors and actuators in smart materials. The piezoelectric effect refers to the phenomenon discovered by Pierre and Paul-Jacques Curie in 1880, namely the appearance of an electric charge on the surface of some crystals when they are mechanically deformed. A year later they discovered that the effect also worked in reverse, i.e. when an electric field was applied to the crystal it changed its shape. This early work concentrated on crystals of materials such as Rochelle salt and quartz, but there are now modern alternatives. The most common are piezoceramics, such as the lead zirconate titanates (PZT) and the piezopolymers, such as the polyvinylidene flourides (PVDF). Piezoceramics are very brittle materials and piezopolymers are very flexible. This makes them suitable for different applications. As a result of their stiffness, the piezoceramics are very rigid and, consequently, convert electrical energy inputs into mechanical movement very efficiently. This makes piezoceramics very suitable for applications as actuators. Being more flexible, piezopolymers offer little resistance to bending and are therefore very suitable for applications as sensors.

Depending on their direction of polarisation, piezoelectric materials can provide either an axial extension or a beam deflection as shown in Fig. 4.9.

elongation of piezoelectric material
+
V
Direction of polarisation
Direction of polarisation
–
construction of piezoelectric material
beam deflects transveresly
+
V
Direction of polarisation
Direction of polarisation
–
if polarised in the same direction the piezoelectric material will elongate (or contract, if the direction of the applied voltage is reversed)

Figure 4.9 Piezoelectric materials giving an axial extension or beam deflection.

Fibre-optic sensors

The simplest form of fibre-optic sensor is shown in Fig. 4.10. The optical fibres are embedded alongside other structural fibres in composite materials. If a crack in the material develops, then the fibre is broken and prevents light being transmitted. If light signals are sent down the fibres, they will not be received by a detector and thus indicate the failure of the component. It is this technology that can be embedded in the composite structures used in aeroplanes in order to monitor them during service.

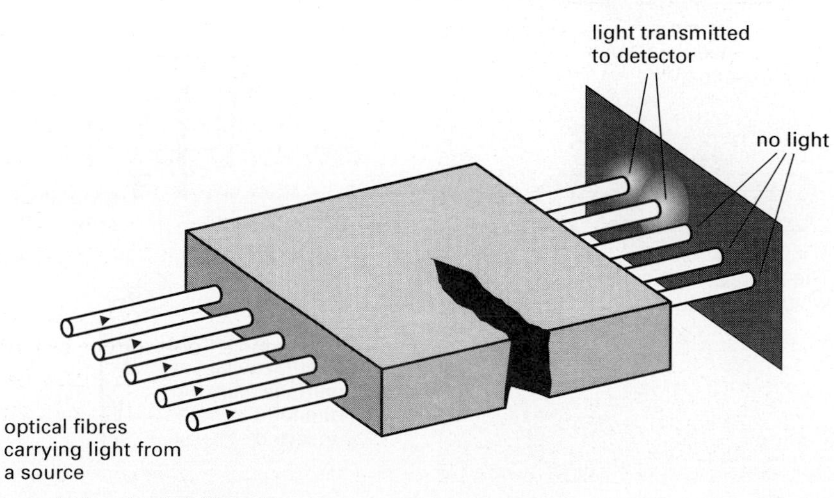

Figure 4.10 Cracked fibres here …

However, researchers have found ways of using optical fibres in much more sophisticated ways. The transmission characteristics of optical fibres can be altered by external stimuli, e.g. strain and heat. These effects have allowed optical fibre systems to be developed that can detect such variables as applied forces and pressure and temperature changes, but it is also possible for optical fibres to detect electric and magnetic fields, X-rays, Gamma rays and changes in density. So, for example, fibre-optic sensors can monitor the moulding and curing of a composite component as well as its performance in service.

Apart from their versatility, optical fibres have many other features that have led to their increased use. They are light and do not generate heat. They do not conduct electricity and do not generate electromagnetic interference. These electrical concerns can be a crucial advantage over conventional electrical wires, where these might otherwise perform similar roles.

Intelligent structures

Having introduced shape-memory alloys and plastics, piezoelectric materials and fibre-optic sensors, it should now be possible for you to imagine what might be meant by the kind of intelligent structures referred to in the article on page 172. One article referred to rotor blades that change their shape depending on the helicopter's pitch. Consider the design of a new generation of sports equipment made from similar materials. It would be possible to embed sensors, so that the force being applied was detected. This could be used to trigger appropriate changes in actuators and change the stiffness of the equipment. So tennis, badminton and squash rackets might be 'programmed' to alter the tension in the strings in different circumstances. The player would also be able to alter the stiffness of the racket during a match as an additional 'tactic'. Similar technology could be employed in adapting the stiffness of suspension systems for vehicles.

4.1.3 Other materials developments

Smart materials are an important new technology, but sustainable development requires us to look again at all our uses of material. The recycling of some materials is already common, e.g. stainless steel, aluminium, glass and paper, and efforts are being made to recycle more polymers. Companies like Smile Plastics Ltd in the UK are continually developing new materials for designers to consider. This will be essential if the mountains of plastic bottles that are collected are not to be burnt in furnaces or dumped in landfill sites. Of course burning the polymers in power generating equipment is potentially an excellent option, in that the energy contained in the polymers (which were made from oil) is recovered. However, if they can be recycled several times before eventually burning them as fuel, then there is an opportunity for further improvements from an ecological perspective. (There is a risk of dioxins being produced if polymer waste is burnt at too low a temperature.) The crucial factor that determines whether the recycling of polymers is economic is the transport costs incurred in collecting the material and transporting it to a processing plant. There must be a concentrated source of the polymer, e.g. collecting bins in an urban area.

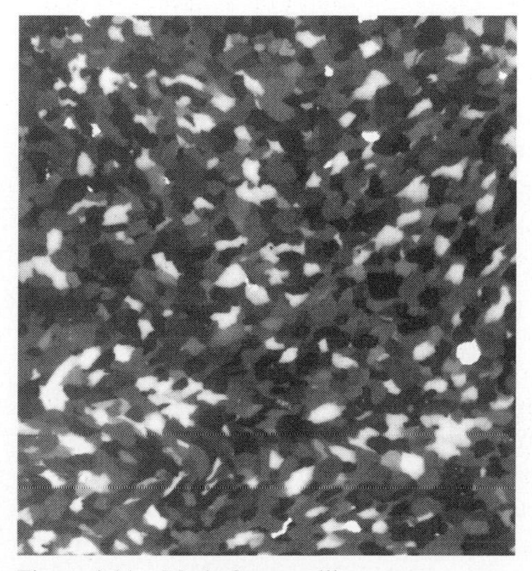

Figure 4.11 Made from wellies.

The new materials can be very striking. They can be made by recycling coffee cups (high impact polystyrene or HIPS), used plastic bottles and packaging (high density polyethylene or HDPE), crisp packets, chopping boards and even wellington boots (plasticised PVC). Fig. 4.11 shows the (multi-coloured) sheet material formed from wellington boots.

The materials challenge the common perception of the way polymers should look and consequently represent an interesting challenge to designers. Examples of the work of Industrial Design and Technology undergraduates at Loughborough University with these new materials can be found at the Department of Design and Technology's website (www.lboro.ac.uk/departments/cd). Materials from Smile Plastics were used in the construction of the Millennium Dome in London, where they also feature in the Design Museum. They have found applications as shelving, doors and cases, and it seems likely that such materials will be used more commonly in the next century.

Not only can materials made from recycled polymers offer striking visual qualities and the opportunity to reuse waste polymers, they can also offer unexpected benefits. Boat and yacht marinas are traditionally constructed from wood, but in order to stop the wood from rotting it must be treated with chemicals that are toxic to much marine life. The result is that the water in marinas is often unable to support significant ecosystems. However, at least in the USA, this is changing. Pylons and boarding have been manufactured by heating and compressing mixed polymer waste to form structural sections. These materials are chemically inert and when they are used for the construction of marinas life is able to return to the water. So the careful consideration of issues surrounding designing for a sustainable future can lead to significant environmental improvements.

4.2 The nature of materials

Everyone involved with designing or improving products must have sufficient knowledge of materials to select, process and finish the materials that best fit the design requirements. An architect aims to design strong buildings that are aesthetically pleasing and resistant to pollution. An aircraft engineer must create a light structure that retains its strength at both high and low temperatures, can be flexed repeatedly and is resistant to the elements.

In choosing a material for a particular application we must consider the internal structure of the material, its properties, and the method of processing and finishing the product. A change in any one of these factors can affect the others and may alter the material's suitability for its task. Therefore it is essential for a designer to have an understanding of the effect and interrelation of a material's structure, properties and the method of processing.

Solid materials can be divided into four main types:

- metals
- ceramics
- polymers
- composite materials.

Within these groups there can be many different types of structure. These will be examined at length in section 4.2.3, but for the moment we can look at the typical compositions, structures and characteristic properties of these broad groups.

4.2.1 Metals and alloys

Metals, either in pure form or combined to give alloys, are *crystalline* structures which have the atoms arranged in a regular pattern within the material. These crystals grow around a central nucleus and expand to form *grains* of crystals within the structure all with their axes of growth in different directions. The crystal structure will generally give them the typical characteristics of metals: high strength, stiffness, ductility, ease of working, good conductivity and toughness, so they are ideal structural materials. Their high ductility and malleability enables them to be used in a variety of forms.

Nonferrous materials

Metals and alloys can be classified as ferrous or nonferrous. Ferrous materials, for example steel, are composed primarily of iron, whereas nonferrous materials are not. Titanium, copper, aluminium, zinc, antimony and magnesium are examples of nonferrous metals and the principal properties and uses of some of these metals are outlined below.

Aluminium is extracted from the ore, bauxite. It is a soft, malleable and ductile metal which has a low density and a low melting point. Its strength and stiffness can be greatly increased by alloying, so that it is suitable for structural applications, but these alloys are more difficult to weld and many aluminium-based products are often joined with rivets. A hard, insoluble oxide layer forms on the surface and makes it corrosion-resistant so it is of great value for cooking utensils and foil wrap in the food industries, where cleanliness is all important. It is also used for lightweight drinks cans.

Malachite (55% copper) or cuprite (88% copper) are the principal oxide ores from which copper is refined. Copper is soft, malleable and ductile making it suitable for drawing out into wire and tube. It has a high melting point and is an excellent thermal and electrical conductor. Consequently it is the main material used in electrical wiring, cables and machines. As copper is easily formed and joined, and has an attractive colour, it has been used as a decorative finish. You may have a tower in your area, such as the Loughborough Carillon, where the upper surface has been sheathed with copper that has turned green over the years as the surface has converted to copper carbonate in the atmosphere. Alloys of copper, such as brass and bronze, are used for bearings and machine parts as well as decorative work.

Silver paper used to wrap sweets is in fact made from tin. The pure metal is very light, soft and malleable. To change these properties and to decrease the cost, tin is usually used in alloy form. Tin cans are made from mild steel plated with tin to give a cheap, strong corrosion-free container for food.

Of the common metals, lead is the densest; it is also soft and malleable. Its weight and its resistance to the effects of sea water means that it is used for keels which act as stabilisers for boats. Lead is alloyed with tin to obtain soft solder. In the past lead has been used for water pipes and the manufacture of paint but now that the harmful long-term effects are realised, its use for these products has been discontinued. We are now also conscious of the harmful effect of lead in the atmosphere from the exhaust of cars that run on leaded petrol.

Ferrous materials

The ferrous metals have been the most important group of metals since the large-scale introduction of iron in the eighteenth century. This is principally because they are strong, rigid and cheap. The many forms of ferrous metals from cast iron to stainless steel mean that they are suitable for a wide variety of purposes and usually cost less than alternative materials. Their properties can also be modified greatly by various forms of heat treatment and mechanical working, which make them even more adaptable.

Iron, the metal from which ferrous metals are derived, is contained in varying quantities in a number of ores. These are widely distributed, making iron one of the most common elements in the Earth's crust. Magnetite, haematite and limonite contain differing amounts of iron and are found in different regions. Small deposits of haematite, a purple-reddish rock yielding up to 60% iron, are found in Britain. Calcium, oxygen, phosphorus, sulphur and silicon are all present in most of the ores. The ores are washed, crushed and smelted in a blast furnace with coke and limestone to reduce these elements to a minimum and leave as pure a form of iron as possible. This is known as *hot metal* in the steel industry. When it is cast into solid form it becomes *pig iron*, which is weak and brittle and needs further processing to convert it into more useful metals.

Figure 4.12 An old section of wrought iron showing the fibre-like structure.

Wrought iron, shown in Fig. 4.12, is made by further refining the pig iron in a puddling furnace until it is almost totally pure iron. It has good tensile strength and is useful for outdoor decorative work, as it is corrosion-resistant. When red-hot it is soft and ductile enough to be hammered into shape. When it is white-hot, pieces can be hammered together to weld them. Traditionally, the village blacksmith was able to form many items for the community at his forge and anvil, but nowadays wrought iron is made in very small quantities as mild steel has to a great extent superseded it.

To produce cast iron the pig iron from the blast furnace is further refined in a much smaller furnace, where the main element, carbon, and the lesser components, silicon, sulphur, phosphorus and manganese, can be controlled to the correct composition.

The proportion of carbon added to the iron and its form of inclusion influence greatly the properties of the cast iron. It can be present in solution as free graphite which is relatively soft. When it is present as long flakes in the iron it forms grey cast iron, which is relatively weak and easy to machine. The presence of the graphite makes the cast iron weaker under tension but allows it to absorb vibrations and resist wear. Small amounts of magnesium or cerium cause the graphite to take up a spherical form, which improves the mechanical properties of the iron.

Carbon can combine with some of the iron to form iron carbide or cementite. This is hard and brittle and so alters the properties accordingly. The white cast iron formed is so hard that it has to be ground. It is usually heat treated to give improved properties so that, where greater strength and shock resistance are necessary, it can be used in place of grey cast iron.

Table 4.3 lists some common metals and their principal applications.

Metal	*Applications*
aluminium	kitchen utensils, cooking foil, cans, engine parts
cast iron	cylinder blocks, vice and machine bodies
copper	electrical wires and components, decorative work, plumbing
lead	soft solder, roof flashing, ships keels, containers
zinc	galvanising and sherardising steel

Table 4.3 Some common metals and their applications

Alloys

An *alloy* is a metal compound produced when a metal is combined with one or more other elements. If there are two alloying elements a *binary alloy* is formed. Three elements will give a *ternary alloy*. Alloying alters the properties of the base metal. Alloying can:

- lower the melting point,
- increase strength, hardness and ductility,
- change the colour (compare the colour of brass and copper),
- give better castings,
- change the electrical and thermal properties,
- change the resistance to corrosion and oxidation.

The change of properties obtained by alloying is clearly demonstrated by the aluminium alloys. High-strength aluminium alloys have a higher tensile strength and a better strength-to-weight ratio than the pure metal but they are not weldable and lose considerable corrosion resistance. Table 4.4 gives some typical alloys and their common applications.

Steel

Steel is the most versatile metal and is also the most common alloy. It is obtained by alloying iron with carbon. Nowadays steels are produced in quantity by three different methods: the *electric arc furnace*, the *basic oxygen process* and the *open hearth furnace*. This last method is the traditional favourite although it is now becoming obsolete as it takes about fifteen times as long as the basic oxygen process. The electric arc furnace method, although as slow as the open hearth furnace, is useful because its charge is mainly scrap steel which it converts to a new steel of the desired composition. More recently, a highly efficient molten jet technique for producing steel has been developed.

Carbon steels may be classified into three main groups:

1 low-carbon steel, containing less than 0.3% carbon,
2 medium-carbon steel, containing 0.3 to 0.6% carbon,
3 high-carbon steel, containing 0.6 to 1.7% carbon.

Approximately 90% by weight of all steel produced is low carbon or mild steel. Mild steel is a cheap material which may need expensive processing. It contains only 0.10% to 0.30% carbon, which means that it cannot be hardened or tempered and so is not a good material to use in products where wear is significant. When the carbon content is less than 0.2% the steel has excellent ductility, which enables it to be deep drawn but makes rapid machining difficult. As the carbon content increases, the ductility decreases, and the strength increases, making it suitable for constructional purposes.

Alloy	*Composition*	*Applications*
copper-zinc		
best brass		
60/40 brass	60% Cu 40% Zn	locks, watch mechanisms and electrical fittings
gilding metal	80–95% Cu 20–5% Zn	jewellery and decorative work
nickel silver	Cu + 13–27% Zn + 10–30% Ni	electroplated nickel silver (EPNS), electrical components
copper-tin		
casting bronze	97% Cu 3% Sn	castings, bells, statues
gunmetal	88% Cu 10% Sn 2% Zn	marine use, bearings, pumps and valves
bell metal	80% Cu 20% Sn	bell castings
aluminium alloys		
Duralumin	Al + 4% Cu + 0.4% Si + 0.5% Mg	pulleys, screws, aircraft parts
LM4	92% Al 3% Cu 5% Si	general purpose sand casting
LM6	88% Al 12% Si	thin-section casting
tin-lead		
pewter	80% Sn 20% Pb	decorative work, tankards and dishes
white metal	6% Sn 76% Pb 18% Sb	machinery bearings
iron alloys		
mild steel	Fe + 0.25% C	structural and engineering work; nails, screws, nuts and bolts
high-carbon steel (tool/cast steel)	Fe + 0.7–1.5% C	hand tools: saws, chisels, files, springs and gauges
high-speed steel	carbon steel + 18% W	high-speed cutting tools, e.g. lathe, miller
stainless steel	carbon steel + 12% Cr	sinks, cutlery, furniture and sterile medical implements

Table 4.4 Some common alloys and their applications

The mild steel billets are hot-rolled into rough bars, rods, flats and sections. This produces *black mild steel* which has a blue-black oxide scale on the surface. It is the simplest and cheapest form of mild steel, available as rods, bars, plate, sheet, wire and tubes. The scale gives protection from corrosion and so mild steel is ideal for outdoor work. The scale will quickly blunt tools used on the steel, making some processing more difficult.

When mild steel bars are drawn through a tungsten carbide die, steel with a smooth, bright surface, a homogeneous structure and an accurate cross-sectional shape is formed. This *bright drawn mild steel* can be cold-worked and forged and is easy to saw, drill, file and machine. As it has no surface oxide scale it should be greased to protect it from rusting in extreme conditions.

The wide range of steels available are obtained by the varying amounts of other elements present, either by deliberate addition during manufacture or natural occurrence in the ore. Each combination of elements produces specific properties to meet the requirements of its function at an economic cost. Carbon is the most important alloying element in steel and is the main contributor to its hardness and tensile strength. The effect of the other elements is shown in Table 4.5.

Element	
manganese	this is the most important alloying element after carbon. It complements the effects of carbon, reduces brittleness, increases toughness and abrasion resistance, promotes hardening.
chromium	used in stainless steel and heat-resistant steels. Increases resistance to wear.
molybdenum	improves strength and toughness of heat-treated steel. It enables nickel-chromium steel to be hardened and tempered satisfactorily.
nickel	increases strength and toughness, especially at low temperatures. It is important in engineering applications as the weight can therefore be reduced. Also used to harden steel at lower temperatures.
vanadium	increases toughness, strength and shock resistance.
tungsten	enables steel to be cut at high speed as it retains hardness and abrasion resistance up to 600°C.
silicon	increases hardness.
cobalt in association with molybdenum and tungsten	essential in high-speed steel.

Table 4.5 Changes effected to steel by alloying elements

4.2.2 Ceramics

Ceramics can be crystalline or *amorphous* (glassy) and are complex combinations of metallic and nonmetallic elements. Both the precious stone ruby and a building brick are examples of a crystalline ceramic. In fact they have the same basic crystal structure of aluminium oxide but the ruby is one large regular crystal and the brick has many randomly joined small crystals. Common salt (sodium chloride, NaCl) is another example of a crystalline ceramic and silicates are the basis for many more, giving us compounds such as magnesium silicate (Mg_2SiO_4) and ferrous silicate (Fe_2SiO_4). Glass is a common example of an amorphous ceramic. Window glass can be made from silica (SiO_2), soda (Na_2O) and calcium oxide (CaO).

Although there is a wide range of ceramics the majority are clay based. Clay is an important ceramic as it can be shaped easily when water is added and is cheap. A disadvantage is that it needs glazing to prevent the permeation of liquids – this would obviously be undesirable in sewage pipes for example. Ceramics made from clay, sand or cement contain silica in the form of silicates. The silicate molecules form into ring, chain or laminar structures giving different properties to the silicates.

Bricks, earthenware, china, porcelain and glass are common ceramics. Another class of ceramics, *refractories*, may be less familiar. A refractory material has a very high melting point (>1700°C) and can withstand large stresses at high temperatures without breaking or deforming. Consequently they are used for lining furnaces, making furnace crucibles and similar applications. Kaolinite (clay) is one of the oldest known refractory materials. Others are alumina (Al_2,O_3), magnesium oxide (MgO) and silica (SiO_2).

Ceramics are usually made by dry-mixing purified materials of the right size and then adding water to obtain the correct consistency for the shaping method that is to be used. This varies from nearly 0% for dry pressing to 25–50% for slip casting. The product is then dried and fired. When a typical ceramic composed of clay, silica and a flux is fired it changes a weak and pliable form into a strong, hard product. The flux assists the firing process by lowering the temperature at which the materials may be fused. In this process, any excess moisture is driven out and then the mixture is brought up to a temperature at which much of the material liquefies causing a bonding of the particles during the subsequent cooling. (See also section 6.2.5.) Sintered ceramics, such as the pure alumina ceramics, are formed by a different method in which they are never brought to a high enough temperature to melt the materials. The bonding in these is dependent on the diffusion of atoms in adjacent particles which forms fewer large pores. This is also the method by which powder metallurgy components are manufactured.

Generally ceramics have poor conductivity, ductility, workability and toughness. They are often brittle and not suitable for structural applications although many have good strength and hardness. However, when virtually all the air is removed from ceramics they lose their brittleness and gain elasticity to the extent that ceramic springs have been made. Other properties, such as electrical insulation, resistance to high temperatures and corrosion, unusual optical and electrical characteristics, make them suitable for a variety of different applications such as pottery, house bricks, cookware, abrasives and magnets (magnadur). Barium titanate which exhibits *piezoelectric* behaviour and can convert sound to electricity is used in sound system transducers.

4.2.3 Polymers

Polymers such as rubber, plastics and adhesives are organic structures. They can be natural, like amber and shellac, modified natural materials, like cellulose and casein, or synthetic, like nylon and polypropylene. Synthetic polymers are produced from gas, coal or oil products. Large molecules are built up by the connection of many base units, the *monomer molecule*, to give a polymer. This process is called *polymerisation* and as the polymer grows, perhaps to several thousand atoms, the melting point increases and it becomes stronger and more rigid. Polymers are lightweight, have good corrosion resistance but relatively low strength and conductivity and do not retain their properties at high temperatures. There are three main types of polymer structure which all react in different ways when heated:

- thermoplastics,
- thermosets,
- elastomers.

Thermoplastics

Thermoplastic articles can be formed at high temperatures, cooled, reheated and reshaped many times without changing their structure and properties irreversibly. The thermoplastic structure is composed of linear chain molecules, sometimes with side branching of the molecules, which have weak, secondary bonding between adjacent chains. Because of their two-dimensional structure thermoplastics are flexible materials but the flexibility decreases with growth of the polymer chain and with increased branching.

When subjected to pressure, thermoplastics can deform in two ways. *Elastic deformation* occurs in which the initially curled long chains stretch out and this deformation is recovered when the pressure is removed. At higher temperatures or pressures, *viscous* or *plastic deformation* occurs in which the secondary bonds weaken and allow the molecular chains to slide over each other. This deformation is not recoverable except by reheating and reforming the article. This weakening of the bonds at higher temperatures means that thermoplastics are easy to cast or form. Common examples of thermoplastics are polyethene, also known as polythene or polyethylene (used in bottles, pipes and packaging), polyamide (nylon, used in fabrics, gear wheels and bearings), polyvinyl chloride (PVC, used in records, cable insulation pipes and guttering), polystyrene (used in flower pots, modelling kits and packaging), polymethyl methacrylate (acrylic or Perspex, used when transparent plastic is needed) and polypropylene (used in plastic chairs, crates and packaging film).

Thermosets

Thermoset polymers are formed by a condensation reaction which forms strong, primary cross-linking bonds between adjacent polymer chains, giving them a rigid three-dimensional structure in which the atoms cannot move easily. Heating helps this cross-linking process and so the polymer hardens as it is heated. This means that thermosets cannot be hot-worked and are more difficult to form than thermoplastics but they do remain hard at quite high temperatures and are, therefore, ideal where resistance to heat is required, such as in cookware and electrical fittings.

When they are subjected to a tensile stress they behave like a brittle metal or ceramic, showing high strength, low ductility, a high modulus of elasticity and poor impact properties compared with other polymers. Common examples of thermosetting polymers are polyester resin (often used with glass fibre reinforcement in boat hulls and furniture), polyurethane (used in upholstery and paint), phenolic resin (Bakelite, used in electrical fittings and pan handles), melamine formaldehyde resin (Formica or Melamine, used in laminated surfaces and moulded products), epoxy resin (Araldite; used as a laminating and bonding agent) and urea formaldehyde resin (used in electrical fittings, adhesives and cavity-wall insulation).

Additives

Most polymers contain additives that give the material special characteristics.

Pigments — give the plastic its colour.

Stabilisers — prevent the environmental deterioration of the plastic. Antioxidants are added to polyethylene and polystyrene and heat stabilisers are necessary for PVC.

Lubricants — Wax or calcium stearate is often added to reduce the viscosity of the molten thermoplastic and make it easier to form.

Plasticisers — are added to make a plastic, such as PVC, less hard and brittle at temperatures of normal use and more easily formed at higher temperatures.

Fillers	can be added to improve the properties or extend the polymer. Carbon added to rubber improves the strength and wear and makes it suitable for use in tyres. Inorganic fibres, such as glass, can strengthen a polymer. Silica, calcium carbonate and clay are used in different polymers to give a greater volume at reduced cost.
Blowing agents	are used to expand some polymers into a plastic foam. Beads of a polymer such as polystyrene are formed round small particles of a blowing agent which will decompose into a gas when heated and expand, so pushing the plastic beads out to form a hollow bead of plastic foam.
Flame retardants	Chlorine, phosphorous, bromine or metal compounds are added to polymers, such as polyester, to reduce the likelihood of combustion which is quite high in an organic material.
Antistatic agents	Polymers, being poor electrical conductors, tend to build up an electrical charge. Antistatic agents improve the surface conductivity by attracting moisture from the surroundings, so reducing the static charge.

Elastomers

When a tensile force is applied to an elastomer such as natural or synthetic rubber, it can deform elastically to an astonishing extent with no permanent damage, e.g. an elastic band. Elastomers have a long polymer chain that is then removed. There may also be some plastic deformation at very high loads when the chains are forced to slide over one another. The addition of sulphur to rubber enbles the polymer chains to cross-link, thus reducing the plastic deformation and making the rubber harder and more rigid.

Table 4.6 shows the common polymers, their trade names, forms, properties and uses.

4.2.4 Composites

Very often two or more materials are combined to obtain different properties than those available in the original substances. This is true of concrete, plywood and fibreglass and these are known as composite materials. Composites are usually formed to increase the strength-to-weight ratio, ductility and temperature and shock resistance. A metal-ceramic composite of cemented carbide is used to make cutting tools that are shock resistant, hard enough to maintain their cutting edge and will continue to produce a quality finish when they become very hot.

Composites can be any combination of metal, ceramic and polymer and can be in particle, fibre or laminar form. Concrete is a mixture of cement and gravel and is a particle composite of two ceramics. GRP or glass fibre reinforced plastic is a ceramic reinforced polymer composite. The polyester resin polymer is brittle with low strength but when fibres of glass are embedded in the polymer it becomes strong, tough, resilient and flexible enough to use for building boat hulls, car bodies, roofing and furniture. Steel reinforced concrete is a ceramic-metal fibre reinforced composite that provides enhanced properties suitable for constructional purposes. Plywood with layers of wood and adhesives is a laminar composite of two polymers whereas chipboard is a particle composite of the same materials.

Timber

Timber is a natural fibre-reinforced polymer that has been used for many centuries. All timbers are essentially composed of a large number of close-packed tubular filaments of *cellulose* bonded together with an organic resin, *lignin*. Fig. 4.13 shows a sketch of a typical cross-section through a tree trunk. This is called a transverse section. Longitudinal sections through XX and YY are called radial and tangential sections respectively.

Plastic	*Trade*	*Different forms*	*Properties*	*Uses*
Thermosetting plastics				
phenolic resin	Bakelite	powder, granules, reinforced laminates	strength, hardness and ridigity; can be produced at low cost; colours limited to black or brown	pan handles, knobs, electrical switch covers, appliance parts
polyester resin	Beetle Orel	liquids, pastes	good surface hardness; can be formed without heat and pressure	light switches, tuning devices, coatings, structural coverings (when reinforced with glass fibres) e.g. boat hulls, car bodies etc.
epoxy resin (epoxide)	Araldite	liquids, pastes	exceptional adhesive qualities with low shrinkage; high strength when reinforced	as a bonding agent, encapsulation, surface coating, laminating
melamine-formaldehyde resin	Formica Melaware	laminates, granules, powder	low water absorption; tasteless, odourless; resists scratching and marking	tableware, buttons, distributor heads, laminated surfaces (e.g. table tops), industrial baking enamel, cookers, refrigerators
polyurethane	Suprasec Daltolac	rigid and flexible foams, coatings	weather resistant even though colour changes; high tear resistance	(rigid) insulating material, floats (flexible) upholstery, mattresses, paint
urea-formaldehyde resin	Aerolite Cascomite	powder, syrup, granules	stiff and hard with good adhesive qualities	adhesives, laminating timber, coating of paper and textiles, electrical fittings
Thermoplastics				
acrylonitrile butadeine styrene (ABS)	Cycolac	powder, granules	excellent impact and scratch resistance; good strength, lightness and durability; high surface finish	kitchenware, clock and camera cases, toys (Lego), crash helmets
polyvinyl chloride (PVC)	Corvic Welvic	(rigid form) powder, pastes (flexible form) powders, pastes, liquids, sheet	strong with good abrasive resistance; low moisture absorption; good chemical resistance	(rigid) pipes, plumbing fittings, corrugated roofing (flexible) packaging, textiles, upholstery
polymethyl methacrylate (acrylic)	Perspex Diakon	sheet, rod, tube	excellent light transmission qualities; hard and rigid; takes colour well	display signs and cases, lenses, dials, furniture, jewellery
polyamide (nylon)	Maranyl Kapton	powder, granules, rod, tube, sheet	very good resistance to temperature extremes; tough and mechanically durable; high chemical resistance	gears, bearings, washers, bristles, textiles, clothing, stockings, upholstery
polyethene or polyethylene	Rigidex	(hard) powder, granules, sheet	good resistance to breakage; withstands low temperatures	(rigid) household wares e.g. buckets, bowls (flexible) bags or food, bottles, electrical cable coating
	Alkathene Visqueen	(soft) powder, film, sheet	transparent, but colours well; (with rubber) impact resistance	food containers, lamp shades, toys, model kits, yogurt cartons
polystyrene	Lustex Styron	powders, granules, sheet expanded foam, beads, slabs	light, buoyant; good insulation	packaging, insulation
polypropylene	Propathene	powders, granules, sheet	higher density and more rigid than polythene; very light; good chemical resistance	crates, chair seats, ropes, plumbing fittings, kitchenware
polytetrafluoro-ethene	Teflon Fluon	powder	resistance to high temperature; strength; hard; good friction qualities	coatings, nonlubricated bearings gaskets, heating cable, plumbers' tape
cellulose acetate	Dexel	powder, film, sheet, rod	hard but tough; can be made flexible	photographic film, packaging, spectacle frames, toothbrush handles

Table 4.6 Common polymers

The *cambium* is the layer that contains living cells that produce new *sapwood*, or *xylem*, towards the centre of the trunk, and new *phloem* (inside layer of bark) cells on the outside. The phloem cells conduct nutrients from the leaves down through the tree and the woody xylem cells transport water and mineral salts up from the roots. Only the new xylem cells close to the cambium are living and the majority of the xylem cells are tubular filaments, composed of the original cell walls, through which liquids can be absorbed. As resin and oils accumulate in the older xylem towards the centre of the tree the wood darkens and can no longer conduct water. This area, which still serves to support the tree, is known as the *heartwood*.

Vascular rays can be seen radiating from the centre of the trunk and these carry water and nutrients between the central *pith* and the outer *cortex*. The *annual growth rings* emanating from the centre indicate the number of seasons of growth and the size of the cells within the ring shows the conditions at the time of growth. A good growing season with plenty of moisture, nutrients and light will be evident in the large-sized cells produced at that time. Trees grown in tropical areas with a continuous growing season will not have any apparent growth rings. Fig. 4.14 shows the annual growth rings in a specimen of yew. The large cells produced in spring and the smaller cells produced later in the growing season can be clearly identified.

Weight for weight, wood can be as strong as steel but steel is about fourteen times as dense. Cellulose, the main structural element of all plants, is contained in the xylem cells and gives trees their strength and stiffness. Cellulose is a long chain condensation polymer of ring-shaped glucose (sugar) molecules with its axis parallel to the direction of the fibre and hence the tree trunk. The cellulose molecule is about 30% crystalline and about 70% amorphous and it is the amorphous section that can

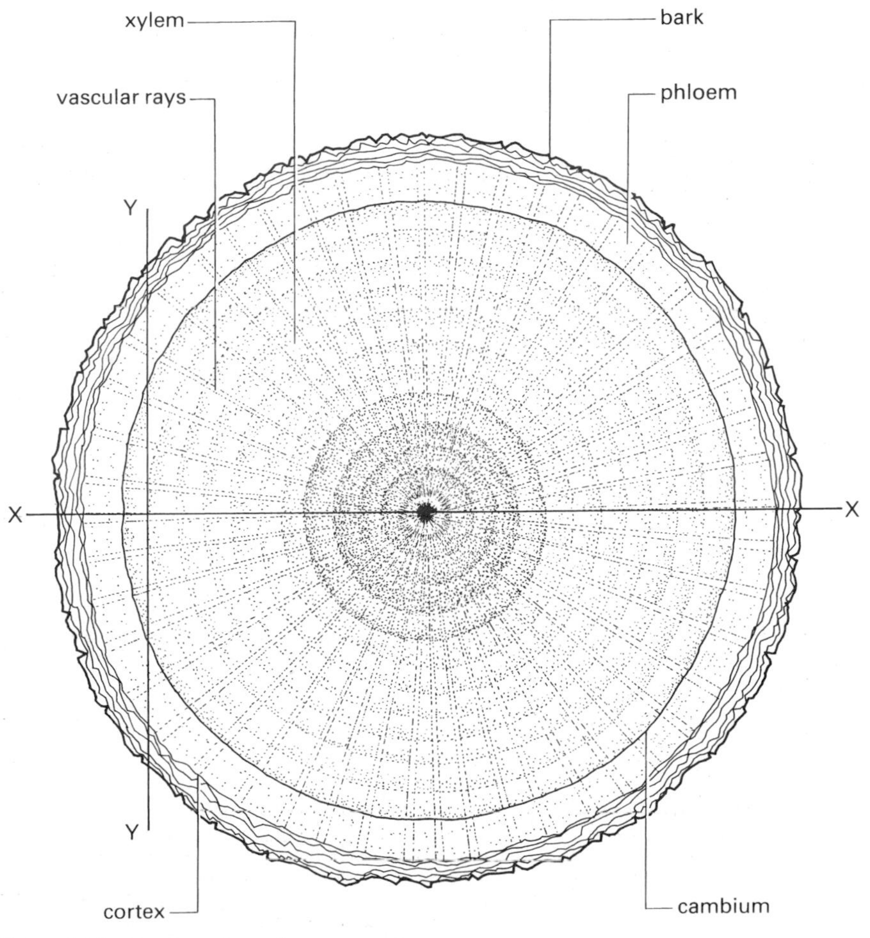

Figure 4.13 Sections through a tree trunk.

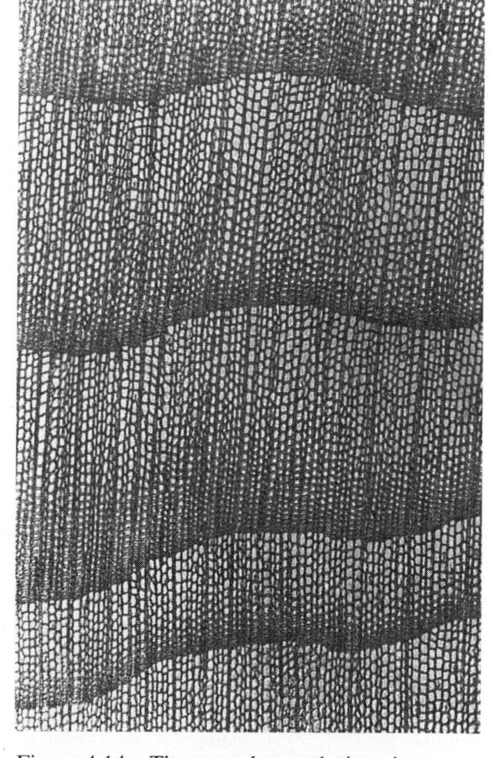
Figure 4.14 The annual growth rings in a specimen of yew showing the large cells produced each spring and the smaller cells produced later in the growing season.

absorb water. Softwoods and hardwoods have a similar structure but the main difference is the thickness of the tubular cell walls. This results in differing densities and mechanical properties for different types of wood. Table 4.7 gives details of various woods, their properties and uses.

Tree	*Places grown*	*Wood colour*	*Properties*	*Uses*
Common hardwoods				
oak	British Isles, Europe, USA, Japan	several species: English – light yellow-brown, with silver grain Japanese – light brown, tinged with grey	English: very strong, tough, durable Japanese: less strong but even-textured; can be worked quite easily	furniture, veneers, panelling; doors, windows, roofs, gates, fencing
beech	Europe, British Isles	white to pale brown	close grain that works and finishes well; hard and strong; but not durable outdoors	functional furniture, toys, tools, turned ware
mahogany	Central and South America, West Indies, West Africa	several varieties: pink to reddish-brown	African (e.g. gaboon, sapele, utile): quite hard and strong; but not easy to finish American: easier to work; gives high finish	indoor work only: furniture, panelling, veneers, pattern making
walnut	Europe, Africa, USA, Japan, Australia	several varieties: mid-brown with dark stripes English: black stripe Australian: dark brown stripe European: pinkish stripe	all highly decorative; vary according to variety. Australian: soft and close grain English: hard; tough; difficult to obtain African: low in weight and strength	superior joinery: furniture, veneers, floor block
teak	India, Burma	golden brown	natural oils make it very durable; hard, strong and fire resistant; works well but quickly blunts tools	fine furniture: chairs, tables, etc; superior joinery: doors, windows, shop fronts, boat-building
obeche	Congo	white to pale straw	very light and soft timber, easily worked; coarse texture; not very durable	general indoor work (often hidden and painted): framework, drawers
balsa	Central and South America	light straw	very light and soft, yet strong for its weight	model-making, rafts, lifebelts, refrigerators
Common softwoods				
European redwood (Scots, pine, red deal)	British Isles, Scandinavia, Russia	pale to reddish-brown	quite strong and hard, with straight grain; works well, but impaired by knots; very durable when preserved	general woodwork: doors, cupboards, shelves etc; building: roofs, floors etc.
Sitka spruce (whitewood)	British Isles, Europe, USA, Canada	cream to light golden yellow, brown	tough but easily worked; straight grain; contains resin pockets but resists splitting	indoor work only: low cost furniture, steps, flooring, packing cases
Parana pine	South America	pale cream to brown with purple	tough with fine grain; shrinks rapidly on drying; also prone to twisting	general indoor woodwork: shelves, cupboards, fitted furniture
western red cedar	Canada	reddish-brown with strong odour	straight grain; soft and weak; works well with sharp tools; natural oils make it outstandingly durable against weather, insects and rot	decorative and protective: panelling, weather boarding, linen chests
Douglas fir (Oregon pine, Columbian pine)	Canada, USA	gold to reddish-brown	strong, with straight grain; slightly resinous; quite durable but needs protection if used outdoors	furniture, plywood doors, windows, frames

Table 4.7 Hardwoods and softwoods

Prepared and processed timber

Timber is manufactured into boards of various types for the building industry and examples of these are illustrated in Fig. 4.15. Chipboard, made by compressing or extruding wood chips with a resin such as urea-formaldehyde is slightly stronger and its strength varies with the density of the board. It is often laminated with a plastic veneer to increase its strength and provide a decorative finish. This enables it to be used for shelving, cheaper furniture and kitchen units. Fibreboard is the next strongest timber board and you may be familiar with two types. Hardboard can be used for facing boards, wallboards and the underlay of a floor on which vinyl is to be laid. Medium density fibreboard (MDF) is used for model-making and light structural items such as speaker cabinets. Fibreboard is made from a pulp of wood or other vegetable fibres which is dried under heat and pressure. For adhesion it relies principally on the natural resin contained in the pulp.

Plywood is made by layering sections of a timber such as gaboon and Douglas fir with the direction of the grain of each layer at right angles to the previous layer. A synthetic resin adhesive is then used to bond the plies of wood and the resulting material is stronger and stiffer than solid timber. A 3-ply board has three layers of timber bonded together. As the top and bottom faces must have the grain going in the same direction to ensure similar properties on each side of the board, there must always be an odd number of plies. Plywood is used for floorboards and furniture. In America the main frame of a house is assembled quickly and easily from plywood although this makes the home vulnerable to the carpenter ant which will eat the

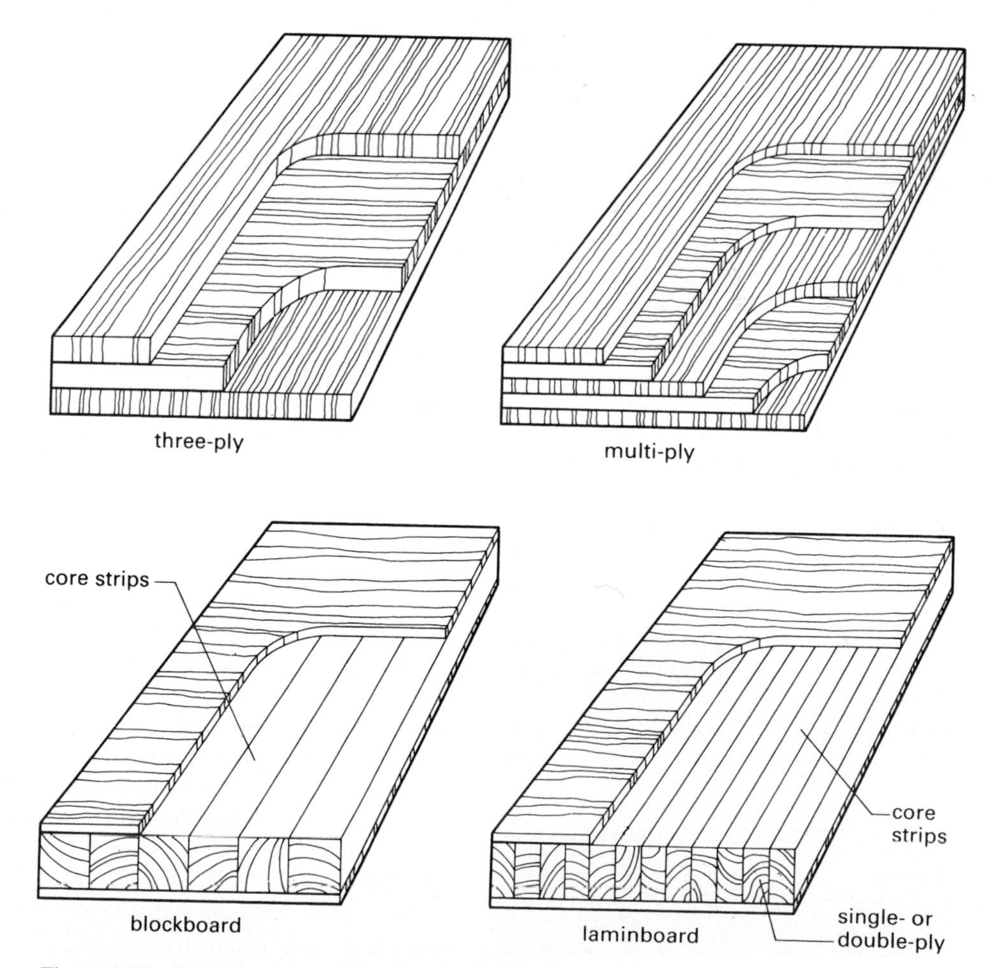

Figure 4.15 Examples of manufactured timber boards.

structure of the house if it is given a chance. Plywood can also be veneered to give it additional strength and versatility of application. Three main grades of plywood are available and they are dependent on the quality of the board, taking into account knots, holes, splits and stains. Each grade has a corresponding cost and therefore each is suitable for a different purpose. Blockboard used in the manufacture of large panels is made by using strips of wood as the core and a layer of plywood on either side. Blockboard with narrow core strips and a high-quality finish is called laminboard. This is used for flush doors and table tops.

Timber is available in several different qualities of finish.

- Rough-sawn timber is the full or nominal size.
- PBS is planed on both sides from the rough-sawn timber and is approximately 3 mm smaller in thickness.
- PAR is planed all round and is about 3 mm smaller in width and thickness.

Timber and boards are also available in several different forms.

- Boards
length: 1.8 m and over
width: 100 mm to 200 mm (softwood) and 150 mm to 300 mm (hardwood)
thickness: 9, 12, 16, 22 mm (PBS)
- Strips
length: 1 m and over
square section 22 × 22, 35 × 35, 47 × 47 mm (PAR)
- Dowel
length: 0.9 m to 2.4 m
circular section diameter: 4, 6, 8, 9, 12, 15, 18, 21, 25, 28, 34, 38 mm
- Plywood
usually 1.5 × 1.5 m or 1.2 × 2.4 m up to a maximum 1.5 × 4.0 m
thickness: 3 to 25 mm
- Blockboard
usually 1.2 × 2.4 m up to a maximum 1.65 × 4.66 m
thickness: 12 to 48 mm
- Chipboard
usually 1.2 × 2.4 m up to a maximum 1.7 × 5.3 m
thickness: usually 9, 12, 18 mm obtainable up to 40 mm
- Hardboard
usually 1.2 × 2.4 m up to a maximum 1.7 × 5.3 m

Figure 4.16 Macrostructure of zinc crystal on a galvanised surface.

4.3 The structure of solids

The structure of a material can be examined at several different levels.

The *macrostructure* is the detail that we can see with the naked eye or a low-powered lens. Examples are: the grain of timber; the surface porosity or cracks in a material; ice crystals on a window; individual crystals of zinc 1 to 2 cm long on a galvanised surface such as a dustbin or many of the railings at Alton Towers, shown in Fig. 4.16; and the presence of different materials such as slag or aggregate.

The *microstructure* is the detail that we can see with an optical microscope with a magnification of about 100. We would be able to identify the size, shape, types and interrelation of crystals in a material. Figure 4.17 shows the microstructure of wrought iron in which the grains can clearly be identified.

Figure 4.17 Microstructure of wrought iron.

The *crystal* or *molecular structure* is the grouping of atoms or molecules that we can see in a crystalline or glassy structure by using an X-ray or electron microscope of magnification 50 000.

The *atomic structure* is that in which we can identify the constitution of individual atoms in a material by use of a field ion microscope or electron probe analyser.

4.3.1 Microscopic examination

Using a microscope

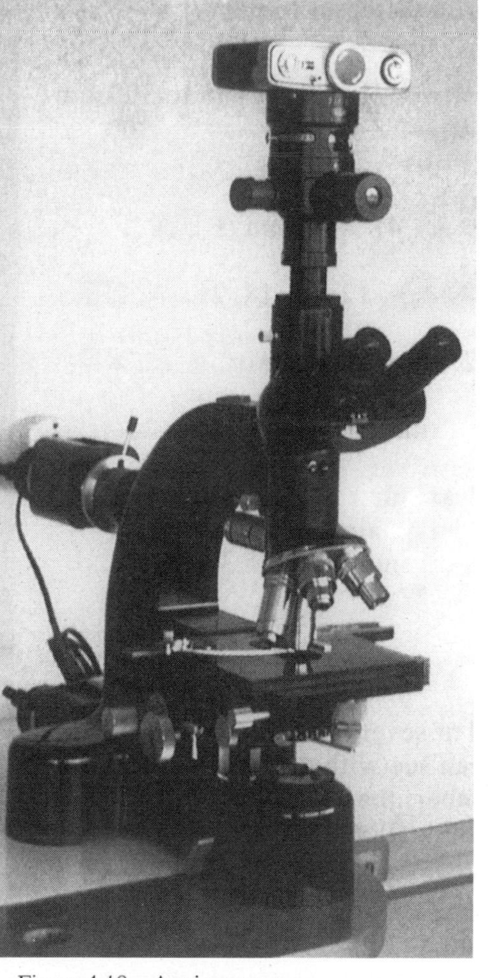

Figure 4.18 A microscope.

A microscope, Fig. 4.18, relies on the combined effect of two complex lenses, the eyepiece and the objective, which together provide large magnification of a small specimen. Usually there will be three or four objective lenses of different magnification on a swivel mounting and some means of illuminating the specimen as this is essential for high-magnification work. The required objective lens can be positioned above the aperture, remembering that the overall magnification will be equal to the product of the magnification of the eyepiece lens and that of the chosen objective.

Great care should be taken when handling the microscope as it is a very expensive instrument. If the specimen is to be mounted in a parallel-sided plastic mount, it may be placed directly on the microscope stage. Otherwise it can be mounted temporarily in Plasticine on a slide, using a ring to ensure that the surface is normal to the microscope stage. The specimen should be pressed onto a clean dry cloth and not the bench top or hard surface.

It is good practice first to bring the specimen very close to the objective lens of the microscope carefully, judging this by sighting from the side, then lower the stage while viewing through the eyepiece until the surface is in focus. This avoids possible damage to the objective caused by pushing the specimen into it. When the surface is in focus, check the adjustments of the illuminator. The polygon produced by the field of view iris should then be adjusted until the edges of the polygon just disappear. The illumination iris should then be closed until the image just starts to darken. If no illumination can be obtained, check the adjustment of the plane glass in the optical tube. This may also be used to centre the image. Before rotating the turret to change the objective, check that the objective lenses will not foul any portion of the specimen and thus cause damage.

Objectives and eyepieces should be cleaned gently only with lens tissue. If an oil immersion objective is used it should be cleaned immediately with xylene. Oil immersion lenses are useful when a magnification as high as $\times$ 400 is required. Oil of cedar wood with a refractive index of 1.5 is commonly used for immersion oil as it has the same refractive index as the lens itself.

Metal macroscopy

To examine metal specimens by eye or a low-powered lens for gross features such as coarse grain structures, forged flow lines, slag and porosity, the surface of the specimen should be ground on 0 or 00 emery paper and then washed. Deep etching should then be carried out in a suitable reagent. The reagents appropriate for different metals are given in Table 4.8.

Metal microscopy

The surface of the metal specimen to be examined should be made smooth by filing and then rubbed on successively finer grinding papers. It is useful to start on a coarse paper (120 grit or 120 G) to ensure the surface is plane and blemishes are removed quickly. The specimen should be rubbed in one direction on the papers until the surface is uniformly scratched and then rubbed at 90° to the former direction until all scratches are removed. The process should be repeated with successively finer papers, 220 G, 320 G, 400 G and finally 600 G. The surface should now be showing a moderate polish and all previous scratch marks should have been removed.

Soft metals and alloys are best ground using water as a lubricant. Steels may be ground dry but care must be exercised to avoid heating the specimen as this may alter the structure. Care must also be taken with alloys containing soft inclusions, since these may be ripped out by hard grinding. Steels for heat treatment should be ground first with 100 G to avoid having to remove a lot of hard material to produce a plane surface.

Reagent	*Specimen*	*Comments*
nital	ferrous metals	good general-purpose etch using 1–4% nitric acid in alcohol
50% hydrochloric acid (HCl)	wrought iron	cold swab or immersion for several minutes
	steel	immerse in boiling reagent for 10–45 min. The higher the carbon content the longer the time
	zinc	cold swab or immersion for one minute
25% Nitric acid (HNO_3)	steel and iron	cold swab or immersion for several minutes
10% ammonium peroxodisulphate (VI)	steel	cold swab of fresh solution
acid-iron (III) chloride solution (25 gm $FeCl_3$ 25 ml HCl 100 ml H_2O)	copper and its alloys	cold swab or immersion this acid-iron (III) chloride etch may also be done in an alcohol base which will give a gentler etch
Tucker's etch (45 ml HCl 15 ml HF 15 ml HNO_3 25 ml H_2O)	aluminium and its alloys	highly corrosive, great care should be taken to avoid contact with liquid or fumes cold swab or immersion for one minute
Marbles reagent copper (II) sulphate (4 gm Cu_2SO_4 20 ml HCl 20 ml H_2O)	stainless steel	

Table 4.8 Etching reagents for metal microscopy

The surface should next be polished on a cloth using alumina (aluminium oxide) as an abrasive. A circular movement may be adopted to try to avoid directionality in the polishing. The whole surface of the specimen should then be examined under a low-powered lens to check that the preparation is satisfactory. If the surface is smooth and free from scratches, it should be washed and degreased with soap or detergent and the surface must not be fingered after this. The specimen should be dried with alcohol and a warm air blower and then lightly etched.

A good general-purpose etch for ferrous specimens is Nital. This is 1–4% nitric acid in alcohol. Acid iron (III) chloride is a good general etch for copper alloys, and this may be carried in either a water or alcohol base although the alcohol base gives a gentler etch. Aluminium responds well to a mixed acid etch containing hydrogen fluoride. However, this substance is highly corrosive and extreme care should be taken when handling it. It should be used in a fume cupboard and care should be taken to avoid any form of contact with the liquid or its gases. Table 4.8 gives more detailed information on the etching reagents.

After lightly etching, examine the whole surface under a low-powered lens to establish whether the etch was sufficient or whether further etching is required. The specimen is invariably improved by polishing and etching several times.

Timber microscopy

Timber samples should be taken as far from the tree pith as possible and notes taken as to whether it is from the branch or trunk (low, middle or top), together with other relevant detail. Then a wide area of end grain should be cleaned with a razor or knife and examined to select a typical section of either heartwood or sapwood. Freehand sections can be cut from dry wood and are useful for rapid and rough examination

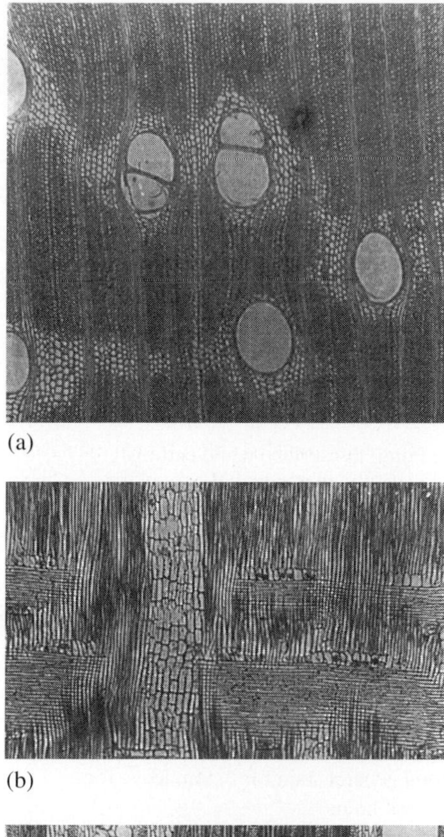

(a)

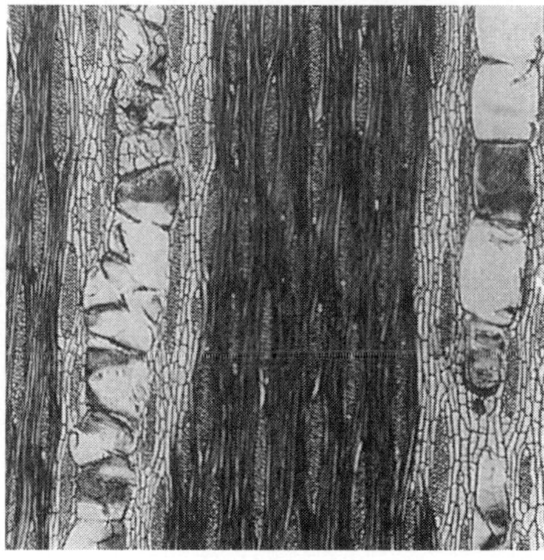

(b)

(c)

Figure 4.20 Microstructure of an African hardwood, iroko.
(a) Tranverse section in which the growth rings are not apparent. The large open cells are the end of the vertical fluid transportation vessel cells, the light areas are the food storage parenchyma cells and the vertical lines are the vascular rays.
(b) Radial section in which the rays are clearly visible as horizontal bands. The pale vertical bands are the parenchyma cells and the dark areas are the fibres.
(c) Tangential section with two columns of vessel cells surrounded by parenchyma and the rays embedded in a matting of fibres.

and either a sledge or a hand *microtome* can be used for finer sections. A good wedge-shaped blade or botanical razor should be used to cut sections about 3 mm square whilst the area is flooded with spirit to ensure that the condition of the specimen is maintained.

The microtome will enable larger, uniformly thin sections of up to 1 cm^2 to be cut. The microtome contains a metal tube in which the specimen block is wedged or clamped to enable a section of suitable thickness to be cleanly sliced free. Fig. 4.19 shows a sledge microtome. The specimen should be positioned with the rays parallel to the direction of travel of the knife which should be at about 10° to the cutting plane. Again the area should be flooded with spirit and several sections should be cut, the thinnest representative sample being chosen. The block will probably require softening before a section is cut. Methods of softening are given in Table 4.9.

Figure 4.19 A sledge microtome.

Some sections need to be stained to obtain contrast between different structures. Some of the most common stains are given in Table 4.10. In some cases where a contrast colour is required two stains may be used. Common combinations are safranin and fast green; safranin and iron alum; haematoxylin; and methyl blue and erythrosin.

The final stained sections should be mounted on a glass slide in a resinous medium such as Canada balsam to exclude air and protect the specimen indefinitely. For temporary work, glycerine jelly may be used as a mounting medium. Fig. 4.20 shows the transverse (endgrain), radial and tangential microstructure of a sample of iroko.

4.3.2 The atomic structure of solids

Atoms and molecules

There are about 110 chemically different types of *atom*, one for each element, and these can bond together in different ways to give us an almost infinite number of materials. The atom is the smallest part into which an element can be chemically divided. All atoms have the same basic structure but they vary in size and mass. Fig. 4.21 shows the constituent parts of the simplest atom, hydrogen.

Each atom has a positively charged nucleus at its centre, composed of positive *protons* and neutral *neutrons*, which together comprise 99% of the mass of the atom.

Method	*Comments*
boiling	Boil until air has all been driven out. Can transfer between boiling and cold water to speed up the process. Many softwoods and some hardwoods require no further treatment.
propane-l,2,3-triol and ethanol	After boiling drop specimen into solution of 50% propane-l,2,3-triol (glycerine) and 50% ethanol (industrial meths) and leave for some time. A 15% solution of propane-1,2,3-triol (glycerine) in ethanol can be used to store specimens.
hydrofluoric acid	Most common method for drastic softening but it is highly corrosive and care should be taken. With 50% commercial acid softening takes about 2 days for oak and 6 for ebony.
ethanoic acid and hydrogen peroxide	A new, quick method which softens resin using a 50% solution. Impregnated wood should be soaked for 45–90 min.
steam	A jet of steam can be directed onto the surface while cutting.
(poly)cellulose ethanoate*	Boil, place in acetone and then into a 12% (poly)cellulose ethanoate solution for 3 days to a few weeks. Results have not been found to be consistently good by this method.

*old name cellulose acetate

Table 4.9 Methods of softening timber specimens

Stain	*Final colour*	*Comments*
safranin	red	good general stain for lignin, leave for several hours
haematoxylins	blue	use for sections to be temporarily mounted in glycerine jelly
light green or fast green	green	good for cellulose
methylene blue	blue	a vital stain (nontoxic to living tissue) which stains lignin
erythrosin	red	stains nonlignified tissue
phenylammonium chloride or phenylammonium sulphate	yellow	temporary stain for lignin
benzene-l,3,5-triol and concentrated hydrochloric acid	red	temporary stain for lignin

Table 4.10 Methods of staining timber specimens

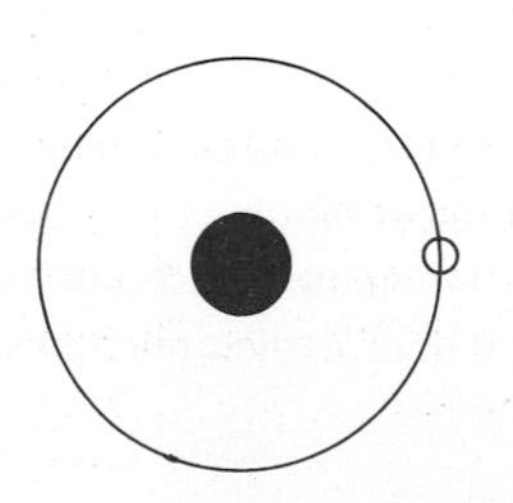

Figure 4.21 The structure of a hydrogen atom.

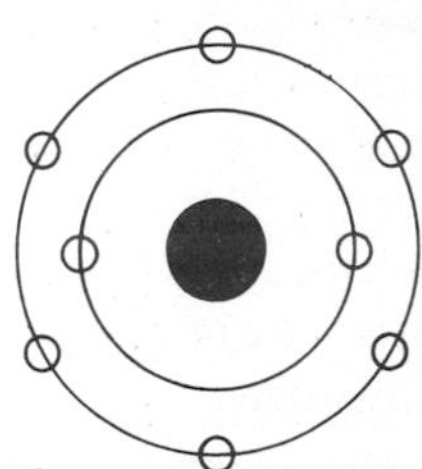

Figure 4.22 The atomic structure of oxygen.

Satellite *electrons* orbit at different distances round the nucleus. They are extremely light, only 0.00054 times the mass of a proton or neutron, but with exactly the right negative charge to balance the charge of one proton. Each atom has equal numbers of electrons and protons to ensure that it is electrically neutral.

Stable electrons associated with a nucleus can only arrange themselves in certain energy levels in shells round the nucleus. Table 4.11 lists some elements from the early part of the atomic table, giving their pattern of electrons. The total number of electrons or protons in a neutral atom is equivalent to its atomic number. For instance the oxygen atom, Fig. 4.22, with an atomic number 8, has eight electrons, two in the full inner shell and six in the outer shell. *Isotopes* are materials whose atoms are of the same element that contain different numbers of neutrons in the nucleus.

Nonmetals always have fairly full energy levels and the electrons are bonded tightly to their nucleus. Metals, however, have free electrons that can easily be detached from the atom leaving a positively charged *ion* of the element. If an atom has the correct number of electrons to fill each of its energy levels it will be a totally unreactive inert gas. Three of them, helium, neon and argon, are shown in Table 4.11.

Atomic number	*Element*	*Symbol*	*Electrons in each energy level or shell*
1	hydrogen	H	1
2	helium	He	2
3	lithium	Li	2.1
4	beryllium	Be	2.2
5	boron	B	2.3
6	carbon	C	2.4
7	nitrogen	N	2.5
8	oxygen	O	2.6
9	fluorine	F	2.7
10	neon	Ne	2.8
11	sodium	Na	2.8.1
12	magnesium	Mg	2.8.2
13	aluminium	Al	2.8.3
14	silicon	Si	2.8.4
15	phosphorus	P	2.8.5
16	sulphur	S	2.8.6
17	chlorine	Cl	2.8.7
18	argon	Ar	2.8.8

Table 4.11 The elements in the early part of the periodic table

Atomic bonding

All solids are held together by the action of their *valence electrons*, which are the electrons in the incomplete outer shell. If a number of atoms come together and distribute their electrons in such a way that they attain or approach full energy levels, then these bonded atoms will have a much lower overall energy and this will be a more stable unit.

If two or more atoms bond together, a *molecule* is formed, which is the smallest particle of a compound that exhibits the properties of the compound. Fig. 4.23 shows molecules of water which are formed from gas atoms and are liquid at room temperature. The atoms and molecules in a material are attracted by the electrical forces exerted by their nuclei and electrons, and this electrical attraction ensures that the volume is preserved if outside conditions remain the same.

If heat or a voltage is applied, the atom becomes *excited* and the electrons absorb energy and move into higher energy levels away from the nucleus. The movement of

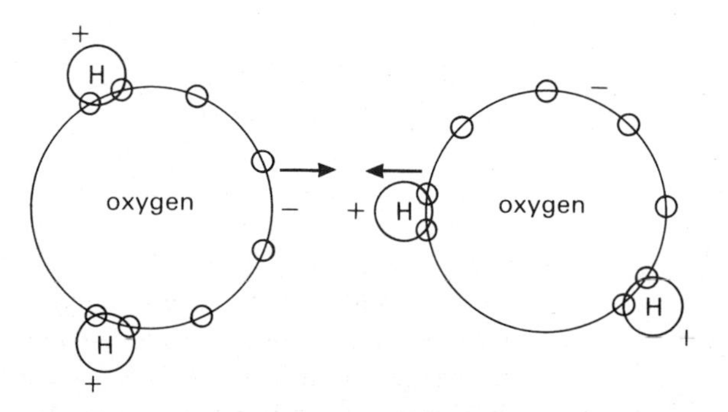

Figure 4.23 The electrostatic attraction between molecules of water.

electrons to lower levels results in a transfer of energy by the emission of radiation, and the change in energy associated with electron movement can be seen with a spectroscope as sharp bands of colour.

Crystalline solids can be grouped broadly into four classifications according to the bonding of the atoms. There are three types of primary atomic bond: *ionic*, *covalent* and *metallic*, and a secondary van der Waals bond. Many materials are bonded by complex combinations of the atomic bonds.

Ionic bonding

Crystals of salt, or sodium chloride, can be seen clearly with an optical microscope; this is a good example of ionic bonding in a crystal. If we consider what happens when metal sodium and nonmetal chlorine atoms are brought together under suitable conditions we can appreciate the important factors of ionic bonding.

Fig. 4.24 shows that the sodium atom has a free electron in its outer shell and the chlorine has a free space in its outer electron shell. The free electron will move over to the chlorine atom and in so doing will release a great deal of energy in the form of heat and combustion. The sodium is now a positively charged ion, the chlorine a negatively charged ion, and the attraction between these ions forms the ionic bond.

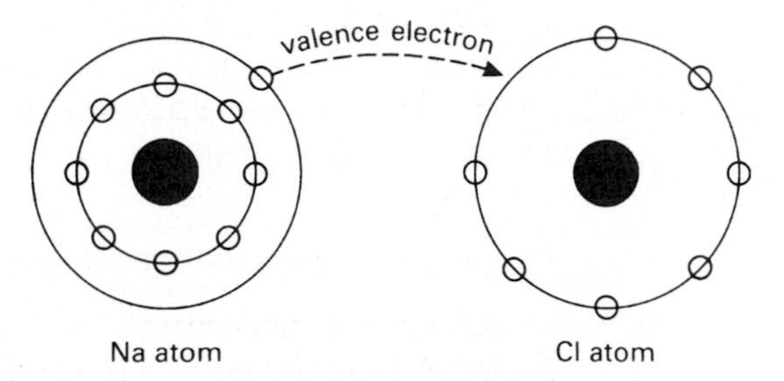

Figure 4.24 The ionic bonding of sodium and chlorine in sodium chloride.

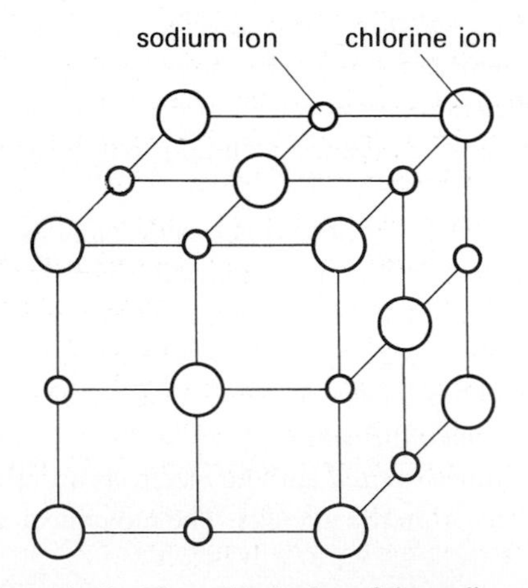

Figure 4.25 The cubic lattice of the sodium chloride crystal.

Fig. 4.25 shows how the opposing charges of these ions hold the crystal together in a cubic lattice. The sodium and chlorine ions will separate easily when dissolved in water, but under stable conditions the electrons stay bonded to their respective ions inside the crystal and it is relatively easy to predict their physical properties. Ionically bonded materials are usually brittle and have poor electrical conductivity because entire ions must move to cause a flow of electrical charge.

Covalent bonding

Covalent bonding occurs in ceramics, glass, wood, oil, sandstone, hydrogen and other mainly organic materials. If two similar atoms with free electrons become linked to share electrons to create a closed outer orbit for each of the atoms, they are said to have covalent bonding. Fig. 4.26 shows two hydrogen atoms with covalent bonding.

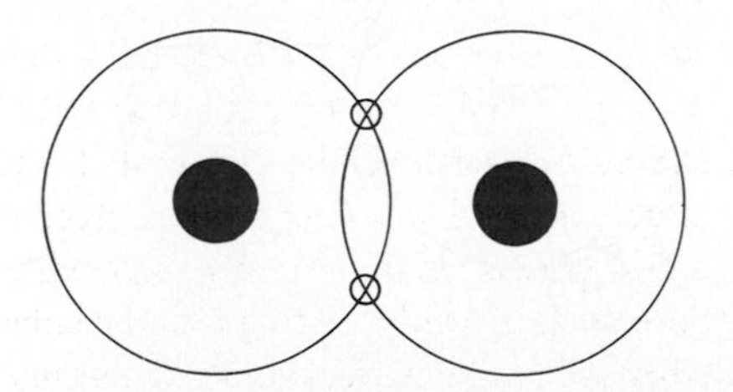

Figure 4.26 Covalent bonding in the hydrogen molecule.

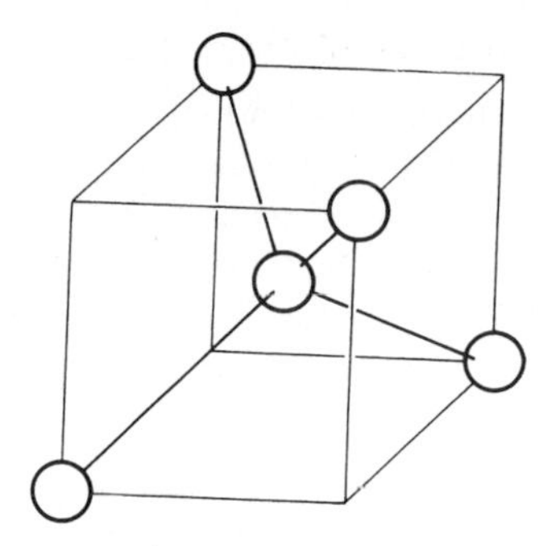

Figure 4.27 The diamond molecule – covalent bonding of four carbon atoms.

An example is the structure of the diamond, Fig. 4.27, which is composed of carbon atoms covalently bonded together. Each carbon atom, which has four valent electrons and can accommodate eight in its outer shell, shares a valency electron with each of its four nearest neighbours so that all the atoms will effectively have complete outer shells.

This bonding is very strong and the cohesive energy of a solid such as a diamond makes it very hard and gives it a high melting point. Once bonded, the atoms or molecules so formed have no electrical charge and no free electrons and so are electrical insulators. The direction of the covalent bonds is fixed and therefore the bonds will break under stress and the material will be brittle.

Metallic bonding

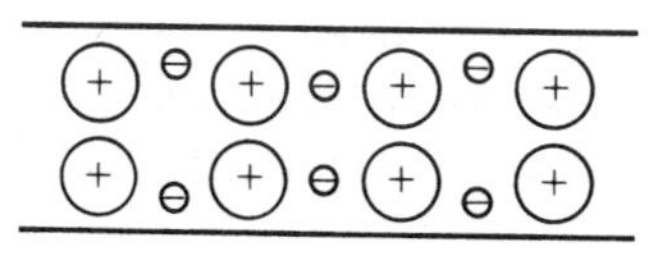

Figure 4.28 Positive ions of metal and free electrons.

Another form of bonding between atoms with valency electrons is metallic bonding. Instead of sharing free electrons with neighbouring atoms, as in covalent bonding, the few valency electrons present in the atom of a metal are free to move through the lattice. Hence a metal is an ordered array of positively charged ions through which the free electrons move in all directions at high speed, as shown in Fig. 4.28. The binding forces that hold a metallic crystal together are the forces between the positive ions and the cloud of moving electrons.

The movement of the free electrons means that metals are good thermal and electrical conductors. The metallic bonds are not fixed in position and can therefore allow metals to deform without the bond breaking. This ductility given by the metallic bond allows useful forming of metals to take place.

Secondary bonds

In addition to the three types of bonding already mentioned, there are also weaker secondary bonds which exist in substances such as water and many ceramics and plastics. The water molecules in Fig. 4.23 have a positive charge near the hydrogen atoms and a concentration of electrons causing a negative charge at the other side of the molecule. This distribution of charge causes attraction between the water molecules. The primary covalent bond between the oxygen and hydrogen atoms in each molecule is very strong but this secondary *hydrogen bond* between the molecules can be broken by heating which results in the water vaporising.

Another type of secondary bond is the *van der Waals bond*. This is formed when the fluctuating electrostatic charge in adjacent atoms of different molecules produces a weak electrostatic force between the molecules. Van der Waals bonding is also often present as a secondary bond between the *long-chain molecules* of polymers. Although the molecules within each polymer chain have strong primary bonds it is the van der Waals bonds that bind the chains together. When the polymer is stretched the van der Waals bonds break easily, allowing significant deformation of the material. They can then easily form again between new neighbouring atoms when the material is released.

Solid structures

The atoms or ions in a solid may be thought of as hard spheres. If we examine the manner in which they interlock to form the final solid, we can appreciate how this structure determines the different characteristics of ceramics, plastics and metals. A structure which has a regular arrangement of atoms repeated in all directions is referred to as *crystalline*. This structure is present in all metals. Where the atoms are arranged randomly, as in glass, the substance is said to be *amorphous*. The atomic structure of a material mainly determines its microstructure and chemical and physical properties.

Amorphous solids

In the simplest solids, the atoms have no regular pattern but are arranged randomly in a given space. Examples of these amorphous solids are those formed by freezing inert gas atoms with closed electron shells such as helium, neon and argon to 4 K.

Where the arrangement of the atoms extends only to each atom's nearest neighbours we have an example of a short-range order material. In a molecule of silica (SiO_2), for example, the silicon atom is covalently bonded to four oxygen atoms in a tetrahedral structure similar to that of the diamond molecule shown in Fig. 4.27 and the molecules of silica are then randomly bonded together. Other examples of amorphous solids are glass and pitch. These amorphous solids are generally unstable and glassy in appearance. The manufacture of amorphous ceramics (*ceramic glasses*) and polymers is easy and inexpensive, and provides us with many useful transparent materials.

Recent developments have been made in producing amorphous metals. When a jet of molten metal is subjected to rapid cooling of more than 10^6 °C/s the normal crystalline structure of the metal does not have time to form and the molecules form a random pattern just like glass. Thin ribbons of *metal glasses* can he produced with good strength, ductility and electromagnetic properties, making them suitable for use in transformers and metal joining.

Long-chain molecules

Some materials form a simple pattern of atoms that is repeated in a long chain to form a molecule rather like a strand of spaghetti. All polymers and some ceramic glasses have long-chain molecules.

Carbon is contained in all organic materials such as wood, coal and oil. It can link up easily with hydrogen, oxygen, nitrogen and chlorine in a vast number of ways. Each combination will give a different material. A carbon atom has four electrons in its outer shell and can link with four other atoms. Hydrogen and chlorine atoms can link with one other atom, and oxygen atoms with two.

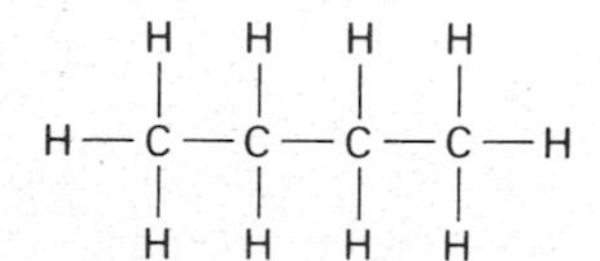

Figure 4.29 Model of the butane molecule.

A simple chain of carbon and hydrogen gives us butane (C_4H_{10}) as shown in Fig. 4.29. Carbon can also form double or triple bonds, using two or three of its links to bond to another carbon atom. By splitting the double bond of ethene (C_2H_4) and joining on another ethene molecule a long-chain of poly(ethene), polythene, is formed, Fig. 4.30.

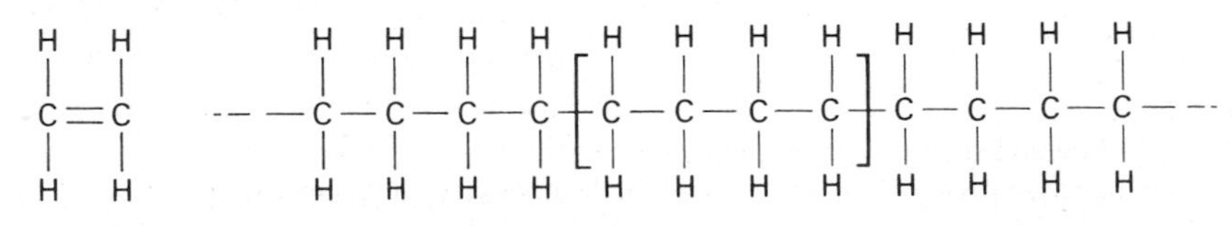

Figure 4.30 (a) Model of ethene molecule.
(b) Model of poly(ethene) (polyethylene) polymer.

In the same way the *monomer*, phenylethene (styrene), can become poly(phenylethene) (polystyrene) and propene can become poly(propene) (polypropylene), shown in Fig. 4.31. These are called *addition polymers* and can have 1000 to 50000 monomers linked together. Fig. 4.32 shows a model of phenylethene (styrene) with its integral benzene ring (C_6H_6). This causes the poly(phenylethene) (polystyrene) molecule to have a benzene ring attached to one side of alternate carbon atoms thus stiffening the polymer chain. Consequently polystyrene is a hard plastic with a fairly high softening point although it is a linear molecule. However, these polymers with bulky atom groups attached to the chain are generally found to be easy to dissolve and are prone to distortion by swelling.

Figure 4.31 3-D model of poly(propene) (polypropylene) polymer.

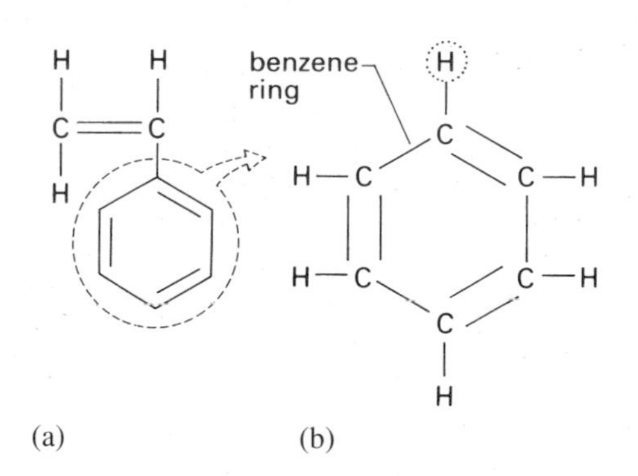

Figure 4.32
(a) Model of phenylethene (styrene) molecule.
(b) Model of benzene ring (C_6H_6) contained in the phenylethene molecule.
(c) 3-D model of the poly(phenylethene) (polystyrene) polymer.

Because these polymer molecules are rather like a plate of spaghetti strands, some can easily be stretched by uncoiling the strands. The long-chain molecules give plastics their ductility and toughness. If strands are stretched still further, they will be found to have some extra give along the line of atoms, but will break easily across the line of atoms.

The structure of long-chain molecules can be altered to give added strength and resistance in two ways. If a hydrogen atom in the chain is removed and replaced by a carbon atom the structure will begin to grow in this new direction and will form a *branched-chain* structure as shown in Fig. 4.33. This will form a material similar to the linear chain polymer but will be softer and have a lower melting point. A *cross-linked* polymer, shown in Fig. 4.34, can be formed when two hydrogen atoms from

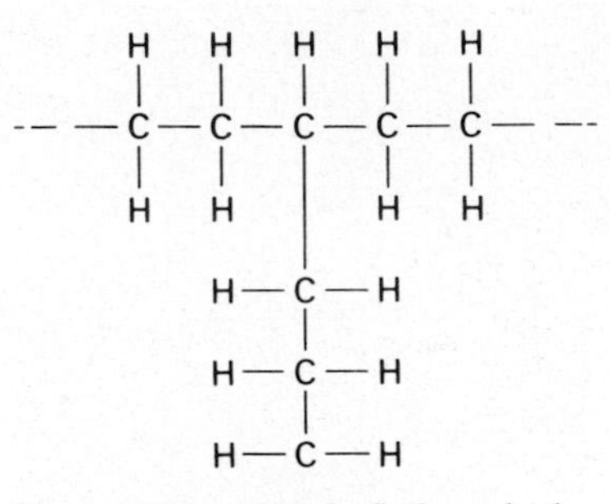

Figure 4.33 Model of a branched-chain polymer.

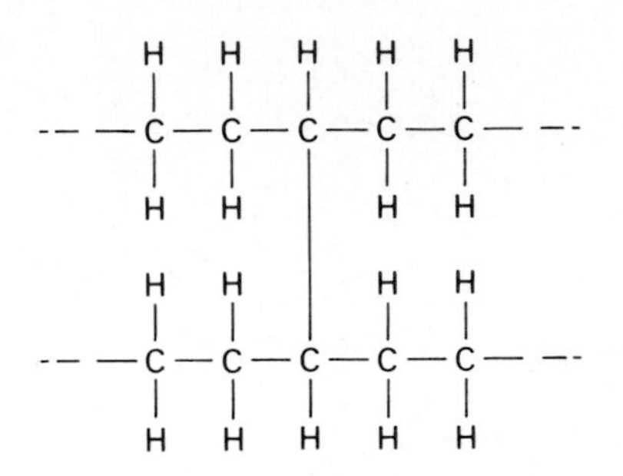

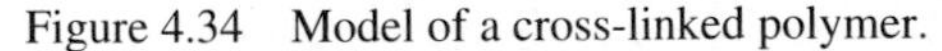

Figure 4.34 Model of a cross-linked polymer.

adjacent long-chain molecules are removed and the carbon atoms become linked. This cross-linking polymerisation has given us polyesters, hard rubbers and thermosetting resins.

Unbranched long-chain solids can form crystalline regions called *spherulites*, especially if cooled slowly. In these regions the molecules are closely packed and give an X-ray diffraction pattern similar to crystals, shown in Fig. 4.35. Poly(ethene) is about 60% crystalline. To add stiffness to rubbers and other elastomers that do not crystallise readily a solid filler such as silica or alumina is added which attaches to the polymer chains and achieves an effect similar to crystallisation.

Nylon is also a long-chain molecule but is a *condensation polymer* formed by two different compounds instead of a repeated single monomer. The two compounds chemically react to form the polymer, with a secondary substance produced as a by-product. *Addition polymers* are those polymers whose structural units are constitutionally identical to the reacting monomers. The best-known examples are those involving vinyl or related monomers, for example polyvinyl chloride PVC (poly(chloroethane)). The model of PVC is shown in Fig. 4.36. *Copolymerisation* entails combining two or more monomers to form a polymer with particular properties. A commercially important example is the copolymer of vinyl chloride (chloroethene) with vinyl acetate.

Each stiffening or strengthening mechanism of long-chain polymers has been examined separately. However, they can be combined to give further modified properties of the polymer material.

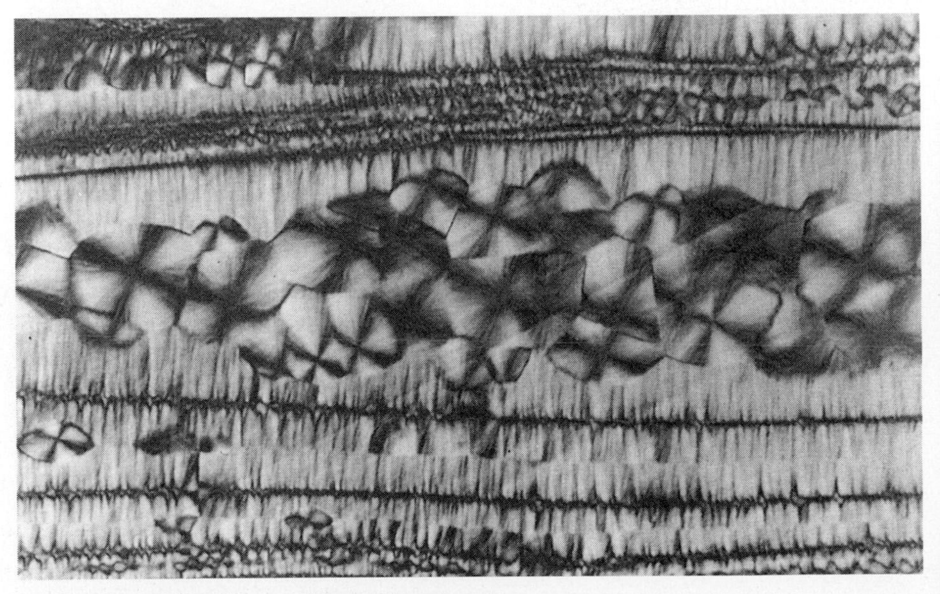

Figure 4.35 Spherulites in a long-chain polymer.

Figure 4.36 3-D model of PVC.

Crystalline solids

Figure 4.37 B.c.c. crystal structure.

Figure 4.38 F.c.c. crystal structure.

Metals, many ceramics and some polymers have a uniform, geometrical arrangement of atoms or ions that is repeated throughout the material and are therefore said to be *crystalline*. This regular 3-D atomic pattern is known as the *space lattice* and the *unit cell* is the smallest unit of the lattice that retains the overall characteristics of the lattice. The *crystal structure* refers to the size, shape and atomic arrangement of the lattice unit cell and varies from one substance to another.

Although there are seven different crystal shapes containing fourteen different lattice structures (*Bravais lattices*), 80% of metals can be classified into the three main types of crystal structure.

The *body-centred cubic* cell (b.c.c.), shown in Fig. 4.37 is formed by an atom at each corner of an imaginary cube with an extra atom at the centre of the cube. B.c.c. metals are generally strong and moderately ductile over a small temperature range. Examples of this crystal structure are tungsten, vanadium, chromium, molybdenum and alpha iron.

The *face-centred cubic* cell (f.c.c.), shown in Fig. 4.38 has an atom at each corner of an imaginary cube and one at the centre of each face of the cube. F.c.c. metals tend to be soft and ductile over a wide range of temperatures. Examples are gold, silver, copper, aluminium, lead, nickel and gamma iron.

The *hexagonal close-packed* cell (h.c.p.), shown in Fig. 4.39 has an atom at each corner of an imaginary right hexagonal prism with an atom at the centre of each hexagonal face and three atoms equally spaced in the centre of the prism. This is the crystal structure for zinc, cadmium, magnesium, cobalt and beryllium. H.c.p. metals are normally relatively brittle.

You will have noticed that iron has appeared in different forms as both a body-centred cubic and a face-centred cubic structure. Materials which can have more than one crystal structure are called *allotropic* or *polymorphic*. Iron has a b.c.c. structure at low temperatures but then transforms to a f.c.c. structure at 910°C. Above 1400°C iron again changes back to a b.c.c. structure. Fig. 4.40 illustrates the transformation of iron with temperature.

Other allotropic metals are titanium (h.c.p. and b.c.c.), zirconium (h.c.p. and b.c.c.) and cobalt (f.c.c. and h.c.p.). Allotropic transformations are produced in the heat treatment of steel and titanium. The metallic bond holding the atoms of any metal will allow it to conform to one of the fourteen simple crystal structures.

Figure 4.39 H.c.p. crystal structure.

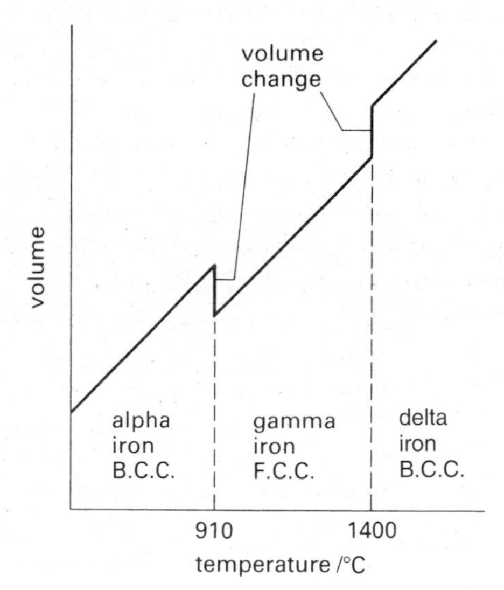

Figure 4.40 The allotropic transformation of iron.

However, nonmetals with different atomic bonding build up more complex crystal structures but may still be allotropic. Carbon can exist as soft graphite or hard diamond. Many ceramic materials, such as silica, undergo allotropic transformations on heating or cooling. A volume change may accompany an allotropic transformation which, if not properly controlled, may cause the material to crack and fail. This volume change is apparent in the transformation of iron as shown in Fig. 4.40.

4.3.3 Crystal growth

Any substance can exist in one of three *states* or *phases*: as a gas, a liquid, or a solid. Each of these states is a homogeneous, physically distinct substance as can be illustrated by the three phases of water: water vapour, water, and ice. The atoms or molecules of a solid are tightly bonded together and the solid consequently has a definite shape and size. In a liquid the bonding forces are not so strong and therefore the atoms or molecules can move, changing the shape of the liquid to fit the container although its volume remains fairly constant. The atoms or molecules of a gas gain sufficient extra energy during the vaporisation process to overcome the atomic bonding forces, thereby enabling them to move freely in the available space. Thus the gas has no fixed shape or volume.

Solidification

Solidification is the transformation of a liquid to a solid material. Fig. 4.41 shows a typical cooling curve for pure molten copper.

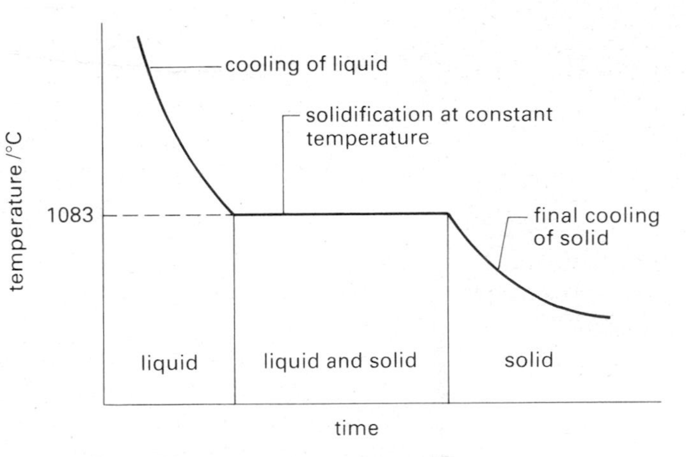

Figure 4.41 The cooling curve for pure copper.

Solidification is achieved by a process of *nucleation* and *growth*. When a pure, molten metal is cooled to just below its freezing temperature minute nuclei of the solid first form in the liquid. In molten tungsten, for example, nine atoms will form a b.c.c. lattice nucleus which will grow as other atoms join the lattice structure. The crystal grows in a tree-like formation called a *dendrite*. The main 'trunk' of the dendrite grows quickly in the direction of the fastest heat loss and then secondary and tertiary branches develop in a geometrical pattern consistent with the lattice structure. Each dendrite continues to grow until it meets other neighbouring dendrites at which point the branches thicken to form a totally solid *grain* of metal, Fig. 4.42. Each grain has the same lattice structure but usually has a different orientation. The *grain boundary* between these lattice regions is a narrow zone, of perhaps three atoms, in which the atoms are not properly spaced because the different orientations of neighbouring grains prevents complete uniformity. The grains and grain boundaries can be seen clearly in the microstructure of wrought iron

formation of nuclei

dendrites formed

dendritic growth

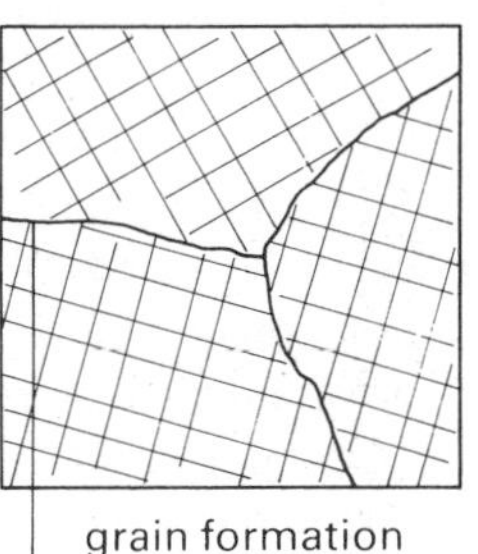

Figure 4.42 Dendritic growth of crystal grains.

in Fig. 4.43. A single crystal will be *anisotropic*, that is it will display different properties in different directions. A *polycrystalline* material will exhibit some anisotropic behaviour if all the crystal grains have a similar orientation, but more commonly it will have a random orientation of the grains and will be *isotropic*, i.e. have similar properties in all directions.

The rate of cooling of a molten metal will determine the grain size. Gradual cooling leads to only a few nuclei being formed and therefore a large grain size, whereas rapid cooling will result in many nuclei and small grain size. When an ingot of metal is cast into a cold metal mould rapid cooling takes place near the surface which causes very small *chill crystals* to form. As the mould is warmed up by the liquid long *columnar crystals* are formed towards the centre of the mould. The liquid at the centre of the mould cools very slowly and so the grains at the centre of the casting are large and are said to be *equiaxed crystals*. The rate of cooling is dependent both on the casting temperature and the temperature surrounding the mould, so that all three types of crystal may not always be present. Figure 4.44 shows the differing crystal structure in aluminium castings made from varying temperatures of molten metal.

Solidification of a pure metal is said to be *homogeneous nucleation*. There is no evidence of dendritic growth in a pure metal under microscopic examination as all the atoms are identical. In normal conditions impurities in the molten material provide the centre for the growth of grains and this is known as *heterogeneous nucleation*. The dendritic growth pattern of heterogeneous nucleation can clearly be seen under microscopic examination.

All engineering materials and alloys will solidify by heterogeneous nucleation. If impurities as low as 0.02% titanium or 0.01% boron are added to liquid aluminium metal or alloy they provide many nuclei for simultaneous growth. The large number of grains formed have many grain boundaries which prevent any defects from progressing through the material. The effect of the impurities in a pure metal can, therefore, be to induce a large number of fine grains which will give a stronger and harder metal than a large grain structure. This introduction of impurities to control the material's grain size and hence its strength is known as *grain size strengthening*. Typical examples of grain size strengthening are the addition of titanium carbide to nucleate aluminium, aluminium carbide in magnesium and tungsten carbide in steel. Again, titanium dioxide dissolved in molten glass enables the crystallisation of Pyroceram which is resistant to shock and thermal stress. The addition of impurities must be carefully controlled as too many impurities introduced into a substance may cause an accumulation at the grain boundaries which will in fact weaken the material.

Lattice imperfections

As the lattice structure builds up from different centres of growth there will be areas where it is not entirely uniform. We have already examined the *surface defects* caused at the grain boundary by the intersection of neighbouring grains. *Point defects* and *line defects* will also inevitably be present in the lattice however carefully the metal has been prepared.

Point defects occur when one point of the crystal lattice is not correctly filled. After solidification from a high temperature or radiation damage it is possible for an atom to be missing from the crystal structure. This is known as a *vacancy*, Fig. 4.45, and causes the bonds of the surrounding atoms to be stretched as they try to fill up the extra space.

An *interstitial defect*, Fig. 4.46, is formed by an extra atom being present in the lattice and causes compression of the surrounding atoms. A *substitutional defect*, Fig. 4.47, is created when impurities or alloying compounds cause an atom to be replaced by an atom of a different substance. These point defects cause tensile or

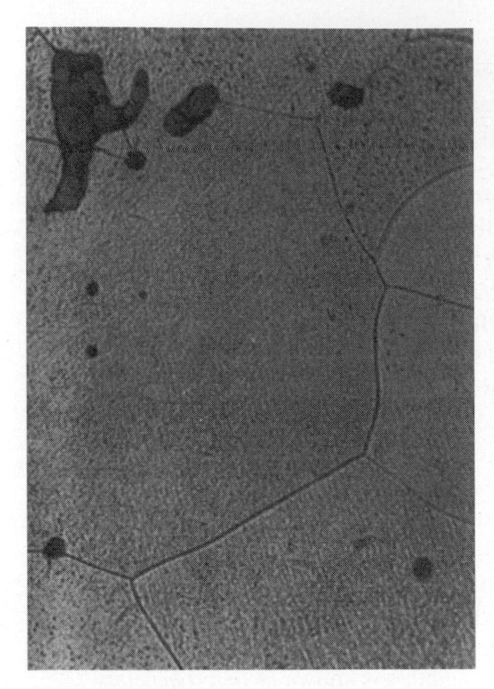

Figure 4.43 The grain structure of wrought iron.

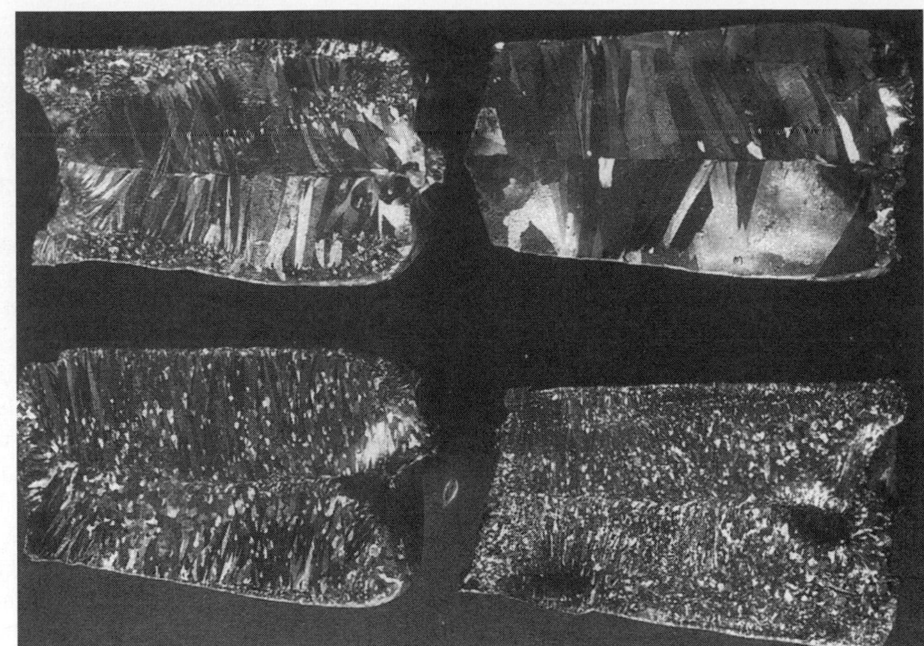

Figure 4.44 Aluminium castings showing the effect of the casting temperature on the type and size of the grain formation.

Figure 4.45 A vacancy point defect.

Figure 4.46 An interstitial point defect.

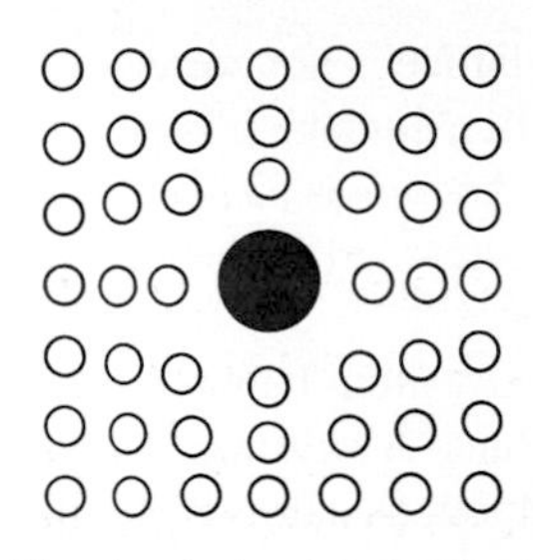

Figure 4.47 A substitutional point defect.

compressive stress fields to form around them. It therefore requires a higher stress to deform the material at the position of these point defects and so they serve to strengthen the material.

Solid solution strengthening is achieved by intentionally introducing interstitial and substitutional defects into a material to produce an alloy with greater strength than the pure metal.

An incomplete line of atoms causes a linear imperfection in the lattice and is termed a *dislocation*. At an *edge dislocation* a partial plane of atoms is present and a *screw dislocation* is where the atoms in an original plane have been displaced so that they form a spiral about a line perpendicular to the original plane of the material. When a shear force is applied perpendicular to the plane of atoms in an edge dislocation, the bonds between the atoms in the next plane can be broken and part of them joined to the original dislocation to form a complete plane of atoms as shown in Fig. 4.48. The next plane of atoms is now incomplete and so the dislocation has effectively moved through the material. This is called *slip* and occurs in metals and some ceramics and polymers. F.c.c and h.c.p. crystals have close-packed planes and a low shear stress is required to cause slip of the dislocations. B.c.c. crystal structures

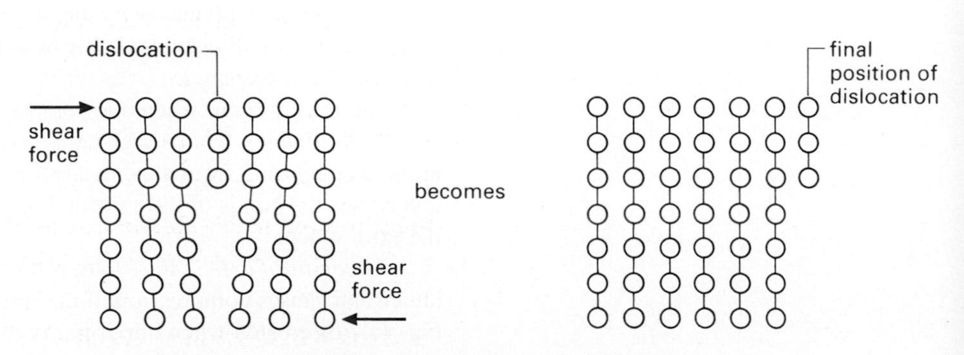

Figure 4.48 Movement of a dislocation through a crystal lattice.

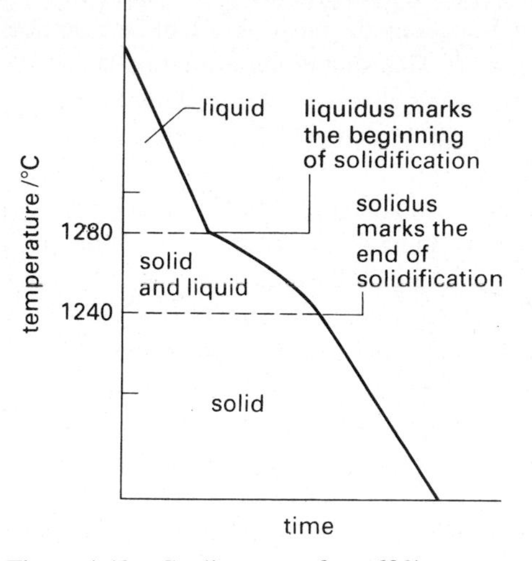

Figure 4.49 Cooling curve for a 60% copper, 40% nickel Ni alloy.

have no close-packed planes and require a greater force to move the dislocations. The b.c.c. crystal lattice materials therefore have greater tensile strength than those of the other crystal structures.

A single dislocation would weaken the material as only a fraction of the lattice bonds would need to be broken to deform the material. However, the more dislocations that are present the more likely it is that one of them will be blocked by a point defect or another line defect and therefore the strength of the material is increased. This is the basis for *strain hardening*. If a material can be worked at ambient temperatures without excessive hardening it is known as *cold-working*. If a material is worked above its recrystallisation temperature it is known as *hot-working*. At the recrystallisation temperature lattice imperfections such as dislocations can disperse and allow the atoms to restructure themselves more regularly. This restructuring process, known as *recrystallisation*, is dealt with in greater detail in section 4.3.5.

4.3.4 Equilibrium structures

Elements may be alloyed in varying compositions which are stated as a percentage by weight. For example a sample of copper-nickel alloy of 60% copper and 40% nickel will contain a weight of copper equivalent to 60% of the total weight of the sample.

The elements to be alloyed are heated together until they form a liquid solution which is then cooled. Unlike the cooling curve for pure copper, shown in Fig. 4.41, where solidification takes place at a constant temperature, the alloy freezing may take place over a range of temperatures. Fig. 4.49 shows the cooling curve for 60% copper and 40% nickel alloy. At 1280°C the solution starts to solidify and until 1240°C a paste of mixed liquid and solid is present. By 1240°C the alloy is completely solid.

We are familiar with substances that have different solubilities in a liquid. For instance, oil is insoluble in water and being less dense, will sit on top of the water and each substance can be clearly identified. On the other hand, if we slowly add salt to a glass of water it will at first dissolve and a single phase solution called brine will be created. However, the continued addition of salt will produce two phases; a saturated brine solution and excess salt deposited on the bottom of the glass. Salt exhibits a *partial solubility* in water. Alcohol is completely soluble in water and, when a solution of alcohol and water is made, only one phase is produced as a result, regardless of the proportions of alcohol and water. This is an example of *total* or *unlimited solubility.*

When a hot, liquid solution of two materials is cooled the resultant alloy can behave in any of these three ways: an *alloy of insoluble solids* is formed when the two materials are completely insoluble; a *partially soluble solid* is formed when there is partial solubility between the elements; and a *solid solution* is formed when only one phase is produced regardless of the relative quantities of the alloying materials.

Solid solution alloys

Solid solutions can be formed in one of two ways.

Substitutional solid solution Atoms of the alloying (solute) element replace the atoms of the basis (solvent) metal in the lattice. This may occur in either a random or an ordered way.

Interstitial solid solution An alloying element with smaller atoms than the solvent metal can squeeze its atoms into the spaces between the parent lattice atoms.

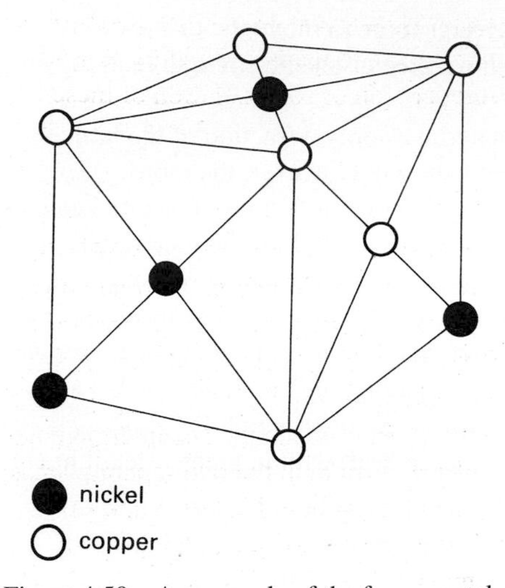

Figure 4.50 An example of the f.c.c. crystal structure of a copper-nickel alloy; the lattice sites are filled randomly by copper or nickel atoms.

Copper and nickel atoms show unlimited solid solubility and are of similar size. They do not separate on cooling but randomly occupy the positions of an f.c.c. lattice forming a substitutional solid solution, Fig. 4.50. This single-phase solution is an alloy with uniform composition and properties, and which has the appearance of a pure metal under the microscope. Other binary alloys that exhibit complete solid solubility are gold-silver and bismuth-antimony.

If we established cooling curve data for all possible compositions of a copper-nickel alloy, such as that contained on Figs 4.41 and 4.49, we could collate it on one *thermal equilibrium diagram* or *binary phase diagram*. This diagram is a temperature-composition graph which shows the degree of solidification taking place during the cooling of an alloy from its hot liquid state. The term 'equilibrium' is used to indicate that solidification has occurred under conditions which have given very slow cooling so that *diffusion* of the alloy is allowed to take place. Diffusion is the movement of atoms through the structure to give an alloy of a uniform composition.

The thermal equilibrium phase diagram for copper-nickel alloy is shown in Fig. 4.51. It is called an *isomorphous* phase diagram because it displays unlimited solid solubility. The freezing point of pure copper is shown to be 1083°C on the left-hand side of the graph. The first and second *arrest points* for 60% copper, 40% nickel are marked on the diagram at 1280°C and 1240°C. The first arrest point is the lowest temperature at which the alloy is completely liquid and the second arrest point is the highest temperature at which the alloy is completely solid. The line joining all the first arrest points for different compositions of copper and nickel gives us the *liquidus*. This indicates the temperature at which an alloy of any composition begins to solidify. The lower arrest points are joined to give the *solidus* or the point at which the alloy becomes completely solid. The area in between the liquidus and solidus is the two-phase region where the material is a mixture of solid and liquid.

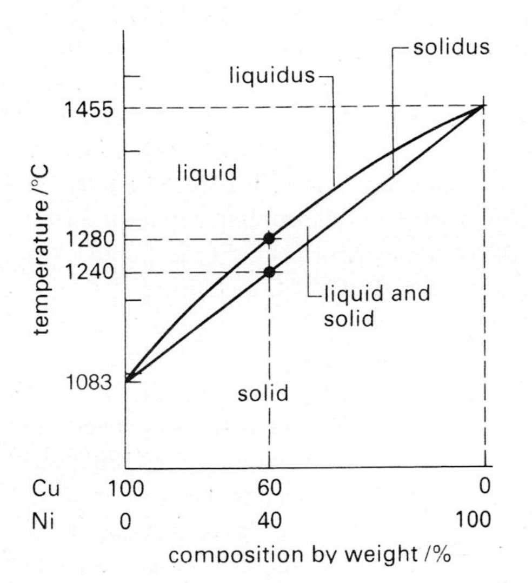

Figure 4.51 The thermal equilibrium phases diagram for copper-nickel alloys.

The ratio of solid to liquid at a particular temperature and composition is given by the *inverse lever rule*. For an alloy of given composition X at a temperature T °C, represented by P in Fig. 4.52, the relative masses of the two phases is given by:

$$\text{mass of liquid } L \times \text{distance MP} = \text{mass of solid S} \times \text{distance NP}.$$

$$\text{Therefore} \quad \frac{\text{mass of liquid } L}{\text{mass of solid } S} = \frac{\text{NP}}{\text{MP}}$$

Many other metallic and ceramic alloy systems show similar characteristics to the copper-nickel alloys. One example of a ceramic alloy which has a similar isomorphous phase diagram is that of nickel oxide and magnesium oxide. Polymers

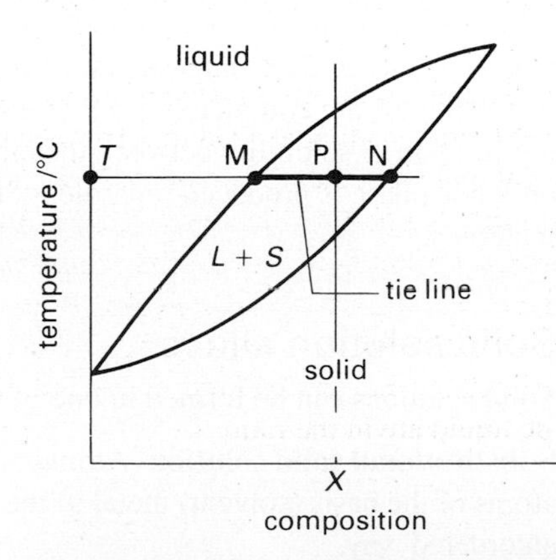

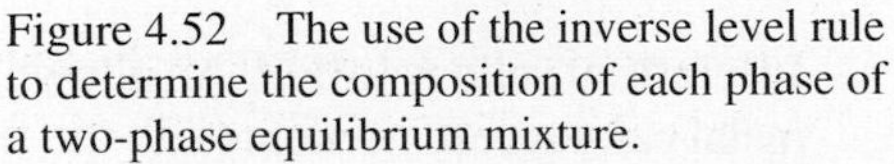
Figure 4.52 The use of the inverse level rule to determine the composition of each phase of a two-phase equilibrium mixture.

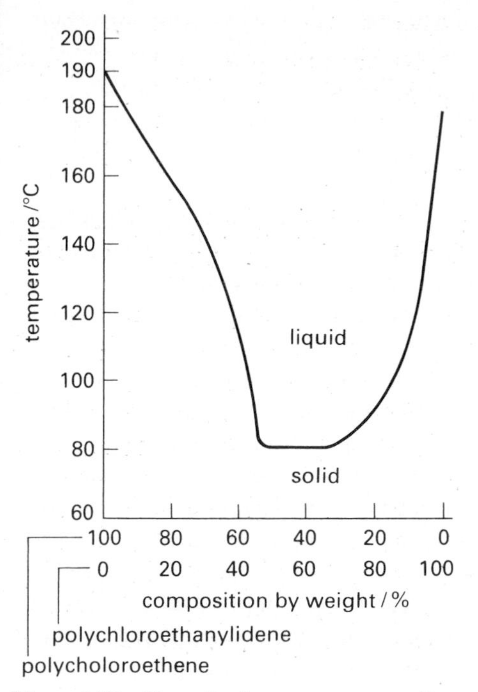

Figure 4.53 The softening temperatures of long polymer chains formed of polychloroethane and polychloroethanylidene.

do not behave in the same way although polychloroethene and polychloroethanylidene molecules, for instance, combine in any ratio to form long polymer chains. A graph showing the softening temperatures of any composition of solid solution of these substances is shown in Fig. 4.53.

Insoluble solid solution alloys

Cadmium-bismuth alloys are soluble in the liquid state but completely insoluble in the solid state for which two separate solid phases are identifiable in the final alloy structure. Other such alloys are those of copper-lead and tin-zinc. An equilibrium diagram can also be constructed for these alloys from cooling curve data but it will look very different from the solid solution alloy diagrams.

Fig. 4.54 shows the equilibrium diagram for cadmium-bismuth. The intermediate region between the liquid and solid phases is clearly visible in the two separate areas; one of liquid solution and solid cadmium and the other of liquid solution and solid bismuth. There is one point, E, where the two arrest points coincide and the two solids form *eutectic* structures at a constant temperature of 140°C with no intermediate stage of mixed liquid and solid.

This is the *eutectic temperature*, which is the lowest melting point and the temperature at which both solid phases are in equilibrium with the liquid phase. The composition of 60% bismuth and 40% cadmium alloy is therefore said to be a *eutectic alloy* which freezes at a constant temperature of 140°C.

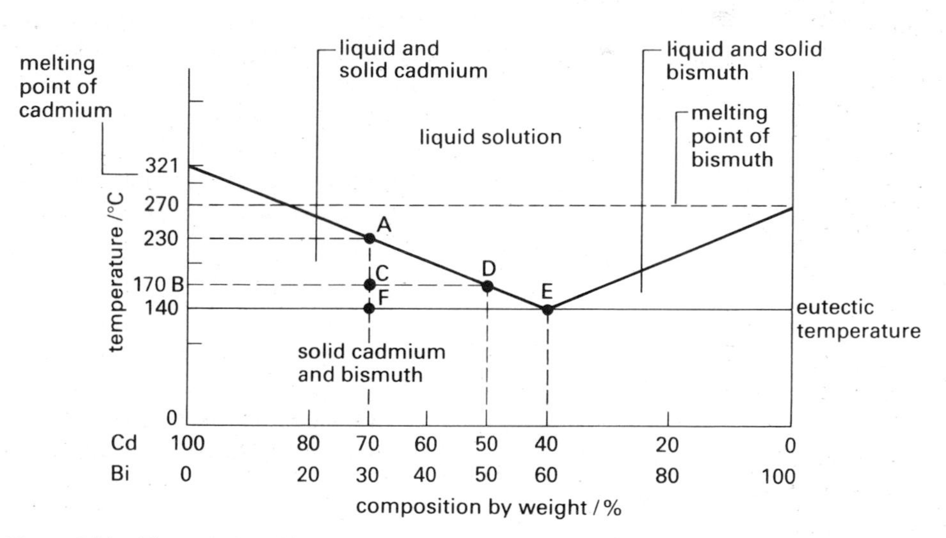

Figure 4.54 The cadmium-bismuth equilibrium phase diagram.

To the left of this eutectic point in Fig. 4.54 it can be seen that there is more cadmium than that needed to form the eutectic alloy structure and similarly to the right there is excess bismuth. A 70% cadmium, 30% bismuth alloy will begin to solidify at point A, found to be 230°C by measurement of a scale diagram, where the excess cadmium will start separating and crystallising. At 170°C the composition of the liquid is given by that of point D, the point on the liquidus at 170°C and corresponds to 50% bismuth, 50% cadmium. Applying the inverse lever rule and using the values for cadmium, it is found that the mass of solid cadmium and that of the liquid are in the ratio

$$\frac{CD}{BC} = \frac{2}{3}$$

This separation continues until the remaining liquid is of eutectic composition; the eutectic alloy structure will form at 140°C by the laying down of crystals of cadmium and bismuth simultaneously. On microscopic examination, the crystals of cadmium can be seen clearly interspersed within the eutectic 60% bismuth, 40% cadmium. A cadmium-bismuth alloy with greater than 60% bismuth will show crystals of bismuth within the eutectic structure.

Partially soluble solids

Copper and zinc exhibit limited solubility in the solid state. If the zinc content is less than 40% an f.c.c. solid solution is formed. If there is more than 40% zinc the excess zinc atoms combine with some of the copper to form a compound and two solid phases are now present; a saturated solution of zinc in copper plus the copper-zinc compound.

Lead and tin are also examples of metals which have limited solid solubility in each other, and so lead-tin alloys form partially soluble solids. As lead is heated, tin can be dissolved in it until a maximum composition of 19% tin is reached at 183°C. If a 10% tin, 90% lead alloy is cooled to between 140°C and 270°C, a single-phase solid is formed. If it is cooled below 140°C a two-phase solid is formed.

The lead-tin thermal equilibrium diagram is shown in Fig. 4.55. It is a combination of the two types of solid solution that we have previously examined and shows areas of solid solubility as well as a eutectic. The liquidus can be identified as the curve CED and the solidus is CBEFD. The line AB is known as the *solvus*, which shows the maximum solubility of tin in lead at the indicated temperature just as the solvus FG shows the maximum solubility of lead in tin.

Lead-tin alloys are the basis for most solders and by examining different compositions of alloy it can be seen that four different solid structures of alloy can be formed. Alloys that contain between 0% and 2% tin, such as that marked V on Fig. 4.55, will pass through a brief intermediate stage of liquid and a single-phase solid solution of tin in lead, α, before becoming completely solid. After the alloy has been cooled to 0°C it is completely composed of the solid solution, α. This composition of alloy shows similar behaviour to copper-nickel alloys.

An alloy with between 2% and 19% tin, for example W on Fig. 4.55, also forms an α + L stage and then solidifies to a single-phase solid solution, α. However, as

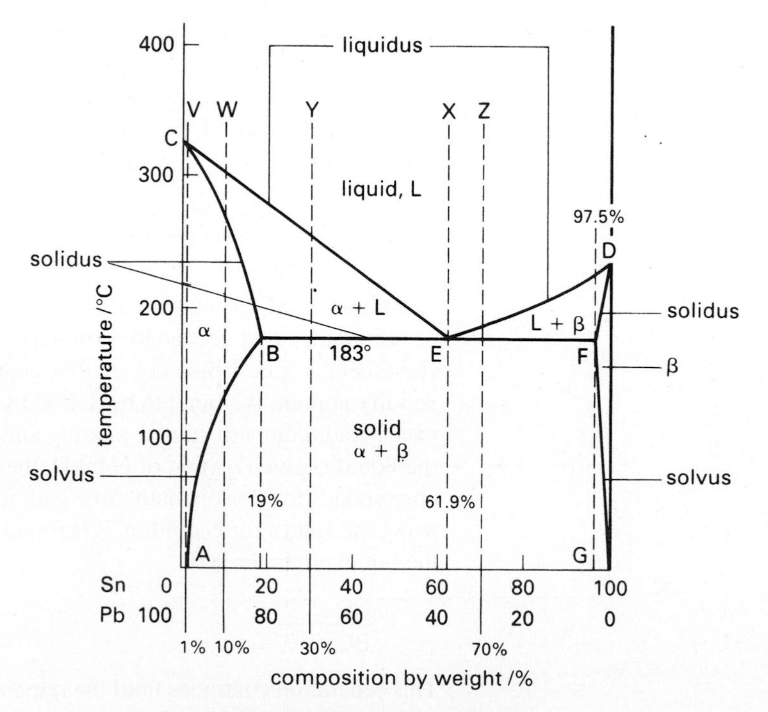

Figure 4.55 The lead-tin equilibrium phase diagram.

the alloy cools past the solvus the solution becomes saturated and a solid state reaction causes a second solid phase, β, to precipitate from the original a phase, leaving a two-phase solid of α + β.

An alloy of 61.9% tin and 38.1% lead, marked X, has the eutectic composition and when it cools to 183°C, the eutectic reaction causes platelets of the two solid solutions α + β to form. The compositions of α and β are found by the values, B and F, at the end of the eutectic line; the composition of α is 19% tin and 81% lead, and that of β is 97.5% tin and 2.5% lead.

Plumbers' solder, represented by the composition at Y in Fig. 4.55, is in the range 19% to 61.9% tin. It begins to solidify at the liquidus temperature of 260°C and has a fairly long intermediate stage in which the pasty consistency of the solid and liquid mixture lasts for some time, while the α solid increases during cooling, which allows the plumbing joint to be fashioned satisfactorily. As the alloy cools to the eutectic temperature, the eutectic platelet structure of α + β is laid down around the existing *primary microconstituent* α regions. Alloys with 19% to 61.9% tin are *hypoeutectic* alloys as they are alloys containing less than the eutectic composition of tin.

An alloy such as Z with between 61.9% and 97.5% tin is a *hypereutectic* alloy that contains more than the eutectic amount of tin. This would consist of the eutectic structure around solid β formations.

Intermediate compounds

In some binary alloys the two elements combine chemically to form an *intermediate phase*. There are two types of intermediate phase. One is where a new compound of fixed composition is formed with a new crystal structure and properties, for example, iron carbide or *cementite* (Fe_3C). In such compounds one element (for example, iron) exhibits much stronger metallic chemical properties than the other (for example, carbon). Other similar intermediate compounds are represented by the formulae $CuAl_2$, Cu_3P, Cu_3Sn, Mg_2Sn and Cu_2Sb. In general these compounds are hard and brittle, appearing as angular particles in the microstructure and contributing hardness to the alloy properties. Where both elements are metal the compound is known as an *intermetallic compound*. The existence of such an intermediate compound is shown by a vertical line on the phase diagram. Fig. 4.56 shows the equilibrium diagram for two metals, aluminium and antimony, which form an intermetallic compound, AlSb, referred to as γ.

It is possible for an intermediate phase of a compound of variable composition to be formed as occurs in the copper-zinc and molybdenum-rhodium alloy systems. Such a phase is represented by the region, γ, in Fig. 4.57.

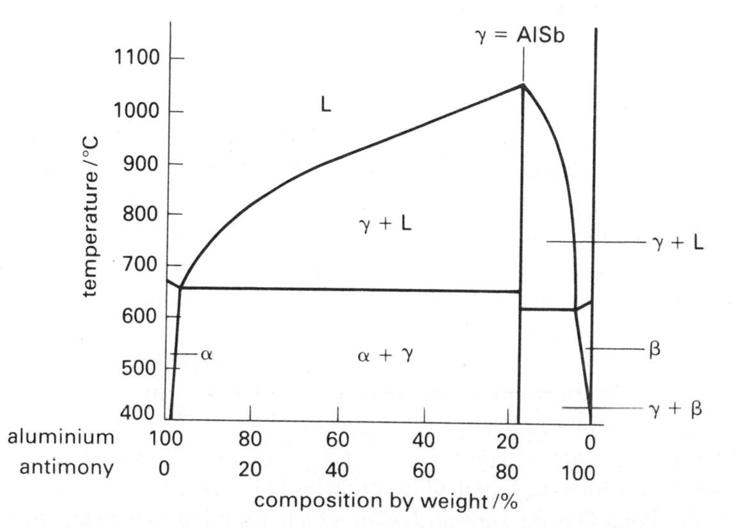

Figure 4.56 The aluminium-antimony equilibrium phase diagram.

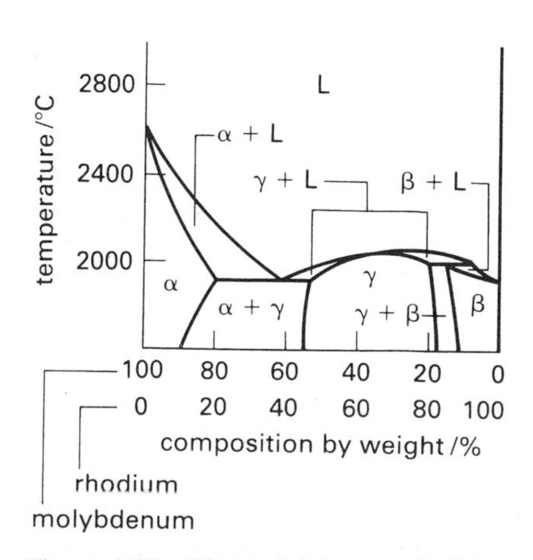

Figure 4.57 The molybdenum-rhodium equilibrium phase diagram.

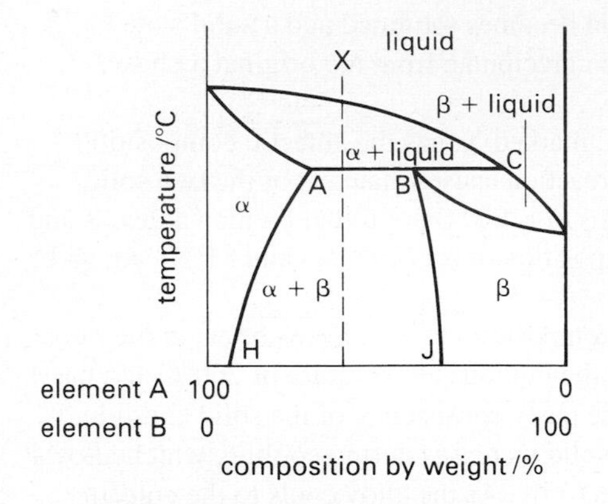

Figure 4.58 A peritectic phase diagram.

Peritectic reactions

In some alloys with partial solid solubility the solid already formed in the cooling of the alloy reacts with the remaining liquid. The final solid formed has an intermediate composition between that of the original solution and that of the liquid. This is a *peritectic reaction* and occurs within the iron-carbon and copper-zinc alloy systems. It is illustrated in a peritectic equilibrium diagram such as Fig. 4.58. This shows how an alloy of composition X initially forms solid α and liquid. Once the peritectic line, ABC, is reached some of the α solid reacts with the liquid to form a new solid, β, and when solidification is complete a two-phase solid exists composed of primary α, of composition A, and secondary β, of composition C.

The iron-carbon phase diagram

The iron-carbon diagram is probably the most important equilibrium diagram of all to consider as plain carbon steels (up to 1.7% carbon) and cast irons (2% to 4% carbon), formed by the addition of carbon to iron, are such vital engineering materials.

Iron itself is allotropic and has three distinct crystal structures formed at *critical temperatures* or arrest points on cooling from its freezing point of 1535°C.

δ-iron is a b.c.c. structure iron stable above 1390°C
γ-iron is an f.c.c. structure iron stable between 1390°C and 910°C
α-iron is a b.c.c. structure iron stable below 910°C

These critical temperatures that mark the change of form are different for the heating and cooling of iron. As the iron changes between the different structures there is a change of volume associated with the structure because the f.c.c. crystals in γ-iron are more densely packed than the b.c.c. crystals in δ-iron and α-iron.

Iron and carbon have only partial solid solubility although they are completely soluble in the liquid phase. The addition of carbon lowers the freezing point of iron and alters the critical temperatures marking the change in crystal structure. The new crystals formed are interstitial solid solutions of carbon atoms in iron. Carbon has only very limited solid solubility in b.c.c. iron but it will form *ferrite* with α-iron at low temperatures and δ-*ferrite* with δ-iron at high temperatures. Ferrite contains up to 0.022% carbon while δ-ferrite can dissolve up to 0.09% carbon. *Austenite* is formed by the solution of up to 2.11% carbon in γ-iron (f.c.c.). The solid solutions of carbon in iron are all soft and ductile but stronger than the pure iron because of solid solution strengthening. When the solubility of carbon in a particular crystal structure is exceeded an intermetallic compound, cementite (6.67% carbon), is formed. This is extremely hard and brittle and is present in all of the commercially produced steels. Varying the amount and size of cementite allows the properties of the steel to be changed.

The equilibrium diagram which illustrates the phases present in iron-carbon alloys up to 6.67% carbon, the carbon content of cementite, is shown in Fig. 4.59. It is not a true equilibrium diagram as the cementite can decompose to graphite and so it is said to be a *metastable* system. Several different reactions such as eutectic, eutectoid and peritectic can be recognised in the phase diagram.

Between points A and B in Fig. 4.59 it can be seen that there is a peritectic reaction where δ-ferrite, containing 0.7% carbon, on cooling to 1490°C reacts with the remaining liquid, containing 0.55% carbon to form a new solid solution, austenite, containing 0.18% carbon. By the time the alloy has cooled to room temperature the solid structure has again changed and so this peritectic reaction has no effect on the final structure and properties.

Point E is the *eutectic point*, at a composition of 4.3% carbon and a temperature of 1130°C, at which the liquid alloy solidifies to a mixture of austenite and cementite. Point G is the *eutectoid point* which is similar to the eutectic except that a solid is cooling and changing structure at this point. Here, at a composition of 0.8%

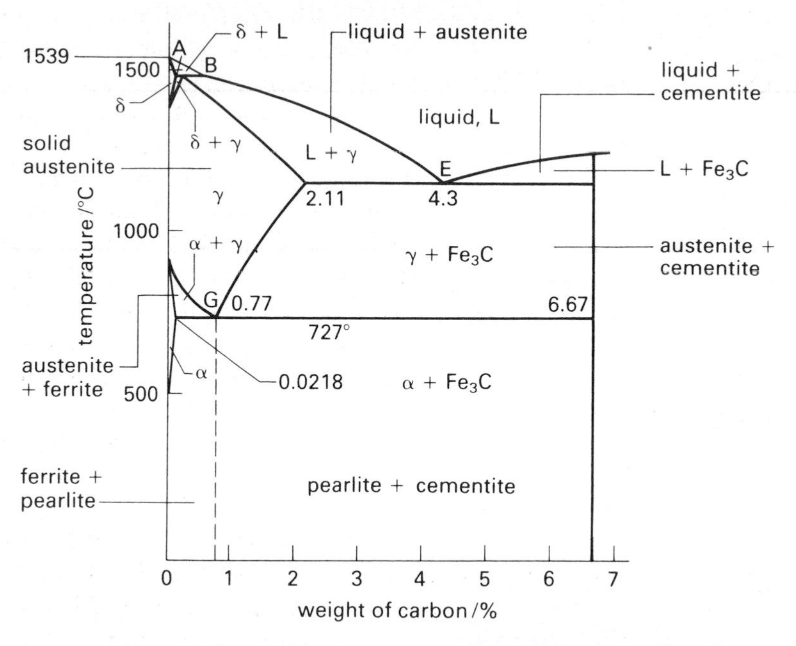

Figure 4.59 The iron-carbon equilibrium phase diagram.

carbon and 723°C, the solid austenite transforms into a fine lamellar structure of ferrite and cementite by a process of diffusion. This new solid has a pearly appearance and is known as *pearlite* and is shown in Fig. 4.60. Most of the austenite disperses to soft ferrite which surrounds the layers of brittle cementite thus causing dispersion hardening. The temperature at which the eutectoid reaction takes place is the same for all plain carbon steels.

Hypoeutectoid steels are those with less than 0.8% carbon and their primary microconstituent is ferrite. *Hypereutectoid steels* have more than 0.8% carbon and have cementite as their primary microconstituent. When a steel is cooled sufficiently it will have one of the following structures.

- Less than 0.006% carbon. Pure ferrite alloy, classed as commercially pure iron.
- 0.006% carbon to 0.83% carbon. Ferrite and pearlite present. The amount of pearlite increases as the carbon content increases.
- 0.83% carbon. Only pearlite present.
- 0.83% carbon to 1.7% carbon. Cementite and pearlite present. The amount of cementite increases as the carbon content increases.

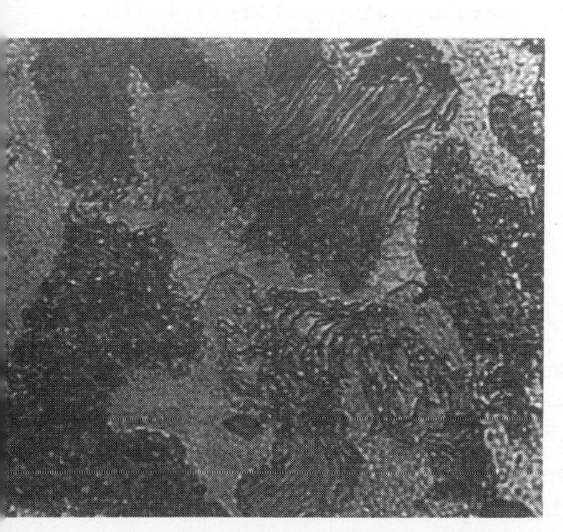
Figure 4.60 Pearlite formations in an EN9 steel.

Commercially produced steels rarely contain more than 1.4% carbon and other elements such as manganese, nickel and chromium are deliberately added to improve the properties of the steel. Elements such as phosphorus and sulphur are also present in the alloy because they occur as impurities in the ore.

Brass equilibrium diagram

Brass is a binary phase solid solution of copper and zinc. Alloys with less than 38% zinc form a single-phase solid solution, α, of zinc in copper which has good properties for cold-working and machining. As the percentage of zinc increases the mechanical properties such as hardness and corrosion resistance improve. The ductility, in fact, increases up to 30% zinc and then it decreases again until at 38% zinc, the limit of the a phase, the ductility is the same as for pure copper. Brasses with 38% to 45% zinc are much better hot-worked. Fig. 4.61 shows the binary phase diagram of brass containing up to 60% zinc.

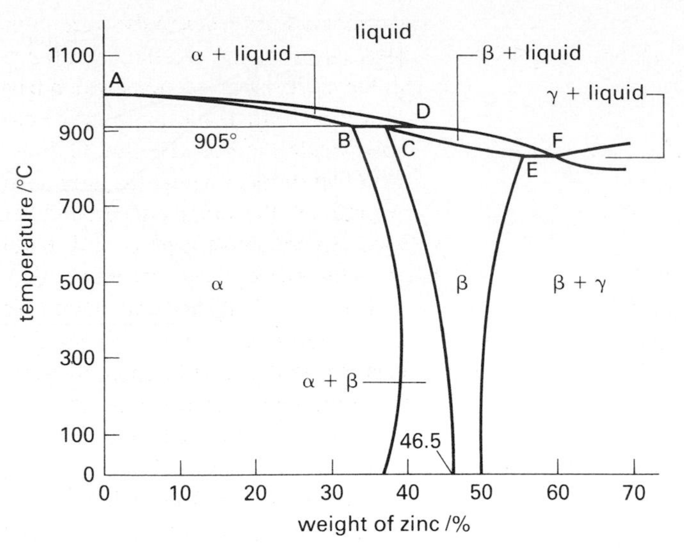

Figure 4.61 Copper-zinc equilibrium phase diagram.

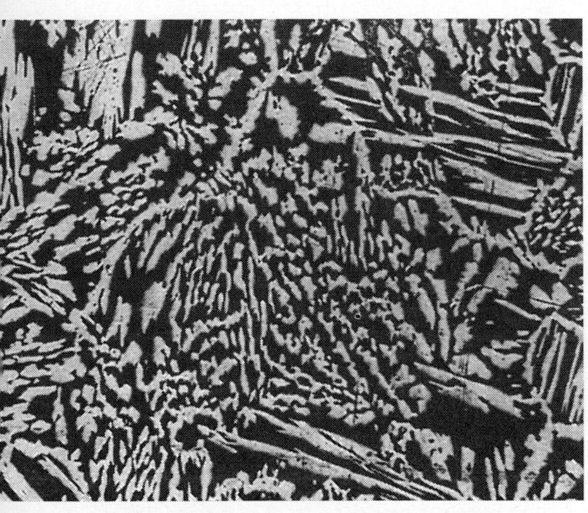

Figure 4.62 The microstructure of brass showing the light α regions and the dark β regions.

For compositions between that of A and B, zinc dissolves in copper to form the solid solution α. This is an f.c.c. structure and the grains that are formed are usually cored (see section 4.3.5 for an explanation of coring). Compositions between B and C solidify initially as a solid solution and then at 905°C they undergo a peritectic reaction to form an α + β phase. As this cools some of the solid changes back to α solid as can be seen by the phase lines from B and C. Compositions between C and D initially form α solid and then at 905°C this changes to form β solid. On cooling further it can be seen that α solid is precipitated to give an α + β solid. Compositions between D and E solidify to a β solid with compositions of less than 46.5% precipitating some *α* solid and those above 50% precipitating some γ phase. Fig. 4.62 shows the microstructure of a typical brass with the light α regions and the dark β areas.

Three-phase binary reactions

When two elements are combined many complex combinations can be produced. The three principal three-phase reactions – the eutectic, the peritectic, and the eutectoid – are identified in Fig. 4.63.

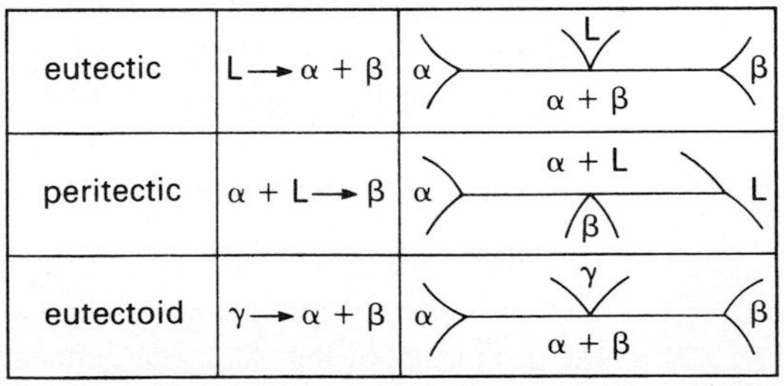

eutectic	L → α + β	α, L, α + β, β
peritectic	α + L → β	α, α + L, β, L
eutectoid	γ → α + β	α, γ, α + β, β

Figure 4.63 The three principal three-phase binary reactions.

4.3.5 Nonequilibrium structures and heat treatment

The imperfections in a crystal structure are critical in determining the properties of the material. These imperfections form as a result of the structural changes which occur as the material solidifies and cools. The solidification temperature and the type of nucleation determine the nature and speed of the grain growth, and hence their orientation and size. As the solid material cools, phase changes will occur, but these are unlikely to proceed to conclusion because the time required for the solid state diffusion to occur is greater than the cooling time between the relevant temperatures.

This sometimes leads to the formation of grains in which the composition varies between the centre and the grain boundaries. This is known as *coring*. Together with all the lattice imperfections which have developed and the results of volume changes associated with phase transformations, the final outcome is a crystal structure which is not in equilibrium. The solid has a multitude of strained regions and if any movement is promoted within the atomic structure these can be relieved with an associated reduction in the stored energy, and hence a movement towards equilibrium. Typically the movement towards equilibrium will result in a significant loss of strength, although there may be improvements in the material's ductility.

The phenomena associated with the formation and control of nonequilibrium structures are a key feature in our ability to manipulate the properties of a material. Together with the addition of alloying elements, they allow the creation of almost any combination of properties and this is particularly true in the case of steels. All common steels are nonequilibrium structures and it is the controllability of these structures and their stability which helps make steel such a remarkable material in the service of man. The usefulness of aluminium alloys in the production of strong, lightweight components is also a result of the control of nonequilibrium structures. These are the two most common examples and they are therefore discussed in more detail below. It is worth noting, however, that similar principles may be applied to many other metals and ceramics.

The main means of promoting the change in nonequilibrium structures is the application of heat and this is discussed in depth here. However, ultrasonic vibrations have also been used to promote the relief of stresses within steel fabrications without the application of heat.

Heat treatment of steels

The iron-carbon phase diagram and the formation of the equilibrium structures of austenite, ferrite, cementite and pearlite were examined in the previous section. If steels with a sufficiently high content of alloying elements are cooled rapidly then the nonequilibrium structures, bainite and martensite, will result. This is illustrated in Fig. 4.64 which shows a schematic version of the time-temperature transformation (t.t.t.) diagram for a eutectoid steel on which three cooling curves, A, B and C, are shown. If the cooling of a steel follows path A and takes about 10 s to cool to 500°C, then there is sufficient time for all the austenite to transform to pearlite. If path B is followed, it takes about 5 s to cool to 500°C, then bainite is likely to form. If however 500°C is reached in about 1 s, as in path C, then the cooling is sufficiently rapid for the austenite to transform to martensite which is a highly distorted form of the austenite f.c.c. structure.

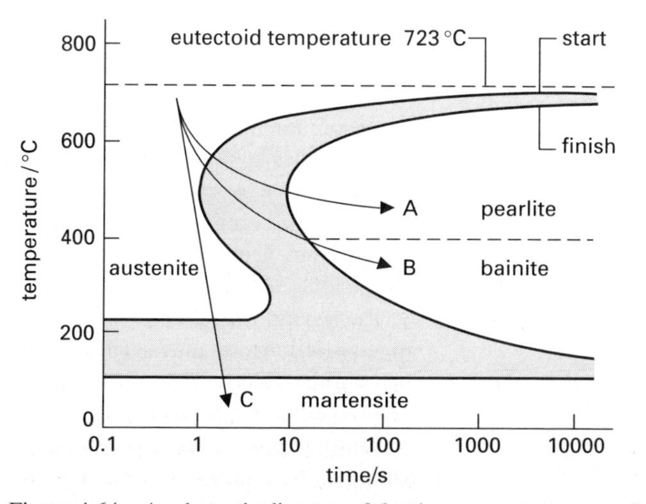

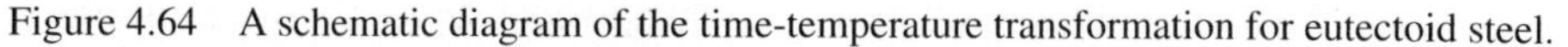

Figure 4.64 A schematic diagram of the time-temperature transformation for eutectoid steel.

Bainite is a highly strained structure and it cannot be transformed to a more useful structure on reheating. Therefore the heat treatment of steels normally involves either cooling rapidly (*quenching*) to promote the formation of martensite, or cooling very slowly so that pearlite results.

Quenching hot material in such a violent manner is only possible for small components. It can also result in cracking because of the high thermal stresses. For these reasons various alloying elements are added to steels in order to promote the formation of martensite and reduce the risk of the formation of bainite. Manganese, chromium, nickel, vanadium and molybdenum all act in this way and are said to increase the *hardenability* when added in small quantities. They do this by moving the 'pearlite nose', as the centre region of the t.t.t diagram is called, to the right, thus allowing martensite to be formed at lower cooling rates.

The martensite that is formed is usually much too hard and brittle to use. It is therefore heated and allowed to decompose partially into more stable, tougher structures. This process called *tempering* is described below.

Annealing

There are two forms of annealing, subcritical annealing and full annealing, which take place at the temperatures indicated in Fig. 4.65. Subcritical annealing is carried out in the temperature range of 650–700°C, which means that there are no phase changes. It is used to remove the effect of work-hardening and this is achieved through the process of recrystallisation. The strains are removed from the atomic structure through the growth of new grains from initiation sites at the grain boundaries. This is shown in Fig. 4.66.

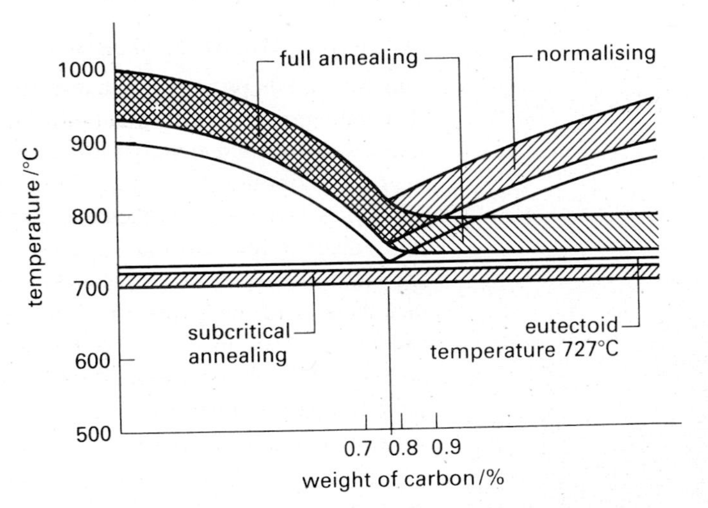

Figure 4.65 The annealing and normalising temperatures of steel.

It is important that the component or assembly being annealed is allowed to soak so that it reaches an even temperature, and it must then be allowed to cool slowly. This is usually done by turning off the oven and waiting until it has cooled before removing the component. For small components, good results can be obtained in the workshop by heating to the required temperature and then surrounding the components with firebricks or sand in order to achieve the required slow cooling.

The second annealing process, full annealing, is used to reverse the formation of nonequilibrium structures. This means that the steel must be heated to allow all of the bainite or martensite structures to transform back to austenite. For a hypereutectoid steel this simply means heating above 723°C as the cementite in the structure does not cause complications. Again it is necessary to allow the components to soak in order to ensure the transformation to austenite is complete and then to cool slowly to give equilibrium structures.

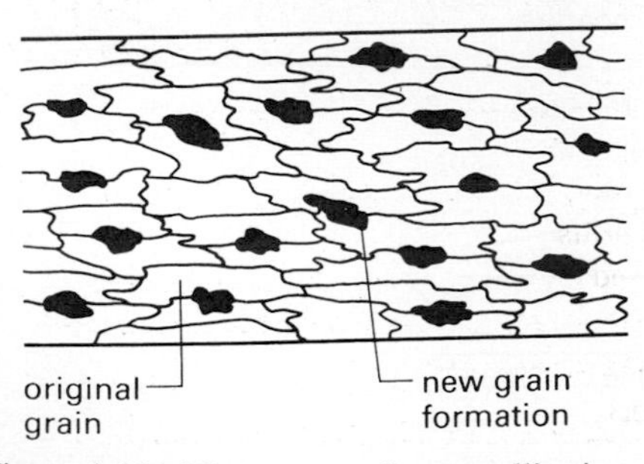

Figure 4.66 The process of recrystallisation.

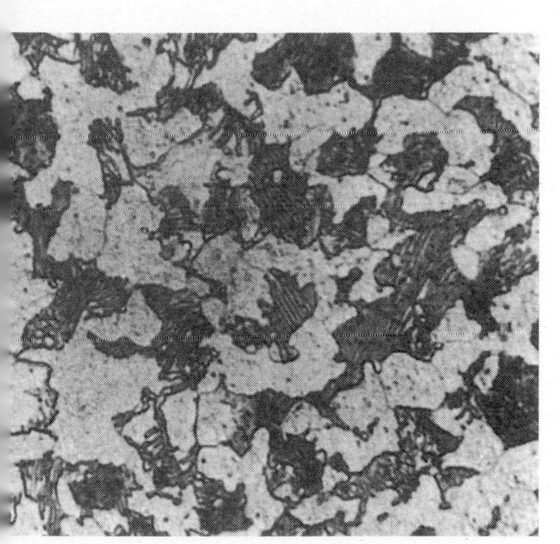

(a)

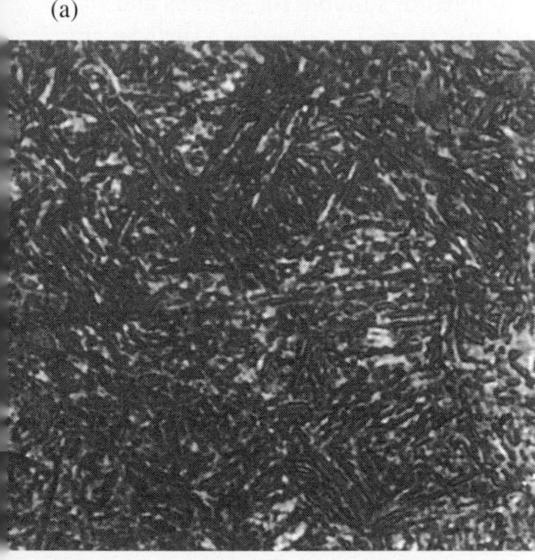

(b)

(c)

Figure 4.67
(a) The microstructure of a 0.7% carbon steel.
(b) The microstructure after it has been quenched.
(c) The microstructure after being normalised.

Normalising

Normalising is a process closely related to annealing, and is used on steels to give a finer grain structure. Figure 4.67 shows the effect of water quenching and normalising on the microstructure of a 0.7% carbon steel. The very slow cooling associated with annealing gives a soft, ductile structure but there is also often excessive grain growth. A faster cooling rate, usually achieved by cooling in air, will give smaller grains and a correspondingly tougher and stronger material. It is essential when normalising for the high-temperature structure to be fully austenite, and consequently hypereutectoid steels must be heated to a higher temperature than for annealing. For hypoeutectoid steels the temperatures for the annealing and normalising will be the same.

Hardening and tempering

As we have already seen, hardening depends on the fast cooling of austenite to produce martensite. Consequently the starting temperatures required to harden or fully anneal steels will be the same. At these temperatures the only phases present are austenitic, and in hypereutectoid steels, cementite. It will be found that the hardening effect on steels with a carbon content of less than 0.3% is negligible, because the atomic structure is insufficiently strained, but above this carbon content fast cooling will give a hard brittle structure. Where the hardenability of the steels is great enough an oil rather than a water quench can be used. This gives a slower cooling rate and consequently less risk of cracking as a result of thermal stresses. This risk can also be reduced by avoiding sharp corners or sudden changes of section wherever possible in the components to be treated.

As with annealing and normalising, it is important to ensure that the component is soaked to ensure that all of the structure reaches the required temperature and that the cooling rate is sufficiently rapid where hardening is required. Heating the components significantly above the normalising temperatures should however be avoided as this merely results in grain growth.

In order to produce useful components it is necessary to temper them. This process is entirely temperature-dependent and allows some of the martensite to transform. The higher the temperature the greater the percentage of martensite transformed and the softer the structure. The temperature is also indicated by the thickness of the oxide layer which causes the component to change colour as it it heated. These colour changes are the common method used in workshops to indicate the tempering temperature. These techniques will give useful results, although the control offered by a furnace is to be preferred where available. Table 4.12 shows typical temper colours apparent at different temperatures and some common applications for this degree of tempering. It should be evident from this table that as the toughness required by the application increases then so does the tempering temperature.

Tempering temperature (°C)	*Colour*	*Typical components*
220	pale straw	cutting edges
230	medium straw	lathe tools
240	dark straw	twist drills
250	brown	taps and dies
260	brownish-purple	press tools, plane irons and woodworking tools
280	purple	cold chisels and axes
300	blue	springs

Table 4.12 Applications of tempering temperatures and their indicators

Case hardening

It has already been mentioned that it is only possible significantly to harden steels with a carbon content greater than 0.3%. Advantage can be taken of this to produce components with a 'soft' inside and 'hard' surface by *carburising* the surface (raising the carbon content) prior to heat treatment. Such components retain the toughness associated with mild steel, whilst gaining the hardness and wear-resistance produced by hardening and tempering. The surface of the components can be carburised by prolonged heating in a carbon-rich environment, during which time carbon will be absorbed to a depth of between 0.5 mm and 1 mm.

The carbon-rich environment can be solid, liquid or a gas. The oldest method is to pack the components in a sealed box surrounded by charcoal and other substances and then heat it for a few hours at 850–900°C. It is possible to leave some areas soft if they are protected by copper plating or a coating of clay. A sodium cyanide bath provides a liquid medium for carburising. This method ensures even heating and carburising, but the fumes and the molten salt are dangerous. This process needs a carefully controlled industrial plant and is therefore not suitable for schools and colleges. It is also possible to use a carbon-rich gas flame to carburise a surface; any of methane, butane and propane are suitable. Once the surface has been carburised the components will need suitable heat treatment.

Proprietary hardening compounds are available for use in school workshops. Typically these are in powder form and can be sprinkled on to small components once they have been brought to red heat. This will result in a hardened layer although not of the depth and quality of that produced by the commercial process.

With alloy steels it is possible to produce hardening by *nitriding* which results from the absorption of nitrogen. The components to be hardened are heated in the presence of ammonia.

Precipitation and age hardening

The process by which martensite is softened by tempering is the precipitation of carbon in the form of carbide (ε-carbide) from the atomic structure. Low-temperature tempering produces a very finely dispersed distribution of ε-carbide in low-carbon martensite. As the tempering temperature increases, the carbon content of the martensite continues to be reduced by further carbide precipitation. A similar process also occurs with nonequilibrium structures in aluminium alloys except that the precipitate hardens and strengthens the structure rather than softening it and making it more ductile as it does in steels. Similarly copper-beryllium alloys may be made as strong as some tool steels by careful control of the precipitation process.

Generally the formation of the precipitate requires the alloy to be reheated and the alloys are said to *age-harden*. The age-hardenable aluminium alloys are known as duralumins and contain varying amounts of elements such as copper, manganese, magnesium, silicon, iron and zinc. The heat-treatment of these alloys begins by *solutionising*, which means heating the alloy until a uniform high temperature structure is obtained. This is analogous to the formation of austenite in steels prior to heat treatment. The alloy is then quenched in order to preserve the high-temperature structure as a supersaturated solution. Precipitation can now be made to follow, which results in the strengthening of the structure in aluminium and copper- beryllium alloys.

Fig. 4.68 shows the critical nature of the reheating temperature for duralumins. If the alloys are heated to the correct temperature then a hard, strong structure will result. This strength is obtained by having just the right dispersion and size of precipitate throughout the structure. If the reheating temperature is too low the critical dispersion of the precipitate is never achieved, although the hardness and strength increase as the size of the precipitate grows. If the reheating temperature is too high then the precipitate size increases and the initial gain in hardness and strength is lost. This is known as *overaging*. Generally the precipitate size required for maximum strength is so small that it cannot be detected by optical means.

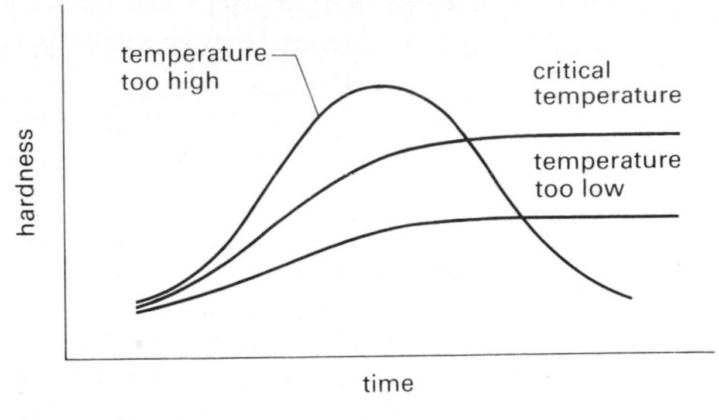

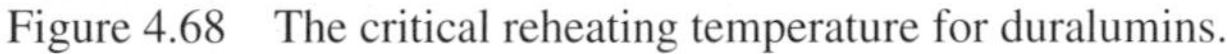

Figure 4.68 The critical reheating temperature for duralumins.

Bibliography

Alexander W and Street A 1976 *Metals in the Service of Man*. Penguin

Askeland D R 1990 *The Science and Engineering of Materials* 2nd SI edn. Chapman and Hall

British Steel Corporation and NCST Trent Polytechnic 1976 *Which Metals*. BST

British Steel Corporation and NCST 1976 *Crystal Structure of Metals*. BST

Davies D J and Oelmann L A 1983 *The Structure, Properties and Heat Treatment of Metals*. Pitman

Desch H E 1973 *Timber – Its Structure and Properties*. Macmillan

Fishlock D 1967 *The New Materials*. Murray

Gordon J E 1991 *The New Science of Strong Materials: or Why You Don't Fall Through the Door*. 2nd edn. Penguin

Greenwood D P 1983 *Modern Design in Plastics*. Murray

Hall M J D 1988 *Design and Plastics*. Hodder and Stoughton

Hennessy L and Smyth L 1985 *Engineering Technology*. Iona Print Dublin

Jane F W 1970 *The Structure of Wood*. Black

Johnson H 1982 *The International Book of Wood*. Artists House

Morton-Jones D H and Ellis J W 1986 *Polymer Products Design, Materials and Processing*. Chapman and Hall

Pettit T 1981 *Appreciation of Materials and Design (Craft Education)*. Arnold

Rollason E C 1973 *Metallurgy for Engineers* 4th edn. Arnold

Slade E 1969 *Metals in the Modern World*. Aldus

Scientific American 1967 *Materials*. Freeman

Stoker A 1984 *Modern Materials*. Cambridge University Press

Assignments

1 (a) Explain the differences between thermosetting plastics and thermoplastics.
(b) Explain the term 'plasticised' in relation to plastics, giving an example of its application and reasons for its use.
(c) Explain why fillers are added to plastics. Give examples of suitable fillers and their value to the final material.
(d) Blowing agents are added to some polymers. Explain what their function is and how they operate to make the final material.

2 Each piece of solid timber is unique in structure whereas most metals and plastics have a more homogeneous structure. Discuss in detail, giving examples, how the structural differences affect the way designers use these materials.

3 Chipboard is but one of a number of manufactured materials. Using two other materials as examples, list for *each* of them three reasons for preferential use by the designer.

4 (a) Explain briefly how metals, polymers and timbers are classified into generic groups. Draw up a table listing examples of materials which belong to these groups.
(b) Micro-examination of a material is often necessary to identify it precisely. Describe, using sketches, the microstructure of TWO of the following:
(i) metal,
(ii) polymer,
(iii) timber.

5 Describe the molecular structure of
(a) a metal (such as aluminium),
(b) a plastics material (such as polyethylene).
Explain how these structures account for the different behaviour of these materials when tensile loads are applied.

6 (a) Name and describe the type of bonding in each of the following substances:
(i) copper;
(ii) diamond;
(iii) sodium chloride.
(b) Explain how the different characteristics of thermosetting and thermoplastic polymers result from their structures.
(c) With the help of examples, give an account of the advantages of the use of plastics materials in engineering over metals.

7 (a) Explain, with the aid of sketches, the basic composition and structure of thermoplastics and thermosetting plastics.
(b) Explain, with the aid of sketches, the process of dendritic solidification in metals and how this affects the final structure of a metal.

8 (a) State the difference between an elastic change and a plastic change.
(b) (i) Describe briefly the arrangement of atoms in the structure of a polycrystalline material such as copper.
(ii) Explain briefly what happens to this arrangement when the material is strained elastically.

9 The properties of metals can be modified by alloying, and the properties of polymers can be modified by copolymerising.
(a) Discuss the similarities and the differences between alloying and co-polymerising.
(b) Choose *either* an alloy *or* a copolymer and compare its properties with the properties of the materials from which it is made.
(c) Name and describe a product for which the alloy or copolymer would be suitable, and explain how the properties of the new material are exploited in the product.

10 (a) Sketch the Iron/Carbon Thermal Equilibrium diagram up to 1.7% carbon and label the important features. Using this diagram explain the different forms of steel at 0.4% carbon, 0.83% carbon and 1.2% carbon.
(b) Using the diagram, explain in structural terms how steel is normalised.

11 (a) Three identical screwdrivers with wooden handles have blades made of a medium-carbon steel. Each blade is subjected to one of the following heat treatments:
(i) Heated to 900°C and allowed to cool very slowly.
(ii) Heated to 900°C and quenched in cold water.

(iii) Heated to 900°C, quenched in cold water, reheated to 400°C and cooled.

How would you expect each to behave when using the maximum torque available to tighten screws by hand?

Explain the reasons for the differences in behaviour.

(b) An aluminium alloy of the duralumin type (i.e. containing about 4% by weight of copper) can be strengthened by the process of precipitation hardening.

Describe the stages in the process and in particular the effects of time and temperature at each stage of the process.

12 (a) Describe how a thermoset plastic can be reinforced and the effect on the properties of that plastic material.

(b) Explain what changes in the microstructure of a metal take place when it becomes work-hardened and detail how the material can be returned to its normal state.

(c) Explain in detail how lead/tin alloys can have significantly different properties because of their differing composition.

(d) Explain what you understand by 'age-hardening'.

5 Material properties and testing

Gold is for the mistress – silver for the maid –
Copper for the craftsman cunning at his trade.
'Good!' said the Baron, sitting in his hall,
'But iron – cold iron – is master of them all.'

(Rudyard Kipling 1865–1936)

In selecting the material for a component it is essential to choose one whose properties meet the requirements of the design. The properties of a material can be divided into its physical properties (such as density, thermal and electrical conductivity) and its mechanical properties (such as hardness, strength and toughness) which indicate how the material behaves under various loads. In this chapter these properties are tabulated for a representative selection of materials. The tables are by no means comprehensive and, for other materials, you should refer to data books such as those listed in the bibliography.

5.1 Physical properties and testing

The physical properties of most interest to a designer are the density and useful working temperature range. The *density* of a substance is its mass per unit volume. It is denoted by ρ and, in SI units, it is measured in kg/m^3. For a ship's keel we would use a dense material such as lead, but for a lightweight camping frame we would use a material of low density such as aluminium. The density of seasoned timber falls within the range 385–835kg/m^3, plastics 900–1400 kg/m^3, stones 2080–3200 kg/m^3 and common metals 2640–11373 kg/m^3.

The density of a composite can be calculated from the proportions of the constituent materials. If a composite is to be two-fifths by volume of element A and three-fifths by volume of element B then the density of the composite ρ_C, is related to the densities of A and B, ρ_A and ρ_B, and is given by the *rule of mixtures*:

$$\rho_C = \tfrac{2}{5}\rho_A + \tfrac{3}{5}\rho_B$$

This can be extended for any number of constituent materials.

The tables 5.1 to 5.3 show these physical properties for metals, alloys and nonmetals. The density and the softening point of polymers or the melting point of the other materials are given. Table 5.1 also gives the boiling point and crystal structure of some common metals. Table 5.4 gives the average density of different timbers at 15% moisture content. The density will vary by approximately 0.5% for a 1% change in the moisture content.

5.1.1 Thermal properties

Heat transfer, heat absorption and their effects on materials are of major concern to the designer. The surface of a saucepan, for instance, must be of a material that will give even, efficient heat transfer to the food. In a refrigeration system, the aim is to prevent the contents from absorbing heat. Colour is an important factor when considering heat absorption as dark colours will absorb heat and light colours will reflect it (see section 5.1.4).

Metal	*Crystal structure*	*Density* (kg/m^3)	*Melting point* (°C)	*Boiling point* (°C)
aluminium	f.c.c.	2700	660	2400
copper	f.c.c.	8960	1083	2580
gold	f.c.c.	19300	1063	2660
iron	b.c.c/f.c.c.	7900	1535	2900
lead	f.c.c.	11300	327	1750
nickel	f.c.c.	8900	1453	2820
silver	f.c.c.	10500	961	2180
tin	diamond/tetragonal	7300	232	2500
titanium	h.c.p./b.c.c.	4500	1680	3300
tungsten	b.c.c.	19300	3380	6000
zinc	h.c.p.	7100	420	907

Table 5.1 Physical properties of metals

Alloy	*Density* (kg/m^3)	*Melting point* (°C)
brass (65% Cu–35% Zn)	8450	927
Dural (5.4% Cu)	2800	640
mild steel	7850	
nichrome (80/20)	8360	

Table 5.2 Physical properties of alloys

Material	*Density* (kg/m^3)	*Melting point (m.p.) or Softening point (s.p.)* (°C)	
brick	1400		
carbon-fibre reinforced plastic	1500		
concrete	2400		
glass	2400–3500	≈1100	(m.p.)
GRP	2000		
mica	2800		
nylon 6	1140	200–220	(m.p.)
Perspex	1200	85–115	(s.p.)
polystyrene	1060	80–105	(s.p.)
polythene	930	65–130	(s.p.)
PVC (plasticised)	1700	70–80	(s.p.)
porcelain	2400	1550	
quartz crystal	2650		
rubber	1100	125	
Terylene filaments	1380		
water	1000		

Table 5.3 Physical properties of nonmetals

Timber	*Density* (kg/m^3)
balsawood	160
spruce	450
Scots pine	513
sapele	641
iroko, walnut, teak	657
ash (European)	705
beech, oak (European)	721
ebony (African)	1025
lignum vitae	1249

Table 5.4 Average density of timbers at 15% moisture content

The thermal expansion of materials and their ability to withstand heat are other characteristics that must be understood if they are to perform correctly at the required temperature. The elements and oven-linings in a cooker show little reaction to the wide temperature fluctuations to which they are subjected, apart from the necessity for the oven-linings to be loosely fitted to allow for the expansion of the metal. Plastics, however, have much less capacity to withstand high temperatures. The washing-up bowl may melt if it is in contact with a hot pan, or a jug may deform if it is placed in boiling water. The melting or softening point of a material indicates the temperatures at which it is not suitable to be used.

The three main thermal properties of materials, *specific heat capacity, thermal expansion and thermal conductivity,* are now examined in more depth and their values for common materials are given in Table 5.5. Formal definitions of these properties will be found in Chapter 12.

Material	*Coefficient of linear thermal expansion (a)* (10^{-6}/K)	*Thermal conductivity (k) at* 0°C (W/mK)	*Specific heat capacity (c) at* 0°C (J/kgK)
Metals			
aluminium	23	235	880
copper	16.7	383	380
brass (65%Cu–35%Zn)	18.5	120	370
Invar	1	11	
iron	12	76	437
lead	29	35	126
magnesium	25	150	
mild steel	11	55	450
silver	19	418	232
tin	6	60	140
Nonmetals			
brick	3–9	0.4–0.8	800
concrete	11	0.1	1100
graphite	2	150	
polyethylene	300	0.3	
polyurethane foam	90	0.05	
polystyrene	60–80	0.08–0.2	1300
porcelain	2.2	0.8–1.85	1100
PVC (plasticised)	50–250	0.16–0.19	
Pyrex glass	3	1.2	
rubber	670	0.15	900
water (at 15°C)	–	–	4210

Table 5.5 Thermal properties of some common materials

Specific heat capacity

The heat capacity of a body is the energy required to raise its temperature by one degree. In SI units it is measured in J/°C or J/K. The *specific heat capacity* of a material is the energy required to raise the temperature of unit mass by one degree and is given by:

$$\text{specific heat capacity} = \frac{\text{heat energy supplied}}{\text{mass} \times \text{temperature rise}}$$

In the case of a gas the value of the specific heat capacity depends on the manner in which it is allowed to expand or contract (see section 12.6.3).

It can be seen from the specific heat capacity values in Table 5.5 that it takes seven times as much heat energy to raise the temperature of aluminium by one degree as it does for the same mass of lead.

Thermal expansion

Thermal expansion results when heat causes the atoms of a substance to vibrate more vigorously and occupy a greater volume. The increase in length, X, depends on the coefficient of linear expansion, α, the original length of the material, L, and the rise in temperature, t°C. The coefficient of linear expansion is defined as the fractional change in length per degree change in temperature. Thus:

$$\text{fractional increase in length } \frac{X}{L} = \alpha \times t$$

and

$$X = \alpha L t$$

The units of α are 'per °C' or 'per K', sometimes written K^{-1}.

Materials with strong atomic bonds have low coefficients of linear expansion and high melting points. Tungsten, diamond and ceramics are examples of materials which have low coefficients of linear expansion whereas lead and polyethylene have high coefficients and expand readily on heating. Small gaps are provided in many bridge structures and railway tracks to allow adjacent sections to expand without causing undue stress and consequent buckling when the temperature rises. Similar precautions should be taken in designing any structure or mechanism which will be subjected to significant temperature changes.

The thermal expansion of materials is used to advantage in the operation of thermostats and flashing indicators where the expansion of a metal at a predetermined temperature causes an electrical contact to be broken. After cooling a few degrees, contact is again made and the unit is switched on, thus creating an automatic switching unit. This is usually achieved by joining two parallel strips of metal which have very different coefficients of expansion, such as brass and Invar. (See Table 5.5 for the values of their thermal expansion coefficients.) This forms a bimetallic strip which will flex as the metals expand at different rates.

The coefficient of thermal expansion can be measured by placing a sample of material in a steam jacket and measuring the change in length accurately with a dial gauge at recorded temperatures.

Thermal conductivity

The transfer of heat through a body by conduction occurs because the faster-moving molecules in the hotter part transfer some of their energy by impacts to adjacent molecules. The ability of a material to conduct heat is measured by its thermal conductivity, k (see section 12.9). In the SI system, the units of k are W/mK or $Wm^{-1}K^{-1}$.

Metals are generally good conductors of heat and materials such as copper and aluminium are therefore used where good heat transfer is essential. Nonmetals tend to be *thermal insulators* and are used to prevent heat transfer between different items. Normally ceramics will show an increase in thermal conductivity as the temperature rises. The low values of thermal conductivity, k, given in Table 5.5, show why brick, polystyrene, porcelain, PVC and rubber are useful insulators. Air is a very good thermal insulator and so materials which trap air are very often used in products. Examples of such use are polyurethane foam as refrigerator insulation, urea-formaldehyde foam as cavity-wall insulation, duck down in duvets, fibreglass wadding in loft insulation and porous building bricks.

5.1.2 Electrical properties

Electrical conductivity

In many applications the electrical properties of a material are of primary importance. Copper wire is chosen for electrical wiring because of its extremely high electrical conductivity despite the fact that it has a high thermal conductivity and expansion coefficient. Its high ductility, shown by the low value of its modulus of elasticity, makes it suitable to extrude and draw into shape. The plastic casing round the copper wiring prevents excessive strain during use but retains flexibility. On the other hand, it is the low electrical conductivity of the PVC which is primarily important in this usage as it insulates between and round the conducting elements.

For a conducting material the e.m.f., V, the current, I, and the resistance, R, are related by Ohm's law. This states that:

$$V = IR$$

This resistance to current flow in a circuit is proportional to the length, l, and inversely proportional to the cross-sectional area, A, of the component and can be given as:

$$R = \frac{\rho l}{A} \quad \text{where } \rho \text{ is the electrical resistivity}$$

or

$$R = \frac{l}{\sigma A} \quad \text{where } \sigma \text{ is the electrical conductivity}$$

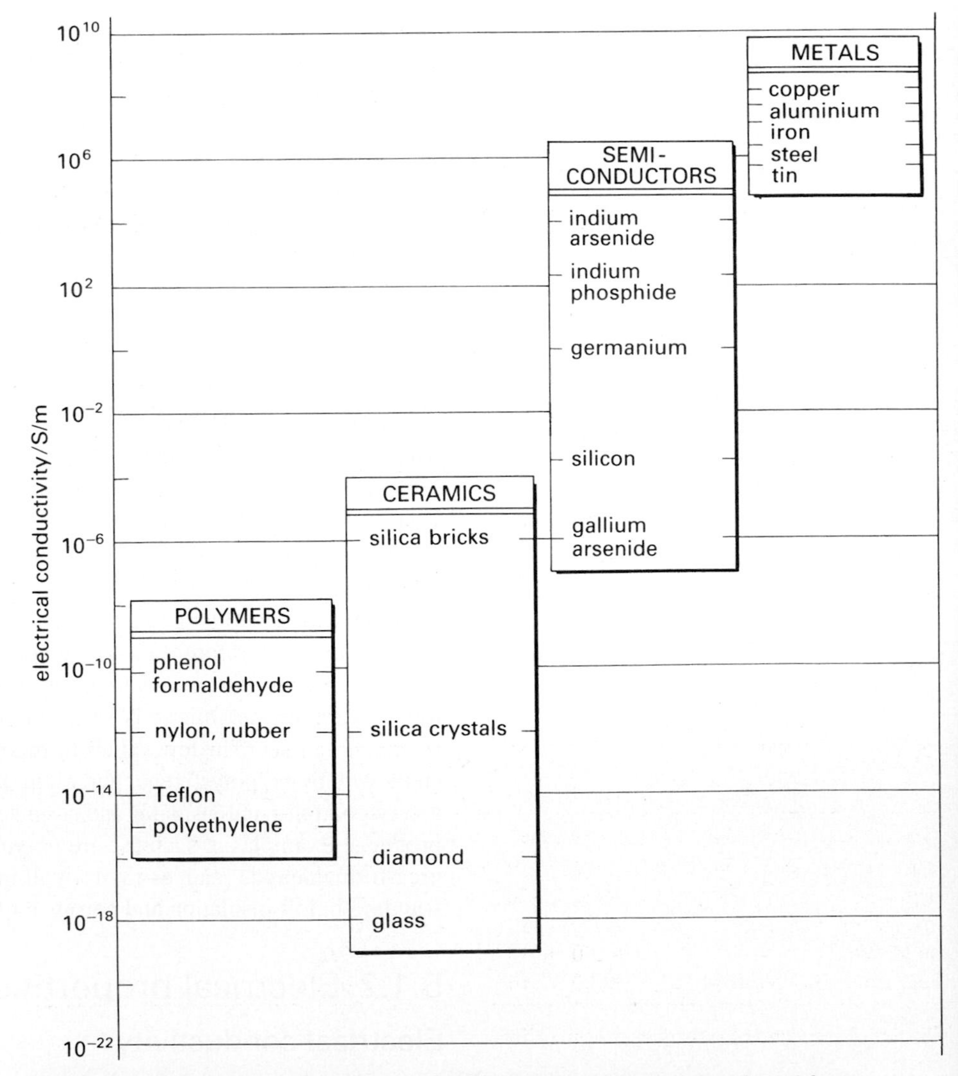

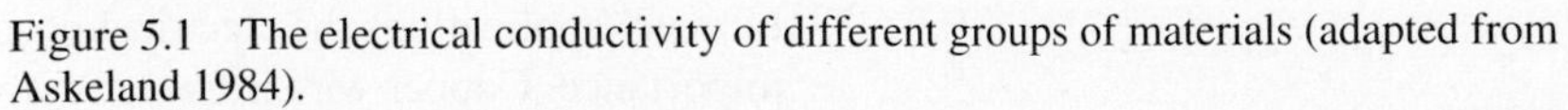

Figure 5.1 The electrical conductivity of different groups of materials (adapted from Askeland 1984).

The SI unit of resistivity is the ohm metre (Ω m), and of conductivity is the siemens per metre (S/m).

It can be seen that the electrical conductivity is the inverse of the electrical resistivity. It is heavily dependent on the molecular structure of the material and Fig. 5.1 shows the widely differing characteristics of the different groups of materials. Electrical conductivity depends on the number of free electrons and their rate of movement through the material and so metals, in which free electrons form the basis of the bonding, have very good electrical conductivity. This is reduced as the temperature is increased or lattice imperfections are introduced by alloying or working. In semiconductors and insulators the covalent bonds must be broken to release electrons to conduct electricity. In ionic bonded compounds entire ions must move through the material to transmit the charge so most ionic materials are electrical insulators.

The best electrical conductors are the metals copper, silver and gold and this is partly explained by the fact that they have only one electron in their valence shell, and this can readily be dissociated from the atom. In a totally pure metal with no lattice imperfections, at absolute zero temperature, the free electrons would have a completely unimpeded path. However, no metal is 100% pure and flaws in the crystal structure cause obstacles that electrons collide with, so slowing their rate of progress through the material. The effect of temperature increase is to cause vibration of the atoms so that the free electrons have an even harder time progressing through the material. Thus, electrical resistivity is proportional to temperature over a large range of values and so the resistance in a light-bulb filament increases as it gets hot. The rate of increase of electrical resistivity is given by the temperature resistivity coefficient. Values of electrical resistivity is at room temperature and the temperature resistivity coefficients for a few common metals are given in Table 5.6. Table 5.7 shows how alloying affects these properties and Table 5.8 gives values of electrical resistivity for some polymers and ceramics.

Material	*Electrical resistivity at* 20°C (Ωm)	*Temperature resistivity coefficient* (10^{-3}/K)
aluminium	27×10^{-9}	4.2
copper	16.8×10^{-9}	4.3
gold	23×10^{-9}	3.9
lead	206×10^{-9}	4.3
silver	16×10^{-9}	4.1
tungsten	55×10^{-9}	4.6

Table 5.6 Electrical resistivity of some metals

Material	*Electrical resistivity at* 20°C (Ωm)	*Temperature resistivity coefficient* (10^{-3}/K)
brass (65% Cu–35% Zn)	69×10^{-9}	1.6
constantan (60% Cu–40% Sn)	490×10^{-9}	≈0.02
Dural (4.4% Cu)	$\approx 52 \times 10^{-9}$	≈2.3
manganin (84% Cu)	440×10^{-9}	≈0
mild steel	$\approx 120 \times 10^{-9}$	≈3.0
nichrome (80% Ni–20% Cr)	1030×10^{-9}	0.18

Table 5.7 Electrical resistivity of some alloys

Material	*Electrical resistivity at* 20°C (Ωm)
alumina	$1\text{–}1000 \times 10^{9}$
dry ground	$0.01\text{–}0.1 \times 10^{6}$
mica	$0.1\text{–}1000 \times 10^{12}$
nylon	$10\text{–}10^{4} \times 10^{9}$
polythene	100×10^{9}
PVC	$10\text{–}10^{4} \times 10^{9}$
quartz	$1\text{–}200 \times 10^{6}$
rubber	$\approx 10 \times 10^{12}$

Table 5.8 Electrical resistivity of some polymers and ceramics

Superconductivity

Superconducting materials are those in which the resistivity falls to zero when they are cooled to a critical temperature and any magnetic field is minimised. There is then no resistance to the flow of current through the materials, and therefore superconductors can carry current without waste heat being produced. All perfect crystals would in fact behave as superconductors at 0 K. Magnetic fields cannot penetrate superconducting materials and this leads to the standard test for superconductors, the Meissner effect: when the superconductor is placed above a magnet it will hover in midair.

It was discovered in 1911 that mercury superconducts at 4 K and soon afterwards lead was found to have a critical temperature of 7.2 K and tin, 3.7 K. Although other superconductors were discovered, it was many years before a material with a critical temperature above 23 K was found. During 1986 a ceramic copper oxide based on the rare-earth element lanthanum was found to have a critical superconducting temperature of 28–40 K. All superconductors up to this date needed to be cooled in liquid helium which is very costly and gave limited applications of these materials. By the end of 1986 the critical temperatures had risen to 77 K so liquid nitrogen, which is readily available, could be used as the coolant and so the possible applications widened. In January and February 1987, other rare-earth ceramic oxide compounds were found to have a critical temperature of 98 K when the samples were sintered to fuse the powdered ingredients without melting them.

By February 1988 many non rare-earth metal oxides based on bismuth, calcium and copper were found with a critical temperature of about 81 K. These were cheaper and more ductile and robust than former superconductors. Since this date other less stable compounds have exhibited superconductivity at temperatures up to about 125 K. If it were possible to find a readily available material that would superconduct at room temperature then the world would change dramatically as most electrical applications would greatly increase their efficiency if high-temperature superconductors were included. It seems possible that the new materials that have already been found could be used to advantage in electric power generation, transmission and storage; high speed rail transport; computers and electronic instruments.

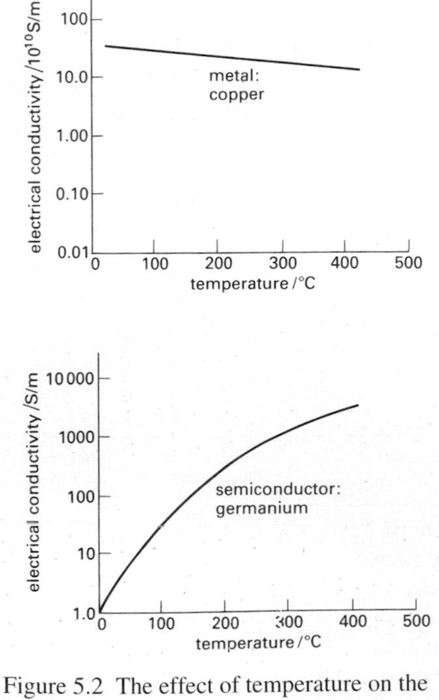

Figure 5.2 The effect of temperature on the conductivity of a metal and a semiconductor.
(a) The variation of the conductivity of copper with temperature.
(b) The variation of the conductivity of germanium with temperature.

Semiconductors

Semiconductors, as their name implies, are materials whose electrical conductivity lies between that of good conductors, such as most metals, and insulators such as polythene. The best-known examples are silicon and germanium but there are many others such as gallium arsenide, cadmium sulphide and indium antimonide. The resistance of semiconductors decreases with increasing temperature. This contrasts with the effect of temperature on the conductivity of a metal, see Fig. 5.2.

In order to control the conductivity of these elements small amounts of impurity atoms are introduced and this is called *doping*. If antimony, which has one more valence electron than silicon or germanium, is used as a doping agent then an *n*-type semiconductor is formed as it has had electrons added and therefore has more negative charge than before. When gallium (a valency of three) is the dopant then a *p*-type semiconductor is formed, with effectively an increased number of absences of electrons or 'holes' which act like positive charges.

P-type and *n*-type semiconducting materials are not used in isolation. They are combined together in different ways, the simplest combination being the *p-n* junction, Fig. 5.3. The *p-n* junction is made by doping a single crystal of silicon in such a way that half of it is *p*-type and the other half is *n*-type.

As soon as such a crystal is created, electrons from the *n*-type materials diffuse into the *p*-type material and fill some of the holes. Also, holes from the *p*-type material diffuse into the *n*-type material and are filled by electrons. This exchange

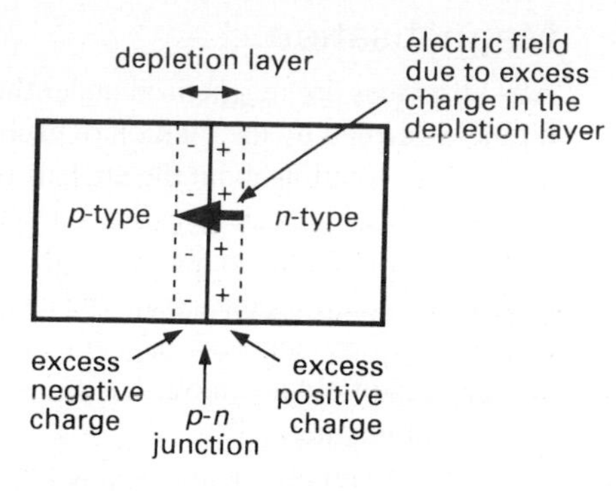

Figure 5.3 The *p-n* junction.

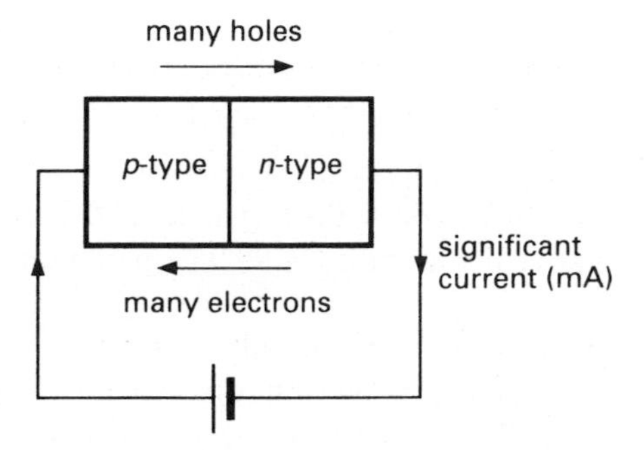

Figure 5.4 The forward biased *p-n* junction.

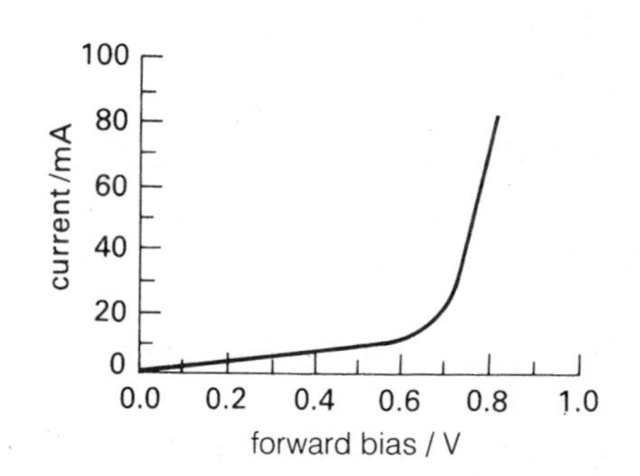

Figure 5.5 Graph to show the forward bias characteristic of a *p-n* junction.

takes place in a very narrow region known as the depletion layer. The *n*-type material in the depletion layer has lost electrons and gained holes so is positively charged; the *p*-type material in the depletion layer has gained electrons and has lost holes so is negatively charged. The diffusion process has therefore established a potential difference across this junction which quickly becomes large enough to prevent any further diffusion. The size of this *contact potential* is approximately 0.6 V for silicon. If a battery is now connected across the *p-n* junction, as shown in Fig. 5.4, then the junction is said to be forward biased.

When the battery p.d. exceeds the contact potential there will be a flow of majority carriers – holes in the *p*-type and electrons in the *n*-type material. This rate of flow is considerable (in the order of milliamp). The graph of this relationship (*forward bias*) is shown in Fig. 5.5.

If the battery is now reversed, the depletion layer widens and the contact potential increases. There is a small flow of minority carriers (usually known as leakage current). The graph of this relationship (*reverse bias*) is shown in Fig. 5.6.

It can be seen that a diode (a *p-n* junction diode) has a low resistance when forward biased but a high resistance when reverse biased.

The increase of conductivity with temperature has made semiconductors an essential part of thermistors where they can be used to sense the temperature and signal when it reaches a predetermined level. Semiconductors are also used in pressure transducers, magnetometers and transistors.

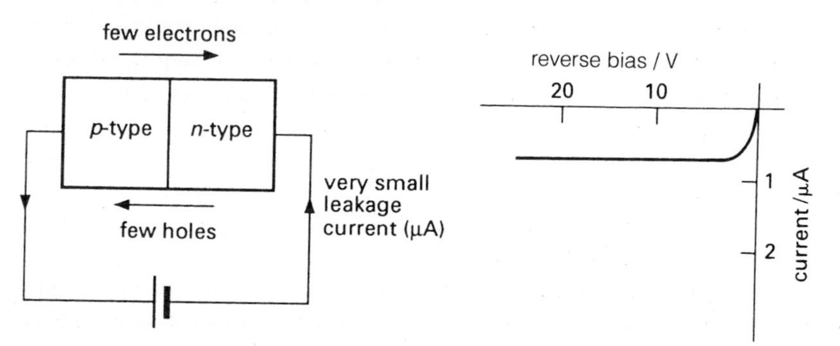

Figure 5.6 Reverse biased *p-n* junction and graph.

5.1.3 Magnetic and dielectric properties

The magnetic properties of a material depend on the ability to be magnetised. This occurs when atoms which have a magnetic field, *dipoles*, are realigned by the influence of a magnetic field. In much the same way the dielectric properties of a material depend on the ability of atoms which have an unbalanced charge to be polarised by an electric field.

Magnetisation

All substances are magnetised under the effect of a magnetic field. A magnetic dipole is formed by the rotation of each electron about its own axis (electron spin) and also its rotation about the nucleus of the atom. This induces a magnetic field in the material which can be enhanced when the dipoles are aligned. Some materials, called *paramagnetic*, magnetise with their dipole axis in the direction of the field and others, with their dipole axis perpendicular to the field are *diamagnetic*. Iron, which shows a very pronounced magnetic effect and retains some residual magnetism when the magnetic field is removed, is *ferromagnetic*. These materials can be used for permanent magnets.

When a magnetic field is applied to a magnetic material lines of magnetic flux are produced. The flux density, measured in tesla (T), is known as the inductance, B, and indicates the degree of magnetisation that can be obtained. This is related to the applied field, H, measured in amp/metre (A/m), by:

$$B = \mu H$$

where μ is the magnetic permeability and is measured in henry/metre (H/m).

The ratio of this magnetic permeability to the permeability measured in a vacuum gives an indication of the degree of magnification of the magnetic field and is known as the relative permeability. The relative permeability of 99.95% pure iron is 180 000.

When a magnetic field is applied to a magnetic material it gradually causes dipoles to align and magnetisation to occur. The magnetisation with respect to an applied magnetic field is shown by path VW in Fig. 5.7, until at W it has reached its *saturated magnetisation*. When the field is removed the flux density now follows path WX. At X the *remanent magnetisation* is the permanent magnetic effect that has been induced in the material.

In order to demagnetise the substance an opposite magnetic field must be applied to remove the magnetic flux by the path XY shown in Fig. 5.8. This is called the *coercive field* and is shown by the value $-H_C$ at the point Y on the figure. If this reverse field is increased still further the point Z would be reached with *reverse saturated magnetisation*. H_C is the field required to remove the reverse saturated magnetisation. The full *hysteresis loop* shown in Fig. 5.8 describes the effect on the magnetic flux density caused by an alternating magnetic field. The power of a magnet is given by the maximum value of the product of *inductance* or magnetic flux density, B, and magnetic field, H. This occurs at point W for the magnet shown in Fig. 5.8. The maximum value of the product $B \times H$ in the second or fourth quadrants gives the energy required to demagnetise the material. This can be obtained by drawing a rectangle of as large an area as possible in these quadrants of the ferromagnetic hysteresis loop diagram.

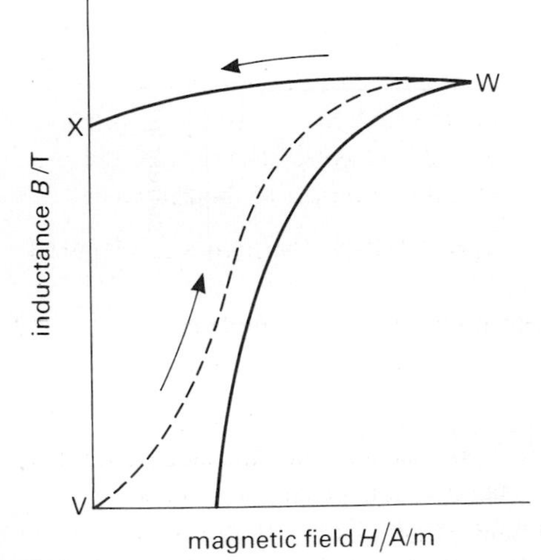

Figure 5.7 The magnetisation of a ferromagnetic material with respect to an applied magnetic field.

Ferromagnets with different characteristics are required for different applications. Fig. 5.9 shows the very different hysteresis loops that are required for electrical, computer and permanent magnet applications.

Ferromagnetic materials are required as cores for electromagnets, motors and transformers and they have a continually cycling alternating field applied to them. They require a high value of saturation magnetisation and small remanent effect. Silicon-iron formed when small amounts of silicon (3–5%) are added to iron produces an excellent ferromagnet with these properties. Amorphous metallic glasses which can be produced on thin tape and then stacked together also have excellent magnetic characteristics.

For computer applications where a magnetic memory store is required, magnetic tapes with a square hysteresis loop, low saturation magnetisation and low coercive field are produced. An 81.5% iron-nickel alloy or ceramic oxide such as magnetite, iron(II) diiron(III) oxide (Fe_3O_4), applied to plastic tape gives the required characteristics.

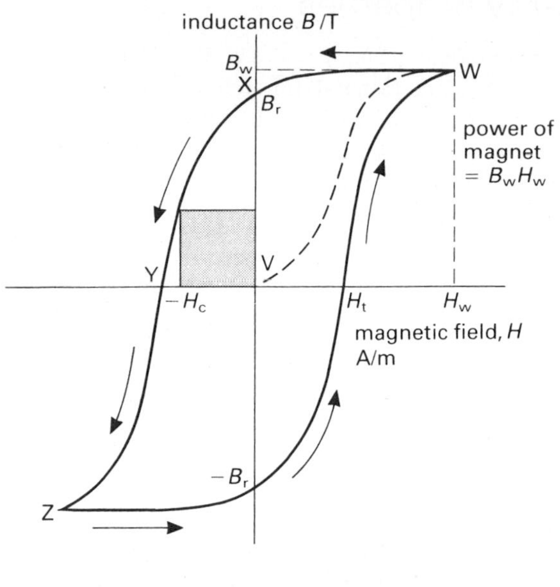

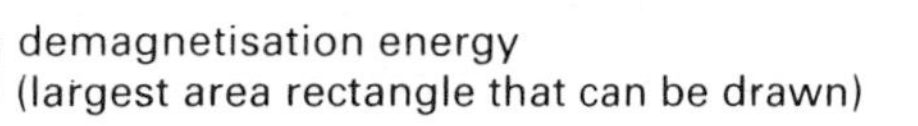

Figure 5.8 The hysteresis loop for a ferromagnetic material.

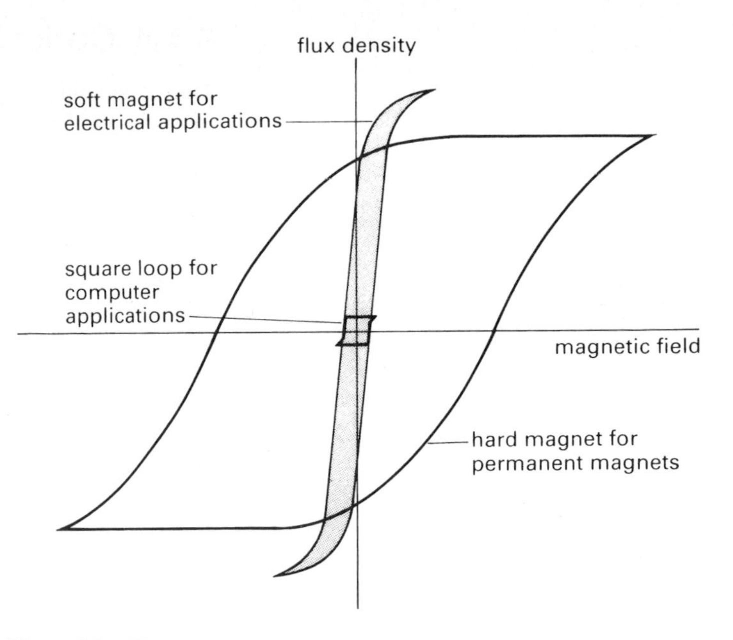

Figure 5.9 The magnetic characteristics required for different applications (adapted from Askeland 1984).

Permanent magnets can be made from iron, nickel and cobalt. Small amounts of impurities can improve the characteristics but more complex metal alloys with a very small grain size give much better results. Alloys such as Alnico are produced by phase transformations to obtain a fine crystal grain size and powder metallurgy is used for intermetallic compounds such as cobalt-samarium.

Common magnetic ceramics with a special crystal structure form the *ferrimagnetic* materials. In magnetite (Fe_3O_4), for example, each metallic ion behaves as a dipole and although different dipoles have varying strengths and alignments there is an overall magnetic effect. They otherwise behave as ferromagnetic materials.

Dielectrics

A dielectric material is used in a capacitor to store an electric charge for later use. When a voltage is applied to the conductor plates the central dielectric must easily polarise but also have high electrical resistance ($10^{11}\Omega$ or greater) to prevent the charge from passing between the plates. There are three sources of dielectrics:

- Liquids of polar molecules such as Al–Al_2O_3 or Ta–TaO. The liquids can be used by themselves or impregnated into paper.
- Polymers and plastic foil such as polyester, polystyrene and cellulose.
- Some ceramics such as mica and glass.

Barium titanate is an example of a *ferroelectric material* which is a dielectric material that retains some degree of permanent polarisation when the electric field is removed. The polarisation of a ferroelectric material with an electric field shows similar characteristics to that of ferromagnetic materials. Fig. 5.10 shows the ferroelectric hysteresis loop. Point W indicates saturation polarisation, X the remanent polarisation, Y the point which shows the coercive electric field needed to remove polarisation, $-E_C$, and Z the point of reverse saturated polarisation.

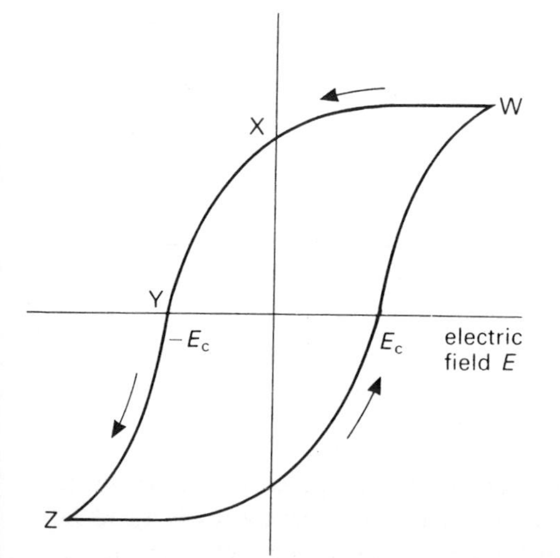

Figure 5.10 The hysteresis loop for a ferroelectric material.

5.1.4 Optical and X-ray properties

The visible light spectrum, ultraviolet radiation and X-rays are all produced by movements of electrons within the atoms of a material. Infrared radiation, radio waves and microwaves have longer wavelengths and are caused by vibrations of the atoms within the crystal structure. Different materials or surface finishes will react very differently to electromagnetic radiation. The rays can be absorbed, reflected or refracted and different effects such as colour, thermal conductivity and fluorescence are produced. Fig. 5.11 shows the relative wavelengths of the different types of electromagnetic rays.

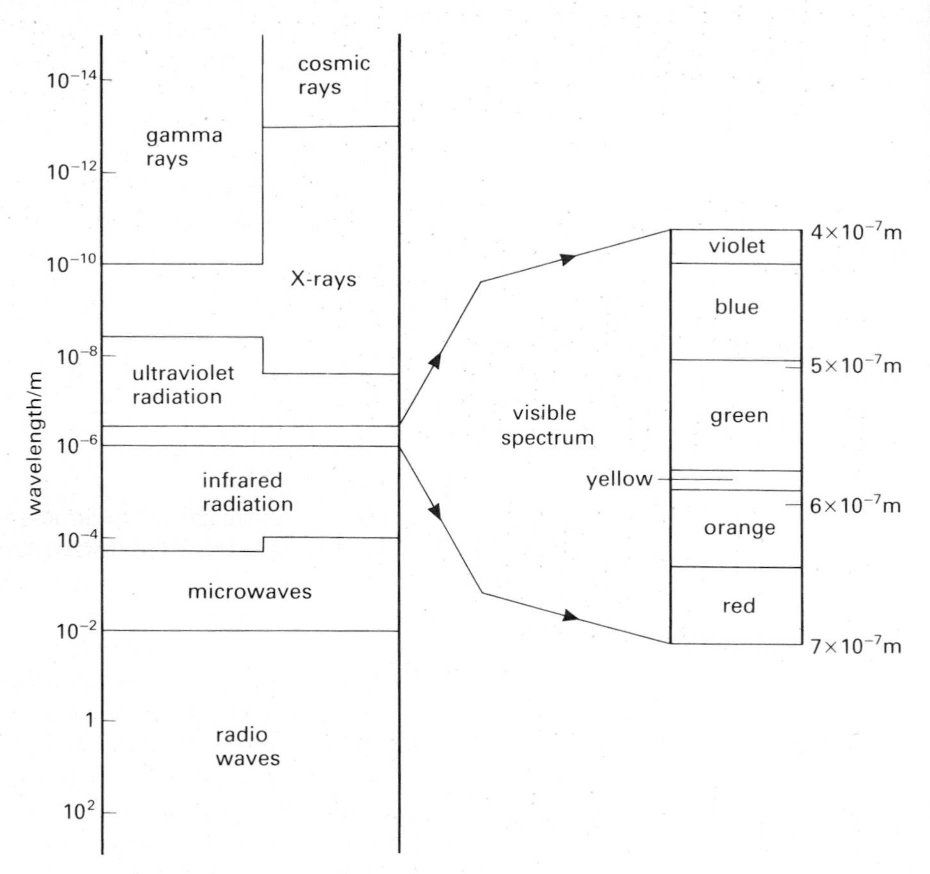

Figure 5.11 The electromagnetic spectrum of radiation.

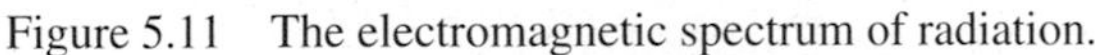

Refraction and reflection

Materials such as glass and clear acrylic are completely transparent and are, therefore, essential in many building applications. Others such as quartz are translucent and allow beams of light to pass through although you cannot see through them. In materials such as these, the speed of light through the material is apparently reduced and a ray of light (beam of photons) changes direction in its passage into and out of the material.

$$n = \frac{c}{v} = \frac{\sin \alpha}{\sin \beta}$$

The value n is the refractive index of the material, c is the speed of light in a vacuum and v is the speed of light in the material. The angle of the incident ray is α and the angle of the refracted ray is β, as indicated in Fig. 5.12.

The denser the material the higher its refractive index will be. Also materials that polarise easily, such as dielectric materials, have a high refractive index.

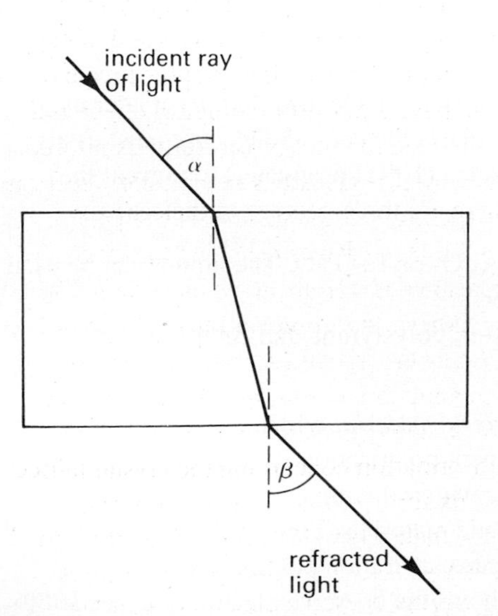

Figure 5.12 The refraction of light.

A smooth, highly polished metal will be a good reflector of light. Some photons are reflected while the rest are refracted through the material. The percentage reflection, R, is related to the index of refraction and those materials with a high index of refraction have a high reflectivity.

$$R_{\text{vacuum}} = \left[\frac{n-1}{n+1}\right]^2 \times 100$$

Colour

Colour is determined by the portion of the spectrum that a material reflects. Copper, for instance, absorbs the short wavelength blue or violet rays whilst reflecting the longer rays with red colourings; hence copper appears reddish. White reflects all the colours in the visible spectrum while black absorbs them. Glass can appear to reflect a wide variety of colours. This is caused by impurities in the material which cause a particular wavelength to be absorbed and therefore a localised colour is reflected. The same effect is seen (but for a different reason) when light is reflected from an oily puddle of water. Metal surfaces which are nonreflective are required for internal components of a camera. This is achieved by using a matt black paint finish. Tropical buildings are often finished with a white paint to reflect the sun's rays.

Photoconduction

A beam of light falling on a semiconductor in a circuit can produce an electric current by causing the increased movement of electrons or vacancies in the atoms of the material. This property is used in photoelectric components such as electronic eyes which can open or close doors when a person passes through a gap interrupting a beam of light which is directed on to a semiconductor. They can also be used to operate automatic lighting triggered by the daylight fading. Solar cells again use the photoconductive property of materials to produce power from the sun's rays.

X-rays

X-rays are produced when high energy electrons are directed on to a material and cause electrons to move from outer shells to the inner shell of an atom or to decelerate the electrons and convert the lost kinetic energy to X-ray emission. X-rays are invisible and travel in straight lines at the speed of light but have a smaller wavelength and a higher energy than light. They can penetrate solids with some scattering of the rays and the degree of penetration depends on the density, thickness and relative molecular mass of the material and the wavelength of the rays. This ability to penetrate solids is utilised in radiography for medical analysis and material testing purposes where the ionising X-ray beam produces a photographic image of the material that it passes through. X-rays can have a harmful biological effect and the amount of radiation that any person is subjected to must be carefully monitored.

If a material is subjected to high energy electrons it gives out an emission spectrum of X-rays. This is composed of a continuous spectrum with characteristic peaks superimposed, as in Fig. 5.13. By measuring these peaks and matching the wavelengths with those of known materials, an unknown material can be identified. The intensity of the characteristic peaks will also indicate the composition of the material.

X-ray diffraction can be used to determine the crystal and atomic structure of materials. Beams of X-rays are directed at an angle on to a sample of the material so that a diffracted ray of the same wavelength as the original can be detected. The angle of diffraction enables a great deal of information concerning the crystal lattice to be determined. William Henry Bragg and his son, William Lawrence Bragg, shared the Nobel Prize for Physics in 1915 for their work on X-ray diffraction and X-ray crystallography. The techniques they developed were used, some 29 years later, in the discovery of deoxyribonucleic acid (DNA), the means by which we inherit our parents' characteristics.

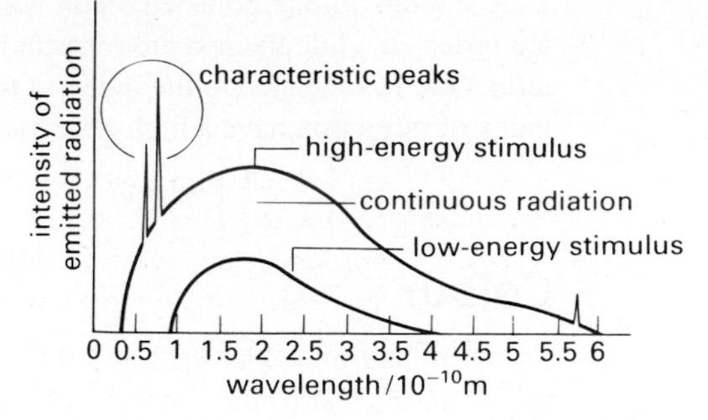

Figure 5.13 The X-ray emission spectrum of molybdenum.

5.2 Mechanical properties and testing

The mechanical properties of a material can be divided into static and dynamic properties.

Static properties

These include strength and hardness, where the test results are to a large extent independent of the rate of loading. The definitions of these static properties are given below.

Strength is the ability to resist a force without breaking. There are three fundamental kinds of loading that a component may be called upon to resist (see sections 8.5.2 and 9.3). These are:

(i) tensile – a stretched wire will need tensile strength;
(ii) compressive – a mine roof-prop will need compressive strength;
(iii) shear – a train coupling bolt needs to resist the equal and opposite forces pulling against each other across the component.

Elasticity is the material's ability to return to its original shape after being deformed. It will depend on the material and the load carried. Rubber is elastic up to its breaking point.

Plasticity is the readiness to deform to a stretched state when a load is applied. The plastic deformation is permanent even after the load is removed. Plasticine exhibits plastic deformation.

Ductility is the ability to be drawn out longitudinally to a reduced cross-section. A ductile material must therefore have high plasticity.

Hardness is the resistance to wear or indentation of a material. It is necessary for engine parts or cutting tools where constant friction causes abrasion and wear.

Malleability is the ability of a material to be stretched in all directions without fracture. A malleable material can be hammered into shape.

Dynamic properties

These include fatigue, creep and impact resistance where the time or rate of loading is significant.

Creep is a slow plastic deformation that can occur when a load is applied for prolonged periods. It can be observed at room temperature in the case of lead but for most alloys it is a high temperature effect and can cause failure at a lower stress than the static test value. It is an important consideration where a component is subjected to high temperatures for long periods, as in the case of gas turbine blades.

Fatigue is the phenomenon by which a material can fail at a lower than normal stress if the load is applied many times. It will generally result in a small crack at the surface which will gradually extend into the material, progressively reducing its ability to carry loads.

Toughness is the ability to withstand sudden loading. It is measured by the total energy that the material can absorb.

Resilience is the ability of the material to absorb energy. It depends on the amount the material extends under load as well as the stress it can bear. For this reason a material such as nylon, that can be stretched considerably, is suitable for tow ropes which have to absorb snatch loads.

Brittleness implies the lack of ductility and toughness. A material that shows no significant plastic deformation before fracture is said to be brittle.

Results from creep and fatigue tests are difficult to relate directly to theoretical predictions and are therefore of empirical importance; they provide an experimental basis for predicting the expected life of a product under prolonged or repeated loads.

5.2.1 Tensile testing

Tensile testing, in which a specimen is loaded to destruction, is one of the most important materials property tests. The results depend on the size and shape of the test piece and specimen sizes should therefore conform to British Standard BS EN 10002 (1990) *Tensile testing of metals.* Specimens should have a central *gauge length* of uniform cross-section, thereby avoiding regions of high stress concentration. Nonuniformity of the specimen may result in premature failure. Tensile specimens can be rectangular or circular in cross-section, as shown in Fig. 5.14. The gauge length should be 5 × (diameter) in the case of round specimens and 5.64 × √(cross-sectional area) for rectangular ones.

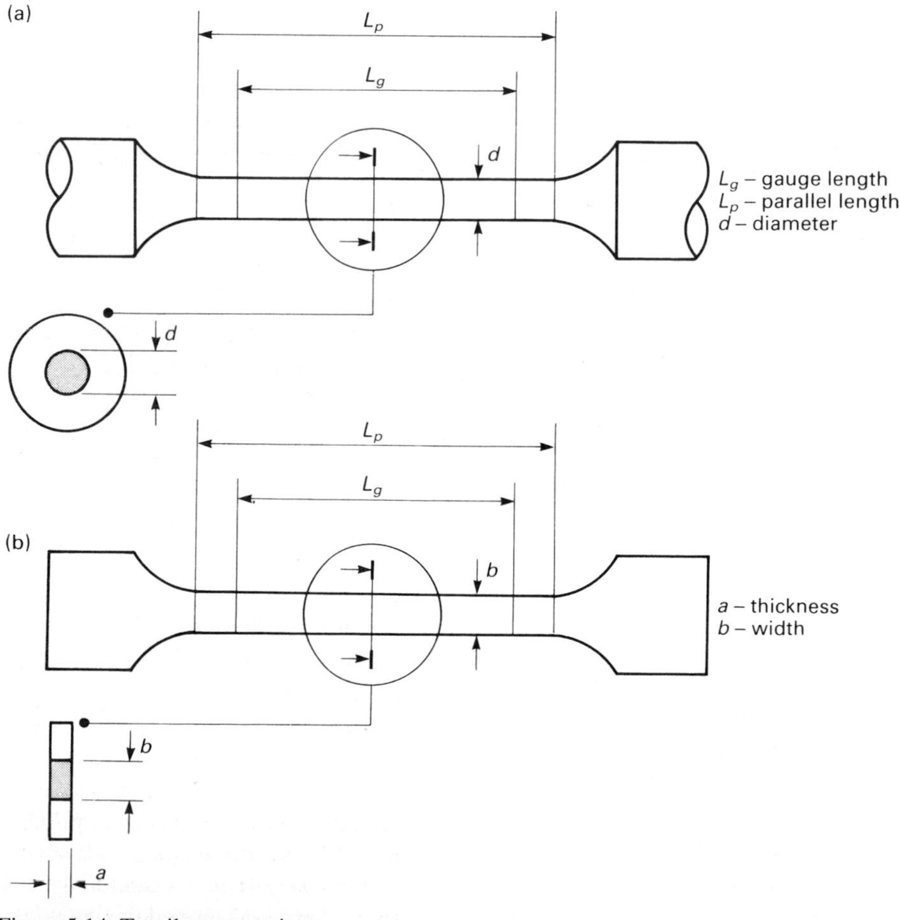

Figure 5.14 Tensile test specimens
(a) Circular cross-section
(b) Rectangular cross-section.

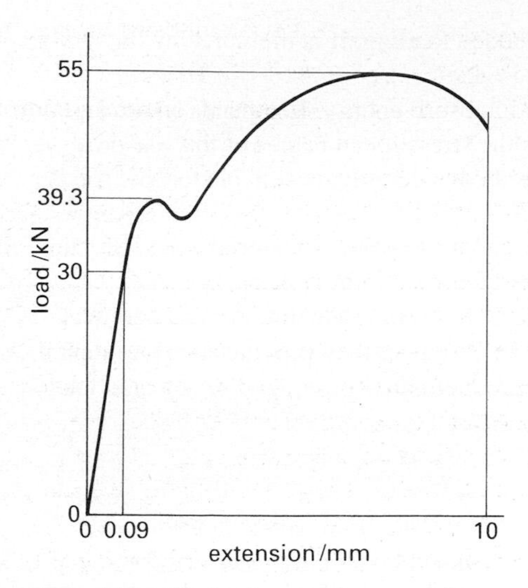

Figure 5.15 The load–extension curve for a low-carbon steel. The initial extension has been exaggerated to show its variation with load more clearly.

Tensile testing machines range from small hand-operated bench-top models (known as *tensometers*) for specimens of a few millimetres' diameter to floor-standing machines that can exert forces of several hundred kilonewton. In the large machines the load is applied through a screw or hydraulic mechanism. The loads and corresponding extensions are recorded and plotted on a load-extension graph, which will yield the tensile properties of the test material.

Tensile properties of low-carbon steel

Fig. 5.15 shows the tensile load–extension curve for a specimen of a constructional steel with about 0.4% carbon content.

We can see that the specimen extended gradually for a time and then went through a transition phase, after which it extended more rapidly until fracture occurred.

Stress, σ, is the load divided by the cross-sectional area (see section 9.2). If it is based on the original cross-sectional area it is known as the nominal stress but, with a ductile material, the area reduces and the real stress may be considerably higher as the fracture point is approached. *Strain*, ε, is the extension divided by the original length of the specimen. Consequently the (nominal) stress–strain curve for the specimen has a similar shape to the load-extension curve.

The stress–strain curve for the low-carbon steel specimen, of 10 mm diameter and 50 mm gauge length, is shown in Fig. 5.16. This has been obtained by dividing the loads by the cross-sectional area of the specimen (25π mm^2) to obtain the stress and dividing the extensions by the original length (50 mm) to obtain the values of strain.

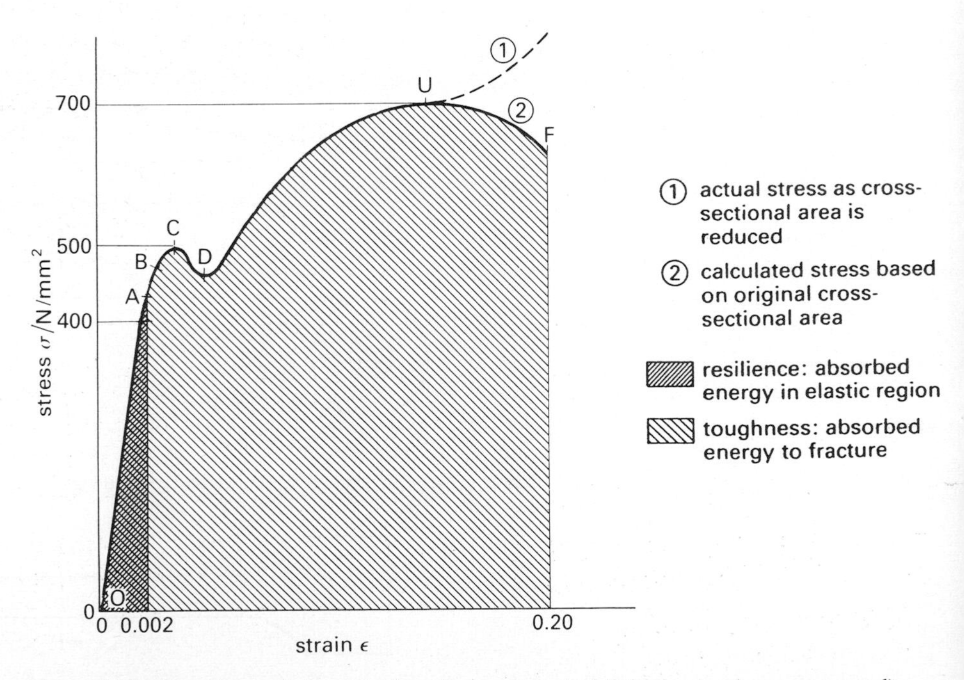

Figure 5.16 The stress–strain curve for a low-carbon steel (initial extension exaggerated).

The key points on the stress–strain curve are marked by the following letters:

O–A indicates the range of values for which the stress is directly proportional to the strain, a result known as Hooke's law. In this region the ratio of stress to strain is termed the *modulus of elasticity* (see section 9.2.2). It is given the symbol E and is often referred to as the Young modulus.

A is the *limit of proportionality*.

O–B indicates the range in which the specimen extends elastically, and will return to its original length if the load is removed.
B is the *elastic limit*, and may be coincident with A for some materials.
C–D marks the region in which the material undergoes internal structural changes before *plastic deformation* and *permanent set* occur. Many dislocations are formed at points of stress concentration and travel through the specimen as observable lines known as *Luders bands*.
C is the *upper yield point* at which the specimen begins to extend appreciably with no increase in stress. This value of stress is known as the *yield strength*.
D is the *lower yield point*.
D–F shows the region of continued plastic deformation. It is the region where *work hardening* takes place.
U marks the *ultimate tensile stress* (or *tensile strength*) and represents the maximum nominal stress that the specimen can endure. Beyond this point the specimen *narrows* or *necks* (the cross-sectional area is reduced) and the *nominal stress* falls.
F indicates the *fracture stress* and is the point at which the material fails.
– – – shows the *real stress* in the specimen based on the reducing cross-sectional area that occurs in the ductile phase.

Mild or low-carbon steels have a large number of uses in engineering, but attention here is focused on two – structural steel members, I-beams for example, and pressed components such as car wings. Until the stress level reaches the elastic limit, Hooke's law is being obeyed and the steel is behaving elastically. In the case of structural members this is the required behaviour, as it would be quite disastrous for a bridge support to grow a little longer every time a vehicle passed over the bridge. If, however, the stress developed by the pressing machine did not exceed the upper yield point the pressed car wings would spring back flat as they left the machine. In this case, it is clearly necessary to produce permanent or plastic deformation. Table 5.9 gives the modulus of elasticity, tensile strength and yield strength for some common metals and alloys. For the mechanical properties of other materials see Tables 5.11 and 5.12 later in the text.

Material	*Modulus of elasticity(E)* (kN/mm^2 or GN/m^2)	*Tensile strength* (N/mm^2 or MN/m^2)	*Yield strength* (N/mm^2 or MN/m^2)
Metals			
aluminium	70	60–160	30–140
copper	124	200–350	47–320
iron (wrought)	195	350	160
iron (cast)	115	140–320	
lead	16	15–18	
tin	47	15–200	9–14
titanium	110	250–700	200–500
tungsten	360	1000–4000	
zinc	97	110–200	
Alloys			
brass (65% Cu–35% Zn)	105	330–530	62–430
Dural (5.4% Cu)	70	230–500	125–450
nichrome (80% Ni–20% Cr)	186	170–900	
steel (mild)	210	480	240
steel (high strength)	210	600	450
steel (ultra high)		2000	1600

Table 5.9 Mechanical properties of some common metals and alloys.

For the design of the structural bridge support we must ensure that the internal stress never exceeds the elastic limit stress. It is common design practice to use the tensile strength of the metal, which can be determined easily, and a *factor of safety* to ensure that the stress stays within safe limits. This is given by:

$$\text{safe working stress} = \frac{\text{tensile strength}}{\text{factor of safety}}$$

The factor of safety will have to take into account:

- the uniformity of the material;
- the type of loading, static or dynamic;
- the effect of failure;
- and the effect of wear or corrosion on the material.

Factors of safety for steel will vary from about 3 for static loads to about 15 for impact loads or 20 where fluctuating loads may cause fatigue failure.

Another characteristic of the material derived from the tensile test is the *percentage elongation*, given by:

$$\text{percentage elongation} = \frac{\text{increase in gauge length}}{\text{gauge length}} \times 100\%$$

This is a measure of the ductility of the material, and clearly this is particularly important when considering pressing or drawing operations. For a mild steel specimen conforming to BS EN 10002 the percentage elongation will be around 28%. The percentage elongation values for a number of materials are given in Table 5.10.

Fractures are classified as either *ductile* or *brittle*. Brittle fracture occurs by the very rapid propagation of a crack after little or no plastic deformation. A ductile fracture is accompanied by considerable plastic deformation and exhibits a 'cup and cone' effect which results from the slow propagation of a crack. Such fractures in

Material	*Elongation (%)*
Metals and alloys	
mild steel	28
copper	60
brass (70% Cu–30% Zn), annealed	70
hard	5
brass (63% Cu–37% Zn), annealed	55
hard	4
aluminium alloy (1.2% Mn), soft	34
hard	4
cast aluminium alloy (3% Cu)	3
Thermoplastic polymers	
polystyrene	1–35
polypropylene	50–600
polyethylene	50–800
PVC, rigid	10–130
plasticised	240–380
nylon	60–300
polyurethane, thermoplastic	up to 700
Thermosetting polymers	
melamine formaldehyde	0.7
urea formaldehyde	1.0
polyester, unsaturated	2.0
casein	2.5–4
polyurethane, thermoset	up to 500

Table 5.10 Typical values of percentage elongation

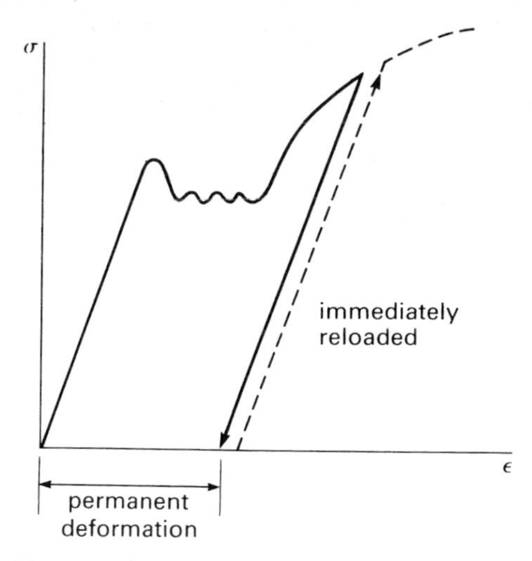

Figure 5.17 An interrupted tensile test with an immediately re-applied load.

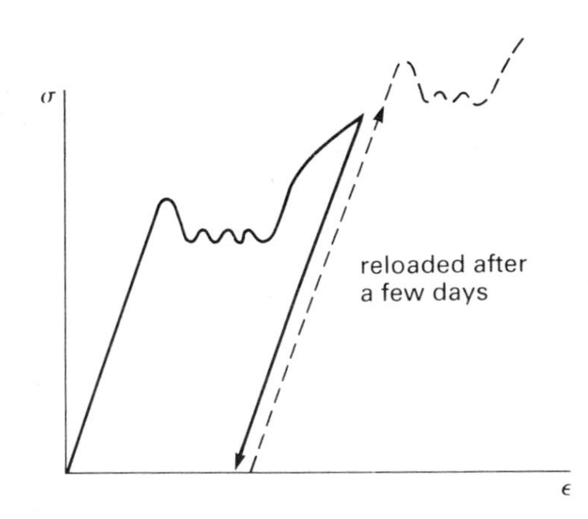

Figure 5.18 A tensile test resumed after an interruption of a few days.

crystalline materials have a characteristic dull, fibrous appearance and obvious necking of the specimen. The rate of loading can have a significant effect on percentage elongation and rapid loading is likely to produce a more brittle fracture.

It is obvious that there is a change in the width of the specimen that accompanies the longitudinal strain. The ratio of this lateral to longitudinal strain is known as the Poisson ratio, v. The two strains are of opposite signs so that an increase in length, for instance, is accompanied by a decrease in width or diameter. The negative sign in the formula below ensures that the value of v is positive.

$$v = -\frac{\text{lateral strain}}{\text{longitudinal strain}}$$

For an incompressible material the volume would remain constant and the lateral strain would exactly compensate for the longitudinal strain. This requires a Poisson ratio of 0.5 but for many materials it is in the region of 0.3.

The percentage reduction of cross-sectional area of a member under tensile load is a measure of the malleability of the material, and shows the degree to which a material can be forged into complex shapes.

$$\text{percentage reduction in area} = \frac{\text{reduction in cross-sectional area}}{\text{original cross-sectional area}} \times 100\%$$

The area under the stress–strain curve represents the work done per unit volume, and hence the energy absorbed before fracture. This is a measure of the *impact strength*, or *toughness* of the material. The large shaded area in Fig. 5.16 shows the toughness of the specimen tested while the shaded area under the elastic strain region indicates the *resilience* of the specimen.

Let us consider what would happen if a specimen were loaded beyond the yield point, unloaded and then the stress re-applied. When the stress drops to zero some permanent deformation has occurred. The specimen will be longer and thinner, because the overall volume is unchanged as plastic deformation results from the relative movement of planes of atoms. If the stress is then re-applied immediately, enough dislocations will be present in the material for the test simply to continue along its uninterruped path as shown in Fig. 5.17. (This graph also shows multiple yield points that are apparent in low-carbon steels.)

However, if the stress is resumed after a few days (or even a few hours at 100°C) these dislocations will have disappeared due to the diffusion of atoms within the solid, and another upper and lower yield point will appear when the new dislocations form. This is illustrated in Fig. 5.18.

Stress–strain curves for other materials

Metals

Fig. 5.19 shows typical stress–strain curves for a variety of metals. Cast iron is obviously a very *brittle* metal exhibiting no plastic deformation before fracture. The curves for low- and high-carbon steels show that an increase of carbon content greatly increases the strength of the metal. However, if we compare the areas under the high-carbon steel and stainless steel curves, they apparently have similar impact strength. It can be seen that stainless steel and copper are ductile metals, exhibiting approximately 60% elongation. They also show only a small area of proportionality and no clear yield point. The strain is shown as a percentage strain which is the actual strain multiplied by 100. This is found frequently on stress–strain curves.

This difficulty in defining the onset of plastic deformation, which is found in the majority of metals, has led to the definition of the *proof stress*. The 0.1% proof stress is found by drawing a line parallel to the elastic slope through the point corresponding to 0.1% strain to intersect with the stress–strain curve. A line drawn horizontally through this point of intersection to intercept with the stress axis will

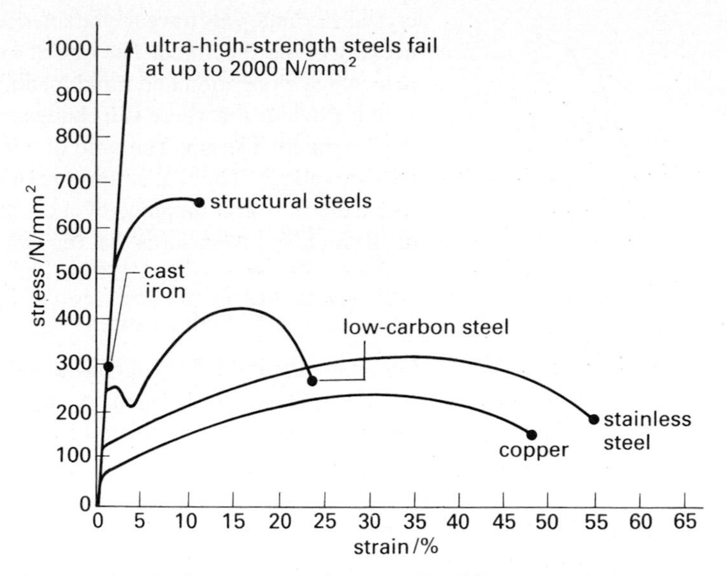

Figure 5.19 Typical stress–strain curves for different metals.

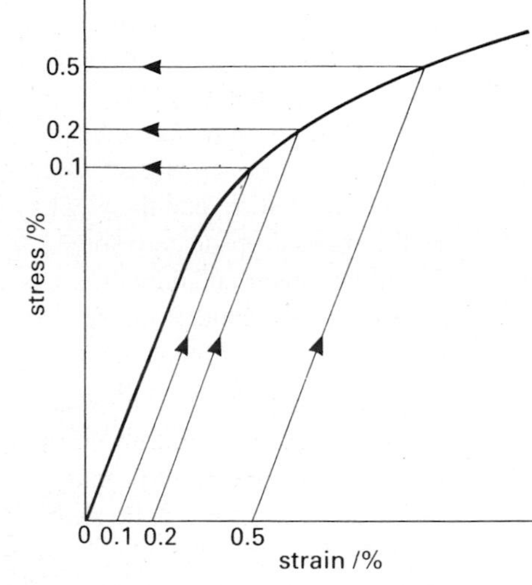

Figure 5.20 Stress–strain curves of ductile metals.

give the 0.1% proof stress, as shown in Fig. 5.20. Tables of properties of engineering materials will commonly give any or all of the 0.1%, 0.2% and 0.5% proof stresses, which are used as design stresses for ductile materials.

The most common uses for copper as an engineering material are as a wire or tubing, and the drawing processes for both depend greatly on its ductility.

Nonmetals

Fig. 5.21 is a typical stress–strain graph for nonmetals. It illustrates the extreme plasticity of thermoplastics and soft rubber, and, by contrast, the brittleness of concrete and a thermoset plastic, Bakelite.

Thermoplastic polymers

The stress–strain curves for thermoplastics are generally nonlinear, but are most remarkable in that the strain at fracture is very large, typically several hundred percent as can be seen from the percentage elongation values in Table 5.10.

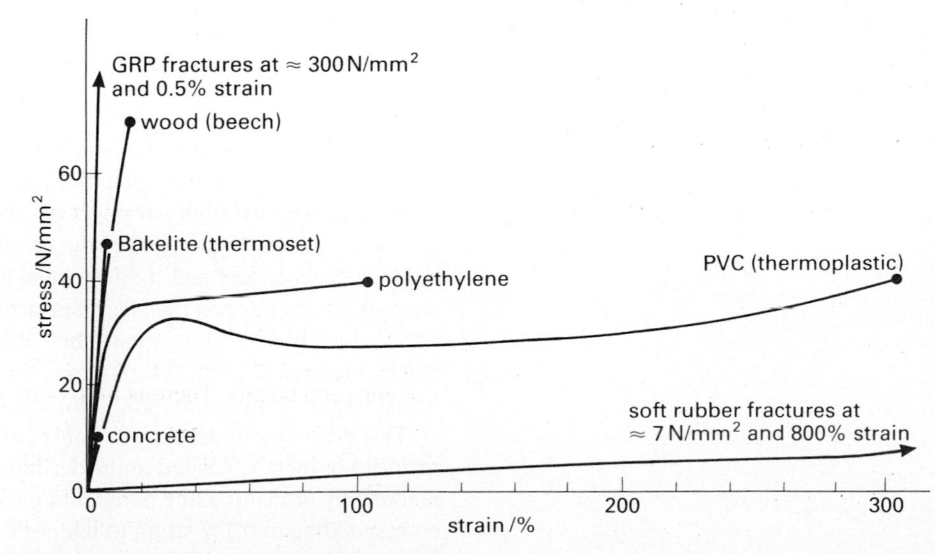

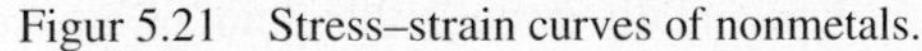
Figur 5.21 Stress–strain curves of nonmetals.

Thermoplastics have a much greater flexibility than metals or thermosetting polymers and this is indicated by lower values of the modulus of elasticity. Table 5.11 includes values of modulus of elasticity for nonmetals and these values can be compared with those for metals and alloys in Table 5.9. Yielding in a thermoplastic, such as polythene, results from the alignment or reorientation of long molecular chains and not as a result of the slipping of layers of atoms as in metals. These chains can be forced apart by *plasticisers* which contribute up to 30% of the weight of the polymer. Fillers and reinforcements are also added to lower the cost, improve the abrasion resistance, strength and colour of the polymer and these additions generally explain why the measured mechanical properties of polymers can vary widely.

Material	*Modulus of elasticity (E)* (kN/mm^2 or GN/m^2)	*Tensile strength* (N/mm^2 or MN/m^2)	*Compressive strength* (N/mm^2 or MN/m^2)
alumina	200–400	140–200	1000–2500
carbon-fibre reinforced plastic	130–190	1000–1400	
concrete (28-day)	10–17		27–55
glass	50–80	30–90	
GRP	40	1100	
nylon-6	1.0–2.5	70–85	50–100
Perspex	2.7–3.5	50–75	80–140
polystyrene	2.5–4.0	35–60	80–110
polythene	0.1–1.0	7–38	15–20
PTFE	0.4–0.6	17–28	5–12
PVC (plasticised)	$\approx$0.3	14–40	75–100
rubber (natural, vulcanised)	$\approx$0.001–1	14–40	

Table 5.11 Mechanical properties of some common nonmetals

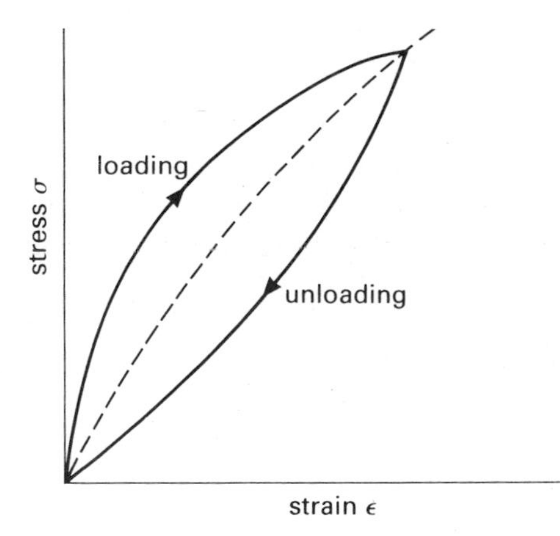

Figure 5.22 Hysteresis stress–strain curve for rubber.

Rubber

Rubbers or elastomers are capable of even greater strains before fracture, typically 800%. The first 600% strain is characterised by a very low elasticity resulting from the uncoiling of the molecules. At higher values of strain the stretching of primary bonds within the molecular chains becomes the dominant mechanism, significantly increasing the modulus of elasticity. One interesting phenomenon exhibited by rubbers, known as hysteresis, is the noncoincidence of the loading and unloading curves under cyclic stressing. Although all the strain is recovered, the loading and unloading relationships between the stress and strain are different, as shown in Fig. 5.22. The area between the paths represents the energy dissipated per cycle and accounts for the ability of rubber to absorb vibrational energy. The amount of energy absorbed can vary considerably and depends on the frequency of the cycle.

5.2.2 Compression testing

The tensile test results for cast iron and concrete show them both to be very brittle. Such materials are not used in tension, but in compression members. For example, cast iron was used for the struts and wrought iron (almost pure and very ductile) for the ties in early steel frameworks. Concrete is used to take compressive loads in columns and beams. There is little point in compression testing extremely ductile materials because the result is the kind of barrelling shown in Fig. 5.23. This is caused by friction at the points of contact with the plate and results in a stress distribution which is extremely difficult to analyse. For suitable compression test materials the length of the specimen should not be greater than three times the diameter in order to counteract the tendency for a compression specimen to buckle.

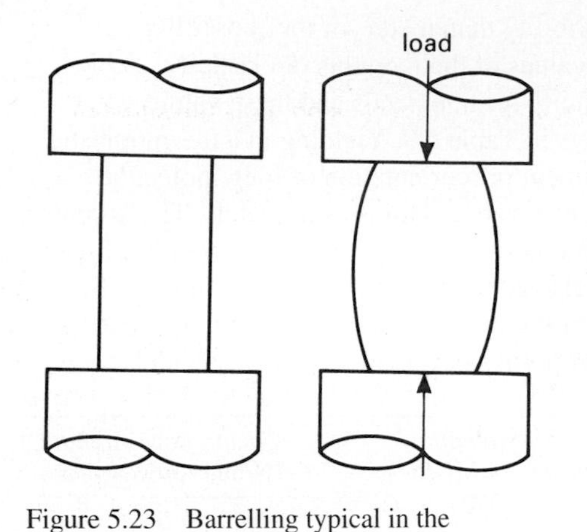

Figure 5.23 Barrelling typical in the compression of ductile materials.

Fig. 5.24 indicates typical tension and compression test results for concrete showing its higher compression failure stress. Table 5.11 gives compression strengths for some nonmetals and Table 5.12 shows average values of modulus of elasticity, tensile strength and compression strength for different timbers.

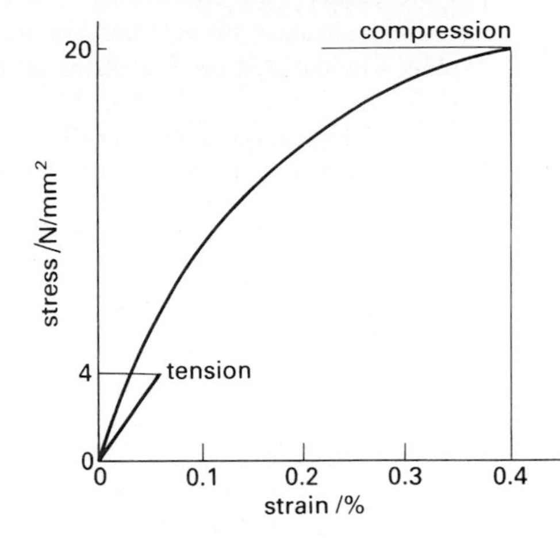

Figure 5.24 Tension and compression test results for reinforced concrete with a dense aggregate.

The first figure is the value for green wood (>18% moisture) and the figure in brackets is for dry wood <18% moisture)

Timber	*Mean modulus of elasticity (E)* (kN/mm^2 or GN/m^2)	*Tensile strength* (N/mm^2 or MN/m^2)	*Compressive strength* (N/mm^2 or MN/m^2)
balsawood	(3.3)	(9.7)	
Parana pine	6.9 (8.3)	6.9 (7.9)	5.2 (6.6)
spruce (European)	5.9 (6.9)	7.6 (10.3)	5.5 (8.3)
teak	11.0 (12.4)	22.1 (26.2)	16.5 (22.1)
iroko	9.0 (10.3)	20.7 (23.4)	15.2 (19.3)
sapele	9.7 (11.0)	19.3 (23.4)	15.9 (23.7)
mahogany (African)	7.9 (8.6)	12.4 (15.2)	9.7 (13.1)
ash (European)	10.0 (11.4)	17.2 (22.8)	9.7 (15.2)
beech (European)	10.0 (11.4)	17.2 (22.8)	9.7 (15.2)
oak (European)	8.6 (9.7)	15.9 (20.7)	9.7 (15.2)

Table 5.12 Mechanical properties of timbers when loaded parallel to the grain

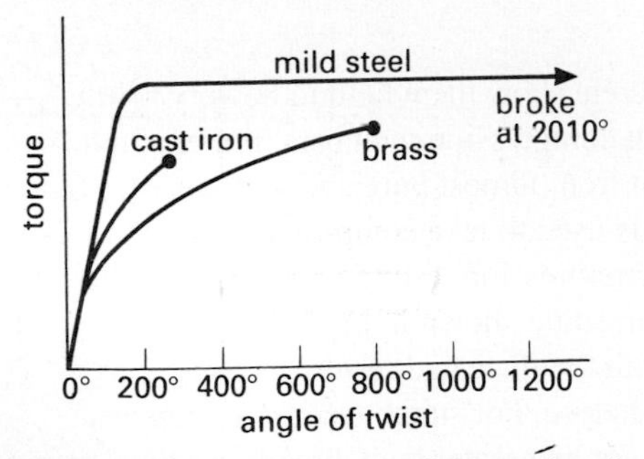

Figure 5.25 A comparison of the results of torsion tests on mild steel, brass and cast iron.

5.2.3 Shear testing

Torsion tests

It is extremely difficult to test a material in pure shear of uniform intensity. An approximation to pure shear occurs in a torsion test, where a twisting couple is applied to an extremely thin-walled cylinder. Torsion results are usually plotted as torque–angle-of-twist curves, which are similar to the stress–strain curves, as shown in Fig. 5.25.

Values of the modulus of rigidity (or shear modulus), the shear yield strength and apparent shearing strength can be obtained from these curves using the torsion equation given in section 9.10.

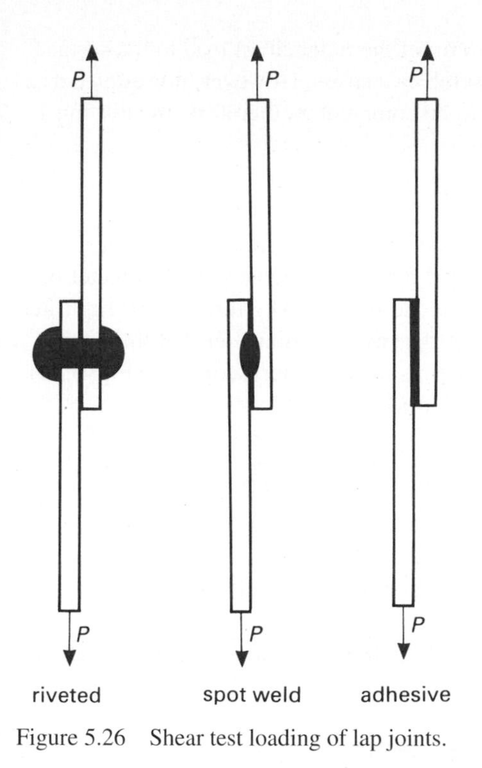

Figure 5.26 Shear test loading of lap joints.

Lap joints

Shear loading applied to lap joints can be tested as in Fig. 5.26.

Unfortunately the offset loads tend to bending as in Fig. 5.27, thus introducing tensile and compressive stresses.

Spot welds are tested in this way to find the ultimate shear strength.

5.2.4 Bending

Bend testing enables an estimate of the ductility of a material to be made relatively easily. Fig. 5.28 shows four forms of this test. Figs 5.28(a) and 5.28(c) show guided bend tests where the radius is controlled while Figs 5.28(b) and 5.28(d) show free bend tests where no former is used to control the radius of the bend. The material is examined for cracks after the test is complete to assess its performance. Bend testing specifications are given in BS 1639.

The Erichsen cupping test and the Olsen cupping test are alternative methods of examining a sheet material's ductility by generating defects by indentation and examining them with a microscope.

5.2.5 Hardness

The hardness tests measure the resistance of a material to indentation. A load is applied by slowly pressing an indenter at right angles to the test surface for a given period of time. The test causes plastic deformation in the material, and all variables affecting deformation therefore affect hardness. For this reason there is a good correlation between hardness and tensile strength for materials which work harden. Due to the plastic deformation, hardness tests are never taken near the edge of a sample or too close to a previous impression.

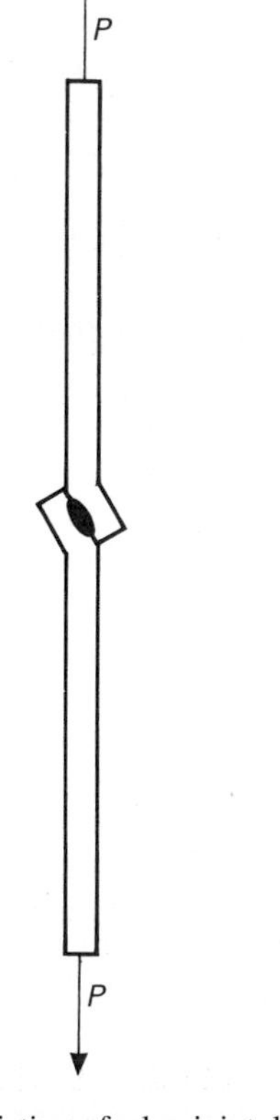

Figure 5.27 Twisting of a lap joint due to the offset loads.

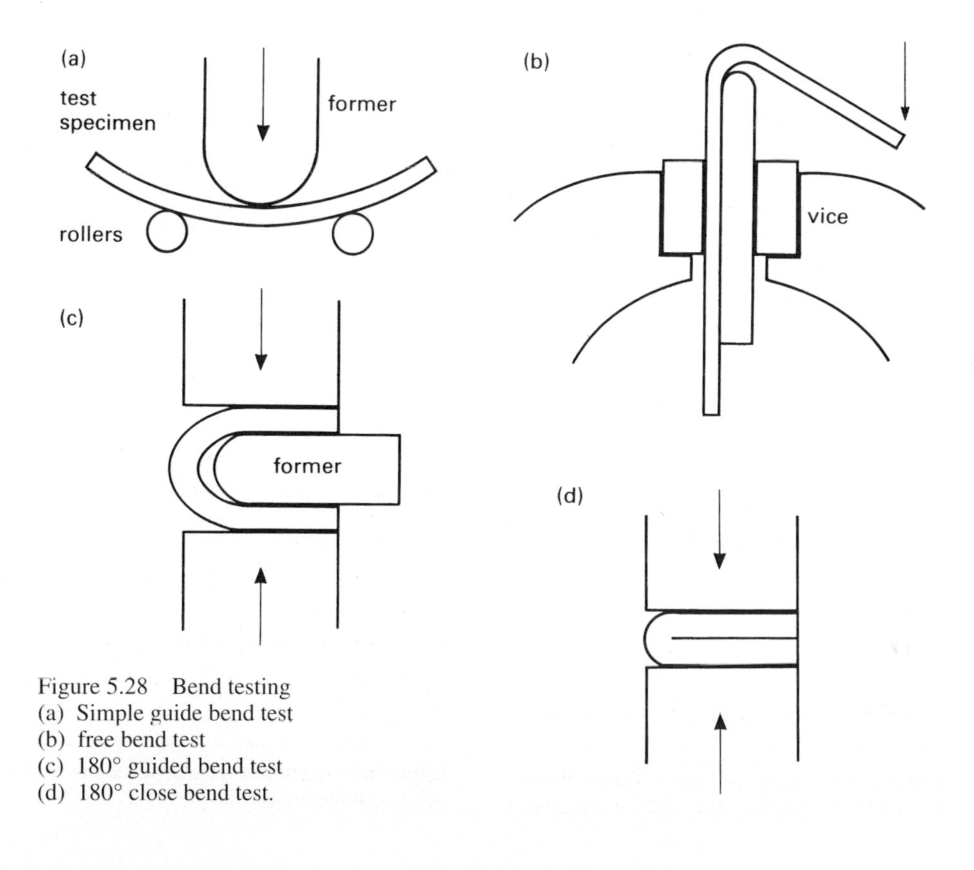

Figure 5.28 Bend testing
(a) Simple guide bend test
(b) free bend test
(c) 180° guided bend test
(d) 180° close bend test.

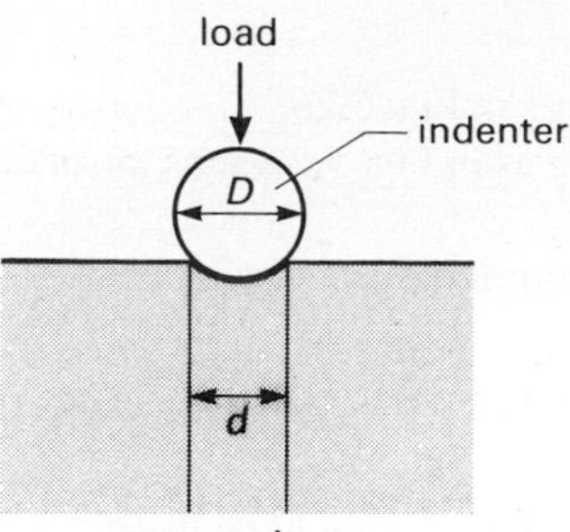

Figure 5.29 Indentation measurement in the Brinell hardness test.

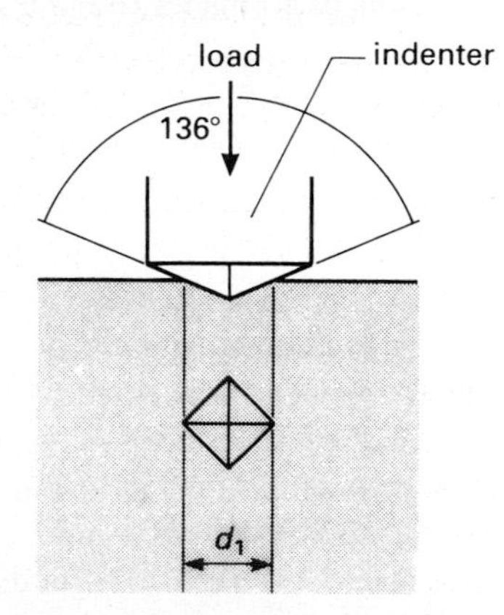

Figure 5.30 Indentation measurement in the Vickers hardness test.

The hardness number is defined as the ratio of the indentation load to the surface area of the indentation, and it therefore resembles a stress. However, it is quoted as a number and is not given units. For values to be comparable, therefore, the test must be carried out under standard conditions.

The Brinell hardness test

The Brinell hardness test uses a steel or tungsten carbide sphere as the indenter which is forced into the test surface under load for fifteen seconds. The diameter of this ball D is usually 10 mm and the standard load F is 3000 kg force, or 500 kg force for soft materials (1 kg force = 9.81 newton). The average diameter d of the indentation made is then measured in millimetres using a microscope (see Fig. 5.29).

The Brinell hardness number HB is then defined as:

$$HB = \frac{F}{\text{surface area of indentation}} = \frac{2F}{\pi D[D - \sqrt{(D^2 - d^2)}]}$$

If the indentation force F is measured in newtons (instead of kg force) this expression must be multiplied by the factor 0.102.

The ratio F/D^2 is standardised in order to obtain reliable results that can be compared for different materials.

It is found that for many steels the tensile strength in MN/m^2 (or N/mm^2) is about three times the Brinell hardness number.

The Vickers and Knoop hardness tests

The Vickers and Knoop hardness tests are both microhardness tests where the indentation is so small that a microscope is required to measure it accurately. Both tests use a square-based diamond pyramid with an included angle of 136° as the indenter, which will give geometrically similar indentations for different loads, as in Fig. 5.30, thus ensuring reliable results without any further standardisation. The Vickers test generally has an applied load of 10 kg force while the Knoop test has an applied load of 0.5 kg force.

The Vickers hardness number (VHN) for a load of F kg is given by:

$$VHN = \frac{F}{\text{surface area of indentation}} = \frac{2F \sin 68°}{d^2} = \frac{1.854F}{d^2}$$

where d is the diagonal length of the indentation in millimetres. If F is measured in newtons, the expression must again be multiplied by 0.102.

The numbers obtained from these tests show the relative hardness of different materials and a rank order can be established from the Knoop values in Table 5.13. Table 5.14 gives the Brinell hardness number for some alloys.

Material	*Knoop hardness*
gypsum	32
silver	60
ferrite	154
copper	163
pearlite	453
nickel	557
martensite	600
quartz	820
cementite	1720
alumina	2100
silicon carbide	2480
diamond	7000

Table 5.13 Knoop hardness values of some common materials

Material	*Brinell hardness*
brass	60–100
steel (low-carbon)	120–140
grey cast iron	150–240
steel (stainless 18/8)	170
steel (med-carbon)	200–250
teel (cutlery)	450–540
steel (tool)	780–800

Table 5.14 Brinell hardness values

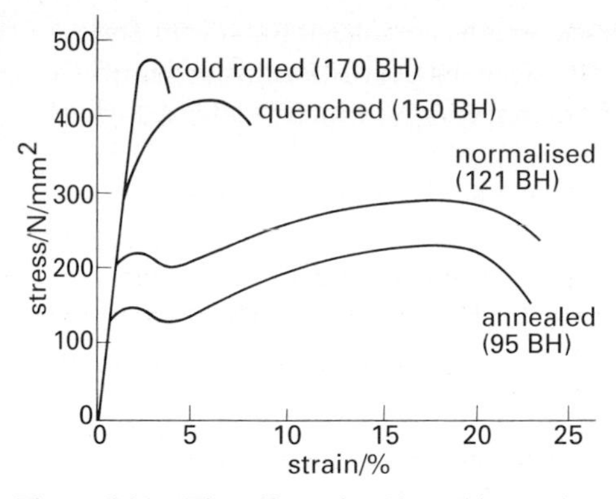

Figure 5.31 The effect of cold-working and heat treatment on the hardness and tensile strength of a steel.

The Rockwell hardness test

A third test for hardness is the Rockwell test, which is less accurate than the Vickers test but useful for rapid routine checks on finished materials as the depth of indentation is converted to a hardness reading and is shown on a dial while the specimen is still under load.

The properties of metals can be considerably altered by work hardening and heat treatment. Fig. 5.31 shows the changes in hardness and tensile strength after heat treating a cold-worked steel.

5.2.6 Toughness and brittleness

Toughness is the resistance of a material to fracture, which depends on the energy absorbed by the fracture process. The amount of energy absorbed during fracture depends on the size of the component, but the amount of energy absorbed *per unit area* of crack is constant for a given material, and this material property is called the toughness.

Tough materials require a lot of energy to break them (e.g. mild steel), usually because the fracture process causes a lot of plastic deformation; a brittle material may be strong but once a crack has started the material fractures easily because little energy is absorbed (e.g. glass). High toughness is particularly important for components which may suffer impact (cars, toys, bikes), or for components where a fracture would be catastrophic (pressure vessels, aircraft).

Rigorous measurements of toughness can be quite laborious to perform, so frequently the toughness of materials is compared using simple impact tests.

Compact tension test

Proper toughness tests use specimens with starter cracks – an example is the compact tension specimen shown in Fig. 5.32.

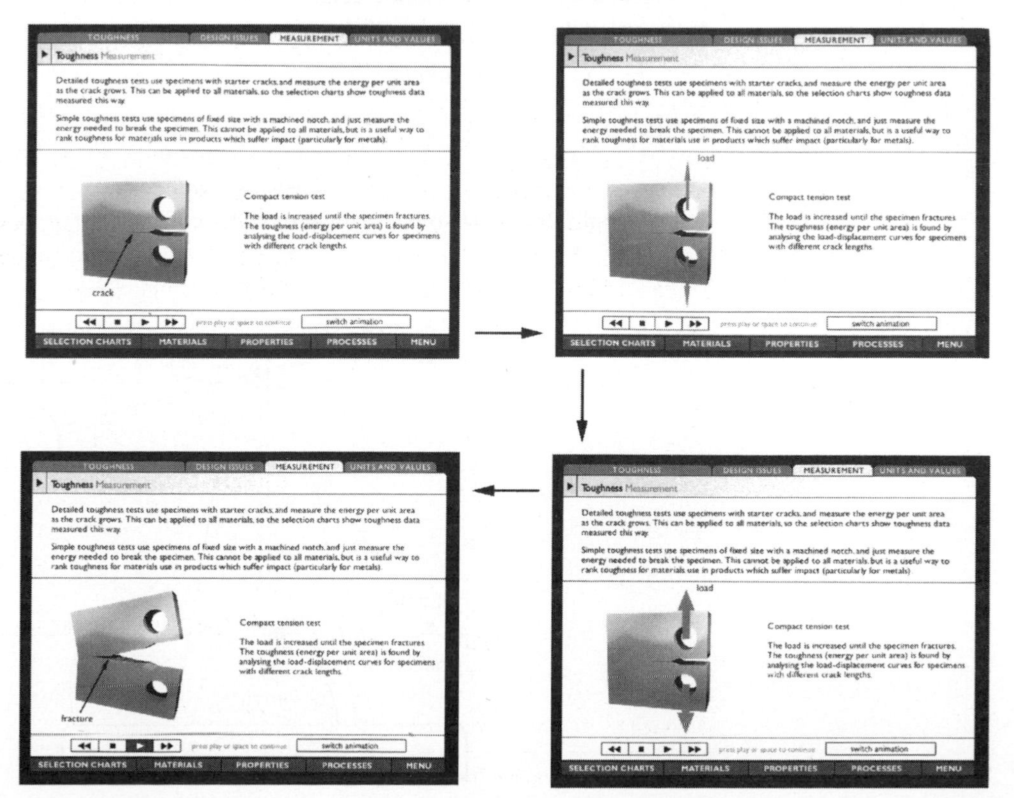

Fig 5.32 Compact tension test specimen here.

Several specimens with different starter crack lengths are loaded to failure. The toughness is found by analysing the load–displacement curves for this series of specimens, since this indicates the available stored energy, which is released when each specimen fractures.

Toughness measured in this way is a genuine material property that is independent of specimen size. The method can be applied to all materials, provided certain limits on specimen size are observed. It has units of energy per unit area, or Joules/m^2 (J/m^2).

Impact tests

Impact tests measure the energy necessary to fracture a standard notched bar specimen using an impulse load. It is a simple test to perform, and by using standard specimen sizes, the energy absorbed provides a ranking of how tough materials are. Given the fixed specimen size, the test cannot be applied to all materials, but is a useful way to rank toughness for materials used in products that suffer impact (particularly for metals). It is not strictly measuring the toughness, since part of the energy goes into general yielding of the specimen and part into the fracture event itself. It therefore falls somewhere in between a true toughness test like the compact tension test, and simply measuring the area under a stress-strain curve, which is sometimes called the 'work of fracture' (per unit volume) for a material.

There are two common kinds of impact test in which a heavy hammer swings in a vertical arc, striking a specimen held in a vice at the bottom of the swing, the Izod test and the Charpy test.

The swing of the hammer fractures the specimen and the loss of height of the hammer from the start to end of the arc gives the energy absorbed as shown in Fig. 5.33.

$$\text{The change in energy} = mg(h_1 - h_2)$$

Figure 5.34 shows how these heights can be obtained from the angles normally measured and a pendulum length L.

$$h_1 = L - L\cos\theta \text{ and } h_2 = L - L\cos\phi$$

Therefore

$$\begin{aligned} h_1 - h_2 &= (L - L\cos\theta) - (L - L\cos\phi) \\ &= L\cos\phi - L\cos\theta \\ &= L(\cos\phi - \cos\theta) \end{aligned}$$

$$\text{and the energy absorbed} = mgL(\cos\phi - \cos\theta)$$

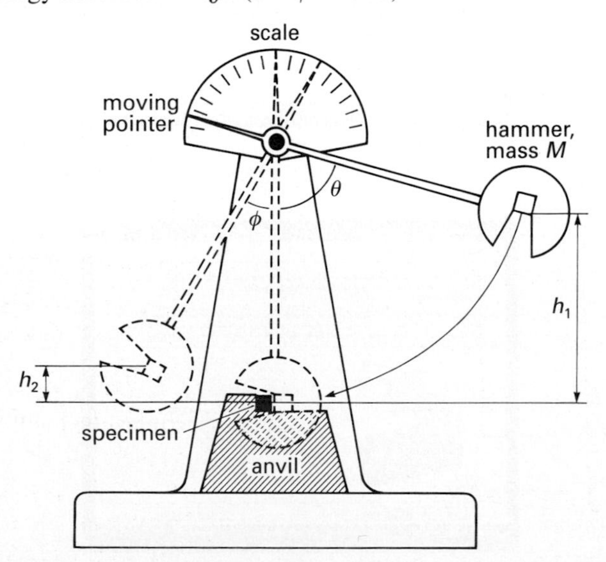

Figure 5.33 Diagram of the standard impact testing machine.

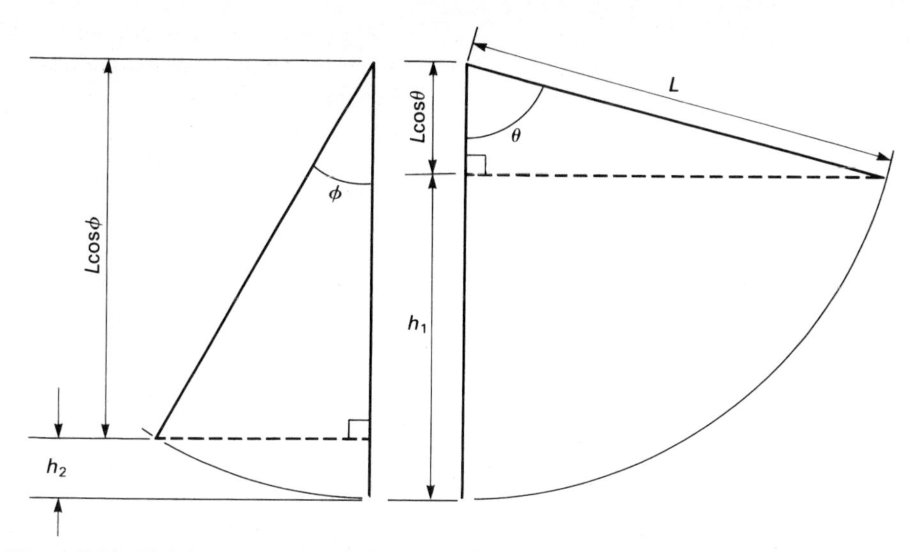

Figure 5.34 Height measurement in impact testing.

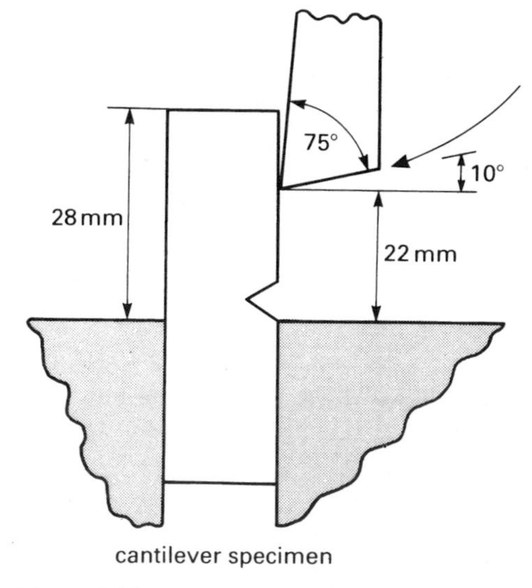

Figure 5.35 Impact test specimen for the Izod test.

The Izod test

The Izod test uses a circular cantilever specimen with a V-notch mounted vertically in the vice as shown in Fig. 5.35.

The Charpy test

The Charpy test uses a simply supported beam specimen mounted horizontally in a vice as shown in Fig. 5.36. The specimen is rectangular in section with a round-bottomed V-notch midway between the supports.

Table 5.15 gives typical values of impact test results showing the energy that different materials absorbed during fracture.

Material	*Impact energy* (J)
cast iron	20
steel (mild, low carbon)	55
steel (0.4% C, 0.9% Mn, 1% Ni) (normalised)	91 (Izod)
steel (0.4% C, 0.9% Mn, 1% Ni,) (tempered)	104 (Izod)

Table 5.15 Typical values of impact energy test results for toughness

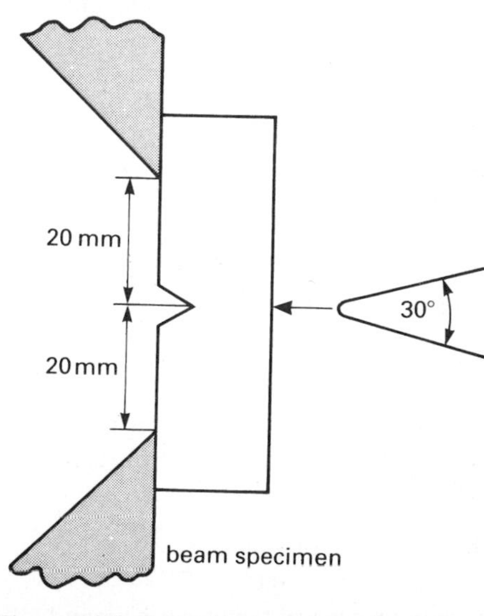

Figure 5.36 Impact test specimen for the Charpy test.

Brittle fracture and the effect of temperature

Brittle fracture occurs by the very rapid propagation of a crack after little or no plastic deformation. In crystalline materials this will occur along preferred cleavage planes and, because of the different orientations of these planes in different grains, the surface will have a shiny granular appearance. Brittle fracture can also proceed along a grain boundary path as a result of embrittling films, for example, sulphides in iron, which collect at grain boundaries. Such films are glass-like and the resulting failure is known as intergranular, rather than transgranular for cleavage planes. Because ductile failure is accompanied by considerable plastic deformation, the energy absorbed will be greater than for a brittle fracture. In body-centred cubic (b.c.c.) metals there is a transition from ductile to brittle behaviour as the metal is cooled, and this results in a transition temperature above which the metal is ductile and below which it is brittle. This is illustrated very clearly in the low-carbon steel shown in Fig. 5.37. As the carbon content increases the effect of temperature and the ductility become less significant.

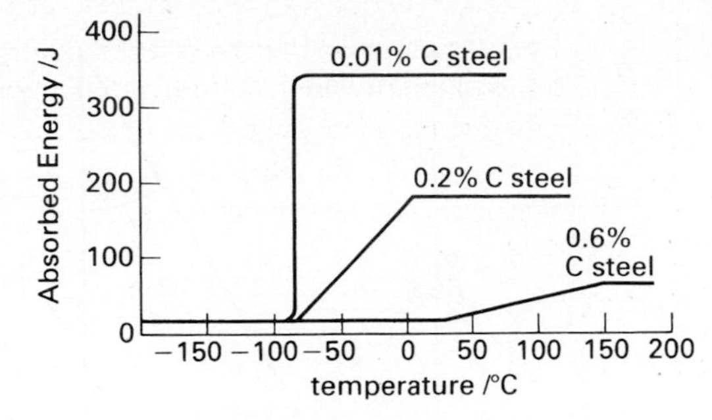

Figure 5.37 Ductile to brittle transition temperatures for steels.

The transition temperature will depend significantly on the shape of the notch because brittle fracture is more likely at higher stress concentrations and is in any case subject to considerable scatter.

In interpreting Izod or Charpy test results, the values obtained for new materials or alloys are compared with those for materials which are known to have given good service in production. The transition temperature indicates a material's likely suitability for low temperature service – in the Alaska oil pipeline, a North Sea oil platform or cryogenic storage vessel perhaps. A fair margin of safety is obviously required in all these cases, because a brittle fracture, perhaps of a platform leg if a ship accidentally collided with it, would have appalling consequences.

5.2.7 Creep

Metals

When forces are applied to a metal, its deformation, whether elastic or plastic, usually takes place very rapidly, and then no further deformation takes place however long the force continues to act. This is not true, though, at temperatures greater than about half of the melting point (in kelvin) of the metal. At these higher temperatures, a metal continues to deform slowly whilst the stress on it is maintained – a process called creep. In machine parts which are stressed at high temperatures, such as turbine blades, creep is a very serious problem. Soft metals such as lead will creep at room temperature, so creep is also a problem in lead pipes and white metal bearings. Creep may occur under static tension, compression, shear, torsion and bending loads.

Polymers

Most thermoplastic materials are subject to creep at room temperatures. There will often be an orientation effect dependent on the direction that the test piece axis runs relative to the rolling or extrusion direction. Plastic water pipes are subjected to constant internal pressure and creep causes them to expand radially, with consequent reduction in the wall thickness and ultimate weakening or failure.

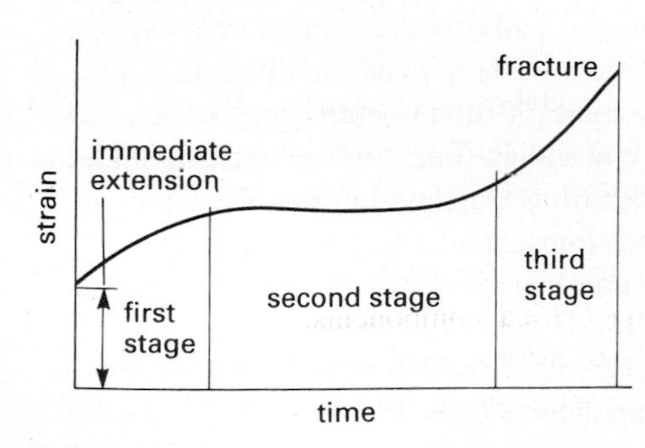

Figure 5.38 A typical creep curve showing the different rates of change of creep in each of the three stages.

Strain–time curves

Creep caused by a static tensile load is determined by measuring the extension with time. After the initial rapid extension it characteristically divides into three regions as illustrated in Fig. 5.38:

1 The first stage, where strain hardening reduces the rate of strain.
2 The second stage, where the strain is approximately constant.
3 The third stage, where necking of the specimen or internal faults cause a rapid increase in the strain.

Creep is both stress and temperature dependent and so greater strain is noticed in all materials when either stress or temperature is increased. If a weight is hung from a thermoplastic hook an initial deflection will be noticed. If this is measured again after the constant weight has been left in place for some time it will be found to have increased. The thermoplastic is exhibiting creep strain and typical curves showing this behaviour are given in Fig. 5.39.

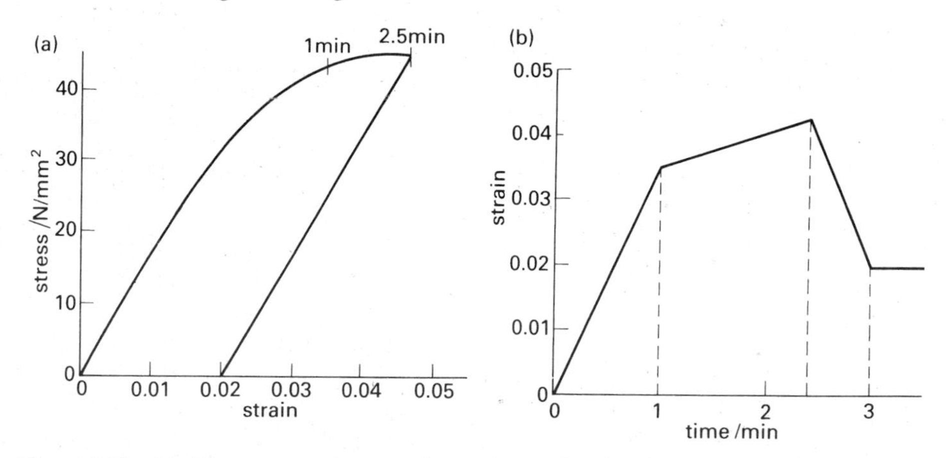

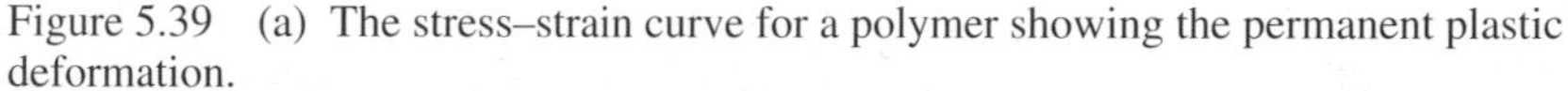

Figure 5.39 (a) The stress–strain curve for a polymer showing the permanent plastic deformation.
(b) The strain–time curve of the polymer showing the initial strain with applied stress and the continued strain while the stress remains constant.

5.2.8 Fatigue

In the middle of the nineteenth century, it was realised that fluctuating loads could be more dangerous to engineering structures than steady ones of the same or even larger magnitude. In 1843, Rankine drew attention to the potential dangers, and about twenty years later comprehensive investigations were reported by Wöhler. It is noteworthy in the light of subsequent events that his work was prompted by the failures of railway wagon axles that had been in service for long periods. The advent of rapidly operating machinery, the motorcar and the aircraft increased the significance of fatigue, and during the twentieth century great efforts were made to understand the phenomenon and to alleviate its drastic effects.

Fatigue failure came to the attention of the general public through the dramatic crash of the first Comet aircraft in 1951 and Nevil Shute's popular novel *No Highway* (1948). Nowadays every aircraft component and other product where safety is paramount is tested to destruction to find its safe working life.

Fatigue testing

With these objectives in mind, fatigue testing machines have been developed to simulate the type of stressing experienced under service conditions. They fall into four main categories.

- Wöhler constructed machines that rotated metal rod specimens loaded as cantilevers (or beams) and this type of machine is still used extensively. As the specimen rotates each point in the material is alternatively in tension and compression, as shown in Fig. 5.40.
- Other bending machines are suitable for testing sheet and strip metal under conditions that occur in leaf springs and electrical components.
- Torsion fatigue machines can twist and untwist test specimens, subjecting the material to the type of stress set up in transmission shafting, splines and torsion bars.
- Direct stress machines can pull and push specimens to stimulate stress conditions in bolts, connecting rods and press tools.

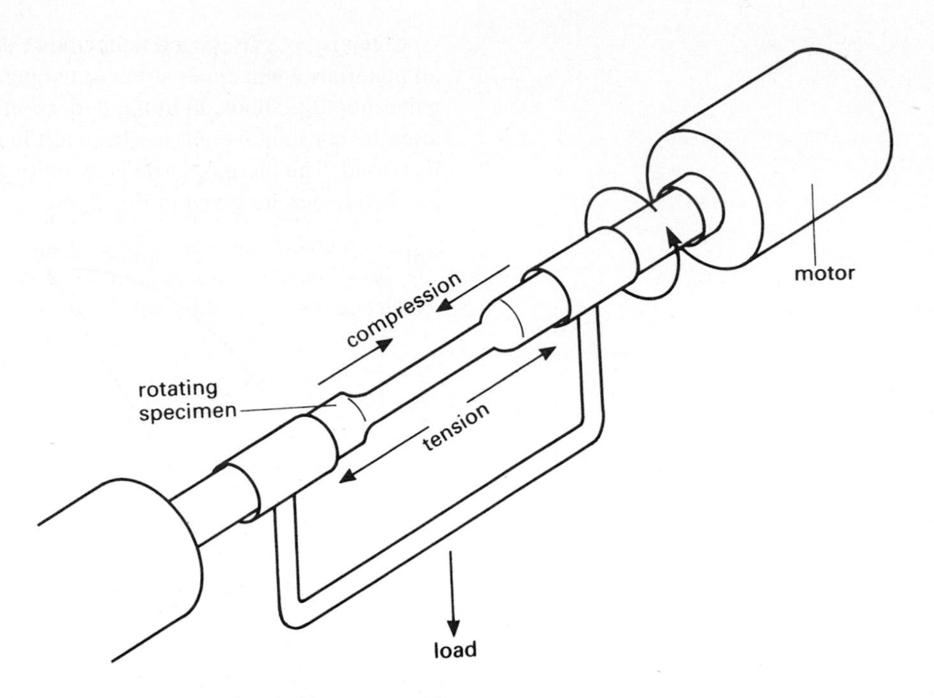

Figure 5.40 A rotating fatigue test machine.

Cylindrical fatigue test specimens should be carefully machined to avoid work-hardening the surface and usually have a diameter between 3 mm and 10 mm. The section to be tested should then be polished longitudinally with fine emery paper until an almost mirror finish is obtained. The specimen should then be set in position very carefully to avoid possible effects of unintentional vibrations. Normally the first test is carried out using a relatively high imposed stress, with failure occurring at 1000 to 10 000 cycles. The next test is carried out using a lower applied stress, requiring an increased number of cycles to cause failure. This technique is continued until a stress level is reached at which failure does not occur even if the number of cycles is as high as 10–100 million. Under these circumstances, the applied stress is known as either the *fatigue strength* or the *endurance limit*.

The pattern of results that can be expected for a ferrous or a nonferrous alloy are shown in Fig. 5.41. With steels, if the stress range can be withstood for 10 million cycles it appears that it can be resisted indefinitely. However, as the diagram shows, there is no corresponding limit for nonferrous metals. Tests have been extended to hundreds of millions of cycles and the safe range continues to reduce.

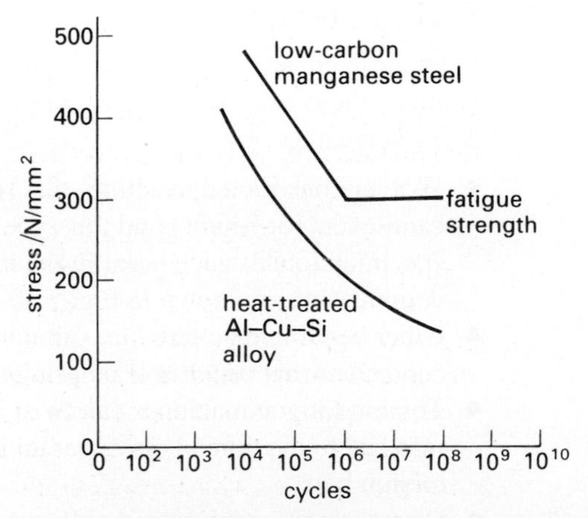

Figure 5.41 The fatigue test results of a low-carbon steel and an aluminium alloy showing the clear failure strength of the steel.

Factors affecting the fatigue strength

Grooves, holes and slots all raise the local stress values. A small oil hole has been found to reduce the fatigue strength of a medium-carbon steel by over 50%. Grey cast iron, however, is quite insensitive to stress concentrations. This behaviour is explained by the fact that the material intrinsically contains many stress raisers in the form of graphite inclusions, so that the addition of another stress raiser, say in the form of a notch, has little effect.

Corrosive environments can also have an adverse influence on fatigue strength and account must be taken of these effects. For instance, the fatigue strength of an aluminium alloy and of mild steel is halved by the presence of tap water. On the other hand, corrosion-resistant materials like titanium or stainless steel are much less affected. Having established a fatigue curve for a particular material, it is suggested that the effects of corrosive elements in the environment and stress concentrations, such as drilled holes and notches, are evaluated.

5.3 Material selection charts

The main classes of engineering materials and the major design-limiting properties have now been described. At this point the task of choosing which materials to use when designing something may seem to imply the need to grapple with huge volumes of material data. Traditionally designers have used extensive handbooks and their own experience to guide the choice of material in design. Electronic databases are making the task of exploring the data more straightforward, but still do not always present material data in a very tangible form or make materials easy to compare.

Here we adopt a graphical approach to presenting material property data, first proposed by Michael Ashby of Cambridge University Engineering Design Centre, and further developed on CD-ROM and the World Wide Web (see Bibliography). The key idea is to plot properties in pairs on *material selection charts* – usually with logarithmic scales, given that most properties extend over several orders of magnitude. Fig. 5.42 shows an example – Young's modulus against density. What does this tell us?

- The properties for a given class of materials (metals, ceramics, polymers, etc.) tend to cluster together.
- The relative values of Young's modulus or density for the different classes can quickly be appreciated (polymers are floppy, most solid polymers sink in water, ceramics are stiffer than metals, and often less dense, etc.).

A 2-D plot enables a clear visual appreciation of the ranges and relative magnitudes of the two properties considered individually, simply by spreading the materials out on the diagram. But there is much more to it than this. Firstly, it is no coincidence that metals, polymers and so on cluster on a Young's modulus–density chart, as these properties reflect the characteristic atomic packing and bonding in each class (Chapter 4). The spread of the data for a given property therefore enhances our appreciation of the underlying physics of the property.

The main advantage of the charts however stems from the fundamental need in engineering design to consider multiple objectives and constraints – light and stiff, strong and cheap, tough and recyclable (or maybe all of these at once!). Material selection in design is therefore a matter of assessing trade-offs between several competing requirements. Selection charts provide insight into these trade-offs by pairing properties which must commonly both be considered, avoiding the need to work with tedious tables of numerical data. This approach is developed in Chapter 7.

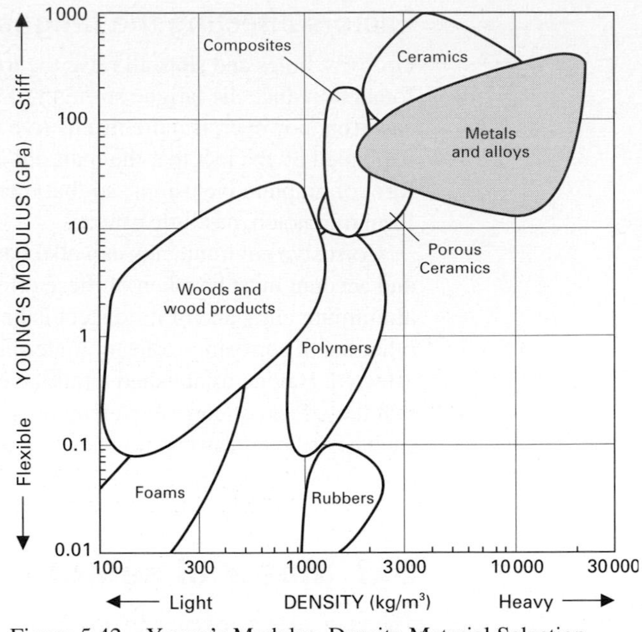

Figure 5.42 Young's Modulus–Density Material Selection Chart, showing the classes of engineering materials.

Fig. 5.43 illustrates two other important properties – strength and toughness. For this chart, 'true' toughness data have been used: G_c, the strain energy release rate/unit area for unstable crack propagation. As noted in section 5.2.6, this is more universally valid and available than measures such as impact energy. Note that on this diagram the 'bubbles' are much bigger, as the plasticity and fracture of materials are sensitive to the material microstructure (which is affected by processing and heat treatment), unlike Young's modulus and density, which only depend on atomic packing and bonding.

The wide ranges of strength and toughness are even more apparent if we populate the material classes on a selection chart with data for materials within a class. Fig. 5.44 shows the strength–toughness chart with a selection of metals illustrated. Note that in general the toughness of a type of alloy falls as its strength is increased.

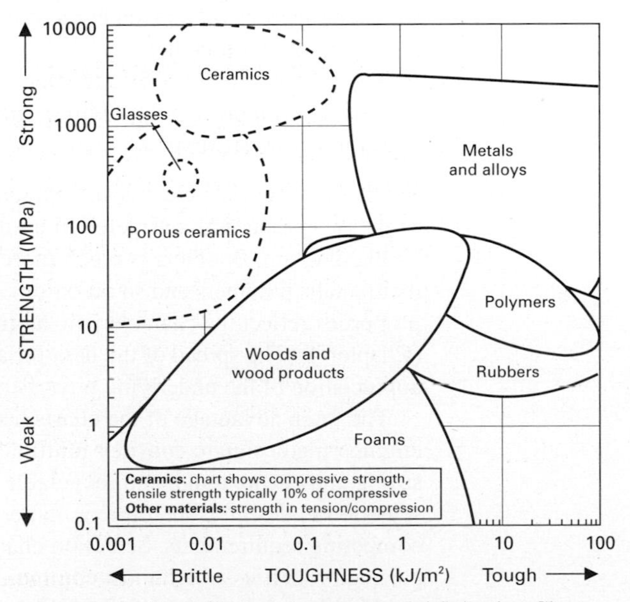

Figure 5.43 Strength–Toughness Material Selection Chart, showing the classes of engineering materials.

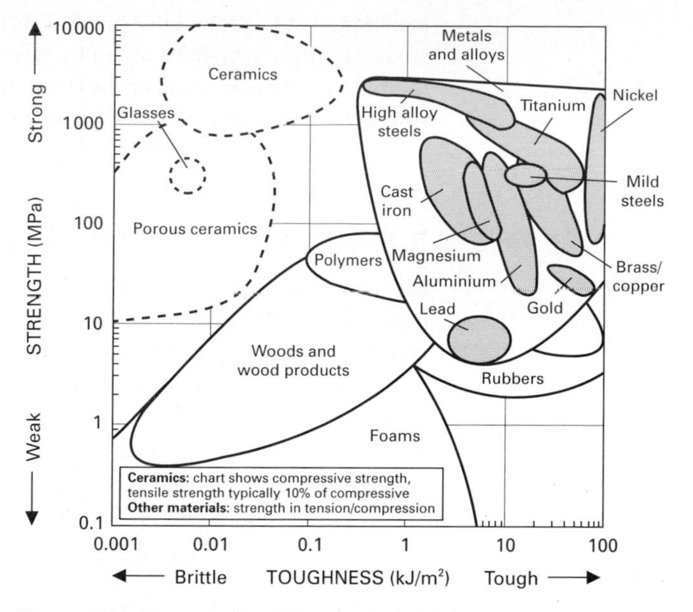

Figure 5.44 Strength–Toughness Material Selection Chart, showing the classes of engineering materials, and a selection of metals.

5.4 Nondestructive testing

Manufactured components and structural members need to be tested to ensure that they will perform in a predicted manner. If the test procedure does not destroy or damage the specimen it will clearly save specimen preparation time, material and cost and is known as *nondestructive testing*. These methods are obviously necessary for items such as an aircraft structure and a nuclear reactor vessel which should be proved safe without causing any damage to them. Surface imperfections can be examined in four different ways: by visual inspection, penetrant tests, magnetic particle tests and electrical tests. Internal structural flaws on the other hand can be tested by radiography or ultrasonic methods. Definitions of these test procedures are to be found in BS 3683.

5.4.1 Visual inspection

Visual inspection can provide useful information about the surface condition of a material. For some components a low-powered microscope may be used to obtain greater magnification but in all visual techniques great care must be taken to ensure no material flaws have been overlooked. An aircraft structure for instance shows the usefulness of visual testing. In aluminium alloys the critical crack length is in excess of 300 mm and so fatigue cracks can be identified visually or perhaps with the aid of a hand lens long before they would cause failure of the structure. If a subcritical crack is found in aluminium alloy a hole is drilled at each end to prevent further propagation of the crack. Low strength steels are also appropriate for visual testing but in high strength steels the critical crack length is less than 0.5 mm and so these steels require the more sensitive ultrasonic or radiographic testing.

5.4.2 Penetrant tests

By applying a low viscosity liquid to the surface of a material and allowing time for it to penetrate, superficial material flaws can be detected. The excess penetrant must be removed and usually a coat of developer, a white powder, is applied to absorb any penetrant that has found its way into surface imperfections and so identify the critical areas. Castings have been tested by this means for many years. They are immersed in oil for several hours, cleaned and then dusted in french chalk. Penetrant tests are suitable for detecting porosity, grinding cavities and shrinkage cavities but are unable to detect very small or very large, open flaws.

5.4.3 Magnetic particle tests

Magnetic particle tests are very sensitive and can detect fine cracks in ferromagnetic material. Any discontinuities in or near the surface distort the magnetic lines of force and the magnification of the discontinuity can be up to 200. Magnets can be used to magnetise the specimen or more commonly an electric current is used. Flaws can be detected which are parallel to the direction of the magnetic lines of force as shown in Fig. 5.45. After magnetisation the component is sprayed with a magnetic flaw detection ink (fine particles of iron(II) diiron(III) oxide in a carrier liquid such as kerosene), examined and demagnetised.

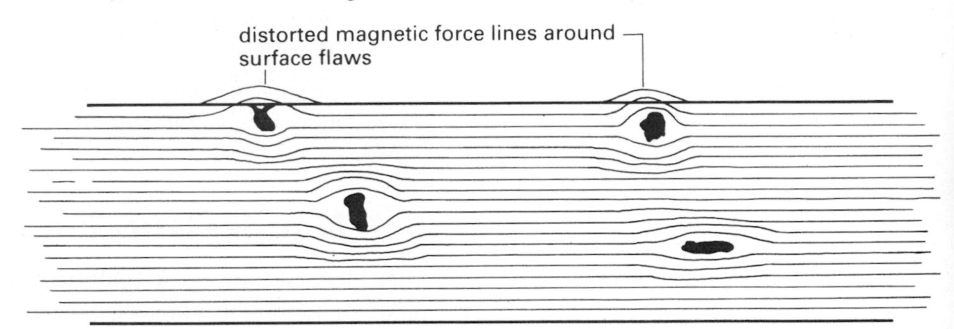

Figure 5.45 Magnetic particle testing.

5.4.4 Electrical tests

When an electric coil is moved close to a conducting surface a magnetic field is produced with electromagnetically induced loops of alternating current in the material called 'eddy currents'. These can then be detected using an eddy current search coil and any changes in electrical pattern will signal a flaw in or near the surface of the material as shown in Fig. 5.46. This is a fast method of testing and is the most widely used nondestructive test for nonferrous wrought metals.

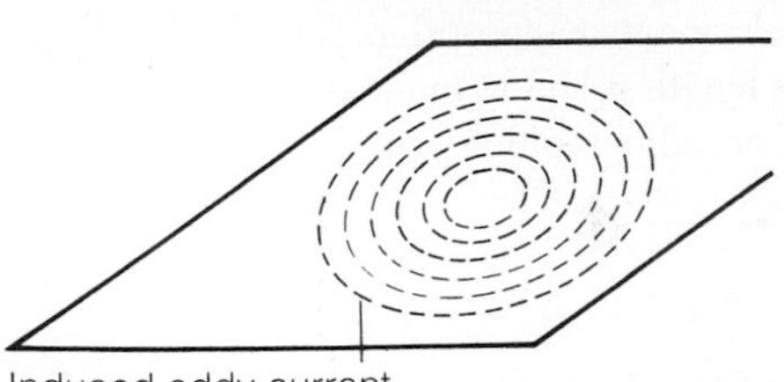

Figure 5.46 Using an eddy current search coil in electrical testing.

5.4.5 Radiography and fluoroscopy

Radiography

When a beam of X-rays (see section 5.1.4) passes through a material on to photographic film an X-ray image is obtained. After washing and drying, the film plate can be viewed on a light screen just as medical X-rays are viewed in a hospital. Due to their different absorption and transmission characteristics defects such as cavities show up as darker areas on the X-ray film. Very fine cracks however are difficult to see and an ultrasonic test method is much more appropriate for these. Radiography gives a permanent record of the defects and is widely used to examine castings for blowholes, cracks, inclusions, porosity and shrinkage cavities despite the high capital and maintenance costs of the equipment necessary for this technique. It is also used to examine butt welds in boilers, oil and gas pipelines and high pressure vessels.

γ-rays can also be used in the same way as X-rays in radiographic testing. γ-rays are emitted when a radioactive element undergoes spontaneous nuclear distintegration. Cobalt-60 is normally used to produce an intense beam of a single wavelength suitable for thick, absorbent materials. Iridium-192 or Caesium-137 is used when a less intense radiation is required. γ-ray radiography is used in preference to X-rays when thick sections (greater than 70 mm thick), are to be examined which would need a very large source of X-ray radiation. γ-ray techniques have the additional advantage that they are portable and do not need an electricity or water supply close at hand as X-rays do. Their disadvantages are that they need longer exposure times and so do not always produce a well-defined image.

Fluoroscopy

Low-energy X-rays directed on to a fluorescent screen of zinc cadmium sulphide are converted to ultraviolet and visible light which are then reflected back to the observer. This can be used in a similar way to the X-ray film technique to obtain a fast, cheap but temporary image of the structure of the component. This method is typically used for fast on-line inspection of low-density materials such as aluminium alloys.

5.4.6 Ultrasonic tests

Sound waves in the frequency range 16–20 kHz are audible to humans while some animals can hear considerably higher frequencies than these. Frequencies of 20 kHz to 1000 MHz are classified as ultrasonic waves. Ultrasonic tests use waves with frequency of 500 kHz to 20 MHz which readily pass through homogeneous solids and liquids. Flaws cause reflection of the ultrasonic vibration which can then be picked up and analysed. Steel plate and bars of thickness in the range 5–30 mm can be tested for inclusions, seams, rolling cracks and laminations with very fine sensitivity, while bars and castings up to 1.5 m diameter can be examined for shrinkage and cracking. Ultrasonic testing is also used to examine for cracks and other defects like porosity and slag inclusion in welds.

When an alternating voltage is applied across a quartz crystal it expands and contracts accordingly, and conversely when it is forced to expand and contract it produces a change in output voltage. This is known as the piezoelectric effect. Piezoelectric crystals can therefore be used to convert an electrical signal to mechanical vibrations and so transmit ultrasonic waves to a material. When operated in reverse it can also receive a transmitted wave if the surface is clean and an oil film is applied to exclude air which would otherwise give rogue results.

The *pulse reflection method* uses an ultrasonic beam transmitted through a material until it is reflected from the back face of the specimen. The time taken for the reflected wave to be received at the starting point should depend on the length of the specimen and the frequency of the wave. If a pulse is received earlier than that expected for the reflected ray, it is an indication that part of the wave has been reflected by an imperfection in the material, see Fig. 5.47. Thus the position of the flaw can be accurately measured. Both the test pulse and the received signal are

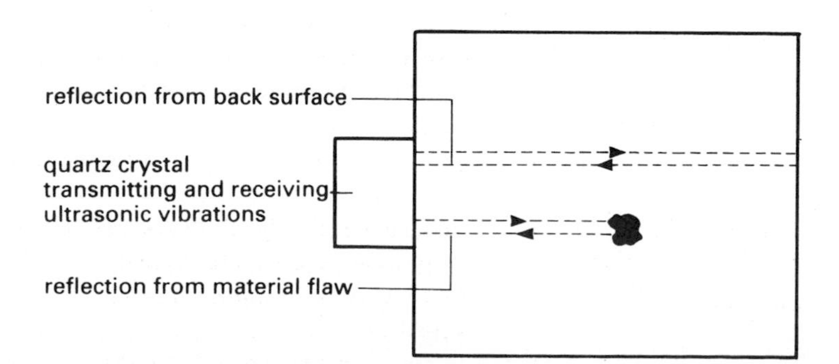

Figure 5.47 Ultrasonic pulse reflecting testing.

usually magnified and shown on a cathode ray oscilloscope. This method can also be used to scan large sheets of material for flaws when they are placed in a large bath of water to exclude the air.

An ultrasonic beam can also be directed through a parallel-sided material and the transmitted beam picked up and examined. This is a *transmission method* and relies on the outgoing and incoming beam being of different intensity if a flaw has reflected some of the ultrasonic wave. Obviously this method needs duplicate equipment with no benefit and so is not as widely used as the pulse reflection method.

British Standards publications

BS EN 10002 (1990) *Tensile Testing of Metals*
BS 131 *Methods for Notched Bar Tests*
Part 1: 1961 *The Izod Impact Test on Metals*
Part 2: 1972 *The Charpy V-notch Impact Test on Metals*
BS 240: 1986 *Brinell Hardness Test and the Verification of Brinell Hardness Testing Machines*
BS 427 *Method for Vickers Hardness Test*
Part 1: 1961 *Testing of Metals*
BS 891 *Method for Rockwell Hardness Test*
Part 1: 1962 *Testing of Metals*
BS 2782: 1970 *Method of Testing Plastics* (Parts 1–11)
BS 5411 *Methods of Test for Metallic and Related Coatings*
Part 6: 1981 Vickers and Knoop Microhardness Tests

Bibliography

Ashby M F 1999 *Materials Selection in Mechanical Design*. 2nd edn. Butterworth–Heinemann
Askeland D R 1996 *The Science and Engineering of Materials.* 3rd SI edn. Chapman and Hall
Cornish E H 1987 *Materials and the Designer*. CUP
Couzens E G and Yarley V E 1968 *Plastics in the Modern World*. Penguin (Pelican)
Desch H E 1980 *Timber – its Structure and Properties.* 6th edn. Macmillan
Everett A 1978 *Materials – Mitchell's Building Construction. Wiley*
Gandhi M V and Thompson B S 1997 *Smart materials and structures*. Chapman and Hall
Hennessy L and Smyth L 1985 *Engineering Technology*. Iona Print Ltd, Dublin
Higgins R A 1976 *Materials for the Engineering Technician*. Hodder and Stoughton
Higgins R A 1977 *Properties of Engineering Materials*. Hodder and Stoughton
Howatson A M, Lund P G and Todd J D 1972 *Engineering Tables and Data*. Chapman and Hall
Lovatt A M, Shercliff H R and Withers J 2000 *Material Selection and Processing*, CD-ROM and supporting booklets, Technology Enhancement Programme (TEP), London
Orgorkiewicz R M 1977 *The Engineering Properties of Plastics.* Oxford University Press
Project Technology 1972 *Handbook 3 Simple Materials Testing Equipment.* Heinemann/Schools Council
Ryan W 1978 *Properties of Ceramic Raw Materials*. Pergamon Press
Schools Council 1981 *Modular Courses in Technology, Materials Technology*. Oliver and Boyd
West G H 1986 *Engineering Design in Plastics*. Plastics and Rubber Institute

Assignments

1 Choose any common product which uses at least two different materials and explain how the properties of the materials help the product to fulfil its function.

2 Thermal insulation is one factor to be taken into account when selecting materials for some situations.

(a) Describe the characteristics of materials which make them good thermal insulators.

(b) Explain how the thermal and other properties would affect your choice of material for *two* of the following situations. In each case suggest suitable materials.

(i) The walls of a house
(ii) An anorak
(iii) A cool box for camping
(iv) A drinking vessel without a handle

3 (a) Describe *four* important properties or characteristics which should be possessed by a material used in the construction of an item of hand-held kitchen equipment.

(b) Name materials which possess these properties or characteristics and describe how the structural components of your chosen artefact might be made.

4 Describe the way in which the variation of an electrical property with temperature is exploited in *one* type of thermometer.

5 Describe the atomic structure of a typical material which could be easily magnetised. Explain briefly the effect of magnetisation on such a material.

6 Sketch a labelled graph of flux density against magnetic field strength for the magnetisation of a piece of soft iron when the magnetic field strength is gradually

(a) increased from zero to a high positive value at which saturation occurs
(b) decreased to zero,
(c) reversed to a high negative value, and
(d) returned to a high positive value.

Show on the diagram the quantities *remanence* and *coercivity*.

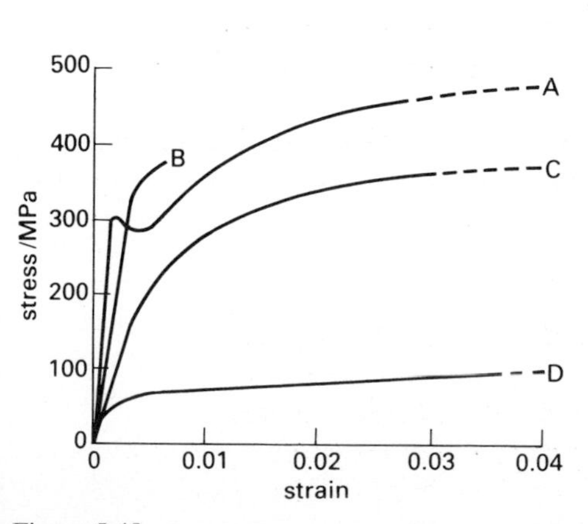

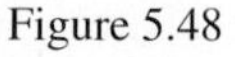
Figure 5.48

7 Figure 5.48 shows the stress–strain curves for pure aluminium, annealed brass, cast iron and mild steel.

(a) Match the curves to the materials.
(b) Calculate the Young modulus of elasticity for material B.

8 (a) A metallurgist may carry out the following tests on samples of a metal. Describe the nature of the properties involved.

(i) a tensile test
(ii) an impact test
(iii) a hardness test

(b) Describe the form of the specimens used for carrying out each of the above tests.

(c) Which of the properties considered in the tests described in (a) would be used in the selection of material for each of the following applications and why?

(d) How would the properties of a medium carbon steel be affected by:
   (i) annealing,
   (ii) heating to 900°C and quenching
   (iii) heating to 500°C after (ii) and cooling to room temperature?

(e) In the manufacture of a cutting tool, for example the blade of a carpenter's plane, from a medium carbon steel;
   (i) what processes listed in (d) might be used,
   (ii) at what stage or stages would they be used,
   (iii) what are the reasons for their use?

(Note: Part (d) is covered in Chapter 4 and (e) in Chapters 6 and 7.)

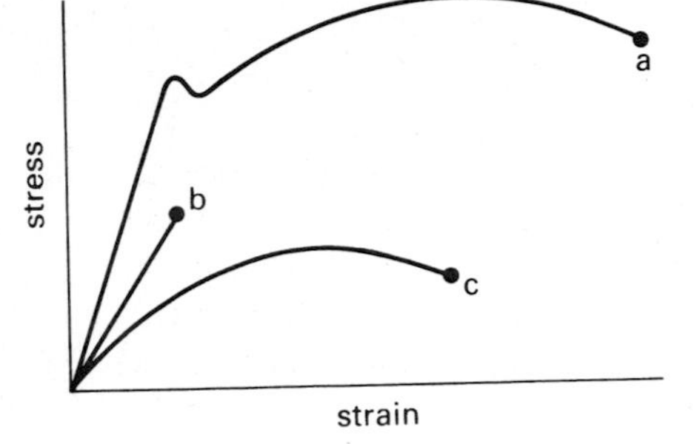

Figure 5.49

9 Figure 5.49 shows stress–strain curves for three common materials.
(a) Which is the most brittle?
(b) Which is the most ductile?
(c) Which has the highest tensile strength?
(d) Explain, briefly, the difference between the elastic and the plastic state when a material is being stressed.

10 Sketch a typical graph resulting from a tensile test on a mild steel specimen. Explain what is understood, in this context, by the terms
(a) elastic,
(b) plastic,
(c) 1% proof stress,
(d) yield stress,
(e) ultimate tensile stress,
(f) elongation.

Select *one* nonferrous metal and describe how the graph for this would differ from that for mild steel.

List and explain briefly other properties which affect the failure of components in service. Give examples of situations in which these properties may be more important than those mentioned in (c), (d), (e) and (f) above.

11 (a) A tensile test specimen made of aluminium alloy has a diameter of 11.3 mm and a gauge length of five times the diameter. In a tensile test to destruction the following results were obtained.

Load (kN)	2	4	6	8	10	12	14	16	18	20	22	24
Extension (μm)	1.6	3.3	5.0	6.6	8.0	10.0	11.5	13.5	18.0	25.0	36.0	57.0

   (i) Plot a load–extension graph.
   (ii) Calculate the Young's modulus for the alloy.
   (iii) Determine the 0.1% proof stress.
   (iv) Explain why it is necessary to determine the proof stress for aluminium alloys.

(b) Show why an understanding of part of the iron–iron carbide diagram is important in the heat treatment of plain carbon steels.

12 Concrete is a material which is strong in compression, yet with structural columns carrying compressive loads the normal practice is to reinforce them.
(a) Suggest two reasons why this practice is considered necessary.
(b) Sketch an arrangement in elevation and plan view showing how the reinforcing material would be incorporated in the column.
(c) Show how the column would fail if the reinforcing was inadequate.

13 Distinguish between the terms *creep* and *fatigue* when they are applied to engineering materials. In the following applications comment on the relative importance of these two phenomena:
(a) an axle of a dumper truck,
(b) a turbine blade in a jet engine,
(c) a tall lightweight communications aerial.

14 (a) In quality control, samples of materials are tested to verify the specification of a batch. Describe, with the aid of sketches, how three of these tests should be carried out in school stating which property is being tested.
(b) Discuss the reasons for using destructive and non-destructive tests.
(c) Explain metal fatigue, and describe two examples where metal fatigue has seriously affected the performance of a product.

6 Materials processing

6.1 A designer's view of manipulating materials

An overview of the materials available to a designer and their properties has been given in Chapters 4 and 5. The importance of considering the manufacturing route as early as possible in the design process has already been emphasised in section 2.5.1. This chapter sets out to give a summary of the key features of manipulating materials from a designer's perspective. Designers will tend to think of extruded sections in aluminium or plastic and rolled steel sections as alternatives and hence these have been included in the same section. Again, designers will often think about die-casting and injection moulding together and it is not unknown for experimental injection mouldings to be made from the same tools as those used for die-casting metals. For the same reason the lamination of wood and GRP (glass fibre reinforced plastic) and the pressure forming of timbers, plastics and metals are linked together.

Information about manipulating materials could be arranged in the traditional manner according to the material classification. From this point of view the machining of metals and wood are very different processes and the forging of metals and the steam bending of timbers would not be seen as related activities. Such a viewpoint has the virtue of emphasising the differences associated with particular material properties, but fails to generate the kind of overview a modern designer needs. An open-minded approach to materials selection depends on being confident that the kind of result required can be achieved in a range of materials and by a number of processes. Equally, cost-effective decisions are only going to be made if the limitations associated with particular materials and processes are well understood. It is a subtle balance of general principles and detailed knowledge which is required and it is this that the arrangement of this chapter is intended to promote.

It is essential that the detailed design of a product is carried out in relation to specific materials and manufacturing processes. Matters like draft angles on castings and the difficulties of dealing with re-entrant features depend critically on the process selected. For example, although a draft angle is required for sand-casting none may be required if shell moulding is used instead. The choice between these two is normally related to the production quantity and hence the detailed design can only take place once the size of the market has been considered. In the early stages of the design process, it is an overview of what can be achieved which is necessary: a much more specific kind of knowledge is required for detail design. Design students should develop an overview of materials processing, and at least know where the specific knowledge they need for detail design can be found.

6.2 Forming techniques

Forming techniques alter the shape of materials in a controlled manner. The control is achieved through the method by which the material is forced into the required position or by the use of control surfaces. Blacksmiths produced very complex shapes by skilfully weighted hammer blows and the use of the flat and curved surfaces of the anvil. The anvil surfaces were not designed for a specific item but served in the manufacture of a variety of products. In contrast GRP mouldings are made by hand in a mould which is made for a specific product and the GRP material is *stippled* into place. (Stippling is a tapping action using the hairs of a soft brush.) Panels for products like cars and aircraft are normally curved in three dimensions and require complex tooling so that the material can be forced into shape using powerful presses.

Getting the material into the correct shape is often only the initial stage of the forming process. Some materials will not stay in position unless they are held there for a period of time. Thermoplastics will need to be held until they have cooled sufficiently, laminated wood whilst the adhesive dries and GRP mouldings whilst a chemical reaction which sets the resin takes place. The difficulty of the holding operation will depend on the magnitudes of the forces involved. GRP will be held against the mould by the force of gravity. Laminated or steamed wood, however, may require quite large clamping forces depending on the thickness of the veneer or timber being used.

More highly stressed components require stronger materials to be formed. This may mean the heating of a metal like steel to make it more workable, or the compaction of powders and their heating to promote bonding. These forming processes – forging and the processing of powders – are described in sections 6.2.4 and 6.2.5.

Casting processes are distinguished from other forming techniques because the material is normally liquid and flows into the mould. These processes are described in section 6.3.

6.2.1 Bending operations, using presses and rotational forming techniques

Bending sheet and tubing

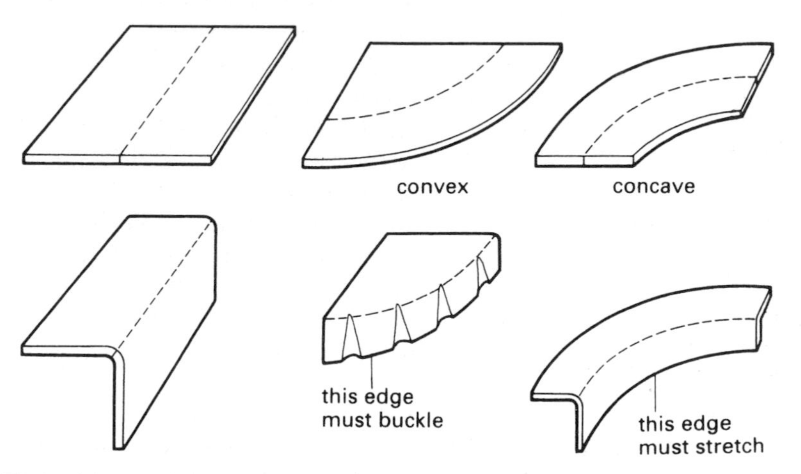

Figure 6.1 Bends in sheet material.

Figure 6.2 Copper trays made with simple tooling.

Fig. 6.1 shows the result of making bends in sheet materials for straight and curved edges with a forming tool (control surface) under the material only. The diagrams would apply equally well to metals or thermoplastics, although the latter would need to be heated in the area surrounding the bend line by a strip heater.

A straight edge will bend without deforming, but a concave edge must stretch and a convex edge will buckle. This is because the outer edge of a convex corner is longer than the bend line, and the outer edge of a concave corner is shorter than the bend line. Products can be found where the stretching and flute-like buckling have been incorporated in the aesthetics of the product, but more often designers will strive to avoid them. Fig. 6.2 shows alternative strategies for the corners of pure

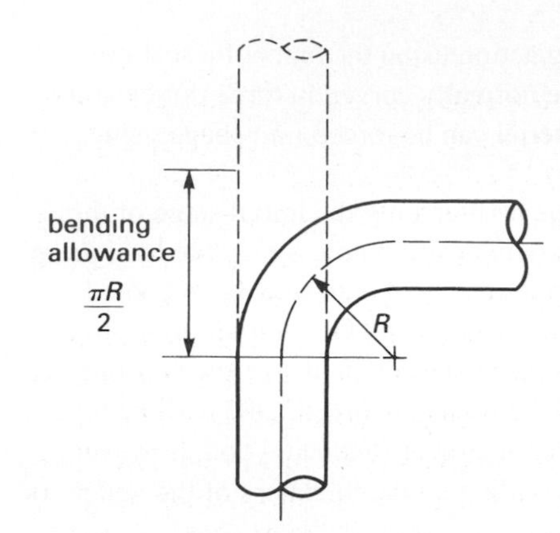

Figure 6.3 Bending tubing.

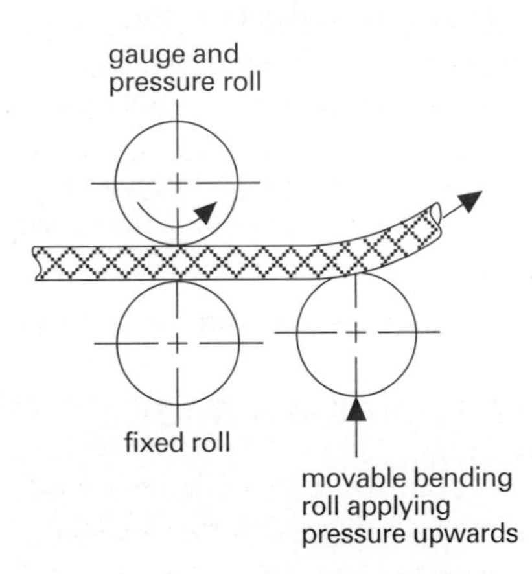

Figure 6.4 Three-roll bending machine.

copper trays which were manufactured with very simple tooling in Zimbabwe. One has the corners cut away and in the other the surplus material has been re-shaped to support cigarettes.

Smooth curves could be produced either by eliminating the buckles once formed or by preventing their formation. Preventing their formation depends on constraining the sheet and forcing it to flow rather than buckle. In metals this means the generation of stresses higher than the elastic limit, so that the material deforms plastically. When sufficiently heated, thermoplastics flow much more freely and, therefore, require much lower stresses. Once formed the buckles can be removed only by spreading the material and trimming it: for metals by hammer blows on to a metal surface and, in plastics, by reheating and forcing out the buckles.

Similar problems occur when a bend is required in tubing. The material on the outer radius must stretch and on the inner radius it must compress. Fig. 6.3 shows a right-angled bend and the length of tubing that will be required, which will be a quarter of the circle circumference (or $\frac{1}{2}\pi R$).

Because of the buckling and stretching of the tube, the minimum radius for successful bending is usually three times the radius of the tubing, although this is dependent on the wall thickness and the material condition. It is essential however that the material is confined on one side by a circular surface and on the other by a roller or sliding former if the tube shape is not to be deformed or crushed. Some bends are made by flattening the tubing before bending in order to avoid the tooling requirements, although this is not always aesthetically acceptable. For one-offs or prototypes of unusual radius the centre of the tube can be filled with sand, a steel spring, rosin, pitch or lead to prevent the tube collapsing.

Bends of large radius are made by using the bending rolls as shown in Fig. 6.4. This technique can be used on sheet, plate, tubing or other sections. The rollers would clearly need to be grooved to take tubing. The material is fed by applying pressure between the gauge and the fixed roll, and rotating the gauge roll. The material is bent by applying a force with the movable bending roll. In order to roll-bend plate, it normally needs to be red hot to reduce the forces involved.

It is not only metals and plastics which must be bent but also timbers, particularly when they are being laminated (see section 6.2.3). Depending on the angle which is needed, the timber selected and its thickness, it may be possible to bend it sufficiently in its natural state. If not, then it must be softened with steam. Timbers have an obvious grain direction, but what is not so apparent is that metals also have a rolling direction. Bending perpendicular to the rolling direction or the wood grain will be successful, whereas bending parallel to the rolling direction or along the grain can result in tearing. This is illustrated in Fig. 6.5.

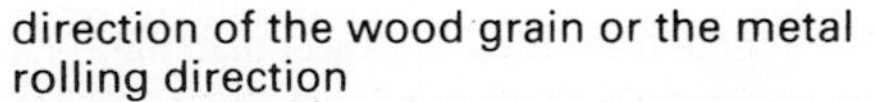

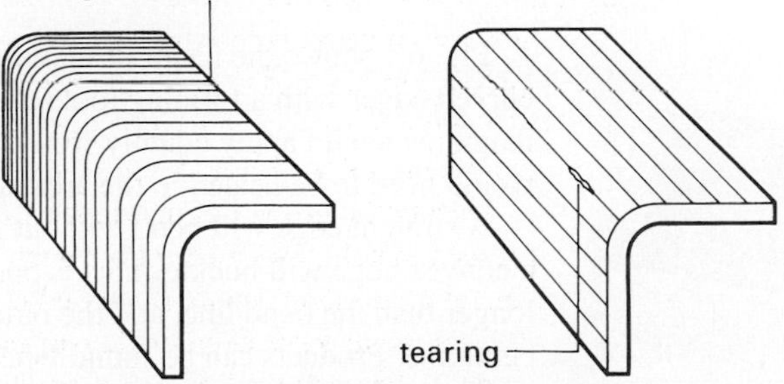

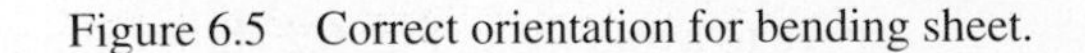

Figure 6.5 Correct orientation for bending sheet.

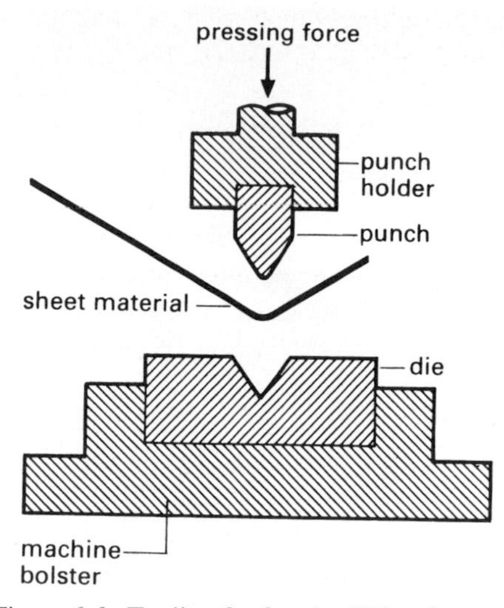

Figure 6.6 Tooling for forming V-bends.

Using presses

'Presswork' is the term usually applied when pressing metals, but wood, glass and plastics are also formed using presses. Metals are pressed at room temperature, although the sheet may well have to be annealed in order to ensure sufficient ductility. Glass must be heated to make it flow more freely before it can be pressed. Both thermoplastic and thermosetting plastics are supplied in sheet form suitable for pressing. The thermoplastics – normally polythene, polypropylene or nylon reinforced with glass or carbon fibre mat – must be heated to soften them prior to pressing. The thermosetting plastics or SMCs (sheet moulding compounds) are based on glass mat reinforced polyester resins and must be pressed with heated tools in order to cure the resin. SMC components have mechanical properties closer to those made of metals, but require a significantly longer cycle time. The thermoplastic sheets are shaped with cold tools and hence have a cycle time of 25–30% of those required for SMC pressings. It is important to note however that the presses for sheet metal, GRP and plastic materials need to have different characteristics. For example, for thermoplastics the press must close rapidly and remain in the closed position until the material has cooled, thermosets must be held whilst the resin cures, but sheet metals can be released immediately. Hence, not only will the force required change with the different materials, but the cycle times must also alter. It may also be necessary to heat or cool parts of the press.

Fig. 6.6 shows the kind of tooling that is required to make a V-bend in sheet material using a press. Similar tooling can be used to produce bends of small radius in sheet.

There are two key components, the punch and the die. For long life, both of these must be manufactured from a very hard material which is resistant to wear and impact loads. Achieving these properties in steels requires them to be hardened and tempered, and consequently it is also necessary that the steel used must not distort when heat-treated. Such tool steels are often highly alloyed and very expensive. For economy, the punch is normally held in a standard punch holder and the die in a standard bolster. Apart from locating the tools these components help the punch and die to withstand the impact forces.

In school and college workshops the force needed to bend the material is normally produced by a toggle mechanism or using a screw thread in one of the commonly available types of hand-operated presses, but industrial presses are power operated. Both in hand-operated and power presses it is common for inertia to be added to store the energy needed for the pressing operation. This is done by using a large steel ball on the handle of fly presses. Fig. 6.7 shows typical large power presses as used in industry. These are from the Rover Group's Swindon plant and programmable robots can also be seen transferring the panels between presses (see also section 6.7).

Form presswork – where the sheet is curved in three dimensions – is believed to have originated in America and come over to the Austin Motor Company in the 1920s. The stiffness of car panels is a result of this complex double curvature and achieving such forms requires complex tooling and the generation of very high stresses. If car panels were simpler shapes they would have to use very much thicker material in order to achieve the same stiffness. Sometimes the depth of the sections is so great that several pressing operations are required. Fig. 6.8 shows side panels manufactured for Rover Group cars.

During the growth of mass manufacture since the 1920s the advantages of using pressed metal components have been exploited in a variety of industries. Now most consumer products contain pressed components. Not surprisingly, significant process developments have been made particularly associated with the key problem areas, namely, the need for large, powerful presses and the tooling costs for complex forms. These advances are described at the end of this section. As would also be expected,

Figure 6.7 Industrial power press.

Figure 6.8 Three-dimensionally curved sheet metal components.

other operations associated with sheet metal components have come to be carried out on the presses, in particular:

- blanking or cutting out the original shape,
- piercing or the punching of small holes,
- nibbling or the punching of overlapping holes,
- rimming or the removal of surplus material after pressing.

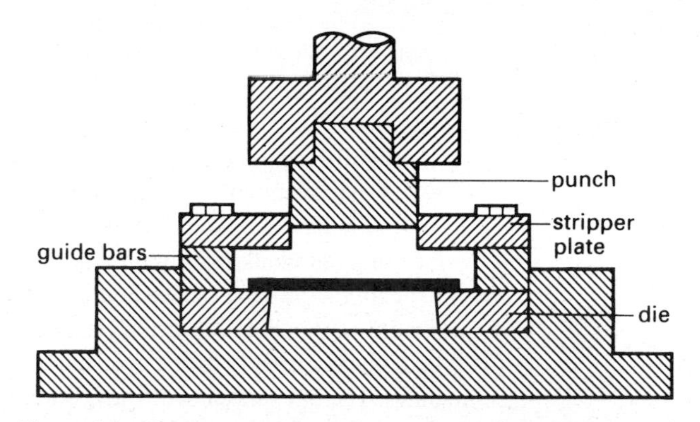

Figure 6.9 Tooling requirements for blanking or piercing.

Figure 6.10 A simple piercing tool.

The tooling required for blanking or piercing is indicated by the layout shown in Fig. 6.9.

For blanking, the punch is the shape and size of the component and the die has a corresponding aperture. In comparison to the tooling required for bending there are two important additions: a stripper plate and the guide assembly. The stripper plate is added so that when the punch is withdrawn the sheet remaining after blanking is forced off should it have become wedged. The guide assembly controls the feeding of the sheet and usually consists of two guide bars and a stop. The sheet position is very important because, if there is too little material left between blanks, the tools can become burred and, clearly, leaving too much material is wasteful. Fig. 6.10 shows simple tooling for piercing operations which could be made up by school or college students.

Successful blanking or piercing operations depend critically on the clearance between the punch and the die and Table 6.1 indicates the values for a number of common sheet materials. Fig. 6.11 shows the normal appearance of the surface of the cut and also the effect of too much or too little clearance.

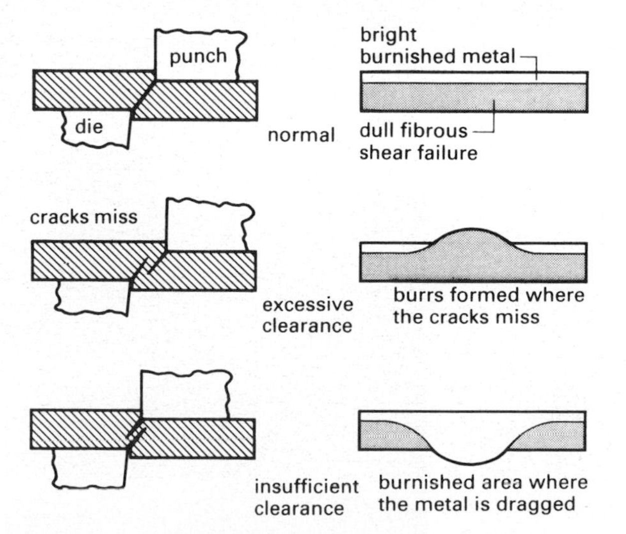

Figure 6.11 The effect of the clearance between the punch and the die on the piercing operation.

Material	*Clearance* (percentage of thickness)
aluminium	3
copper	4–5
brass	5
stainless steel	5
other steels	5–10

Table 6.1 Clearance required between blades for shear cutting

Initially the material is cut as a result of plastic deformation and failure at the tool edges, but eventually the remaining material will be thin enough to fail in shear. This kind of behaviour should be observed when metal is cut on guillotines or when carrying out blanking, nibbling or piercing operations. The cut surface directly beneath the moving blade should have a burnished appearance and the remainder should be much duller. If the clearance is too great then the shearing action does not take place correctly as the cracks miss one another. The metal is then torn, forming burrs. The cut surface will show a wavy line between the burnished and unburnished areas. If there is too little clearance a double shearing action occurs. With the correct clearance the shear plane forms between the two blade edges. With too little clearance the angle formed by the line joining the blade edges and the material surface will be near to 90° and too great for the material to shear along it. Consequently two shear planes form at steeper angles and the material tears between them. It is then dragged between the blades and the resulting cut surface shows burnished areas. There will be a consequential increase in the power consumption.

If necessary, the power required for a blanking operation can be reduced by putting a slope on to the die so that the cut is progressive. This is known as putting shear on the die and is illustrated in Fig. 6.12.

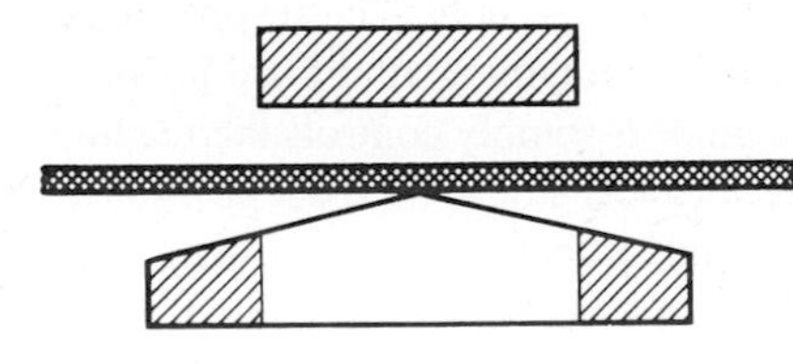

Figure 6.12 Adding shear to the die.

Fig. 6.13 shows pressed steel components for a sewing machine which have been made to a high degree of accuracy and washers which everyone tends to take for granted.

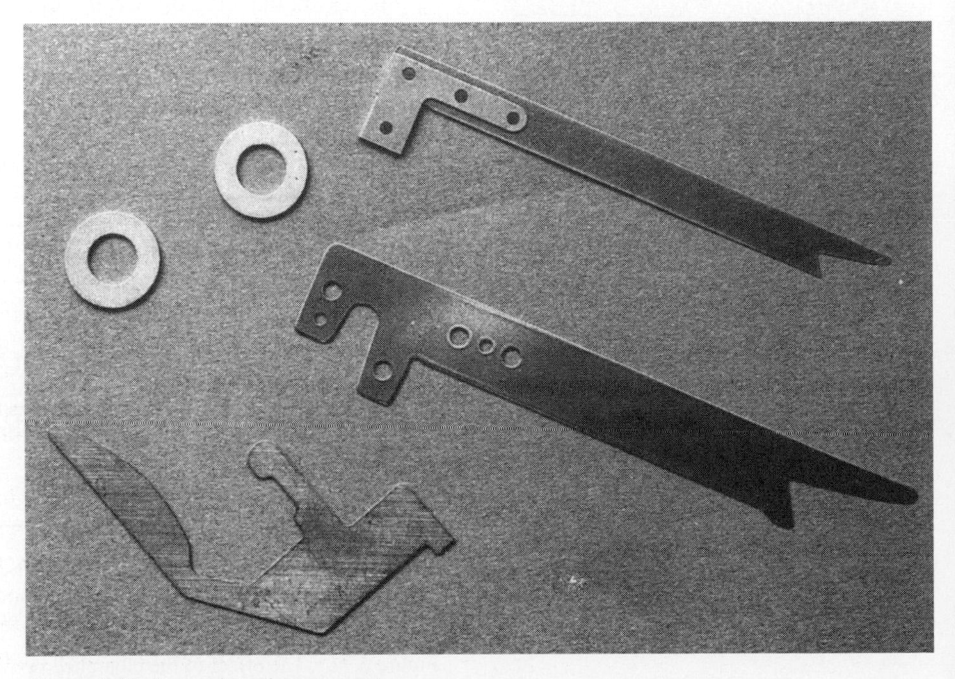

Figure 6.13 Pressed steel components.

The manufacture of a washer requires both a blanking and a piercing operation and these are normally combined in what is known as a progression tool as shown in Fig. 6.14. As the strip is fed a hole is initially pierced and then the blanking punch produces the circular shape.

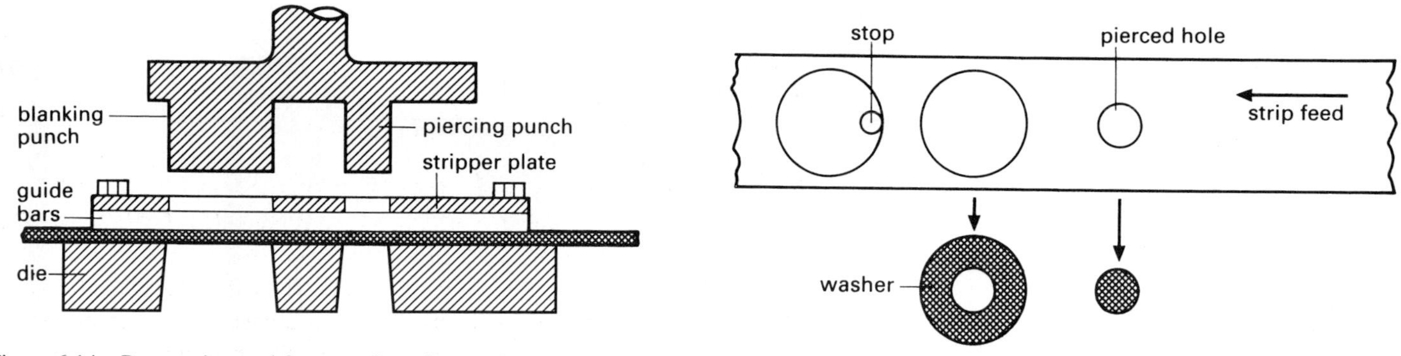

Figure 6.14 Progression tool for manufacturing washers.

Nibbling is the punching of overlapping holes. It has the advantage over cutting on a guillotine of not distorting either of the cut edges. For this reason it is often used to cut out shapes in sheet metals, particularly where the production quantity is small or the components very large.

It should be remembered that the cost of a sheet metal component is associated with the total amount of sheet metal consumed, which will include the scrap which remains after blanking. This is because the scrap is usually of comparatively little value. Fig. 6.15 shows an example of the way in which a designer should aim to fit components on to standard sheet sizes in order to minimise wastage.

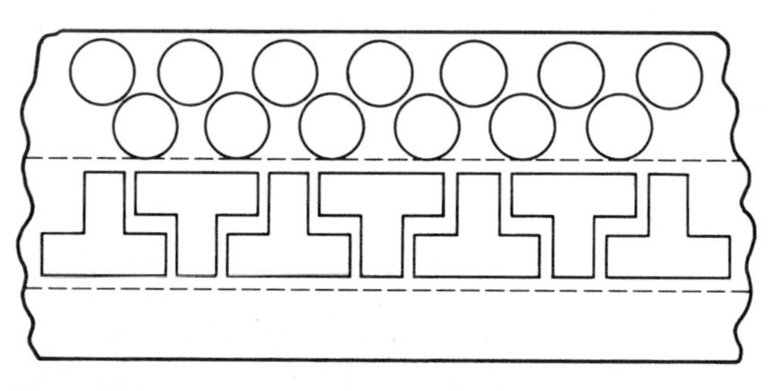

Figure 6.15 Economy in blanking operations.

Deep drawing

The deep drawing process is illustrated in Fig. 6.16 which shows the production of a simple cylindrical form. This process is the basis for the production of most three-dimensionally curved pressings.

The blank is held by a pressure ring which provides sufficient force to prevent the sheet from wrinkling, but not enough to prevent it from drawing into the tool. All the surfaces on the ring, die and punch are highly finished to minimise friction and normally a lubricant is also used.

The depth which can be drawn in a single stroke depends on many factors, including the sheet material, its tensile strength and the tool design, but it is generally limited to about 60% of the outside diameter. Greater depths can be obtained by re-drawing operations, but it may be necessary to anneal the material first. Deep drawing results in considerable work hardening and, sometimes, associated grain growth. It is the resultant material properties, in particular the ductility and tensile strength, which will determine whether annealing is necessary.

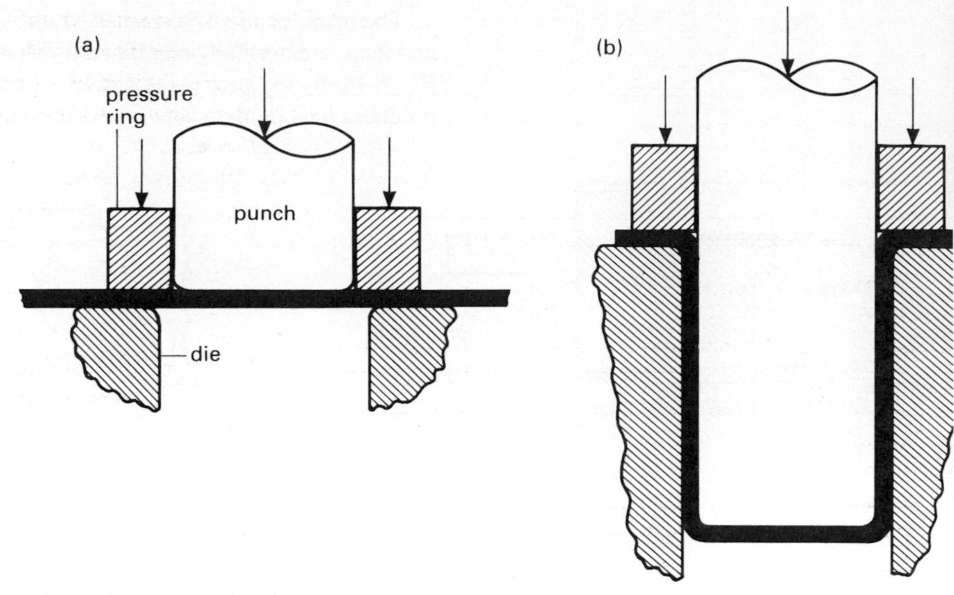

Figure 6.16 Deep drawing.

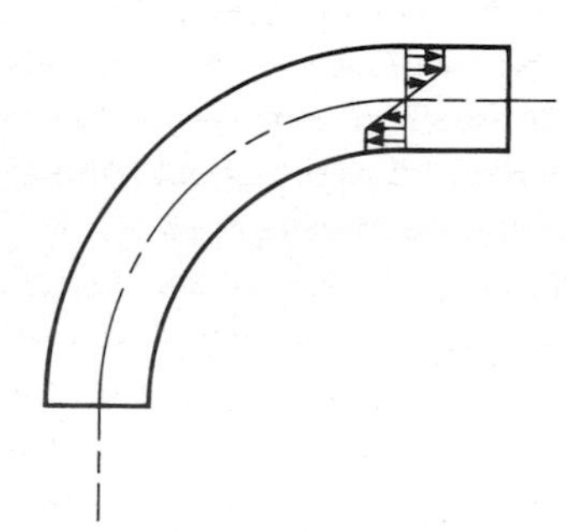
Figure 6.17 Stresses generated during bending.

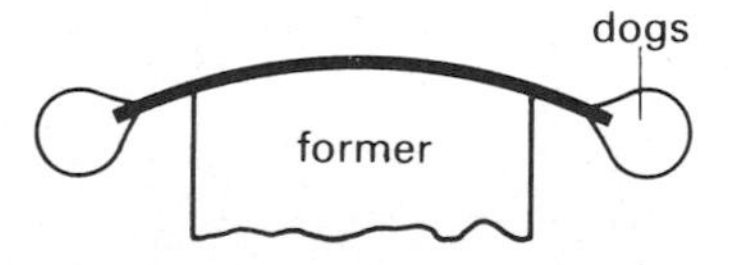

Figure 6.18 Stretching a material during bending to reduce springback.

Designing for presswork

The most significant design issues in basic pressworking operations are related to the material properties. There are two properties which can cause problems: the material's elasticity which will result in *springback* and the grain structure or rolling direction. Springback occurs because when the loading is removed the material will unload elastically, even after it has been plastically deformed (see section 5.2.1). It should also be recognised that when a bend is made the material on the neutral axis is not stretched as shown in Fig. 6.17.

Clearly, some of the material close to the neutral axis on either side is never plastically deformed and will want to return to its original position. The outer plastically deformed material will return elastically to its new position. The problem of springback can be minimised by making as much of the material as possible deform plastically, that is, producing as high a stress as possible in the outer fibres. A very small amount of springback is inevitable, however, but this can be allowed for in the tool design by overbending the material. Typically a right angle will result from overbending sheet steel by about 2°. Springback can be reduced by bending the material whilst it is under tension so that the stress levels are higher and a greater proportion of the material is plastically deformed. One way of stretching the material is the use of 'dogs' as shown in Fig. 6.18.

The use of bottoming dies which squeeze the material and force it to flow rather than air-bend tools which rely on three point bending will also reduce springback. A comparison of air-bend and bottoming tools is shown in Fig. 6.19. Bottoming dies will clearly require the use of more force and power, but give a sharper bend radius.

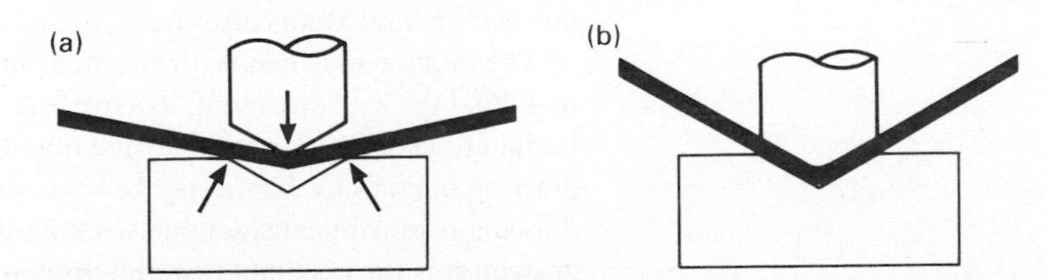

Figure 6.19 (a) Air bend and (b) bottoming dies.

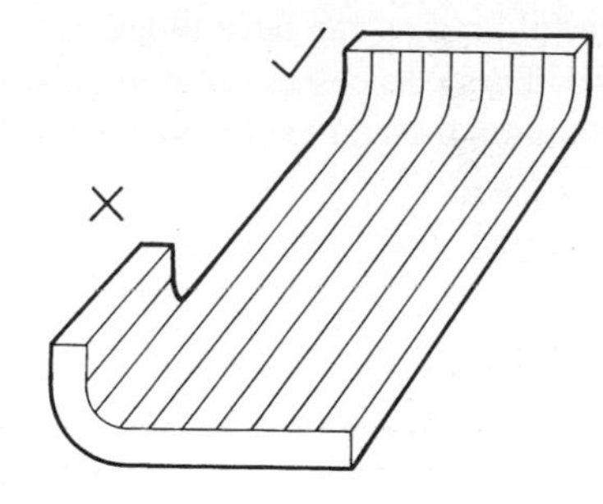

Figure 6.20 Component requiring bends perpendicular to one another.

As there is less material near the neutral axis springback will normally be less of a problem with hollow sections than with solid rods and bars.

The difficulty of bending parallel to the rolling direction means that components like the one shown in Fig. 6.20 requiring bends at right angles to one another could be very troublesome.

It also means that the direction of the bend line must be thought about when the blanking operation is carried out as indicated in Fig. 6.21.

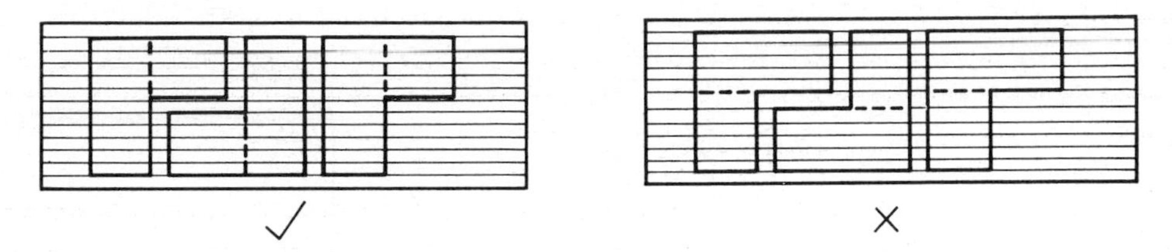

Figure 6.21 Correctly orientating components for blanking.

Process developments

The cost of making matching male and female tools and the capital investment in powerful presses can only be justified for very large production quantities. Development shops and small batch manufacturers need only small numbers of a particular pressing. All companies are, of course, interested in reducing the capital needed for manufacture and consequently there is a commercial pressure for simplified forms of the pressing process. One process which overcomes the need for a female tool is *hydroforming* (or fluid forming) as illustrated in Fig. 6.22.

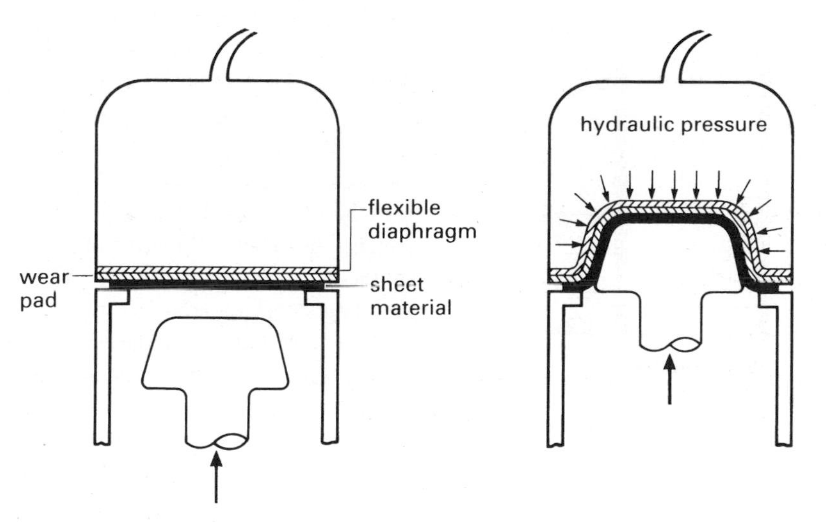

Figure 6.22 Hydroforming.

This can be used with most metals and thermoplastics and can reduce tooling costs to a fraction of conventional deep drawing tool costs (5% has been claimed). The disadvantage of the process is that it is relatively slow and therefore uneconomic for large quantities. For small quantities, soft materials or where dimensions are not critical the punch can be made of aluminium, brass, hardwood or plastics, but for large quantities or pressing stronger materials then cast iron or steel would normally be selected. Ideas similar to hydroforming have been pursued by toolmakers for many years in development shops where they need to improvise. Forming can be carried out on materials like rubber and polyurethanes, and temporary dies can sometimes be made from metals with low melting points, like lead and aluminium. The temporary die is produced by casting round the punch.

The large capital investment needed for powerful presses can be avoided by employing very high rates of loading the material. These processes are normally known as high energy rate forming. The most common uses an explosive charge immersed in water as indicated in Fig. 6.23.

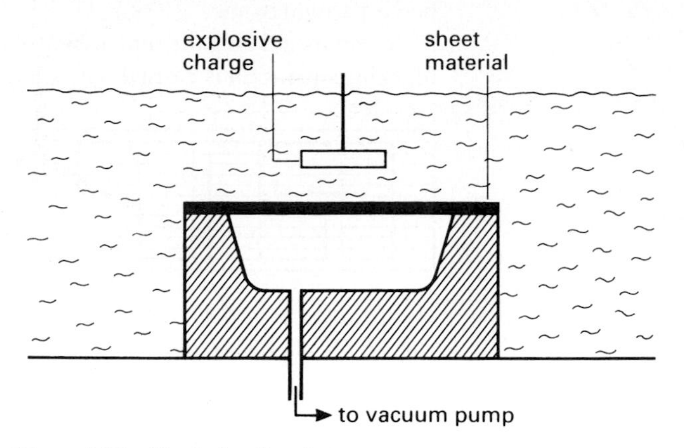

Figure 6.23 Explosive forming.

The shock waves generated by the explosion deform the material so fast that it does not have sufficient time to work-harden. Because of the speed of the deformation, air will become trapped between the sheet and the die giving a dimpled surface. In order to avoid this, the air must be evacuated from behind the sheet material before the charge is detonated. If the air is not exhausted then the process is known as 'free-forming'; when a vacuum has been created under the sheet first it is known as 'bulkhead forming'.

The major advantages of the process are the low capital cost, potentially as little as 1% of that for a hydraulic press, and the simple tooling. There is no punch and the dies are cheap as they can be made from materials like epoxy-faced concrete, plaster and GRP (with support). There are, however, other advantages like the absence of springback, high accuracy and being able to form materials of different thicknesses with the same dies. It is also possible to form very large components with dimensions of several metres in a single piece. The only significant disadvantages are the requirements for safety and security in handling the explosives and the need for a large working area.

As with spinning (see the next section) the use of preformed shapes is a great aid to the explosive forming process. Fig. 6.24 indicates how a preform is used.

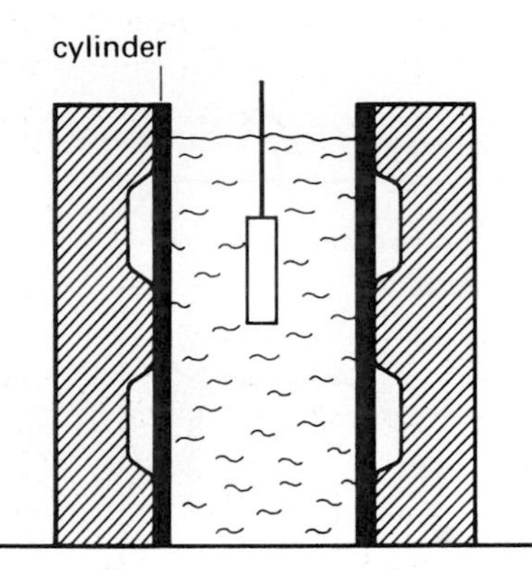

Figure 6.24 Use of a preform.

Rotational forming techniques

Shapes which are symmetrical about a central axis can be produced by forcing the material on to a rotating former. Although they are sometimes used in combination, these rotational forming techniques can be thought of in one of three categories: spinning, shear-forming and flow-forming.

Spinning

Fig. 6.25 indicates the tooling needed to spin a simple component from sheet metal. Forming the shape will require the tool to make a number of passes to force the material on to the former – exactly how many will depend on the component's curvature, the material's ductility, the wall thickness and the force which can be exerted. Hand spinning is a highly skilled operation as the avoidance of puckering, or wrinkles in the outside diameter, requires considerable judgement and the use of back support for the sheet. The force which can be exerted is limited by the strength of the operator. On powered spinning machines, the force which can be exerted is

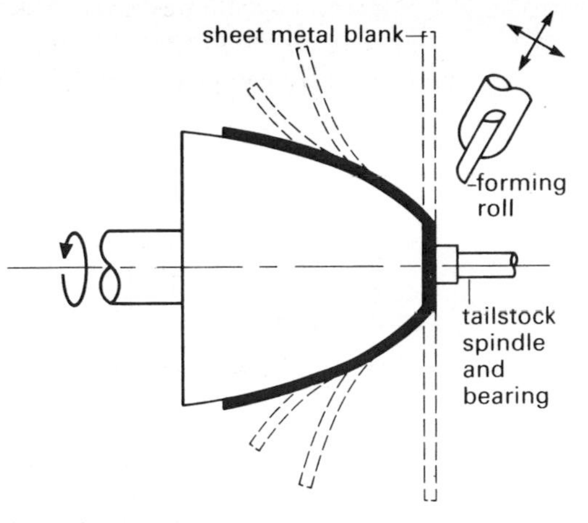

Figure 6.25 Spinning.

limited ultimately by the strength of the material. Although Fig. 6.25 shows the material spinning on to a rotating former it is equally possible to spin 'outwards', typically on to a freely rotating roller.

Spinning is at its most useful when producing shapes which cannot readily be pressed or when finishing off pressed components. Fig. 6.26 shows shapes which would be difficult to press, although they could be produced by explosive forming. Whichever way they were produced would necessitate the use of segmented tooling so that the former could be extracted. The sheet material for these components would normally be preformed into a cylinder or a cylindrical cup, so that the major part of the forming operation is completed prior to spinning.

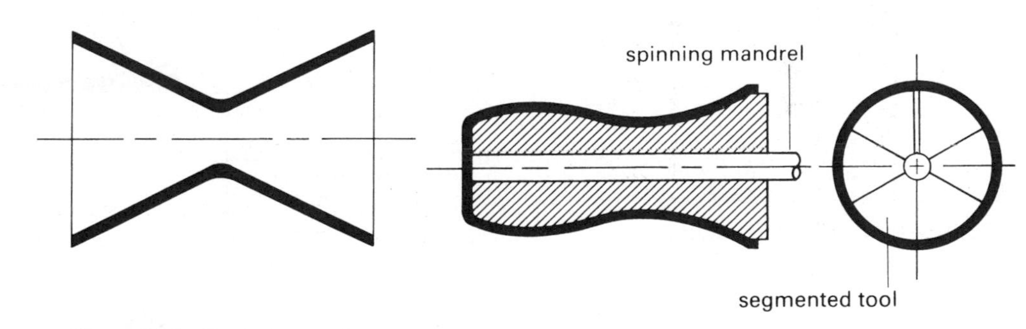

Figure 6.26 Typical spun shapes.

Shear-forming

Shear-forming (sometimes known as cold-flow turning) differs from spinning only in that, as the material is forced on to the former, it is also thinned down. This requires much larger forces to be exerted because the material must be made to deform plastically. Shear-formed components typically have a thick base and the wall thickness then tapers down as shown in Fig. 6.27.

Because of the high forces involved the machines must be very robust and the tools are usually hydraulically operated rollers. There are significant advantages to the process because of the low tool cost and the excellent finish resulting from the tool burnishing the outside surface. It has found applications in the manufacture of many products, for example, beer barrels, gas cylinders and spin dryers.

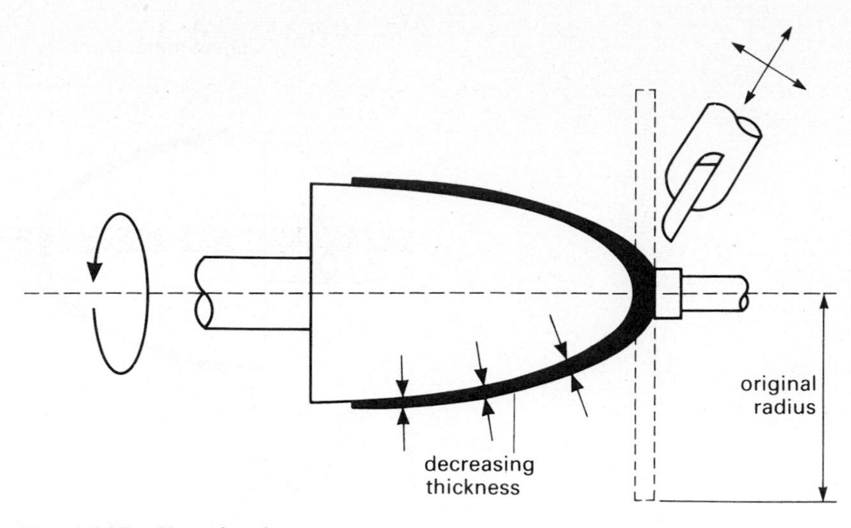

Figure 6.27 Shear-forming.

Flow-forming

Flow-forming differs from shear-forming in that the component has parallel walls and begins as a preformed, thick-walled, cylindrical cup, rather than a circular blank. Fig. 6.28 shows the tooling configuration normally used. As the forces involved are very great, it is quite common for a number of tools to be working the material simultaneously in order to make them easier to balance.

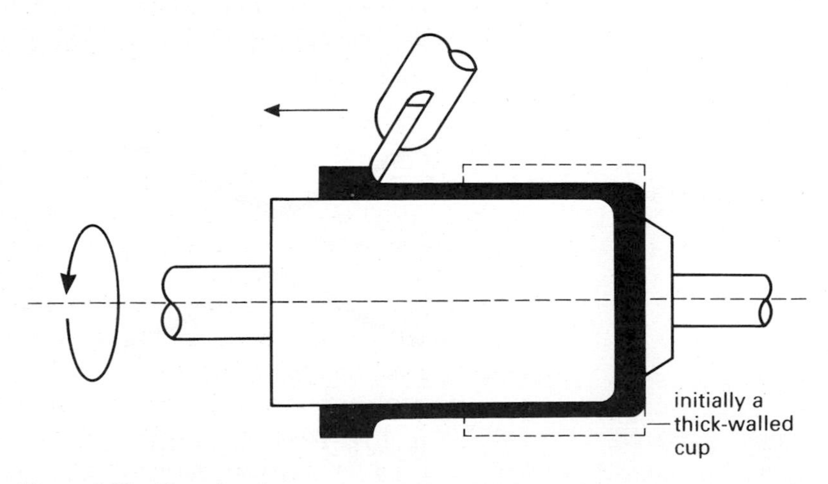

Figure 6.28 Flow-forming.

6.2.2 Vacuum forming and blow-moulding

Vacuum forming and blow-moulding are manufacturing processes which are very closely related to the sheet forming techniques described in section 6.2.1. Both these processes rely on the use of pressure forces to bring sheet material into contact with a mould. In vacuum forming this is achieved by sucking the sheet material down on to a former; in blow-moulding by raising the pressure to force the material against the mould. The tool costs are much lower than for presswork and it is possible to get a high-quality finish, which is difficult when pressing SMCs, for example. One disadvantage is that the stress levels required to deform the material cannot be too high and this limits the materials which can be used and the shapes that can be generated.

Vacuum forming

Fig. 6.29 shows the equipment normally used for vacuum forming. The sheet material is clamped in position, heated so that it softens and a vacuum then draws the sheet material over the mould.

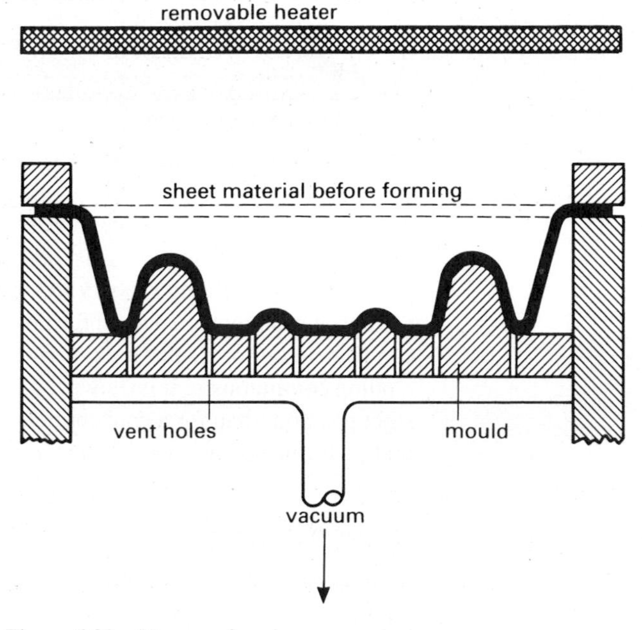

Figure 6.29 Vacuum forming.

In order to allow the shape to form, all the corners must be rounded and small vents located in positions on the mould where the air might become trapped. It is also necessary to taper the mould – typically 5° on male and 2° on female features. It is often thought that only thermoplastic sheet can be vacuum formed, but certain alloys of aluminium, zinc, stainless steel and titanium can be thermoformed at relatively low temperatures and stresses. These are known as superplastic alloys. Generally these materials are stiffer and more heat-resistant than thermoplastics, and could be used for car panels, machine covers etc. The alloys available are, however, rather expensive and with large production quantities to justify the tool cost conventional presswork is likely to prove cheaper. Wood fibre mouldings are also made for applications like car interiors and furniture in a very similar process to the vacuum forming of plastics, although the tool cost is much higher.

The moulds can be made from any material which can withstand the comparatively low temperatures and stresses produced – wood, plaster, cast aluminium and epoxy resins are all possibilities. Metal tooling will give longer life of course, but will take much more time to manufacture. Whichever material is used it should be finished as well as possible to avoid marks resulting in blemishes to the outer surface.

Drape-forming

Shapes which have very deep features or tight radii can be difficult to produce with vacuum forming, mainly because of the thinning of the sheet which occurs as the material is drawn down. Drape-forming seeks to increase the capabilities of the process by forcing the mould into the softened sheet using a ram before the vacuum is applied. When the mould is pushed into the heated sheet it drapes over it, partly forming the shape. When the vacuum is applied the sheet is sucked on to the mould, but the thinning in deep sections is reduced.

Vacuum-bag presses

Fig. 6.30 shows a typical application of a vacuum-bag press – holding sheets to be laminated in the required shape whilst the adhesive sets.

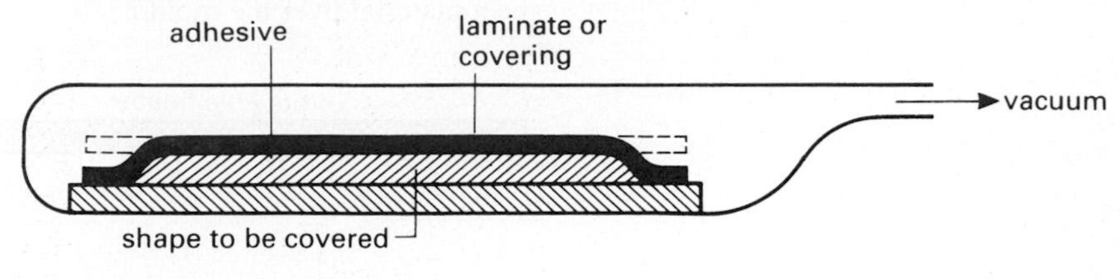

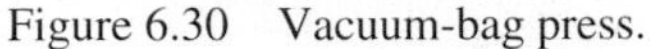

Figure 6.30 Vacuum-bag press.

The former and sheet materials are placed inside the bag which is then sealed and the air extracted. This process is most often used when processing wood and, in particular, when gluing veneers and other finishes to base materials. Although it is similar in principle to vacuum forming, the important difference is that the vacuum can be held on for a substantial period by simply sealing the bag, rather than running a pump continuously. It is, however, more difficult to set up and hold the sheet in the right position, than for a conventional vacuum forming operation. Heating sheet material prior to, or after, forming would also cause significant difficulties.

Blow-moulding

Blow-moulding uses compressed air to form the softened sheet. As the material will naturally form into a smooth bubble shape the tooling needed may be very simple, perhaps just a yoke and restraining plate as indicated in Fig. 6.31.

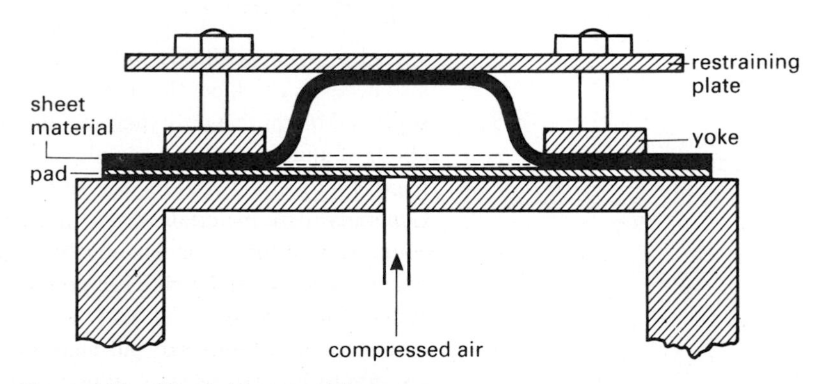

Figure 6.31 Simple blow-moulding.

Generally the sheet material will cool rapidly on contact with the cool mould or restraining plate, but the compressed air is often deliberately warmed to prevent it cooling prematurely. Blow-moulding is, like vacuum forming, normally associated with thermoplastics, but its origins can be found in the blowing of glass. Glassware is usually made by the kind of two-stage process shown in Fig. 6.32. The glass is heated to red heat and a gob of the liquid glass is then pressed to produce a blank. The blank is then transferred to a mould of the shape required.

When making bottles, a liquid glass gob is dropped into a preliminary (or parison) mould, the neck is formed and a thick-walled preform is blown. The *parison*, as it is now called, is then transferred to a finishing mould where the final shape is blown. This process was adapted for forming thermoplastic bottles from a cylindrical extruded preform. This technique is illustrated in Fig. 6.33. The similarity of the blow-moulding of glass and thermoplastics to the explosive forming or spinning of metals from a preformed cylinder is evident.

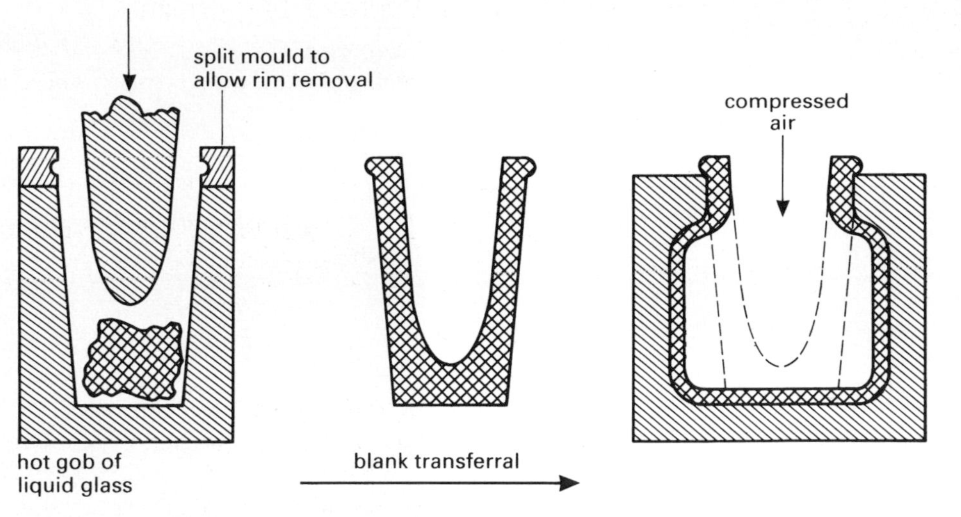

Figure 6.32 Manufacturing glass bottles.

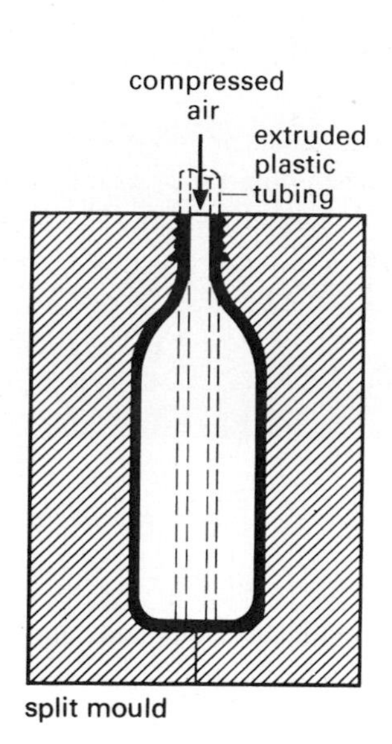

Figure 6.33 Blow-moulding plastic bottles.

The blow-moulding process is used a great deal commercially in the manufacture of a wide variety of plastic bottles and containers. The most important plastics used for blow-moulding are polyolefins, for example, polyethylene, which has been used in the manufacture of a range of products, like surfboards and car fuel tanks. One of the major advantages of the process is that there is less restriction on the kind of shape which can be generated than with metal pressings and this gives the car designer freedom to fit the tank round other assemblies and the opportunity to use space very efficiently.

6.2.3 GRP moulding, laminating and steam-bending timber

Laminated materials are produced in sheet form by stacking the material in layers with a suitable adhesive or bonding agent and applying pressure, and sometimes heat, in a flat press. Laminated mouldings can also be produced by pressing the materials into a mould. Lamination is a very effective way of combining the properties of different materials and also achieving uniform strength where this is directional in the individual sheets. (See also section 4.2.4.)

Prototype GRP mouldings

For prototypes and small batch manufacture, and certainly in school and college workshops, the kind of moulding indicated by Fig. 6.34 is the right approach.

This shows the three stages in producing a GRP moulding. First, a plug or shape to mould round must be produced. In this case MDF was chosen, but other woods, plaster or clay could also have been used. This is then covered with a release agent, normally a wax emulsion, and a female GRP moulding is then made around the plug. The GRP moulding is built up from layers of glass mat and resin which is stippled or forced into the gaps between the fibres using the ends of the hairs of a brush. Although glass mat is most commonly used, carbon fibre reinforcement is now also available. This is more expensive, but will give a better strength-to-weight ratio and greater stiffness because of the higher modulus of elasticity. Having made the female mould, which will be smooth inside but comparatively rough outside, the final product can be made by building up layers of GRP inside it. The final product – in

Figure 6.34 The stages in producing a GRP moulding.

this case a mechanic's trolley – will now have a smooth surface outside and a rough interior. The recesses the designer has included in the moulding are for light fittings which would replace the inspection lights conventionally used under vehicles.

Fig. 6.35 shows a far more technically demanding application of GRP moulding. The requirement for the prototype aerobic weight was for a hollow cavity which could be filled with sand. In order to manufacture this a male and female mould had to be taken from the plug to produce the two halves of the moulding. These were then bonded together and filled to give the necessary weight.

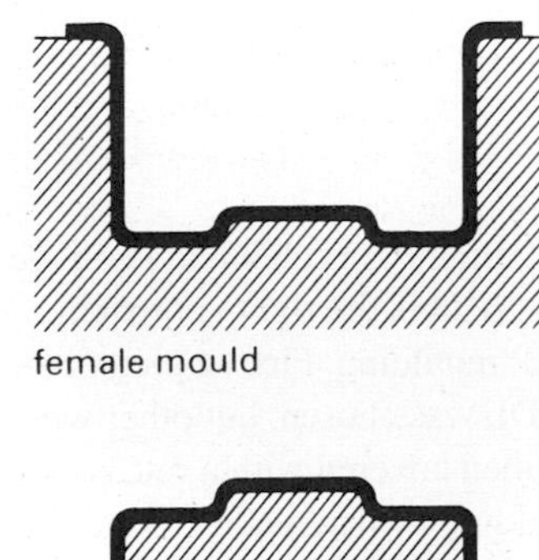

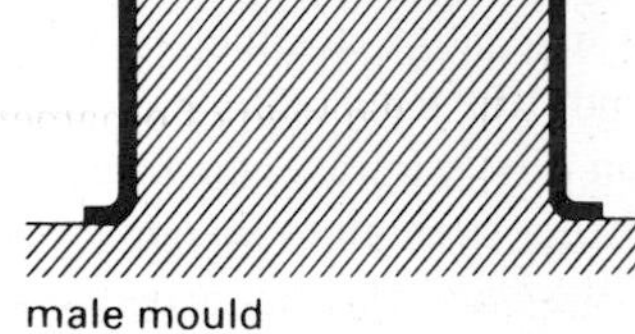

Figure 6.35 Complex GRP mouldings.

Laminating and steam-bending timber

Because of the grain structure of timber it can only be bent significantly in one direction, namely, at right angles to the grain flow. Bending along the grain will tend to cause splitting at quite a low curvature. Consequently laminated timber is best bent across the grain, as shown in Fig. 6.36.

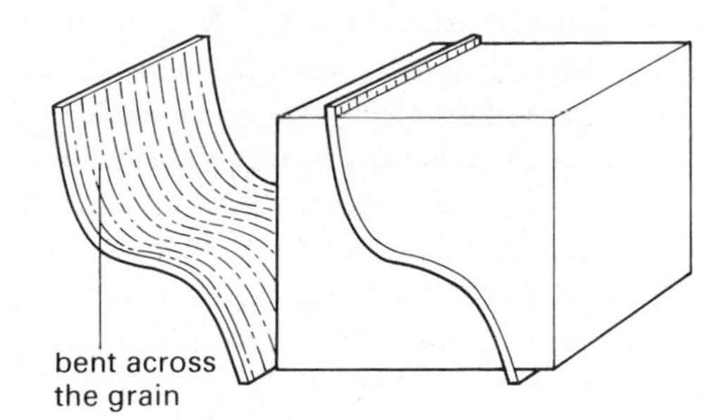

Figure 6.36 Laminating timber.

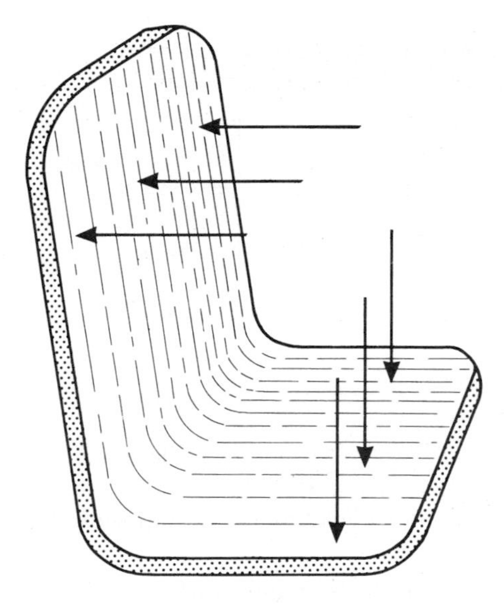

Figure 6.37 Laminated chair back.

This will, of course, lead to a great difference in the strength properties of the completed lamination, and this must be taken into account by the designer. If a chair back is laminated as shown in Fig. 6.37 then when the occupier sits down and leans back the stresses generated will lie along the grain and not cause problems.

If strength is required both along and across the grain, then it is possible to laminate sheet material with the grain in alternate layers perpendicular to one another. It is also possible to bend plywood with the grain in the front and back veneers lying perpendicular to the bend. It should be remembered when ordering plywood that the first dimension gives the length along the grain – i.e. 2440 × 1220 mm indicates a normal board and 1220 × 2440 mm indicates a cross-grained board. The designer should consider carefully which is the best choice, and be sure to use it economically. It is important that the designed shapes fit conveniently on to standard sheets to avoid waste, in just the same way as sheet metal blanking must be carefully planned.

Clearly the preparation and bending of a large number of timber strips and the manufacture of the former takes considerable time and it would be much quicker if timber of the required size could be bent directly. This can often be achieved if the wood is first softened with steam. The bending strength of wood is very dependent on its moisture content because, as it dries, it initially loses water from the lumens and later from the cell boundaries (see section 4.2.4) which increases their stiffness. Immersing the timber in steam can reverse this process and make the timber more workable. If the timber is held in the bent position it will tend to settle to this shape as it dries. The drying timber is, however, prone to twisting and bending and this leads to a lack of accuracy and repeatability. Improved precision is a major advantage of laminating timber in comparison with steam-bending.

6.2.4 Forging

Forging is associated industrially with the manufacture of strong metal components and traditionally with decorative metalwork. Blacksmithing is one of the oldest metalworking crafts and the blacksmith's work used to embrace both these fields of activity. They produced swords and ploughshares as well as the kind of ornamental ironwork which enhances many old buildings. Because the work of the blacksmith is so highly skilled and the making of ornamental ironwork so labour-intensive, it has moved out of the general product design area and become a very specialist activity. In industrial terms, the hot forging of copper alloys and the cold forging of steels are probably the most significant applications.

Hot forging

The oldest technique for forming steel is by hammering it against an anvil, which has different shaped flat and curved surfaces. This is also known as 'open-die forging'. Obtaining the desired shape is very skilled and repeatability is difficult to achieve. Industrially the need for highly skilled craftsmen has been removed and the repeatability improved by using specially shaped dies in order to help create the desired form. The hot metal is placed in the die and then forced into it by a power-driven hammer. Many products where considerable strength is needed, like crankshafts and spanners, are made by this process. With a specially made die the process is known as 'closed-die', 'drop' or 'die forging'.

The tools must be manufactured by machining or casting and are very expensive to produce. One of the reasons for this is the need to use very highly alloyed steels in order to prevent the heat lost to the tools causing them to wear too quickly under the impact loads. For some components it may be necessary to use a succession of dies in order to achieve the required form – each stage bringing the forging closer to its final dimensions. Fig. 6.38 shows the kind of tooling typically required for a closed-die forging operation.

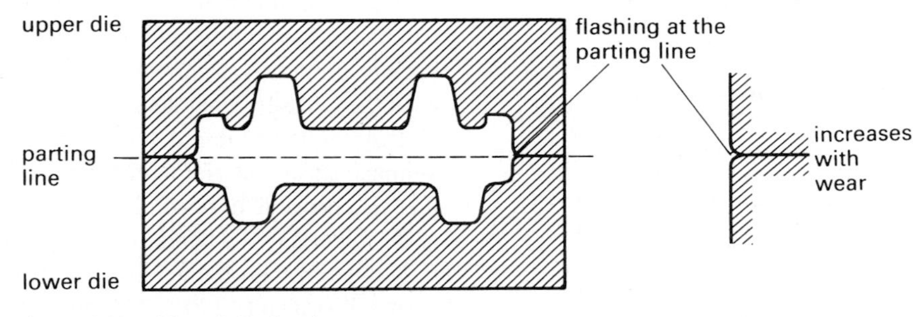

Figure 6.38 Closed-die forging.

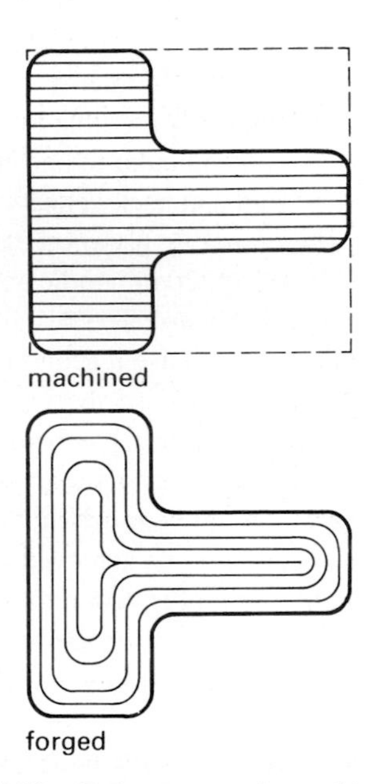

Figure 6.39 Grain structure in machined and forged components.

For the hot forging of steels considerable *draft* (taper) is required because of the surface oxidation. On an internal feature this is generally 10° but on an external surface only 7° is needed. Nonferrous materials would require lower draft angles. The flash is eventually removed with a trimming die.

The significant advantage of forged components is the improvement in their grain structure which can result from hot working. Fig. 6.39 shows a comparison of a machined and a forged component.

In the machined component all the grain fibres still lie in the direction of rolling: in the forged component the fibres tend to follow the outside contour. Fig. 6.40 shows the kind of grain structure that is produced by hot forging a blank for a gear.

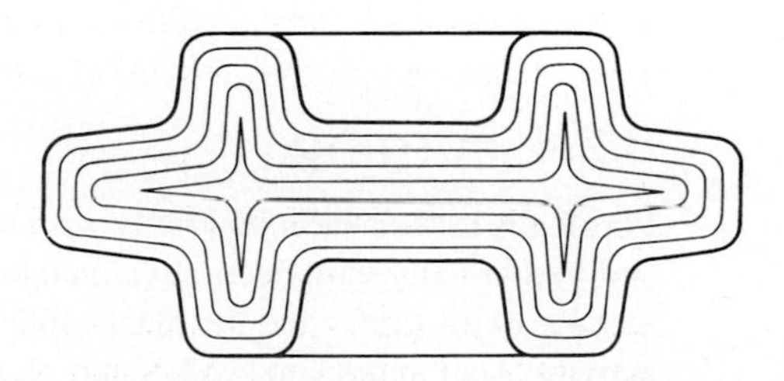

Figure 6.40 Grain fibres in a forged gear block.

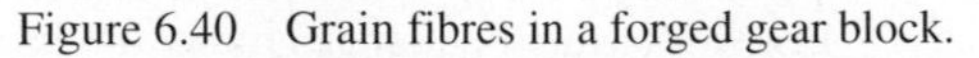

Ideally the loading should be perpendicular to the grain fibres in order to give optimum strength and forging comes closer to achieving this than other manufacturing processes. Fig. 6.41 shows the forces acting on the teeth which can be

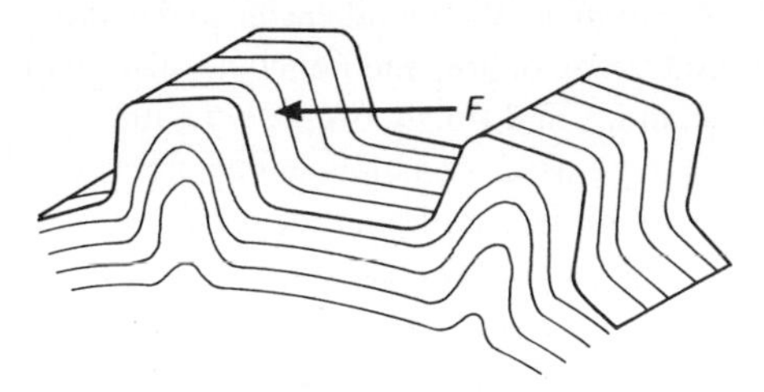

Figure 6.41 Force on a gear tooth.

seen to be perpendicular to the grain fibres. The torque must then be transmitted through the gear body to the shaft and again the forged blank provides the right grain fibre orientation.

Apart from the manufacture of strong steel components like crankshaft and gear blanks, the most common application of hot working is in the closed-die forging of copper alloys. This is also known as hot stamping, particularly if brass is being used. Before the introduction of plastic components most plumbing hardware was made by this process, especially because of the absence of porosity.

Cold working

Many of the components which are commonly used in assembling products will have been made by the cold working of steel. Rivets, bolts and screws are available at low cost and with good working properties because this technique is used. There are two important processes: cold-roll forming and cold heading.

Cold-roll forming

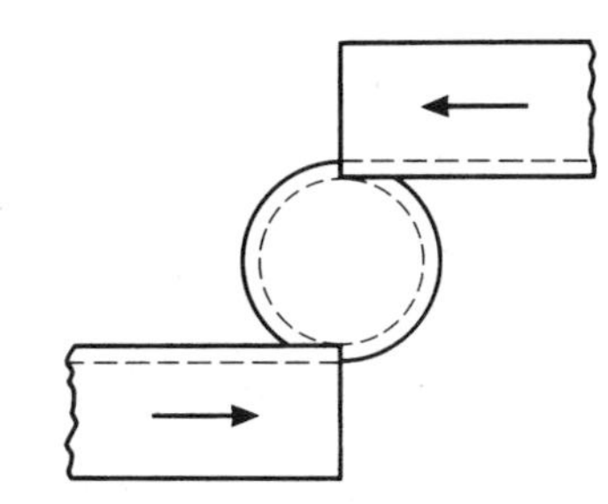
Figure 6.42 Cold-roll forming.

Fig. 6.42 shows the basic cold-rolling technique. This is used to put threads, splines, serrations, knurling and other grooves and indentations on to steel or other materials.

The component to be formed is mounted between centres and the two tools are then moved towards each other. When they make contact with the component it rotates, so that the pattern is produced all round the circumference. Some machines are built to take rotating circular dies instead of the horizontal tools shown in Fig. 6.42. Cold-roll forming results in much stronger teeth than machining because the grain fibres follow the thread profile rather than being cut. The surface is also work-hardened in a rolled thread which improves their wear resistance. This is illustrated in Fig. 6.43.

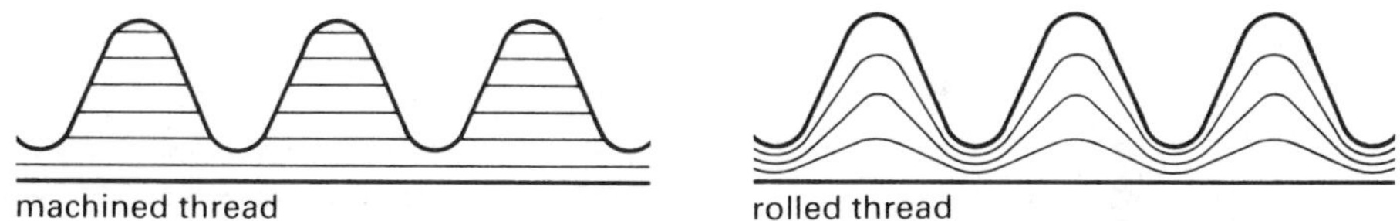

Figure 6.43 Grain flow in machined and rolled threads.

Cold heading

Cold heading machines are used to create the heads on rivets, screws, bolts and similar components. Material is fed to the machine from a coil, sheared to length and then transferred to the die. A punch then produces the required head form, normally in a single blow. For very difficult head forms a progressive series of punches may be used, and more than one blow may be needed for hard alloy steels. Fig. 6.44 indicates the tooling requirements for the cold heading process.

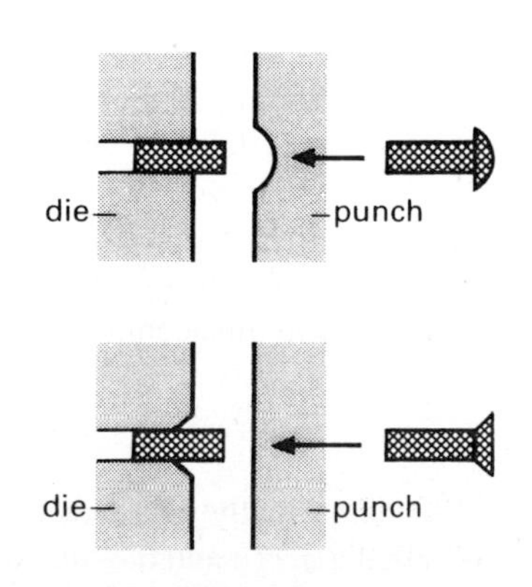

Figure 6.44 Cold heading.

Process developments

Although hot forgings give good material properties, they are generally rather inaccurate and have a poor finish. In order to overcome these difficulties the possibility of using a very high deformation rate, as with the explosive forming of sheet materials, has been investigated. The new high-energy rate forging processes are based on the projection of the punch at very high speeds as a result of gas ignition. The component metal is effectively injected into the die giving far greater detail, improved accuracy and a much better finish. Webs as thin as 0.25 mm have been produced in the more forgeable metals.

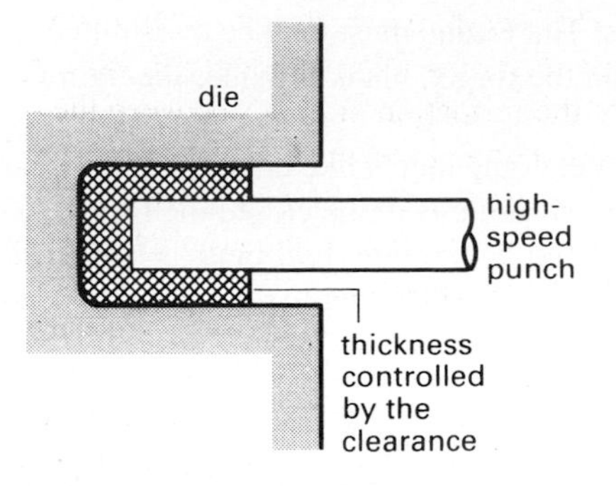

Figure 6.45 Impact extrusion.

Using a high rate of loading rather than heat to make the metal easier to deform plastically means that there is no high temperature oxidation and hence no scaling or flash formation. It is this that improves the accuracy and finish, because the metal reliably fills the die. It also allows the process to be used on metals which are difficult to forge conventionally, for example, niobium, titanium, tungsten, zirconium etc. As there is no heat entering the die the tool life is also much greater. The high rate of loading and lack of oxidation also mean that draft angles can be very low.

Impact extrusion

A technique related to forging at a high rate of loading is impact extrusion (also see section 6.2.6). This process is used on metals such as aluminium and lead and is shown schematically in Fig. 6.45.

A cold slug of metal is struck by a punch moving at about 25 m/s which causes the metal to become plastic and flow up the sides of the punch. The clearance between the punch and the die determines the thickness of the extrusion. Products such as cigar and thin-walled cans, are produced by this technique.

6.2.5 Using powders

The transformation of powders into components usually depends on two key stages:

- the compaction of the powder to produce a strong 'green' compact of the required shape;
- the application of heat to promote the bonding of the particles (although this is not required with some ceramic materials which set through a cementation reaction).

The compaction of the powder is made much easier if it can be heated or blended to help it flow, but dry powders will not generally flow very easily. Thermoplastics can be made to flow by heating and plasticising them and ceramics by blending the powders with water or a resin. With dry powders it is necessary to avoid thin sections or sharp corners; even stepped diameters and chamfers can cause difficulties.

Powders are used to form components from metals, ceramics and, with mixtures of materials, to produce composite structures. A mixture of tungsten and silver powders is used for electrical contacts to provide wear resistance and conductivity. Cutting tool inserts are often made from a hard material in a metallic matrix – tungsten carbide cemented in cobalt is commonly used. As the tungsten carbide particles at the surface lose their sharpness the cutting forces increase and the particles pull out of the cobalt matrix exposing new, sharp-edged particles. By varying the amount of the cobalt used, the ease with which the particles pull out can be altered making them suitable for different applications (for example, finishing tools have little cobalt so that the particles come away easily and the tool remains sharp). Cobalt-samarium magnets are also made using powder metallurgy and provide an alternative to ferrous alloys.

Because historically they were the first group of materials to be formed using powders the processing of ceramics is described first. Powder metallurgy and the compression moulding of plastics can then be seen in relation to this technique.

Forming ceramics

Ceramic powders with a controlled particle size are blended, usually with water, and then compressed into moulds. The resulting shape is called a *green* and has sufficient strength to allow it to be transferred either to racks for drying or to a furnace for firing. Where water has been used to blend the ceramic particles this evaporates during the initial drying stage and significant dimensional changes result. Heating after the evaporation of the water promotes the formation of bonds between the particles by partial or complete vitrification (melting to form a glass) or sintering. Where a liquid resin such as sodium silicate or Portland cement has been used to blend the particles, a chemical reaction known as 'cementation' causes the bonds to form. These three bonding processes are described opposite.

Vitrification

During the firing of a ceramic vitrification or melting often occurs. Flux materials in the clay react with other substances to produce a liquid phase at the grain surfaces. This liquid helps to fill the pores and hence to reduce porosity. The amount of vitrification which occurs is related to the firing temperature – higher temperatures giving greater vitrification. The eventual structure of ceramics bonded in this way is shown in Fig. 6.46. This kind of bonding due to a glass phase is known as a 'ceramic bond'.

The enamels and glazes which are used on the surface of clay vessels are made up from clays that vitrify easily. Consequently they serve to reduce porosity as well as providing protection and decoration.

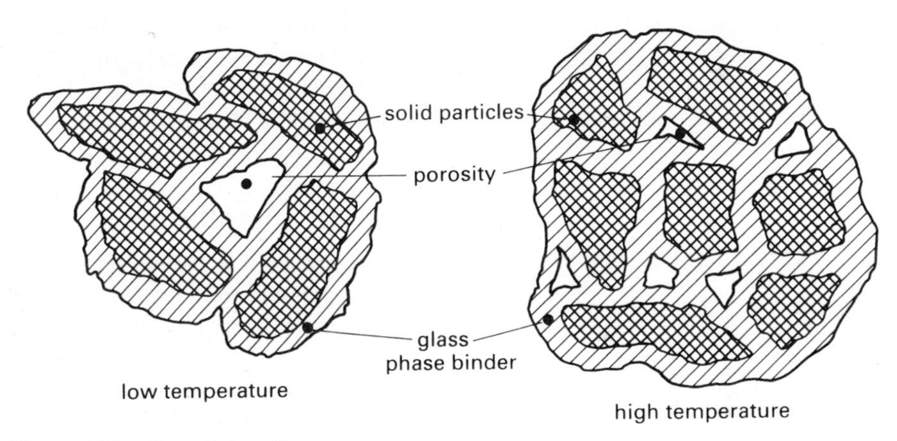

Figure 6.46 Ceramic bonding.

Sintering

When the powdered material is compacted the particles are brought into contact at numerous positions, but there is no actual join. They are held together largely by surface tension forces. As the material is heated atoms in the particles diffuse towards the contact zone creating bridges and eventually filling the pores. This process, which occurs only in the solid phase, is known as 'sintering' and is illustrated in Fig. 6.47.

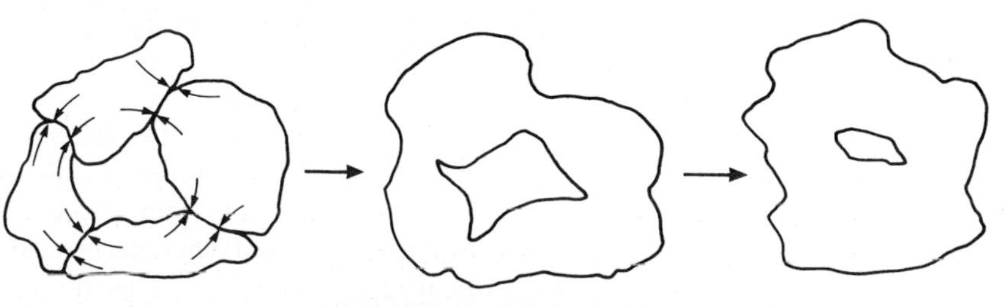

Figure 6.47 Sintering bonds formed by solid state diffusion.

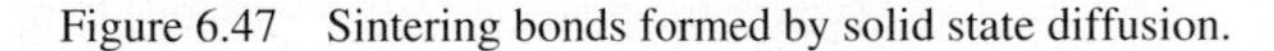

The diffusion process, the movement of atoms within the solid, is different depending on the type of bonding. In metals and alloys, the atoms can move to any nearby vacancy or interstitial site. In ionic materials, the diffusing ion can only enter sites having the same charge and to reach these it must get past regions of opposite charge. Activation energies are therefore high and the diffusion rates are low for ionic materials compared to metals. In polymers the molecules must diffuse between long chains and this is easier in amorphous polymers which have no long-range order in comparison to polymers where the chains form a glassy (or crystalline) pattern.

Cementation

The most common example of a cementation reaction is the setting of Portland cement, but there are several others like the setting of plaster of Paris, calcium aluminate, aluminium phosphate and sodium silicate cements. In all these cases a reaction occurs which produces a solid bridge between the particles. Normally this reaction is with water and the cement sets as it dries. In some cases an additional substance is involved, however, like carbon dioxide which is necessary to set sodium silicate cement.

The resulting ceramics are very porous and permeable and hence may be used as ceramic filters. They are also used to make moulds and cores for metal castings, because they have a strong, rigid structure and their permeability allows the gases generated in the casting process to escape. Clearly, if these gases were trapped they would result in large pores in the casting.

Powder metallurgy

Most metals can be formed by compaction and sintering, but not aluminium because of cold pressure welding. Metal powders are produced with particle diameters from 0.001 mm to 0.3 mm using a numher of processes. Brittle metals are crushed and ground, some metals are atomised by a jet of gas operating on the liquid metal (in the same way as petrol is atomised in a carburettor) and some metal powders are produced from chemical reactions.

Once formed, the metal powders are compressed to about 50% of their volume under very high pressure. The component can be made from a single metal or powders can be mixed, which is particularly useful for combinations that do not form alloys or for mixing ceramics and metals. The greens are transferred to an oven and heated – for metals, to 70–90% of their melting temperature. When heated the particles join by the sintering process in exactly the same way as ceramics.

Fig. 6.48 shows shapes which can be easily formed using powder metallurgy.

It is important to avoid thin sections into which the powder will not easily flow and sharp corners from which the powder will break away. Chamfers should be at an angle greater than 30° and there could be difficulties with uneven density of the component if the ratio of the length to the diameter exceeds about 2.5:1.

It must be possible to withdraw the component in a straight line, and consequently side holes and undercut features are not possible. Gear teeth, straight knurling and serrations are all easily achieved however. A small draft angle (0.5°) is necessary to aid ejection – and to prevent the die from becoming scored.

Components made from powder metallurgy are now used in a variety of products. Typically gears and other mechanical parts are made by this process, and, because the pore size can be controlled through the sintering time and temperature, porous components are also produced. This allows filters and self-lubricating bearings (where the pores are filled with oil) to be produced.

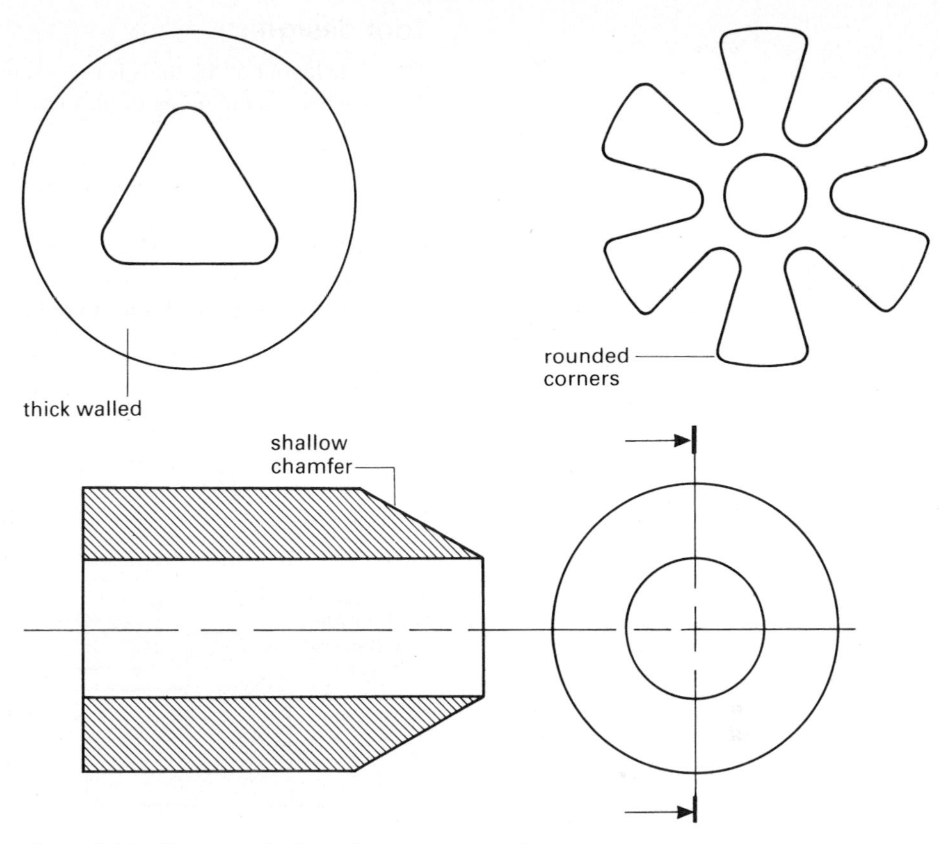

Figure 6.48 Shapes easily formed by powder metallurgy.

Amongst the advantages of the powder metallurgy process are the avoidance of any material wastage, the very small draft angle required, the very high production rate (up to 1000 per hour) and the avoidance of machining, except for side holes and undercuts. The components are also very accurately manufactured and the accuracy can be made even more precise by *coining*. Coining is a cold forging process by which the sintered component is finally sized by a high-powered press. It gets its name from the pressing of coins in the mint where patterns and lettering are produced by cold forging. The significant disadvantage of the process is the high tooling costs which means that it only becomes economic for large production quantities of about 10 000–20 000.

Compression moulding of plastics

Although there are significant differences between the compression moulding of plastics and the pressing of ceramic and metal powders, there are sufficient similarities to justify discussing them together. In particular the tool design for the three processes is very similar, as discussed below. Compression moulding is used to form thermosetting plastics such as phenol-formaldehyde and urea-formaldehyde. In its conventional form a measured amount of powder is placed into the mould cavity where it is heated and plasticised. It is then compressed into the required form by a punch which may also be heated. The raw material is sometimes compressed into tablets, or preplasticised, that is, heated outside the mould in order to reduce the cycle time. In this form the process has a cycle time very close to that of injection moulding.

Items like electric light and power sockets, knobs and plastic products which must resist temperature increases are all made by this process. Generally the properties and finish of compression-moulded products are very good.

Tool design

The basic layout of the tools for pressing ceramic powders, powder metallurgy and the compression moulding of plastics is very similar and there are several common features, namely:

- a female mould or die which must contain the initial powder volume,
- a punch which compresses the powder and produces the required form,
- a very shallow draft angle,
- an ejector mechanism of some kind. Contraction on cooling may be sufficient for some thermosetting plastics to reduce this requirement significantly.

Fig. 6.49 shows the kind of tooling which would be necessary to produce components by any of these processes.

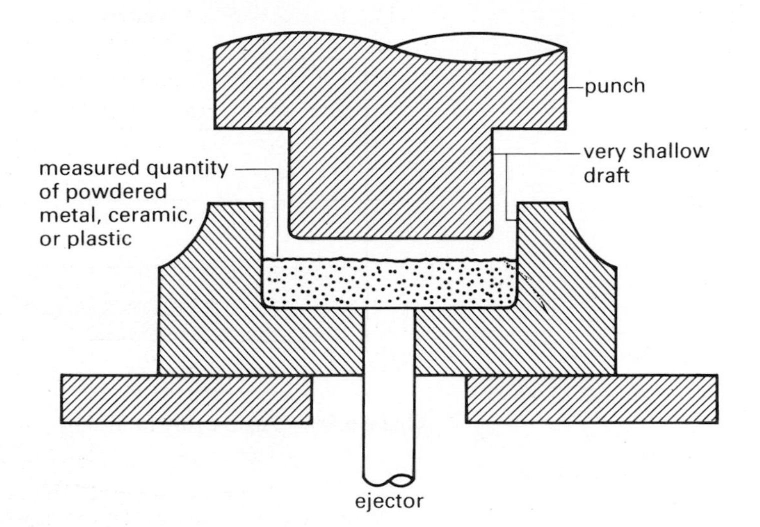

Figure 6.49 Tooling requirements for compression moulding.

6.2.6 Extruded, drawn and rolled sections

Industrial designers make a great deal of use of extruded and rolled sections. Complicated forms are used in products like double glazing and simple shapes, like tee and angle sections, are commonly found in shelving systems and furniture. The two processes are described first and then their advantages and disadvantages and potential applications are compared. Drawing can be thought of as the opposite of extrusion – the metal being pulled rather than pushed through a die or a series of dies. It is therefore included here for easy comparison.

Extrusion

The basic extrusion process is illustrated in Fig. 6.50.

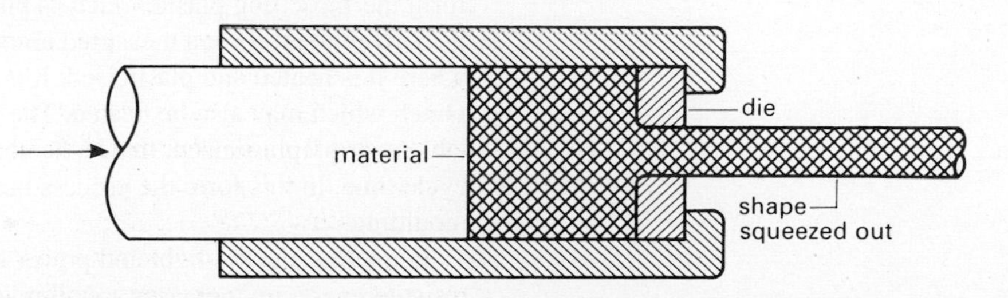

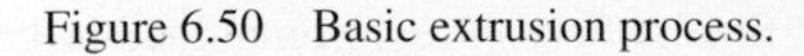
Figure 6.50 Basic extrusion process.

The material is forced through a die which has the required cross-section in much the same way that toothpaste is squeezed from a toothpaste tube. Many different materials can be extruded. Metal extrusions can be produced by either hot or cold working, and aluminium and copper alloys are the most commonly used. Cast iron is extruded hot to produce steam pipe, and steel can now be extruded hot in specially built machines. The extrusion of thermoplastics is basically the same as that of metals except that temperature control is vital. The newly formed extrusion is normally water-cooled by allowing it to sag into a water bath. Blended powders, whether purely ceramics or metals or mixtures, can be easily extruded. Composites are also extruded, typically plastics and metal foils as used in decorative trims or plastics over metal inserts in order to improve their stiffness.

One of the most commonly extruded products is tubing. The two alternative techniques for producing this are shown in Fig. 6.51.

Fig. 6.51(a) shows a mandrel attached to a ram which results in tubing being formed after the material has passed through the die. Fig. 6.51(b) shows the mandrel (or torpedo) attached with spiders to the die. The material parts as it flows past the spiders and then recombines under the pressure in the die cavity. Plastic-coated wire is produced by feeding a cold drawn copper wire through the centre of the die as a very thin-walled plastic tubing is extruded. The coated wire then passes through a water bath in order to cool the plastic. The wire is produced at several hundred metres per minute and wound on to large diameter drums.

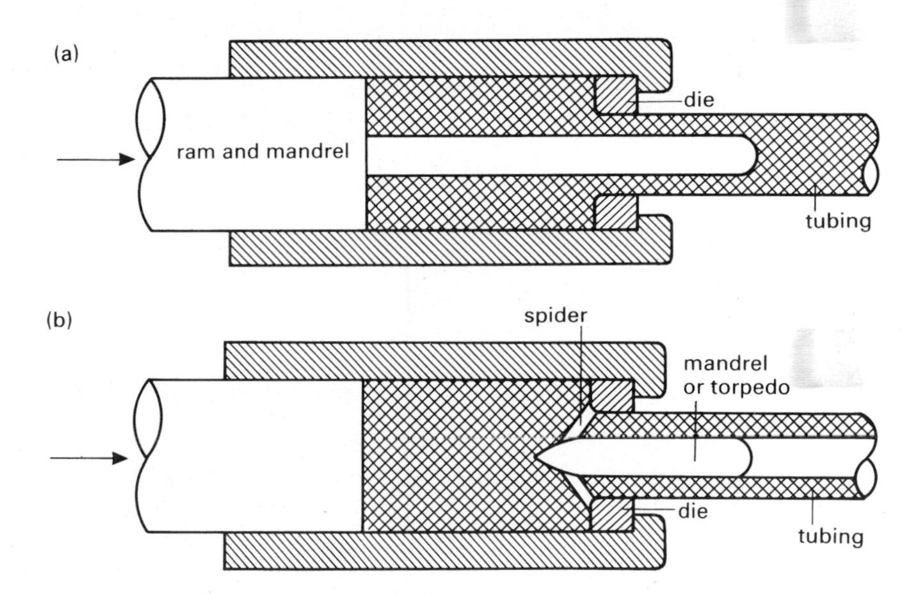

Figure 6.51 Two approaches to extruding tubing.

One of the significant differences in the extrusion of different materials is the means by which the pressure is generated. A helical screw is used with thermoplastics and they are heated as they progress towards the die (see injection moulding in section 6.3.3). Metal rods and powders are often fed using an extrusion wheel as shown in Fig. 6.52. This is known as the 'conform continuous extrusion process'.

The conform process is sometimes modified to produce large numbers of components from powder materials as shown in Fig. 6.53. This is effectively a continuous pressing operation.

Thermoplastic materials can be formed into sheet by extrusion through a slit die, although this is not possible with metals. Very thin-walled plastic tubing can be made by blowing up extruded tubing like a balloon as it leaves the die, again a technique which has not proved possible with metals.

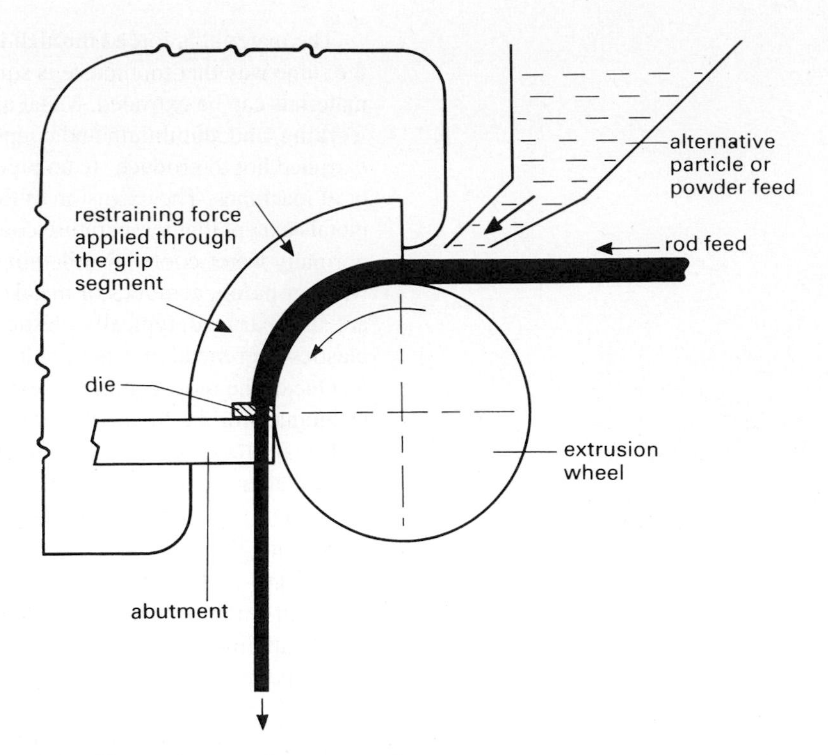

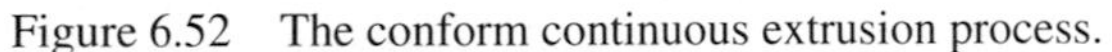

Figure 6.52 The conform continuous extrusion process.

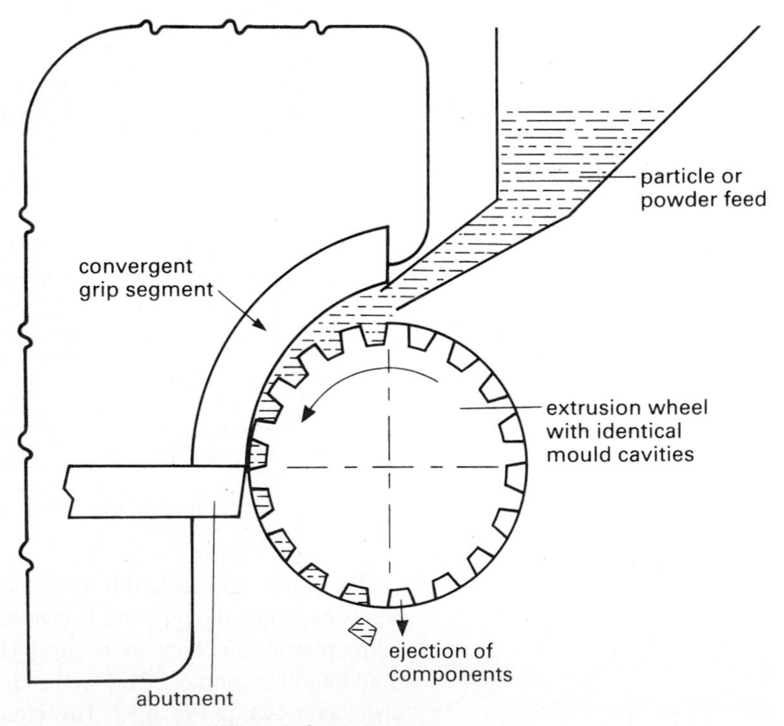

Figure 6.53 The modified conform process.

Although one of the potential advantages of the extrusion process is the ability to distribute material in the cross-section where it is most advantageous from a design point of view, this option is limited by the need to keep fairly uniform sections. The extrusion of metal will result in an even grain structure providing it all passes through the die at the same speed. However, the metal will tend to flow faster in the middle away from the friction with the die in the same way that a stream flows faster away from the bank. Thus, if the thickness of the section varies greatly, all the metal will not pass through the die at the same speed and the section properties will be adversely affected. Plastics extrusions need to be of uniform thickness for similar reasons. Differences in the flow of polymers can lead to differential cooling and warping of the section.

Fig. 6.54 shows a range of shapes which could be produced easily by extrusion. In developing these shapes the designers have striven to keep the sections as uniform as possible. Symmetrical sections are generally easier to produce and transitions between different thicknesses should be as smooth as possible.

Figure 6.54 Cross-sections of shapes which could be easily extruded.

Drawing

Rods, wires and metal tubing are normally produced by drawing the material through a series of dies of gradually reducing diameter. This is shown schematically in Fig. 6.55.

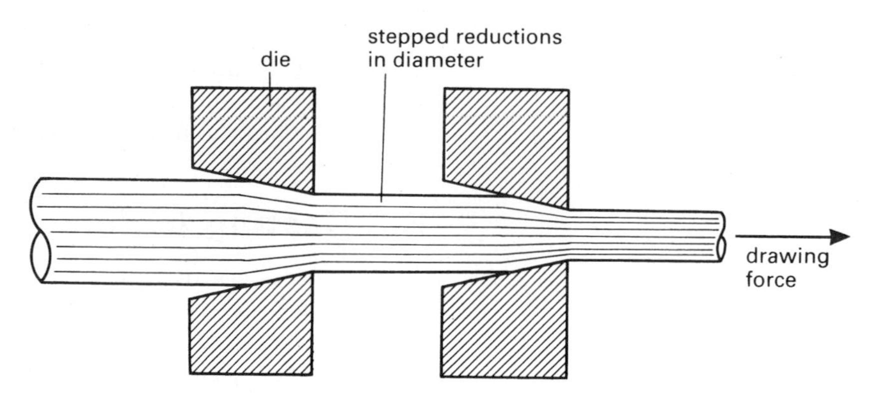

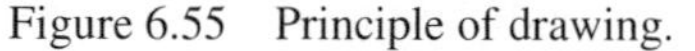

Figure 6.55 Principle of drawing.

The amount by which the diameter can be reduced at each stage is limited by the force which can be exerted in pulling the material and its ductility. The material will tend to work-harden as it is cold worked. If tubing is being produced then the initial section from which the tubing is drawn would be formed by the extrusion process. For example, copper tubing is produced by extruding a heated copper billet and cold drawing to near the required size. It is then annealed before the final drawing operation is carried out. This results in just sufficient hardening for the tubing to remain undamaged during distribution, but in sufficient ductility to allow the formation of bends.

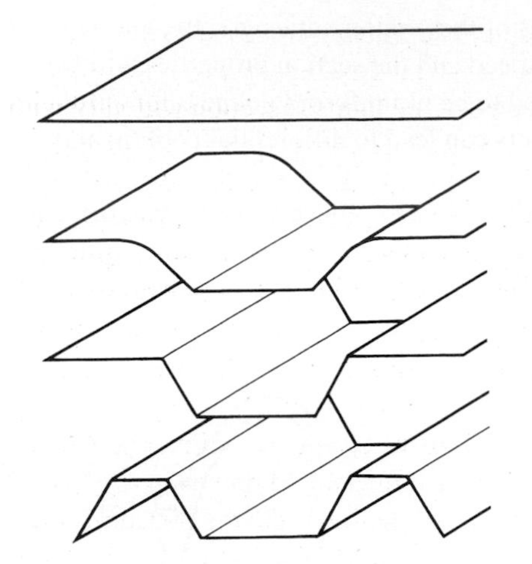

Figure 6.56 Stages in rolling a section.

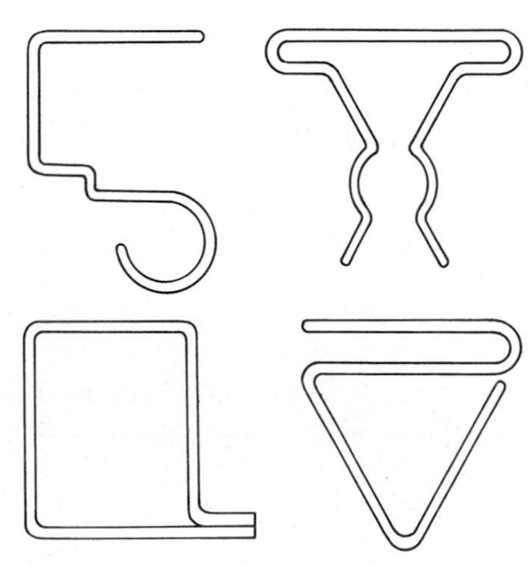

Figure 6.57 Typical rolled section.

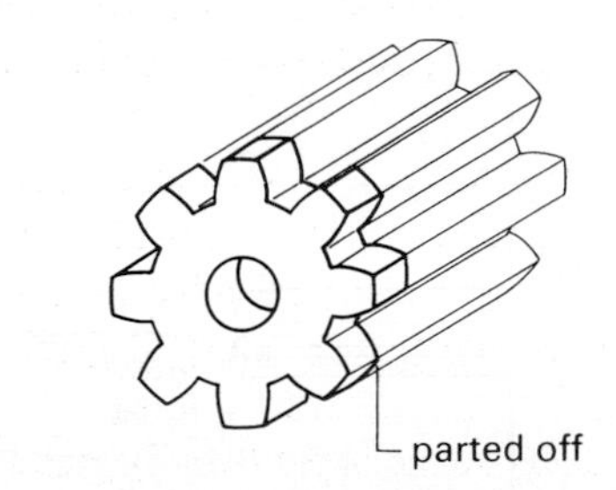

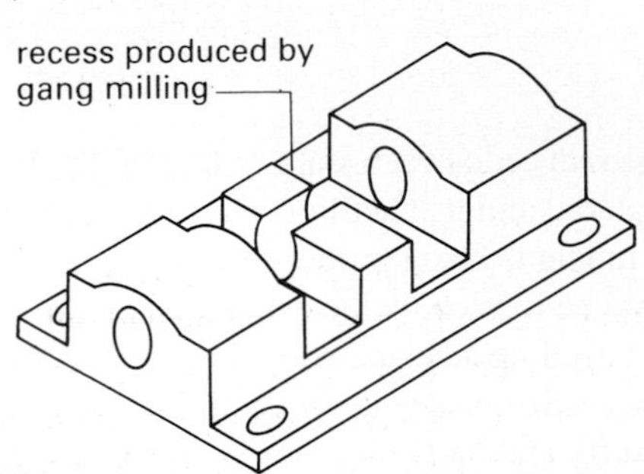

Figure 6.58 Extruded (a) gears and (b) bolt body.

Rolled sections

Sections with design possibilities similar to some of the extruded shapes shown in Fig. 6.54 can be formed progressively from sheet materials using a series of rollers. Metals like steels, aluminium and copper alloys can all be formed by this technique and it is also possible to use metals which have been precoated with zinc or plastic materials. Fig. 6.56 indicates the kind of progression it would be necessary to carry out in order to form the section.

Similar sections can also be produced on a brake press where each bend is made in sequence. This requires a different order in the forming operations but has a similar end result.

The choice between forming the section using rollers or by presswork is primarily associated with the production quantity. The time and money spent in manufacturing and setting up the rolls would need to be recovered. Fig. 6.57 shows some typical rolled sections.

Clearly most of these are open sections. Producing a closed section requires a seam welding operation. Although this could usually be carried out quite easily using resistance seam welding (see section 6.5.2), it is an additional operation with consequential costs.

Comparing rolled and extruded sections

Roll-formed sections are generally thinner than extrusions and consequently cheaper. Rolled sections are also available in sheet steel, and steel cannot be extruded in such small thicknesses. As steel is cheaper than aluminium or copper alloys, rolled steel sections should be considered very seriously if the application might allow their selection.

The extrusion process provides many unique design opportunities however. For example re-entrant angles and undercuts in the cross-sectional shape, some variation in section thickness allowing material to be placed where it can contribute most to performance and thin-walled tubing of large diameter are all possible in extruded forms. Designers can also exploit the extrusion process in producing gears or the bolt body as shown in Fig. 6.58.

Gears are produced from the extruded section shown in Fig. 6.58(a) by a parting-off operation. Producing the bolt bodies shown in Fig. 6.58(b) requires parting off, gang milling and drilling operations. In both cases the designer has greatly simplified their production by using an extruded section.

6.3 Casting techniques

A large range of materials can be cast, that is, poured into a mould of the required shape, in which they become solid. Plastics are cast using a fluid monomer which is poured into the mould where it finishes polymerising. Plastic film is produced by pouring plastic resin on to a moving belt. Ceramics are made into a slurry – a runny mixture of particles and fluid – and poured into moulds. A hollow shape is often produced by pouring off the excess liquid from the centre after the shape has begun to set. The same technique is used with metals to make hollow spouts for domestic products from pure aluminium and was once used to make lead soldiers. When plastic products are made in a similar way from PVC paste it is known as 'slush moulding'.

The casting of metals is one of the oldest techniques known. Around 600 BC very accurate bronze castings were produced by the lost wax technique. In this process the shape is first made in wax. It is then surrounded by a ceramic material and heated so that the wax melts out, leaving the required cavity. Equally, some forms of die-casting have been known for thousands of years. Here the cavity is formed by the two halves of a split mould which register together. The molten metal is then poured in and the product removed after it has cooled sufficiently, hence allowing the mould to be re-used.

Some basic design principles apply to all casting processes, and also to the injection and transfer moulding of plastics. In all these processes it is necessary to design for easy fluid flow. The acceptable size of the thinnest section will vary with the size of the casting and the materials, perhaps 10–20 mm on large steel castings and 1 mm on zinc die-castings and plastic injection mouldings. It is necessary to avoid abrupt changes of section and to ensure that thin sections are fed from thick ones. The thicker regions will cool more slowly and hence material can be drawn into the thinner areas as they contract. Shrinkage holes or cracks can otherwise result. Large volumes of material will tend to result in a poor surface finish because of the uneven shrinkage. Such surfaces need to be broken up visually by adding some kind of texture or pattern.

Because they are fundamental to understanding the casting techniques developed for quantity manufacture the basic sand-casting and full-mould processes are described first. Quantity manufacture and related techniques like the injection and transfer moulding of plastics are then discussed.

6.3.1 Sand-casting and the full-mould process

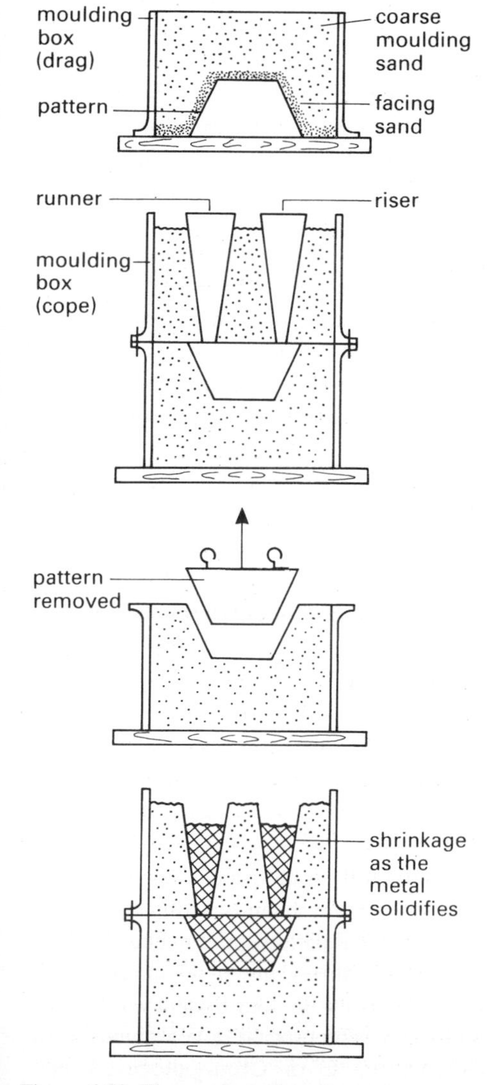

Figure 6.59 The sand-casting process.

Damp sand will retain its shape when compressed and this is the basis of the sand-casting technique, which can be used with any metal. The 'green sand' used in the basic process is simply ordinary moulding sand (or Mansfield sand) with no chemical additives. An alternative, Petrobond, is sand-blended with oil so that it does not dry out. The sand next to the molten metal is, however, burnt and must be discarded after use. More stable moulds can be produced if the sand is bonded by a cementation reaction or a resin (see section 6.2.5). Normally a special pattern is used to create the cavity, but the need for this can be avoided by making impressions in the sand on the foundry floor using tools, templates or existing products to create the shape. If a cavity which is fully surrounded is to be created, however, then a pattern will need to be made up. If the product has sand on all sides then more even cooling will take place and there can be better control of shrinkage, distortion and the surface finish. In its conventional form, the pattern is extracted by making it in two pieces and using a split box. When the sand has been formed round the pattern the box is opened and the two halves of the pattern are removed from their respective sides of the box. This conventional form of the sand-casting process is shown in simplified form in Fig. 6.59.

Many materials could be used to make the patterns. Yellow pine is the traditional choice, but hardwoods, epoxy resins, rubber and metal could also be used. The patterns must have a good finish in order to be removed easily from the sand and are best covered with a special pattern-makers' paint. Even so there must be a small draft angle on the pattern so that it can be removed.

In the first stage of the process high-quality facing sand is put over the pattern and the remainder of the moulding box (the drag) is filled with coarse moulding sand. This is then compacted with a ram and the excess sand removed from the top to leave it level (*strickling*). The drag is then inverted and the other half of the moulding box (the cope) is located on top. A runner is added to create the tapered hole through which metal will be poured and a riser to allow the metal to flow through the mould cavity. In order to extract the pattern the moulding box is split open and the casting poured once the box has been re-assembled.

In the full-mould process the whole pattern is buried in the sand and it cannot therefore be removed intact. The pattern could be made from wax and melted out, but this is normally only used if the wax is itself first cast into the required shape. This is a very useful way of making a large quantity of precise castings of a difficult form (see section 6.3.2). For single castings it is more usual to make the pattern from polystyrene which is burnt out by the molten metal when it is poured, rather than melted out in advance. The full-mould process is shown in Fig. 6.60.

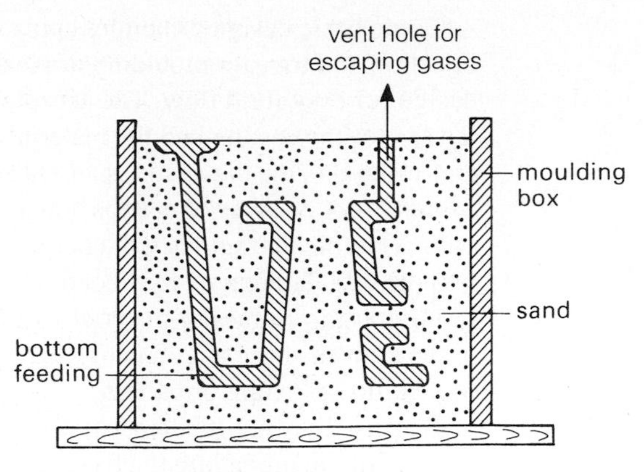

Figure 6.60 The full-mould process.

Burning out the polystyrene results in the formation of a lot of gas and consequently it is usual to feed the mould from the bottom so that the gases can escape more easily. If this is not done the castings can be very porous. As the pattern does not have to be withdrawn from the mould, no draft angle is required.

Generally the accuracy achievable with either sand-casting or the full-mould process is not very good. For sand-casting ± 1.5 mm on smallish castings and ± 6 mm on a one-metre length is usual commercial practice and rather worse with the full-mould process. The major advantages of sand-casting are that it is economic for any production quantity, because of the low tooling cost, and that it can be used to produce castings of almost any size. The disadvantages lie in the comparatively slow production time, caused by the slow cooling rate and the poor accuracy and finish. Without careful design the *fettling* associated with the removal of the runner and riser and any flashing at the split line will tend to make the appearance even worse.

A hollow cavity can be formed inside a sand-casting by placing a sand core in the required position. The core must be made separately and the sand particles bonded by one of the processes described in section 6.2.5. The core must be supported adequately in the mould or there is a danger of its being swept away when the metal is poured. Leaving a long core unsupported is very poor design practice. Steel pins can be used but it is much better if the long core is supported by bearing surfaces in the sand mould.

The major advantage of the full-mould process is the absence of any restrictions on the shapes which can be produced. The absence of a draft angle and the ease of dealing with undercut features mean that difficult shapes can be made quickly. The finish is, however, poor because of the problems in shaping the polystyrene. This is normally cut with hot wires or machined, and the blocks are cemented together. Runners and risers, heads and gates can all be obtained preformed from polystyrene as standard items. Another difficulty is that the fumes produced when cutting the polystyrene or during casting can be dangerous and it is essential to take appropriate safety precautions.

6.3.2 Quantity manufacture

Both the conventional sand-casting process and the full-mould process have been developed for quantity manufacture. The production rate for sand-casting can be increased by mechanising all the operations using metal patterns and pneumatic ramming equipment. Fully automated sand-casting facilities now make many products where there is a large enough quantity to justify the tooling and set-up costs. The bodies for bench vices are made this way. The process development work has also resulted in shell-moulding (also known as the 'C' or 'Cronig process') and various forms of die-casting. In shell-moulding, a split metal pattern is made in the shape of the required metal product, and a temporary female mould formed around it using sand mixed with a thermosetting resin. In die-casting, a split metal female mould is manufactured and the casting produced inside it. These processes are described below. The full-mould processes have been developed to use expendable patterns made in permanent moulds from expanded polystyrene foam or wax. These processes are described in more detail in the following sections.

Shell-moulding

The shell-moulding process is illustrated in Fig. 6.61. Each half of the split metal tool is heated and clamped to the top of a dump box. This is then inverted so that the pattern becomes surrounded by sand which has been mixed with a thermosetting resin, such as phenol-formaldehyde. The heat from the tool melts the resin and a casing about 10 mm thick is formed round the tool. The tool and casing are separated using ejector pins and the casing baked in an oven for about a minute to cure the resin. Finally the two half shells are clamped together and the metal is poured in.

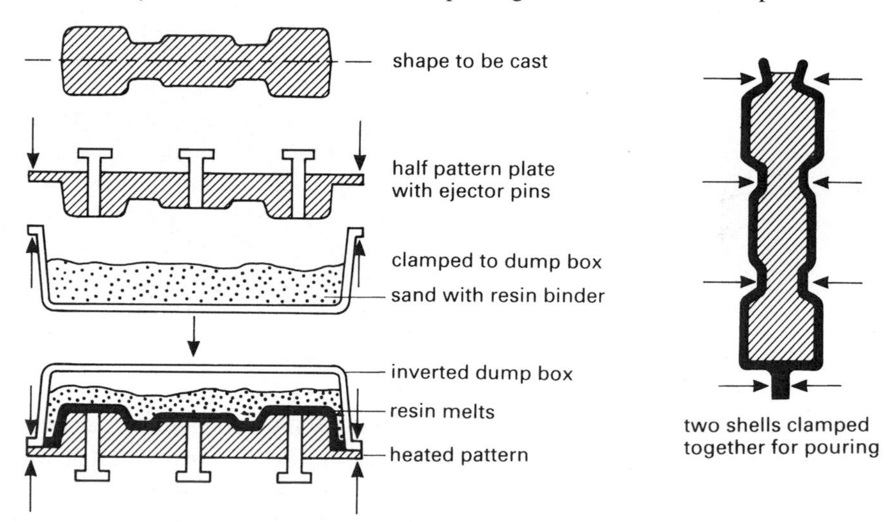

Figure 6.61 The shell-moulding process.

Apart from a decrease in the production time the shell-moulding process does offer other advantages. Generally it is more accurate and gives a better surface finish than sand-casting. There is very little draft required because the tool and the resin-bonded sand tend to separate as they cool and the resin sets. The tool cost is also quite low as it is only necessary to make the male pattern and hence it is economic for production quantities of as few as five hundred. It is particularly useful for producing ferrous castings, which are not easily die-cast. The low mass of the shell also avoids the surface chilling and hardening associated with sand-casting because it heats up more rapidly. This allows better metal flow during pouring and can improve the machining characteristics of the casting. The major disadvantages are the high cost per casting of the resin and the time required to assemble the shells. Many products – for example, brake drums, engine parts and bells – are made by the shell-moulding process.

Moulds for casting polyester resins can be produced in a silicone rubber material (*silastomer*). The technique, which is very similar to shell-moulding, is shown in Fig. 6.62.

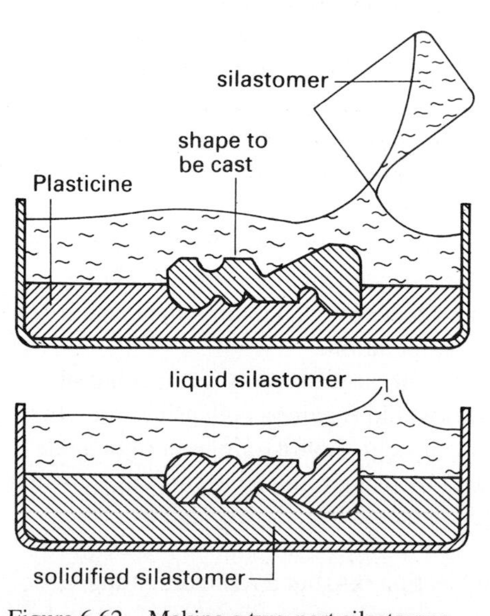

Figure 6.62 Making a two-part silastomer mould.

The object is initially half buried in Plasticine and the first layer of silastomer is added. When this has set, the assembly is inverted, the Plasticine is removed and a second layer of silastomer added. After the object has been removed the shells are clamped together and the casting resin is poured in. As the shell is flexible and rubber-like, it can be removed without damage and therefore used many times, unlike the sand shells used for metals. This is an excellent technique for producing plastic copies of scientific specimens like bones and examples of defects in metals etc.

Die-casting methods

Die-casting differs from sand-casting in that a metal mould is used rather than a sand mould. It differs from shell-moulding in that the tooling produced must be the female moulds or the inverse of the shape required. This has significant implications for product designers because, when they are considering the feasibility and difficulty of producing the tooling, they must think about the inverse of the shape of the product. The same problem must be faced when designing shapes to be made by injection moulding. The die-casting process exists in four primary forms – high pressure die-casting with hot and cold chambers, low pressure die-casting and gravity die-casting (also known as permanent mould casting). The choice between the hot and cold chamber methods depends very largely on the material being used. The hot chamber method can be used with zinc and low melting-point alloys which are kept molten and injected by a plunger when required. Molten aluminium tends to react with steel and cannot therefore be cast by this method. In the cold chamber method molten metal is transferred to the machine between castings which tends to slow the production time. It is, however, necessary for metals with higher melting points, such as aluminium, brass and magnesium alloys. Centrifugal action is also sometimes used to help the molten metal to flow into the mould.

In all these process variants the shape of the cavity and the general design considerations are similar. The differences lie in the tool size and the material it must be made from in order to withstand the temperatures and pressures involved. From the product designer's point of view the key differences are the materials which can be cast with each of the processes, the economic production quantity, the achievable accuracy and the surface finish. These aspects are discussed for each of the process variants after a general description of the issues associated with the tool design.

Tool design

The important feature of the die-casting process is that the mould which moves to allow the casting to be taken out must be withdrawn in a straight line. This imposes similar restrictions on shape to those found when using powders. Lettering and holes can be easily formed on surfaces which are facing the tool, but side holes and other features require special arrangements. In components formed using powder metallurgy these would be produced by machining or coining after sintering. This approach could also be taken with die-cast components but retractable cores and removable inserts built into the die are other possibilities. Fig. 6.63 shows the simplified layout of a die-casting tool.

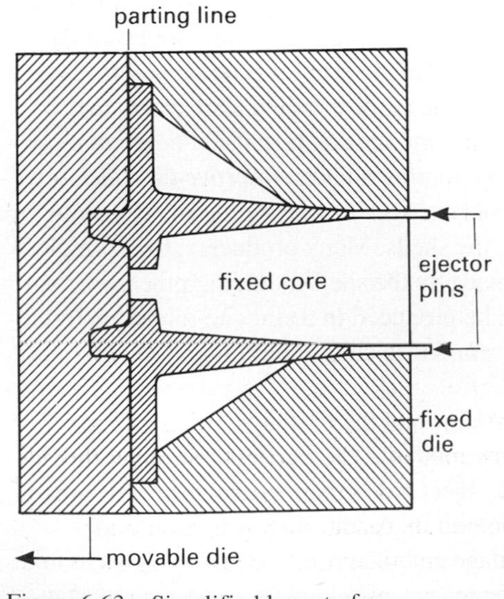

Figure 6.63 Simplified layout of a die-casting tool.

It is important that there is a draft angle (or taper) on the cores so that they can be removed easily. The allowable length of a cored hole is limited to two or three times the diameter depending on the metal being cast. The parting line does not have to be at right angles to the direction of the tool movement or in a single plane, but the tool costs will be significantly reduced if it is. Small radii help to ease the removal of the casting, but it is still necessary to make provision for ejector pins. Generally, as there are no flash, runners or risers the surface finish is excellent and consequently the ejector pins must act on a sufficiently large and preferably an unimportant area in order to avoid an unnecessary extra machining operation.

Normally, it is only possible to deal with re-entrant features by using collapsible or retractable cores. The design in Fig. 6.64(a) would require a collapsible core, but this can be avoided if it is redesigned as shown in Fig. 6.64(b). Holes at right angles to the direction the die moves can be produced by using retractable cores as shown in Fig. 6.64(c), although the necessity for them can sometimes be avoided by clever design as shown in Fig. 6.64(d).

Designing for machining is the subject of section 6.4, but a reminder of the key issues is given here. It is important to consider access and the direction of the tool path, the clearances necessary for tool approach and overrun and how the component

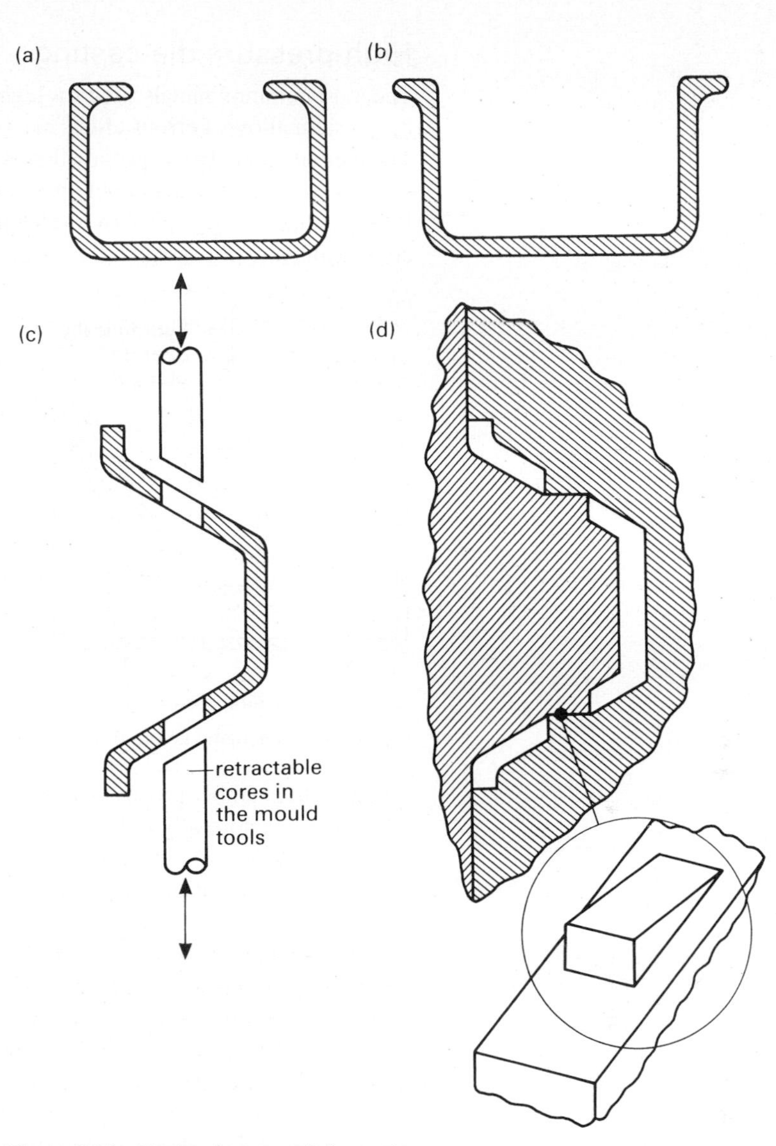

Figure 6.64 Design features of die-castings.

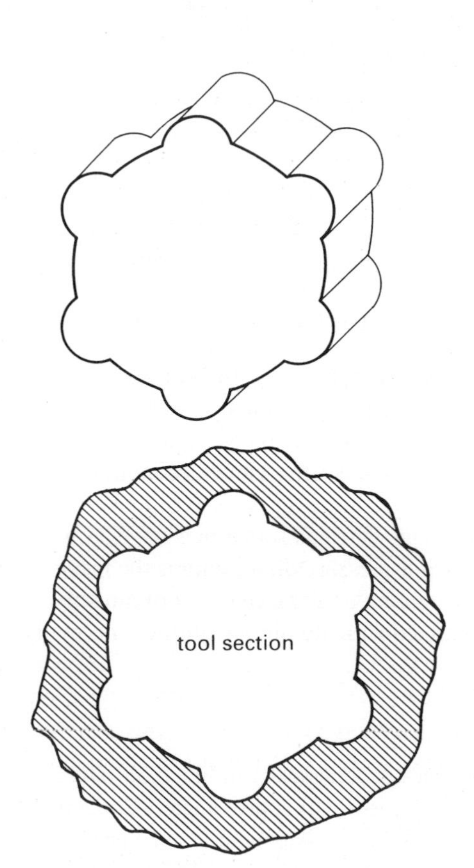

Figure 6.65 Knob designed for die-casting.

Figure 6.66 Knob designed for machining.

is to be located and clamped during machining. Fig. 6.65 shows the shape of a finger knob as it would be designed for die-casting. The corresponding tool section shows how grooves would be machined into the outer surface of the cavity.

A similar shape could be used for extruded sections and on moulded handles. Fig. 6.66 shows how a knob would be made by machining from solid bar – effectively the inverse of the shape for die-casting.

The cost of the tool will depend on its size and complexity, the cooling necessary, the nature of the joint line and the use made of retractable cores. Generally the tool will need to be larger with higher-pressure machines, increasing tool cost. With rapid production rates and alloys which melt at higher temperatures it may be necessary to water-cool the dies to ensure reasonable life.

Many features are best produced by using inserts in die-castings. Stronger threads can be obtained by incorporating steel bolts or threaded rods, bearings by inserting a brass or phosphor bronze bush and curved holes by casting-in preplaced tubing. Localised areas of increased strength or stiffness can be produced by using steel reinforcement. The positioning of such inserts must also be taken into account during the tool design.

High-pressure die-casting

The most common metals used for high-pressure die-casting are zinc, aluminium and magnesium alloys. Ferrous alloys can be die-cast but this is less commonly done. The dies are made from special alloy steels and normally weigh several tonnes. Cycle times are typically about one minute, although this could be longer on very large machines. Fig. 6.67 shows schematically the difference between the hot and cold chamber methods.

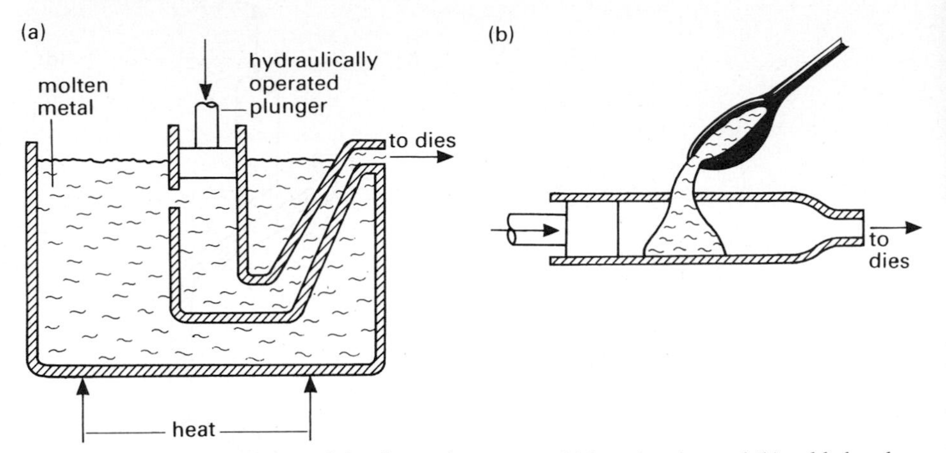

Figure 6.67 Schematic layout of the die-casting process (a) hot chamber and (b) cold chamber.

The cold chamber process operates at a pressure of between 14 MPa and 70 MPa and is normally only economic for production quantities greater than 20000. Because of the lower pressures – typically 2.5–3.5 MPa – and consequentially smaller tooling associated with the hot chamber process, this can be economic for quantities as low as 10000. The finish obtained with the process is excellent and only requires polishing before painting or plating. Zinc castings sometimes have a finish good enough for plating without polishing. High-pressure die-casting is the most accurate of the commercial casting processes giving a tolerance of approximately ± 0.05 mm on a 25 mm dimension. The tolerance with aluminium alloys is normally a little worse, say ± 0.07 mm.

Low-pressure die-casting

The low-pressure die-casting process is illustrated schematically in Fig. 6.68. The molten metal is forced up into the die by air at a pressure of approximately 2.8–5.6 kPa (4–8 p.s.i.). As with high-pressure die-casting, complex components can be produced to close tolerances and with a good surface finish. The cycle time is longer but the tooling is much cheaper to produce, so that much shorter production runs are potentially economic. All sizes of components can be produced, a well-known example being the aluminium beer casks which have superseded the previous wooden barrels.

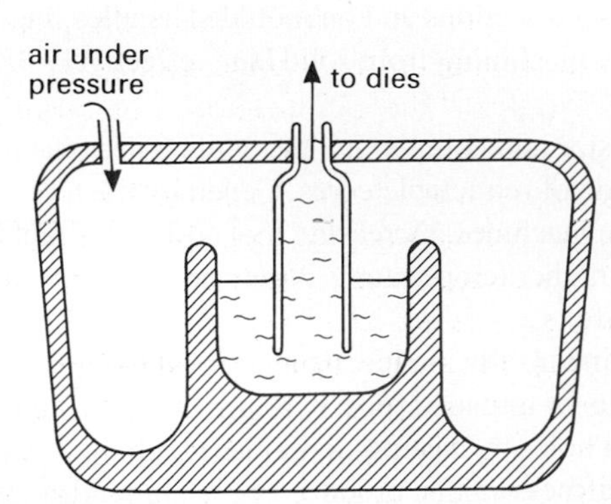

Figure 6.68 Schematic layout of the low-pressure die-casting process.

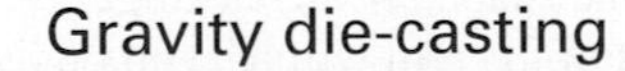

Gravity die-casting

Forms of gravity die-casting (or permanent mould casting) have been known for thousands of years. The metal flows into the mould simply under the influence of gravity. It solidifies and the mould, which is often hinged, is then opened to remove the product. Because of the much lower pressures involved, the required clamping force and the tool dimensions to give adequate stiffness are much reduced. Economic production quantities can therefore be as low as 500–1000. The process can be used with a range of aluminium, copper and magnesium alloys, but it is not suitable for zinc. Some gravity die-casting is done with ferrous materials, but this is a rare and specialist process. Particular grades of nylon can also be die-cast successfully into polished aluminium moulds. For metals, the moulds are usually made from machined cast iron.

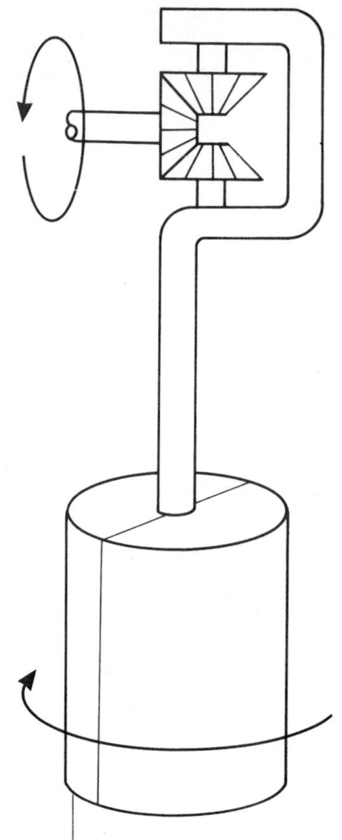

Figure 6.69 Rotational casting.

As a result of the lower operating pressure, it is not possible to produce sections much thinner than about 4 mm thick. For the same reason, the surface finish is not as good as pressure die-castings, but the permanent mould does result in significant improvement over sand moulds. A draft angle of several degrees is necessary in order to ensure that the component is easy to free and to give adequate die life. The accuracy obtained with gravity die-casting is approximately ±0.25 mm between points parallel to the parting line, but the tolerance on lengths which cross the parting line needs to be much higher.

Centrifugal methods

Centrifugal action can be used in two ways in casting processes. It can either be used to help to force the metal into the die by simply rotating the whole assembly, or it can be used to produce hollow products in metals and plastics. Accelerations of up to $60g$ can be produced by rotating the die, but as this results in pressures much less than those in conventional die-casting sand or plaster moulds can be used. This form of die-casting is not commonly used, however.

Metal pipes are produced by pouring the metal on to a rotating mould. The centrifugal effect causes the metal to flow to the outside of the mould resulting in the formation of a hollow cylinder. Hollow plastic products can be similarly formed from PVC paste. A measured quantity is placed in the die cavity and the assembly is then spun about two axes as shown in Fig. 6.69.

The mould is heated which causes the plastic to fuse. Once cooled the die is split open to withdraw the completed casting. Typical examples of products formed this way are footballs and moulded furniture.

Expendable pattern processes

Expendable pattern processes have been developed in a number of forms and are typically used either for the precision casting of steels or for producing shapes which are difficult to machine. The expendable patterns are normally made from wax or expanded polystyrene, but are also made out of thermoplastics and low melting-point metals. They are produced in moulds and then processed either individually or in a group. If they are to be processed as a group, then they are built up on a central runner to form a 'tree' as shown in Fig. 6.70.

The moulds are processed by one of two techniques: flask moulding or the ceramic shell investment process. The choice of process depends largely on the size of the casting. Small individual castings or trees are produced using flask moulding. Larger castings or trees use the ceramic shell technique.

In *flask moulding* (also known as the 'block mould process') the patterns are submerged in a ceramic slurry contained in a metal flask. The whole assembly is heated so that the ceramic sets and the wax melts out to be re-used. The plaster mould is then heated to a higher temperature to remove any remaining wax and reduce the water content. The metal is poured in with the mould still hot so that there is no time for water to be re-absorbed. Pouring the metal into a mould at a high temperature means that the cooling time is very long. There is, therefore, significant grain growth and a limit to the achievable toughness.

The *ceramic shell investment process* is very similar to shell-moulding. The expendable pattern is dipped alternately into a ceramic slurry and dry, fine particles of a refractory material known as 'stucco'. The ceramic layer is built up to a thickness of between 5 mm and 12 mm, the thickness necessary to give sufficient strength depending on the weight of the casting. The shell is then air dried before being heated to remove the wax. As in flask moulding, the shell is baked at a higher temperature and pouring is done while it is still hot. The cooling is more rapid because it is a shell rather than a solid mould. The finish and tolerances obtainable from the ceramic shell process are excellent, but it is comparatively costly because of the tooling and component costs associated with expendable patterns.

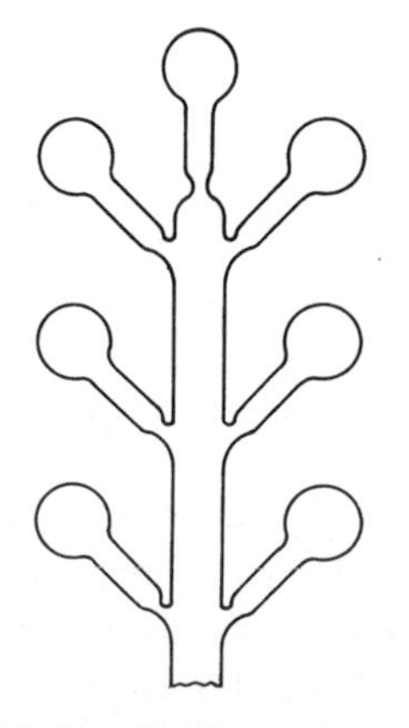

Figure 6.70 Completed tree for casting ball joints.

A recent development of the full mould process using expanded polystyrene foam patterns is the *magnetic moulding process*. As with wax patterns, the polystyrene moulds may be processed either individually or in trees formed by cementing them together. They are then given a special coating and surrounded by a free-flowing magnetic moulding material. This contains ferrous particles of between 0.5 mm and 1 mm grain size, but vibrations may be necessary to aid its compaction round the pattern. The assembly is placed in a magnetic field to hold the moulding material in position and the metal is poured in, burning out the pattern. When the casting has cooled, the magnetic field can be removed and the moulding material drawn off for re-use. The cost of the mould for the expanded polystyrene patterns means that the process is only economic for production quantities of a few hundred or more.

Other casting developments

The use of silicone rubber shell-moulds for casting polyester resins has already been described (see shell-moulding) and this allows the precise reproduction of difficult forms – for example, those with some undercut features. A similar technique, called the *Shaw casting process*, has been developed for use with metals. For this technique a specially formulated ceramic slurry is poured over the pattern and allowed to gel for two or three minutes. At this stage it is like a hard rubber and can be sprung away from the pattern if there are slight undercuts. The mould is then ignited, which results in a permeable, rigid structure. The moulds are poured after they have cooled, which allows tough, fine-grained castings to be produced. The process is very accurate because there is no expansion or contraction during setting, but is quite expensive because of the special materials needed.

If plaster is used instead of sand the surface finish of the resulting casting will be much improved. There is, however, a further advantage of casting in plaster moulds, namely their rigidity. This allows the use of a flexible pattern, which can be manipulated out of the mould. The moulds are formed in the same way as for expendable mould castings: immersion in a ceramic slurry, removing the pattern, baking and pouring hot. The advantages of the technique are the simple tooling requirements and the high accuracy which can be obtained, but it can only be used with low melting point alloys and the materials used are very expensive in comparison with pressure die-casting. It tends to be used for making prototypes and the tooling for other processes, like vacuum forming, blow-moulding and rotational casting.

6.3.3 Injection and transfer moulding of plastics

Injection moulding is by far the most common process employed by designers for plastic products. The main reason for this is that the dimensions and shapes can be accurately controlled and the process is very reliable in production. A schematic illustration of the industrial process is shown in Fig. 6.71.

The thermoplastic is fed from the hopper and is forced through the heating units by an Archimedean screw. When sufficient plastic has collected in the shot chamber the screw stops rotating and acts as a plunger forcing the plastic into the mould. When sufficient cooling has taken place the mould can be opened and the product removed.

This type of equipment can be used only with thermoplastic materials. Transfer moulding is a hybrid of injection moulding and compression moulding which can be used with thermosetting plastics. A schematic illustration of the transfer moulding process is shown in Fig. 6.72.

The powder or pellets are placed in a cavity which is connected to the mould by a *sprue*. The material is heated and injected into the mould.

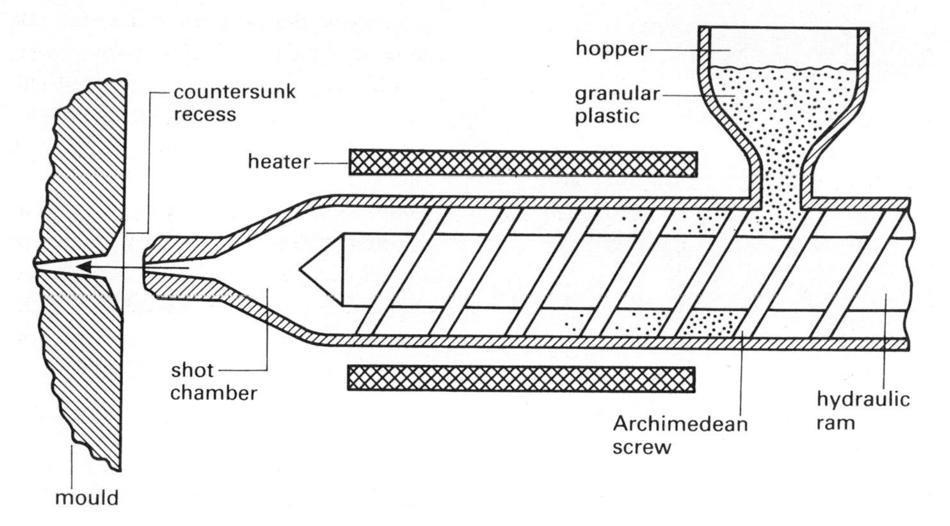

Figure 6.71 Schematic illustration of the injection moulding process.

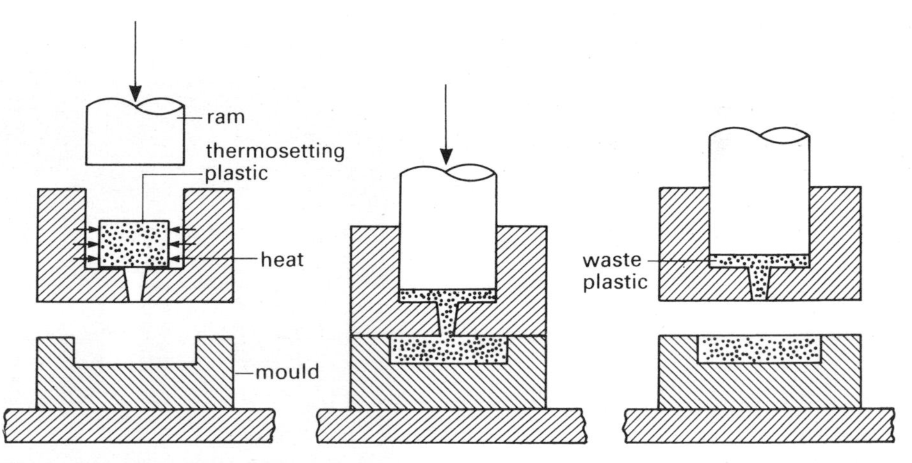

Figure 6.72 Transfer moulding.

There is a third variation of the injection moulding process known as the *RIM* (reaction injection moulding) *process*. The RIM process starts with the basic chemicals from which the plastic is made and hence eliminates the stage of forming powder or pellets. The chemicals are mixed and then injected. They react under the heat and pressure to produce a rigid plastic moulding.

The similarity of all three of these processes to high-pressure die-casting should be apparent. The designer faces the same difficulties in considering the feasibility of the tooling. It is the inverse of the required shape which is needed in the mould tool. Fig. 6.73 shows two designs for injection mouldings intended to act as connectors for everyday articles, drinking straws and lolly sticks. These were intended to be used in construction kits for primary school pupils.

Fig. 6.73(a) shows the production of triangular holes to grip the straws. This might at first be considered a difficult operation, until it is realised that all that is required is to file a circular insert to give a triangular cross-section. Fig. 6.73(b) shows the use of rectangular inserts to create the rectangular holes for the lolly sticks.

Fig. 6.74 shows the way in which these processes can be used commercially. Virtually all the components for the alarm clock – the casing, mechanism and even the hands – have been manufactured by injection moulding. A pressed steel spring can be seen holding the plastic gears in place.

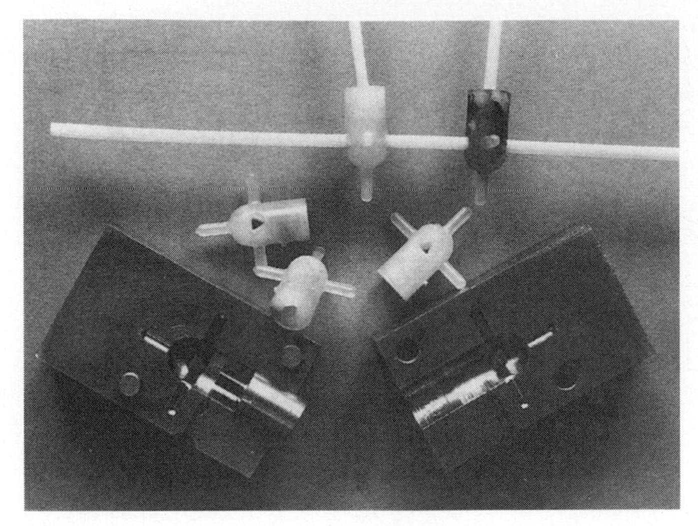

(a)

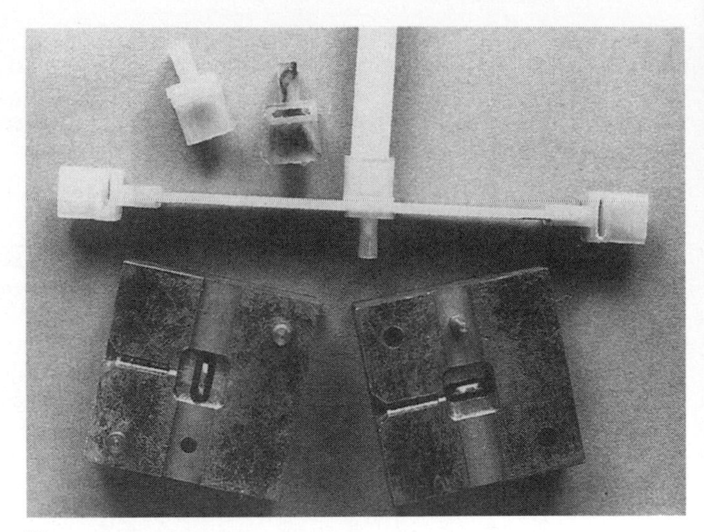

(b)

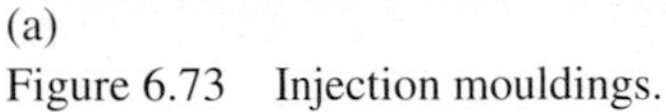

Figure 6.73 Injection mouldings.

Figure 6.74 Injection moulded alarm clock components.

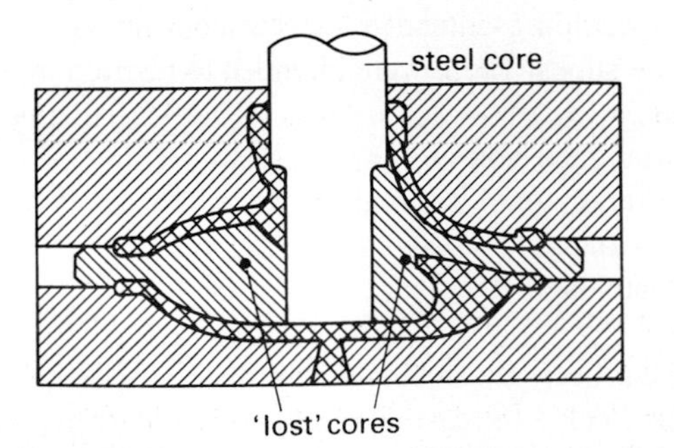

Figure 6.75 The use of 'lost' cores with injection moulding.

Injection moulding can be used to make very small and very large products. One of the smallest must be a gearwheel used to drive the seconds hand of a wrist watch which has a diameter of 1.3 mm and a mass of only 0.00056 g. Among the largest are a 37 kg garbage container, the body of the Sinclair C5 and the hull of a sailing dinghy which have all been made by injection moulding. The potential of the process is, however, still being developed. Water-meter housings would have once been made in cast iron or hot stamped in brass, but injection moulding plastics is now an option. Fig. 6.75 shows how the required shape can be achieved.

Some of the internal cores are steel components – the others being *lost cores* made from low melting point alloys. After the plastic has cooled the metal lost cores are melted out by using a high-frequency field to induce eddy current heating. Pump impellers are produced by the same method.

It is also possible to produce multi-component injection mouldings. Common examples are ball joints and the push-button pads used in telephones. The push-button pads are made with a twin mould. The numbers or lettering are injected in a bold coloured plastic in the first mould. The buttons and the remainder of the pad are then injected round them in the second mould. Similarly ball joints can be produced by injecting plastic round a steel or plastic ball. A steel ball would normally be preheated to provide the necessary clearance when the ball contracts on cooling. With a plastic ball the clearance needed can be obtained from the shrinkage of the injected plastic material.

6.4 Wasting techniques

Almost any shape can be cut out of a solid mass of any material by some technique. If designers choose to use such a method in order to create their product, then their role is to facilitate the task so that it becomes cost-effective. Every effort must be made to minimise the amount of material which needs to be removed and to simplify its removal. This means, wherever possible, designing around standard sections, conventional machinery, standard tooling and workholding assemblies. The design of special jigs and fixtures as part of the production process is quite common, but there must be a very clear justification.

When shapes are produced by removing material from a solid mass this is generally known as a 'wasting technique'. In many cases it may be possible to use or recycle the material removed, but certainly the swarf produced from machining steels is difficult to reprocess. Wasting should not, however, be thought of only in terms of the cutting of metals. Wood and plastics can be machined, clay jiggered, ceramics strickled, stone carved and metals can be shaped by electrical, chemical and thermal methods. Ultrasonic machining can be carried out on gems, ceramics and sintered carbides as well as on the most brittle metals.

The problems associated with designing for a wasting technique can be considered in two areas: firstly, how the material is going to be removed and, secondly, how the workpiece is to be held, the tool manipulated and provided with power. These latter problems relate to the design of jigs, fixtures and machines. Although special jigs and fixtures might well be made up in order to produce quite a small quantity of components, building a special machine of any complexity can only be justified for very high production volumes. It is clearly essential for designers to develop an awareness of the capabilities of existing machines.

6.4.1 Methods of material removal

The methods available for removing materials can be categorised into four groups: mechanical, electrical, chemical and thermal methods. Mechanical methods are by far the most common and are described first. Electrical methods find applications in working difficult materials and chemical methods in dealing with large components. Methods relying entirely on heat are used to cut shapes from metals and plastics and are important production processes.

Mechanical methods

You will probably be familiar with mechanical cutting using wedge-shaped tools and perhaps form tools, but you should remember that abrasive particles, whether bonded in a solid or flowing in a fluid, also have a cutting action. The wedge shape is the basis for the cutting action of the majority of hand tools, for example, chisels and saws, as well as that of HSS (high-speed steel) lathe tools and drill bits where they employ a positive rake. Under some circumstances it is possible to use a form tool which has a profile matching the required shape, but the potential size of the cutting force is very high. The cutting action of abrasive particles is the basis for the action of grinding wheels and ultrasonic machining, as well as being the foundation of the inserts used in modern machine tools.

Wedge-shaped cutting tools

The essential features of a wedge-shaped cutting tool are shown in the cold chisel illustrated in Fig. 6.76.

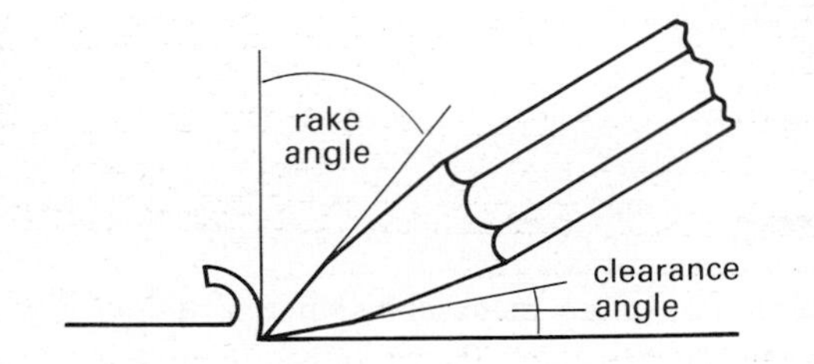

Figure 6.76 Rake and clearance angles.

The clearance angle is necessary to prevent the cutting edge rubbing along the metal surface. The actual clearance angle is less critical than the rake angle as this determines the cutting efficiency. If a cold chisel is manipulated to lower and raise the rake angle, the tool will be found respectively to dig in and come away from the workpiece. Efficient cutting depends on the rake angle being correct. With powered machinery, the rake angle will determine the magnitude of the forces needed for a given cut, and hence the power consumption and tool life.

Fig. 6.77 shows how the same wedge angle appears on hacksaw blades, shaper tools, lathe tools and drill bits.

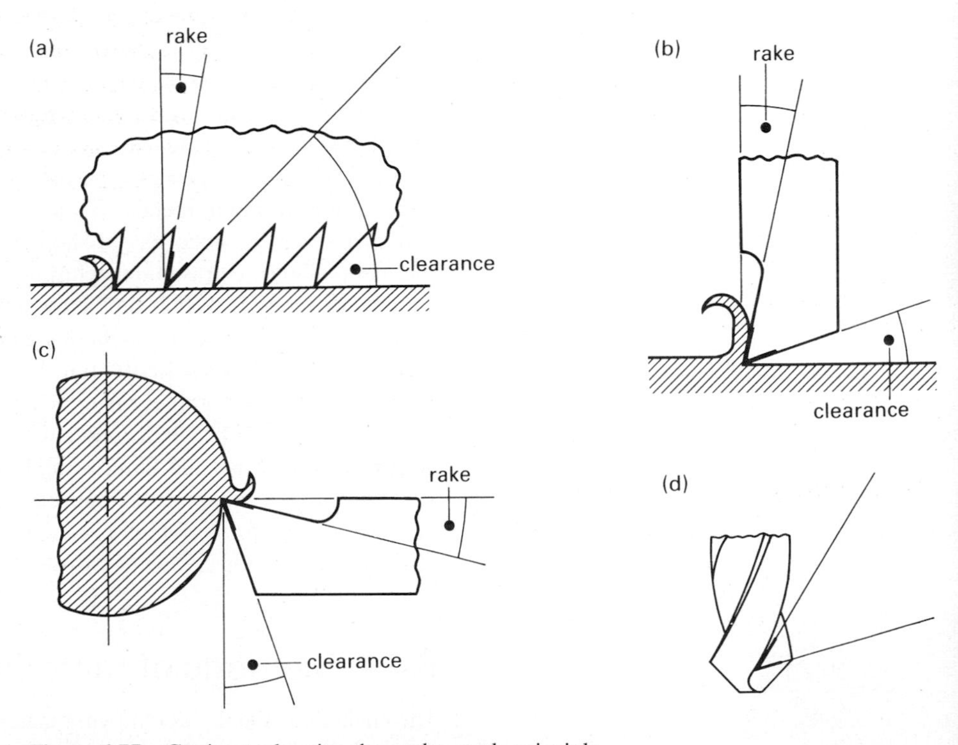

Figure 6.77 Cutting tools using the wedge angle principle
(a) hacksaw
(b) shaping tool
(c) lathe tool
(d) drill.

The cutting action is similar in all these cases, the material flowing away from the cutting edge in a thin ribbon. Sometimes this ribbon will shear under the action of the cutting forces, and sometimes the swarf is continuous. In this case it can be difficult to control and *chipbreakers* are then introduced to break it deliberately. These normally consist of a step ground parallel to the cutting edge across the face of the tool over which the material flows. It is sometimes necessary to provide a *secondary clearance* angle on the wedge shape, where the relative movements of the tool and workpiece after cutting would bring them into contact. Fig. 6.78 shows this for a rotating workpiece, a lathe boring operation, and for a rotating tool, a side and face milling cutter. In the latter case it is necessary to provide side clearance on the tool. This is also the case when a parting off tool is used to cut a groove. These side clearance angles are also illustrated in Fig. 6.78.

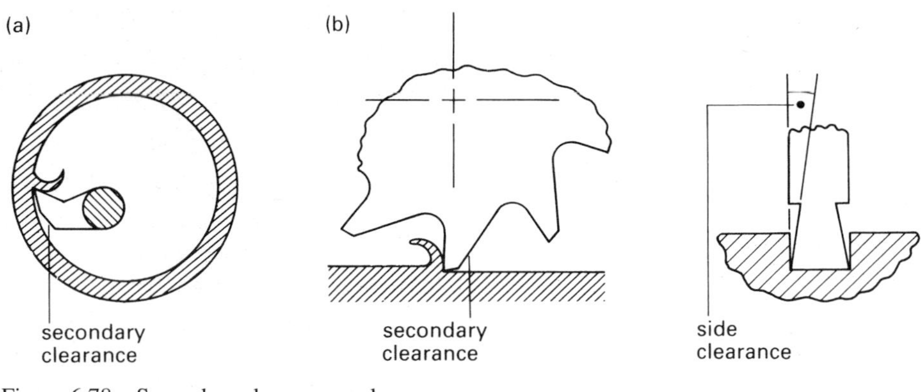

Figure 6.78 Secondary clearance angles on
(a) a boring bar
(b) a side and face milling cutter and
(c) a parting off tool.

The primary cutting force acts at right angles to the cutting edge and, in order to reduce its effect, the cutting edge is often turned away from the workpiece as shown in Fig. 6.79. This is known as 'oblique cutting'. A component of force is produced acting towards the toolholder which is much easier for the tool to withstand. When orthogonal cutting using a knife tool as shown in Fig. 6.79(a) there will be substantial bending loads.

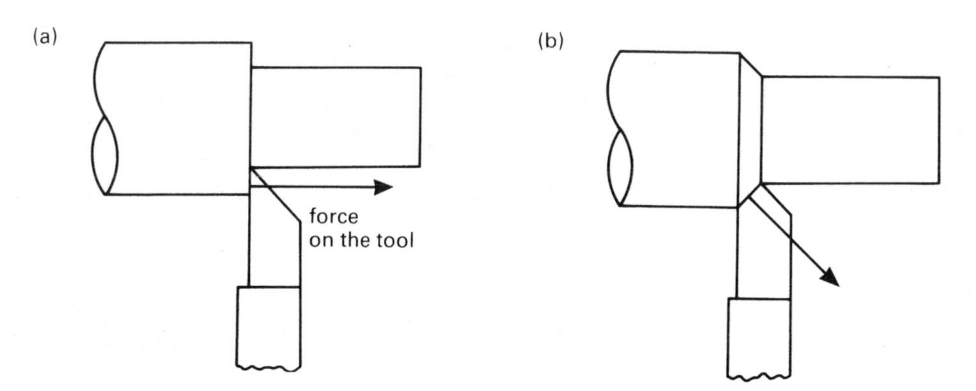

Figure 6.79 Forces on lathe tools.

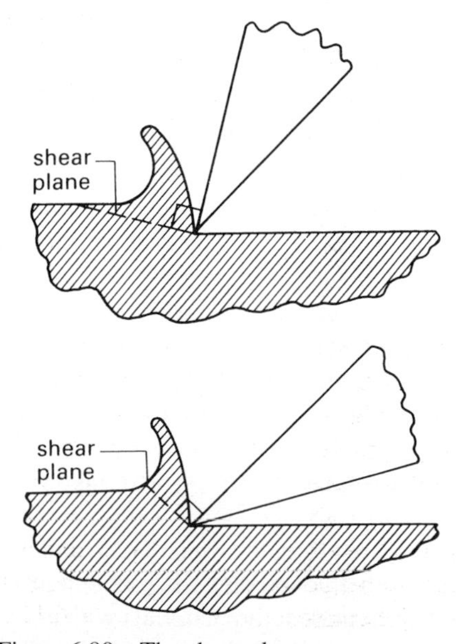

Figure 6.80 The shear plane.

In general, as the rake angle is increased, the cutting force and the power consumed will be reduced. This is because the shear plane is at right angles to the tool surface and its length will decrease as it approaches the thickness of the cut. This is illustrated in Fig. 6.80.

It might initially seem logical to use the maximum rake angle possible, but unfortunately it would soon be discovered that the tool wear becomes more rapid as the rake angle increases. Hence the angle chosen is a compromise which gives the most economic combination of cutting efficiency, power consumption and tool life. It will also be found that the cutting action will be most efficient at a particular cutting speed. For steel this is 25–30 m/min depending on its composition, for copper approximately 60 m/min and for brass a little less (about 50 m/min for free-cutting material). The cutting action is significantly influenced by the temperature, which increases with the cutting speed. It is sometimes necessary to use a cutting fluid in order to remove sufficient heat to ensure an adequate tool life, but flushing away the swarf, lubricating the cutting action and protecting the newly cut surface are equally important roles.

One difficulty which can occur is the build-up of material on the cutting edge, particularly if there is some chemical affinity between the tool material and the material being cut. This, of course, blunts the tool and, when the cutting force increases, the material built up eventually breaks away. Some of the resulting particles adhere to the swarf and some to the workpiece giving it a very rough appearance. This problem can occur between steel and certain aluminium alloys and cannot easily be resolved without using a carbide-tipped tool.

Tipped tools

Carbide tips are made from a mixture of carbide powders and cobalt by the standard powder metallurgy technique (see section 6.2.5). They are very much harder than HSS or carbon steel and retain their hardness well at high temperatures as shown in Fig. 6.81.

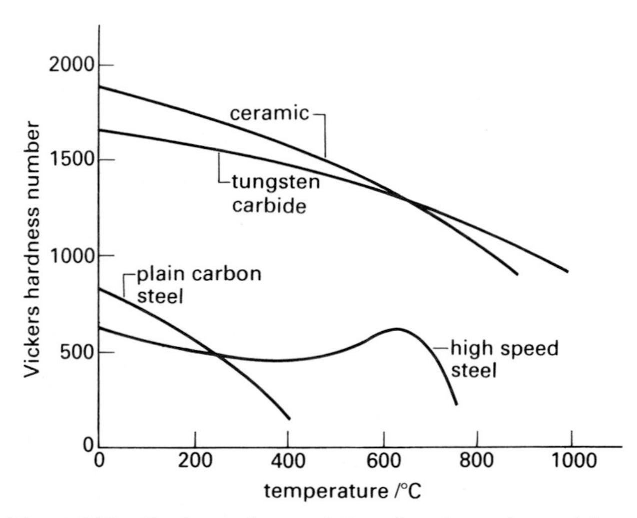

Figure 6.81 Hardness characteristics of cutting tool materials.

From a product designer's point of view their primary significance lies in their ability to machine materials which can prove difficult using conventional techniques. This means that most materials can be machined at a cost. Tungsten carbide tools are most suitable for machining brittle materials where there is no metallic adhesion: cast iron, wood, stone, glass etc. Mixed carbide tools, particularly those containing titanium carbides, are best suited to machining steels, where there is usually a difficulty of a built-up edge forming on a tungsten carbide tool.

From the machinist or tool designer's perspective, carbide tipped tools present a different set of problems. The tools, although they retain their hardness, tend to be brittle and have little resistance to impact loads. Vibrations and shock loads must therefore be avoided. The tools also operate best at much higher cutting speeds. With a positive rake tool, finishing speeds may be up to several hundred metres per minute. Fig. 6.82 shows the effect of using a negative rather than a positive rake with a carbide-tipped tool.

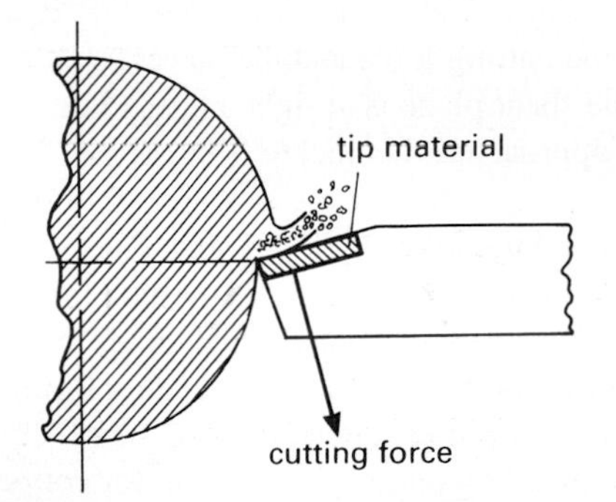

Figure 6.82 Using a negative rake.

The direction of thrust is altered and the steel shank offers the tip much greater support. This technique was originally developed to help overcome the problem of material building up on the cutting edge when machining ductile materials with a positive rake, but was soon discovered to have other advantages. More heat is carried away by the waste material and hence coolants are not needed. At high speeds the heat helps to soften the chip and hence assists in its removal. Negative rake cutting is therefore carried out at speeds two or three times those used with a positive rake and is most effective for machining ductile steels.

Tipped cutting tools are available in even harder materials than sintered carbides. Diamond tips can be used for producing a superfine finish on difficult materials – for example, cast aluminium pistons or some plastics. As they are very brittle, they must be used at very low cutting forces and care must be taken to avoid shock loads. The diamonds are usually brazed to a steel toolholder. Ceramic tips are also available, typically made from aluminium oxide which, although it is very hard, cannot easily be brazed. They are therefore manufactured in a form which can easily be clamped mechanically to the toolholder. Both diamond and ceramic tips need to be used at very high speeds and consequently the machines used must be very powerful. As it is important to avoid vibration and chatter, the machines must also be very robust.

Form tools

The use of a tool which matches the required shape means that the cutting action is not associated with a point, but with a much longer edge. Clearly the magnitude of the forces generated could now be very much greater and it is only possible to cut soft materials or use very fine feeds with a long cut profile. Fig. 6.83 shows a variety of situations where form tools can be employed. Fig. 6.83(a) shows a screw-threading tool being fed directly into a workpiece at right angles to it. The tool is normally fed at a slight angle to avoid contact on the back cutting edge, but the operation is sometimes carried out as indicated. Producing hemispherical ends and industrial wood turning are other common applications.

(a)

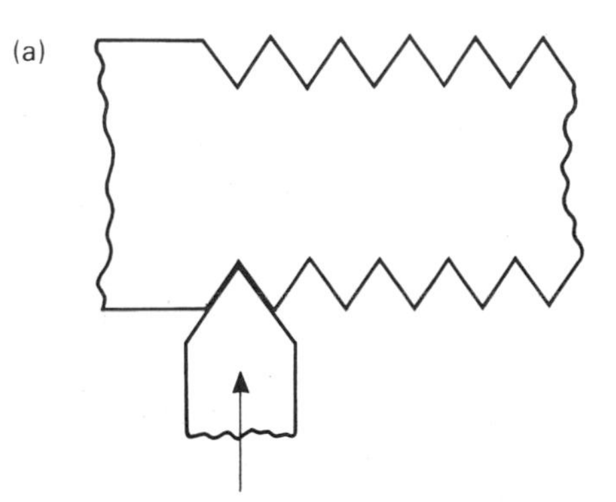

(b)

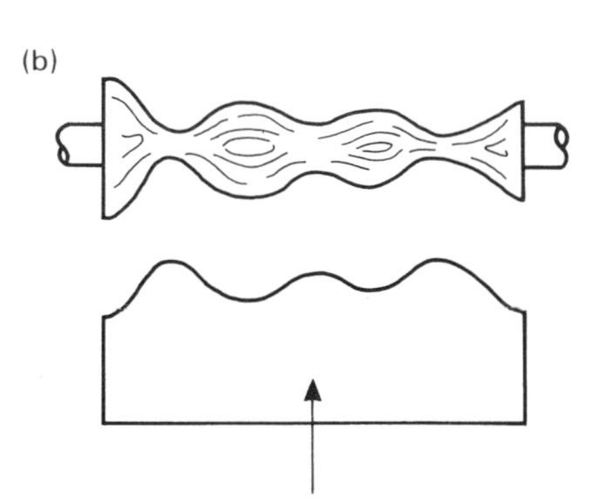

(c)

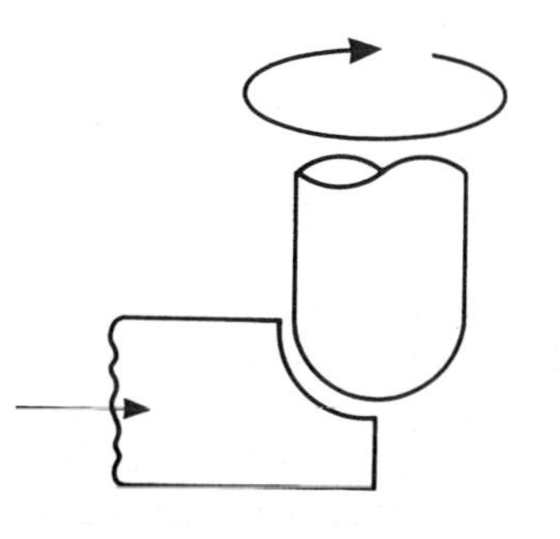

(d)

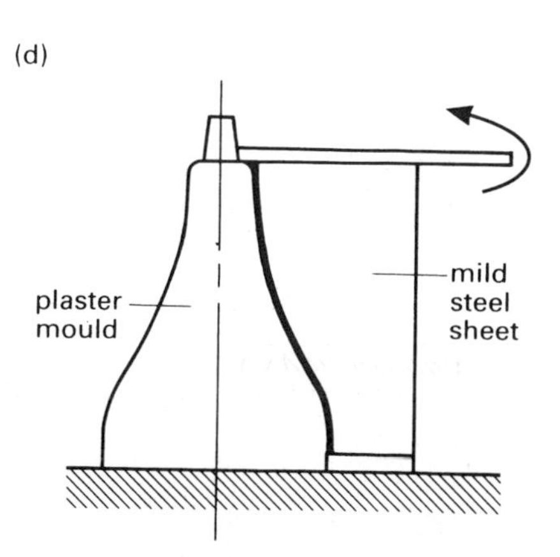

Figure 6.83 Using form tools.

Soft materials like plaster and clay can be shaped easily and the form tool only needs to be supported by hand. With clay this is called 'jiggering' and with plaster 'strickling'. The latter process is commonly employed in producing axially symmetrical formers for GRP moulding. Small spherical surfaces and threads are sometimes produced on metals in this way, but the tool restraint must be very rigid and feeds very small.

Grinding wheels

Grinding wheels consist of abrasive particles bonded together in a matrix. The abrasive particles are normally either silicon carbide or aluminium oxide, of which emery and corundum are naturally occurring impure forms. The size of the particles can be graded from quite a coarse-sized grit to very fine 'flour' grades, which are typically used by jewellers. The bond can be of many forms: a vitrified clay is the most common, but vulcanised rubber, synthetic resin, sodium silicate and shellac are also used. The more flexible bonds are used for slitting discs and fine finishing. The bond is chosen so that when the cutting edge on the abrasive particle becomes blunt it is torn from the wheel surface allowing new sharp particles to take up the cutting. This means that the wheel selected for any application must be carefully chosen to be right for the material and finish required. Grinding can give a very fine surface finish, but represents an additional operation and consequently it should only be specified with good reason. Designers must consider carefully whether or not the finish 'as cast' or 'as machined' is adequate for their purpose.

Ultrasonic machining

Fig. 6.84 shows another way of cutting using abrasive particles – ultrasonic machining.

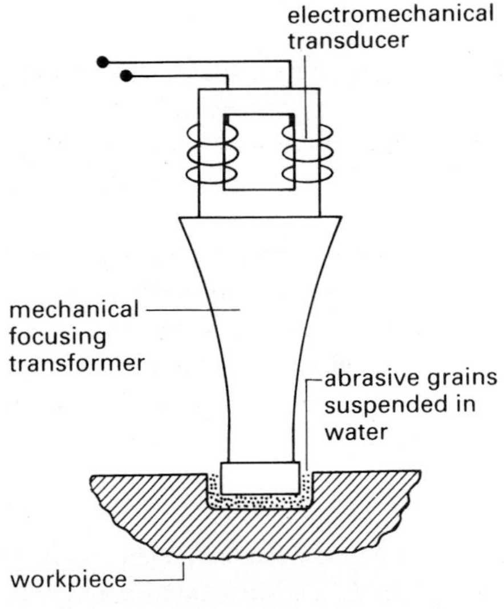

Figure 6.84 Ultrasonic machining.

A high-frequency mechanical oscillation is produced using an electromechanical transducer and these vibrations are then focused to intensify their effect. A tool of the required shape is fixed to the mechanical amplifier so that it vibrates in a direction perpendicular to the workpiece surface. Abrasive particles in a slurry are pumped under the tool and impacted into the surface, which is gradually removed. Both the amplitude of the vibrations and the grit size are about 0.02–0.07 mm.

Ultrasonic machining can be used on almost any material and is particularly useful for working on the very hard tool steels used for extrusion, press and forge tools. It can also find application with very brittle materials like glass, ceramics and precious stones. Ultrasonic machining can give an excellent surface finish and dimensional accuracy of ± 0.005 mm. The maximum penetration speed that can be expected is around 20 mm/min.

Electrical methods

There are two primary electrical machining methods: arc-discharge machining (or spark erosion) and electrochemical machining.

Arc-discharge machining

Arc-discharge machining works as a result of the eroding effect of arcs (or sparks) formed between the electrode and the workpiece. The arcs are generated at voltages from 20–500 V and frequencies from 1000–2000 Hz. The whole of the workpiece assembly must be immersed in a dielectric liquid, such as paraffin, so that the arc discharges are quenched rapidly. Sometimes the liquid is fed down the centre of a hollow electrode in order to wash away the swarf. Electrodes are made in the shape required – usually from brass. They are made slightly undersize as the machined area is a little greater than the electrode dimensions. However, with the correct allowances, tolerances of about 0.01 mm are achievable. Fig. 6.85 shows a typical arrangement.

The major advantage of the arc-discharge method is its ability to machine very hard materials like sintered carbides and hardened tool steels. It is an ideal way of producing

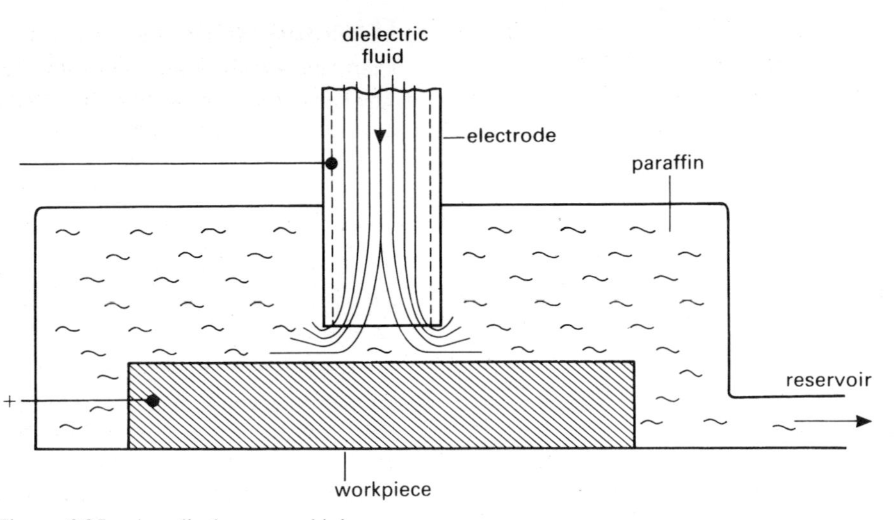

Figure 6.85 Arc-discharge machining.

holes in drawing and extruding dies, nozzles and similar components made from very hard, wear-resisting materials. The surface finish is dependent on the energy content of the discharge and, hence, the rate of metal removal. With a low current level and high frequency the surface finish can be very good. The major disadvantages of the process are the fact that it can only be used on materials which conduct electricity and the high rate of electrode wear. For small-diameter holes the weight of the electrode consumed can be even greater than the weight of material removed.

Electrochemical machining

Electrochemical machining depends on electrolytic action. A current is passed from a copper cathode to the workpiece through an electrolyte – normally salt water. The voltage is kept low enough to prevent arcs from occurring, and consequently the machining action depends on the metal dissolving at the anode. Hydrogen is liberated at the cathode.

The arrangement of an electrochemical machining tool is very similar to that of an arc-discharge machine. The electrode is again produced in the shape required, but this time usually from copper. The process has similar advantages to arc-discharge machining, but can give a finer surface finish. Also the electrode is not worn away by the electrolytic action. The disadvantage is that very high power levels have to be employed to achieve acceptable rates of metal removal.

Chemical methods

You may already be familiar with etching processes in one form or another – perhaps associated with the production of prints or circuit boards. The technique depends on masking those areas which are needed and exposing the remainder of the surface to chemical attack. This approach can be particularly useful in industry because of its ability to deal with large areas and with both sides of a component simultaneously. There are, of course, also no tooling costs. It was originally developed in the aircraft industry for use with aluminium alloys, but can be equally effective with magnesium and titanium. Different etchants and working temperatures are required for each material, but a removal rate of about 1 mm depth per hour is generally achievable.

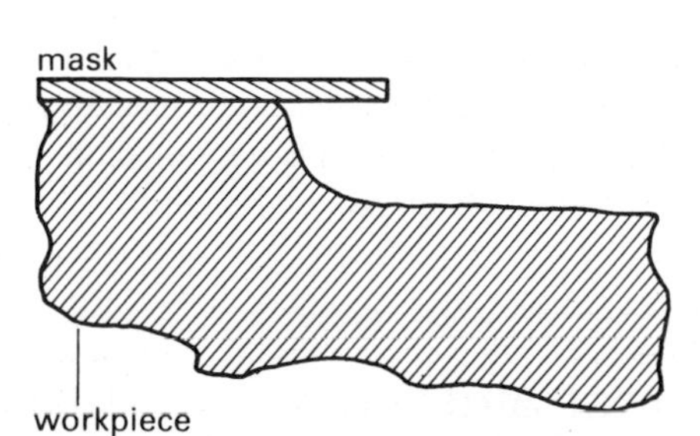

Figure 6.86 Undercutting in chemical machining.

The tolerance which can be achieved depends primarily on the accuracy of the masking operation. The undercutting which will occur is illustrated in Fig. 6.86. Typically the undercut will extend to between 1 and 1.5 times the depth of the etching, but this is slightly variable and results in tolerances of around ± 1 mm on length dimensions. On thicknesses an accuracy of about ± 0.10 mm can be expected as well as a good surface finish.

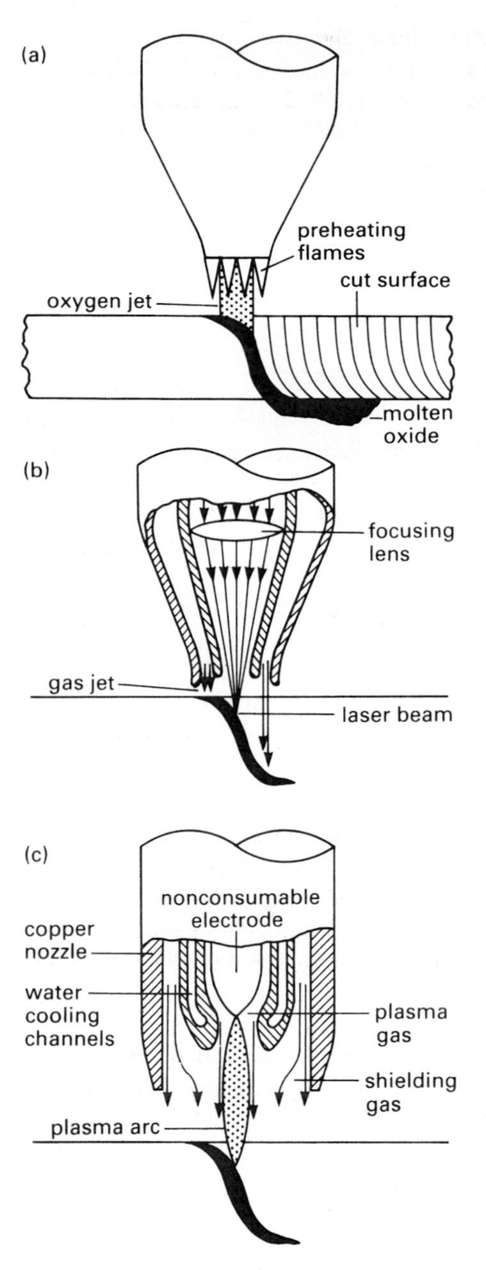

Figure 6.87 High-energy heat sources for thermal cutting
(a) oxyacetylene
(b) laser
(c) plasma arc.

Thermal methods

Thermal methods of removing material depend on melting or vaporising the material they work. There are two common types of process, cutting with high-energy heat sources and cutting with hot wires, and these are described below.

High-energy heat sources

There are three primary heat sources which are used for cutting materials: a mixture of oxygen and ethyne (acetylene) gases, lasers and plasma arcs. These are illustrated schematically in Fig. 6.87. All three processes are used to cut shapes from flat sheet or plate.

Oxyacetylene cutting is used on steels and depends for its operation on the exothermic reaction that occurs between steel and pure oxygen above 875°C, which is the ignition temperature of iron. This is why the process is commonly known as 'burning'. Once the preheating flames have heated the top surface sufficiently, the oxygen jet is turned on and the exothermic reaction which occurs helps to keep the oxide formed molten. This oxide is very fluid and it flows freely from the cut under the action of the oxygen jet. If the tool is traversed at the correct speed a continuous cutting action will he sustained. The process can be used on steel plate from 6 mm to 150 mm thick.

When laser-cutting steels, a jet of pure oxygen is used to blow away the molten material. Clearly the heat generated by the exothermic reaction will help the laser to heat the material and the jet action again helps to blow the oxide from the cut. Lasers are, however, also used to cut other materials such as nonferrous metals, gasket materials and textiles. For these applications a jet of pure nitrogen is used to remove the material instead of oxygen. The cut width can be less than 0.5 mm and consequently there will be less distortion than with oxyacetylene cutting, but it is limited to thicknesses of about 20 mm.

The use of plasma arcs for cutting materials is a comparatively recent innovation. The process was developed in the late 1950s for cutting stainless steel and aluminium. An arc is struck between the nonconsumable electrode and the workpiece (see section 6.5.2) and this is then constricted by a flow of gas plasma – typically argon or hydrogen. This process can cut materials up to 150 mm thick, but it has never really been considered as a competitor to the oxyacetylene cutting of steels because of the cost, the poorer finish, the fumes and the noise. Recent developments using air as the shielding gas and injecting water into the plasma arc have however reduced costs and improved the squareness of the cut. It is also possible to operate the whole process underwater which reduces the noise to virtually background levels and also alleviates the fume problem.

Hot-wire cutting

Materials like polystyrene can be cut effectively using a hot wire. This is heated by a low voltage supply and the current is limited to prevent the temperature of the wire rising towards red heat which would result in the formation of styrene fumes. If operated correctly the hot wire will leave a very smooth cut surface. Although the process is again primarily used to cut shapes from flat sheet, it can also be used in a more sculptural way to shape polystyrene. This is likely to be the primary method used in producing one-off expendable polystyrene moulds for use with the full mould casting process (see section 6.3.1).

6.4.2 Machine designs and restraint

Mechanical power is associated with a force and a movement; electrical power with a voltage and a current. It follows that the design of machines for electrical methods of material removal is concerned with the location of the workpiece and the maintenance of the correct gap between the electrodes which, typically, determines the voltage. The design of machines for mechanical methods of material removal depends on providing a relative movement between the tool and the workpiece and sufficient restraint of both the tool and the workpiece to withstand the cutting forces involved. The implications for the product designer of these relative movements, the standard tooling and the restraint requirements for mechanical material removal methods are discussed below.

Relative workpiece and tool movement

The relative movement necessary for developing mechanical cutting power can be generated through either a linear or a rotary action. A linear relative movement is used on planers and shapers – the tool being stationary on planing machines and the work being stationary on a shaper. Many machines use a rotary movement. On lathes the tool is fixed and the workpiece is rotated: on millers, grinders and drills the tool is rotated and the workpiece restrained. In order to cut a surface or contour it is necessary to reposition the cutting tool, either in steps by some form of indexing, or continuously. Simple machines will use steps in a particular direction or about a particular axis: more complex machines will be able to move or turn about several axes simultaneously.

Designers must ensure that they are aware of the types of machine available and how flat and curved surfaces can be created. Efficient machining operations must minimise the number of operations and machine settings and provide for the approach and the overrun of the tool. Parallel surfaces only require one dimension, say a depth setting, to be altered. Nonparallel surfaces will require a more complex alteration to the machine settings. It may be a question of altering the angle of a rotating table on a milling machine, or changing the position of a taper turning attachment on a lathe before the cut depth is set. The shape of the surface which is needed is also very important. Clearly flat, cylindrical or conical surfaces can be produced by moving the tool in a single direction. Unless the use of a form tool is feasible, spherical, elliptical or similar shapes require a point tool to move in a curve. This can be achieved by using a mechanism to force the tool to follow a template or by driving two axes simultaneously under computer control. Clearly such operations are more costly than those for the production of simple surfaces.

Standard tooling

Designers need to have a thorough knowledge of the sizes and capabilities of standard tooling. Lathe tools working on the outside surface of a component present few restrictions, but when boring an internal diameter, a hole large enough for the boring bar must be present. This can be a significant restriction because deep holes with limited access might mean using a small-diameter boring bar with little stiffness. The consequential vibrations and tool chatter would result in a poor surface finish. The stiffness of milling cutters is also an important issue. Thin large-diameter cutters for horizontal millers need to be supported by thick washers and only shallow cuts are possible with small-diameter vertical milling tools.

Many years ago, matching pairs of male and female threads were made with tools which were ground by individual craftsmen. The result was slight variations in the thread angles, which meant that if a nut or bolt was lost a special replacement had to be manufactured. Preformed screw-cutting tools are now available for the standard thread forms, and there are significant reductions in both the initial and replacement costs if designers select these screw threads. With good reason designers can, of course, specify a nonstandard thread, but they should be aware of the consequential penalties.

THE LIBRARY
GUILDFORD COLLEGE
of Further and Higher Education

Restraint

Workpieces are likely to be held in a machine vice, a chuck, a collet or between centres. In any of these cases the designer needs to make suitable provision. Machine vices can be used in two ways, either on their own or in conjunction with a V-block. Fig. 6.88 shows some of the most common configurations.

The arrangement shown in Fig. 6.88(a) cannot be used safely on a rough surface and a circular rod or knife-edge is inserted between the workpiece and jaw in order to give more reliable contact. In components designed with this form of restraint in mind it is important to provide sufficient material to grip and also to avoid conical or tapered forms. In order to hold cylindrical objects a V-block is usually inserted to give three-point contact. Fig. 6.89 indicates how the form of a component can easily be altered to simplify restraint. In the original form the sharp edge between the two conical surfaces would inevitably be damaged by the clamping forces and there would be a significant risk of slip. The method shown in Fig. 6.88(c) is used to clamp components directly to the tables of drills and milling machines.

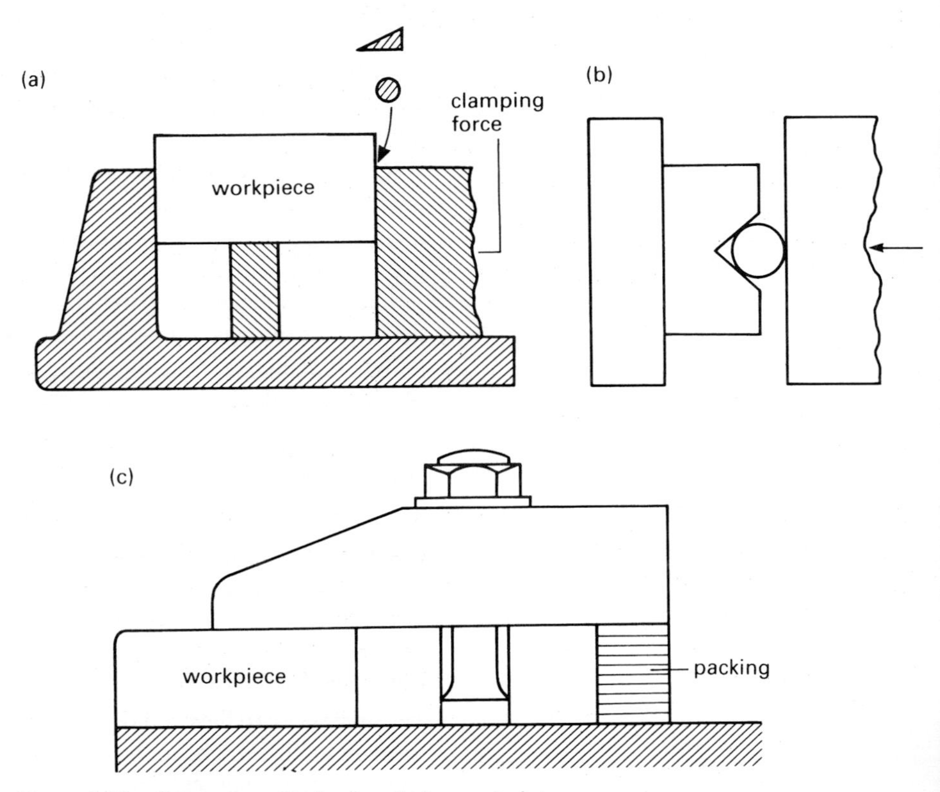

Figure 6.88 Common methods of workpiece restraint.

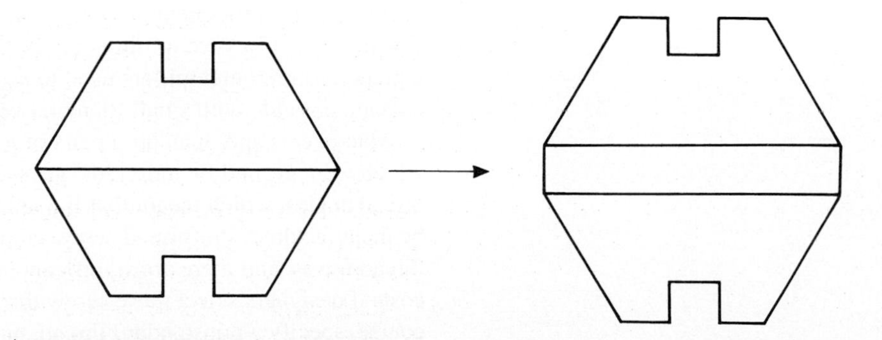

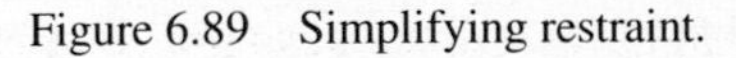

Figure 6.89 Simplifying restraint.

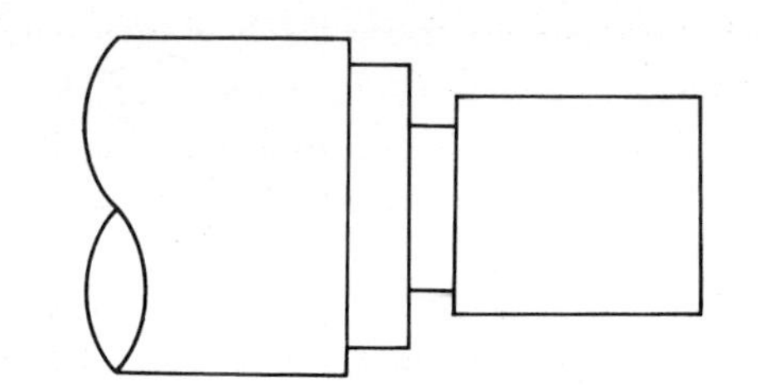

Figure 6.90 Turned component.

Chucks can have three or four jaws and these may close independently or together. It is therefore possible to hold most shapes, but there are two potential difficulties. Firstly the surface which is gripped can be damaged and secondly there must be sufficient material (approximately 20 mm) in the chuck in order to hold it properly. Components like the one shown in Fig. 6.90 would, therefore, be machined whilst still on the rod and then cut (or parted) off. Similarly if a component is held between centres, either the holes in the ends must be tolerated or they must be cut off and discarded as shown in Fig. 6.91.

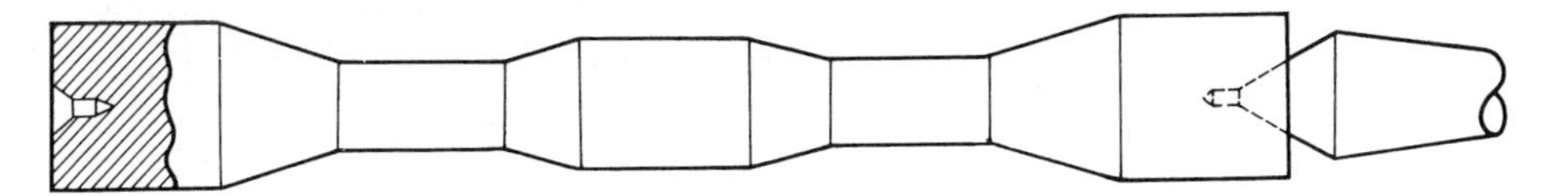

Figure 6.91 Holding work between centres.

6.5 Fabrication techniques

Previous sections of this chapter have been concerned with the manufacture of a component in a single piece. For a variety of reasons it can be advantageous to produce a component in a number of parts and then to join them together. This is known as 'fabrication'. Some of the reasons that fabrication can be the most effective approach are listed below:

- a combination of materials with different properties may be required,
- the mould tools for an injection moulding or casting may be significantly simplified,
- substantial weight savings are possible in large castings because there is a limit on the minimum section size which will allow the molten metal to flow,
- thicknesses, widths or lengths can be built up where material of sufficient size is not available,
- much of the material removal associated with a wasting technique can be avoided,
- standard components can be incorporated into the design.

Whatever the reasons for employing a fabrication technique, the particular difficulties lie in the joining of the parts. These may have been produced by any of the processes described earlier in this chapter. Fundamentally there are three approaches to joining components: mechanical methods, thermal methods and adhesives. The materials processing aspects of these three methods and some of the associated design considerations are discussed below.

6.5.1 Mechanical joints

The classical methods of joining materials mechanically were developed by craftsmen many centuries ago. Woodworkers developed a large variety of joint configurations for use in different circumstances, some allowing movement as the wood expands and contracts with changing humidity, some being combined with adhesives and others incorporating a mechanical wedge action. Fig. 6.92 shows a selection of some common wood joints.

From a designer's viewpoint, the key requirements are to ensure that the joints selected have appropriate mechanical properties, such as strength and stiffness, and can be produced using the available woodworking tools. These may be hand tools or power tools like circular saws and routers.

Early metalworkers developed riveting methods and forge welding techniques, but also adopted some of the joint configurations from the woodworkers' practice. Perhaps the best known example of this is the joining of the cast iron sections at

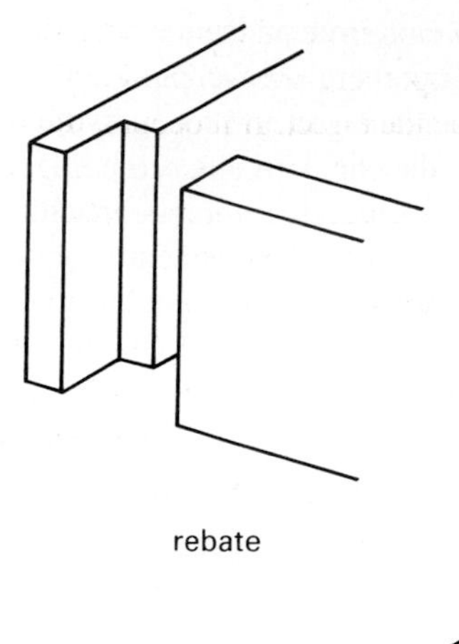

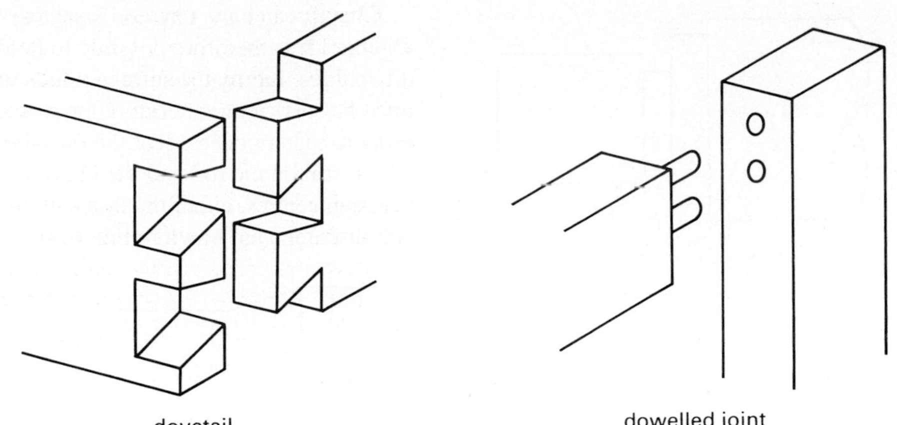

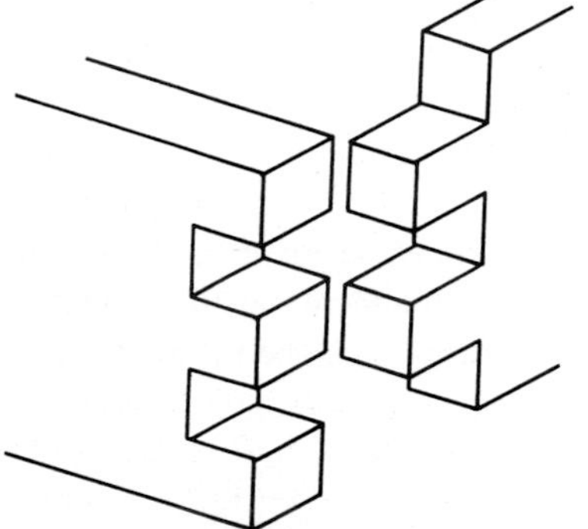

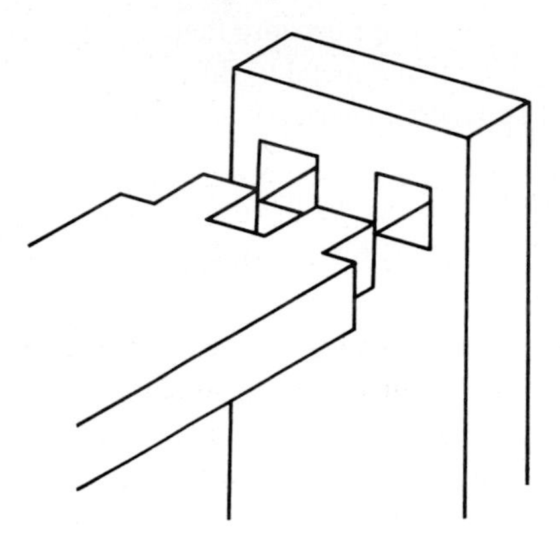
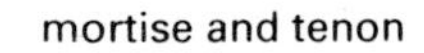

Figure 6.92 Some common wood joints.

Ironbridge. The bridge was completed in 1780 and its builders used dovetails and mechanical wedges in its fabrication. Clearly there have been many developments since then, and Fig. 6.93 shows a selection of the modern mechanical fasteners available to designers.

Most of the fasteners require only the production of holes or grooves in the metal, but it is sometimes necessary to machine flats for bolts or rivet heads. Mechanical fasteners are under continuous development and improved designs are constantly coming on to the market, particularly for knock-down furniture and self-assembly products.

The major advantage of mechanical fasteners is their independence of the materials being joined. Parts can be produced from metal, plastic, wood or ceramic and the joining process does not require any mixing of the materials or adhesion between the surfaces. There is also no change in the mechanical or other properties of the components as a result of the joining process. The same cannot be said of most thermal methods or adhesives. However, even though mechanically joined components are not in direct contact it is possible for dissimilar materials to react chemically in the presence of a suitable electrolyte. Unprotected steel in close proximity to aluminium, for example, can result in corrosive action (see section 6.6.1).

6.5.2 Soldering, brazing and welding

Soldering, brazing and welding are thermal joining methods which make it possible to produce joints directly between components, that is, without any fasteners such as springs, clips or screws. There are a large number of welding processes which are discussed later in this section, but fundamentally they can be divided into two categories: those in which the materials being joined (the parent materials) are melted and those which occur in the solid phase. When the parent materials melt, it is known as *fusion welding*. Bonds in solid phase welding processes develop between the atoms at the interface of the two components as they are brought into close proximity at raised temperature and pressure. Soldering and brazing also take place without the parent materials being melted, but in these two processes a filler material is melted between the parts being joined and forms a bridge between them. Fusion welding is fundamentally different from all the others in that the parent materials are mixed in the molten state. The fusion welding of dissimilar materials thus depends on what happens when they are mixed as liquids and then resolidify. Because it depends on melting, fusion welding is confined to metals and thermoplastics. Brazing is, of course, also used with ceramics (for example, tipped lathe tools).

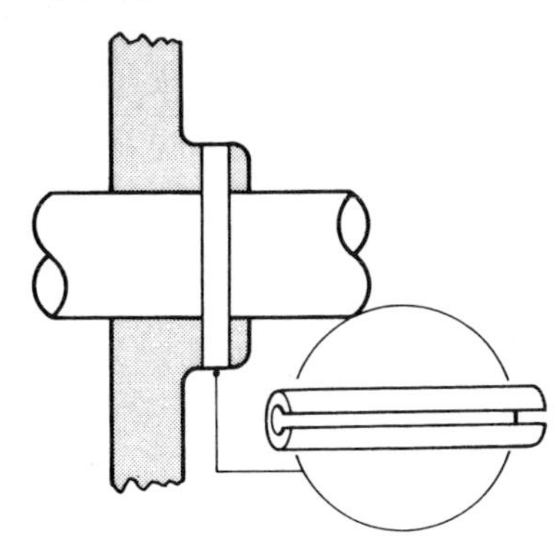

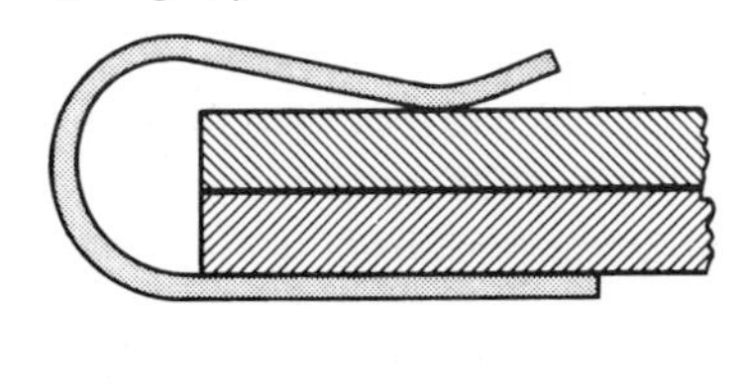

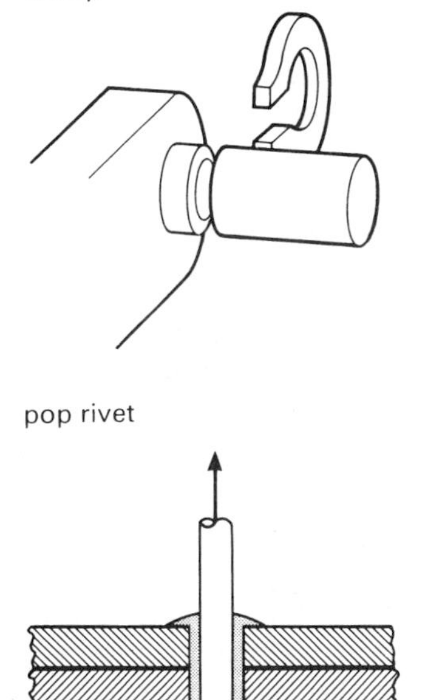

Figure 6.93 Mechanical fasteners.

Carrying out any of these processes requires the parent materials to be heated. Heat is often deliberately used to alter the material properties (see section 4.3.5) but, when using thermal joining methods this is more of a side effect. If it occurs, the change in properties will only take place close to the joint where the temperature has been raised sufficiently and, consequently, there will be a variation in properties across the section. Clearly this is much more likely to occur when welding or brazing, which uses fillers that melt at a higher temperature, than with soldering, where the filler normally melts at only a few hundred degrees. Generally the result is a softened or annealed zone either side of the joint, but there can also be regions of considerable hardness resulting from rapid cooling. Such variations in properties can only be removed by heat-treating the entire assembly after the joining operations are complete.

Soldering

The term 'soldering' should only be used when the filler material melts below 450°C. The molten filler metal is drawn into the gap between the materials by capillary action, in which a liquid is drawn into a narrow gap by surface tension. Fig. 6.94 shows a number of joint types intended to promote this capillary action.

These joints could be used for soldering or brazing (which uses fillers which melt above 450°C). The capillary action depends on the molten filler flowing on the parent material surface and this is critically affected by its cleanliness. If the material is contaminated then the filler will not 'wet' (or flow on) the surface. It is therefore essential that the components to be joined are cleaned mechanically before heating and chemically by a flux during heating. The flux can both remove oxides from the surface and prevent their reforming. The mechanical cleaning is normally carried out by wire brushing or using emery cloth on the critical surfaces before assembly. These surfaces should also be degreased using a suitable solvent because fluxes do not remove oil or grease. The flux (for soft soldering this is usually zinc chloride) is normally applied first and the components then brought up to the necessary temperature before the solder is applied. Solder pastes are, however, also available where the solder has been produced in a fine form and made up in a suspension with the flux. Solder wire is normally used for electrical work in which the flux is contained within a metal tubing as a core.

The solder composition will not only affect its strength and other properties, such as the electrical conductivity, but also the manner in which it melts and the temperature at which it does so (see section 4.3.4). A filler of the eutectic composition melts and resolidifies at a single temperature, and therefore rather suddenly. At all other compositions there is a temperature range over which the alloy solidifies. During this temperature range the alloy is a pasty mixture of solid and liquid particles. It is in this semisolid state that the solder is 'wiped' by plumbers – shaping the joint before it solidifies. The selection of filler alloys is discussed in more detail in the section on brazing, as well as joint design to give the required strength.

The heat for soldering can be from a gas flame or soldering iron as the temperatures needed are quite low. Fig. 6.95 shows the joint configuration used for electrical work.

The soldering iron is applied on one side of the copper wire and when the temperature has been raised sufficiently the flux-cored wire is touched down on the other. The flux rapidly removes the oxide and the solder then runs through the joint.

Brazing

Brazing is similar in operation to soldering and, with appropriate filler materials, it can be used on nearly all metals and many ceramics. It can also be used on the complex composite materials produced by sintering powders.

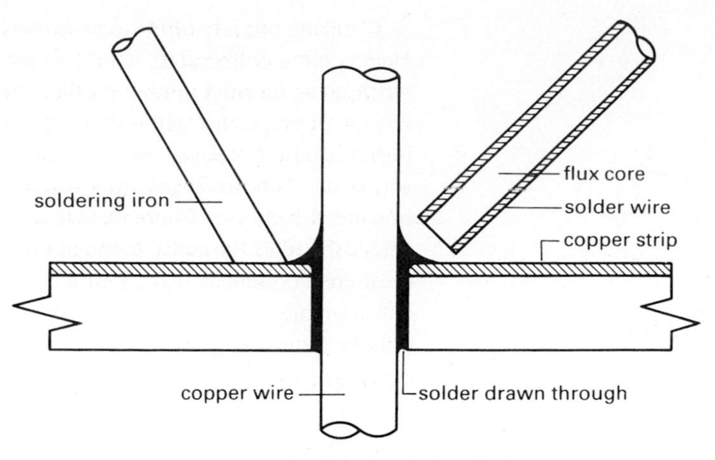

Figure 6.95 Arrangement used for electrical soldering.

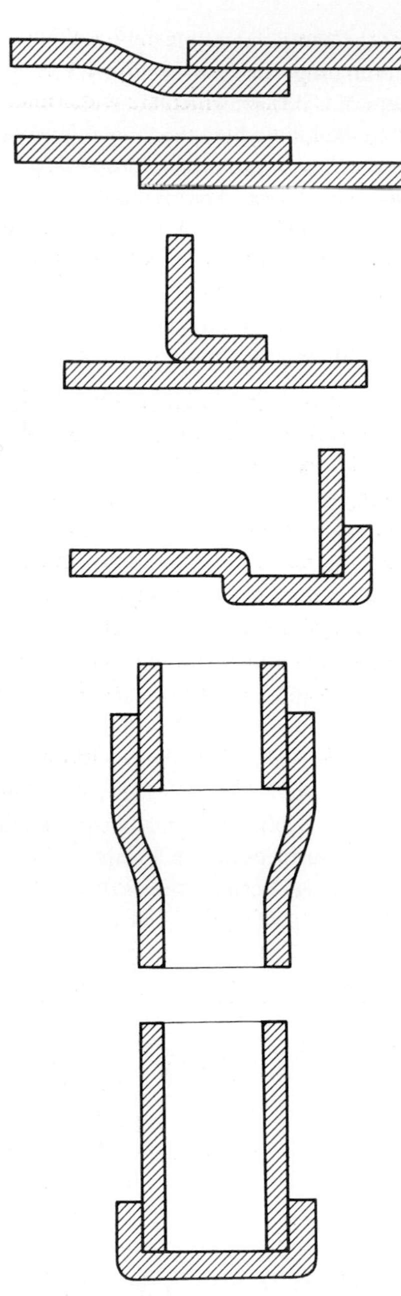

Figure 6.94 Joint configurations suitable for soldering and brazing.

The melting temperature of the filler materials is much higher than those used in soldering. This has the advantage that the filler alloys are of much higher strength, but makes the operation more critical for two particular reasons:

- The higher temperature means that the heating time is longer and hence the flux has more oxide to remove. It is possible for it to become spent or exhausted and, consequently, it is not only essential to select the correct flux but also to use plenty of it. Vacuum brazing is a good alternative.
- Different thermal expansion in the components is more likely to close the gaps and thus prevent capillary action taking place. The recommended joint gap is normally between 0.05 mm and 0.20 mm.

Filler alloys

Filler alloys for brazing are often known as 'hard' or 'silver solders'. The latter name indicates their most common constituent, silver, which is normally alloyed with copper and zinc to give a range of alloys with different melting ranges and strengths. About 20% cadmium is also often included to give more fluid alloys with lower melting points, but as cadmium is toxic the range of applications is limited and care must be taken to avoid overheating the filler. If there are a number of components to be brazed together and this cannot be completed in a simultaneous operation, then a series of fillers with decreasing melting temperatures can be used to make the joints in sequence. This is quite common practice in manufacturing jewellery. There is also a range of copper alloys with phosphorus as a major alloying element. Aluminium alloys require the use of special aluminium alloys as fillers.

Melting behaviour is the most important characteristic in selecting a filler alloy, eutectic alloys giving rapid flow. If, however, fillet joints, wide gaps or poor fit-up are likely to be encountered, then a wide melting range can be advantageous. The pasty stage will help the filler to form bridges across any gaps in the joint. Clearly alloys with higher silver contents will be more expensive. For every filler alloy there will be a proprietary flux which is active at the appropriate temperature for use with it. Some flux residues can be difficult to remove and joints should be designed with this in mind. The fluxes used on aluminium are actually corrosive and will damage the parent materials if the components are not effectively cleaned.

Joint strength

The joint strength depends on the strength of the filler alloy and correct execution of the brazing operation, but equally it is important that the designer ensures there is sufficient joint area. In general, the maximum tensile strength for a joint is developed when the overlap is four times the thickness of the thinnest component, but this varies with the parent materials. For example, with copper alloys the maximum strength is developed with overlaps of 2–2.5 times. Overlaps which are wider than necessary can result in poor filling with voids and variable joint properties. Overlaps should not normally exceed 20 mm for this reason, but if greater strength is necessary then the use of preformed or preplaced filler materials might give sufficient control. It is possible to purchase aluminium alloys with the filler materials already clad to the surface which obviously eliminates difficulties associated with flow of the filler.

Welding

There is a large range of welding processes in commercial use (see Houldcroft 1977) and a full discussion of these is beyond the scope of this book. Almost all possible sources of energy are used in welding processes, for example, explosives, electric arcs, burning gases (oxyacetylene and oxyhydrogen), exothermic chemical reactions, lasers and electron beams. The welding processes have been developed to gain control of the energy sources, and to utilise them to produce efficient joints. Welding processes can be broadly separated into two categories: those which rely on the parent materials melting (fusion welding) and those which take place with the materials still in the solid phase. The following notes are intended to illustrate the range of techniques available, and also briefly to discuss those most commonly used in schools and colleges.

Soldering and brazing use lap joints as shown previously in Fig. 6.94. Similar joint types would also be used with adhesives, but most welded joints are either butt joints or fillet welds. These are shown in Fig. 6.96.

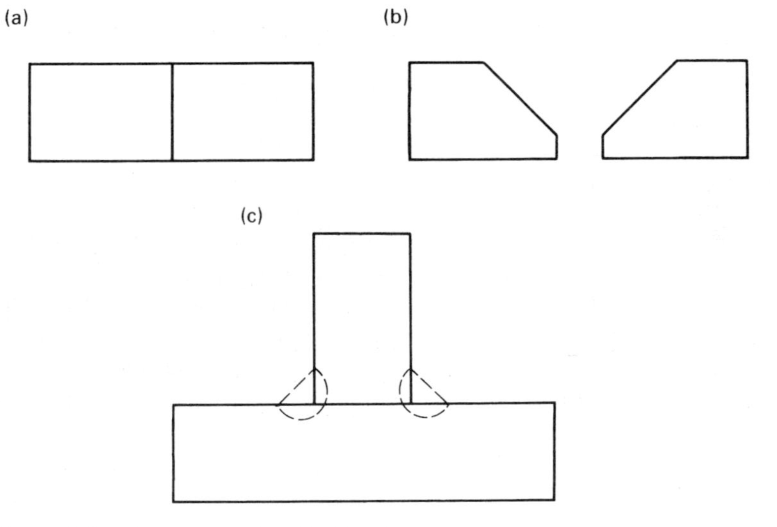

Figure 6.96 Basic types of welded joint
(a) butt joint (b) joint with a V-preparation (c) fillet welds.

It is of course possible to weld thermoplastics as well as metals, but a much lower heat input is required. It is also necessary to use wider joint preparations to ensure adequate penetration in butt joints.

Fusion welding processes

Oxyacetylene welding

Oxyacetylene welding is one of the most commonly available techniques in schools and colleges and is typically used to produce single pass welds in steel sheet of 0.5–4 mm thickness. (Strictly the term 'sheet' should only be applied to material of thickness 3 mm or less, after which it becomes plate.) Above this thickness a multipass procedure becomes necessary and in this form the process can be used to weld thicknesses of up to approximately 12 mm. It would also be necessary to put a V-preparation on the edges of a butt weld in order to ensure sufficient penetration.

Arc welding

The most common form of arc welding uses a metal electrode with a flux covering. The arc is struck between the electrode and the workpiece and this produces a very intense heat source as shown in Fig. 6.97.

The metal electrode melts and droplets are transferred across the welding arc to help form the weld bead. The flux is specially formulated to generate a gas shield to protect the molten weld pool and a glassy slag which solidifies over the weld bead, protecting it whilst it cools. The slag can be chipped away when it is cold.

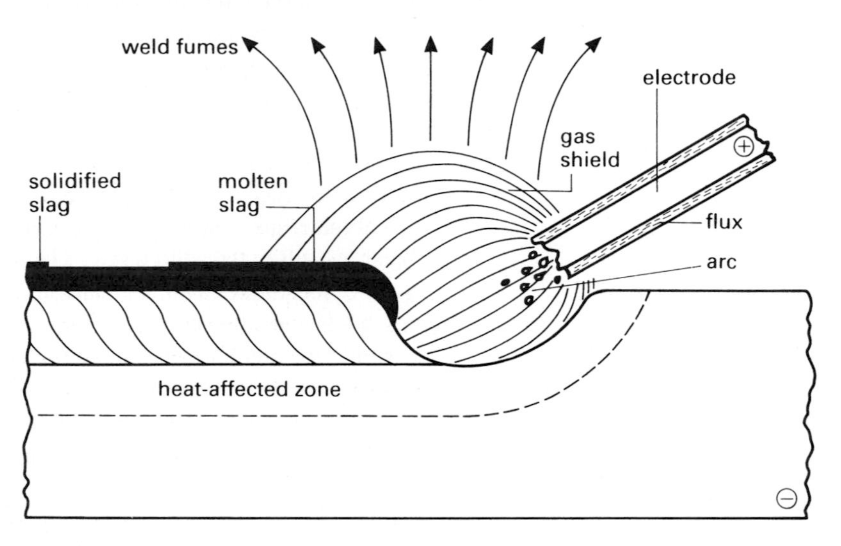

Figure 6.97 Schematic illustration of a metal arc.

There are many variations on this fundamental arc welding process – some developed to make the process more economic and others to increase the range of applications. The metal electrodes are usually supplied in short lengths, but it is possible to obtain coils of flux-cored wire. The process works in a very similar way to manual metal arc welding, but the flux is now on the inside rather than the outside. The flux-cored wire is fed through a welding gun, thus allowing more continuous operation. At very high currents, the arc becomes too intense for manual operation and so an automated version, using a continuous wire feed, has been developed with the arc completely submerged beneath powdered flux. This is shown in Fig. 6.98. A version of this process – submerged arc welding – using a strip electrode rather than a wire is used for cladding steels with very hard cobalt alloys for wear resistance or with stainless steel for corrosion protection. In recent years, the solid wire version has also found application using low currents, because of certain metallurgical and process advantages.

The tenacious oxide film which forms on aluminium and its alloys means that arc welding fluxes are just as corrosive and difficult to formulate as brazing fluxes. As aluminium is commercially a very important material, this provided the incentive to develop an alternative welding method. It was discovered soon after the Second World War that if inert gases like argon and helium were used to shield the arc and weld pool instead of a flux, then the cathodic action of the arc on the workpiece effectively removed the aluminium oxide. This discovery led to the development of the MIG (metal inert gas) and TIG (tungsten inert gas) welding processes which are now commonly used for welding aluminium. The MIG process uses a continuous bare metal electrode which is supplied in coils and, although developed primarily for

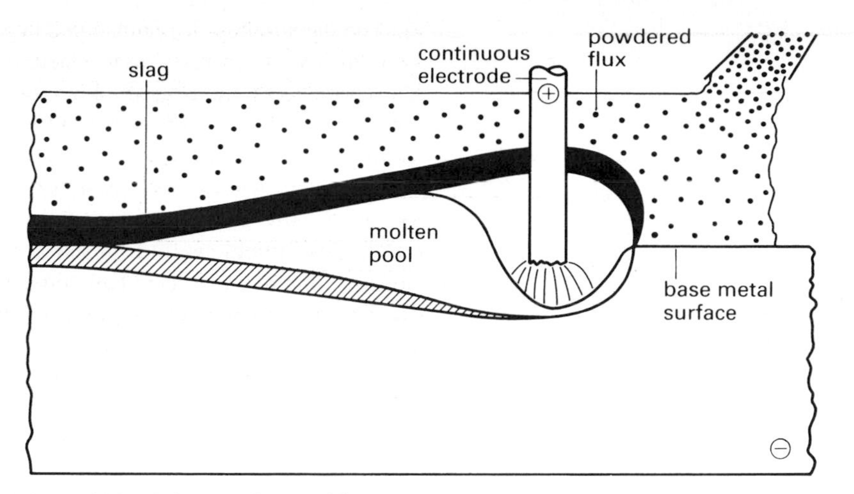

Figure 6.98 Submerged arc welding process.

use on aluminium, variations of the process are now used on steels and other metals and alloys. It is also now more properly known as *gas metal arc welding* (or GMAW). In the TIG process a nonconsumable tungsten electrode replaces the metal electrode, and consequently filler metal additions are made independently if they are needed. This means that TIG is a highly controllable welding process, and it is therefore commonly used in critical or difficult welding operations – like the root run in pipe joints. TIG is now more properly known as *gas tungsten arc welding* (or GTAW).

Resistance welding

Although less commonly available in schools and colleges, resistance welding processes are vital in industry. The commonest form is resistance spot welding as shown in Fig. 6.99.

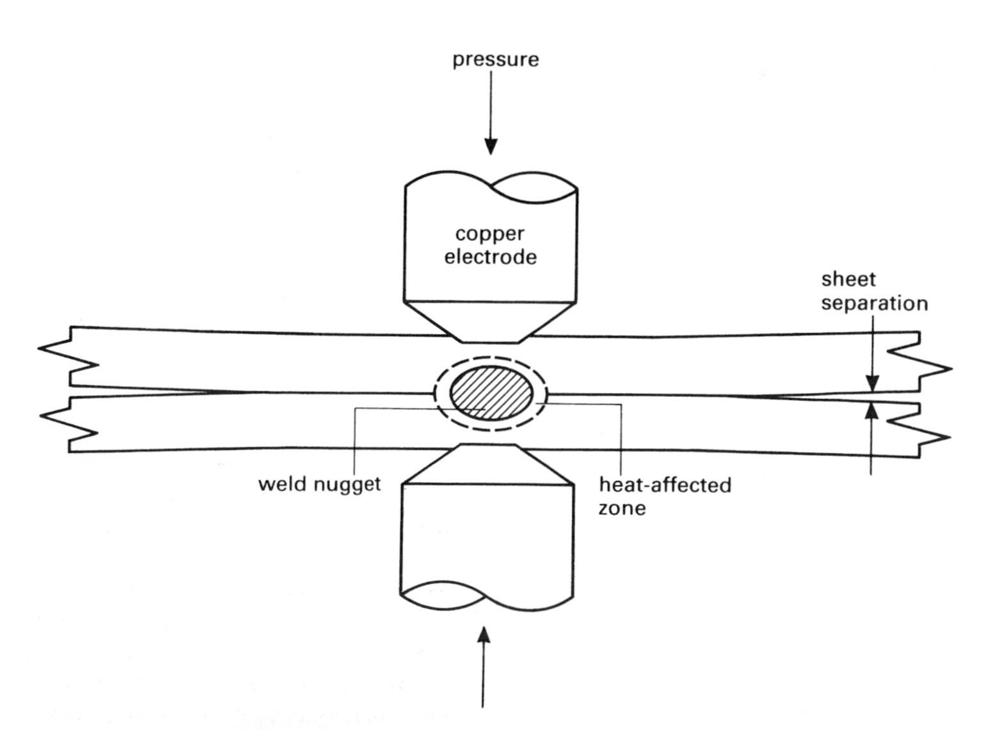

Figure 6.99 Resistance spot welding.

The sheet materials are clamped between copper electrodes and, when a current is passed, heat is generated where there is most resistance, that is, at the interface. The electricity takes the form of a low-voltage, high-current pulse of short duration. The pressure is held on after the current pulse in order to forge the sheets together as the weld solidifies. The end result is a weld nugget slightly wider than the electrode tip diameter and surface indentation where the electrode pressure was applied. The process is most commonly used for welding sheet steel, particularly for the production of vehicles and consumer durables. The difficulty in getting resistance spot welding to work effectively on aluminium sheet is one of the primary reasons why aluminium vehicles are not mass produced. In order to function, very high currents are required and the electrodes wear too rapidly. They must be removed and reshaped every 1000 or so welds making automated production uneconomic.

If instead of single cylindrical electrodes copper wheels are used, then a series of spot welds can be made by sending a number of pulses through the wheels as the electrodes rotate. These spot welds can either be intermittent or made to overlap in order to produce a seam weld. Either variation is known as resistance seam welding. Figure 6.100 shows the process schematically as well as the configuration used for producing seam welds in tubing.

The resistance welding process is used on numerous sheet steel products like canisters and petrol tanks where a sealed joint is required.

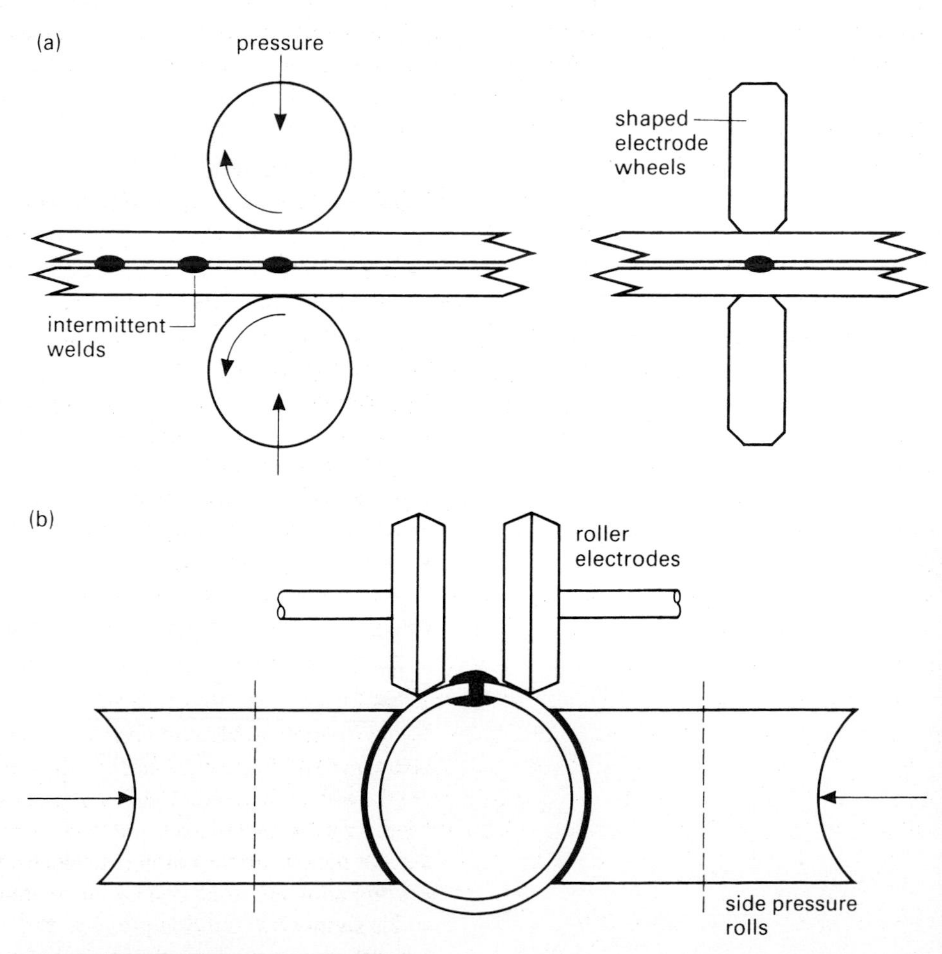

Figure 6.100 Resistance seam welding
(a) schematic layouts
(b) configuration for welding tubing.

High-energy beam processes

Generally fusion welding processes result in considerable distortion as a result of the uneven contraction during cooling as well as a wide heat-affected zone. Both these problems can be reduced by using a high-energy beam which results in a narrower weld bead. Electron beams and laser beams are both used for welding in production. Electron beam welding has the disadvantage that it can only be carried out in a vacuum, because otherwise the power of the beam is lost in collisions with molecules in the air. It can weld much greater thicknesses than lasers, but these can be operated in air because the light is not dispersed.

Solid phase processes

Forge welding

Forge welding is the oldest welding process and for several thousand years was the only one. It was used by blacksmiths for joining iron, and one of the most impressive examples of its use – the Iron Pillar of Delhi – dates from AD 310. The iron is heated to over 1000°C, and at this temperature the metal is in a softened state and the surface oxides are fluid. The metal is then hammered together forcing out the molten oxides and bringing the iron parts into sufficiently intimate contact for a bond to form. The components can, of course, be rolled, squeezed or otherwise brought together and the use of hammer blows is not essential.

Successful welding depends primarily on the removal of the surface oxides. The oxide formed on carbon steels is less fluid and requires substances like silica sand, fluorspar or borax to be sprinkled on the heated components to make the oxide more fluid. Alloy steels which contain elements giving solid refractory oxides cannot be forge welded.

Friction welding

Friction welding is a modern development of forge welding. It had been used for welding thermoplastics in many countries, but it was in the USSR that a practical method of friction welding metals was introduced just after the Second World War. One of the components to be joined is held stationary and the other is rotated in contact with it. The heat generated raises the temperature in the region to be joined, and when the temperature is high enough the components are forced together.

The rotary action is a very efficient way of removing the oxides from the surfaces and also flattening out any irregularities. The result is that the surfaces are brought close together everywhere and a very reliable joint is formed. As it is a solid phase process, welds can be produced between both similar and dissimilar metals, for example, copper to aluminium. Friction welding has become a vital joining process in industry, and is used in the manufacture of twist drills to join the high-speed steel ends to the softer carbon steel shanks, as well as in the automotive industry, in for example the production of rear axle casings.

Cold bonding processes

Forge welding and friction welding both require the materials to be heated. There are also other solid phase processes which use a combination of heat and pressure, like roll-bonding where sheets of base material are clad with skins of other metals by squeezing the heated sheets between heavy rollers. Mild steel clad with stainless steel or nickel for corrosion resistance and aluminium alloys ready bonded with brazing alloy are produced by this method. It is however also possible to produce welds simply by applying pressure and without first heating the materials. In space, the surfaces are so clean that components tend to weld together without any pressure being applied at all. Cold bonding can be demonstrated by hammering copper or aluminium sheets together with a punch. Bonds will be formed between the sheets where the punch acts. The oxides break up in a brittle way under deformation allowing the freshly exposed metal surfaces to join.

Alternatively, cold bonding can be achieved by using a very high impact speed for one of the components. Explosive charges can be used for this purpose as can the electromagnetic forces generated by high currents flowing in magnetic coils.

One of the common applications of the cold bonding process is indicated by Fig. 6.101 which shows how aluminium cans are sealed by cold bonding.

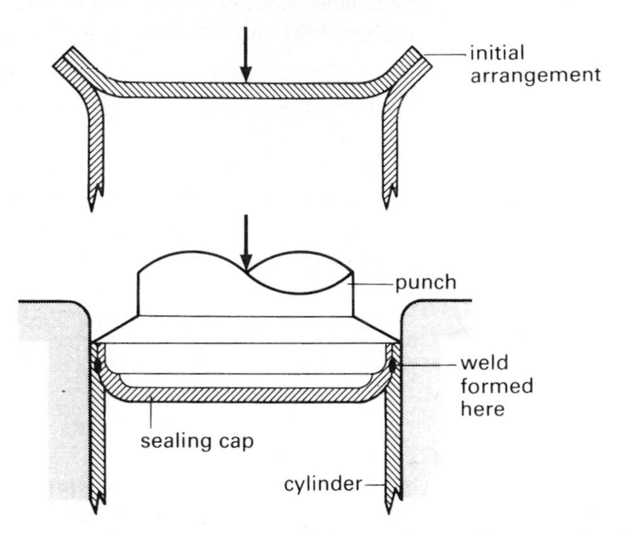

Figure 6.101 Sealing of aluminium cans by cold pressure bonding.

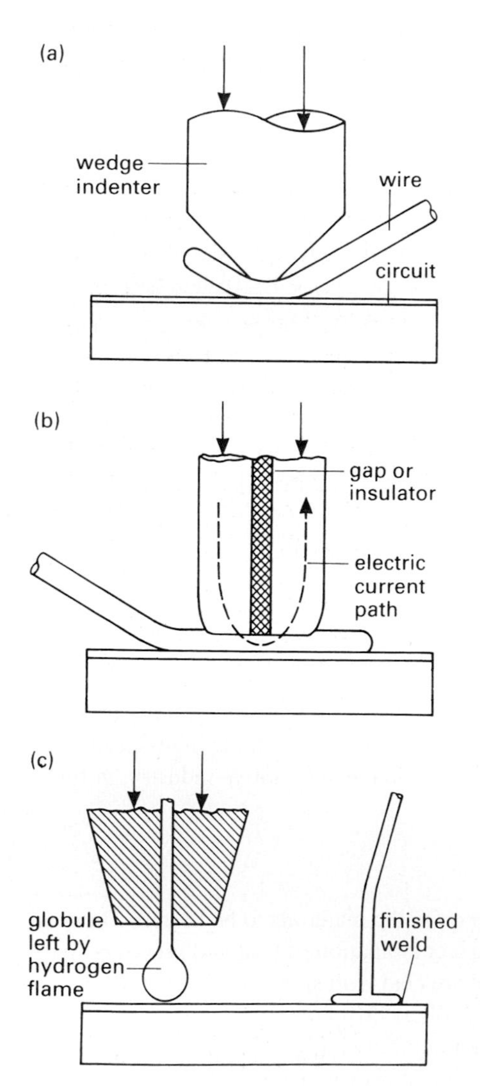

Figure 6.102 Three methods of thermocompression bonding
(a) wedge bonding
(b) parallel gap bonding
(c) ball bonding.

Thermocompression bonding

A process closely related to cold bonding has been developed for joining wires to circuits. The wire is heated to a temperature of about 200°C and is forced against the surface of the circuit board. Three variations of this process are shown in Fig. 6.102.

With wedge bonding, either the wedge indenter or the sheet may be heated. In the parallel gap method, the electrode is heated by electrical resistance heating. In ball bonding, the wire is parted by a hydrogen flame leaving a heated globule on the end. This carries the heat to the joint area. Many examples of these joint types will be found in industrially produced electronic circuits.

Cladding (or roll-bonding)

Steel can be protected by forming a sandwich of steel between sheets of a corrosion-resistant metal like nickel or stainless steel. The sheets are heated and passed through rollers where significant deformation results in a solid phase weld forming between the sheets. This process is also used to clad aluminium alloys with a brazing alloy which melts at a lower temperature, and was once used to produce Sheffield plate, which was how copper used to be clad with silver. Nowadays electroplating would be used (see section 6.6.1).

Ultrasonic welding

The head used for ultrasonic machining has already been shown in Fig. 6.84. A similar arrangement can be used to weld materials. The probe introduces vibrations between the surfaces which disrupt the surface oxides and raise the temperature through friction. When pressure is applied this allows a bond to be formed. Ultrasonic welding is generally used for smaller components and can be used to weld a variety of metals and thermoplastics. Good results can be obtained when welding either similar or dissimilar metals.

6.5.3 Adhesives

Adhesives of one form or another were in use in even the earliest civilisations. Animal and vegetable glues were used in ancient Egypt for applying wood veneers to furniture and gold leaf to decorate artefacts. Natural adhesives are typified by glues, waxes, casein gums, carbohydrates such as starch, rubber and its derivatives and inorganic substances like silicates and bitumen. Generally these are nontoxic substances and hence present few hazards in manufacturing the joints or in the final product. In contrast their modern synthetic counterparts tend to be toxic substances requiring care in their use, although they are generally safe when cured. Most thermoplastic materials can be welded, but the only method available for thermosets is adhesive bonding.

Modern adhesives are characterised in a number of ways: as structural or nonstructural, for interior or exterior use and as thermoplastic or thermosetting. Although all of these categorisations offer some useful insights, none of them is completely satisfactory. The eight groups shown in Table 6.2 cover those in general use.

The use of any of these adhesives requires clean and suitably prepared surfaces. Metal surfaces should be degreased and roughened with an abrasive. Abrasive papers

Type	*Curing method*	*Notes*
anaerobic acrylic resins	action of a catalyst out of contact with the air	One type, cyanoacrylates, cure by reacting with moisture on the surfaces.
modified phenolic resins	application of heat and pressure	These resins provided the first successful adhesives for metal–metal, metal–wood and metal–plastic joints. The basic phenol-formaldehyde resin is typically mixed with nylon, neoprene or polyvinyl butyrate. These are thermosetting.
epoxy-based adhesives	action of a hardening compound on the epoxy resin, although some one-part epoxy resins are cured by heat	Generally these are supplied in twin packs. Also capable of producing joints between most materials.
polyurethane adhesives	reaction between two compounds	Fast acting.
modified PVC dispersions	action of heat	
rubber-based adhesives	evaporation of a solvent	
PVA adhesives	removal of water from the emulsion	Used to bond porous materials such as wood or concrete. PVA is a thermoplastic material.
hot melts	action of heat	Special polymers used to produce quick joints for lightly loaded structures.

Table 6.2 Common adhesive categories

or grit blasting are both suitable. The surfaces of plastics and glass should be similarly cleaned and roughened before bonding. Specific surface treatments prior to applying the adhesive can also increase the bond strength for some metals. Mild steel may be treated in dilute sulphuric acid or with chromium salts. Aluminium alloys give greater bond strength when anodised (see also section 6.6.2) but not sealed. Copper alloys are best etched in nitric acid.

The actual mechanism of adhesive bonding is not yet fully understood. The strongest joints are made through a chemical reaction – a purely physical bond giving less strength. Some theories of bonding are based on intermolecular attraction associated with the formation of covalent bonds or van der Vaals forces between the adhesive and the metal (see section 4.3.2). Others emphasise the physical interlocking of the surfaces through the adsorption of molecules or the diffusion of molecules into the material.

The adhesive is applied to the cleaned and prepared surfaces and the adhesive must then be cured. The components must be held in place until the adhesive has gained sufficient strength. This may take a few minutes or up to a day at normal temperatures. Increasing the temperature through infrared panels, ovens or heating pads will reduce the curing time significantly. Temperatures up to 180°C may be necessary to cure one-part epoxy resins. Accelerators can also be used to reduce curing times, but there will be some loss of strength.

The strength of adhesive joints

Figure 6.103 indicates the stress distribution associated with a lap joint.

This kind of stress distribution will be found in the overlap region however the joint is formed and is the fundamental reason why increasing the width of the overlap does not result in a proportional increase in the joint strength. Increasing the length of the joint, of course, does increase the joint strength proportionally. Loading of the form shown in Fig. 6.103 is shear loading and the maximum average shear stress will obviously depend on the width of the overlap as well as the particular adhesive and the materials being joined. Typical shear strengths are in the range 30–60 N/mm^2. One other important variable is the thickness of the adhesive – the strength decreasing as the thickness of the glue increases. Generally the optimum thickness is between 0.05 mm and 0.15 mm.

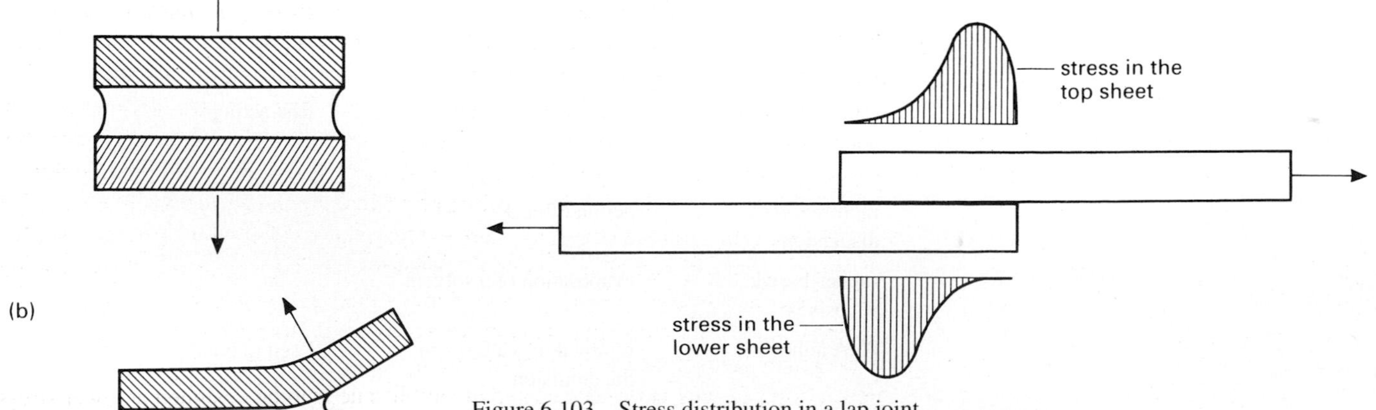

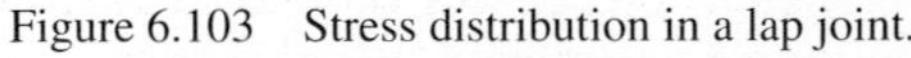

Figure 6.103 Stress distribution in a lap joint.

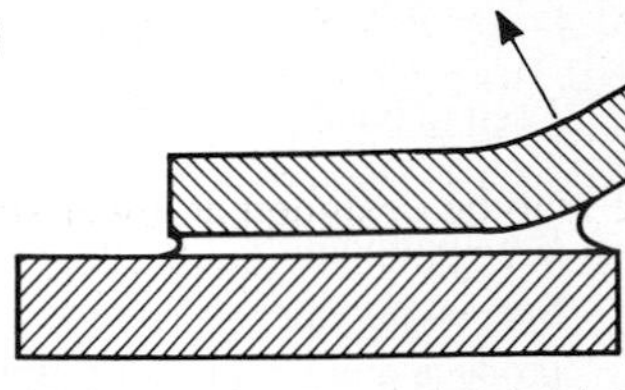

Figure 6.104 Adhesives joints loaded in
(a) tension
(b) peel.

Adhesively bonded joints are strongest either in shear or in compression. They can also be loaded in tension and peel as shown in Fig. 6.104, but the joints are weaker in these modes and they should be avoided wherever possible.

Advantages and disadvantages

Adhesives have some specific advantages and disadvantages in relation to other joining methods. Amongst the advantages are:

- more uniform stress distribution because of the absence of sudden discontinuities,
- the ability to bond dissimilar materials and composites,
- a separate sealing operation can sometimes be avoided,
- there is reduced risk of galvanic corrosion between dissimilar metals because the adhesive joint will act as an insulator,
- no metallurgical damage when joining metals as would be associated with significant heat inputs (see section 6.5.4).

Amongst the disadvantages are:

- joints must be held firmly during curing,
- strength tends to fall off sharply above 80–100°C,
- adhesives are often expensive, with a limited shelf life.

6.5.4 Selection of a joining method

There are a number of factors which may influence the selection of the best joining method, but the following are the most significant:

- the degree of permanency required,
- the necessary strength of the joint,
- the stiffness needed,
- the type of loading, particularly whether static or fatigue,
- the effect of heat on the materials involved,
- the appearance of the product round the joint,
- the environment, particularly potential corrosion sites.

It is clearly crucial to know whether or not the assembly must be easily dismantled. If it must, then nonpermanent fasteners must be used. It is important to have a clear view of the required strength of the joint. If all the members are critically stressed, then the joints must be capable of carrying an equal load, but this is not always the case. The structural members in furniture generally carry very small stresses and the joints do not need to be produced on the basis of the members being stressed to their limit. Stiffness is often more crucial in furniture joints; for example, some give is desirable in chair joints to allow all four legs to carry loading on uneven floors.

A knowledge of the magnitudes and types of the loading is crucial, particularly whether they are static or fatigue loads. Fatigue failure starts from a stress concentration and this will exist at any discontinuity. The discontinuity might be the excess metal above the plate surface for a weld bead (the reinforcement), a hole for a bolt or a rivet or a screw thread cut on to a shaft. Any of these could provide the initiation site for a fatigue crack. Under fatigue loading the attractions of adhesive joints with little stress concentration are evident, but welded joints dressed level with the plate surface would give even better performance. The removal by heat treatment of the residual stresses present in welded assemblies as a result of the thermal stresses generated on cooling also improves fatigue performance by reducing the average stress level. Mechanical fastening methods like bolts and rivets generally have poorer fatigue performance and assemblies need to be designed at lower stress levels. Anything which might vibrate loose must also be secured – normally by a pin or lock washer.

For any joining method involving significant heat input there is a likelihood of a softened zone forming adjacent to the weld. A recrystallised zone may form in work-hardened material and an overaged zone in age-hardened alloys (see section 4.3.5).

Table 6.3 shows the melting points and approximate recrystallisation temperatures for several metals. (The recrystallisation temperature is approximately 0.4 × the melting temperature in Kelvin.)

Alloys recrystallise at higher temperatures than pure metals but it is clear that, when welding, brazing and hard soldering, the metal close to the joint will be heated above the recrystallisation temperature. There will therefore be a region of lower tensile strength adjacent to the joint. For this reason, the high-strength aluminium alloys used in aircraft are riveted or adhesively bonded.

Metal	*Melting point* (°C)	*Recrystallisation temperature* (°C)
lead	327	below room temperature
zinc	420	below room temperature
aluminium	660	150
magnesium	650	200
copper	1085	200
iron	1538	450

Table 6.3 The melting points and recrystallisation temperatures of some metals

Clearly if heat is to be applied to the final assembly, or generated in use as with brake pads, it is important to avoid using thermoplastic adhesives. A thermosetting adhesive can be used for temperatures up to a few hundred degrees Celsius, but in order to withstand higher temperatures it is necessary to produce a brazed or welded joint. It is also worth noting that the thermal coefficients of expansion for most adhesives are very different from those of most metals, as are their moduli of elasticity, and these differences may be the cause of significant difficulties when the assembly is heated or loaded.

Equally crucial, however, are other factors like the appearance of the product round the joint and the environment in which the joint is used. Methods like spot welding leave characteristic deformation of the surfaces and both bolts and rivets may be unacceptable. The obtrusive nature of either of these can, of course, be much reduced by countersinking the heads, although this necessitates an extra manufacturing operation. Jointing methods often bring dissimilar metals into contact resulting in potential problems with electrochemical corrosion (see section 6.6.1). Adhesives are often insulators and hence reduce this effect.

6.6 Corrosion and finishing processes

It is not only to give protection against corrosion or the environment that finishing processes are undertaken, but for a whole host of other reasons as well. There are technical reasons like the improvement in wear resistance and fatigue strength. Probably more significantly for the product designer, there are a variety of factors associated with the visual qualities of the product. It may be that aesthetic and decorative issues are central, perhaps identification as when red or yellow are used for telephone boxes, perhaps camouflage or safety – bright colours being more easily seen. There are also factors associated with the product use which might be crucial, like the reflection or adsorption of radiant heat, insulation from heat or electricity, antifouling, water repellent or anticondensation finishes. Whatever the finish required, the process by which it is applied is almost certain to have implications for the design of the product.

As corrosion is one of the major influences over the finishing of metals and the electrochemical cell is not only the major corrosion mechanism but also the basis for electroplating, these are briefly discussed before describing finishing processes.

6.6.1 Corrosion and the electrochemical cell

Students will already be aware that corrosion is very rapid in damp environments, and particularly near the sea, but much reduced in dry regions. Vehicle bodies tend to last for great lengths of time in Africa or India – scratches remaining bright for months after they are made, unlike the unsightly brown rust which soon appears in European environments. These phenomena result from the need for an electrolyte – a solution containing positive and negative ions – for a metal to corrode by an electrochemical method, which is the predominant route. The rate at which metals corrode will be different in different electrolytes. The salt solution associated with the sea and the dilute sulphuric acid associated with acid rain are particularly effective.

Essentially what happens is that a small battery is set up, either between two components or between two regions on a component – the electrolyte conducting the associated current flow. An electrochemical cell is said to have been established and the detailed behaviour of such cells is described below. It is also possible for a material simply to dissolve in a corrosive liquid and in this case it is referred as chemical rather than electrochemical corrosion. Examples of chemical corrosion are the dezincification of brass, when the zinc dissolves in aqueous solutions at high temperatures, and the graphitic corrosion of grey cast irons, when the iron dissolves in water or soil leaving graphite flakes connected by corrosion products. It is this which is often the cause of leakage or failure of buried cast iron gas or water pipes. Polymers can also suffer chemical corrosion if they come into contact with solvents with a similar structure, for example, polystyrene in benzene.

At elevated temperatures metals, ceramics and polymers can all be destroyed by reactions with oxygen. Metals form oxides and polymers form cross links which radically alter their properties. It is also possible for materials to be damaged by exposure to radiation or bacteria. The corrosion or degradation of materials costs millions of pounds every year and everything that a designer can do in a cost-effective manner to reduce these losses is a vital contribution to the economy and the conservation of resources.

The electrochemical cell

Fig. 6.105 shows the four key components of the electrochemical cell.

- The anode which gives up electrons and corrodes through forming positively charged ions, known as anions, which go into solution.
- The cathode which receives electrons. These recombine at the cathode surface with anions in the solution.
- A connection between the anode and the cathode which allows electrons to flow.
- An electrolyte which allows the circuit to be completed by transporting ions from the anode to the cathode.

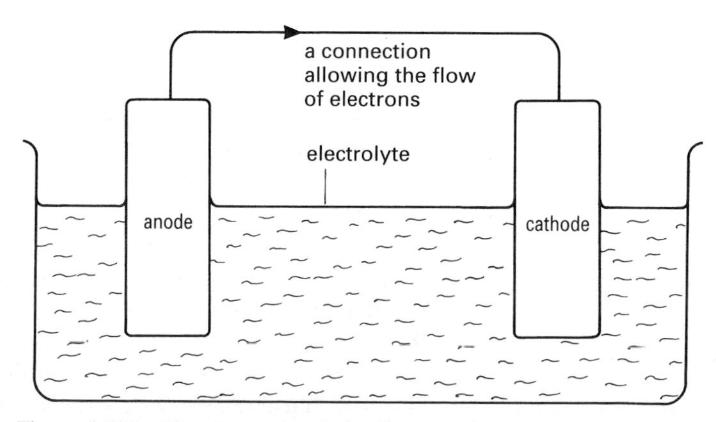

Figure 6.105 The electrochemical cell.

		Electrode potential (V)
anode ↑		
	sodium	− 2.71
	magnesium	− 2.37
	aluminium	− 1.66
	titanium	− 1.63
	manganese	− 1.63
	zinc	− 0.76
	chromium	− 0.74
	iron	− 0.44
	cadmium	− 0.40
	nickel	− 0.25
	tin	− 0.14
	lead	− 0.13
	hydrogen	0.00
	copper	+ 0.34
	silver	+ 0.80
	platinum	+ 1.20
	gold	+ 1.50
cathode ↓		

Table 6.4 Standard electrode potentials – the electrochemical series.

The same basic arrangement can produce either of two important effects: the anode dissolves – electrochemical corrosion – or the cathode is covered with anions which adhere to the surface – electroplating.

Electrochemical corrosion

All metals have a tendency to corrode in the sense that they will release electrons to form anions. The tendency of a metal to give up its electrons can be found by measuring the voltage generated when it is connected to a standard electrode in a standard electrolyte. The electrode is known as a calomel electrode and measures the electromotive force (e.m.f.) generated with respect to hydrogen which is taken as zero. This is known as the standard electrode potential. A table of these values, also known as the electrochemical series, is shown in Table 6.4.

When two of these elements are connected together in the presence of an electrolyte, the element with the greater tendency to lose electrons (that is, higher up the table) becomes the anode and the other element (that is, the one lower in the table) becomes the cathode. The element higher up the table, the anode, corrodes. Fig. 6.106 shows the result of using a steel screw to secure the handle to an aluminium saucepan lid. The corrosion of the aluminium where the different metals were in contact can clearly be seen.

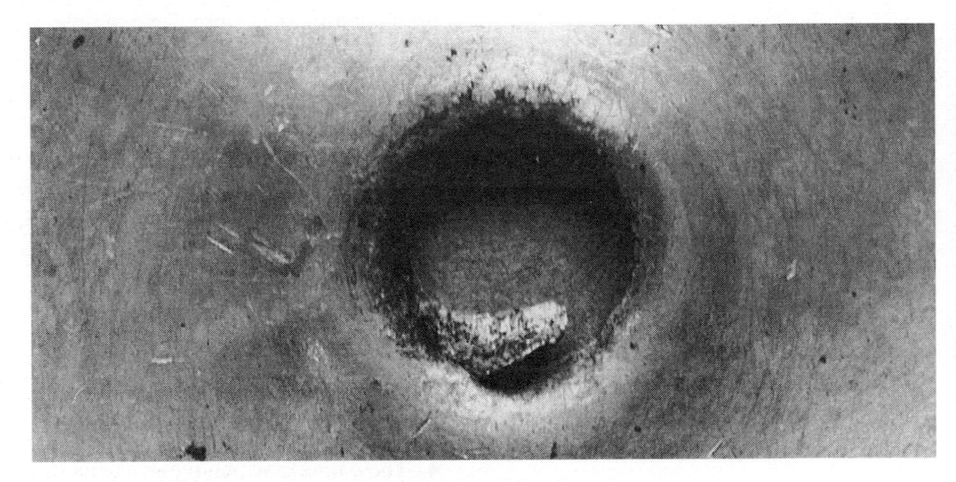

Figure 6.106 Corrosion between an aluminium lid and steel screw.

anodic ↑	
	magnesium
	magnesium alloys
	zinc
	galvanised steel
	aluminium
	cadmium
	Duralumin
	low-carbon steel
	cast iron
	stainless steel (active)
	lead-tin solder
	lead
	tin
	brass
	aluminium bronze
	copper
	cupronickel
	nickel
	silver
	stainless steel (passive)
	titanium
	graphite
	gold
	platinum
cathodic ↓	

Table 6.5 The electrochemical series in seawater

The electrochemical series does not always accurately predict the behaviour of different alloys in a particular electrolyte, because of the variety of ways in which electrochemical cells can be formed. Table 6.5 shows the electrochemical series which shows the anodic or cathodic tendencies of metals in seawater.

There are a variety of circumstances under which an electrochemical cell can lead to corrosion and three of the more common ones are described below.

Composition cells result from the connection of dissimilar metals. These may be components made from different metals like copper and iron or simply different phases of the same alloy – for example, ferrite is anodic with respect to cementite. Equally the grain boundaries are often a different composition from the interior of the grains and consequently intergranular corrosion occurs.

Stress cells result from the existence of regions of differing atomic stresses within the material. The most highly stressed, or high-energy, regions act as anodes and the less stressed regions act as cathodes. This is why stress relieving can improve corrosion resistance and also why the rate of growth of fatigue cracks can be accelerated by corrosion.

Concentration cells result from differences in composition within the electrolyte. The region of the metal in contact with high concentration acts as the cathode and the region of metal in contact with a low concentration acts as the anode. Fig. 6.107 shows such a concentration cell for iron.

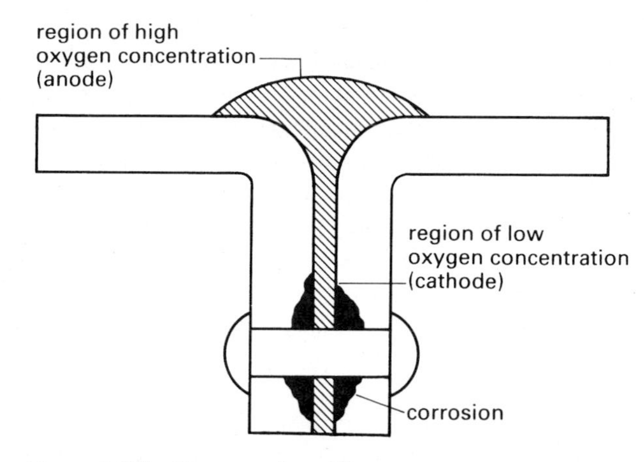

Figure 6.107 Concentration cell.

The corrosion occurs at the tip of the crack where there is little oxygen. This is why, once rusting begins, all oxide must be removed from the metal surface before repainting. Otherwise the deposited rust will shield the underlying iron creating regions of low oxygen levels. Corrosion will therefore continue under the rust deposits in the anodic regions.

Waterline corrosion is a similar phenomenon. The metal above the waterline is exposed to oxygen and is, therefore, cathodic. The oxygen concentration decreases as the depth beneath the surface increases, producing anodic regions. The major corrosion occurs just below the waterline.

Electroplating

Components are connected as the cathode in an electrochemical cell in order to electroplate them. The anode consists of a piece of the metal which is to be plated on to the component. A voltage up to about 15 V is applied between the anode and the cathode with the result that the anode dissolves to form anions and electrons flow to the cathode. At the cathode the electrons will seek to combine with the anions requiring the lowest potential which will be the lowest one in the electrochemical series. As H^+ ions are present in aqueous solutions, only those metals below hydrogen can be plated on to components through this mechanism. The most common are copper, silver, platinum and gold. If the H^+ ions recombine at the cathode, hydrogen gas is given off and no plating occurs.

A simple aqueous solution is not normally used in commercial operation. Nickel and chromium are two of the metals most often plated on to components, although often on top of another metal like copper. The same principles apply to the deposition of these metals, but the electrolyte used will be a much more complex formulation. As they are above hydrogen in the electrochemical series, the cathode reaction would otherwise tend to produce hydrogen gas. Nonmetals like ABS thermoplastics can be electroplated if they are first sprayed with a conducting coating.

Successful electroplating requires flat surfaces and the avoidance of sharp edges. This means that all corners must be radiused and recesses avoided. The polishing costs associated with the preparatory work can represent a large proportion of the costs of the total plating operation.

6.6.2 Protecting against environmental damage

There are a number of general approaches to the problem of minimising corrosion. Probably the most important is careful design to avoid water traps – simply putting in a drain hole may prevent the corrosion beginning. It is also good practice to seal crevices and joints which are not continuously welded. Equally, bringing dissimilar metals into contact, for example by using steel bolts with aluminium components, should be avoided if possible, but if unavoidable insulating washers can stop electrochemical cells occurring. It is also possible to reduce the effect of the environment, for example, by adding corrosion inhibitors to antifreeze. Electrochemical corrosion can be avoided by using a sacrificial anode which will corrode in preference to the component. The metal highest in the electrochemical series will be the anode and hence corrode first. Zinc and magnesium are used to protect steel pipelines, ships and offshore drilling platforms in this way.

It is also possible to stop electrochemical corrosion by applying an e.m.f. of the opposite polarity to that generated by the electrochemical cell as shown in Fig. 6.108.

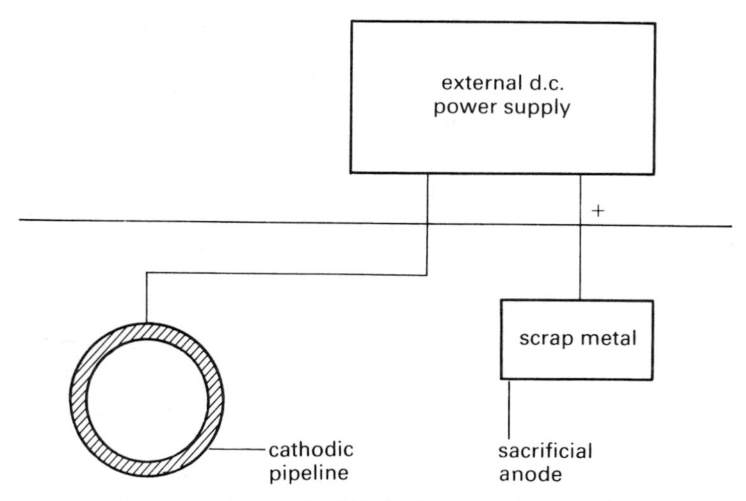

Figure 6.108 Protection of a buried pipeline by an impressed e.m.f.

This ensures that the component to be protected is the cathode and will not therefore corrode. The other lead from the power supply is connected to some scrap metal which acts as the anode and corrodes.

The finishes which are applied to protect against corrosion generally relate to either metals or woods and these are discussed separately below.

Metal finishes

If it is to stop an electrochemical cell forming, the finish must provide a barrier between the metal and the environment which is providing the electrolyte. The barrier can either be an oxide of the metal itself or an applied barrier.

Metal oxides

Some metal oxides adhere very strongly to the base metals. Aluminium oxide on pure aluminium is a well-known example, although the situation is complicated by some alloying additions. Once formed, the oxide layer effectively prevents further corrosion taking place. If it is scratched or damaged it will also re-form almost immediately – on aluminium in a small fraction of a second. The oxide layer is so thin that it is transparent and therefore does not affect the high polish which aluminium can have. The thickness of the oxide layer can be increased by anodising where the aluminium component is made the anode in an electrochemical cell.

Oxygen is formed at the anode which results in the anode thickness growing from approximately 0.00001 mm to 0.01 mm in about 30 min. The oxide layer is porous which enables it to absorb dyes and take a variety of coloured finishes. Many of the colours tend to fade in strong light and are not therefore suitable for outdoor use, but, by using suitable dyes, silver, gold, copper, bronze and brass can be simulated. Because the anodised oxide layer is porous it should be sealed to give maximum corrosion resistance. Anodising is not generally used on the alloys used for casting.

The chromium additions to steel, which make it stainless, work in a similar way. Complex chromium oxides form on the surface of the steel and it is these which provide the corrosion resistance. If all the chromium which can diffuse to the corrosion area becomes used up, then the corrosion resistance will be lost – the stainless steel is then said to be active. This is why it appears twice in Table 6.5, once as passive and once as active.

The oxide which is formed by iron when it is heated also adheres to the surface, although less strongly than chromium oxides. This means that a process like blueing, which is the common workshop technique for finishing steels, provides some corrosion resistance because of the oxide layer formed. Blueing is carried out by heating the steel to around 300°C and then quenching in oil. The oxides formed by zinc, lead, copper and nickel also adhere strongly to the base metal giving good corrosion resistance. The copper oxide formed tends to react with atmospheric pollutants forming a green patina of sulphates and carbonates.

Applied barriers

There is a very large range of materials which are applied to metals in order to form a barrier to the environment and prevent corrosion. These coatings can be applied by brushing, rolling or spraying them on to the surface or by dipping the components into tanks of the coating material. Generally other metals are applied by electroplating, but they can also be sprayed on by using a special gun which includes a heating system or applied by dipping the components into a heated tank containing the molten metal. It is also possible to dip components into solid powders by fluidising the powder with compressed air. The air is passed through a membrane into the powder creating a fluid-like suspension of the powder in air. An exhaustive discussion of all of the possible materials which can be used is beyond the scope of this text, but the following notes on some of the more common possibilities serve to show the range available. Before any finish is applied, it is essential to have a very clean metal surface often obtained by pickling (or cleaning) the components in acid – although other methods like abrasion and degreasing can be used.

Zinc is most commonly associated with the galvanising process, which involves dipping the fluxed components into a bath of molten zinc (at 450–460°C). Window frames, dustbins, buckets, railings, nuts, bolts and screws are all finished in this way. If the zinc coating is damaged and an electrochemical cell established, the zinc will corrode in preference to the steel because it is higher in the electrochemical series. No corrosion of the underlying steel will occur until all the zinc is exhausted. Zinc is also applied by electroplating and sprayed on to large structures before painting.

Tin is most commonly applied to steels during production by passing the steel sheet through molten baths of tin – producing tinplate. The sheet steel passes through a layer of flux on the surface of the tin as it enters the molten bath which is at a temperature of 315–320°C. Tinplate is used to produce food cans, and will generally form an effective barrier. It does however react with some acids and sulphur, which is sometimes added as a preservative, and in these cases the inside of the can needs to be lacquered. If the tin layer is damaged at any point and the underlying steel is exposed, the steel will form the anode and corrode first because tin is below iron in the electrochemical series. It is for this reason that the edges of cut tinplate should be sealed with solder.

Vitreous enamel is basically a borosilicate glass and is used not only because of the protection it offers, but also because of its aesthetic qualities and the hard, durable finish. Small particles of the glass (known as 'frits') are suspended in water and the components are dipped into the suspension. An even coating of the glass is obtained which is then fired in an oven. Different colours are produced by adding metal oxides to the glass.

Plastic coatings are applied to metal components in a similar way to vitreous enamels. The components to be coated are heated and dipped into a fluidised bed of plastic powder suspended in pressurised air. A uniform layer is obtained which fuses on to the component as a result of the heat the component contains, although further heating in an oven is likely to be necessary to produce a smooth, even coating. A variety of plastics are used, for example, PVC which is low-cost and corrosion-resistant and often used for street furniture, polythene which is low-cost and nontoxic and used on domestic equipment, PTFE for its nonstick qualities and many others like epoxy, nylon and polyesters.

Painting is one of the traditional methods of finishing metals, but its nature has changed with the development of paint technology. Paint is not now associated only with oil-based paints, but also with modern polymers and resins. The paints may be set by polymerisation at room temperature, through a thermosetting reaction or by the evaporation of a volatile solvent. (Paints which dry by evaporation are known as 'lacquers'.) A variety of materials are used as the basis of paints, for example, cellulose which produces a hard, resistant surface, vinyls which are excellent for outdoor use, urethanes which give a hard, durable finish, polyesters which are extremely hard and can be buffed to a high gloss finish, as well as neoprene, epoxy and alkyds.

Hot tar dipping used to be the ancient method of protecting iron and proved to be extremely effective. At least part of the reason was its ability to withstand impacts without chipping. There are a variety of similar rubber-like finishes available today which are applied like paints and can withstand repeated impacts; for example, for the undersealing of vehicles.

Wood finishes

One of the major difficulties in using wood in outdoor applications is its ability to absorb water. This causes the wood to expand and can lead to its degradation. Wood can also be subject to attack by bacteria and fungi if they are able to gain access. Clearly one approach to protecting timber is, therefore, to provide a barrier between the wood and its environment. Another approach is to impregnate the internal structure of the wood with a preservative. These alternatives are discussed briefly below.

Applied barriers

As for metals, a range of new, modern paint materials have been developed. Many paints can now be used on either wood or metal, but it is important that the wood is first sealed. If this is not done the paint will tend to be absorbed into the interior of the wood and not provide an effective barrier. The other aspect which makes wood very different from metal is the aesthetic quality of the natural wood surface. It is often desirable for this to be left visible and consequently a range of transparent varnishes and lacquers are now also available.

Internal preservatives

Particularly for wood which is to be used outside, for example as posts to support overhead cables or in harbour and bridge constructions, it is important to ensure the sustained long-term life of the wood. The most effective method is pressure treatment which is carried out in large vessels capable of accommodating timber up to 40 metres long. Once treated, the timber can last fifty years in all weather conditions. It is particularly important to use internal preservatives where there is likely to be erosion so that the new wood exposed will not be vulnerable to attack.

When pressure treating wood, water is first driven from the structure by circulating hot oil round the vessel. When all the water has been removed the preserving fluid is run in and the pressure is then slowly increased until the required amount of fluid has been injected. The preservative penetrates mainly along the fibre direction particularly into the sapwood. The heartwood is often impervious and only a little penetration occurs across the grain, because this requires the preservative to pass through the cell walls. Creosote is still the most commonly used, but this is not appropriate if the wood is subsequently to be painted. Before it is removed from the vessel, the timber is steam dried.

6.6.3 Other finishing processes

Protection against the environment is not the only reason that finishes are applied. Simply improving the aesthetic qualities may be the only reason and many of the methods discussed in section 6.6.2 could be used for this purpose. However other factors might be just as significant. The following notes indicate some of the possibilities.

Hammer finishes are normally produced by an aluminium powder suspended in a solvent – the 'hammer blow' effect resulting from a chemical reaction. This is a cheap, effective way of hiding surface imperfections and is often used on cast components.

Flock spraying provides a soft textural covering for hard materials. This can be used to improve the acoustic properties.

Nonslip finishes are often produced by adding an abrasive like pumice to a varnish or paint in order to provide grip, for example, on the decks of boats.

Antique finishes are sometimes required on metal objects. This can be achieved for steel components by grit blasting and then clear lacquering the resulting surface. Antique silver, copper or bronze effects can be obtained by electroplating the components, chemically oxidising and then lightly abrading the surfaces before polishing.

Phosphating is now carried out on a variety of galvanised steel, zinc and aluminium components to prepare them for painting. This is achieved by immersing them in appropriate chemicals. Phosphating not only improves the corrosion resistance, but provides an excellent surface for subsequent painting. It is also possible to seal phosphated components as an alternative to painting.

6.7 Computer-aided manufacture (CAM)

There are some industrial manufacturing activities that humans neither enjoy nor are efficient at repeating on a continuous basis. This has resulted in a long history of attempts at mechanisation either to eliminate some of the drudgery of shopfloor production work or to improve its efficiency. The first machine tools to be brought under automatic control were developed in America in the early 1950s following pioneering efforts by the J C Parsons Corporation. It was the production of templates for helicopter blades which had prompted this company to attempt to control a jig boring machine automatically. These early machines used punched tapes to store the machines' instructions in what is still known as a *part program*. They were said to be *numerically controlled* (NC) and hence the machines were known as NC machine tools. Such machines can still be found giving sterling service in industry and over the years have been found to offer significant advantages some of which are listed below:

- elimination of operator error,
- reduced labour costs,
- longer tool life due to the consistent use of optimum feeds and speeds,
- improved flexibility in producing different designs,
- greater predictability allowing more accurate costings.

The most significant disadvantage had been the need for large mainframe computers to produce and store the part programs. These tended to be very expensive and only large corporations could afford the capital investment, although some schemes were set up for companies to share a central facility by renting computer time. The computer revolution of the twentieth century has radically changed this situation, because it is now possible for each machine tool to be provided with the necessary computing power cheaply. However, the changes associated with the advent of the microprocessor go much further than simply the control of individual machine tools. The emphasis is now on the development of *flexible manufacturing systems* (FMS) and on *computer integrated manufacture* (CIM) which link CAD (*computer-aided design*) and CAM effectively. These two areas – the control of individual machine tools and recent developments in FMS and CIM – are discussed below, followed by a brief note about choosing to use CAM.

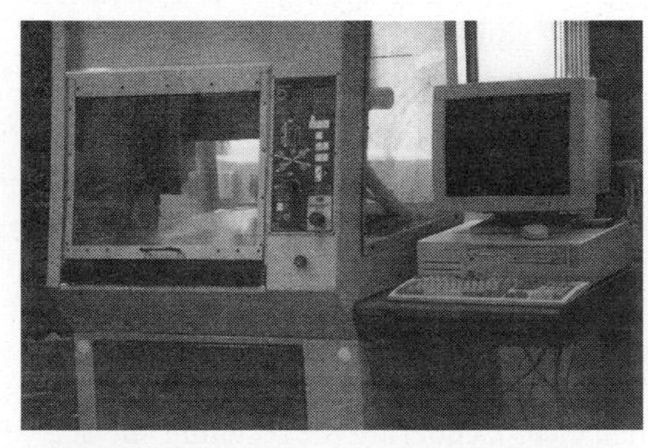

Fig 6.109

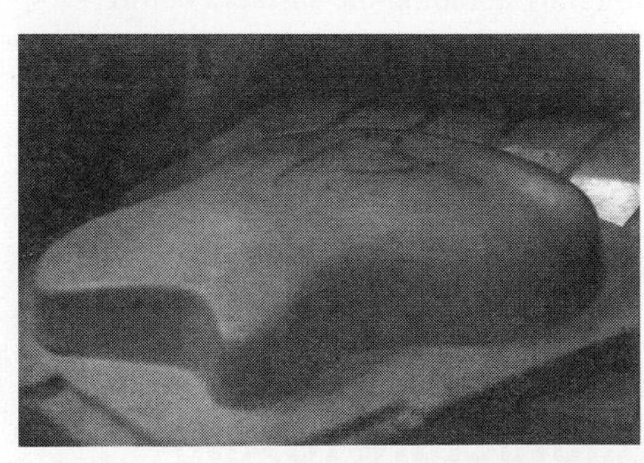

Fig 6.110

6.7.1 Individual machine tools

The simplest machine tool to control would be one with only one degree of freedom (see section 10.1.1). A drill or a resistance welding head could be such a machine. A lathe has two degrees of freedom because the tool can move both along and across the lathe bed. A milling machine has at least three degrees of freedom because the bed can move in two directions and the tool can move at right angles to the bed. A milling machine can have even more degrees of freedom if the machining head can also rotate about one or more axes. The fundamental ideas associated with implementing CAM have been outlined below in relation to lathes and this is followed by a discussion of the more complicated multi-axis machines. However, it is important to realise that lathes are just one of a number of machine tools available. Fig. 6.109 shows a CNC router and 6.110 a wooden mould for a GRP guitar back that has been manufactured using it.

Lathes

Automatic lathes were used in industry during most of the twentieth century – the early versions being controlled by special cams which were made up for each component. The setting of these machines was a highly skilled and time-consuming operation. Fig. 6.111 shows the kind of CNC (*computer numerically controlled*) lathe which could now be available in student workshops and Fig. 6.112 a typical smaller training lathe.

Using this kind of machine tool depends on being able to enter the required component shape and the associated machining instructions and tooling requirements through the computer keyboard – a much less skilled and time- consuming activity than setting up their earlier mechanical counterparts.

Programming CNC machines

In order to help both users and suppliers of machine tools an international system of standard codes has been developed to describe machining operations. Table 6.6 shows the codes used on Boxford's CNC lathes and Table 6.7 the codes used on their milling machines. The full lists are very extensive and cover every conceivable machining operation and eventuality.

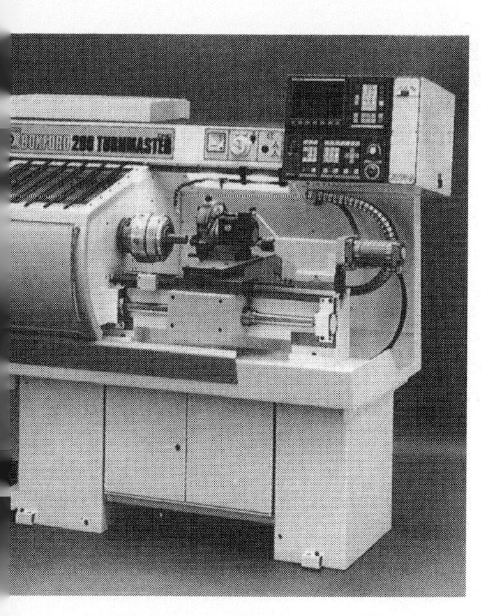

Figure 6.111 CNC lathe.

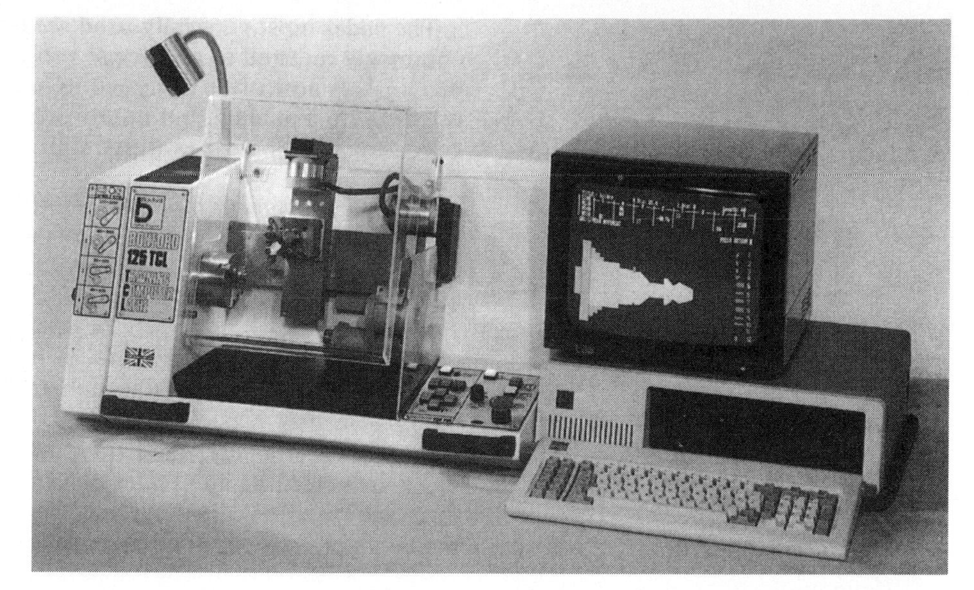

Figure 6.112 Training computer lathe.

G00	rapid movement, point-to-point
G01	linear movement
G02	circular interpolation, clockwise
G03	circular interpolation, counterclockwise
G70	imperial units selected
G71	metric units selected
G81	outside diameter (parallel) turning cycle
G82	facing/grooving cycle
G83	peck drill cycle
G84	thread cycle
G90	absolute programming selected
G91	incremental programming selected
M02	end of program
M03	start spindle forwards
M04	start spindle reverse
M05	spindle stop
M06	tool change
M08	coolant on
M09	coolant off
M39	close air chuck
M40	open air chuck

Table 6.6 Selected codes used to describe turning operations

G00	rapid movement
G01	linear interpolation (straight line cutting)
G02	circular interpolation, clockwise
G03	circular interpolation, counterclockwise
G04	time dwell
G25	jump to block number
G26	return from jump
G28	mirror image
G53	cancel change of datum
G57	set new datum position
G70	imperial units
G71	metric units
G79	point-to-point milling
G81	drilling cycle
G84	tapping cycle
G85	boring cycle
G86	PCD drilling cycle
G87	dish milling cycle
G88	rectangle milling cycle
G89	circle milling cycle
G90	absolute programming
G91	incremental programming
M02	end of program (single quantity)
M03	start spindle forward
M04	start spindle reverse
M05	stop spindle
M06	tool change
M08	coolant on
M09	coolant off
M30	end of program (repeat)
M39	close automatic chuck
M40	open automatic chuck
M43	subroutine create
M44	subroutine terminate
M45	subroutine call

Table 6.7 Selected codes used to describe milling operations

The codes most generally used are G-codes and entering these codes is commonly referred to as G-code programming. The user interface on most machines is now often designed to help in entering these codes without reference to a manual, but notes on the general meaning of some of the G-codes are given below as illustrations.

G00 Rapid movement, point-to-point. This instruction would cause the tool to move rapidly from one point to another, and hence is used to position the tool prior to machining. The start and end positions are specified by two coordinates (typically X across and Z along the bed).

G02 Circular interpolation, clockwise rotation. This instruction will cause the tool to move in a circular path and is used to turn fillets and small radii.

G81 Outside diameter (parallel) turning cycle. This instruction will result in the lathe turning a cylinder parallel to the machine axis. It will be necessary to specify the final diameter and the length of the cut (X and Z coordinates), the depth of each cut and the associated chuck speed and feed rate.

G82 Facing cycle. This instruction is concerned with movements across the lathe bed (in the X direction) and hence is used for facing off. It can also be used for undercutting or similar operations.

G90 Absolute programming selected. This means that all dimensional movements are given relative to a fixed reference point or datum.

G91 Incremental programming selected. This means that all dimensions are given as movements relative to the last position of the tool.

These G-codes representing the instructions for tool movements are combined with codes like M03 (to start the spindle forwards) and M06 (to change tools) to make up a block. A set of these blocks then makes up a part program. Once the blocks are entered there is normally a facility incorporated in the software to show the part program in a graphical representation on the computer screen. This process is known as emulation and allows the programmer to see exactly what has been entered before material is actually cut (see also Fig. 6.119).

Clearly the continuous operation of a machine tool depends on being able to change or replace worn or broken tooling without loss of accuracy. Consequently the tooling used on computer-controlled machine tools is often qualified tooling. This is tooling manufactured to a high tolerance such that it can be replaced, or its tip replaced, in the certain knowledge that positional accuracy of the cutting edge will be maintained. Qualified tooling should conform to ISO 1832 which specifies that the tool tip dimensions from three datum faces is within ± 0.08 mm.

Manual part programming is becoming a less important skill because many manufacturing software applications can now generate tool paths directly from the CAD model. However, it is still important that operators understand and can troubleshoot machining programs when necessary.

Multi-axis machines

Fig. 6.113 shows a programmable die sinker from the Rover Group's Swindon plant. This forms part of the company's computer-integrated engineering system and automatically machines the dies used to press out the body panels. Fig. 6.114 shows a similar, but much smaller, milling machine which could be available in a school or college workshop.

It is not however only computer-controlled milling machines which are used in the Swindon plant. Fig. 6.7 shows programmable robots being used to transfer panels between the presses and Fig. 6.115 shows a 5-axis laser cutter which is used to trim the panels.

Figure 6.113 Programmable die sinker at Rover.

Figure 6.114 CNC vertical milling machine.

Figure 6.115 Computer-controlled laser cutter.

A machine for this purpose needs the extra two degrees of freedom associated with the head movement to maintain the correct angle for laser cutting as the head goes round the edge of the panel.

The first robots used were generally employed in pick and place operations – one of the earliest being used by General Motors for unloading a die-casting machine. Nowadays robots are also commonly used for operations like welding and paint spraying. Clearly with more axes of rotation there are increased mechanical complexities, but the more major problems are associated with programming the machines. The programmers can enter instructions through a number of methods which are outlined below:

point-to-point The robot can be moved from one key point to another and at each the readings on the robot sensors are recorded.
calculated trajectories The operator can instruct the robot to move in linear, circular or more complex paths.
recorded trajectories The operator goes through the sequence of movements required and the robot samples data points as the motion is executed.

The strategy of recording trajectories has significant advantages in applications like paint spraying where the movements required are very complex. The computing power necessary for either point-to-point or calculated trajectories can become very large with multi-axis machines. For applications like straightforward welded joints the point-to-point or calculated trajectory methods are likely to be more successful, however, because they will give greater accuracy and smoothness than an operator can achieve.

The major components of a robot system are normally supplied by a manufacturer, but there is often interesting project work to be done in the design of end effectors. This is the name given to any device attached to the end of a robot's 'arm' or wrist, that is designed to perform a specific function such as gripping small, hot, fragile or soft components – placing delicate electrical devices, for example, or transferring hot metal.

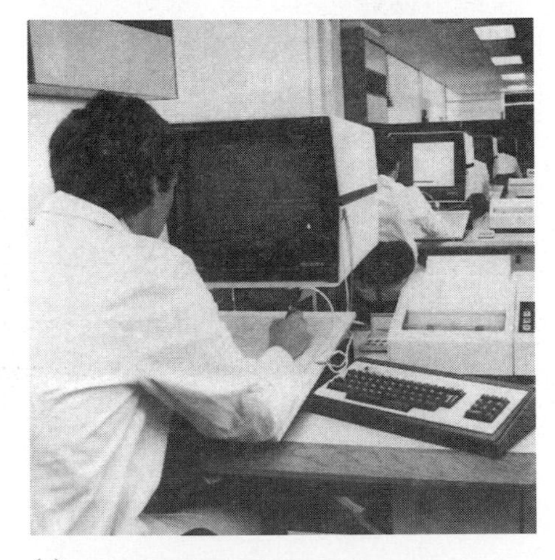

(a)

(b)

Figure 6.116
(a) Traditional CAD office
(b) CIM in the textile industry.

6.7.2 Computer-integrated manufacture

The workplace of a designer working in a traditional CAD office might look something like the one shown in Fig. 6.116(a). However, CIM is also applied in industries like textile design, as in Fig. 6.116(b).

The data the designer enters can be used directly in the manufacturing operations via computer links. Ultimately all of the movements and processing of raw materials, semifinished and finished components can be planned, executed, inspected and monitored using computer systems. The achievement of such a goal would, however, require the outlay of large sums of money and different companies may progress towards it in ways which best suit their state of development and their products. It is not necessary to do everything at once and elements like flexible work cells and automated materials handling equipment can be introduced independently.

Fig. 6.117 shows a flexible work cell with a robot specially developed to operate with the Boxford CNC machines shown previously. Such a unit is able to produce a variety of components unlike traditional dedicated tooling and consequently could form part of a flexible manufacturing system. It is important to note that the modern approach to quality management regards each cell as responsible for its own quality and hence its own inspection. Even when the output of the cell is destined to another area of the factory – an internal customer – the cell's operators are required to ensure that all the components it supplies are of the required quality. Nowadays inspection is more than just a part of the final inspection process. Such a 'right first time' approach is the essence of TQM or Total Quality Management.*

In order to develop this cell into a computer-integrated manufacturing system, a number of elements still need to be added – most notably automated inspection and materials handling equipment. Fig. 6.118 shows a modular system developed by TQ International to train students in the principles of computer-integrated manufacture.

Not only are the key manufacturing elements here, but also a network of computer workstations allowing the remote development of part programs. The designer can use simulations of their programs like the ones shown in Fig. 6.119 during the development of the part program. A computer controlling the overall cell operation will implement the programs on the machine tools when time is available.

*see section 1.4.3

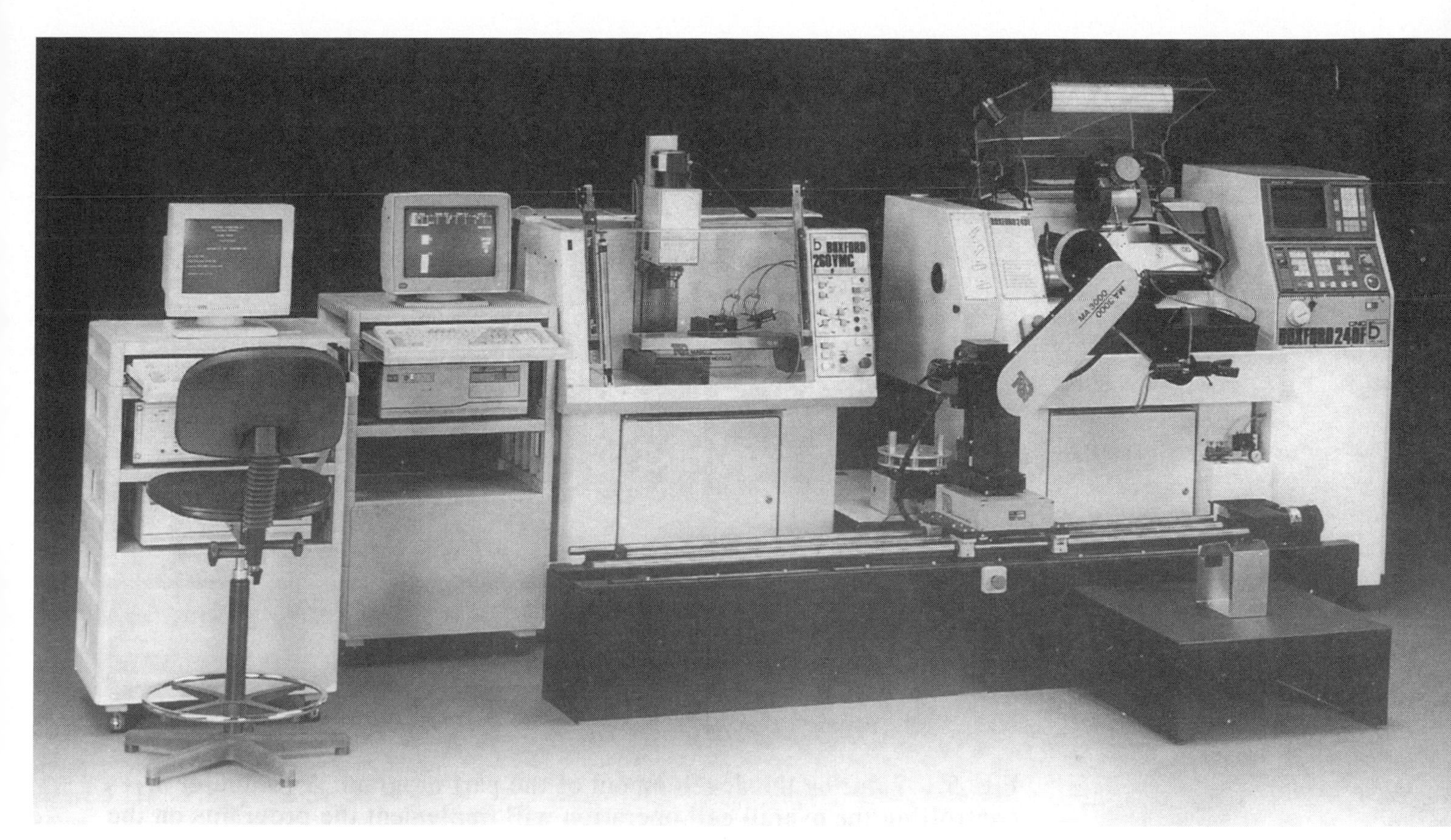

Figure 6.117 A flexible work cell.

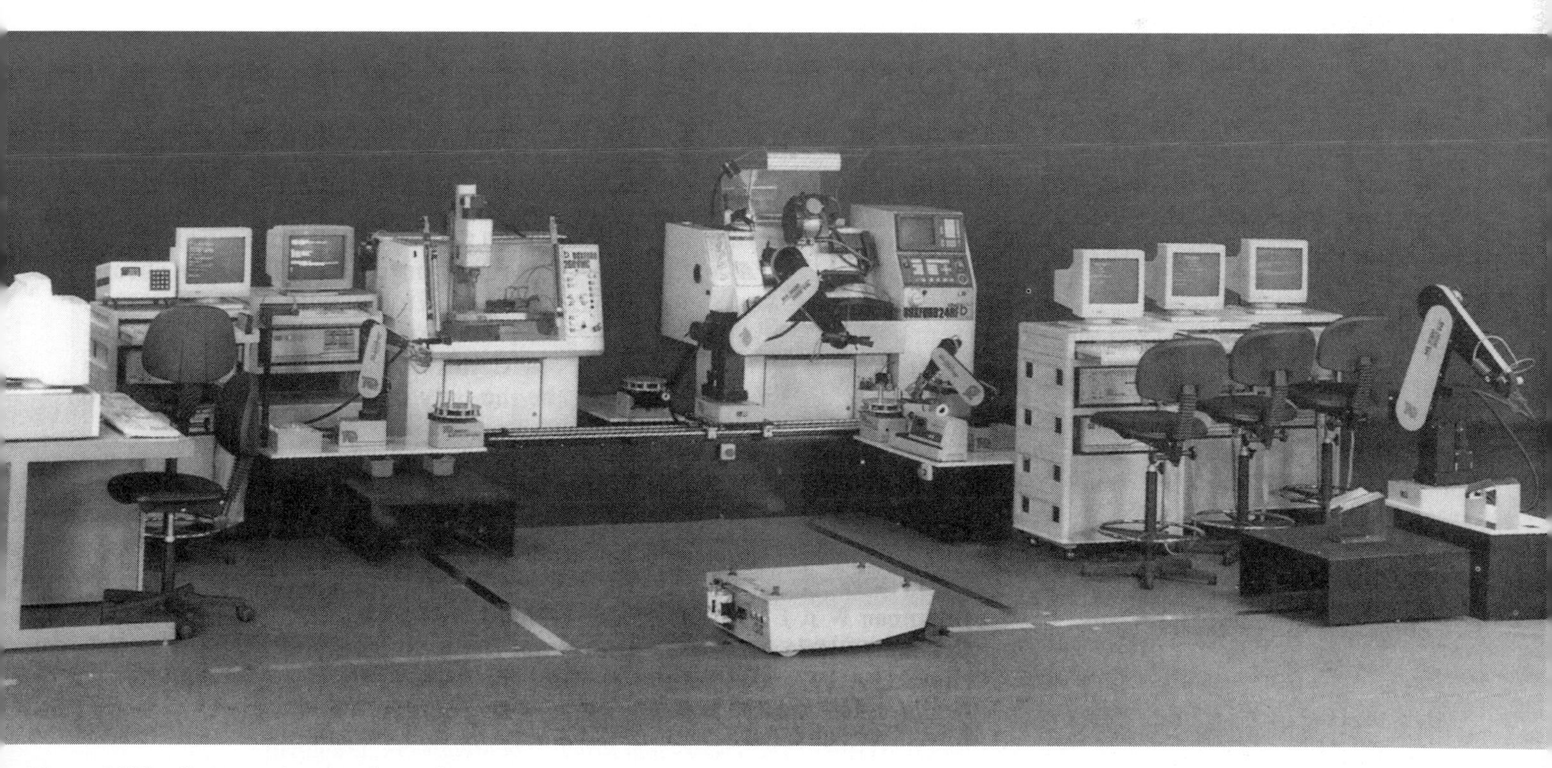

Figure 6.118 Computer-integrated manufacture.

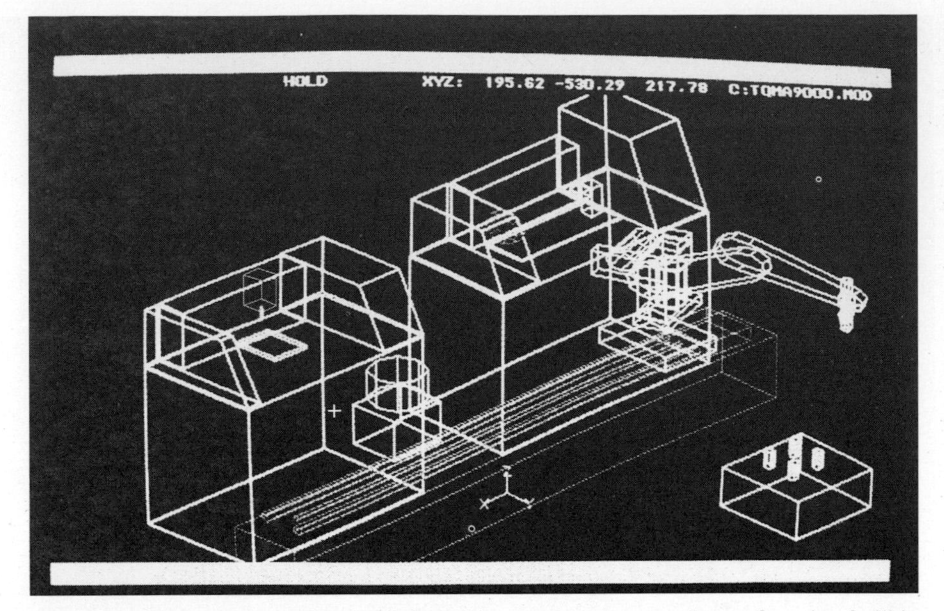

Figure 6.119 Computer screen simulation of the computer-integrated manufacturing facilities.

6.7.3 Choosing to use CAM

Deciding to use CAM presents exactly the same kind of difficulties as the selection of any other manufacturing process. It is a question of looking at the set-up (or fixed) costs and deciding whether any consequential savings in the cost of each component produced (the variable costs) justify the investment (see section 7.2). There is likely to be a breakeven quantity above which CAM is more economic than manual operation and equally a production quantity above which specially developed machinery or tooling will prove cheaper.

Generally CAM is thought to compete effectively with production quantities from about 10 to 1000. However, there are circumstances where it is economic to produce only a few components by CAM. For some highly complex shapes CNC machining may, in any case, be the only readily available production method.

The decision to use CAM or CAD/CAM is becoming ever easier to justify. Integrated software is being produced with shorter learning curves and training materials (e.g. CD-ROMs) are becoming more sophisticated. The day will soon arrive when CAD/CAM becomes an effective option for many 'one-off' components that would 'automatically' be made manually today.

Bibliography

Beadle J D 1972 *Plastics Forming*. Macmillan
Chapman W A J 1972 *Workshop Technology Parts 1–3*. Edward Arnold
Gorham C 1975 *Corrosion Attack and Defence*. BSC and NCST
Hicks G A 1975 *Design and Technology: Metal*. Wheaton
Houldcroft P T 1977 *Welding Process Technology*. Cambridge University Press
Jordan M 1985 *Creative Design and Technology*. Longman
Kempster M H A 1984 *Engineering Design III*. Hodder and Stoughton
Kenyon W 1982 *Welding and Fabrication Technology*. Pitman
Love G 1974 *The Theory and Practice of Metalwork*. Longman
Love G 1981 *The Theory and Practice of Woodwork* 4th edn. Longman

Marden A 1987 *Design and Realisation*. Oxford University Press
Millet R 1977 *Design and Technology: Plastics*. Wheaton
Millet R and Storey E W 1974 *Design and Technology: Wood*. Wheaton
Morton-Jones D H and Ellis J W 1986 *Polymer Products Design, Materials and Processing*. Chapman and Hall
Timings R L and Savage T E 1983 *Manufacturing Technology Checkbooks (1–4)*. Butterworth
West G H 1985 *Manufacturing in Plastics*. Plastics and Rubber Institute
Wingfield L E 1979 *Essential Information for Product Design*. RCA internal publication.

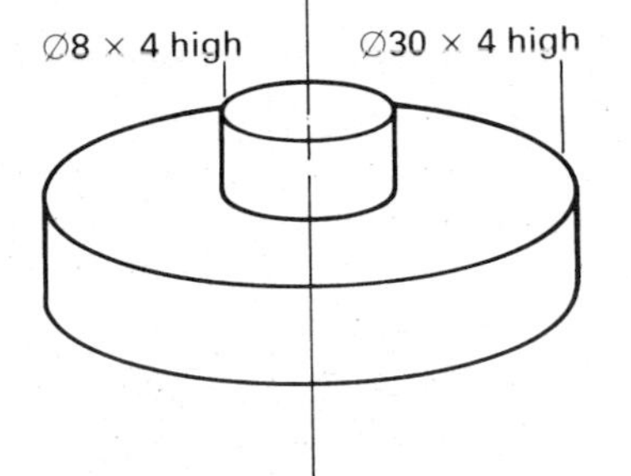

Figure 6.120

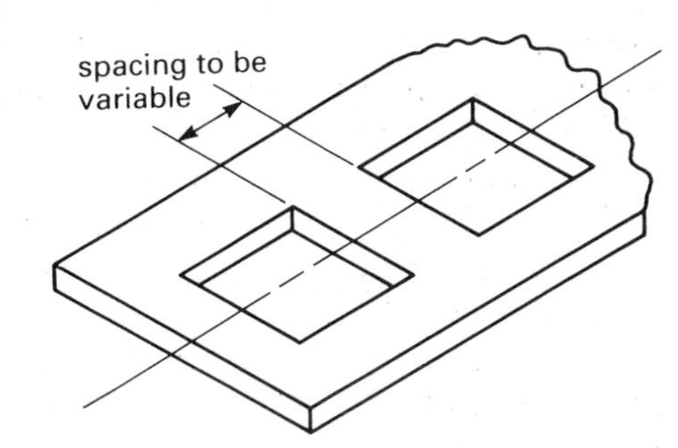

Figure 6.121

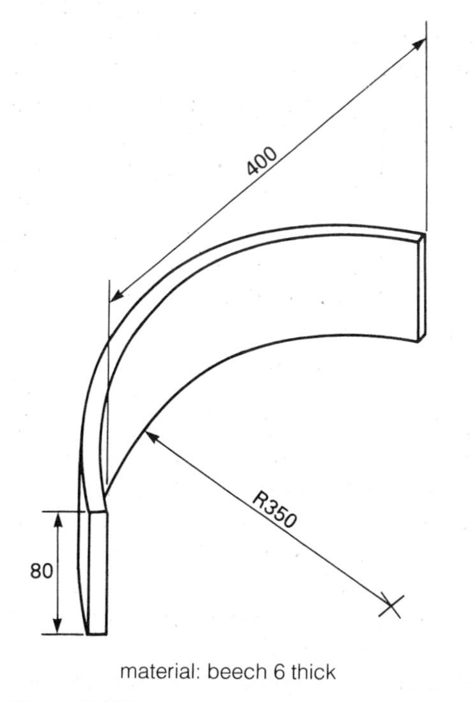

Figure 6.122

Assignments

1 (a) (i) What are the advantages and disadvantages of steaming timber in order that it can be bent to a predetermined curve?
(ii) How does the steaming enable the bending to take place easily?
(iii) Draw a diagram of a simple steaming apparatus you have used or seen.
(b) The rockers of a modern rocking chair are to be made of laminated strips of beech that have been steamed, glued and clamped around a former to produce the desired curve.
(i) Why is lamination a good choice for chair rockers?
(ii) Devise a test that could be undertaken to determine the shortest steaming time which would produce pliable strips supple enough to be bent to the required curves with minimum effort.
(iii) What problems would you be watching for in manufacturing the two rockers in this way?

2 Special tools, including jigs and moulds, are used a great deal in manufacturing.
(a) Select *two* of the following, and using sketches and notes, describe how each could be produced with special reference to the features of the jigs, moulds or special tools required.
(i) An injection moulded plastic disc as shown in Fig. 6.120.
(ii) 10 mm square holes on a central axis of a 30 mm wide strip of 18s.w.g copper for a bracelet as shown in Fig. 6.121.
(iii) A laminated chair back as shown in Fig. 6.122.
(b) Describe briefly how the special tools and/or the manufacturing process used in your answers would differ for small batch production and mass production.

3 Using a series of diagrams and sketches, describe in thorough detail *one* of the following plastic conversion processes. Amplify your answer by listing at least *three* differing products which would be made by the method you have selected, explaining why it had been chosen as the most appropriate technique for that product:
(a) vacuum forming;
(b) injection moulding;
(c) rotational or slush moulding;
(d) blow moulding.

4 Using notes and sketches describe a manufacturing process for four of the following:
(a) 'I' section universal steel beam.
(b) Plastic hosepipe from plastic granules.
(c) Brass contacts used in electrical switches from brass strip.

(d) One off shooting stick handle from molten aluminium alloy.
(e) Copper wire using copper rod.
(f) Section for greenhouse using aluminium alloy.
(g) Curved laminated beam using wood.

5 Metals can be moulded by the following processes:
(i) pressure die-casting;
(ii) sand casting;
(iii) lost wax casting.
(a) For each process select a product that would he suitable for production by the process, describe the product, and explain why the process is particularly suitable.
(b) In each case, identify the skills involved and where the major costs are incurred.

6 Material may be removed by the use of:
(a) heat;
(b) a shearing process;
(c) a chemical action;
(d) a wedge shaped cutting tool. For each of these methods give *one* example of its appropriate use, and explain why the process is suited to the material and the workpiece.

7 Control of the relationship between the tool and the workpiece is important in shaping materials.
(a) Describe three different ways of shaping materials to a circular cross-section. For each, explain how the shaping is controlled and, giving examples, explain what factors would determine which method to use in a particular situation.
(b) Describe three different ways of creating a flat surface. For each, explain how the shaping is controlled and, giving examples, explain what factors would determine which method to use in a particular situation.

8 Adhesives are now widely used to join wood, metals and plastics to themselves and to each other.
(a) For each combination of materials (six in all), name a suitable adhesive, stating in each case, why it is particularly appropriate. You should refer to the adhesive type or composition and not merely to its trade name.
(b) (i) Discuss your own use of adhesives in solving a particular design or manufacturing problem.
(ii) What steps did you take to ensure the adhesive(s) used was totally suitable for this particular purpose?

9 Figure 6.123 illustrates the following five situations:
(a) Joining the treads of a wooden step ladder.
(b) Attaching a stainless steel anchorage point to a GRP boat-deck.
(c) Constructing a permanent framework for a stool out of 25 mm square teel tube of 1 mm wall thickness.
(d) Attaching a cast aluminium alloy wall light fitting to a brick wall.
(e) Joining a brass tube to the ceramic base of a table lamp.

For any three situations:
(i) draw a suitable method of joining, showing any necessary changes to the components; and
(ii) explain why the method you have chosen is appropriate, with reference to the stress in the joint, the materials involved, and any other factors.

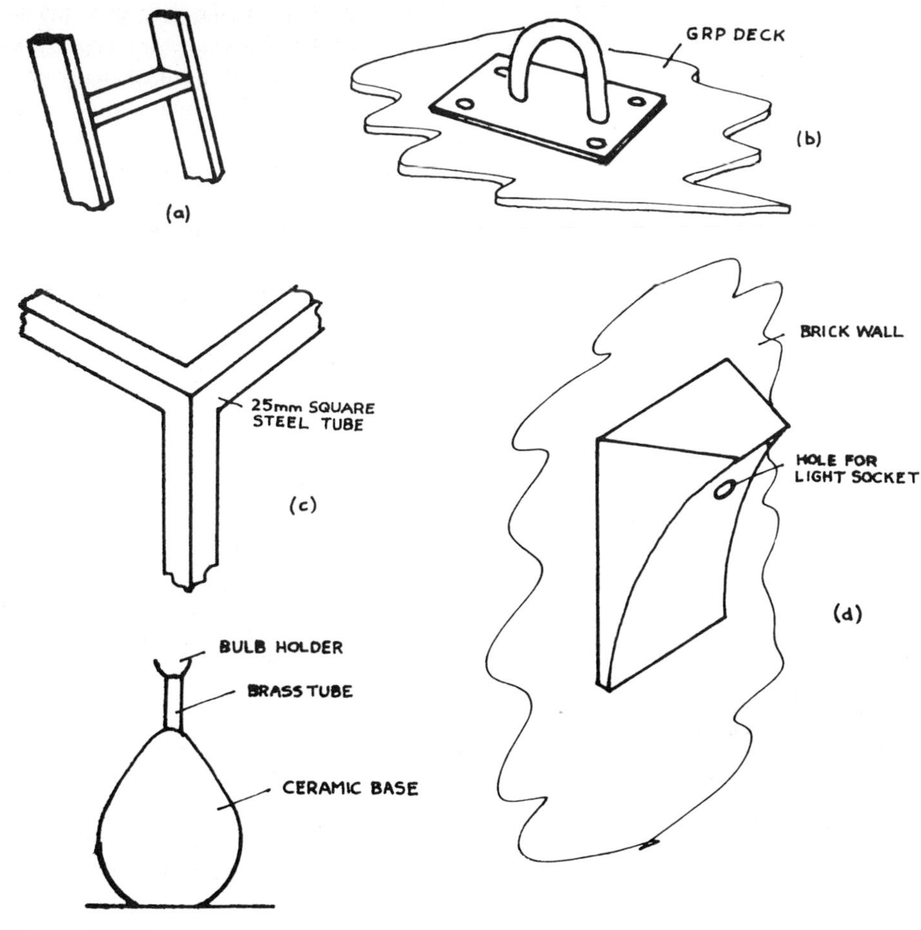

Figure 6.123

10 Select two products which are fabricated from metal:
– one involving mainly thermal joining;
– one involving, predominantly, mechanical fasteners.
(a) Sketch each product.
(b) For each product, explain why the joining process used is the most appropriate for the efficient construction and safe use of the product.
(c) For each product, discuss the effectiveness of the finishing processes used.

11 Corrosion is a major problem that has to be taken into consideration in the design and manufacture of a motor car. Most corrosion takes place in the wings, sills and at the bottom of the doors.
(a) Explain why corrosion takes place in each of the areas identified above.
(b) Suggest how the effects of corrosion could be reduced by appropriate action taken
(i) at the design stage
(ii) once the car is in everyday use.

12 Corrosion of some metals is a very serious problem confronting industry.
(a) How are metal finishes used to improve the appearance of products?
(b) Explain and compare the effects on tinplate and zinc-coated mild steel when the protective layer is damaged and the material is exposed to a damp environment.
(c) To improve the protection of aluminium, it is anodised.
Explain this process and the enhanced properties aluminium derives from it.

13 (a) Name *two* widely differing methods of surface finishing wood and *two* other methods of surface finishing metals. Describe a situation in which each named finish would be highly appropriate.
(b) For each situation given in your answer to (a)
(i) explain why the method of surface finishing is particularly suited to the material,
(ii) discuss any implications for further maintenance of the surface finish,
(iii) discuss briefly the economic implications of this method of surface finishing.

14 When selecting a suitable finish for an article a designer must consider many factors, for example:
(i) the material from which the article is made;
(ii) the environment in which the article will eventually be placed;
(iii) the visual effect.
Select *either*, (a) a sailing dinghy with its fittings *or*, (b) garden equipment, and discuss the interdependence of the factors listed.

15 (a) Explain what the initials CNC stand for and describe its principle of operation.
(b) Software is now available that will automatically produce CNC data from a drawing done on a CAD package. Explain the basic principles being used here to produce CNC code.
(c) Explain what 'emulation' is when referring to software for a CNC lathe and describe the advantages of using it.
(d) Servo motors are often used in place of stepper motors.
(i) Explain the differences between the two types of motor.
(ii) Explain with the aid of diagrams how a CNC machine knows where the cutter is in relation to the work with both of these motors.

16 (a) (i) Explain the terms 'part programming' and 'manual data input (MDI)'.
(ii) Discuss the advantages and disadvantages of MDI.
(b) Produce a program in any format with which you are familiar to machine and part off the engineering component shown in Fig. 6.124. The stock size is 25 mm dia. and the material is free cutting aluminium alloy.
(c) In the batch production of components from the same stock size, collet chucking can be an advantage. Briefly explain the principle of collet chucking and discuss the advantages and disadvantages of this form of work holding.

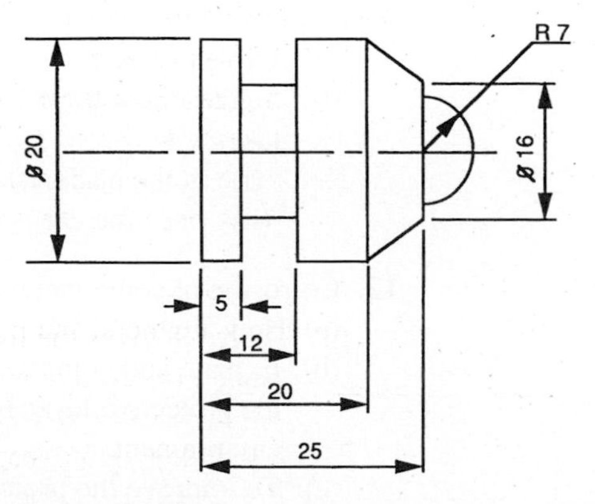

Figure 6.124

17 Figure 6.125 shows a sheet of aluminium alloy 5 mm thick which is mounted securely and with adequate clearance on the bed of a CNC milling machine.
(a) Using the given table as a guide compile a co-ordinate table for the points 0 to 6 enabling the 8 mm diameter cutter to produce the given profile.
(b) List and explain in detail the appropriate ISO codes necessary to program the cutter movement.
(c) Explain the advantages and disadvantages of manufacture by a CNC machine compared to manufacturing by other types of machine tools.

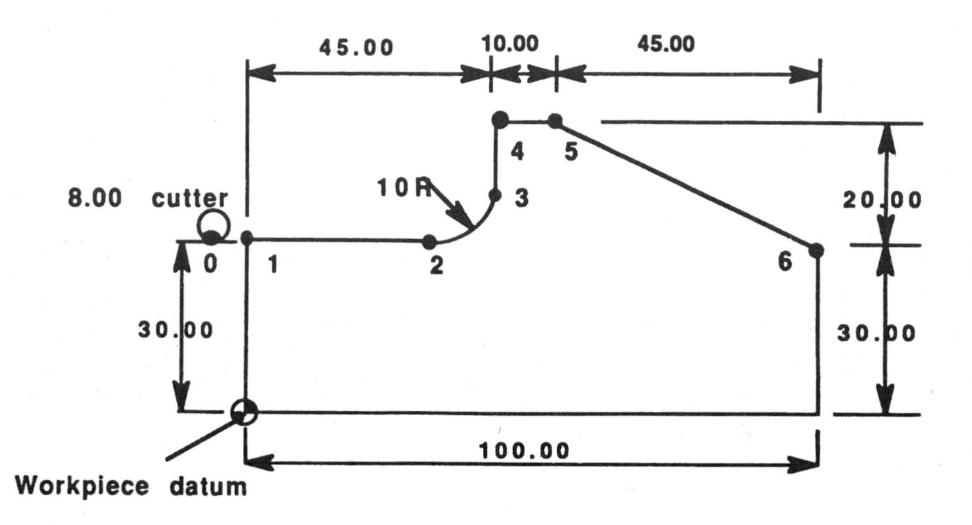

point	Co-ordinates from datum point X	Y
0	-5.00	+34.00
1		
2		
etc.		

Figure 6.125

18 (a) Draw a simple diagram to represent a CNC milling machine and indicate on it the directions of X+, Y+ and Z+.
(b) Describe the factors which limit the size of components that can be produced on such a machine.
(c) Name a material that is suitable for machining with a small CNC milling machine that might be found in a school Design and Technology department. State, giving your reasons in full, why your selected material is suitable.
(d) Table 6.8 gives the listing of a program for a CNC milling machine. Given that the material to be machined is 150 mm square and 6 mm thick with the datums for X and Y 10 mm outside its edge and Z set to the top surface.
(i) on the graph paper, plot the centre path of the cutter using a thin hain line,
(ii) on the graph paper, number the significant positions of the cutter,

(iii) given that the tool number 1 is an 8 mm diameter endmill, and tool number 2 is a 3 mm diameter endmill, draw using a thickcontinuous line the final milled profile around the centre line you have drawn,

(iv) add explanations of what each step of the program does in the remarks column of Table 6.8.

N	G	M	X	Y	Z	F	R	S	T	C	REMARKS
10	90										
20	71										
30		06							1		
40		03						3000			
50	00		0	0	2						
60	00		85	-27							
70	01				-1.5	100					
80	01		67.5	-45							
90	01		124	-103							
100	01		90	-69							
110	01		70.5	-89							
120	01		62	-81							
130	01		81.5	-61							
140	01		86	-65							
150	01		66	-85							
160	00				2						
170	00		53.5	-89							
180	01				-1.5	100					
190	01		39.5	-103							
200	00				2						
210	00		62.5	-97.5							
220	01				1.5	100					
230	01		22.5	-137.5							
240	00				2						
250	00		100	-56							
260	82				3.5	100	16			1	
270	80										
280	00		-30	30	40						
290		05									
300		06							2		
310		03						3000			
320	00		70.5	-28	2						
330	01				-1.5	100					
340	01		104.5	-28							
350	00				2						
360	00		114.5	-28							
370	82				3.5	100	11.5			1	
380	80										
390	00		-30	30	40						
400		05									
410		02									

Table 6.8

28 Many modern manufacturing industries use Flexible Manufacturing Systems (FMS).

(a) Explain the term Flexible Manufacturing Systems with particular reference to its 'flexibility'.

(b) Within a flexible manufacturing system describe the role of:

(i) manufacturing cells;

(ii) automatic guided vehicles;

(iii) automatic machine tools;

(iv) buffer storage.

(c) Explain why quality control needs to be carried out within a flexible manufacturing system and suggest how this might be achieved

7 Material and process selection

The selection of appropriate materials and manufacturing processes to make a product or system presents issues of ever-increasing complexity. The number of materials available has grown rapidly in response to the efforts of materials scientists – adding new alloys, plastics, ceramics and complex composites to the naturally occurring woods and rubbers and the materials developed by previous generations. Fig. 7.1 shows the foam and plywood composite used for flooring in caravans and boats, and the GRP (glass fibre reinforced plastic) and aluminium honeycomb composite used in aircraft.

Figure 7.1 Composite flooring materials.

Also, production engineers are continually improving existing manufacturing procedures and developing new ones. The exploitation of computers in controlling manufacturing operations and new power sources like the laser for welding and cutting are recent examples. Such innovations are changing the nature of our factories and workplaces, bringing demands for new intellectual talents rather than the traditional manual skills. Decisions about manufacturing methods therefore have social as well as economic implications.

New materials bring with them beneficial properties which can be exploited, but also difficulties in their use which would not have been found with their traditional counterparts. For example, connections to a honeycomb composite must be made in a way very different from the approach used with solid, homogeneous materials. The loads must be carefully spread to avoid the adverse effects of stress concentrations. Similarly, the need for careful control and the fumes associated with setting resins mean that manufacturing a GRP boat is a very different operation from making a traditional wooden craft. Of course the GRP has the advantage of needing no surface finishing of the hull or repeated annual painting.

Designers will have all of these issues in their minds, but perhaps the primary concern for them is the relationship between materials and processes, and the form of the product. One way of observing this relationship is to look carefully at several different designs which fulfil the same function but are made from different materials.

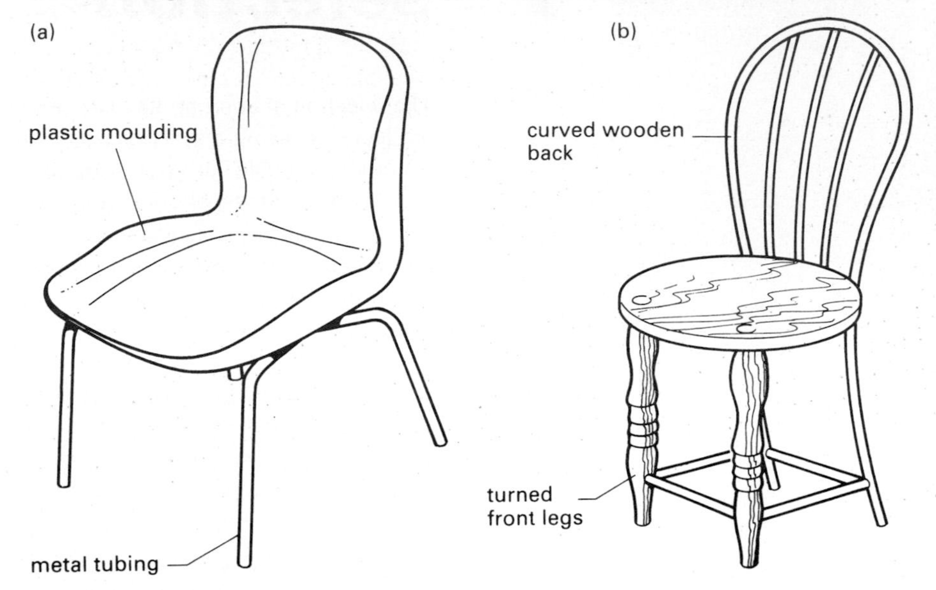

Figure 7.2 Classic chair designs.

Fig. 7.2 shows some sketches of classic chair designs. The influence of the material on the product form is evident but, equally, the nature of the manufacturing process can be seen to have been accommodated in the designer's concept. The tubular steel has been curved, both to exploit an easy manufacturing route and to give good aesthetic qualities. The form of the plastic seat shows how a simple moulding can be functional, easy to manufacture and visually pleasing. The front wooden legs have been turned to give a good appearance with limited manufacturing effort and the chair back bent to exploit this property of wood. All of these design features show the effective synthesis of functional and aesthetic issues in the materials and processes selected. To understand this relationship further, consider Fig. 7.3 which shows two possible responses to the idea of using a polymer concrete for chair construction.

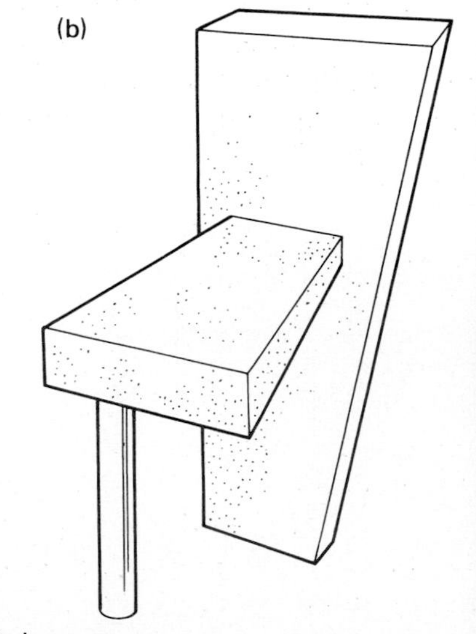

Figure 7.3 Concept sketches for polymer concrete chairs (a) suitable for casting (b) for fabrication from sheet.

This is no longer so absurd a concept as it might once have been, because polymer concrete can now be produced with tensile strengths of about 50 MN/m^2. Apart from its comparatively low cost, the good finish which can be obtained with suitable additions and the interesting sculptural forms which might be possible through casting are potential advantages. On the other hand its cold feel and high density could be significant disadvantages, although not if used for a city centre bench in a hot climate for example. Weight might give stability, but transportation costs could be very much increased. The design indicated by Fig. 7.3(a) shows the kind of form which might be expected if casting were adopted as the method of construction. The freely curved shapes which are possible give plenty of opportunity for the synthesis of aesthetic, functional, ergonomic and manufacturing considerations. Fig. 7.3(b) shows a more unexpected alternative – utilising the material's easy availability as sheet. (There are significant advantages in producing polymer concrete as sheets in controlled factory conditions.) Designs of this form would clearly have much in common with wooden chairs made from planks or boards and the designer might well expect to be able to transfer much of the experience gained from the service performance of traditional designs. It may be possible to anticipate areas which require strengthening or effective jointing methods without undertaking major testing. Such intuitive feel might significantly shorten the design development process bringing savings in both time and money.

In considering the issues surrounding the design of a polymer concrete chair the designer would need to consider the relationship of the form with both the material selected and the manufacturing route, and bring these together with a perception of the service requirements. For such well-known products as chairs the service requirements are felt to be intuitively understood, but for other products it is easier to transfer clearly expressed analysis to a design in a new material.

Both analysis and intuition should, however, be regarded as valid means of transferring experience. Product analysis was introduced in Chapter 1 from the point of view of assessing the market, product lifetimes, utility and so on. For the purposes of understanding material and process selection, it is useful to develop further aspects of product analysis, to give the maximum opportunity for learning from existing designs. The current goal of product analysis is to understand the important materials, processing and economic decisions which are required before any product can be manufactured.

The first task in product analysis is to become familiar with the product! What does it do? How does it do it? Here we are considering mainly the mechanical (and possibly electrical or optical) requirements, though it is also important to consider the ergonomics, how the design has been made user-friendly and any marketing issues which all have an impact on design decisions.

An important initial distinction needs to be drawn between products which are essentially one *component*, and whole *systems* which are made up of assemblies of many components. We can begin product analysis by looking at a whole system – consider the example of a bicycle:

- What is the function of a bicycle?
- How does the function depend on the type of bike (e.g. racing, about-town, child's bike)?
- What are the different components of the bicycle, and how do they fit into the overall product design?
- How is it made to be easily maintained?
- What should it cost?
- What should it look like (colours etc.)?
- How has it been made comfortable to ride?

To understand the materials and processes used, we need to think at the level of each component (frame, wheels, pedals, forks, etc.). It is important to start with an awareness of where a component fits into an overall system design – e.g., to appreciate the importance of weight, cost or safety.

At the component level we can analyse the function in more detail and draft a design specification for each part. A useful checklist of questions to ask could be:

- What are the requirements of each part (mechanical, electrical, aesthetic, ergonomic, etc.)?
- How many of each part are going to be made?
- What is the function of each component, and how do they work?
- What is each part made of and why?
- What manufacturing methods were used to make each part and why?
- Are there alternative materials or designs in use and can you propose improvements?

These questions act as a general guide – clearly the requirements need to be customised to the product. For example, for a drink container, the design requirements might look something like:

- provide a leak-free environment for storing liquid
- comply with food standards and protect the liquid from health hazards
- for fizzy drinks, withstand internal pressurisation and prevent escape of bubbles
- provide an aesthetically pleasing view or image of the product
- if possible create a brand identity
- be easy to open
- be easy to store, transport and be disposed of after use
- be cheap to produce for volumes of 10,000+.

With this as background, the materials and manufacturing processes used for different drink containers can be set in the context of the product function and the cost level which the market would allow.

Analysis of many products will reveal a tremendous range of alternatives in current use – both in the materials used, and the ways in which they are processed. For example, bicycle frames are made of steel, aluminium, titanium, magnesium, CFRP and wood, and even plastics have been tried. Choosing the materials and processes means balancing many requirements against one another. Furthermore, imagine that a range of new alloys or plastics or composites comes onto the market. Designers who had already analysed the reasons for the use of existing materials would be in a much better position to evaluate their potential than someone who relied on intuition alone. Clearly there is no single correct material or process to use in all circumstances, but it is vital to take a systematic approach to weighing up the options.

7.1 Selecting an appropriate material

Generally, the first stage of design (or product analysis) is to consider the materials. The materials are important because they provide the 'look' of the component as well as its behaviour. The three main things to think about when choosing materials are (in order of importance):

1. Will they meet the performance requirements?
2. Will they be easy to process?
3. Do they have the right 'aesthetic' properties?

The processing aspects of materials are discussed in section 7.2. It is important to note that experienced designers aim to make the decisions for materials and processes together to get the best out of selection – material and process selection are closely coupled activities. Of course, materials are not only chosen for technical reasons. They must also give the product visual qualities, e.g. providing associations and a sense of quality. Material selection must be sensitive to people's senses and emotional responses. It is important to try to learn to take these into account at the same time as the more technical issues, but, by their very nature, these matters are very difficult to quantify. This chapter provides a visual approach to taking the technical issues into account and, in so doing, is providing every opportunity to deal with the expressive functions of materials simultaneously.

As discussed in the previous section, most products need to satisfy several performance requirements, e.g. they must be cheap, or stiff, or strong, or light, or perhaps all of these things. It is usually possible to identify the key objectives we wish to maximise or minimise – e.g. we may require minimum cost or weight. Other requirements will just be specified as acceptable limits or constraints on the design – e.g. a specified allowable deflection at the working load. Consideration of these performance requirements will determine which material properties are important. Chapter 5 covers ways of determining the mechanical and physical properties of materials. Designers must be sure to recognise the relationship between test and service conditions, and to base selection on the correct properties – e.g. tensile test data are not sufficient for fatigue loading, and room temperature strength is not appropriate for high temperature design when creep may be an issue. The important point is that part of the designer's responsibility is to ensure that the material data used are valid. In practice, handbooks or manufacturers' data are rarely sufficient to ensure the performance of the product – some level of special testing can be expected (particularly for difficult properties such as wear and corrosion).

Having identified the dominant properties for the problem in hand, one way of selecting the best materials would be to look up values for the properties in tables, but this is likely to be a time-consuming and dull activity. A better way was introduced in Chapter 5, in which two material properties are plotted for all possible materials on a *material selection chart*. As shown below this greatly simplifies the consideration of multiple design requirements, and also ensures that no materials are overlooked.

Material selection charts

To illustrate the use of material selection charts, consider choosing materials for a rucksack frame. For this product the material must be stiff, light, strong, tough and cheap – in fact, this is a very common set of requirements for many structural components (usually with the objective of minimising weight or cost). We therefore need to find information about the Young's modulus (also known as the modulus of elasticity, E), density (ρ), strength, toughness and cost. It is unlikely that the cheapest material will also be the stiffest, lightest, or strongest, so we are going to have to make some judgements about how these properties trade off against each other.

First consider Young's modulus and density, which are plotted in the material selection chart in Fig. 7.4. Note that the scales are logarithmic, to cover the very wide ranges of each property, and that a range of values is shown for each material since each 'bubble' contains many individual materials with the same generic name. Initially we will consider generic material classes which look promising for stiffness at low weight. On this requirement alone, we can see what falls towards the top left corner of Fig. 7.4: woods, composites, some metals, ceramics.

Some of these materials clearly sound absurd for a rucksack frame. For example, the E-ρ chart suggests that ceramics offer a good compromise between high stiffness and low density. On these grounds alone they do, but we need to consider all the major requirements – so what about strength and toughness? Fig. 7.5 shows these two properties, and it is clear that though ceramics are strong in compression, they are less strong in tension than metals and have insufficient toughness for a product which will suffer impact, like a rucksack frame. Woods, too, are not very strong, so we will limit further consideration to metals and composites.

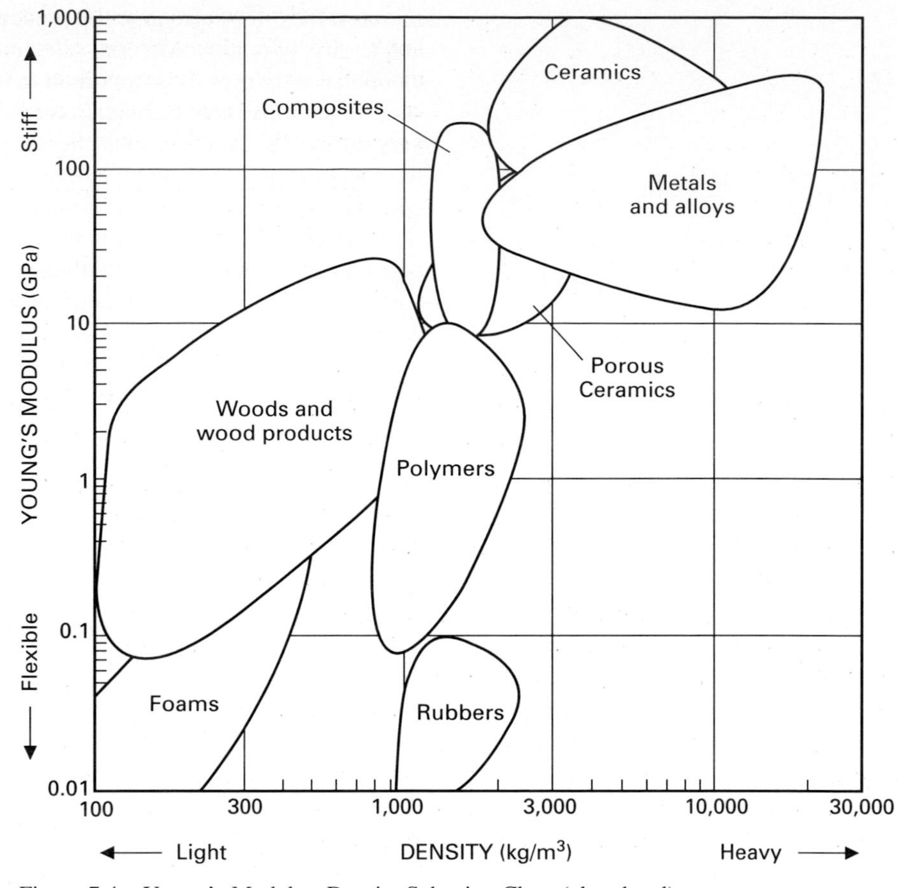

Figure 7.4 Young's Modulus–Density Selection Chart (class level).

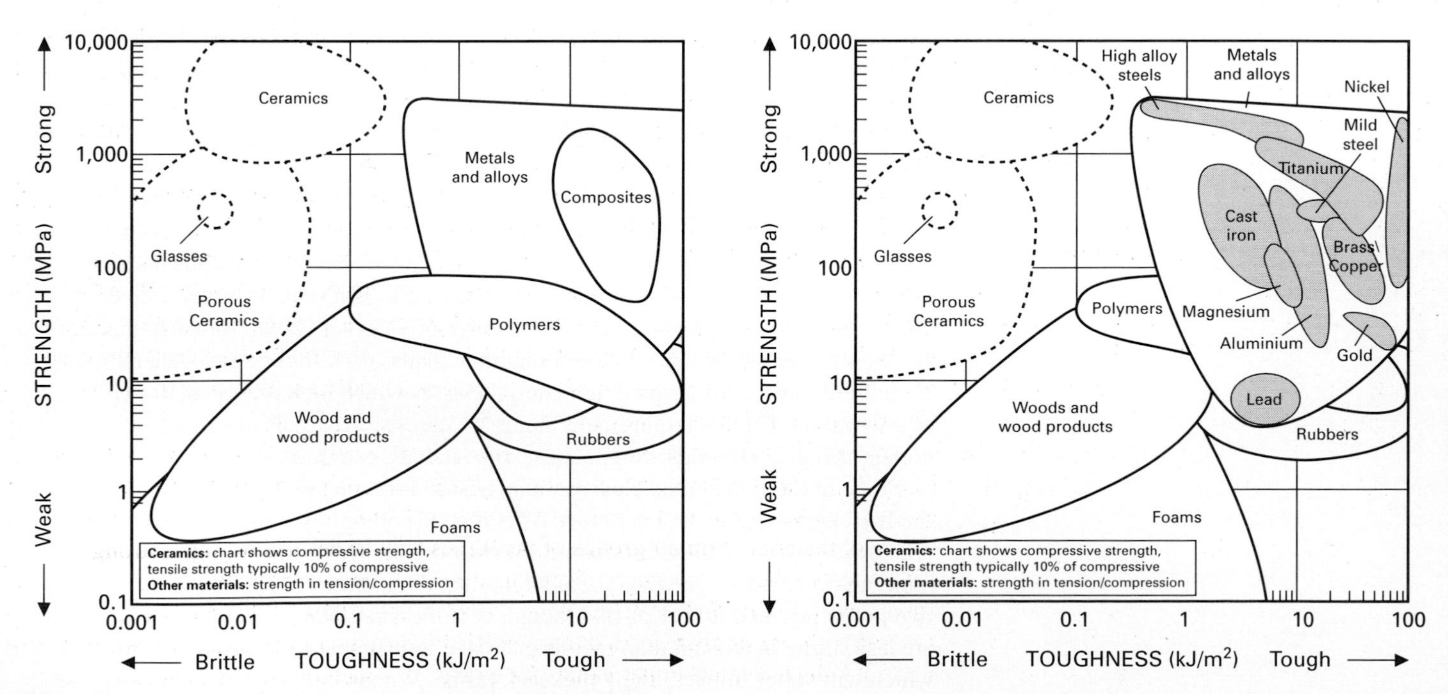

Figure 7.5 (a) class level; (b) class level and selected materials.

Having identified two material classes of interest, we can 'zoom-in' and look in more detail at the different options within these classes. Fig. 7.6 shows the Young's modulus–density chart, now with data for various metals and composites. Note that each material still covers a range, as there are many variants of each, but that the ranges are more tightly defined.

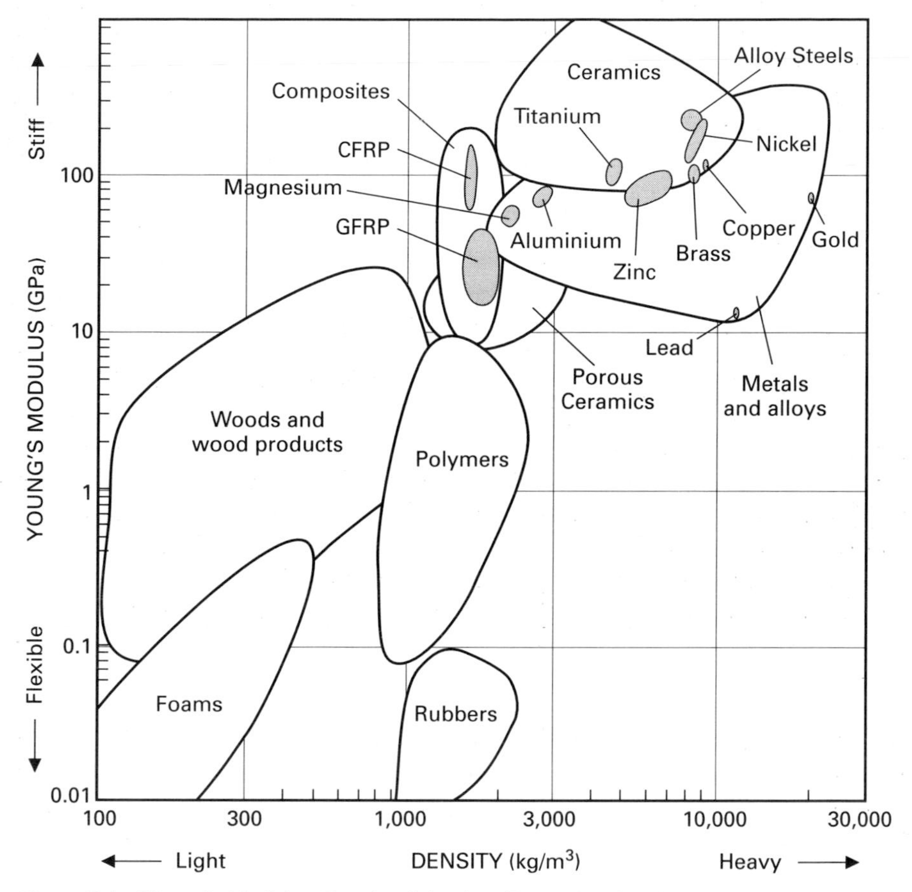

Figure 7.6 Young's Modulus–Density Selection Chart (class level, and selected metals and composites).

From this chart it appears that aluminium alloys might be a good choice: although they are less stiff than steel, they are a lot lighter. Magnesium alloys also look attractive, offering similar performance to composites. Of course we should also check the strength and toughness of the materials, and might now also consider the cost. Fig. 7.7 shows a strength against cost selection chart. It is important to realise that cost here is a raw material cost, and that manufacturing costs will be added on top of this. These added costs will vary depending on the process (see section 7.3), so raw material cost should only be used as a first indicator. From the chart, the cost of aluminium alloys, magnesium alloys and steels are all attractive compared to CFRP – but GFRP could compete more strongly.

Trading off stiffness or strength against weight is a very common design scenario. Rigorous analysis of material selection for light, stiff components shows that it is useful to consider combinations of properties, such as E/ρ or $\sqrt{E}/\rho$, to optimise the choice of material. Similar groups of properties may be defined for light, strong components – these groups are referred to as *performance indices*. While not universally valid for lightweight design, the specific stiffness, E/ρ, and specific strength, σ/ρ, are frequently used to give a first indication of material performance. Table 7.1 shows typical upper values for a range of structural materials. It is interesting to note that the woods and metals have very similar values of E/ρ.

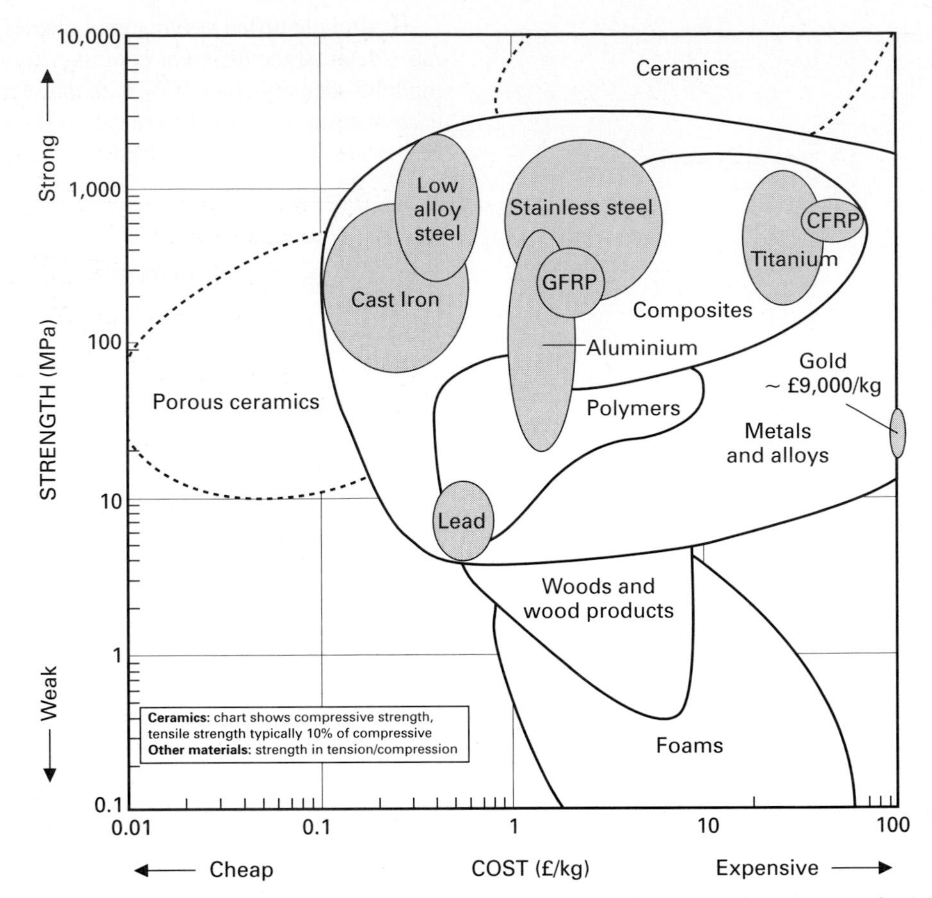

Figure 7.7 Strength–Cost Selection Chart (class level, and selected metals and composites).

Material	*Tensile strength* σ (MN/m^2)	*Young's modulus* E (GN/m^2)	*Density* ρ (kg/m^3)	σ/ρ in SI units	E/ρ
balsawood	10	3.3	144	0.069	0.023
spruce	35	9	500	0.070	0.018
steel	500	210	7800	0.064	0.027
stainless steel	980	200	7855	0.124	0.025
aluminium	90	70	2700	0.033	0.026
aluminium alloy	500	70	2810	0.178	0.025
GFRP†	250	48	1800	0.620	0.017
CFRP*	800	189	1500	0.533	0.067

†Glass fibre reinforced plastic
*Carbon fibre reinforced plastic

Table 7.1 Combining mechanical and physical properties

We may also plot these specific properties on a selection chart, as in Figure 7.8. The clustering of the metals around a similar value of E/ρ is apparent. The chart also illustrates that strength can vary widely for a given material, while Young's modulus has a very narrow range, since the bubbles are very elongated and parallel to the specific strength axis.

The selection charts enable us to identify promising classes of materials, down to the level of a sub-class of alloys, polymers or composites. It is important to leave the options quite wide at this point, as there are many other factors to consider. The most important are manufacturing issues, considered in the next section, but another aspect for structural components is the shape. Structural design is considered in detail in later chapters. For present purposes it is sufficient to note that the

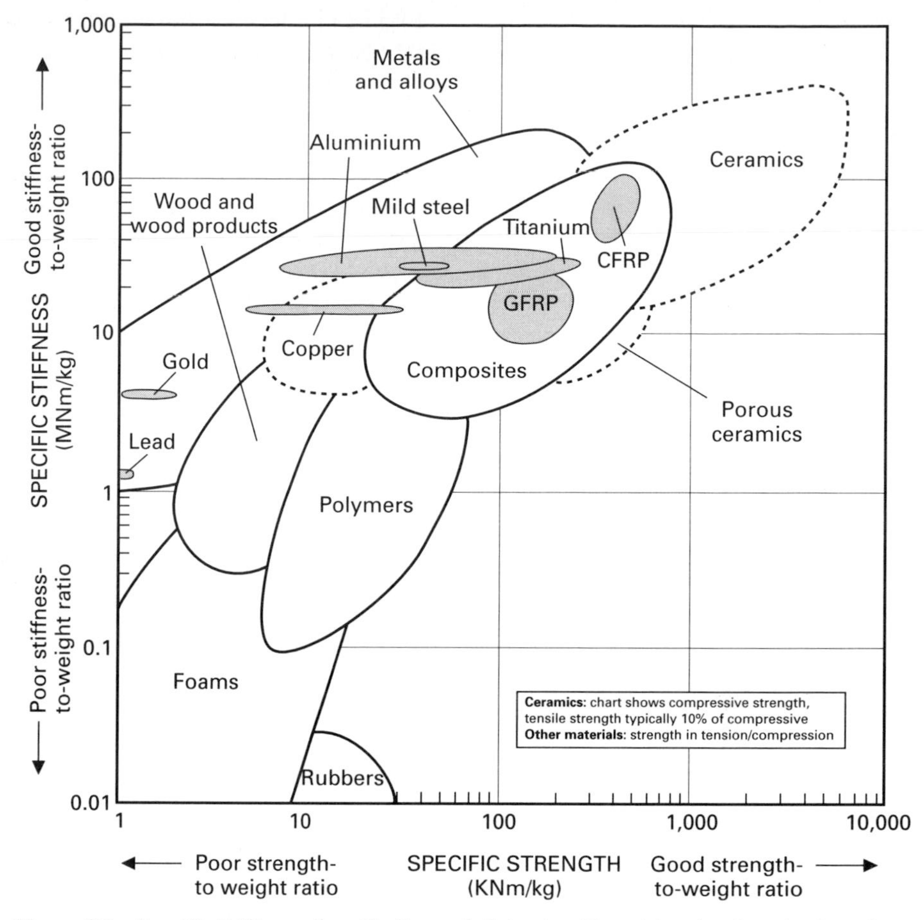

Figure 7.8 Specific Stiffness–Specific Strength Selection Chart (class level, and selected metals and composites).

performance of materials can be greatly improved by shaping, particularly for loading in bending or torsion. For example, steels are rolled into I-beams, and aluminium is extruded into hollow sections and tubes. Shaping a section adds rigidity in bending or torsion without increasing the weight (or, alternatively, a given stiffness requirement can be met with a lower mass of material). This influences material selection because some material classes can be shaped much more efficiently than others. On the whole, metals can be shaped very efficiently, but woods and polymers have to be used in solid and more inefficient shapes. Composites lie somewhere in between – so some of their apparent advantage on specific stiffness and strength (Fig. 7.8) is recovered by the metals. Detailed comparison of different materials in different shapes would be needed in the later design stages. It is sufficient for initial material selection to note that the competition between structural materials will be altered somewhat in stiffness or strength limited design, if shaping the materials is an option.

7.2 Selecting the manufacturing method

Having established some kind of product concept and provisionally chosen a material, the designer's next task is to develop a clear view of the manufacturing options. It is often possible to make comparable plastics products by vacuum forming or injection moulding. Vacuum forming limits the radii of corners and the depth of features but only requires simple tooling. Injection moulding gives more freedom in the product form but requires expensive tooling. Similarly, comparable forms can sometimes be created in metal by casting processes, forging, machining or fabrication, each with very different tooling requirements. There are, therefore, normally several ways a product can be manufactured from a material and the detailed design requirements will be different depending on the process selected.

The selection of a suitable process to manufacture a component is not a straightforward matter. There are many factors which need to be considered, for example, the size and shape of the component, the material to be processed and the accuracy required on dimensions. Whilst all processes have slightly different capabilities, there is often a large overlap – for many components there are a large number of processes which would do the job well enough. Conversely, there are some processes which are very dedicated – either to a type of material (such as composite forming), or to making particular sorts of shape (such as extrusion to make long thin sections). It is also necessary to consider what primary manufacturing steps may be needed (mostly to do with making the basic shapes of the components) and what secondary processes could be involved (dominated by joining and assembly, but including things like heat treatment or surface engineering). Finally, and most importantly, processing is where value is added to raw materials, so the manufactured cost becomes an important part of the design decision making.

To deal with this complexity, it is helpful to take a reasonably systematic approach – though, as with material selection, there is plenty of room for experience and intuition. As a first step, we can distinguish between two major stages in selecting a suitable process:

- **Technical suitability:** can we make the product with the chosen material and can we make it to the required level of quality?
- **Economics**: if we can make it, can we make it cheaply enough?

The first thing to check as far as technical suitability is concerned is the match between the material and the possible processes. In product analysis and in a lot of design work, the material to be processed is often known (at least down to the level of the class of material) before processing needs to be decided. This makes life a little easier, but it is important not to refine the choice of materials too far before asking about shaping and joining issues, which may effectively eliminate the chosen material at a late (and expensive) stage in the design. A convenient way to check material–process compatibility is first to classify materials and processes into their main groups, and then to draw up a table showing the viable combinations. Processes may be split up into:

- *metal shaping:* e.g. forging, rolling, casting processes
- *polymer shaping:* e.g. blow moulding, vacuum forming
- *composite forming:* e.g. hand lay-up
- *ceramic processing:* e.g. sintering
- *machining:* e.g. grinding, drilling
- *joining*: e.g. welding, adhesives, fasteners.

Material classification has been illustrated via the selection charts – we might consider one level below the initial classes of metals, polymers, ceramics, composites – e.g. carbon steels, stainless steels, aluminium alloys and so on. Table 7.2 shows a *material–process compatibility* table for a subset of materials (two classes of polymers, and woods) to determine which of the relevant processes are suitable.

These tables show whether a particular material–process combination is *routine*, *difficult* or *unsuitable* – 'difficult' processes may be worth considering if the options are few, but come with a warning that there may be technical (and therefore cost) implications. Using this table we can usually narrow down our choice of processing options, though it is true to say that for metals and thermoplastic polymers most of the shaping, machining and joining processes are all suitable. To discriminate between processes, we therefore turn to other technical issues. The most important to consider are:

- Can the process make something this **size**? For instance, you cannot easily die-cast an engine block – it is too big.
- Is the process suitable for the **shape** we need? For example: tubes are long and thin so ideal for extrusion but not casting; and you cannot blow-mould a telephone case because of all the holes in it!
- Will we get the **dimensional accuracy** and **finish** we want? Both dimensional tolerance and *surface roughness* are strongly influenced by which process is used, for instance sand casting is poor for both whilst die-casting is very good.
- How good will the **quality** be? This is the most difficult problem to address and usually there is little that can be said without actually trying it, or relying on past experience! However, we can sometimes make rough generalisations – for instance sand castings can often be *porous* and so might not be as strong as other casting processes or forging.

+ : routine *? : difficult* *✗ : unsuitable*		*Polymer*		*Wood*
		Thermoplastic (e.g. ABS, PE)	*Thermoset (e.g. epoxies)*	*(e.g. pine)*
Polymer Shaping	Polymer extrusion	+	✗	n/a
	Compression moulding	+	+	n/a
	Injection moulding	+	?	n/a
	Blow moulding	+	✗	n/a
Machining	Milling	+	✗	+
	Grinding	✗	✗	+
	Drilling	+	?	+
	Cutting	+	?	+
Joining	Fasteners	+	+	+
	Solder/braze	✗	✗	✗
	Welding	+	✗	✗
	Adhesives	+	+	+

Table 7.2 Example Material–Process Compatibility Table

This kind of information can only be obtained by reading descriptions of the various processes and then making informed decisions, and of course experience helps. The main concept to grasp is that aspects of the design, the material and the process can be closely integrated. Extrusion is suitable for making long, thin prismatic shapes, and it is also very suitable for aluminium alloys and thermoplastics. Forging and casting lead to very different microstructures in a given material – so the product properties depend not just on the alloy but also on the process route.

Making decisions based on the considerations above is essentially a screening activity – that is, we are eliminating processes which definitely cannot provide the design requirements, or building up our knowledge of possible problems. This approach works well for both the primary shaping processes, and for secondary steps such as joining, and provides a short list of candidate processes. The final step is to consider the most discriminating aspect of all, and to compare the costs of the various options.

Note that there are many costs involved in the making and selling of a product, including research, advertising, packaging, distribution and manufacturing. For different products, the importance of each contribution will vary. Note that the cost is not the same as the price – the difference is the manufacturer's profit! Here we are only interested in the *manufacturing* cost – the other costs are not likely to be affected much by our choice of process. The designer will, of course, need to be just as aware of the implications of the other influences on the product's eventual price – but keeping manufacturing costs down is everybody's concern.

Estimating process costs

Any production operation will have associated with it costs which will be incurred whatever the quantity produced, and other costs which will only be incurred when components are actually made. These are known respectively as fixed and variable costs. It is also useful to divide variable costs into the cost of the material used, and the costs associated with running the equipment. Basic manufacturing cost generally therefore has three main elements: material, startup and running costs, defined as follows.

Material cost

The material cost per component depends on the size of the component. We may assume that (for a given component) essentially the same amount of material is used for all processes:

Material cost per part = constant
(*same value for all processes*)

Startup cost

All new products have one-off startup costs, such as special tools or moulds which have to be made. This cost only occurs once, so it is shared between the total number of components made – the 'batch size':

Startup cost per part = one-off cost ÷ batch size
(*gets less for bigger batches and is different for each process*)

Running cost

Many manufacturing costs will be charged at an hourly rate, such as the cost of energy and manpower. In addition the capital cost of the machine must be 'written off' over several years, which can also be regarded as an hourly cost – the same would apply if a machine was rented instead. The share of this hourly running cost per part depends on how many parts are made per hour, the production rate:

Running cost per part = hourly cost ÷ production rate
(*constant, but different for each process*)

Each process has a range of values for *one-off costs*, *hourly costs* and *production rate* – these values can sometimes be obtained from data sheets, but manufacturers are often reluctant to disclose this kind of information (or may not actually know!). To estimate the manufactured cost, a particular value must be chosen from each range depending on what item is to be made. Factors in this choice include component size and complexity, but choosing sensible values needs some experience.

We can now find the total cost per part by summing the material cost, the startup cost and the running cost. Since the startup cost per part depends on the 'batch size', we can plot the total cost as a function of batch size, as in Fig. 7.9. The figure shows that the total cost falls with batch size, levelling off at very large batches where the startup contribution is negligible and the cost is effectively made up of the material and running costs only.

Fig. 7.9 shows how the cost varies with batch size for one process. If we repeat the calculation for another process, the two cost curves can be plotted together and the costs compared. Consider two casting processes – sand-casting and die-casting. Generally sand-casting is only used for small batches and die-casting for large batches – plotting the cost curves for these processes should show us why this is the case. Table 7.3 shows typical data for the contributions to cost in each process, to make a small, simple part. Then assuming a material cost of 10p per part, we calculate and superimpose the total cost for each process against batch size, as in Fig. 7.10.

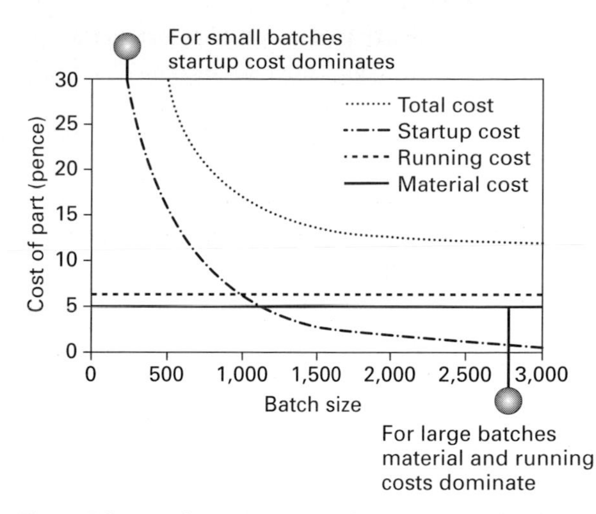

Figure 7.9 Total cost per part, and the three contributions to the cost, as a function of batch size.

	Sand-casting	*Die-casting*
One-off cost	£100	£2,000
Hourly cost	£30/hour	£35/hour
Production rate	100 parts/hour	500 parts/hour

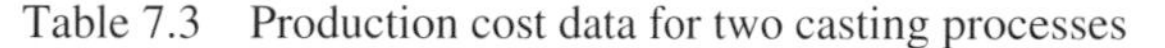

Table 7.3 Production cost data for two casting processes

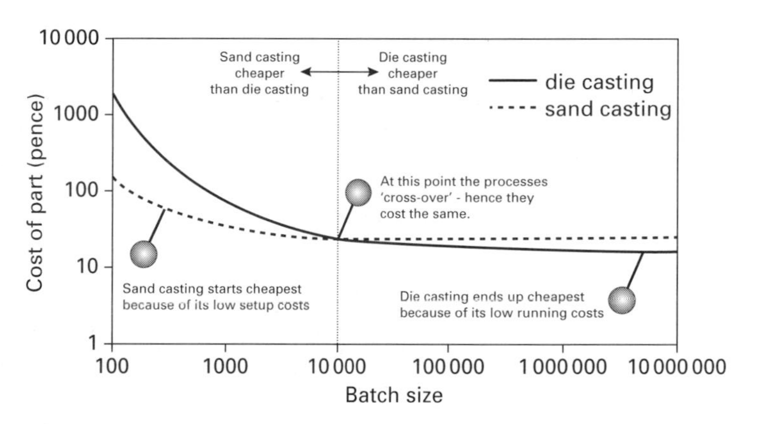

Figure 7.10 Total cost per part against batch size, comparing sand and die-casting.

We can see immediately that:

- Sand-casting is cheaper than die-casting if 1000 parts are to be made.
- Die-casting is cheaper than sand-casting if 100 000 parts are to be made.
- Sand-casting and die-casting cost the same if around 10 000 parts are to be made.

This procedure gives good insight into the way process costs compare, and indicates that many factors determine which will be the cheaper of two processes. In practice, there may be lots of processes to compare, and the cost curves are often rather less well defined than in Fig. 7.10 (due to uncertainty in the input data). A useful concept is therefore the 'economic batch size', which is the range of batch size for which a given process is normally found to be competitive. Data for this range is based on knowledge of cost comparisons of the type shown in Fig. 7.10 for many processes, but is largely based on production experience. Sand-casting, for instance, might have an economic batch size of 1–500 and die-casting 20 000+. This differs from the result when comparing just the two processes, since there are other processes which are competitive for the mid-size batches.

Some care must be taken when using economic batch size data, because it does not always take account of how the competitiveness of a process will change depending on the shape or complexity of the component. Forging, for instance, is typically economic for batches over 10 000 if the components are small, but this could fall to 500+ if the components are large (because there are fewer competing processes for large components). Some judgement often has to be used, based on quite detailed knowledge about the process, but the economic batch size is a convenient concept to give a first order indication as to whether a process will be economic.

Note also that we have assumed in our costing that the manufacturer has no extra reasons for choosing a process other than simple economics. This is often not the case and other factors may also affect the choice of the most economic process:

- machinery that has already been paid for
- special deals with suppliers
- problems finding trained operators
- environmental considerations.

For the purposes of product analysis or preliminary design, we can initially neglect these problems, but ultimately they can be very important and cannot be ignored.

Table 7.4 lists production processes under the headings of individual items, small-batch and quantity manufacture. Some indication is given of economic batch size limits (upper limits for small batch, and lower limits for quantity manufacture). As noted above, this table is for guidance only – process economics is one of the most complex aspects of design and manufacturing.

Individual items	*Small batch*	*Quantity manufacture*
sand-casting – wood patterns	sand-casting – metal patterns (500)	mechanised sand-casting
manual machining	shell-moulding (500)	gravity die-casting (1000)
rolling	full-mould casting (500)	high-pressure die-casting
open-die forging	expendable pattern casting (500)	cold chamber (20 000)
simple 2-D presswork	spinning	hot chamber (10 000)
laminating timber	capstan machining	low-pressure die-casting
steam bending timber	vacuum forming	sintering (10 000)
routing	blow-moulding	injection moulding (20 000)
manual welding	drape-forming	compression moulding (5000)
manual soldering		transfer moulding
manual brazing		rotary casting
		deep drawing (20 000)
		hydroforming (10 000)
		closed-die forging
		extrusion
		automatic machining
		impact extrusion
		production welding
		production soldering
		production brazing
		cold heading
		cold-roll forming
		shear forming
		flow forming
		NC machining

Figures quoted are typical limits on economic batch size. Actual quantities may vary considerably.

Table 7.4 Production processes

7.3 Environmental factors

The environmental impact of manufacturing, using and disposing of products has emerged as a major issue in recent years. This is prompted by increasing social concern about permanent ecological damage, particularly global warming and the greenhouse effect, and this can be expected to have an increasing influence on the decisions made by designers. The environmental impact of a product is a difficult thing to measure, as it must include the whole 'life cycle' of the product – extracting the raw material, manufacturing the product, and what happens when the consumer has finished with it. There are three alternatives for dealing with products which are to be scrapped:

- **disposal** (bury it or burn it)
- **reuse** (collect it intact, clean it up and use it again)
- **recycle** (collect it and recover the material which is used again).

These choices influence the designer in choosing the materials and processes to use in the product.

Design for disposal should ideally use sustainable, biodegradable materials and low energy manufacturing processes. This will minimise the impact of the continuous production and disposal cycle. Designs based on non-sustainable materials should also use as little material as possible. Good examples of this approach are provided by thin paper cartons and polythene packaging for food.

In design for reuse, the energy and resources associated with production and ultimate disposal are less important, but instead designs must be robust and long lasting. Now the environmental cost of recovering the products and making them fit for reuse is important, as well as the average number of times the product is re-used before disposal. Examples of good design for re-use are glass bottles for beer and milk.

Designers increasingly design for recycling. Here material resources are conserved and the energy required for primary extraction saved, but collection and reprocessing costs have to be considered. The most familiar examples of recycling are newspapers, aluminium and steel cans and glass bottles. For complex systems such as cars, the speed with which the product can be dissassembled and the materials identified becomes important.

The economics of recycling and reuse can also be strongly influenced by legislation. Some governments set targets for recycling fractions, and enforce these by mandatory deposit or collection systems. For example, until recently only glass bottles were allowed in Denmark, with deposits used to encourage recovery for reuse – plastic and aluminium drinks containers were prohibited. In Germany it is mandatory for supermarkets to collect waste packaging – but in fact a lot of this material still ends up in landfill as there is no parallel market which can use it cost effectively.

A great deal of research is going on into finding ways of measuring the environmental damage caused by products over their life cycle. Measuring environmental impact can involve factors such as:

- harmful emissions when the product is being made or used (e.g. tons of CO_2 produced)
- the costs of transporting materials
- the costs of landfill
- the energy returned if the material can be incinerated in a power station.

A useful first indicator of the environmental impact of a product is to consider the energy which goes into extracting and purifying the raw material used in the product. This data can also be displayed conveniently in a material selection chart of energy content and raw material cost, as in Fig. 7.11.

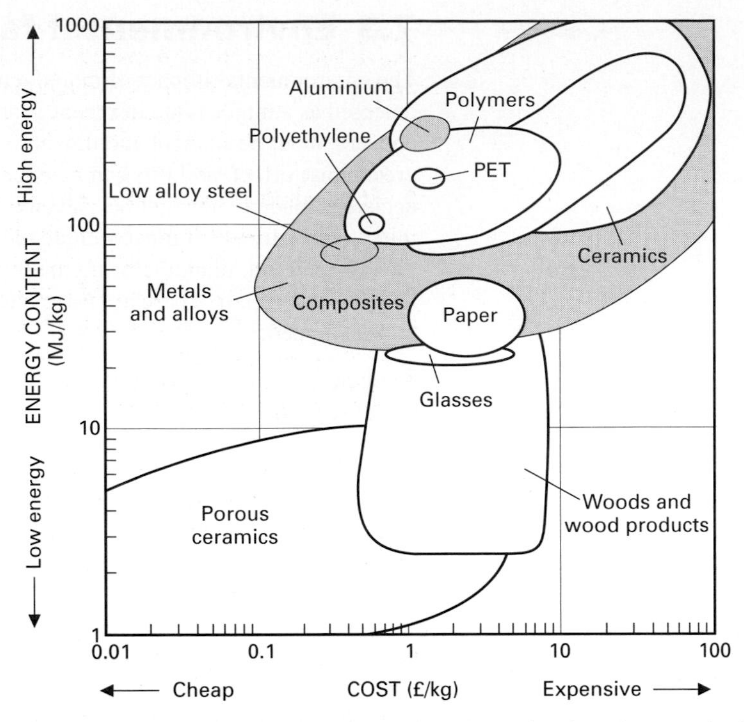

Figure 7.11 Energy Content–Cost Selection Chart (showing classes and selected materials).

In this chart, the materials lie in a broad band from bottom left to top right. This shows that the energy needed to make the material is a significant factor in determining its cost. We might expect that more effort will go into recovering expensive (energy-intensive) materials. Is this the case? A further selection chart shown in Fig. 7.12, shows how much of a material it is currently feasible to recycle, as a fraction of the total production, again plotted against material cost.

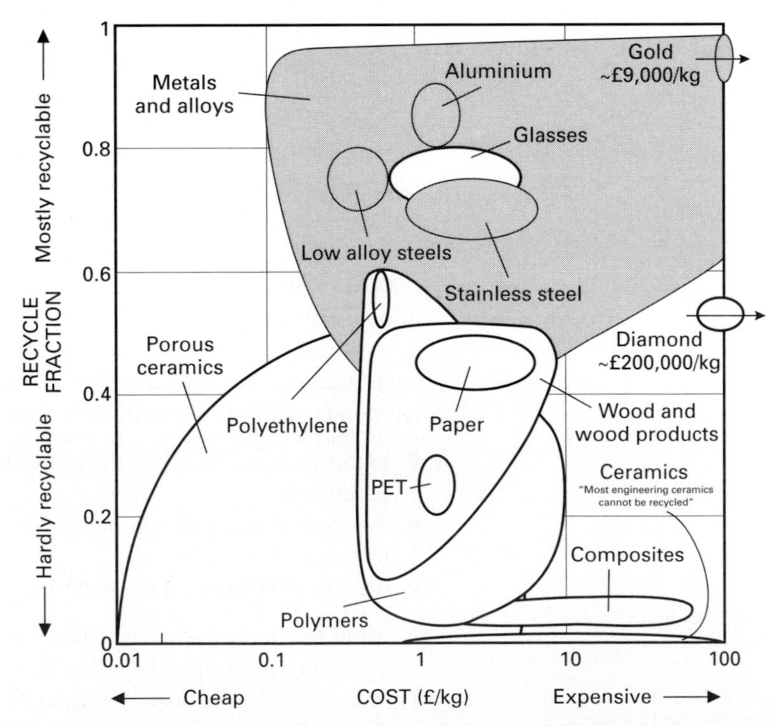

Figure 7.12 Recycle Fraction – Cost Selection Chart (showing classes and selected materials).

On the whole we see that expensive materials are the most recycled – e.g. diamonds and gold are recycled if at all possible! The picture is a little more complex than this however, e.g. it is worth noting the following:

- Aluminium is very energy intensive to produce from ore, but as it is easy to remelt, it is particularly cost effective to recycle.
- Although thermoplastics can be easily recycled once separated, the bulkiness of scrap polymer products like drink bottles means that very large volumes have to be collected, which is rarely economic.
- Even if they can be collected, mixed thermoplastics are difficult to separate and it is probably more economic to burn the material to produce energy.
- In spite of their higher cost, composites are difficult to recycle because the fibre and matrix cannot easily be separated, and ceramics cannot effectively be recycled at all.

The influence of the environment on design is, rightly, here to stay. Ultimately the responsibility of the designer is to reach decisions which achieve the service requirements in the most cost-effective manner, but increasingly the cost will include some evaluation of the environmental penalty.

7.4 An overview

The integrated nature of material and process selection is best illustrated by Fig.7.13. There are many factors leading to the selection of a material – the most important being the properties and cost of the material, the design and service requirements,

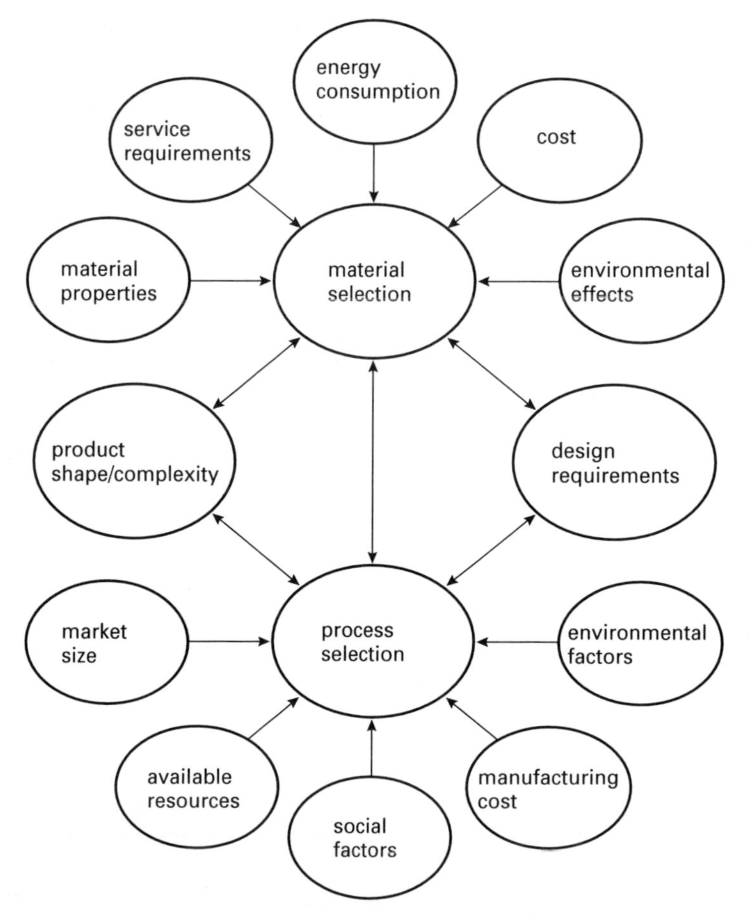

Figure 7.13 An integrated overview of material and process selection.

and the impact on the environment, such as energy consumption in producing the raw material, and what happens to the product at the end of its life. Many of these are closely coupled to the manufacturing process. Increasingly designers must have the skill to evaluate both material and process together earlier and earlier in the design process, due to the pressure to accelerate design cycles and reduce costs. Material and process selection are therefore at the heart of modern design methodologies such as concurrent engineering and design for manufacture.

Bibliography

Ashby M F 1999 *Materials Selection in Mechanical Design,* 2nd edition, Butterworth–Heinemann.

Lovatt A M, Shercliff H R and Withers P J 2000 *Material Selection and Processing,* CD-ROM and supporting booklets. Technology Enhancement Programme (TEP), London. World Wide Web: http://www-materials.eng.cam.ac.uk/mpsite/

Assignments

1 When designing and making your final project, the material(s) from which it was made would have been chosen to meet a detailed specification.
   (a) What were the major criteria you used in the selection and why were they so important?
   (b) Was there any clash between material(s) suitable for the end product and the method(s) of manufacture available to you? If so, how were these resolved? If not, why did you think this was so?
   (c) Assume that your design is to be taken up by a manufacturer for mass production. Recommend what changes might be necessary with respect to the choice of materials and the methods of manufacture. Justify your recommendations.

2 The Governing Body of a school has recommended that display stands should be provided in the entrance hall to show pupils' work.
   The boards will be 2000 × 1500, and 40 thick. They will be double-sided with surfaces that can be re-used and will accept staples and pins. The board will have an edging to give a good finish.
   (a) Identify the criteria to select materials for the display board; this should include the central sandwich of the board, the covering and edging.
   (b) Choose suitable materials and explain how they meet the criteria in (a) above.
   (c) Describe how one of these boards could be made in a school workshop, identifying any major difficulties.
   (d) Explain how the mass-manufacturing of these boards may differ in the materials used and how the manufacturing process may vary.

3 Choose two items of tableware, which will perform a similar function but which are made from different material(s), for example, a plastic teaspoon and a stainless steel teaspoon. For the pair of items you have chosen, explain:
   (a) how the aesthetic and structural qualities of the material(s) used relate to the requirements of the product;
   (b) the processes involved in making the product and how these relate to the properties of the material(s); and
   (c) the possible influence of economic and life cycle considerations on the choice of material(s).

4 Rowing oars are almost always made out of wood or fibre composites. Use the material selection charts to identify light, stiff materials, and explain why these are not preferred materials for oars. Which of these two classes of material would be chosen for the oars in: (a) a pleasure boat in a park; (b) an Olympic racing shell?

5 Use the material selection charts showing strength to explain why steels are used for such a diverse range of engineering products. Distinguish between cast iron, mild steel, alloy steels and stainless steels, and give examples of products made from each.

6 On the specific stiffness – strength chart, the bubbles for the metals and alloys tend to be elongated parallel to the strength axis. By considering the physical origins of Young's modulus and strength in these materials, explain why this is so.

7 Explain why bike frames are made from steel, aluminium alloy, titanium alloy and carbon fibre composites. Why are carbon fibre composites and titanium generally only found in performance racing bikes. Discuss the practicality of making a bycycle frame out of a polymer.

8 The dominant material in car bodies is steel, but there is now fierce competition from aluminium and glass fibre composites. Use the energy content and recycle fraction selection charts to compare how these materials compete with steel in a life cycle analysis of the car.

9 Under the following headings, discuss briefly the relative advantages and disadvantages of timber, metal and plastic window-frames. Your answer should refer to one specific named material from each of the three groups of materials:
  (i) manufacturing methods employed;
  (ii) durability and maintenance;
  (iii) aesthetic factors.

10 Cast iron, PFTE, carbon steel, beech, PVC, copper, ABS and melamine formaldehyde are all used in kitchens.
Select *five* of these materials and for each:
  (i) give one example of its use in this situation;
  (ii) explain how its properties make it suitable for that use.

11 Timber, metal, plastics, clay and concrete can all be used for flooring. For each of these materials select a flooring situation and explain what properties make that material suitable for that situation.

12 The production of an artefact in an alternative material can lead to significant changes to its form.
Choose an example where this has occurred and explain the changes.

13 Fig. 7.14 on the following page illustrates some of the components of a portable electric drill. Select six components used in such a drill which require different manufacturing techniques. Describe a method of manufacture for the components in your list, stating in each case the material you would use and the properties that make it suitable.

Figure 7.14

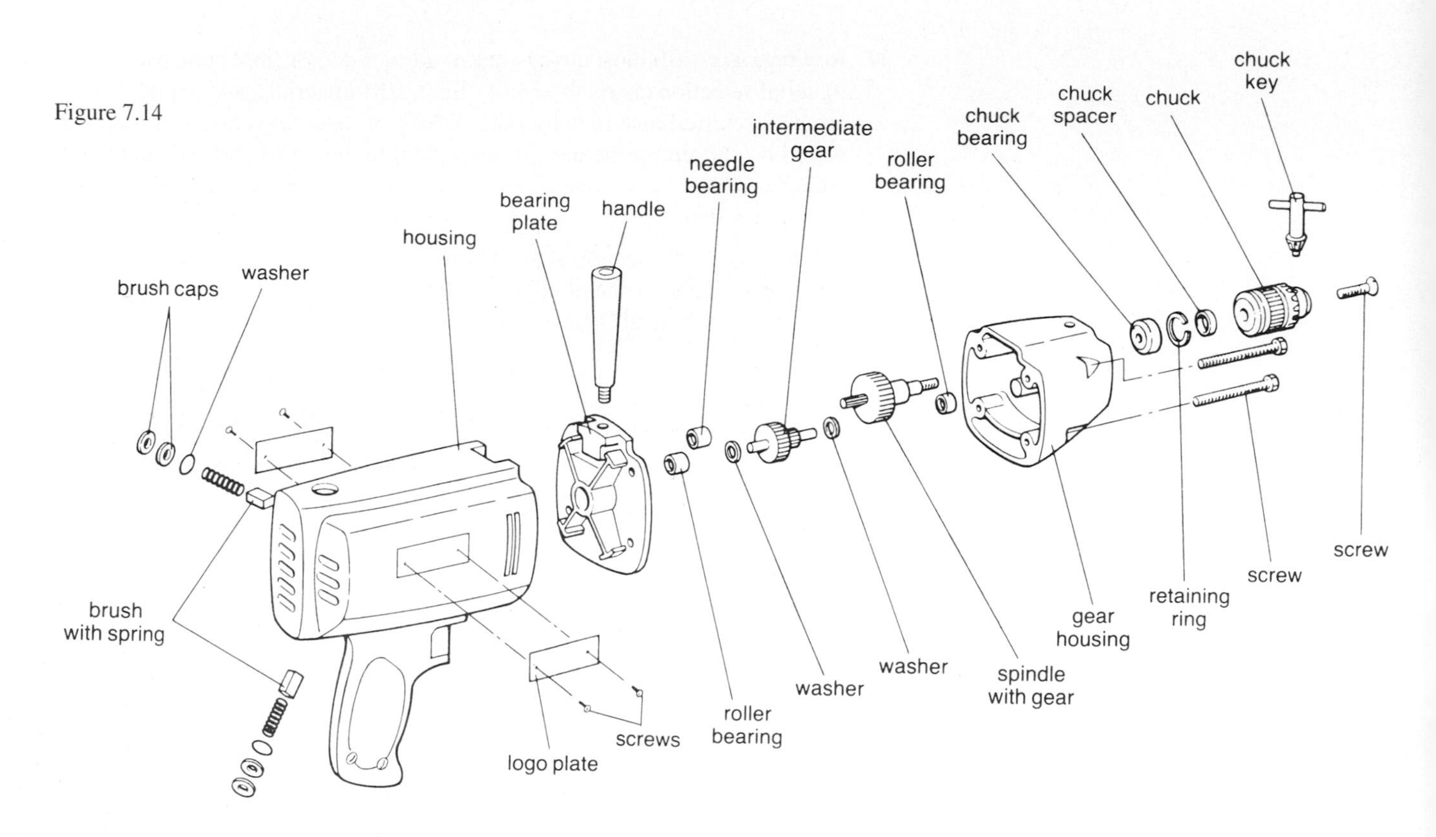

14 It is often possible to produce a component by differing manufacturing methods.

Either State what factors should be considered and how they could influence the decision whether to make a curved chair back in wood either by laminating or by steam bending.

or State what factors should be considered and how they could influence the decision to make the tailstock of a small model-making lathe by casting or by fabrication.

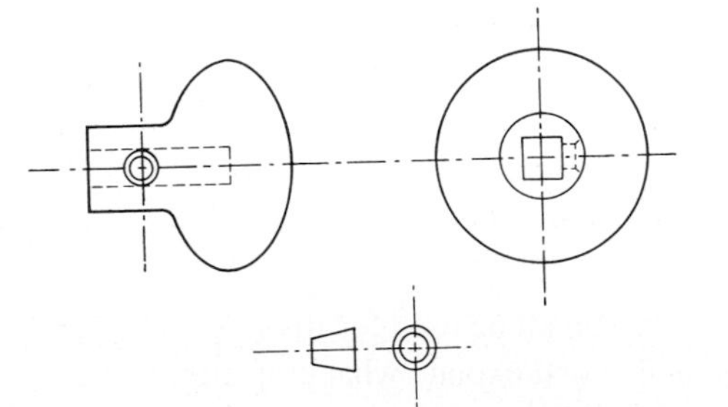

Figure 7.15

15 (a) Door knobs, of the type illustrated in Fig. 7.15, can be purchased in hardware shops made in wood, metal, plastic or ceramic. Name six specific examples of materials, including at least one from each of the four classes, which might be used for such knobs.

(b) Describe in outline, with the aid of flow charts or otherwise, the manufacturing process for a door knob made from each class of material.

(c) Discuss the advantages and disadvantages of each material you name in (a) for this application.

16 (a) Designing and making a one-off product is very different to producing many thousands. Using, as an example, a project you have made, discuss and sketch the changes which would be needed to make it viable for mass production. Your answer should compare your one-off product with a similar mass produced one in terms of:

(i) choice of material,
(ii) shaping and forming,
(iii) joining and assembly,
(iv) applying finishes, and
(v) evaluating and testing.

(b) Discuss the disadvantages of mass production of products.

17 (a) Explain, with the aid of diagrams where necessary, the processes of:
(i) sand casting,
(ii) sintering,
(iii) continuous casting
giving typical examples of the commercial use of each and, in each case, state why the preferred process is appropriate.
(b) A component, as illustrated in Fig. 7.16, is to be made in a low-carbon steel. Suggest appropriate manufacturing processes for the production of
(i) a single component,
(ii) a single batch of 1000,
(iii) an on-going requirement of 1 million per year.
(c) List in detail the steps of each manufacturing process selected in (b).

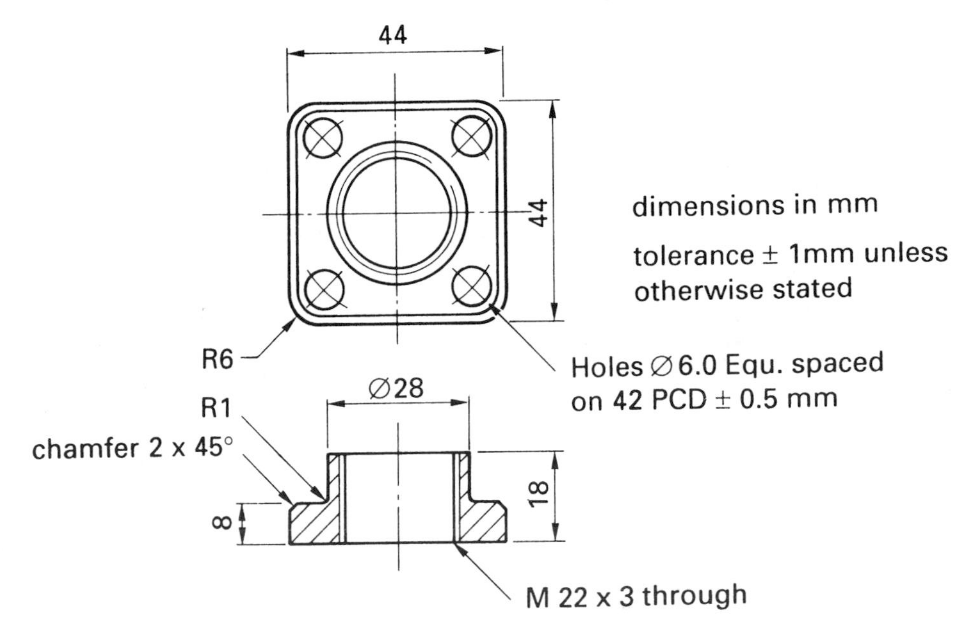

Figure 7.16

18 A small plastic bucket is to be made by injection moulding or rotational moulding. The designer wishes to know which process is likely to be cheaper for different batch sizes. The total cost per part is the material cost plus startup cost and running cost (as defined on page 358). Use the data in Table 7.5 to find the running cost for each process, and then the startup cost for batch sizes of 1000 and 50 000. Hence, calculate the total cost per part for both processes at these batch sizes, and recommend which process to use in each case.

Process	*Material (£)*	*One-off cost (£)*	*Hourly cost (£/hr)*	*Production rate (parts/hr)*
injection mouding	0.25	5000	20	120
rotational moulding	0.20	1000	20	40

Table 7.5 Cost data for two moulding processes.

SECTION **3**

Systems at Rest

The structure of every object must be designed so that it does not break, twist, deform or collapse when loads are placed on it. The photographs below show four different types of structure:

a) a static structure – a tent
b) a static structure with some moving parts – a gym machine
c) a movable structure – a computer workstation
d) a moving structure – a bicycle.

a) tent

b) exercise machine

c) computer workstation

d) bicycle

Different types of structure

The framework for each of these can be designed using the same techniques. The length, cross-section, type of material and method of joining the components of each framework must be analysed carefully to ensure a good, safe design.

- Chapter 8 introduces us to different types of structure, the loads that they carry and the analysis of the internal and external forces of frameworks.
- Chapter 9 explains how to determine the length and cross-section of different types of components to enable them to carry these internal forces.

The computer software associated with these chapters, which is listed below, is available for you to download from Longman's website http://www.longman.co.uk
Chapter 8: Computer program for frame analysis
Chapter 9: Computer program for Shear Force (SF) and Bending Moments (BM)
Computer program for section properties

8 Structures

8.1 What is a structure?

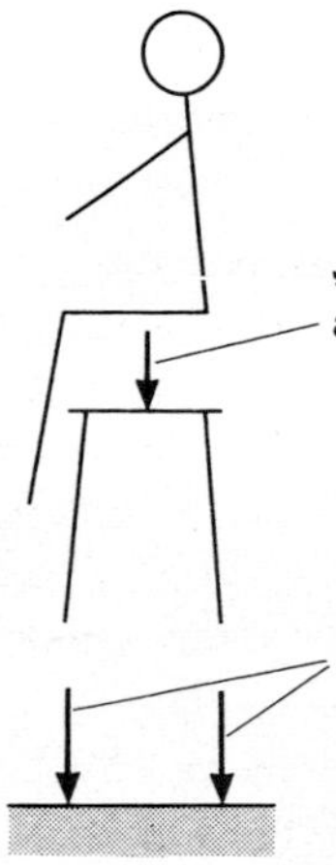

weight of person
acting on stool

weight
transmitted
to floor

Figure 8.1 Transmission of weight to the floor.

A *structure* may be defined as a body that can resist applied forces without changing its shape or size, apart from the deformations due to the elasticity of the materials from which it is made. In this book we shall be concerned mainly with manufactured structures but there are many examples in nature. The elastic deformations can be substantial, as in the flexing of an aircraft wing during take-off or the bending of tall trees in a strong wind but in most structures they are small compared with the overall dimensions. Provided it is not permanently strained the structure returns to its original shape when the forces are removed.

A *mechanism*, in contrast, is designed so that there are precise relative movements between its components (see Chapter 10). The term *machine* is often used for objects such as vehicles and workshop equipment that are combinations of structures and mechanisms.

In order to design safe, efficient structures we need to understand the effects of forces, and the study of forces at rest is given the name *statics*. The term *dynamics* is used for the study of forces on bodies in motion. The word *mechanics* covers both.

Examination questions on forces and structures can usually be solved by graphical methods. These are covered in the first part of this chapter and you should make your own scale drawings when working through the examples. Suitable scales are suggested in the solutions. Later sections (8.6–8.11) show how numerical results can be obtained by calculation.

8.1.1 Purposes of structures

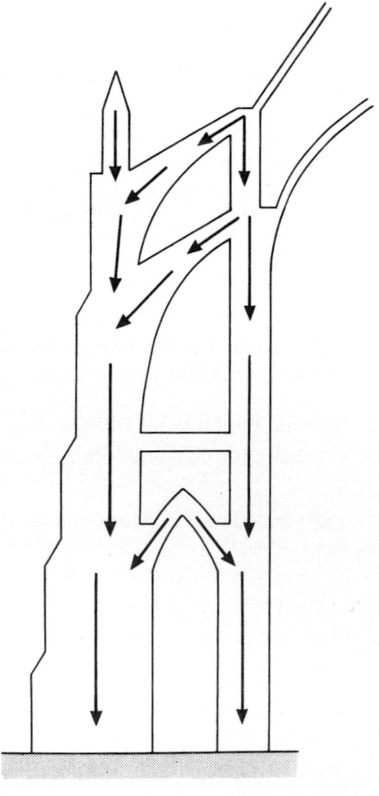

Figure 8.2 Transmission of forces through a buttressed cathedral wall.

The function of a structure is to transmit forces from one place to another. If you sit on a stool, it transmits your weight to the floor (Fig. 8.1) and if you walk across a footbridge over a stream, it transmits your weight to the banks. In a structure with slender components, such as the buttressed cathedral wall shown in Fig. 8.2, the transmission of the forces can be visualised as flowing along the axes of these components.

Objects that are structures can have additional purposes. The walls of a house form a structure for transmitting various weights to the ground but (along with the roof) they also provide shelter from the elements. Almost every object we can think of from a soap bubble to the hull of a bulk carrier is a structure that is transmitting forces from one place to another. Even a work of art such as a sculpture is a structure in that it has to support its own weight. Almost all designing involves aesthetic, as well as technical, considerations and we should aim to give our structures an attractive appearance so as to enrich our environment.

8.1.2 Equilibrium, stability and strength

A structure at rest (or one that is moving at a constant velocity) is said to be in *equilibrium*. This means that the forces acting on it just balance and have no *resultant*. In the case of a chair, for example, the downward force due to the combination of its own weight and that of the person sitting on it must be exactly balanced by the upward forces exerted by the floor (Fig. 8.3(a)). These balancing forces are called *reactions*.

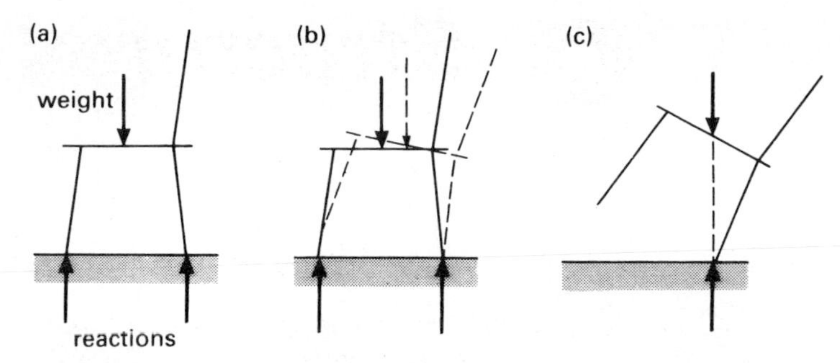

Figure 8.3 (a) Chair at rest (b) Stable equilibrium (c) Unstable equilibrium.

If the structure returns to its original position after being disturbed, the equilibrium is said to be stable. Taking the example of the chair (Fig. 8.3(b)), this will be the case if the weight acts within the area whose corners are the points where the legs touch the floor. Suppose, however, the person sitting on the chair leans back (Fig. 8.3(c)) until the weight is acting in the vertical plane through the points where the back legs touch the floor. Although the reactions at these points can still balance the weight, the slightest disturbance can cause the chair (and the sitter) to topple. In this case the equilibrium is said to be *unstable*.

In addition to ensuring that a structure is in stable equilibrium we must design it so that every component is strong enough to withstand the forces it has to transmit. If one of the chair legs breaks under the weight of the person sitting on it he or she will again end up on the floor. The present chapter is concerned mainly with the equilibrium of structures. The strength of structural components is considered in Chapter 9.

8.2 Types of structures

Structures can be divided into three broad categories according to the way in which they transmit and resist the forces acting on them.

Mass structures

These are solid structures such as gravity dams (Fig. 8.4) that resist the applied forces by virtue of their own weight.

Figure 8.4 Mass structure – the Kariba dam.

Framed structures

A second type is one in which bars (usually straight) are joined together at their ends to form a framework. They are sometimes called 'skeletal structures'. If the bars all lie in the same plane they are termed 'plane frames' (Fig. 8.5(a)) but if they extend in three dimensions we use the name 'space frames' (Fig. 8.5(b)). The strength of frames depends upon their ability to retain their shapes under the action of the external forces.

It is often necessary to cover a space frame with some form of sheet material. For example, a tall modern building (Fig. 8.6) is likely to consist of a steel frame together with a cladding of some other material (it can even be glass). In this case the frame is designed to carry all the weights and other forces acting on the building, and the cladding does not add to the strength.

Shells

In contrast to the last example a shell transmits the forces imposed on it through the sheet material of which it is made. Boilers, balloons and the domes of buildings are all examples of shell structures. The strength of shells is considered in Chapter 9.

(a)

(b)

Figure 8.5
(a) Plane frame – roof truss
(b) Space frame – British Rail station, Reading.

A noteworthy development in the last half century has been a change in the type of structure used for cars, railway carriages and other vehicles. Previously these were designed as frames covered with sheet materials. Nowadays they are usually designed as shells with the forces being transmitted by the sheet material forming the skin. This form is known as *monocoque construction.* The shell is often reinforced with stiffeners that help to preserve its shape and thereby add to its strength.

8.3 Types of load

Figure 8.6 A framed structure with cladding – the Hong Kong and Shanghai Bank.

The forces applied to structures are usually referred to as *loads* and the term is also used for the forces being transmitted by individual components within structures. In many cases the loads acting on a structure are the weights of objects it is supporting, including itself. For some structures, such as a large motorway bridge, this self-weight may be far larger than any of the other loads. On the other hand, some structures are capable of carrying many times their own weight. Since the weights of objects act towards the centre of the Earth they can be regarded as parallel vertical loads.

To calculate the weight of an object use:

$$\text{Weight} = \text{mass} \times g$$

where g is the acceleration due to gravity. The value of 9.81 is a widely used approximation for g but you may be given a slightly different value in examinations (see Appendix 2).

In contrast, an orbiting satellite or space station may be in a state of 'weightlessness'. When in space its structure has no weights to carry and needs little strength. This advantage can only be obtained if the structure is assembled in space. Otherwise it will have to be designed to bear the weights that are present when it is on the surface of the Earth and the large forces that can arise when it is launched.

Loads and forces are measured in newton (abbreviated to N) and practical values are often given in kilonewton (kN). Sometimes we are told the mass of an object being supported by a structure and it is then necessary to calculate the corresponding load. Suppose a 60 kg person stands on a table. What load is the table carrying? On the surface of the Earth a mass of 1 kg has a weight of 9.81 N (1 kg $\times$ 9.81 m/s^2). Hence:

$$\text{load} = 60 \times 9.81\ \text{N} = 588.6\ \text{N}$$

8.3.1 Concentrated and distributed loads

For some purposes, the weight of an object carried by a structure can be considered as a force acting at a single point, its *centre of gravity* (Fig. 8.7(a)). Such loads are called 'point' or 'concentrated loads'. Other loads such as the aerodynamic lift on an aircraft wing (Fig. 8.7(b)) may be spread over a large area. We call these 'distributed loads'. The term 'uniformly distributed load' (UDL) is used when the load is spread evenly over the surface.

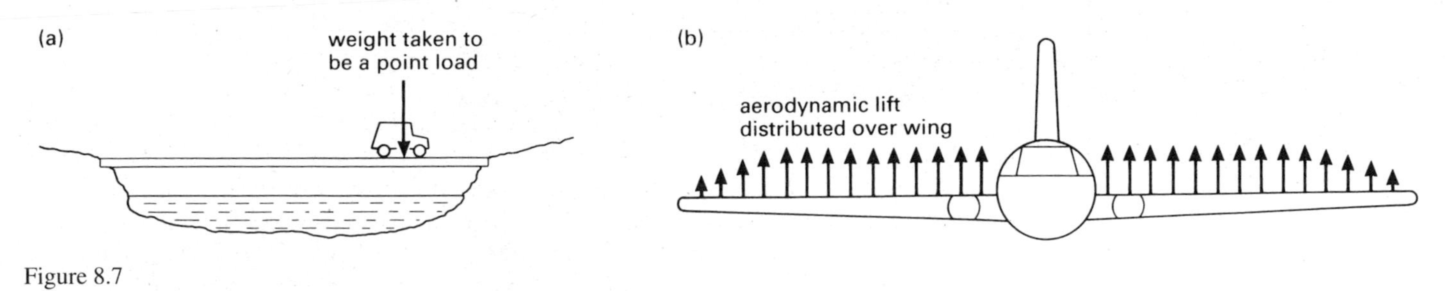

Figure 8.7

8.3.2 Static and dynamic loads

When you are answering examination questions on structures, assume that the loads have been gradually applied unless you are told otherwise. Such loads are called 'static loads'. A structure that is subjected to impact loads or accelerations may experience much larger forces than if it were at rest under static loads.

When a high-speed vehicle crosses a bridge, for example, the structure has to bear its weight almost instantaneously. This is called a 'suddenly applied load', and it can be shown that the effective force is twice as much as for the same weight gradually applied. Civil engineers often refer to static and suddenly applied loads as 'dead' and 'live loads' respectively.

If a structure is accelerated or decelerated violently, as in an aircraft pulling out of a dive or a car crashing, the effective loads can be several times the weights involved and they will not necessarily act vertically downwards. The structures may therefore have to be designed to withstand forces in other directions. The forces arising from accelerations are known as 'inertia' loads.

Some structures often have to carry the forces arising from the pressure of a liquid or gas. Buildings have to resist wind loads, dams are subjected to hydrostatic pressure and gas cylinders and boilers have to withstand the pressures of the fluids they contain. These pressures are examples of distributed loads and they act at right angles to the surfaces in contact with the fluid.

8.4 Properties of forces

Suppose a block is resting on a horizontal table and is subjected to a single applied force. If the force acts downwards, as in Fig. 8.8(a), it will have no visible effect on the block provided it is strong enough not to break. If it is a horizontal force, however, its effect depends upon the point at which it is applied. If it acts near the bottom of the block (Fig. 8.8(b)) it may cause sliding: if it acts near the top the block may topple (Fig. 8.8(c)).

The effect of a force on a structure therefore depends on three things:

- its magnitude (we measure this in newton),
- its direction (usually measured anticlockwise from a horizontal datum),
- the point at which it is applied.

Together, the last two define a line along which the force acts. This is called its *line of action*.

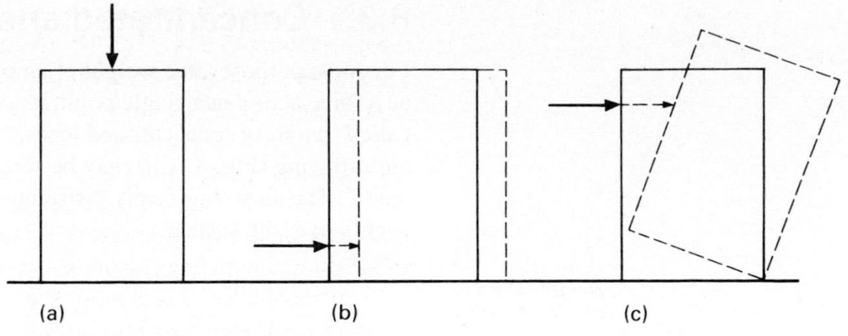

Figure 8.8 Single applied force acting on block
(a) downward (b) causing sliding (c) causing toppling.

Quantities, such as force, that possess magnitude and direction are called *vectors*. Other examples are velocity, acceleration and momentum. A vector quantity can be represented by a straight line drawn in the same direction as itself with a length proportional to its magnitude. Fig. 8.9(a) shows two loads acting at a point on a framework. One has a magnitude of 7 kN and acts vertically downwards, and the other is a 3 kN load acting at 30° to the horizontal. They are represented by OA and OB in the vector diagram, Fig. 8.9(b). Suggested scale: 1 cm = 1kN.

If a parallelogram is constructed with OA and OB as adjacent sides the diagonal OC represents the resultant, that is, the single force which has the same effect as the 7 kN and 3 kN forces acting together. This construction for finding the resultant is known as the *parallelogram of forces*. By measurement of the diagonal, the resultant in the present case is found to be 6.08 kN at an angle of 25.3° to the vertical.

If the second vector is drawn as AC (Fig. 8.9(c)) instead of OB, the resultant is given by the line from O to C. A triangle is therefore sufficient to determine the resultant of two forces.

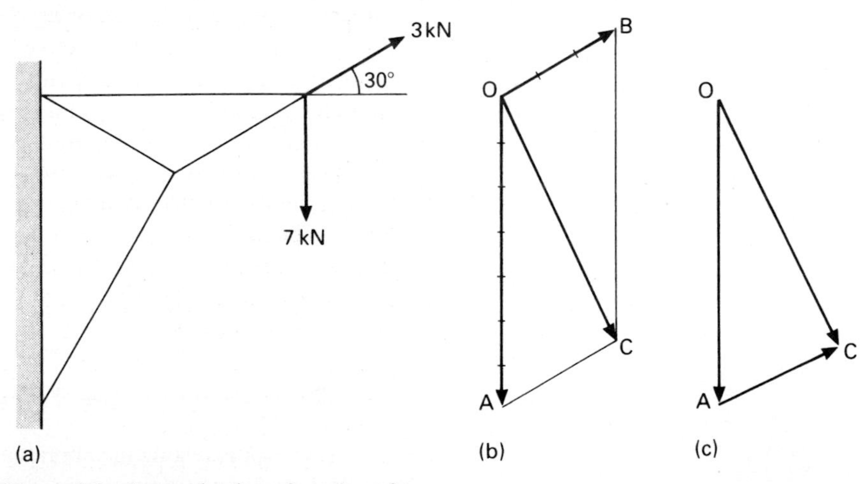

Figure 8.9 Determination of resultant force
(a) two loads acting at a point (b) parallelogram of forces (c) triangle of forces.

8.4.1 Equilibrium of concurrent forces

If the lines of action of a set of forces lie in the same plane they are said to be *coplanar*. If all the forces pass through the same point they are called *concurrent*.

To achieve equilibrium with only two forces, they must be equal in magnitude, opposite in direction and act along the same line. They may act towards each other like the forces exerted by the jaws of a vice on a workpiece or away from each other like the pulls on the rope in a tug-of-war contest. These are examples of *compression* and *tension* respectively.

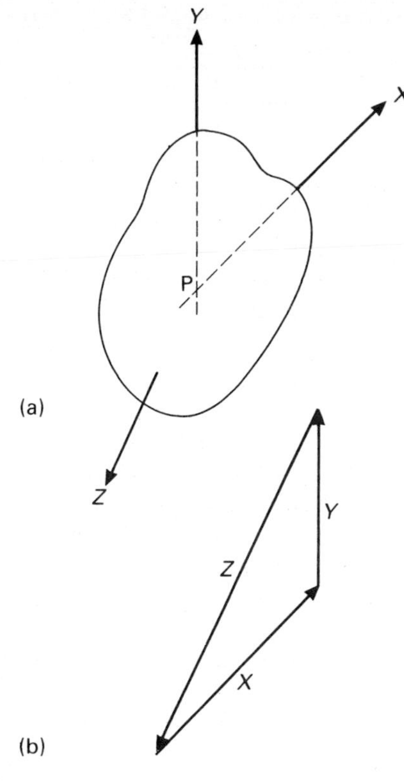

Figure 8.10

Suppose next that the body shown in Fig. 8.10(a) is in equilibrium under the action of three forces X, Y and Z. The lines of action of any two of them, say X and Y, will intersect at some point P (unless they are parallel) and, by using the parallelogram of forces, they can be replaced by their resultant which will also pass through P. The force system is then reduced to two forces and these must be equal and opposite, and act along the same line. For equilibrium therefore the third force Z must act through P and be equal and opposite to the resultant of X and Y.

If lines are drawn to represent X and Y taken in order (Fig. 8.10(b)) then, for equilibrium, the line representing Z must complete a triangle.

The Principle of Concurrency states that if three nonparallel forces are in equilibrium they must be concurrent. The forces can be represented in magnitude and direction by the sides of a triangle taken in order. Parallel forces are considered in section 8.8.

In tackling problems it is important to start with a clear diagram showing the forces that are to be taken into account. As the solution to the following example shows, it is helpful to have a drawing from which other details have been removed.

EXAMPLE 8.1 A straight bar AB, 4 m long, is hinged to a vertical wall at the end A. It is maintained in a horizontal position by a rope attached to the bar at a point C, 3 m from A, and to the wall at a point D, 5 m vertically above A. Find the tension in the rope and the magnitude and direction of the hinge reaction when a mass of 100 kg is suspended from B. Neglect the weights of the bar and rope.

SOLUTION A pictorial view of the problem is given in Fig. 8.11(a). The required answers can be found by considering the equilibrium of the bar AB. The bar and the forces acting on it are picked out in Fig. 8.11(b). The forces are:

(a) the weight of the 100 kg mass acting vertically downwards at B. This amounts to $100 \times 9.81 = 981$ N.
(b) the tension in the rope. Although its magnitude is unknown to start with, it acts through the point C and its direction is towards D.
(c) the reaction at the hinge A. At the beginning its magnitude and direction are both unknown but its line of action must pass through A.

The lines of action of (a) and (b) intersect at the point E and therefore the third force (c) must also act through this point. By measurement, its line of action is found to be 22.6° to the horizontal. The triangle of forces can now be drawn (Fig. 8.11(c) or (d)). Suggested scale: 1 cm = 100 N. Force (a) is set out first because its magnitude and direction are both known. The triangle is completed by setting out lines in the directions of the other two forces and locating the point where they intersect. Note that there are two possible ways of doing this depending on the sequence in which they are taken.

By measurement of the triangle of forces the results are:

tension in the rope = 1530 N
hinge reaction = 850 N at 22.6° to the horizontal

8.4.2 Static friction

Suppose (Fig. 8.8(b)) a block rests on a horizontal table and a gradually increasing horizontal force is applied near its base. If it remains at rest the forces acting on it must be in equilibrium. These are shown in Fig. 8.12(a) and comprise:

- the weight of the block W acting vertically downwards,
- the perpendicular (or normal) reaction of the table N vertically upwards,
- the applied force P from left to right,
- a resisting frictional force F between the table and block acting from right to left.

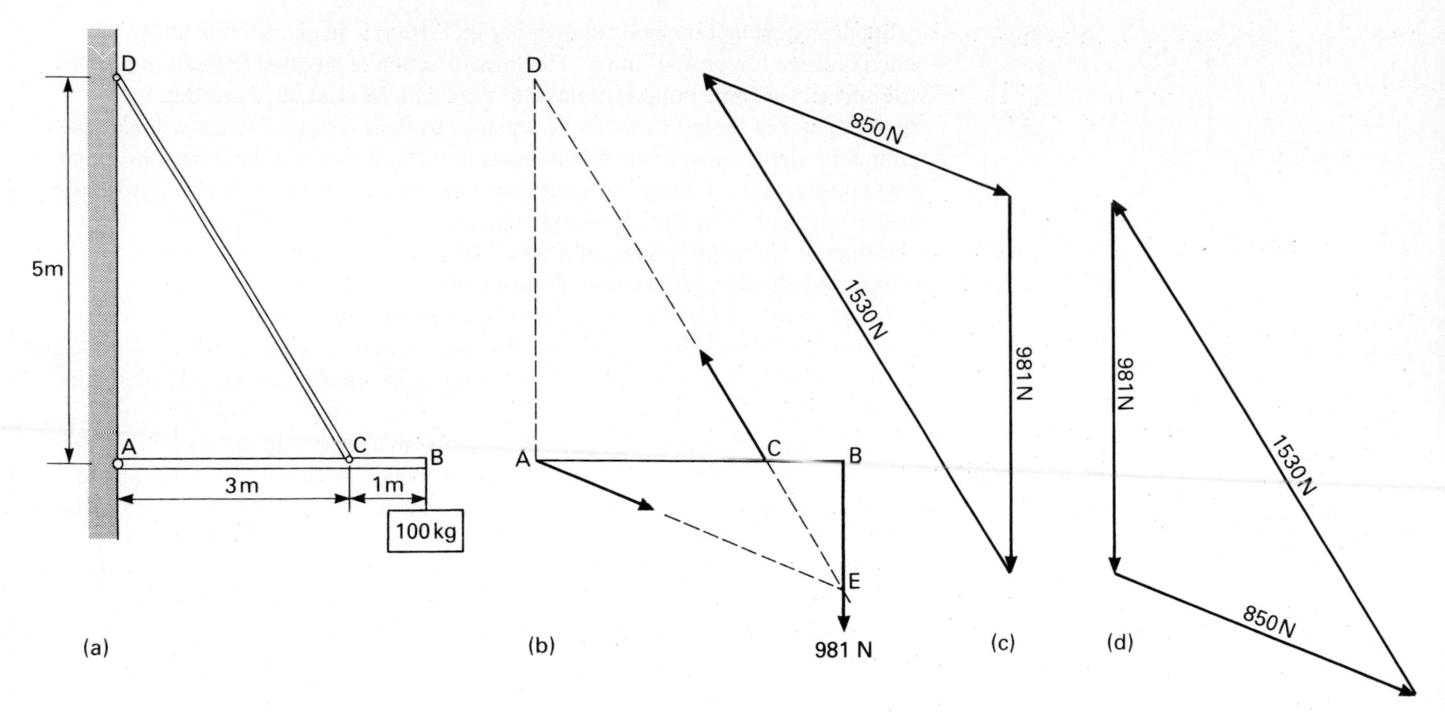

Figure 8.11

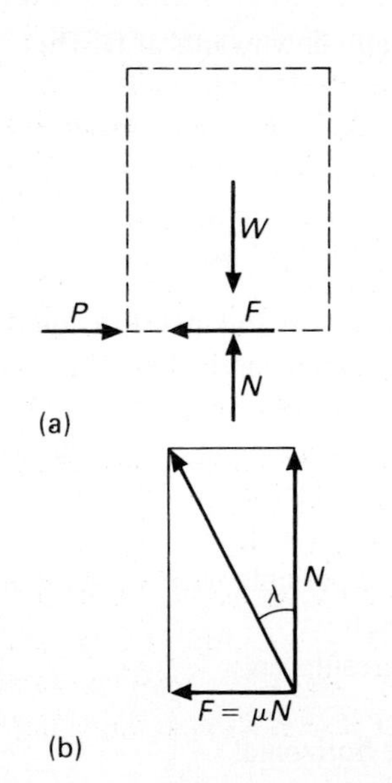

Figure 8.12 Block at rest on horizontal table
(a) forces in equilibrium
(b) resultant reaction.

For equilibrium, $N = W$ and $F = P$ (see 8.6.6). As the applied force P increases so does the friction force F up to a maximum value that depends upon the surfaces of the block and table. If P is increased beyond this amount, friction is overcome and the block slides.

It is found by experiment that this maximum value of F is roughly proportional to the normal reaction N. The constant of proportionality is usually denoted by μ and thus:

$$F = \mu N$$

The constant μ is called the *coefficient of limiting friction* and its value will be given in questions where it is needed. The resultant reaction can be determined by a parallelogram of forces (Fig. 8.12(b)) which in this case is a rectangle. The maximum inclination that the resultant can make with the normal to the surface is called the *angle of friction* and is denoted by λ. By trigonometry,

$$\tan \lambda = F/N = \mu N/N = \mu$$

A smooth surface is one for which μ and λ are zero. It is unable to offer any frictional resistance; the only reaction it can provide is normal to the surface.

EXAMPLE 8.2 A lightweight ladder, 4 m long, stands on rough ground and leans against a smooth vertical wall. The foot of the ladder is 1.2 m from the base of the wall. To what height can a person climb the ladder if the coefficient of friction between the ground and the ladder is 0.2? Neglect the weight of the ladder in comparison with that of the person.

If the person's mass is 65 kg what are the reactions at the wall and the ground in this limiting case?

SOLUTION This is another example of three forces in equilibrium. Suppose the points of contact at the ends of the ladder are A and B as shown in Fig. 8.13(a) and the person has climbed to a point C. It is usual to denote the reactions by R_A and R_B.

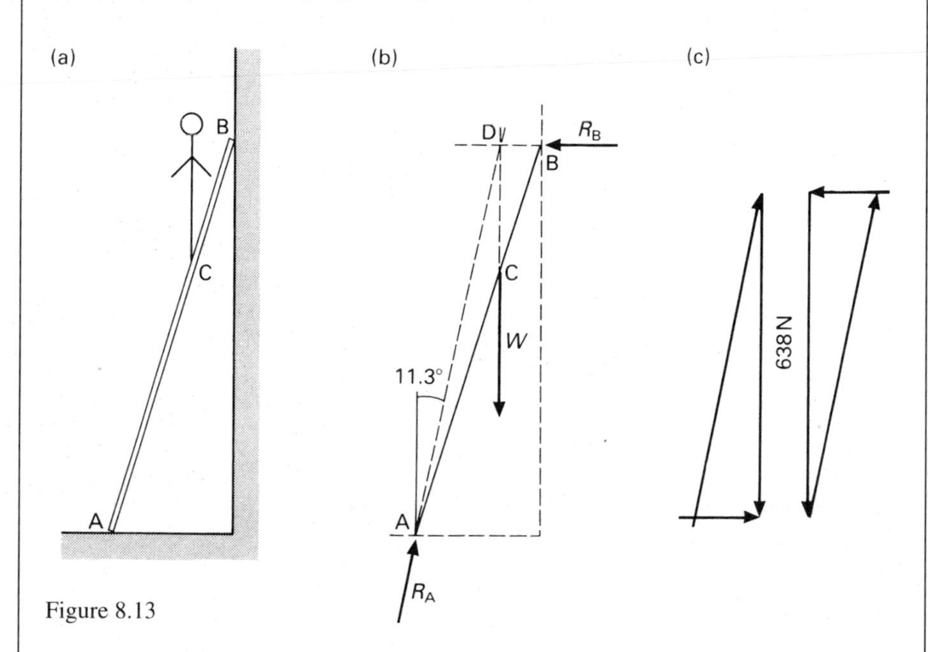

Figure 8.13

Since the wall is smooth, R_B is normal to it and therefore horizontal. The weight of the person W is a vertical force and its line of action will intersect that of R_B at some point D. The third force R_A must pass through the same point and it therefore acts along the direction AD as shown in Fig. 8.13(b). The maximum angle that R_A can make with the normal to the ground is the angle of friction λ. Since

$$\mu = \tan \lambda$$

then

$$\lambda = \tan^{-1} \mu = \tan^{-1} 0.2 = 11.3°$$

Using this result and a scale drawing (Fig. 8.13(b)) the highest point C to which the person may climb before the ladder slips is 2.54 m from A.

The triangle of forces can now be drawn (Fig. 8.13(c)). Suggested scale: 1 cm = 100 N. The weight of the person (65 kg × 9.81 m/s^2 = 638 N) is set out first as a vertically downwards force and the triangle is completed by adding sides in the directions of the reactions R_A and R_B. As in the previous example there are two possible configurations. By measurement of the triangle the reactions are found to be:

$$R_A = 651 \text{ N} \quad \text{and} \quad R_B = 128 \text{ N}$$

8.4.3 Polygon of forces

The method of vectors that leads to the triangle of forces can be extended to any number of concurrent coplanar forces. If the vectors representing the forces form a polygon that closes then the forces are in equilibrium. If the diagram does not close it can be used to find the resultant.

EXAMPLE 8.3 Find the resultant of the four concurrent forces shown in Fig. 8.14(a). What additional force is needed to produce equilibrium?

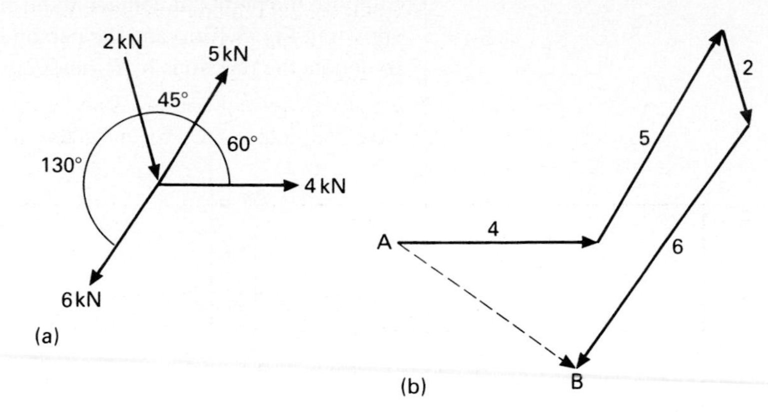

Figure 8.14
(a) Four concurrent forces (b) Vector polygon of forces.

SOLUTION The vector polygon is shown in Fig. 8.14(b). Suggested scale: 1 cm = 1 kN. Each force is represented by a vector line with the arrowhead of one becoming the starting point for the next. If this rule is followed the forces may be taken in any order. In the present case the polygon does not close. Therefore the forces are not in equilibrium and their resultant is represented by the broken line in the direction AB. By measurement it is found to be 4.37 kN downwards to the right at an angle of 35° to the 4 kN force.

The force required to produce equilibrium (sometimes called the *equilibrant*) must be equal and opposite, that is, 4.37 kN upwards to the left and at the same angle to the 4 kN force. On the diagram it is represented by the broken line in the direction BA.

8.4.4 Components and resolved parts

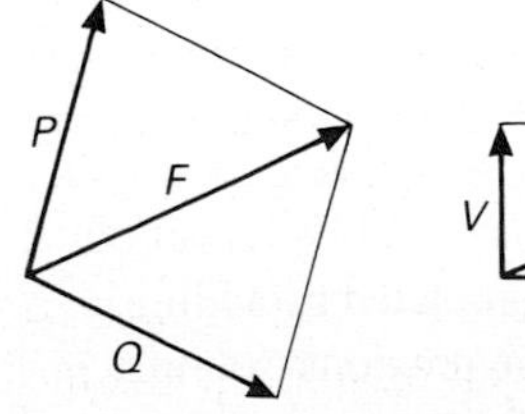

Figure 8.15

So far we have used the parallelogram of forces to find the resultant of two given forces. The construction can be used in reverse to find two forces that together have the same effect as a single given force. Suppose the given force is F (Fig. 8.15(a)). If its vector is the diagonal of a parallelogram the containing sides represent two forces P and Q that together can replace F. They are called the components of F and they may be chosen to act in any desired directions.

In many examples it is useful to have the components at right angles and they are then called the resolved parts of the given force. Often the chosen directions are horizontal and vertical as in Fig. 8.15(b) and the resolved parts are then denoted by H and V. If the force F makes an angle θ with the horizontal as shown, then, by trigonometry, the resolved parts are $H = F \cos \theta$ and $V = F \sin \theta$.

Note that a force has no component in a direction at right angles to itself since $\cos 90° = 0$.

8.4.5 External and internal forces

In analysing the forces acting on a structure it is helpful to think of it as being suspended in space. We show the line of action of each force on a drawing known as a free body diagram from which unnecessary details have been removed. As well as the weight, and wind and other loads, the set of forces must include the reactions on the structure where it is in contact with fixed supports such as a wall or the ground. We can use this diagram to examine the equilibrium of the structure.

This is illustrated by Fig. 8.11 and Fig. 8.13. In each case the elements of the problem are shown in (a) and the corresponding free body diagram in (b). Each example involves three forces and their directions are used to draw the triangle of forces from which their magnitudes can be found.

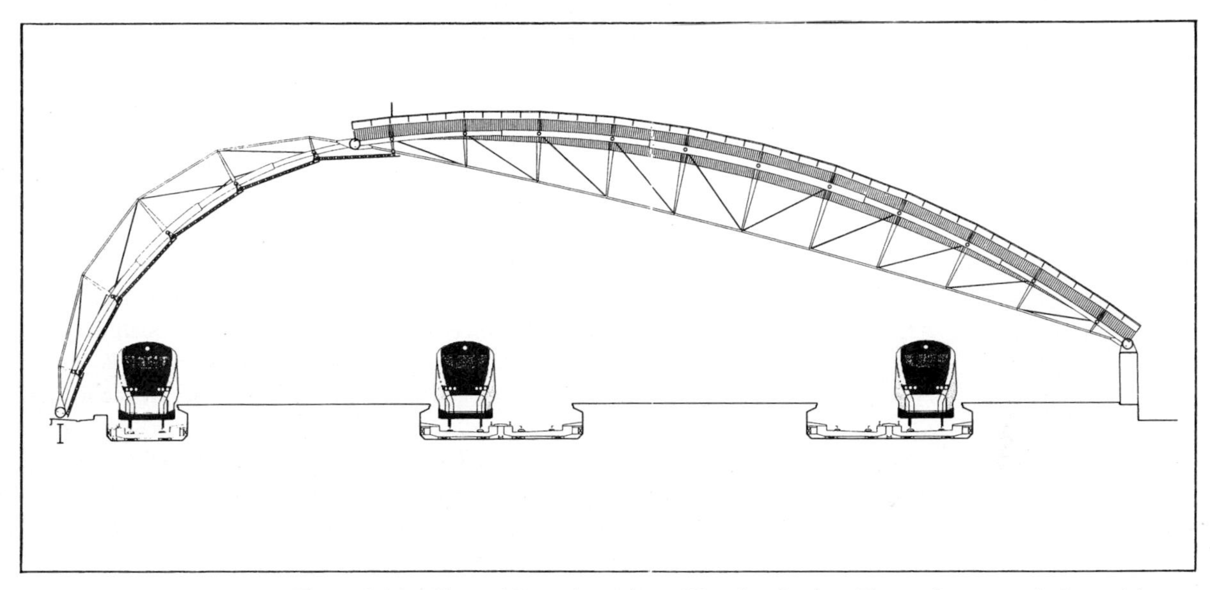

Figure 8.16 Channel Tunnel terminus, Waterloo Station. The roof structure is formed from a three-pin arch with the 'centre pin' located to one side.

In many practical cases the structure contains many components and we need to examine one section at a time. Fig. 8.16 shows one of the arches supporting the roof of the Channel Tunnel rail terminal at Waterloo Station. It consists of two frames hinged (or pinned) to vertical columns at their ends and to each other where they meet. The conditions for equilibrium can be applied to the whole structure or to any section of it that we choose.

If we consider the structure of Fig. 8.16 as a whole, the free body diagram must include the reactions acting on it at the two supports. We can also consider the equilibrium of each of the two frames separately; in this case the reaction between them at the top hinge is then an external force for each. This principle is used in the solution to Example 8.4.

We can take the idea further and apply it to individual bars, to selected portions of a frame or to small elements of the material of which it is made. In Chapter 9 the principle is also used for relating the force acting on a bar to the internal stresses within the material.

EXAMPLE 8.4 A bridge spans a stream between banks of different heights as shown in Fig. 8.17(a). The main structure consists of two quadrants AC and CB hinged to the abutments at A and B and to each other at the crown C. (This configuration is known as a *three-pinned arch*.) The radii of the quadrants are 10 m and 5 m respectively.

A vertical load of 20 kN is applied to the arch at a point 6 m horizontally from A. Find, due to this load,

(a) the reactions at A and B in magnitude and direction,
(b) the horizontal and vertical components of these reactions.

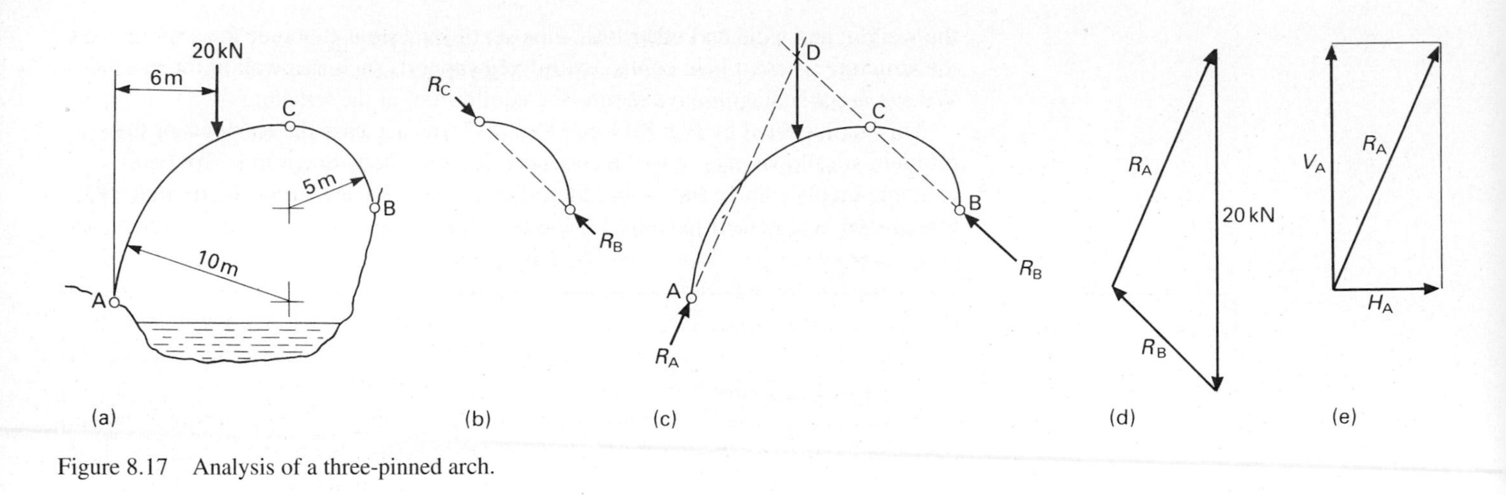

Figure 8.17 Analysis of a three-pinned arch.

SOLUTION Each quadrant (as well as the complete arch) is in equilibrium under the action of the forces acting on it. The smaller quadrant CB has only two forces acting on it (Fig. 8.17(b)), the hinge reactions R_C and R_B. These must be equal and opposite, and act along the same line. Hence R_B acts in the direction BC.

The complete arch is subjected to three external forces (Fig. 8.17(c)), the applied load of 20 kN together with the reactions R_A and R_B. (R_C is now an internal force.) The lines of action of the 20 kN load and R_B meet at D, and it follows that the third force R_A passes through the same point. Its direction is therefore AD.

Once the directions of the forces are known, the triangle of forces can be drawn (Fig. 8.17(d)). Suggested scale: 1 cm = 2 kN. The 20 kN load is drawn first and the triangle is completed by two sides in the directions of R_A and R_B.

(a) By measurement of the triangle of forces the reactions are:

$$R_A = 15.2 \text{ kN at } 67^\circ \text{ to the horizontal}$$
$$R_B = 8.5 \text{ kN at } 45^\circ \text{ to the horizontal}$$

(b) The horizontal and vertical components of R_A are found by making its vector the diagonal of a rectangle as shown in Fig. 8.17(e). By measuring the sides the results are found to be:

$$V_A = 14.0 \text{ kN and } H_A = 6.0 \text{ kN}$$

A similar construction for B gives:

$$V_B = 6.0 \text{ kN and } H_B = 6.0 \text{ kN}$$

8.5 Plane frames

The basic principles of two-dimensional frames can easily be demonstrated using the flat strips found in Meccano and similar construction kits. Take five strips of various lengths, lay them on a table and join them end to end to make a closed assembly as shown in Fig. 8.18(a). Use a single bolt and loose nut at each joint to allow movement; this is called a *pin joint*. Hold one of the strips still (the bottom one is chosen in the diagram) and investigate the possible movements of the others. Repeat the process with four and three strips as in Fig. 8.18(b) and (c).

With five strips there is no pattern to the movement. With four, movement is possible and with a distinct pattern; each point on each member follows a unique

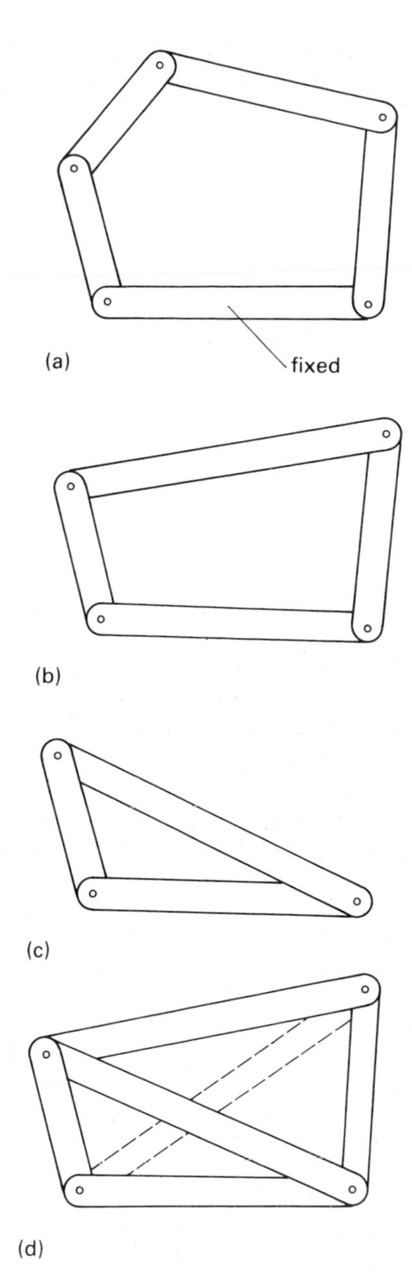

Figure 8.18 Two-dimensional frames
(a) five bar
(b) four-bar chain
(c) three bar
(d) triangulated frame.

path called its *locus*. By putting the point of a pencil through one of the intermediate holes in the strip its locus can be drawn. With three strips no movement is possible – apart from any slack in the joints.

An assembly of five bars joined in this way has no practical application. The four-bar case is an example of a mechanism and is called a four-bar chain.

The triangle (Fig. 8.18(c)) is the basis of many practical structures and it can be extended to any number of joints. For example the four-bar case of Fig. 8.18(b) can be converted to a structure by adding a diagonal as in Fig. 8.18(d). Structures built up from triangles are often referred to as *pin-jointed triangulated frames* and are said to be *just stiff* or *perfect*. The forces in the bars of such frames can be determined using the methods of statics such as the polygon of forces. Problems are of this kind are said to be statically *determinate*.

Starting with a triangle, it is clear that each extra joint requires two additional bars if it is to be just stiff. So for 3, 4, 5, 6,... joints the required numbers of bars are 3, 5, 7, 9,... respectively. If j is the number of joints, the required number of bars b is given by:

$$b = 2j - 3$$

In practice many frames have more bars than are needed for them to be just stiff. For example a second diagonal member could be added to a frame with four joints, as shown by the broken lines in Fig. 8.18(d). The frame is then described as *overstiff* or *redundant*. This does not mean that the extra bars are unnecessary; the designer may have specified them to give the structure more strength. However, redundant frames require each member to be a good fit. If one of the diagonals of the frame in Fig. 8.18(d) was the wrong length and was forced into position it could impose forces in the frame even before the external loads were applied. The analysis of redundant frames is beyond the scope of this book.

8.5.1 Types of support

Fig. 8.19(a) shows a pin-jointed frame with four members that is attached to a vertical wall. A frame that projects from a wall in this way is said to be *cantilevered*. If each of the supports A and B is also pinned (or hinged) it can provide a reaction in any direction perpendicular to the axis of the pin. The number of joints, including these supports, is 4 and, using the formula obtained above, the number of bars needed to make the frame just stiff is:

$$b = 2j - 3 = 2 \times 4 - 3 = 5$$

This is achieved in the present example because the wall acts as a bar in addition to the members AD, AC, BC and CD. A simpler way is to distinguish between the pinned supports A and B and the free joints C and D. The required number of bars is then twice the number of free joints. This means that with two free joints four bars are required and, since four are present, the frame is just stiff or perfect.

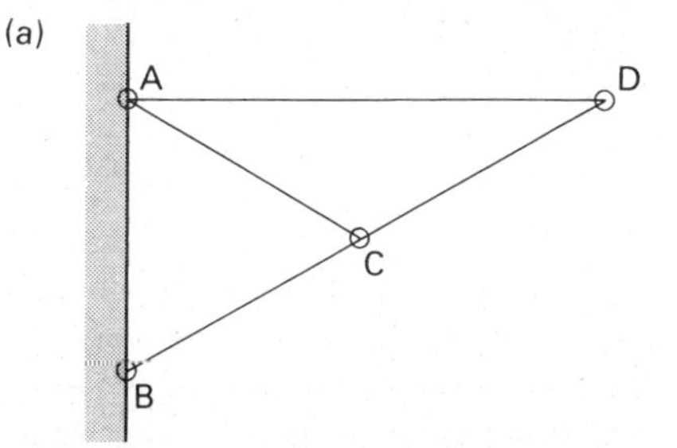

Figure 8.19 (a) Cantilevered pin-jointed frame

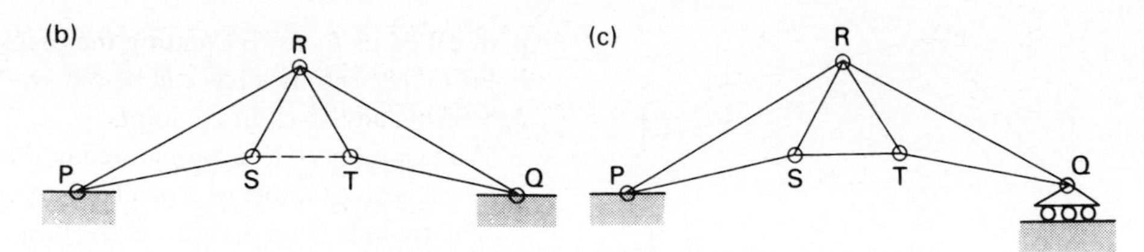

Figure 8.19 (b) Frame pinned at both ends (c) Roller support at one end.

Fig. 8.19(b) shows a frame pinned at its ends. It represents a structure called a roof truss that is commonly used for supporting the roof of a building. This frame has three free joints and therefore needs six members to be just stiff. The inclusion of a bar between S and T (shown by the broken line) would make a total of seven and the frame would be redundant. It would not then be possible to find the internal forces using the force and link polygons. Suppose, however, one of the supports Q is not pinned but consists of a frictionless roller on which the frame rests (Fig. 8.19(c)). The reaction at this support must then be normal to the surface on which the frame rests and the forces in the bars can be found by the methods given earlier. Roller supports of this kind can be seen in many large bridge structures.

8.5.2 Internal forces in plane frames

Pin-jointed frames are usually designed so that the external loads are carried at the joints. As a result there are effectively two forces acting on each bar, one at each end. For the equilibrium of the bar, these must be equal and opposite, and act along the axis of the bar. There are two possibilities. One is that the external forces are pulling on the ends of the bar tending to stretch it. This is called *tension* and the bar is termed a *tie* or *tie-bar* (Fig. 8.20(a)).

The other possibility is that the external forces are pushing inwards on the ends of the bar tending to shorten it. This is called *compression* and the bar is then termed a *strut* (Fig. 8.20(b)).

Note that the internal forces, which are the forces exerted by the bar on the joints, are in the opposite direction to the external forces. A tie-bar pulls inwards on the joints and a strut pushes outwards on them. This is the way the forces will be shown on diagrams for the rest of this chapter.

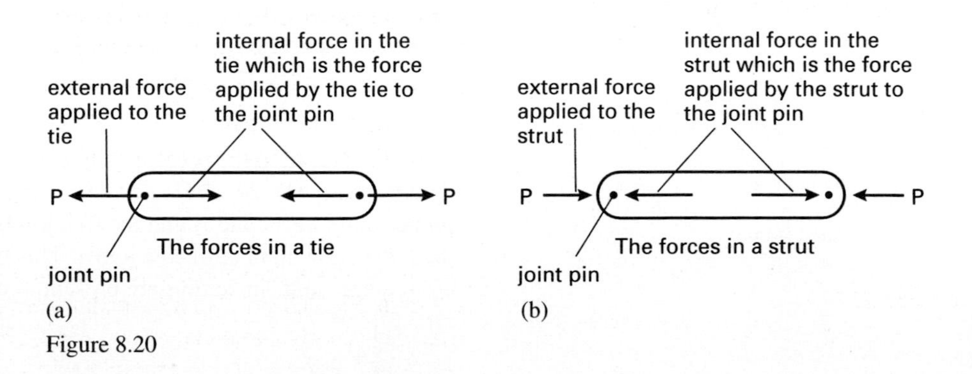

Figure 8.20

The forces in all the bars of a perfect pin-jointed frame can be found by considering the equilibrium of each of the joints using a triangle or polygon of forces. Since the directions of all the forces are known the magnitudes of two of them can be found each time this is done. The results can then be used in drawing the triangle or polygon for other joints. It is convenient to use Bow's notation for lettering the forces and the separate triangles and polygons can be joined together in a single diagram as the next two examples show.

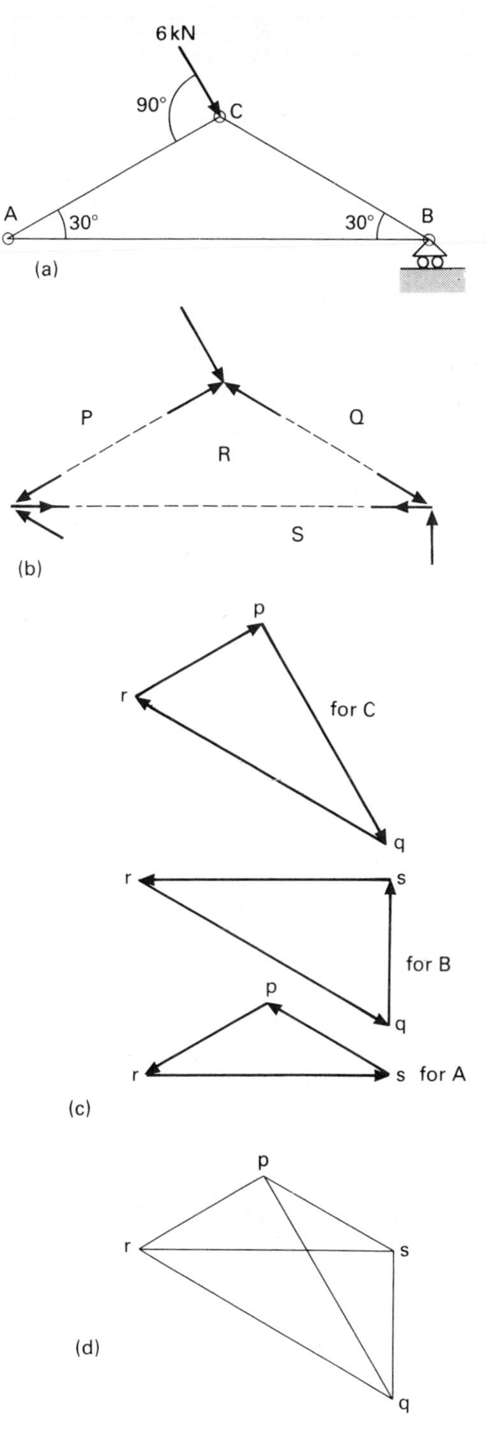

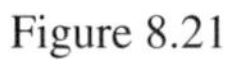
Figure 8.21

EXAMPLE 8.5 Fig. 8.21(a) shows a pin-jointed triangle ABC that carries an inclined load of 6 kN at C. The frame is pinned to a support at A and rests on a roller support at B. Find:

(a) the force in each of the bars,
(b) the magnitude and direction of the reactions at A and B.

SOLUTION The forces acting at each joint are shown in Fig. 8.21(b). At the beginning the only known force is the external load of 6 kN at C. Following Bow's notation the spaces between the forces are given capital letters, P, Q, R and S.

The triangles of forces are shown in Fig. 8.21(c). Suggested scale: 1 cm = 1 kN. The first triangle to be drawn is that for C. The 6 kN load is set out as pq and the point r is located by adding sides qr and rp parallel to the bars CB and AC respectively. The arrow on pq is in the direction of the external load and the arrows on the other two vectors follow in clockwise order. These arrows can be added to Fig. 8.21(b) at the joint C and in opposite directions at the other ends of the same bars.

The triangle for B can now be drawn. Since the support is a frictionless roller the reaction at B must act vertically upwards. The vector qr, representing the force in CB is copied from the first triangle, the only difference being the direction of its arrow. The other two sides are drawn vertically and horizontally since these are the directions of the reaction at B and the force in AB. The reaction is represented by the vector qs and the force in AB by sr. Note that the arrows on this triangle follow an anticlockwise order. The arrow on sr shows that the bar AB is pulling to the left at joint B as shown in Fig. 8.21(b). An arrow in the opposite direction is then added to the same bar at its other end A.

Two of the forces at A are now known in magnitude and direction and the triangle for this joint can be drawn. The arrows follow an anticlockwise order and the reaction at A is represented by sp.

Since the arrows on AC and BC are pushing outwards at the joints these bars are in compression. The arrows on AB are pulling inwards at the ends and this bar is therefore in tension. By measurement of the triangles of forces the answers are found to be:

(a) Force in AB (represented by rs) = 6.0 kN (tension),
Force in AC (represented by pr) = 3.46 kN (compression),
Force in BC (represented by qr) = 6.93 kN (compression),

(b) R_A (represented by sp) = 3.46 kN upwards to the left at 30°,
R_B (represented by qs) = 3.46 kN vertically upwards.

The work can be shortened by combining the three triangles of forces in a single diagram, Fig. 8.21(d). No intermediate measuring is required and each vector is drawn once only. Note that arrows cannot be shown because their order can be clockwise at one joint and anticlockwise at another. At each joint the direction of one force will be known and the others can be found by following it round the corresponding triangle or polygon. The use of Bow's notation and the combined vector diagram is the standard graphical method for the analysis of plane frames.

EXAMPLE 8.6 All the bars shown in the pin-jointed Warren girder of Fig. 8.22(a) are the same length. Find the forces in all the members due to the given loads, stating whether they are struts or ties.

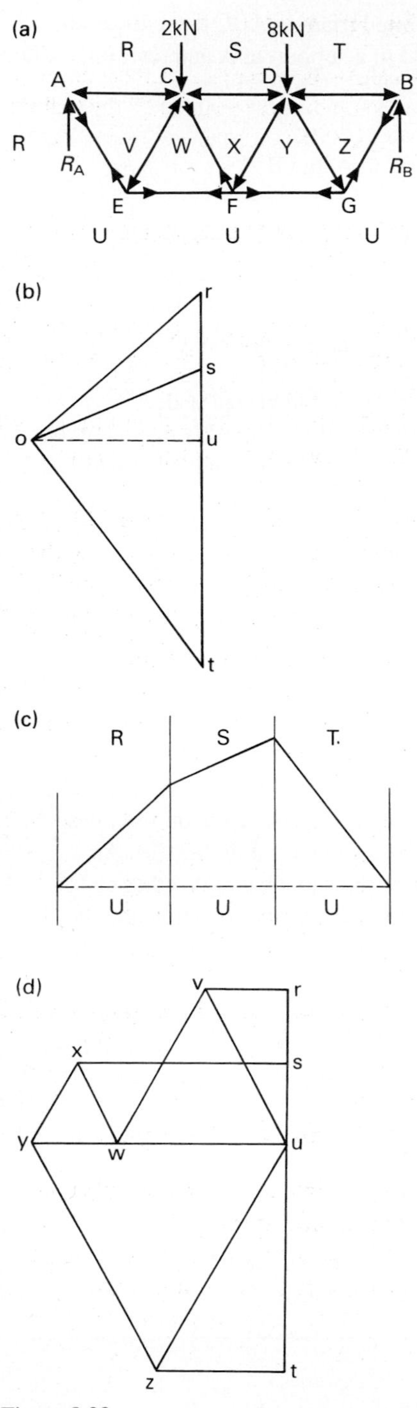

Figure 8.22

SOLUTION A Warren girder is a frame consisting of identical equilateral triangles with all the internal angles being 60°. This form of frame has been widely used for bridge structures in many parts of the world.

The supports A and B cannot both be pinned or the frame would be redundant, having five free joints and eleven bars. It is assumed, therefore, that one of them is a frictionless roller. In the present case it does not matter which because all the external loads are vertical.

The spaces between the forces are given upper case (capital) letters RSTUVWXYZ in accordance with Bow's notation. There are four spaces between the external loads and it is useful to repeat the letters R, T and U to make clear which forces they separate.

The applied loads of 2 kN and 8 kN are represented by the vectors rs and st drawn in Fig. 8.22(b). Suggested scale: 1 cm = 1 kN. A pole o is chosen and the rays or, os and ot give the directions of the sides of the funicular polygon, Fig. 8.22(c). In the present example, this can be started at any point on one of the lines of action of the external forces. (Note, however, that when the direction of one of the reactions is unknown the polygon can only be started at the corresponding joint.)

The closing link of the funicular polygon, shown by a broken line, gives the direction of the remaining ray ou and this locates the point u on the force line. The reactions R_A and R_B are represented by ur and tu respectively and, by measurement, they are found to be 4 kN and 6 kN.

The force diagram can now be drawn, Fig. 8.22(d). For clarity it is separated from the polar diagram but it starts with the same force line rsut and the two can be combined. It is built up as follows.

1 Choose a joint for which there are no more than two unknown forces, say A.
2 Identify the spaces that meet at this joint (R, V and U).
3 Locate the corresponding points on the force diagram. At this stage u and r are already present and the remaining point v is found by drawing vectors uv and rv parallel to the bars that separate the spaces U/V and R/V.
4 For the triangle or polygon of forces just completed, the direction of at least one arrow will be known and the others follow in sequence. The reaction R_A represented by ur acts upwards and therefore rv acts from right to left and vu downwards to the right.
5 Put arrows on the space diagram near the joint to show the directions of these forces.
6 Put opposing arrows at the other ends of the same bars.

If these steps are repeated for the remaining joints the complete force diagram, Fig. 8.22(d), will be obtained and the directions in which the forces in the bars act on the joints will be seen. As shown in Fig. 8.20(b) internal forces pulling inward indicate tension and internal forces pushing outward mean compression. The magnitudes of the forces are obtained by measurement of the force diagram and the results are tabulated below, T indicating tension (tie) and C compression (strut):

Bar	Vector	Force (kN)	T or C
AC	rv	2.31	C
CD	sx	5.77	C
DB	tz	3.46	C
BG	uz	6.93	T
GF	uy	6.93	T
FE	uw	4.62	T
EA	uv	4.62	T
EC	vw	4.62	C
CF	wx	2.31	T
FD	xy	2.31	C
DG	yz	6.93	C

8.6 Mathematical analysis of forces

In industry, graphical methods for frame analysis have been replaced almost entirely by calculations. Pocket calculators and personal computers have removed the tedium of the arithmetic and answers can be found quickly and accurately.

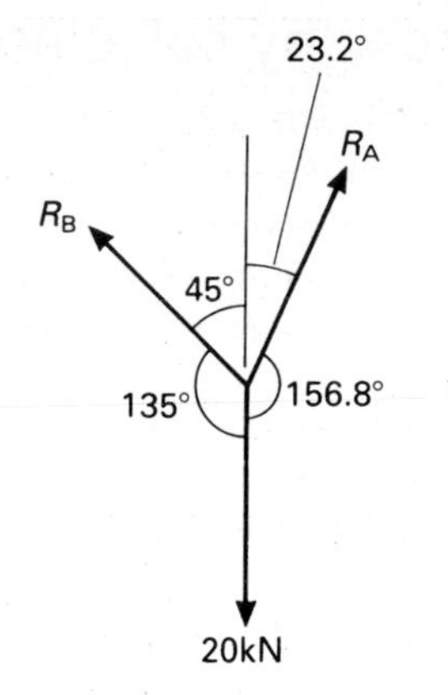

Figure 8.23

8.6.1 Mathematical solution of three forces in equilibrium

For three forces in equilibrium we can use Lami's theorem to find any unknown forces. This is simply the application of the sine rule for the solution of triangles to the triangle of forces. It states that if three forces are in equilibrium, each is proportional to the sine of the angle between the other two. Fig. 8.23 shows the forces acting on the arch structure of Example 8.4. They are the applied load of 20 kN together with the reactions R_A and R_B.

R_B makes an angle of 45° with the vertical and R_A acts in the direction of AD. Fig. 8.17(c) shows that the point D is 14m vertically above A and 6m horizontally to the right of it. Thus R_A makes an angle $\tan^{-1}(14/6)$ or 66.8° with the horizontal and 23.2° with the vertical. The angles between the forces are shown in Fig. 8.23 and, applying Lami's theorem:

$$\frac{R_A}{\sin 135°} = \frac{R_B}{\sin 156.8°} = \frac{20}{\sin 68.2°}$$

Thus

$$R_A = 20 \times \frac{\sin 135°}{\sin 68.2°} = 15.23 \text{ kN}$$

and

$$R_B = 20 \times \frac{\sin 156.8}{\sin 68.2°} = 8.49 \text{ kN}$$

Compare these answers with the results you obtained graphically in the solution to Example 8.4.

8.6.2 The resultant of two right angled forces

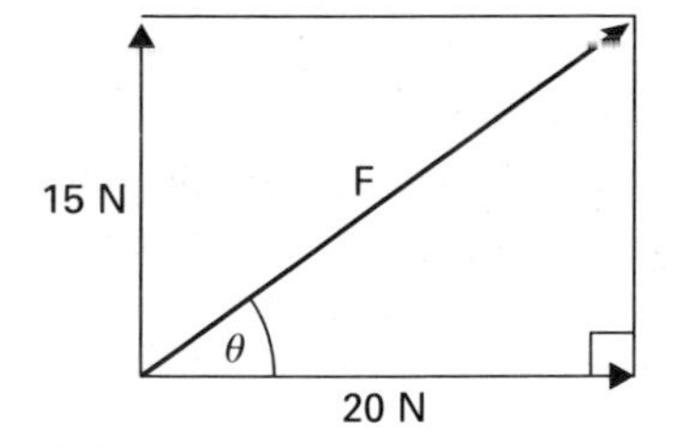

Figure 8.24

If we wish to find the resultant of two forces acting at right angles to each other we need to use a combination of trigonometry and Pythagoras' theorem.

If an object has two perpendicular forces of 15 N and 20 N acting on it, as shown in Fig. 8.24, we can calculate the magnitude and direction of the resultant force acting on the object as follows:

Using Pythagoras' theorem

$$F^2 = 15^2 + 20^2$$

Therefore

$$F^2 = 625 \text{ and so } F = 25 \text{ N}$$

From the diagram

$$\tan\theta = \frac{15}{20} \quad \text{therefore } \theta = 36.9°$$

Therefore the resultant of the given 15 N and 20 N forces is a force of 25 N at 36.9° to the 20 N force.

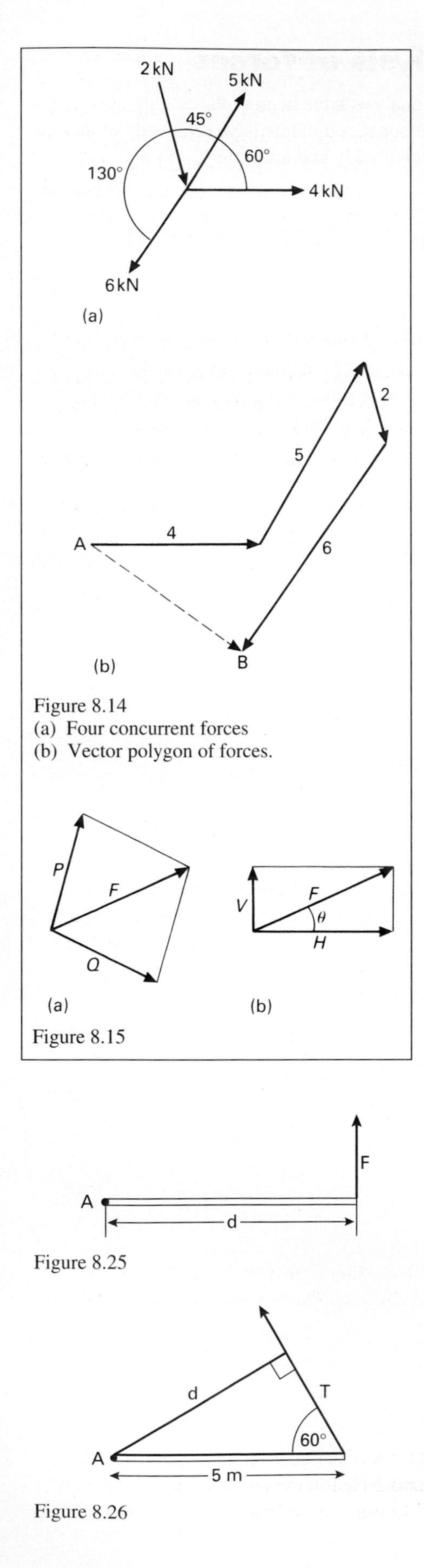

Figure 8.14
(a) Four concurrent forces
(b) Vector polygon of forces.

Figure 8.15

Figure 8.25

Figure 8.26

8.6.3 Resultant of concurrent forces by calculation

The component of a resultant in a given direction is equal to the sum of the components of all the separate forces in the same direction. As shown in section 8.4.4, the horizontal and vertical components (or resolved parts) of a force F which makes an angle θ with the horizontal are $F \cos \theta$ and $F \sin \theta$. For a set of forces the horizontal and vertical components of the resultant (H and V) can be written:

$$H = \Sigma F \cos \theta \quad \text{and} \quad V = \Sigma F \sin \theta$$

The resultant itself corresponds to the diagonal of a rectangle whose sides are H and V as in Fig. 8.15(b). Its magnitude, by Pythagoras, is $\sqrt{(H^2 + V^2)}$ and it makes an angle $\tan^{-1} (V/H)$ with the horizontal datum. Note that the second result gives two possible angles between 0° and 360° and the correct one is chosen by examining the signs of H and V.

Take the values in Example 8.3 (Fig. 8.14) and let the direction of the 4 kN force be the horizontal datum. Then the forces of 4, 5, –2 and 6 kN make angles (anticlockwise) of 0°, 60°, 105° and 235° respectively with the datum. The force of 2 kN is regarded as negative because it is pushing inwards to the point of intersection.

Using the results above for H and V the horizontal and vertical components of the resultant are:

$$H = 4 \cos 0° + 5 \cos 60° - 2 \cos 105° + 6 \cos 235° = 3.576 \text{ kN}$$
$$V = 4 \sin 0° + 5 \sin 60° - 2 \sin 105° + 6 \sin 235° = -2.517 \text{ kN}$$

The magnitude of the resultant is therefore:

$$\sqrt{(H^2 + V^2)} = \sqrt{[3.576^2 + (-2.517)^2]} = 4.373 \text{ kN}$$

and its line of action makes an angle with the horizontal of:

$$\tan^{-1} (V/H) = \tan^{-1}[(-2.517)/3.576] = -35.1° \text{ or } 144.9°$$

Since H is positive and V is negative the correct answer is −35.1° and the resultant acts downwards to the right.

8.6.4 Moment of a force

We know from everyday experience that the turning effect of a force on a body depends on the point at which it is applied (see Fig. 8.8(b) and (c)). We define the moment of a force about a point as its magnitude multiplied by the perpendicular distance of the point from its line of action. This distance is called the *moment arm*. The moment of a force about a point on its own line of action is zero because the moment arm is zero. In the SI system the unit of moment is the newton metre (Nm).

Fig. 8.25 shows the force, F, acting perpendicular to a bar of known length, d. The moment of F about A is given by F.d. If F was known to be 82 N and d was 4 m then:

the moment of the force about A is $4 \times 82 = 328$ N m

When the force, T, acts at an angle to the bar, as shown in Fig. 8.26, the moment of the force about A is the product of T and the perpendicular distance from A onto the line of action of the force, d. If the length of the bar shown is 5 m then, from the geometry, d = 5 sin 60° = 4.33 m.

Therefore the moment of T about A is $5 \sin 60° \times T = 4.33$ T

For a force, T, equal to 50 N then the moment of T about A is 216.5 N.

An alternative method is to say that the components of T are T cos 60° along the bar and T sin 60° perpendicular to the bar (Fig. 8.27). Now T cos 60° passes directly through A and so its moment arm is zero. Therefore the moment of T cos 60° about A must be zero. Also the moment of T sin 60° must be T sin 60° × 5, as this component is at right angles to the bar. This gives us the same result as before as T sin 60° × 5 = 5 sin 60° × T.

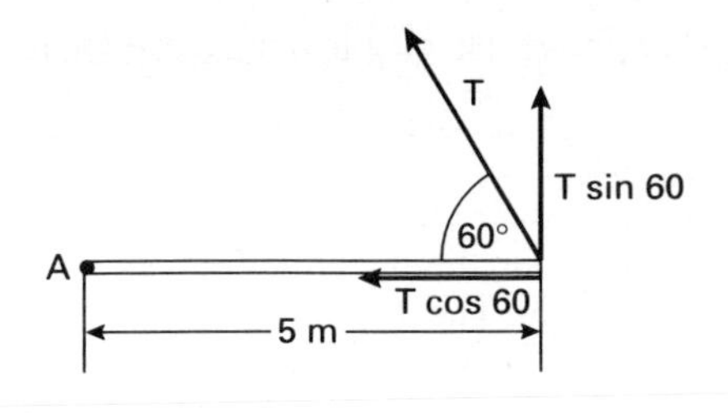

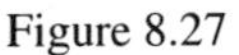
Figure 8.27

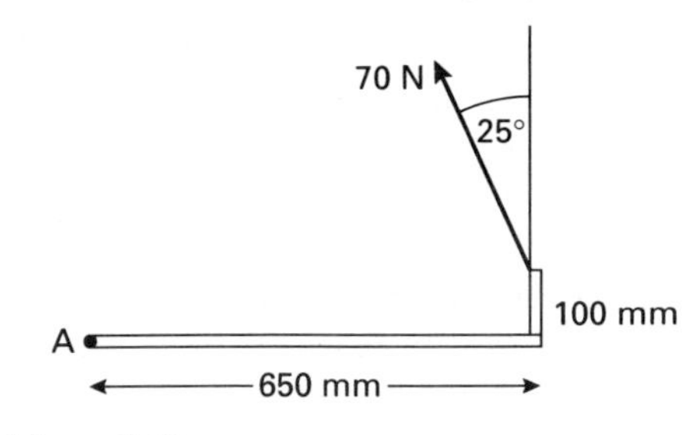

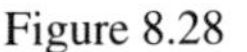
Figure 8.28

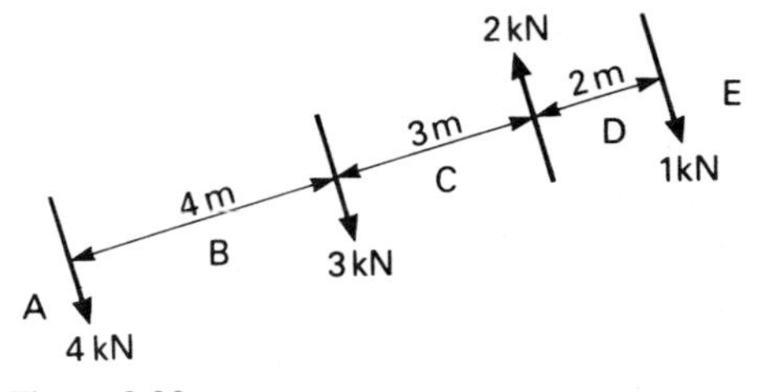

Figure 8.29

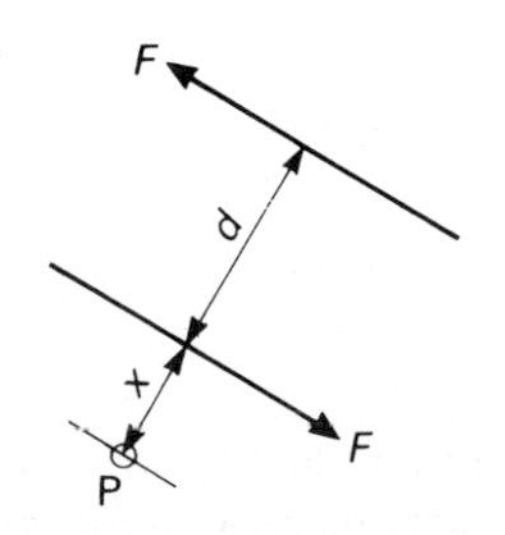

Figure 8.30

The force may not always be acting at one end of a straight bar. A cranked lever, for instance, carries an offset force. In Fig. 8.28 a cranked lever carries a force of 70 N as shown. To calculate the moment of the 70 N force about A, it is much easier to consider the moments of the components of the force about A than to calculate the perpendicular distance from A onto the 70 N force. One component of the force is 70 cos 25° acting vertically 650 mm from A, causing an anticlockwise moment about A. The second component is 70 sin 25° acting horizontally 100 mm from A, also causing an anticlockwise moment about A.

Therefore, with the length in metres,

$$\text{the moment of the force about A} = (70 \cos 25°) \times 0.650 + (70 \sin 25°) \times 0.100 = 44.20 \text{ N m}$$

The sum of the moments of a set of forces about a given point equals the moment of their resultant about the same point. This result is useful in finding the line of action of the resultant of non-concurrent forces. It is important to distinguish between clockwise and anticlockwise moments when making calculations; one is chosen to be positive and the other as negative.

Consider the parallel forces shows in Fig. 8.29. The magnitude of the resultant is $4 + 3 - 2 + 1 = 6$ kN and it acts parallel to the existing forces and in the same direction as the 4 kN force.

Suppose its line of action is above that of the 4 kN force and a distance x m from it. Then, taking moments about a point on the line of action of the 4 kN force with clockwise as positive,

$$\text{moment of resultant} = \text{sum of moments of given forces}$$

and in kN m units,

$$6x = 4 \times 0 + 3 \times 4 - 2 \times 7 + 1 \times 9 = 7$$

from which

$$x = 7/6 = 1.167$$

The resultant therefore acts parallel to the 4 kN force and 1.167 m from it.

If a set of forces is in equilibrium the sum of their moments about a given point is zero. This result can be used to find the reactions at the supports of a structure. Take Example 8.7 (Fig. 8.22). The lengths of the members are not specified but they are all equal and it is convenient to take each one as 1 m.

At the beginning of the calculation both reactions are unknown, but if moments are taken about one of the supports its reaction is eliminated because it has zero moment arm. To find R_B, therefore, take moments about A. With clockwise as positive,

$$2 \times 1 + 8 \times 2 - R_B \times 3 = 0$$

from which

$$R_B = 6 \text{ kN}$$

R_A can now be found by subtracting R_B from the total downward load but taking moments about support B provides a useful check on the working. By either method:

$$R_B = 4 \text{ kN}$$

8.6.5 Couple

A pair of equal forces acting along parallel lines of action, but in opposite directions, is called a *couple* or *torque* (Fig. 8.30). Suppose each force is F and the perpendicular distance between them is d. Choose a point P as shown. Its distance from the nearer force is x and from the other is $(x + d)$. The resultant moment about P, taking anticlockwise as positive, is:

$$F(x + d) - Fx = Fd$$

Hence the moment of a couple is equal to the magnitude of one of the forces multiplied by the perpendicular distance between them. The distance x does not appear in the result, which is therefore the same for all positions of P.

The earlier sections of this chapter show that the resultant of a system of forces in two dimensions can be found by scale drawing or calculation. Its magnitude and direction can be determined either by the polygon of forces or by calculating the sum of the resolved parts of the forces. The line of action can be found by the funicular polygon or by taking moments.

There is one exception. The system may reduce to a couple for which there is no single resultant force.

If the forces are in equilibrium the force and funicular polygons both close. The sum of the components in any given direction will be zero and the sum of the moments about any given point will also be zero.

8.6.6 Equilibrium of coplanar force

If an object is in equilibrium, there is no net force in any direction and no net moment about the centre of gravity. In two dimensions the object can be said to have three degrees of freedom. Therefore we can find three equations relating the forces and moments acting on the object and use them to solve for three unknown forces. The equations are obtained by considering the balance of the forces in any two directions and the moment of the forces about any point in space. It is very often helpful to use vertical and horizontal as the directions, and any point through which some of the unknown forces act as the point about which to take moments. Therefore the *Conditions of Equilibrium* mean that:

Σ Vertical forces = 0
Σ Horizontal forces = 0
Σ Moments about any point (M_O) = 0

This is the most common approach but it is also possible to take moments about any three different positions in space to obtain the three equations or sum the forces in any two other perpendicular directions.

If we have only one unknown, it is necessary to find just one equation and any of the above may be used.

EXAMPLE 8.7 A uniform bar of length 3 m and weight 40 N acting at the centre of the bar is hinged at one end. It is supported in a horizontal position by a string attached to the other end, making an angle of 60° with the bar as shown in Fig. 8.31. Calculate T, the tension in the string, and the reaction at the hinge.

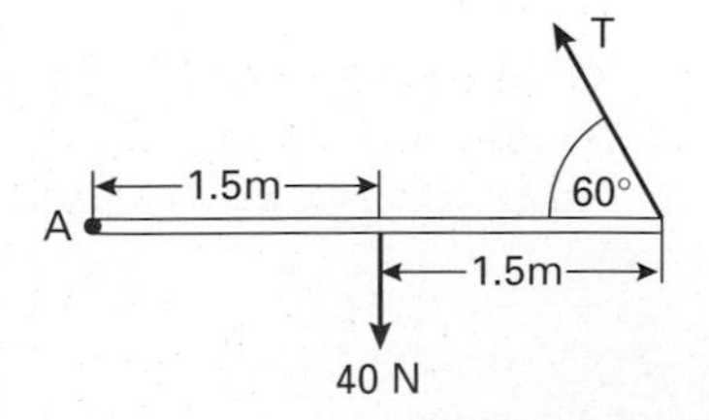

Figure 8.31

SOLUTION The easiest appoach to find T is to consider the moments about A. This is because the hinge reaction passes through this point and therefore has no moment about A. The string provides an anticlockwise moment about A which is equal to $3 \times T \sin 60°$ N m. The weight of the bar provides a clockwise moment about the bar which is equal to 40×1.5 N m.

If the bar is in equilibrium then

Σ MA = 0 i.e. Anticlockwise moments = Clockwise moments

Therefore $3\ T \sin 60° = 40 \times 1.5$ and so $T = \dfrac{60}{3 \sin 60°} = 23.09$ N (1)

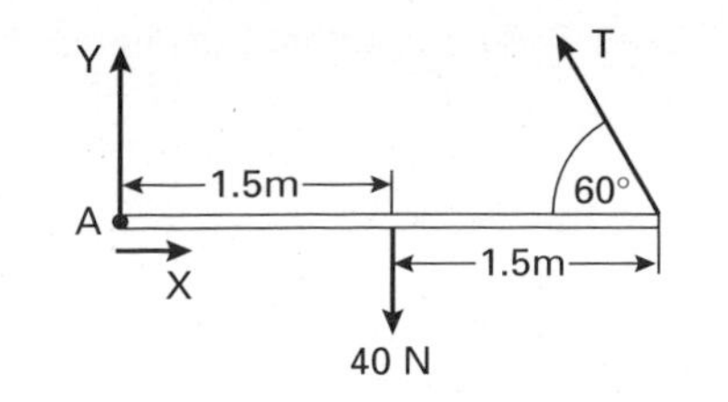

Figure 8.32

To find the magnitude and direction of the hinge reaction we could consider its horizontal and vertical components acting at A, which we have called X and Y respectively (Fig. 8.32). These could then be combined to find the reaction R at an angle θ.

Σ Horizontal forces $X = T \cos 60°$
therefore $X = 0.5$ (2)
Σ Vertical forces $Y + T \sin 60° = 40$
therefore $Y + 0.866\,T + 40$ (3)

Previously, taking moments about A (equation (1)) gave $T = 23.09$ N

Substituting for T in equation (2) gives us $X = 0.5 \times 23.09$
therefore $X = 11.54$ N

Substituting for T in equation (3) gives us $Y + 0.866\,(23.09) = 40$
therefore $Y = 20.01$ N

Now let us combine X and Y to find the magnitude and direction of the reaction at the hinge.

From Fig. 8.33 $\text{Tan}\,\theta = \dfrac{Y}{X} = \dfrac{20.01}{11.54} = 1.734$
therefore $\theta = 60.0°$

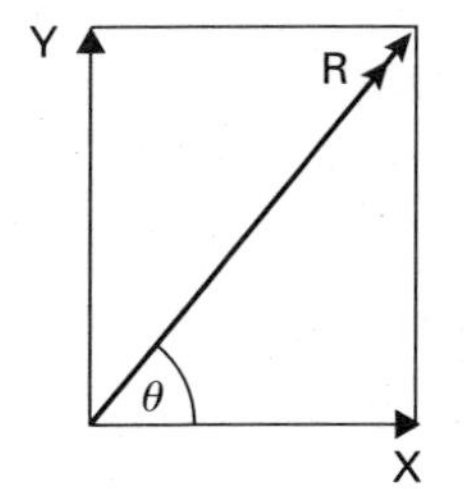

Figure 8.33

By Pythagoras' theorem $R^2 = X^2 + Y^2$, so $R^2 = 11.54^2 + 20.01^2$

therefore $R = \sqrt{533.6} = 23.1$ N

(or using $X = R \cos \theta°$, $R = 11.54 = 23.1$N)

The tension in the string is therefore 23.1 N and the reaction of the hinge is 23.1 N at 60° with the bar, vertically upwards.

8.7 Mathematical analysis of frameworks

The loads acting on a framework will cause external reactions at the supports of the framework. It is necessary to calculate these external reactions before considering the internal forces of the framework.

8.7.1 Establishing external reactions of a plane frame

If we consider the framework shown in Fig. 8.34, which is simply supported at both ends, the left-hand and right-hand external reactions are shown by R_L and R_R respectively. In this particular case the loads are all vertical, so that we can obtain two equations that will yield the values of R_L and R_R.

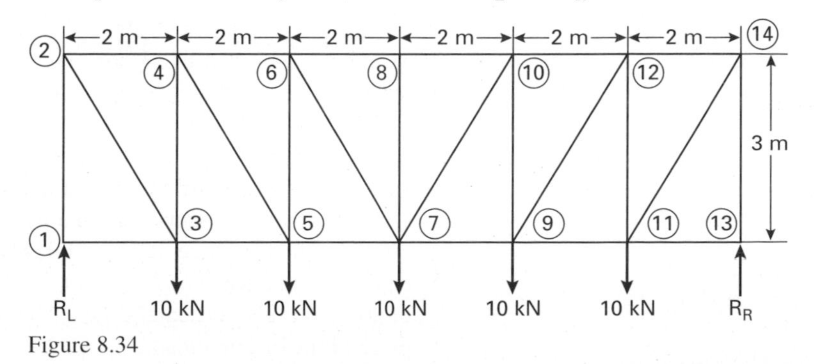

Figure 8.34

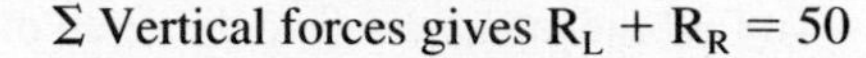

Σ Vertical forces gives $R_L + R_R = 50$ (1)

Taking moments about joint (1) gives

$$10 \times 2 + 10 \times 4 + 10 \times 6 + 10 \times 8 + 10 \times 10 = R_R \times 12$$

so $\quad 20 + 40 + 60 + 80 + 100 = 12 \times R_R$

therefore $\quad R_R = \dfrac{300}{12} = 25 \text{ kN}$

Substituting R_R into equation (1) gives $R_L = 25$ kN

(This is perhaps obvious as the structure is symmetrical and therefore the 50 kN load must be divided equally.)

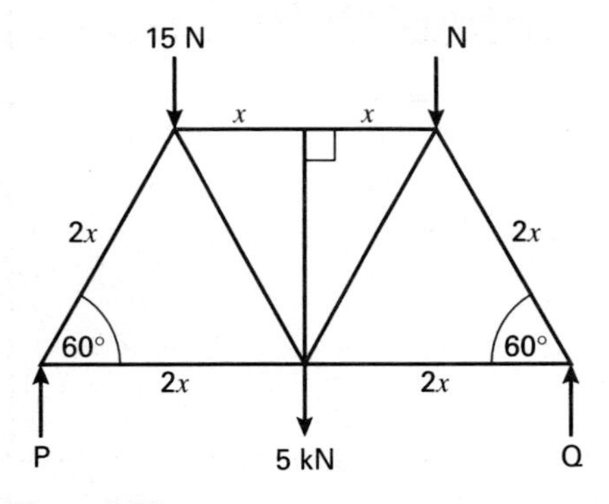

Figure 8.35

EXAMPLE 8.8 The framework shown in Fig. 8.35 has vertical but non-symmetrical loading. Calculate the value of the support reactions, P and Q.

SOLUTION Σ Vertical forces gives $P + Q = 40$ (1)

Taking moments about the left hand joint $15(x) + 5(2x) + 20(3x) = Q(4x)$

Dividing through by x gives $15 + 10 + 60 = 4 \times Q$

Thus $\quad Q = \dfrac{85}{4} = 21.25 \text{ N}$

Substituting for Q into equation (1) gives $\quad P = 18.75$ N

EXAMPLE 8.9 The framework shown in Fig. 8.36 is simply supported at A, has a fixed suport at B and the loads are at 60° to the horizontal as shown. If the vertical reaction at A is represented by P and the vertical and horizontal components of the reaction at B are represented by Q and R respectively, calculate the values of P, Q and R.

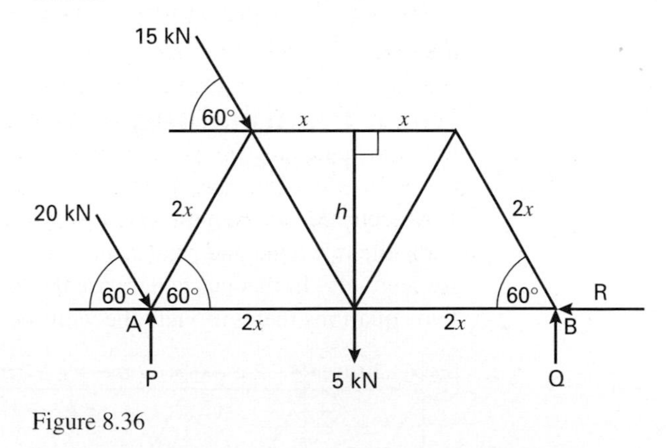

Figure 8.36

SOLUTION First consider the components of the 15 kN and 20 kN loads. The 15 kN load will be equivalent to a force of 15 cos 60° kN horizontally to the right as well as a force of 15 sin 60° kN vertically down. The 20 kN load will be equivalent to a force of 20 cos 60° kN horizontally to the right as well as a force of 20 sin 60° kN vertically down. These could be drawn onto the diagram in place of the original loads.

Σ Vertical force gives $P + Q = 15 \sin 60° + 20 \sin 60° + 5$ (1)
therefore $P + Q = 35.311$ kN

Σ Horizontal force gives $R = 15 \cos 60° + 20 \cos 60°$
therefore $R = 17.5$ kN

If we take moments about A, the 20 kN force and the reactions P and R will all pass through this point and will therefore have no moment about A. The 15 kN and 5 kN forces will provide a clockwise moment about A and Q will give an anticlockwise moment.

The vertical height of this frame, h, is given by $\sqrt{3}.x$ from the geometry of the equilateral triangles.

so $15 \cos 60° \times h + 15 \sin 60° \times x + 5 \times 2x = Q \times 4x$

substituting for h gives $15 \cos 60° \times \sqrt{3}.x + 15 \sin 60° \times x + 5 \times 2x \times Q \times 4x$

(dividing through by x) and simplifying gives $35.981 = Q \times 4$

so $Q = 8.995$ kN

substituting this into equation (1) gives $P = 26.316$ kN

If a single force was needed to replace Q and R at the right-hand support then this could be found by using Pythagoras' theorem and its direction found by trigonometry, as in 8.6.2.

8.7.2 Establishing internal reactions of a plane frame

Once the external forces on a framework in equilibrium have been determined, there are a number of approaches to establishing the internal forces. There are two analytical methods: the method of resolution at the joints and the method of sections; and one graphical method: Bow's notation.

We could use Bow's notation to find the forces graphically in the internal members of any framework but drawing this accurately takes time. In practice, a sketch of the Bow's force polygon is sufficient to identify the critical members. The forces can then be identified by an analytical method or the computer program for frame analysis (FRAMES) given on www.longman.co.uk There are many possible approaches and it is up to you to find the one that suits you best; this may vary according to the complexity of the frame. Table 8.1 shows the preliminary preparation, method of analysis and validation that are required by each of these methods.

Main analysis by	*Medium*	*Preliminary analysis needed*	*Validate by*
Methods of joints	Analytical	None	Double checking last joint
Bow's scale drawing	Graphical	None	Closed polygon which double checks last joint
Method of sections	Analytical	Bow's sketch or similar	Bow's sketch or similar
FRAMES	Computer	None	Any analytical or graphical check

Table 8.1

8.7.3 Method of resolution at the joints

The forces in the bars of a pin-jointed plane frame can be calculated by resolving in two directions at each joint, horizontal and vertical being convenient in most cases. Each joint, therefore, gives rise to two equations whose terms are the resolved parts of the forces in the bars. It is usual to denote each force by F with a suffix, written as a subscript, identifying the bar. For example the force in the bar CD is called F_{CD}. If tensile forces are taken as positive then each bar can be regarded as pulling outwards from the joints at its ends. When the equations are solved, negative results will indicate compression.

If the frame is supported at its ends it is first necessary to calculate the reactions but if it is cantilevered from a wall this is not generally necessary.

EXAMPLE 8.10 Fig. 8.37(a) shows a pin-jointed cantilever frame carrying a vertical load of 5 kN at C. AB is horizontal, AD and BD are inclined at 60° to the horizontal and the remaining members are at 30° to the horizontal.

Find the forces in all the bars stating whether they are tensile or compressive.

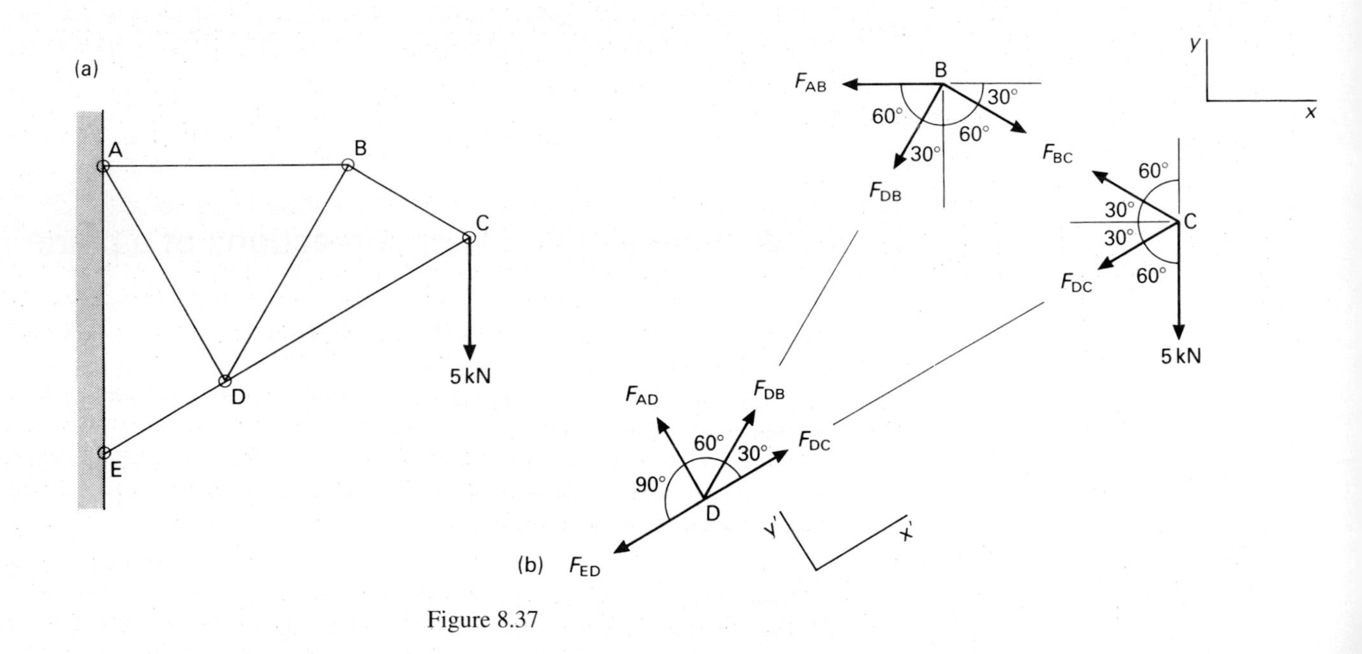

Figure 8.37

SOLUTION The forces acting at the joints B, C and D are shown in Fig. 8.37(b). For B and C it is convenient to resolve horizontally and vertically as indicated by the x and y axes. In the usual way these are taken as positive to the right and positive upwards. For D the work is simplified if the axes are taken in the DC and DA directions, x' and y'.

A force may be acting at an angle greater than 90° to the positive direction of the axis. For example, at joint C the force F_{BC} acts at an angle of 150° to the positive direction of x. This will lead to a negative term in the equation. Alternatively, the force can be resolved through 30° giving a component acting from right to left and the sign changed afterwards. The result is the same because cos 150° equals −cos 30°. With forces in kilonewton and angles in degrees the equations arising from resolving at the three joints are:

$$
\begin{array}{llllll}
\text{C/x:} & -F_{BC}\cos 30^\circ & -F_{CD}\cos 30^\circ & & & = 0 \\
\text{C/y:} & F_{BC}\cos 60^\circ & -F_{CD}\cos 60^\circ & -5 & & = 0 \\
\text{B/x:} & F_{BC}\cos 30^\circ & -F_{AB} & & -F_{DB}\cos 60^\circ & = 0 \\
\text{B/y:} & -F_{BC}\cos 60^\circ & & & -F_{DB}\cos 30^\circ & = 0 \\
\text{D/x':} & F_{CD} & -F_{DE} & & +F_{DB}\cos 30^\circ & = 0 \\
\text{D/y':} & F_{AD} & & & +F_{DB}\cos 60^\circ & = 0
\end{array}
$$

These six simultaneous equations are solved for the forces in the six bars.

From C/x, $F_{BC} = -F_{CD}$ and, substituting this result in C/y, the first two results are $F_{CD} = -5$ and $F_{BC} = 5$.

With these values B/y gives $F_{DB} = -2.887$ and B/x gives $F_{AB} = 5.774$. Finally, equations D/x' and D/y' give the results $F_{DE} = -7.5$ and $F_{AD} = 1.443$. The negative values indicate compression and the results are collected in the following table.

Bar	*Force* (kN)	*T or C*
AB	5.774	T
BC	5.000	T
CD	5.000	C
DE	7.500	C
AD	1.443	T
DB	2.887	C

8.7.4 The method of sections

The method of resolution at the joints (or the corresponding computer program) analyses one joint at a time. In the case of a frame supported at its ends, such as a roof truss or a bridge girder, we have to start at an end joint. It can then take many steps to reach the centre of the frame where the bars may be carrying the largest loads. We can avoid this and go straight to the members in the middle of the frame by using the *method of sections*.

We divide the frame into two by a section that cuts through the members whose forces are required. We then consider the equilibrium of that part of the frame to one side of the section by the usual methods of resolving forces and taking moments, see Fig. 8.38(b). The forces in the bars that have been cut become additional external loads on the part of the frame under consideration.

EXAMPLE 8.11 The roof truss shown in Fig. 8.38(a) is hinged at one end and rests on a roller support at the other. It carries a vertical load of 2 kN at each of the upper chord points. Determine the forces in the bars a, b, c and d, stating whether they are tensile or compressive.

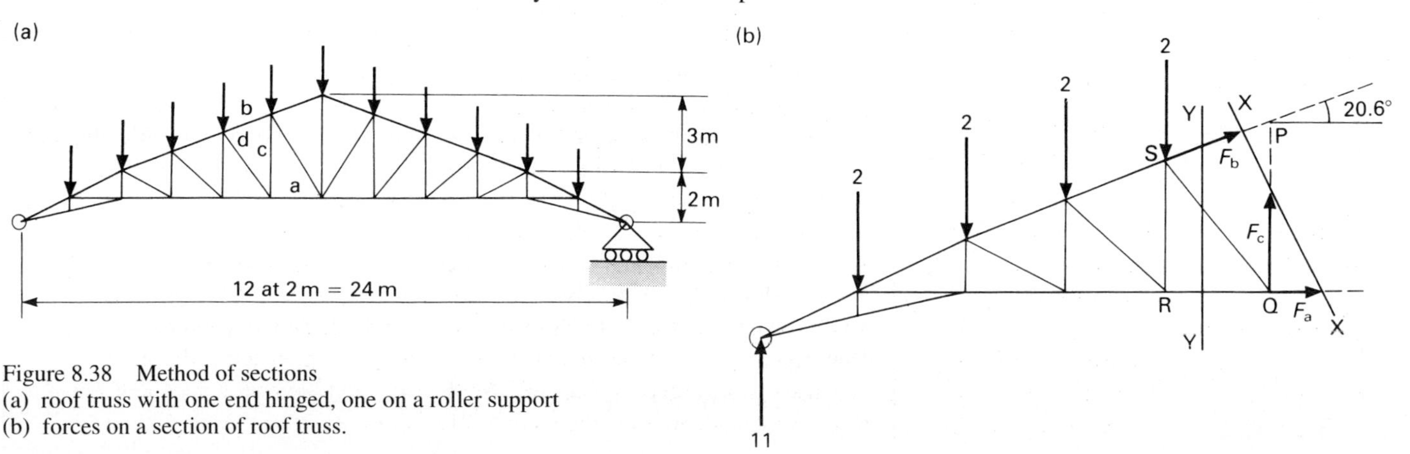

Figure 8.38 Method of sections
(a) roof truss with one end hinged, one on a roller support
(b) forces on a section of roof truss.

SOLUTION The bars along the top of the frame are often called the upper chord members and the corresponding joints are known as the upper chord, or upper panel, points. The total downward load on the frame is 22 kN, symmetrically distributed. The reaction at each end is therefore 11 kN vertically upwards.

The choice of section depends on the bars whose forces are required. The bars a, b and c can be dealt with by taking the section XX as shown in Fig. 8.38(b). Let the forces in these bars be F_a, F_b and F_c, positive if tensile and pulling outwards in the directions of the arrows. The part of the frame to the left of XX is therefore in equilibrium under the action of these forces together with the four downward loads of 2 kN each and the reaction of 11 kN upwards at the left hand.

Each unknown force can be found by resolving or taking moments and we can select the method in each case so as to simplify the work. For the force F_a it is convenient to take moments about the point P. It does not matter that P is outside the part of the frame under consideration and it has the advantage that F_b and F_c both act through it. These two forces therefore have zero moments about P. From the dimensions given in Fig. 8.38 the moment arm for F_a (PQ in Fig. 8.38(b)) equals 1 m + 3 × 3/4 m = 3.25 m.

Taking anticlockwise as positive, and working in kilonewton and metre units, moments about P give the equation:

$$3.25F_a + 2 \times 2 + 2 \times 4 + 2 \times 6 + 2 \times 8 - 11 \times 10 = 0$$

and

$$F_a = (110 - 4 - 8 - 12 - 16)/3.25 = 21.54$$

Although F_b can be found by taking moments about Q it is easier to resolve horizontally. The upper chord members of the inner panels rise a total of 3 m vertically in a horizontal distance of 8 m and their angle with the horizontal is therefore $\tan^{-1}(3/8) = 20.6°$. The force F_b therefore has horizontal and vertical components of $F_b \cos 20.6°$ and $F_b \sin 20.6°$.

Resolving horizontally (this eliminates all the vertical loads on the frame) and taking positive to the right,

$$F_a + F_b \cos 20.6° = 0$$

and

$$F_b = -21.54/\cos 20.6° = -23.0$$

The force in bar c can be found with the same section by resolving vertically. Taking upwards as positive the equation is:

$$F_c + F_b \sin 20.6° - 2 - 2 - 2 - 2 + 11 = 0$$

and

$$F_c = -(-23.0) \sin 20.6° + 8 - 11 = 5.08$$

With the lettering for the panel points shown in Fig. 8.38(b), bar d is SQ and its angle with the horizontal is:

$$\tan^{-1}(SR/RQ) = \tan^{-1}(2.5/2) = 51.3°.$$

To find the force in this bar, F_d, a new section is needed and YY is convenient. This cuts the bar RQ whose force is not called for. To eliminate it from the equation we resolve vertically upwards and, taking tensile as positive in the usual way,

$$F_b \sin 20.6° - F_d \sin 51.3° - 2 - 2 - 2 - 2 + 11 = 0$$

and

$$F_d = [(-23.0) \sin 20.6° - 8 + 11]/\sin 51.3° = -6.50$$

Collecting results,

Bar	*Force* (kN)	*T or C*
a	21.54	T
b	23.00	C
c	5.08	T
d	6.50	C

8.8 Designing structures

Many examination questions on structures are analytical. A pin-jointed frame is given, the external loads are specified and the question asks for the forces in some or all of the members. Designing a structure is different. The designer has to estimate the loads it will carry, choose the materials from which it will be made and decide the shape and form of construction.

Design codes

In large structures such as bridges, high rise buildings and airliners where there is concern for public safety the designer has to follow codes of practice. These are drawn up by national or international standards organisations and they specify the loads to be allowed for and the strength that can be assumed for the materials to be used.

Load factor

If we are designing a hi-fi storage unit we first have to work out the weights of all the components, records, tapes and compact discs it will carry. Then, to provide a margin of safety, we design the structure to carry a greater load than we would expect under normal conditions. The ratio of the design load to the expected load is called the 'load factor'. In many cases it corresponds to the factor of safety defined in Chapter 9 but there are instances where the two are different. Design codes specify the load factors to be used for large structures.

Form and materials

At an early stage in the design process decisions have to be taken about (a) the general form of the structure and (b) the materials of which it will be made. In some cases the general shape and size of the structure will be the same whatever materials are used. A chair frame can be made from solid timber, laminated wood, tubular metal or a modern plastic material but its dimensions and overall form are governed largely by the function it has to perform.

In contrast there can be great differences between the shapes of bridges built with different materials. Materials such as brickwork and masonry have a reasonable strength in compression but are very weak in tension (see Chapter 5). A bridge made from such materials must therefore be in the form of an arch (Fig. 8.39(a)) in which each block presses against those next to it so that they all carry compressive loads.

On the other hand, cables and chains can withstand tension but offer no resistance to compression. Bridges using such components are therefore made in the form of suspension cables (Fig. 8.39(b)). The deck of the bridge is hung from these cables which take up a shape corresponding to the funicular polygon arising from the vertical loads. The cables, in turn, are supported by piers or towers near the ends of the bridge and these are in compression.

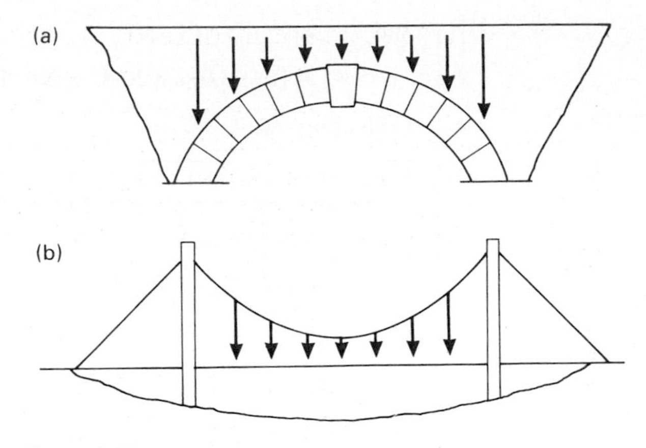

Figure 8.39
(a) Compression in an arched bridge
(b) Tension in cables of a suspension bridge.

The history of bridges provides many examples of how the design of structures is related to the materials available. The rail and road bridges over the Forth (Fig. 8.40) cross the river near to each other but their shapes are entirely different. The high tensile steel used in the cables of the modern road bridge was not available at the end of the nineteenth century when the rail bridge was built. In projects of this kind the choice of materials and form is usually made on the grounds of cost.

Figure 8.40 Contrasting structural forms – Forth rail and road bridges.

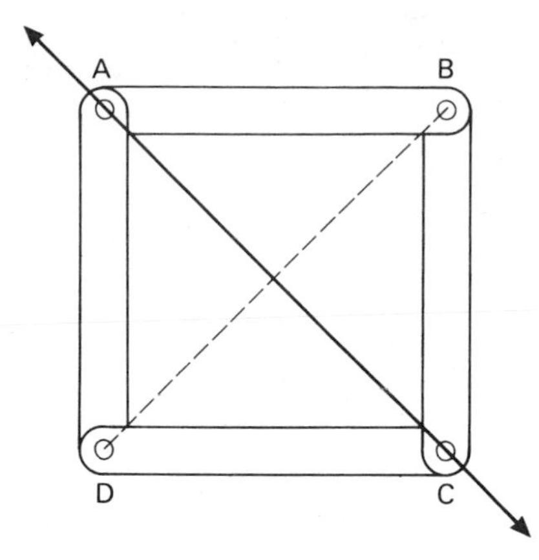

Figure 8.41 Square pin-jointed frame.

Redundant structures

In practice many structures have more members than are required to meet the stiffness criterion given in section 8.5. Provided the bars can be fitted together without strain these redundancies increase the strength of the structure and in some circumstances they can lead to a reduction in weight.

Consider the square pin-jointed frame ABCD in Fig. 8.41. With four joints the number of bars required for stiffness is given by

$$\begin{aligned} b &= 2j - 3 \\ &= 2 \times 4 - 3 \\ &= 5 \end{aligned}$$

Only one diagonal is required but suppose two are provided in the form of wires. When outward loads are applied at A and C the wire AC will be in tension and will stretch a little. This causes BD to go slack so that the frame effectively has only five members. If, instead, the forces are applied at B and D (or the forces at A and C are reversed) it is BD that is in tension and AC that goes slack. The frame can therefore withstand both types of loading without having a diagonal bar designed for compression which would be heavier. This double bracing was extensively used in early biplanes because the loading on the wings during flight was opposite to that when the aeroplane was on the ground.

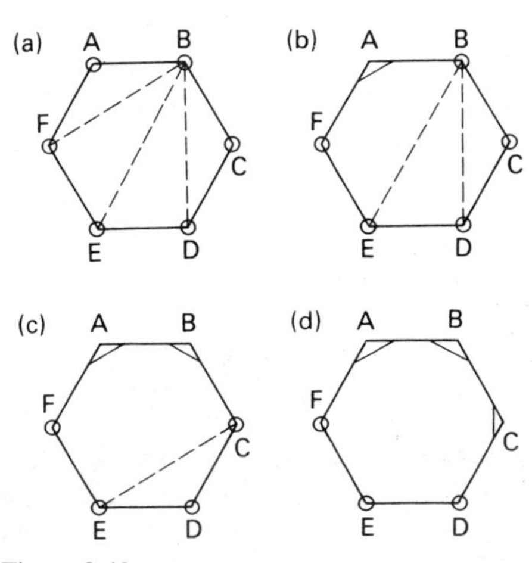

Figure 8.42

Stiff joints

Throughout this chapter it has been assumed that each joint of the frame consisted of a single, frictionless pin. This idealisation is far removed from most practical joints which are designed to prevent any relative movement of the bars. To achieve this stiffness, timber structures use means such as slotted joints (mortise and tenon), adhesives and stiffening brackets, and the joints in metal structures are welded or riveted at several points.

In general such joints strengthen the frame so that designs based on the assumption of frictionless pins will have a margin of safety. On the other hand, many structures depend on the joints being stiff. A chair or table could not stand up if its legs were merely attached by pinned or hinged joints.

The effect of stiff joints is shown in Fig. 8.42. Suppose the hexagonal frame ABCDEF is pinned at every joint (Fig. 8.42(a)). The formula for stiffness shows that the total number of bars required is give by:

$$\begin{aligned} b &= 2j - 3 \\ &= 2 \times 6 - 3 \\ &= 9 \end{aligned}$$

In addition to the six perimeter members three internal bars are therefore required. One possibility is shown by the broken lines BD, BE and BF. If, on the other hand, the joint A is made stiff (Fig. 8.42(b)) then B and F are fixed relative to each other and do not need to be connected by a bar. In this case only eight bars are required.

If joints A and B are both made stiff (Fig. 8.42(c)) only one internal bar is required, making seven in all, and with three stiff joints A, B and C the frame is just stiff without internal bars. Each stiff joint is therefore equivalent to one extra bar.

The joints where frames are attached to a wall or the ground can also be made stiff. Instead of the bar being pinned at the support it can be built in as shown in Fig. 8.43. If it projects from a wall (Fig. 8.43(a)) it is called a *cantilever*; this is considered in more detail in Chapter 9.

Fig. 8.43(b) shows a frame with two members built into the ground as in the case of goal posts. In considering the stiffness of this arrangement the ground counts as one member so that, with two uprights, a crossbar and two fixed joints, there are effectively six members altogether. With four pin joints, only five bars are required for stiffness. As a two-dimensional structure, therefore, the frame is redundant.

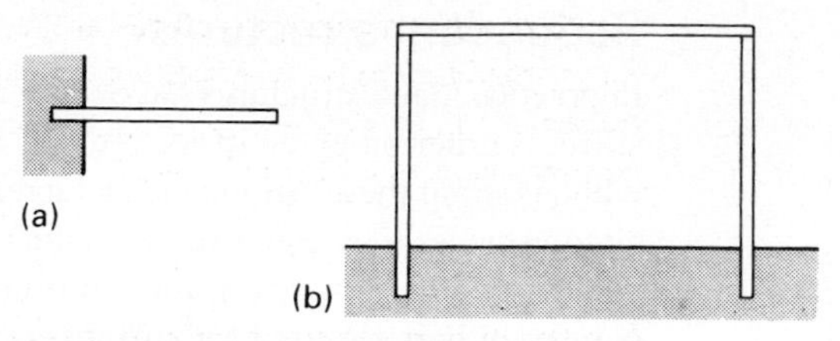

Figure 8.43
(a) Bar built into the support
(b) Frame built into the ground.

Three-dimensional structures

Although we live in a three-dimensional world, the study of structures is concerned to a large extent with two-dimensional frames. There is some justification for this in that many practical structures consist of two or more vertical plane frames that are braced laterally to keep them upright. The roof structure of a modern house is often made up from several similar trusses that are supported by the walls and are joined by longitudinal members called purlins.

Bridge structures, too, consist of two parallel girders, arches or suspension cables with the carriageways in between (Fig. 8.44).

Although vertical loads can be carried by two-dimensional structures of these kinds, some provision must be made for stiffness and strength in the third dimension. This can be done by ensuring that there are triangles or stiff joints in the plane perpendicular to that of the parallel frames.

Figure 8.44 Three-dimensional structure formed from two parallel plane frames – Runcorn Bridge.

Three-dimensional structures are not all assemblies of two-dimensional frames and some have to be analysed using the general conditions for equilibrium in three dimensions. This entails resolving forces in three directions rather than two and is beyond the scope of this book.

Safety

It is the designer's responsibility to ensure that structures are safe throughout their working lives. The shape and dimensions of the structure and its components, the materials chosen for its construction, the stiffness of its joints, in the case of a frame, and the method of attachment to the ground or other support all affect the safety of a structure. The designer must think in three dimensions and structures that are strong enough as vertical plane frames can collapse sideways if they are not supported against lateral loads.

A special problem arises in the case of large civil engineering structures. The designer of a bridge can make sure that it will be safe under all foreseeable loads when it is completed, but problems may arise during construction. When it is partially built the loads on members may be very different from those it will eventually carry and there have been spectacular failures in partially completed structures. Figs 8.45 and 8.46 show a bridge under construction and in its completed form. Note the temporary structures (known as 'falsework') supporting the bridge during construction.

As in all design every detail requires careful consideration and the safety of a structure depends on the strength of each of its components. This is the subject of the next chapter.

Figure 8.45 Foyle Bridge, Londonderry, under construction.

Figure 8.46 The completed Foyle Bridge.

Bibliography

Blundell A, Hawkins R, Luddington D 1981 *Modular Courses in Technology: Structures*. Oliver and Boyd

Croxton P C L and Martin L H 1987 *Solving Problems in Structures,* vol 1. Longman

Gauld B J B 1991 *Structures for Architects* 3rd edn. Longman

Gordon J E 1991 *Structures or Why Things Don't Fall Down*. Penguin

Hannah J and Hillier M J 1999 *Mechanical Engineering Science* 3rd edn. Longman

Morgan W 1978 *The Elements of Structure* 2nd edn. Longman

Reynolds T J and Kent L E 1973 *Introduction to Structural Mechanics* 6th edn. Longman

Rich S 1992 *Structures with Materials*. Stanley Thornes

Shirley-Smith H 1964 *The World's Great Bridges* 2nd edn. Phoenix House

Assignments

1 A block of 200 kg mass rests on an inclined plane of 15° slope as shown in Fig. 8.47. A rope is attached to the block and pulled with a tension T at an angle of 30° to the slope.

(a) If the block is at rest when the tension is 1 kN what is the magnitude and direction of the resultant reaction R?

(b) If the tension is gradually increased and the block just begins to move when $T = 1.2$ kN, what is the coefficient of limiting friction and the corresponding angle of friction?

(c) What is the least tension required to overcome friction if the rope is parallel to the plane?

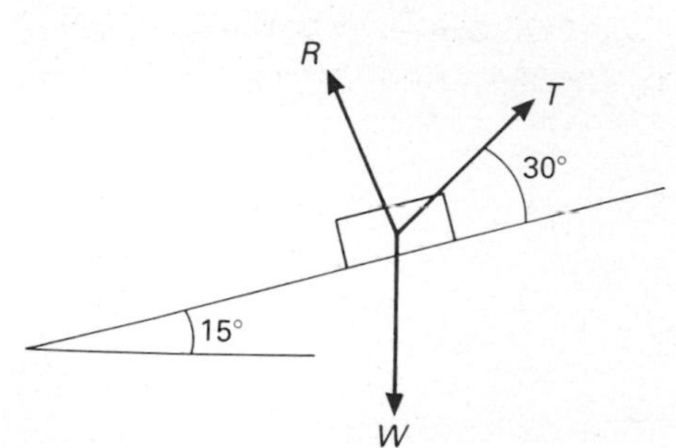

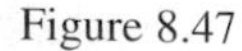

Figure 8.47

2 (a) Explain clearly what is meant by the term *conditions of equilibrium*. State the conditions which must be satisfied for a body to be in equilibrium under the action of a number of forces in the same plane.

(b) Three structural members (A, B and C) form part of a bridge truss and are tied together in a bracket as shown in Fig. 8.48. The force in member A is 150 kN and it is required that the force in member C should act horizontally. Calculate the *magnitude* of the forces in members B and C.

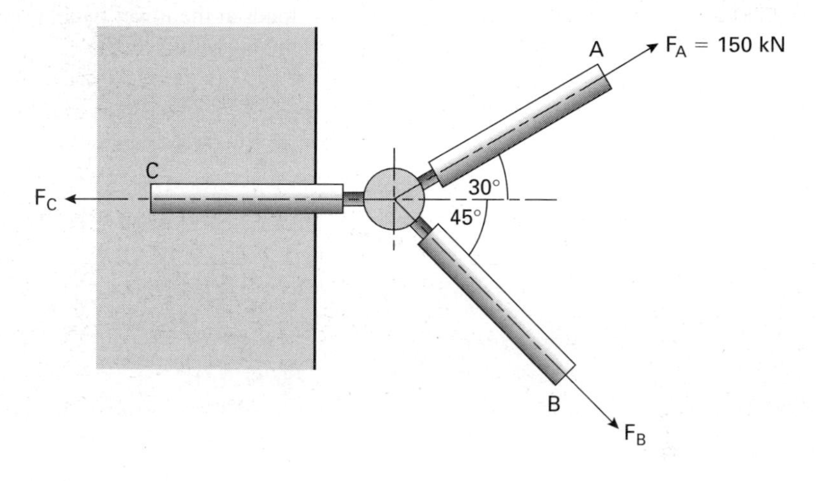

Figure 8.48

3 The sides of a square plate ABCD are each 1 m long. Forces of 2, 3, 5 and 8 kN act along the sides AB, BC, CD and DA respectively. Find the magnitude, direction and line of action of the resultant.

4 Examine each of the two-dimensional, pin-jointed assemblies shown in Fig. 8.49 for stiffness, distinguishing between mechanisms, perfect frames and redundant frames. The bars are pinned at the circled joints.

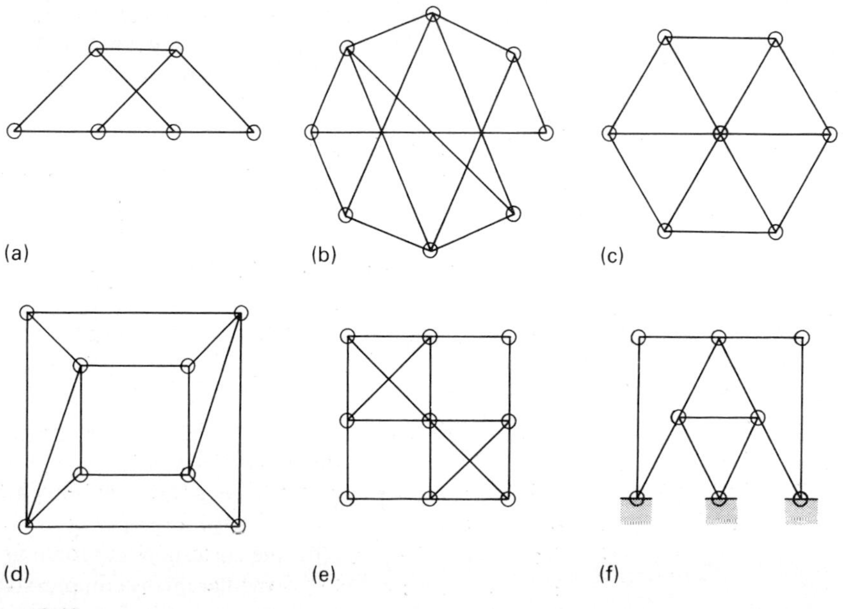

Figure 8.49

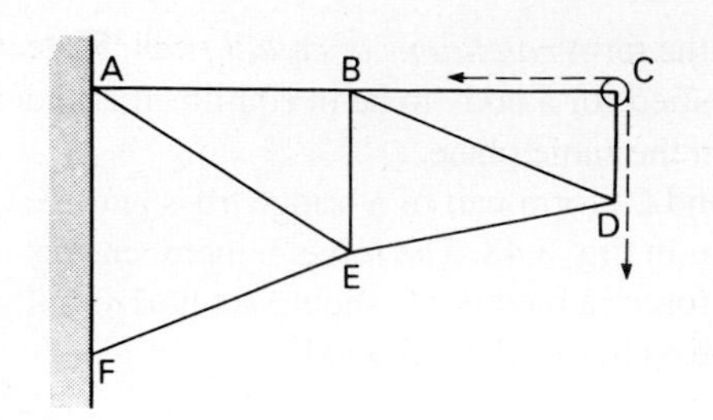

Figure 8.50

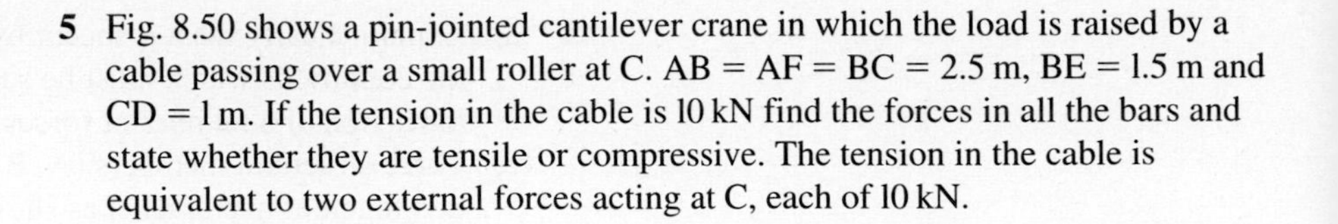
5 Fig. 8.50 shows a pin-jointed cantilever crane in which the load is raised by a cable passing over a small roller at C. AB = AF = BC = 2.5 m, BE = 1.5 m and CD = 1 m. If the tension in the cable is 10 kN find the forces in all the bars and state whether they are tensile or compressive. The tension in the cable is equivalent to two external forces acting at C, each of 10 kN.

6 The pin-jointed frame of Fig. 8.51 is supported at its ends and carries vertical loads at the lower panel points. The sloping members are all inclined at 45° to the horizontal.

Find (a) the reactions at A and E,

(b) the forces in the bars, stating whether they are tensile or compressive.

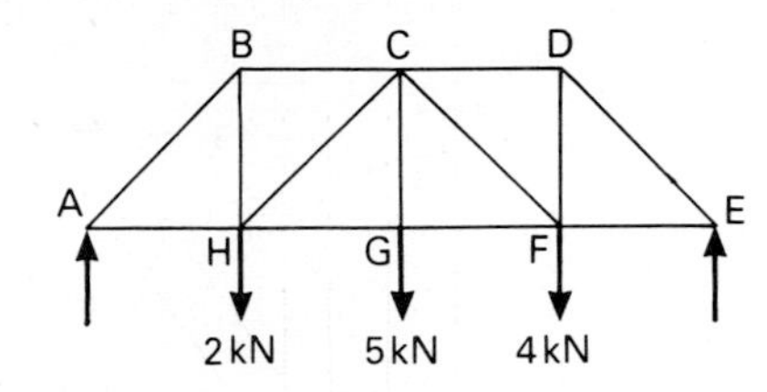

Figure 8.51

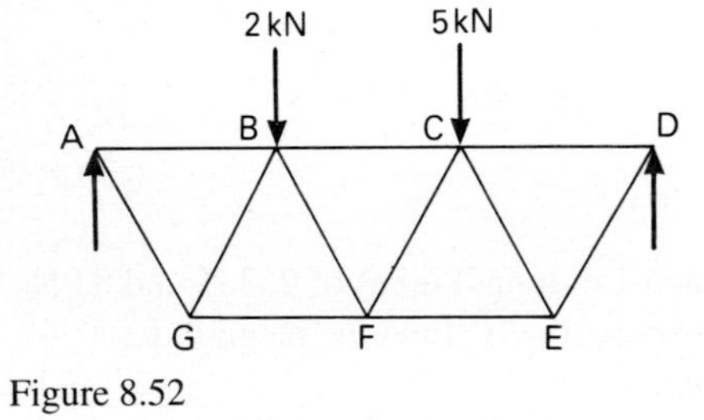

Figure 8.52

7 All the bars in the pin-jointed frame of Fig. 8.52 have the same length. The frame carries vertical loads of 2 kN and 5 kN at B and C.

Find (a) the reactions at A and D,

(b) the forces in all the bars, stating whether they are tensile or compressive.

8 Each panel of the pin-jointed plane frame shown in Fig. 8.53 is 3 m wide and 4 m high. Determine, for the loading shown, the forces in the bars a, b, c and d, stating whether they are tensile or compressive.

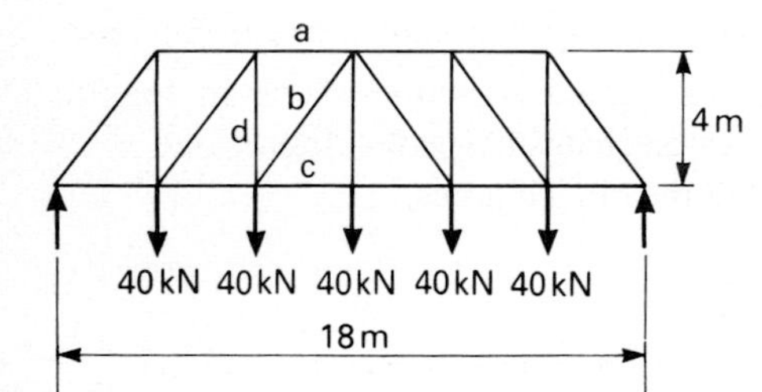

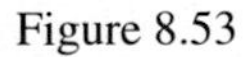
Figure 8.53

9 The diagram (Fig. 8.54) shows a pin-jointed frame, comprising seven members of equal length, supported by a hinge at A and a roller at E. Under a particular loading condition the external loads are represented by a horizontal force 25 kN at B and a vertical force 300 kN at C.

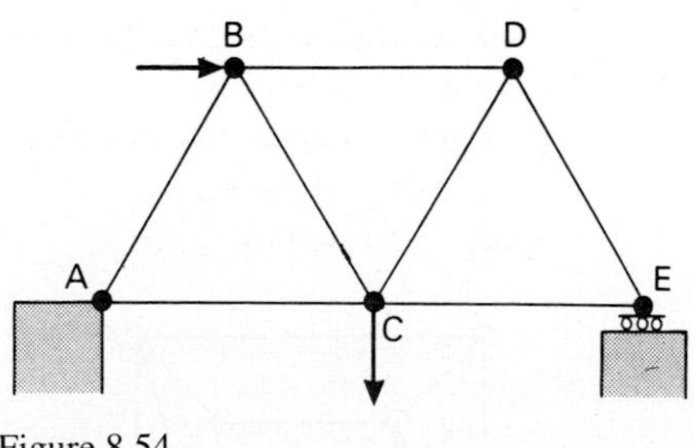

Figure 8.54

Determine

(a) the magnitude and direction of the reactions at A and E arising from the external forces,

(b) the corresponding force in each member of the framework indicating whether this is compressive or tensile.

(Solution by a graphical method is acceptable)

10 The diagram (Fig. 8.55) shows a small pin-jointed frame to which a vertical load F is applied. AB is vertical. ABC = 90°.
Derive expressions, in terms of F and the lengths x and y, for the forces in members AC and BC.

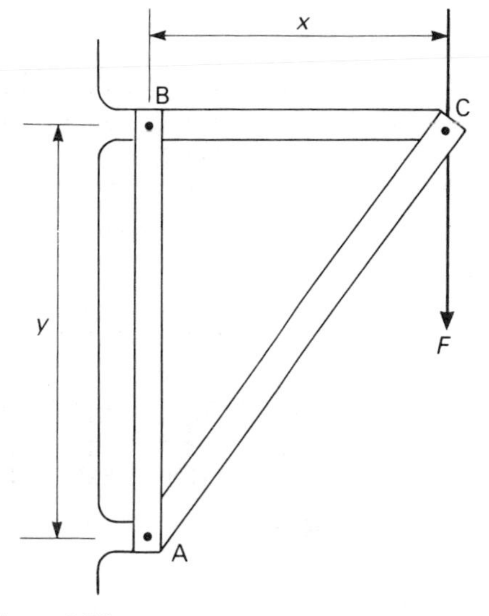

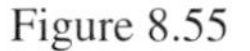
Figure 8.55

11 The simple pin-jointed framework shown in Fig. 8.56 supports a bank of floodlights of weight 0.80 kN. The wall exerts a horizontal reaction force against the roller at E. You may neglect the weight of the framework.

(a) Find the magnitude of the horizontal force exerted by the wall at E.

(b) Find the force in each member of the framework, stating whether the force is tensile or compressive.

(c) Determine the magnitude and direction of the force exerted by the framework on the bracket which holds the hinge at A.

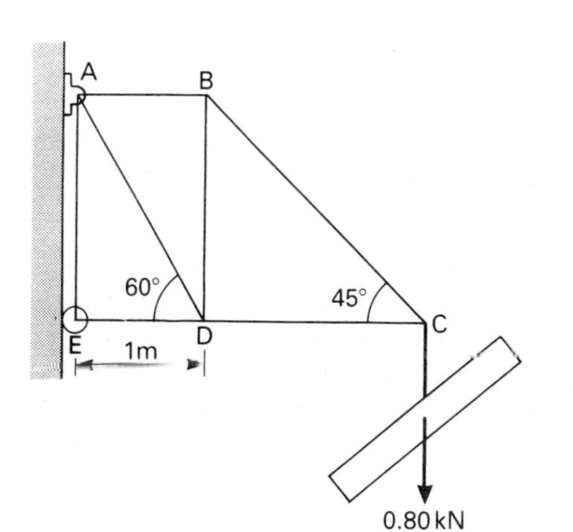

Figure 8.56

12 The shear leg derrick shown in Fig. 8.57 is used to raise loads vertically by means of a pulley system attached to point D. Determine the magnitude and type of force produced in the legs BD and CD and in the backstay AD when raising a load of 8kN.
HINT: This is an example of a space frame. The forces in the members BD and CD have their resultant in the direction ED. This resultant, together with the force in AD and the 8 kN load, form a set of three coplanar forces in equilibrium.

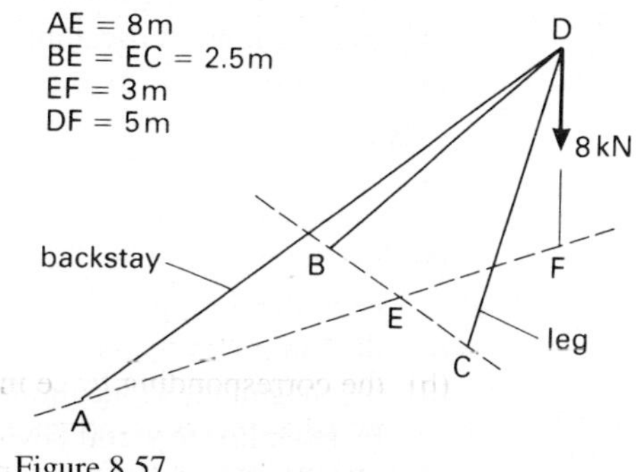

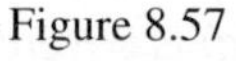
Figure 8.57

13 Fig. 8.58 shows the elevation of two pin-jointed steel frameworks, to each of which a gantry rail is attached. A bogie can be mounted on the rails providing approximately 7.5 metres of horizontal movement.
The bogie has a winch and 4 wheels mounted on two axles set at 0.75 metres apart. The maximum carrying capacity is 2000 kN, supported equally by the four wheels.

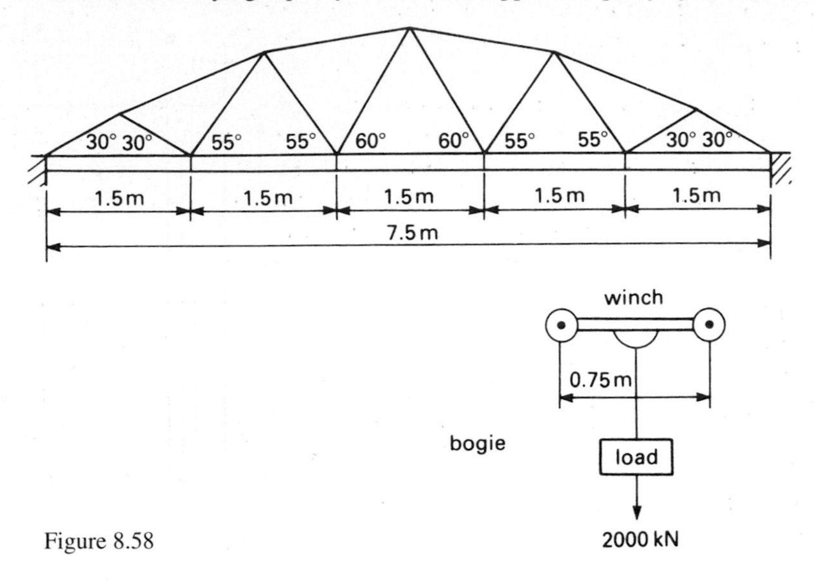

Figure 8.58

Ignoring the mass of the structure and bogie

(a) Select the position in which the bogie will apply maximum stress to the supporting frames.

(b) (i) Determine the members subject to maximum tension and compression force as a result of the applied load. State the magnitudes of these forces.

(ii) If the bogie is in the position to give the maximum forces, which member is subject to the minimum force? Explain how the force in this member will change if the bogie is moved from the position selected in (a).

(c) If the extension of any of the members forming the lower horizontal section of the framework is not to exceed 1.0 mm, determine the minimum cross sectional area of steel required for these members.
[E for steel = 200 GN/m^2.]

NOTE: Part (c) is covered in Chapter 9. Assume that the rail is in sections simply supported at the joints of the lower horizontal section of the framework.

14 A series of pin-jointed roof trusses, loaded as shown in Fig. 8.59, are mounted on 5 m high walls.
Two adjacent trusses also support a water tank (dimensions 4 × 1 × 0.25 m) which rests on two beams suspended on 3.5 m long steel rods.
The tank is filled to capacity and emptied regularly. In order to avoid fracturing pipe connections, vertical movement of the tank relative to the trusses must not exceed 2 mm.
Ignoring the mass of the framework and the mass of the tank, determine:

(a) the nature, direction and magnitude of the reactions supporting one of the roof trusses when the tank is full of water.

(b) the nature and magnitude of the forces in the two members meeting at the right-hand support.

(c) the minimum diameter of the rods supporting the tank.

(d) the magnitude of the turning moment acting on the base of the left-hand wall.
[E for mild steel 200 GN/m^2. 1 m^3 of water has a mass of 1000 kg.]
NOTE: Part (c) is covered in Chapter 9.

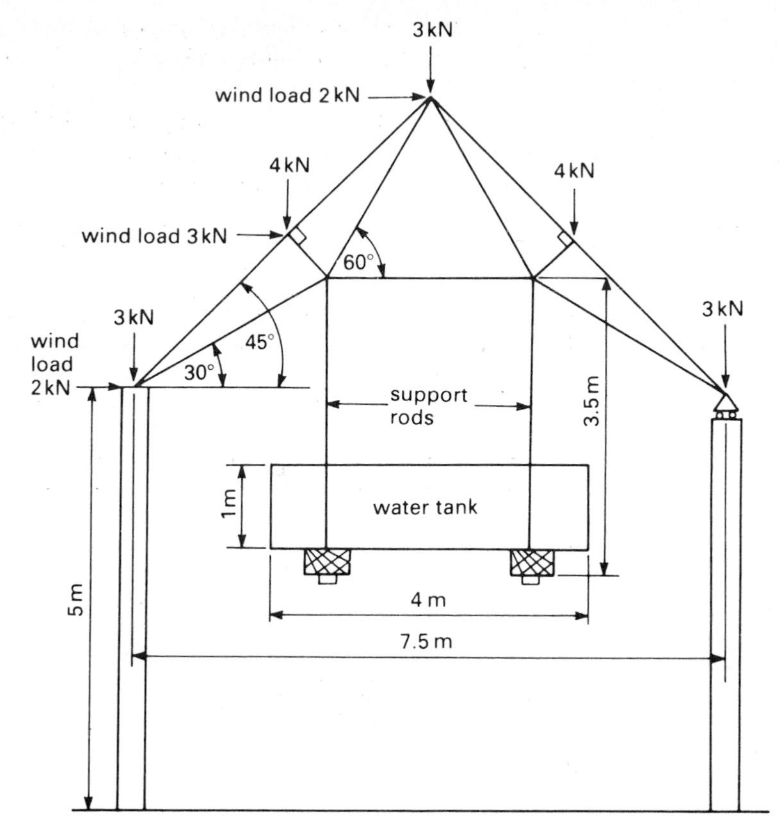

Figure 8.59

15 Fig. 8.60 shows the outline design of one of two frameworks forming the sides of a bridge. The roadway is supported each side by vertical cables attached to the pin-jointed arched frame. The mass of the roadway and passing vehicles apply a load of 80 kN to each supporting cable as shown in the figure.

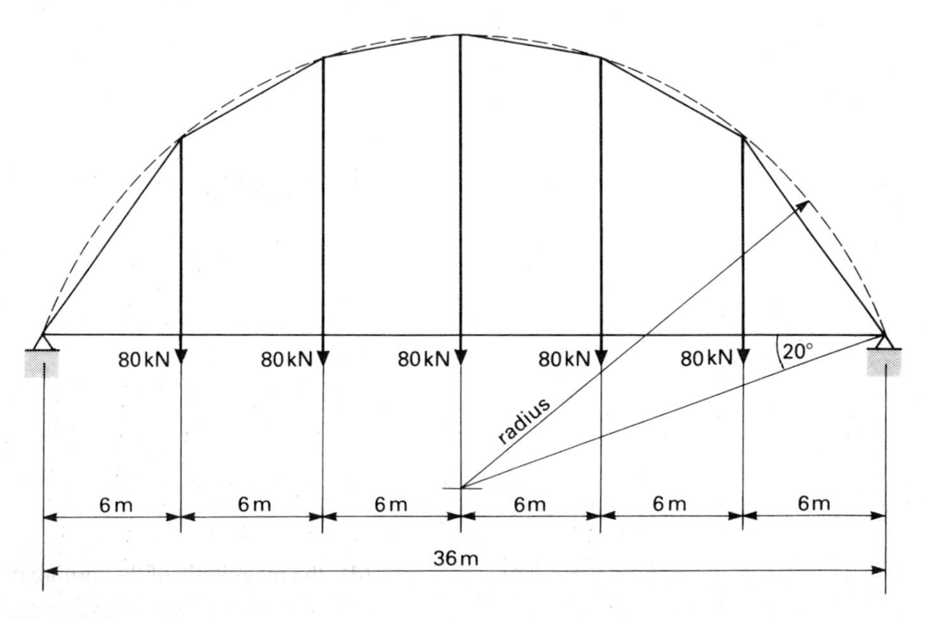

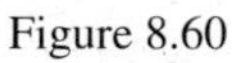

Figure 8.60

Disregarding any forces originating from the mass of the framework and supporting cables:

(a) Add sufficient members to the existing design to create a rigid structure.

(b) State the nature and magnitude of the forces acting on the members you have added to the framework.

(c) Identify which individual member(s) are subject to the greatest load. State their nature and magnitude(s).

(d) Traditionally stone bridges have been built by forming a compressive arch. Identify the factors which enable such a bridge to support a load. Explain why a pin-jointed framework similar to the one shown in the figure is unable without additional members to support a load.

(e) Describe the stability and the rigidity problems that are likely to occur during the construction of your final design. Describe any additional support that may be required whilst construction is in progress.

9 Stress analysis

9.1 The strength of structural components

The previous chapter was concerned with the equilibrium of structures and how the forces in the bars of a frame can be found by drawing or calculation. As well as checking for equilibrium the designer must make sure that each component of the structure is strong enough to carry the forces acting on it. Forces in structures are often called *loads*.

The strength of a component will depend upon the material of which it is made and also its shape and size. The properties of materials are the subject of Chapter 5; the present chapter deals with the stresses within a component and how they are related to its shape and the loads acting on it. This subject is called *stress analysis* or *strength of materials*. In engineering courses it is often included in a broad subject called *mechanics of solids*.

9.2 Stress and strain

In everyday language we often use the words 'stress' and 'strain' to mean the same thing. In mechanics, however, they have distinct and precise meanings.

Suppose a tie-bar, length L and cross-sectional area A, is acted upon by equal and opposite forces, each P, as shown in Fig. 9.1(a). In this chapter the arrows show the forces acting on the bar as in Fig. 8.20(a). Note that the load being carried by the bar is P, not $2P$. Take a section such as XX and consider the equilibrium of that part of the bar to one side as in Fig. 9.1(b). To balance the end force P there must be an equal opposite force distributed over the section XX.

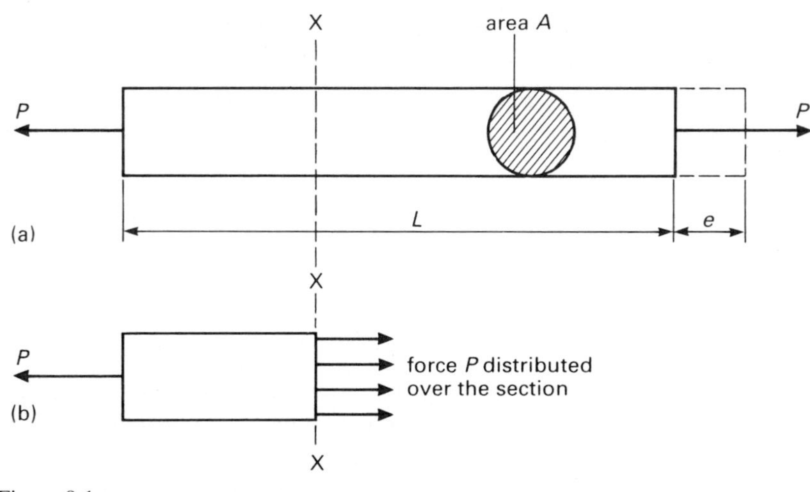

Figure 9.1

The ratio of the force to the cross-sectional area is called the *stress* and is denoted by σ. Thus:

$$\text{stress } \sigma = \frac{\text{load } P}{\text{area } A}$$

The same definition applies when the bar is in compression. Tensile and compressive stresses are called *direct* stresses and many formulae can be used for both. Suppose the load P causes the bar of Fig. 9.1(a) to stretch by an amount e. The ratio of the increase in length to the original length is called the *strain* and is denoted by ϵ. Thus:

$$\text{strain } \epsilon = \frac{\text{change in length } e}{\text{original length } L}$$

The same definition applies in compression and where a distinction is necessary it is usual to treat tensile strains as positive and compressive strains as negative.

9.2.1 Units of stress and strain

Since stress is defined as a force divided by an area, its units in the SI system are *newton per square metre*. This is written N/m^2 or N m^{-2}. The name *pascal* (Pa) has been given to this unit and you must be prepared for all three forms in books and examination questions. The first method (N/m^2) emphasises that stress is a ratio and is used throughout the book.

Practical values often run into millions. The working stress for one type of steel might be 150 000 000 N/m^2 and this can be written as 150×10^6 N/m^2, 150 MN/m^2 or 150 MPa, where the prefix M stands for mega (see Appendix 2). Another possibility is to measure the area in square millimetres and to quote the stress in N/mm^2. This is effectively the same as MN/m^2 because 1m^2 = 10^6 mm^2 and values will be identical in the two systems.

From its definition, strain is the ratio of two lengths and is therefore a pure number with no units. Practical values are very small and are sometimes quoted in millionths for which the name *microstrain* is used.

EXAMPLE 9.1 A steel wire 6 mm diameter and 8 m long extends 5 mm under a pull of 3.6 kN. Calculate the stress and strain in the wire.

SOLUTION Working in metres and newtons,

$$\text{cross-sectional area} = \pi d^2/4 = \pi(0.006)^2/4 \text{ m}^2 = 28.27 \times 10^{-6} \text{ m}^2$$

$$\begin{aligned}\text{stress} = \text{load/area} &= 3600/(28.27 \times 10^{-6}) \text{ N/m}^2 \\ &= 127.32 \times 10^6 \text{ N/m}^2 \text{ (or Pa)} \\ &= 127.32 \text{ MN/m}^2 \text{ (or MPa)}\end{aligned}$$

$$\begin{aligned}\text{strain} = \text{extension/original length} &= 0.005/8 \\ &= 0.000\,625\end{aligned}$$

9.2.2 Elasticity

For many materials it is found that the change in length of a bar is proportional to the force causing it (see section 5.2.1). This result is known as *Hooke's law* but engineers usually quote the equivalent relationship 'stress is proportional to strain'. This statement is almost exactly true for many materials up to a stress called the *limit of proportionality* (see Fig. 5.16).

For a material that obeys Hooke's law the ratio of stress to strain is a constant. It is called the *modulus of elasticity*, the *linear elasticity*, or the *Young modulus* and is denoted by E:

$$\text{modulus of elasticity } E = \frac{\text{stress } \sigma}{\text{strain } \epsilon}$$

Since strain is a ratio of two lengths and has no units, modulus of elasticity has the same basic units as stress, that is N/m^2 or Pa. Practical values are very large and are usually expressed in giganewton per square metre, GN/m^2, or gigapascal, GPa (see Appendix 2). You may also find this unit expressed as kN/mm^2 but this does not affect the numerical value. In this book the form GN/m^2 is used.

EXAMPLE 9.2 A railway signal of the semaphore type is operated from the signal box by a wire 800 m long and 5 mm diameter. The tension in the wire needed to move the signal arm is 900 N. If a movement of 150 mm is required at the signal post end of the wire what is the corresponding movement at the signal box end? Take $E = 206\ GN/m^2$.

SOLUTION Cross-sectional area $= \pi d^2/4 = \pi \times 0.005^2/4\ m^2$
$= 19.6 \times 10^{-6}\ m^2$

stress $\sigma =$ load/area $= 900/(19.6 \times 10^{-6})\ N/m^2 = 45.8 \times 10^6\ N/m^2$

since $E =$ stress/strain then

strain $\epsilon = \sigma/E = (45.8 \times 10^6)/(206 \times 10^9) = 0.000\,223$

Also $\epsilon =$ extension/original length and therefore

extension $= \epsilon L = 0.000\,223 \times 800\ m = 0.178$ m or 178 mm

Movement required at signal box end $= 150 + 178 = 328$ mm

9.3 Shear stress and strain

Suppose a rectangular block is firmly fixed at the base and has a force F acting along its top face as shown in Fig. 9.2(a). At a section such as PQRS there is a tendency for sliding to take place between the upper and lower parts of the block. Shear stress is the force F divided by the area PQRS and it is given the symbol τ. If the area is A, then:

$$\text{shear stress } \tau = \frac{\text{shear force } F}{\text{area } A}$$

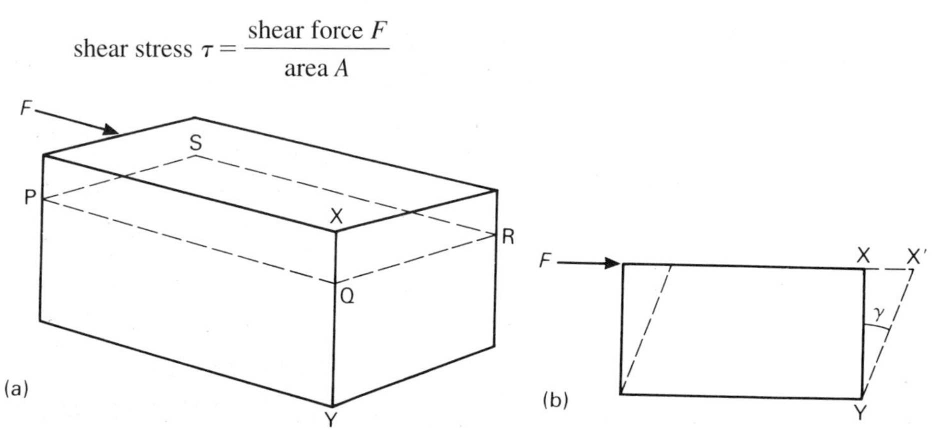

Figure 9.2
(a) Shear force acting on top surface of fixed block
(b) Shear strain.

Note that the area resisting shear is parallel to the force, whereas for direct stress it is perpendicular to the line of action of the force. Shear stress has the same units as direct stress.

Due to the shear force F, a vertical edge such as XY becomes X′Y. The distortion is exaggerated in Fig. 9.2(b). *Shear strain* is defined as the ratio XX′/XY and for small displacements this is equal to the angle γ in radian. Shear stress is proportional to shear strain for many materials and the ratio between them is called the *modulus of rigidity* or shear modulus. It is denoted by G and has the same units as the modulus of elasticity. Thus:

$$\text{modulus of rigidity } G = \frac{\text{shear stress } \tau}{\text{shear strain } \gamma}$$

Suppose a joint in a structure has to transmit a force F by means of a rivet as shown in Fig. 9.3. The rivet is said to be in *single shear* (Fig. 9.3(a)) when the tendency for shearing to take place occurs at one section only (ab). If the joint is made so that the shearing effect occurs at two sections (pq and rs in Fig. 9.3(b)) the rivet is said to be in *double shear*. The area resisting shear is twice as much in the case of double shear and, for the same load F, the stress is halved.

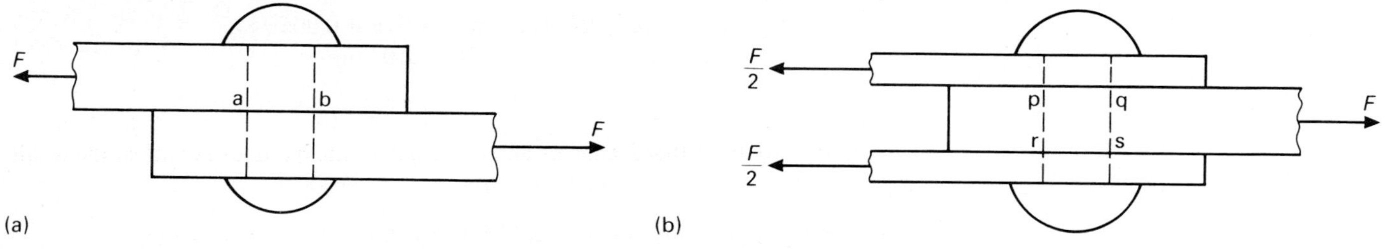

Figure 9.3

EXAMPLE 9.3 Fig. 9.4(a) shows a timber structural member that is loaded in tension by close-fitting steel pins 35 mm diameter. Calculate the shear stress in the timber and steel if the load in the member is 15 kN and the pins are in double shear. The timber thickness is 12 mm.

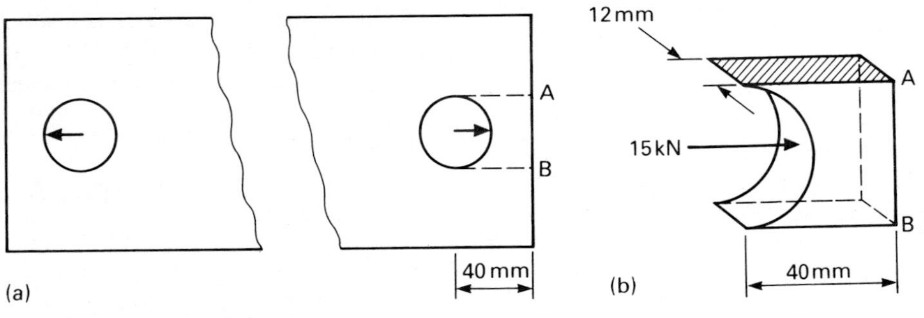

Figure 9.4
(a) Timber member loaded in tension by steel pins
(b) Block indicating planes of shear failure.

SOLUTION Shear failure in the timber would occur along planes through A and B parallel to the axis of the member. The block shown in Fig. 9.4(b) would be pushed out of the bar and the area resisting shear is two rectangles each 40 mm × 12 mm. Hence, working in newtons and metres,

$$\begin{aligned}\text{shear stress in timber} &= \text{shear force/area resisting shear}\\ &= 15\,000/(0.040 \times 0.012 \times 2)\ \text{N/m}^2\\ &= 15.6 \times 10^6\ \text{N/m}^2 \ \text{ or } \ 15.6\ \text{MN/m}^2\end{aligned}$$

For the steel pins,

$$\begin{aligned}\text{area in shear} = 2 \times \pi d^2/4 &= 2 \times \pi \times 0.035^2/4\ \text{m}^2\\ &= 0.00192\ \text{m}^2\end{aligned}$$

$$\begin{aligned}\text{shear stress} = \text{load/area} &= 15\,000/0.001\,92\\ &= 7.80 \times 10^6\ \text{N/m}^2 \quad \text{or} \quad 7.80\ \text{MN/m}^2\end{aligned}$$

9.4 Thermal stresses

It was shown in section 5.1.1 that the free expansion of a bar, length L, when heated through a temperature increase t, is αLt, where α is the coefficient of linear expansion. Its units are 'per °C' or 'per K', and this is written K^{-1}. The same formula will give the contraction in the bar when its temperature falls by an amount t.

If the ends of the bar are restrained so that it cannot expand or contract, a stress is induced in the bar. This stress corresponds to a change in length equal and opposite to that due to the temperature change. With the usual symbols,

$$\text{stress } \sigma = \text{strain } \epsilon \times E = \frac{\text{change in length}}{\text{original length}} \times E$$

$$= \frac{\alpha Lt}{L} \times E = \alpha tE$$

For a fall in temperature this stress will be tensile, and for a rise in temperature, compressive. Note that L cancels and the length of the bar does not affect the result.

EXAMPLE 9.4 A welded length of steel railway track is laid at 18°C. Determine the stress in the rails at 2°C if all contraction is prevented. For steel, $\alpha = 12 \times 10^{-6}\ °C^{-1}$ and $E = 207\ GN/m^2$.

SOLUTION The fall in temperature is 18 – 2 = 16°C and, using the result obtained above,

$$\text{stress } \sigma = \alpha tE = (12 \times 10^{-6}) \times 16 \times (207 \times 10^9)\ N/m^2$$
$$= 39.7 \times 10^6\ N/m^2 \quad \text{or} \quad 39.7\ MN/m^2$$

Since the temperature change is a fall this stress is tensile.

9.5 Bars of two materials

Some structural components are formed from two materials that are joined in such a way that they expand or contract together. Many civil engineering structures are constructed of concrete reinforced with steel bars or mesh; other examples are given in Chapters 4 and 7. This form of construction can lead to strong, economical structures and has the advantage of shielding the steel from corrosion and fire.

When a load is applied to the component it will be shared between the two materials. If they expand or contract together they will have the same strain. However, if they have different moduli of elasticity their stresses will differ. These two principles – load sharing and equal strains – can be used in various ways to solve numerical problems.

EXAMPLE 9.5 A short concrete column, 200 mm square, is reinforced by four steel bars in the corners as shown in section in Fig. 9.5. The modulus of elasticity for steel is fifteen times that for concrete. What load can the column bear if the total cross-sectional area of the steel bars is 4000 mm² and the stress in the concrete must not exceed 4 MN/m²?

What area of steel is required if the load is increased to 500 kN, the stresses being unchanged?

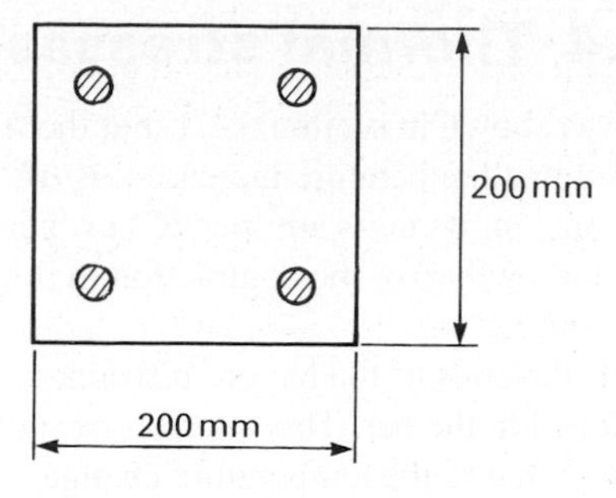

Figure 9.5

SOLUTION Let the stresses in the steel and concrete be σ_s and σ_c respectively, A_s and A_c the corresponding areas and E_s and E_c the values of the moduli of elasticity. Then, since strain = stress/E and the strain is the same for both materials, we can write:

$$\sigma_s/E_s = \sigma_c/E_c \quad \text{or} \quad \sigma_s = (E_s/E_c)\sigma_c$$

The ratio of the moduli is given as fifteen and this result becomes

$$\sigma_s = 15\sigma_c$$

With the value of σ_c given in the question, the stress in the steel is

$$\sigma_s = 15 \times 4\text{MN/m}^2 = 60\text{MN/m}^2$$

Also, working in millimetres,

$$\text{area of concrete} = 200^2\ \text{mm}^2 - 4000\ \text{mm}^2 = 36\,000\ \text{mm}^2$$

Using these results and converting to newton and metre units,

$$\begin{aligned}\text{safe load} &= \text{load in concrete} + \text{load in steel}\\ &= (\sigma_c \times A_c) + (\sigma_s \times A_s)\\ &= (4 \times 10^6 \times 36\,000 \times 10^{-6})\ \text{N} + (60 \times 10^6 \times 4000 \times 10^{-6})\ \text{N}\\ &= 384 \times 10^3\text{N} \quad \text{or} \quad 384\ \text{kN}\end{aligned}$$

If A_s mm^2 is the required area of steel then the area of concrete becomes $(40\,000 - A_s)$ mm^2 or $(40\,000 - A_s) \times 10^{-6}$m^2. With the new figures,

$$(\sigma_c \times A_c) + (\sigma_s \times A_s) = 500 \times 10^3$$

or

$$[4 \times 10^6 \times (40\,000 - A_s) \times 10^{-6}] + [60 \times 10^6 \times A_s \times 10^{-6}] = 500 \times 10^3$$

from which

$$A_s = 6071\ \text{mm}^2$$

9.5.1 Thermal stresses in composite bars

It was shown in section 9.4 that a temperature change can induce stresses in a bar if it is prevented from expanding or contracting. A bar of two materials can be stressed by a temperature change even if not constrained in this way. The method of finding the stresses is best explained by a numerical example.

EXAMPLE 9.6 A compound bar, 2 m long, consists of three bars, two of copper and one of steel, each of rectangular section 20 mm × 5 mm. The bars are securely fixed together with the steel in the middle to form a compound bar of rectangular section 20 mm wide and 15 mm thick.

Find the stresses in the steel and copper due to a temperature rise of 120°C.

For steel: $E = 207$ GN/m^2; $\alpha = 12 \times 10^{-6}\,°\text{C}^{-1}$
For copper: $E = 108$ GN/m^2; $\alpha = 18 \times 10^{-6}\,°\text{C}^{-1}$

SOLUTION Suppose that all expansion in the bar takes place at the right-hand end as shown in Fig. 9.6. If the copper and steel were free to expand separately, the copper would lengthen more than the steel as indicated by the broken lines.

Using the formula for thermal expansion:

$$\text{free expansion of copper} = \alpha_c Lt = 18 \times 10^{-6} \times 2 \times 120 \text{ m} = 0.004\,32 \text{ m or } 4.32 \text{ mm}$$

$$\text{free expansion of steel} = \alpha_s Lt = 12 \times 10^{-6} \times 2 \times 120 \text{ m} = 0.002\,88 \text{ m or } 2.88 \text{ mm}$$

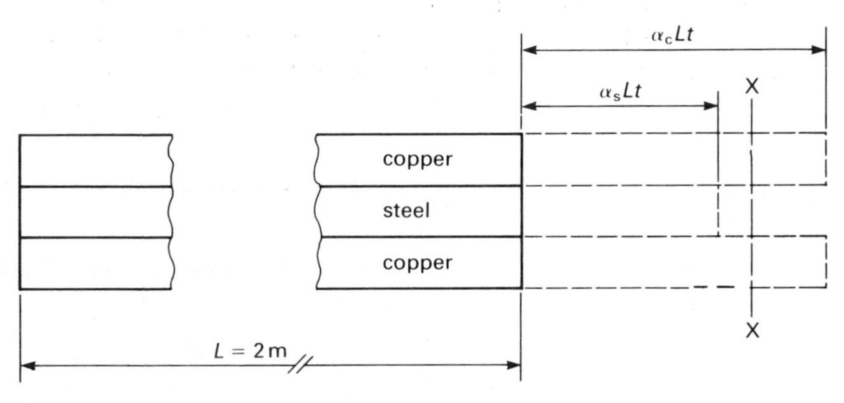

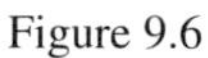

Figure 9.6

Since the two materials are securely joined, the steel is stretched and the copper compressed to some intermediate position XX (Fig. 9.6). Let σ_s be the tensile stress in the steel and σ_c the compressive stress in the copper, A_s and A_c the corresponding cross-sectional areas. For equilibrium the forces (but not the stresses) in the two materials must be equal and opposite. This gives the equation:

$$\sigma_s A_s = \sigma_c A_c$$

and, because the area of the copper is twice that of the steel,

$$\sigma_s = 2\sigma_c \qquad [i]$$

Using E_s and E_c for the moduli of elasticity we can write

$$\text{extension of the steel due to } \sigma_s = L \times \text{strain} = \frac{L\sigma_s}{E_s}$$

$$\text{and shortening of the copper due to } \sigma_c = \frac{L\sigma_c}{E_c}$$

It is sufficiently accurate in these expressions to take L as 2 m, because the free temperature expansions are very small in comparison.

From Fig. 9.6 it is clear that the difference in the free expansions is equal to the extension of the steel plus the shortening of the copper. With the previous results this leads to the equation:

$$\frac{L\sigma_s}{E_s} + \frac{L\sigma_c}{E_c} = (0.004\,32 - 0.002\,88)$$

and, using the numerical values given in the question,

$$\frac{2\sigma_s}{207 \times 10^9} + \frac{2\sigma_c}{108 \times 10^9} = 0.001\,44 \qquad [ii]$$

Results [i] and [ii] are simultaneous equations for σ_s and σ_c. The solution is $\sigma_s = 76.10 \times 10^6$ MN/m^2 and $\sigma_c = 38.05 \times 10^6$ MN/m^2. Thus

stress in steel $= 76.10$ MN/m^2 (tensile)
stress in copper $= 38.05$ MN/m^2 (compressive)

Although the length of the bar was given in the question the results would be the same for any length since L is a factor in every term of equation [ii].

The method used in this solution assumes that the compound bar remains straight and this is ensured by a symmetrical arrangement of the two materials. If there were only one bar of copper, for example, bending would occur when the temperature changed. This effect has a practical application in some forms of thermostat but the theoretical analysis is complex.

9.6 Beams and cantilevers

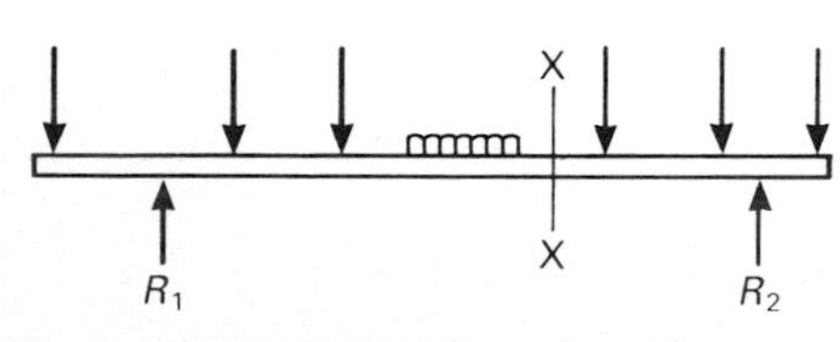

Figure 9.7 Beam carrying point and uniformly distributed loads.

A beam is a bar that carries external forces inclined to its axis. Many beams are horizontal and the loads they carry are weights acting vertically downwards. In examination questions assume (unless you are told otherwise) that beams are horizontal and straight, and the external forces on them are vertical.

Fig. 9.7 shows a beam carrying *point* (or concentrated) *loads* and a *uniformly distributed load* (UDL). Point loads are denoted by arrows and a UDL is usually shown by a symbol ⌒⌒⌒⌒. A common example of a UDL is the weight of the beam itself.

A *simply* (or *freely*) supported beam is one that rests on knife-edges. These are supports that provide upward forces of the point load kind but do not prevent the beam from rotating at the supports. The beam in Fig. 9.7 is resting on two such supports. Beams with three or more supports are beyond the scope of this book.

The set of forces acting on a beam at rest must satisfy the laws of equilibrium explained in Chapter 8. In particular, by taking moments about one point of support the reaction at the other can be calculated. The reactions can also be obtained graphically using the funicular polygon.

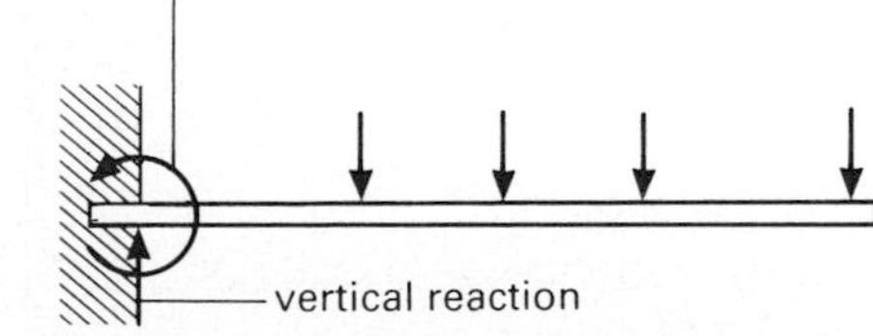

Figure 9.8 Cantilever.

A *cantilever* is a beam that is supported at one end only (Fig. 9.8) as in the case of a diving board over a swimming pool. The support cannot be of the knife-edge type or the whole cantilever would rotate about it. In addition to a vertical upward force, the support must provide a moment equal and opposite to the sum of the moments of all the loads about the supported end. It is called the *fixing moment*.

9.7 Shearing force and bending moment

To investigate the equilibrium of a beam, all the forces acting on it (including reactions) must be considered. If we take only the forces on one side of a section, such as XX in Fig. 9.7, they will not normally be in equilibrium. Those to the left will have a resultant equal and opposite to the resultant of those to the right.

The *shearing force* at a section of a beam is the algebraic sum of the forces (including reactions) on one side or the other. If the resultant force to the left of the section is upward, that to the right will be equal but downward. Shearing force is abbreviated to SF. It is the lateral force that the beam has to resist at the section.

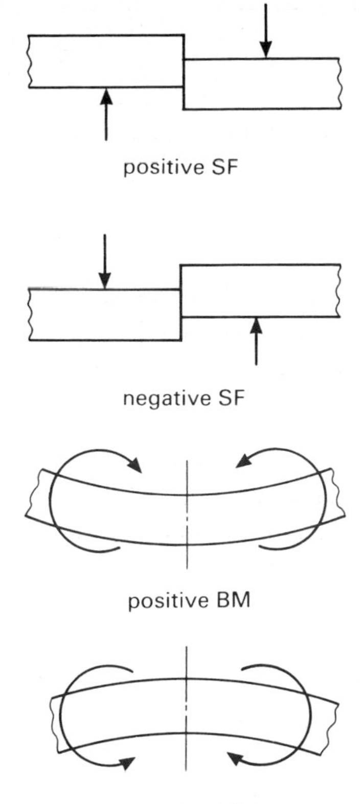

Figure 9.9 Sign convention for shearing forces and bending moments.

The *bending moment* at XX is defined as the sum of the moments of all the forces on one side of the section about a point in that section. The total amount for all the forces on the left will be equal and opposite to that for forces to the right. Bending moment is abbreviated to BM. It is the moment the beam has to resist in bending at the section.

9.7.1 Sign convention for SF and BM

The signs used in this book are shown in Fig. 9.9. Shearing in which the resultant force is upwards to the left of the section and downwards to the right will be considered positive.

Bending moments will be taken as positive if the resultant moment of the forces on the left is clockwise, and the resultant moment of those on the right is anticlockwise. These make the beam concave upwards and are called *sagging* bending moments. If the moment is anticlockwise on the left and clockwise on the right the beam will become convex upwards and the bending moment is called *hogging*.

You may find different conventions in other books but, apart from the signs, the answers will be the same whatever system is used. To avoid ambiguity it is best to state left-up or left-down for shearing force and sagging or hogging for bending moment.

EXAMPLE 9.7 A beam ABCDE, 6m long, is simply supported at B and D and carries point loads of 80 kN and 10 kN at C and E respectively.
AB = 1 m, BC = 2 m, CD = 1 m, DE = 2 m. Neglecting the beam's own weight,

(a) find the shearing forces and bending moments at P, Q and R which are 2 m, 3.5 m and 5 m from A,
(b) draw to scale the shearing force and bending moment diagrams,
(c) locate the point of contraflexure.

SOLUTION The beam and its loading are shown in Fig. 9.10(a). If you take moments about B you will find that the reaction at D is 70 kN. By subtracting this value from the total downward load, or by taking moments about D, the reaction at B is found to be 20 kN.

(a) For P (Fig. 9.10(b)), there is only one force to the left, the reaction of 20 kN at B. This acts upwards and the shearing force is therefore (+)20 kN. Its moment about P is $20 \times 1 = 20$ kNm and this is also positive being clockwise on the left. Hence:

SF at P = 20 kN left-up
BM at P = 20 kNm sagging

For Q (Fig. 9.10(c)), there are two forces to the left, 20 kN upwards and 80 kN downwards. The shearing force at Q is therefore $20 - 80 = -60$ kN.
Taking clockwise as positive the moment of these forces about Q is $(20 \times 2.5) - (80 \times 0.5) = 10$ kN m. Therefore:

SF at Q = 60 kN left-down
BM at Q = 10 kNm sagging

For the section at R, there are three forces to the left (Fig. 9.10(d)) and it is easier to take the one force to the right (Fig. 9.10(e)). This is 10 kN downwards on the right, which is equivalent to left-up, and the shearing force is therefore (+)10 kN. The moment of this force is $10 \times 1 = 10$ kNm clockwise on the right, which is negative and therefore hogging. Hence:

SF at R = 10 kN left-up
BM at R = 10 kNm hogging

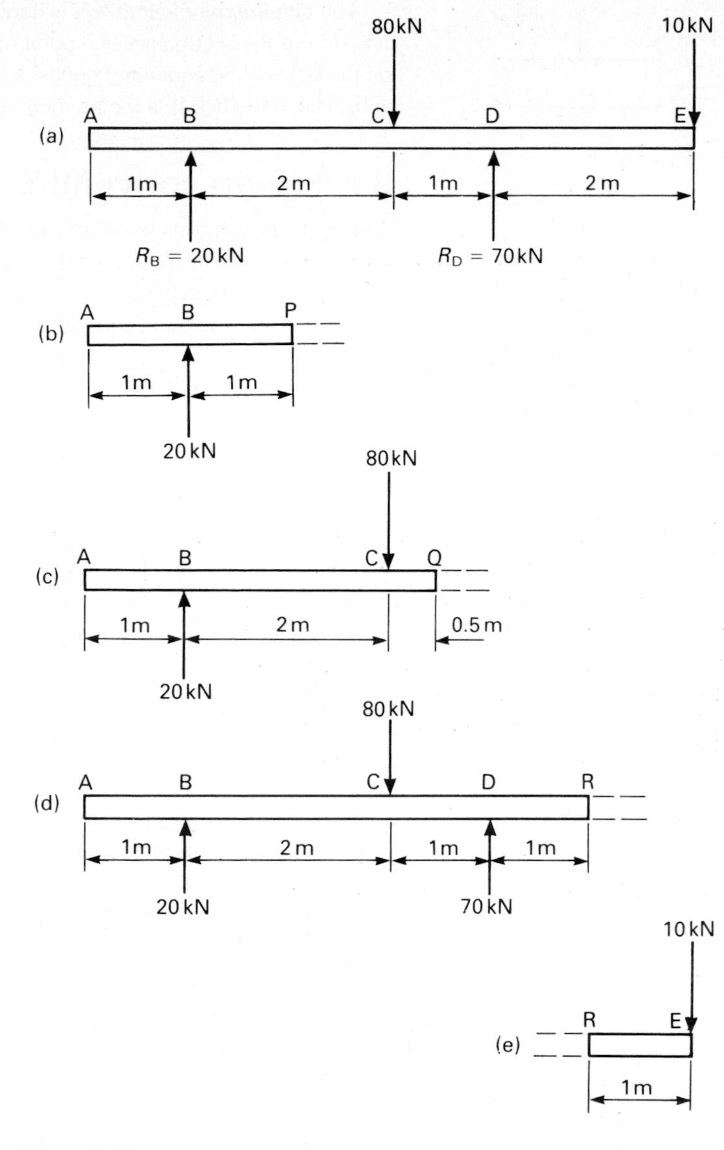

Figure 9.10

(b) Shearing force and bending moment diagrams are graphs showing the variation of these quantitites along the beam. They are drawn below a diagram of the beam so that the principal values can be related to the corresponding sections. Fig. 9.11 shows the diagrams for the beam in the question, the numerical values being kilonewton for SF and kilonewton metre for BM. Suggested scales: horizontal distance,
1 cm = 0.5m; SF, 1cm = 10 kN; BM, 1 cm = 10 kNm.

For sections between A and B there is no force to the left. The SF and BM are therefore zero for this part of the beam.

Between B and C there is one force to the left, the reaction of 20 kN at B. The SF is therefore constant at +20 kN. The moment arm of this force increases from zero at B to 2 m at C. The BM therefore increases linearly from zero at B to +40 kNm at C passing through the value of +20 kNm calculated earlier for the point P.

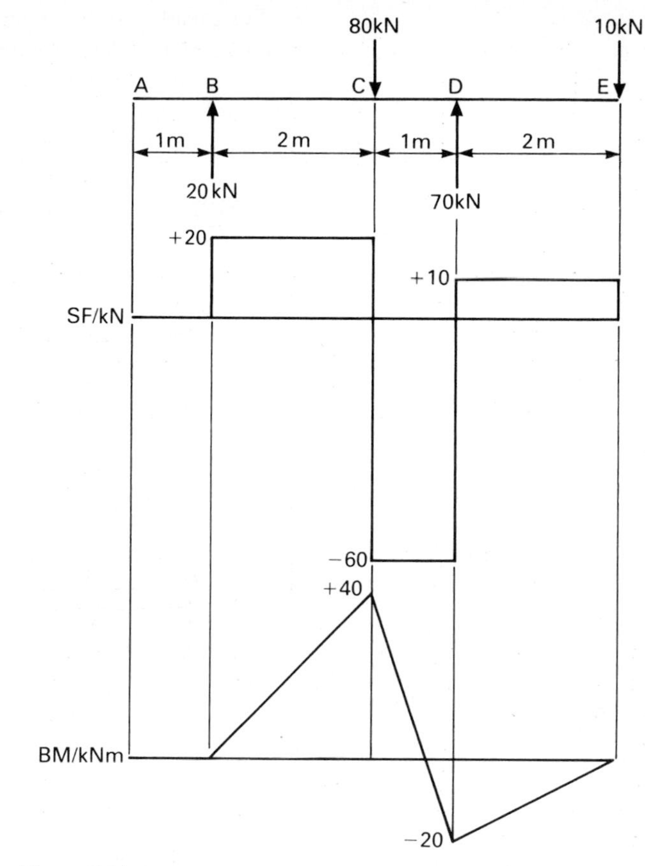

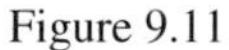

Figure 9.11

Between C and D there are two forces to the left. The SF is constant at – 60 kN, the value for point Q, and the BM again varies in a straight line. It is +40 kNm at C and −20 kNm at D passing through +10 kNm at the point Q.

For the portion DE it is easier to consider the one force to the right than the three to the left. The SF is constant at +10 kN and, starting from the right-hand end, the BM varies linearly from 0 at E to −20 kNm at D.

In all cases of point loads the SF diagram consists of horizontal steps and the BM diagram is a series of straight lines, the slope changing each time a load is passed.

(c) A point of *contraflexure* (or *inflexion*) is one at which the BM passes through zero. It is a section at which the bending changes from sagging to hogging. From the BM diagram (Fig. 9.11) it is between C and D. By measurement or calculation, it is found to be 3.67 m from A.

At a point of contraflexure the beam is required to resist shearing but no bending. It could, therefore, be hinged at such a point but it should be remembered that its position will change if the loading is altered.

9.7.2 SF and BM diagrams for uniformly distributed loads

If a beam carries a uniformly distributed load over the whole or part of its length then the total amount can be regarded as acting at its centre of gravity when taking moments to find the reactions. Since the distribution is uniform this will be the midpoint of the load.

For calculating the SF and BM at a selected section, take that part of the UDL to one side or the other as acting at its own midpoint.

For the portion of the beam that is carrying the UDL the SF diagram is a sloping straight line and the BM diagram is a curve (a parabola). It is therefore necessary to calculate the value of the BM at several points and plot a graph.

EXAMPLE 9.8 A beam ABC, 10m long, is simply supported at A and B. AB = 6 m; BC = 4 m. The beam carries a uniformly distributed load of 20 kN/m between A and B and a point load of 18 kN at C. Draw the shearing force and bending moment diagrams for the beams, state the greatest sagging and hogging bending moments and find the point of contraflexure.

SOLUTION The loading is shown in Fig. 9.12(a) and it is convenient to work in kilonewton and metre units. The total UDL is $20 \times 6 = 120$ kN and its centre of gravity is 3 m from A. Taking moments about A with anticlockwise as positive

$$6R_B - 120 \times 3 - 18 \times 10 = 0$$

from which

$$R_B = 90$$

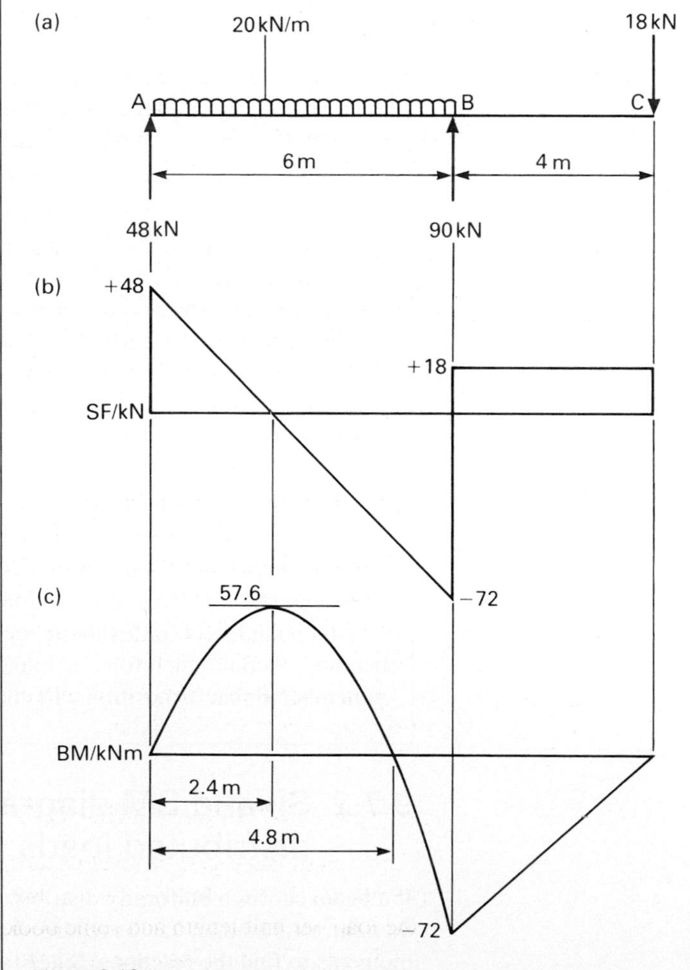

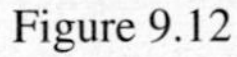

Figure 9.12

In a similar way, moments about B give $R_A = 48$.

Between A and B it is convenient to take intervals of 1 m. For a section immediately to the right of A the only force on the left is 48 kN upwards. Hence the SF is +48 kN and, since the force is effectively acting at zero distance, the BM is zero.

At 1 m from A there are two forces to the left, 48 kN with a moment arm of 1 m, and the 1 m length of the UDL. This exerts a force of 20 kN downwards and its centre of gravity is 0.5 m from the section. Hence the SF is 48 – 20 = +28 kN and the BM is 48 × 1 – 20 × 0.5 = +38 kN m.

At 2 m from A the UDL to the left of the section totals 40 kN and its centre of gravity distance is 1 m. For this section therefore the SF is 48 – 40 = +8 kN and the BM is 48 × 2 – 40 × 1 = 56 kN m.

If the calculations are continued the following table is obtained:

Distance from A	(m)	0	1	2	3	4	5	6
SF	(kN)	48	28	8	−12	−32	−52	−72
BM	(kNm)	0	38	56	54	32	−10	−72

(It is a coincidence that the SF and BM values at B are equal; this is not a general result.)

Between B and C the SF is constant at +18 kN and the BM varies in a straight line from −72 kNm at B to zero at C. The diagrams are shown in Fig. 9.12(b) and (c). Suggested scales: 1 cm = 1 m of horizontal distance; 1 cm = 20 kN of shearing force; 1 cm = 20 kNm of bending moment.

The greatest hogging (negative) bending moment occurs at B and equals 72 kN m.

A useful general result is that the SF passes through zero at a section where the BM has a maximum or minimum value, and this provides a quick method of locating the greatest sagging BM in the present example. By measurement of the SF diagram (Fig. 9.12(b)), or by calculation, the SF is zero at a section 2.4 m from A. Using this value and taking forces on the left,

$$\text{greatest sagging BM} = 48 \times 2.4 - 20 \times 2.4 \times 1.2 = 57.6 \text{ kN m}$$

By measurement of the BM diagram (Fig. 9.12(c)) or by calculation the point of contraflexure, where the BM passes through zero, is 4.8 m from A. To the left of this section the BM is sagging (positive BM) and to the right it is hogging (negative BM).

9.7.3 SF and BM diagrams for cantilevers

It is not necessary to find the vertical reaction and fixing moment (Fig. 9.8) before calculating the SF and BM for a section of a cantilever. If the forces on the free end side of the section are considered, the reaction and fixing moment at the support do not appear in the calculations.

If the cantilever carries downward forces only, the bending moment will be hogging (negative) throughout.

9.7.4 Standard cases of SF and BM

Fig. 9.13 shows four cases that occur frequently in design work. In each one, L is the length and W the total downward load. For the two UDL cases, $W = wL$ where w is the load per unit length and some books quote the results in terms of w.

The results for point loads apply to a central load on a beam and an end load on a cantilever, and the greatest bending moments will be smaller for loads applied at other points. The formulae given in the table therefore represent the worst cases and designs based on them will be on the safe side if the loads move to other points.

Note that a beam supported at its ends is stronger than a cantilever and that the UDL causes a smaller BM than if the same load were concentrated at the midpoint of the beam (or the free end in the case of the cantilever).

The deflection formulae are explained in section 9.9.1.

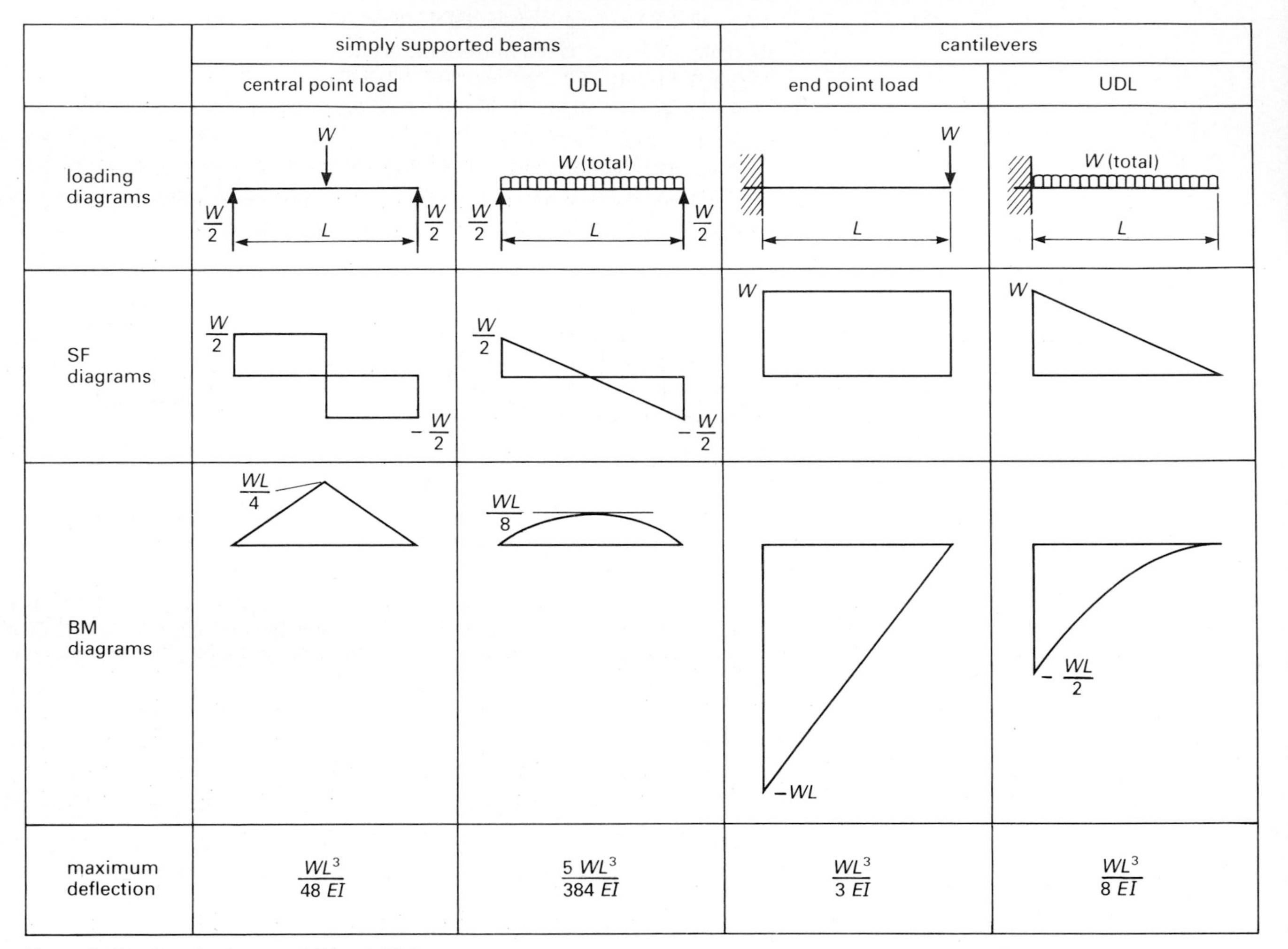

Figure 9.13 Standard cases of SF and BM.

9.8 The theory of simple bending

In designing beams and cantilevers we need to calculate the stresses caused by bending. To develop a theory for calculating bending stresses, a number of assumptions must be made. The following set leads to the equation that is used for most practical design work.

1. The material of the beam is uniform throughout.
2. Each cross-section is symmetrical about the plane of bending.
3. The beam is straight and unstressed before bending.
4. The loads are applied in the plane of bending.
5. Plane cross-sections before bending remain plane after bending.
6. The material of the beam obeys Hooke's law and the modulus of elasticity is the same in tension and compression.
7. The resultant force normal to the cross-section is zero.

These assumptions are used in the next two sections.

9.8.1 Bending stresses

The nature of the stresses in a loaded beam can be demonstrated with a rectangular piece of sponge or foam material. Draw some parallel lines on one face as shown in Fig. 9.14(a). Bend the material as shown in Figs 9.14(b) and (c) and measure the new distances between these lines. With sagging moments (Fig. 9.14(b)) the lines become closer together at the top and further apart at the bottom. With hogging the reverse effect is obtained.

This experiment indicates that bending causes longitudinal stresses in the material and that these stresses vary across the section from top to bottom. There is an intermediate surface which is unstressed and this is called the neutral surface. Its line of intersection with a cross-section of the beam is called the *neutral axis* of that section and is denoted by n.a. on diagrams. If the section is symmetrical the neutral axis is midway between the top and bottom surfaces.

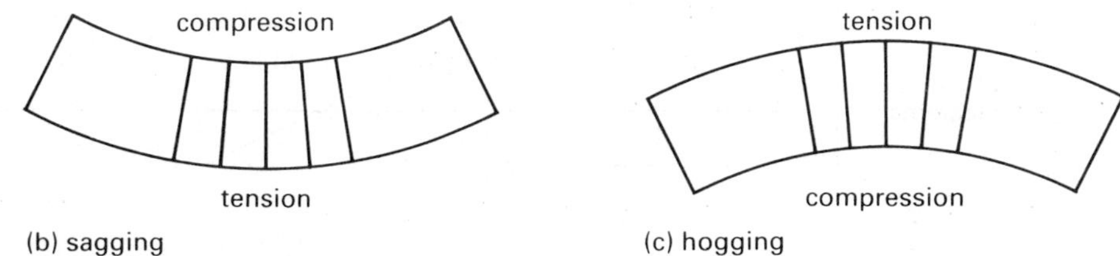

(a) before bending (b) sagging (c) hogging

Figure 9.14 The nature of stresses in a loaded beam.

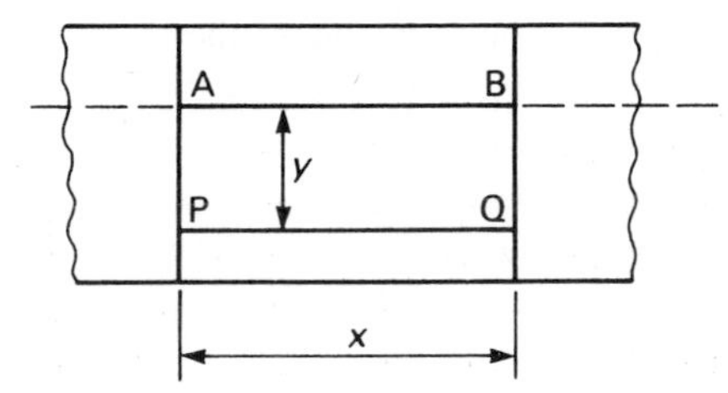

(a) before bending

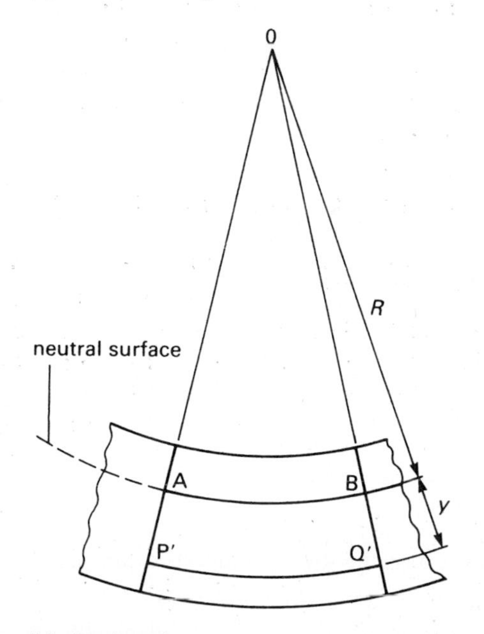

(b) after bending

Figure 9.15

Suppose a beam that is straight before loading (Fig. 9.15(a)) bends to an arc of a circle (Fig. 9.15(b)). Cross-sections that are parallel before bending will tilt and intersect at a line through O.

Consider two thin, longitudinal layers of material, one in the neutral surface, such as AB, and the other a distance y below it, such as PQ. Before loading they have the same length x. After bending, AB is unchanged because it is not stressed but PQ is stretched to P'Q'. From the geometry of Fig. 9.15(b), the lengths of P'Q' and AB are proportional to their radii from O, R and $R + y$ respectively. Thus:

$$\mathrm{P'Q'} = \frac{R+y}{R}\,\mathrm{AB} = \frac{R+y}{R}\,x = \left(1 + \frac{y}{R}\right)x = x + \frac{yx}{R}$$

and the strain along P'Q' is

$$\frac{\text{extension}}{\text{original length}} = \frac{\mathrm{P'Q'} - \mathrm{PQ}}{\mathrm{PQ}} = \frac{(x + yx/\mathrm{R}) - x}{x} = \frac{y}{\mathrm{R}}$$

The corresponding stress is therefore

$$\sigma = E \times \text{strain} = E \times \frac{y}{R}$$

This result is usually written

$$\frac{\sigma}{y} = \frac{E}{R}$$

At a given section the bending stress is therefore proportional to y, the distance from the neutral axis and the greatest values will occur at the top and bottom edges of the section. One will be tensile and the other compressive. For a symmetrical section they will be numerically equal.

In most examples the radius R varies from one section to another. It is not then the radius of a circular arc but is called the radius of curvature at the section.

EXAMPLE 9.9 A steel strip of rectangular cross-section 2.5 mm thick is initially straight and unstressed. Calculate the maximum stress in the steel if it is bent round a drum 3 m radius. $E = 206\ \text{GM/m}^2$.

SOLUTION The cross-section is symmetrical and the neutral axis is therefore 1.25 mm from the top and bottom edges. When the strip is bent round the drum the radius to the neutral axis is 3 m plus 1.25 mm, making 3.001 25 m.

The maximum stress will occur at the edges of the strip where the value of y is 1.25 mm or 0.001 25 m. Using the result obtained in section 9.8.1 this stress is:

$$\sigma = \frac{yE}{R} = \frac{0.001\,25 \times (206 \times 10^9)}{3.001\,25} = 85.8 \times 10^6\ \text{N/m}^2 \quad \text{or} \quad 85.8\ \text{MN/m}^2$$

This stress will be compressive on the inner (concave) surface and tensile on the outer (convex) surface of the strip.

Note that the thickness of the strip is so small compared with the radius of the drum that it would have been sufficiently accurate to take R as 3 m.

9.8.2 Bending equation

Fig. 9.16(a) shows the cross-section of a beam, symmetrical about a vertical axis but not about a horizontal axis. The neutral axis is not midway between the top and bottom edges and it is shown later that it passes through the centroid of the section.

Fig. 9.16(b) illustrates the distribution of stress across this section due to a sagging bending moment. It is a straight line variation, with compressive stress at the top, tensile stress at the bottom and zero at the neutral axis.

To find the normal force and the moment on the whole section consider first a thin slice, area A and distance y from the neutral axis, as shown in Fig. 9.16(a).

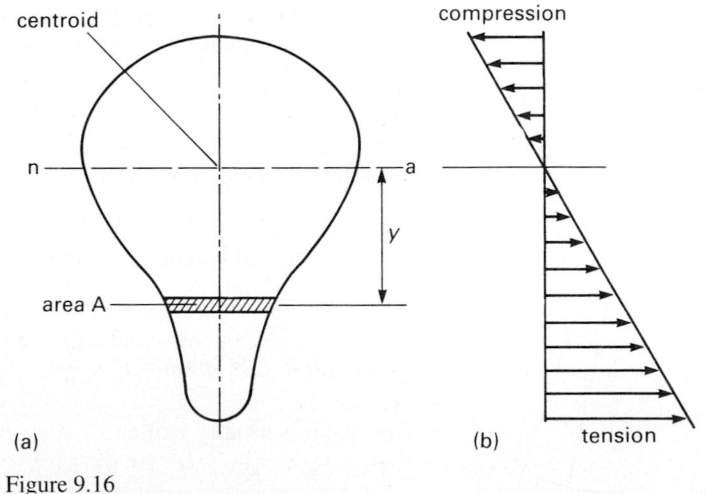

Figure 9.16
(a) Cross-section of a beam
(b) Distribution of stress across the section.

From the result obtained in the previous section the stress on this slice can be written as:

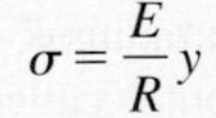

$$\sigma = \frac{E}{R}y$$

Using this result the force F on the slice can be written:

$$F = \text{stress } \sigma \times \text{area } A = \frac{E}{R}Ay$$

and the moment M of this force about the neutral axis is:

$$M = \text{force } F \times \text{moment arm } y = \frac{E}{R}Ay^2$$

The quantities Ay and Ay^2 are called the *first and second moments of area* respectively. The total force and total moment for the cross-section are the results of adding all such terms and these results can be expressed as integrals. Since the first moment is an area multiplied by a distance its units are cubic metres (m^3). Second moment is an area multiplied by the square of a distance and its units are therefore m^4.

The total first moment for a given figure can be calculated by regarding the total area as concentrated at its *centroid*. This is equivalent to considering the whole weight of a body as acting at its centre of gravity.

Assumption 7 in section 9.8 states that there is no resultant longitudinal force in the beam and, for this to be so, the total first moment must be zero. It follows that the neutral axis must pass through the centroid of the cross-section. With y having positive and negative values on opposite sides of the axis the first moment of the portion above the axis is just balanced by that of the portion below.

The second moment, on the other hand, is not zero about this axis (or, indeed, any other) because y^2 is always positive, whatever the sign of y. Second moment of area is denoted by I and the total moment M can therefore be written:

$$M = \frac{E}{R}I \quad \text{or} \quad \frac{M}{I} = \frac{E}{R}$$

The quantity M is the total moment arising from the stresses on the section. It is sometimes called the *moment of resistance* and it is equal to the bending moment acting at the section. The calculation of second moment of area I is explained in section 9.8.4. It is related to the moment of inertia that occurs in the dynamics of rotating bodies (Chapter 11), and some authors (and some examiners) use the name *moment of inertia* for both.

The last result is combined with the relationship obtained in the previous section to obtain the complete bending equation:

$$\frac{\sigma}{y} = \frac{M}{I} = \frac{E}{R}$$

This double equation is the basis for all later work in the bending of beams and the meaning of each symbol, and the corresponding units, should be carefully noted. The variables are:

σ = a stress in N/m^2 at a distance y m from the neutral axis
M = bending moment (or moment of resistance) in Nm
I = second moment of area about the neutral axis in m^4
E = modulus of elasticity in N/m^2
R = radius of curvature of the neutral surface in m

Numerical values of I are usually very small in the basic unit of m^4. In examples they may be given in mm^4 or cm^4. Multiply by 10^{-12} or 10^{-8} respectively to convert them to the basic unit.

EXAMPLE 9.10 A horizontal beam, simply supported at its ends, is formed from a rolled steel T-section of the dimensions shown in Fig. 9.17. What is the maximum uniformly distributed load (including its own weight) that the beam can carry over a span of 5 m if the tensile stress must not exceed 160 MN/m^2? $I_{xx} = 253$ cm^4.

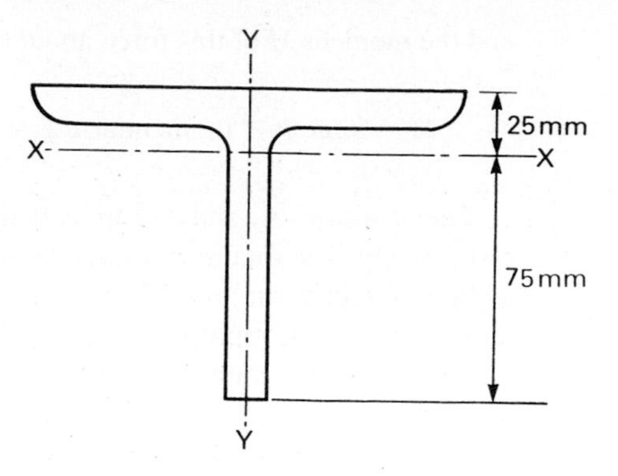

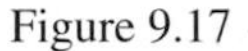
Figure 9.17

What is the maximum compressive stress in the beam under these conditions?

SOLUTION The horizontal and vertical axes through the centroid are usually lettered XX and YY as shown in Fig. 9.17. Since the beam is sagging the maximum tensile stress will occur at the bottom edge of the section, 75 mm below the n.a. (XX). From the bending equation,

$$\frac{M}{I} = \frac{\sigma}{y} \quad \text{or} \quad M = \frac{\sigma I}{y}$$

In newton and metre units, $\sigma = 160 \times 10^6$ N/m^2, $I = 253 \times 10^{-8}$ m^4 and $y = 0.075$ m. With these values the maximum permissible bending moment is:

$$M = \frac{(160 \times 10^6) \times 253 \times 10^{-8}}{0.075} = 5397 \text{ N m}$$

As shown in Fig. 9.13 the maximum bending moment for a UDL on a beam, simply supported at its ends, is $WL/8$ at midspan. With the present values,

$$\frac{WL}{8} = 5397 \quad \text{or} \quad W = \frac{5397 \times 8}{L} = \frac{5397 \times 8}{5}$$

$$= 8636 \text{ N (total)} \quad \text{or} \quad 1727 \text{ N/m}$$

The maximum compressive stress will occur at the top of the midspan section. This is 25 mm from the neutral axis and, since the bending stress is proportional to the distance from this axis,

$$\text{maximum compressive stress} = \frac{25}{75} \times (160 \times 10^6) = 53.3 \times 10^6 \text{ N/m}^2$$

$$\text{or} \quad 53.3 \text{ MN/m}^2$$

9.8.3 Efficient beam sections

If you take a flat sheet of A4 paper and support it along its shorter edges it will sag and collapse under its own weight. If you fold it as shown in Fig. 9.18(a), however, it can support many times its own weight. This strengthening effect has many practical applications such as the use of corrugated sheet materials (Fig. 9.18(b) and (c)) for the roofs of sheds, garages and industrial buildings.

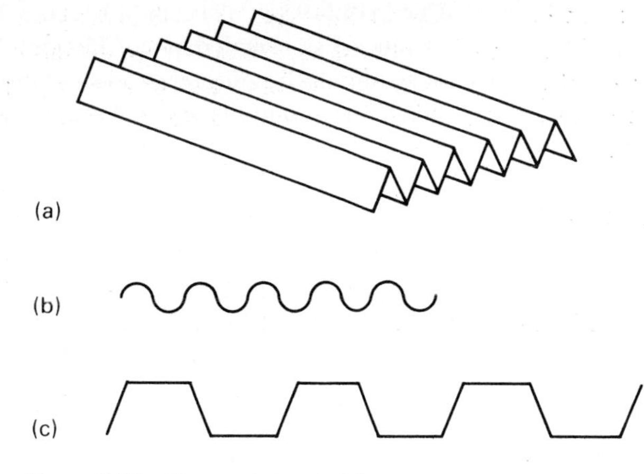

Figure 9.18 Corrugated materials

This demonstration shows that the strength of a beam depends upon the shape of its cross-section, and not simply its area as in the case of a bar loaded in tension or compression. Fig. 9.16(a) shows that the contribution to the moment of resistance made by an element of area A depends on its distance from the neutral axis. As y increases the stress increases and this leads to a greater force on a given area. Furthermore, the moment of the force about the neutral axis is proportional to y. Many beam sections are therefore designed so that most of the area is at a considerable distance from the neutral axis. Fig. 9.19(a) shows a section that is widely used in structural steelwork. It consists of two flanges joined by a thin web. It is produced by a rolling process and for this reason the flanges are tapered and the corners rounded. It is often referred to as an I-section or rolled steel joist (RSJ). A range of standard dimensions has been adopted for these sections and (in the UK) they are described as universal beams or British standard beams (BSBs).

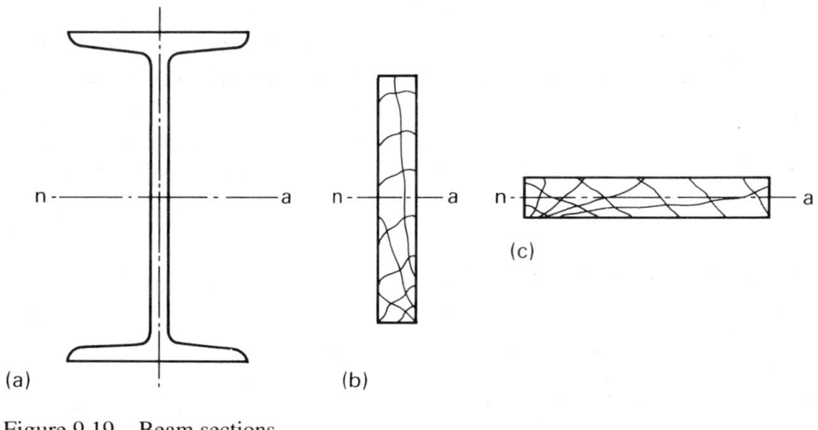

Figure 9.19 Beam sections
(a) a rolled steel joist or I-section
(b) timber floor joist
(c) horizontal timber board.

To obtain the greatest possible strength in bending the beam must be correctly placed relative to the plane in which the loads are applied. Suppose a timber beam of rectangular cross-section is carrying vertical loads. It will be able to carry greater loads with the longer sides of the section vertical as in Fig. 9.19(b) than if it is placed with the longer edges horizontal as in Fig. 9.19(c). It is for this reason that floor joists are placed with the longer sides of the cross-section vertical.

EXAMPLE 9.11 A beam of I-section has rectangular flanges each 120 mm by 10 mm and an overall depth of 150 mm. What is the greatest permissible bending moment if the bending stress must not exceed 80 MN/m^2, assuming that the stress in each flange is uniform and that the effects of the web may be neglected?

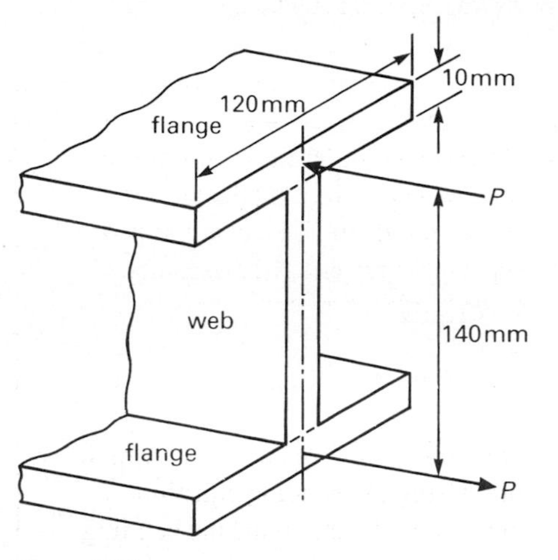

Figure 9.20

SOLUTION If the stress in each flange is uniform there will be a resultant force P equal to the stress multiplied by the flange area and acting at the midpoint of the flange as shown in Fig. 9.20. This force will be tensile for one flange and compressive for the other, the two having parallel lines of action 140 mm apart and forming a couple. Working in newton and metre units, the force in each flange is:

$$P = \text{stress} \times \text{area} = (80 \times 10^6) \times (120 \times 10^{-3}) \times (10 \times 10^{-3})$$
$$= 96\,000 \text{ N}$$

and, since the moment of a couple equals one of the forces multiplied by the distance between them, the greatest permissible bending moment is:

$$M = 96\,000 \times (140 \times 10^{-3}) = 13\,440 \text{ N m}$$
$$\text{or } 13.44 \text{ kN m}$$

9.8.4 Calculation of the second moment of area

The value of I, the second moment of area of the cross-section, is needed for most calculations on bending stress. In some problems, such as Example 9.10, its value is given in the question. In the practical design of steelwork values are obtained from published tables. In other cases it must be calculated from the dimensions of the section and Fig. 9.21 shows three results that are required frequently.

For a rectangle, width b and depth d, about an axis through the centre parallel to the side b (Fig. 9.21(a)) the value is $bd^3/12$. For the same rectangle about an axis along the side b (Fig. 9.21(b)) the value is $bd^3/3$. For a circle about a diameter (Fig. 9.21(c)) $I = \pi d^4/64$.

All three formulae contain the fourth power of length and if the cross-section dimensions are given in millimetres the result will be in mm^4. It will then be necessary to divide by 10^{12} to obtain the value in the basic SI unit of m^4.

The rectangle formulae are used in calculating second moments for I- and T-sections, and sections in the form of angles and channels. Such calculations ignore the tapering of flanges and corner fillets.

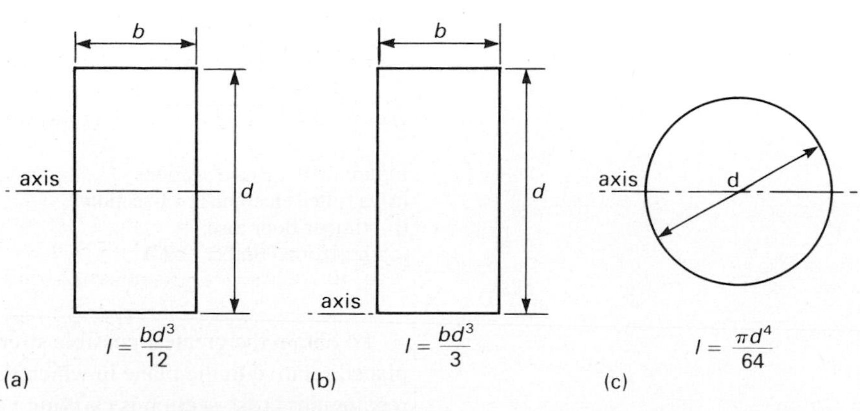

Figure 9.21 The second moment of area
(a) rectangle, axis through centre
(b) rectangle, axis along side b
(c) circle, axis through diameter.

EXAMPLE 9.12 Find the greatest permissible bending moment for the I-section beam of Example 9.11 allowing for the variation of bending stress across the section and the contribution of the web. Take the maximum bending stress as 80 MN/m^2 and the web thickness as 12 mm.

What is the result if the section is used with the web horizontal?

SOLUTION Fig. 9.22(a) shows the dimensions of the section in millimetres. The value of the second moment I can be calculated by considering the cross-section as a rectangle 120 mm wide and 150 mm deep with two rectangular cutouts (shown by hatching) each 54 mm wide and 130 mm deep. The central axis of each of these coincides with the neutral axis and the formula $bd^3/12$ can be applied to all three. Working in millimetre units,

$$I = \frac{120 \times 150^3}{12} - 2 \times \frac{54 \times 130^3}{12} = 13.98 \times 10^6 \text{ mm}^4$$

The maximum bending stress occurs at the top and bottom edges for which the distance y from the neutral axis is 75 mm. Using the bending equation $M/I = \sigma/y$ and converting all quantities to newton and metre units the bending moment is:

$$M = \frac{\sigma I}{y} = \frac{(80 \times 10^6) \times (13.98 \times 10^6 \times 10^{-12})}{(75 \times 10^{-3})}$$

$$= 14\,910 \text{ N m} \quad \text{or} \quad 14.91 \text{ kN m}$$

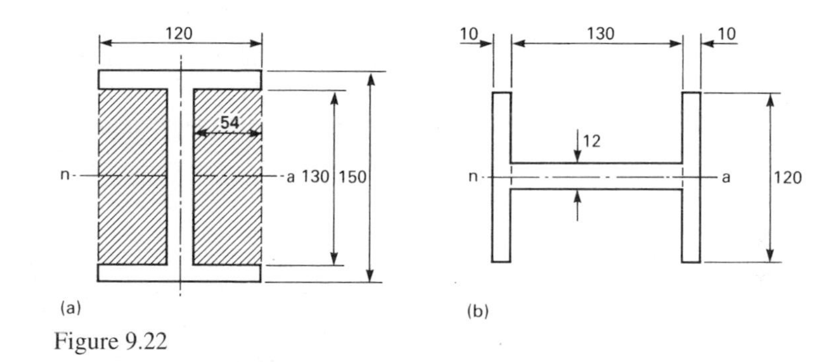

Figure 9.22

Compare this answer with the approximate result obtained in Example 9.11.

If the beam is used with the web horizontal, as shown in Fig.9.22(b), the cross-section can be divided into three rectangles each having a central axis that coincides with the neutral axis. The total second moment can therefore be found by adding the values for the two flanges and the web obtained with the formula $bd^3/12$. The required dimensions are shown on the diagram (in millimetres) and the second moment of area is:

$$I = \frac{10 \times 120^3}{12} + \frac{130 \times 12^3}{12} + \frac{10 \times 120^3}{12} = 2.899 \times 10^6 \text{ mm}^4$$

The distance y from the neutral axis to the top and bottom edges of the section is now 60 mm and the bending moment is therefore:

$$M = \frac{\sigma I}{y} = \frac{(80 \times 10^6) \times (2.899 \times 10^6 \times 10^{-12})}{(60 \times 10^{-3})} = 3865 \text{ N m}$$

$$\text{or} \quad 3.865 \text{ kN m}$$

Compare this result with the previous one; the permissible bending moment is very much smaller with the web horizontal.

EXAMPLE 9.13 A beam of I-section has unequal flanges. The upper flange is 30 mm by 15 mm and the lower flange is 80 mm by 15 mm. The web thickness is 10 mm and the overall depth is 90 mm.

Calculate the maximum span over which the beam can carry a uniformly distributed load of 2 kN/m (including its own weight) if the tensile and compressive bending stresses are not to exceed 35 MN/m^2 and 70 MN/m^2 respectively.

SOLUTION The section is shown in Fig. 9.23, all dimensions being in millimetres. The neutral axis passes through the centroid of the section; suppose its distance from the bottom edge is y mm as shown. The first moment of area of the section about the bottom edge will equal the sum of the first moments of the two flanges and the web. The centroid distances from this edge for the lower flange, web and upper flange are 7.5, 45 and 82.5 mm respectively. Thus:

$$y \times (\text{total area}) = (\text{total first moment})$$

or, working in millimetres

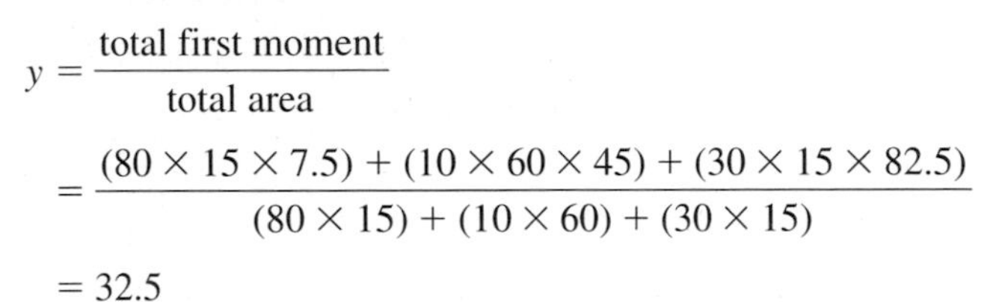

$$y = \frac{\text{total first moment}}{\text{total area}}$$

$$= \frac{(80 \times 15 \times 7.5) + (10 \times 60 \times 45) + (30 \times 15 \times 82.5)}{(80 \times 15) + (10 \times 60) + (30 \times 15)}$$

$$= 32.5$$

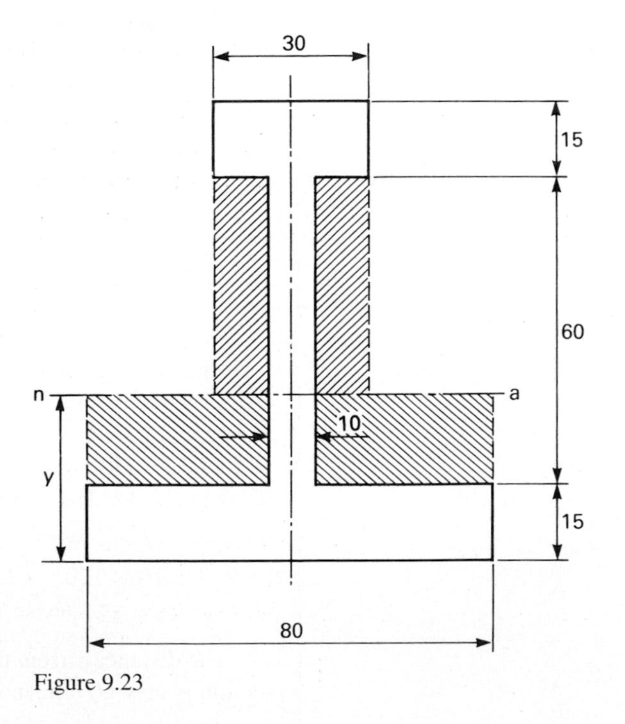

Figure 9.23

The neutral axis is therefore 32.5 mm from the bottom edge and 57.5 mm from the top edge.

The second moment of area I can be found by considering rectangles and rectangular cutouts (shown by hatching) each having one edge on the neutral axis. The formula $bd^3/3$ can be used for all of them.

For the portion of the section above the neutral axis the rectangle is 30 mm wide and 57.5 mm deep and the two cutouts are each 10 mm wide and 42.5 mm deep. For the portion below the neutral axis the rectangle is 80 mm wide and 32.5 mm deep, the cutouts each being 35 mm wide and 17.5 mm deep. Working in millimetre units the second moment of area about the neutral axis is:

$$I = \frac{30 \times 57.5^3}{3} - 2 \times \frac{10 \times 42.5^3}{3} + \frac{80 \times 32.5^3 \times}{3} - 2 \times \frac{35 \times 17.5^3}{3}$$

$$= 2.180 \times 10^6 \text{ mm}^4$$

Two limiting stresses are given and neither must be exceeded. If the beam rests on simple supports at its ends the bending moment will be sagging throughout and the greatest tensile stress will occur at the bottom of the section. Its distance y from the neutral axis is 32.5 mm, the permitted stress is 35 MN/m^2 and, converting all quantities to newton and metre units, the bending moment is:

$$M = \frac{\sigma I}{y} = \frac{(35 \times 10^6) \times (2.180 \times 10^6 \times 10^{-12})}{(32.5 \times 10^{-3})}$$

$$= 2348 \text{ N m}$$

The corresponding bending moment for the permitted compressive stress of 70 MN/m^2 at the top edge (y = 57.5 mm) is:

$$M = \frac{\sigma I}{y} = \frac{(70 \times 10^6) \times (2.180 \times 10^6 \times 10^{-12})}{(57.5 \times 10^{-3})}$$

$$= 2654 \text{ N m}$$

To satisfy both stress limits the smaller of these two results (2348 Nm) must be used. As shown in Fig. 9.13 the maximum bending moment due to a uniformly distributed load on a beam, simply supported at its ends is $WL/8$ where W is the total load and L the span. In the present example the load is given as 2 kN/m and for a span of L m the total load is $W = 2000\,L$ N. Using this result the maximum bending moment is:

$$\frac{2000\,L \times L}{8} = 2348 \quad \text{or} \quad L^2 = \frac{2348 \times 8}{2000} = 9.392$$

$$\text{and } L = \sqrt{9.392} = 3.065$$

The greatest possible span is therefore 3.065 m.

9.8.5 Section modulus

In design work it is the maximum value of the bending stress (σ_{max}) that is of interest. This occurs at the top or bottom edge of the section where the distance from the neutral axis is greatest. If this distance is denoted by y_{max} then, from the bending equation $M/I = \sigma/y$, the bending moment can be written:

$$M = \left(\frac{I}{y_{max}}\right)\sigma_{max}$$

The quantity I/y_{max} is called the *section* modulus and it is usually denoted by Z. It is a geometric property of the section and in the SI system it has the units cubic metres (m^3). Practical values are often given in cm^3 or mm^3 and these must be divided by 10^6 or 10^9 respectively to obtain m^3. If the section is symmetrical about the neutral axis the value of Z will be the same for the top and bottom edges. Otherwise two values are required.

Once the values of Z are known for a section the moment and maximum stress can be related by the equation:

$$M = Z\,\sigma_{max}$$

This result is the bending equivalent of the relationship

force = area × stress for a bar in tension or compression.

EXAMPLE 9.14 Fig. 9.24 shows the cross-section of a beam, the dimensions being in millimetres. Calculate:
(a) the section moduli,
(b) the greatest compressive and tensile stresses on the section when the beam is subjected to a sagging bending moment of 2 kN m.

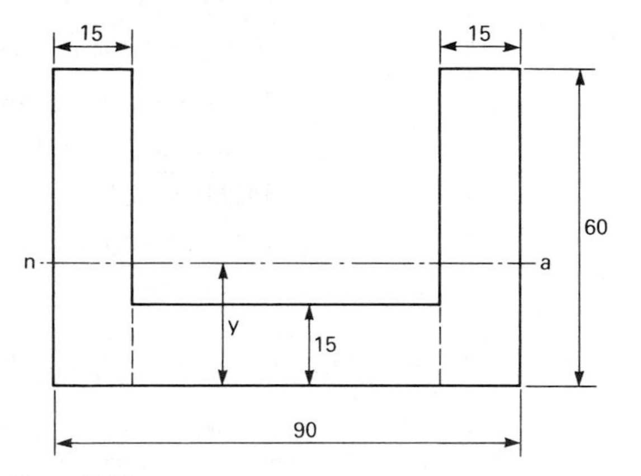

Figure 9.24

SOLUTION To find the position of the neutral axis, it is convenient to divide the section into three rectangles as shown by the broken lines. The end rectangles are each 15 mm wide and 60 mm deep with a centroid 30 mm from the bottom edge of the section. The middle rectangle is 60 mm wide and 15 mm deep with a centroid 7.5 mm from the bottom.

Working in millimetre units and considering first moments of area about the bottom edge, the height of the centroid of the whole section is:

$$y = \frac{(15 \times 60 \times 30) + (60 \times 15 \times 7.5) + (15 \times 60 \times 30)}{(15 \times 60) + (60 \times 15) + (15 \times 60)} = 22.5 \text{ mm}$$

The neutral axis is therefore 22.5 mm from the bottom edge and 37.5 mm from the top.

The second moment of area about the neutral axis can be found as in the previous solution by taking rectangles and a rectangular cutout each having one edge on the neutral axis. There are two rectangles above the neutral axis, each 15 mm wide and 37.5 mm deep. Below the neutral axis there is one rectangle 90 mm and 22.5 mm deep and a cutout 60 mm wide and 7.5 mm deep. Using the formula $bd^3/3$ in each case the total second moment is:

$$I = 2 \times \frac{15 \times 37.5^3}{3} + \frac{90 \times 22.5^3}{3} - \frac{60 \times 7.5^3}{3}$$

$$= 860\,625 \text{ mm}^4$$

(a) At the top edge $y_{max} = 37.5$ mm and the section modulus is

$$Z = \frac{I}{y_{max}} = \frac{860\,625}{37.5} = 22\,950 \text{ mm}^3$$

At the bottom edge $y_{max} = 22.5$ mm and

$$Z = \frac{860\,625}{22.5} = 38\,250 \text{ mm}^3$$

Note Although the values of I for the various rectangles can be added or subtracted, the values of Z for the section can only be found from the overall value of I.

(b) Since the bending moment is sagging, stresses are compressive above the neutral axis and tensile below. The greatest compressive stress occurs at the top edge and using $M = Z\sigma_{max}$ it is given by:

$$\sigma_{max} = \frac{M}{Z} = \frac{2 \times 10^3}{22\,950 \times 10^{-9}}$$

$$= 87.15 \times 10^6 \text{ N/m}^2 \text{ or } 87.15 \text{ MN/m}^2$$

Similarly, the greatest tensile stress (at the bottom edge) is:

$$\sigma_{max} = \frac{2 \times 10^3}{38\,250 \times 10^{-9}}$$

$$= 52.29 \times 10^6 \text{ N/m}^2 \quad \text{or} \quad 52.29 \text{ MN/m}^2$$

9.9 Deflection of beams and cantilevers

So far in this chapter the bending equation has been used for calculations on bending stress. It is also the basis for the theory of beam deflections.

Suppose a straight horizontal uniform cantilever is built into a wall at the left-hand end A (Fig. 9.25(a)) and B is the position of the free end before the loads are applied. Consider the effect on the position of B of the bending of a very short length of the cantilever, dx, distance x from B. (In the diagram the distortion is

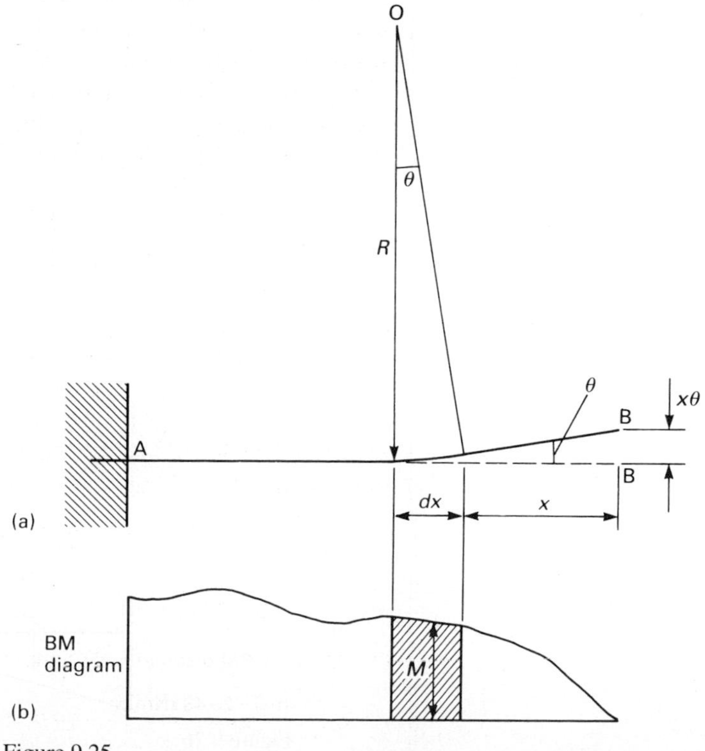

Figure 9.25

greatly exaggerated.) This portion bends to an arc, radius R, that makes an angle θ (radians) at the centre O. The length of an arc is $R\theta$ and therefore:

$$R\theta = dx \quad \text{or} \quad \theta = dx/R$$

The length x of the cantilever on the free end side of this arc turns through the same angle and causes B to move to the position B′. Hence:

$$BB' = x\theta = xdx/R$$

The radius R and bending moment M are related by the bending equation $M/I = E/R$ from which $R = EI/M$. Substituting this result in the previous equation the deflection may be written:

$$BB' = \frac{xMdx}{EI}$$

The quantities M and dx are the height and width of the hatched area in Fig. 9.25(b) and the product $M.dx$ therefore represents the area of the BM diagram under the portion dx of the cantilever. Furthermore $xM.dx$ is the moment of this area about B. If this analysis is extended to the whole cantilever the total deflection at the free end is given by the first moment of area of the complete BM diagram about B, divided by EI. This result can be written:

$$\text{deflection at free end} = \frac{1}{EI} \times \left(\begin{array}{c}\text{first moment of area of BM} \\ \text{diagram about the free end}\end{array}\right)$$

The symbol for deflection at a specified point is δ; when it is a variable, at a distance x along the beam or cantilever, it is usually denoted by y.

EXAMPLE 9.15 Calculate the deflection at the free end of a cantilever, 6.0 m long, when a mass of 500 kg is suspended there. Take $E = 206$ GN/m^2 and I for the cross-section = 30 000 cm^4.

SOLUTION The cantilever is shown in Fig. 9.26(a). The mass of 500 kg causes a load at the free end of 500 × 9.81 = 4905 N. The greatest value of the bending moment (at the fixed end) is 4905 × 6 = 29.43 kN m. It is a hogging moment and therefore negative according to the sign convention used in this chapter but the sign can be ignored in this solution.

The complete BM diagram (Fig. 9.26(b)) is a triangle.

$$\begin{aligned}\text{area of BM diagram} &= \tfrac{1}{2} \times \text{base} \times \text{height} \\ &= 0.5 \times 6 \times 29.43 \\ &= 88.29 \text{ kN m}^2\end{aligned}$$

$$\text{centroid distance from free end} = \tfrac{2}{3} \times \text{length} = \tfrac{2}{3} \times 6 = 4 \text{ m}$$

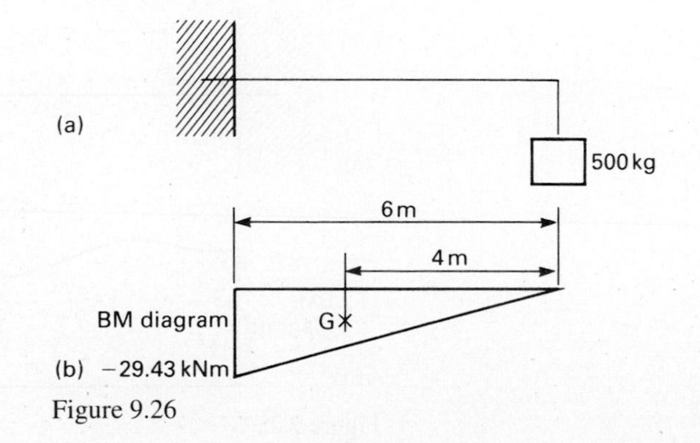

Figure 9.26

Working in newton and metre units, the required deflection is:

$$\delta = \frac{1}{EI} \times \text{(first moment of BM diagram about free end)}$$

$$= \frac{(88.29 \times 10^3) \times 4}{(206 \times 10^9) \times (30\,000 \times 10^{-8})}$$

$$= 0.005\,71 \text{ m} \quad \text{or} \quad 5.71 \text{ mm}$$

9.9.1 Deflection formulae

The definition of a beam or cantilever varies along its length and it is the maximum value that is usually required. It is therefore convenient to have formulae for the four standard cases of loading shown in Fig. 9.13. These can be derived from the BM diagrams using the areas and centroid positions given in Fig. 9.27 for the triangle and parabola.

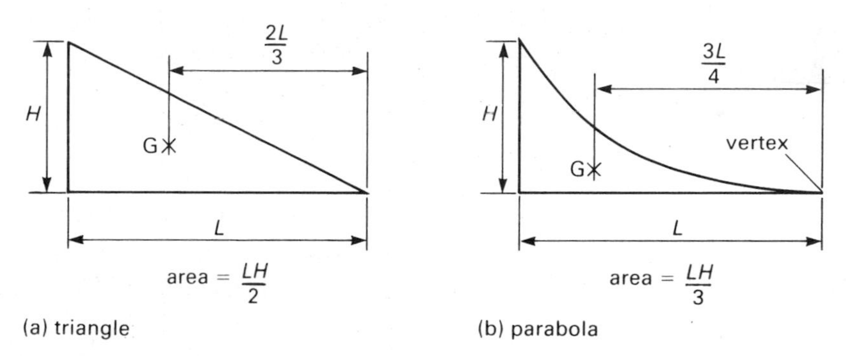

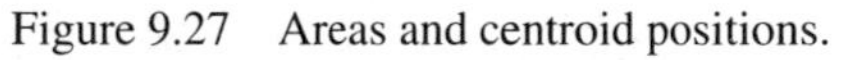
Figure 9.27 Areas and centroid positions.

It is easier to start with the cantilever cases and then use the results to find the formulae for beams. Although the bending moments for the cantilevers are negative (hogging), the areas of the BM diagrams are treated as positive in this analysis. The four standard cases are the following.

Cantilever with end point load

The BM diagram (Fig. 9.13) is a triangle with a length L and a maximum height WL. Its centroid is $2L/3$ from the free end. The maximum deflection occurs at the free end and, using the result obtained in the previous section,

$$\delta = \frac{1}{EI} \times \text{(first moment of area of BM diagram about the free end)}$$

$$= \frac{1}{EI} \times \text{(area of BM diagram} \times \text{centroid distance)}$$

$$= \frac{1}{EI} \times \left[\left(\frac{1}{2}WL \times L\right) \times \frac{2}{3}L\right]$$

$$= \frac{WL^3}{3EI}$$

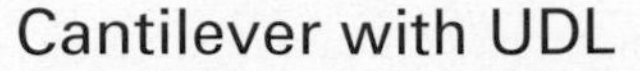

Cantilever with UDL

The BM diagram is a parabola (Fig. 9.13) with the vertex at the free end and a maximum value of $WL/2$. Using the results given in Fig. 9.27(b), the deflection at the free end is:

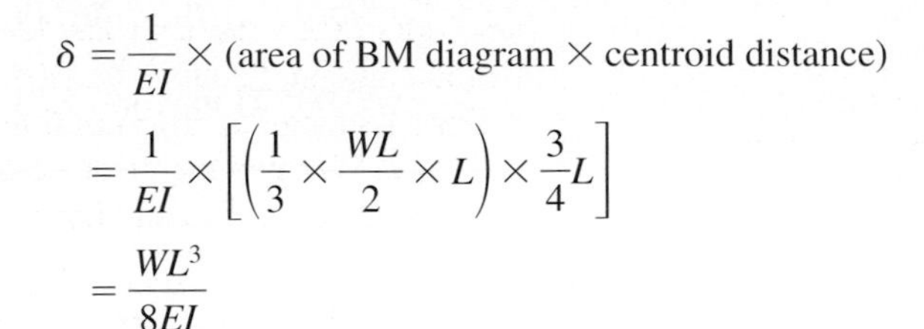

$$\delta = \frac{1}{EI} \times \text{(area of BM diagram} \times \text{centroid distance)}$$

$$= \frac{1}{EI} \times \left[\left(\frac{1}{3} \times \frac{WL}{2} \times L \right) \times \frac{3}{4} L \right]$$

$$= \frac{WL^3}{8EI}$$

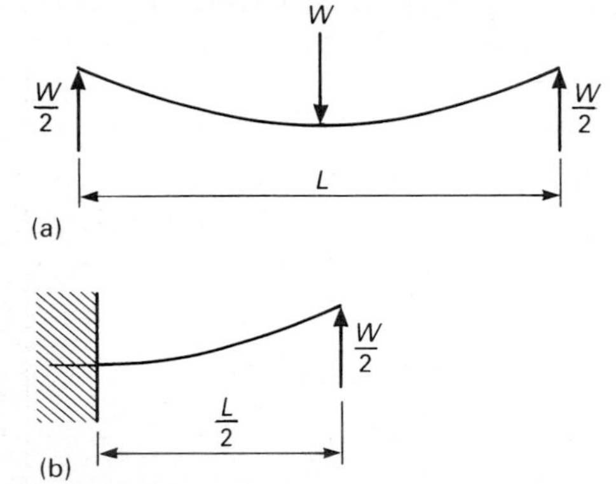

Figure 9.28
(a) Beam with central point load
(b) Equivalent cantilever with upward load at free end.

Beam with central point load

The deflected shape of the beam is shown in Fig. 9.28(a) (much exaggerated). The maximum deflection occurs at midspan. Imagine a cantilever of half the length with an upward load of $W/2$ at the free end (Fig. 9.28(b)). Its deflection curve would be the same as that for the right-hand half of the beam and the upward deflection at its free end would be equal to the downward deflection at the midpoint of the beam. We can therefore use the cantilever formula $WL^3/3EI$ with L replaced by $L/2$ and W by $W/2$. The required deflection is:

$$\delta = \frac{(W/2)(L/2)^3}{3EI} = \frac{WL^3}{48EI}$$

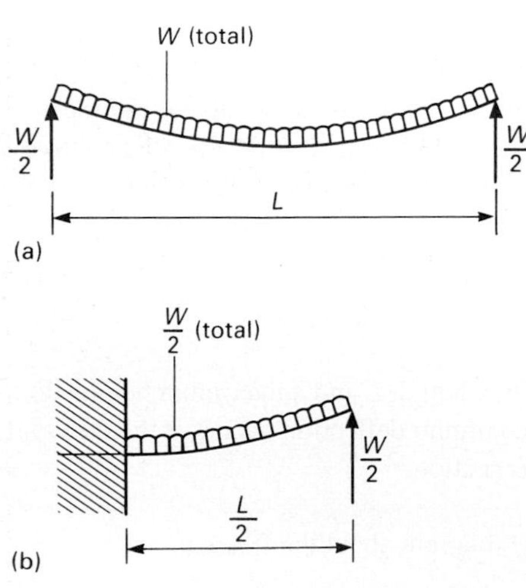

Figure 9.29
(a) Beam with uniformly distributed load
(b) Equivalent cantilever.

Beam with uniformly distributed load

The beam is shown in Fig. 9.29(a) and the equivalent half-length cantilever in Fig. 9.29(b). For the cantilever the upward deflection at the free end is the same as for the previous case. In addition there would be a downward deflection due to the UDL. Its value can be found by replacing L by $L/2$ and W by $W/2$ in the formula $WL^3/8EI$ obtained above for a cantilever with a UDL. The resultant deflection at the free end of the equivalent cantilever, and hence the midpoint deflection of the beam is, therefore:

$$\delta = \frac{WL^3}{48EI} - \frac{(W/2)(L/2)^3}{3EI} = \frac{WL^3}{EI} \left(\frac{1}{48} - \frac{1}{128} \right)$$

$$= \frac{WL^3}{EI} \left(\frac{8-3}{384} \right)$$

$$= \frac{5WL^3}{384EI}$$

The formulae for the four standard loading cases are collected for reference in the table of Fig. 9.13. They apply only to uniform beams whereas the other information in the table may be used for beams in which the cross-section varies.

EXAMPLE 9.16 A wooden shelf, simply supported at its ends, is required to carry a 20-volume encyclopaedia, each volume being 45 mm thick and weighing 2 kg. Determine the thickness of timber required under the following conditions:

width of shelf 240 mm,
maximum permissible tensile stress 30 MN/m^2,
maximum permissible compressive stress 12 MN/m^2,
deflection not to exceed 3 mm,
modulus of elasticity 10 GN/m^2.

SOLUTION This problem is an example of simple beam design with limiting values of both stress and deflection.

With 20 volumes, each 45 mm thick, the span L is $20 \times 45 = 900$ mm or 0.9 m. (In practice the shelf might be made slightly longer to allow easy handling of the volumes.) The load may be regarded as uniformly distributed, the total amount being $W = 20 \times 2 \times 9.81 = 392.4$ N.

For a rectangular cross-section, width b and depth d, the second moment of area I is $bd^3/12$ and the distance from the neutral axis to the most stressed fibres is $y_{max} = d/2$ at the top and bottom edges. Hence the section modulus is:

$$Z = \frac{I}{y_{max}} = \frac{(bd^3/12)}{(d/2)} = \frac{bd^2}{6}$$

From Fig. 9.13 the maximum bending is:

$$M = \frac{WL}{8} = \frac{392.4 \times 0.9}{8} = 44.15 \text{ N m}$$

With a symmetrical cross-section, the bending stresses at the top and bottom edges are numerically equal. From the data given in the question the compressive stress is the limiting one and therefore:

$$\sigma_{max} = 12 \text{ MN/m}^2 = 12 \times 10^6 \text{ N/m2}$$

Substituting these values in the relationship $Z = M/\sigma_{max}$ we obtain:

$$\frac{bd^2}{6} = \frac{44.15}{12 \times 10^6} = 3.679 \times 10^{-6}$$

But $b = 240$ mm $= 0.240$ m and therefore:

$$d^2 = \frac{3.679 \times 10^{-6}}{b} = \frac{3.679 \times 10^{-6} \times 6}{0.240} = 91.98 \times 10^{-6}$$

from which

$$d = \sqrt{(91.98 \times 10^{-6})} = 0.009\,59 \text{ m} \quad \text{or} \quad 9.59 \text{ mm}$$

The maximum permitted deflection is $\delta = 3$ mm $= 0.003$ m and, from Fig. 9.13, the appropriate deflection formula is $\delta = 5WL^3/384EI$. Hence:

$$I = \frac{5WL^3}{384E\delta} = \frac{5 \times 392.4 \times 0.9^3}{384 \times (10 \times 10^9) \times 0.003} = 124.2 \times 10^{-9} \text{ m}^4$$

But $I = bd^3/12$ and therefore

$$\frac{bd^3}{12} = 124.2 \times 10^{-9}$$

or

$$d^3 = \frac{12 \times 124.2 \times 10^{-9}}{b} = \frac{12 \times 124.2 \times 10^{-9}}{0.240} = 6.21 \times 10^{-6}$$

and

$$d = \sqrt[3]{(6.21 \times 10^6)} = 18.4 \times 10^{-3} \text{ m} \quad \text{or} \quad 18.4 \text{ mm}$$

To meet both the stress and deflection requirements the minimum shelf thickness is 18.4 mm, the deflection being the limiting factor.

9.10 Struts

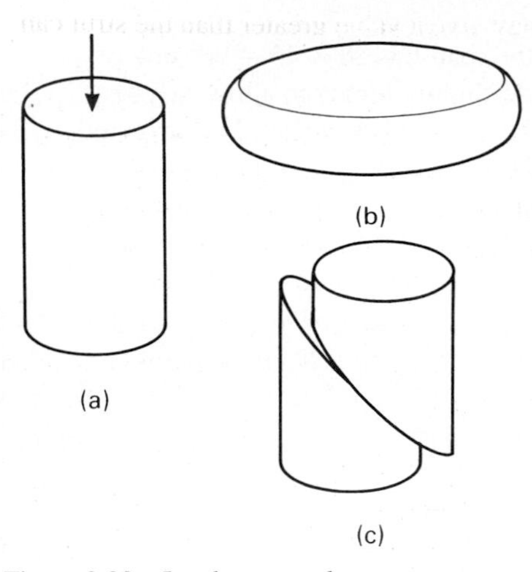

Figure 9.30 Load on a stocky strut.

The strength of a bar in tension is generally unaffected by its length; for a given material it depends only on its cross-sectional area. In compression, however, the failure loads may be very different for short and long bars of the same cross-section and material. You can demonstrate this by compressing thin rods, say 5 mm diameter, of steel or wood, 10 mm and 1 m long.

A strut that is short in comparison with its cross-sectional dimensions is called *stocky*. One that is comparatively long is termed *slender*.

Suppose a stocky strut in the form of a round bar (Fig. 9.30(a)) is loaded until it fails. If it is made from a soft material, such as aluminium, it will be squeezed to a much larger cross-section as shown in Fig. 9.30(b). A brittle material, such as cast iron, may fracture suddenly along a plane at approximately 45° to the axis as shown in Fig. 9.30(c). When a slender strut is loaded it may bow as shown in Fig. 9.31. If the load is increased the bowing will increase until the strut buckles. Although the load is compressive the failure occurs in bending.

If the strut were perfectly straight to start with and the load acted precisely along the axis bowing would not begin but there is inevitably some initial crookedness or eccentricity of load that causes bending. There is a critical value of the end load at which this bending increases until the strut collapses, an effect known as *elastic instability*.

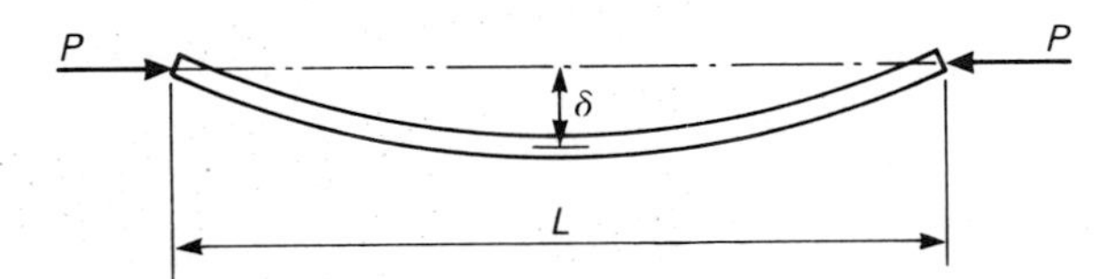

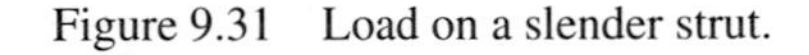

Figure 9.31 Load on a slender strut.

Suppose (Fig. 9.31) the length of the strut is L, the end load is P and the bowing has a maximum deflection at the midpoint of δ. The crucial question is – what value of P will maintain the strut in this bowed form? An approximate answer can be found by assuming the strut bends to the same curve as a beam with a central point load.

From the table of Fig. 9.13 the deflection at the centre of a beam due to a point load W is:

$$\delta = \frac{WL^3}{48EI}$$

The bending moment at the midpoint of the strut is $P\delta$ and for the beam it is $WL/4$. If the strut and beam bend to the same curve they will have the same deflections and bending moments at midspan. Equating the moments and substituting for the deflection δ we have:

$$P\delta = \frac{WL}{4} \quad \text{or} \quad P\frac{WL^3}{48EI} = \frac{WL}{4}$$

and, on arrangement,

$$P = \frac{12EI}{L^2}$$

This value of P is called the *critical load* for the strut. If it is exceeded, any bowing will become progressive and the strut will collapse. It can be shown by a more rigorous analysis that the strut bends to a sine wave and that the critical load is given by:

$$P = \frac{\pi^2 EI}{L^2}$$

This result is known as the *Euler critical* (or *crippling*) load. The formula must be used with caution. Although it gives reasonable results for very slender struts it does not apply to stocky struts. If L is small it may give a value greater than the strut can withstand in axial compression. The formula is then invalid. In practice struts are designed using formulae that allow for both effects – bowing and direct compression.

If it is not restrained the strut will bend in the plane for which it is weakest. This corresponds to the minimum value of I for the cross-section. If the section is rectangular, for instance, the strut will bend with the neutral axis parallel to the longer side. For this reason circular sections have advantages for struts.

EXAMPLE 9.17 An alloy tube has inner and outer diameters of 50 mm and 60 mm respectively. Calculate the crippling load for a 4 m length of this tube by the Euler formula. Modulus of elasticity, 70 GN/m^2.

What is the failure load for a short length of the tube if the crushing stress for the alloy is 60 MN/m^2?

SOLUTION Using the formula for I given in Fig. 9.21(c), the second moment of area for the cross-section is:

$$I = \frac{\pi}{64} \times 60^4 - \frac{\pi}{64} \times 50^4$$

$$= 329\,376 \text{ mm}^4$$

Converting all values to newton and metre units and substituting in the Euler formula the crippling load is:

$$P = \frac{\pi^2 EI}{L^2} = \frac{\pi^2 \times (70 \times 10^9) \times (329\,376 \times 10^{-12})}{4^2}$$

$$= 14\,220 \text{ N} \quad \text{or} \quad 14.22 \text{ kN}$$

The short length of tube fails by axial compression at the crushing stress. In millimetre units the cross-sectional area is:

$$A = \frac{\pi}{4}(60^2 - 50^2) = 863.9 \text{ mm}^2$$

Converting to newton and metre units the failure load is:

$$P = \sigma A = (60 \times 10^6) \times (863.9 \times 10^{-6})$$

$$= 51\,830 \text{ N} \quad \text{or} \quad 51.83 \text{ kN}$$

9.10.1 Struts with built-in ends

The Euler formula given in the previous section applies to struts whose ends are pin-jointed or hinged. This type of end fixing is called position fixed, direction free. For other end fixings the crippling load formula has to be modified. Fig. 9.32 shows four possibilities in increasing order of strength. The length of the strut is L in each case and the formula is obtained by considering the length of the equivalent pin-jointed strut.

1 One end built-in, one completely free

This case is shown in Fig. 9.32(a) and can be described as one end position and direction fixed, one end position and direction free. The strut bends as one half of a pin-jointed strut of length $2L$. Replacing L by $2L$ in the Euler formula, the crippling load becomes:

$$P = \frac{\pi^2 EI}{(2L)^2} = \frac{\pi^2 EI}{4L^2}$$

2 Both ends pin-jointed

The case with both ends position fixed and direction free (Fig. 9.32(b)) was the basis of the Euler formula and the cripping load is therefore:

$$P = \frac{\pi^2 EI}{L^2}$$

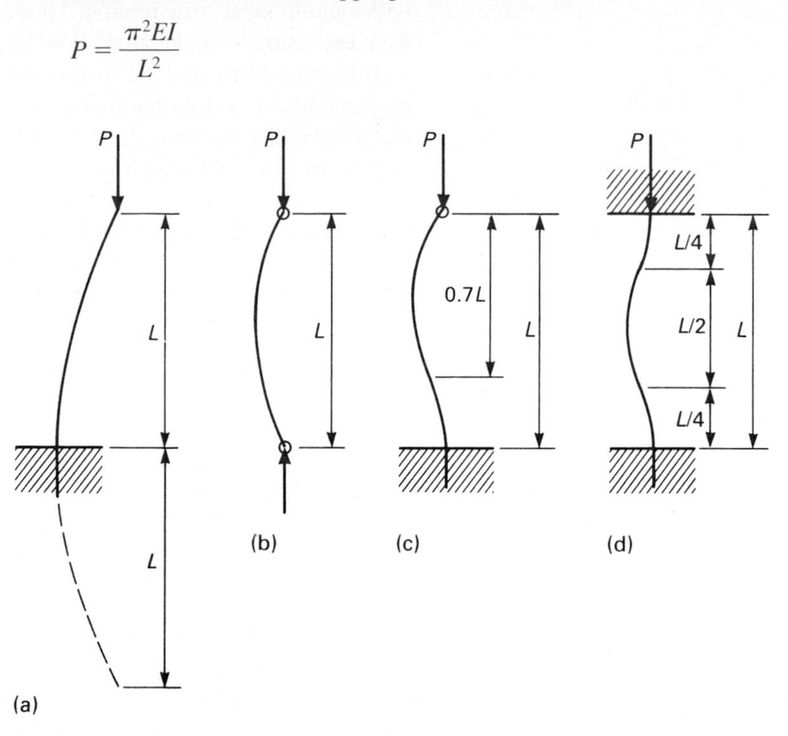

Figure 9.32 Struts with built-in ends.

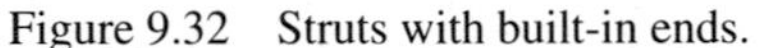

3 One end pinned, one built-in

This arrangement (Fig. 9.32(c)) can be described as one end position fixed and direction free, the other position and direction fixed. It can be shown that the length of the equivalent pin-jointed strut is approximately $0.7L$ and, substituting this value for L in the basic Euler formula, the crippling load becomes

$$P = \frac{\pi^2 EI}{(0.7L)^2} = \frac{2\pi^2 EI}{L^2} \quad \text{(approximately)}$$

4 Both ends built-in

Fig. 9.32(d) shows a strut with both ends position and direction fixed. It bends in such a way that there are points of contraflexure at the quarter-span points. The length of the equivalent pin-jointed strut is $L/2$ and the crippling load is now:

$$P = \frac{\pi^2 EI}{(L/2)^2} = \frac{4\pi^2 EI}{L^2}$$

Cases 2 and 4 are sometimes referred to as ends free and ends fixed respectively. The description one end free, one end fixed is ambiguous because it could refer to 1 or 3.

A comparison of the results shows that the crippling loads in cases 1, 3 and 4 are respectively one quarter, twice and four times that given by the original Euler formula. For the tubular strut of Example 9.17 the crippling loads in cases 1, 3 and 4 would be 3.56 kN, 29.44 kN and 56.88 kN. The last of these results is not valid since it is greater than the failure load by axial compression calculated in the solution to Example 9.17, namely 51.83 kN.

9.11 Torsion of circular shafts

Suppose a circular bar is fixed at one end and has a bracket attached to the other at right angles to its axis (Fig. 9.33). If a couple is applied to this bracket, the bar will twist and there will be a shearing effect at each point in the material.

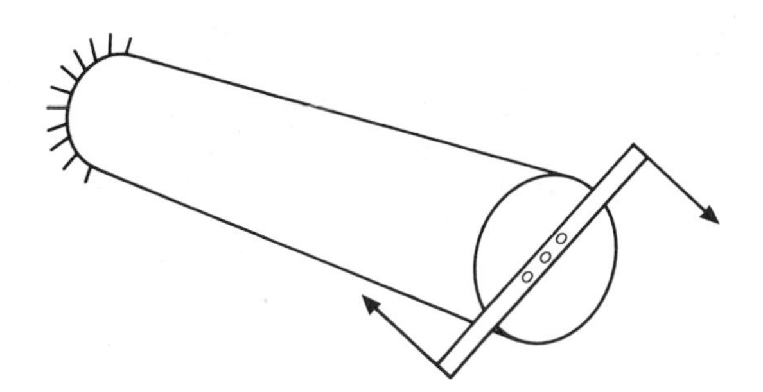

Figure 9.33

The applied couple is usually called the *twisting moment* or *torque* and its magnitude, the angle of twist and the shear stress are related by an equation that corresponds to the bending equation for beams. The theory of simple torsion depends on the following assumptions:

1 The material is uniform throughout.
2 The cross-section is circular.
3 The shaft is initially straight and unstressed.
4 The axis of the twisting moment is the axis of the shaft.
5 The material of the shaft obeys Hooke's law in shearing.
6 The radii of each cross-section remain straight after twisting.

The notation is shown in Fig. 9.34. Suppose the length of the shaft is L and that a straight line AB parallel to the axis at radius r becomes the helix AB′ when the shaft is twisted. The angle of this helix (B′AB) is the shear strain γ (Fig. 9.34(a)). If the corresponding shear stress is τ then, since stress/strain = modulus, $\gamma = \tau/G$ where G is the modulus of rigidity.

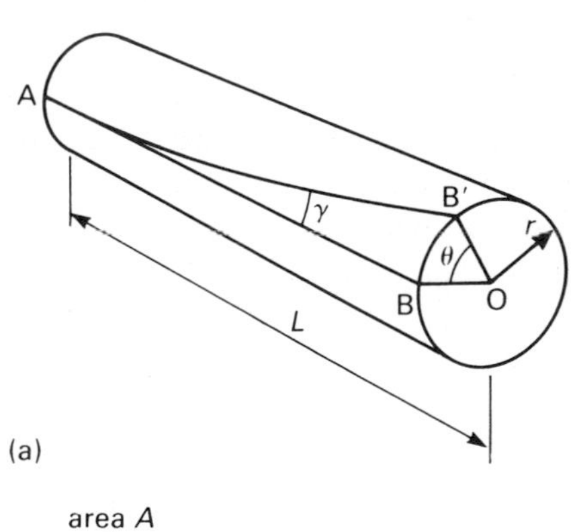

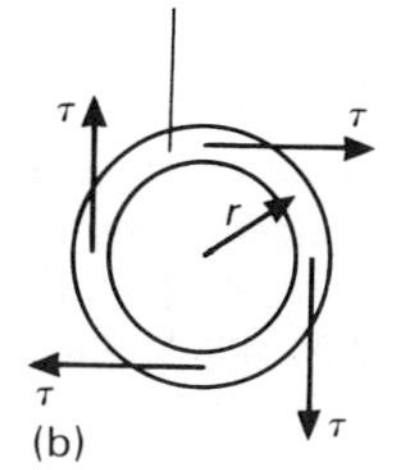

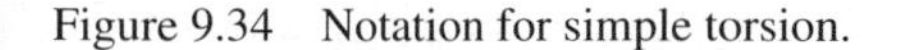

Figure 9.34 Notation for simple torsion.

The arc length BB′ can be expressed in two ways. Since θ is a small angle in radians,

$$BB' = L\gamma = L\tau/G$$

Also, if θ is the angle by which one end twists relative to the other, it can be written:

$$BB' = r\theta$$

By equating these two results,

$$r\theta = L\tau/G$$

or

$$\frac{\tau}{r} = \frac{G\theta}{L}$$

To find the twisting moment or torque consider a small ring (or *annulus*) at radius r (Fig. 9.34(b)). From the last result the stress can be written:

$$\tau = \frac{rG\theta}{L}$$

and the corresponding tangential force on this ring is:

$$\tau \times \text{area } A = \frac{rAG\theta}{L}$$

This force is acting at a distance r from the axis and therefore produces a moment about it of:

$$\text{force} \times \text{radius} = \frac{rAG\theta}{L} \times r = \frac{G\theta}{L} \times r^2 A$$

The quantity r^2A is called the *polar second moment of area* for the ring. The total polar second moment for the whole cross-section is denoted by J (or sometimes I_p). The total twisting moment or torque T is therefore given by:

$$T = \frac{G\theta}{L} \times J \quad \text{or} \quad \frac{T}{J} = \frac{G\theta}{L}$$

Combining this result with the previous one gives the torsion equation:

$$\frac{\tau}{r} = \frac{T}{J} = \frac{G\theta}{L}$$

There are obvious similarities between the torsion and bending equations. The polar second moment of area J is sometimes called the polar moment of inertia. For a circular section, diameter d, $J = \pi d^4/32$, twice the value of I about a diameter. The units of J and I are the same, m^4 in SI with practical values often given in cm^4 or mm^4.

9.11.1 Polar section modulus

In torsion, as in bending, the greatest stress occurs at the greatest distance from the axis of the shaft. If r_{max} is the radius to the outer surface and τ_{max} is the corresponding shear stress then, from the torsion equation,

$$T = \frac{J}{r_{max}} \tau_{max} = Z_p \tau_{max}$$

where $Z_p = J/r_{max}$ and is called the polar section modulus. Its units are cubic metres in the SI system with practical values often given in cm^3 or mm^3.

9.12 Strain energy

When a tensile force P is gradually applied to a bar it causes an extension e as shown in Fig. 9.1(a). The point of application of the force moves in the same direction as P and work is done (see section 11.1). Provided the stress is within the elastic limit (see section 5.2) this work can be absorbed and stored by the material of the bar. It can be recovered when the force is removed. It is known as strain energy or resilience and is usually denoted by U.

Within the limit of proportionality the graph of force against extension is a straight line and the work done is given by the area of the triangle shown in Fig. 9.35. It can be calculated as the average value of the applied force multiplied by the extension. Thus:

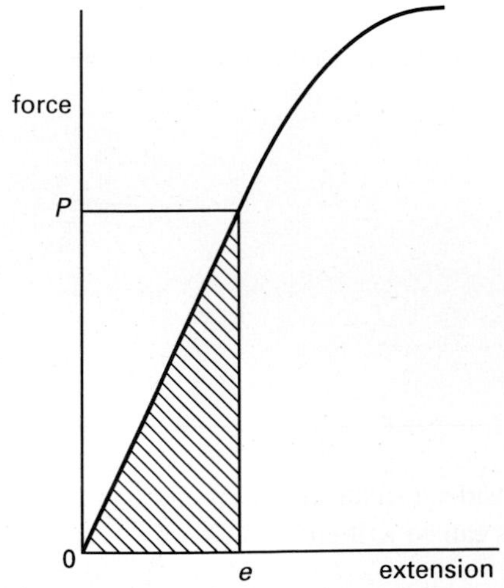

$$U = \frac{1}{2} Pe$$

Figure 9.35

Using the notation of sections 9.2 and 9.2.2, $P = \sigma A$, $e = \epsilon L$ and $E = \sigma/\epsilon$. With these results,

$$U = \frac{1}{2}(\sigma A)(\epsilon L)$$

$$= \frac{1}{2}(\sigma A)(\sigma/E)L$$

and

$$U = \frac{\sigma^2}{2E} \times \text{(volume of the bar)}$$

A similar result applies to a bar in compression and there are corresponding formulae for bending. In each case the strain energy is given by an expression of the form:

$$U = \frac{\sigma^2}{kE} \times \text{(volume of the material)}$$

where σ is the maximum stress in the material, E is the modulus of elasticity and k is a numerical constant that depends on the type of loading. It follows that a material with a low modulus of elasticity may absorb as much energy as a stronger one with a high modulus (see Tables 5.9 and 5.11). This is an important consideration in the design of components such as tow ropes that have to absorb the energy of impact loads.

Table 9.1 gives the value of U in terms of the volume of material for uniform bars and wires of rectangular and circular cross-sections when used as beams, cantilevers and springs. For materials in shear and torsion, the shear stress τ and modulus of rigidity G replace the direct stress σ and modulus of elasticity E. All the results are given as factors to be multiplied by the volume. This explains why a long tow rope in tension or a long beam in bending can absorb more energy than a short one.

Mechanical springs are designed so that they can withstand large distortions without the elastic limit being exceeded. Although straight bars in bending, torsion or tension are used as springs, there are two groups that are particularly useful in practice; those consisting of a length of wire wound in a helix or spiral and those known as leaf springs which consist of several rectangular plates strapped together. Examples of both kinds are used in vehicle design.

Uniform bar in tension or compression:	$\frac{\sigma^2}{2E}$
Uniform block in shear:	$\frac{\tau^2}{2G}$
Beam simply supported at its ends, with a central point load or cantilever with an end point load:	
rectangular cross-section	$\frac{\sigma^2}{18E}$
circular cross-section	$\frac{\sigma^2}{24E}$
Circular shaft in torsion or a helical spring of round wire with axial load:	$\frac{\tau^2}{4G}$
Helical spring with axial torsional moment:	
wire of rectangular cross-section	$\frac{\sigma^2}{6E}$
wire of circular cross-section	$\frac{\sigma^2}{8E}$

Table 9.1 Strain energy factors

9.13 Thin shells

If a sealed vessel, such as a boiler or gas cylinder, contains a fluid under pressure its wall will be stressed in tension. If the wall is thin in comparison with the diameter of the vessel the stresses can be estimated from simple formulae.

Suppose a cylindrical vessel contains a gas or liquid at a pressure P. At each point in the wall there will be a tensile stress acting in a tangential direction. This is called the *hoop* or *circumferential stress*. Its value can be found by considering the equilibrium of half the cylinder (Fig. 9.36). Let L be the length, d the inner diameter, t the wall thickness and σ_H the hoop stress. It is assumed that the hoop stress is constant through the wall and that the effect of the ends of the cylinder can be neglected.

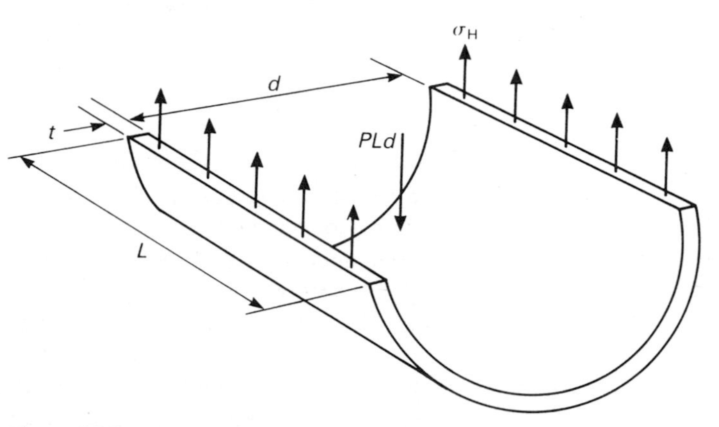

Figure 9.36

It is shown in Chapter 13 that the resultant force on a curved surface subjected to a uniform pressure is equal to the force produced by the pressure on the projected area. In the present case the pressure is P and the projected area is a rectangle length L and breadth d. The resultant downward force (Fig. 9.36) is therefore PLd.

The hoop stress σ_H acts on two rectangles each length L and breadth t as shown in Fig. 9.36. The total upward force is therefore $2\sigma_H Lt$. For equilibrium the upward and downward forces must be equal. Thus:

$$2\sigma_H Lt = PLd$$

and

$$\sigma_H = \frac{Pd}{2t}$$

The cylinder is also being stretched in the direction of its axis due to the pressure acting on the ends. The stress in this direction is called the *longitudinal stress* (σ_L) and, as shown in Fig. 9.37, it acts on a ring of diameter d (internal) and width t. The area of this ring is $2\pi r t$ where r is the mean radius. Since the wall thickness is small compared with the diameter it is sufficiently accurate to take d as the mean diameter and the area becomes $\pi d t$.

The force due to the longitudinal stress is therefore $\sigma_L \pi d t$.

The force on the end of the cylinder due to the pressure P is $P(\pi d^2/4)$ and for equilibrium this equals the force arising from the longitudinal stress. Therefore

$$\sigma_L \pi dt = P(\pi d^2/4)$$

and

$$\sigma_L = \frac{Pd}{4t}$$

The longitudinal stress is therefore half the hoop stress.

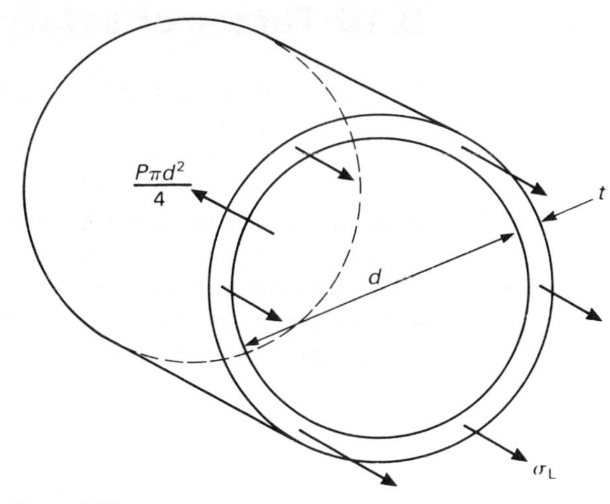

Figure 9.37

In the case of a spherical shell the stress is the same in all tangential directions. The analysis is the same as for the longitudinal stress of the cylinder since this would be unchanged if the ends of the cylinder were hemispherical. The expression $Pd/4t$ can therefore be used for the stress in a thin-walled sphere.

Fluid pressures are often given in terms of a unit called a *bar*. A bar is defined as 10^5 N/m² or 100 kN/m² and 1 bar is approximately equal to atmospheric pressure at sea level.

The pressure P should be taken as the difference between the internal and external pressures acting on the wall of the vessel. If the external pressure is atmospheric then the required value of P is the gauge pressure (see section 13.1.2). If P is given as an absolute pressure then the atmospheric pressure (1 bar) must be subtracted from it.

EXAMPLE 9.18 A thin cylindrical shell, 1.2 m diameter and 20 mm wall thickness, is subjected to an internal gauge pressure of 40 bar. Calculate the hoop and longitudinal stresses in the wall.

Find also the greatest capacity of a spherical shell that can withstand the same pressure with the same maximum stress and wall thickness.

SOLUTION In newton and metre units the pressure P is $40 \times 10^5 = 4 \times 10^6$ N/m². Hence, the hoop and longitudinal stresses (both tensile) are:

$$\sigma_H = \frac{Pd}{2t} = \frac{(4 \times 10^6) \times 1.2}{2 \times (20 \times 10^{-3})} = 120 \times 10^6 \text{ N/m}^2 \quad \text{or} \quad 120 \text{ MN/m}^2$$

$$\sigma_L = \frac{Pd}{4t} = \frac{(4 \times 10^6) \times 1.2}{4 \times (20 \times 10^{-3})} = 60 \times 10^6 \text{ N/m}^2 \quad \text{or} \quad 60 \text{ MN/m}^2$$

For the spherical shell the stress is $\sigma = Pd/4t$ in every tangential direction and, using the permissible value of 120 MN/m², the greatest diameter is:

$$d = \frac{4t\sigma}{P} = \frac{4 \times (20 \times 10^{-3}) \times (120 \times 10^6)}{4 \times 10^6} = 2.4 \text{ m}$$

The corresponding radius is 1.2 m and, using the formula for the volume of a sphere, the capacity is:

$$V = \frac{4\pi r^3}{3} = \frac{4\pi (1.2)^3}{3} = 7.238 \text{ m}^3$$

9.14 Factor of safety

It is the designer's responsibility to ensure that every component of a structure can withstand the forces acting on it. The first step is to decide on a safe stress for the material of the component. This is usually called the *working stress* and it is obtained by dividing the failure stress by a factor called the *factor of safety*. In examination questions, the factor of safety will be given whenever it is required and for practical design it will be specified in the code of practice.

The failure stress is obtained by experiment (see Chapter 5). If 'failure' is taken to mean the onset of permanent strain then the yield point is used. In some circumstances failure is considered to be fracture and the ultimate strength is taken as the failure stress.

The value of the factor of safety will depend upon the type of loading to be expected. If the structure carries only static loads a low value such as two may be used. If there are suddenly applied or impact loads a higher value must be taken.

An alternative to using a factor of safety is to increase the load acting on a component by a *load factor*.

$$\text{Load factor} = \frac{\text{design load}}{\text{normal maximum load}}$$

In simple tension or compression it will not matter whether the factor is used to reduce the stress or increase the load but in complex loading cases the load factor and factor of safety methods lead to different results.

9.15 Design considerations

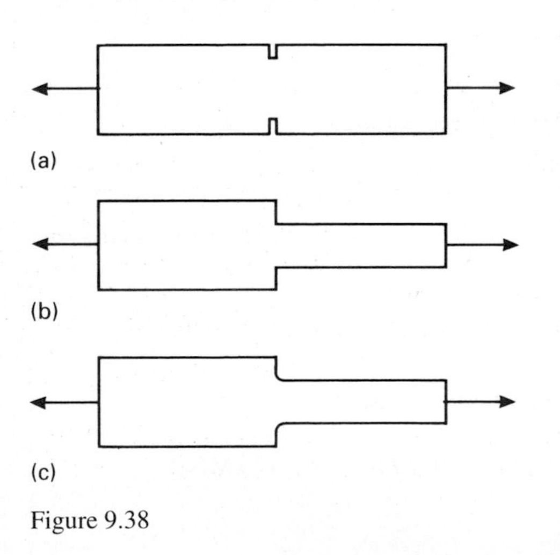

Figure 9.38

This chapter has introduced a number of equations and formulae that can be used for deciding the shapes and dimensions of structural components under simple systems of loading. Theoretical results exist for almost every type of loading that can occur and details will be found in the specialised books on the subject. In many cases the main difficulty is anticipating the mode of failure. Where is failure most likely to occur? Will it be by tension, compression, shear, bending, torsion or elastic instability? Will there be impact loads, stress concentration, fatigue or creep effects? If these questions can be answered correctly the component can be designed to prevent failure.

It must be remembered that all the formulae depend on assumptions, often quite drastic ones. In simple tension or compression, for instance, we assume that the load acts precisely along the axis of the bar; in practice it may be applied eccentrically. In addition, we take the force at each cross-section to be spread evenly over the whole area. This is reasonable for a uniform bar but a small notch (Fig. 9.38(a)) will lead to a variation in stress across the section with very high values at the bottom of the notch itself. This phenomenon is called *stress concentration*. It also occurs where there is a sudden change in cross-section (Fig. 9.38(b)) but its effects can be mitigated by the use of fillets (Fig. 9.38(c)).

The limitations of the bending and torsion equations must also be taken into account. In deriving the bending equation it was assumed that each cross-section of the beam is symmetrical about the plane in which the loads are applied. The torsion equation given in this chapter only applies to bars of circular section. Furthermore it has been assumed that the beams and shafts were initially straight.

A further complication arises when a component carries two or more types of loading simultaneously. A bar may be able to carry a certain axial load, a bending moment and a torque if they are applied one at a time, but it may fail when these loadings are applied simultaneously. Failure in such cases may occur on a plane inclined to the cross-section and this indicates that stresses can arise in directions that are different from those of the applied loads. This failure on an inclined plane can be observed by twisting a stick of blackboard chalk and examining the fracture.

Among other things, the designer has to specify the material from which each component is to be made. Many materials are not equally strong in tension and compression and the designer must therefore consider both tensile and compressive stresses. Concrete, for instance, has little strength in tension and slender steel rods offer little resistance to compression. On their own, neither is suitable for beams and cantilevers. In combination, however, they provide a very useful composite material – reinforced concrete. If the steel rods are embedded in a concrete beam near the tension face, the bending moments are resisted by tension in the steel and compression in the concrete. For sagging bending moments the reinforcing bars are needed near the bottom of the section but for hogging, as in the case of cantilevers, they should be near the top. The benefits of reinforced concrete can be enhanced by prestressing, in which the steel rods or wires are put under tension when the beam is cast. This has the effect of compressing the concrete before the external loads are applied; when the beam is in use, the concrete contributes to the tension through a reduction in compression.

A full treatment of all these topics will be found in more advanced books on stress analysis, solid mechanics and strength of materials.

The designer must consider other factors that affect the strength of components. If a bar is subjected to a load many times it may fail at a lower stress than if the load were applied only once, an effect known as *fatigue*. If it operates at a high temperature strain may become progressive at a constant load, a phenomenon called *creep*. Fatigue and creep testing are described in Chapter 5.

Bibliography

Benham P P and Crawford R J 1987 *Mechanics of Engineering Materials.* Longman

Case J, Chilver A H and Ross C T F 1999 *Strength of Materials and Structures* 4th edn. Arnold

Gordon J E 1991 *Structures: or Why Things Don't Fall Down.* Penguin

Iremonger M J 1982 *Basic Stress Analysis.* Butterworths

Jackson J H and Wirtz H G 1983 *Statics and Strength of Materials.* Schaum's Outline Series. McGraw-Hill

Megson T H G 1987 *Strength of Materials for Civil Engineers.* Arnold

Roark R J 1989 *Formulas for Stress and Strain* 6th edn. McGraw-Hill

Ryder G H 1969 *Strength of Materials* 3rd edn. Macmillan

Assignments

1 (a) Show, with the aid of sketches, the principles involved in creating stiffness in structures made from thin sheet material.

(b) Give three different examples of products in which rigid structures have been created from thin sheet material. For each explain how the required stiffness is achieved and how this relates to the application.

2 (a) Calculate the stress and extension for a vertical steel wire, 4 mm diameter and 6m long, when a mass of 200kg is suspended from it. For steel take $E = 207\text{GN/m}^2$.

(b) A square, rigid, horizontal plate hangs from four vertical steel rods each 8 mm diameter and 5 m long, one at each corner. The plate carries a load, symmetrically placed, of 20 kN (including its own weight). The rods are all rigidly fixed at their upper ends at exactly the same level. Calculate the stress and extension of each. Modulus of elasticity, 207 GN/m^2.

(c) If two of the rods in part (b), (diagonally opposite each other) are initially 1 mm shorter than the others calculate the stresses in each.
Hint First calculate the force required to stretch the shorter rods until they are the same length as the others. The remaining load is shared equally between all four rods.

3 A load of 2 kN is to be raised at the end of a wire. If the stress in the wire must not exceed 80 MN/m^2, what must be the minimum diameter?

4 The link shown in Fig. 9.39 is loaded in tension by a pin, 30 mm diameter, passing through the hole. Calculate the tensile stresses on the sections XX and YY and the shearing stress on the pin (assuming it is in double shear).

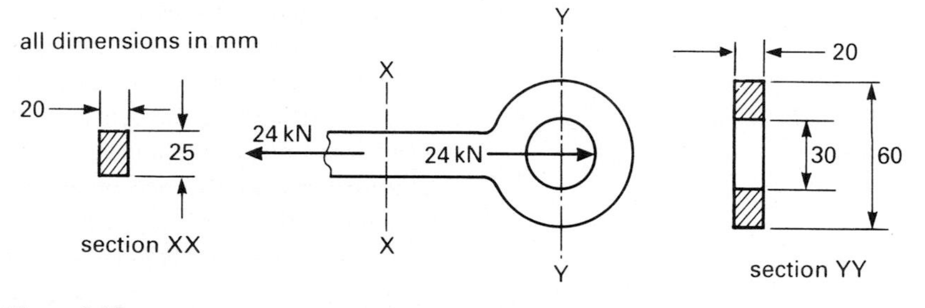

Figure 9.39

5 A brass tube, 500 mm long, has outside and inside diameters of 48 mm and 36 mm respectively. It is held between two stops exactly 500 mm apart. Find the stress in the tube if its temperature is raised 60°C and the distance between the stops is unchanged. Modulus of elasticity $E = 95$ GN/m^2; coefficient of linear expansion $\alpha = 19 \times 10^{-6}\,°C^{-1}$. What force does the tube exert on the stops?
What would be the answers if the distance between the stops increased by 0.5 mm?

6 (a) ABCDE is a simply supported beam 10 m long. AB = 2 m, BC = 4 m, CD = 3 m, and DE =1 m. Calculate the reactions for each of the following loading systems:
(i) The beam is supported at A and E, and carries concentrated loads of 30, 50 and 20 kN at B, C and D respectively.
(ii) The beam is supported at B and D and carries concentrated loads of 20, 80 and 35 kN at A, C and E respectively.
(iii) The beam is supported at A and D, there is a uniformly distributed load of 20 kN/m between B and D, and a concentrated load of 40 kN at E.

(b) Draw to scale the shearing force and bending moment diagrams for the loading systems of questions (i) to (iii).

7 A beam 12 m long is simply and symmetrically supported over a span of 8 m. It carries a uniformly distributed load throughout its length (including its own weight) of 20 kN per metre. Make well-proportioned sketches of the SF and BM diagram stating the principal values. Locate the points of contraflexure. How does the maximum BM compare with that for the same beam simply supported at its ends?

8 A steel beam spanning 8 m is simply supported at its ends and carries a central load of 130 kN. The permissible bending stress is 165 MN/m².

(a) Select a suitable section for the beam from the following table of I-sections (rolled steel joists or universal beams) neglecting the beam's own weight in the first instance.

(b) Calculate the weight of the beam you have chosen and check whether the permissible stress would be exceeded. If so, select another section to suit.

(c) An approximation to the shearing stress in a beam can be made by assuming that the whole of the shearing force at any section is carried by the web with a uniform shearing stress. Use this approximation to estimate the maximum shearing stress in the beam. Take the depth of the web as the same as the depth of the section.

Extract from *Universal Beam Tables*:

Size (min) depth × breadth	*Mass per metre (kg)*	*Section modulus Z for XX axis (cm³)*	*Web thickness (mm)*
686 × 254	152	4364	13.2
610 × 229	125	3217	11.9
533 × 210	101	2293	10.9
457 × 191	82	1610	9.9
406 × 152	67	1155	9.3
356 × 171	57	894	8.0
305 × 127	42	530	8.0
254 × 102	25	265	6.1

9 (a) The sections shown in Fig. 9.40 are to be used (upright as shown) for horizontal beams. Calculate for each the position of the centroid, the second moment of area about the neutral axis and the section moduli for the bottom and top edges.

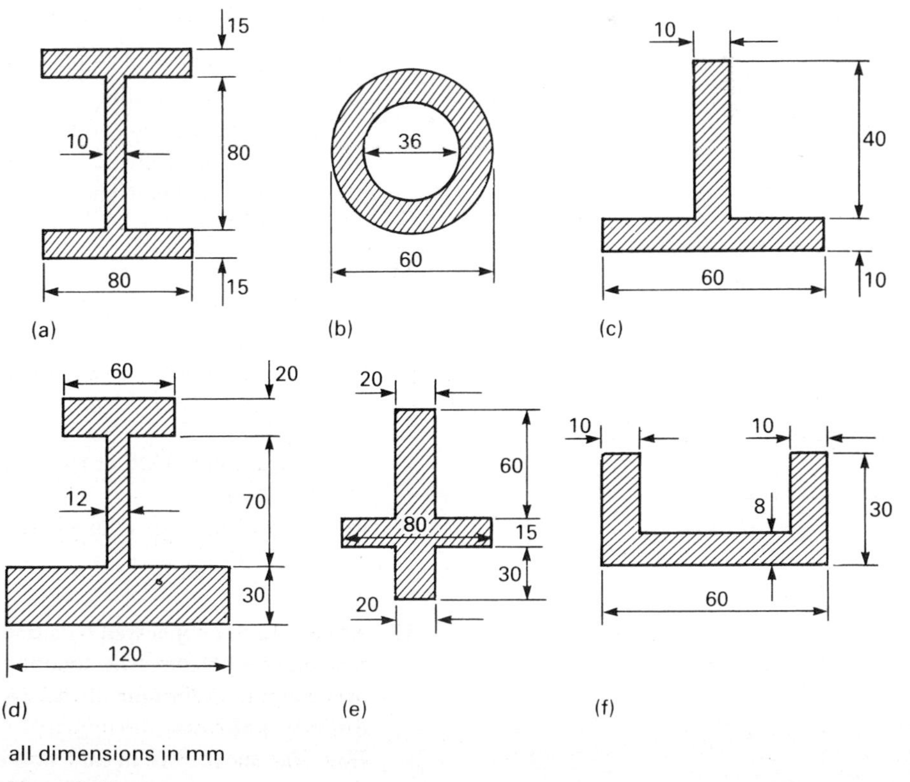

Figure 9.40

(b) In the following bending problems the cross-sections are those of part (a) (Fig. 9.40).
  (i) Calculate the moment of resistance for the section of Fig. 9.40(a) if the maximum bending stress is 120 MN/m^2. Check the result by the approximate method of Example 9.11, taking the flange stress as 120 MN/m^2.
  (ii) A cantilever 4m long has the section shown in Fig. 9.40(b). Find the maximum bending stress and maximum deflection due to a point load of 600 N at the free end. $E = 206$ GN/m^2.
  (iii) A cantilever 5 m long has the section shown in Fig. 9.40(c). Find the maximum stress and maximum deflection due to a uniformly distributed load of 60 N/m over its whole length. $E = 206$GN/m^2.
  (iv) A cast iron beam 2 m long, having the section shown in Fig. 9.40(d), is simply supported at its ends. What is the maximum central load it can carry without the tensile and compressive bending stresses exceeding 10 MN/m^2 and 20 MN/m^2 respectively? What are the actual stresses under these conditions?
  (v) A beam 4 m long and having the section shown in Fig. 9.40(e) is simply supported at its ends and carries a uniformly distributed load of 1.2 kN/m throughout. Find the maximum tensile and compressive bending stresses due to this load and the deflection at midspan. $E = 206$GN/m^2.
  (vi) Over what span can a beam, simply supported at its ends and having the section shown in Fig. 9.40(f), carry a central point load of 400 N if the bending stress is not to exceed 80 MN/m^2?

10 Determine the crippling load by the Euler formula for the following slender struts, using the equivalent pin-jointed length. Take the modulus of elasticity $E = 206$GN/m^2.
(a) A circular rod, 30 mm diameter and 1.25 m long, with hinged ends.
(b) A bar 0.75 m long and of rectangular cross-section, 75 mm by 25 mm, built-in at one end, and direction and position free at the other.
(c) A tube 3 m long with inner and outer diameters of 30 mm and 40 mm, built-in at one end and position fixed, direction free at the other.
(d) A structural steelwork member in the form of a universal beam, 10 m long and built-in at both ends. The relevant second moment of area is 38 700 cm^4.

11 Select diameters (to the nearest millimetre) of round steel bar for each of the following torsion conditions, taking the modulus of rigidity G as 80 GN/m^2.
(a) To transmit a torque of 1 kNm with a maximum shear stress of 30 MN/m^2.
(b) To transmit 5 kNm with a twist of 1° in a length of 6m.
(c) To twist through 5° in a length of 3 m with a maximum shear stress of 40 MN/m^2.
(d) To transmit 50 kW at 800 rev/min with a maximum shear stress of 25 MN/m^2.
(e) To twist through 4° in a length of 4 m when transmitting 60 kW at 30 rad/s.

12 Two shafts are connected by a flange coupling having six bolts on a pitch circle diameter of 300 mm. The coupling is required to transmit 800 kW at a speed of 500 rev/min. Determine the necessary bolt diameter if the shear stress is not to exceed 40 MN/m^2.
Hint The shear force in each bolt is equal to the torque divided by the radius of the bolt centres and the number of bolts.

13 A high tensile steel bolt, 8 mm diameter, is used to clamp together two large pieces of machinery, total thickness 62.5 mm. After the bolt is fitted so that the nut is touching the surface of the material the nut is tightened by turning through 40°. The thread pitch is 1.25 mm. The Young modulus of elasticity for the material is $200 \times 10^9\ \text{Nm}^{-2}$. Find the tensile force in the bolt.

14 The uniform steel bench, shown in Fig. 9.41, rests on a support at A and C. The mass of the bench top is 120 kg and that of a machine tool fixed at B is 80 kg.

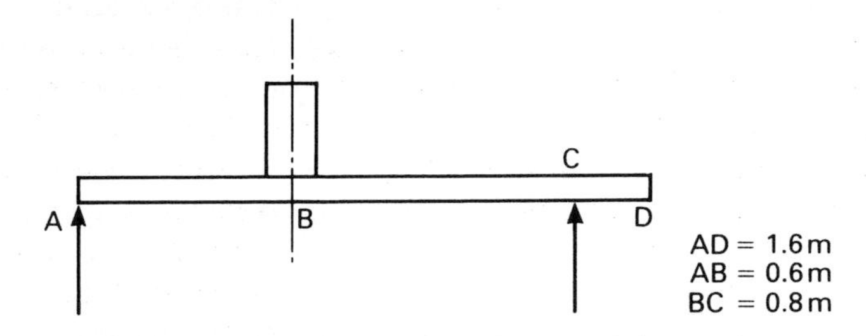

Figure 9.41

Sketch a shear force diagram for the bench top giving the important values of shear force. Indicate the point at which the bench experiences the maximum shear force.

15 A steel tube 60 mm diameter and 50 mm inside diameter is embedded vertically in concrete and used as a clothes line post. Two lines are attached to the post and can be assumed to exert horizontal tensions in the directions shown in Fig. 9.42.

Determine the common level h above the ground at which the two lines can be attached to the post so that the maximum stress due to the bending does not exceed 140 N/mm² for the tensions given.

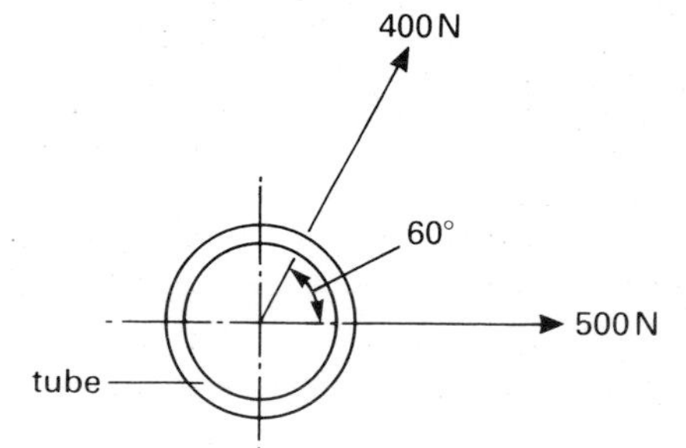

Figure 9.42

16 Overhead electric power transmission cables consist of aluminium and steel wires twisted together as a hawser.

In such a cable the total cross-sectional area of the aluminium wire is 600 mm² and it is subject to a tensile load of 25 kN.
[Take the Young modulus for steel to be 200GN/m² and that for aluminium to be 70GN/m².]

(a) If the stress in the aluminium under the loading must not exceed 30 N/mm², calculate:
  (i) the total cross-sectional area of the steel wires;
  (ii) the stress in the steel wires;
  (iii) the extension of the cable over a length of 105 m.

(b) Explain the purpose of using such a compound cable for overhead transmission.

17 In Fig. 9.43 AB is a simply supported concrete floor joist which has to take the load of vertical walls at B and D. Neglecting the weight of the joist, calculate the size and position of the maximum shear force in the joist.

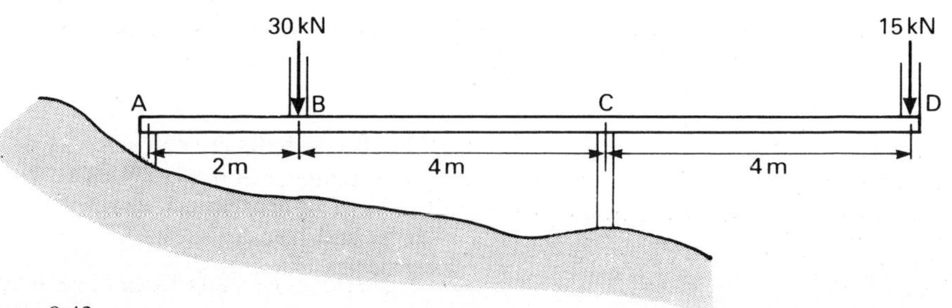

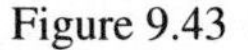

Figure 9.43

18 (a) Explain the significance of *factor of safety* in engineering design.

(b) Fig. 9.44 shows a connecting rod subject to a maximum tensile force of 150 kN. The ultimate tensile strength of the material of the rod is 70×10^6 N/m^2 and the ultimate shear stress for the material of the pin is 58×10^6 N/m^2.

Assuming a factor of safety of 5, calculate the minimum diameters of:

(i) the rod;

(ii) the pin.

(c) The pin of the coupling is made from 1.6% C steel. Describe how the correct heat treatment of quenching and tempering will alter the structure and physical properties of the steel.

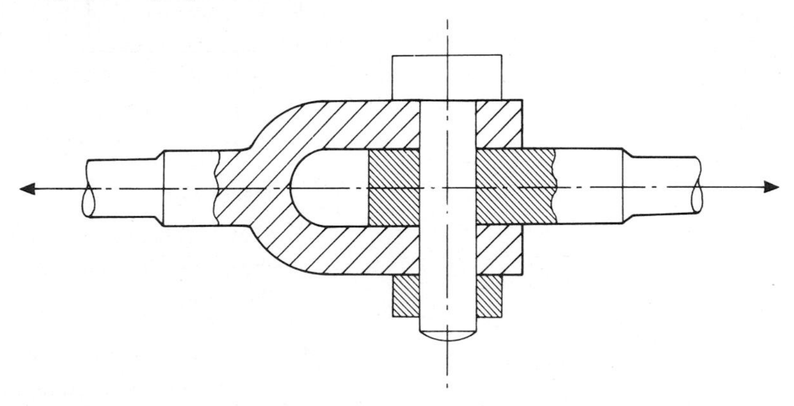

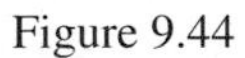

Figure 9.44

SECTION 4 Mechanical systems in motion

Consider the conversion of energy when riding a bicycle. If we were to list some of the possible forms of energy and work needed during a typical bicycle journey we might come up with:

- human energy (calories) is converted to pedal rotations
- rotary motion of the pedals is converted by the gear mechanism to rotary motion of the drive wheel
- rotary motion of the drive wheel is converted to forward velocity of the bike
- work is done to produce this forward motion against the road resistances
- work is done to overcome the air drag and any incline of the road.

If we considered a cycle which has a 2-stroke petrol engine mounted on it to assist the cyclist we would also have to add:

- the use of chemical energy to power the bicycle and the thermodynamics of the engine.

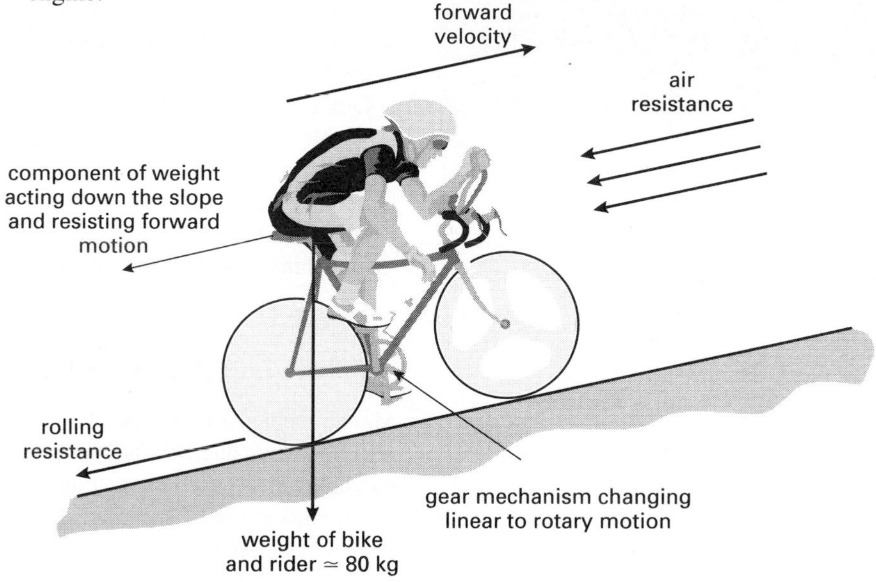

Section 4 introduces all the concepts connected with energy, work, power and thermodynamics.

- Chapter 10, 'Mechanisms and motion', introduces us to levers, mechanisms and machines, and gearing systems as well as linear and rotary motion.
- Chapter 11, 'Energy and dynamics', explains work, power, and energy, for both linear and rotary motion. It introduces us to different sources of energy and to Newton's laws of motion.
- Chapter 12 introduces thermodynamics and its application to the internal combustion engine, refrigerators and heat pumps.
- Chapter 13, 'Fluid dynamics', explains the concept of fluid pressure, buoyancy and fluid flow. It looks at water turbines, model testing, lift and drag forces and aerofoils.

The computer software associated with this section, which is listed below, is available for you to download from Longman's website http://www.longman.co.uk

Chapter 10: Computer program for piston velocity and acceleration

10 Mechanisms and motion

10.1 Machines and mechanisms

Figure 10.1 A food mixer.

A *machine* is a combination of moving (and fixed) mechanical components that transmits or modifies the action of a force or torque to do useful work. Machines such as the food mixer, Fig. 10.1, the washing machine, food processor, electric sander or drill are familiar to us all. An *engine* is a machine that converts a natural energy source to the output power that is required for a particular application.

Whereas a machine is primarily concerned with the transmission of forces, a *mechanism* is principally concerned with movement. A mechanism is a system that transforms one kind of motion to another. It may consist of a single component or a combination of the fundamental components – the lever, cam, screw, pulley, gear and ratchet. The sewing machine, mechanical clock and vehicle linkages are everyday examples of mechanisms.

Other common examples of mechanisms are to be found in children's toys, especially where the rotation of the wheels causes other parts of the toy to move. Mechanisms can be identified in all machines; the piston-crank mechanism in the reciprocating engine, and the quick return mechanism of a shaping machine, for example. It is the function of these mechanisms to provide the correct positioning of the machine's components at each stage in the sequence of its operation.

10.1.1 Degrees of freedom

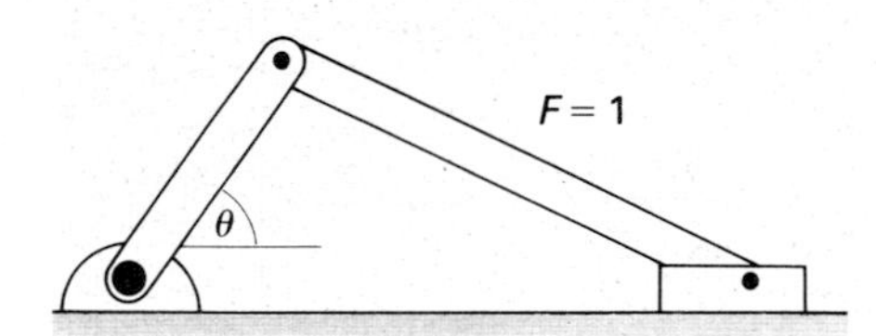

Figure 10.2 A crank and slider mechanism

Many mechanisms are complex combinations of levers, gears and linkages moving relative to one another and these are often difficult to analyse. The analysis of mechanisms is based on the concept of *degrees of freedom*.

All motion can be regarded as a combination of translation and rotation. By considering the three dimensions of space, a translation can be defined in terms of a movement along one or more of the three axes. The rotation can also be defined as a rotation about one or more of the axes. Thus a maximum of six separate types of movement can be identified at any instant in time. These possible movements are the degrees of freedom of a body and a mechanism can have a maximum of six degrees of freedom.

Fig. 10.2 shows a representation of a crank-slider mechanism in which the slider is constrained to move along one path only and therefore its number of degrees of freedom, F, is said to be 1. When the angle, θ, is known the position of every part of the mechanism is defined. When $F = 1$, the output motion is predictable from a known input and the linkage is a mechanism.

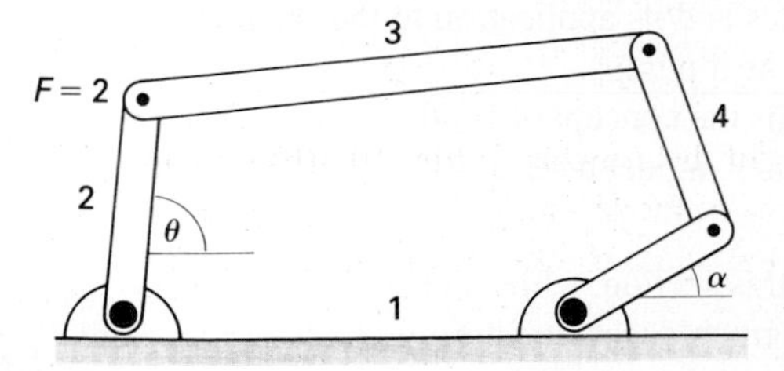

Figure 10.3 A five-bar chain linkage

The five-bar chain shown in Fig. 10.3, where one of the linkages is a fixed surface, has two degrees of freedom and two independent variables such as θ and α must be established to identify the position of every other part of the linkage. Its motion cannot therefore be predicted and it is not a mechanism. With four links there is one degree of freedom and this constitutes a mechanism; with only three, the triangle, there is no possible relative movement and we have a structure (see Chapter 8, section 8.5 and Fig. 8.18).

10.2 Levers and linkages

10.2.1 Simple lever

Fig. 10.4(a) shows a crowbar being used to force a bicycle tyre free of the rim. This can be considered to be a rigid *lever* pivoted at the fulcrum. The effort applied at one end lifts a load placed at the other, as shown in Fig. 10.4(b). The crowbar and also the scissors in Fig. 10.5 are examples of *class 1 levers* in which the pivot or *fulcrum* is between the load and the effort. The lever will magnify a force if the load point is closer to the fulcrum than the effort. When using scissors it is noticeable that the greatest cutting force is obtained when the material to be cut is placed as near to the pivot as possible.

(a)

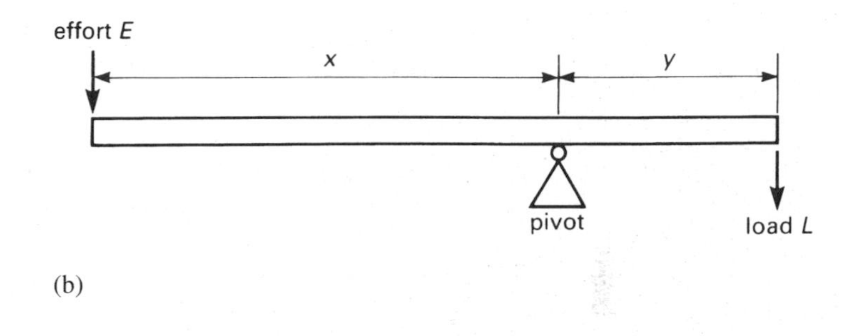

(b)

Figure 10.4 The class 1 lever
(a) a crowbar being used to remove a bicycle tyre.
(b) the forces in a class 1 lever.

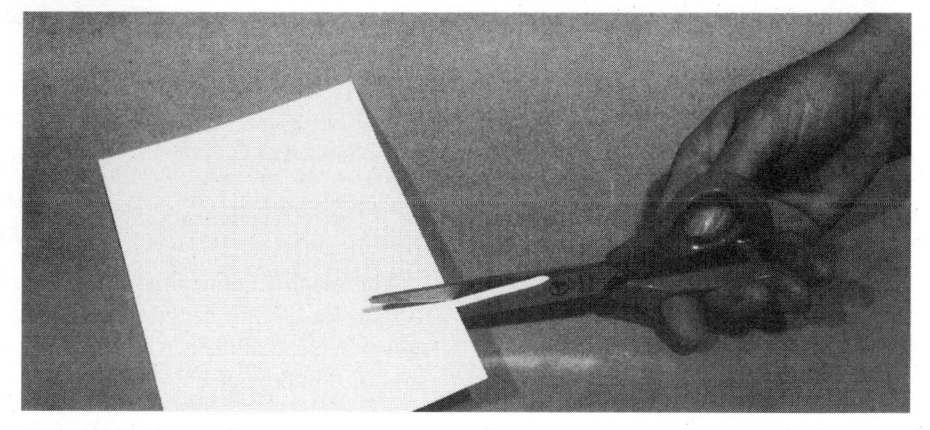

Figure 10.5 Scissors, a double lever mechanism.

The *mechanical advantage* (MA) is an expression of the amplification of the effort and is defined as:

$$\text{mechanical advantage} = \frac{\text{load}}{\text{effort}} \quad \text{or, in symbols, MA} = \frac{L}{E} \qquad (1)$$

By taking moments about the fulcrum of the crowbar in Fig. 10.4(b) we can express the mechanical advantage in terms of the distances x and y. For equilibrium, the anticlockwise moment due to E is equal to the clockwise moment of L. Thus:

$$Ex = Ly \quad \text{and} \quad \text{MA} = \frac{L}{E} = \frac{x}{y}$$

It follows that the greater the ratio of x to y, the greater the mechanical advantage. Note that MA may be greater or less than 1, depending on the values of x and y.

This analysis ignores the frictional forces and, if these are allowed for, the value of MA is reduced. In practice the mechanical advantage of a machine should be obtained experimentally by measuring test values of effort and load. Alternatively, if the efficiency of the machine is known or can be estimated, the mechanical advantage can be determined.

If MA is greater than 1, the lever magnifies the force. However, the work input must be equivalent to the work output and therefore (since work = force × distance) the effort must move a greater distance than the load. The *velocity ratio* of a machine (VR) is the ratio of distance moved by the effort to the distance moved by the load and can be obtained either by calculation or graphical means. Thus:

$$\text{velocity ratio} = \frac{\text{distance moved by effort}}{\text{distance moved by load}} \quad \text{or} \quad \text{VR} = \frac{d_E}{d_L} \qquad (2)$$

where d_E and d_L are the distances moved by the effort and load.

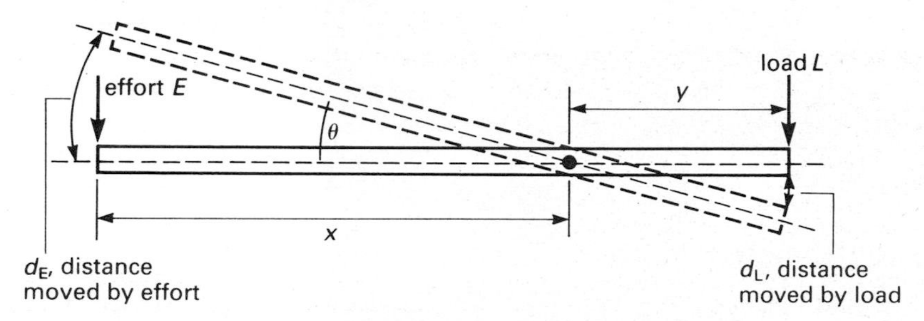

Figure 10.6 Rotation of a crowbar and its effect on the loads.

Fig. 10.6 shows the crowbar after it has moved through an angle θ. The distance moved by an effort that always acts at right angles to the bar is given by the arc length $x\theta$ and the distance moved by the vertical load is given by the vertical distance $y \tan \theta$. For a small angle θ (in radians) the distance moved by the load can be approximated to $y\theta$.

$$\text{Hence for small angles} \quad \text{VR} = \frac{x\theta}{y\theta} = \frac{x}{y}$$

This amplification of movement that occurs in the lever is often utilised in design, as in the case of the car park barrier shown in Fig. 10.7. The control mechanism moves one end of the barrier through a small distance, which is then greatly amplified to provide enough room for a car to pass through.

(a)

(b)

Figure 10.7 Amplification of movement in a car park barrier.

The *mechanical efficiency* η is the ratio of the work output to the work input. Thus:

$$\text{mechanical efficiency } \eta = \frac{\text{work out}}{\text{work in}} = \frac{Ld_L}{Ed_E}$$

Substituting from equations (1) and (2) yields the efficiency in terms of the mechanical advantage and velocity ratio. In symbols,

$$\eta = \frac{\text{MA}}{\text{VR}} \quad \text{or, as a percentage,} \quad \eta = \frac{\text{MA}}{\text{VR}} \times 100\%$$

When a load is raised by a lifting machine, it will remain securely in its new position if the efficiency of the machine is less than 50%. However, if the efficiency is greater than 50% the potential energy gained by the load, when it is raised, is greater than the work needed to overcome the frictional forces and the machine will run back to its starting position. This is called *overhauling* and obviously must be prevented for the safe operation of the machine. A ratchet mechanism or brake is usually used for this purpose.

The nutcrackers shown in Fig. 10.8 are an example of a *class 2 lever*. The load and effort are both on the same side of the fulcrum with the effort further away from the pivot to produce a mechanical advantage greater than 1.

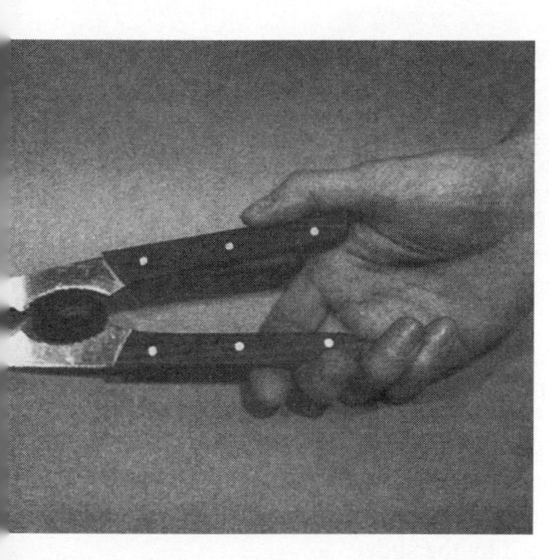

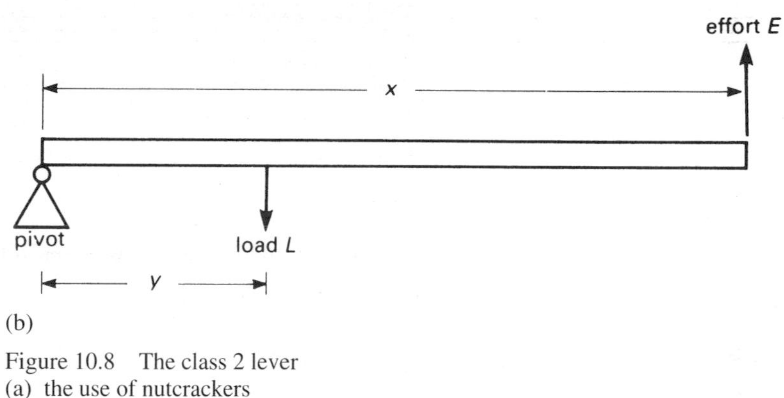

(b)

Figure 10.8 The class 2 lever
(a) the use of nutcrackers
(b) the forces in a class 2 lever.

The tweezers shown in Fig. 10.9 are an illustration of a *class 3 lever*. Both the load and effort are on the same side but the load is further from the fulcrum, giving a mechanical advantage less than l. Tweezers are therefore used for delicate work where the output force is less than the applied effort.

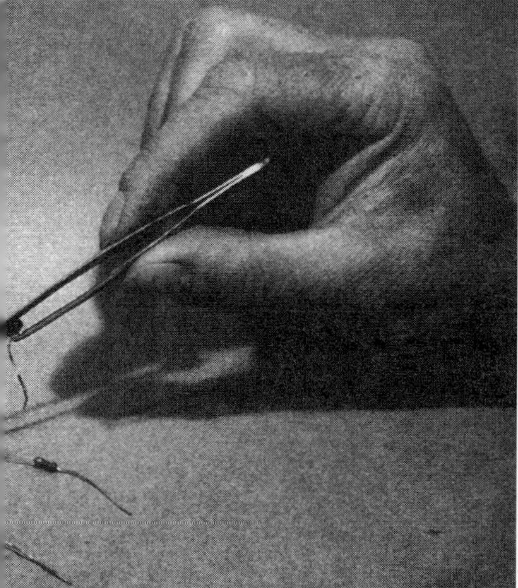

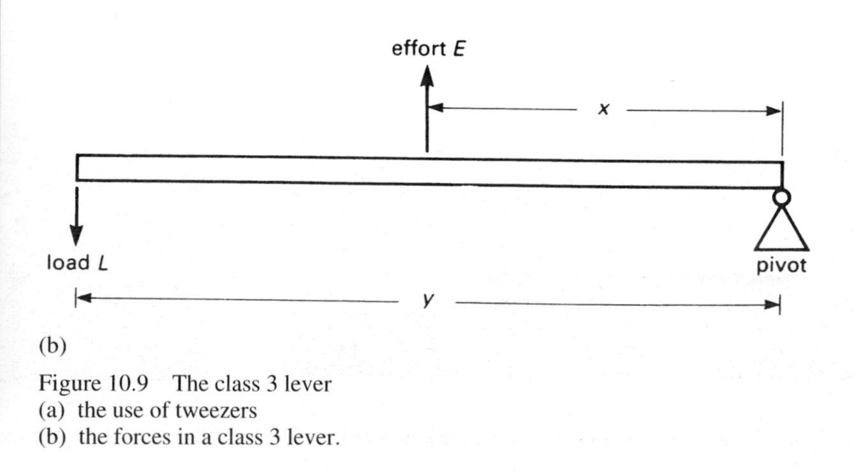

(b)

Figure 10.9 The class 3 lever
(a) the use of tweezers
(b) the forces in a class 3 lever.

10.2.2 The bell crank lever

The *bell crank lever* is an angled class 1 lever for non-parallel forces as shown in Fig. 10.10. Its name dates from the time when it was used to ring a bell to summon the servants. Bell crank levers are also found in mechanically operated railway signalling and weighing machines. A *double bell crank* illustrated in Fig. 10.11 can provide two output signals from the one input signal.

Figure 10.10 The bell crank lever used in a lorry accelerator linkage.

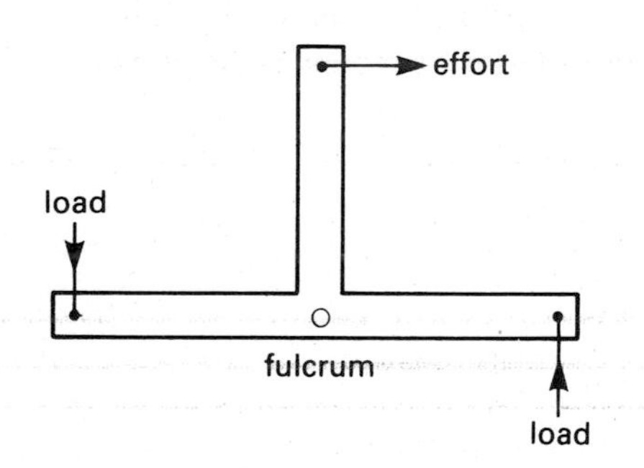

Figure 10.11 The double bell crank lever.

10.2.3 The toggle mechanism

Toggle mechanisms are used to produce very large clamping forces. The mechanism consists of two links pivoted together at one end. One end is then pivoted to a fixed surface while the other is free to move in a straight line. When an effort is applied to the central pivot a large force can be effected by the slider.

The forces acting on a toggle clamp are represented in Fig. 10.12 and, by considering the equilibrium of these forces, an expression for the output clamping force F can be obtained. The sum of the vertical forces is equal to zero. Therefore:

$$2N = E \quad \text{or} \quad N = \frac{E}{2}$$

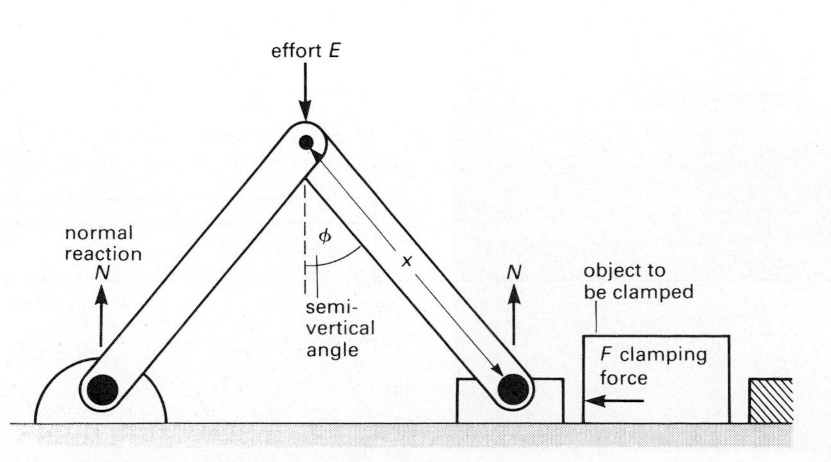

Figure 10.12 The forces acting in a toggle mechanism.

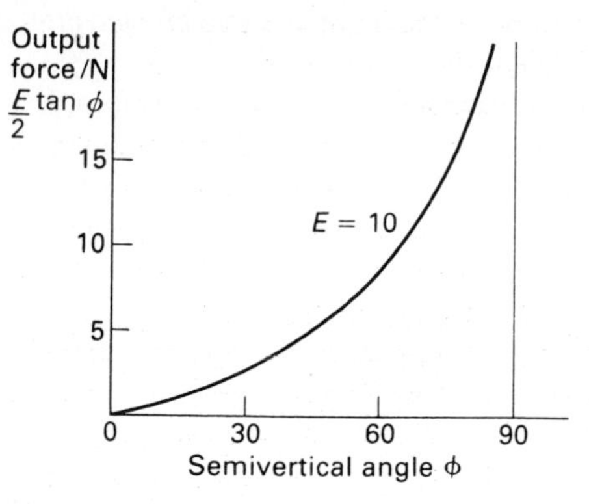

Figure 10.13 The theoretical output of a toggle mechanism.

Figure 10.14 The toggle used to fix a folding pushchair.

The perpendicular distances from the effort pivot to the lines of action of the clamping force F and right-hand vertical reaction N are $x \cos \phi$ and $x \sin \phi$. Therefore, taking moments about the effort pivot for the right-hand link

$$Fx \cos \phi = Nx \sin \phi$$

Substituting for $N = E/2$ and dividing through by $x \cos \phi$, we have:

$$F = \frac{E}{2} \tan \phi$$

Therefore the clamping force increases proportionally to the tangent of the semivertical angle. This thoretical variation of the clamping force with ϕ for an effort of 10 N is shown in Fig. 10.13.

Since $\tan 90° = \infty$, it would appear that the clamping force is infinite when the two bars are in a straight line and $\phi = 90°$. It is true that the greatest clamping force is achieved in this position but, in practice, frictional forces and the self-weight of the toggle, which were not considered in this theory, would be present and would reduce the clamping effect.

Most parents will have used toggle mechanisms to erect folding pushchairs or secure pram hoods, Fig. 10.14, and, in a workshop, toggle clamps are extremely useful for securing work that is being machined, pressed, formed or drilled.

10.2.4 Linkages

Linkages are often used to change the direction of motion of a component. The simplest of these is the *reverse motion linkage* in which the input direction of motion is reversed. The output movement can either be equal to the input as in Fig. 10.15(a) or amplified by providing an off-centre pivot as shown in Fig. 10.15(b), in which the output movement is smaller than the input. The input and output force will be affected as well as the movement, but in a mechanism the motion is the prime concern.

Fig. 10.16 shows a *rotary linkage*. This type produces a reverse motion rotary output.

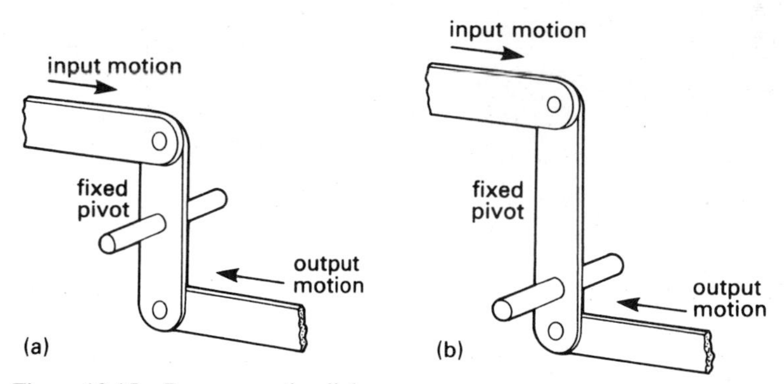

Figure 10.15 Reverse motion linkages
(a) centre fixed pivot (b) off-centre fixed pivot.

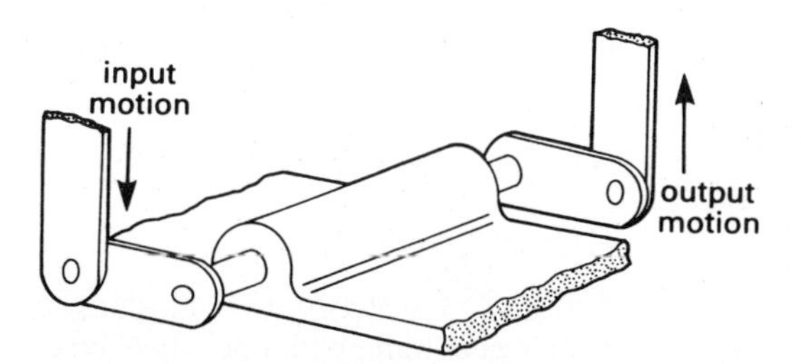

Figure 10.16 A rotary linkage.

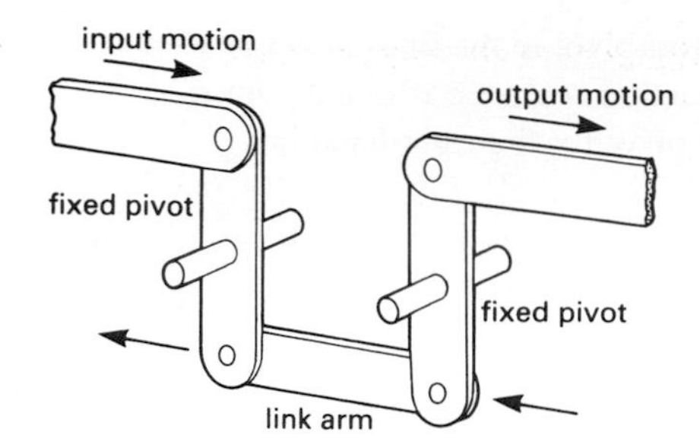

Figure 10.17 A push-pull linkage.

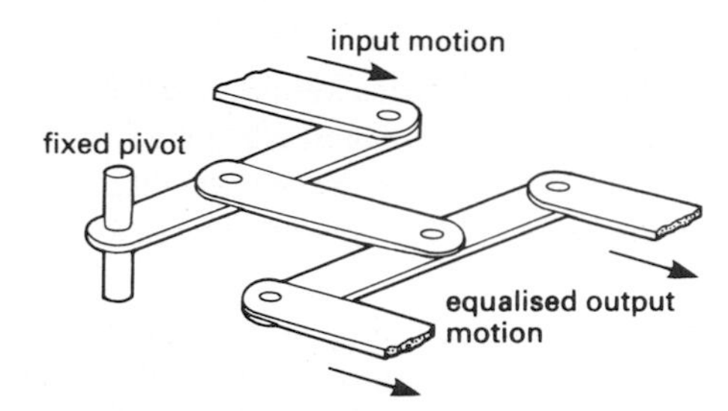

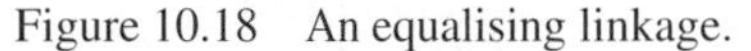

Figure 10.18 An equalising linkage.

To obtain a linear output in the same direction as the input an extra pivoted arm and linkage are required. Fig. 10.17 shows this *push-pull linkage*.

When it is necessary to transmit the input movement to two output positions, a more complex linkage is required. If the two outputs are to be equal the pivot arm is placed connecting the centre of the input and output bars and an *equalising linkage* is created, as shown in Fig. 10.18. Outputs of differing proportions can be obtained by moving the position of the pivot arm.

A parallel rule, used by draughtsmen and navigators, is a good example of a *parallel motion linkage*. Opposite sides of this linkage will always remain parallel but can be placed at different distances from each other. Consequently it is a good mechanism for keeping objects horizontal as in the trays of a tool box, Fig. 10.19.

Figure 10.19 Parallel hinges on a tool box.

Figure 10.20 The compound gear train on a lathe control mechanism showing the spur gears.

10.3 Rotary motion

The *gear wheel* is the basic mechanism which transfers and transforms rotational motion and torque. Gears can often be seen on the side of lathes and are used to control the lathe speed. A car gearbox enables the driver to meet the speed and torque requirements of the terrain with the available engine power. Large gears are used in the mechanisms that operate cranes, winches and lock gates. The *spur gear* is the most common type of gear and an example can be seen in the lathe mechanism shown in Fig. 10.20.

Where a smooth, quiet transmission at high rotational speeds is required *helical gears,* which have each tooth inclined at a slight angle to the axis of rotation and forming part of a helix, are used. The gears in a car gearbox, shown in Fig. 10.21, are usually helical since quiet, efficient transmission is required in this application. They are also well lubricated to reduce friction and increase mechanical efficiency.

Figure 10.21 Helical gears on a car gear box mainshaft.

10.3.1 Gear ratio

Simple gear train

Fig. 10.22 shows two meshed spur gears of different sizes. This is an example of a simple gear train, with one wheel on each shaft. The smaller, 9-toothed *pinion* will have to perform two revolutions for each revolution of the larger, 18-toothed wheel

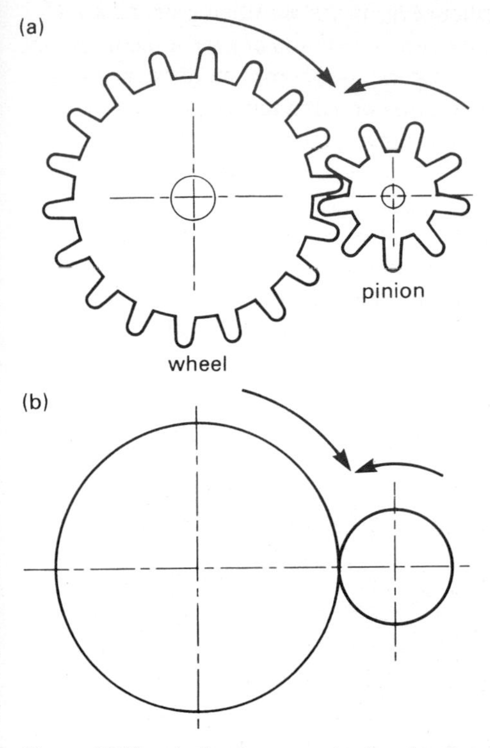

Figure 10.22 A simple gear train showing the direction of rotation of the meshed spur gears.

and so when the wheel is used as the *driver* (input) gear the output motion will be faster than the input. In this case:

$$\text{VR} = \frac{\text{angular movement of wheel (effort)}}{\text{angular movement of pinion (load)}} = \frac{1}{2}$$

or

$$\text{VR} = \frac{9}{18} = \frac{\text{number of driven teeth}}{\text{number of driver teeth}}$$

The velocity ratio for a gear system is usually referred to as the gear ratio and the above value would be quoted as a gear ratio of 1: 2. It equals the ratio of the speeds of the driving and driven gears. In other books you may find velocity or gear ratio defined as the ratio of the driven gear speed to that of the driver. To avoid confusion the gear ratio should be clearly specified.

EXAMPLE 10.1 Find the gear ratio of a simple gear train where the drive gear has 15 teeth and the driven gear has 45 teeth.

SOLUTION

$$\text{The gear ratio} = \frac{\text{number of driven teeth}}{\text{number of driver teeth}} = \frac{45}{15} = \frac{3}{1} \quad \text{or} \quad 3:1$$

Compound gear train

In a compound train at least one shaft carries a compound gear, that is two wheels which rotate at the same speed, Fig. 10.23. A practical example is the gear train shown in the lathe mechanism of Fig. 10.20. The advantage of a compound gear train is that it can produce a high gear ratio without the disproportionate gear sizes that would be necessary in a simple gear train.

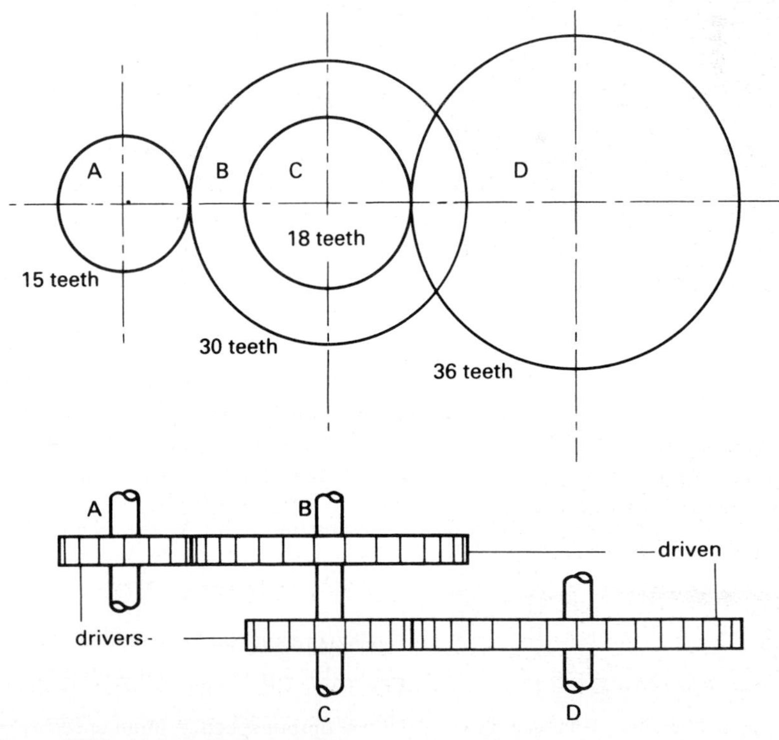

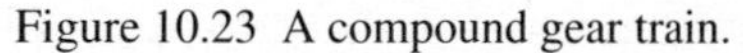
Figure 10.23 A compound gear train.

Now consider that, in the compound train of Fig. 10.23, the drive gear, A, has 15 teeth and an angular speed of 240 rev/min. For one revolution of gear A the meshing gear B (30 teeth) will rotate half a revolution. Gear C is mounted on the same shaft as gear B and will therefore also rotate half a turn. For half a turn of gear C ($\frac{1}{2} \times 18 = 9$ teeth), gear D (36 teeth) will rotate through one quarter of a revolution. We can write:

$$\text{gear ratio} = \frac{\text{distance moved by input}}{\text{distance moved by output}} = \frac{1}{1/4} = \frac{4}{1} \quad \text{or} \quad 4:1$$

Generally, for a compound gear train, in which A and C are the driver gears, and B and D are the driven gears (as in Fig. 10.23),

$$\text{gear ratio} = \frac{\text{product of numbers of teeth in driven gears}}{\text{product of numbers of teeth in driver gears}}$$

Hence, with the figures for the gear train of Fig. 10.23

$$\text{gear ratio} = \frac{30 \times 36}{15 \times 18} = \frac{4}{1} \quad \text{(as before)}$$

10.3.2 Gear wheel speed

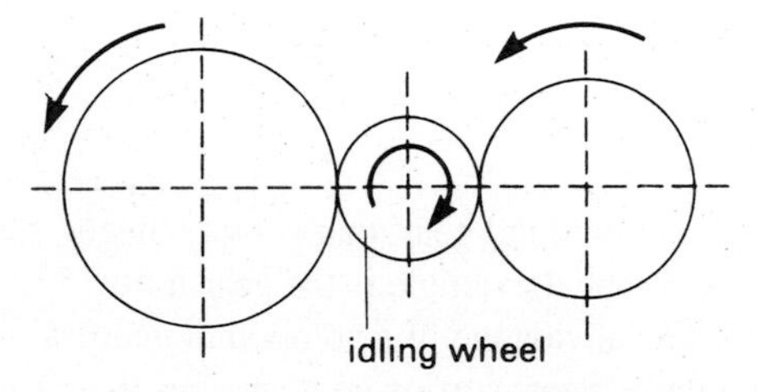

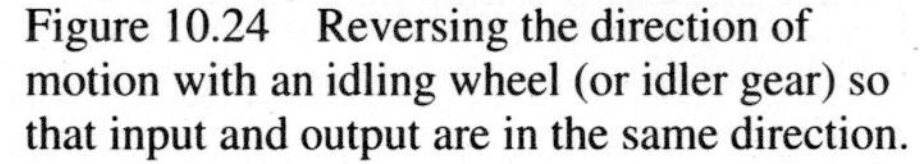
Figure 10.24 Reversing the direction of motion with an idling wheel (or idler gear) so that input and output are in the same direction.

If the pinion in Fig. 10.22 is used as the drive wheel and the larger wheel as the driven gear, the output shaft speed will now be slower than the input. If an intermediate gear of any size is placed between the wheel and pinion, as in Fig. 10.24, it will cause the input and output shafts to rotate in the same direction with no effect on the speed of the driven gear. It is called an *idler gear* or *idling wheel*.

If we look again at Fig. 10.23, wheel B will obviously take twice as long as A to turn through a full revolution and therefore the speed of B must be half the speed of A, that is, 120 rev/min (r.p.m.). To find a relationship between the angular speed of a gear and its number of teeth we must use the fact that the rate at which the teeth pass the point of contact must be the same for each gear. The number of teeth per minute passing any point is given by the product of its rotational speed, N in rev/min, and the number of teeth, t. So for a simple train consisting of two gear wheels A and B:

$$N_B t_B = N_A t_A \quad \text{or} \quad N_B = N_A \times \frac{t_A}{t_B}$$

Hence,

$$\text{speed of driven gear} = \text{speed of driver} \times \frac{\text{number of driver teeth}}{\text{number of driven teeth}}$$

Applying this result to gears C and D.

$$N_D = N_C \times \frac{t_C}{t_D}$$

For the compound gear train in Fig. 10.23 we also know that gears B and C rotate at the same speed. Hence $N_B = N_C$ and the last result becomes:

$$N_D = N_B \times \frac{t_C}{t_D}$$

Substituting for N_B the relationship derived above for gears A and B:

$$N_D = N_A \times \frac{t_A}{t_B} \times \frac{t_C}{t_D}$$

This result can be extended for any number of compound gears and the general result for any compound train may be written:

$$\text{output speed} = \text{input speed} \times \frac{\text{product of numbers of driver teeth}}{\text{product of numbers of driven teeth}}$$

EXAMPLE 10.2 Find the angular speed of gear D of the compound gear train in Fig. 10.23.

SOLUTION

$$\text{output speed} = \text{input speed} \times \frac{\text{product of driver teeth}}{\text{product of driven teeth}}$$

Hence,

$$N_D = 240 \times \frac{15}{30} \times \frac{18}{36} = 60 \text{ rev/min}$$

The speed of gear D is 60 rev/min.

10.3.3 Power and torque

It is shown in Chapter 11 (section 11.7) that the power transmitted when a torque T Nm is acting on a body rotating at ω rad/s is $T\omega$ in watts. For a gear train, the power input is equal to the product of the driving torque and the angular speed of the drive shaft. Assuming no power losses, this available power will be the same for each gear in the train. Therefore torque and angular speed are inversely related for any gear and, if the speed is increased, the available torque will decrease. Consequently when we are accelerating from rest or moving up steep hills on a bicycle or in a car we must use the slow, start-up gears in order to achieve a high torque.

Suppose for a simple gear train with driving gear A and driven gear B, T_A and T_B are the torques, and ω_A and ω_B are the angular speeds in radian per second. Then, since the power is the same for A and B:

$$T_B \times \omega_B = T_A \times \omega_A \quad \text{or} \quad T_B = T_A \times \frac{\omega_A}{\omega_B}$$

In words,

$$\text{output torque} = \text{input torque} \times \frac{\text{angular speed of driver gears}}{\text{angular speed of driven gears}}$$

Also, as the number of teeth of a gear is inversely related to its angular speed then, for a constant input, the more teeth there are in the driven gear the larger the output torque will be.

Since $\omega_B t_B = \omega_A t_A$ the ratio of the speeds ω_A/ω_B equals t_B/t_A and the torque on gear B becomes:

$$T_B = T_A \times \frac{t_B}{t_A}$$

Therefore

$$\text{output torque} = \text{input torque} \times \frac{\text{number of driven teeth}}{\text{number of driver teeth}}$$

and, for a compound gear train

$$\text{output torque} = \text{input torque} \times \frac{\text{product of numbers of driven teeth}}{\text{product of numbers of driver teeth}}$$

EXAMPLE 10.3 Find the power supplied by the motor to the 15-teeth drive gear A to rotate it at an angular speed of 240 rev/min with a torque of 300 Nm. Also, if gear A is simply meshed with a second gear B, which has 30 teeth, find the torque on gear B.

SOLUTION

$T_A = 300$ N m

input speed $= 240 \times 2\pi/60$ rad/s

and

power $= T\omega$

input power $= T_A\omega_A = 300 \times 240 \times 2\pi/60 = 7540$ W $= 7.54$ kW

torque on gear B, $T_B = T_A \times \dfrac{t_B}{t_A} = 300 \times \dfrac{30}{15} = 600$ N m

EXAMPLE 10.4 Find the output torque on gear D of the compound gear train in Fig. 10.23 if there is an input torque of 300 Nm on gear A.

SOLUTION

$$\text{output torque} = \text{input torque} \times \frac{\text{product of driven teeth}}{\text{prodict of driver teeth}}$$

Therefore

$$\text{the output torque} = 300 \times \frac{30 \times 36}{15 \times 18} = 1200 \text{ N m}$$

10.3.4 Power transmitted by rotating shaft

The power P transmitted by a rotating shaft is given by:

$$P = T\omega$$

where T is the torque or twisting moment in newton metre and ω is the angular speed in rad/s. If the speed is given in rev/min (r.p.m.) it must be multiplied by $2\pi/60$ to obtain ω.

EXAMPLE 10.5 What is the minimum diameter of a round steel bar if it is to transmit a torque of 1.2 kN m with a maximum shear stress of 50 MN/m^2?

SOLUTION From the torsion equation given in section 9.10 $T/J = \tau/r$ we have:

$$\frac{J}{r} = \frac{T}{\tau}$$

If d is the required diameter then the polar second moment of area $J = \pi d^4/32$ and the radius to the outer surface is $r = d/2$. Substituting these expressions,

$$\frac{(\pi d^4/32)}{(d/2)} = \frac{T}{\tau}$$

and, working in newton and metre units,

$$\frac{\pi d^3}{16} = \frac{T}{\tau} = \frac{1.2 \times 10^3}{50 \times 10^6}$$

from which

$$d^3 = \frac{1.2 \times 10^3 \times 16}{50 \times 10^6 \times \pi} = 122.2 \times 10^{-6}$$

and

$$d = \sqrt[3]{(122.1 \times 10^{-6})} = 0.0496 \quad \text{or} \quad 49.6 \text{ mm}$$

EXAMPLE 10.6 What power can be transmitted by a hollow shaft, 100 mm external and 75 mm internal diameters, at a speed of 800 rev/min if the shear stress is limited to 25 MN/m^2?

SOLUTION The value of J for a hollow section is the difference between the values based on the outer and inner diameters. Working in millimetres,

$$J = \frac{\pi}{32} \times 100^4 - \frac{\pi}{32} \times 75^4 = 6.711 \times 10^6 \text{ mm}^4$$

The maximum shear stress occurs at the outer surface for which $r = 50$ mm. Working in newton and metre units and using the relationship $T/J = \tau/r$, the torque that can be transmitted is:

$$T = \frac{\tau J}{r} = \frac{(25 \times 10^6) \times (6.711 \times 10^6 \times 10^{-12})}{(50 \times 10^{-3})}$$

$$= 3356 \text{ Nm}$$

The rotational speed is 800 rev/min and, converting this to rad/s

$$\omega = \frac{2\pi \times 800}{60} = 83.78 \text{ rad/s}$$

The power transmitted is therefore:

$$P = T\omega = 3356 \times 83.78 = 281.2 \times 10^3 \text{ W} \quad \text{or} \quad 281.2 \text{ kW}$$

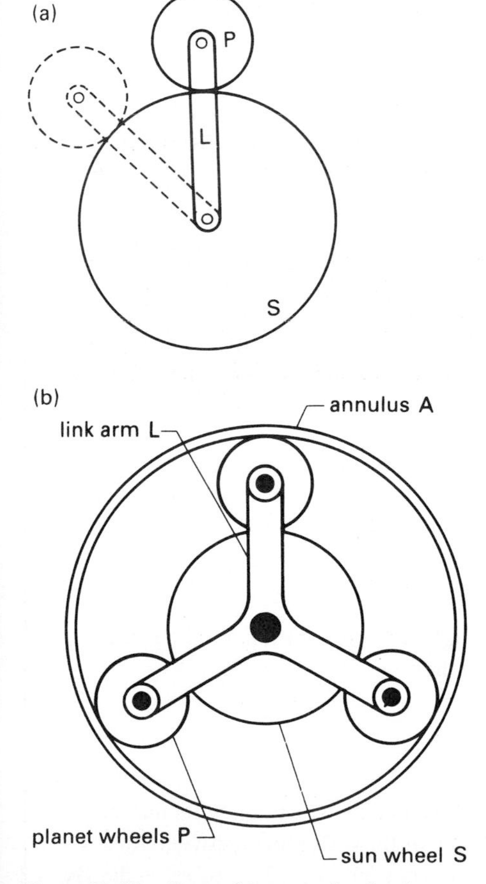

Figure 10.25 General layout of a simple epicyclic gear train
(a) a 1-planet epicyclic gear
(b) a 3-planet epicyclic gear.

10.3.5 Epicyclic gear train

So far, we have looked at simple and compound gear trains in which each gear turns about a fixed axis. If however, the axis of one gear can turn about the axis of another the train is described as *epicyclic*. Fig. 10.25(a) shows two meshing gears, P and S, whose centres are linked by an arm L. At each end of the arm, the gear and link can turn independently about the same axis.

When the link L is fixed, we have an ordinary gear train in which P can drive S or vice versa. If, on the other hand, S is fixed and the arm L rotates, as indicated by the broken lines in Fig. 10.25(a), the gear becomes an epicyclic train since the axis of P turns about S.

The gears S and P were termed the *sun and planet* by James Watt who devised the epicyclic train for converting reciprocating motion to rotary motion after his previous idea, the crank and connecting rod, had been taken and patented by a rival.

In modern practice, the elements S, P and L are contained within a ring or *annulus* A as shown in Fig. 10.25(b). The planet gears mesh with the annulus, which has internal teeth, and the central sun wheel, which has external teeth. The motion is the same for any number of planet wheels but, in practice, there are usually two or three. The most common arrangement is for the annulus to be fixed, providing bearing support, and the link frame, with an arm for each planet, to be connected to the output shaft.

Epicyclic gear trains are the basis for most automatic gear boxes, some steering mechanisms and some cordless drills and screwdrivers (Fig 10.26). The major advantages are that a high-velocity ratio is possible and the fact that it acts as a bearing as well as a transmission system. By fixing different parts of the epicyclic gear train, different gear ratios can also be obtained.

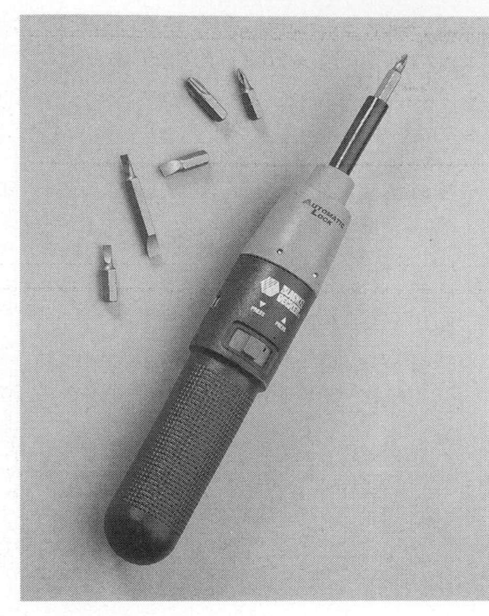

Fig 10.26 A cordless screwdriver

Three methods are in general use for analysing epicyclic gear trains – tabular, graphical and analytical – and each makes use of the velocity ratio obtained when the link L is fixed. The tabular method is widely considered easiest to follow and will be used here. The number of planet wheels does not affect the results and it is only necessary to consider one of them. It is essential to adopt a sign convention for rotations and clockwise will be taken as positive, anticlockwise negative.

In some numerical examples one of the elements will be specified as being at rest – the annulus, for example. The general procedure for this case is, in the notation of Fig. 10.25(b):

1 Imagine the whole assembly to be locked and turned clockwise through one revolution. This has the effect of giving one revolution (+1) to every element (S, P, A and L).
2 Next apply a rotation of –1 to the gear that is specified as being at rest. This has the effect of cancelling the rotation given to it in 1. Imagine the link L to be fixed and calculate the rotations of the other elements arising from this rotation of –1, using the numbers of teeth on the gears.
3 For each element add the results obtained in 1 and 2.

If no element is specified as being at rest the procedure is modified as follows. In step 1 give every element a rotation of $+a$ instead of +1. In step 2, apply a rotation of $+b$ to any wheel (instead of –1 to the one at rest). Obtain the results in terms of a and b and form two equations containing a and b, using the speeds given for two of the elements.

The next two examples illustrate these procedures.

EXAMPLE 10.7 If, in the notation of Fig. 10.25(b), the sun S has 65 teeth and the annulus A has 125 teeth, what is the speed of the link L when A is fixed and S rotates at 250 rev/min?

SOLUTION Note that the number of teeth on the planet P is not required since it is, in effect, an idler between S and A. The various rotations are set out in the table below, using the steps given above. In step 2, the ratio of the number of revolutions of S to that of A equals the inverse ratio of their numbers of teeth; the sign is changed because, with L fixed, they move in opposite directions.

	L	A	S
1 Lock assembly. Give +1 to all	+1	+1	+1
2 Fix L. Give –1 to A	0	–1	$+\frac{125}{65} = 1.923$
3 Add 1 and 2	+1	+0	+2.923

Hence, for 1 revolution of L, S turns through 2.923 revolutions in the same direction. If the speed of S is 250 rev/min then

$$\text{speed of L} = \frac{250}{2.923} = 85.5 \text{ rev/min}$$

If the speed of the planet P is required we must first calculate its number of teeth. From Fig. 10.25(b) the diameter P is half the difference between the diameters of A and S. Hence the number of teeth on P is (125 – 65)/2 = 30. By using this value and adding a column to the table for P, its speed can be determined.

EXAMPLE 10.8 Suppose, in Example 10.7, A is not fixed but rotates at 90 rev/min in the opposite sense to the rotation of S. What is then the speed of L?

SOLUTION The revised table is shown below. In step 1 all elements are given $+a$ instead of $+1$. In step 2, A is given $+b$ instead of –1, the resulting value for S being $-1.923b$ instead of 1.923.

	L	A	S
1 Lock assembly. Give $+a$ to all	$+a$	$+a$	$+a$
2 Fix L. Give $+b$ to A	0	$+b$	$-\frac{125}{65}b = -1.923b$
3 Add 1 and 2	a	$a+b$	$a - 1.923b$

The expressions in step 3 represent the speeds of L, A and S. Since the values for A and S are given as –90 rev/min and 250 rev/min respectively we can form two simultaneous equations as follows:

$$a+b = -90 \qquad [i]$$

$$a - 1.923b = 250 \qquad [ii]$$

Subtracting [ii] from [i]

$$2.923b = -340 \text{ and } b = -116.3$$

Substituting this result in [i],

$$a = -90 - b = -90 + 116.3 = 26.3$$

From the table the resultant speed of L is $+a$. Hence the link L turns at 26.3 rev/min in the same direction as the sun S.

10.3.6 Transmission through a right angle

The gear trains that we have considered so far all have parallel input and output shafts but in many cases the output shaft may need to be perpendicular to the input shaft.

Fig. 10.27 shows how *bevel gears* turn the motion through a right angle in the operation of a hand drill or manual food whisk. When the two bevel gears are the

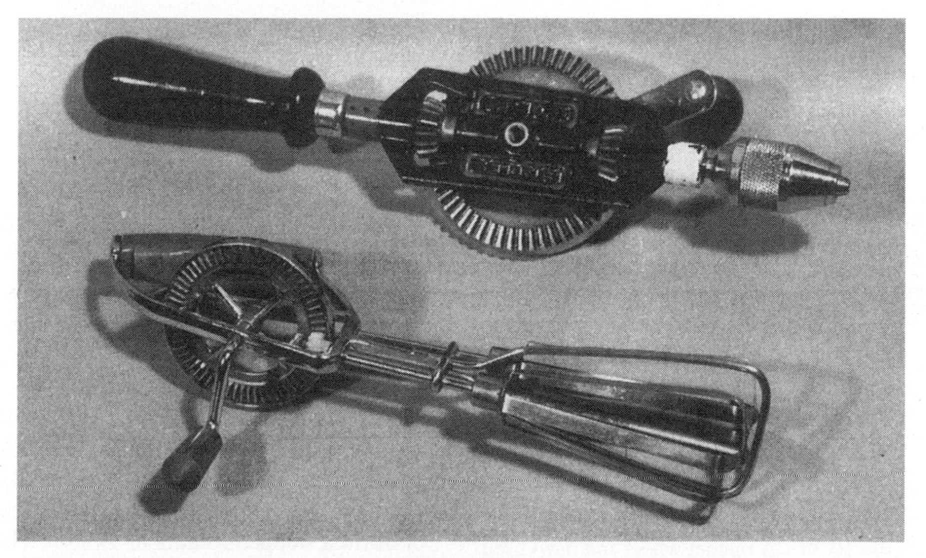

Figure 10.27 Bevel gears in a hand drill and whisk.

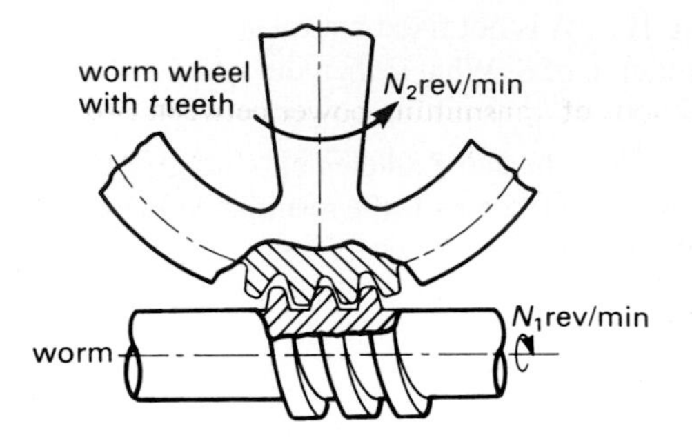

Figure 10.28 The worm and wheel mechanism.

same size, with the same number of teeth, they are referred to as *mitre gears*. Bevel gears, like spur gears, can be helical and they are again used for smooth transmission at high rotational speeds.

When it is necessary to turn the drive through ninety degrees and obtain a very high gear ratio the *worm* and *worm wheel* shown in Fig. 10.28 can be used. The worm is a screw thread with one continuous gear tooth which engages with the worm wheel, which is a helical gear arranged to mesh with a rotating screw thread. One revolution of the worm moves the wheel one tooth, and hence the ratio of the angular velocities is given by the number of teeth on the wheel. Hence,

$$\frac{\text{angular speed of worm}}{\text{angular speed of wheel}} = \text{the number of teeth in wheel}$$

and therefore

$$\text{gear ratio} = \text{the number of teeth in wheel}$$

Because of the large reduction in the speed of drive that is obtained with a worm and wheel, a very high output torque is produced. Fig. 10.29 shows how this is used to advantage in the tightening of guitar strings to tune them. It is also necessary for the tuning mechanism to be less than 50% efficient in order to prevent the strings from slackening when the tuning peg is released (overhauling).

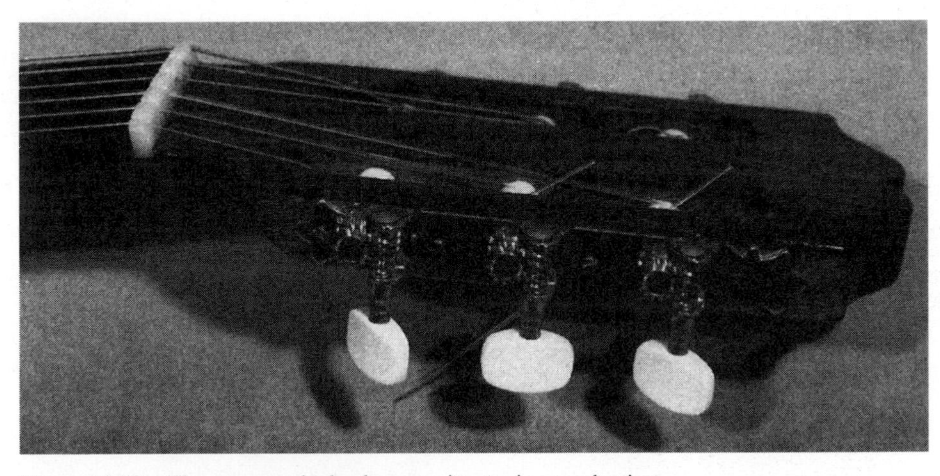
Figure 10.29 The worm and wheel on a guitar tuning mechanism.

10.3.7 Ratchet mechanisms

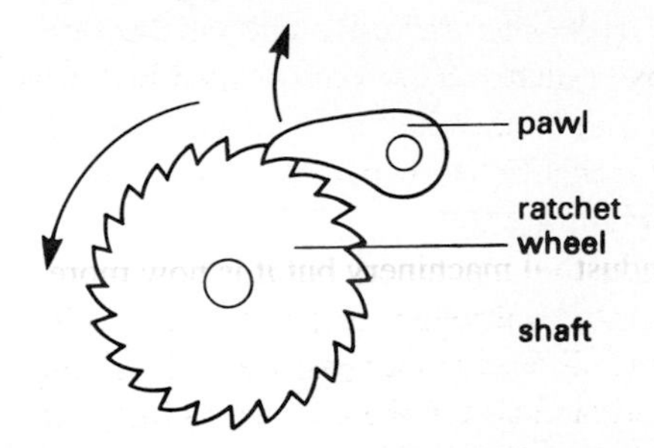

Figure 10.30 A ratchet wheel and pawl used to arrest motion.

A *ratchet mechanism*, shown in Fig. 10.30, consists of a wheel with saw-shaped teeth, called a ratchet, which engages with an arm called a *pawl*. The specially shaped teeth allow rotation in one direction only and this mechanism can be used to prevent motion in the reverse direction.

In many lifting devices, a ratchet device is used to prevent the load running back down once it has been lifted to its new position. In an alternative use of the device, the ratchet is free to move and an oscillating input is applied to the pawl. In this situation, the ratchet will be advanced each time that the pawl contacts it and so will perform a stepped rotational motion. This mechanism is used to regulate mechanical watches and clocks, where it is known as the *escapement*.

10.3.8 Chain and belt drives

Apart from gears there are two main methods of transmitting power between two shafts – *chain drives* and *belt drives*, Figs 10.31 and 10.32. Belts are the cheapest of the three and gears are the most costly to manufacture, assemble and align. The choice of the type of transmission depends mainly on four factors:

- the distance between the shafts,
- the required performance,
- the cost,
- the maintenance of the system.

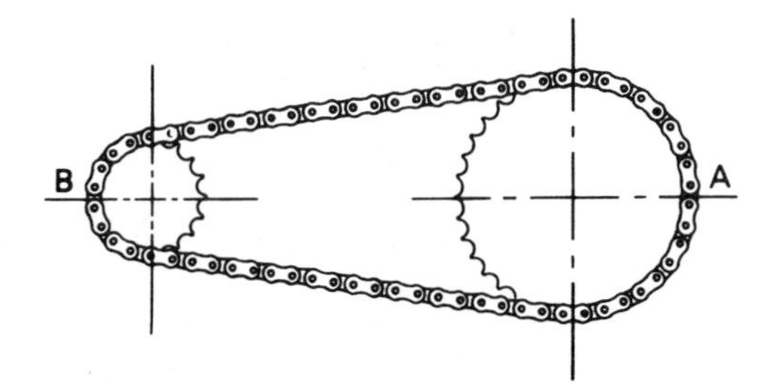

Figure 10.31 Bicycle chain drive.

(a)

(b)

Figure 10.32
(a) Belt drive on a sewing machine
(b) Different belt drive arrangements.

If the centre distance is large, a belt drive is appropriate but over small distances gears are more practical. If precise performance and accurate timing over a large distance are required a flat belt, which will tend to slip, will not prove satisfactory and a more expensive means must be employed such as a chain or toothed-belt drive. These factors explain the use of a chain, gears or toothed belt to drive the camshaft of a car, but a belt, usually a *V-belt* to increase the friction between the belt and the pulley, is used to drive the fan and alternator or dynamo.

Belt drives were once commonplace in industrial machinery but it is now more common for the primary distribution of power to be through electrical cables which then power individual motors. Belt drives are still used in many household machines such as the washing machine or food processor and also in the computer printer. In many cases a stepper motor is used in conjunction with the belt transmission to achieve a smoother drive.

The power available in a belt drive will depend on the tensions on the drive and return sides of the belt. These in turn depend upon how much of the circumference is in contact with the belt. This is called the *angle of wrap* and it can be increased by using a jockey wheel or crossing the belt as shown in Fig. 10.33. The pulley diameter and angular speed of a belt drive are inversely proportional if there is no slip in the belt.

So for a belt drive

$$\text{angular output speed} = \text{angular input speed} \times \frac{\text{diameter of input pully}}{\text{diameter of output pully}}$$

With no slipping, the linear tangential speed v of the two pulleys in a belt drive must be the same.

$$\text{Hence } v_B = v_A$$

where v_B and v_A are the speeds of the driven pulley, B, and the drive pulley, A. Since $v = r\omega$ the rotational speeds of C and D are related by:

$$r_B\omega_B = r_A\omega_A$$

where r_B and r_A are the radii of B and A. If the rotational speeds are measured in rev/min rather than rad/s the result becomes:

$$r_BN_B = r_AN_A$$

where N_B and N_A are the rotational speeds in revolutions per minute of B and A.

A chain drive has toothed wheels and therefore has the same relationship for gear ratio and output speed as a simple gear train.

So for a chain drive

$$\text{gear ratio} = \frac{\text{the number of teeth in the driven sprocket}}{\text{the number of teeth in the driver sprocket}}$$

and the angular speeds are related by:

$$\text{Output speed} = \text{input speed} \times \frac{\text{number of teeth in driver sprocket}}{\text{number of teeth in driven sprocket}}$$

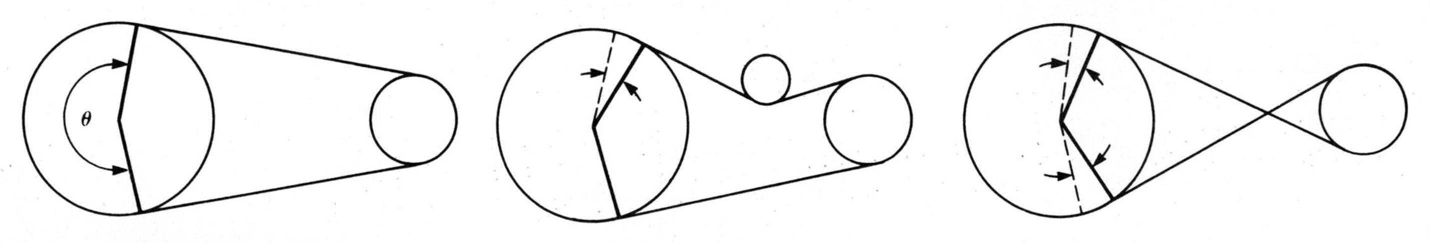

Figure 10.33 Increasing the angle of wrap of a belt.

10.4 Mechanisms for converting rotary to linear motion

10.4.1 Crank mechanisms

Fig. 10.34 shows a *crank mechanism*. The centre of the crank handle moves in a circle and the path traced by a moving point of the mechanism is known as the *locus*. When the handle is turned a torque is applied to the crankshaft. The crank is used in a winch mechanism for raising a heavy weight and for operating a screw jack to raise a car.

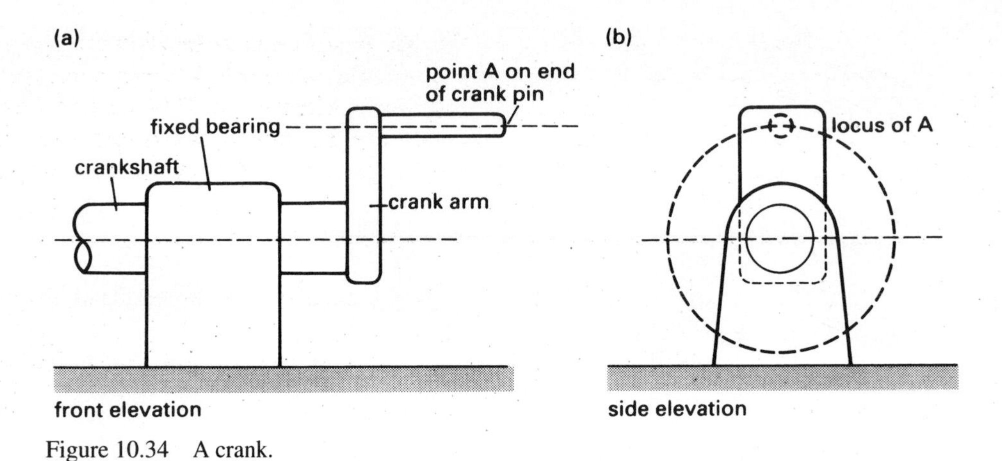

Figure 10.34 A crank.

Fig. 10.35 shows a representation of a *crank and slider* mechanism which can either have a rigid crank arm or a handle on a crank wheel that can be rotated. When the crank is rotated the slider moves back and forth in its frame. Fig. 10.36 shows the locus of the component parts of the crank and slider. The locus of the crank pin is continuous and circular, that of the slider is linear and the locus of the point B on the *connecting rod* is a combination of the two, giving an elliptical path. This mechanism is useful where a large linear movement is required, the total length being equal to twice the length of the crank. Large forces can be exerted on both the outward and inward directions. An example of a crank and slider mechanism is to be found in a car windscreen wiper.

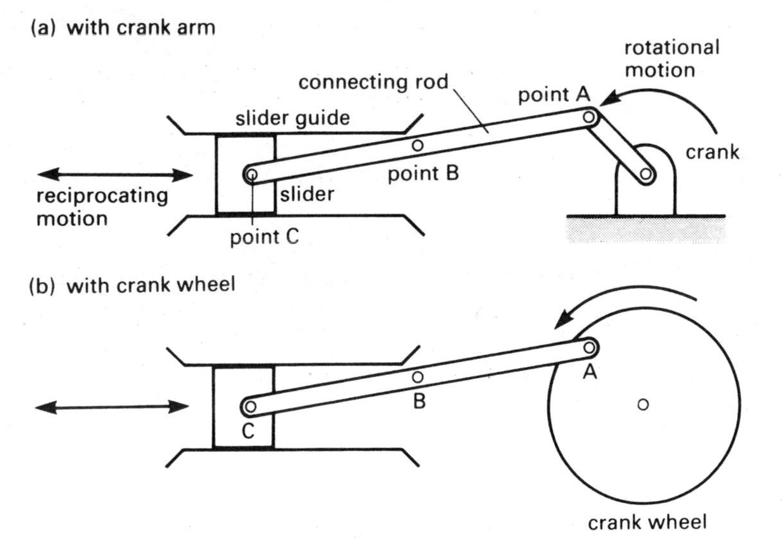

Figure 10.35 The crank and the slider mechanism.

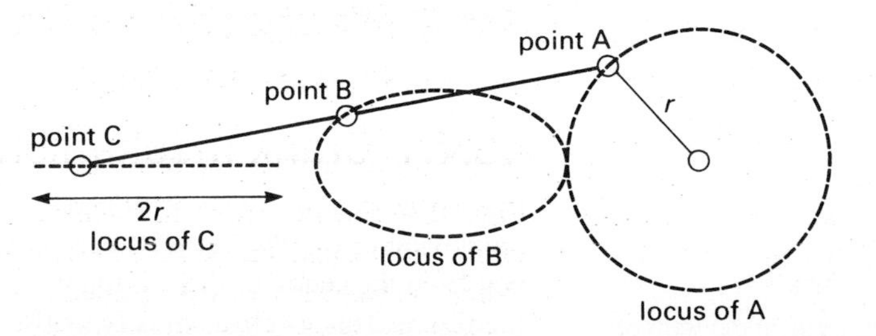

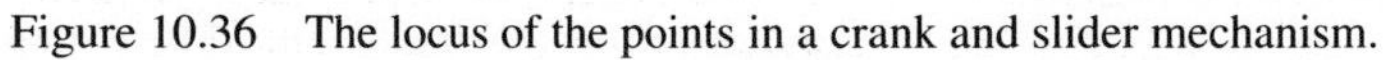
Figure 10.36 The locus of the points in a crank and slider mechanism.

(a)

(b)

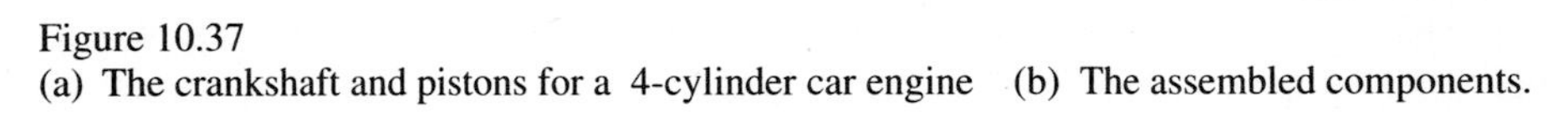

Figure 10.37
(a) The crankshaft and pistons for a 4-cylinder car engine (b) The assembled components.

The motion of the slider is reciprocating and its speed varies. The velocity can be derived from the displacement–time graph described below.

In a car engine a crankshaft is joined to the pistons by connecting rods (con-rods) and this mechanism transforms the linear piston movement into the rotation of the crankshaft which can drive the wheels of the car. The crankshaft of a car engine is shown in Fig. 10.37. This is one example of several crank and slider mechanisms being used to convert linear to rotational motion. In the internal combustion engine, the steam engine and the air compressor, which are all reciprocating engines, the length of the con-rod is usually between three and six times the length of the crank.

If the position of point C on the slider is plotted against time the *displacement diagram* or *displacement–time curve* for the crank and slider mechanism is obtained. The displacement diagram for a crank and slider with a stroke of 50 mm is shown in Fig. 10.38. If it were necessary to find the speed of the slider at any point in the motion, the gradient at the corresponding point on the displacement curve would give the required value. The mathematical relationships between displacement, speed and acceleration are given in section 10.6.

An *eccentric* is a crank mechanism produced when the shaft of the wheel is positioned off-centre and the connecting rod is connected to the centre of the wheel as in Fig. 10.39. The eccentric was used in the valve control in steam engines as it can produce powerful forces in both directions for relatively small linear movements. The length of travel of the slider is equivalent to twice the distance between the centre of rotation and the drive pivot of the connecting rod.

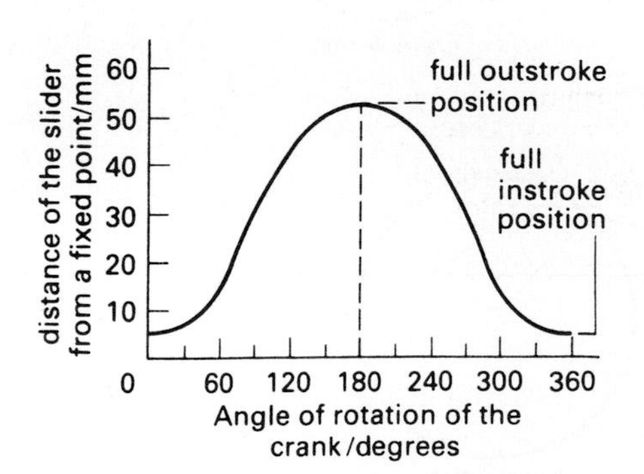

Figure 10.38 The displacement diagram of the crank and slider.

A *peg and slot* mechanism can be used when a reciprocating motion with a different speed and force for each direction of its movement is required. The peg and slot mechanism shown in Fig. 10.40 will produce a slow outward motion with a large pushing force and a light, fast return. To obtain a slow, large pulling motion and a light, fast push the direction of rotation of the peg should be reversed. The range of movement of the slider can be very large and will depend on both the distance of the peg from the centre of the driving wheel and the ratio of the distances *a*:*b* shown on the figure. This mechanism is typically used in machine tools. An illustration of a quick-return mechanism used in a shaping machine is given in Fig. 10.41.

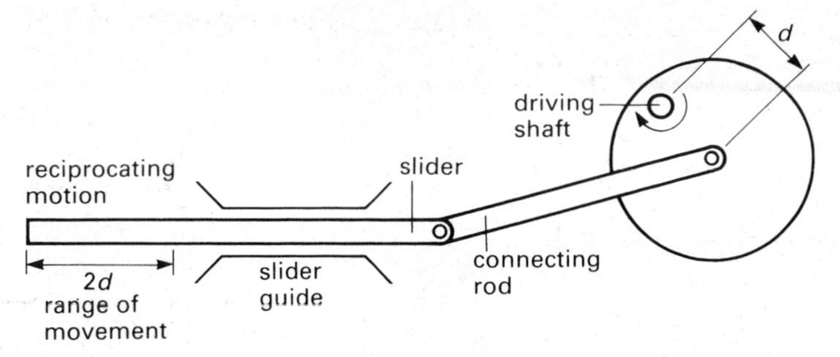

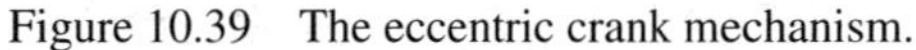
Figure 10.39 The eccentric crank mechanism.

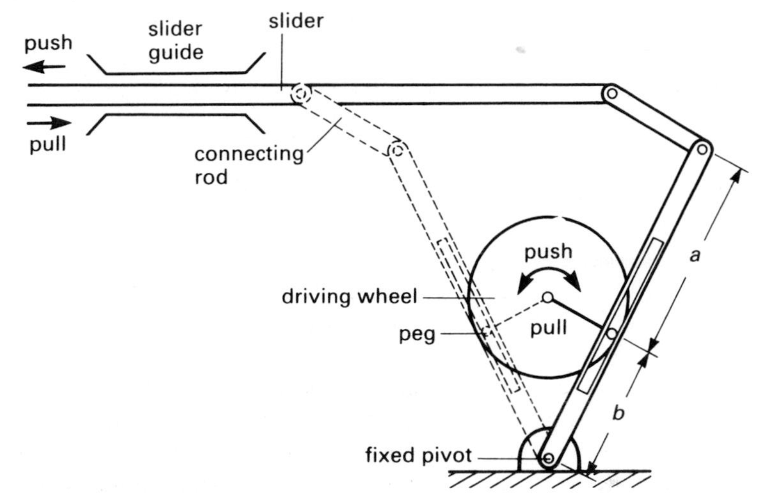

Figure 10.40 The peg and slot mechanism.

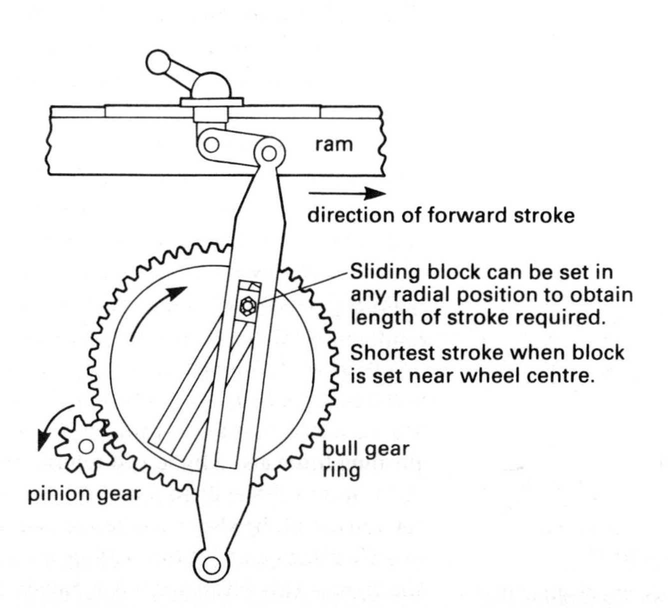

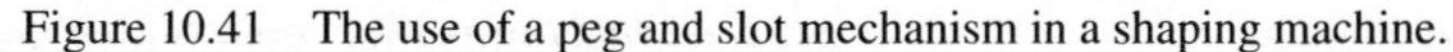
Figure 10.41 The use of a peg and slot mechanism in a shaping machine.

10.4.2 The cam and follower mechanism

In the internal combustion engine another method of converting rotational to linear motion is also employed. Instead of a driving wheel or crank a specially shaped *cam* is used to control each valve of the engine and as it rotates a *follower* is kept in contact with the surface and follows the cam profile.

Fig. 10.42 shows the *cam and follower* mechanism. The follower may be kept in contact with the cam either by gravity when in the upright position or the action of a spring. The cam is used for small linear movements and the distance that the slider moves is given by the distance $d_2 - d_1$ in the figure. It can be seen that the follower will stay stationary or *dwell* when the cam surface is circular and will move out when the cam profile protrudes from this. On the out-stroke of the slider, considerable force can be exerted but the in-stroke force will be determined by the strength of the spring or the self-weight of the follower and slider.

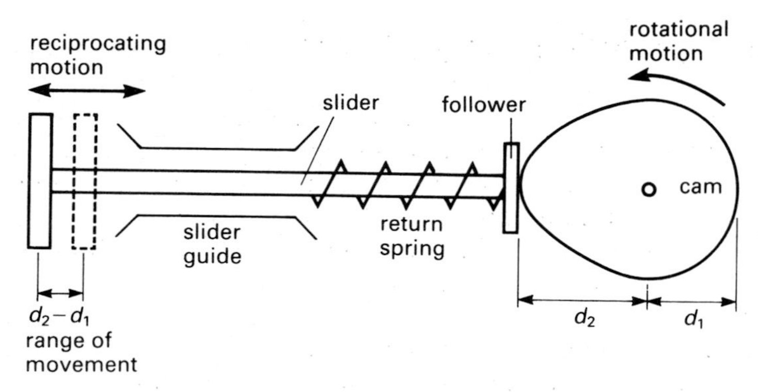

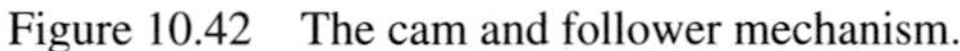
Figure 10.42 The cam and follower mechanism.

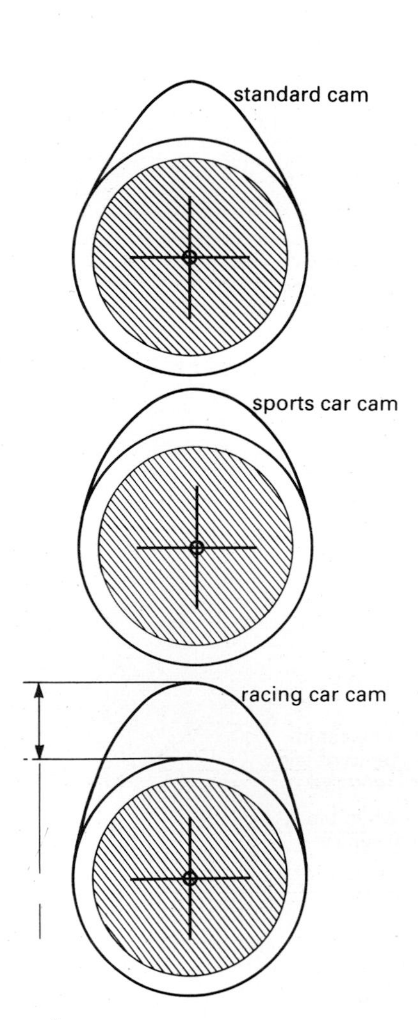

Figure 10.43 Profile of cams controlling the valves in different car engines.

The *pear-shaped cam* is typical of that used for engine valve control where the valve is the follower and is directly operated by the cam. Slight differences in the cam profile will produce different engine characteristics and so a high-performance sports or racing car will have different shaped engine cams from those of a family saloon. Fig. 10.43 shows the cam profiles for three different engines.

The exact cam shape required to produce a particular motion will depend on the type of follower that is used and Fig. 10.44 shows some different types of follower. The followers that have a small area of contact such as the point, knife and edge followers will experience significant wear during the repetitive operation of the mechanism. Others, such as the angled foot, roller and sliding and oscillating followers, can be used for rotation of the cam in one direction only. The sliding yoke follower is designed to be used with an eccentric cam which is always the same width and would not be effective with other designs of cam.

A cam and follower will usually be designed to produce the desired output motion and the cam is normally a flat metal plate cut to a smooth curve so that the linear motion transmitted is also smooth and continuous. The displacement–time curve of the required output motion shows the necessary positioning of the follower at each point in time and so can be used to determine the cam profile. The vertical displacement at any point of the cycle is projected onto the line of the cam follower and rotated to the appropriate radial line of the cam. The direction of this rotation will be opposite to the required direction of rotation of the cam and when all the plotted points are connected the cam profile will be obtained.

Typical displacement diagrams with the corresponding projected cam profiles necessary for a simple knife follower are shown in Figs 10.45 to 10.48.

Fig. 10.45 shows the pear-shaped cam used in engine valve control. The follower is stationary for about half a revolution of the cam and rises and falls symmetrically in each of the remaining quarter revolutions.

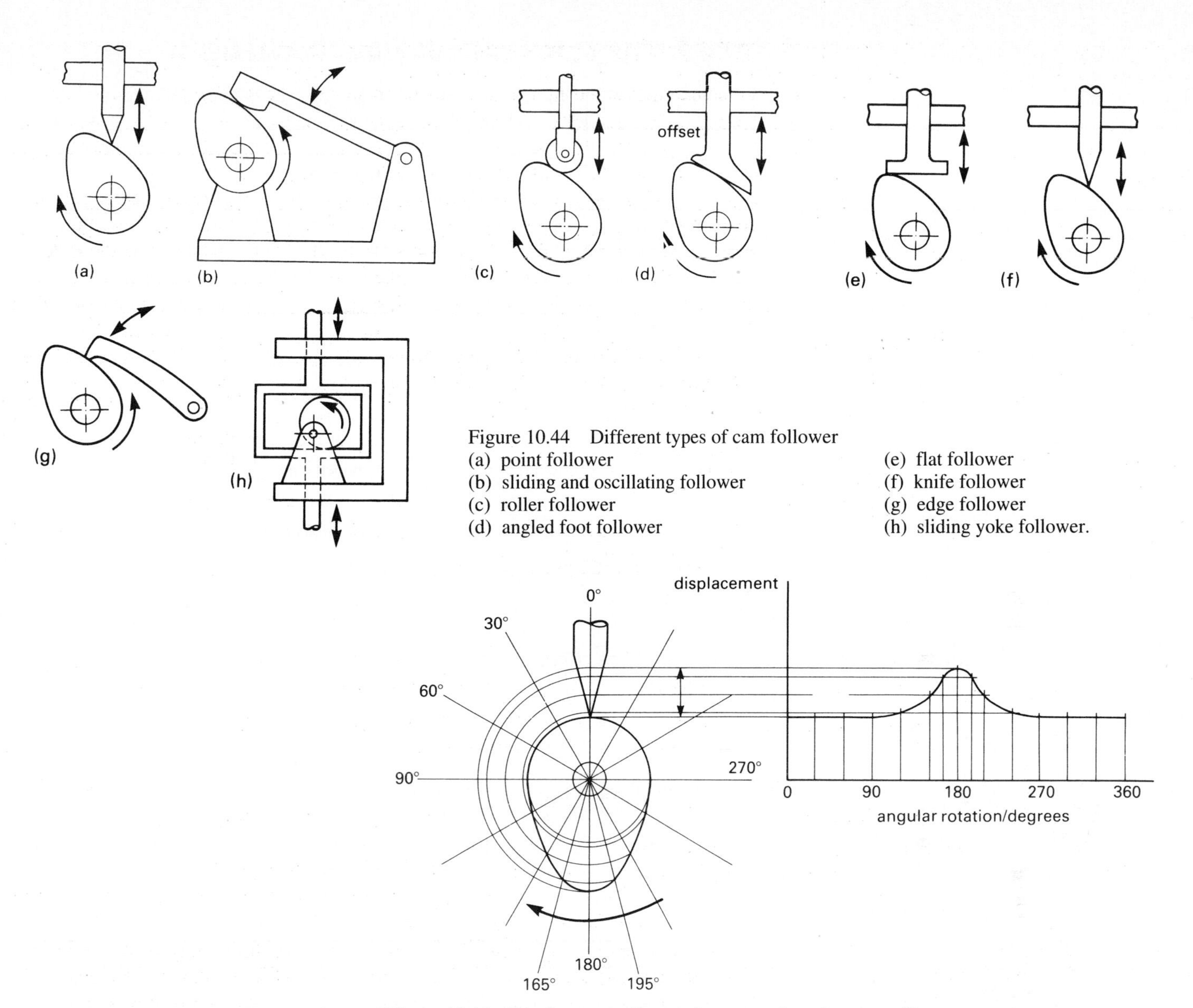

Figure 10.44 Different types of cam follower
(a) point follower
(b) sliding and oscillating follower
(c) roller follower
(d) angled foot follower
(e) flat follower
(f) knife follower
(g) edge follower
(h) sliding yoke follower.

Figure 10.45 Displacement diagram for a pear-shaped cam profile.

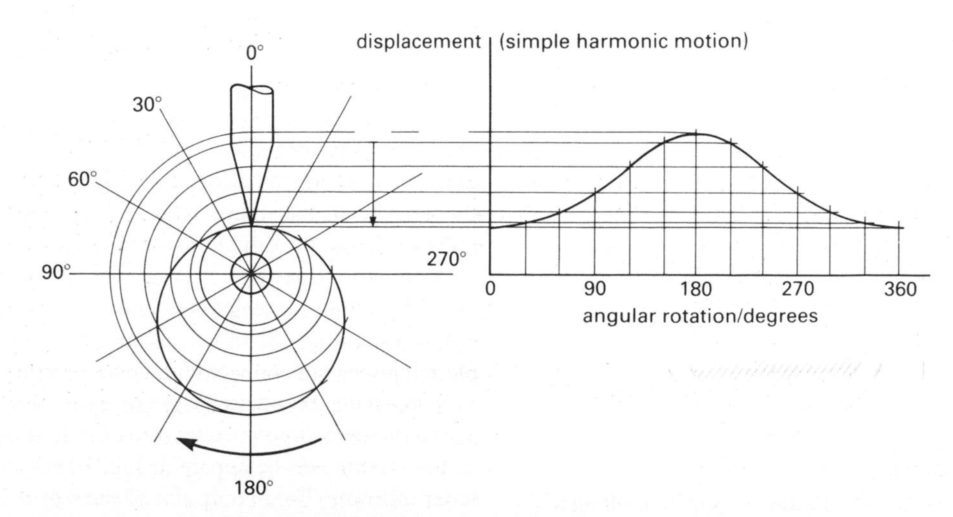

Figure 10.46 Displacement diagram for an eccentric cam.

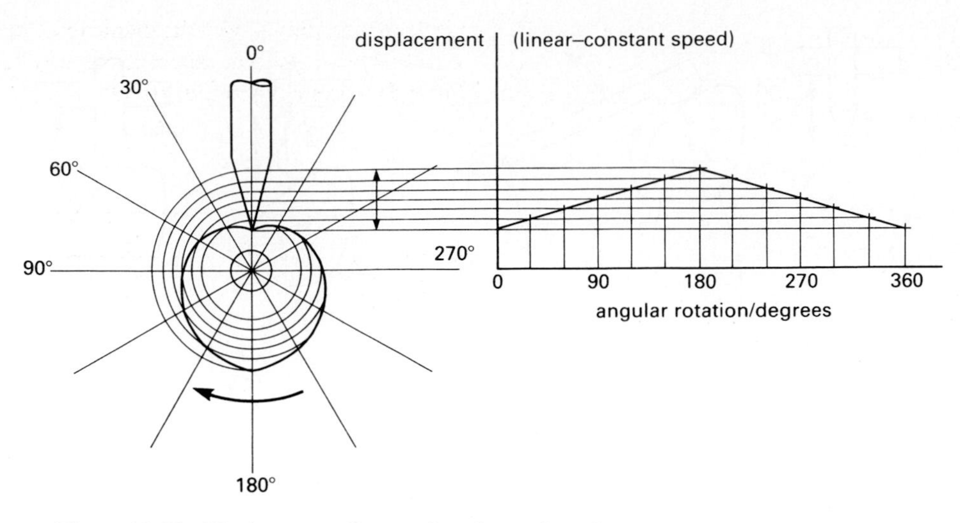

Figure 10.47 Displacement diagram for a heart-shaped cam.

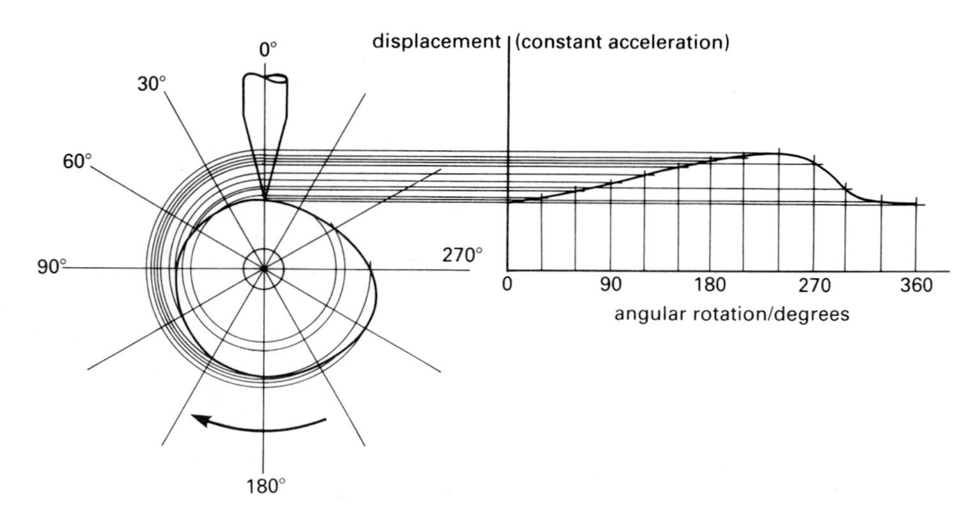

Figure 10.48 Displacement diagram for a uniform acceleration cam.

The *eccentric cam* shown in Fig. 10.46 is a circular cam with an offset centre of rotation. This mechanism has very similar characteristics to the eccentric crank mechanism. It produces simple harmonic motion and is used in pumps. The rise and fall phases of the motion have similar characteristics and each accounts for half a revolution of the cam.

The *heart-shaped cam* can be seen on the spool-winding mechanism of an old hand-operated sewing machine with long cylindrical bobbins. It is used for this purpose as it gives a uniform velocity of the follower and so will wind the spool evenly. The displacement diagram for the heart-shaped cam in Fig. 10.47 is a straight line for each of the directions of movement, indicating constant speed of motion for each of these phases.

Where uniform acceleration and deceleration are required the *uniform acceleration cam* shown in Fig. 10.48 will achieve this function. The displacement curve is a second order function which will yield a uniformly increasing or decreasing gradient (velocity) and a constant second derivative of the displacement (acceleration).

If a different type of follower is used on these same cams, slight changes in the output motion will be apparent. Fig. 10.49 shows the required modifications for a roller follower. The pivot point of the roller, in which the direction of movement is normal to the surface of the cam, is projected on the appropriate radial line and a circle is then drawn with the diameter of the roller. If the roller is also offset, then a series of tangents to the appropriate offset circle must be used instead of the radial lines. This is shown in Fig. 10.50.

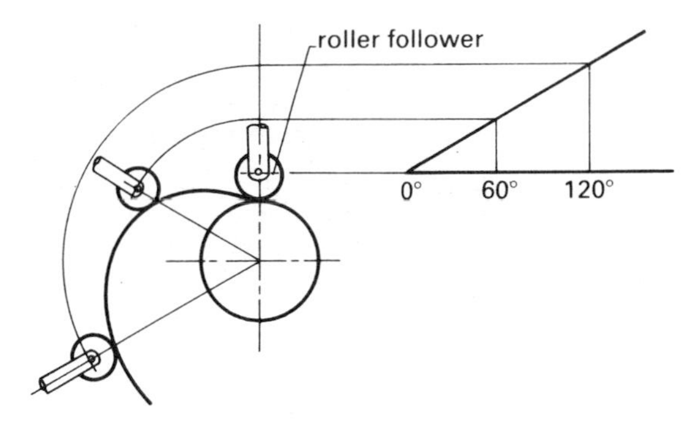

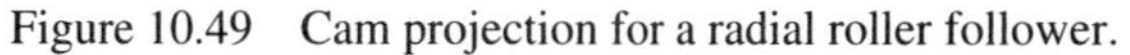
Figure 10.49 Cam projection for a radial roller follower.

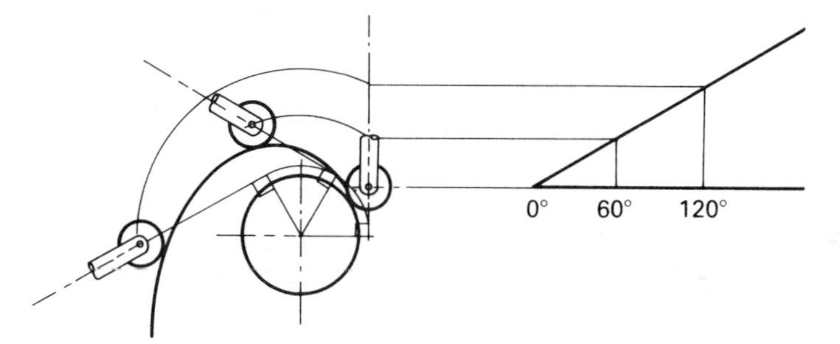

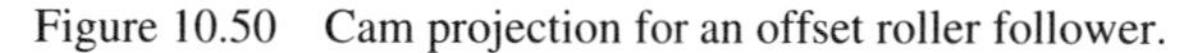
Figure 10.50 Cam projection for an offset roller follower.

10.4.3 The rack and pinion

The *rack and pinion* mechanism shown in Fig. 10.51 is another method of converting rotary to linear motion. It differs from those previously examined in that the motion of the rack is not reciprocating for continuous rotation of the drive. There would be continuous movement of the rack at constant speed in one direction for each revolution at uniform angular speed of the pinion. Therefore the pinion would have to be reversed in order to retract the rack. The rack and pinion mechanism will transmit large forces and is used in some car steering mechanisms and in some lifting equipment.

The rack and pinion can be operated in several ways. If the axis of the wheel is held stationary, the drive can either be applied to the wheel, which will result in the linear movement of the rack, or to the rack, which will then give a rotary output to the wheel. Alternatively, the rack may be held still and the wheel free to move, in which case the wheel will rotate and its centre will also progress parallel to the rack.

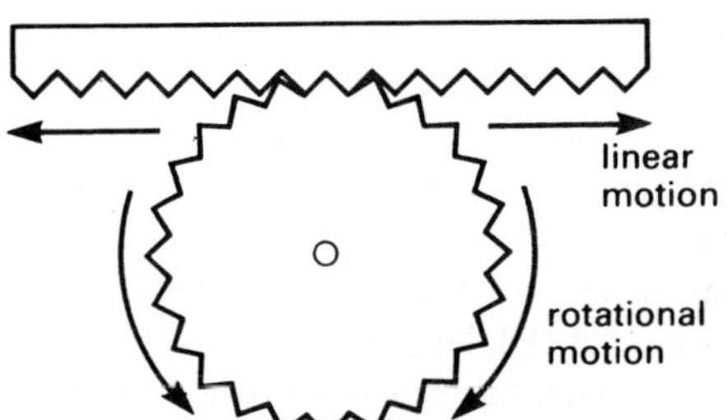

Figure 10.51 The rack and pinion.

10.5 Simple lifting machines

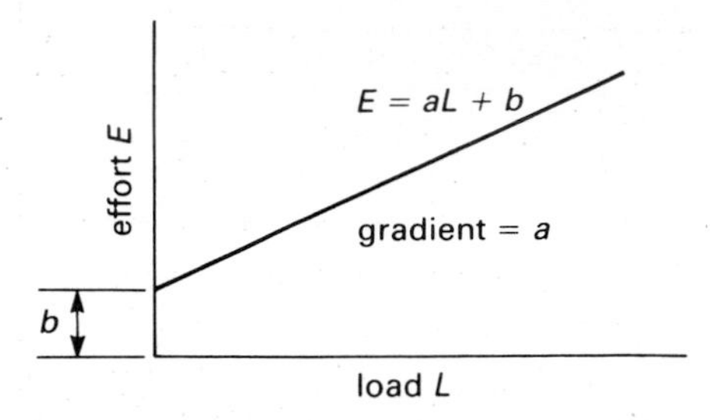

Figure 10.52 Effort against load graph for a simple lifting machine.

From the earliest times, people have needed to raise weights greater than they could lift unaided. Consequently, they developed lifting machines with a large mechanical advantage to lift heavy weights more easily. The simplest of these is based on the lever such as the crowbar described in section 10.2.1.

When the values of effort are plotted against load for a lifting machine, the graph usually approximates to a straight line as in Fig. 10.52. The effort intercept, b, shows the effort needed to overcome the friction in the machine before any load can be lifted and the gradient depicts the rate of increase of effort with applied load. The equation of the straight line is given by:

$$E = aL + b$$

where E is the effort,
L is the load,
a is the slope of the line,
and b is the effort intercept.

This equation is known as the *law of the machine* and if the values of a and b are known the effort required to lift a given load can be calculated.

EXAMPLE 10.9 The results for a series of measurements of load and corresponding effort used on a lifting machine are given below. The effort moves through 1.885 m while the load is lifted 19 mm.

Load (N)	72	210	330	425	600
Effort (N)	3	5	7	8.5	11

(a) Plot the characteristic of effort against load and determine the law of the machine. Find the effort needed to raise a load of 1 kN.
(b) Find the velocity ratio of the machine, plot the efficiency against load curve and find the efficiency for a load of 500 N. Will the machine overhaul at this load?

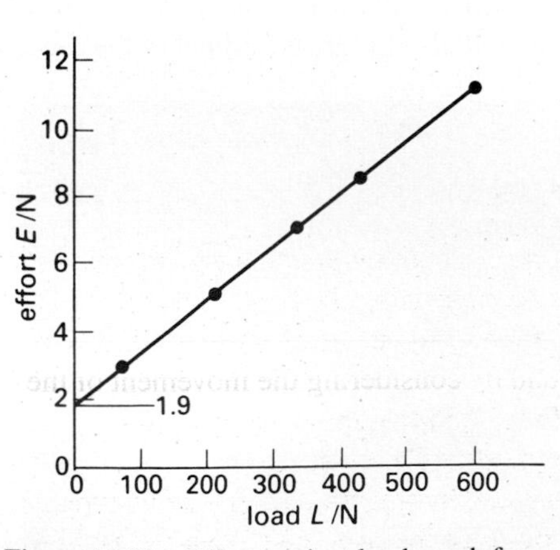

Figure 10.53 Effort against load graph for Example 10.9.

SOLUTION

(a) Plotting the values of effort against load for this machine gives us the graph shown in Fig. 10.53. The effort intercept is 1.9 N and the gradient is found to be 0.0155 from measurement of the graph. Therefore, the law of the machine is $E = 0.0155L + 1.9$.

The effort needed to raise a load of 1 kN is given by putting $L = 1000$ N in the law of the machine. This gives:

$$E = 0.0155 \times 1000 + 1.9 = 17.4\ \text{N}$$

(b) The velocity ratio is the same for all values of the load since it depends only on the geometry of the machine. Working in millimetres it is given by:

$$\text{VR} = \frac{\text{distance moved by effort}}{\text{distance moved by load}} = \frac{1885}{19} = 99.21$$

On the other hand, the mechanical advantage and efficiency both vary with the load and must be calculated for each value. They are given by

$$\text{MA} = \frac{\text{load}}{\text{effort}} \quad \text{and} \quad \eta = \frac{\text{MA}}{\text{VR}}$$

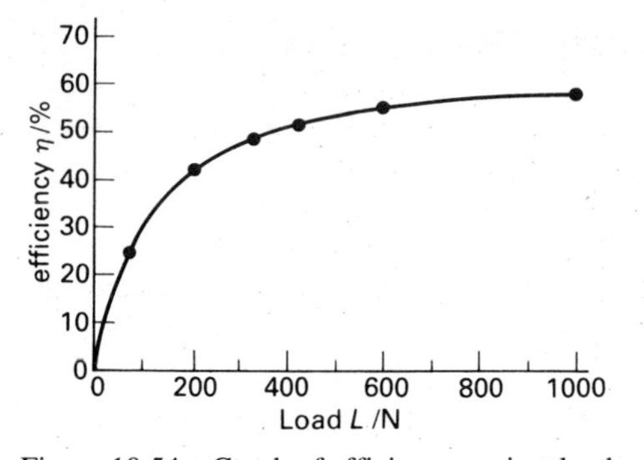

Figure 10.54 Graph of efficiency against load for Example 10.9.

The calculated values of mechanical advantage and efficiency are given in the following table:

Load (N)	72	210	330	425	600	1000
Effort (N)	3	5	7	8.5	11	17.4
MA	24	42	47.1	50	54.6	57.5
η(%)	24.2	42.3	47.5	50.4	55	58

Plotting these values of efficiency against load gives us the curve in Fig. 10.54. From the graph the efficiency for a load of 500 N is found to be 52.5%. This value of efficiency (>50%) means that the machine would overhaul and would need to be secured after the load has been raised.

Alternatively, by calculation, when $L = 500$,

$$E = 0.0155L + 1.9 = 0.0155 \times 500 + 1.9 = 9.65$$

Hence

$$\eta = \frac{\text{MA}}{\text{VR}} = \frac{L}{E \times \text{VR}} = \frac{500}{9.65 \times 99.21} = 0.523 \quad \text{or} \quad 52.3\%$$

10.5.1 Limiting efficiency

It can be seen from the curve in Fig. 10.54 that the efficiency of a machine increases with increasing load and approaches an upper limit or *limiting efficiency*. Since the efficiency can be expressed as:

$$\eta = \frac{\text{MA}}{\text{VR}} = \frac{L}{E \times \text{VR}} = \frac{L}{(aL + b) \times \text{VR}} = \frac{1}{(a + b/L) \times \text{VR}}$$

and the term b/L decreases with increasing load then the limiting efficiency is given by:

$$\eta_{\text{LIM}} = \frac{1}{a \times \text{VR}}$$

EXAMPLE 10.10 Find the value of limiting efficiency for the machine quoted in Example 10.9.

SOLUTION

$$\eta_{\text{LIM}} = \frac{1}{a \times \text{VR}} = \frac{1}{0.0155 \times 99.21} = 0.650 \quad \text{or} \quad 65.0\%$$

10.5.2 Simple pulley systems

Fig. 10.55 shows three pulley systems, each with a continuous rope wound round the pulley sheaves giving 1, 2 or 3 sections of rope supporting the load. Pulley systems such as these are often used as lifting machines and can have a very high efficiency. The velocity ratio of each system can be found by considering the movement of the effort and the load and it can be seen from Fig. 10.55 that the velocity ratio is equivalent to the number of rope sections supporting the load.

The arrangement of pulleys in Fig. 10.56 is different from that in the previous example but, by again examining the effort distance for a set load distance, we find that the velocity ratio of this system is 4.

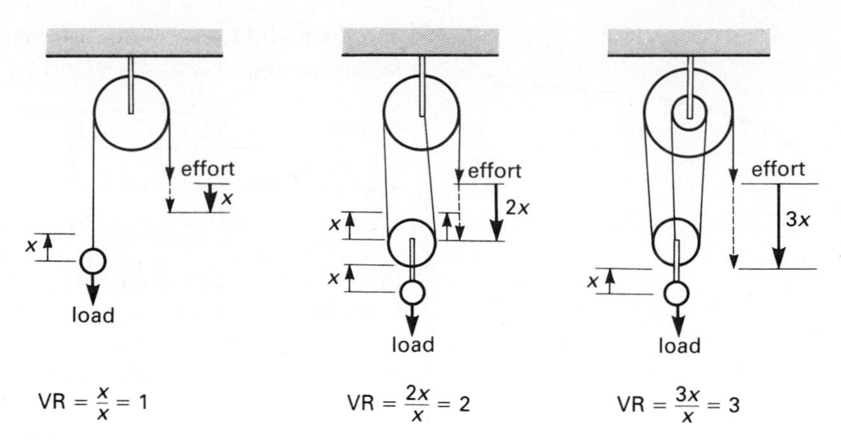

Figure 10.55 Pulley lifting devices.

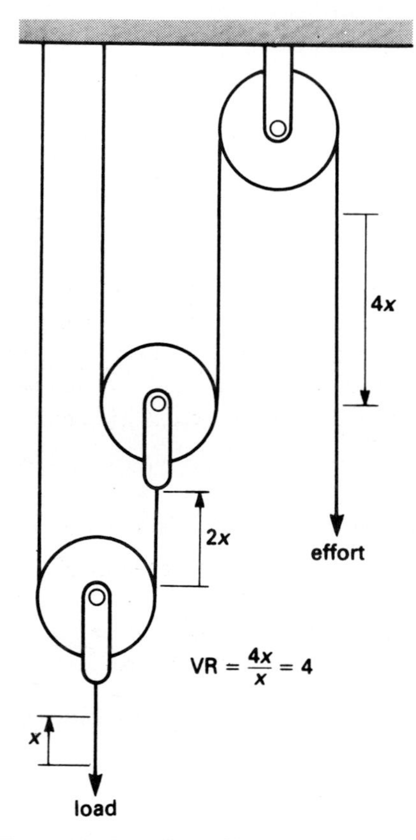

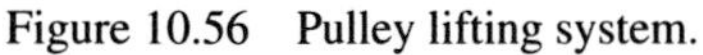

Figure 10.56 Pulley lifting system.

10.5.3 Weston differential pulley

The *Weston differential pulley* is used to lift very heavy loads. It is a toothed pulley system in which two pulleys of different sizes are locked together on the same shaft and turn as one. A continuous chain passes round each of these pulleys and the pulley supporting the load as shown in Fig. 10.57.

If pulley A is rotated through one complete revolution then the effort will move a distance $2\pi R$. This will also cause $2\pi R$ of the chain to be taken up by the right-hand side of pulley A and $2\pi r$ to be let out from the left-hand side of pulley B. Consequently the chain supporting the load will be taken up by $2\pi(R - r)$ and the load will rise by a distance equal to half this value.

$$\text{VR} = \frac{\text{distance moved by effort}}{\text{distance moved by load}} = \frac{2\pi R}{\pi(R - r)} = \frac{2R}{R - r}$$

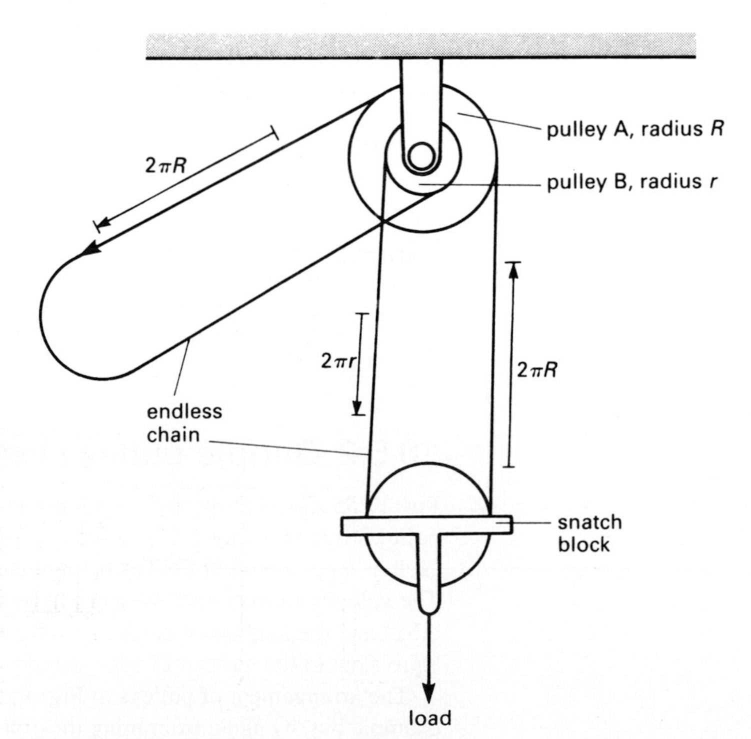

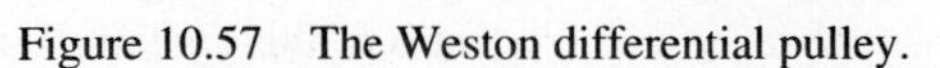

Figure 10.57 The Weston differential pulley.

EXAMPLE 10.11 A Weston differential pulley with radii of 150 mm and 140 mm is to be used to lift a load of 7 kN. If the efficiency is 45% calculate the effort required.

SOLUTION Using the last result, the velocity ratio is given by:

$$\mathrm{VR} = \frac{2R}{R - r} = \frac{2 \times 150}{150 - 140} = \frac{300}{10} = 30$$

As before,

$$\eta = \frac{\mathrm{MA}}{\mathrm{VR}} = \frac{L}{E \times \mathrm{VR}} \quad \text{and} \quad E = \frac{L}{\eta \times \mathrm{VR}}$$

Substituting the numerical values, the effort is:

$$E = \frac{7000}{0.45 \times 30} = 519\,\mathrm{N}$$

10.5.4 Wheel and axle

The *wheel and axle* shown in Fig. 10.58 is a development of the early windlass which was used to lift heavy weights by winding a rope round a pole. The effort wheel is of greater radius than the load drum and both are mounted on the same axle. When effort is applied to the wheel a mechanical advantage is obtained and for one turn of the effort wheel the load drum will also rotate by one revolution. Hence:

$$\mathrm{VR} = \frac{\text{distance moved by effort}}{\text{distance moved by load}} = \frac{2\pi R}{2\pi r} = \frac{R}{r}$$

This is also the velocity ratio for an ungeared winch in which a load drum of radius r is rotated by means of a crank of length R.

When a gear train is incorporated into the wheel and axle system then the mechanical advantage can be greatly increased. The velocity ratio becomes:

$$\mathrm{VR} = \frac{RG}{r}$$

where R is the effort wheel radius,
r is the load drum radius,
G is the gear ratio when the drive wheel is connected to the effort drum.

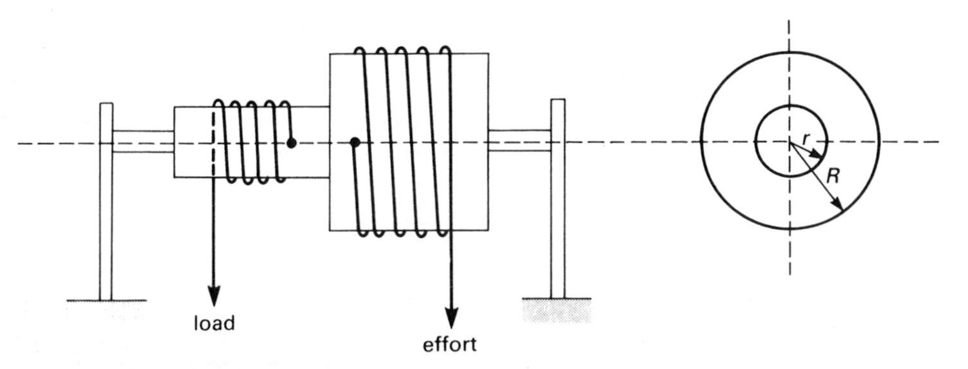

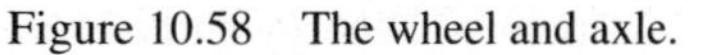

Figure 10.58 The wheel and axle.

EXAMPLE 10.12 When an effort of 600 N is applied at a crank radius of 25 mm the geared hoisting mechanism of a crane can lift a load of 2.55 kN on a drum of 100 mm diameter. The effort crank drives a compound gear train and is directly linked to a gear of 20 teeth. The other drive gear has 15 teeth and the two driven gears have 40 and 75 teeth. Calculate the velocity ratio of the hoist and its efficiency at this load.

SOLUTION The gear ratio is given by:

$$G = \frac{\text{product of driven teeth}}{\text{product of driver teeth}} = \frac{40 \times 75}{20 \times 15} = 10$$

and the velocity ratio is therefore:

$$\text{VR} = \frac{RG}{r} = \frac{25 \times 10}{50} = 5$$

Working in newtons the mechanical advantage is:

$$\text{MA} = \frac{\text{load}}{\text{effort}} = \frac{2550}{600} = 4.25$$

and the efficiency of the hoist at a load of 2.55 kN is therefore:

$$\eta = \frac{\text{MA}}{\text{VR}} = \frac{4.25}{5} = 0.850 \quad \text{or} \quad 85.0\%$$

10.5.5 Screw jack

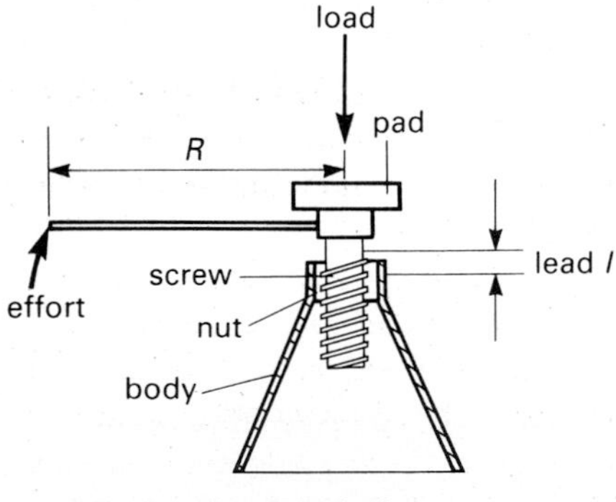

Figure 10.59 The screw jack.

The *screw jack* consists of a screw running inside a nut as shown in Fig. 10.59. When an effort, applied to the handle at a distance R from the centre of the screw, is moved through one complete turn the screw thread will advance by the *lead* distance. For a single start thread this will be equivalent to the *pitch* of the thread, the distance between two adjacent crowns of screw thread. For a multiple start thread the lead will be equal to the pitch multiplied by the number of starts. So for one turn of the effort handle:

distance moved by the effort $= 2\pi R$

distance moved by the load = the lead of the screw, l

$$\text{Velocity ratio} = \frac{\text{distance moved by effort}}{\text{distance moved by load}} = \frac{2\pi R}{l}$$

The frictional forces in a screw jack are very high and so its efficiency is usually considerably less than 50%. If the frictional forces can be calculated an alternative method of finding the efficiency may be appropriate. The efficiency can also be expressed as:

$$\eta = \frac{\text{work out}}{\text{work in}} = \frac{\text{useful work done}}{\text{useful work done} + \text{work done against friction}}$$

$$= \frac{\text{potential energy gained by load}}{\text{potential energy gained by load} + \text{work done against friction}}$$

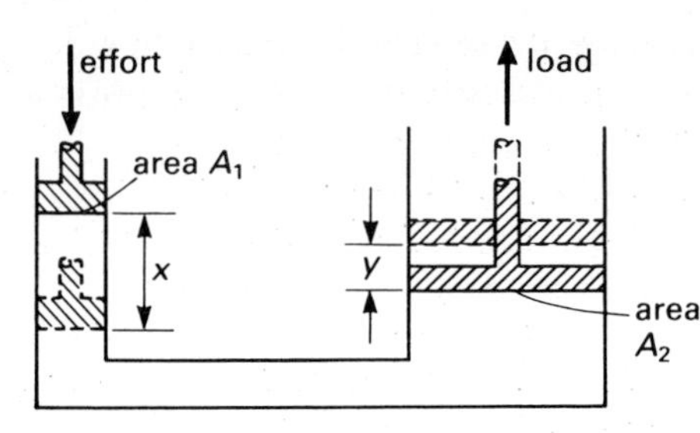

Figure 10.60 The hydraulic lift.

10.5.6 Hydraulic lift

The *hydraulic lift* shown in Fig. 10.60 is a lifting machine that uses fluid pressures. In this lift the amount of fluid displaced by the effort piston must be equal to the amount of fluid displaced at the load piston. So the volumes A_1x and A_2y must be the same.

$$\text{Therefore}\quad \text{VR} = \frac{x}{y} = \frac{A_2}{A_1}$$

The area of the load piston is substantially larger than that of the effort piston and so a small effort applied to a hydraulic jack will lift a large load.

10.6 Speed, velocity and acceleration

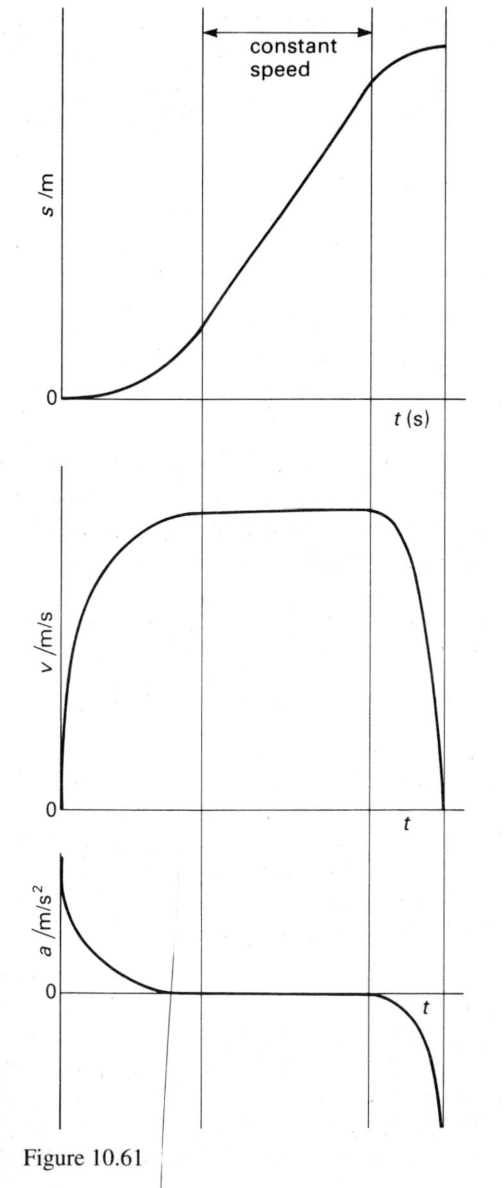

Figure 10.61

All mechanisms involve movement. The designer of a mechanism must therefore understand how its components move relative to one another. The simplest kind of motion is that in a straight line. Suppose a slider or cutting tool starts from rest, moves along a straight line and then comes to rest again. It is convenient to illustrate the motion by graphs of the distance travelled (or *displacement*) s, the speed v and the acceleration a as in Fig. 10.61.

They are all plotted with time t as a base and their general shapes could also apply to the motion of a complete vehicle such as a train travelling between two stations. The middle portion of each graph represents a part of the motion for which the speed is constant. On the displacement graph it is a sloping straight line, on the speed graph it is a horizontal line and over the same part of the movement the acceleration is zero. Note that the acceleration is negative during the slowing down phase; in other words, it is *deceleration*. At the start of the motion ($t = 0$) the displacement and speed are both zero but there is a finite acceleration from the outset.

A distinction is made between the terms *velocity* and *speed*. Velocity is a vector possessing magnitude and direction; speed refers to the magnitude only. The difference is important for motion along a curved path; even if the speed is constant there will be an acceleration due to the change of direction.

For motion in a straight line the slope at a point on the displacement graph represents the velocity. Slope does not mean the angle of the tangent to the curve as it appears on the paper but what it represents in terms of the values on the axis. In the same way, the slope of the velocity graph gives the acceleration. In the language of the calculus we say that velocity is the rate of change of displacement with time and acceleration is the rate of change of velocity with time. These results can be expressed by the differentials:

$$v = \frac{\mathrm{d}s}{\mathrm{d}t} \quad \text{and} \quad a = \frac{\mathrm{d}v}{\mathrm{d}t} = \frac{\mathrm{d}^2s}{\mathrm{d}t^2}$$

Furthermore, the area under the velocity curve between two points in time represents the distance covered in that period. In terms of the integral calculus:

$$s = \int v\,\mathrm{d}t \quad \text{and} \quad v = \int a\,\mathrm{d}t$$

In many examples, the velocity–time graph is particularly useful; the slope represents the acceleration and the area below the curve gives the distance travelled.

EXAMPLE 10.13 A cam is designed so that, for part of its motion, the follower moves with constant acceleration. If its velocity increases from 3 m/s to 5 m/s over a distance of 16 mm, find:

(a) the time taken,
(b) the acceleration,
(c) the distances covered in the two halves of the period.

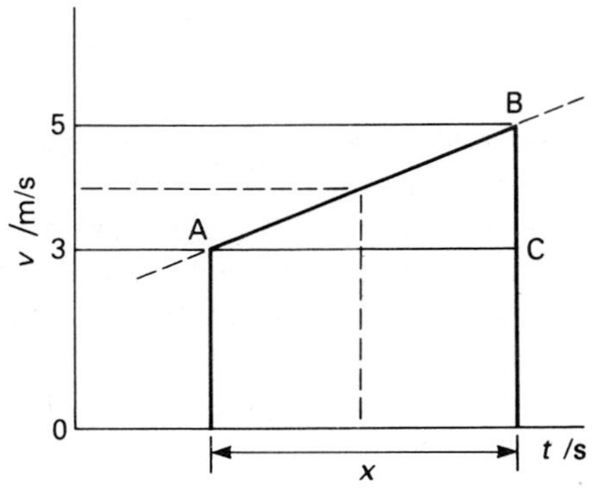

Figure 10.62

SOLUTION The velocity–time diagram is shown in Fig. 10.62. Since the acceleration is constant it is a sloping straight line for the part that is of interest in the present example.

(a) Suppose the time taken is x seconds as shown in the diagram. Then the distance is represented by the area of a trapezium with a base x and average height 4 m/s. Working in SI units,

$$4x = \frac{16}{1000} \quad \text{or} \quad x = \frac{4}{1000} = 0.004 \text{ s}$$

(b) The acceleration is represented by the slope BC/AC. Therefore

$$a = \frac{5-3}{0.004} = 500 \text{ m/s}^2$$

(c) After half the time has expired the velocity is given by the midpoint of AB, that is 4 m/s. Using the rule for the area of a trapezium, as before,

$$\text{distance covered in first } 0.002 \text{ s} = 0.002 \times \frac{(3+4)}{2}$$
$$= 0.007 \text{ m} \quad \text{or} \quad 7 \text{ mm}$$

Similarly,

$$\text{distance covered in second } 0.002 \text{ s} = 0.002 \times \frac{(4+5)}{2}$$
$$= 0.009 \text{ m} \quad \text{or} \quad 9 \text{ mm}$$

10.6.1 Uniform acceleration in a straight line

Motion with a constant acceleration occurs so frequently that it is convenient to have formulae linking the distance covered, time taken, acceleration and velocity. They are expressed in terms of the acceleration a but can be used for deceleration by giving negative values to a. The other symbols are:

t = time of acceleration,
s = distance covered during this period,
u and v = initial and final velocities.

With this notation the increase in velocity is at and we can put:

final velocity = initial velocity + increase in velocity

$$\text{or } v = u + at \tag{3}$$

The distance covered is given by (average velocity) × (time) and, in symbols

$$s = \frac{(u+v)}{2} \times t \tag{4}$$

or, substituting for v from equation (3),

$$s = \frac{(u+u+at)}{2} \times t$$
$$= ut + \tfrac{1}{2}at^2 \tag{5}$$

From equation (3), $t = (v - u)/a$ and, substituting this in (4), the distance is given by:

$$s = \frac{(u+v)}{2} \times \frac{(v-u)}{a} = \frac{v^2 - u^2}{2a}$$

from which

$$v^2 = u^2 + 2as \qquad (6)$$

If u and two of the four quantities, v, a, s and t are given, the other two can be found using equations (3) to (6).

The most common example of constant acceleration is motion under gravity near the surface of the Earth. If air resistance is neglected it may be assumed that an object moving freely has an acceleration g vertically downwards. For most purposes the value of the acceleration due to gravity g (on the surface of the Earth) may be taken to be 9.81 m/s^2 but you may find slightly different values, such as 9.8 or 10, in some examination papers.

Away from the Earth the value of g decreases, being inversely proportional to the square of the distance from the centre of the Earth.

EXAMPLE 10.14 Water from a fountain rises vertically to a height of 8m. What is the velocity at exit from the nozzle?

SOLUTION At the highest point the vertical velocity is zero. The formula to use is the one that includes the two velocities, the acceleration and the distance. This is equation (6).

$$v^2 = u^2 + 2as$$

Taking positive velocity and acceleration as upwards, we have $v = 0$,

$$a = -g = -9.81 \text{ and } s = 8 \text{ in SI units. Therefore,}$$

$$0 = u^2 - 2 \times 9.81 \times 8 = u^2 - 157.0$$

and the initial velocity

$$u = \sqrt{157.0} = 12.53 \text{ m/s}$$

EXAMPLE 10.15 A cricket ball is thrown with an initial velocity of 25 m/s at an angle of 40° to the horizontal across a level field. Find, neglecting air resistance,

(a) the time of flight,
(b) the greatest height during the flight,
(c) the horizontal distance covered.

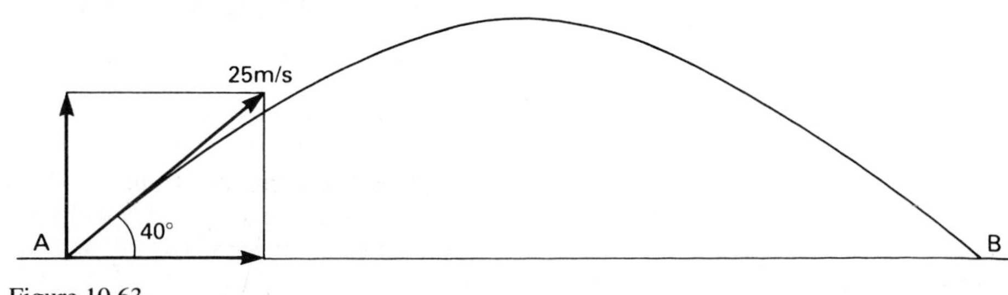

Figure 10.63

SOLUTION Fig. 10.63 shows the path of the ball and the direction of its initial velocity (at A). The initial velocity can be replaced by its horizontal and vertical components in the same way as a force (see Chapter 8). Therefore:

horizontal component of initial velocity $= 25 \cos 40° = 19.15$ m/s
vertical component of initial velocity $= 25 \sin 40° = 16.07$ m/s

The horizontal and vertical motions can be considered separately using these values.

(a) The height of the ball at time t is given by $s = ut + \frac{1}{2}at^2$, where u is the initial vertical velocity. The time of flight is that taken for the ball to return to ground level ($s = 0$). Putting $u = 16.07$ and $a = -g = -9.81$, the equation becomes

$$0 = 16.07t - 0.5 \times 9.81t^2$$

or

$$0 = t(16.07 - 0.5 \times 9.81t)$$

One solution is $t = 0$ and this corresponds to the start of the motion. The other is given by:

$$0 = 16.07 - 0.5 \times 9.81t$$

from which

$$t = 16.07/(0.5 \times 9.81) = 3.28$$

The time of flight is therefore 3.28*s*.

(b) At the highest point, the vertical velocity is zero. Therefore, using $v^2 = u^2 + 2as$ and putting $v = 0$, $u = 16.07$ and $a = -9.81$, we obtain:

$$0 = 16.07^2 - 2 \times 9.81s$$

or

$$s = 16.07^2/(2 \times 9.81) = 13.16 \text{ m}$$

The greatest height during the flight is therefore 13.16m.

(c) The horizontal velocity is constant and the distance covered, AB, is therefore equal to (velocity) × (time).

$$\text{distance AB} = 19.15 \times 3.28 = 62.81 \text{ m}$$

10.7 Angular motion

Consider a pulley rotating about an axis O as shown in Fig. 10.64. (It is usual to take anticlockwise movements as positive.) When the pulley turns through an angle θ, a point P on the circumference moves to a new position P′. If the angle is measured in radians, the length of the arc PP′ is $r\theta$. At the same time, a body suspended from a wire wound round the pulley will fall a distance s. If the wire is effectively wound at the same radius r then s = PP′ and we can put:

$$s = r\theta \quad \text{or} \quad \theta = s/r$$

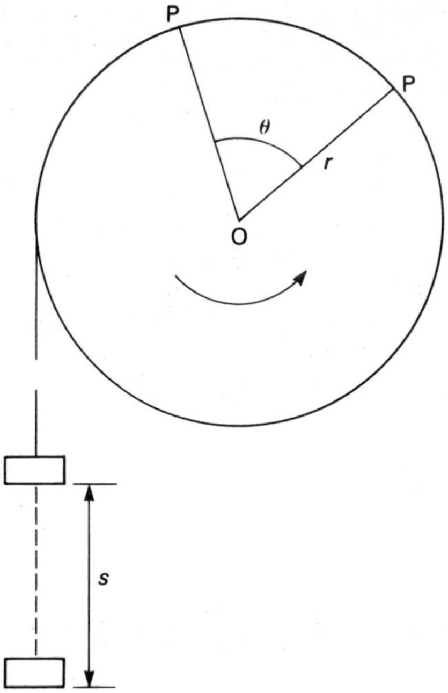

Figure 10.64 Pulley rotating about an axis.

The angular velocity ω (rad/s) and angular acceleration α (rad/s^2) are related to the equivalent linear values in the same way. Thus:

$$v = r\omega \quad \text{or} \quad \omega = v/r$$

and

$$a = r\alpha \quad \text{or} \quad \alpha = a/r$$

Equations 10.3 to 10.6 derived in section 10.6.1 for constant linear acceleration have their equivalent in angular motion. Distance s is replaced by angular displacement θ, the initial and final velocities u and v by the initial and final angular velocities ω_1 and ω_2, and acceleration a by angular acceleration α. The results become:

$$\omega_2 = \omega_1 + \alpha t$$

$$\theta = \frac{(\omega_1 + \omega_2)}{2} \times t = \omega_1 t + \tfrac{1}{2}\alpha t^2$$

$$\omega_2^2 = \omega_1^2 + 2\alpha\theta$$

Angular velocity is often quoted in rev/min for which the abbreviation r.p.m. is widely used. Since 1 revolution = 2π rad, it is necessary to multiply such values by $2\pi/60$ to obtain rad/s.

EXAMPLE 10.16 A car, whose wheels (with tyres) are 550 mm diameter, is travelling along a straight road at 75 km/h. P is a point on one of the wheels at a radius of 150 mm from the axis O. Find:

(a) the angular velocity of the wheel.
(b) the resultant velocity of P at the instant when the wheel has turned through 60° from the position for which P is at its highest point.

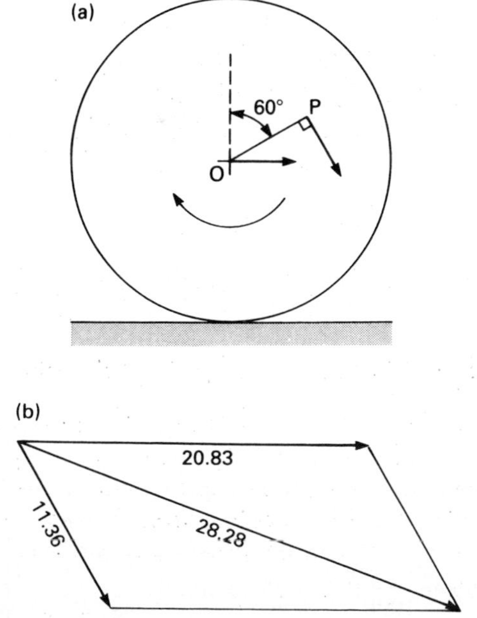

Figure 10.65

SOLUTION

(a) In SI units, the forward speed of the car is $75 \times 1000/(60 \times 60) = 20.83$ m/s. The forward speed of the car and the angular velocity of the wheel are related in the same way as the suspended object and the pulley of Fig. 10.64. Since $v = \omega r$ the angular velocity is given by:

$$\omega = v/r = 20.83/0.275 = 75.75 \text{ rad/s}$$

(b) If the wheel were rotating about a fixed axis with this angular velocity then P would have a linear velocity of

$$\omega r = 75.75 \times 0.150 = 11.36 \text{m/s}$$

in a direction at right angles to OP. At the instant given in the question, this direction is at 60° to the horizontal as shown in Fig. 10.65(a). However, O has a velocity of 20.83 m/s horizontally. The parallelogram of velocities for P is shown in Fig. 10.65(b). By measurement of a scale drawing, or by calculation, the resultant velocity of P (represented by the diagonal) is:

28.28 m/s at 20.4° to the horizontal, downwards to the right

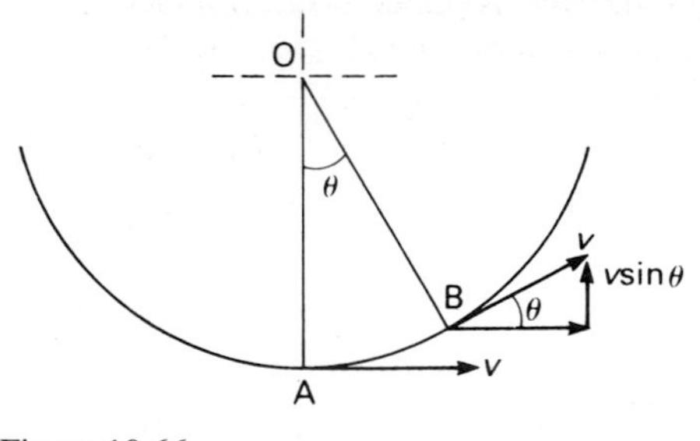

Figure 10.66

10.7.1 Centripetal acceleration

Consider a particle moving round a circle radius r at a constant speed v. Although the magnitude of its velocity is constant the direction is changing all the time so there must be an acceleration. Fig. 10.66 shows two points A and B on the circle that subtend an angle θ at the centre O. When the particle is at A, the direction of its velocity is tangential to the circle, and this velocity has no component in the direction AO.

Similarly at B, there is no component of the velocity in the radial direction BO. However, as the triangle of velocities shows, the velocity at B has a component in the direction AO; it is $v \sin \theta$. Between A and B therefore there is an increase in velocity of this amount in the direction AO.

Provided θ is measured in radians the length of the arc AB is $r\theta$ and, at speed v, the time taken to cover this distance is:

$$t = \frac{r\theta}{v}$$

Hence the average acceleration in the direction AO during this time period t is:

$$a = \frac{\text{increase in velocity}}{\text{time}} = \frac{v \sin \theta}{t} = \frac{v \sin \theta}{(r\theta/v)} = \frac{v^2}{r} \times \frac{\sin \theta}{\theta}$$

As θ becomes smaller the fraction $\sin \theta/\theta$ becomes closer and closer to 1. This can be checked with a scientific calculator. For example, with $\theta = 0.2$ rad, the ratio is 0.9933, to four places of decimals, and with $\theta = 0.1$ rad, it is 0.9983. In mathematical language, the ratio tends to the value 1 as θ tends to zero.

Therefore, the instantaneous value of the acceleration is:

$$a = \frac{v^2}{r}$$

and, since $v = \omega r$, it can also be written:

$$a = \frac{(\omega r)^2}{r} = \omega^2 r$$

Although the magnitude of this acceleration remains constant its direction is always towards the centre and therefore changing. Also, it is always at right angles to the direction of the velocity. It is called *centripetal* acceleration.

EXAMPLE 10.17 A stunt motor cyclist rides along a track that incorporates a vertical circular loop of 5 m radius as shown in Fig. 10.67. What is the minimum speed at which the machine must be travelling at the top of the loop if it is to remain in contact with the track?

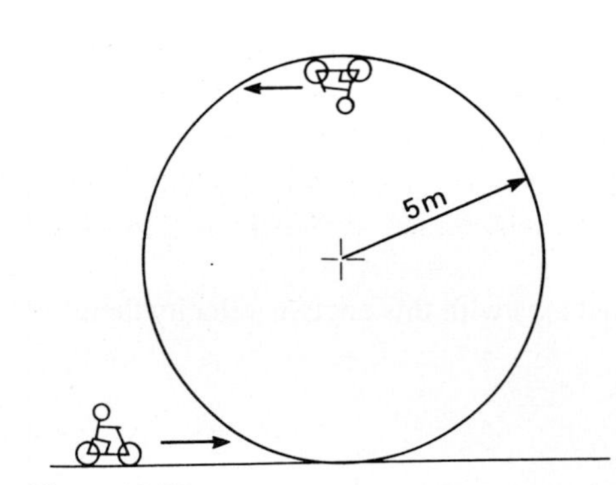

Figure 10.67

SOLUTION At the top of the loop the gravitational acceleration g must not be greater than the centripetal acceleration if the machine is to remain in contact with the track. Therefore, the minimum speed is given by:

$$\frac{v^2}{r} = g \quad \text{or} \quad v = \sqrt{gr}$$

Putting $g = 9.81$ m/s^2 and $r = 5$ m, minimum velocity is:

$$v = \sqrt{(9.81 \times 5)} = 7.00 \text{ m/s} \quad \text{or} \quad \frac{7.00 \times 3600}{1000} = 25.2 \text{ km/h}$$

10.8 Damped and forced vibrations

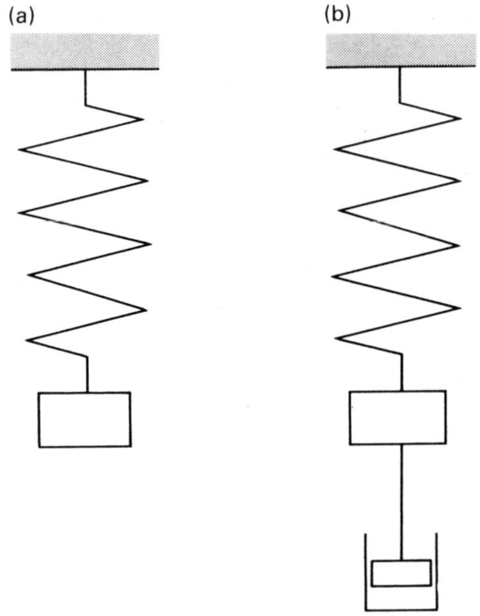

(a) (b)

Figure 10.68

If a mass at the end of a vertical spring, Fig. 10.68(a), is pulled down from its equilibrium position and released, it will oscillate or vibrate. If no driving force is applied to the system during the motion, the vibrations are said to be *free*. In a vacuum the motion would continue indefinitely and the graph of displacement against time would be a sine curve as shown in Fig. 10.69.

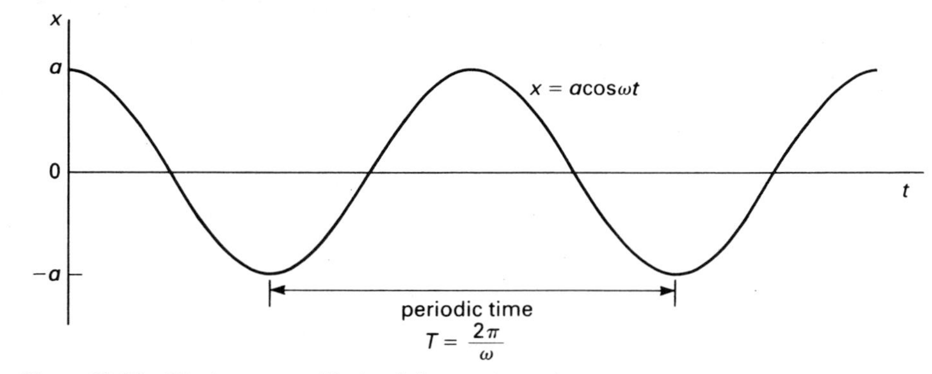

Figure 10.69 Displacement with simple harmonic motion.

In practice the vibrations will die away naturally because of air resistance. This *damping*, as it is called, can be increased artificially. One method is shown diagrammatically in Fig. 10.68(b); the mass is connected to a piston that moves up and down in a cylinder containing oil, a device known as a *dashpot*. The viscosity of the oil produces a resisting force on the piston which is proportional to its velocity. The effect on the motion can be seen in Fig. 10.70 which shows how successive amplitudes become smaller and smaller. Damping can be produced by other devices, such as rubber mountings. Car designers use a combination of springs and shock absorbers to produce damped vibrations and achieve a comfortable ride.

A *forced* vibration is one in which a force is applied to the vibrating mass during the motion. If a young child sitting on a playground swing is given a single push the oscillations will gradually die out. Experience shows that the amplitude can be maintained or even increased by applying a small force each time the child is at the highest point of the motion.

A similar effect occurs with spring systems that support rotating machines. If a motor or engine is not perfectly balanced it applies a force to its mountings that varies in the form of a sine wave. This applied force is called the *excitation* force. The vibrations will persist as long as the excitation force is applied. The amplitude will depend on the magnitude of the force and the frequency at which it is applied.

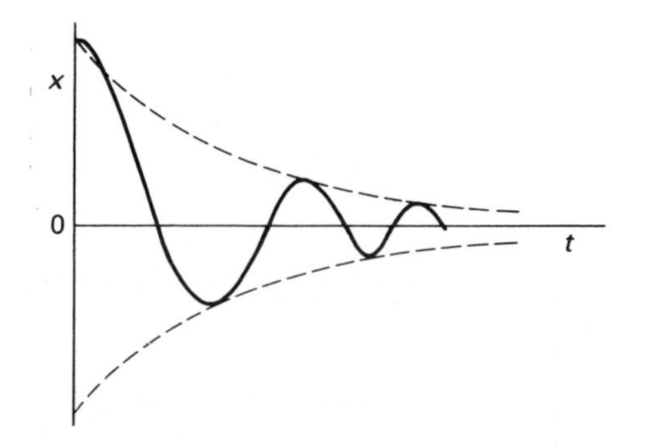

Figure 10.70 Damping effect on oscillations.

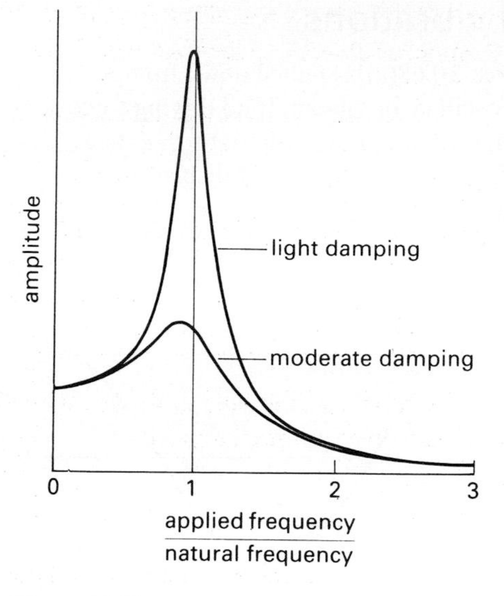

Figure 10.71

Fig. 10.71 shows some possible variations of amplitude as the frequency is increased from zero up to and beyond the natural frequency of free vibrations. As the diagram shows, there can be a large increase in amplitude as the applied frequency approaches the natural frequency. The size of the maximum amplitude depends on the level of damping. If there were no damping at all the amplitude would theoretically become infinite. When the applied frequency approaches that of the natural frequency the system experiences large oscillations, and is said to be in *resonance*. The high noise level that can occur in cars at particular engine speeds is an example of this effect.

10.9 Summary of formulae

The main results and formulae required for linear motion and rotational motion are collected in the following table (10.1).

	Linear motion			*Rotary motion*	
	term	*units*	*linear-rotary links*	*term*	*units*
Displacement	s	m	$s = r\theta$	θ	rad
Initial velocity	u	m/s		ω_1	rad/s
Final velocity	v	m/s	$v = r\omega$	ω_2	rad/s
Acceleration	a	m/s^2	$a = r\alpha$	α	rad/s^2
Equations of motion	$v = u + at$ $s = \frac{1}{2}(u + v)t$ $v^2 = u^2 + 2as$ $s = ut + \frac{1}{2}at^2$			$\omega_2 = \omega_1 + \alpha_1$ $\theta = \frac{1}{2}(\omega_1 + \omega_2)t$ ${\omega_2}^2 = {\omega_1}^2 + 2\alpha\theta$ $\theta = \omega_1 t + \frac{1}{2}\alpha t^2$	

Table 10.1 Summary of formulae

10.10 Bicycle analysis

When riding a bicycle, the drive wheel is turned by the pedals and, in rotating, it moves a certain distance along the ground. Therefore we have a clear link between linear and rotary motion. By analysing a simple bicycle journey, we can get a better understanding of many of the topics covered in this chapter. We can see the relationship between linear and rotational motion and the contribution that the gearing makes to power and torque requirements.

10.10.1 Bicycle power and torque

From 10.3.3 we know that:

power = torque × angular velocity

or $P = T\omega$

If we produce a steady power and assume no losses, then $T_{out}\omega_{out}$ at the rear wheel must be constant. In order to climb a hill or overcome wind resistance we will need to increase the torque on the rear wheel (T_{out}). If we do this, ω_{out} must therefore decrease. This means that we need to reduce speed in order to get the necessary force to climb a hill or battle against a headwind.

The rotational speed of the rear wheel is the same as the speed of the rear cog as they are mounted on the same axis. As we saw in 10.3.2, for a simple gear system (linked by a chain):

$N_{in}.t_{in} = N_{out}.t_{out}$ where N is the speed in rev/min (r.p.m.)
and t is the number of teeth in the gear wheel

To increase the torque output N_{out} must drop and so t_{out} must rise. Therefore, if we change to a larger gear wheel with a greater number of teeth on the back wheel, the speed will fall and the torque available will rise.

10.10.2 Bicycle gearing

Let the front chainwheel of the bicycle be gear A and the rear cog gear B. For a bicycle the gear ratio, g_r, is usually given by

$$\frac{\text{number of teeth in chainwheel (driver)}}{\text{number of teeth in cog (driven)}} = \frac{t_A}{t_B}$$

The continental gear development, g_d, is often used as a measure of the gearing of a bicycle and bikes can be compared very easily using this measurement.

$$g_d = \frac{t_A}{t_B} \times \pi \times d$$

As $\pi \times d$ is the circumference of the wheels, the gear development measures how far along the ground you would travel for 1 complete turn of the pedals. With a front chainwheel of 50 teeth, a rear cog of 16 teeth and a wheel diameter of 0.7 m the continental gear development for this bike and combination of gears is

$$g_d = \frac{50}{16} \times \pi = 0.7 = 6.87 \text{ m/rev}$$

If we examine all the possible gear combinations of a particular 16 speed bicycle with a front chainwheel of 50 or 42 teeth and a rear cog of 11, 13, 16, 18, 20, 23, 26 or 28 teeth we find that the gear development ranges from 3.30 to 10.00 (Table 10.2)

Rear cog, B / *Front chainwheel, A*	28	26	23	20	18	16	13	11
50	3.93	4.23	4.78	5.50	6.11	6.87	8.46	10.00
42	3.30	3.55	4.02	4.62	5.13	5.77	7.10	8.40

Table 10.2 Continental gear development for a 16 speed bicycle

To find the possible speed in each gear we need to know our pedal cadence, or comfortable speed of pedalling. This is found to be between 60 and 80 r.p.m. If a cyclist has a pedal cadence p_c, in one minute he will travel $g_d \times p_c$ metres. Therefore the speed, v, in m/s is

$$g_d \times \frac{p_c}{60} \text{ m/s}$$

For the 50/16 gear combination again we find that the speed at a pedal cadence of 60 r.p.m. is 6.87 m/s and for 80 r.p.m.

$$\text{speed} = \frac{6.87 \times 80}{60} = 9.16 \text{ m/s}$$

Table 10.3 shows the range of speeds possible for each gear combination. It is obvious from this that there is an overlap in the gears and it is not necessary or helpful to use them all. From 10.10.1 we would expect the largest rear cog to give us the greatest torque and the lowest speed and the smallest rear cog to give us the lowest torque and the highest speed. This is shown to be true and means that the 42/28 combination is the lowest gear and the 50/11 combination is the highest gear.

Front chainwheel, A \ Rear cog, B	28	26	23	20	18	16	13	11
50	3.93–5.24	4.23–5.64	4.78–6.37	5.50–7.33	6.11–8.15	6.87–9.16	8.46–11.28	10.00–13.30
42	3.30–4.40	3.55–4.74	4.02–5.35	4.62–6.16	5.13–6.84	5.77–7.70	7.10–9.47	8.40–11.20

Table 10.3 Speed range for each gear of a 16 speed bicycle

10.10.3 Analysing a bicycle journey

The information below describes a simple bicycle journey by an experienced cyclist on a straight road from which a velocity–time graph can be drawn. From the information and the graph the total linear and angular distance, the maximum linear and angular velocity and the maximum linear and angular acceleration and retardation can all be found.

> The cyclist began by accelerating uniformly for the first 30 metres and stopped accelerating after 10 seconds. For the next 3 s the cyclist travelled at uniform velocity. The cyclist then accelerated uniformly again, travelling a further 14 m in 2 s. The cyclist then travelled at uniform velocity for 10 s. Finally, the cyclist braked uniformly to rest taking 5 s to do so. The cyclist used a 50/26 gear combination for the first 9 s of the journey before changing up to a 50/20 combination.

In order to plot a velocity-time graph, we need to calculate the velocity at the critical points in the journey. It helps to divide the journey into sections and draw a sketch to show the type of motion in each section, Fig. 10.72.

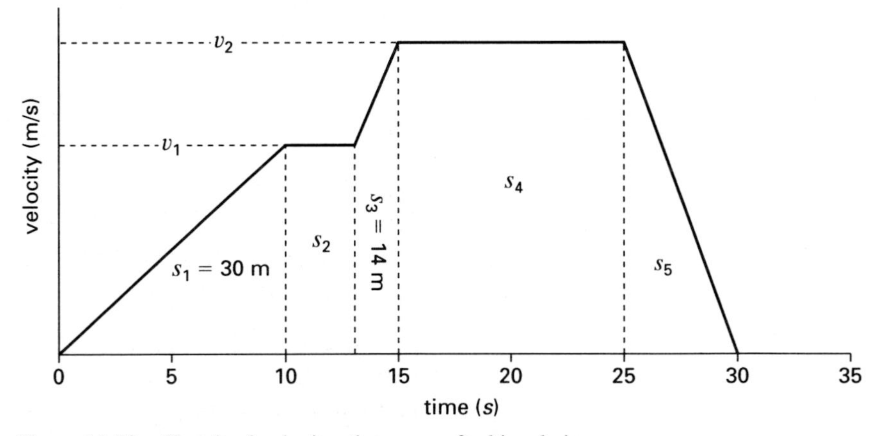

Figure 10.72 Sketch of velocity–time curve for bicycle journey

We know that the initial linear distance travelled, s_1, is 30 m. The linear speed at the end of the initial acceleration phase, v_1, is given by

$$s_1 = \tfrac{1}{2}(u + v_1)t \quad \text{where } s_1 = 30,\ u = 0 \text{ and } t = 10 \quad \text{(see 10.6.1 and 10.9)}$$

$$\text{therefore} \quad v_1 = \frac{60}{10} = 6 \text{ m/s}$$

For the third stage of the journey $s_3 = 14$, $t = 2$, the initial speed, $v_1 = 6$ and the final speed $= v_2$.

$$\text{using} \quad s_3 = \tfrac{1}{2}(v_1 + v_2).\text{t} \quad \text{we find that } v_2 = \frac{14 \times 2}{2} - 6 = 8 \text{ m/s}$$

Now we can plot AB on the graph from the origin to 6 m/s after 10 s (Fig. 10.73). The cyclist now travels at constant speed for 3 seconds so BC can then be drawn horizontally for the next 3 seconds on the graph. C can then be drawn to D which is the point 15 s after the start of the journey when he is travelling at a speed of 8 m/s. DE can then be drawn horizontally for the next 10 seconds and EF drawn back to the x axis to represent the braking to rest.

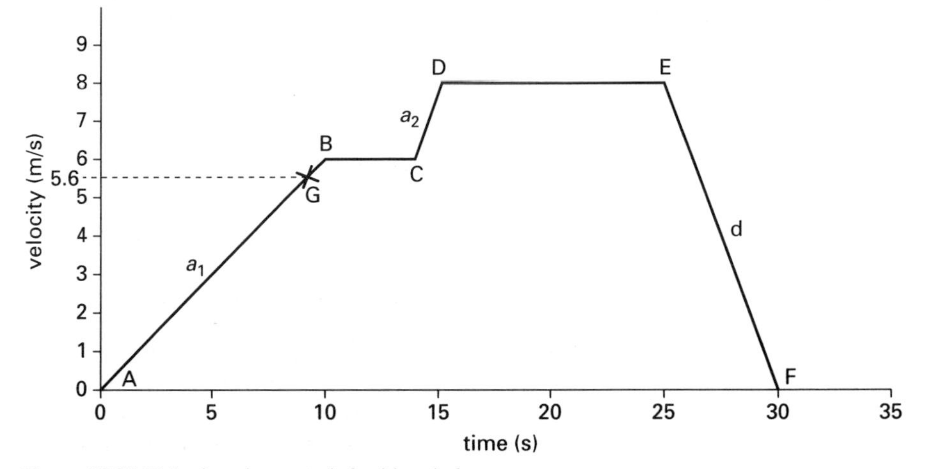

Figure 10.73 Velocity–time graph for bicycle journey

If we wish to find the total distance travelled on this journey we need to find the distances for each stage of the journey.

s_1 is 30 m and s_3 is 14 m.

For section 2 at constant speed

$$s_2 = v_1.t_2 = 6 \times 3 = 18 \text{ m}$$

For section 4 at constant speed

$$s_4 = v_2.t_4 = 8 \times 10 = 80 \text{ m}$$

And for section 5 decelerating to rest

$$s_5 = \tfrac{1}{2}.v_2\, t_5 = \tfrac{1}{2} \times 8 \times 5 = 20 \text{ m}$$

therefore the total distance travelled is 30 + 18 + 14 + 80 + 20 = 162 m

From 10.7 we know that the linear and angular distances are related by $s = r\theta$ and so the total angular distance turned by the wheel is given by θ, where

$$\theta = \frac{s}{r} = \frac{162}{0.35} = 463 \text{ radians}$$

As there are 2π radians in each revolution, this means that the wheels have fully rotated

$\dfrac{463}{2\pi}$ or 73.7 times during the whole journey.

We may also wish to find the acceleration or deceleration of the bike at different stages. The acceleration, a_1, in the first stage (AB) is given by

$$\frac{v_1}{10} = \frac{6}{10} = 0.6 \text{ m/s}^2$$

The second acceleration, a_2, in the stage (CD) is given by

$$\frac{v_2 - v_1}{2} = \frac{8-6}{2} = \frac{2}{2} = 1 \text{ m/s}^2$$

therefore the maximum acceleration is 1 m/s^2.

The deceleration, d, is given by

$$\frac{v_2}{5} = \frac{8}{5} = 1.6 \text{ m/s}^2$$

From the graph we can see that the maximum speed of the bike is 8 m/s. The maximum angular speed of the wheel, ω, is given by

$$v_2 = r\omega \quad \text{where } r \text{ is } 0.35 \text{ m}$$

$$\text{therefore} \quad \omega = \frac{8}{0.35} = 22.9 \text{ rad/s}$$

The maximum angular acceleration of the wheel is given by α where

$$a = r\alpha \quad \text{and} \quad a = 1 \text{ m/s}$$

$$\text{therefore} \quad \alpha = \frac{1}{0.35} = 2.86 \text{ rad/s}^2$$

The angular deceleration on the final part of the journey is

$$\frac{1.6}{0.35} = 4.57 \text{ rad/s}^2$$

It is also possible to find the pedal cadence just before and just after changing gear on the initial stage (represented by G on Fig 10.73). In the first gear combination of 50/26 the gear development is 4.23 and

$$\text{velocity, } v = 4.23 \times \frac{p_c}{60}$$

$$\text{and so } p_c = \frac{v}{4.23} \times 60$$

The speed at the point of changing gear, v is given by

$$v = a.t = 0.6 \times 9 = 5.4 \text{ m/s}$$

$$\text{therefore } p_c = \frac{5.4}{4.23} \times 60 = 76.6 \text{ r.p.m.}$$

This is approaching the maximum comfortable pedal cadence of 80 r.p.m.. After changing to a gear combination of 50/20 The gear development is now 5.50 and the new pedal cadence is

$$p_c = \frac{5.4}{5.5} \times 60 = 58.9 \text{ r.p.m.}$$

This is a low cadence and therefore the cyclist could gradually increase this to increase his speed further.

At the end of the energy chapter (11.10) we will resume the bicycle analysis and look at the torque and power available on this journey.

Bibliography

Bailey O, Pickup R, Lewis R and Patient P 1981 *Modular Courses in Technology: Mechanisms*. Oliver and Boyd

Chironis N P 1965 *Mechanisms, Linkages and Mechanical Controls*. McGraw-Hill (for reference only)

Hannah J and Hillier M J 1995 *Applied Mechanics* 3rd edn. Longman

Hannah J and Stephens R C 1984 *Mechanics of Machines – Elementary Theory and Examples* 4th edn. Edward Arnold

Humphrey D and Topping J 1971 *A Shorter Intermediate Mechanics* SI edn Ch 1–2, 7–8. Longman

Rich S 1991 *Mechanisms.* Stanley Thornes

Assignments

1 The diagram in Fig. 10.74 shows a lever, where X is the load, Y the effort and F the fulcrum.

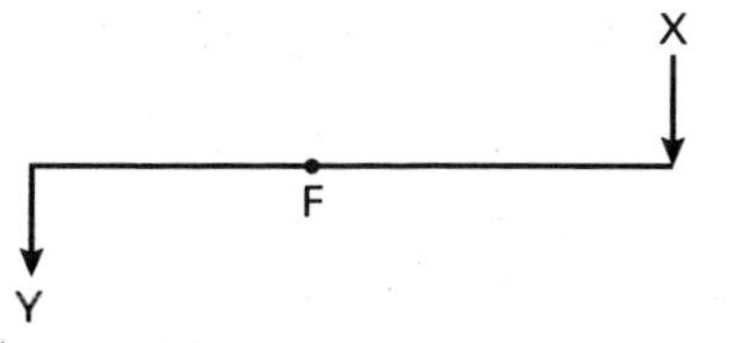

Figure 10.74

(a) The mechanical advantage of this lever is

A $X + Y$

B $X - Y$

C $\frac{X}{Y}$

D $\frac{Y}{X}$

(b) Moving F closer to X

A reduces the velocity ratio

B increases the velocity ratio

C decreases efficiency

D decreases the mechanical advantage.

2 Two jaws of a gripping device are required to open and close a distance of approximately 20 mm and exert a force 20 times greater than that available to operate the device.

Describe three different means of achieving this, the main principles involved, and in each case explain how the relationship between the forces is achieved.

3 (a) An engine develops a torque of 150 Nm at the flywheel at a speed of 2500 rev/min. The constant mesh gears of the gearbox (Fig. 10.75) through which the engine is driving have 20 teeth and 35 teeth, the second gear mainshaft pinion has 30 teeth and the 'layshaft' gear which meshes with it 25 teeth.

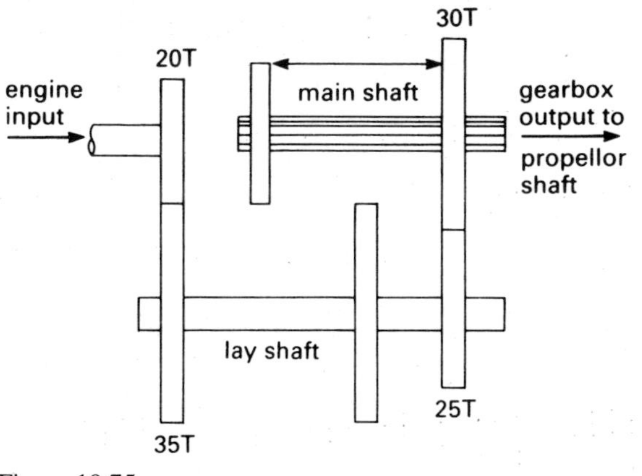

Figure 10.75

Calculate;

(i) the speed of the propeller shaft.

(ii) the torque on the rear wheels given that the rear axle rotates at one fifth the speed of the propeller shaft.

(b) A system of pulleys (Fig. 10.76) consists of two blocks each containing three pulleys, one block being fixed and the other moveable. If an effort of 10N raises a load of 50N, calculate the efficiency of the system.

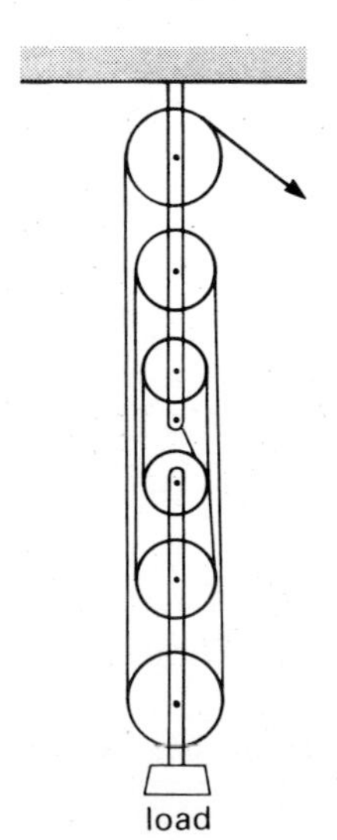

Figure 10.76

4 A clockwork motor used in a child's toy car has sufficient energy in the spring to rotate 20 times when released. On the same shaft as the spring is a gear wheel with 60 teeth. This gear drives a gear wheel with 25 teeth. On the same shaft as the 25 teeth gear wheel is another gear with 40 teeth; this drives a gear wheel on the axle of the toy which has 20 teeth. The wheels on this axle are 35 mm diameter.

Assuming no slip, how far will the toy travel?

5 (a) With the aid of sketches give details of two ways of coupling two shafts together to enable motion and torques to be transmitted.

(b) Sewing machines, internal combustion engines and workshop pillar drills all use belts and pulleys to transmit power and motion.

(i) For each of the three above mentioned cases, specify a type of pulley and belt which is most suitable and give reasons why that type of belt is used.

(ii) Using sketches, show how any one of the pulleys would be located on its particular shaft.

(c) A simple treadle type sewing machine is to be motorised by the addition of a 240 volt electric motor using belt and pulley drive. If the machine makes one stitch for each revolution of the machine pulley and the motor runs at 845 rpm

(i) State the velocity ratio if the motor pulley diameter is 24 mm and the machine pulley diameter is 78 mm.

(ii) How long would it take (in seconds) to sew a seam of 780 mm if the stitch is 2 mm?

6 (a) Describe, with the aid of sketches and/or appropriate graphical symbols, the following types of gear giving an application for each.

(i) Bevel gears.

(ii) Rack and pinion.

(b) A simple mechanical winch is shown in Fig. 10.77. It consists of a single start worm attached to a winding crank and a 35 tooth wheel attached to a

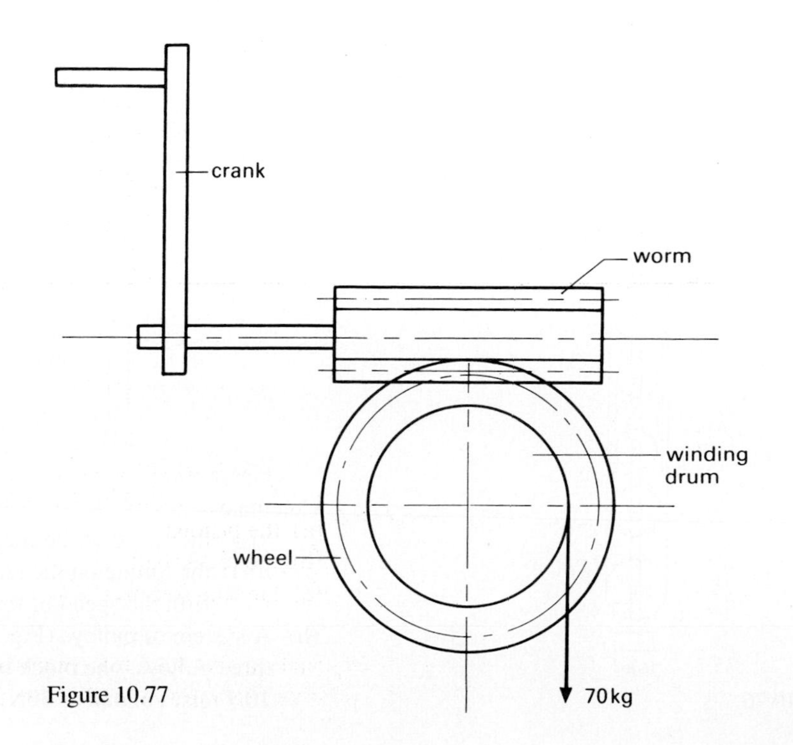

Figure 10.77

winding drum. The crank has a radius of 200 mm and the winding drum has a diameter of 250 mm. An effort of 50 N is needed to raise a load of 70 kg. Determine the efficiency of the mechanism at this load.
(Assume that $g = 10$ m/s^2.)

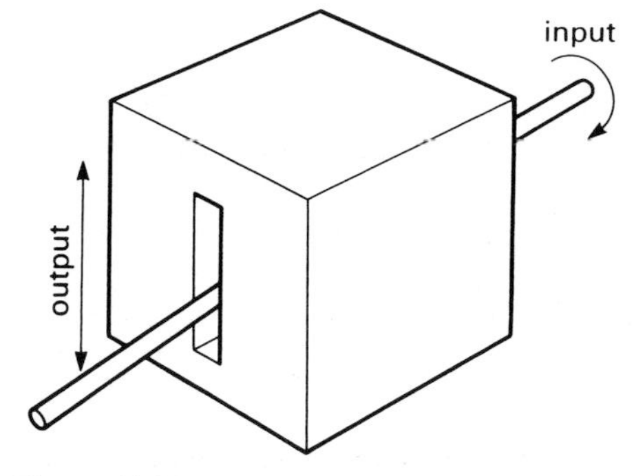

Figure 10.78

7 (a) A screw thread can be visualised as an inclined plane wrapped around a cylinder. Explain, using sketches, what the term 'pitch' means in relation to a thread and then for a screw jack develop an expression for its velocity ratio.
(b) The cutting tool on a lathe is moved horizontally by a screw which rotates at 360 rev/min. The pitch of the screw is 2.5 mm and the force required at the cutting tool is 600 N. Calculate:
(i) the linear speed of the tool in mm/s;
(ii) the power output at the tool;
(iii) the power input to the screw. (Assume an efficiency of 45%.)
(c) Using diagrams to illustrate your ideas, devise a mechanism to go into the 'black box' shown in Fig. 10.78 which will satisfy the input and output requirements. The input is rotating and the output has to be an up and down motion moving through an angle of 45 degrees.

8 (a) Friction can reduce the efficiency of mechanisms. Using a bicycle as an example, discuss two ways in which the friction can be reduced.
(b) A bicycle with only a front caliper type brake is particularly inefficient at braking in wet weather. Explain why this is the case.
(c) Using sketches describe a braking system for a bicycle which is more efficient than a caliper for both front and back wheels.
(d) Fig. 10.79 shows an ordinary Derailleur gear system for a bicycle.
(i) What is the purpose of the jockey wheels in the system?
(ii) Calculate the mechanical advantage when cycling in second gear if the bicycle is 70 per cent efficient.
(iii) Discuss, with the aid of sketches, why alternative drive systems to a chain and sprocket are very hard to achieve successfully on a bicycle.

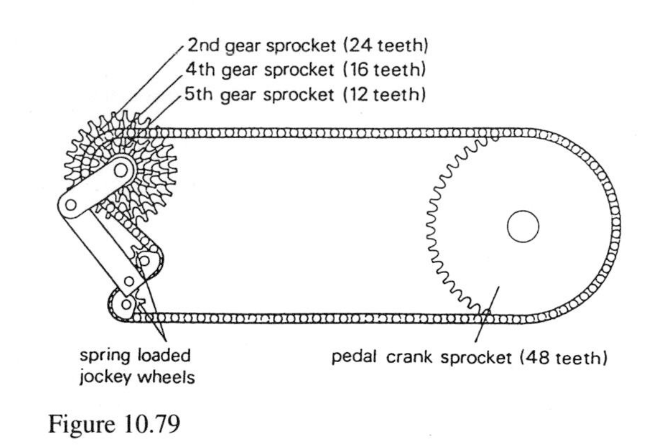

Figure 10.79

9 Using sketches describe the form of motion of each of the following mechanisms found in an internal combustion engine:
(a) the pistons,
(b) the crankshaft,
(c) the cam follower.

10 Make a sketch of any common cam profile, with its associated follower, and give a specific example of its use.

11 The drawing Fig. 10.80(a) and (b) shows the arrangement used in a motor car engine to operate the valves. It consists of a cam, push rod, rocker, valve and spring.

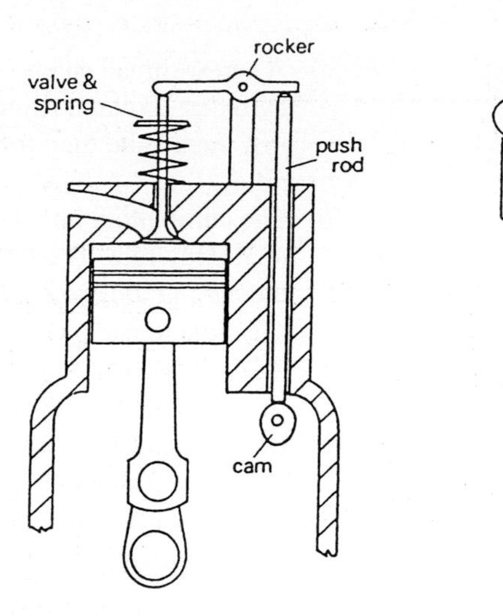

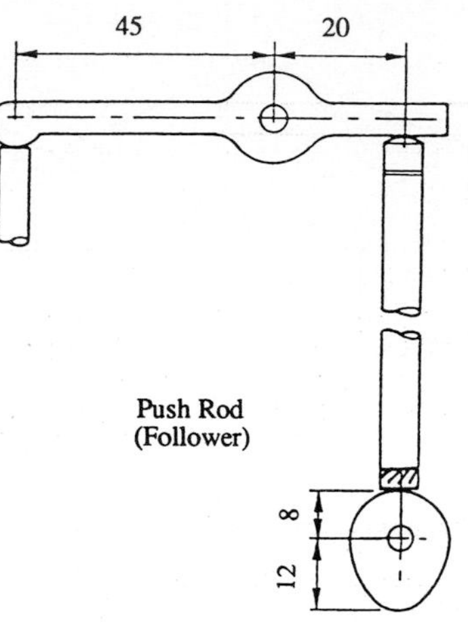

Figure 10.80

(a) Calculate:

(i) the maximum lift of the valve;

(ii) the force experienced in the push rod if the spring, when compressed, exerts a force of 200 N.

(b) The performance of the engine is to be improved to achieve greater acceleration.

(i) What modifications could be made to the valves (inlet and exhaust) and the valve spring to help achieve this improvement?

(ii) How does the shape of the cam and follower (push rod end) influence the engine's performance?

(c) Lubrication is vital to the working on an internal combustion engine.

(i) Give two reasons why lubrication is important;

(ii) Explain the meaning of '20W50' on a 5 litre can of oil;

(iii) Describe, with the aid of sketches, how oil may be distributed to the rockers and valves of an engine.

12 (a) Describe three different types of mechanism which could be used to enable a person to raise a 400 kg weight on to a platform 1 metre high without the assistance of an external energy source. The only source of energy is that provided by the operator.

(b) In each case:

(i) providing your own estimates of any necessary data, explain, using diagrams and calculations, how the mechanical advantage is achieved; and

(ii) explain what factors, other than the mechanical principle, would influence the design of a device which could be used for raising the weight.

13 The arrangement to raise or launch yachts at a marina dock is shown in Fig. 10.81. The block and tackle employed has a velocity ratio of 6.

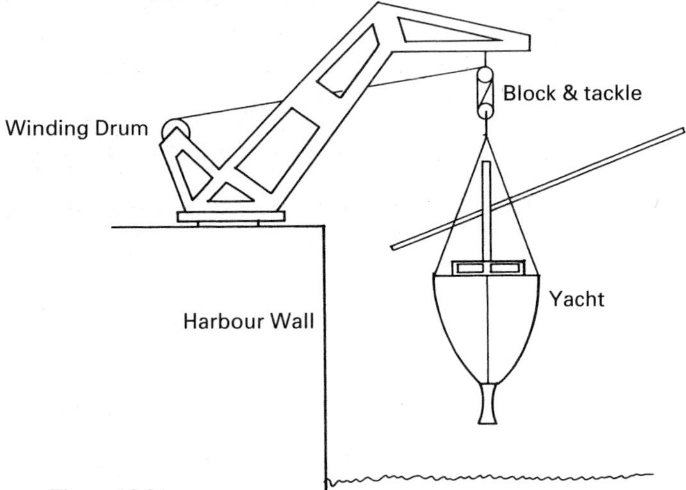

Figure 10.81

(a) (i) Draw a suitable block and tackle arrangement which will give the required velocity ratio.
(ii) If the winding drum diameter is 750 mm and has a stabilised rotational velocity of 25 rev/min, calculate the time for a yacht to be lowered 5 m to the water's surface.

(b) As the winding drum is powered by a large electric motor, a power failure when raising or lowering would be very serious in its outcome. In the event of such a failure:
(i) describe, with the aid of sketches, a system which would automatically slow the rate of fall of a yacht during a launch;
(ii) describe, with the aid of sketches, a system which could prevent a yacht falling back into the water when it was being raised.

14 The speed of a lorry increases from 12 km/h to 48 km/h in a distance of 200 m with constant acceleration. Find:
(a) the acceleration,
(b) the time taken,
(c) the speed in km/h after half this time,
(d) the speed when the lorry has covered half the distance.

15 A load is lowered on a rope which passes round a circular drum of 0.4 m radius. If the load starts from rest and moves with constant acceleration, dropping 30 m in 20 s, what is the angular acceleration of the drum?

16 (a) It is often necessary on machines such as a tractor or four wheel drive vehicle to damp down any vibrations or reduce the effects of shock to which a coupled pair of rotating drive shafts are subjected.
(i) What is the name of the coupling most suited to this application?
(ii) Using sketches, describe one example of the type of coupling you have identified and give two practical applications.

(b) The gearbox on a tractor has a ratio of 8.5:1 for first gear. The differential ratio is 4.5:1 and the rear tyre diameter is 1.5 m
(i) Calculate the drive ratio between the engine and the rear axle in first gear.
(ii) Calculate the road speed in km/h for a constant engine speed of 2500 rev/min in first gear.

(c) Describe, with the aid of sketches, one suitable method which could be used for transmitting steering motion to the wheels of the tractor.

11 Energy and dynamics

11.1 Definitions and units

In everyday conversation the words *force*, *energy* and *power* are often used with similar meanings but in science and technology they are distinct numerical quantities each with its own units.

Force is the quantity which causes, or tends to cause, a change in the position or state of a body. As explained in section 8.1.2 if a body remains at rest (or moves at a constant velocity), the forces acting on it are in equilibrium and have no resultant. Force is a vector quantity and is measured in *newton*, N.

Suppose a force F (Fig. 11.1) causes a body to move a distance x in the direction of action. The force is said to do *work*. Work is a means of transferring energy by the movement of the point of application of a force through a distance in the direction of the force. Provided it is constant, the amount of work done or energy transferred is equal to the force multiplied by the distance moved, Fx. Work is a scalar quantity.

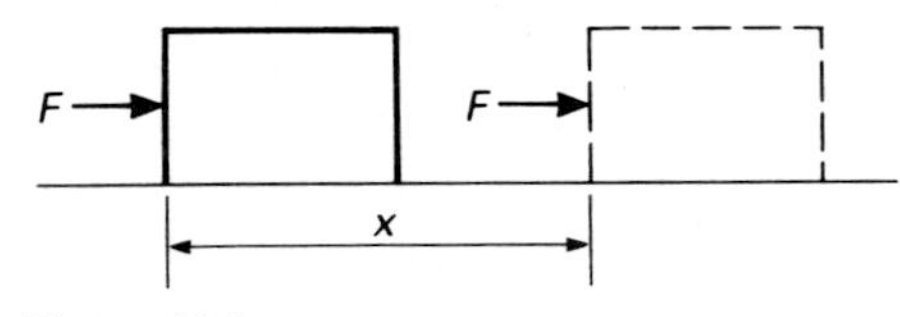

Figure 11.1

From the definition of work its unit could be written newton metre (N m) but this is given the special name *joule* and is denoted by J. (N m is reserved for the unit of moment.) The joule is a very small unit of work and practical values are often given in kilojoule (kJ), megajoule (MJ) or gigajoule (GJ).

There are many instances in our daily lives of work being done in this technical sense and we frequently use machines to help us. For personal transport we can use a pedal cycle or a petrol-engined moped. We can cut the grass on a lawn by pushing a hand mower or we can use one with a petrol engine or electric motor. In these examples the engines or motors do similar amounts of work to ourselves but more quickly. We therefore judge motors and engines not by the total amount of work done but by the *rate* at which they do work. This is called the *power* and in SI units it is the number of joules per second. This unit is given the name *watt* and is denoted by W. The multiple units kilowatt (kW) and megawatt (MW) are often used for practical values of power. Power, like work, is a scalar quantity possessing magnitude but not direction.

Suppose the force F in Fig. 11.1 moves through the distance x in time t at a constant speed v, then, by definition, the power P is

$$P = \text{rate of doing work} = \frac{\text{work done}}{\text{time taken}} = \frac{Fx}{t} = Fv$$

since speed is equal to distance divided by time. If the speed is changing, the last expression gives the instantaneous value of the power.

The unit *horsepower*, introduced by James Watt, is widely used. It is abbreviated to hp, and 1 hp = 746 W. The metric horsepower (equal to 735 W) is also used in some countries.

EXAMPLE 11.1 What is the minimum energy transferred to 40 gallons of water when it is raised to a height of 15 metres? Take 1 gallon = 4.55 litres.

What is the effective power output of a pump that can achieve this in 25 seconds?

SOLUTION The mass of 1 litre of water may be taken as 1 kg. Hence the mass of 40 gallons is $40 \times 4.55 = 182$ kg. The force F required to raise this mass is equal to the weight of the water and therefore:

$$F = 182 \times 9.81 = 1785.4 \text{ N}$$

$$\text{Work done} = \text{force} \times \text{distance} = 1785.4 \times 15 = 26\,780 \text{ J} \quad \text{or} \quad 26.78 \text{ kJ}$$

$$\text{power} = \frac{\text{work done}}{\text{time taken}} = \frac{26\,780}{25} = 1071 \text{ W} \quad \text{or} \quad 1.071 \text{ kW}$$

EXAMPLE 11.2 A car travels at a uniform speed along a level road, the total frictional resistance being 350 N. Find

(a) the work done in overcoming the frictional resistance over a distance of 600 m,
(b) the corresponding power if the speed is 72 km/h,
(c) the power required if the car has a mass of 900 kg and is climbing a slope of 1 in 20 at the same speed.

SOLUTION

(a) work done = force × distance

$$= 350 \times 600 = 210 \times 10^3 \text{ J} \quad \text{or} \quad 210 \text{ kJ}$$

(b) converting to metre and second units the speed is

$$v = \frac{72 \times 1000}{3600} = 20 \text{ m/s}$$

$$\text{power} = \text{force} \times \text{speed} = 350 \times 20 = 7000 \text{ W} \quad \text{or} \quad 7 \text{ kW}$$

(c) Gradients are often expressed in the form of a ratio rather than an angle (Fig. 11.2). The ratio may be taken to be the sine of the angle (as indicated in the diagram) or its tangent. With small angles the difference between the results is negligible. In addition to the frictional resistance, the force F now has to overcome the component of the weight down the slope, $W \sin \theta$. Thus

$$F = \text{frictional resistance} + W \sin \theta = 350 + (900 \times 9.81) \times (1/20) = 791.4 \text{ N}$$

$$\text{power} = \text{force} \times \text{speed} = 791.4 \times 20 = 15828 \text{ W} \quad \text{or} \quad 15.83 \text{ kW}$$

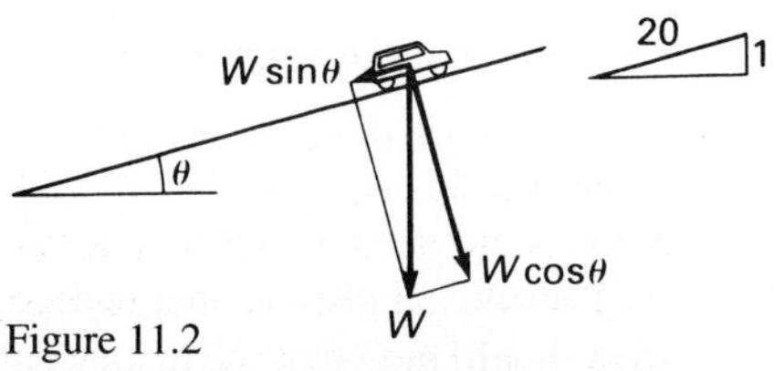

Figure 11.2

EXAMPLE 11.3 A small airliner achieves a top speed of 300 knots in level flight when the engines are generating 5000 hp. (1 knot = 0.5148 m/s.)
Find
(a) the drag at this speed,
(b) the thrust horsepower required at 250 knots assuming that the drag is proportional to the square of the speed.

SOLUTION
(a) Converting the speed and power to SI units,

$$\text{speed } v = 300 \times 0.5148 = 154.4 \text{m/s}$$
$$\text{power } P = 5000 \times 746 = 3.73 \times 10^6 \text{ W}$$

If the airliner is flying at its top speed there is no acceleration and the drag D is equal to the thrust. Hence:

$$D = \frac{\text{power } P}{\text{speed } v} = \frac{3.73 \times 10^6}{154.4} = 24\,160 \text{ N} \quad \text{or} \quad 24.16 \text{ kN}$$

(b) If the drag is proportional to the square of the speed then its value at the lower speed is:

$$D = \left(\frac{250}{300}\right)^2 \times 24\,160 = 16\,780 \text{ N}$$

In SI units the new speed is 250 × 0.5148 = 128.7 m/s and

$$\text{power required} = 16\,780 \times 128.7 = 2.16 \times 10^6 \text{ W}$$

$$\text{or} \quad \frac{2.16 \times 106}{746} = 2895 \text{ hp}$$

EXAMPLE 11.4 During the expansion stroke of an internal combustion engine the force exerted on the piston by the gas in the cylinder varies with the piston displacement as follows:

Piston displacement (mm)	0	5	10	20	40	60	80
Force (kN)	20	11.83	8.14	4.81	2.47	1.58	1.15

Estimate the work done on the piston during the stroke.

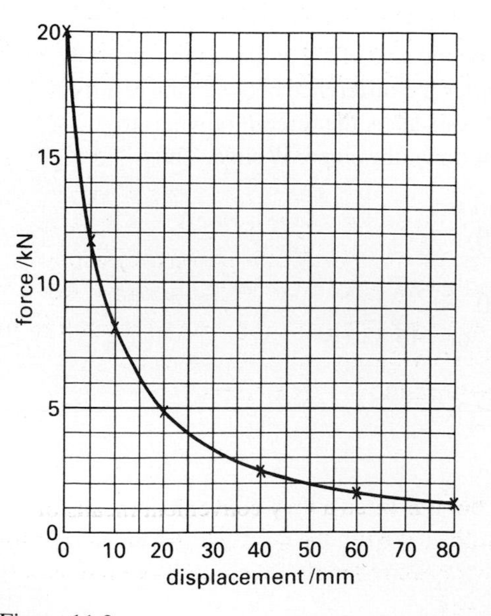

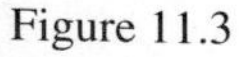
Figure 11.3

SOLUTION When the force varies, as in the present example, the expression for work done (force × distance) can only be applied to a small distance or displacement dx. For this small displacement, the work done is $F\,\mathrm{d}x$, and the total for the whole stroke is represented by the integral $\int F\,\mathrm{d}x$. The limits of integration correspond to the beginning and end of the stroke.

The variation of force with displacement is shown in Fig. 11.3 and the required integral is given by the area under the curve. In the diagram one square represents 1 kN (1000 N) vertically and 5 mm (0.005 m) horizontally. Therefore:

$$\text{area of one square represents } 1000 \text{ N} \times 0.005 \text{ m} = 5 \text{ N m or } 5 \text{ J}$$

The number of squares under the graph is approximately 63 and hence:

$$\text{work done during the stroke} = 63 \times 5 = 315 \text{ J}$$

The result can also be obtained using the trapezoidal rule or Simpson's rule but you should note that the displacement increments given in the question are not all equal.

11.2 Energy

In order for work to be done by machines or human beings, they need to possess energy. It is measured in the same units as work, namely joule (J), with practical values often being expressed in kilojoule (kJ), megajoule (MJ) or gigajoule (GJ). Energy, like work, does not possess direction and is a scalar quantity. It can take several forms and may be converted from one form to another.

Kinetic energy

This is the energy that a body possesses because of its motion. The greater the speed at which it is moving the greater its kinetic energy. The energy in this case is equal to the work done by the body in coming to rest. A hammer possesses kinetic energy as it strikes a nail and this energy enables it to do work driving the nail into a piece of wood. Linear and angular motion both give rise to kinetic energy.

Potential energy

This is the energy that a body possesses because of its position in a particular type of field: gravitational, electric, magnetic or nuclear. Gravitational potential energy for example equals the amount of work that the body can do as it falls towards the Earth under the pull of gravity. Potential energy, like kinetic, can be used to do useful work. In a hydroelectric generating station, for instance, water falls from a higher level to a lower one and does work on the turbines.

Strain energy

When a force is applied to a structure and causes it to distort, the point of application of the force is changed and work is done. This can be stored in the structure as strain energy or resilience (see section 9.12). If the force is removed this energy may be recovered and converted to other forms. In the operation of a mechanical pinball machine, for instance, strain energy is stored in the spring and then released in the form of kinetic energy or potential energy of the ball.

Heat

All substances consist of molecules which are in motion and between which there are forces of attraction. We can think of the molecules as possessing kinetic energy and potential energy corresponding to the two forms described above. In everyday language we call this molecular energy heat. In thermodynamics it is termed *internal energy* and the word 'heat' is used to mean the energy that is transferred from one body to another by conduction, convection or radiation.

Chemical energy

Fuels such as oil and coal contain the elements carbon, hydrogen and sometimes sulphur, each of which combines with oxygen when the fuel is burned, releasing energy in the process. The amount of energy available from the fuel is called its *calorific value* and it is usually quoted for 1 kg. In the case of gaseous fuels it may be given in terms of volume. The energy released during combustion may be used directly for heating purposes or it may be converted to other forms.

Electrical energy

Electricity is not a form of energy itself but it is often a very convenient means of transferring energy from one body to another. We use the term 'electrical energy' for the energy being transferred. The energy transferred in this way can be stored in electrochemical batteries as potential energy.

11.2.1 Units of energy

The recommended SI unit, the joule (J), should be used in all technical calculations. However, for historical reasons, several other units are still used and it is often necessary to convert values to joules.

Kilowatt hour

The electricity industry uses this unit for measuring the energy supplied to its customers. It is abbreviated to kW h and represents the energy that corresponds to a power of 1 kW operating for 1 hour. Since 1 h = 3600 s and 1W = 1J/s then 1kW h =1000 × 3600J = 3.6 MJ.

British thermal unit and the therm

These are heat units based on the pound and Fahrenheit scale of temperature. The British thermal unit (abbreviated to Btu) was defined as the heat required to raise the temperature of 1 lb of water by 1°F. For most purposes it is sufficiently accurate to take 1 Btu = 1055 J = 1.055 kJ.

The gas industry previously used the therm as a unit of heat. This is defined as 100 000 Btu and therefore 1 therm = $10^5 \times 1055$ J = 105.5×10^6 J = 105.5 MJ. The industry now uses the kW h and 1 therm = 29.3 kW h. This enables a direct comparison to be made between gas and electricity prices but it must be remembered that heating appliances are not 100 percent efficient.

Calorie

The relationship between heat and energy was not fully understood until the middle of the nineteenth century. By this time the unit calorie had been widely adopted by scientists for quantities of heat. It is abbreviated to cal and it was originally defined as the heat required to raise 1 g of water by 1°C but this is not a precise definition. By international agreement in 1956 the value of the calorie was fixed as 4.1868 J. Nowadays it is used mainly in the form kilocalorie (kcal) for the energy values of food. You will find it on the packaging of many foods.

In summary:

1kWh = 3.6 MJ
1Btu = 1.055 kJ
1 therm = 105.5 MJ
1 cal = 4.1868 J

EXAMPLE 11.5 Find the energy cost in pence per gigajoule for each of the following:

(a) Coal, with a calorific value of 30 MJ/kg, costing £2.32 for a 20 kg bag.
(b) Gas costing 38p per therm.
(c) Electricity costing 5.32p per unit (kW h).
(d) Petrol, with a calorific value of 51.4 MJ/kg, costing 36p per litre. Density of petrol, 0.9 kg/l.

SOLUTION (a) Since the calorific value is 30 MJ/kg the mass required for 1 GJ is 1000/30 = 33.33 kg. If a 20 kg bag costs £2.32 then the cost per kilogram is 2.32/20 = £0.116 = 11.6p. Hence:

cost per gigajoule = 11.6 × 33.33 = 386.6p

(b) Since 1 therm = 105.5 MJ, the number of therms required is $(1 \times 10^9)/(105.5 \times 10^6) = 9.479$. Hence:

cost per gigajoule = 9.479 × 38 = 360.2p

(c) 1 kW h = 3.6 MJ and therefore the number of units required is $(1 \times 10^9)/(3.6 \times 10^6) = 277.8$. Therefore,

cost per gigajoule = 277.8 × 5.32 = 1478p

(d) Mass of petrol required = $(1 \times 10^9)/(51.4 \times 10^6) = 19.46$ kg and with the given density, number of litres required = 19.46/0.9 = 21.62. Hence:

$$\text{cost per GJ} = 21.62 \times 36 = 778.3\text{p}$$

11.2.2 The principle of conservation of energy

This is generally stated in the form:

Energy cannot be created or destroyed: it can only be converted from one form to another.

It means that an increase in one system must be exactly matched by decreases in others. This idea is used to tackle a wide range of problems in mechanics and thermodynamics.

Some systems involve several energy transfers. Consider the processes that enable us to use power tools in the workshop or at home. In a conventional power station using fossil fuels, the chemical energy of the fuel is released as heat during combustion. This generates steam in the boilers which does work on the turbine blades. The turbines in turn drive electrical generators and some of the original energy is conveyed by electricity to where it is required in factories and homes.

This energy enables machine tools, powered DIY tools and household appliances such as vacuum cleaners to exert forces and do work. This step is not the end of the sequence because we know from experience that the tools and workpieces heat up as work is done.

If we could measure the amounts of all the energy involved in this sequence we would find that the energy released in the combustion of the fuel could be exactly accounted for by the amounts appearing elsewhere. If no energy has disappeared why cannot we use it again and do more work with it? The explanation is that it becomes dispersed in the atmosphere as heat, raising the temperature of our surroundings by a very small amount. It is shown in Chapter 12 that a large difference is needed for the efficient conversion of heat into mechanical energy. Heat that is available at only a slightly higher temperature than the environment is called low-grade heat.

11.2.3 Conversion efficiency

When the energy possessed by a body is converted from one form to another, the total quantity is unchanged – a consequence of the principle of conservation of energy. In many cases, however, we may not obtain all the energy in the form we require. Some energy is said to be lost in the process; it does not disappear but it is not available to us. We define the *efficiency* of the process, denoted by η, as the ratio of the useful output to the input. That is,

$$\text{efficiency } \eta = \frac{\text{useful output}}{\text{input}}$$

An important example of energy transfer is when energy is provided by heat to a heat engine which supplies useful energy by doing work. In this case the useful output is the work done by the engine and the input is the energy obtained from the combustion of the fuel. Thus:

$$\eta = \frac{\text{work done}}{\text{heat supplied}}$$

The efficiency of a heat engine is limited by theoretical and practical considerations. It is shown in section 12.5.3 that there is a theoretical limit to the proportion of heat that can be converted to work, depending on the upper and lower temperatures during the operating cycle. On the practical side, there are losses due to friction in the moving parts. The overall efficiency of a power station, for example, is about 0.3 and the power plant contains many refinements to achieve even this figure. In contrast, the human body is remarkably efficient as a heat engine and we can convert as much as 0.7 of the energy we derive from our food into work.

EXAMPLE 11.6 A petrol engine uses 4.6 kg of fuel per hour when developing 25 hp. If the calorific value of the fuel is 43.2 MJ/kg what is the overall efficiency? (Take 1 hp = 0.746 W.)

SOLUTION Using the conversion factor given at the end of section 11.2.1,

work done per hour = 25 hp/h = 25 × 2.685 MJ = 67.1 MJ
heat supplied per hour = 4.6 × 43.2 =198.7 MJ

$$\text{efficiency } \eta = \frac{\text{work done}}{\text{heat supplied}} = \frac{67.1}{198.7} = 0.338$$

11.2.4 Potential and kinetic energy

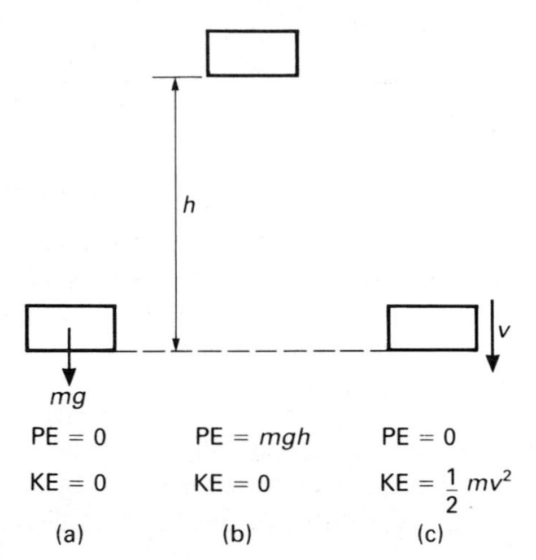

Figure 11.4 Potential and kinetic energy of a body.

Fig. 11.4(a) shows a body of mass m resting on a floor or the ground. Take the potential energy (PE) of the body as being zero at this datum. The kinetic energy (KE) is also zero since the body is not moving. Next suppose that the body is raised to a height h as shown in Fig. 11.4(b), and is again at rest. The force of gravity acting on the body is mg, its weight, and the work done in raising it is (force × distance) and this equals mgh. The body has gained potential energy equal to the work done on it. Therefore,

$$\text{gain in PE} = mgh$$

If the body is released and falls freely back to the original level (Fig. 11.4(c)) it will achieve a speed v. This is an example of uniform acceleration in which the initial velocity $u = 0$, the acceleration $a = g$ and the distance $s = h$. Using the equation that relates speeds and distance, we obtain

$$v^2 = u^2 + 2as = 2gh \quad \text{from which} \quad gh = v^2/2$$

If there is no resistance to the motion, no work is done and only two forms of energy are involved – potential and kinetic. By the principle of conservation of energy the total (PE + KE) must remain constant. Hence the increase in kinetic energy must be equal to the decrease in potential energy. Furthermore the body was released from rest and its initial kinetic energy was zero. Therefore,

$$\text{KE at speed } v = \text{decrease in PE} = mgh$$

But $gh = v^2/2$ and, substituting this in the last expression,

$$\text{kinetic energy} = mv^2/2 \text{ or } \tfrac{1}{2}mv^2$$

A wide range of examples can be solved using the principle that the total energy (potential + kinematic) remains constant. If work is done on or by the body, this must be allowed for in the calculations.

EXAMPLE 11.7 Two masses of 7 kg and 5 kg are connected by a light string passing over a light, frictionless pulley. The system is released from rest. Find
(a) the speed of the masses when they have moved 1.5 m,
(b) the time taken for this distance,
(c) the acceleration of the masses.

SOLUTION The question mentions some of the idealisations that are often made in examples of this kind. The pulley and string are assumed to be so light that their kinetic energy can be ignored. Furthermore no work is done against friction. The system is shown in Fig. 11.5.

(a) loss of PE in 7 kg mass = mgh = 7 × 9.81 × 1.5 =103.0 J
gain of PE in 5 kg mass = 5 × 9.81 × 1.5 = 73.57 J
net loss of PE = 103.0 – 73.57 = 29.43 J

If v m/s is the speed of the masses when they have moved 1.5 m then

gain in KE of 7 kg mass = $\frac{1}{2}mv^2 = \frac{1}{2} \times 7 \times v^2 = 3.5v^2$
gain in KE in 5 kg mass = $\frac{1}{2} \times 5 \times v^2 = 2.5v^2$

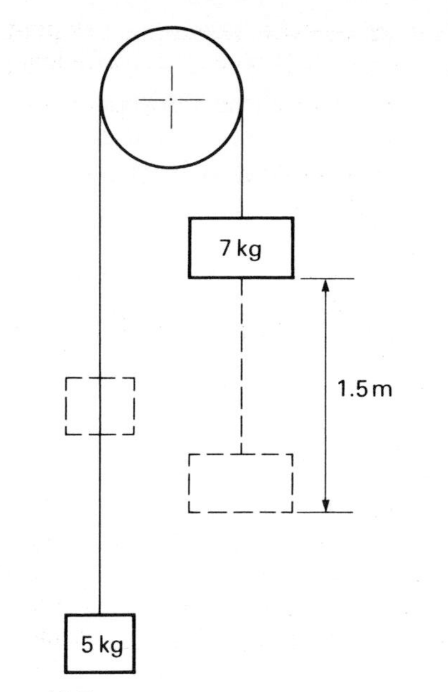

Figure 11.5

Note that the total KE of the system is found by adding these two amounts even though one mass is moving downwards and the other upwards. This follows from the fact that energy is a scalar quantity and does not possess direction.

Equating the gain in KE to the loss of PE gives:

$$3.5v^2 + 2.5v^2 = 29.43 \text{ from which } v^2 = 29.43/6 = 4.905$$

and the required speed is $v = \sqrt{4.905} = 2.215$ m/s

(b) the average speed during the motion is 2.215/2 = 1.107 m/s and therefore:

$$\text{time taken} = \text{distance/average speed} = 1.5/1.107 = 1.355 \text{ s}$$

(c) $\text{acceleration} = \text{increase in speed/time} = 2.215/1.355 = 1.635 \text{ m/s}^2$

An alternative solution based on the relationship between force and acceleration is given in Example 11.14. The solution to Example 11.19 shows how the mass of the pulley can be taken into account.

EXAMPLE 11.8 A car whose mass is 1100 kg starts from the rest and reaches a speed of 90 km/h while covering a distance of 450 m in 40 s. The road is not level and climbs 20 m over this distance. The frictional resistance of the road is 1/30 of the weight of the car. (Note that the acceleration is not constant.) Calculate:
(a) the gain in potential energy,
(b) the gain in kinetic energy,
(c) the work done in overcoming friction,
(d) the average power exerted.

SOLUTION

(a) Gain in PE $= mgh = 1100 \times 9.81 \times 20 = 215.8$ kJ

(b) Converting the final speed to metre per second units,

$$v = \frac{90 \times 1000}{3600} = 25 \text{ m/s}$$

Therefore,

$$\text{gain in KE} = \tfrac{1}{2}mv^2 = \tfrac{1}{2} \times 1100 \times 25^2 = 343.7 \text{ kJ}$$

(c) Frictional force = weight/30 = $mg/30$

$$= 1100 \times 9.81/30 = 359.7\text{N}$$

Therefore, work done against friction $= \text{force} \times \text{distance} = 359.7 \times 450 = 161.9\text{kJ}$

(d) The total work done by the engine is the sum of these three results.

Hence,

$$\text{work done} = 215.8 + 343.7 + 161.9 = 721.4 \text{ kJ}$$

and, average power developed $= \text{work done/time} = (721.4 \times 10^3)/40 = 18\,040 \text{ W or } 18.04 \text{ kW}$

11.3 Sources of energy

As explained in section 11.2.2, the form of energy may be changed several times in a series of processes such as the generation, transmission and use of electricity. Although no energy disappears its final form is low-grade heat which is not available at a high enough temperature to be converted to a mechanical form. For practical purposes this energy is lost.

More energy is therefore needed all the time to provide the power and heat for transport, factories, schools, hospitals and our homes. Fig. 11.6 shows the world energy supply from 1967 to 1992.

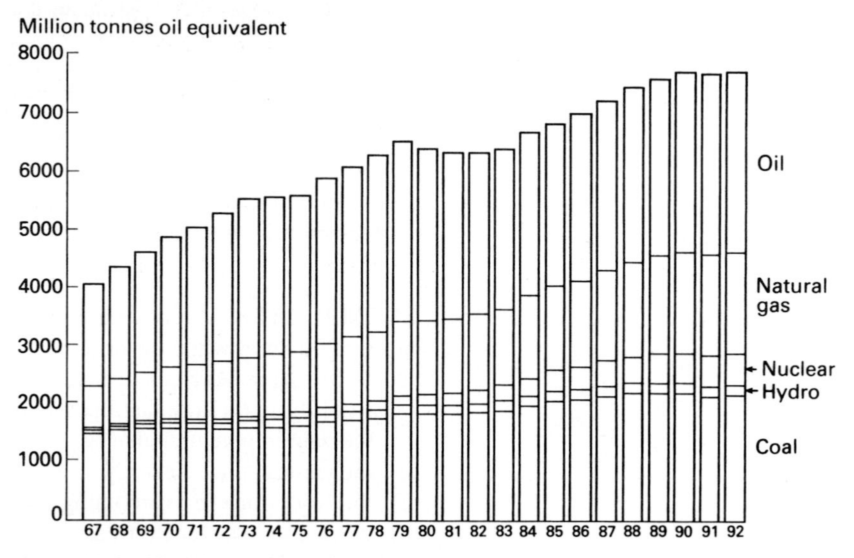

Figure 11.6 World energy supply.

This diagram is taken from the *BP Statistical Review of World Energy*. The vertical scale represents energy consumption with the various sources expressed as amounts that are equivalent to millions of tonnes of oil. Note the levelling out of consumption in the years 1990–92 following a near doubling in the previous twenty years.

The fuels listed on the diagram are known as primary sources and the consumer may receive energy in other forms, notably from the electricity supply. It is estimated that about one-third of the primary energy goes into electricity generation. Transport accounts for another quarter, coming almost entirely from oil. Synthetics are fuels such as ethanol and methanol that can be derived from plant material.

A large proportion of our energy needs is met by burning fuels and releasing the heat of combustion. The increasing demand for energy in recent years has led to a growing concern about fuel supplies. There are several reasons for this.

1 The Earth's reserves of fossil fuels such as coal, oil and natural gas are limited. They have taken millions of years to form but they may run out in a comparatively short period. Some estimates put the deposits of coal at 200–300 years and of oil at only 20–30 years.
2 Although exploration continues for new deposits of oil and coal, they are likely to become increasingly difficult to exploit and therefore more expensive.
3 The production of oil may become the monopoly of a small group of countries. This could lead to great fluctuations in price and supply.
4 The burning of more fossil fuels aggravates the problems of atmospheric pollution, global warming and the greenhouse effect.

As a result, energy is now more than a scientific and technical matter. It has become a social, environmental, political and economic issue. A small country with extensive oil deposits can become very rich; a threat to oil supplies can lead to an international crisis.

A rapid increase in oil prices in the 1970s led to a national campaign for energy conservation. This is not to be confused with the scientific principle of the conservation of energy and the term 'fuel conservation' might have been more apt.

The problems associated with fossil fuels have created interest in alternative sources of energy.

Solar energy

Enormous amounts of energy reach the outer atmosphere of the Earth from the Sun. The quantity reaching the surface depends upon the ozone layer, dust, clouds and latitude. Even in temperate latitudes it can amount to 500 W/m^2. This energy is readily available for producing hot water and space heating using simple solar panels. Installation costs are high, however, and the energy available decreases in winter when the requirement is greatest. Much higher temperatures are needed for power generation and these can be achieved by focusing the Sun's rays with concave mirrors.

Solar cells convert the Sun's energy directly into electricity. These have been used to power small experimental cars through electric motors of about 1 hp. Photoelectric cells are already widely used in light meters and solar-powered calculators where very small currents are sufficient but they are not yet economic for power generation.

Ocean thermal energy

The Sun causes the surface temperatures of oceans to be considerably higher than that at great depths. This temperature difference could be used in a thermodynamic cycle to generate electricity.

Tidal energy

Seawater can be used to drive turbines in a barrage built across a bay or estuary using the potential energy that exists when the levels are different on the two sides. In its simplest form, such a scheme could only generate electricity for part of the day but constant generation is possible using two basins. One tidal generating scheme has been built on a river estuary in northern France.

The use of tidal energy is not new. The tidal mill at Woodbridge in Suffolk had been working for 200 years when it closed in 1957 and the first such mill on the site was built in about 1170.

Hydroelectric power

In mountainous countries substantial amounts of power can be generated from the potential energy of water stored by dams. Although the construction costs can be very high the direct cost of production is very small.

Wave energy

The vertical motion of the sea can be utilised to generate power in several ways. One system uses the relative movements of floating rafts to convert the wave energy. Very large amounts of energy are theoretically available, particularly in the winter when the requirement is greatest.

Geothermal energy

We know that very high temperatures exist within the Earth and in some places heat in practical quantitites is available at the surface. Hot springs have been used at least since Roman times. There are installations in several countries which use geothermal energy to generate electricity.

Wind energy

Wind power has been harnessed for many centuries and is now considered seriously for the generation of electricity. We can make a rough estimate of the power available from a windmill by considering the kinetic energy of the air approaching the circle swept out by the blades. If:

d = diameter of blades in m,
v = speed of the wind in m/s,
ρ = density of air in kg/m^3
m = mass of air per second in kg/s

then

$$\text{volume of air per second} = (\text{area of blade circle}) \times (\text{wind speed})$$
$$= \frac{\pi}{4} d^2 v$$
$$\text{mass of air/s } m = \rho \frac{\pi}{4} d^2 v$$

The kinetic energy per second is given by the formula $\frac{1}{2}mv^2$ and, in the present case, this gives the power P. Thus:

$$P = \frac{1}{2} \times \rho \frac{\pi}{4} d^2 v \times v^2 = \frac{\pi}{8} \rho d^2 v^3$$

This is a theoretical result and we have to allow for the efficiency of conversion. With good design this can reach 0.6 and, taking the sea level density of air as 1.23 kg/m^3, the result is approximately

$$P = 0.3 d^2 v^3$$

The largest windmills (or aerogenerators) in use at present have a diameter of about 50 m and an output of roughly 1 MW.

Biomass energy

Plants can yield energy in several ways and, unlike fossil fuels, are renewable. Wood, for example, is still an important fuel in many countries and, under favourable conditions, a copse of about one hectare (10000m^2) can produce sufficient timber to meet the energy requirements of an average house on a continuous basis.

Perhaps the best way of producing wood fuel on a large scale is through coppicing. This involves the regular cropping of fast-growing trees like willow or poplar. The crop is then dried, chipped and burnt to provide power or heat. The roots remain and regrowth occurs each year. Wood produced in this way or waste farm straw could be used in both large and small scale energy generation schemes. Waste from forestry operations and woodland management can be used and collected for wood fuel. Also the waste from woodworking industries could be burnt to supply heat for the factories; this would both provide energy and remove the need for waste disposal.

Another means of obtaining biofuel is to ferment plant material to produce bio-ethanol which is a substitute for petrol derived from crops such as sugar cane, maize, wheat, straw and woods such as poplar, willow and perennial grasses. It is also possible to obtain ethanol from the waste from many industrial and municipal organisations, such as paper mills and household waste. During their growth, the plants remove carbon dioxide from the air and convert it into starch, cellulose and hemicellulose. Fermentation of the plants then converts the carbohydrates into ethanol. By-products of fermentation can be further processed to produce methane-rich biogas, another energy resource as well as fertiliser and high-nitrogen animal feed. In Brazil ethanol production accounts for over 20% of the country's fuel needs. The first (and only, as far as we are aware) European bio-ethanol plant, running on wheat, is situated in Sweden and there are research programmes and pilot plants in the UK.

Bio-diesel is a diesel substitute, which is derived from oilseed rape crops, which are ideally suited to the climate in the UK. The oil can be easily extracted from the oilseed crops and can be blended and burnt in diesel engines. The addition of ethanol or methanol to the oil increases the efficiency of the bio-diesel. The by-products from the production of bio-diesel include glycerine, a high value substance used in the pharmaceutical, cosmetic and other industries; crushed rapeseed, an animal feed; and rape straw, a combustible fuel. There are bio-diesel plants in Austria, France and Italy.

One of the main benefits of biomass fuels is that they do not contribute to the 'greenhouse' effect. The burning of wood fuel and ethanol releases the same amount of carbon dioxide back into the air as the plants removed during their growth. Ethanol is a clean-burning fuel and does not produce fuel toxins that are produced by burning traditional fuels such as benzene and gasoline. Therefore there can be a reduction in the accumulation of greenhouse gases for biomass fuels. The other advantages are that the fuels are biodegradable; they are a sustainable and renewable energy resource; and reduce our reliance on fossil fuels.

Advantages and disadvantages of alternative energy sources

Although these alternative sources of energy can reduce our dependence on fossil fuels they have their own limitations and disadvantages. Most of them involve large construction costs and these may outweigh the savings in fuel costs. Furthermore, they are not all suitable for every country. Solar energy, for example, requires a favourable climate and hydroelectric power is only possible in mountainous regions. Winds and tides vary from season to season and day to day so their energy is not always available when it is needed.

One argument put forward for most of these alternative energy sources is that they do not cause pollution nor damage the environment in other ways. There may be objections, however, to the installations needed to harness the energy. A tidal barrage, for instance, may cause a major ecological change in its locality. A group of aerogenerators, or *wind farm*, may be regarded as unsightly as well as noisy.

11.3.1 Energy storage

The great advantage of fossil fuels is that they can be stored until their energy is required. In particular, they enable vehicles to carry their supplies of energy with them.

Most alternatives to fossil fuels depend upon the generation and transmission of electricity for making energy available where it is required. Electricity is a highly efficient means of transmitting energy but it has two main disadvantages.

One is that electricity cannot be stored. The amount generated has to be matched to demand and this can vary greatly during the day and from one time of year to another. Batteries can store a limited amount of electrochemical energy but it could take up to a hundred car batteries to store a day's supply of energy for the average house in winter.

Pumped storage is one means of coping with the fluctuating demand for electricity. Water from a lake is pumped to a reservoir at a high level during the off-peak period when there is spare generating capacity. Then, during the peak period the water is allowed to fall and drive turbines as in a hydroelectric scheme. There is a large pumped storage installation at Dinorwic in North Wales.

The second major drawback in using electricity as a means of transferring energy is that the consumer must be connected to the supply.

A promising solution is to use hydrogen as a fuel. It could be produced by the electrolysis of water at power stations and then made available in fuel cells wherever the energy is required. Fuel cells are highly efficient over a range of power outputs and have the added advantage that the only product of combustion is water.

11.3.2 Nuclear energy

For most practical purposes we can rely on the following principles:

- conservation of energy,
- conservation of mass.

The first of these has been explained and used in the earlier sections of this chapter. The second principle states that mass cannot be created or destroyed and this result is used in calculations on chemical reactions. When fuel is burned, for example, the total mass of the products of combustion is taken to be equal to the mass of fuel and oxygen involved in the reaction.

Einstein showed that these two principles are not independent and that mass and energy are equivalent. If a mass m kg disappears in a process the energy E J released is given by Einstein's equation:

$$E = mc^2$$

where c m/s is the velocity of light.

Since c has a very large value (3×10^8 m/s) it is clear that enormous amounts of energy are released from small reductions in mass.

In *nuclear fission*, atoms are caused to disintegrate and mass is lost in the process. In early experiments, the energy required to split the atom, as it was popularly called, was greater than the energy released. In a chain reaction, the energy released from the initial disintegration is sufficient to continue the process and this was achieved in the first atomic bombs in 1945. In nuclear power stations the process takes place in a *reactor* and is controlled by a *moderator* such as graphite or heavy water so that the energy release takes place at a manageable rate. The heat generated raises the temperature of a liquid or gas coolant which passes through a heat exchanger. Here it heats water to form steam which in turn drives turbines as in a conventional power station.

There is much concern about the safety of reactors because of radiation hazards from accidents and the disposal of radioactive waste.

Nuclear fusion is the process of joining atoms to form new ones, again with a loss of mass. It is the means by which energy is released in the Sun and in the hydrogen bomb. It is a difficult process to control and it is likely to be many years before power for peaceful purposes is available from nuclear fusion.

11.4 Energy in the home

In the UK the use of energy in the home accounts for about 30 percent of the national total. The Watt Committee on energy has suggested the figures shown in Table 11.1 for the annual energy requirements of an average house. The units have been converted from kW h (used in the report) to megajoules (MJ).

Fabric loss means the flow of energy out of the house through the walls and roof by conduction and radiation. Ventilation refers to the energy carried away by the air flowing through the house by convection. As the table shows, these two items account for over 60 percent of the energy used in the house.

There are several ways of reducing these losses including loft insulation, double glazing and cavity-wall insulation. New houses have to meet minimum standards of insulation and they incorporate some or all of these means. Improving the standard of insulation in older houses can prove costly and loft insulation usually offers the best savings for the costs involved. In the absence of cavities, the insulation of walls can be increased by fixing purpose-made boards to the inner or outer surfaces.

Ventilation losses can be cut down by reducing the ventilation rate (measured as the number of air changes per hour). This can cause problems, however. Washing, cooking and the breathing and perspiration of the occupants releases about 7 kg of water vapour each day. This amount may be doubled by clothes drying and the use of free-standing paraffin heaters. If there is insufficient ventilation, some of the vapour will condense on windows. Moisture may also be deposited in rooms that are at lower temperatures than the living rooms.

	MJ	MJ	
Space heating			
fabric loss	32 400		
ventilation	19 800		
		52 200	
High and low temperature use			
cooking	7920		
hot water	18 000		
refrigeration	1170		
deep freeze	3600		
		30 690	
Nonthermal use			
television	1340		
lights	1190		
vacuum cleaner	110		
spin dryer	70		
		2710	
Total requirement			85 600

Table 11.1 Annual energy requirements of an average house

We also need adequate ventilation to ensure a supply of fresh air and to remove body odours. Ventilation loss can be reduced without these problems by the process of ventilation heat recovery. In this, the warm, moist air from the kitchen and bathroom is drawn through a heat exchanger where it is used to help raise the temperature in the hot water system. The equipment is difficult to install in existing houses.

11.5 Mass, weight and momentum

The *mass* of a body was defined by Sir Isaac Newton as the 'quantity of matter' in it. Quantity here does not mean volume, and mass should be considered as a fundamental physical property. It can be thought of as the tendency of a body to resist the change in velocity caused by an external force. Mass is therefore said to be a measure of inertia. It has magnitude but not direction and is therefore a scalar quantity. The SI unit is the kilogram (kg).

Weight is a more familiar concept being the force with which the Earth attracts a body. To emphasise that weight is a force it is expressed in the force unit newton (N).

Space travel has emphasised the distinction between mass and weight. If a body is taken to the Moon its mass is unchanged but its weight is much less than on Earth. This reduction in weight could be detected by a spring balance which measures the gravitational force on the body but not by scales that work on the principle of the lever, because the balancing weights would be affected in the same way as the body. Away from the Earth, astronauts experience weightlessness but they still possess mass.

If a body is moving without rotation so that each particle has the same velocity then its *momentum* equals its mass multiplied by its velocity. This is sometimes referred to as linear momentum to distinguish it from the angular momentum associated with rotation.

Momentum has the same direction as the velocity of the body and it is a vector quantity. There is no special name for its unit and in SI it is expressed as kg m/s.

The quantity momentum is used in solving problems in which two bodies collide. If no external force is involved the total momentum of the bodies in a given direction is the same before and after collision. This principle is known as the *conservation of momentum*. This principle can be deduced from one of Newton's laws and is confirmed by experiment.

EXAMPLE 11.9 A nail of 25 g mass is driven into a piece of wood by a hammer of mass 1 kg. If the hammer is moving with a velocity of 5 m/s when it strikes the nail, what is the velocity of both immediately after impact?

SOLUTION Before impact the nail is at rest and its momentum is zero. The total momentum before impact is therefore that of the hammer.

$$\text{initial momentum} = \text{mass} \times \text{velocity} = 1 \times 5 = 5 \text{ kg m/s}$$

After impact the hammer and nail move together, their combined mass being 1 kg + 25 g = 1.025 kg. If their common velocity is v then, equating the momentum after impact to that before,

$$1.025v = 5 \quad \text{from which } v = 5/1.025 = 4.878 \text{ m/s}$$

EXAMPLE 11.10 A disabled ship whose mass is 4000 tonnes is under tow by a tug of 500 tonnes mass. At one stage in the operation when the towing cable is slack the tug is moving forward at 3 m/s and the ship is drifting in the opposite direction at 0.2 m/s. What is the common velocity immediately after the cable becomes taut?

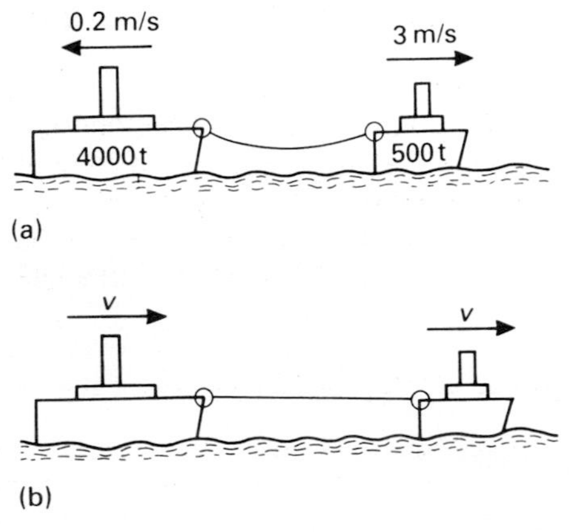

Figure 11.7

SOLUTION Figures 11.7(a) and 11.7(b) show the ships and the relevant data before and after the cable becomes taut. Since the initial velocities of the two vessels are in opposite directions a sign convention is needed. Take velocities to the right as positive. It is convenient to keep the masses in tonnes, the combined mass of the two vessels being 4500 tonnes. If v is their common velocity we have, equating the final momentum to the total initial momentum,

$$4500v = 500 \times 3 - 4000 \times 0.2 = 700$$

and

$$\text{common velocity } v = 700/4500 = 0.156 \text{ m/s}$$

11.5.1 Newton's laws of motion

In designing vehicles, machine tools, household appliances and other products we often need to carry out calculations on force and motion. The methods we use are based on Newton's laws of motion, first published in 1686. These laws cannot be proved in a strict sense but their truth has been confirmed by many experiments and they remain the basis for most practical work.

It is only when very high velocities are encountered – comparable to the speed of light – that Newton's laws fail to give accurate results. The discrepancy was first noticed in the case of certain astronomical observations and the differences were explained by the Theory of Relativity put forward by Einstein in 1905. This theory has had far-reaching consequences in the field of nuclear physics and astronomy.

We can take Newton's laws to be accurate for our purposes and they can be stated as follows.

1 *A body continues at rest or at a constant speed in a straight line unless it is acted on by an external force.*

2 *When an external force acts on a body its rate of change of momentum is proportional to the force and takes place in the direction of the force.*

3 *To every action there is an equal and opposite reaction.*

The first law is a statement of what happens in the absence of an external force. It is easy to accept that a force is needed to cause a body to move when it is at rest. What is less obvious is that a body in motion will continue at a constant velocity for ever if no force is acting on it. Everyday experience shows that a car or train moving along the level will slow down if its power is cut off. The explanation is that there is always a frictional force acting as a brake. The more we can reduce friction and air resistance the closer we come to achieving the constant velocity predicted by the first law.

The second law is a precise relationship between force and momentum. If a force F acts on a body of mass m and increases its velocity from u to v in time t, the momentum increases from mu to mv, the change of momentum is $(mv - mu)$ and the rate of change of momentum is $(mv - mu)/t$. The second law states that F is proportional to this expression, Thus:

$$F = k\frac{(mv - mu)}{t} = km\frac{(v - u)}{t}$$

where k is a constant.

But $(v - u)/t$ is the acceleration a and therefore:

$$F = kma$$

It is convenient to choose a unit for force so that k is equal to 1. In SI this unit is the newton, denoted by N. Thus 1 N is defined as the force which causes an acceleration of 1 m/s^2 in a body of 1 kg mass. This leads to the equation:

$$F = ma$$

This result, that force is equal to mass multiplied by acceleration, can be used for solving a wide range of problems.

On the surface of the Earth a body falling freely has a downward acceleration of 9.81 m/s^2, usually denoted by g. The force producing this acceleration is its weight, the force of gravity. Hence, a force equal to the weight of a 1 kg mass causes an acceleration of 9.81 m/s^2 in a mass of 1 kg. The weight of 1 kg is therefore equal to 9.81 N.

The third law was mentioned in Chapter 8 in connection with the reactions of the supports holding up structures. It is also the starting point for deducing the principle of conservation of momentum.

EXAMPLE 11.11 What force acting on a mass of 20 kg will accelerate it from 2 m/s to 7 m/s in 8 s?

SOLUTION From the equation $v = u + at$ the acceleration is:

$$a = (v - u)/t = (7 - 2)/8 = 0.625\text{m/s}^2$$

The required force is therefore:

$$F = ma = 20 \times 0.625 = 12.5 \text{ N}$$

EXAMPLE 11.12 What is the effective force needed to propel a car of 750kg mass up a slope of 1 in 10 and accelerate it from 40–60km/h in 6s if the frictional resistance is taken as 1/40 of the weight and assumed constant?

How long would it take to achieve the same increase in speed down the slope with the same propulsive force and resistance?

SOLUTION The required force is made up of three parts.

(1) The force F_1 needed to overcome the frictional resistance. This is given as a proportion of the weight (mg). Therefore:

$$F_1 = \frac{1}{40} \times 750 \times 9.81 = 184 \text{ N}$$

(2) The force F_2 needed to overcome gravity. As shown in the solution to Example 11.2, the component of the weight down the slope is $W \sin \theta$ and $\sin \theta$ may be taken as the gradient (1 in 10 in this case). Therefore:

$$F_2 = \frac{1}{10} \times 750 \times 9.81 = 736 \text{ N}$$

(3) The force F_3 required to accelerate the car. Working in metre and second units the initial and final velocities are:

$$u = 40 \times \frac{1000}{3600} = 11.11 \text{ m/s}$$

and

$$v = 60 \times \frac{1000}{3600} = 16.67 \text{ m/s}$$

The acceleration is therefore:

$$a = (v - u)/t = (16.67 - 11.11)/6 = 0.927 \text{ m/s}^2$$

and

$$F_3 = ma = 750 \times 0.927 = 695 \text{ N}$$

Hence the total force required is:

$$F = F_1 + F_2 + F_3 = 184 + 736 + 695 = 1615 \text{ N}$$

With motion down the slope the propulsive force is assisted by gravity. Hence the total force available to overcome the frictional resistance and accelerate the car is 1615 + 736 = 2351 N. If the frictional resistance is unchanged the force available for acceleration is 2351 – 184 = 2167 N. From $F = ma$ the acceleration is now:

$$a = F/m = 2167/750 = 2.89 \text{ m/s}^2$$

and, rearranging the equation $v = u + at$, the required time is:

$$t = (v - u)/a = (16.67 - 11.11)/2.89 = 1.92 \text{ s}$$

EXAMPLE 11.13 A parcel of 50 kg mass stands on the floor of a lift. Find the reaction of the floor when the lift has an acceleration of 3 m/s² (a) upwards, (b) downwards.

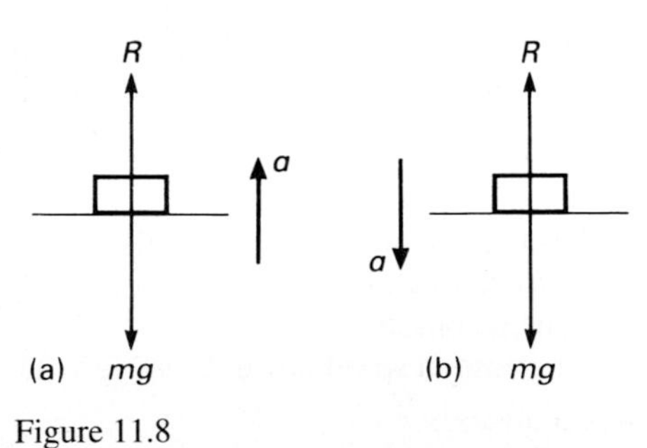

Figure 11.8

SOLUTION The forces and accelerations for the two cases are shown in Figs 11.8(a) and (b). Note that the resultant force in each case must be considered in the same direction as the acceleration.

(a) The resultant upward force is $R - mg$. Substituting in $F = ma$, we have:

$$R - mg = ma$$

or

$$R = m(a + g) = 50(3 + 9.81) = 640.5\text{N}$$

(b) Since the acceleration is now downwards the resultant force must also be considered downwards. Thus:

$$mg - R = ma$$

and

$$R = m(g - a) = 50(9.81 - 3) = 340.5\text{N}$$

EXAMPLE 11.14 Two masses of 7 kg and 5 kg are connected by a light string passing over a light, frictionless pulley. Find the acceleration of the system and the tension in the string.

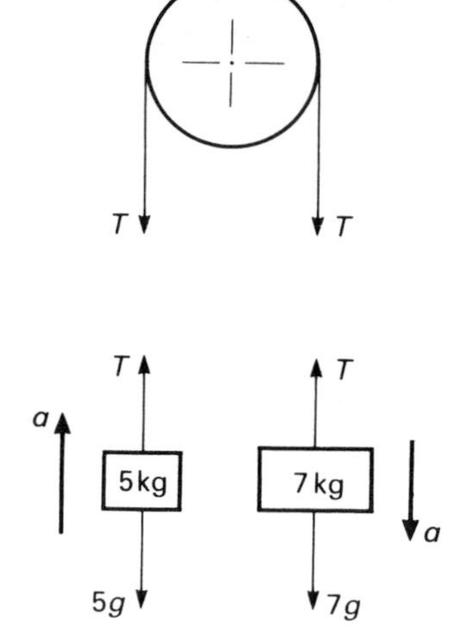

Figure 11.9

SOLUTION The forces acting on the masses and their acceleration are shown in Fig. 11.9. Since the pulley is light and frictionless, no torque is required to accelerate it and the tension in the string T is the same on both sides. It must also be assumed that the string does not stretch since any extension would affect the motion. An equation of motion can be written down for each mass using the relationship $F = ma$.

For the 5 kg mass, $T - 5g = 5a$
and, for the 7 kg mass, $7g - T = 7a$

These are simultaneous equations for a and T. Adding them gives:

$$7g - 5g = 5a + 7a$$

from which,

$$a = 2g/12 = 9.81/6 = 1.635\text{m/s}^2$$

Substituting this result in the first equation, the required tension is:

$$T = 5a + 5g = 5(1.635 + 9.81) = 57.2\text{ N}$$

Although the figures in this example are the same as for Example 11.7 the method of solution is different. Many problems in dynamics can be solved using either the force and acceleration relationship or the conservation of energy principle. Both methods should be learned.

EXAMPLE 11.15 A pile driver of 800 kg mass falls freely through a distance of 2 m before striking the pile whose mass is 400 kg. After impact the two masses move together, the resistance being 500 kN. Determine:
(a) the loss of kinetic energy at impact,
(b) the retardation after impact,
(c) the penetration per blow.

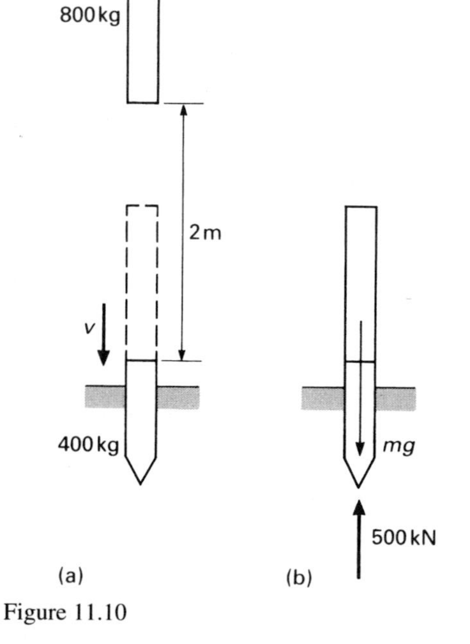

Figure 11.10

SOLUTION Piles are vertical bars used to provide foundations for buildings and other structures. They may be formed of concrete cast in holes bored in the ground or, as in the present case, made of steel and driven in. Fig. 11.10(a) shows the pile and pile driver before impact. If the pile driver falls freely its initial velocity $u = 0$ and its acceleration $a = g$. The distance fallen s is 2 m and, using the equation for uniform acceleration, the velocity v just before impact is given by:

$$v^2 = u^2 + 2as = 0 + 2 \times 9.81 \times 2 = 39.24$$

and

$$v = \sqrt{39.24} = 6.264\text{ m/s}$$

Before impact, only the pile driver is moving and therefore:

$$\text{initial momentum} = mv = 800 \times 6.264 = 5011\text{ kg m/s}$$

After impact, the pile and pile driver move together. Their combined mass is 1200 kg. Suppose their velocity immediately after impact is v_1. Then:

$$\text{momentum after impact} = 1200v_1$$

By the principle of conservation of momentum this result equals the momentum just before impact. Hence:

$$1200v_1 = 5011$$

and

$$v_1 = 5011/1200 = 4.176\text{m/s}$$

(a) KE immediately before impact $= \frac{1}{2}mv^2 = \frac{1}{2} \times 800 \times 6.264^2 = 15\,695$ J

KE immediately after impact $= \frac{1}{2} \times 1200 \times 4.176^2 = 10\,463$ J

Loss of KE at impact $= 15\,695 - 10\,463 = 5232$ J or 5.232 kJ

(This energy is converted to heat and other forms.)

(b) The forces acting after impact are shown in Fig. 11.10(b). The resultant upward force F is given by:

$$F = 500 \times 10^3 - mg = 500 \times 10^3 - 1200 \times 9.81 = 488\,200 \text{ N}$$

This is a retarding force and, using the relationship $F = ma$,

$$a = F/m = 488\,200/1200 = 406.8 \text{ m/s}^2$$

(c) The distance moved in coming to rest can be found in two ways. The initial velocity $u = 4.176$ m/s, the final velocity $v = 0$ and the acceleration $a = -406.8$ m/s^2, the minus sign indicating a retardation. The equation for uniform acceleration is:

$$v^2 = u^2 + 2as$$

and, on rearrangement, the distance to come to rest is:

$$s = \frac{v^2 - u^2}{2a} = \frac{0 - 4.176^2}{2 \times (-406.8)} = 0.0214 \text{ m} \quad \text{or} \quad 21.4 \text{ mm}$$

An alternative method of solution is to use energy and work. The kinetic energy immediately after impact is all used in doing work against the resisting force since the final velocity is zero. The work done against a resistance F over a distance x is Fx. The effective resistance $F = 488\,200$ N (allowing for the weight of the pile and pile driver) and therefore:

$$Fx = \text{decrease in KE}$$

or $x = (\text{decrease in KE})/F = 10\,463/488\,200$
$= 0.0214$ m (as before).

11.6 Motion of a vehicle on a curved track

Suppose a car is travelling at a speed v round a circular level road of radius r (Fig. 11.11(a)). Although the magnitude of the velocity is constant, its direction is continually changing and the car therefore has an acceleration. It was shown in Chapter 10 that this acceleration (known as the centripetal acceleration) is v^2/r towards the centre.

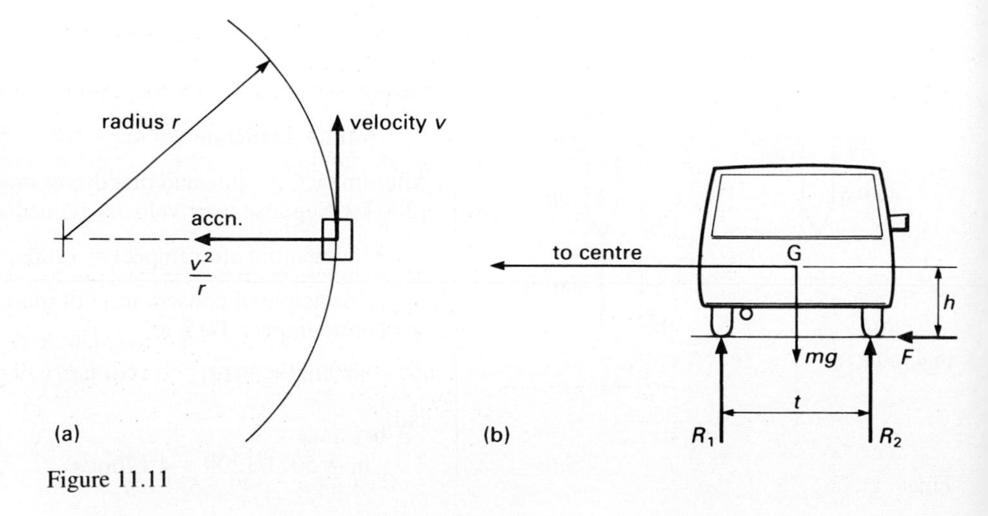

Figure 11.11

We know from experience that two kinds of mishaps can occur. The car may skid or overturn. To investigate these possibilities we have to consider the forces shown in Fig. 11.11(b). Suppose the mass of the car is m and its centre of gravity G is height h above the road. Let the track of the car (the distance between the inner and outer wheels) be t and assume that G is on the vertical centreline. R_1 is the total reaction for the inner wheels and R_2 for the outer. Since there is an acceleration towards the centre there must be a force acting on the car in this direction. It is called the *centripetal force*. The car is not in equilibrium and the radial forces do not balance. However, it is sometimes convenient to imagine a force acting radially outwards which would produce equilibrium. It is equal and opposite to the centripetal force and is called the *centrifugal force*. Its value is mv^2/r or $m\omega^2 r$. The use of centrifugal force is not recommended and it is avoided in this book.

Skidding

The vertical forces are in equilibrium since there is no acceleration in this direction. The total vertical reaction $(R_1 + R_2)$ is therefore equal to the weight of the car, mg. If the coefficient of limiting friction is μ the maximum frictional force is:

$$F = \mu(R_1 + R_2) = \mu mg$$

If the car is to keep to its circular path, this force must be sufficient to achieve the acceleration v^2/r. Using the relationship force = mass × acceleration we obtain:

$$F = mv^2/r$$

Substituting the value of F above, this becomes:

$$\mu mg = mv^2/r \quad or \quad \mu = v^2/gr$$

This is the smallest value of the coefficient of friction that will prevent skidding. It is a common misunderstanding that when skidding occurs the car slides radially outwards but Newton's first law shows that in the absence of a lateral force the car would continue straight ahead in the direction of the velocity v (Fig.11.11(a)).

Overturning

Suppose there is sufficient friction to keep the car to its circular path. Then $F = mv^2/r$ as before. If the car is on the point of overturning then the inner wheels will start to lift and the reaction R_1 will be zero. The total weight is then being carried on the outer wheels and $R_2 = mg$.

At this instant the moments of the forces about G just balance. Taking anticlockwise as positive,

$$Fh - R_2(t/2) = 0$$

With the values obtained above for F and R_2, this equation becomes:

$$m(v^2/r)h - mg(t/2) = 0$$

and

$$h = gtr/2v^2$$

If the height of the centre of gravity exceeds this amount the vehicle will begin to overturn.

In the case of a two-wheeled vehicle (Fig. 11.12) the rider must lean inwards at an angle θ so that the vertical reaction R will provide a moment about the centre of gravity G.

As before,

$$F = mv^2r \quad \text{and} \quad R = mg$$

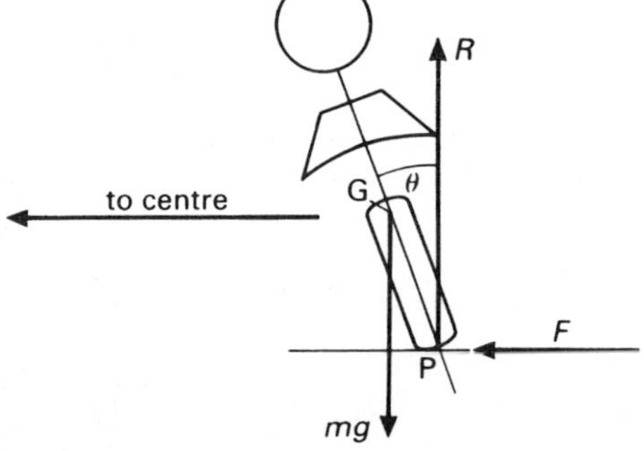

Figure 11.12

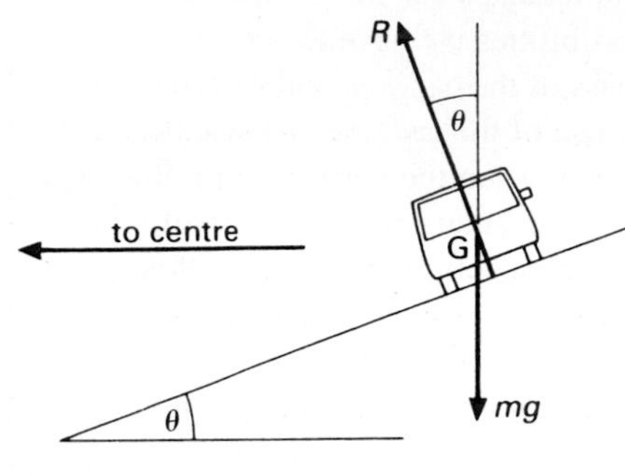

Figure 11.13

Let P be the point of contact with the road. Then taking moments about G, anticlockwise positive,

$$R \times GP \sin\theta - F \times GP \cos\theta = 0$$

Substituting for R and F,

$$mg \sin\theta - m(v^2/r) \cos\theta = 0$$

from which

$$\tan\theta = v^2/gr$$

This result gives the angle to which the rider must lean. (It is assumed that there is sufficient friction to prevent skidding.)

The tendency to skid can be eliminated by banking the road or track. Suppose (Fig. 11.13) θ is the angle at which no lateral force is required at the point of contact. The reaction R, perpendicular to the slope, has components $R \cos\theta$ vertically upwards and $R \sin\theta$ towards the centre.

Using force = mass × acceleration the equation of motion in the radial direction is:

$$R \sin\theta = m(v^2/r)$$

and resolving vertically

$$R \cos\theta = mg$$

Dividing the first of these equations by the second leads to the result:

$$\tan\theta = v^2/gr$$

This is identical to the previous result. Therefore the angle of banking that will eliminate side forces is the same as that to which the rider of a two-wheeled vehicle must lean on a level road. It is also equal to the angle of friction needed to prevent skidding on a level track.

The angle given by this result applies to one speed only. If the road or track is banked to this angle then higher speeds will require an inward force and lower speeds an outward one. On the early racing tracks for cars or two-wheeled vehicles, such as Brooklands, the banking was increased towards the outside so that a range of speeds could be accommodated. On a railway, however, a higher speed than the chosen one will result in an outward thrust on the outer rail, and a lower speed will cause an inward thrust on the inner rail.

EXAMPLE 11.16 A car is rounding a curve of 40 m radius on a level road. The track of the wheels is 1.3 m and the centre of gravity is 0.8 m above the ground and midway between the wheels. Find:

(a) the maximum possible speed if the car is not to overturn,

(b) the least coefficient of friction between the road and tyres which will prevent skidding at this speed,

(c) the angle to which the road should be banked if there is to be no tendency to slide at this speed,

(d) the side thrust required if the car rounds the banked road at a speed 20 per cent greater. The mass of the car is 900 kg.

SOLUTION The formulae of the last section are assumed; if they are not memorised, solutions in examinations must be obtained from first principles.

(a) In metre units, $r = 40$, $t = 1.3$ and $h = 0.8$

Using the result $h = gtr/2v^2$, the maximum speed is given by:

$$v^2 = gtr/2h = 9.81 \times 1.3 \times 40/(2 \times 0.8) = 318.8$$

and

$$v = \sqrt{318.8} = 17.86 \text{ m/s}$$

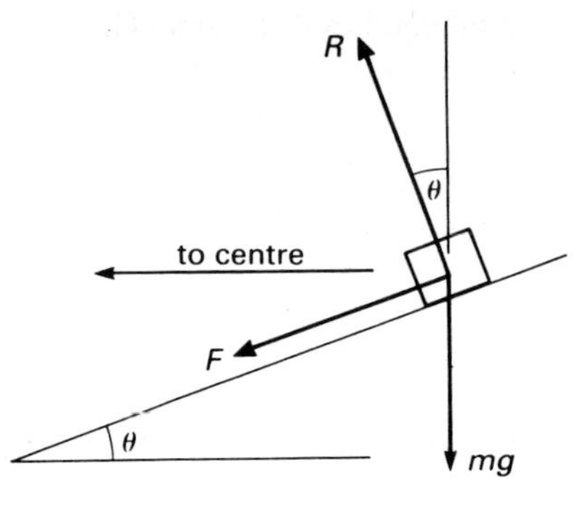

Figure 11.14

(b) The required coefficient of friction is:

$$\mu = v^2/gr = 17.86^2/(9.81 \times 40) = 0.813$$

(c) The angle of banking for which there is no tendency to skid is given by:

$$\tan\theta = v^2/gr = 0.813 \text{ and } \theta = 39.1°$$

(d) With a 20 per cent increase the speed is $1.2 \times 17.86 = 21.43$ m/s and, since this is greater than the value for which the road is banked there will be an inwards sideways force F acting on the car as shown in Fig. 11.14.

The components of R and F radially inwards are $R\sin\theta$ and $F\cos\theta$ respectively. Hence, applying the relationship

force = mass × acceleration, we have:

$$R\sin\theta - F\cos\theta = mv^2/r$$

or, dividing through by $\sin\theta$

$$R + F\cos\theta = mv^2/r\sin\theta \qquad [i]$$

The corresponding vertical components are $R\cos\theta$ upwards and $F\sin\theta$ downwards. Resolving vertically,

$$R\cos\theta - F\sin\theta = mg$$

or, dividing through by $\cos\theta$,

$$R - F\tan\theta = mg/\cos\theta \qquad [ii]$$

Subtracting [ii] from [i] to eliminate R,

$$F\cos\theta + F\tan\theta = \frac{mv^2}{r\sin\theta} - \frac{mg}{\cos\theta}$$

Substituting the numerical values for the various quantities,

$$F \times 1.231 + F \times 0.8125 = \frac{900 \times 21.43^2}{40 \times 0.631} - \frac{900 \times 9.81}{0.776} = 4998$$

from which

$$F = 2450 \text{ N}$$

11.7 Work and power in rotary motion

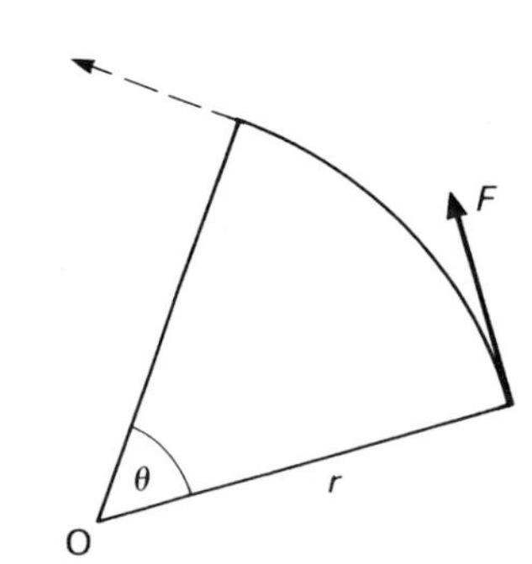

Figure 11.15

Suppose a body turns through an angle θ radians about a point O when acted on by a force F as shown in Fig. 11.15. Let r be the moment arm (or lever arm) of the force. The point of application of the force moves round the arc of a circle radius r whose length is $r\theta$. Hence, the work done or energy transferred (force × distance) is $Fr\theta$. However, Fr is the moment or torque T and we can therefore write:

$$\text{work done} = T\theta$$

If the torque varies, the useful energy transferred is given by the area under the graph of T against θ. (Compare with Example 11.4 and Fig. 11.3.) If the work $T\theta$ is done in time t the power P is given by:

$$P = \text{rate of doing work} = T\theta/t = T\omega$$

where ω is the angular speed in rad/s and is equal to θ/t.

EXAMPLE 11.17 A flywheel is mounted on a horizontal shaft 50 mm diameter which runs in plain bearings for which the coefficient of friction is 0.05. If the combined weight of the flywheel and axle is 120 kg how much work is done against friction when the flywheel makes 500 revolutions?

What is the power expended in overcoming friction when the flywheel rotates at 1200 rev/min?

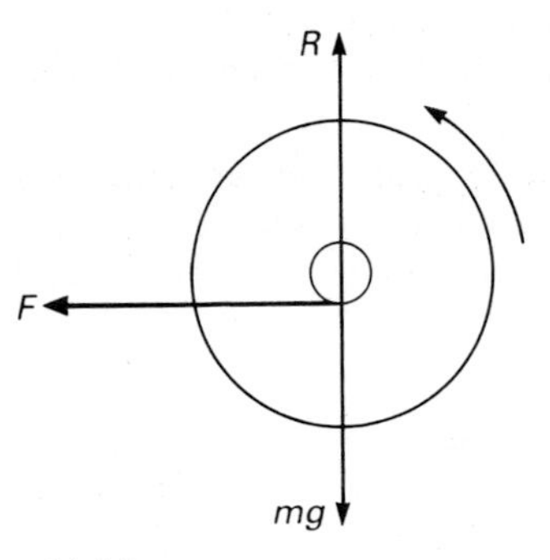

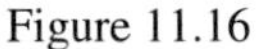

Figure 11.16

SOLUTION The vertical reaction R between the bearing and shaft is equal to the weight of the flywheel and shaft as shown in Fig. 11.16. Hence the friction force is:

$$F = \mu R = \mu mg = 0.05 \times 120 \times 9.81 = 58.86 \text{ N}$$

This acts at a radius of 25 mm and the torque is therefore:

$$T = Fr = 58.86 \times 0.025 = 1.471 \text{ Nm}$$

The angle turned through during 500 revolutions is $2\pi \times 500 = 3142$ rad and hence:

$$\text{work done} = T\theta = 1.471 \times 3142 = 4622 \text{ J} \text{ or } 4.622 \text{ kJ}$$

At 1200 rev/min the angular speed is $2\pi \times 1200/60 = 125.7$ rad/s and the power is:

$$P = T\omega = 1.471 \times 125.7 = 184.9 \text{ W}$$

11.7.1 Kinetic energy of rotation

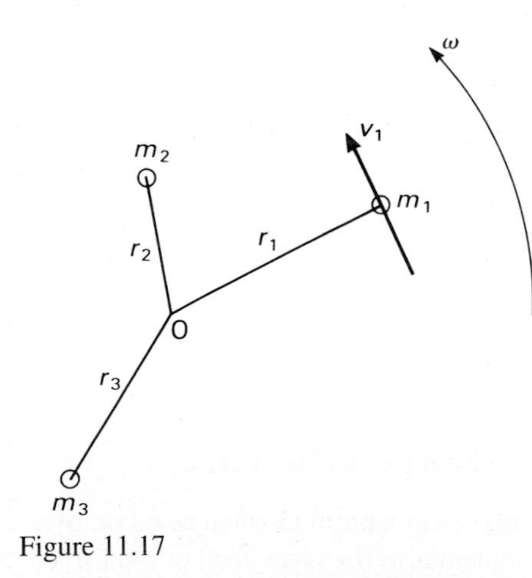

Figure 11.17

Fig. 11.17 shows three very small masses (or particles) m_1, m_2 and m_3, rotating about a point O with the same angular speed ω (in radian units). Suppose m_1 is at a distance r_1 from O and that its tangential velocity is v_1. As shown in Chapter 10, $v_1 = \omega r_1$ and the kinetic energy of this mass can be written:

$$\text{KE} = \tfrac{1}{2} m_1 v_1^2 = \tfrac{1}{2} m_1 (\omega r_1)^2 = \tfrac{1}{2} m_1 r_1^2 \omega^2$$

A similar result is obtained for the other masses and the total kinetic energy for the system may be written:

$$\text{KE} = \tfrac{1}{2} m_1 r_1^2 \omega^2 + \tfrac{1}{2} m_2 r_2^2 \omega^2 + \tfrac{1}{2} m_3 r_3^2 \omega^2$$
$$= \tfrac{1}{2} (m_1 r_1^2 + m_2 r_2^2 + m_3 r_3^2) \omega^2$$

The expression in brackets can be extended for any number of particles and its value depends only on the masses and their distances from O. Once it has been calculated it can be used to find the kinetic energy for any rotational speed ω. It is called the *moment of inertia* and is denoted by I. The kinetic energy can therefore be written:

$$\text{KE} = \tfrac{1}{2} I \omega^2$$

This result should be compared to the expression $\frac{1}{2} mv^2$ for linear motion. The units of I are kg m^2. Moment of inertia is related to the quantity, second moment of area, used in stress analysis and hydrostatics, the difference being that it is the result of multiplying mass (rather than area) by the square of the distance.

For a particle, mass m, at radius r the moment of inertia is:

$$I = mr^2$$

In the case of a finite object or body, the value of I is the sum of the values for all the particles of which it is made. It is convenient to define an equivalent radius at which the whole mass of the body would have to be concentrated to give the same value of I. It is called the *radius of gyration* and is denoted by k. Therefore for a body of mass m the moment of inertia can be written:

$$I = mk^2$$

The value of k will be given in numerical examples whenever it is required. For a thin circular rim rotating about the centre, k is equal to the radius of the rim. For a uniform solid disc rotating about an axis through its centre perpendicular to the disc $k^2 = d^2/8$ where d is the external diameter.

When a body is moving forward and rotating at the same time, as in the case of a wheel rolling along a road, the total kinetic energy is the sum of the amounts due to the linear and rotational motions. These are called the kinetic energy of translation and kinetic energy of rotation.

EXAMPLE 11.18 A railway wagon has a total mass of 4 tonnes of which the wheels and axles account for 600 kg. The radius of the wheels to the point of contact with the rails is 400 mm and the radius of gyration of each pair of wheels and their axle is 180 mm. Calculate the kinetic energy of the truck when it is travelling at 10 km/h.

If the truck is loose shunted at this speed how far will it travel up a slope of 1 in 150 before coming to rest? Ignore frictional resistance.

SOLUTION Converting to SI units the speed of the wagon is:

$$v = \frac{10 \times 1000}{3600} = 2.778 \text{ m/s}$$

$$\text{KE of translation} = \tfrac{1}{2}mv^2 = \tfrac{1}{2} \times 4000 \times 2.778^2$$
$$= 15\,435 \text{ J}$$

Also, $v = \omega r$ and the angular speed of the wheels is:

$$\omega = v/r = 2.778/0.4 = 6.945 \text{ rad/s}$$

Since all the wheels rotate at the same speed the moment of inertia can be calculated to include them all. Therefore,

$$I = mk^2 = 600 \times 0.18^2 = 19.44 \text{ kg m}^2$$

and the kinetic energy of rotation is:

$$\text{KE} = \tfrac{1}{2}I\omega^2 = \tfrac{1}{2} \times 19.44 \times 6.945^2 = 469 \text{ J}$$

$$\text{Total} \quad \text{KE} = 15\,435 + 469 = 15\,904 \quad \text{or} \quad 15.904 \text{ kJ}$$

In climbing the slope the wagon gains potential energy mgh and loses its kinetic energy. If the distance travelled in coming to rest is s, then the vertical height h it gains is $s/150$. Therefore:

$$mgh = 15\,904$$

or

$$4000 \times 9.81 \times (s/150) = 15\,904$$

and

$$s = (15\,904 \times 150)/(4000 \times 9.81)$$
$$= 60.8 \text{ m}$$

EXAMPLE 11.19 Repeat Example 11.7 taking the mass of the pully into account. The pulley has a mass of 4 kg and a radius of gyration of 120 mm, and the radius at which the string passes round it is 150 mm.

SOLUTION Referring to Fig. 11.5, the change in potential energy is not affected by the mass of the pulley because it remains at the same level. Hence, decrease in PE = 29.43 J as before. If the final speed of the masses is v then the corresponding angular speed of the pulley is given by:

$$\omega = v/r = v/0.15 = 6.667v \text{ rad/s}$$

The moment of inertia of the pulley is:

$$I = mk^2 = 4 \times 0.12^2 = 0.0576 \text{ kg m}^2$$

and its final kinetic energy of rotation is:

$$\text{KE} = \tfrac{1}{2}I\omega^2 = \tfrac{1}{2} \times 0.0576 \times (6.667v)^2 = 1.28v^2 \text{ J}$$

(a) Adding the energy of rotation of the pulley to the kinetic energy of the masses that was calculated in the earlier solution,

$$\text{total KE} = 3.5v^2 \times 2.5v^2 + 1.28v^2 = 7.28v^2 \text{ J}$$

and, equating this result to the decrease in potential energy,

$$7.28v^2 = 29.43 \quad \text{or} \quad v^2 = 29.43/7.38 = 4.043$$

Therefore,

$$v = \sqrt{4.043} = 2.011 \text{ m/s}$$

(b) time taken = distance/average speed $= 1.5/(2.011/2)$
$= 1.49$ s

(c) acceleration of masses = increase in speed/time
$= 2.011/1.49 = 1.35 \text{ m/s}^2$

The results show that the inertia of the pulley has reduced the acceleration and final speed, and increased the time taken.

11.7.2 Flywheels

The angular kinetic energy of a rotating body depends on its moment of inertia I, which equals mk^2, where m is its mass and k is the radius of gyration. If we want to store a large amount of energy in this way we must therefore increase k as far as possible. This is achieved by a flywheel, which is made in the form of a disc or rim, so that the mass is distributed away from the axis of rotation.

The store of angular kinetic energy in a flywheel can be used to drive machines, a familiar example being push-and-go toys. The principle has also been used in some experimental full-size vehicles.

The most common use of flywheels, however, is to reduce the fluctuations of speed in machinery caused by the variation in the driving torque. In a single-cylinder, four-stroke, reciprocating engine, for example, the energy gained by the flywheel during the power stroke enables it to maintain the speed during the exhaust, induction and compression strokes.

Suppose I is the moment of inertia, and ω_1 and ω_2 are the greatest and least speeds during the cycle. Then the greatest change in the angular kinetic energy is given by:

$$\text{change in KE} = \tfrac{1}{2}I\omega_1^2 - \tfrac{1}{2}I\omega_2^2 = \tfrac{1}{2}I(\omega_1^2 - \omega_2^2)$$

The expression in brackets is the difference of two squares and it can be written $(\omega_1 - \omega_2)(\omega_1 + \omega_2)$. Furthermore, it is usual to assume that the speed fluctuates equally above and below the mean value. Hence we can put $\omega = (\omega_1 + \omega_2)/2$ and the result becomes:

$$\text{change in KE} = I\omega\,(\omega_1 - \omega_2)$$

The ratio of the change in speed to the mean speed is sometimes called the *coefficient of fluctuation of speed*, and is denoted by c. Thus:

$$c = \frac{\omega_1 - \omega_2}{\omega} \quad \text{and} \quad \omega_1 - \omega_2 = c\omega$$

Substituting this result,

$$\text{change in KE} = I\omega^2 c$$

EXAMPLE 11.20 A flywheel has a radius of gyration of 200 mm and is to run at a mean speed of 1500 rev/min. Determine its mass if the speed is not to fluctuate by more than 1 per cent when the fluctuation of energy is 240 J.

SOLUTION The coefficient of fluctuation of speed c is $1/100 = 0.01$ and, in SI units, the mean speed is:

$$\omega = \frac{2\pi}{60} \times 1500 = 157.1 \text{ rad/s}$$

Hence, by rearranging the last result,

$$I = \frac{\text{fluctuation of energy}}{\omega^2 c} = \frac{240}{157.1^2 \times 0.01} = 0.972 \text{ kg m}^2$$

But $I = mk^2$ and k is given as 200 mm = 0.2 m. Thus, the required mass of the flywheel is:

$$m = I/k^2 = 0.972/0.2^2 = 24.3 \text{ kg}$$

11.7.3 Angular momentum, torque and angular acceleration

In section 11.7 it was shown that the expression $\frac{1}{2}mv^2$ for the kinetic energy of translation has its counterpart $\frac{1}{2}I\omega^2$ for the kinetic energy of rotation. Momentum, too, has both linear and angular forms. In linear motion it equals mv and in angular motion it is given by $I\omega$. The principle of conservation of momentum can also be applied to angular momentum. If two bodies, rotating independently about the same axis, are locked together by a clutch, their common velocity can be found by equating the initial and final values of the angular momentum.

The result $F = ma$ for the force required to produce a linear acceleration also has a corresponding relationship in angular motion. The torque T required to cause an angular acceleration α in a body having a moment of inertia I about the appropriate axis is given by:

$$T = I\alpha$$

EXAMPLE 11.21 Find the torque required to accelerate a flywheel of mass 60 kg and radius of gyration 250 mm uniformly from 600 rev/min to 2000 rev/min in 6s.

SOLUTION Converting the rotational speeds to radian and second units

$$\text{initial speed } \omega_1 = \frac{2\pi \times 600}{60} = 62.8 \text{ rad/s}$$

$$\text{final speed } \omega_2 = \frac{2\pi \times 2000}{60} = 209.4 \text{ rad/s}$$

$$\text{angular acceleration} = (\omega_2 - \omega_1)/t = (209.4 - 62.8)/6 = 24.43 \text{ rad/s}^2$$

$$\text{moment of inertia } I = mk^2 = 60 \times 0.25^2 = 3.75 \text{ kg m}^2$$

Therefore, the required torque is:

$$T = I\alpha = 3.75 \times 24.43 = 91.6 \text{ N m}$$

11.8 Examples of simple harmonic motion

Simple harmonic motion in a straight line can be defined as the motion of a particle in which its acceleration towards a fixed point in the line is proportional to its distance from it. Since force = mass × acceleration, the force on the particle must also be proportional to the distance from the fixed point. There are many practical examples in which this occurs, at least approximately.

Spring oscillations

Figure 11.18 shows a light spring from which a mass m is suspended. If the spring obeys Hooke's law then its extension is proportional to the force producing it. The ratio of force to extension is called the *stiffness* of the spring and it is denoted by S. Thus:

$$S = \text{force/extension}$$

Stiffness has the units newton per metre. Let AA be the equilibrium position of the mass. In this position the tension in the spring equals the weight of the mass and it has no acceleration. If the mass is pulled down and released it will oscillate. When it is a distance x below the equilibrium position the additional force acting on it is, from the definition of stiffness, Sx. This force acts towards the equilibrium position and, using force = mass × acceleration,

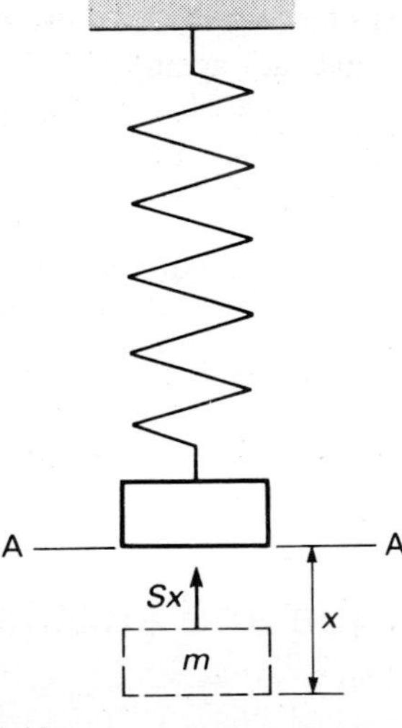

Figure 11.18

$$Sx = ma \quad \text{or} \quad a = Sx/m$$

This result shows that the acceleration a is proportional to the distance x from the equilibrium position AA. The motion is therefore simple harmonic and the constant of proportionality ω^2 is equal to S/m. Thus:

$$\omega = \sqrt{(S/m)}$$

and the period of oscillation is given by:

$$T = \frac{2\pi}{\omega} = 2\pi\sqrt{\frac{m}{S}}$$

This analysis ignores the mass of the spring. It can be allowed for by adding one-third of its value to m.

The vibrations of cantilevers and beams are further examples of simple harmonic motion. If the mass m is attached to the free end of a cantilever or the midpoint of a beam, the stiffness S in the formula for T can be found from the deflection formulae given in Fig. 9.13. In the case of the cantilever the deflection is $WL^3/3EI$ and the stiffness (force/deflection) is therefore $3EI/L^3$. To allow for the mass of the cantilever add 33/140 of it to m; for the beam the corresponding fraction is 17/35.

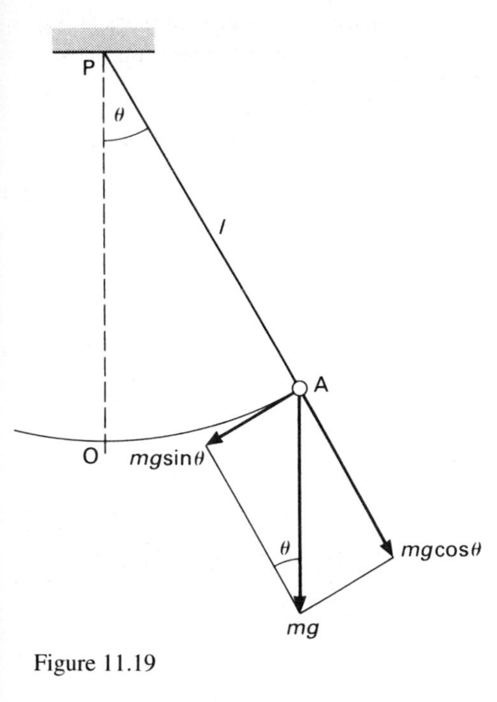

Figure 11.19

Simple pendulum

Suppose a small body (called a *bob*), of mass m, is suspended from a point P at the end of a light string, length l, as shown in Fig. 11.19 and is set oscillating in a vertical plane. The equilibrium position of the bob is at the point O vertically below P.

At the instant when the string makes an angle θ radians with the vertical line PO the displacement x of the mass from O is the arc length OA and this equals $l\theta$. Therefore:

$$x - l\theta \quad \text{and} \quad \theta = x/l$$

The weight of the bob, mg, has a component $mg \sin \theta$ acting tangentially and this is the force accelerating it towards its equilibrium position O. For small angles in radians, $\theta = \sin \theta$ approximately. For example, when the angle is 14° the difference between θ rad and $\sin \theta$ is about 1 per cent. The accelerating force can therefore be taken as $mg\theta$ approximately. Substituting in the relationship force = mass × aceleration, we have:

$$mg\theta = ma \quad \text{from which} \quad a = g\theta = gx/l$$

The acceleration a is therefore proportional to the displacement x and the motion is therefore simple harmonic. The constant of proportionality ω^2 is g/l and $\omega = \sqrt{(g/l)}$. The periodic time T is given by:

$$T = \frac{2\pi}{\omega} = 2\pi\sqrt{\frac{l}{g}}$$

This result can be used so as to obtain an experimental value of g. If a pendulum is made from a heavy bob and a light string, and its period of oscillation timed, a value of g can be found from this relationship. To achieve reasonable accuracy, the value of T should be found by timing a large number of swings, at least twenty. One complete oscillation consists of two swings – across and back again.

The term 'simple' is used to indicate that the mass of the bob can be assumed to be concentrated at a point. Otherwise the pendulum is described as compound and the analysis is more complicated.

EXAMPLE 11.22 A body is carefully hung from a helical spring and produces an extension of 56 mm. The body is then pulled down a further 35 mm and released. Calculate, for the ensuing motion,
(a) the periodic time,
(b) the maximum velocity,
(c) the maximum acceleration.
Neglect the mass of the spring itself.

SOLUTION If the mass is m kg its weight is $9.81m$ N. Working in SI units the stiffness S is given by:

$$S = \text{force/extension} = 9.81m/0.056 = 175.2m \text{ N/m}$$

The ensuing motion is simple harmonic with an amplitude a of 35 mm. The constant of the motion ω is given by:

$$\omega = \sqrt{(S/m)} = \sqrt{(175.2m/m)} = 13.24 \text{ s}^{-1}$$

(a) Periodic time $T = 2\pi/\omega = 2\pi/13.24 = 0.475$s

(b) Using the result obtained in Chapter 10 the maximum velocity is

$$v = \omega a = 13.24 \times 0.035 = 0.463 \text{ m/s}$$

(c) Maximum acceleration = $\omega^2 a = 13.24^2 \times 0.035 = 6.135$m/s^2

EXAMPLE 11.23 Calculate the length of a seconds pendulum, that is, a simple pendulum that beats seconds.

SOLUTION A seconds pendulum is one in which each swing takes 1 s so that the periodic time T is 2 s. Using the formula for a simple pendulum,

$$T = 2\pi\sqrt{(l/g)} \quad \text{or, by squaring,} \quad T^2 = 4\pi^2 l/g$$

Substituting $g = 9.81$ and $T = 2$, the required length is:

$$l = T^2 g/4\pi^2 = 2^2 \times 9.81/(4\pi^2) = 0.994 \text{ m}$$

11.9 Summary of formulae

The main results and formulae required for linear motion and rotary motion are collected in the following table.

	Linear motion	*Rotary motion*
displacement	x	θ
velocity	v	ω
acceleration	a	α
work done	Fx	$T\theta$
power	Fv	$T\omega$
momentum	mv	$I\omega$
potential energy	mgh	mgh
kinetic energy	$\frac{1}{2}mv^2$	$\frac{1}{2}I\omega^2$
force/acceleration	$F = ma$	$T = I\alpha$

Table 11.2

Also $I = mk^2$ where k is the radius of gyration.

These results, together with the two conservation principles (momentum and energy) provide the basis for solving a wide range of problems in dynamics.

11.10 Bicycle analysis

In 10.10 we analysed the linear and rotational speed, acceleration and distance travelled in the different phases of a bicycle journey. Now with the experience of Chapter 11 we can analyse the energy, torque and power requirements on this journey. In addition to the previous information, we also know that the mass of the bike and rider together is 90 kg and the rolling resistance can be taken as 0.05 N/kg.

11.10.1 Bicycle braking

If a bicycle travels on level ground with no wind resistance the rider would have to exert a force to keep the speed constant. This is because, although it is a very efficient machine, the rolling resistance, which is a measure of the resistance to motion due to the frictional contact of the tyres with the road, results in energy losses. If the cyclist allowed himself to come to rest naturally without braking, how far would he travel? The resistance to motion, R, is given by the mass of the rider and bike multiplied by the rolling resistance.

$$\text{therefore} \quad R = 90 \times 0.05 = 4.5 \text{ N}$$

The work done in coming to rest must be equivalent to the energy lost. The cyclist was travelling at 8 m/s just before this so the energy lost is given by the kinetic energy of

$$\tfrac{1}{2}mv^2 = \tfrac{1}{2} \times 90 \times 8^2 = 2880 \text{ J}$$

The work done in stopping is given

$$Rs = 4.5 \times s, \text{ where } s \text{ is the distance travelled}$$

$$\text{therefore} \quad 4.5 \times s = 2880 \text{ and so } s = 640 \text{ m}$$

This means that without brakes it would take the cyclist 640 m to stop from 8 m/s under the stated conditions.

The actual stopping distance was 20 m and the resistance to motion was the previous value, R, as well as the braking force, F_B. Again using the fact that the energy lost is equal to the work done

$$2880 = (4.5 + F_B)\,20 \text{ and so } F_B = 144 - 4.5 = 139.5 \text{ N}$$

This is the braking force applied by the brake pads to the tyres and is obviously a larger force than we use to grip the brake lever, so there must be a mechanical advantage on the braking system. You can investigate this by measuring the distance that the brake lever moves compared with the distance that the brake pad moves.

11.10.2 Bicycle power requirements at constant speed

It is possible for an experienced adult cyclist to maintain about 75 watts for long periods without tiring. Much higher outputs can be maintained for short periods. For example, a fit, experienced cyclist may be able to produce 300 W for up to 5 minutes. However, over prolonged periods, untrained cyclists can maintain a power of only 35 to 40 W and a speed of about $3\frac{1}{2}$ m/s on the level with no wind.

We know that

$$\text{power} = Fv$$

Therefore the power required to overcome rolling resistance, P_R, is equal to the rolling force, R multiplied by v. At a speed of 10 m/s this gives

$$P_R = 4.5 \times 10 = 45 \text{ W}$$

At 5 m/s this would be 22.5 W as it is a linear relationship with the speed, v.

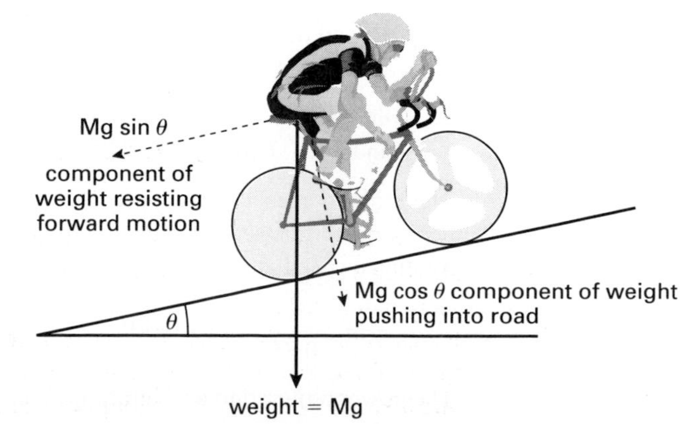

Figure 11.20 The components of weight of a bicycle when hill climbing

If there was also a slope on the road of 2.5° the cyclist would have to expend extra power to climb the hill. In Fig. 11.20 we can see that the extra resistance to motion when climbing a hill is equal to $mg \sin \theta$. Therefore the power needed to climb the hill, P_H is given by

$$P_H = mg \sin \theta \, v$$

For a hill of 2.5° at 10 m/s

$$P_H = 90 \times 9.81 \times \sin 2.5° \times 10 = 385.1 \text{ W}$$

This is obviously far too high for us, so could we go up this hill at 5 m/s?

$$P_H = 90 \times 9.81 \times \sin 2.5° \times 5 = 192.6 \text{ W}$$

This is still high. For a comfortable prolonged power of about 50 W we could go up this hill at only 1.3 m/s which explains why our speed falls so rapidly when we start hill climbing. In fact, by far the biggest power requirement is to overcome the wind resistance (see 13.11) and we have not considered this at all.

The total power requirements for an incline and the rolling resistance from our figures are shown in the following table.

Power (W)	Speed 1 m/s	3 m/s	5 m/s	10 m/s
to overcome the rolling resistance	4.5	11.25	22.5	45
to climb a hill of 2.5°	38.5	115.5	192.6	385.1
total power (3 sig figs)	43.0	127	215	430

Table 11.3

We can see from this that an inexperienced cyclist will be able to climb this hill at about 1 m/s with no wind resistance, while a more experienced cyclist will be able to go up the hill at no more than 5 m/s in still conditions. If there is any wind resistance these speeds will obviously be reduced further.

11.10.3 Bicycle power requirements during acceleration

Let us first look at the torque, T, and power, P_w, necessary to rotate the drive wheel during the second acceleration phase. Let the mass of the wheel, m_w, be 1.5 kg. From 11.7.1 we can see that

$$T = I\alpha \quad \text{and} \quad I = m_w k^2$$

During the phase of maximum acceleration we found that $\alpha = 2.86$ rad/s^2 (See 10.10.3).

Nearly all of the mass of the wheel is on the rim and so the radius of gyration of the wheel, k, can be taken as approximately equivalent to the radius, r or 0.35 m.

therefore $I = 1.5 \times 0.35^2 = 0.184$ kg m^2
and $T = 0.184 \times 2.86 = 0.525$ N m

Therefore it takes a torque of 0.525 Nm to rotationally accelerate each wheel. At the beginning of this acceleration, the linear speed, v_1, is 6 m/s. This means that the rotational speed,

$$\omega \text{ is } \frac{6}{0.35} = 17.14 \text{ rad/s}$$

The power required to accelerate each wheel, P_w, is therefore

$$P_w = T\omega = 0.525 \times 17.14 = 9 \text{ W}$$

The power to accelerate is larger than this as the bicycle and rider need a linear acceleration too. The force needed to provide this acceleration is given by

$F = Ma$ where $M = 90$ kg and $a = 1$ m/s^2 from the second stage of the journey

therefore $F = 90 \times 1 = 90$ N

The force needed to overcome the rolling resistance is still 4.5 N and so the total force needed to accelerate, F_T is 94.5 N. This means that the power needed to provide the linear acceleration, P_L is

$$P_L = F_T \times v = 94.5 \times 6 = 567 \text{ W}$$

This means a total power requirement of 567 + 2(9) W or 585 W. This is high but is only required for 2 s. The person who did this journey was a very fit, experienced cyclist who could produce 300 W for several minutes of constant cycling.

11.10.4 Dissipated energy

In 11.10.1 we found that the linear kinetic energy of the bike and cyclist at 8 m/s = 2880 J. Now we could also include the kinetic energy of both wheels.

$$KE_{wheels} = 2 \times \tfrac{1}{2} I\,\omega^2 \quad \text{where } I = 0.184 \text{ kg m}^2 \quad \text{and} \quad \omega = 22.9 \text{ rad/s.}$$

therefore $KE_{wheels} = 96.5$ J and the total KE = 2880 + 96.5 = 2976 J

If the cyclist were to hit a wall or another vehicle and come to rest very quickly, this energy would have to be absorbed suddenly. If the cyclist was stopped in only 1 m, the effective force that he would be hit with, F_{stop}, is given by:

$$\text{Total KE} = F_{stop} \times 1$$

This means that $F_{stop} = 2976$ N

The measurement of the stopping force in newtons may not mean too much, but if we divide this number by the mass of the bike and rider we obtain the sudden deceleration acting on them. If we divide this by by 9.81 (the acceleration due to gravity) we get the energy in terms of 'g' force or multiple of the normal acceleration due to gravity. It is possible for the brain to resist about an 80g force before becoming irreparably damaged.

$$F_{stop} = 2976 \text{ N} \quad \text{deceleration} = \frac{2976}{90} = 33.1 \text{ m/s}^2 = 3.37 \text{ g}$$

Looking at the table below, we can see the type of 'g' forces acting for stopping distances of between 0.5 m and 10 m when travelling at 8 m/s. It may surprise you to find that you would experience over 3 g if you stopped within a metre from a speed of 8 m/s. If you were to stop suddenly in 0.5 m you would experience nearly 7 g. Imagine if you were on a motorbike and hit a wall where the speed could be as much as 30 m/s. If you stopped within 0.5 m this could result in a force of over 90 g.

Stopping distance	10 m	2 m	1 m	0.5 m
'g' force	0.33 g	1.67 g	3.33 g	6.74 g

On impact the front wheel of the bike will absorb some of the energy. Crash helmets are also designed to absorb the energy of impact. If we look at the maximum acceleration that the brain can withstand, 80 g, we find that for our cyclist this is:

$$80 \times 9.81 = 784.8 \text{ m/s}^2$$

This gives us a force of 784.8 × 90 N = 70632 N
If this were to occur at a speed of 8 m/s the stopping distance, d_{stop}, is given by:

$$2976 = 70632 \times d_{stop}$$

and so $d_{stop} = 0.042$ m

This means that if you were to stop suddenly within 42 mm, and the bike did not absorb any of the energy of impact, you would experience the full 80 g force. Consequently bicycle helmets are designed to absorb much of this impact. Have a look at a cycle helmet and a motor cycle helmet and examine the material and the thickness of padding in the helmet. This is designed to crush and absorb the impact in order to protect the head. How far do you think it could be compacted? How much of the energy of impact do you think it will absorb?

Bibliography

BP Statistical Review of World Energy 1993
British Petroleum various publications. BP Educational Service
Drabble G E 1986 *Elementary Engineering Mechanics*. Macmillan
Foley G 1992 *The Energy Question*. 4th edn. Penguin.
Hannah J and Stephens R C 1984 *Mechanics of Machines* (*Elementary Theory* and Examples) 4th edn. Arnold
Humphrey D and Topping J 1971 *A Shorter Intermediate Mechanics* SI edn. Longman, Ch 4–5, 9
Patterson W C 1982 *Nuclear Power*. Penguin
Sage J *et al* 1980 *Modular Courses in Technology*: *Energy Resources*. Oliver and Boyd
Shell Briefing Service various publications. Shell International Petroleum Company
Watt Committee on Energy 1979 *A Warmer House at Lower Cost*. Institution of Mechanical Engineers

Assignments

Note: 1 Mg (megagramme) is equivalent to 1000 kg or 1 tonne.

1 A railway locomotive exerts a tractive effort of 150 kN at a speed of 75 km/h. Calculate the work done over a distance of 1.4 km and the power developed.

2 The engines of a jet aircraft develop a total thrust of 90 kN. What power is developed at a speed of 900 km/h?

3 Tabulate the energy values of packaged foods in joule and calorie units using the data given on wrappings.

4 Try and obtain the fuel bills for a house over a one-year period. Estimate the energy used annually, expressing your result in kWh, GJ and therms.

5 (a) A van of 2500 kg mass travels along a horizontal road against a resistance equal to 1/20 of its weight. Find the greatest speed attainable if the maximum effective output from the engine is 40 kW.
(b) What is the maximum acceleration attainable by the same van up an incline of 1 in 10 when its speed is 15 km/h, assuming the same power output and resistance?

6 A construction site hoist consists of two cages connected by a cable passing over a winding drum. The full cage of 3000 kg mass is raised while the empty cage of 1000 kg is lowered. The drum of 2000 kg mass has an effective diameter of

2.5 m to the cable and a radius of gyration of 1.0 m. The cages are accelerated from rest at the rate of 1.5 m/s^2. Determine, neglecting friction in the drum bearing and cage guides,

(i) the tensions in the cables on the two sides of the drum,
(ii) the driving torque required on the drum,
(iii) the distance moved by the cages in the first 3 s from rest.

7 A flywheel and its shaft together have a mass of 6000 kg and a radius of gyration of 1.2 m, the shaft diameter being 125 mm. The system is rotating at 120 rev/min and its weight is supported in bearings for which the coefficient of friction may be taken as 0.02. Calculate:

(i) the power required to overcome friction,
(ii) the kinetic energy of the flywheel and shaft,
(iii) the number of revolutions it will make in coming to rest if the power is cut off.

8 (a) Define *force*, *work*, *power* and *torque*.
(b) Give equations relating work to force and relating power to torque, defining all symbols.

9 (a) Explain what is meant by the following terms:
(i) kinetic energy,
(ii) a flywheel.
(b) An experimental vehicle is being developed using a flywheel disc as an energy source. The disc has a mass of 750 kg and has a radius of 1.0 m. When running it revolves at 20 rev/s.
(i) How much energy is stored in the flywheel?
(ii) If the vehicle runs for one hour, before coming to rest, what is the average power available?
(c) How could the energy stored in the flywheel be considerably increased without increasing the mass?
(d) There are many problems associated with flywheel power as an energy source. Identify and discuss three such problems.

10 (a) Solar panels are used in many locations to provide hot water. Draw a fully annotated diagram to illustrate the basic design of a solar panel which could be used for heating domestic hot water.
(b) A community transport mini-bus has the following design specification:

fuel tank capacity	100 litres
fuel energy content	12 kW h/litre
diesel engine efficiency	42%

Calculate the amount of lost or wasted energy if a journey of 600 km takes place at an average fuel consumption of 10 km/litre.
(c) Approximately 25% of the United Kingdom's oil consumption is used for transport. Discuss strategies which could reduce this figure.

11 The internal combustion engine of a car develops a maximum power of 36 kW; the car's transmission has an efficiency of 70%. The car has a mass of 800 kg and the resistance to its motion is to be taken as constant and equal to 500 N.
(a) Describe the energy changes that take place in the car's engine.
(b) Calculate the greatest speed that the car can travel:
(i) on a level road;
(ii) up a 1 in 8 incline.
(c) If the car is allowed to freewheel from rest down the 1 in 8 incline, find its speed after travelling a distance of 200 m.

12 (a) A car of mass 800 kg accelerates uniformly from rest on a straight horizontal road for a distance of 120 m, reaching a speed of 15 m/s. It maintains this speed as it enters a horizontal bend of radius 200 m.

(i) Calculate the initial accelerating force acting on the car.
(ii) What is the sideways force on the car as it rounds the bend?
(iii) What is the minimum coefficient of friction between the tyres and the road surface if side-slip is not to occur as the car rounds the bend?
(iv) With the help of a diagram of the forces acting on the car, show that the sideways force can be made zero by suitable banking, and finding the angle of banking needed at this speed.

(b) Rounded humps ('sleeping policemen') are used on roads as a means of preventing excessive speed. Describe and explain what happens if an attempt is made to drive over these at high speed.

13 A spring system has a rate (or stiffness) 20 kN m^{-1} between loads 0 and 1 kN, and 50 kN m^{-1} at greater loads.
Sketch a load–extension graph and find
(a) the change in extension when the load increases from 500 N to 1.5 kN,
(b) the energy stored when the load is 1.5 kN.

14 Three similar railway trucks, each of mass 2 Mg, are linked together and are moving at a constant speed, 8 km h^{-1} on a horizontal straight track. They meet another similar truck moving in the opposite direction at 1.5 kmh^{-1}.
Find the resulting speed of the four trucks if they link together automatically on contact.

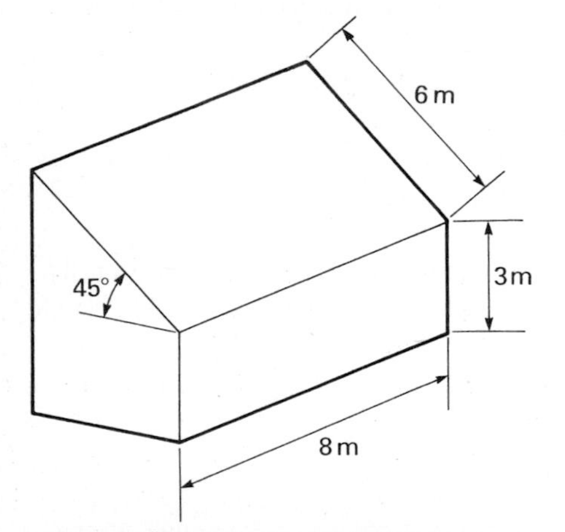

Figure 11.21

15 A conservatory is attached to the south facing wall of a building as shown in Fig. 11.21. The glass of which the wall and roof are made transmits 60% of the energy received. The solar flux is 500 W m^{-2} at noon when the sun is at 30° to the horizontal.
Determine the rate at which energy is gained by the conservatory.

16 An aircraft, launched from a ship by catapult (Fig. 11.22), has mass 15 Mg. The acceleration of the driven member of the catapult, of mass 250 kg, is uniform and the aircraft is launched after travelling 50 m in 1.5 s. The aircraft engine is producing a net thrust 10 kN at 5° to the direction of motion.

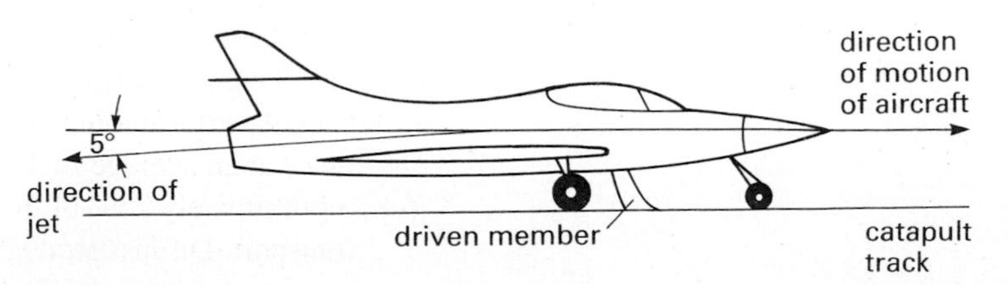

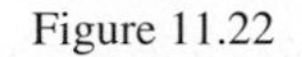

Figure 11.22

Determine
(a) the acceleration of the driven member of the catapult,
(b) the speed of the aircraft at the moment of launch,
(c) the force on the attachment between driven member and aircraft,
(d) the force which would need to be applied to the driven member to reduce its speed to zero in a further 3 m of the track after the launch point,
(e) the maximum power required at any position in the travel of the catapult.

17 (a) Using simple sketches and line diagrams explain how the energy in a raised water reservoir can be converted to the various energies required for domestic and industrial use. Name the form of energy at each conversion.

(b) Water is flowing at 15 m^3 per minute from a reservoir situated at an elevation of 100 metres above a hydroelectric power station. If the overall efficiency of the plant is 60% what will the electrical power output be measured in kW? (Assume density = 1000 kg/m^3 and gravity = 10 m/s^2.)

(c) Calculate the energy stored in a 50 A h (amp hour) 12 volt car battery. How long could this battery maintain the operation of four 55 watt lamps, assuming that the lights were 12 volt types and that the battery maintains its full output voltage until it is completely discharged?

12 Thermodynamics

12.1 What is thermodynamics?

Thermodynamics deals mainly with the transfer of energy by work and heat. It is also concerned with heat transfer (by conduction, convection or radiation, caused by a temperature difference) and chemical reactions that involve heat such as the combustion of fuels.

The machines that do work using energy transferred by heat are sometimes called *heat engines*. They include internal combustion engines (petrol and diesel), gas turbines and steam turbines. Machines that convert work into heat include refrigerators and heat pumps. In all these machines there is a *working substance* (a gas or vapour) that is compressed and expanded, undergoing pressure and temperature changes in the process. Some examples are:

Machine	*Working substance*
air compressor	air
internal combustion engine	mixture of air and fuel vapour
steam turbine	water vapour (steam)
refrigerator	ammonia or a commercial chlorine compound such as Freon

This chapter begins with a review of the physical properties that are used in thermodynamics and the terminology associated with engines, refrigerators and heat pumps. It then concentrates on the properties and behaviour of the working substance. The practical design of engines, turbines and refrigerators also requires a knowledge of materials, structures and mechanisms.

The theory of thermodynamics is now well understood and we have powerful machines for the conversion of heat and work, but these achievements have taken many years. Some of the scientists and engineers who have made important contributions are mentioned in this chapter, with their dates.

12.1.1 Temperature and its measurement

The temperature of a body is a measure of its ability to transfer energy to other bodies in a heating process. If two bodies are at different temperatures (and are not insulated from one another), energy will flow naturally from the hotter to the cooler. If they are at the same temperature, there will be no such flow and they are said to be in *thermal equilibrium*.

Temperature is measured by the effect it has on some physical property of a substance. The general name for instruments that measure temperature is *thermometer* but other terms such as *thermocouple* and *pyrometer* are used for certain specialised devices. Most instruments depend on one or other of the following properties.

Volume In general, substances expand when their temperature rises and this is the principle on which many thermometers are based. Many household thermometers depend on the expansion of a column of mercury (or sometimes alcohol) in a glass tube. The temperature range of a mercury-in-glass thermometer can be extended upwards by introducing nitrogen above the mercury. Other liquids can be used for temperatures lower than the range covered by mercury.

Gas pressure If a gas is contained in a vessel of fixed volume its pressure will vary with temperature.

Electrical resistance The electrical resistance of a wire increases with temperature. Nickel and copper are suitable for temperatures up to the boiling point of water. Platinum is suitable for much higher temperatures and gives high accuracy.

Electrical potential If two wires of different materials are joined at their ends and the junctions are at different temperatures, a current flows through the wires. This device is called a thermocouple. The combinations of metals include copper and constantan (a copper-nickel alloy), iron and constantan, platinum and a platinum-iridium alloy, and platinum and platinum-rhodium. The highest temperatures for a thermocouple are obtained with the last of these combinations.

Thermocouples and electrical resistance thermometers can be linked to microprocessors for the recording of results or for controlling other devices.

Brightness and colour The brightness and colour of a body changes at high temperatures and can be used as a measure of temperature. Instruments that work on this principle are usually called *pyrometers*.

12.1.2 Scales of temperature

In order to give numerical values to temperatures it is necessary to select some physical phenomena such as the freezing and boiling of substances that always occur at the same temperature. The most familiar choices are the freezing and boiling of water (at standard atmospheric pressure). On the Celsius scale, these fixed points are given the values 0 and 100 degrees, abbreviated to °C. On the Fahrenheit scale the same points are given the values 32°F and 212°F. The difference between the boiling points is therefore 100° on the Celsius scale and 180° on the Fahrenheit. Therefore a *change* of 1°C equals a change of 1.8°F.

The physical properties on which thermometers depend do not all vary with temperature in the same way. If the thermometers that use different properties are calibrated so that they agree at certain temperatures they may give differing readings at other temperatures.

The lowest possible temperature is called the *absolute zero* and is the same for all substances. It is the temperature at which all molecular activity ceases (apart from a possible vibration from which the energy cannot be extracted). On the Celsius scale its value is approximately −273.2°C.

The Kelvin scale takes this point as zero and uses the same intervals as Celsius but the unit is called a Kelvin not a degree. For most purposes the Kelvin value can be found with sufficient accuracy by adding 273 to the Celsius value. Kelvin temperatures are denoted by K without the degree symbol (°). Note that a *change* of 1 K equals a change of 1°C.

Lord Kelvin (1824–1907) devised a thermodynamic scale of temperature which is independent of other physical properties but the practical measurement of temperature still depends on such properties.

EXAMPLE 12.1 Place the following in order of increasing temperature:
(a) melting point of lead, 327°C,
(b) boiling point of glycerin, 556°F,
(c) boiling point of mercury, 630 K.

SOLUTION It is convenient to convert (a) and (b) to the Kelvin scale.
(a) 327°C = (327 + 273)K = 600 K.
(b) Fahrenheit temperatures can be converted to Celsius in several ways. One is to calculate the number of degrees above the freezing point of water. In the present case this gives (556 − 32) = 524 in Fahrenheit degrees and 524/1.8 = 291 in Celsius. Since the freezing point of water is 0° on the Celsius scale the given temperature is:

$$556°\text{F} = 291°\text{C} = (291 + 273)\ \text{K} = 564\ \text{K}$$

On comparing the three Kelvin values the required order is seen to be (b), (a) and (c).

12.2 Physical quantities

Some of the physical quantities used in the study of thermodynamics have been introduced in earlier chapters. They are used here with the same symbols and units as before, and include:

mass m in kilogram (kg)
volume V in cubic metre (m^3)
density ρ in kilogram per cubic metre (kg/m^3)
force in newton (N)
absolute pressure p in newton per square metre (N/m^2), or Pa and sometimes bar
energy and work in joule (J)
power in watt (W)

Further quantities are needed in thermodynamics. They include:

Temperature On the Celsius scale this is measured in °C and denoted by t. On the Kelvin scale it is given the unit kelvin (K) and is denoted by T.

Heat Nowadays this term is used only to mean energy transferred from one body to another because of a temperature difference. It is denoted by Q and has the usual unit of energy, joule (J).

Internal energy This is the energy possessed by a body through the motion of its molecules. It is denoted by U and again has the units of energy, joule (J).

Specific heat capacity In SI units this is the heat required to raise 1 kg of a substance by 1°C or 1 K. It is denoted by c and therefore the heat required to raise a mass m from temperature T_1 to T_2 is:

$$Q = mc\,(T_2 - T_1)$$

This result assumes that the specific heat capacity is a constant. In the SI system c has the units joule per kilogram kelvin (J/(kg K)). Practical values are often given in kilojoule per kilogram kelvin (kJ/(kg K)).

Latent heat Substances can exist in three phases, solid, liquid and gas. The term *fluid* is used to include both liquid and gas.

If a solid substance is heated to the temperature at which it melts, more heat is needed to convert it to a liquid at the same temperature. This heat which causes the change of phase without increasing the temperature is called the *latent heat of fusion*. If further heat is transferred to the substance (now in its liquid form), the temperature will again rise. If the process is reversed and the liquid is cooled to the temperature at which it solidifies or freezes, the same latent heat is given out during the change of phase from liquid to solid.

A similar phenomenon occurs at the phase change from liquid to gas. The heat required to cause this change is called the *latent heat of evaporation*. The reverse process, the phase change from gas to liquid, is called *condensation*.

Latent heat is expressed as energy per unit mass and therefore has the units joule per kilogram. Practical values are usually expressed as kilojoule per kilogram. The word 'latent' is used because the heat transfer takes place without a change in temperature. The term *sensible heat* is sometimes used for the heat associated with a temperature change.

The temperature at which evaporation and condensation takes place, and the value of the latent heat, vary considerably with the pressure. Published tables give the values for water/steam and the substances used as refrigerants.

The word *vapour* is used to mean a substance at or just above its condensation point. It is possible for a substance to exist as a mixture of its liquid and vapour, an important practical example being water and steam in a boiler, usually called *wet steam*.

EXAMPLE 12.2 A platinum ball weighing 88 g is taken from a furnace at a temperature of 650°C. It is quickly placed in a vessel containing 500 g of water at 15°C. Find the resulting temperature assuming that all the heat given out by the ball is received by the water.

Specific heat capacities (assumed constant): water 4.187 kJ/(kg K)
platinum 0.145 kJ/(kg K)

SOLUTION Suppose the resulting temperature is t °C. Then, working in kilogram and kilojoule,

$$\text{heat received by the water} = mc\,(T_2 - T_1) = 0.5 \times 4.187 \times (t - 15)\ \text{kJ}$$

$$\text{heat given out by the ball} = 0.088 \times 0.145 \times (650 - t)\ \text{kJ}$$

Equating these results,

$$0.5 \times 4.187 \times (t - 15) = 0.088 \times 0.145 \times (650 - t)$$

from which

$$t = 18.8$$

The resulting temperature is therefore 18.8°C.

EXAMPLE 12.3 How much heat must be extracted from 1.7 kg of water at 20°C to change it to ice at −5°C?

Take specific heat capacity of water 4.187 kJ/(kg K)
specific heat capacity of ice 2.114 kJ/(kg K)
latent heat of fusion of ice 334 kJ/kg

SOLUTION There are three stages in the calculation. Working in kilojoule, the heat extracted is:

$$\text{for cooling the water to freezing point} = mc \times (\text{fall in temperature}) = 1.7 \times 4.187 \times 20 = 142.4$$

$$\text{for phase change to ice at } 0°\text{C} = 1.7 \times 334 = 567.8$$

$$\text{for cooling ice from } 0°\text{C to } -5°\text{C} = 1.7 \times 2.114 \times 5 = 18.0$$

$$\text{Total heat extracted} = 142.4 + 567.8 + 18.0 = 728.2\ \text{kJ}$$

Fig 12.1 2-stroke engine on a scooter

12.3 Internal combustion engines

Heat engines have been the main sources of power for industry and transport for many years. Most can be classified under the following headings:
(a) reciprocating steam engines,
(b) steam turbines.
(c) reciprocating internal combustion engines,
(d) gas turbines.

In (a) and (c) the power is generated as the working substance expands behind a piston in a cylinder. In most reciprocating engines the forwards and backwards motion of the piston is changed to rotary motion by a connecting rod and crank mechanism (see Chapter 10). In (b) and (d) the working substance passes through the turbine blades giving rotary motion directly.

In (a) and (b) the heat released by burning the fuel is used to produce steam in a boiler; the steam then enters the cylinder or turbine at a high pressure. In (c) the fuel is burned within the cylinder; hence the term 'internal combustion'. In (d) the fuel is burned in a combustion chamber.

The reciprocating internal combustion engine has been refined over many years and it offers many advantages, particularly in vehicles, see Fig. 12.1. It has a higher efficiency than the steam engine and a good power/weight ratio. It can operate economically over a very wide range of sizes. The internal combustion engine is almost universal in road vehicles and is now widely used on railways. The reciprocating internal combustion engine was crucial to the development of the aeroplane until the advent of gas turbines and jet propulsion. It is also widely used in ships. Its portability and speed of startup make it useful when the demand for power is intermittent. For example, the large-scale generation of electricity depends on steam turbines or hydroelectric schemes but the internal combustion engine is often used for standby generation or where a mains electricity supply is not available.

Fig. 12.2(a) shows the main features of one cylinder of an internal combustion engine in diagrammatic form. In this illustration the cylinder centre line is horizontal and the piston moves along this line. In practice the cylinder is often placed

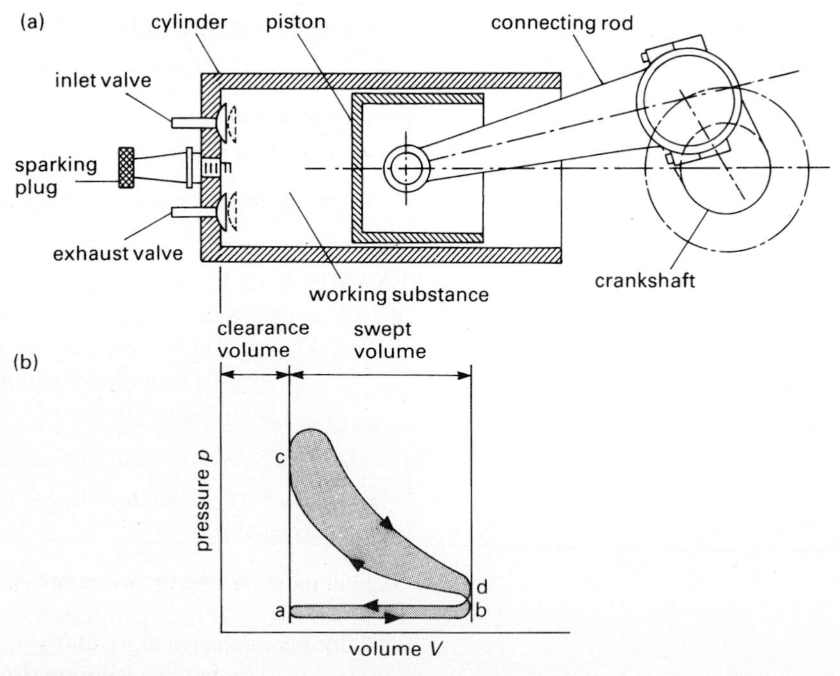

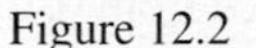

Figure 12.2
(a) Diagram of one cylinder of an internal combustion engine
(b) The four-stroke cycle.

vertically above the crankshaft and the extreme positions of the piston are referred to as *top dead centre* and *bottom dead centre*. The distance travelled by the piston between these two positions is called the *stroke*. During this movement the crank turns through half a revolution so that the stroke equals twice the crank radius. One revolution of the crankshaft corresponds to two strokes.

The operation of a reciprocating engine depends on a set of processes and events that is repeated over and over again. The sequence is called the *cycle of operation*. In many engines the steps are repeated every four strokes of the piston (two revolutions of the crankshaft) and this arrangement is known as the *four-stroke cycle*.

To understand how the energy of combustion is converted to work it is helpful to consider the changes in pressure and volume of the working substance and Fig. 12.2(b) shows in general terms the variation of these quantities with one another. With the notation of Fig. 12.2 the four strokes are, in the case of a petrol engine:

1 Suction (or induction) stroke (a to b) The inlet valve is open, the piston moves from left to right and a petrol–air mixture is drawn into the cylinder at a pressure slightly below atmospheric.

2 Compression stroke (b to c) Both valves are closed, the piston moves from right to left and the mixture is compressed, causing its pressure and temperature to rise. At the end of the stroke the fuel is ignited by the sparking plug and there is a further rapid rise in pressure and temperature.

3 Power (or expansion) stroke (c to d) Both valves remain closed, the piston moves from left to right and work is done on it by the expanding gases. The pressure and temperature fall during this stroke.

4 Exhaust stroke (d to a) The exhaust valve is open, the piston moves from right to left and the burnt gases are pushed out of the cylinder, the pressure being slightly above atmospheric.

The volume of the working substance is a minimum when the piston is at the left-hand end of the stroke (top dead centre). This is called the *clearance volume*. The increase in volume during the power stroke is called the *swept* or *displacement volume*. The swept volume in litres (or cm^2) is often quoted to indicate the power of an engine. It is often referred to as the *cubic capacity*.

The maximum volume during the cycle is equal to the (clearance + swept) volume. The ratio of the maximum volume to the minimum is called the *compression ratio* and is usually denoted by r. Thus

$$\text{compression ratio} = \frac{(\text{clearance} + \text{swept})\ \text{volume}}{\text{clearance volume}}$$

$$= 1 + \frac{\text{swept volume}}{\text{clearance volume}}$$

The efficiency of an engine increases as the compression ratio is increased. However, an increase in compression ratio results in higher temperatures during the compression stroke and, in the case of petrol engines, this can lead to spontaneous combustion of the fuel so that burning begins before the spark occurs. This gives rise to violent pressure waves, causing knocking or *detonation*. For this reason the compression ratio of petrol engines is limited to about ten. The temperature at which self-ignition occurs can be raised by adding small amounts of tetraethyl lead to the fuel but there are objections to this on environmental grounds and cars are now designed to run on unleaded petrol.

The limitation on compression ratio is overcome in diesel or compression-ignition engines. These follow the same cycle of operation, but with one modification. During the suction stroke, air alone is drawn into the cylinder rather than an air–fuel mixture. The air is compressed during the next stroke and at the end the fuel (oil) is injected into the cylinder by a high-pressure pump. The high temperature of the air causes the oil to ignite spontaneously. Modern diesel engines have compression ratios of over twenty but the resulting high pressures require the cylinder to be very strong and therefore heavy.

Some engines (petrol and diesel) operate on a *two-stroke cycle* in which the processes and events described earlier are completed in two strokes of the piston rather than four. This is achieved by opening the exhaust valve before the end of the power stroke. The burnt gases are largely cleared from the cylinder by the end of the stroke, a process called *scavenging*. To allow time for this to be completed or for the introduction of the air–fuel mixture in the case of the petrol engine, the valves remain open during the early part of the next stroke. They are then closed and compression begins.

The cycle is therefore completed in two strokes or one revolution. In practice two-stroke engines do not develop twice as much power as four-strokes of the same size running at the same speed. Nevertheless they can produce up to about 80% more power, albeit at a lower efficiency.

EXAMPLE 12.4 The bore and stroke of a four-cylinder car engine are 73.0 mm and 77.4 mm respectively. Calculate the cubic capacity of the engine and, if the compression ratio is 9.5, the clearance volume per cylinder.

SOLUTION Working in centimetre units,

$$\text{swept volume of one cylinder} = \frac{\pi}{4} \times 7.30^2 \times 7.74$$

$$= 323.9 \text{ cm}^3$$

$$\text{With four cylinders, cubic capacity} = 323.9 \times 4$$

$$= 1296 \text{ cm}^3 \quad \text{or} \quad 1.296 \text{ litre}$$

If the compression ratio is 9.5, the swept volume is 8.5 times the clearance and therefore, for one cylinder,

$$\text{clearance volume} = 323.9/8.5 = 38.1 \text{ cm}^3$$

12.3.1 Internal combustion engine development

The first successful internal combustion engine was built in 1876 by Nikolaus Otto (1832–1891) and the four-stroke cycle of operation is often known as the Otto cycle. Early engines had large cylinders (often only one) and ran at low speeds (typically 200 rev/min). It was Gottlieb Daimler (1834–1900) who pioneered the development of smaller engines running at higher speeds (up to 1000 rev/min). His work made the motor car a practical proposition and it is fitting that his name is still associated with luxury cars. Today's car engines run at 4000 rev/min or more and these speeds, together with high compression ratios, produce far more power than early engines of the same swept volume. The compression-ignition engine was patented by Rudolf Diesel (1858–1913) in 1893, and oil engines (four-stroke and two-stroke) are usually referred to as diesels. Sir Dugald Clerk (1854–1932) is credited with the invention of the two-stroke engine, as a means of getting round the patents held by Otto.

Single-cylinder petrol engines are still widely used in mopeds and motorcycles but there are advantages in having several cylinders, particularly when higher powers are needed. By having the cylinders firing in sequence a more uniform torque on the crankshaft is obtained than is possible with one cylinder. Furthermore, combustion is less efficient in large cylinders and some designers consider that a swept volume of about 300–400 cm^3 per cylinder is an optimum. Large engines therefore require more cylinders and it is usual to take an even number (2, 4, 6, etc.). However, 3- and 5-cylinder engines have proved successful.

With a single carburettor it is impossible to supply all the cylinders with the correct ratio of air to fuel. For this reason multicylinder engines sometimes use two or more carburettors. The power obtained from an engine can also be increased by *supercharging*. This is a means of pumping more air–fuel mixture (or air alone in the case of a diesel engine) into the cylinder than would be drawn in by normal suction. This, in turn, allows more fuel to be burnt per cycle, thus increasing the power developed.

Diesel engines generally cost more than petrol engines of the same power but use less fuel. They are therefore preferred in large vehicles and where continuous running is required. The higher compression ratios of diesel engines lead to higher noise levels and greater vibration than for a petrol engine. In small cars, therefore, petrol engines predominate but many car manufacturers have tackled the noise and vibration problems and now offer diesel alternatives.

12.3.2 Indicated power

Instruments of several kinds are used to record the variation of cylinder pressure with volume and they are known as *indicators*. The output from an indicator has the general appearance shown in Fig.12.2(b) and is known as an *indicator diagram*. The area of this diagram is a measure of the work done on the piston in one cycle of operation. It was shown in Chapter 11, Example 11.4, that the work done on a piston by an expanding gas is represented by the area under a graph of pressure against volume. In the case of Fig. 12.2(b), the work done during the power (or expansion) stroke is given by the area under the curve from c to d. In the same way the work done by the piston on the gas during the compression stroke is given by the area under the curve from b to c. This is negative work and therefore the net amount of work for the two strokes taken together is given by the upper shaded area.

The shaded area at the bottom of the diagram represents the work done during the suction and exhaust strokes. This is negative work since the pressure during the exhaust stroke is slightly higher than for the suction stroke. This area must therefore be subtracted from the previous one.

It is useful to have an equivalent constant pressure which would, in one stroke, give the same amount of work as is done in one complete cycle. It is called the *mean effective pressure* (m.e.p.) and it can be found from an indicator diagram by dividing the net shaded area by the length of the base.

Once this pressure is known the work done on the piston per cycle and the corresponding power are easily calculated. Suppose, in the appropriate SI units,

P = mean effective pressure
A = cross-sectional area of the cylinder
L = stroke
N = number of working strokes per second

Then, average force on piston during power stroke $= P \times A$
work done per cycle = force × distance $= (P \times A) = L$
corresponding power = work done per second $= (P \times A \times L) \times N$

This is the power developed within the cylinder and is called the *indicated power*. The formula is usually remembered as *PLAN* and allowance must be made for the number of cylinders. In a two-stroke engine, N equals the number of revolutions per second but in a four-stroke it is half that number.

EXAMPLE 12.5 The bore and stroke of a six-cylinder four-stroke marine oil engine are 600 mm and 900 mm respectively. During a test on the engine at 200 rev/min a number of indicator diagrams were taken with the following results:

mean area of diagrams	7.81 cm^2
length of diagrams	7.62 cm
pressure scale	600 kN/m^2 per centimetre of height

Calculate the mean effective pressure and the indicated power.

SOLUTION Working in centimetre units in the first instance,

$$\text{main height of diagrams} = \frac{7.81}{7.62} = 1.025 \text{ cm}$$

Hence, with the given scale factor,

$$\text{mean effective pressure} = 1.025 \times 600 = 615 \text{ kN/m}^2$$

For each cylinder, N (working strokes/s) $= 200/(2 \times 60) = 1.667$

Allowing for six cylinders and working in basic SI units,

$$\text{indicated power} = (PLAN) \times 6$$

$$= (615 \times 10^3) \times \left(\frac{900}{1000}\right) \times \frac{\pi}{4}\left(\frac{600}{1000}\right)^2 \times 1.667 \times 6$$

$$= 1.565 \times 10^6 \text{ W} \quad \text{or} \quad 1.565 \text{ MW}$$

12.3.3 Brake power and engine efficiency

The power delivered at the shaft of an engine is less than the indicated power because of friction in the mechanism. It is measured by instruments called *dynamometers*. Some types absorb the power while measuring it and these are known as *absorption dynamometers* or *brakes*. Others, called *transmission dynamometers*, measure the power without absorbing it and these are used when it is necessary to test an engine in service, such as an aircraft engine in flight.

Rope or Prony brake

The earliest and simplest form of brake is one in which wooden blocks are pressed against the rim of the flywheel as shown in Fig. 12.3. The work done against friction is converted to heat.

The rope carries a weight A at one end, the other being attached to a spring balance B. If the readings of A and B are W and S newtons respectively, and the effective diameter of the rope is d m then the brake torque $T = (W - S)d/2$.

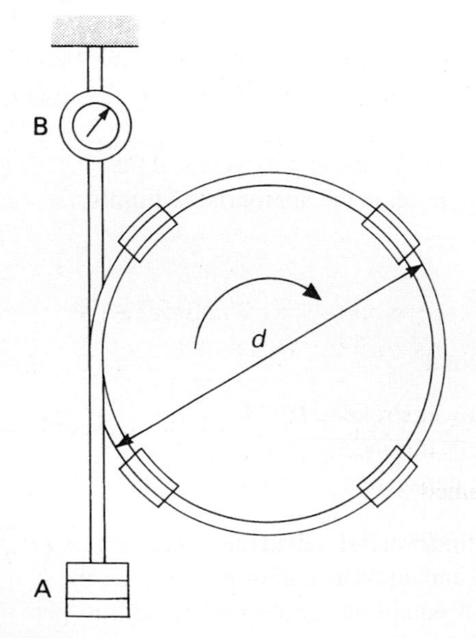

Figure 12.3 Rope brake.

Froude hydraulic brake

The engine is used to drive a wheel carrying several vanes in a freely mounted box containing water. The torque transmitted to the box is measured by a lever system.

Electrical dynamometer

The engine is coupled to the armature of a generator whose case is freely pivoted. The torque is balanced by weights or spring balances on a load arm and the electrical power generated is dissipated as heat in resistances.

Torsionmeter

This is an example of a transmission dynamometer. It is attached to a shaft transmitting the power and measures either the angle of twist over a known length or the shearing stress at the surface (using an electrical resistance strain gauge). The torque being transmitted can then be found using the torsion equation given in Chapter 9.

In all cases the brake power can be calculated from the expression power $= T\omega$ where T is the torque measured by the dynamometer and ω is the angular velocity in radian per second.

The power delivered at the shaft of an engine is called the *brake power* and practical values are still frequently quoted in horsepower units (bhp). The term 'indicated horsepower' (ihp) will also be encountered.

A useful quantity in comparing engines of different cubic capacities is the *brake mean effective pressure*. This is the equivalent pressure that gives the brake power when used in the *PLAN* formula.

Friction power

This is defined as (indicated power – brake power). Several different efficiencies are used in assessing the performance of internal combustion engines. They include:

Indicated thermal efficiency This is the ratio of the indicated power to the power equivalent of the heat available from the fuel. It is given by the fraction:

$$\frac{\text{indicated power (W)}}{\text{rate at which fuel is used (kg/s)} \times \text{calorific value (J/kg)}}$$

Brake thermal efficiency This is defined in a similar way, substituting brake power for indicated power.

Mechanical efficiency This is the ratio of brake power to indicated power.

Air standard efficiency This is the efficiency of an ideal engine operating on the theoretical Otto cycle and using air as the working substance. It is shown in section 12.8.2 that it is given by the formula:

$$1 - \frac{1}{r^{\gamma - 1}}$$

where r is the compression ratio and γ is the ratio of the principal specific heats of the working substance (usually taken as 1.4 for air). See section 12.6.3.

Relative efficiency The thermal efficiency of a real engine is always less than the air standard, and relative efficiency is defined as:

$$\frac{\text{indicated thermal efficiency}}{\text{air standard efficiency}}$$

Another measure of efficiency is *specific fuel consumption*. This is defined as the fuel used in a given time (usually per hour) divided by the power output.

EXAMPLE 12.6 A single-cylinder four-stroke oil engine has a bore and stroke of 127 mm and 203 mm respectively, and a compression ratio of 15 to 1. During a 30-minute test the following results were obtained:

fuel used: 675 g of oil having a calorific value of 41 900 kJ/kg
engine speed: 420 rev/min
mean effective pressure: 758 kN/m^2
brake load: 178 N at an effective radius of 0.8 m

Calculate:

(a) the indicated, brake and friction powers,
(b) the indicated and brake thermal efficiencies,
(c) the mechanical efficiency,
(d) the air standard efficiency based on the ideal Otto cycle and the (indicated) relative efficiency (take $\gamma = 1.4$),
(e) the specific fuel consumption.

SOLUTION

(a) Since the engine is operating on a four-stroke cycle the number of power strokes/s = 420/(60 × 2) = 3.5. Hence:

$$\text{indicated power} = PLAN$$
$$= (758 \times 10^3) \times \left(\frac{203}{1000}\right) \times \frac{\pi}{4}\left(\frac{127}{1000}\right)^2 \times 3.5$$
$$= 6822\ \text{W} \quad \text{or} \quad 6.822\ \text{kW}$$

The brake torque T = force × lever arm = 178 × 0.8 = 142.4 N m

and

rotational speed = $2\pi \times 420/60 = 44.0$ rad/s

Therefore:

brake power = $T\omega$ = 142.4 × 44.0 = 6266 W or 6.266 kW

friction power = (indicated − brake) power = 6822 − 6266
= 556 W or 0.556 kW

(b) Fuel used per second = 675/(30 × 60) = 0.375 g and, converting to basic SI units,

$$\text{indicated thermal efficiency} = \frac{\text{indicated power}}{\text{fuel used/s} \times \text{calorific}}$$
$$= \frac{6822}{(0.375/1000) \times (41\,900 \times 1000)}$$

(c) Mechanical efficiency = brake power/indicated power
= 6266/6822
= 0.918

(d) Using the formula given above,

$$\text{air standard efficiency} = 1 - \frac{1}{r^{\gamma - 1}} = 1 - \frac{1}{15^{1.4-1}}$$
$$= 0.661$$

$$\text{relative efficiency} = \frac{\text{indicated thermal efficiency}}{\text{air standard efficiency}} = \frac{0.434}{0.66}$$
$$= 0.657$$

(e) The fuel used per hour is 675 × 2g = 1350 g or 1.35 kg. Thus,

$$\text{Specific fuel consumption} = \frac{\text{fuel used/h}}{\text{brake power}} = \frac{1.35}{6.266}$$
$$= 0.215\text{kg/kWh}$$

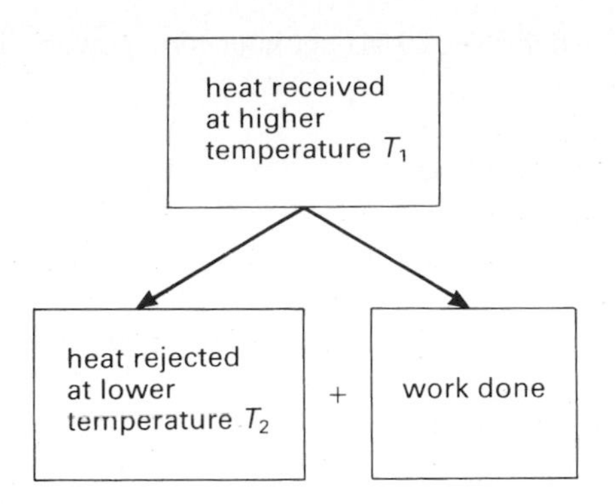

(a) heat engine

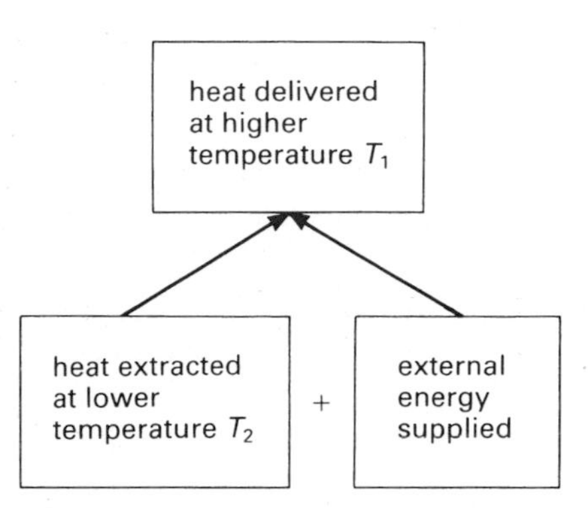

(b) refrigerator or heat pump

Figure 12.4

12.4 Refrigerators and heat pumps

The thermal efficiency of an engine was defined in the previous section as the proportion of the heat available from the fuel that is converted to useful work. The rest of the heat released from the fuel is given out to the surroundings; it is said to be heat rejected by the engine. All our experience shows that an engine can only work if the heat is supplied at a higher temperature than that at which it is rejected. This idea is examined in more detail in section 12.5.1.

Fig. 12.4(a) shows a simple heat balance for an engine. Since energy cannot disappear, the work done and heat rejected are together equal to the heat supplied.

In a refrigerator the transfers take place in the opposite directions, Fig. 12.4(b). Heat is extracted at a low temperature, external energy has to be supplied to drive a compressor or other device and heat is rejected at a higher temperature. This rejected heat warms the air round the refrigerator – as we know if we put our hand over it.

The efficiency of a refrigerator is measured by its *coefficient of performance* and this is defined as the ratio:

$$\frac{\text{heat extracted}}{\text{external energy supplied}}$$

Note that this ratio can have values greater than 1.

A *heat pump* is a refrigerating machine in which the heat delivered at the higher temperature is used to warm a building. The heat may be extracted in an area of the building where a low temperature is required, such as a food store, or from the external surroundings. The performance of a heat pump is given by the ratio:

$$\frac{\text{heat delivered}}{\text{external energy supplied}}$$

Since the heat delivered is the sum of the external energy and the heat extracted, Fig. 12.4(b), this performance ratio can be written as:

$$\frac{\text{external energy supplied} + \text{heat extracted}}{\text{external energy supplied}} = 1 + \frac{\text{heat extracted}}{\text{external energy supplied}}$$

This means that the performance ratio as a heat pump equals

1 + (coefficient of performance as a refrigerator)

EXAMPLE 12.7 A refrigerator has a coefficient of performance of 2.85. How many kilograms of ice at 0°C can be produced from water at 15°C by one megajoule of electricity?

Take specific heat capacity of water: 4.187 kJ/(kg K),
latent heat of fusion of ice: 334 kJ/kg.

SOLUTION It is convenient to work in kilojoules. For m kg of water/ice the sum of the sensible and latent heat to be extracted is

$$(4.187 \times 15 + 334)m = 396.8m \text{ kJ}$$

The external energy supplied is 1 MJ = 1000 kJ.

Since the coefficient of performance equals

(heat extracted)/(external energy supplied)

we can write: $\dfrac{396.8m}{1000} = 2.8$ from which $m = 7.18$

Thus, 7.18 kg of ice can be produced per megajoule of electricity.

EXAMPLE 12.8 A domestic heat pump can deliver 25 MJ per hour for a power input 3 kW.
Calculate
(a) the heat extracted per hour at the lower temperature,
(b) the performance ratio of the heat pump,
(c) its coefficient of performance as a refrigerating machine.

SOLUTION
(a) The power input is equivalent to 3 kJ per second. Thus:

$$\text{external energy supplied per hour} = (3 \times 10^3) \times 3600$$
$$= 10.8 \times 10^6\,\text{J} \quad \text{or} \quad 10.8\,\text{MJ}$$
$$\text{heat extracted per hour} = 25 - 10.8 = 14.2\,\text{MJ}$$

(b) performance ratio = heat delivered/external energy supplied
= 25/10.8 = 2.315

(c) coefficient of performance as a refrigerator
= heat extracted/external energy supplied
= 14.2/10.8 = 1.315

12.5 Reversibility

The theory of thermodynamics is based on a number of fundamental concepts and laws. These are of great help in understanding the energy transformations that take place in an engine but engineers had developed successful steam engines long before these ideas were fully understood. (The early history of aviation is another example in which practice was ahead of theory.)

One of these basic ideas is *reversibility*. We know instinctively that there are many processes in everyday life that cannot be reversed. If we make a film of breaking an egg, or peeling and eating a banana, and run it backwards, the effect is strange and slightly amusing because we know that it is impossible to reverse such processes.

A reversible process in the thermodynamics sense is one in which the energy transformations could be reversed and the working substance (a fluid) could be returned precisely to its original state. Suppose in Fig. 12.2(a) the piston is at the left-hand end of the stroke and the crankshaft is at rest. Suppose also that the fluid expands behind the piston and does work on it at the expense of its internal energy. This work can be used to speed up the flywheel which then possesses kinetic energy. If the process were reversible, the kinetic energy would be sufficient to compress the fluid on the return stroke back to its original pressure, volume, temperature and internal energy. This cannot happen in practice for several reasons.

There is always mechanical friction in the system: at the bearings, between the piston and cylinder wall, and between the moving parts and the atmosphere. Furthermore, during the expansion of the fluid, heat is transferred through the walls of the cylinder. We know from all our experience that the energy associated with friction and the heat dissipated to the atmosphere cannot be used to help compress the gas on the return stroke. These are examples of *external irreversibility*.

There are also energy transfers within the fluid that cannot be reversed. Due to the movement of the piston the pressure of the fluid will vary from one point to another causing internal motion and fluid friction. The energy absorbed in this way is not returned during the compression stroke. This is an example of *internal irreversibility*.

To achieve reversibility several conditions would have to be met:

- no mechanical or fluid friction,
- no temperature change during the transfer of heat,
- no transfer of heat when the temperature is changing.

A process that takes place at constant temperature is termed *isothermal*. A process in which there is no heat transfer is called *adiabatic*; if it is adiabatic and reversible it is called *isentropic*.

12.5.1 Laws of thermodynamics

Until the nineteenth century, heat and its relationship with other forms of energy were not properly understood. In the metric system different units (the calorie and the erg) were used for heat and work. The experiments carried out over many years by James Prescott Joule (1818–1889) showed that heat and work are equivalent. This result is the basis of the *first law of thermodynamics*. In its original form it stated that there was a fixed relationship between units of heat and units of work. The conversion factor was known as *Joule's equivalent* and in SI units its value is 4.187 J/cal. However we now measure both heat and work in joules and so the first law can be stated as:

The net energy supplied by heat to a body is equal to the increase in its internal energy and the energy output because of work done by the body.

The *second law of thermodynamics* has been stated in many ways. It was first put forward in 1850 by Rudolf Clausius (1822–1888) in the form

It is impossible for a self-acting machine unaided by an external agency to convey heat from one body to another at a higher temperature.

This means that the natural flow of energy is always from the hotter body to the colder. We can only cause energy to flow in the other direction by supplying external energy as in the case of a refrigerator or heat pump. The second law, like the first, cannot be proved but it is borne out by all our observations and no exceptions to it have ever been found.

A result of the second law is that some of the heat supplied at the higher temperature is rejected at the lower temperature. This means that no engine, actual or theoretical, can be 100 per cent efficient; the energy transferred by a heat process and used to do work is always less than the energy it receives.

So far it has been assumed that if two bodies at the same temperature are brought together there will be no transfer of energy between them, that is, they are in thermal equilibrium. It is something that we know instinctively which is confirmed by all our experience. Nevertheless, it is a profound idea and amounts to another law of thermodynamics. Logically it should precede the first and second laws because it describes the outcome when there is no temperature difference and no transfer of energy. The statement is therefore called the *zeroth law of thermodynamics*. (In the same way, Newton's first law is a statement about the outcome when there is no external force and his second law describes what happens if an external force is present.)

12.5.2 Reversible engine

The word 'reversible' refers to the thermodynamic conversions; it does not mean the ability to run backwards in the mechanical sense. A reversible engine is a theoretical concept but it leads to important results in terms of the efficiency of practical engines.

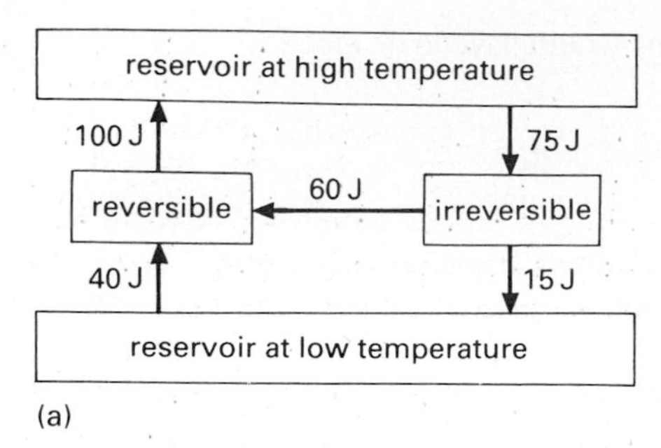

(a)

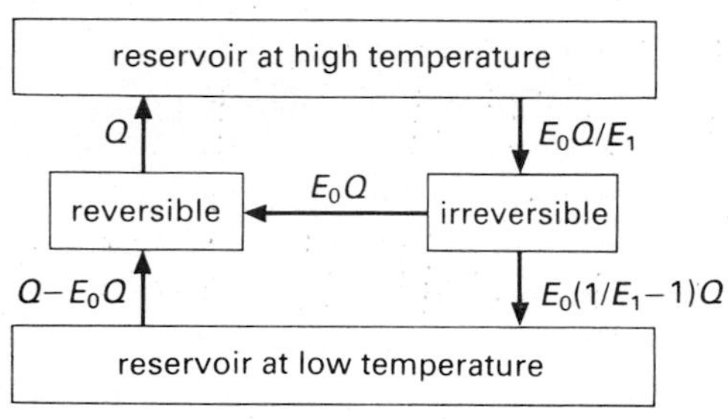

(b)

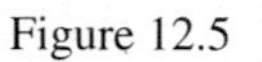
Figure 12.5

It can be proved that no engine can be more efficient than a reversible one. This is done by first imagining the opposite to be true and showing that this leads to an impossibility. This method of argument is often used as a basis for a scientific or mathematical proof.

Suppose there are two bodies, one at a high temperature and one at a low temperature, from which heat can be extracted or to which it can be delivered. These are called *reservoirs*, Fig. 12.5.

It is convenient to work with numbers in the first instance and the outcome will be the same whatever values are chosen. Suppose the reversible engine has an efficiency of 0.6 and the other, irreversible engine has a greater efficiency, say 0.8.

Then, if the reversible engine receives 100 J from the reservoir at the higher temperature, it converts $100 \times 0.6 = 60$ J to mechanical work and rejects $100 + 60 = 40$ J at the lower temperature. Since it is reversible it could operate as a refrigerator or heat pump between the same temperature limits with the same quantities of energy being transferred or converted in the opposite direction. For an input of 60 J of mechanical energy it extracts 40 J at the lower temperature and delivers a total of 100 J at the higher temperature. This is illustrated on the left-hand side of Fig. 12.5(a).

Next suppose that the other, more efficient, engine drives the first one and operates between the same temperature limits. To deliver 60 J of mechanical energy with an efficiency of 0.8 it requires $60/0.8 = 75$ J from the hotter reservoir and will reject $75 - 60 = 15$ J to the colder one. These values are given on the right-hand side of Fig. 12.5(a).

The diagram shows that at the lower temperature more heat is being extracted than delivered, the net amount being $(40 - 15) = 25$ J. The same net amount is being delivered at the higher temperature. In other words, this self-acting system is conveying energy from the reservoir at the lower temperature to the one at the higher temperature. This contravenes the second law of thermodynamics and is impossible.

The argument can be generalised using symbols. Suppose the reversible engine has an efficiency of E_0 and receives energy Q from the reservoir at the higher temperature. It converts E_0Q to mechanical work and the balance $(Q - E_0Q)$ conveyed to the reservoir at the lower temperature. Since the engine is reversible it could operate as a refrigerator or heat pump between the same temperature limits with the same quantities of energy being transferred in the opposite direction. The amounts are shown on the left-hand side of Fig. 12.5(b).

If the irreversible engine has a higher efficiency E_1 then, to drive the first, it has to provide E_0Q of mechanical energy and therefore needs E_0Q/E_1 from the reservoir at the higher temperature. The difference between these amounts,

$$(E_0Q/E_1 - E_0Q) \text{ or } E_0(1/E_1 - 1)Q,$$

is rejected to the reservoir at the lower temperature. These results are shown on the right-hand side of Fig. 12.5(b).

Using the quantities shown on Fig. 12.5(b), the net energy received by the reservoir at the higher temperature is

$$Q - E_0Q/E_1 = (1 - E_0/E_1)Q$$

and it is easily shown that the same amount is given out by the reservoir at the lower temperature. If E_1 is greater than E_0, therefore, energy is flowing from the lower temperature to the higher. It follows that E_1 cannot be greater than E_0.

Another way of expressing the argument is to say that the system is a perpetual motion machine, another impossibility. Either way it is clear that *no engine can be more efficient than a reversible one.*

12.5.3 Carnot cycle

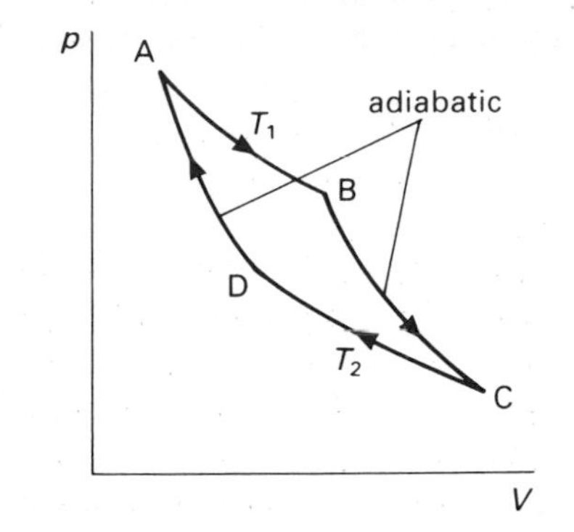

Figure 12.6

Although a reversible engine can only exist in theory it is an important idea in that it would have the highest possible efficiency of any engine. To be reversible, its cycle of operation would have to consist of reversible processes. In section 12.5 it was shown that only two types of process can be thermodynamically reversible – isothermal and reversible adiabatic (isentropic). A cycle of operation based on such processes was devised in about 1820 by Sadi Carnot (1796–1832). All heat transfer takes place at constant temperature for which there is no change in internal energy. The four-stage sequence is represented by the pressure–volume diagram shown in Fig. 12.6 and explained below. In the notation of Figs 12.4(a) and 12.5, the sequence is:

(a) A–B the fluid receives heat from the hot reservoir at constant absolute temperature T_1 (isothermal process),
(b) B–C the fluid expands, without heat being transferred, to a final absolute temperature T_2 (isentropic process),
(c) C–D heat is rejected to the cold reservoir at constant temperature T_2 (isothermal),
(d) D–E the fluid is compressed, without heat being transferred, to the original temperature T_1 (isentropic).

It can be shown that, for this cycle, the ratio (heat rejected)/(heat received) is T_2/T_1 and this is the same for all working substances. Thus if Q is the heat received at the higher temperature,

$$\text{heat rejected} = Q(T_2/T_1)$$

and

$$\text{heat converted to work} = Q - Q(T_2/T_1) = Q(1 - T_2/T_1)$$

Therefore,

$$\text{Carnot efficiency} = \frac{\text{heat converted to work}}{\text{heat received}}$$

$$= \frac{Q(1-T_2/T_1)}{Q} = \frac{T_1 - T_2}{T_1}$$

It is not possible for a practical engine to approximate to the Carnot cycle since two of the processes would have to be very slow to approach isothermal conditions, and the other two would need to be very fast to be approximately adiabatic.

The same quantities of heat and work would apply (in the reverse directions) to an ideal refrigerator or heat pump (Fig. 12.4(b)). For a refrigerator:

$$\text{ideal coefficient of performance} = \frac{\text{heat extracted}}{\text{external energy supplied}}$$

$$= \frac{Q(T_2/T_1)}{Q(1 - T_2/T_1)} = \frac{T_2}{T_1 - T_2}$$

For a heat pump:

$$\text{ideal performance ratio} = \frac{\text{heat extracted}}{\text{external energy supplied}}$$

$$= \frac{Q}{Q(1 - T_2/T_1)} = \frac{T_1}{T_1 - T_2}$$

The three results are measures of ideal performance and all practical machines will have lower values.

EXAMPLE 12.9 An ideal heat pump takes in heat from a reservoir at 5°C and delivers it to a reservoir at 75°C. The work input to the heat pump is from a reversible heat engine which takes heat from a reservoir at 1000°C and rejects heat to the reservoir at 75°C.
Calculate
(a) the thermal efficiency of the engine,
(b) the performance ratio of the heat pump.
How much heat must be supplied from the reservoir at 1000°C if 20 MJ is supplied to the reservoir at 75°C?

SOLUTION
(a) For the reversible heat engine the Kelvin temperatures at which heat is received and rejected are respectively

$$T_1 = 1000 + 273 = 1273 \text{ K and } T_2 = 75 + 273 = 348 \text{ K}.$$

Hence,

$$\text{ideal efficiency} = \frac{T_1 - T_2}{T_1} = \frac{1273 - 348}{1273} = 0.727$$

b) For the pump the upper temperature $T_1 = 348$ K and the lower temperature $T_2 = 273 + 5 = 278$ K. Therefore,

$$\text{ideal performance ratio} = \frac{T_1}{T_1 - T_2} = \frac{348}{348 - 278} = 4.97$$

Suppose the heat supplied to the engine at 1000°C is Q, then from the result of (a),

work done $= 0.727Q$
and heat rejected to reservoir at 75°C $= Q - 0.727Q = 0.273Q$

For the heat pump the external energy supplied is $0.727Q$ and, using the result found in (b),

heat delivered to reservoir at 75°C $= 0.727Q \times 4.97 = 3.613Q$
Adding results, the total heat delivered at 75°C $= 0.273Q + 3.613Q$
$= 3.886Q$

This result corresponds to the figure of 20 MJ given in the question. Therefore, $3.886Q = 20$ and $Q = 20/3.886 = 5.147$ MJ.

12.6 Properties and functions

Some of the physical quantities used in the study of thermodynamics were introduced in section 12.2. The present section is concerned with the properties of the working substance. These include its pressure, temperature, volume and internal energy; other properties will be introduced later.

Under given conditions the pressure and temperature of a fluid are independent of the mass but other properties such as volume and internal energy are proportional to the mass of fluid present. For these it is often convenient to quote the values for 1 kg (unit mass) and these are known as *specific volume*, *specific internal energy* and so on. The total values for a mass m are sometimes called *extensive properties* and the specific values are known as *intensive properties*. To make a distinction in the symbols, total values are usually given capital (upper-case) letters, V for volume and U for internal energy, and specific values are denoted by lower-case letters such as v and u. Pressure p and Kelvin temperature T will be the same whether we are dealing with total quantities or unit mass (1 kg).

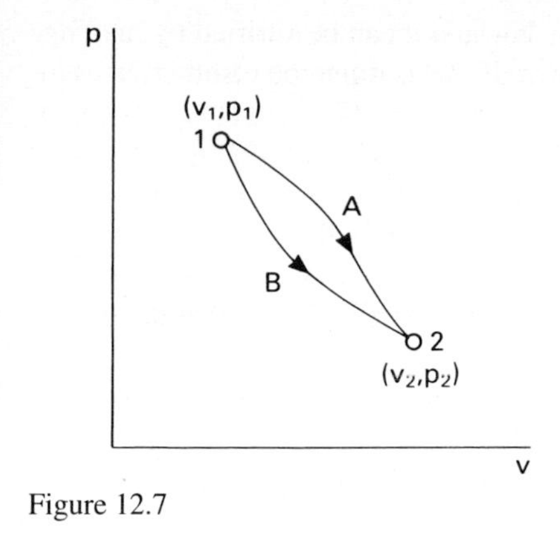

Figure 12.7

The thermodynamic condition of a fluid is called its *state*. For each state, it will have particular values of pressure p, temperature T, specific volume v and specific internal energy u. If two or more states have to be considered, it is convenient to distinguish between them by using number subscripts (suffixes). In the first state the values would be p_1, v_1, T_1 and u_1; in the second state p_2, v_2, T_2 and u_2.

If two independent properties of the fluid are specified, its state is completely defined. For example, if the value of p and v are known, the values of T and u are settled. It is often helpful to show two of the properties by graphical coordinates; pressure p and specific volume v are often the most convenient (see Fig. 12.7).

The diagram shows two states represented by coordinates (p_1, v_1) and (p_2, v_2). The values of p, v, T and u depend only on the position of the point on such a diagram and they are called *point functions*. They are not affected by the path taken between one point and another. In moving from state 1 to state 2, therefore, the values of the properties at the second point will be the same whether path A, path B or any other path is followed.

In contrast the work done W and heat Q received or rejected will be different for different paths. These quantities are sometimes called *path functions*.

12.6.1 Gas laws

The relationships between the pressure, volume and temperature of a gas were investigated by a number of scientists in the seventeenth and eighteenth centuries and it is interesting to note that different experimenters made the same discoveries at about the same time without being aware of each other's work.

The earliest result relates pressure and volume. It may be stated as follows:

If the temperature of a given mass of gas remains constant its volume V varies inversely with its absolute pressure p.

In symbols this may be written:

$$V = \text{constant} \times \frac{1}{p} \quad \text{or} \quad pV = \text{constant}$$

The value of the constant will be different for different temperatures.

The result was discovered by Robert Boyle (1627–1691) and the Frenchman Edme Mariotte (1620–1684). In English-speaking countries it is known as *Boyle's law*.

The effect of temperature was discovered about one hundred years later by two Frenchmen, Jacques Charles (1746–1823) and Joseph Gay-Lussac (1778–1850) independently of each other. The relationship is usually known as *Charles's law* and it can be expressed in two ways:

If the volume of a given mass of gas remains constant, its pressure p is proportional to the absolute temperature T.

If the pressure of a given mass of gas remains constant, its volume V is proportional to the absolute temperature T.

In symbols, these results may be written:

$$\text{For constant } V, \quad p = \text{constant} \times T \quad \text{or} \quad \frac{p}{T} = \text{constant}$$

$$\text{For constant } p \quad V = \text{constant} \times T \quad \text{or} \quad \frac{V}{T} = \text{constant}$$

Note that the value of the constant will be different in the different equations.

In many examples all three properties (p, V and T) will change as the gas changes from one state to another and it is convenient to use the single relationship:

$$\frac{pV}{T} = \text{constant} \quad \text{or} \quad pV = \text{constant} \times T$$

This is sometimes called the Boyle–Charles law and it can be verified by making each of the properties p, V and T constant in turn. If this is done the result reduces to the previous equations.

The combined Boyle–Charles law can be illustrated by a series of graphs (Fig. 12.8). To make the arithmetic easy, suppose that p, V and T are measured in units that are chosen so that the values of each property are in the range 0–8. (For temperature, the range 0–800 would be more realistic and, in any case, the gases would liquefy as the temperature neared absolute zero.) Suppose, too, that the mass of gas is such that the constant has a value of 2.

We can therefore write:

$$\frac{pV}{T} = 2 \quad \text{or} \quad pV = 2T$$

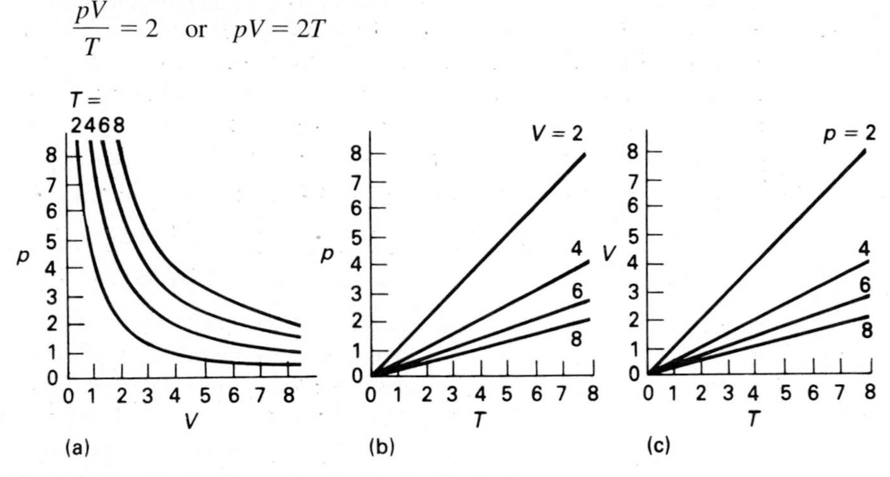

Figure 12.8 Graphs illustrating the Boyle–Charles law.

If we fix the value of one property we can draw a graph showing the variation between the other two. For example if we put $T = 2$ the equation becomes $pV = 4$. In that case, if $p = 1$ then $V = 4$, if $p = 2$ then $V = 2$ and if $p = 4$ then $V = 1$. Using these results and extending the range of values we obtain the following table:

p	0.5	1	2	4	6	8
V	8	4	2	1	0.667	0.5

This table is represented by the lowest curve of Fig. 12.8(a). The other curves in this diagram are obtained by changing the value of T to 4, 6 and 8 respectively.

If we return to the equation $pV = 2T$ and now fix the value of V we obtain an equation linking p and T. When $V = 2$ the equation is $p = T$, when $V = 4$ the equation is $p = T/2$ and so on. Each of these equations gives a straight line and Fig. 12.8(b) shows four possibilities.

Finally, by taking fixed values for p, we obtain equations linking V and T. Again they give straight lines, and Fig. 12.8(c) illustrates four possibilities. The three diagrams of Fig. 12.8 illustrate Boyle's law and the two versions of Charles's law. If p, V and T are taken as three perpendicular axes in three dimensions (Fig. 12.9) the equation $pV = 2T$ is represented by a curved surface. Straight lines representing fixed values of V and curves representing fixed values of T are shown and these form a grid on the surface. (Lines representing fixed values of p could be added.) If the surface were viewed in the direction of the T axis, the curves would appear as Fig. 12.8(a) and, if it were viewed in the direction of the V axis (towards the origin), the lines would appear as in Fig. 12.8(b).

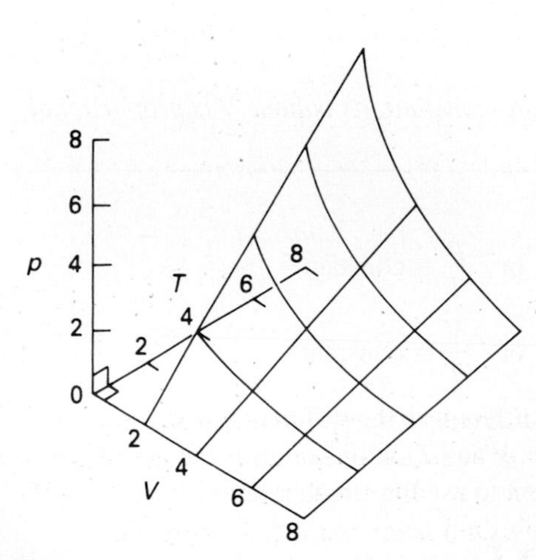

Figure 12.9

A gas that follows Boyle's and Charles's laws exactly is called a *perfect* or *ideal gas*. Although no actual gas is perfect, air and other gases are nearly so at the temperatures encountered in internal combustion engines and gas turbines. In many calculations, therefore, the laws give answers that are sufficiently accurate for practical purposes. This is not, however, true for steam and other vapours. In calculations on steam turbines and refrigerators the properties have to be obtained from published tables.

EXAMPLE 12.10 An oil engine has a compression ratio of 18. At the beginning of the compression stroke the cylinder contains air at a pressure of 1 bar and temperature 20°C. Determine the temperature at the end of the compression stroke if the pressure is then 45 bar. (Assume that combustion has not begun.)

SOLUTION Since the mass of air is constant we can use the combined Boyle-Charles law (pV/T = constant). Let 1 and 2 refer to the conditions at the beginning and end of compression respectively. Then:

$$\frac{p_1V_1}{T_1} = \frac{p_2V_2}{T_2} \quad \text{or} \quad T_2 = \frac{T_1p_2V_2}{p_1V_1}$$

Since we are concerned only with ratios the units of pressure and volume need not be consistent provided the initial and final values of each are measured in the same units. The ratio of pressures $P_2/P_1 = 45/1$ and, since the compression ratio is 18 to 1, the ratio of volumes $V_2/V_1 = 1/18$. The absolute temperature at the beginning of compression is $T_2 = 273 + 20 = 293\text{K}$.
Substituting all these values, the final temperature is:

$$T_2 = \frac{293 \times 45 \times 1}{1 \times 18} = 732.5 \text{ K}$$

$$= 732.5 - 273 = 459.5°\text{C}$$

12.6.2 Characteristic equation of ideal gas

For a given amount of gas, the pressure p, volume V and absolute temperature T are related by the combined Boyle–Charles law which states that pV/T is constant. For different amounts of the gas at the same pressure and temperature the volume V will be proportional to the amount and we can then write:

$$\frac{pV}{T} = \text{constant} \times (\text{amount of the gas})$$

The constant in this result is known as the *gas constant* and is denoted by R. However, engineers and scientists differ in the way they specify the amount of gas and this leads to two forms of the equation and two forms of the units for R. The examination questions in the assignments at the end of this chapter use the engineering approach.

Engineering form of the characteristic equation

If the amount of gas is taken to be its mass m, the last result becomes:

$$\frac{pV}{T} = Rm$$

or

$$pV = mRT \tag{1}$$

This result is the engineer's version of the *characteristic equation of a gas*. It is also known as the *equation of state*, the *perfect gas equation* or the *ideal gas law*. The constant R has the units of pV/mT. In the SI system these will be $(\text{N/m}^2 \times \text{m}^3)/(\text{kg K})$, which reduces to J/(kg K), or J $\text{kg}^{-1}\,\text{K}^{-1}$. In examination questions where it is required the value of R will be given. For air it is about 287 J/(kg K).

For an ideal gas the value of R in these units is approximately $8300/M$ where M is its relative molecular mass. Thus for nitrogen (molecular mass 28.02) we can take $R = 8300/28.02 = 296$ J/(kg K).

Scientific form of the characteristic equation

Chemists and physicists are concerned with atoms, molecules, electrons and other elementary particles. These *entities*, as they are called, are all very small and we should be involved in very large numbers if we specified practical amounts of gas by the numbers of atoms or molecules they contained. Scientists therefore use another unit called the *mole*, for which the symbol is mol. This is defined as follows:

---

1 mole is the amount of substance which contains as many elementary entities as there are atoms in 12 g of carbon-12.

Note that the definition is based on a mass of 12 g or 0.012 kg. Careful experiments have shown that 1 mole corresponds to 6.022×10^{23} entities. The amount of substance, measured in moles, is denoted by the symbol n. The relative molecular mass M of a compound equals the sum of the relative atomic masses of its constituent elements. The molecular mass is the same number with the units of gram per mole (g mol^{-1} or g/mol). The units are usually converted to kilogram per mole (kg mol^{-1} or kg/mol) in the SI system. For instance, since the relative atomic masses of carbon and oxygen are 12.01 and 16.00 respectively, the molar mass of carbon dioxide (CO_2) is $12.01 + 2 \times 16.00 = 44.01$ g/mol. That is, 1 mole of CO_2 has a mass of 44.01 g.

In symbols,the amount of substance measured in moles $n = m/M$ where m is the mass of gas and M is its relative molecular mass.

It is found that different gases containing equal amounts of substance occupy the same volume at the same temperature and pressure, a result known as Avogadro's law. This means that the value of n will be the same for all perfect gases if the values of p, V and T are the same.

It follows that the combined Boyle–Charles law leads to the result

$$\frac{pV}{T} = \text{constant} \times n$$

where the constant has the same value for all gases. This constant is called the molar gas constant or simply the *gas constant* and it is also denoted by R. Thus:

$$pV = nRT \qquad (2)$$

This result should be compared with Equation 1. The definitions of R are different in the two cases, but related. If we replace n by m/M Equation 2 becomes:

$$pV = \frac{mRT}{M} \qquad (3)$$

Now compare Equation 3 with Equation 1. The gas constant R in Equation 1 will have a different value for every gas but the molar gas constant R in Equation 2 and Equation 3 has the same value for all gases. It is approximately 8.3 J mol^{-1} K^{-1} (or J/(mol K)).

It will always be clear from the units which definition of R is intended. If it is given in Jkg^{-1} K^{-1} (or J/(kg K)), it is the value for use in Equation 1. If its units are J(mol^{-1} K^{-1}) (or J/(mol K)), then it is the molar gas constant and it should be used in Equation 2.

You will find a detailed treatment of the mole and the molar gas constant in books on physical chemistry such as those by MacGregor or Atkins, Clugston, Frazer and Jones (see Bibliography).

EXAMPLE 12.11 A cylinder contains argon (relative molecular mass 39.95) at a pressure of 30 bar. What is the density of the gas if the temperature is 20°C?

SOLUTION This example will be used to illustrate both approaches to the characteristic equation and its application. The solutions to later examples are based on the engineering form of the gas law $pV = mRT$.

(a) Using Equation 1. In this case the value of R is 8300/M in SI units. Thus:

$$R = \frac{8300}{39.95} = 207.8 \ \text{J/(kg K)}$$

Since 1 bar = 10^5 N/m^2, the pressure $p = 30 \times 10^5 = 3 \times 10^6$ N/m^2. The absolute temperature $T = 273 + 20 = 293$ K. Density is the ratio of mass to volume (m/V) and is denoted by ρ. By rearranging Equation 1, the density is

$$\rho = \frac{m}{V} = \frac{p}{RT} = \frac{3 \times 10^6}{207.8 \times 293} = 49.3 \ \text{kg/m}^3$$

(b) Using Equation 2. By rearranging this form of the ideal gas equation, the amount of substance is given by:

$$n = \frac{pV}{RT}$$

In this formula R is the molar gas constant with the value 8.3 J mol^{-1} K^{-1}. The pressure p and absolute temperature T have the same values as before so that:

$$n = \frac{3 \times 10^6 \times V}{8.3 \times 293} = 1233.6 \, V \ \text{mol}$$

Substituting $n = m/M$ gives:

$$\frac{m}{M} = 1233.6 \, \text{V}$$

This result would apply to all ideal gases at these values of pressure and temperature. Rearranging and substituting $M = 39.95$ for argon, the density is:

$$\rho = \frac{m}{V} = 1233.6 \, M$$

$$= 1233.6 \times 39.95 = 49\,300 \ \text{g m}^{-3}$$
$$= 49.3 \ \text{kg m}^{-3}$$

as before.

12.6.3 Specific heat capacities of a gas

Specific heat capacity c was defined in section 12.2 as the heat required to raise 1 kg of a substance by 1°C (or 1 K). Its units are J/(kg K).

In the case of a gas the specific heat capacity depends on the way in which the gas expands or contracts. Two special values are frequently required and these are called the *principal specific heat capacities*.

Constant volume

Suppose the gas is contained in a vessel of fixed volume. As heat is supplied to the gas its temperature and pressure, but not volume, will increase. The *specific heat capacity at constant volume*, c_v, is defined as the heat required to raise 1 kg by 1°C (or 1 K), the volume being unchanged.

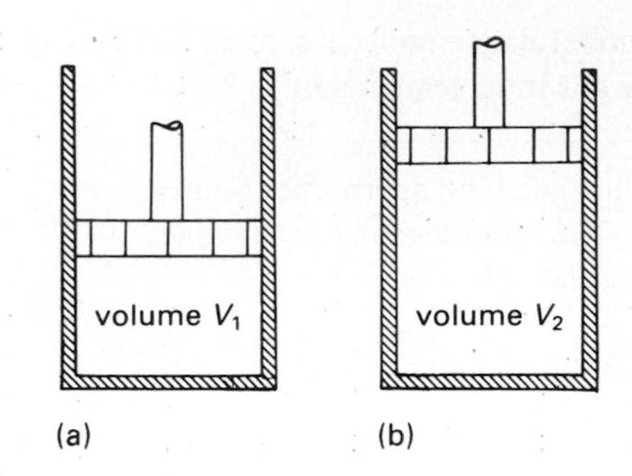

Figure 12.10 Gas expanding at constant pressure.

Constant pressure

Next consider a cylinder with its axis vertical and open at the top, the gas being trapped below a frictionless piston, as shown in Fig. 12.10(a). As heat is supplied the pressure will remain constant, since it depends only on the weight of the piston, but the volume will increase, Fig. 12.10(b). The *specific heat capacity at constant pressure*, c_p, is the heat required to raise 1 kg by 1°C (or 1 K) when the pressure remains constant.

In this case, work is done by the gas in raising the piston and the heat required at constant pressure is therefore greater than at constant volume. In the case of air the values of c_p and c_v are about 1005 J/(kg K) and 718 J/(kg K) respectively.

12.6.4 Joule's law and internal energy

It can be shown theoretically that, for an ideal gas, *the internal energy depends only on its temperature*. This result was obtained experimentally by Joule and is known as Joule's law. It leads to reasonably accurate results in the case of real gases but cannot be applied to vapours.

When a gas is heated in a closed vessel of fixed volume, it does no work and, since it remains at rest, it possesses no external kinetic energy. The heat supplied under these conditions is therefore all accounted for by an increase in its internal energy U. If the initial and final states are denoted by 1 and 2, the increase in internal energy is therefore:

$$U_2 - U_1 = \text{heat supplied at constant volume} = mc_v(T_2 - T_1)$$

or, in terms of specific internal energy,

$$u_2 - u_1 = c_v(T_2 - T_1)$$

Although this result has been derived under conditions of constant volume, it applies to all cases since, by Joule's law, the change in internal energy depends only on the change in temperature.

12.6.5 The gas constant and principal specific heat capacities

Suppose heat is supplied to a gas under constant pressure conditions as illustrated in Fig. 12.10. During the heating process the gas expands and work is done on the piston. If the pressure of the gas is p (constant) then this work is given by:

$$\begin{aligned} W &= \text{force} \times \text{distance} \\ &= p \times \text{piston area} \times \text{distance moved by piston} \\ &= p(V_2 - V_1) \end{aligned}$$

From the engineering form of the characteristic equation $pV = mRT$, we have $pV_2 = mRT_2$ and $pV_1 = mRT_1$

Hence the work done may be written:

$$W = pV_2 - pV_1 = mR(T_2 - T_1)$$

Since the pressure remains constant the heat added is:

$$Q = mc_p(T_2 - T_1)$$

and, from the result of the previous section, the increase in internal energy is:

$$U_2 - U_1 = mc_v(T_2 - T_1)$$

In this process the heat supplied is partly accounted for by the increase in internal energy and partly by the work done on the piston. Therefore,

$$Q = (U_2 - U_1) + W$$

and, substituting the expressions obtained above,

$$mc_p(T_2 - T_1) = mc_v(T_2 - T_1) + mR(T_2 - T_1)$$

Dividing through by m and $(T_2 - T_1)$ the equation becomes:

$$c_p = c_v + R \quad \text{or} \quad R = c_p - c_v$$

This is a general result and if two of the three quantities in the equation are given, the third may be calculated. It must be remembered that the gas constant R has the meaning and units used in Equation 1. The quantities R, c_p and c_v all have the units J/(kg K). In the case of oxygen, for example, the principal specific heat capacities c_p and c_v are respectively 918.2 and 658.6 J/(kg K) and its gas constant is therefore:

$$\begin{aligned} R &= 918.2 - 658.6 \\ &= 259.6 \text{ J/(kg K)} \end{aligned}$$

EXAMPLE 12.12 Suppose the cylinder of Fig. 12.10 contains 0.1 kg of gas (below the piston) at a temperature of 20°C and pressure 200 kN/m^2. If the gas is heated to 100°C at constant pressure, find:
(a) the heat supplied,
(b) the change in internal energy,
(c) the work done on the piston,
(d) the final density of the gas.
Take $c_p = 1040$, $c_v = 744$ J/(kg K).

SOLUTION

(a) The heat supplied is:

$$Q = mc_p(T_2 - T_1) = 0.1 \times 1040 \times 80 = 8320 \text{ J} \quad \text{or} \quad 8.32 \text{ kJ}$$

(b) The change in internal energy is:

$$\begin{aligned} U_2 - U_1 = mc_v(T_2 - T_1) &= 0.1 \times 744 \times 80 \\ &= 5952 \text{ J or } 5.952 \text{ kJ (increase)} \end{aligned}$$

(c) Since the heat supplied is greater than the increase in internal energy the difference equals the work done on the piston. Hence:

$$W = Q - (U_2 - U_1) = 8.32 - 5.952 = 2.368 \text{ kJ}$$

This result could also be found by first calculating the initial and final volumes and then using the constant pressure result $W = p(V_2 - V_1)$.

(d) The engineering form of the gas constant is given by:

$$R = c_p - c_v = 1040 - 744 = 296 \text{ J/(kg K)}$$

With this value, the final density is:

$$\rho_2 = \frac{m}{V_2} = \frac{p_2}{RT_2} = \frac{200 \times 10^3}{296 \times (273 + 100)} = 1.811 \text{ kg/m}^3$$

12.7 Steady flow processes

The principle of conservation of energy applies to all transformations and processes (apart from those associated with nuclear energy). In Chapter 11 the principle was applied to forms of mechanical energy (kinetic and potential) with examples on vehicles and other solid bodies. In these problems the changes in temperature are usually small and thermal effects can be ignored. In thermodynamics, however, we must include the forms of energy associated with changes in pressure, volume and temperature.

It is convenient to think in terms of a *system* enclosed by a *boundary*. A *nonflow* system is one in which the working substance does not cross the boundary (but energy, in the form of heat Q and work W, does). An example of a nonflow system is that of a gas in a cylinder as shown in Fig. 12.10.

In a *flow system* the working substance also crosses the boundary as shown in Fig. 12.11. A *steady flow process* is one in which the flow rates and working substance properties do not change with time.

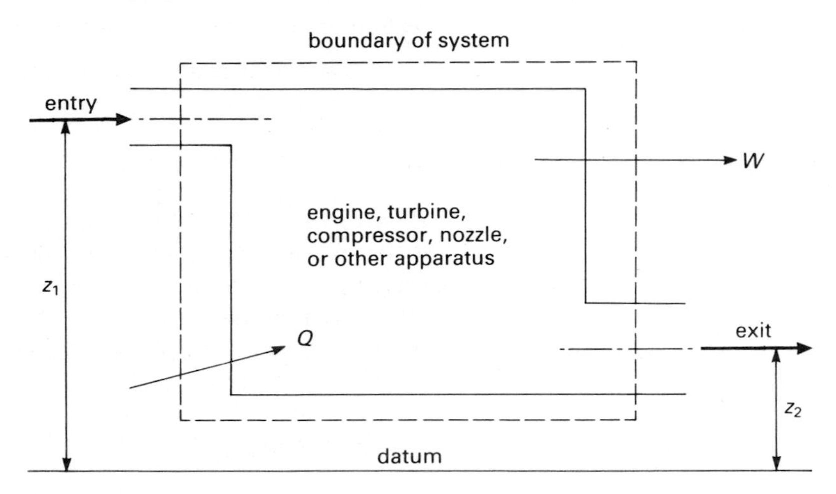

Figure 12.11 A flow system.

It is usual to regard Q as positive for heat flowing into the system and W as positive for work done by the system. Hence the energy gained by the system is $Q - W$. If E is the total energy stored in the working substance (kinetic, potential, internal, etc.) and the initial and final states are denoted by 1 and 2 then, by the principle of conservation of energy.

$$\text{initial stored energy} + \text{energy added} = \text{final stored energy}$$

or

$$E_1 + (Q - W) = E_2$$

This applies to flow and nonflow systems, the difference being in the items to be included in E. In a nonflow system the only stored energy that need be considered is the internal energy U. Hence the result becomes:

$$U_1 + (Q - W) = U_2$$

or

$$Q = (U_2 - U_1) + W$$

This can be remembered as:

$$(\text{heat supplied}) = (\text{increase in internal energy}) + (\text{work done})$$

In a steady flow system other forms of energy must be taken into account. Suppose a mass m of the working substance flows through the system. The volume of this mass corresponding to the pressure p and temperature T at a given point in the system is, say, V. Suppose also that its velocity at the same point is denoted by c and its height above the datum by z. (These symbols are chosen because v is now used for specific volume and h is required for another quantity to be introduced shortly.)

The following forms of energy are now taken into account:

- Potential energy. As shown in Chapter 11, this equals mgz.
- Kinetic energy. Taking c for velocity this is $\frac{1}{2}mc^2$.
- Internal energy U.
- Flow energy. If the pressure at a boundary is p then the work done in pushing a volume V across the boundary into the system is pV. This form of energy is generally called *flow work*.

If the conditions at entry and exit to the system, Fig. 12.11, are denoted by 1 and 2 then, as before, the principle of conservation of energy gives:

$$E_1 + Q = E_2 + W$$

and, allowing for the four forms of stored energy listed above,

$$mgz_1 + \tfrac{1}{2}mc_1^2 + U_1 + p_1V_1 + Q = mgz_2 + \tfrac{1}{2}mc_2^2 + U_2 + p_2V_2 + W$$

If the equation is applied to unit mass (1 kg) then $m = 1$ and U and V are replaced by u and v, the specific internal energy and specific volume. It can then be written:

$$gz_1 + \tfrac{1}{2}c_1^2 + u_1 + p_1v_1 + Q = gz_2 + \tfrac{1}{2}c_2^2 + u_2 + p_2v_2 + W$$

This result is known as the *steady flow equation*. It will be used again in the next chapter. In thermodynamics examples the change in height is often negligible and the first term on each side of the equation can therefore be omitted. For the flow of a liquid through a pipe, however, this is often an important term.

12.7.1 Enthalpy

The quantities U, p and V are all point functions. For a given state of the working substance, their values are fixed. It follows that the expression $U + pV$ is also a point function and it is convenient to give it a separate name and symbol. It is called *enthalpy* and it is denoted by H. Specific enthalpy (the enthalpy per kilogram) is given the symbol h.

$$\text{Thus } H = U + \text{pV} \quad \text{and} \quad h = u + pv$$

With this notation the steady flow equation can be written:

$$gz_1 + \tfrac{1}{2}c_1^2 + h_1 + Q = gz_2 + \tfrac{1}{2}c_2^2 + h_2 + W$$

For a perfect gas at absolute temperature T, the internal energy U and specific internal energy u reckoned from absolute zero are, from Joule's law,

$$U = mc_vT \text{ and } u = c_vT$$

Also, from the characteristic equation in its engineering form, $pV = mRT$ and the enthalpy H is therefore given by:

$$H = U + \text{pV} = mc_vT + mRT = mT(c_v + R)$$

But, as shown in section 12.6.5, $R = c_p - c_v$, and the expression in brackets becomes $(c_v + c_p - c_v) = c_p$. Therefore,

$$H = mc_pT \text{ and } h = c_pT$$

These results are sufficiently accurate in most cases for calculating the change in enthalpy of a real gas. They cannot, however, be used for steam and other vapours; for these, the value of enthalpy must be obtained from tables.

EXAMPLE 12.13 Air enters a water-cooled compressor at 102 kN/m^2 and 20°C. At exit its temperature is 110°C. The heat carried away by the cooling water is 7 kJ per kilogram of air. The air velocities at inlet and outlet are 10 m/s and 16 m/s respectively.

Calculate the power required to drive the compressor if the air flow at inlet is 6 m^3/min and the mechanical efficiency of the compressor is 0.9. Take c_v and c_p for air as 718 and 1005 kJ/(kg K) respectively.

SOLUTION The potential energy term in the steady flow equation may be ignored and, for 1 kg of air, it becomes:

$$\tfrac{1}{2}c_1^2 + h_1 + Q = \tfrac{1}{2}c_2^2 + h_2 + W$$

and, on rearranging, we have

$$W = \tfrac{1}{2}(c_1^2 - c_2^2) + (h_1 - h_2) + Q$$

Taking the three terms on the right-hand side separately,

$$\tfrac{1}{2}(c_1^2 - c_2^2) = \tfrac{1}{2}(10^2 - 16^2) = -78 \text{ J/kG}$$

Using the ideal gas result for enthalpy.

$$(h_1 - h_2) = c_p(T_1 - T_2) = 1005\,(20 - 110)$$
$$= -90.45 \times 10^3 \text{ J/k} \quad \text{or} \quad -90.45 \text{ kJ/kg}$$

The heat Q is negative if it is extracted from the air. Hence, in the present case,

$$Q = -7 \text{ kJ/kg}$$

Collecting results and working in kJ/kg units,

$$W = -0.078 - 90.45 - 7 = -97.53 \text{ kJ/kg}$$

Since the answer is negative work is done *on* the air.

The mass flow can be calculated from the characteristic equation. The gas constant $R = c_p - c_v = 1005 - 718 = 287$ J/(kgK) and, using the inlet conditions,

$$m = \frac{pV}{RT} = \frac{(102 \times 10^3) \times 6}{287 \times (273 + 20)} = 7.278 \text{ kg/min} \quad \text{or} \quad 0.1213 \text{ kg/s}$$

Allowing for the mechanical efficiency of 0.9.

$$\text{power required} = \frac{\text{rate at which work is done on the air}}{0.9}$$
$$= \frac{(97.53 \times 10^3) \times 0.1213}{0.9}$$
$$= 13.14 \times 10^3 \text{ W} \quad \text{or} \quad 13.14 \text{ kW}$$

12.8 Laws of expansion and compression

For a given mass of an ideal gas the combined Boyle–Charles law states that pV/T is a constant. This relationship is illustrated in Figs 12.8 and 12.9. If the gas changes from one state to another (1 to 2) the initial and final values are related by the equation:

$$\frac{p_1 V_1}{T_1} = \frac{p_2 V_2}{T_2}$$

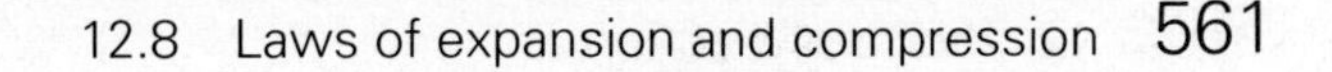

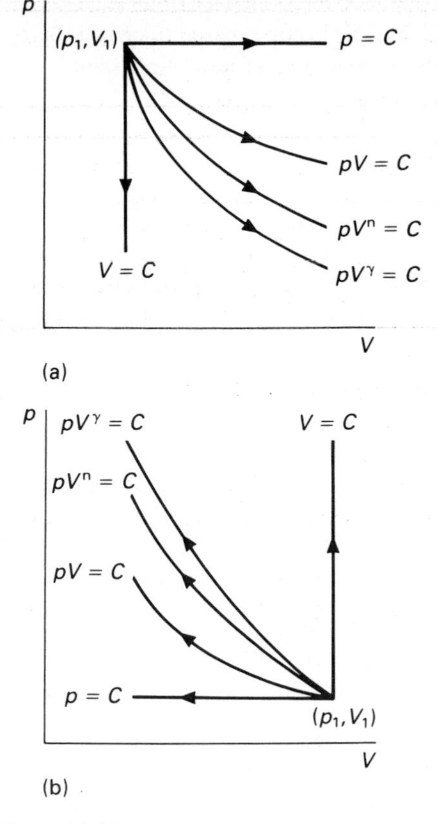

Figure 12.12

Given five of the quantities in this equation the sixth can be calculated. In some examples the ratio of initial and final values is given (e.g. V_1/V_2) instead of the separate values.

However, two states can be joined by any number of paths as indicated in Fig. 12.7. They are usually defined by equations relating pressure and volume and there are five processes of an ideal gas that are of special interest. These are illustrated in Fig. 12.12(a) and (b). In each case the ratio V_1/V_2 is denoted by r, and C is a constant.

Constant pressure $p = C$

In the case $p_1 = p_2$ and the Boyle–Charles law becomes

$$\frac{V_1}{T_1} = \frac{V_2}{T_2} \quad \text{or} \quad T_2 = T_1\left(\frac{V_2}{V_1}\right) = T_1/r$$

Isothermal $T = C$

This is the condition for Boyle's law $pV = C$.

$$p_1V_1 = p_2V_2 \quad \text{or} \quad p_2 = p_1\left(\frac{V_1}{V_2}\right) = rp_1$$

Reversible adiabatic (isentropic)

It can be shown that for the expansion or compression of an ideal gas in which no heat is transferred ($Q = 0$) the pressure and volume are related by:

$$pV^{\gamma} = C$$

where the index $\gamma = c_p/c_v$ the ratio of the principal specific heat capacities. In this case,

$$p_1V_1^{\gamma} = p_2V_2^{\gamma} \quad \text{and} \quad p_2 = p_1\left(\frac{V_1}{V_2}\right)^{\gamma} = p_1r^{\gamma}$$

If this result is substituted in the Boyle–Charles law the temperatures are related by:

$$T_2 = T_1\frac{p_2V_2}{p_1V_1} = T_1\frac{r^{\gamma}}{r} = T_1r^{\gamma-1}$$

For air, the index γ is about 1.4.

Polytropic

Many practical processes can be represented by an equation of the form:

$$pV^n = C$$

In most cases the index n lies between 1 (the value for an isothermal process) and γ, the isentropic index.

The expressions for p_2 and T_2 can be found by substituting n for γ in the results for the previous case.

Constant volume $V = C$

In this case there is no compression nor expansion. $V_1 = V_2$ and the Boyle–Charles law becomes:

$$\frac{p_1}{T_1} = \frac{p_2}{T_2}$$

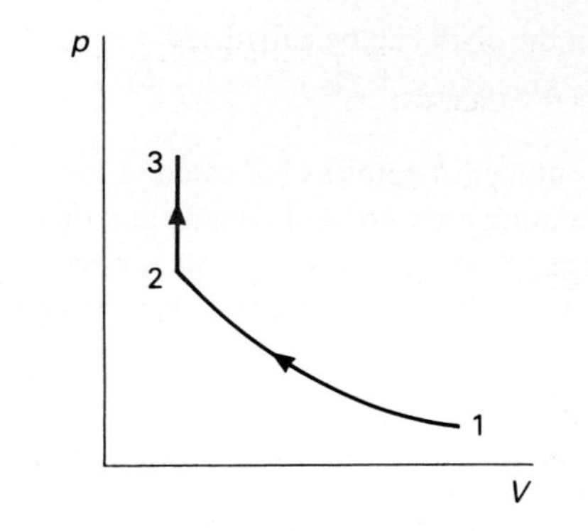

Figure 12.13 Ideal cycles in an engine.

EXAMPLE 12.14 A cylinder, fitted with a piston, contains 0.1 kg of air at a pressure of 110 kN/m^2 and temperature 25°C. The air is compressed to one-ninth its volume to the law $pV^{1.3}$ = constant and it then receives 8 kJ of heat at constant volume. Find the pressure and temperature of the air

(a) at the end of compression,
(b) after the constant volume heating.
Take c_v for air as 720 J/(kg K).

SOLUTION The two processes are shown on the p–V diagram of Fig. 12.13. Let 1 and 2 denote the states at the beginning and end of compression, 3 the state at the end of the constant volume heating.

(a) Since $pV^{1.3} = C$ the pressure at the end of compression is:

$$p_2 = p_1\left(\frac{V_1}{V_2}\right)^{1.3} = 110 \times 9^{1.3} = 1914\,\text{kN/m}^2 \quad \text{or} \quad 1.914\,\text{MN/m}^2$$

From the Boyle–Charles law $p_1V_1/T_1 = p_2V_2/T_2$ and the required temperature is:

$$T_2 = \frac{T_1p_2V_2}{p_1V_1} = \frac{(273 + 25) + (1.914 \times 10^6)}{(110 \times 10^3) \times 9} = 576\,\text{K} \quad \text{or} \quad 303°\text{C}$$

(b) For the constant volume heating, $Q = mcv\,(T_3 - T_2)$ and

$$(T_3 - T_2) = \frac{Q}{mc_v} = \frac{8 \times 10^3}{0.1 \times 720} = 111$$

From which

$$T_3 = T_2 + 111 = 576 + 111 = 687\,\text{K} \quad \text{or} \quad 414°\text{C}$$

Again using the Boyle–Charles law, $p_2V_2/T_2 = p_3V_3/T_3$, and noting that $V_2 = V_3$, we have:

$$p_3 = \frac{p_2V_2T_3}{V_3T_2} = \frac{(1.914 \times 10^6) \times 687}{576} = 2.283 \times 10^6\,\text{N/m}^2$$

$$\text{or} \quad 2.283\,\text{MN/m}^2$$

12.8.1 Nonflow work

Suppose a perfect gas expands behind a piston in a cylinder. If the motion of the piston is fully resisted the work done by the gas can be found by considering the changes in pressure and volume. Four processes are of special interest.

Constant pressure

It was shown in section 12.6.5 that if the pressure remains constant the nonflow work is:

$$W = p(V_2 - V_1)$$

Polytropic

If the pressure varies this last result can only be applied to an infinitesimal increase in volume dV. For this increase the work done is pdV and for the complete process it is given by the integral:

$$W = \int_{V_1}^{V_2} pdV$$

In the polytropic case $pV^n = C$ and $p = C/V^n = CV^{-n}$. Also, the integral of x^n is $x^{n+1}/(n+1)$ (except when $n = -1$), and the result becomes:

$$W = \int_{V_1}^{V_2} CV^{-n}\,dV = \left[\frac{CV^{-n+1}}{-n+1}\right]_{V_1}^{V_2} = \frac{CV_2^{1-n} - CV_1^{1-n}}{1-n}$$

But $C = p_2V_2^n$ and also $p_1V_1^n$. Using these results in turn.

$$W = \frac{p_2V_2 - p_1V_1}{1-n} = \frac{p_1V_1 - p_2V_2}{n-1}$$

This result can be used for all values of n except 1 (the value for an isothermal process, see below).

Reversible adiabatic (isentropic)

For this process the law of expansion is $pV^\gamma = C$ and, substituting γ for n in the last result, the nonflow work is given by:

$$W = \frac{p_1V_1 - p_2V_2}{\gamma - 1}$$

The same expression can be obtained independently using the energy conservation principle. Since no heat is transferred, $Q = 0$ and the work done is equal to the decrease in internal energy. Thus:

$$W = mc_v(T_1 - T_2)$$

Rearranging the engineering form of the characteristic equation $pV = mRT$, we have $mT = pV/R$ and so

$$W = \frac{c_v(p_1V_1 - p_2V_2)}{R} = \frac{c_v(p_1V_1 - p_2V_2)}{c_p - c_v} = \frac{p_1V_1 - p_2V_2}{\gamma - 1}$$

since $R = c_p - c_v$ and $c_p/c_v = \gamma$

The agreement between this result and the one found by integration is an indirect proof that the isentropic law is $pV^\gamma = C$.

Isothermal

If the temperature is constant the pressure and volume follow Boyle's law $pV = C$, from which $p = C/V$. In this case the nonflow work is given by:

$$W = \int_{V_1}^{V_2} p\,dV = \int_{V_1}^{V_2} \frac{C}{V}\,dV$$

The integral of $1/x$ is $\ln x$ and $(\ln a - \ln b) = \ln(a/b)$. The result is therefore

$$W = [C \ln V]_{V_1}^{V_2} = C(\ln V_2 - \ln V_1) = p_1V_1 \ln\left(\frac{V_2}{V_1}\right)$$

Note that ln is the natural logarithm, that is, to the base e. On most calculators and computers this is shown as LN but some machines use LOG for natural logarithms and LOG10 for common logarithms (to the base 10).

EXAMPLE 12.15 A cylinder fitted with a piston contains 1.5 kg of a perfect gas at an initial pressure and temperature of 1.05 MN/m^2 and 260°C respectively. The gas expands to four times its initial volume and the pressure is then found to be 198 kN/m^2.

If the expansion follows the law pV^n = constant, find:

(a) the work done by gas,

(b) the heat transferred to the gas during the expansion.

Take $R = 188$ J/(kg K); $c_v = 479$ J/(kg K).

SOLUTION From the characteristic equation in the form $pV = mRT$ the initial volume is:

$$V_1 = \frac{mRT_1}{p_1} = \frac{1.5 \times 188 \times (273 + 260)}{1.05 \times 10^6} = 0.1431 \text{ m}^3$$

With a fourfold increase in volume the final volume is

$$V_2 = V_1 \times 4 = 0.1431 \times 4 = 0.5724 \text{ m}^3$$

The law of expansion is $pV^n = C$ and therefore $p_1V_1{}^n = p_2V_2{}^n$, from which:

$$\left(\frac{V_2}{V_1}\right)^n = \frac{p_1}{p_2} \quad \text{or} \quad 4^n = \frac{1.05 \times 10^6}{198 \times 10^3} = 5.303$$

Taking logs,

$$n \ln 4 = \ln 5.303$$

and

$$n = \ln 5.303/\ln 4 = 1.2034$$

In this calculation, either natural or common logarithms may be used since it is a ratio that is required.

(a) Substituting these results in the formula for nonflow work during a polytropic expansion, the required answer is:

$$W = \frac{p_1V_1 - p_2V_2}{n-1} = \frac{(1.05 \times 10^6) \times 0.1431 - (198 \times 10^3) \times 0.5724}{1.2034 - 1}$$

$$= 181.5 \times 10^3 \text{ J} \quad \text{or} \quad 181.5 \text{ kJ}$$

(b) In section 12.8 it was shown that, for a polytropic process $T_2 = T_1 r^{\gamma - 1}$ where r is the ratio V_1/V_2. In the present case $r = \frac{1}{4}$ and

$$T_2 = (273 + 260) \times (0.25)^{0.2034} = 402.3 \text{ K} \quad \text{or} \quad 129.3°\text{C}$$

The increase in internal energy is:

$$U_2 - U_1 = mc_v(T_2 - T_1) = 1.5 \times 479 \times (129.3 - 260)$$
$$= -93.9 \times 10^3 \text{ J} \quad \text{or} \quad -93.9 \text{ kJ}$$

(the negative sign indicating a decrease). From the energy conservation equation for a nonflow process, the heat received by the gas is:

$$Q = (U_2 - U_1) + W = -93.9 + 181.5 = 87.6 \text{ kJ}$$

12.8.2 Cycles of operation

A number of ideal cycles have been suggested for the changes of state that occur in an engine and they are usually illustrated by p–V diagrams. Four important cases are shown in Fig. 12.14. They all assume that the working substance is an ideal gas and each consists of four processes. Two of these processes are reversible adiabatic (isentropic) changes in which no heat is received or rejected by the gas ($Q = 0$). The cycles differ from one another in the way in which the gas receives and rejects heat. The efficiency of a cycle is defined, in the usual way, as the proportion of the heat

(a) Carnot

(b) Otto (constant volume)

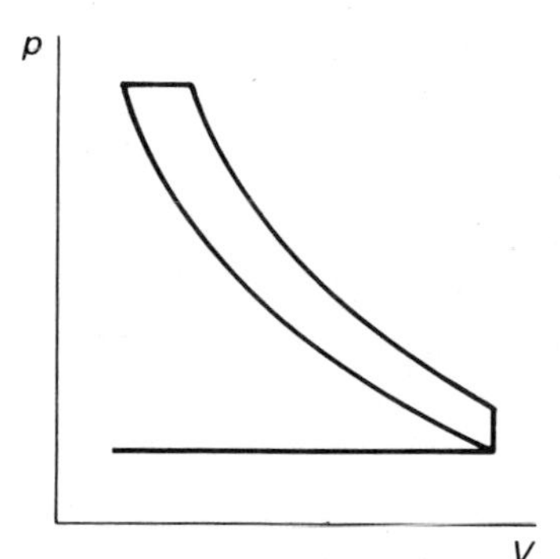

(c) Diesel (constant pressure)

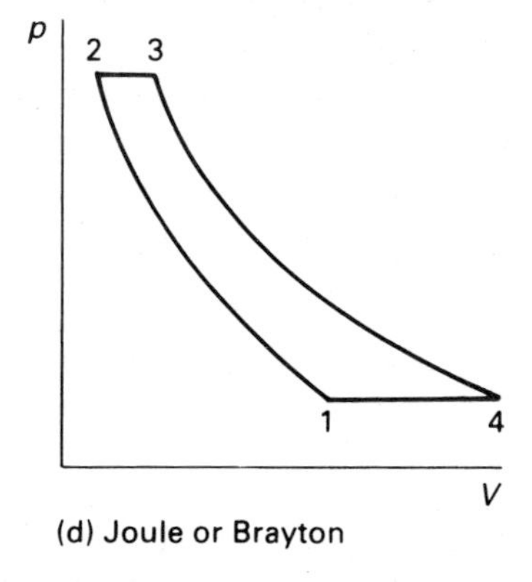

(d) Joule or Brayton

Figure 12.14

received by the gas that is converted to work. Furthermore the heat converted to work is the difference between the heat received and the heat rejected. Hence, for all such cycles, the efficiency η is:

$$\eta = \frac{\text{heat received} - \text{heat rejected}}{\text{heat received}} = 1 - \frac{\text{heat rejected}}{\text{heat received}}$$

Carnot cycle

The four processes were described in detail in section 12.5.3 and are shown again in Fig. 12.14(a).

Otto or constant volume cycle

This is shown in Fig. 12.14(b). It assumes that the induction (suction) and exhaust strokes take place at the same constant pressure and effectively cancel one another. The supply of heat to the gas is considered to take place instantaneously (2 to 3) and at constant volume. Similarly it is supposed that the heat is rejected at constant volume (4 to 1). The expansion and compression strokes are adiabatic ($Q = 0$). The cycle efficiency is therefore:

$$\eta = 1 - \frac{\text{heat received}}{\text{heat rejected}} = 1 - \frac{mc_v(T_3 - T_2)}{mc_v(T_4 - T_1)} = 1 - \frac{T_3 - T_2}{T_4 - T_1}$$

The expansion and compression strokes have the same adiabatic law and the same volume ratio r. Therefore $T_4 = T_3 r^{\gamma - 1}$ and $T_1 = T_2 r^{\gamma - 1}$. The efficiency is therefore:

$$\eta = 1 - \frac{T_3 - T_2}{T_3 r^{\gamma - 1} - T_2 r^{\gamma\;1}} = 1 - \frac{1}{r^{\gamma - 1}}$$

This result is called the *air standard efficiency* and is used as the basis of comparison for the efficiency of internal combustion engines.

Diesel or constant pressure cycle

In this cycle the supply of heat to the gas takes place at constant pressure, see Fig. 12.14(c). In practice the combustion process is so rapid that the standard efficiency based on the Otto cycle is often used as the basis of comparison.

A dual combustion cycle is sometimes considered, in which the supply of heat takes place partly at constant volume and partly at constant pressure.

Brayton or Joule cycle

In this cycle, Fig. 12.14(d), the reception and rejection of heat takes place at constant pressure. The processes in gas turbines approximate to this cycle.

In terms of the temperatures T_1, T_2, T_3 and T_4 the efficiency is the same as for the Otto cycle, above. However, it is usual to express the result in terms of the pressure ratio. Suppose $p_2/p_1 = r_P$ and, as before, $V_1/V_2 = r$. From the isentropic law,

$$p_2 V_2^{\gamma} = p_1 V_1^{\gamma} \quad \text{and} \quad \frac{p_2}{p_1} = \left(\frac{V_1}{V_2}\right)^{\gamma} \quad \text{or} \quad r_p = r^{\gamma}$$

Thus $r = r_p^{1/\gamma}$ and the efficiency can be written

$$\eta = 1 - \frac{1}{r^{\gamma - 1}} = 1 - \frac{1}{(r_p^{1/\gamma})^{\gamma - 1}} = 1 - \frac{1}{r_p^{1 - 1/\gamma}}$$

12.8.3 Summary of results

All the results that follow can be applied to unit mass (1 kg) by putting $m = 1$ and replacing volume V, internal energy U and enthalpy H by the specific values v, u and h.

For a steady flow process the principle of conservation of energy leads to the equation:

$$mgz_1 + \tfrac{1}{2}c_1^2 + H_1 + Q = mgz_2 + \tfrac{1}{2}c_2^2 + H_2 + W$$

where

$$H = U + pV$$

For a nonflow process this reduces to:

$$Q = (U_2 - U_1) + W$$

For a perfect gas, the Boyle–Charles law gives:

$$\frac{p_1V_1}{T_1} = \frac{p_2V_2}{T_2}$$

The engineering form of the characteristic equation is:

$$pV = mRT$$

in which the gas constant R corresponds to unit mass (1 kg) of gas. In SI units (J/(kg K)) the value of R is approximately $8300/M$ where M is the relative molecular mass.

The characteristic equation can also be written:

$$pV = nRT$$

in which R is the molar gas constant in J mol^{-1} K^{-1} and n is the amount of substance in moles.

If R, c_p and c_v are in the same units,

$$R = c_p - c_v$$

and

$$\gamma = c_p/c_v$$

The changes in internal energy and enthalpy are given by:

$$U_2 - U_1 = mc_v(T_2 - T_1)$$

and

$$H_2 - H_1 = mc_p(T_2 - T_1)$$

For an isothermal process $T_2 = T_1$ and the last two results become zero.

The following table summarises the formulae for various processes with ideal gases:

Process	*Constant volume*	*Constant pressure*	*Isothermal*	*Reversible adibatic/ isentropic*	*Polytropic*
$p - V$ equation	$V = C$	$p = C$	$pV = C$	$pV^\gamma = C$	$pV^n = C$
$\frac{p_2}{p_1} =$	$\frac{T_2}{T_1}$	1	$\frac{V_1}{V_2}$	$\left(\frac{V_1}{V_2}\right)^\gamma$	$\left(\frac{V_1}{V_2}\right)^n$
$\frac{T_2}{T_1} =$	$\frac{p_2}{p_1}$	$\frac{V_2}{V_1}$	1	$\left(\frac{V_1}{V_2}\right)^{\gamma-1}$	$\left(\frac{V_1}{V_2}\right)^{n-1}$
nonflow work	0	$p(V_2 - V_1)$	$p_1V_1 \ln\left(\frac{V_2}{V_1}\right)$	$\frac{p_1V_1 - p_2V_2}{\gamma - 1}$	$\frac{p_1V_1 - p_2V_2}{n - 1}$

12.9 Heat transfer

Note that several symbols used earlier in this chapter have new meanings when used in heat transfer theory. It will always be clear from the context which meaning is intended.

Whenever two bodies (or two parts of the same body) are at different temperatures there is a tendency for energy to be transferred between them in the form of heat Q. The name *heat exchanger* is used for any device which is designed to transfer heat rapidly from one substance to another. The transfer may be achieved by the direct mixing of different fluids or it may occur through a wall separating them.

Heat transfer may take place in three ways: *conduction*, *radiation* and *convection*. Conduction is the transfer of energy from faster moving molecules to adjacent slower ones by impact. Suppose heat is conducted through a flat plate, area A and thickness x, as shown in Fig. 12.15(a), T_A and T_B being the temperatures of the two surfaces.

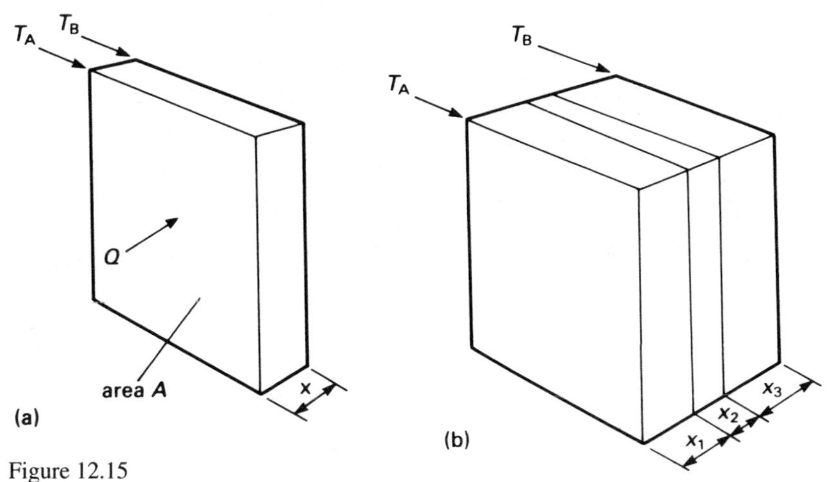

Figure 12.15

If the temperatures are steady and the whole heat flow is perpendicular to the surface of the plate then it is found that the heat Q conducted per second is

- proportional to the area A,
- proportional to the temperature difference $(T_A - T_B)$
- inversely proportional to the thickness x.

We can therefore write:

$$Q = \frac{kA(T_A - T_B)}{x}$$

where k is a constant for the material of the plate. It is called the *thermal conductivity* for the material.

From the definition it will have the units of $Qx/A(T_A - T_B)$ that is (energy per unit time × length)/(area × temperature)

In the SI system the units of k are:

$$\frac{(\text{J/s}) \times \text{m}}{\text{m}^2 \times \text{K}} = \text{W/(m K)} \quad \text{or} \quad \text{W m}^{-1}\,\text{K}^{-1}$$

The fraction x/kA is called the thermal resistance. If it is denoted by R (not to be confused with the gas constant) then the heat flow rate is given by:

$$Q = \frac{T_A - T_B}{R}$$

This result is similar in form to the relationship between current, voltage drop and electrical resistance in a d.c. electrical circuit. The electrical circuit is said to be an *analogue* of the thermal conditions (see Chapter 15) and the flow of heat in complex cases is often determined by measurements made on corresponding electrical circuits. In particular, the total thermal resistance for a composite plate, such as the one in Fig. 12.15(b), equals the sum of the resistances for the separate plates of which it is formed, in the same way that the total electrical resistance of several resistors in series is found by adding their separate values.

EXAMPLE 12.16

(a) The wall of a building is 225 mm thick and is constructed of bricks having a thermal conductivity of 0.7 W/(m K). Calculate the heat loss per square metre of wall if the inner and outer surface temperatures are 20°C and 0°C respectively.
(b) A wall board 6 mm thick ($k = 0.05$ W/(m K)) is added to the inside surface and a 25 mm thick rendering ($k = 0.8$ W/(m K)) is added to the outside. Find, for the same overall temperature difference, the percentage reduction in heat loss.

SOLUTION

(a) If $A = 1\ \text{m}^2$ and $x = 225\ \text{mm} = 0.225\ \text{m}$, the heat loss is given by:

$$Q = \frac{kA(T_A - T_B)}{x} = \frac{0.7 \times 1 \times 20}{0.225} = 62.2\ \text{W}$$

(b) If the wall board, brickwork and rendering are denoted by 1, 2 and 3 respectively the total thermal resistance is:

$$R = \frac{x_1}{k_1 A} + \frac{x_2}{k_2 A} + \frac{x_3}{k_3 A} = \frac{1}{A}\left(\frac{x_1}{k_1} + \frac{x_2}{k_2} + \frac{x_3}{k_3}\right)$$

the area A being the same (1 m²) for each material.

In SI units:

$$R = \frac{0.006}{0.05} + \frac{0.225}{0.7} + \frac{0.025}{0.8} = 0.473$$

The heat loss per square metre is now:

$$Q = \frac{20}{0.473} = 42.3\ \text{W}$$

The reduction is therefore (62.2 − 42.3) = 19.9 W, or 32.0 per cent of the original value.

In the last example, the surface temperatures were given. These may be very different from the temperatures of the surroundings because of radiation and convection effects. Often there is a rapid change in temperature through the thin film of fluid adjacent to the surface. Figure 12.16 shows a typical variation in the case of a composite wall. The flow rate through a surface film can be expressed as:

$$Q = hA \times (\text{temperature fall through the film})$$

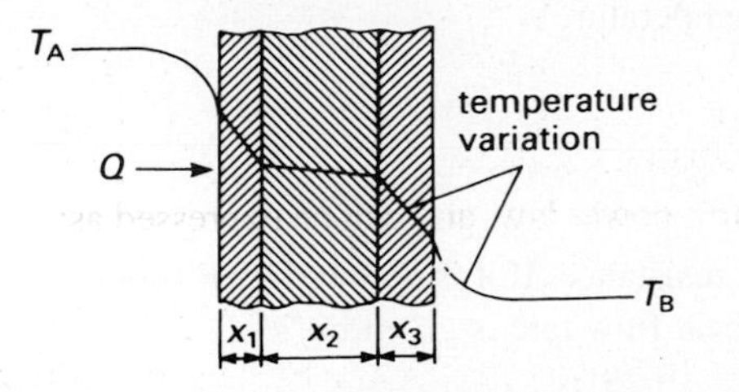

Figure 12.16 Temperature variation through a composite wall.

where A is the surface area and h is called the *coefficient of surface heat transfer* or *film coefficient*. Do not confuse this use of the symbol h with specific enthalpy. Here, the units of h are those of Q/AT, giving W/(m² K²) (or W m^{-2} K^{-2}) in the SI system.

The expression for heat transfer by conduction can be modified to allow for surface film effects by adding terms to the expression for thermal resistance. If the wall has three components, as shown in Fig. 12.16, with thermal conductivities k_1, k_2 and k_3, and surface heat transfer coefficients h_1 and h_2, the heat flow can be written:

$$Q = \frac{A(T_A - T_B)}{\frac{x_1}{k_1} + \frac{x_2}{k_2} + \frac{x_3}{k_3} + \frac{1}{h_1} + \frac{1}{h_2}}$$

where T_A and T_B are now the temperatures of the surroundings on the two sides of the wall.

The denominator of this fraction represents the thermal resistance. Its reciprocal is called the *transmittance*, *overall conductance* or *overall coefficient of heat transfer*. It is denoted by U (another example of a symbol having two meanings in thermodynamics) and its units are $W/(m^2\ K)$ or $W\ m^{-2}\ K^{-1}$. The overall heat transfer coefficient is often referred to simply as the U-value. Once it is known for a particular wall or partition the heat flow can be obtained in terms of the temperatures of the surroundings by the equation:

$$Q = UA(T_A - T_B)$$

EXAMPLE 12.17 A furnace wall consists of two layers of brickwork each 225 mm thick. The inner layer is built from refractory bricks and the outer one from insulating bricks. The thermal conductivities are 0.86 W/(m K) and 0.26 W/(m K), respectively. The coefficient of heat transfer for the outer surface is 1.12 $W/(m^2\ K)$.

Find the heat loss per square metre of surface area if the temperature of the inner surface is 1100°C and of the surroundings is 20°C.

SOLUTION Since the temperature of the inner surface is given, only one coefficient of surface heat transfer is required, that for the outer surface. Denoting the two kinds of brickwork by 1 and 2, the overall heat transfer coefficient is given by:

$$\frac{1}{U} = \frac{x_1}{k_1} + \frac{x_2}{k_2} + \frac{1}{h} = \frac{0.225}{0.86} + \frac{0.225}{0.26} + \frac{1}{1.12} = 2.02$$

and

$$U = 0.495\ W/(m^2\ K)$$

Putting $A = 1$, the rate of heat loss is:

$$Q = UA(T_A - T_B) = 0.495 \times (1100 - 20) = 534.6\ W$$

12.9.1 Radiation

The transfer of heat by radiation takes place through electromagnetic waves of the same kind as light and radio waves. In general, some of the radiant heat striking a body is reflected, some passes through and some is absorbed. A *black body* is defined as one that absorbs *all* the radiant heat striking it. It is also the ideal *radiator*. In 1879, Stefan suggested from an analysis of experimental results that the amount of radiation is proportional to the fourth power of the Kelvin temperature and in 1884 Boltzmann came to the same conclusion on theoretical grounds. The result is known as the Stefan–Boltzmann law, or simply the fourth power law, and can be expressed as:

$$Q = \sigma A T^4$$

where T is the absolute temperature of the body, A is the radiating area and σ is a factor known as the Stefan–Boltzmann constant. In SI units its value is about $57.0 \times 10^{-9}\ W/(m^2\ K^4)$.

If the energy radiated from one black body at temperature T_1 is absorbed by another at temperature T_2, the heat flow between them per square metre of surface is:

$$Q = \sigma(T_1^4 - T_2^4)$$

The radiation from an actual body is less than that from the ideal black body and, to allow for this, the Stefan–Boltzmann law is written:

$$Q = \epsilon\sigma AT^4$$

where ϵ is a coefficient (less than 1) called the *emissivity*.

12.9.2 Convection

The third mode of heat transfer is convection. A fluid is heated by radiation or conduction and then moves to a cooler body or space and heats it, again by radiation or conduction. *Free convection* occurs when the fluid moves of its own accord because of differences in density arising from temperature variations. In *forced convection* the fluid is blown or pumped by some external means.

There are no simple rules for calculating the heat transfer by convection, and practical design relies on experimental results and empirical formulae.

Bibliography

Atkins P W, Clugston M J, Frazer M and Jones R A Y 1988 *Chemistry: Principles and Applications*. Longman

Eastop T D and McKonkey A 1993 *Applied Thermodynamics for Engineering Technologists* 5th edn. Longman

Joel R 1996 *Basic Engineering Thermodynamics in SI Units* 5th edn. Longman

MacGregor J 1993 *Higher Chemistry*. Longman

Rogers G F C and Mayhew Y P 1992 *Engineering Thermodynamics* 4th edn. Longman

Spalding D B and Cole E H 1973 *Engineering Thermodynamics* 3rd edn. Arnold

Assignments

1 A single cylinder air compressor has a piston area 8000 mm^2 and piston stroke 100 mm. Calculate the mass of air which would occupy the swept volume at the free air conditions 15°C and 101.5 kPa.

If 85 per cent of this mass is delivered on each revolution, find the mass flowrate for a compressor speed 1200 revolution $minute^{-1}$.
(For air $R = 0.287$ kJ kg^{-1} K^{-1}.)

2 A domestic oven has a uniform thickness of insulation, 25 mm, on all its walls. The total surface area of the walls is 1.1 m^2 and the conductivity of the insulation is 0.04 W m^{-1} K^{-1}.

Find the energy lost through the walls in 20 minutes when the wall surfaces are at 220°C inside and at 40°C on the outside.

3 Explain what is meant by a *black body*.

A body which may be regarded as black is at a temperature 200°C. Energy is radiated to the surroundings, at temperature 20°C, at a rate P. Express, in terms of P, the rate at which the body would radiate energy to the same surroundings if its temperature were increased to 300°C.

4 A heat pump takes energy from a river which is at 5°C and uses this to evaporate the working fluid. This is then compressed and passed into a heat exchanger in which it condenses, giving energy to the air in the building at 20°C. The fluid then expands through a valve to complete the cycle.

Find the maximum rate at which energy could be given to the building if the power input to the compressor is 15 kW.

State why this would not be achieved in practice.

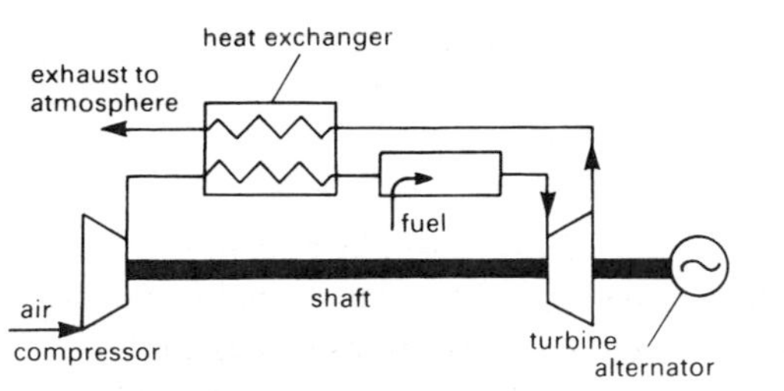

Figure 12.17

5 An industrial gas turbine plant (Fig.12.17) is used to generate electricity. It incorporates a heat exchanger, which may be assumed to be completely insulated, in which the exhaust from the turbine is used to raise the temperature of the compressed air.

The following data were obtained in a test:

Air : fuel ratio by mass	80:1
Heating value of fuel	44 MJ kg^{-1}
Air temperature at inlet to compressor	20°C
Air temperature at outlet from compressor	430°C
Gas temperature at inlet to turbine	1000°C
Gas temperature at outlet from turbine	530°C
Fuel mass flowrate	0.2 kg s^{-1}

Assume that the specific heat capacity, c_p, of air and gas is 1.05 kJ kg^{-1} K^{-1} throughout.

Find
(a) the power required by the compressor,
(b) the shaft power output to the alternator,
(c) the overall efficiency of the plant,
(d) the temperature of gas exhausted to atmosphere.

6 Give an expression for the relationship between mass m, pressure p, volume V and temperature T for an ideal gas. Indicate clearly the units in which each of these properties is measured for the expression to be consistent.

A mass 50 g of an ideal gas is initially at ambient temperature 17°C and pressure 100 kPa. In two separate experiments it is compressed in a cylinder to a pressure 4000 kPa under different conditions:
(a) with the cylinder fully insulated,
(b) slowly and with no insulation of the cylinder.

Find, in each case, the final gas temperature and the work done on the gas during the compression process. In case (b) find also the heat transfer. (For the gas:

the specific heat capacity at constant pressure, $c_p = 0.82$ kJ kg^{-1} K^{-1},

and

the specific heat capacity at constant volume, $c_v = 0.63$ kJ kg^{-1} K^{-1})

7 Explain the following terms used in heat transfer, indicating carefully the differences between them:
(a) conduction,
(b) convection,
(c) radiation.

State which method or methods of heat transfer is involved in each of the following devices, justifying in each case the answer given.
(i) A soldering iron.
(ii) An infrared paint-drying oven.
(iii) A fan heater.
(iv) A central heating system 'radiator'.
(v) A solar water-heating panel.

8 A power plant receives energy by heat transfer from a source at 1050°C at a rate 3 MW. It transmits energy by heat transfer to a region at 35°C.
Determine the maximum power output obtainable and give two reasons why an actual plant would not achieve this output.

9 A small portable drill is driven by a turbine supplied from a factory compressed air main at 600 kPa and 40°C. The turbine discharges air at 100 kPa and −23°C.
Find
(a) the isentropic efficiency of the air turbine,
(b) the power output for an air flowrate 25 kg h^{-1}.
(For air $c_p = 1.005$ kJ kg^{-1} K^{-1}, $\gamma = 1.4$.)

10 (a) The *U*-value of a steel-clad factory roof is 6.7 W m^{-2} K^{-1}. The average outside temperature in winter is 8°C and the average temperature in the roof space is 22°C. The winter period lasts for 35 weeks. During this period the factory is heated 12 hours per day, 6 days per week. The heating system has an efficiency of 75% and the cost of fuel is 0.5p MJ^{-1}.
Calculate:
(i) the rate of loss of energy per m^2 through the roof in winter,
(ii) the energy loss per m^2 through the roof each winter,
(iii) the cost of the fuel which must be burned each winter to make good the energy lost through each m^2 of the roof.
(b) The roof is to be insulated with panels of glass fibre insulation. The panels are available in several thicknesses; the corresponding *U*-values and installation costs are given in the following table:

thickness	mm	0	50	75	100
U-value	Wm^{-2} K^{-1}	6.7	0.55	0.39	0.30
installation cost	£m^{-2}	0	7.65	8.42	10.28

Determine the thickness which results in the lowest total cost per m^2 of roof over an eight-year period, assuming the cost of the fuel remains constant.
(c) Would your answer to (b) have been greater, the same or lower, if the evaluation period had been:
(i) three years?
(ii) thirty years?
Justify your answers.

11 (a) (i) Sketch a typical indicator diagram for one cylinder of a four-stroke petrol engine. Label the strokes and show with an X where ignition would occur.
(ii) Explain in clear, nontechnical language what is happening in each of the strokes.
(b) The following data were obtained in a test on a single-carburettor, four-cylinder, four-stroke petrol engine:

engine speed	3000 rev min^{-1}
area of indicator diagram	0.214 kJ
output shaft torque	58 Nm
calorific value of fuel	43.7 MJ kg^{-1}
density of fuel	740 kg m^{-3}
rate of fuel consumption	50 cm^3 in 26.8 s

Use the above information to determine:

(i) the input power,
(ii) the indicated power,
(iii) the brake power,
(iv) the mechanical efficiency,
(v) the brake thermal efficiency of the engine.

Why are your answers to (ii) and (iv) likely to be only approximate?

12 A geothermal power station generates electricity at a rate of 50 MW in a region where the maximum available temperature is 700 K. A nearby river at 300 K is to be used for cooling purposes.

(a) Calculate the Carnot efficiency of the power station.
(b) If the power station were to operate at its Carnot efficiency, at what rate would energy be rejected to the river?

13 *Fluid mechanics*

13.1 Properties of fluids

A fluid is defined as a material that offers no permanent resistance to change of shape. It flows to match the shape of the vessel containing it. The definition applies to liquids and gases.

A liquid forms a free horizontal surface but a gas fills the whole space to which it has access. Although the shape of a liquid adapts to that of its container, its volume can only be changed to a limited extent. The volume of a gas can be readily changed. Liquids are almost *incompressible*; gases are highly *compressible*.

The general principles and laws of mechanics that were introduced in earlier chapters for solids apply equally well to fluids. For example, the forces acting on a liquid at rest are governed by the laws of equilibrium given in Chapter 8. Momentum and energy have the same meanings for fluids as for solids, and Newton's laws are as valid for fluids as for solids. There is a close connection between fluid mechanics and thermodynamics, and the steady flow equation, based on the principle of conservation of energy, is widely used in both subjects. They are sometimes studied together under the name *thermofluids*.

For fluids, as for solids, it is convenient to divide the work into statics (forces at rest) and dynamics (forces in motion). The prefix 'hydro-' is used for liquids (not just water) and 'aero-' for gases. This gives four branches altogether: hydrostatics, hydrodynamics, aerostatics and aerodynamics.

13.1.1 Density

Density is defined as mass divided by volume. It is denoted by ρ and its SI units are kg/m^3. The density of fresh water may be taken as 1000 kg/m^3 and this will not be greatly affected by changes in pressure and temperature. The densities of other liquids are often quoted as ratios of that of water. This ratio is called *relative density* (or specific gravity). The relative density of oil, for example, is about 0.9; for mercury, it is approximately 13.6.

The density of a gas depends on its temperature and pressure; for an ideal gas it can be found from the characteristic equation, see section 12.6.2. The density of air at sea level and 15°C is about 1.23 kg/m^3.

Density ρ (mass/volume) is the reciprocal of specific volume v (volume/mass) that was introduced in Chapter 12.

13.1.2 Pressure

If a fluid is at rest it exerts the same pressure in all directions. This pressure acts everywhere at right angles to the surface in contact with the fluid.

The basic SI unit of pressure is N/m^2 or Pa. For many purposes this is a very small unit and pressures are often quoted in bar, particularly in meteorology. The bar is defined as 100 kN/m^2, which equals 10^5 N/m^2 (or 10^5 Pa) and it is approximately the value of atmospheric pressure at sea level. Pressures are also given in terms of an equivalent column of liquid. Consider a point y, Fig. 13.1, at a depth h below the surface of the liquid. If the density of the liquid is ρ and the cross-sectional area of the column is A, then the volume of liquid above y is Ah and its mass is ρAh. The weight of this liquid is ρgAh and therefore the pressure at y due to the liquid is:

$$p = \frac{\text{force}}{\text{area}} = \frac{\rho gAh}{A} = \rho gh$$

The height h is sometimes called the *pressure head*, or simply the *head*, and pressure may be given in terms of head. We speak of an atmospheric pressure of 760 mm of mercury or a household gas supply pressure of 100 mm of water.

A simple device for measuring pressure is the *U-tube manometer* shown in Fig. 13.1. It depends on the principle that all points in a liquid at the same level have the same pressure. The pressure at y in the right-hand column is therefore equal to the pressure at x in the left-hand column. Hence the required pressure is given by the column of liquid above y.

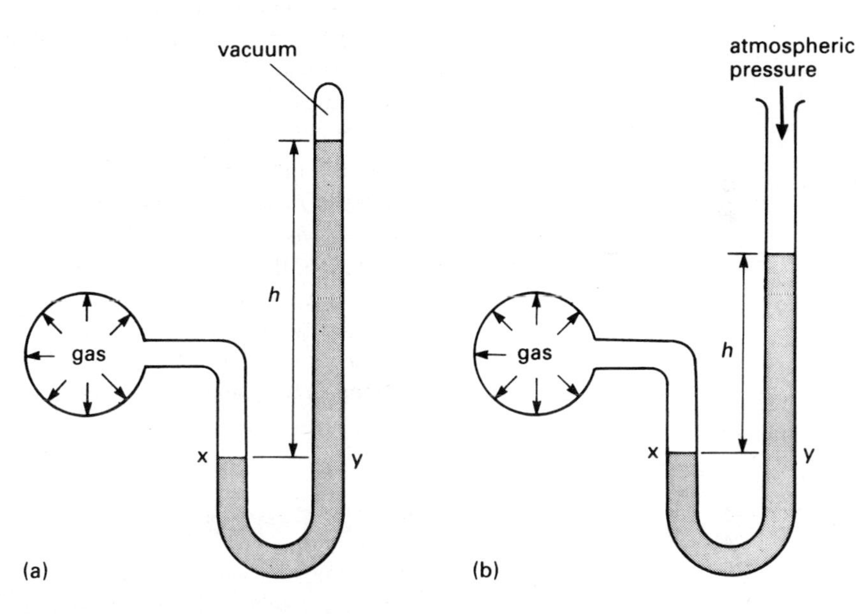

Figure 13.1 The U-tube manometer (a) absolute pressure (b) gauge pressure.

If the right-hand column is sealed (Fig. 13.1(a)), and there is a vacuum above the liquid, the value will be the *absolute pressure*. Atmospheric pressure is measured in this way using a mercury barometer.

If the right-hand column is open to the atmosphere, Fig. 13.1(b), the column of liquid gives the *gauge* value. Absolute and gauge values are related by:

$$\text{absolute pressure} = \text{gauge pressure} + \text{atmospheric pressure}$$

If the required pressure is below atmospheric (sometimes known as a *partial vacuum*) the level in the right-hand column in Fig. 13.1(b) will be below that in the left-hand column.

Make sure you use gauge and absolute pressures correctly. The stresses in the walls of a pressure vessel (Chapter 9) depend on the difference between the internal and external pressures. The external pressure is normally atmospheric and gauge pressure must therefore be used in the formulae. On the other hand, the gas laws (Chapter 12) require absolute values of the pressure. In fluid mechanics it is usually the *change* in pressure that is required and this will be the same whether the separate values are gauge or absolute.

EXAMPLE 13.1 The pressure at the top of a mountain is found to be 0.7 bar. Express this in metres of water. What would be the reading of a mercury barometer at this point? Relative density of mercury, 13.6.

SOLUTION The density of water may be taken as 1000 kg/m^3. Working in basic SI units the equivalent head h is given by:

$$\rho gh = 0.7 \times 10^5 \quad \text{or} \quad h = \frac{0.7 \times 10^5}{\rho g} = \frac{0.7 \times 10^5}{1000 \times 9.81} = 7.14 \text{ m}$$

The corresponding head for mercury is 7.14/13.6 = 0.525m or 525 mm.

13.1.3 Properties of the atmosphere

Suppose p_0 is the pressure of the atmosphere at sea level and p the value at an altitude h. If the density were constant the difference in pressure would be given by the equation:

$$p_0 - p = \rho gh$$

In fact, the density ρ decreases with altitude and this result should not be used for heights of more than a few hundred metres. To allow for the variation in density, we have to consider a small difference in height dh and then find the total change by integration. If this is done the result becomes:

$$p_0 - p = \int_0^h \rho g \, \mathrm{d}h$$

To find the value of this integral the density ρ must first be expressed in terms of h. See Assignment 2.

13.2 Force on an immersed surface

If a rectangular tank with a level base is filled with liquid, the resultant force on the base is simply the total weight of liquid in the tank.

There is also an outward thrust on each of the vertical sides. To find its value consider a rectangle, width b and depth d, with one edge in the surface of the liquid as shown in Fig. 13.2(a). The pressure variation is triangular as shown in Fig. 13.2(b), increasing from zero at the surface to a maximum at the bottom. The total force F acting on the rectangle is given by:

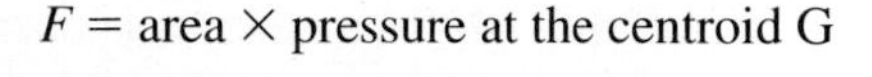

$$F = \text{area} \times \text{pressure at the centroid G}$$

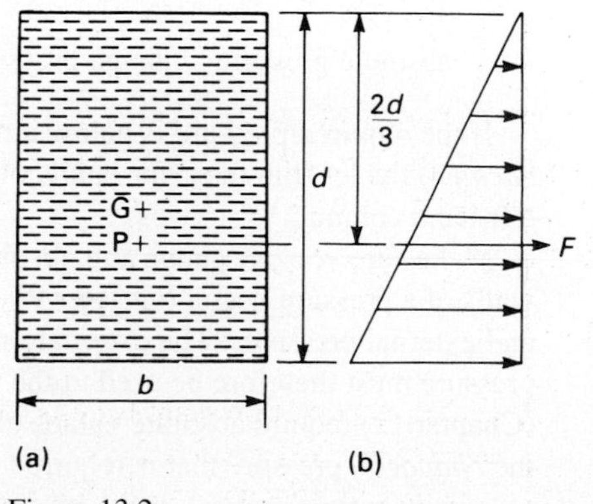

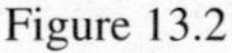

Figure 13.2

In the case of the rectangle, the area is bd and G is at a depth $d/2$. The total force is therefore:

$$F = bd \times \rho gd/2 = \tfrac{1}{2}\rho gbd^2$$

The point, P, at which the resultant acts is called the *centre of pressure*. In the present case its position corresponds to the centre of gravity of the pressure triangle, two-thirds of the depth from the surface.

If the rectangle is not vertical, the resultant force is still obtained by multiplying its area by the pressure at its centroid. Note that this pressure is based on the *vertical* depth of the centroid and is no longer equal to $d/2$. The centre of pressure is still $2d/3$ from the surface, but is measured down the plane of the rectangle (not vertically).

EXAMPLE 13.2 A rectangular hole in the vertical side of a tank is 2.5 m wide and 1.5 m deep, and is closed by a cover plate. Find the magnitude of the force on this plate and its centre of pressure if the tank is filled with water to a level 2 m above the top edge of the hole.

Density of water, 1000 kg/m^3.

SOLUTION The rectangular plate is shown as ABCD in Fig. 13.3(a). The result $2d/3$ for the depth of the centre of pressure only applies to a rectangle with one edge in the surface. The answer can therefore be obtained by considering two rectangles ABEF and DCEF, each with one edge EF in thc surface. If the force on the second rectangle is taken as negative, their resultant will be the required force on ABCD.

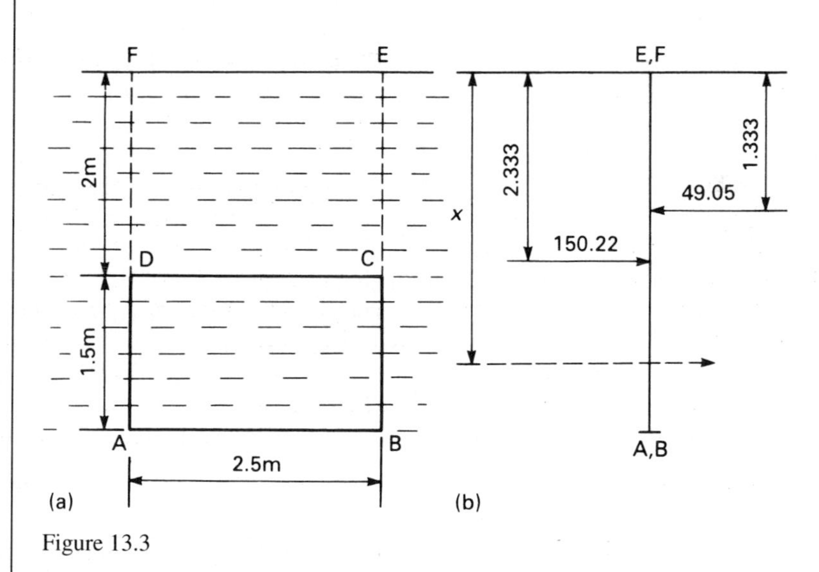

Figure 13.3

For ABEF the centroid (G) is at a depth 3.5/2 = 1.75 m, and the pressure at this point is $\rho gh = 1000 \times 9.81 \times 1.75 = 17.17$ kN/m^2. Hence the force on ABEF is given by:

$$\begin{aligned} \text{force} &= \text{area} \times \text{pressure at centroid} \\ &= 3.5 \times 2.5 \times (17.17 \times 103) \\ &= 150.22 \times 10^3\ \text{N} \quad \text{or} \quad 150.22\ \text{kN} \end{aligned}$$

The centre of pressure for this force is two-thirds of the total depth from the surface, ie $(2/3) \times 3.5 = 2.333$ m.

The corresponding results for DCEF are 49.05 kN and 1.333 m. The two forces and their lines of action are shown in kilonewton and metre units in Fig. 13.3(b). The magnitude of the resultant force R on the plate is the difference between the separate magnitudes. Therefore:

$$R = 150.22 - 49.05 = 101.17 \text{ kN}$$

Suppose R acts at a depth x from the surface (EF). The moment of the resultant R about EF will equal the sum of the moments of the two separate forces, one being regarded as negative. Therefore:

$$101.17 \times x = 150.22 \times 2.333 - 49.05 \times 1.333 = 285.08$$

and

$$x = \frac{285.08}{101.17} = 2.818 \text{ m}$$

or 0.818 m below the top edge of the plate.

The magnitude of the force acting on ABCD can be found directly using the formula (area) × (pressure at its centroid) but the two-thirds rule only works for a rectangle with one edge in the surface.

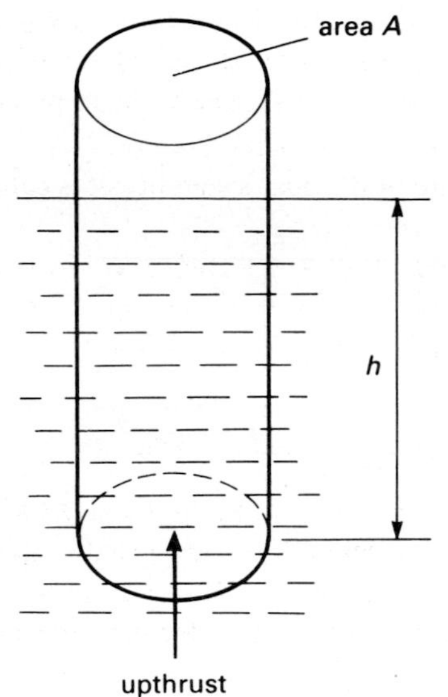

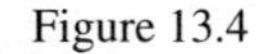
Figure 13.4

13.3 Buoyancy

The pressure at a point in a liquid at rest acts equally in all directions, including upwards. It therefore causes an upthrust on the underside of a body immersed in it. Suppose a uniform cylinder of cross-sectional area A is lowered in a liquid of density ρ to a depth h as shown in Fig. 13.4.

The pressure at the bottom of the cylinder is ρgh and it acts on an area A. The upthrust is therefore ρgAh. But Ah is the volume of the cylinder immersed in the liquid and this equals the volume of the liquid displaced. The mass of liquid displaced is ρAh and its weight is ρgAh. It follows that the *upthrust is equal to the weight of liquid displaced*. This is the famous principle of Archimedes. Since the upthrust reduced the apparent weight of the body it can also be expressed as 'the apparent loss of weight of a body immersed in a liquid is equal to the weight of liquid displaced'.

The result is not restricted to objects of uniform cross-section but applies to a body of any shape.

EXAMPLE 13.3 Calculate the available lift of a balloon containing 500 m^3 of helium at an altitude of 1000 m.

Density of air at 1000 m = 1.15 kg/m^3; density of helium relative to air = 0.14.

SOLUTION Archimedes principle applies to gases as well as liquids. Since the balloon displaces 500 m^3, we have:

$$\begin{aligned} \text{upthrust} &= \text{weight of air displaced} \\ &= (\text{density of air} \times (\text{volume}) \times g \\ &= 1.15 \times 500 \times 9.81 \\ &= 5640 \text{ N} \quad \text{or} \quad 5.64 \text{ kN} \end{aligned}$$

The weight of the same volume of helium is 0.14 times this result. Therefore,

$$\begin{aligned} \text{available lift} &= \text{upthrust} - \text{weight of helium} \\ &= 5.64 - (0.14 \times 5.64) \\ &= 4.85 \text{ kN} \end{aligned}$$

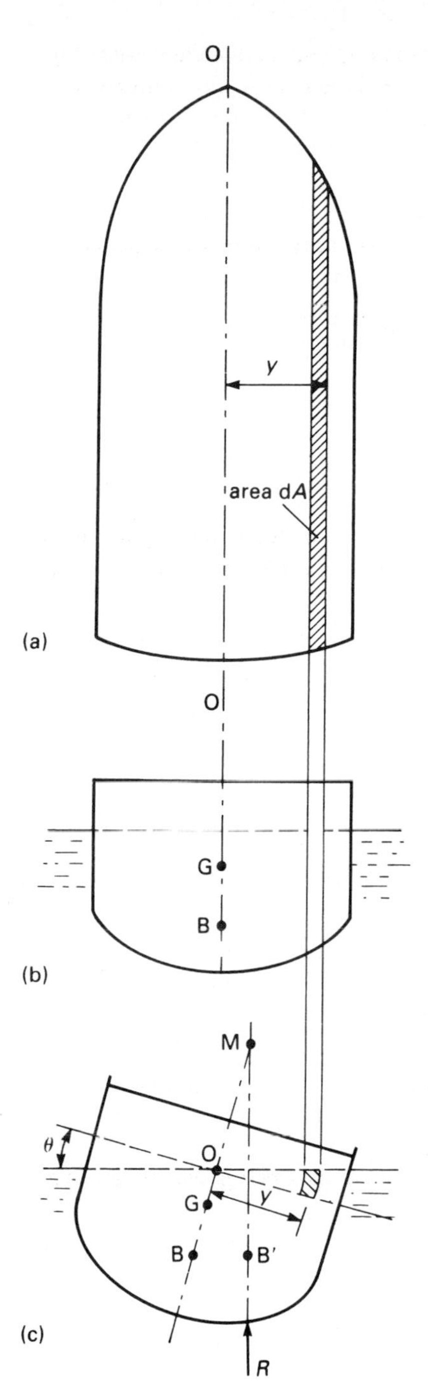

Figure 13.5

13.3.1 Equilibrium of floating bodies

If a ship or other body is floating at rest in water, it is in equilibrium under the forces acting on it. In particular, the upthrust (or buoyancy) is equal to its weight. The equilibrum is stable if it returns to its initial position after being disturbed. For most ships the main concern is for stability in *rolling*, that is, rotation about a longitudinal axis. Stability in *pitch* (motion about a lateral or transverse axis) and in *yaw* (motion about a vertical axis) can be important in some cases.

Fig. 13.5(a) and (b) show the waterline plan and a vertical cross-section of a boat in its equilibrium position. If W is the weight of the ship and R the total upthrust due to the displacement of the water, $R = W$. Also if V is the volume of the boat below the waterline then the weight of water displaced, and hence the upthrust, is

$$R = \rho g V$$

Let G be the boat's centre of gravity and B the *centre of buoyancy*, that is, the centre of gravity of the volume of displaced water. This is the point through which the upthrust acts. In the equilibrium position, Fig. 13.5(b), points B and G lie on the same vertical line.

Now consider what happens if the ship tilts about O through a small angle θ radians, as shown in Fig. 13.5(c). The upthrust R and weight W are still equal but the shape of the volume of displaced water is different and the centre of buoyancy is at a new position B′. The point M at which the vertical through B′ cuts the axis of symmetry through B, G and O is called the *metacentre*. If M is above the centre of gravity, as it is shown in Fig. 13.5(c), the moment of the force R about G is tending to restore the boat to its upright position and the equilibrium is stable. If, on the other hand, M is below G, the moment of R tends to increase the angle of tilt and the boat is *unstable*. The height GM is therefore a measure of the boat's stability; it is called the *metacentric height* and is positive when M is above G.

The position of M is found by considering the restoring moment caused by the movement of the centre of buoyancy from B to B′. The line of action of the force R has moved by the distance BB′ and, for small angles, this is equal to $\mathrm{BM} \times \theta$. With the result $R = \rho g V$ obtained above, this gives:

$$\text{restoring moment} = R \times \mathrm{BB}' = \rho g V \times (\mathrm{BM} \times \theta) = \rho g V\theta \times \mathrm{BM}$$

This moment can also be obtained in terms of the waterline section, Fig. 13.5(a). Consider a thin strip of area $\mathrm{d}A$, distance y from the centre line. As a result of the tilting this moves downwards in an arc as shown in the section view, Fig. 13.5(c). The length of this arc is θy and therefore the volume displaced by the movement of $\mathrm{d}A$ is $\theta y\,\mathrm{d}A$. For this element of area therefore,

$$\text{mass of water displaced} = \rho\theta y\,\mathrm{d}A$$

The corresponding weight of water displaced, and therefore the resulting upthrust, is $\rho g\theta y\,\mathrm{d}A$. The moment of this force about O is $\rho g\theta y\,\mathrm{d}A \times y$ and, by integration, the total for the whole section is

$$\text{restoring moment} = \rho g\theta \int y^2\,\mathrm{d}A = \rho g\theta I$$

where I is the second moment of area of the waterline section about the axis OO. Equating the result to the previous one for the restoring moment,

$$\rho g\,V\,\theta \times \mathrm{BM} = \rho g\,\theta I \quad \text{or} \quad \mathrm{BM} = \frac{I}{V}$$

The second moment of area I is the same quantity that occurs in the bending equation (see Chapter 9) and its SI unit is m^4. For a rectangle or circle the value of I can be calculated using the formulae given in Fig. 9.21. For the streamlined shapes that occur in ships the waterline section may have to be divided into small elements and the value of I determined for each. One method is to divide the section into longitudinal strips like the one in Fig. 13.5(a) and then to use the computer program for section properties on www.longman.co.uk. It will also be necessary to allow for the variation in section below the waterline in the calculation of volume V.

EXAMPLE 13.4 The plan view of a barge, Fig. 13.6, is a rectangle 10m long by 2.4 m wide together with a semicircular nose of 1.2 m radius, giving an overall length of 11.2 m. It has a flat bottom and vertical sides and ends. The barge is loaded until its total mass including the contents is 15 tonnes. The load is distributed so that the vessel is level, and its centre of gravity is 0.85 m above the bottom. Find:
(a) its draught in fresh water,
(b) the metacentric height.

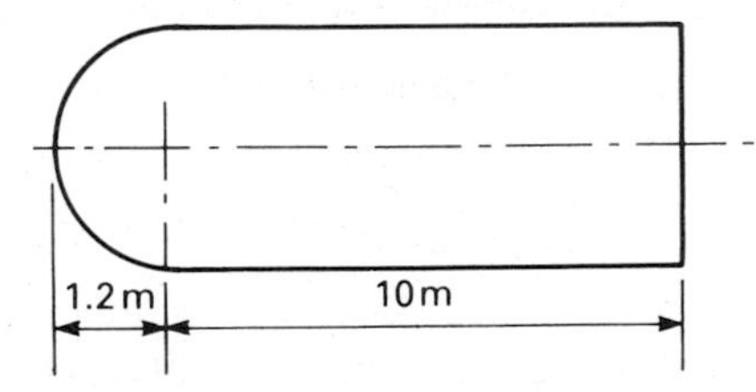

Figure 13.6

SOLUTION

(a) The total area in plan is:

$$A = b \times d + \pi r^2/2 = (10 \times 2.4) + (\pi \times 1.2^2/2) = 26.26 \text{ m}^2$$

If V is the volume of the barge below the waterline then, equating the weight of water displaced ($\rho g V$) to the weight of the barge, we have

$$1000gV = 15\,000g$$

and

$$V = 15\,000/1000 = 15 \text{ m}^3$$

The draught h, that is the depth to which the vessel is immersed, is:

$$h = V/A = 15/26.26 = 0.571 \text{ m}$$

(b) Since the barge has the same cross-section at every depth the centre of buoyancy B is at $h/2$ from the bottom. Also using the formulae given in Fig. 9.21, the second moment of area of the waterline section about the longitudinal axis is:

$$I = \frac{bd^3}{12} + \frac{1}{2} \times \frac{\pi}{64} d^4 = \frac{10 \times 2.4^3}{12} + \frac{\pi \times 2.4^4}{2 \times 64} = 12.33 \text{ m}^4$$

Using the lettering of Fig. 13.5(c), the height of the metacentre M above the centre of buoyancy B is:

$$\text{BM} = I/V = 13.33/15 = 0.822 \text{ m}$$

and the height of M above the bottom is:

$$\text{BM} + (h/2) = 0.822 + (0.571/2) = 1.108 \text{ m}$$

Since the centre of gravity G is 0.75 m above the bottom the metacentric height is:

$$\text{GM} = 1.108 - 0.75 = 0.358 \text{ m}$$

The equilibrium is stable because M is above G.

13.4 Force exerted by a jet

Throughout human history water and air in motion have been used as sources of energy. The turbines used in modern hydroelectric schemes have their origins in the waterwheels that powered corn mills in the middle ages and the early factories of the industrial revolution; and the latest aerogenerators depend on the same principles as traditional windmills.

The force exerted by a jet or stream of fluid is given by Newton's second law. In dealing with solids (Chapter 11) it is often convenient to use the form 'force = mass × acceleration' but for fluids it is useful to express it in a slightly different way. If a mass m has a velocity v then its momentum is mv and, from the second law, the force is given by:

$$\text{force} = \text{rate of change of } mv$$
$$= \frac{\text{change of } mv}{\text{time}}$$

If the mass flow rate is constant then mass/time is constant and the result can be written:

$$\text{force} = (\text{mass/second}) \times (\text{change in velocity})$$

Suppose a jet of water with a velocity v and cross-sectional area A strikes a flat plate at right angles and that the direction of flow after impact is parallel to the plate as shown in Fig. 13.7. If the plate is fixed, as in Fig. 13.7(a), the velocity after impact (in the original direction) is zero. The change in velocity is therefore v. Also, the volume of water striking the plate per second is Av and its mass is ρAv, where ρ is the density. With these results the expression for force becomes:

$$\text{force on the plate} = (\rho Av) \times v$$
$$= \rho Av^2$$

Next, suppose that the plate is not fixed, Fig. 13.7(b), but is moving in the same direction as the jet with a velocity u, u being less than v.

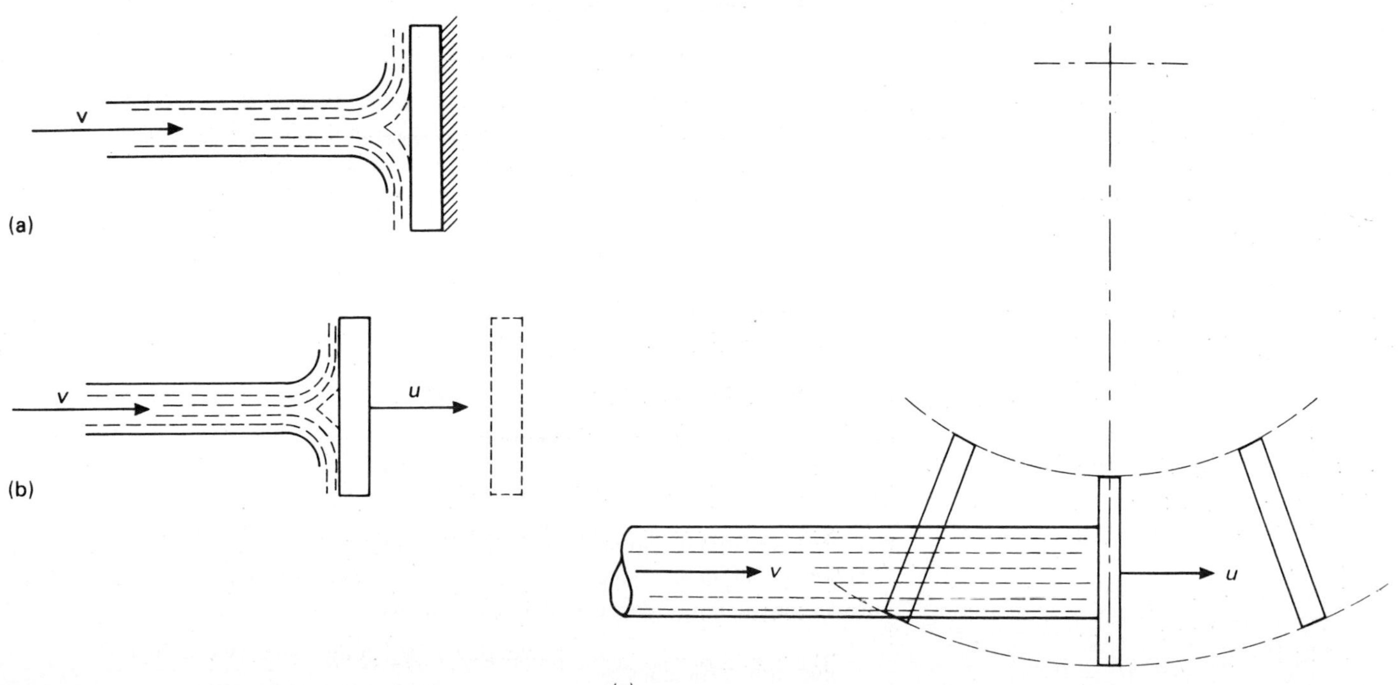

Figure 13.7

The mass of water striking the plate per second is now $\rho A(v - u)$. After impact, it moves with the plate at velocity u so that its change in velocity is $(v - u)$. Using these results,

$$\begin{aligned}\text{force on the plate} &= (\text{mass per second}) \times (\text{change in velocity})\\ &= \rho A(v - u) \times (v - u)\\ &= \rho A(v - u)^2\end{aligned}$$

This result assumes that the jet velocity v remains constant, however far the plate moves. Of more practical interest is the waterwheel shown in Fig. 13.7(c), in which a set of plates (or vanes) are attached to a wheel. If the diameter of the wheel is large it can be assumed that the jet strikes each plate at right angles. The mass striking the plates per second is ρAv and the change in velocity is $(v - u)$. Hence:

$$\text{force on the plate} = \rho Av(v - u)$$

A wheel in which the jet strikes the vanes at the bottom, as in Fig. 13.7(c) is described as *undershot*. An *overshot* waterwheel is one in which the jet strikes the vanes at the top.

EXAMPLE 13.5 An undershot waterwheel has flat radial vanes at an effective diameter of 4 m. The stream of water issues from a nozzle of 150 mm diameter with a velocity of 30 m/s. If the wheel is rotating at 80 rev/min, find:
(a) the force on the vanes,
(b) the power developed.

SOLUTION

(a) The cross-sectional area of the jet is:

$$A = \frac{\pi}{4} \times \left(\frac{150}{1000}\right)^2 = 0.0177\ \text{m}^2$$

The mass of water striking the vanes per second is:

$$\rho Av = 1000 \times 0.177 \times 30 = 530\ \text{kg}$$

The tangential velocity of the vanes is:

$$u = \omega r = \frac{2\pi \times 80}{60} \times 2 = 16.76\ \text{m/s}$$

$$\begin{aligned}\text{Hence, force on the vanes} &= (\text{mass per second}) \times (\text{change in velocity})\\ &= 530 \times (30 - 16.76)\\ &= 7019\ \text{N} \quad \text{or} \quad 7.019\ \text{kN}\end{aligned}$$

(b) The power developed is:

$$\begin{aligned}P &= \text{work done per second}\\ &= \text{force} \times \text{velocity of vanes}\\ &= 7019 \times 16.76\\ &= 117.6 \times 103\ \text{W} \quad \text{or} \quad 117.6\ \text{kW}\end{aligned}$$

13.4.1 Hydraulic efficiency

The efficiency of a waterwheel can be calculated by taking the work done on the vanes as the output and the kinetic energy of the jet as the input. It is convenient to consider the quantities per second and, on this basis,

$$\text{hydraulic efficiency } \eta = \frac{\text{power developed at the vanes (work done/s)}}{\text{kinetic energy of the jet/s}}$$

With the results and notation of the previous section the expressions become, in the case of a waterwheel with flat vanes,

$$\text{power} = \text{force} \times \text{velocity} = \rho Av(v-u) \times u = \rho Auv(v-u)$$

The mass of water per second (m) is ρAv and therefore:

$$\text{kinetic energy of jet per second} = \tfrac{1}{2}mv^2 = \tfrac{1}{2}\rho Av \times v^2 = \tfrac{1}{2}\rho Av^3$$

Dividing the output by the input,

$$\text{hydralic efficiency } \eta = \frac{\rho Auv(v-u)}{\frac{1}{2}\rho Av^3} = \frac{2u(v-u)}{v^2}$$

It is interesting to use this expression to see how the efficiency changes when the velocity of the wheel varies. To simplify the arithmetic, take a jet velocity (v) of 10 m/s and a range of vane velocities (u) from 0 m/s to 10 m/s. The following table shows the results:

u (m/s)	0	1	2	3	4	5	6	7	8	9	10
η	0	0.18	0.32	0.42	0.48	0.50	0.48	0.42	0.32	0.18	0

The table shows that the maximum hydraulic efficiency in the flat vane case is 0.5 and it is achieved when the vane velocity is half the jet velocity. This general result can also be derived by calculus.

The overall efficiency is less than the hydraulic efficiency because of mechanical friction losses in the same way that the overall (or brake) thermal efficiency of a heat engine is less than the indicated thermal efficiency (see Chapter 12).

EXAMPLE 13.6 Find the hydraulic efficiency for the waterwheel of Example 13.5.

SOLUTION The mass of water striking the vanes is 530 kg/s and its initial velocity is 30 m/s. Hence:

$$\text{kinetic energy of jet} = \tfrac{1}{2}mv^2 = \tfrac{1}{2} \times 530 \times 30^2$$

$$= 238.5 \times 10^3 \text{ J/s} \quad \text{or} \quad 238.5 \text{ kW}$$

The power developed is 117.6 kW and therefore:

$$\text{hydraulic efficiency} = \frac{\text{power developed at the vanes}}{\text{energy of jet per second}}$$

$$= \frac{117.6}{238.5} = 0.493$$

13.4.2 Water turbines

The efficiency of a waterwheel can be increased by using curved vanes. If they are shaped as shown in Fig. 13.8, the jet is turned through a greater angle than before and the change in velocity in the direction of motion is increased. This results in a greater change of momentum and hence a greater force on the vanes. In practice these vanes, or buckets as they are sometimes called, have a shape that is symmetrical about the plane of rotation (as in the diagram) so that there is no longitudinal force on the shaft or bearings. A turbine of this kind is called a *Pelton wheel*. It is described as an *impulse* turbine. The water emerges from the jet at atmosphere pressure and its energy is entirely in the form of kinetic energy.

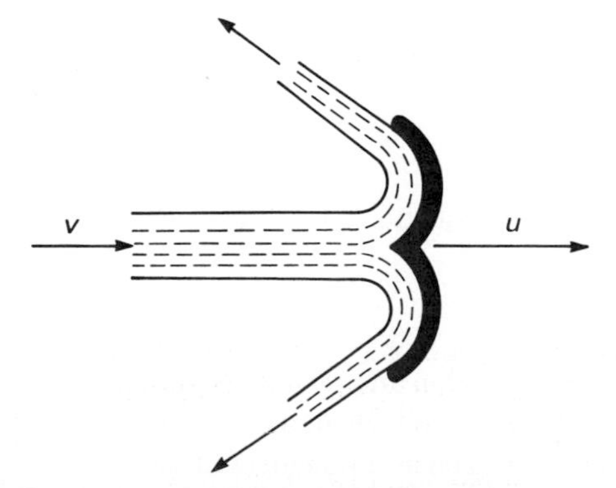

Figure 13.8 A Pelton wheel or impulse turbine.

An alternative is the *pressure* or *reaction* turbine in which the water enters the machine under pressure and only part of its energy is in the form of kinetic energy. One example is the *Francis turbine* in which the water passes through a ring of fixed vanes to strike a rotating ring of blades known as the impeller. Unlike the Pelton wheel, this turbine is full of water and all the blades are in contact with the water continuously.

The rotating ring of blades, or *runner*, can be placed downstream from the guide vanes so that the water turns through a right angle and approaches the runner axially. The blades on the runner are twisted as in an aircraft propeller to allow for the changes in velocity at different radii. To achieve high efficiencies over a range of loads variable pitch blades are used and axial flow machines of this kind are known as *Kaplan turbines*.

All three types have been developed to a high level of efficiency (greater than 0.9). The Pelton wheel is capable of using large pressure heads at moderate turbine speeds. On the other hand, it cannot use more than two jets on a single wheel efficiently and this restricts the volume of water that can be handled. Reaction turbines can deal with large volumes of water because it enters the machine all round the circumference. They are not suited to low powers at high pressure heads since the volume is then small.

13.5 Bernoulli's equation

We can apply the principle of conservation of energy to fluids as well as solids, provided we allow for the appropriate forms of energy. Two forms – potential and kinetic – were introduced in Chapter 11 and two more – internal energy and flow work – in Chapter 12. All four were included in the steady flow equation given in section 12.7 in the form:

$$mgz_1 + \tfrac{1}{2}mc_1{}^2 + U_1 + p_1V_1 + Q = mgz_2 + \tfrac{1}{2}mc_2{}^2 + U_2 + p_2V_2 + W$$

The subscripts 1 and 2 were used for the conditions at entry and exit to the system. In the flow of a fluid they can represent two points in a stream of fluid or two cross-sections of a pipe carrying a liquid. In thermodynamics, the change in potential energy is usually very small and the first term on each side of the equation may be omitted. However, for the flow of a liquid through an inclined pipe the potential energy terms must be included.

In the case of a liquid the temperature change is usually negligible and the internal energy terms may be left out. Assuming no heat transfer ($Q = 0$) and no work done ($W = 0$) the equation simplifies to:

$$mgz_1 + \tfrac{1}{2}mc_1{}^2 + p_1V_1 = mgz_2 + \tfrac{1}{2}mc_2{}^2 + p_2V_2$$

Liquids can be regarded as incompressible for most purposes and the volume of a given mass will remain the same. Hence V_1 and V_2 are equal and both can be replaced by one volume V. If we divide through the equation by V the first two terms on each side will contain m/V and this is the density ρ. It, too, is constant for an incompressible fluid and the equation can be written:

$$\rho gz_1 + \tfrac{1}{2}\rho c_1{}^2 + p_1 = \rho gz_2 + \tfrac{1}{2}\rho c_2{}^2 + p_2$$

Now that the symbol V has disappeared from the equation we can use the more familiar v for velocity and the result can be written:

$$\rho gz + \tfrac{1}{2}\rho v^2 + p = \text{constant}$$

This result is generally known as Bernoulli's equation after Daniel Bernoulli (sometimes spelt Bernouilli) who lived from 1700 to 1782. He derived it by considering force and momentum but it can be regarded as a statement of the principle of conservation of energy. The three terms represent potential energy, kinetic energy and flow work respectively.

It does not allow for compressibility of the fluid and cannot be used for air at speeds approaching that of sound. Furthermore, it does not take into account any friction losses in pipes or the shock losses that occur with a sudden enlargement or contraction of a pipe.

Each term in the expression has the units of pressure, newton per square metre or pascal. For some calculations it is convenient to have the result in a slightly different form. The equation assumes incompressibility so ρ is a constant and, dividing through by ρg, it becomes:

$$z + \frac{v^2}{2g} + \frac{p}{\rho g} = \text{constant}$$

Each of the three terms now has the units of length (metre). In this form they are referred to as heads and are known as the *potential head*, *velocity head* and *pressure head*. The constant is called the *total head* and is denoted by H. Bernoulli's equation can therefore be stated in words as:

potential head + velocity head + pressure head = total head (constant)

13.5.1 Equation of continuity

If a liquid is flowing along a tapering pipe, as shown in Fig. 13.9(a), its velocity will change from one section to another. Under steady conditions, the mass flowing across each section per second will be the same. Assuming that the pipe is running full and the liquid is incompressible, the volume flow rate Q (in cubic metre per second) will also be constant.

With a velocity v and cross-sectional area A the flow rate is vA and, using the symbols given in the diagram,

$$Q = v_1A_1 = v_2A_2$$

This result is called the equation of continuity for incompressible flow. In numerical examples it can be used to find the velocity at one section when it is known for another. The two velocities can then be substituted in Bernoulli's equation to give the pressure difference between the two sections.

The principle of continuity can be applied to branching pipes, Fig. 13.9(b). If the flow divides into two pipes, as shown, the flow rate in the main pipe will equal the sum of the flow rates in the two branches. In symbols,

$$v_1A_1 = v_2A_2 + v_3A_3$$

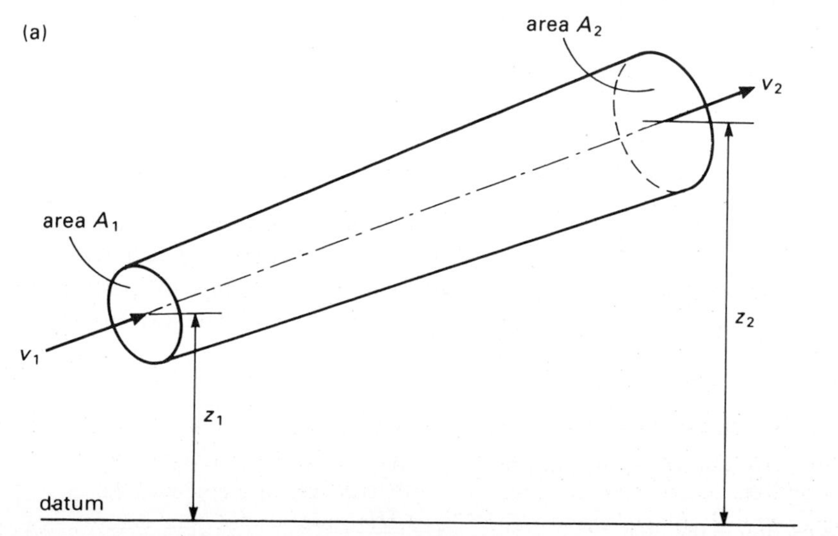

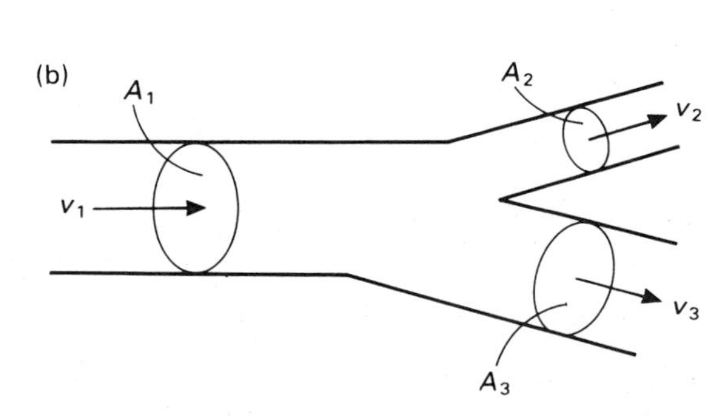

Figure 13.9 Flow along (a) a tapering pipe (b) a branching pipe.

EXAMPLE 13.7 The cross-section of a circular pipe, which is running full of water, expands gradually from a diameter of 0.2 m at one section to 0.3 m at another. The second is 6 m above the first. If the velocity at the first section is 5 m/s, find:
(a) the volume flow rate,
(b) the velocity at the second section,
(c) the pressure difference between the sections. Give the answer as a head of water and also in newton per square metre.

SOLUTION It is convenient to use subscripts 1 and 2 for the two sections as shown in Fig. 13.9(a).
(a) The volume flow rate Q is given by:

$$Q' = v_1A_1 = 5 \times \frac{\pi}{4} \times 0.2^2 = 0.1571 \text{ m}^3\text{/s}$$

(b) From the equation of continuity, $v_1 A_1 = v_2A_2$, and the second velocity is:

$$v_2 = v_1 \times \frac{A_1}{A_2} = 5 \times \frac{(\pi/4) \times 0.2^2}{(\pi/4) \times 0.3^2} = 2.222 \text{ m/s}$$

(c) To obtain the answer as a pressure head Bernoulli's equation is used in the form:

$$z_1 + \frac{v_1^2}{2g} + \frac{p_1}{\rho g} = z_1 + \frac{v_2^2}{2g} + \frac{p_2}{\rho g}$$

Rearranging the equation and noting that the difference in level, $z_2 - z_1$, is 6m, the pressure difference, expressed as a head, is given by:

$$\frac{p_1}{\rho g} - \frac{p_2}{\rho g} = z_2 - z_1 + \frac{v_2^2}{2g} - \frac{v_1^2}{2g} = 6 + \frac{2.222^2 - 5^2}{2 \times 9.81} = 4.977 \text{ m}$$

The pressure difference is therefore 4.977 m of water, being greater at the first section. Converting this to newton per square metre gives:

$$p_1 - p_2 = \rho g \times 4.977 = 1000 \times 9.81 \times 4.977$$
$$= 48.8 \times 10^3 \text{ N/m}^2 \quad \text{or} \quad 48.8 \text{ kN/m}^2$$

13.5.2 Losses in pipes

Bernoulli's equation states that the total energy of the fluid in motion is the same at each section or point along its path. In practice there will be losses of energy due to shock and friction. (As explained in Chapter 11 the term 'loss' does not mean that energy has disappeared; it has merely been transferred, and is not available for useful purposes.)

The losses depend on the velocity v and it is convenient to express each one in terms of the velocity head $v^2/2g$. The head lost is therefore written as $k \times (v^2/2g)$ where k is a constant that depends on the type of loss. Suppose a pipe of constant cross-section connects two reservoirs as shown in Fig. 13.10, with the exit from the pipe being below the surface in the lower reservoir.

Then:

for a sharp exit to a reservoir $k = 1$

for a sharp entrance from a reservoir $k = 0.5$

for friction in the pipe $k = \dfrac{4fL}{d}$

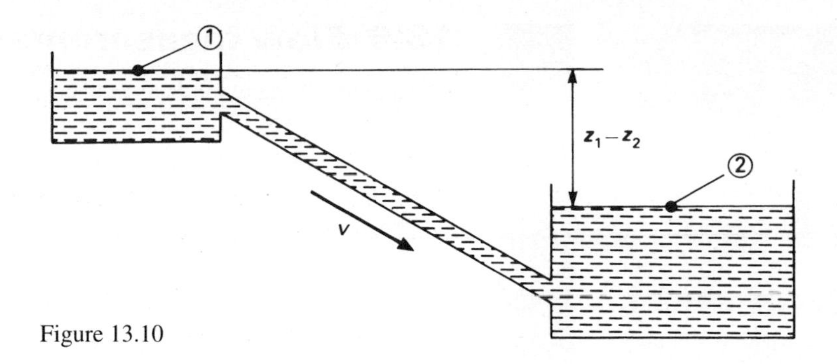

Figure 13.10

where L is the length of the pipe, d is the diameter and f is a resistance coefficient that depends on the material and roughness of the pipe, and the type of flow. This result for friction loss is known as the *Darcy formula*. The entry and exit losses can be greatly reduced by making the junctions of the pipe with the reservoir rounded or bell-mouthed. For a pipe discharging to the atmosphere the loss is negligible.

EXAMPLE 13.8 Two large tanks are connected by a pipe 150 mm diameter and 40 m long. Allowing for a sharp entrance and exit, and assuming a resistance coefficient $f = 0.008$, find the flow rate in the pipe when the difference in level is 3 m.

SOLUTION Let the subscripts 1 and 2 refer to points in the water surfaces of the tanks as shown in Fig. 13.10. Using the pressure head form of Bernoulli's equation and allowing for pipe losses,

$$z_1 + \frac{v_1^2}{2g} + \frac{p_1}{\rho g} = z_2 + \frac{v_2^2}{2g} + \frac{p_2}{\rho g} + \text{pipe losses}$$

For a large tank the velocity at the surface is negligible. Hence $v_1 = 0$ and $v_2 = 0$. Also since the surfaces are at atmospheric pressure we can put $p_1 = p_2$. With these values two terms on each side of the equation disappear leaving the simple result:

$$z_1 - z_2 = \text{pipe losses}$$

In terms of the velocity in the pipe (v), the formulae for shock and friction losses are:

$$\text{loss at entry to pipe} = 0.5 \times (v^2/2g)$$
$$\text{loss at exit from pipe} = 1 \times (v^2/2g)$$

$$\text{friction loss} = \frac{4fL}{d} \times (v^2/2g) = \frac{4 \times 0.008 \times 40}{0.15} \times (v^2/2g)$$
$$= 8.533 \times (v^2/2g)$$

Collecting results the equation becomes:

$$z_1 - z_2 = (0.5 + 1 + 8.533) \times (v^2/2g) = 10.03 \times (v^2/2g)$$

But $z_1 - z_2$ is given as 3 m and therefore:

$$v^2 = \frac{2g(z_1 - z_2)}{10.03} = \frac{2 \times 9.81 \times 3}{10.03} = 5.868$$

from which the velocity $v = \sqrt{(5.868)} = 2.422$ m/s

The flow rate is therefore

$$Q = vA = 2.422 \times \frac{\pi}{4} \times 0.150^2 = 0.0428 \text{ m}^3/\text{s}$$

Fig 13.11 Petrol pump

13.6 Flow measurement

Instruments of many kinds have been developed to measure the velocity and flow rate of fluids, Fig. 13.11. Some are used to measure the flow in pipes and channels, others to measure wind speed or the speed of aircraft in flight. Most are based on the principle of conservation of energy in the form of Bernoulli's equation. Almost always, the actual velocity or flow rate is slightly different from the theoretical value and the instruments are therefore calibrated by experiment.

13.6.1 Venturi meter

The flow in a pipe can be determined by measuring the change in pressure caused by a reduction in the cross-section. One device that works on this principle is shown in Fig. 13.12. It is inserted in the pipeline at a convenient position and consists of three sections – a converging cone, a short parallel throat and a longer expansion section.

Pressure tappings are made in the main pipe and at the throat, and tubes lead to the arms of a U-tube manometer or other type of pressure gauge. Note that the fluid fills the arms of the manometer above the liquid that is used to measure the pressure difference. The diverging section is usually made with a much smaller angle of taper than the converging cone. The instrument is known as a *Venturi meter*.

If it is used in a horizontal pipeline, as shown in the diagram, the potential energy is the same at the throat as in the main pipe. Only two forms of energy – kinetic and pressure – are involved in changes. At the throat section, the velocity and kinetic energy are increased and the pressure is therefore reduced. This means that the level of the liquid is higher in the arm of the manometer that is connected to the throat.

The actual flow rate (or discharge) is less than the theoretical value, which can be written:

$$\text{actual discharge} = C_d \times \text{theoretical discharge}$$

where C_d is a constant for the instrument known as the *coefficient of discharge*. With careful design its value can be as high as 0.98.

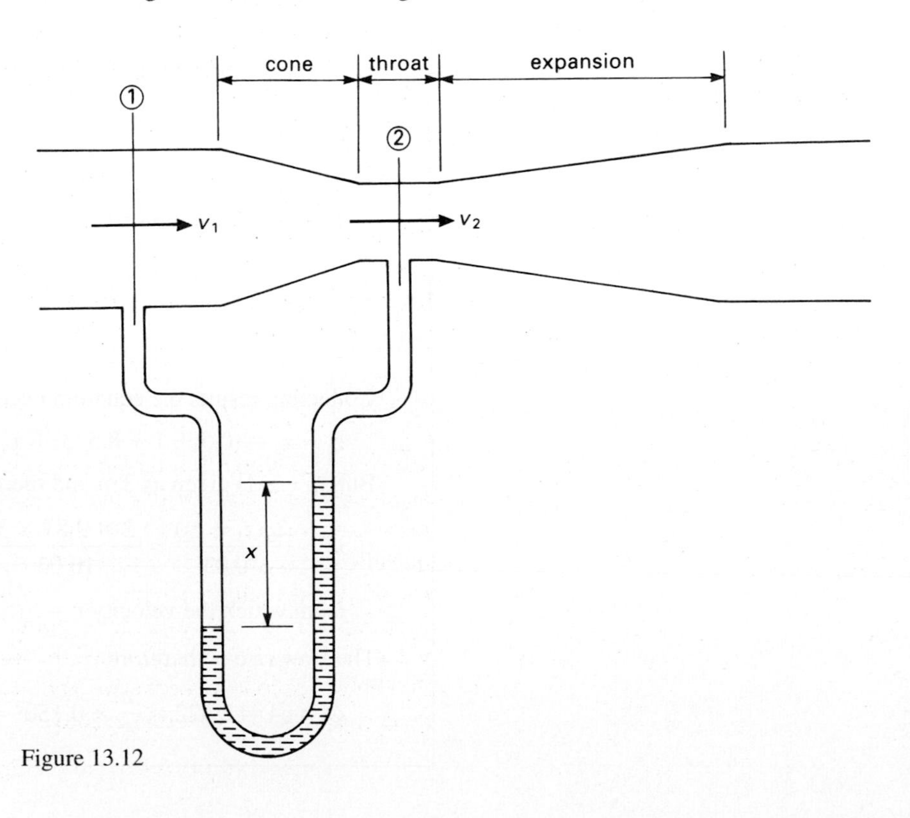

Figure 13.12

EXAMPLE 13.9 In a Venturi meter used for measuring the flow of oil through a horizontal pipeline, the pipe diameter is three times the diameter at the throat. Connections are made from the main pipe and the throat to a vertical U-tube containing mercury. Neglecting energy losses, calculate the velocity of the oil in the main pipe if the difference in levels of the mercury in the U-tube is 300 mm.

Densities (relative to water): oil 0.81; mercury 13.6.

SOLUTION Let the subscripts 1 and 2 refer to the main pipe and the throat as shown in Fig.13.12. Since the pipe is level, $z_1 = z_2$, and Bernoulli's equation becomes:

$$\frac{v_1^2}{2g} + \frac{p_1}{\rho g} = \frac{v_2^2}{2g} + \frac{p_2}{\rho g}$$

and this can be rearranged to give:

$$\frac{v_2^2 - v_1^2}{2g} = \frac{p_1 - p_2}{\rho g}$$

The right-hand side of this equation is the pressure difference in terms of a column of the liquid flowing through the pipe. In the U-tube, the difference in levels (x) is 300 mm of mercury. Expressing this pressure in terms of a column of liquid:

$$\frac{p_1 - p_2}{\rho g} = 300 \text{ mm of mercury}$$

$$= 300 \times 13.6 \text{ mm of water}$$

$$= \frac{300 \times 13.6}{0.81} \text{ mm of oil} = 5.037 \text{ m of oil}$$

Also, since the pipe diameter is three times the throat diameter, $A_1 = 9A_2$ and from the equation of continuity ($v_1A_1 = v_2A_2$), the velocity v_2 is given by:

$$v_2 = v_1 \times A_1/A_2 = 9v_1$$

Substituting these results on the two sides of the equation,

$$\frac{(9v_1)^2 - v_1^2}{2g} = 5.037$$

and

$$v_1^2 = \frac{5.037 \times 2 \times 9.81}{81 - 1} = 1.235$$

The required velocity is therefore $v_1 = \sqrt{(1.235)} = 1.111$ m/s

13.6.2 Orifice

We can use Bernoulli's equation to estimate the velocity of a jet emerging from a small circular hole or orifice in a tank, Fig. 13.13(a). Suppose the subscripts 1 and 2 refer to a point in the surface of the liquid in the tank, and a section of the jet just outside the orifice. If the orifice is small we can assume that the velocity of the jet is v at all points in this section.

The pressure is atmospheric at points 1 and 2 and therefore $p_1 = p_2$. In addition the velocity v_1 is negligible, provided the liquid in the tank has a large surface area. Let the difference in level between 1 and 2 be h as shown, so that $z_1 - z_2 = h$. With these values, Bernoulli's equation becomes:

$$h = v^2/2g \quad \text{from which} \quad v = \sqrt{(2gh)}$$

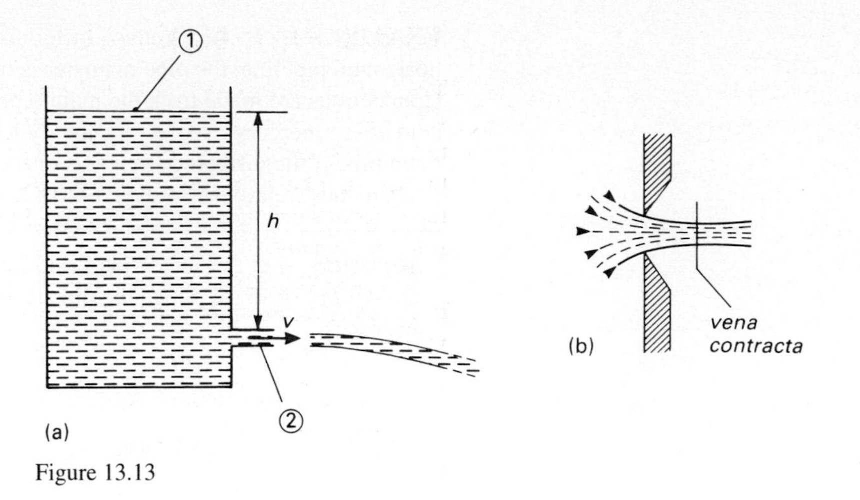

Figure 13.13

This result is known as Torricelli's theorem. If the area of the orifice is A the theoretical discharge is:

$$Q(\text{theoretical}) = vA = A\sqrt{(2gh)}$$

The actual discharge will be less than this. In practice the liquid in the tank converges on the orifice as shown in Fig. 13.13(b). The flow does not become parallel until it is a short distance away from the orifice. The section at which this occurs has the Latin name *vena contracta* (*vena* = vein) and the diameter of the jet there is less than that of the orifice. The actual discharge can be written:

$$Q(\text{actual}) = C_d A\sqrt{(2gh)}$$

where C_d is the coefficient of discharge. Its value depends on the profile of the orifice. For a sharp-edged orifice, as shown in Fig. 13.13(b), it is about 0.62.

EXAMPLE 13.10 A water tank discharges through a sharp-edged orifice, 30 mm diameter, near the base. In a calibration test, the water level was kept constant at a height of 1.5 m above the centreline of the orifice and 70 kg of water were collected in 30 s. Find the coefficient of discharge.

SOLUTION The theoretical discharge velocity is:

$$v = \sqrt{(2gh)} = \sqrt{(2 \times 9.81 \times 1.5)} = 5.425 \text{ m/s}$$

The corresponding discharge (or flow rate) is:

$$Q(\text{theoretical}) = vA = 5.425 \times \frac{\pi}{4} \times 0.03^2 = 0.003\,83 \text{ m}^3/\text{s}$$

Taking the density of water as 1000 kg/m³ the actual discharge is:

$$Q(\text{actual}) = \frac{70}{30 \times 1000} = 0.002\,33 \text{ m}^3/\text{s}$$

The coefficient of discharge is therefore:

$$C_d = \frac{Q(\text{actual})}{Q(\text{theoretical})} = \frac{0.002\,33}{0.003\,83} = 0.608$$

13.6.3 Pitot-static tube

The Venturi meter can be used for measuring air flow as well as the flow of liquids. It is confined to flow in pipes because it depends on having a throat section in which the speed of the fluid increases and the pressure decreases. One method of finding the speed of an unconfined flow such as the wind is to measure the pressure change that occurs when the stream is brought to rest. A point where this occurs is called a *stagnation point*.

Suppose the pressure and velocity of the undisturbed stream of air are p and v. Let p_0 be the pressure at the stagnation point (where the velocity is zero). With these values, and ignoring changes in level, Bernoulli's equation becomes:

$$\frac{p}{\rho g} + \frac{v^2}{2g} = \frac{p_0}{\rho g}$$

or

$$p + \tfrac{1}{2}\rho v^2 = p_0$$

On rearranging this result the velocity is given by:

$$v = \sqrt{\left[\frac{2(p_0 - p)}{\rho}\right]}$$

The pressure difference $(p_0 - p)$ can be measured by an instrument called a *Pitot-static tube*, Fig. 13.14. It consists of two parallel tubes and is placed facing the flow of air. One tube (the Pitot) is open at the end and the other (the static) is closed at the end but has holes in its side wall. In one version, developed at the National Physical Laboratory, the tubes are arranged concentrically, as shown in Fig. 13.14(a) and (b), but they can be separated as in Fig. 13.14(c).

In the original National Physical Laboratory version, the inner (Pitot) tube had inner and outer diameters of approximately 4 mm and 5 mm, the corresponding values for the other (static) tube being about 7 mm and 8 mm. It had three rows of seven holes, each 1 mm diameter, in the outer wall. There is a stagnation point at the entrance to the Pitot tube and the pressure in this tube is therefore equal to p_0. From the difference in pressure in the two tubes $(p_0 - p)$, the velocity can be determined.

As with other flow measuring instruments, Pitot-static tubes need to be calibrated.

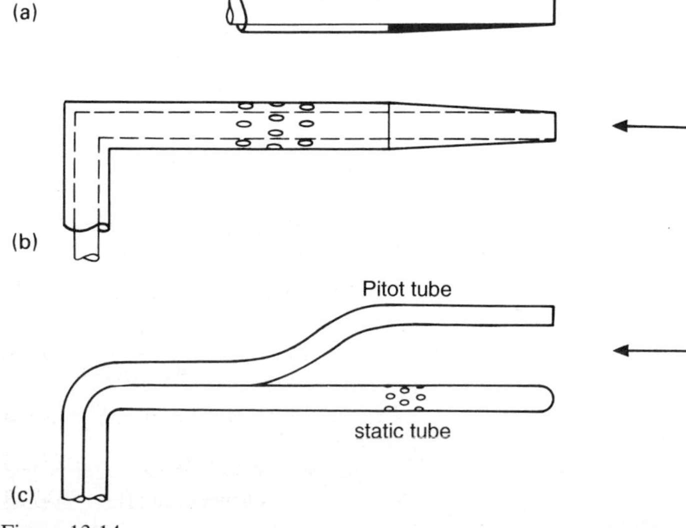

Figure 13.14

The Pitot-static tube has been used for many years to measure the speed of aircraft in flight. If the instrument is calibrated at ground level it does not give the true airspeed at altitude because of the change in density. This is not altogether a disadvantage because stalling depends on the indicated, rather than the true, airspeed.

Pitot-static tubes on aircraft suffer from errors of two kinds. There are position errors due to the airflow being affected by the rest of the aeroplane and the fact that the tube is inclined to the airstream except at one speed. There is also a compressibility error at high speeds because the formula for v applies only to incompressible flow.

An advantage of both the Venturi meter and the Pitot-static tube is that they require no power supply. A Pitot tube can be mounted on the centre line of a pipe with the static pressure being tapped at the wall and this requires fewer modifications to the pipe than a Venturi meter.

The accurate measurement of low airspeeds in a laboratory can be made by a *hot-wire anemometer*. In this a wire is placed in the airstream and heated electrically, the heat carried away being a measure of the velocity.

Wind speed is measured by anemometers that consist of flat vanes or hemispherical cups mounted at the ends of radial arms. The speed of rotation is a measure of the airspeed.

13.6.4 Pressure definitions

The following terms are in general use.

Static pressure This is defined as the pressure at a point on a body moving with the fluid. It is given by p in Bernoulli's equation.

Dynamic pressure This is the increase in pressure which arises when a stream of fluid is brought to rest. Using the symbols of the previous section it is $(p_0 - p)$ and, for incompressible flow, it is equal to $\frac{1}{2}\rho v^2$.

Total head is the sum of the static and dynamic pressures. It is equal to p_0 and is the pressure measured in the Pitot tube. For this reason it is sometimes called the Pitot head.

Reference pressure is specified as half the product of the density and the square of the velocity. It is therefore equal to the dynamic pressure when the flow is incompressible. However, when compressibility effects are significant, a different expression is required for dynamic pressure. Reference pressure is usually given the symbol q.

13.7 Nozzles

The speed at which water emerges from a hose is increased by discharging it through a nozzle. Take three planes perpendicular to the flow as shown in Fig. 13.15, the first in the main pipe or hose, the second at the nozzle exit and the third in the jet after leaving the nozzle. Since the area of the section is smaller at 2 than at 1, the velocity at 2 is greater. From Bernoulli's equation an increase in velocity results in a fall in pressure.

At section 3 the pressure is reduced to atmospheric and this further reduction leads to a still higher speed. As a result the cross-section of the jet is smaller than that of the nozzle itself.

The acceleration of the water is accompanied by a reaction force on the nozzle. By Newton's second law this force is equal to the rate of change of momentum and it can be calculated by multiplying the mass flow of the water by its increase in velocity.

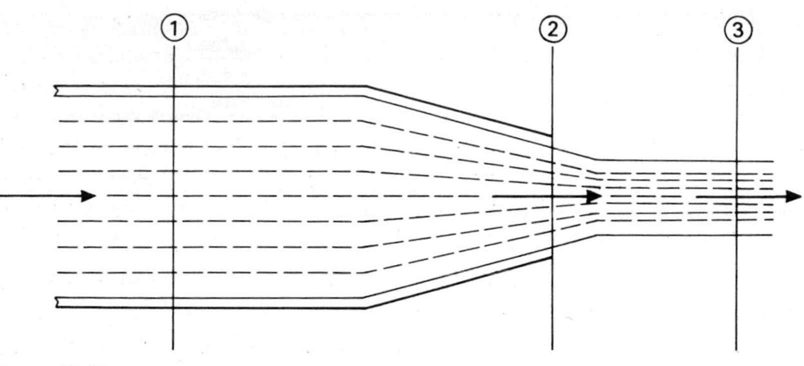

Figure 13.15

13.8 Viscosity

A fluid does not offer permanent resistance to change of shape. *Viscosity* is the property by which it *tends* to resist such a change. Treacle, for instance, is highly viscous but the viscosity of air is comparatively small. When a solid body is at rest in a stationary fluid, the force between them is perpendicular to the surface of the body at all points. If there is relative movement between them the body experiences a force tangential to its surface. This is called *skin friction*.

13.8.1 Laminar and turbulent flow

Suppose a motorway has several lanes of traffic each moving at its own steady speed and that each car and lorry stays in its own lane. The motion of a fluid that follows this pattern is called *laminar*; the individual particles of fluid move along parallel paths but the velocity changes from one path to another. If, on the other hand, the flow is unsteady so that particles change speed and move across the general direction of flow (like vehicles weaving from one lane to another on the motorway) the flow is called *turbulent*.

Laminar flow is also known as *viscous flow* because it occurs when the forces of viscosity are predominant. The factors that decide whether the flow is laminar or turbulent are explained in section 13.9.6.

13.8.2 Coefficient of viscosity

Consider two parallel plates distance y apart, as shown in Fig. 13.16, one being stationary and the other moving with velocity u parallel to it. Slipping does not occur between a surface and the fluid immediately in contact with it so the velocity varies from zero at the stationary plate to the value u at the moving one. If the flow is laminar, it can be proved that the velocity variation is a straight line as shown in the diagram.

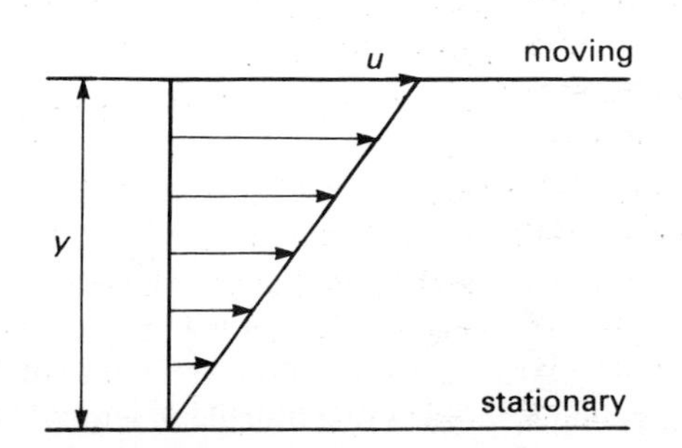

Figure 13.16 Velocity variation of laminar flow between parallel plates.

The tangential force between the fluid and the body is proportional to the area of contact and the ratio force/area is called the *viscous shear stress*. It corresponds to the shear stress that arises in solids and it is denoted by the same symbol τ. It has been found by experiment that τ is proportional to the velocity u and inversely proportion to the distance y between the plates.

Therefore:

$$\tau = \text{constant} \times \frac{u}{y}$$

The ratio u/y is known as the *velocity gradient*. The constant is called the *coefficient of viscosity* and is denoted by the symbol μ. It is also known as the *dynamic* (or *absolute*) *viscosity*.

The units of μ are those of stress (force/area) divided by velocity gradient (velocity/distance) and in the SI system this gives:

$$\frac{(\text{N/m}^2 \times \text{m})}{(\text{m/s})} = \text{N s/m}^2 \quad \text{or} \quad \text{kg/(m s)}$$

since the force unit newton can be replaced by kg m/s^2. Numerical values will be the same in both forms of the unit. The *poise*, *g*/(cm s), is still sometimes in use as a unit of dynamic viscosity (see Appendix 2).

The viscosity of a fluid is sensitive to changes in temperature. For liquids the viscosity decreases as the temperature rises; we describe this by saying that liquids become thinner when they are heated. The variation with temperature can lead to lubrication problems in high performance engines and the manufacturers of motor oils have made strenuous efforts to develop engine oils that are equally suitable for winter and summer conditions. In contrast, the viscosity of gases increases as the temperature rises.

An alternative means of specifying viscosity was adopted by the American Society of Automobile Engineers (SAE) for lubricating oils. The SAE system uses a simple numerical scale with a normal range of 10 to 60 and relates to the viscosity at a temperature of 99°C, 10 being the thinnest oil and 60 the thickest. Higher values have been added for the very heavy oils used in transmissions and gears. In addition, a range of 'W' numbers (usually 5W, 10W and 20W) has been introduced to show that such oils have an acceptable viscosity at –18°C and are therefore suitable for winter motoring. Multigrade oils therefore carry numbers such as 5W–30, 10W–40 and 20W–50.

EXAMPLE 13.11 An oil, for which $\mu = 0.05$ kg/(m s), completely fills the space between two parallel, flat plates, 6 mm apart. One plate is fixed and the other moves parallel to it at 3 m/s, the velocity distribution between them being linear. Assuming the flow to be laminar, find the viscous shear stress on each plate.

SOLUTION The viscous shear stress is:

$$\tau = \mu \times \frac{u}{y} = 0.05 \times \frac{3}{0.006} = 25 \text{ N/m}^2$$

13.8.3 Kinematic viscosity

There are a number of formulae in fluid mechanics that include the ratio (coefficient of viscosity μ)/(density ρ). This ratio is given the name *kinematic viscosity* and the Greek symbol ν. Thus:

$$\nu = \frac{\mu}{\rho}$$

In the SI system its units are those of μ (kg/(m s)) divided by those of density ρ (kg/m^3), giving m^2/s. The mass unit kilogram has cancelled and ν has the units of area divided by time. The absence of mass, and hence force, from the result is the reason for the term 'kinematic'. Dynamic implies force. The unit cm^2/s, called the *stoke*, is still sometimes used (see Appendix 2).

13.9 Dimensional analysis

If you examine the formulae for the areas of various geometric figures – rectangle, triangle, trapezium, circle – you will find that they all involve the product of two lengths, or some equivalent such as the square of a length. For a rectangle whose sides are b and d the area is $b \times d$; for a triangle, base b and height h, it is $bh/2$; and for a circle, radius r, it is πr^2. However, if two lengths are added, or one is subtracted from another, the result is still a length. For example, the area of a triangle in terms of its sides a, b and c is:

$$\sqrt{[s(s-a)(s-b)(s-c)]}$$

where s is the semiperimeter. The expression within the square root is therefore the product of four lengths and this is equivalent to the product of two lengths when the square root is taken.

This idea that one physical quantity (such as area) is related to others (such as length) in a precise mathematical way is useful in checking equations. It can also be used to determine the arrangement of variables in a formula and sometimes enables us to tackle a problem for which a complete theoretical solution does not exist. The method, which is known as *dimensional analysis*, can be applied to many branches of technology but it is especially valuable in the field of fluid flow.

13.9.1 Primary and secondary quantities

In using dimensional analysis, it is first necessary to choose a few physical quantities that are independent of one another. These are called *primary quantities*; other quantities that can be expressed in terms of these are called *secondary quantities*. The combination of primary quantities that is equivalent to a given quantity is called its *dimensions*. Area, for instance, is said to have the dimensions of (length)^2 because it is obtained by multiplying two lengths together.

In mechanics the primary quantities are normally taken as mass, length and time and their dimensions are written M, L and T. Square brackets [ ] are used to mean 'the dimensions of'. Velocity v, for example, has the dimensions of length divided by time, and acceleration a has the dimensions of velocity divided by time. These results are written:

$$[v] = \mathrm{L/T} \quad \text{or} \quad \mathrm{LT^{-1}}$$

and

$$[a] = \mathrm{(L/T)/T} \quad \text{or} \quad \mathrm{LT^{-2}}$$

The dimensions of other quantities can be found by seeing how they are related to those for which the dimensions are known. The dimensions of force are those of (mass × acceleration) and of stress and pressure are those of (force/area). With some quantities the dimensions cancel out. For example, strain is equal to one length divided by another and it has no dimensions. Such quantities are said to be *dimensionless*.

Table 13.1 shows the dimensions of the primary quantities and some of the secondary quantities used in mechanics. Note that angle is dimensionless since it is the ratio of arc length to radius.

Quantity	*Dimensions*
mass	M
length	L
time	T
angle	—
area	$\mathrm{L^2}$
volume	$\mathrm{L^3}$
first moment of area	$\mathrm{L^3}$
second moment of area	$\mathrm{L^4}$
velocity	$\mathrm{LT^{-1}}$
acceleration	$\mathrm{LT^{-2}}$
angular velocity	$\mathrm{T^{-1}}$
angular acceleration	$\mathrm{T^{-2}}$
frequency	$\mathrm{T^{-1}}$
force	$\mathrm{MLT^{-2}}$
stress and pressure	$\mathrm{ML^{-1}T^{-2}}$
moment of torque	$\mathrm{ML^2T^{-2}}$
modulus of elasticity	$\mathrm{ML^{-1}T^{-2}}$
momentum	$\mathrm{MLT^{-1}}$
energy and work	$\mathrm{ML^2T^{-2}}$
power	$\mathrm{ML^2T^{-3}}$
density	$\mathrm{ML^{-3}}$
coefficient of viscosity	$\mathrm{ML^{-1}T^{-1}}$
kinematic viscosity	$\mathrm{L^2T^{-1}}$

Table 13.1 Dimensions of physical quantities

EXAMPLE 13.12 The speed of sound in a fluid is given by the formula $\sqrt{(\gamma p/\rho)}$ where γ is the ratio of the principal specific heats, and p and ρ are the pressure and density. Check the formula dimensionally and use it to calculate the speed of sound in air at sea level, taking $\gamma = 1.4$, $p = 102\ \mathrm{kN/m^2}$ and $\rho = 1.23\ \mathrm{kg/m^3}$.

SOLUTION The ratio of the specific heats γ is dimensionless and hence

$$[\sqrt{(\gamma p/\rho)}] = \sqrt{\left(\frac{\mathrm{ML^{-1}T^{-2}}}{\mathrm{ML^{-3}}}\right)} = \sqrt{(\mathrm{L^2T^{-2}})} = \mathrm{LT^{-1}}$$

Since the dimensions of velocity are $\mathrm{LT^{-1}}$ the formula is dimensionally correct. With the numerical values given in the question,

$$\text{speed of sound} = \sqrt{(1.4 \times 102 \times 10^3/1.23)}$$
$$= (116.1 \times 10^3) = 340.7\ \mathrm{m/s}$$

13.9.2 Deriving formulae by dimensional analysis

In earlier chapters, formulae have been obtained using principles such as the laws of equilibrium, Newton's laws and the conservation of energy. Some of the results can also be obtained by the method of dimensions. More importantly, this method enables us to analyse certain problems that cannot be solved by other means.

As a first example, consider the formula for the periodic time of a simple pendulum. This was derived in section 11.8 using the principles of mechanics. Imagine, for a moment, that we do not know the result and that we wish to find it by dimensional analysis.

The first step is to list the quantities (or *variables*) that might affect the periodic time t. The periodic time is denoted by t to avoid confusion with the dimension T. If the bob is regarded as a particle we do not have to consider its size, and, if we ignore the mass of the string, our list might include:

mass of bob,	m
length of the string,	l
gravitational acceleration,	g

In a complete analysis we could include the density and viscosity of the air and the amplitude of the pendulum's swing. However, if we restrict the analysis to these three variables, we say in mathematical language that the periodic time t is a function of m, l and g. In symbols this is written:

$$t = \phi\{m,l,g\}$$

where$\phi\{\ \}$ means a function of.

Suppose next that the formula contains the variables m, l and g raised to various powers. This can be written

$$t = k \times m^a l^b g^c$$

where a, b and c are powers at present unknown, and k is a numerical constant. The values of these powers must ensure that the dimensions are the same on the two sides of the equation. Since k is a numerical constant, it is dimensionless and, equating dimensions,

$$[t] = [m^a l^b g^c]$$

Substituting the primary dimensions for each quantity, this becomes

$$\mathrm{T} = \mathrm{M}^a \mathrm{L}^b (\mathrm{LT}^{-2})^c = \mathrm{M}^a \mathrm{L}^{b+c} \mathrm{T}^{-2c}$$

The powers of each dimension M, L and T must be the same on the two sides on the equation. On the left-hand side the powers of M and L are zero and the power of T is 1. Therefore:

for M,	$a = 0$
for L,	$b + c = 0$
for T,	$-2c = 1$

From these questions, $a = 0$, $c = -\frac{1}{2}$ and $b = -c = \frac{1}{2}$.

Since $a = 0$ the mass m does not appear in the formula, which can now be written:

$$t = k \times l^{1/2} g^{-1/2}$$

or

$$t = k\sqrt{\frac{1}{g}}$$

This result should be compared with the one obtained in section 11.8. Dimensional analysis has produced a result of the correct form but it cannot give the value of the constant k. It has eliminated the mass m but it cannot show whether an essential variable was omitted from the original choice. Its success depends, therefore, on selecting the variables correctly at the beginning.

13.9.3 Model testing

In fluid dynamics, there are many problems for which complete theoretical solutions do not exist. The designers of aircraft and ships have to depend on experimental results obtained from models tested in wind tunnels or ship tanks. Car models, too, are tested in wind tunnels to develop shapes that give good performance. The effect of wind on tall buildings and suspension bridges is also investigated using models.

Dimensional analysis helps the designer to predict the behaviour of the full-scale vehicle or structure from the results obtained on the model.

As an example of the method, suppose that we wanted to find the periodic time of a simple pendulum 9 m long and that we did not know the formula. It would be inconvenient to test a 9 m pendulum in a school or college laboratory but we could test a model, a 1 m pendulum, for instance. Can we predict the period of the 9 m pendulum from the value of the 1 m model?

Let the subscripts F and M stand for full-scale and model. Then, from the result obtained by dimensional analysis in the previous section:

$$t_F = k\sqrt{\frac{l_F}{g}}$$

and

$$t_M = k\sqrt{\frac{l_M}{g}}$$

In the present case, the gravitational acceleration g is the same for the model and the full-scale pendulum. Dividing the first result by the second, both g and k cancel out leaving the result:

$$\frac{t_F}{t_M} = \sqrt{\frac{l_F}{l_M}}$$

But $l_F = 9$ m and $l_M = 1$ m. With these values the ratio becomes:

$$\frac{t_F}{t_M} = \sqrt{\frac{9}{1}} = 3 \quad \text{and} \quad t_F = 3t_M$$

The periodic time of the long pendulum is therefore three times that of the model. If you time a 1 m pendulum (take 10–20 oscillations to ensure an accurate result) you will find that its periodic time is about 2 s. The period of the 9 m pendulum is therefore 6 s.

Although this example merely confirms the theoretical result given by the formula $2\pi\sqrt{(l/g)}$, the technique of combining dimensional analysis with model testing to predict full-scale performance has important applications in vehicle design.

13.9.4 Aerodynamic and hydrodynamic forces

Consider the problem of finding the drag force D (or resistance) on a vehicle due to its motion through the air or through water. (The analysis will be very similar for other forces such as the lift on an aircraft wing.) To use the results of model tests, the variables to be included in the dimensional analysis must first be selected. General experience suggests the following:

- l, a representative length as a measure of size and scale,
- v, the velocity,
- ρ, the density of the fluid
- μ, the dynamic viscosity,
- a, the velocity of sound, since this is a measure of compressibility,
- g, the gravitational acceleration, which affects the formation of waves by a ship.

To simplify the analysis suppose that the vehicle is a car or aircraft, and that viscosity and compressibility can be neglected. The variables μ, a and g can therefore be omitted and the drag D can be written:

$$D = \phi\{l,v,p\} = k \times l^a v^b \rho^c$$

where a, b and c are unknown indices and k is a dimensionless constant.

Equating dimensions,

$$[D] = [l^a v^b \rho^c]$$

or

$$\begin{aligned} \text{MLT}^{-2} &= \text{L}^a(\text{LT}^{-1})^b(\text{ML}^{-3})^c \\ &= \text{M}^c\text{L}^{a+b-3c}\text{T}^{-b} \end{aligned}$$

Equating the powers of the three primary dimensions,

$$\begin{aligned} &\text{for M,} & c &= 1 \\ &\text{for L,} & a + b - 3c &= 1 \\ &\text{for T,} & -b &= -2 \end{aligned}$$

From the the last equation $b = 2$ and, substituting in the second,

$$a = 1 - b + 3c = 2 + 3 = 2$$

With these values the formula for D becomes

$$D = k \times l^2 v^2 \rho$$

It is often convenient to express the result in terms of the reference pressure $\frac{1}{2}\rho v^2$. To compensate for the $\frac{1}{2}$ we replace k by a drag coefficient C_D which is equal to $2k$. (For lift, a lift coefficient C_L is used.) Furthermore, l^2 has the dimensions of area and it is convenient to use a representative area A instead of a length. When dealing with the drag on a car it is taken as the projected area perpendicular to the direction of motion. In the case of an aircraft it is the plan area of the wing.

With these modifications the formula becomes:

$$D = C_D \tfrac{1}{2} p v^2 A$$

EXAMPLE 13.13 Wind tunnel tests on a quarter-scale model of a new car give a drag of 5.80 N at a wind speed of 15 m/s. Estimate the full-scale drag on the car at 120 km/h.

If the projected area of the car is 1.81 m^2, calculate the drag coefficient, taking the density of air to be 1.23 kg/m^3.

SOLUTION Using F and M for the full-scale and model conditions, the drag formula can be writtten:

$$D_F = C_D \tfrac{1}{2} \rho v_F^2 A_F$$

and

$$D_M = C_D \tfrac{1}{2} \rho v_M^2 A_M$$

Dividing the first result by the second gives

$$\frac{D_F}{D_M} = \left(\frac{v_F}{v_M}\right)^2 \frac{A_F}{A_M}$$

In SI units the full-scale velocity is

$$v_F = 120 \times \frac{1000}{3600} = 33.33 \text{ m/s}$$

Also, if the model is quarter-scale, the full-scale area will be sixteen times that of the model. Hence $A_F / A_M = 16$ and:

$$\frac{D_F}{D_M} = \left(\frac{33.33}{15}\right)^2 \times \frac{16}{1} = 79.0$$

The full-scale drag is therefore:

$$D_F = 79.0 \times D_M = 79.0 \times 5.80 = 458.2 \text{ N}$$

Using the full-scale figures, the drag coefficient is:

$$C_D = \frac{D}{\frac{1}{2}\rho v^2 A} = \frac{458.2}{\frac{1}{2} \times 1.23 \times 33.33^2 \times 1.81} = 0.371$$

13.9.5 Reynolds number and dynamic similarity

If the analysis of the previous section is extended to allow for viscosity, the formula for drag D becomes:

$$\begin{aligned} D &= \phi\{l, v, \rho, \mu\} \\ &= k \times l^a v^b \rho^c \mu^d \end{aligned}$$

and the dimensional equation is:

$$[D] = [l^a v^b \rho^c \mu^d]$$

or

$$\begin{aligned} \text{MLT}^{-2} &= \text{L}^a(\text{LT}^{-1})^b(\text{ML}^{-3})^c(\text{ML}^{-1}\text{T}^{-1})^d \\ &= \text{M}^{c+d}\text{L}^{a+b-3c-d}\text{T}^{-b-d} \end{aligned}$$

Equating the powers of the primary dimensions,

$$\begin{aligned} &\text{for M,} & c + d &= 1 \\ &\text{for L,} & a + b - 3c - d &= 1 \\ &\text{for T,} & -b - d &= -2 \end{aligned}$$

With three equations and four indices (a, b, c and d) there is no unique solution. However, three of the indices (a, b and c) can be expressed in terms of the fourth (d). From the first and third equations,

$$c = 1 - d$$

and

$$b = 2 - d$$

THE LIBRARY
GUILDFORD COLLEGE
of Further and Higher Education

Substituting these results in the second equation,

$$a = 1 - b + 3c + d = 1 - (2 - d) + 3(1 - d) + d = 2 - d$$

The drag D can now be expressed as:

$$D = k \times l^{2-d} v^{2-d} \rho^{1-d} \mu^{d}$$

or, collecting factors that involve the index d,

$$D = k \times l^2 v^2 \rho \times (l^{-d} v^{-d} \rho^{-d} \mu^{d})$$

The expression in the brackets can be written $\left(\frac{\rho v l}{\mu}\right)^{-d}$. Although the analysis does not lead to a numerical value for d, it shows that the drag D involves some function of the expression $\rho vl/\mu$. The form of this function is not known, and the result can be written as:

$$D = k \times l^2 v^2 \rho \times \phi\left\{\frac{\rho v l}{\mu}\right\}$$

The expression $\rho vl/\mu$ occurs frequently in the study of fluid flow. It is called *Reynolds number* and is given the symbol Re. It is named after Osborne Reynolds who first discovered its significance in relation to flow in pipes. Since μ/ρ equals ν, the kinematic viscosity, we can write:

$$Re = \frac{\rho v l}{\mu} = \frac{vl}{\nu}$$

and the expression for drag D can be written:

$$D = \rho v^2 l^2 \times \phi\{\mathrm{Re}\}$$

since the constant k can be considered as part of the function.

If the dimensions for each quantity are substituted in the expression for Re, they cancel out. Re is therefore a dimensionless number. As before we can introduce the factor $\frac{1}{2}$ to obtain the reference pressure and replace l^2 by an area A. Dividing through by the expression $\frac{1}{2}\rho v^2 A$ then gives C_D on the left-hand side. This leads to the result:

$$C_D = \phi\{Re\}$$

The importance of this result is that C_D is a function of a single quantity Re. For a body of given shape, variations in velocity, scale, fluid density and viscosity can all be accommodated in a single graph. Furthermore, since C_D and Re are dimensionless numbers the scale of the graph will be the same in all systems of units, provided they are self-consistent. If experimental results fall on a single curve, the original choice of variables (l, v, ρ, and μ) is correct. If not, some other quantity (such as compressibility) may be important.

In wind tunnel testing, it is essential that the model has the same shape as the full-scale vehicle. This is called *geometric similarity* and must apply even to the roughness of the surface if results are to be strictly comparable.

Dynamic similarity relates to the fluid motions and the pressures they produce. If the flow patterns are identical, the motions are dynamically similar. Dynamic similarity cannot exist unless geometric similarity is first achieved.

The condition for two flows to be dynamically similar is that the Reynolds number is the same for both (unless there are compressibility effects). Since $Re = \rho vl/\mu$, a decrease in size (represented by l) must be compensated by an increase in $\rho v/\mu$. Normally, the velocity of the model cannot be made much greater than that of the actual vehicle because of the increase in power required to drive the tunnel and the onset of compressibility effects. Nor is it possible to greatly reduce the viscosity μ. The remaining quantity is the density ρ. This can be increased either by using a compressed air wind tunnel or by changing to another fluid, water instead of air, for example.

13.9.6 Critical Reynolds number for pipe flow

If the flow of fluid through a pipe is laminar, each particle will move parallel to the centre line. Its velocity will depend on its radial position, being a maximum at the centre and zero at the wall. It can be proved that the variation is parabolic as shown in Fig. 13.17 and, if v is the mean velocity, the maximum is $2v$. Furthermore, the tangent to the velocity curve at the wall cuts the centre line at a value of $4v$. The velocity gradient at the wall is therefore $(4v)/(d/2) = 8v/d$.

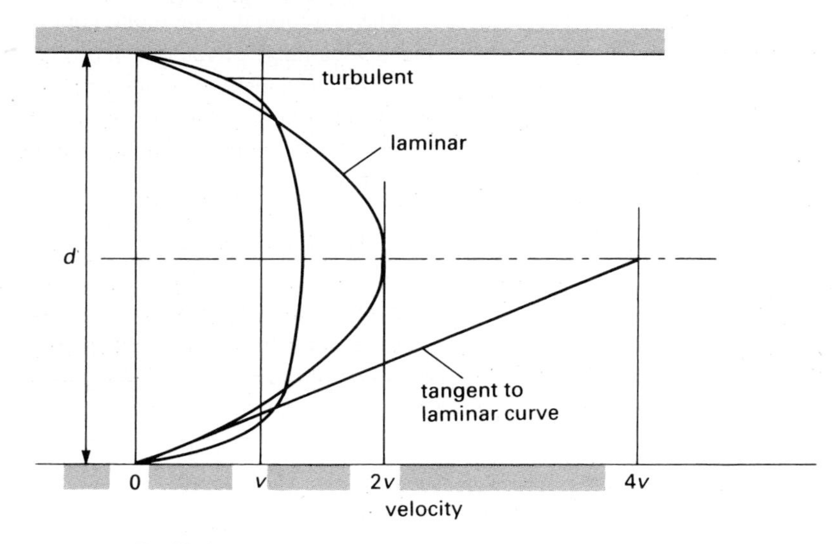

Figure 13.17 Variation of flow velocity in a pipe.

The viscous shearing stress at the wall is therefore:

$$\tau = \mu \times \text{velocity gradient} = \mu \times \frac{8v}{d}$$

and the corresponding force for a length L of the pipe is:

$$F = \tau \times \text{surface area} = \mu \times \frac{8v}{d} \times \pi d L = 8\pi\mu v L$$

This force is provided by the pressure difference over this length of pipe. Therefore:

$$\text{loss of pressure} = \frac{\text{force } (F)}{\text{cross-sectional area}} = \frac{8\pi\mu v L}{\pi d^2/4} = \frac{32\mu v L}{d^2}$$

It was shown in section 13.5.2 that the friction loss, in terms of the friction coefficient f, is

$$\text{friction loss} = \frac{4fL}{d} \times \frac{v^2}{2g} \quad \text{(expressed as a head)}$$

$$= \frac{4fL}{d} \times \frac{v^2}{2g} \times \rho g \quad \text{(in units of pressure)}$$

Equating the two results,

$$\frac{4fL}{d} \times \frac{v^2}{2} \times \rho = \frac{32\mu v L}{d^2}$$

and

$$f = \frac{16}{(\rho v d/\mu)} = \frac{16}{Re}$$

since the denominator is the Reynolds number for the flow, with the diameter d as the representative length.

If the flow is turbulent, the velocity distribution is very different, as shown in Fig. 13.17. Note that the variation relates to time-average values since the flow is now unsteady. For a smooth pipe the maximum is about 1.22 times the mean. No simple theory exists for the friction loss with turbulent flow but an analysis of experimental results for smooth pipes gives the following formula:

$$f = \frac{0.0791}{Re^{1/4}}$$

The two formulae are plotted in Fig. 13.18. A logarithmic scale is taken for *Re* so that a wide range of values can be accommodated. If Reynolds numbers below 500 are to be considered it is convenient to take a logarithmic scale for *f* also. This has the effect of making both graphs into straight lines. For rough pipes the turbulent graph will be similar in shape to the one given but with higher values of *f*.

For the flow to be laminar, the Reynolds number *Re* (based on mean velocity and pipe diameter) must normally be less than 2100. This is called the *critical Reynolds number* and for higher values the flow is turbulent. With very careful experimentation, laminar flow may be maintained up to $Re = 20\,000$ but the flow is unstable above $Re = 2100$ and becomes turbulent if disturbed.

In practice, there is a transition region near the critical Reynolds number in which the value of *f* increases with *Re*.

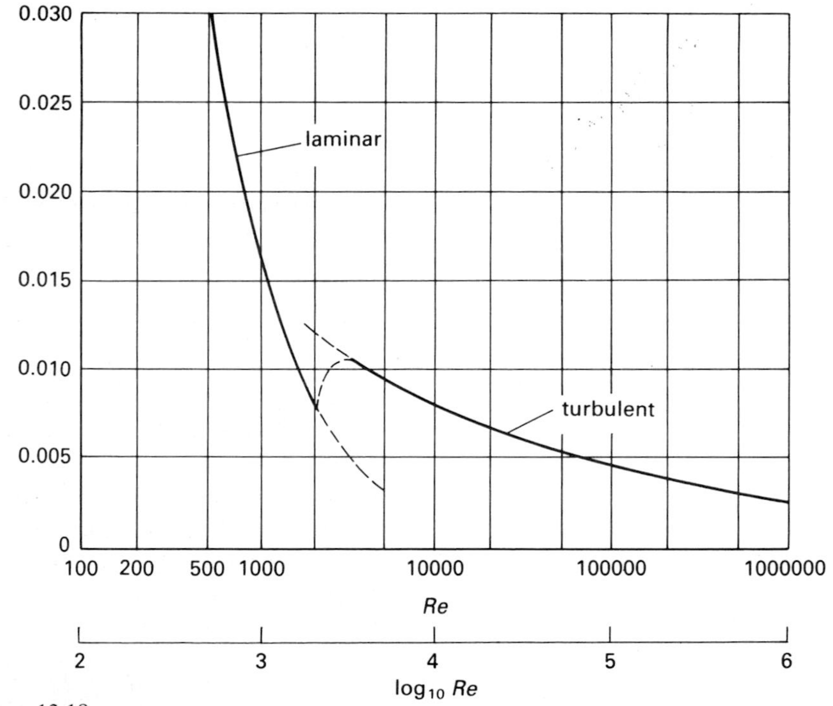

Figure 13.18

13.9.7 Mach and Froude numbers

At high speeds the forces acting on an aircraft, such as the drag *D*, are affected by compressibility. Pressure waves move through the air with the speed of sound and, to allow for compressibility, this velocity is included in the dimensional analysis. It is found that *D* is a function of v/a where v is the velocity of the aircraft and a is the velocity of sound. The ratio v/a is known as the Mach number *Ma*. Allowing for both viscosity and compressibility, the drag formula can be written:

$$D = \rho v^2 l^2 \phi\{Re, Ma\}$$

For complete dynamic similarity, the Reynolds and Mach numbers of the model should both have the same values as the full-scale machine. It is not generally possible to arrange this. At supersonic and high subsonic speeds, where compressibility effects are predominant, the correct Mach number is preferred.

In contrast, a large part of the drag on a ship is due to the wave motion brought about by its progress through the water. Since the formation of waves depends on gravitational forces, the acceleration g due to gravity is included in the dimensional analysis together with the representative length l, the velocity v and the density ρ. The drag can therefore be written:

$$D = k \times l^a v^b \rho^c g^d$$

The dimensional equation is:

$$[D] = [l^a v^b \rho^c g^d]$$

from which

$$\mathrm{MLT}^{-2} = \mathrm{L}^a(\mathrm{LT}^{-1})^b(\mathrm{ML}^{-3})^c(\mathrm{LT}^{-2})^d = \mathrm{ML}^{a+b-3c+d}\mathrm{T}^{-b-2d}$$

Equating the primary dimensions in turn,

$$\begin{aligned} &\text{for M,} & c &= 1 \\ &\text{for L,} & a + b - 3c + d &= 1 \\ &\text{for T,} & -b - 2d &= -2 \end{aligned}$$

From these equations, $c = 1$, $b = 2 - 2d$ and $a = 2 + d$. The drag can be written:

$$D = k \times l^2 v^2 \rho \times (l^d v^{-2d} g^d) - k \times l^2 v^2 \rho \times \left(\frac{v^2}{gl}\right)^{-d}$$

The quantity v^2/gl is dimensionless. For dynamic similarity the value of v^2/gl must be the same for the model and the full-scale ship. Its square root $v/\sqrt{(gl)}$ is called the Froude number Fr (though some authors apply the name to v^2/gl itself). The formula for wave-making drag is therefore:

$$\mathrm{D} = \rho v^2 l^2 \phi\{Fr\}$$

Since g is for practical purposes a constant, it is the ratio v^2/l that must be kept constant to ensure dynamic similarity. This is convenient because it means the velocity for the model will be less than that of the full-scale ship.

EXAMPLE 13.14 The wave-making resistance of a racing power boat at a speed of 50 knots in sea water is to be estimated from experiments in a fresh-water tank on a model one-tenth full size. Determine the best speed for the test and calculate the full-scale wave-making resistance if the model drag at this speed is found to be 8.7 N. Take the density of sea water as 1024 kg/m³. 1 knot = 0.5148 m/s.

SOLUTION For dynamic similarity the ratio v^2/l should be the same for the model and full-scale boat. Using M and F for model and full-scale, $l_M = 0.1 l_F$. Also, in SI units, $v_F = 0.5148 \times 50 = 25.74$ m/s. Hence:

$$\left(\frac{v_F}{v_M}\right)^2 = \frac{l_F}{l_M} = 10$$

and

$$v_M = v_F/\sqrt{10} = 0.3162 v_F$$

The best speed for the test is therefore:

$$v_M = 0.3162 \times 25.74 = 8.14 \text{ m/s}$$

For the same Froude number, the ratio of full-scale to model drag is

$$\frac{D_F}{D_M} = \frac{\rho_F v_F^2 l_F^2}{\rho_M v_M^2 l_M^2} = \frac{1024 \times 10 \times 10^2}{1000 \times 1 \times 1} = 1024$$

The full-scale wave-making resistance is therefore

$$D_F = D_M \times 1024 = 8.7 \times 1024 = 8909 \text{ N} \quad \text{or} \quad 8.909 \text{ kN}$$

13.9.8 Model structures testing

Dimensional analysis is of particular importance in the design of aircraft, ships and cars where model testing is often the only means of predicting full-scale performance. Models can also be used to forecast the behaviour of structures under static and dynamic loads.

Suppose the deflection of a complex framed structure under static loading is to be estimated from tests on a geometrically similar model. This similarity must include the cross-section of the members and the nature of the joints; if the full-scale structure is to have pin joints then so must the model.

Let the deflection at the point of interest be d when a force F is applied at this or some other point. The resulting formula will apply to all points in the structure. From experience of structural analysis, the deflection can be expected to be a function of the force F, the modulus of elasticity of the material E and the size, given by a representative length l. It does not matter which length is chosen because the final formula will depend only on ratios. With the usual functional notation, this gives:

$$d = \phi\{l,E,F\} = k \times l^a E^b F^c$$

where k is a dimensionless constant and a, b and c are powers that are initially unknown. The dimensional equation is therefore:

$$[d] = [l^a E^b F^c]$$

The deflection d is a length so that its dimensions are simply L. From Table 13.1 the dimensions of E are $ML^{-1}T^{-2}$ and those of F are MLT^{-2}. Substituting these results the dimensional equation becomes:

$$L = L^a(ML^{-1}T^{-2})^b(MLT^{-2})^c$$

and, equating the powers of the primary dimensions,

$$\begin{aligned} &\text{for M,} \quad & b + c &= 0 \\ &\text{for L,} \quad & a - b + c &= 1 \\ &\text{for T,} \quad & -2b - 2c &= 0 \end{aligned}$$

Although there are three equations for a, b and c, they are not independent since the first and third both give the result $b = -c$. By itself, therefore, dimensional analysis cannot determine the form of the result. However, Hooke's law can be expected to apply to an elastic structure and this means that the deflection is proportional to the force. Hence c, the power of F in the original equation, is 1 and it follows that $b = -c = -1$. With these values the remaining power, a, can be found from the dimensional equation for L.
Thus:

$$a = 1 + b - c = 1 - 1 + 1 = 1$$

Using these values of a, b and c the deflection d can be expressed as

$$d = k \times l^{-1}E^{-1}F = k \times \frac{F}{lE}$$

To predict the full-scale deflection from the value measured on the model let r and m refer to the real and model structures respectively. Then:

$$d_r = k\frac{F_r}{l_r E_r} \quad \text{and} \quad d_m = k\frac{F_m}{l_m E_m}$$

Dividing the first equation by the second,

$$\frac{d_r}{d_m} = \frac{F_r l_m E_m}{F_m l_r E_r}$$

This result was used in the testing of the trailer chassis described in section 2.4.4. Previously L was used for length but has been replaced here by l as, in dimensional analysis, L represents the dimension of length as shown in Table 12.1.

If a one-tenth scale model is made, in aluminium alloy, of a steel structure the ratio E_m/E_r would be about 1/3. Suppose a force is applied to the model 1/100th of that expected on the real structure, then:

$$\frac{d_r}{d_m} = \frac{100 \times 1 \times 1}{1 \times 10 \times 3} = 3.333$$

and the full-scale deflection would be 3.333 times that of the model.

The dynamic behaviour of structures is also investigated with models. In 1940 wind loads on a newly opened suspension bridge at Tacoma, Washington, USA, caused torsional oscillations that led to its collapse – an example of forced vibrations producing large amplitudes as described in section 10.8. Since that time, the designs for new suspension bridges have been subjected to rigorous wind tunnel testing.

13.10 Aerofoils

The aerodynamic resistance to a body moving through the air is called the drag D. The force perpendicular to the motion is called the lift L. An aerofoil is a body so shaped that the lift may be considerably greater than the drag. The wings, tailplane and fin of an aircraft are all examples of aerofoils.

13.10.1 Aerofoil definitions

A tapered aerofoil is shown in plan in Fig. 13.19(a). The plan area (usually denoted by S) is used as the representative area for force coefficients and the span is denoted by b.

Fig. 13.19(b) shows a section (to a larger scale). The line joining the centres of the arcs which form the leading and trailing edges is the *chord line*. The *chord* (or *chord length*) c is the length of this line between its intersections with the section profile.

The mean chord is the total area S divided by the span b. *Aspect ratio* is the span b divided by the mean chord (S/b). It is given the symbol A and therefore:

$$A = b/(S/b) = b^2/S$$

Taper ratio is tip chord divided by root chord (or, by some authors, the reciprocal of this fraction).

Wing loading is the total weight of the aircraft divided by the wing area. Its units are therefore N/m^2.

In wind tunnel testing, the model is at rest and v is taken as the velocity of the undisturbed airflow – the velocity away from the model. For an aircraft in flight, v is its velocity relative to the air.

The angle made by the chord line with the direction of the relative airflow is called the *angle of incidence* α (or sometimes the angle of attack).

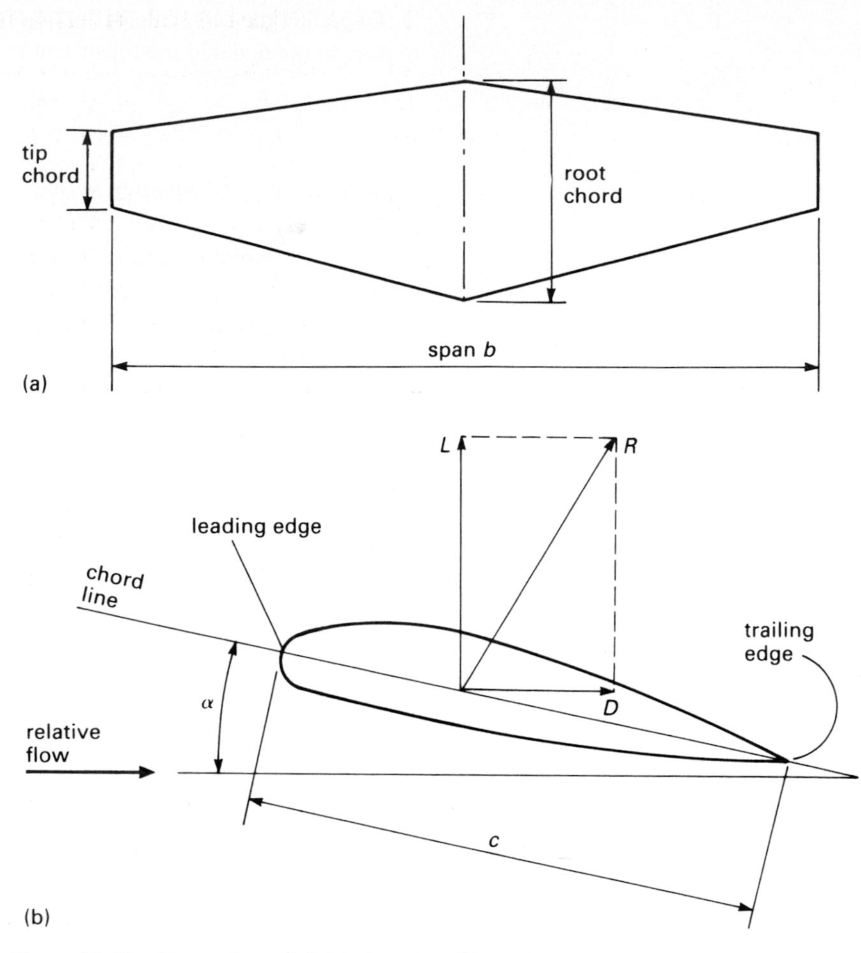

Figure 13.19 Tapered aerofoil (a) plan view (b) section.

EXAMPLE 13.15 When a rectangular aerofoil 2 m span and 0.3 m chord was tested in a wind tunnel at 30 m/s the resultant force was 320 N at an angle of 6° to the lift axis. Determine the lift and drag coefficients, taking $\rho = 1.23 \text{ kg/m}^3$.

SOLUTION The lift, drag and resultant forces are shown in Fig. 13.19(b). Resolving the resultant force perpendicular and parallel to the relative airflow gives:

$$L = 320 \cos 6° = 320 \times 0.9945 = 318.3 \text{ N}$$

and

$$D = 320 \sin 6° = 320 \times 0.1045 = 33.45 \text{ N}$$

The reference pressure $q = \frac{1}{2}\rho v^2 = \frac{1}{2} \times 1.23 \times 30^2 = 553.5 \text{ N/m}^2$

The area $S = 2 \times 0.3 = 0.6\text{m}^2$ and the lift is given by $L = C_L qS$, where C_L is the lift coefficient. Therefore:

$$C_L = \frac{L}{qS} = \frac{318.3}{553.5 \times 0.6} = 0.958$$

Similarly the drag coefficient is

$$C_D = \frac{D}{qS} = \frac{33.45}{553.5 \times 0.6} = 0.101$$

13.10.2 Lift and drag characteristics

Although there has been much progress in the theory of aerofoils, designers still depend to some extent on the results of wind tunnel experiments. Fig. 13.20 presents some test results and shows how the lift and drag coefficients C_L and C_D vary with the angle of incidence α. In this test the aerofoil was extended to the walls of the tunnel to prevent flow round the tips. This corresponds to an infinite span and infinite aspect ratio and is described as a two-dimensional aerofoil.

It is usual to plot C_L and C_D against the angle of incidence α. The section used in this test was asymmetrical and C_L is zero at a small negative angle of incidence called the *no-lift angle*. As α is increased, the lift coefficient increases almost linearly up to a maximum value which, in the present case, is about 1.74 at an angle of 17°. At this angle the aerofoil is said to *stall* and beyond it the value of C_L is reduced. The gradient of the graph of C_L against α is called the *lift slope*. The theoretical value for thin aerofoils of infinite aspect ratio is 2π; for the present test it is about 5.7.

The value of C_D is small and, for clarity, $10C_D$ is plotted. At low angles of incidence the variation is small with a minimum in the region of the no-lift angle. There is a gradual increase in C_D up to the stall beyond which it rises rapidly.

Since the lift L and drag D are based on the same area their ratio is the same as C_L/C_D. This ratio is plotted (to a different scale) on the same axes and the maximum value is approximately 90. With a finite aspect ratio this value would be reduced.

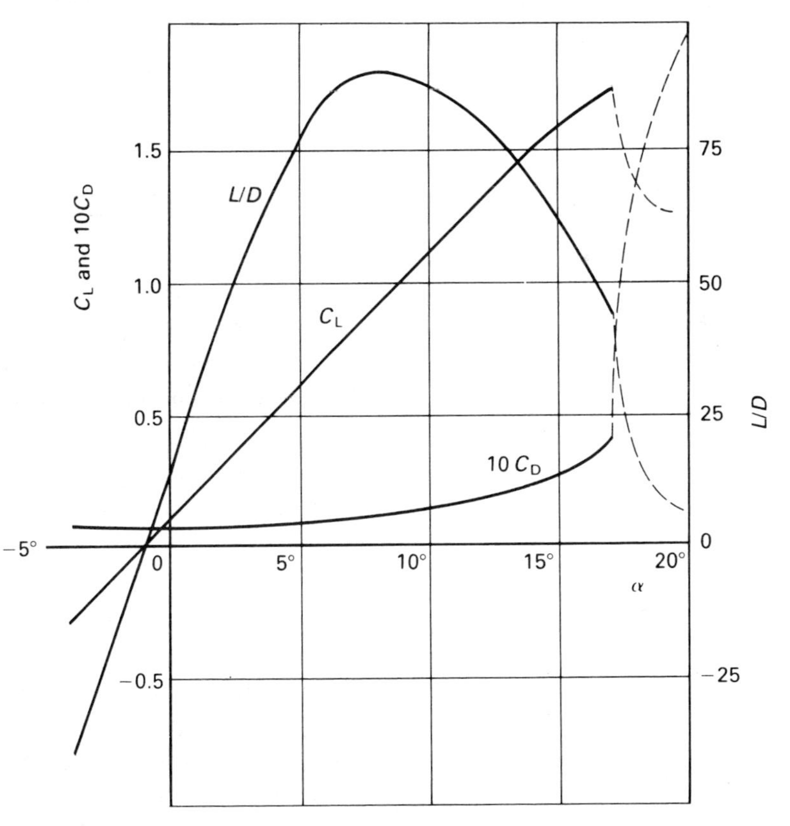

Figure 13.20

13.10.3 Effect of thickness and camber

Fig. 13.21 shows the profile of an asymmetric aerofoil section. The centre line is defined as the curve which is everywhere equidistant from the upper and lower surfaces in directions perpendicular to itself. Suppose the maximum thickness of the

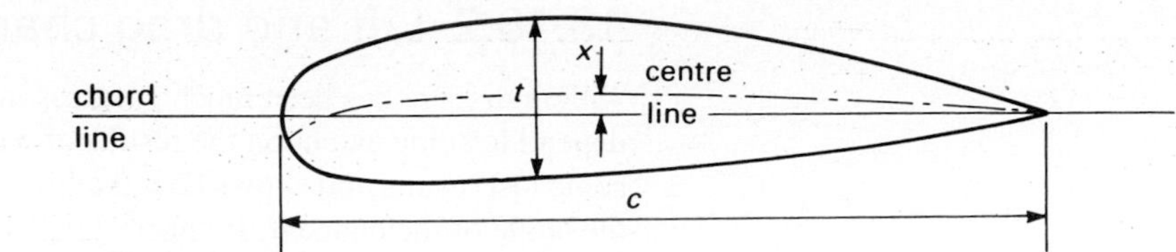

Figure 13.21 Profile of an asymmetrical aerofoil section.

section perpendicular to the chord line is t and the maximum height of the centre line above the chord line is x. With these dimensions, the following definitions are made:

$$\text{thickness ratio} = \frac{t}{c} \text{ and centre line camber} = \frac{x}{c}$$

Thickness and camber both affect the maximum lift coefficient. The values depend on the details of the profile but, in general, the maximum C_L increases with thickness ratio up to a certain value and falls away for thicker sections. It also increases with the camber of the section. The minimum drag coefficient increases steadily with thickness.

The early aerofoil sections developed in the government laboratories of the UK, USA and Germany had their maximum thickness at approximately 0.3 of the chord from the leading edge. Modern laminar flow aerofoils, which may have a minimum drag less than two-thirds as much, usually have their maximum thickness further aft.

The maximum C_L and minimum C_D are also affected by Reynolds number. For aerofoils, Re is based on chord length c. The maximum C_L increases steadily with Re in the range 10^5 to 10^7. In contrast, the minimum C_D decreases rapidly up to $Re = 10^6$ beyond which the variation is small. If Re is greater than 10^6 the results are generally taken to be applicable to flight conditions.

13.10.4 Three-dimensional aerofoils

When an aerofoil is placed in an airstream, a stagnation point occurs at or near the leading edge and the flow divides at this point. If the aerofoil is cambered as in Fig. 13.21, the air passing along the upper surface has further to travel to reach the trailing edge than that passing along the lower surface. Conditions are similar for a symmetrical aerofoil if it is inclined to the airstream. In both cases the higher velocities along the upper surface result in lower pressures in accordance with Bernoulli's equation.

This difference in velocity between the upper and lower surface is equivalent to a combination of a parallel air stream and a clockwise rotational velocity, Fig. 13.22(a). A rotating mass of fluid is called a *vortex* and the pattern of flow round an aerofoil section corresponds to that of a vortex superimposed on a parallel air stream.

Fig. 13.22(b) shows a front view of a finite aerofoil (to a smaller scale) and, as indicated, there is a tendency for the pressure on the upper and lower surfaces to equalise by flow round the tips. The rotation that is set up persists and proceeds downstream. As a rough approximation, the flow round a finite aerofoil can be pictured as a U-shaped or horseshoe vortex consisting of long wing-tip vortices joined by a cross vortex in place of the wing, Fig. 13.22(c).

The generation of the wing-tip vortices requires a continuous supply of energy and, in consequence, the drag coefficient of a finite wing is greater than that of a two-dimensional aerofoil of the same section. The additional drag is known as the *induced drag* and under ideal conditions the increase in the drag coefficient is $C_L^2/\pi A$, where A is the aspect ratio. In practice this quantity is increased by an induced drag factor.

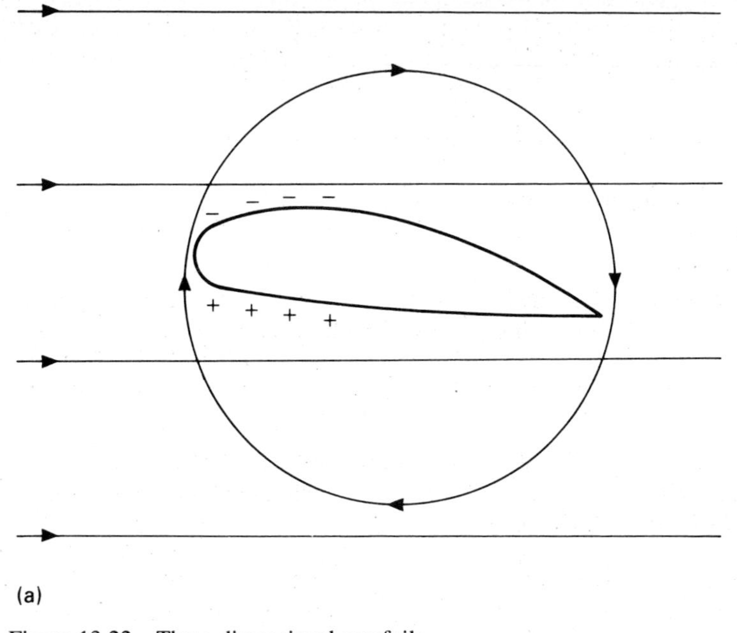

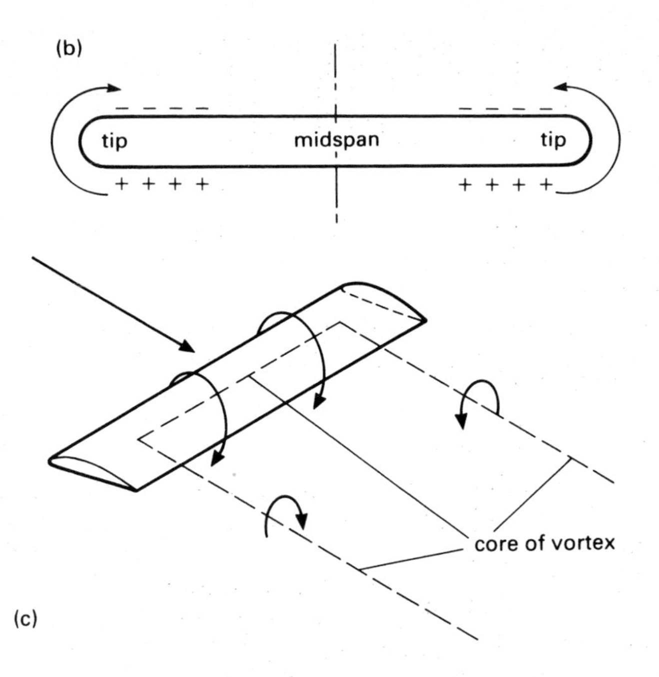

Figure 13.22 Three-dimensional aerofoils.

13.10.5 Aircraft control surfaces

The main wings, the tailplane and the fin of an aircraft are all examples of aerofoils. The rear portions of these aerofoils are separated and attached by hinges so that they can be rotated to control the aircraft in flight. Fig. 13.23 shows the general arrangement of control surfaces on a conventional aeroplane but the detailed shapes and sizes vary considerably. The *elevator* is the hinged rear portion of the tailplane and its movement controls the pitch of the aircraft, its angle of incidence. The *rudder* is the corresponding movable part of the fin and controls the yawing. The *ailerons* are hinged to the wings near the tips and have differential movement (one up and one down) to control the aircraft in roll, its motion about the longitudinal axis. In tailless aircraft, *elevons* are used; these combine the functions of elevators and ailerons.

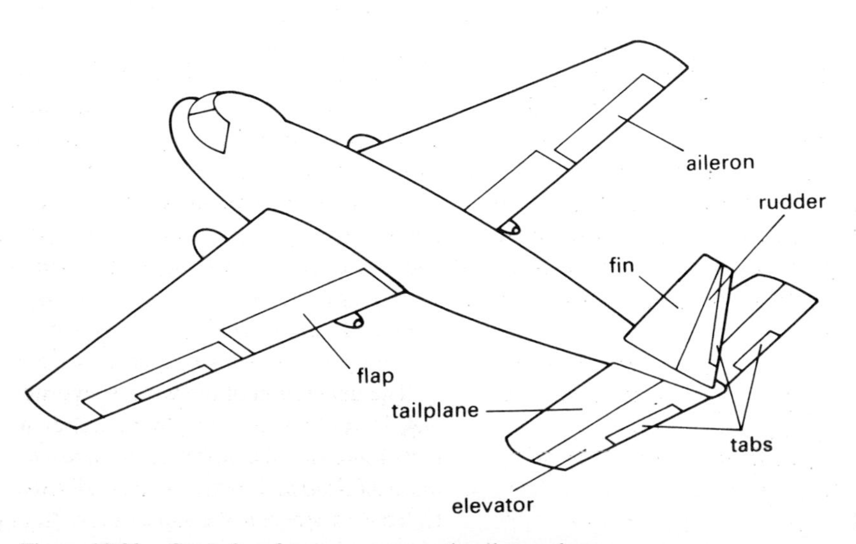

Figure 13.23 Control surfaces on a conventional aeroplane.

In large aeroplanes, the forces needed to move the control surfaces are so great that they are power operated. In smaller aircraft, they can be operated by the pilot's efforts through mechanical linkages. The force required of the pilot can be reduced by the use of *tabs*, small hinged surfaces attached to the main control surfaces.

Fig. 13.24 shows the variation of normal pressure on an aerofoil from the leading to the trailing edge due to changes in the angle of incidence, the control angle and the tab angle. The curves were obtained from wind tunnel tests and give the values of the pressure coefficient c_p. This is the resultant of the pressures on the upper and lower surfaces divided by the reference pressure q. The horizontal axis shows the distance x from the leading edge as a fraction of the chord c.

The aerofoil section (shown at the top of the diagram) was symmetrical with a thickness ratio 0.09 (9%). Curve a was obtained when the aerofoil was placed at an angle of incidence of 5°. The curve shows that most of the lift is due to the pressures near the leading edge.

Curve b gives the corresponding result for a control surface, hinged at $0.3c$ from the trailing edge, when it was rotated through 5°. Note the peak value that occurs at the hinge. The pressure acting on the control surface produces a moment about the hinge which must be balanced by the pilot's force or the power operating system.

Curve c is the result of a 5° rotation of a tab, hinged at $0.06c$ from the trailing edge. The effect of tab movement is the same as that of a small control surface. Tabs are used to counteract the hinge moments of the main control surface. If a tab is given a rotation opposite to that of the main control, the resulting pressures provide a moment about the control surface hinge opposing that due to the movement of the control itself.

Tabs are of several types. A *trimming tab* is one that can be rotated independently of the main control and it is used to balance the control hinge moment due to the incidence and control angle. When a control surface is trimmed no force is required except to move it from this balanced position, to which it should return when released.

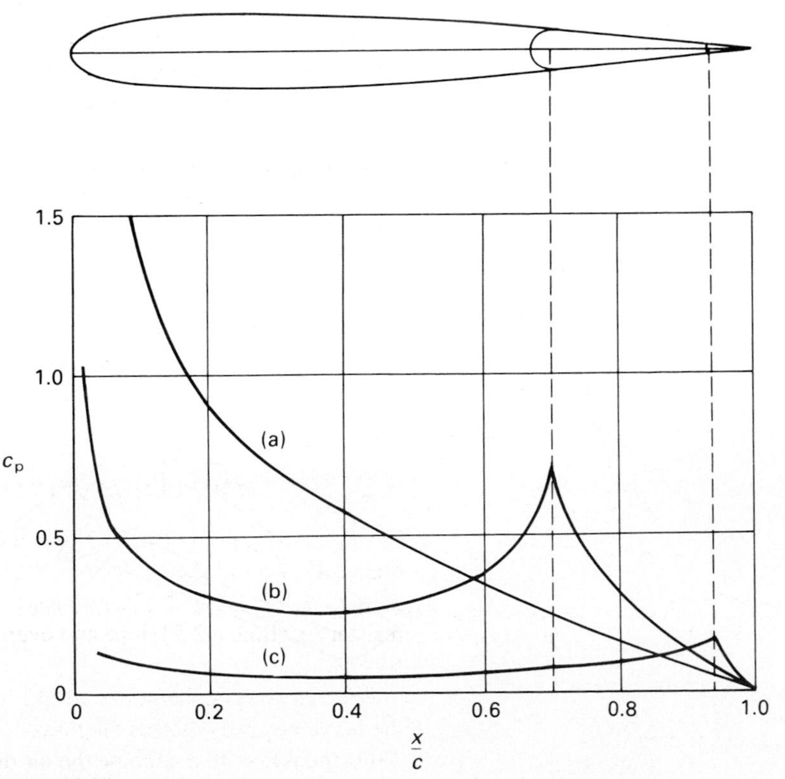

Figure 13.24

Tabs can also be used to reduce the forces needed to rotate the control surface itself. The tab can be geared to the main control so that any rotation of the control is accompanied by an opposite rotation of the tab. This is called a *balance tab.*

A *spring tab* is one in which the pilot's effort is applied partly to the main control and partly to the tab. A spring is included in the circuit so that the tab gives more help as the required effort increases. In the *servo tab* the pilot operates the tab alone and this produces aerodynamic forces which, in turn, rotate the main control.

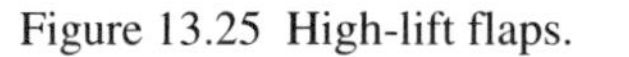

Figure 13.25 High-lift flaps.

13.10.6 High-lift flaps

The general definition of a *flap* is that it is any portion of an aerofoil that can be adjusted in flight to alter the lift. The term therefore includes control surfaces (elevators, rudders and ailerons) but it is applied particularly to those devices that are used to increase the lift during take-off and landing. They are located at the inboard ends of the main wings as shown in Fig. 13.23.

The control surface shown in Fig. 13.24 is known as a *plain flap*. Some other designs are shown in Fig. 13.25. Fig. 13.25(a) is the *split flap* in which only the lower part rotates. A modification of the split flap is the *zap flap* in which it moves both rearwards and downwards. In the *leading-edge slat*, Fig. 13.25(b), the nose portion of the aerofoil moves forwards and downwards forming a slot. This type is not confined to the inboard part of the wing because it does not interfere with the operation of the ailerons.

Single and double *slotted flaps*, Fig. 13.25(c) and (d), open one or more air passages between the upper and lower surfaces of the aerofoil. The *extension* or *Fowler flap* shown in (e), moves backwards thus increasing the effective area of the aerofoil. The *auxiliary aerofoil flap*, shown in (f), is permanently positioned behind the main aerofoil and can be rotated to a high angle of incidence.

The general effect of all these devices is to increase the camber and hence produce a higher maximum lift coefficient. In one set of two-dimensional tests a split flap with a chord of 0.2c increased the maximum C_L from 1.6 to 2.35. In other experiments a wing, whose maximum lift coefficient was 1.13, was fitted with various types of flap. This value became 2.95 with a single extension flap, 3.57 with a double flap and 4.44 with a double flap and leading-edge slat.

It must be remembered that all these results apply to flaps extending over the whole span and the increases will be smaller for practical wings. Flaps also cause substantial increases in drag; this may be an advantage during landing but necessitates an increase in power during take-off. Flaps also lead to structural and mechanical complications.

Nevertheless, modern aircraft use complex flap systems to achieve the low speeds needed for safe take-off and landing.

13.11 Bicycle dynamics – overcoming air drag

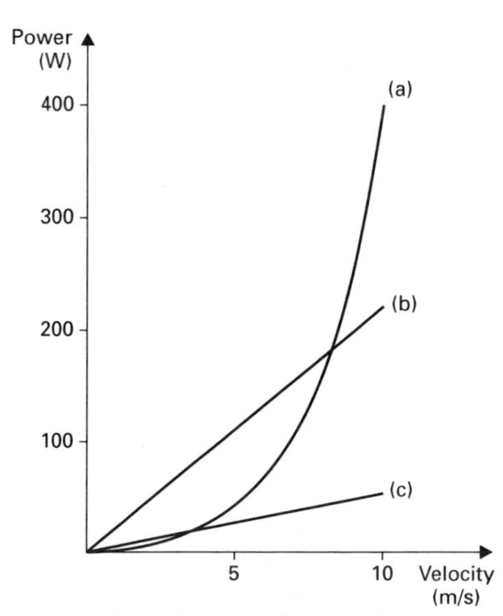

Figure 13.26 Power requirements for a bicycle journey.
(a) power to overcome drag
(b) power to climb slope of about 2.5°
(c) power to overcome rolling resistance.

In Chapters 10 and 11 we examined the dynamics of a bicycle journey but neglected one of the most important factors, the power to overcome the air resistance when there is no wind. Fig. 13.26 shows typical power required to overcome the rolling resistance, climb a 2.5° slope and overcome the air drag when travelling at different speeds.

We can see that the power for rolling resistance and hill climb vary linearly with the bicycle speed whereas the power for air drag does not. At speeds of less than 2 m/s the power to overcome the air drag is similar to that for the rolling resistance but after about 5 m/s it becomes very significant.

In 13.9.4 we found that the drag on a vehicle, D, is given by

$D = C_D \frac{1}{2}\rho v^2 A$ where C_D is the drag coefficient and can be taken as 1.0 for a sports bike
ρ is the density of air and is 1.23 kg/m^3
v is the speed of travel in still air
and A is the frontal area and can be taken as 0.4 m^2 for a sports bike

The power to overcome this drag, $P_{AR} = Dv$ [Power = Fv]

And so $P_{AR} = C_D \frac{1}{2} \rho v^3 A$

Therefore for our cyclist this gives us

$$P_{AR} = 1.0 \times \tfrac{1}{2} \times 1.23 \times 0.4 \times v^3 = 0.246\, v^3$$

The following table shows the power required to overcome the air resistance at different speeds.

Speed (m/s)	1	3	5	8	10
Power to overcome the air resistance (W)	0.246	6.64	30.75	126	246

We can see from these values that the power to overcome the air drag increases sharply and most of us would have trouble cycling at more than 5 m/s for any distance.

Consider the previous values that we found for the power requirements of our bicycle in 11.10 and now include the figures for air drag to obtain total power requirements when travelling in no wind. Now consider the difference that a following or head wind of 10 m/s or 25 m/s would make to you.

Bibliography

Douglas J F 1969 *An Introduction to Dimensional Analysis for Engineers.* Pitman
Douglas J F 1986 *Solving Problems in Fluid Mechanics* vol. 1. Longman
Kermode A C 1995 *Mechanics of Flight* 10th edn. Longman
Ward-Smith A J 1984 *Biophysical Aerodynamics and the Natural Environment.* Wiley

Assignments

1 A rectangular tank, 3 m long and 2 m wide, contains water to a depth of 1.5 m. Calculate the thrust on each side of the tank and the base.

2 A tank containing water has a rectangular opening 3 m wide by 2 m high in one of its vertical sides, the upper and lower edges being horizontal. The opening is covered by a plate carried on a horizontal hinge at the top. What horizontal force is required at the bottom edge of the cover plate to prevent it from opening when the water level inside the tank is 1 m above the hinge?

3 A timber plank, with a relative density of 0.8 and a rectangular cross-section 200 mm by 120 mm, floats in water with its longitudinal axis horizontal. Find the depth to which it is immersed and whether it is stable if (a) the 200 mm side is horizontal, (b) the 120 mm side is horizontal.

4 Two points on the surface on an aircraft fuselage differ in level by 2 m. Air of density 1.23 kg/m^3 flows steadily past the fuselage, the velocities at the upper and lower points being 100 m/s and 80 m/s respectively. Find the difference in pressure between the two points assuming Bernoulli's equation applies.
What percentage error is caused by neglecting the difference in level?

5 A circular pipe is running full of water and at one section, where the diameter is 0.3 m, the velocity is 4 m/s. Calculate the velocity at another section where the diameter is 0.2 m.
If the second section is 10 m lower than the first what is the pressure difference between the two sections (neglecting friction losses)?

6 What pressure differences will occur in a Pitot-static tube placed in an air stream having a velocity of (a) 10m/s, (b) 100 m/s? Take the air density as 1.23 kg/m^3.
If the pressure difference is measured by a U-tube containing water and the difference in level is h show that the velocity v is given by:

$$v = \text{constant} \times \sqrt{h}$$

Find the value of the constant if h is measured in millimetre and v is to be in metre per second.

7 A vertical cylindrical tank has at the bottom a 50 mm diameter sharp-edged orifice for which the discharge coefficient is 0.6. If water enters the tank at a constant rate of 0.01 m^3/s, find the depth of water above the orifice for which the level in the tank becomes steady.

8 Two pipes of 50 mm and 100 mm diameter, each 100 m long, are connected in parallel between two reservoirs whose difference of level is 10 m. Find
(a) the flow in cubic metre per second for each pipe,
(b) the diameter of a single pipe 100m long which would give the same total flow as the two given pipes.
Neglect entry and exist losses and use the Darcy formula with $f = 0.008$ in each case.

9 The flow of a gas in a duct is to be investigated using a scale model. Show, by dimensional analysis, that the power P required to propel a fluid of density ρ and dynamic viscosity μ through the duct at velocity v may be written

$$P = \rho v^3 l^2 \phi\{Re\}$$

where Re is the Reynolds number based on the representative length l and ϕ means 'a function of'.
If the flow of gas is to be simulated by means of water in a one-quarter scale model and the full-scale gas velocity is expected to be 12 m/s find
(a) the corresponding water velocity in the model for dynamically similar conditions,
(b) the power required in the full-scale duct if that of the model is 500 W. Take μ for water as 60 times μ for the gas, ρ for water as 750 times ρ for the gas.

10 Fig. 13.27 shows a roadside warning sign which is used to inform drivers about road surface temperatures. A sensor embedded in the road detects temperature and the signal produced by the sensor is processed and displayed on the sign.

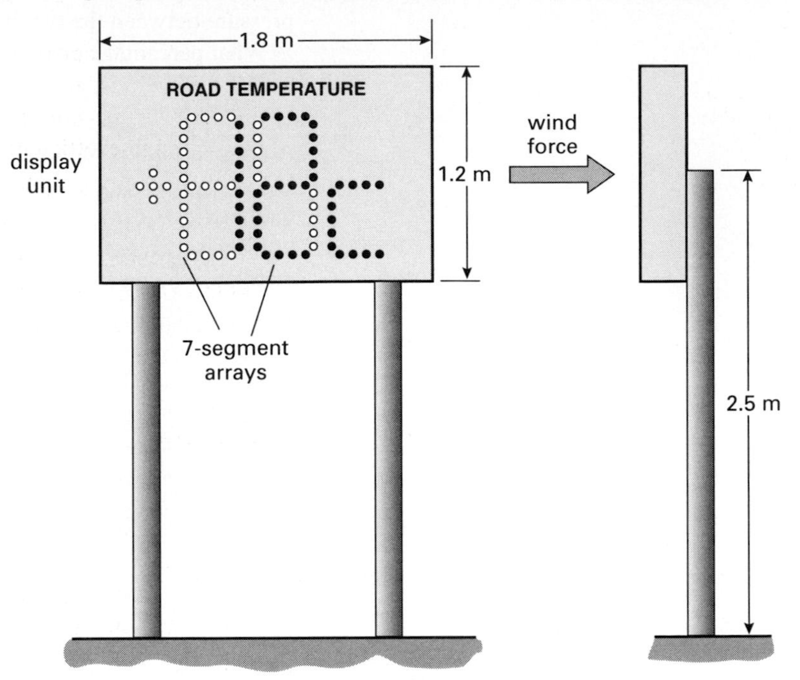

Figure 13.27

The road sign is supported above ground level using two tubular mild steel posts of outside diameter 150 mm. The supports are to be designed to deflect by no more than 5 mm when a force produced by a wind speed of 55 m/s, is applied directly to the front surface of the sign.

Fig. 13.28 shows the graph of wind speed/wind pressure which is used in designing structures such as the road sign.

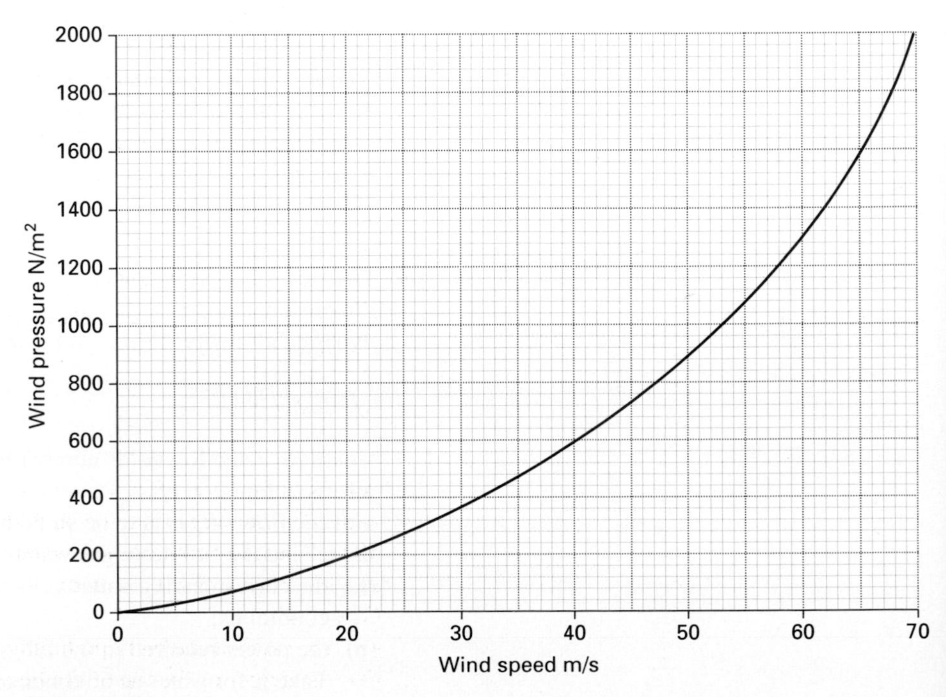

Figure 13.28

(a) The wind force acting on a structure can be calculated from the formula:

$$F_W = PAC_d$$

Where
F_W – Wind force;
P – Wind pressure;
A – Area of structure;
C_d – Design coefficient applied to the structure. In this application $C_d = 1.75$.

Calculate the wind force acting on the sign at a wind speed of 55 m/s. (Ignore the effect of wind pressure on the support legs.)

It can be assumed that the wind force acting on the road sign creates a point-load at the top end of the posts.

(b) Calculate the wall thickness of the mild steel tube required for the posts. (Ignore the weight of the sign.)

11 A small airliner has a wing area of 90 m² and a total mass of 26 100 kg. The drag and lift coefficients are related by:

$$C_D = 0.0196 + 0.0484C_L{}^2$$

The aircraft is in steady horizontal flight at a speed of 200 knots and an altitude of 5000 m where the air density is 0.74 kg/m³. (1 knot = 0.5148 m/s.)
Find, for these conditions,
(a) the wing loading in newton per square metre,
(b) the reference pressure,
(c) the lift coefficient,
(d) the corresponding drag coefficient,
(e) the total drag,
(f) the thrust power required.

12 State *two* methods of measuring air flow in a pipe, giving *one* advantage and *one* disadvantage of each method.

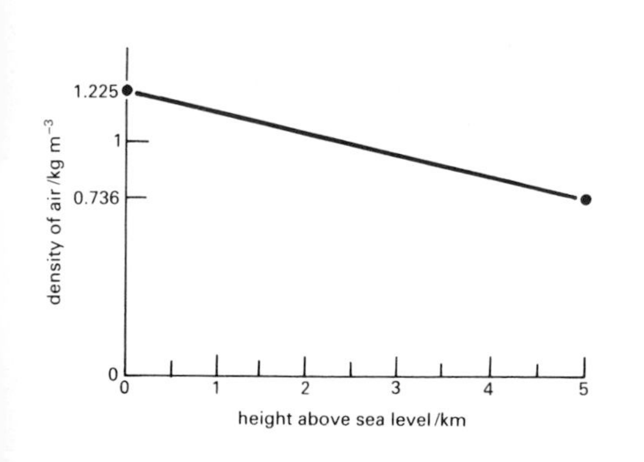

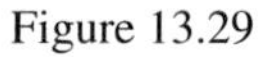
Figure 13.29

13 The density of air at various heights above sea level is given in a table published by the International Civil Aviation Organisation. A graph drawn from part of this table is shown in Fig. 13.29.

Assuming the pressure at sea level to be 101.3 kPa, calculate the pressure at 5000 m above sea level.

14 A venturi is inserted into a pipe in order to measure liquid flow rate. Sketch such a venturi, indicating the direction of flow, and show against its length plots of
(a) velocity, and
(b) pressure for
the liquid.

15 A hollow buoy is made, as a cylinder closed at both ends, from steel plate for which the mass per unit area is 800 kgm^{-2}. It has a diameter 12 m and height 50 m. The buoy contains sea water up to a height 15 m from the base and floats in the sea with its axis vertical.
Calculate
(a) the mass of the buoy including the water contained,
(b) the height of the top of the buoy above sea level when floating. (Density of sea water = 1028 kgm^{-3})

16 Write the Bernoulli equation for the flow of fluid in a pipe, explaining the meaning of each term.
A pipe of inside diameter 120 mm is fitted with an orifice plate having orifice diameter 50 mm to enable the flow of water in the pipe to be measured. As a check on this a Pitot-static tube is inserted as shown in Fig. 13.30.
The mass flow rate, in kg s^{-1}, is given by the equation:

$$\dot{m} = 0.62A\ \sqrt{(2\rho\Delta p)}$$

where A is the orifice area in m^2
ρ is the density in kg m^{-3}
Δp is the pressure difference in Pa

When the orifice meter pressure difference is 13.5 kPa, find
(a) the mass flow rate,
(b) the velocity of water in the pipe,
(c) the difference in head in the paraffin manometer connected to the Pitot-static tube by pipes containing air.
(Density of paraffin, $\rho = 800 \text{kgm}^{-3}$.)

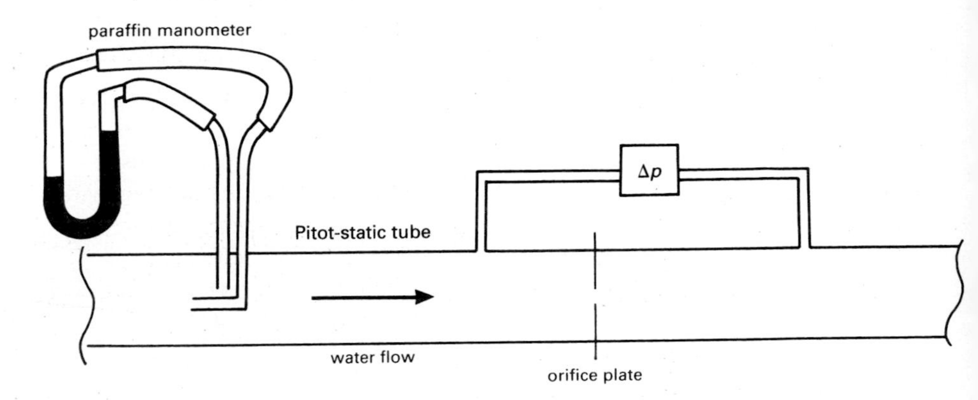

Figure 13.30

17 A Pelton wheel is one type of water turbine which may be used in a hydroelectric power station.
Water is stored by a dam so that its level is 292 m above the turbine. Water flows from this level through a pipeline and into a nozzle of effective outlet diameter 53 mm. The pressure loss caused by friction in the pipeline is 0.9 MPa. Water from the nozzle strikes the blading, a section through which is shown in Fig. 13.31, and is turned, leaving at the angle shown.
The mean diameter of the turbine wheel is 0.8 m.
Determine
(a) the pressure at entry to the nozzle,
(b) the velocity of water at exit from the nozzle and the mass flow rate,
(c) the force exerted on the blading, and the torque applied to the turbine wheel if, for test purposes, the wheel were locked to prevent rotation.

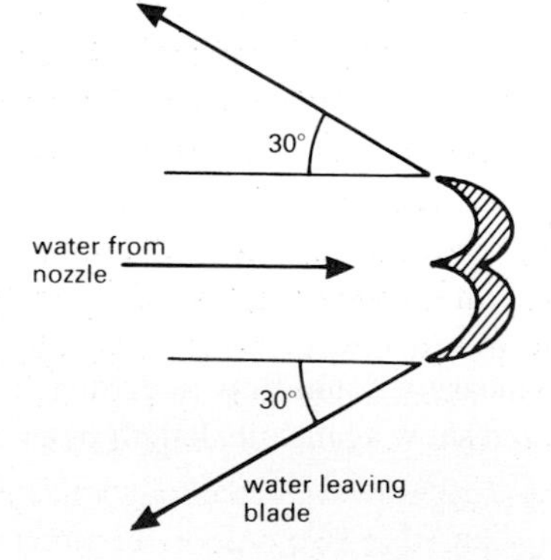

Figure 13.31

18 Name any one type of water turbine and briefly describe the *principle* on which it operates. You are not asked to write a full description of the machine.

19 (a) State what is meant by
(i) laminar and
(ii) turbulent flow.
(b) For each type of flow outline *one* situation in which it occurs.

14 Electronic systems

14.1 Distinction between electricity and electronics

To the lay person electricity and electronics appear to be one and the same thing. To the technologist they are different but they have some characteristics in common. Electricity is a means of transferring energy whereas electronics uses electrical signals to convert and process information.

Information from the outside world is converted into an electrical form (the signal), is processed and is then converted it back to another form. Electronics techniques provide the means of processing the electrical signal. A study of electricity is therefore needed to understand electronics.

Some textbooks refer to electronics as the controlling of electrons but since electric current is itself defined as the rate of flow of electrons this is not a helpful distinction.

14.2 The systems approach

The study of electronics has changed in recent years. Previously, circuits were analysed in terms of the behaviour of components such as resistors and capacitors. However, this component-centred approach is not the way electronics engineers work in practice. Devices are rarely considered on their own in the first instance. It is more usual to decide what function part of a system is supposed to perform and to select the combination of components that will achieve this task. Firstly, there are certain standard functions and, secondly, there are alternative ways of putting components together to achieve a particular function.

From this view has grown a *systems approach* to the study of electronics. It has been recognised that there are many definable functions in electronic systems. These functional building blocks are known as subsystems. All subsystems can be categorised as input subsystems, output subsystems or processing subsystems and they can be connected in a systems diagram, as shown in Fig. 14.1.

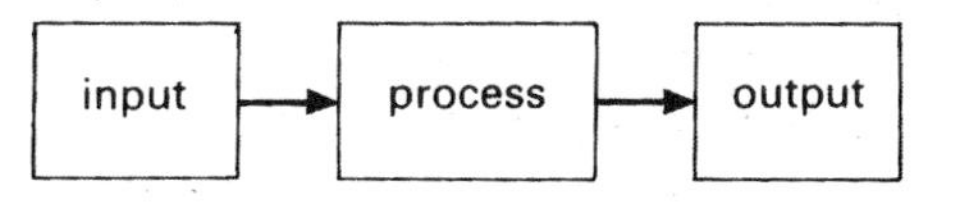

Figure 14.1 Simple systems diagram.

It should be noted that the power supply is not shown in a systems diagram – it is always assumed to be present. The arrows on the lines connecting the subsystems show the direction of the information flow. This information is in the form of an electrical signal. The different forms of signal are examined in section 14.3.

Input subsystems usually convert information from the outside world into an electrical form. A few, such as the pulse generator, generate a signal independently from the environment. Input subsystems usually consist of two or more components connected in a particular way. A list of input subsystem functions would include:

reference signal,
light sensor,
temperature sensor,
position sensor,
magnetic field sensor,
sound sensor,
pressure sensor,
continuous-pulse generator,
light-operated switch,

temperature-operated switch,
position-operated switch,
magnetically-operated switch,
sound-operated switch,
pressure-operated switch.

The different types of signal produced by these subsystems will be examined in section 14.3 whilst some circuits to achieve these functions will be suggested in section 14.4.

Processing subsystems take the signal from an input subsystem and modify it and/or combine it with other signals. Possible processing functions are:

current amplification,
voltage amplification,
voltage comparison,
inverter or NOT,
AND, OR, NAND, NOR, EOR gates,
latch, flip-flop, memory,
timer,
counter,
pulse stretcher,
square pulse producer.

Most of these functions are examined in more detail in Chapters 15 and 16.

Output subsystems take the signal after it has been processed and convert it back into another form of energy. You should be able to identify the main form of output energy for each item in this list.

motor,
solenoid,
buzzer,
loudspeaker,
lamp,
seven-segment display,
relay.

Output subsystems are considered in more detail in section 14.5.

The systems approach will be used in all chapters involving electronics in this text. We will start with the function, examine alternative ways of achieving the function, explore principles exemplified by the various configurations and finally consider the devices. It should be stressed that this covers exactly the same ground as other approaches to electronics – it is just that the order is different.

The important consideration is how you should begin to study electronics. It is tempting to begin to study the components but it is much better to begin by studying systems that perform useful tasks. Most of the systems kits on the market have printed material which gives problem-solving examples and a sound starting point for further study. Much of this work should be completed before starting an advanced level course. For example, the Unilab Alpha range of equipment is supported by two publications, entitled *An Introduction to the Systems Approach Using Alpha* and *The Alpha Resource*. Together they form an excellent introduction.

14.3 Types of signal

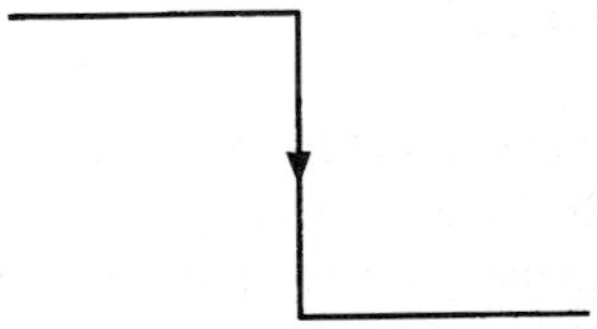

Figure 14.2 Signal trace.

Most books define signals as being either analogue or digital. This is undoubtedly correct but the distinction between analogue and digital signals is not as obvious as it appears. Merely examining the signal does not help, as can be seen in Fig. 14.2.

Is this signal analogue or digital? Well, it could be either!

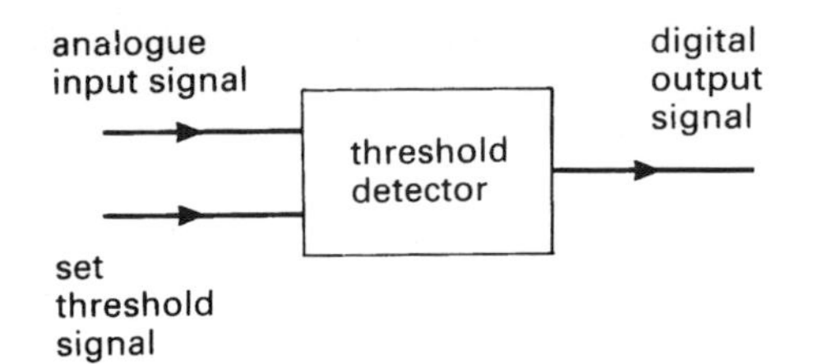

Figure 14.3 Block diagram of threshold detector.

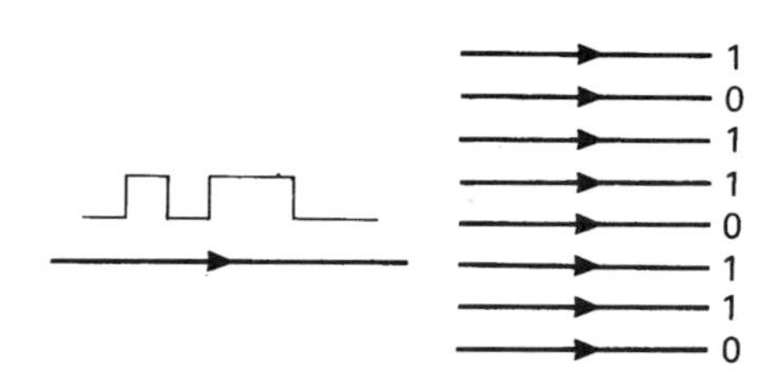

Figure 14.4 Serial and parallel digital signals.

The crucial question is to ask what information it is conveying. It could be showing that the light level has now passed some pre-set value in which case it would be digital. On the other hand it might be showing that a bottle top has been placed on top of the light sensor. This would be the response from an analogue sensor.

An *analogue signal* is an analogy with a physical property. There will be a mathematical relationship between the change in the physical property and the change in the signal. It need not be a simple linear relationship. The signal should be capable of having any value between a maximum and a minimum. It does *not* have to be smooth – it *could* be smooth if that was the analogy.

A *digital signal* is an encoded signal. The simplest form of digital signal is a binary or two state signal. It is relatively easy to convert an analogue signal into a binary signal with a threshold detector (Fig. 14.3).

More complex digital signals can take the form of a pattern of pulses along a single wire (a serial signal) or a pattern of binary digits on a set of wires (a parallel signal) as in Fig. 14.4.

Analogue (A) and digital (D) signal processing are dealt with separately in the next two chapters. There is a special section on D to A and A to D conversion in section 16.5.3.

14.4 Input sensors – analogue and digital

This section looks at a range of input sensors and the transducers needed to create the signal. It first considers how to provide a reference signal and how to convert an analogue signal into a two-state digital signal. It cannot be totally comprehensive because new devices are constantly appearing.

An input transducer is defined as a device that converts energy from a non-electrical form into an electrical form.

All the circuits are shown operating on a single supply rail – from 0 V to +V. It would be possible to operate them on split rail supplies if necessary.

Whenever alternatives are encountered it is vital to consider what criteria are to be used in making the choice. Just because you know a particular circuit well may not be good grounds for choosing it. The criteria may include:

- what value power supply is needed?
- what is the cost?
- how much space does it take up?
- what kind of signal does it produce or require?

As you gain more experience you will be able to add to this list. Remember that the 'best' alternative may be considered to be the one that fits together with the other subsystems, making a total system which meets the original specification at the lowest cost.

14.4.1 Provision of a reference signal

The simplest form of *reference signal* is derived from a potential divider arrangement. (For the theory of the potential divider see section 20.4.) The first diagram of Fig. 14.5 shows a simple two-resistor potential divider whilst the second shows a potentiometer. The third diagram shows a combination of resistors and a potentiometer.

The two-resistor arrangement is very cheap but not adjustable. The only way to change the reference signal is to change one or both of the resistors. It is difficult to predict the exact value of the reference signal due to the tolerance of the resistors. (For an explanation of tolerance see section 14.7.1.)

A potentiometer is more expensive but it is easy to change the value. The

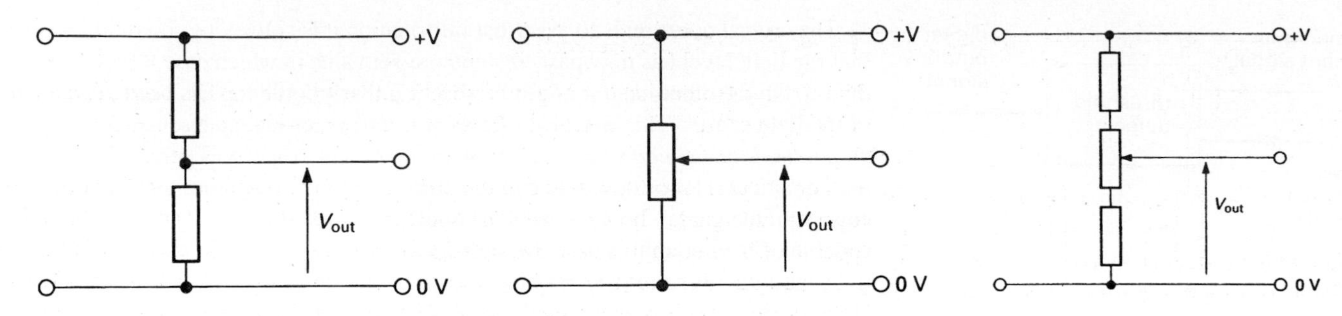

Figure 14.5 Alternative means of providing a reference signal.

sensitivity of the potentiometer will be low if it is a high value; that is, small movements of the slider may produce large changes of signal.

The use of two fixed resistors and a potentiometer allows for an increased sensitivity. With all potential divider systems the current flowing through the resistors should be kept small or else heating will occur. If any significant amount of current is taken from this system then the reference signal will change. Potential dividers are generally used when a reference signal is required by a high impedance input such as on an op-amp. (For an explanation of impedance see section 20.6.)

An alternative method is to use a *zener diode* and resistor in a potential divider, Fig. 14.6.

It should be noted that the zener diode is reverse biased. The resistor value must be chosen carefully so that the maximum power rating of the zener diode is not exceeded. This arrangement gives a very accurate reference signal which will allow some current to be taken off to the next stage. The values of zener diodes available are limited but they can be connected in series to give otherwise unobtainable values. In practice it is more usual to accept a fixed reference signal of whatever value and adjust the other signal which is being compared to the reference.

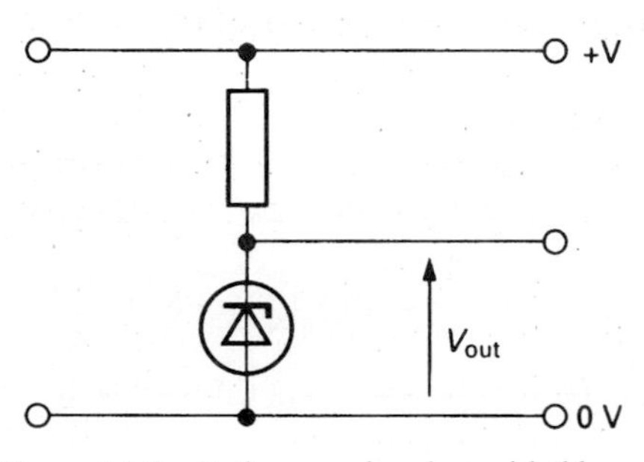

Figure 14.6 Reference signal provided by zener diode arrangement.

14.4.2 Converting an analogue signal into a two-state digital signal

The basic operation is shown in Fig. 14.7.

More details of circuits that will process the analogue signal to achieve this effect can be found in section 15.4.1.

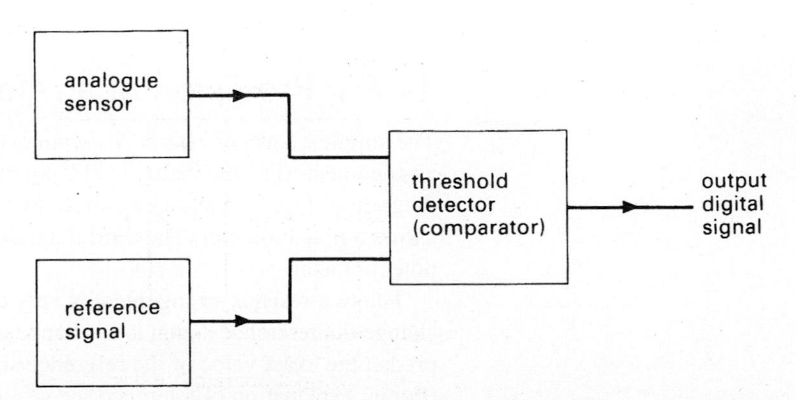

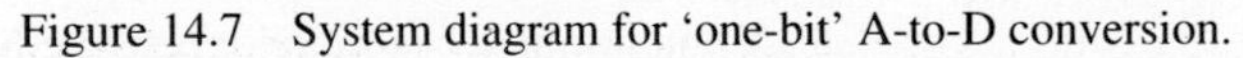

Figure 14.7 System diagram for 'one-bit' A-to-D conversion.

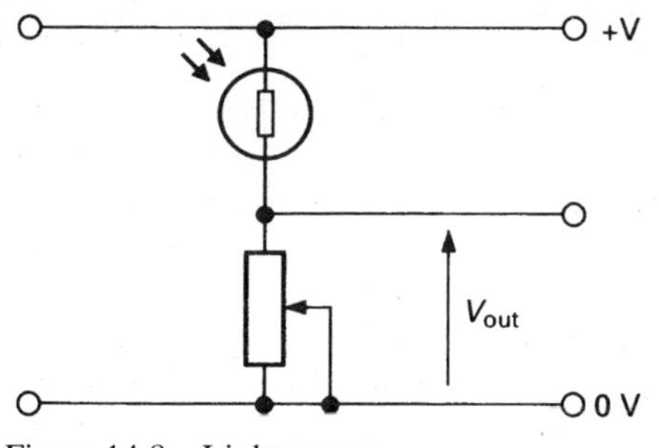

Figure 14.8 Light sensor.

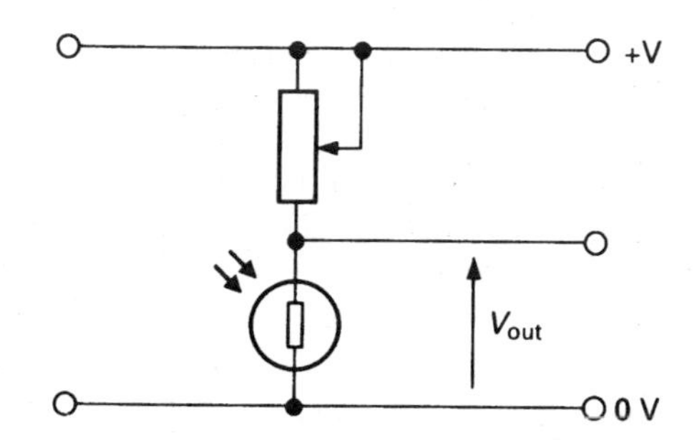

Figure 14.9 Alternative arrangement for light sensor.

14.4.3 Light sensors

The most common sensing system for light intensity involves the *light dependent resistor* or LDR. This is usually incorporated in a potential divider as shown in Fig. 14.8.

The use of a variable resistor in the other arm of the potential divider allows the level of the signal to be adjusted. The resistance of the LDR increases as the light level decreases so in this case the signal level will rise as it gets dark. The LDR can experience a wide range of resistance change – from about 100 ohm in very bright light to several megohms in total darkness. It should also be noted that it has an exponential response. The most common type of LDR used is the ORP12.

Changing the position of the LDR and variable resistor allows the signal to change in the opposite direction, Fig. 14.9.

In this arrangement the signal level will rise as it gets light. The LDR is relatively expensive, quite large and rather slow in its response to rapid changes of light level. Its response to different wavelengths of light is very similar to that of the human eye.

Alternative light sensors are based upon the photodiode or the *phototransistor*. There are special sensors available that are particularly sensitive to infrared illumination. This can be particularly useful if ambient lighting is a problem. Digital signals are available from opto-switching devices. These can be free standing or combined with a transmitter – the reflective opto-switch and the slotted opto-switch are two examples.

Examples of these circuits are shown in Fig. 14.10.

Details of the cost of the different transducers, their size as well as their electrical characteristics and requirements, can be found in manufacturers' catalogues and data sheets. A list of these can be found at the end of this chapter.

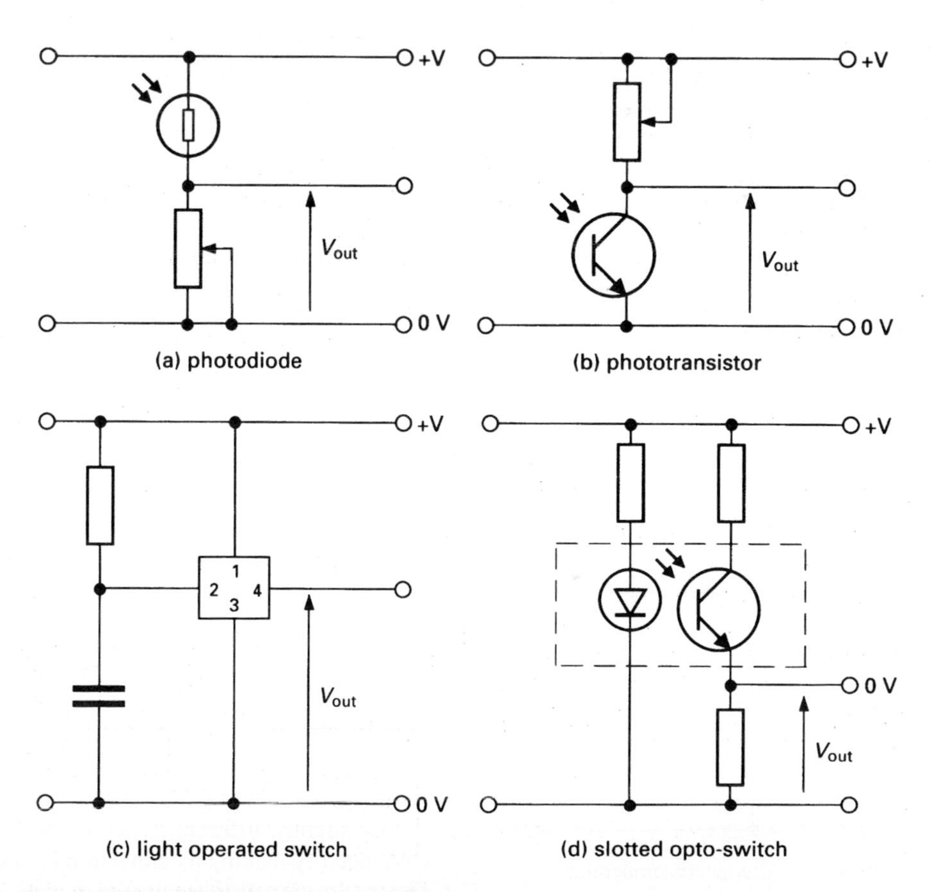

Figure 14.10 Alternative light-operated sensors.

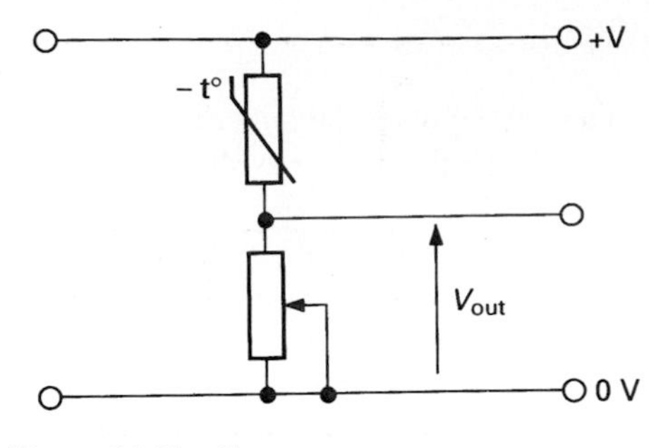

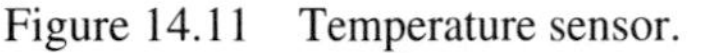
Figure 14.11 Temperature sensor.

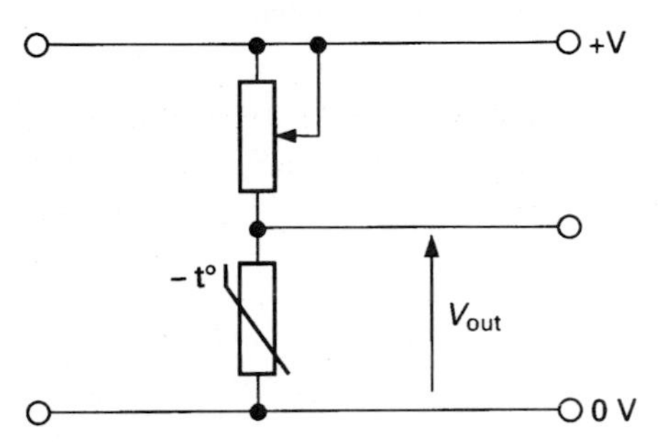

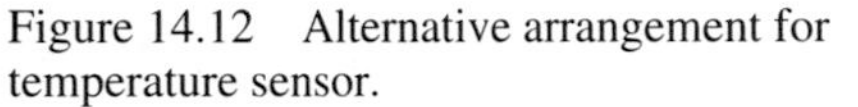
Figure 14.12 Alternative arrangement for temperature sensor.

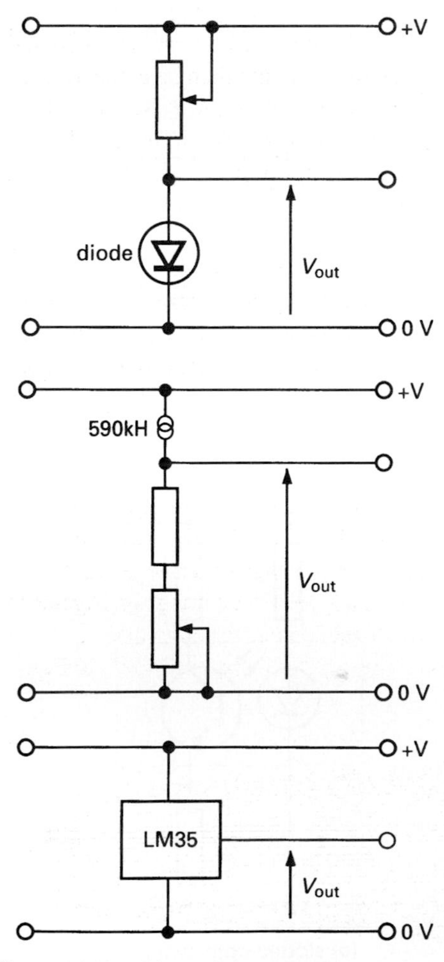

Figure 14.13 Alternative temperature sensors.

14.4.4 Temperature sensors

The most common temperature sensing system involves a temperature-dependent resistor (*thermistor*) used in a potential divider, Fig. 14.11.

This is a very similar arrangement to light sensing and, like the light sensor, the positions of the thermistor and variable resistor can be reversed to produce a signal that moves in the opposite direction when the same temperature change occurs.

The thermistor shown is a negative temperature coefficient device (n.t.c.) and is by far the most common. Its resistance decreases as the temperature increases. It is possible to buy p.t.c. (positive temperature coefficient) devices. There are many different types of thermistor – both different shapes and sizes as well as different values of resistance change over different temperature ranges. The manufacturers' catalogues and data sheets give more details.

The value of the variable resistor should be chosen to ensure that the signal will have the correct range of values over the required temperature range. For example, let us suppose we wish to sense a temperature at about 25°C. The thermistor chosen has a resistance of about 4.7 kΩ at this temperature. (The numerical coding for resistors is explained in section 14.7.1.) If the signal we require is to be about half the supply voltage then a fixed value resistor of 4.7 kΩ would put the signal in approximately the right place. A variable resistor of about twice this value, that is, 10 kΩ, would allow us to adjust this signal and would be using the variable resistor with its wiper in about the centre position, which is good practice.

It should be noted that thermistors are non-linear devices but it may be possible to select one that is reasonably linear over a limited temperature range. Alternatively the signal must be processed to produce a linear output or the output must be calibrated.

Another common temperature transducer is a *thermocouple*. This is known as an active transducer since it generates its own e.m.f. rather than the passive transducer which has to have an external e.m.f. applied to it. The thermocouple, which consists of two dissimilar metals joined together, produces very small signals with a low impedance. The signal is reasonably linear over a range of temperatures.

Alternative temperature-sensing subsystems are based on semiconducting devices. The resistance of most semiconductor devices is affected by temperature and a simple temperature sensor can be made with a single diode and resistor combination. Alternatively, there are special devices that produce linear outputs, for example, 1 microvolt per kelvin, but at an increased cost. Some of these circuits are shown in Fig. 14.13.

There is a device called a temperature-operated switch but it is not very versatile since its switching point cannot be controlled. It operates on the principle of the bimetallic switch.

14.4.5 Position sensors

There are two types of position sensors: one type answers the question 'Am I there?' whilst the other type answers the question 'Where am I?'

The first type are often known as *limit switches* and produce a digital output. They can be light-operated, pressure-operated (for example, microswitches) or magnetically-operated switches, and are usually arranged in one of the two configurations shown in Fig. 14.14, depending on whether the signal change required is high-going (a rising edge) or low-going (a falling edge) when the switch is closed.

One slightly different switch is the *tilt switch*, which is usually a sealed container with mercury making a connection between two electrodes when the switch is in the upright position. It is connected in exactly the same way as the microswitch.

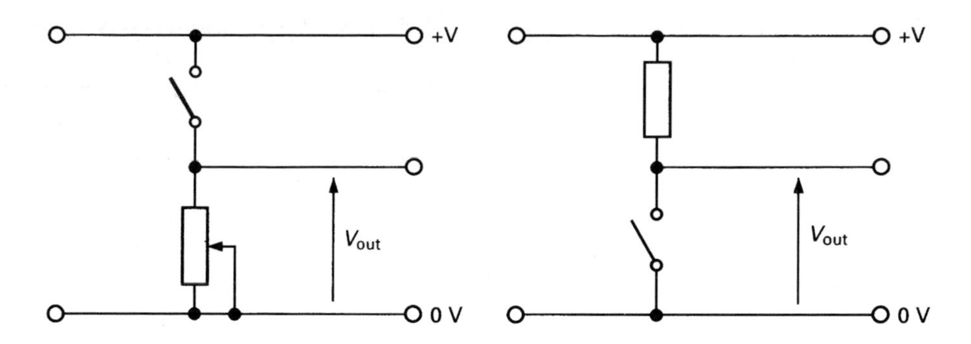

Figure 14.14 Alternative switching arrangements.

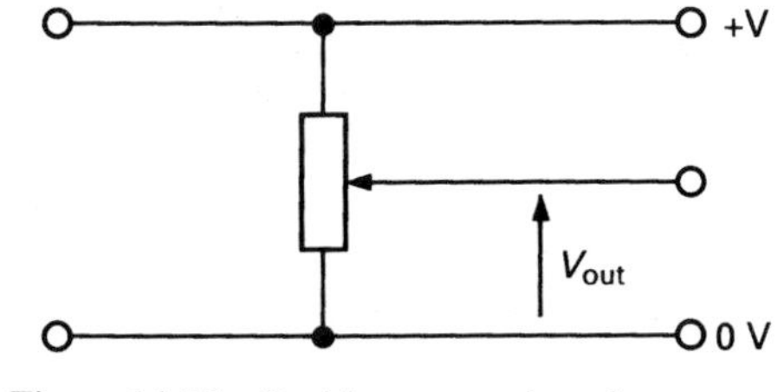

Figure 14.15 Position sensor based on a potentiometer.

The second type of sensor is usually either an analogue signal produced from a potentiometer (Fig. 14.15) coupled to the moving object, or a digital encoded signal from a disc with black and white sections (or clear and opaque) rotating in a beam of light.

Electrically, the resistive type is always connected in the same way. The potentiometer can be rotary or linear (slider) type. The multiturn rotary potentiometers are particularly useful for this application since they allow several full rotations before reaching the limits of their travel.

The digital type of position sensor can be quite a complex system. It is often linked to a microcomputer or microprocessor because the signal can be difficult to process. The simplest form consists of a disc that is divided into segments, Fig. 14.16.

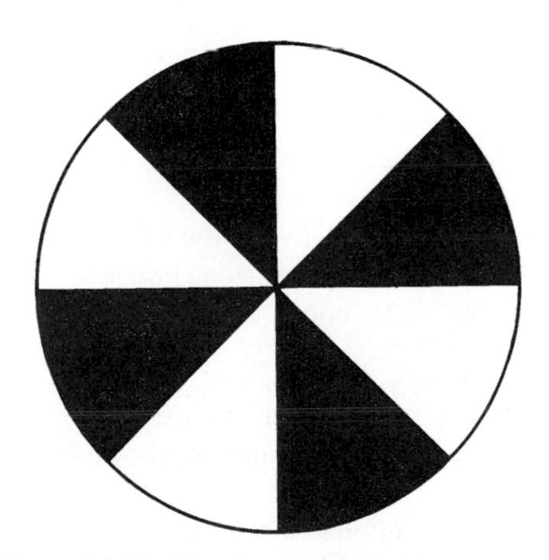

Figure 14.16 Disc for digital position sensor.

This will rotate in a slotted opto-switch or in front of a reflective opto-switch and will produce a chain of pulses as it rotates. These pulses could then be counted and the count would be a measure of the position. The sensitivity of the device is not high. The segments cannot be made too small or the opto-switch will get conflicting information. It cannot easily cope if the direction changes. If it moves a distance and then returns the counter will show that it has moved twice the distance – not that it is back where it started.

A development of this is to use a binary encoded disc, Fig. 14.17.

This requires four opto-switches – one for each of the concentric rings. This will give 16 unique positions, from 0000 to 1111, and the direction moved is no longer important. It can, however, still give problems if the final position is on the junction of two sectors.

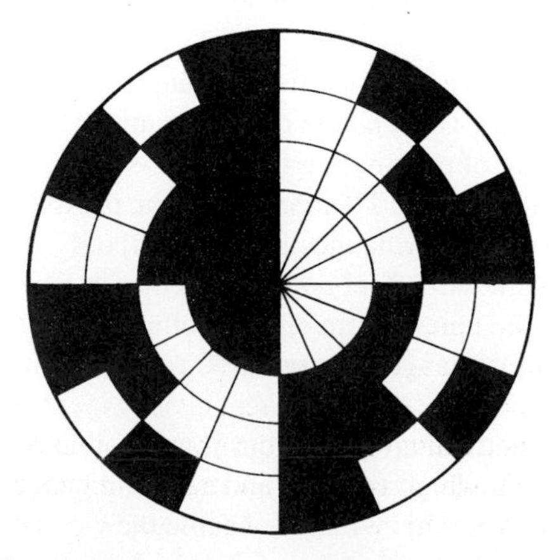

Figure 14.17 Binary encoded disc.

14.4.6 Strain sensors

The sensing of strain is almost exclusively achieved with a subsystem based on a strain gauge. The output signal produced is very small and can be easily swamped by other changes that may be taking place – in particular a change of temperature. For this reason the strain gauge is often incorporated in a Wheatstone bridge arrangement. The function of this is to convert changes in resistance to changes in voltage.

The strain gauge consists of a very thin wire arranged on a pad as shown in Fig. 14.18.

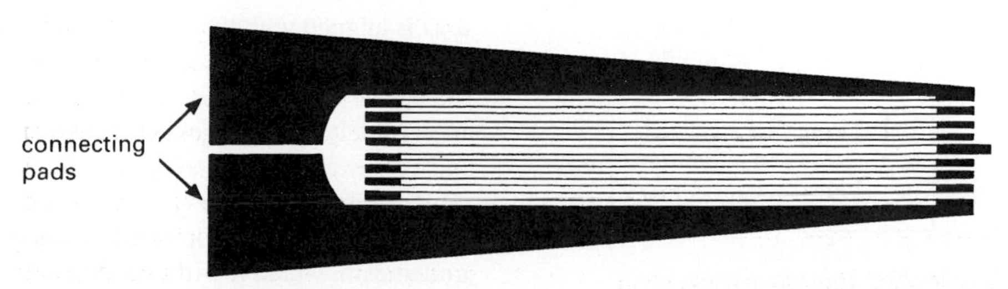

Figure 14.18 Strain gauge.

The long axis is termed the active axis and the short axis is termed the passive axis. The strain should cause changes to the active axis. If the pad experiences strain then the wire will be deformed. The resistance of the wire is affected by its length and cross-sectional area. When the wire is deformed both these properties will change producing a change of resistance. The relationship between the resistance, R, the length of the wire, l, and its cross-sectional area, a, is given by:

$$R = \frac{\rho l}{a}$$

where ρ is the resistivity of the material (a constant value). See section 5.1.2.

However it is more usual to use the gauge factor to calculate the resistance change. The gauge factor, G, is the fractional change in the resistance of the gauge, $\Delta R/R$, divided by the fractional change in the length of the gauge, $\Delta l/l$, along its active axis. Thus:

$$G = \frac{\Delta R/R}{\Delta l/l}$$

But $\Delta l/l$ is the strain, ϵ, of the body to which the strain gauge is attached. Therefore:

$$\Delta R/R = \epsilon G$$

or

$$\Delta R = R\epsilon G$$

If, for instance, the gauge resistance is 120 Ω and the gauge factor is 2.1 then, for a strain of 7×10^{-4} the expected change in resistance would be $120 \times 7 \times 10^{-4} \times 2.1 = 0.176$ Ω.

If the strain gauge was incorporated in a simple potential divider the voltage change produced would be very small. Some theory of the Wheatstone bridge is given in section 20.4. The Wheatstone bridge, Fig. 14.19, converts this resistance change to a voltage change.

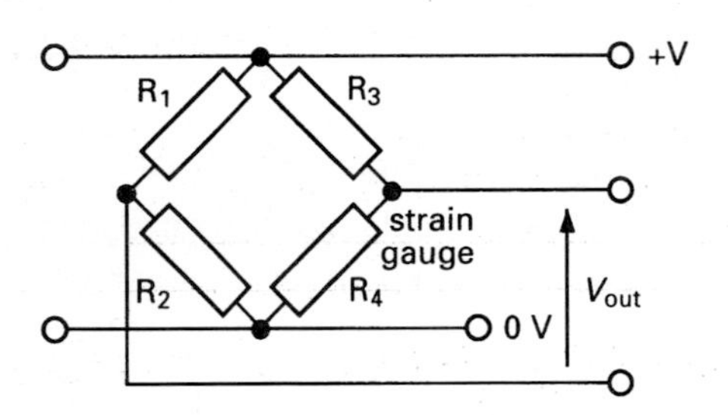

Figure 14.19 Strain gauge in Wheatstone bridge arrangement.

If $\Delta R/R = \epsilon G$ and $V_{out} = V_s \Delta R/4R$ (see section 20.4) we can now say that:

$$V_{out} = V_s \epsilon\, G/4$$

This shows that the output voltage signal is directly proportional to strain.

Note that this arrangement gives two output signals for processing. We are interested in the difference between the two signals so a difference amplifier will often be the next subsystem. Details of difference amplifiers can be found in section 15.3.2.

Unfortunately, when the deformation takes place there is often a change of temperature in the material undergoing the strain. This change of temperature can also affect the resistance of the gauge. One way of eliminating this effect is to include a second strain gauge aligned at 90° to the first. This means that its passive axis is aligned with the active axis of the first device. Changes in strain will only affect the first gauge but changes in temperature will affect both. When the difference between the two readings is taken the temperature effect is eliminated. It is possible to buy gauges containing two separate gauges pre-set at 90°. The circuit for this arrangement is shown in Fig. 14.20.

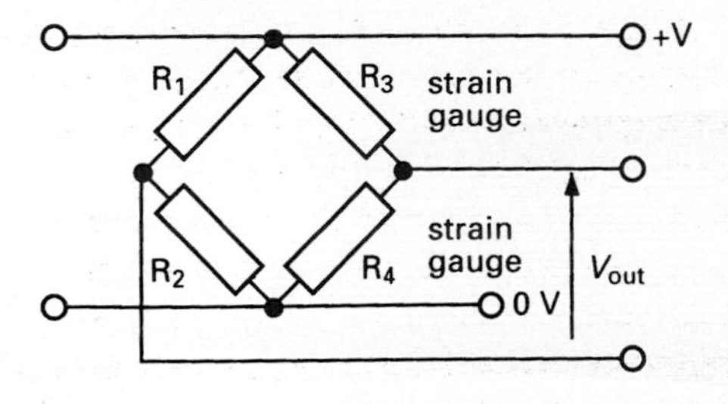

Figure 14.20 Two strain gauges in Wheatstone bridge arrangement.

If the strain occurs in a cantilever (of symmetrical cross-section) it is possible to put one gauge on the top surface, where it will undergo tension, and a second gauge underneath, where it will experience compression. This will give double the resistance change and also eliminate the temperature effect.

14.4.7 Magnetic flux density sensor

The strength of a magnetic field is traditionally measured by moving a search coil across the field lines and measuring the deflection produced on a ballistic galvanometer. Modern electronics has provided us with the alternative, as shown in Fig. 14.21.

Two outputs are available from this subsystem. One produces a signal that rises with increasing field strength whilst the other produces a signal that falls as the field strength rises.

A digital signal is usually obtained from a reed switch arranged in a potential divider, Fig. 14.22. In this arrangement the reeds will close together when the field strength becomes large enough and the signal will go from low to high.

The components can be exchanged to produce a signal that will fall when the magnetic field strength is greater than the predetermined value. This value is not adjustable. Although reed switches are quite cheap they are also quite large and fragile. A semiconductor Hall effect switch achieves the same effect but is more expensive. It is approximately the same size as a transistor and is connected as shown in Fig. 14.23.

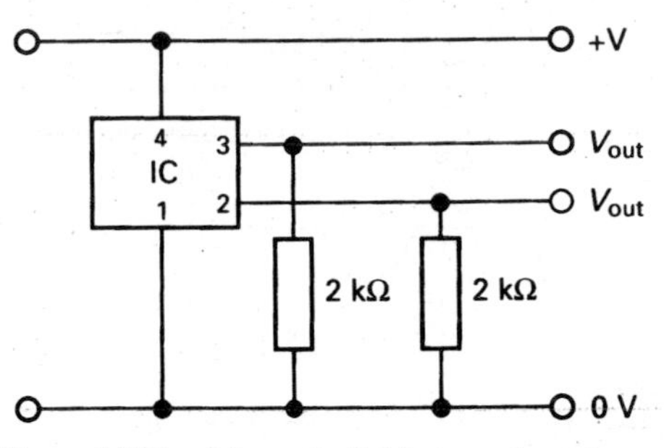

Figure 14.21 Magnetic field sensor based on linear Hall effect IC.

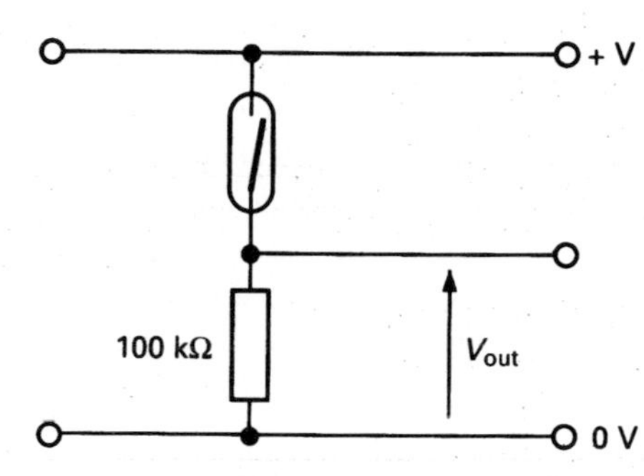

Figure 14.22 Reed switch used as a digital magnetic field detector.

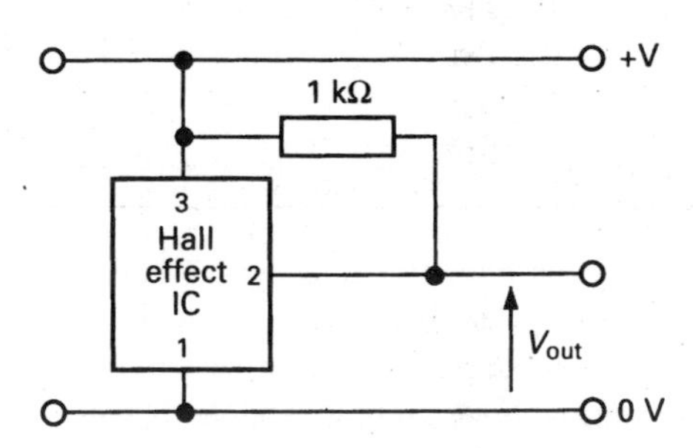

Figure 14.23 Digital magnetic flux density sensor based on Hall effect IC.

14.4.8 Sound sensors

Sound sensing is achieved with a *microphone*. Some microphones are active transducers whilst others are passive. The carbon microphone is a passive transducer and can be connected into a simple potentiometer circuit, Fig. 14.24. Changes in sound level produce resistance changes in the carbon.

Crystal microphones, on the other hand, are active transducers and generate an e.m.f. when they receive sound, Fig. 14.25. This has a high output impedance and would normally send its signal directly to an amplifier that is capable of amplifying this signal, that is, one with a high input impedance.

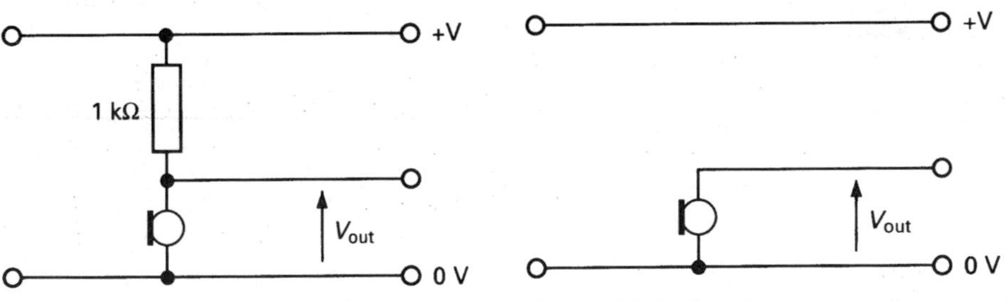

Figure 14.24 Sound sensor based on carbon microphone.

Figure 14.25 Sound sensor based on crystal microphone.

14.4.9 Pressure sensors

The simplest form of pressure sensor is given by some form of switching arrangement that will give a two-state output. The switch can be a simple push-to-make type or, for greater sensitivity, a microswitch may be employed. There are some special arrangements, known as pressure pads, which are used in intruder detection systems. Their function is exactly the same as a push-to-make switch. The arrangement of components is shown in Fig. 14.26.

The signal will rise if the switch is pressed. The components can change position, Fig. 14.27, if a low signal is required when the switch is pressed.

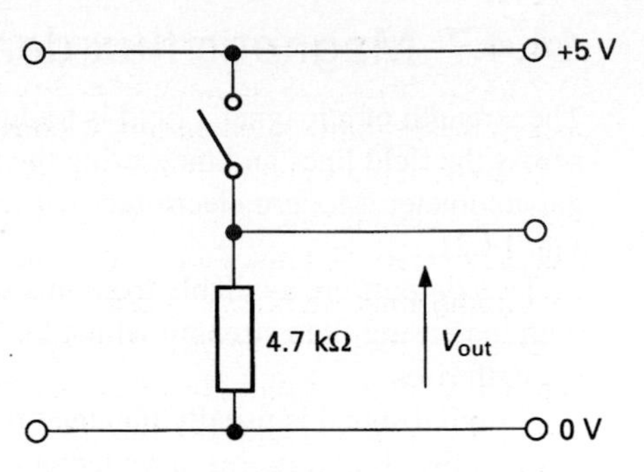

Figure 14.26 One form of pressure sensor.

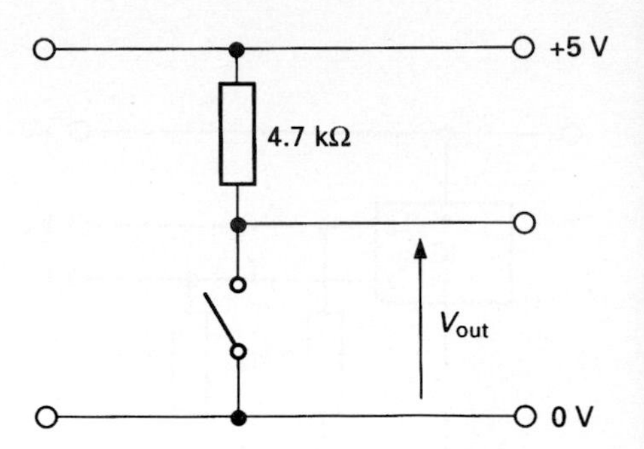

Figure 14.27 Pressure sensor giving alternative signal when pressure is detected.

14.5 Output transducers

At the output of an electronic system there must be some means of converting the electrical signal back into some other form of energy. The device that performs this conversion is known as an *output transducer*. Almost all output devices require significant currents to operate whilst most signals have a low electrical current. It is very often necessary to amplify the current just before the output transducer. Details of current amplification subsystems can be found in Chapters 15 and 16. As well as amplifying the current it may be necessary to code or decode the signal in a particular way. This may often be found in conjunction with the current amplification subsystem. A good example is the decoder-driver required for the seven-segment light emitting diode (LED).

There are two possibilities for passing current through output transducers. The current may flow from the positive power supply, through the output transducer and back into the driver, whose output is low. This is known as *sinking current*. The other option is for the current to come from the driver, which now has a high output, pass through the output transducer and go on to the 0 V power supply. This arrangement is known as *sourcing current*. Both options are represented diagrammatically in Fig. 14.28.

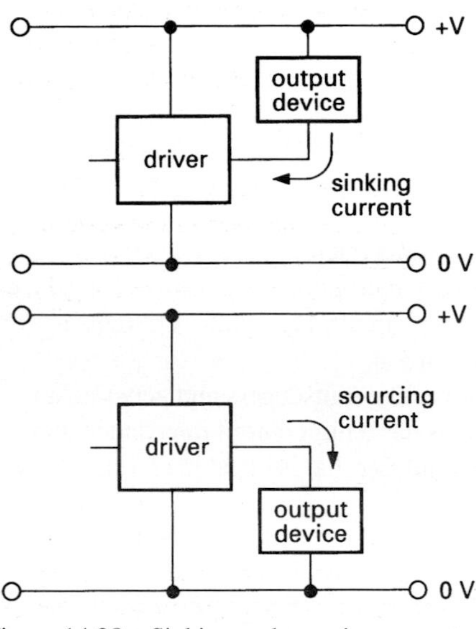

Figure 14.28 Sinking and sourcing current.

14.5.1 Motors – d.c. and stepping

The d.c. motor converts the electrical signal into rotational kinetic energy. Details of the construction of d.c. motors can be found in most physics textbooks. (See the bibliography to Chapter 20.) Before interfacing these devices it is necessary to know the working voltage and the maximum current drawn by the motor. This will enable the correct choice of driver to be made.

Some d.c. motors tend to be noisy – this is particularly true of cheap motors. The noise referred to is not sound but electrical noise. Inside the motor there is a part called the commutator which rotates against conductors called brushes. The commutator is made up of a series of separate sectors with insulation between them and, as the brushes pass from one sector to another, there is a switching of current. This rapid switching causes voltage spikes to appear on the power lines and this can disturb the working of the rest of the circuit. If electronic methods of suppression (usually a resistor in series with the motor and a capacitor across the terminals of the motor) are not successful it may be necessary to use a separate power supply for the motor and switch it on through a relay. This would not help if the circuit was designed to vary the motor speed in a manner proportional to the strength of the signal.

The *stepping motor* (also known as a *stepper motor*) requires pulses to drive it. It converts the signal into rotational kinetic energy but has the added possibility of moving a known amount for each pulse received. The most usual stepping motor is a four-phase device and this requires four signals to activate it. (One method, using a special driving integrated circuit, is considered in section 16.3.3.)

Each time the stepping motor receives a pulse the rotor advances by a known amount. This amount is different for different motors but a common value is 7.5°. If, with this value, 48 pulses are sent to the motor it will rotate through one complete revolution, since $48 \times 7.5° = 360°$.

Stepping motors are relatively low torque devices so they are often geared down to produce more torque. They cannot rotate very quickly compared to d.c. motors and, if they are geared down, that speed will further decrease. They are also large and heavy compared to d.c. motors. See section 16.3.3 for details of driving circuits.

14.5.2 Solenoids

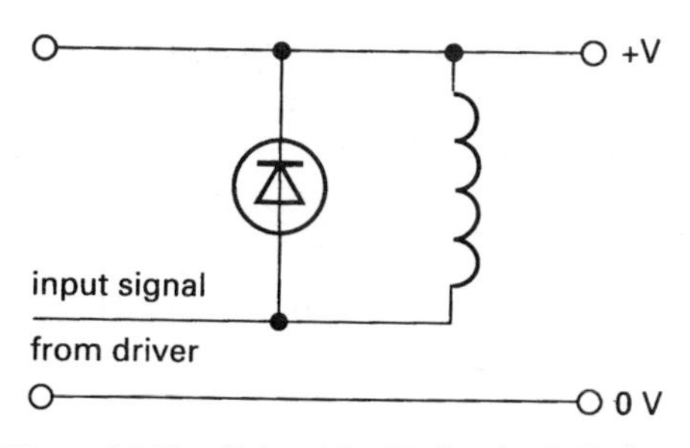

Figure 14.29 Solenoid with flywheel diode.

As with all output devices, the first consideration is to make sure the working voltage and maximum safe current are compatible with the rest of the system. A *solenoid* consists of a core of iron positioned in a coil. When a current flows through the coil, it becomes a magnet, causing the iron core to move. It converts the electrical signal into linear kinetic energy. A *flywheel diode* needs to be incorporated into the subsystem. This is because the solenoid is an inductive device, and when the magnetic field is turned off it generates a large back e.m.f. which would cause considerable damage to unprotected components in the driving part of the system. The flywheel diode allows the energy to be dissipated by providing an alternative path for the current. The circuit is shown in Fig. 14.29. A solenoid is the moving part of a relay (14.5.5) and devices such as solenoid valves, are used in washing machines and other pneumatic/hydraulic systems

14.5.3 Loudspeakers and buzzers

Both these devices convert the electrical signal into sound. A loudspeaker must receive an alternating signal and the sound it produces will be at the same frequency as the alternating signal. The impedance of the speaker is usually quoted along with its power rating. Typical impedance values are 8 ohm and 64 ohm and the power will range from 100 mW up to several tens of watts.

Most buzzers require a steady signal and they will produce a note at a fixed, predetermined frequency. The sound output power is often quoted in decibels for a certain distance from the device. Buzzers contain an oscillator circuit and usually have the polarity marked on the terminals. They must be connected in the circuit the correct way round or else the oscillator may be permanently damaged.

14.5.4 Lamps, LEDs and seven-segment displays

Lamps are relatively easy to drive. It should be remembered that the switch-on current will be significantly higher than the current when the filament is at its working temperature. This is due to the rise in resistance as the filament warms up. A lamp will be rated according to the maximum safe voltage and the current that will flow at this voltage. Thus a lamp rated at 6 V, 0.06 A can be seen to have a resistance of 100 ohms when it is working at normal brightness.

Light emitting diodes or LEDs are polarised devices and must be connected the right way round in a circuit. Since the maximum voltage to be applied across an LED is of the order of 2 V they almost always have a resistor connected in series. The value of the resistor is calculated from the voltage of the power supply and the current required by the LED.

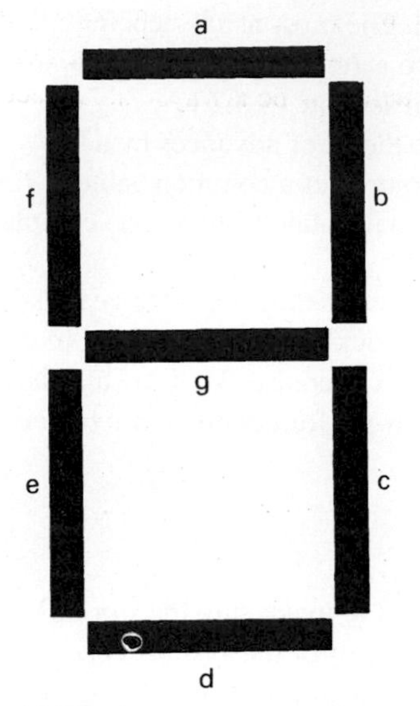

Figure 14.30 A seven-segment light emitting diode.

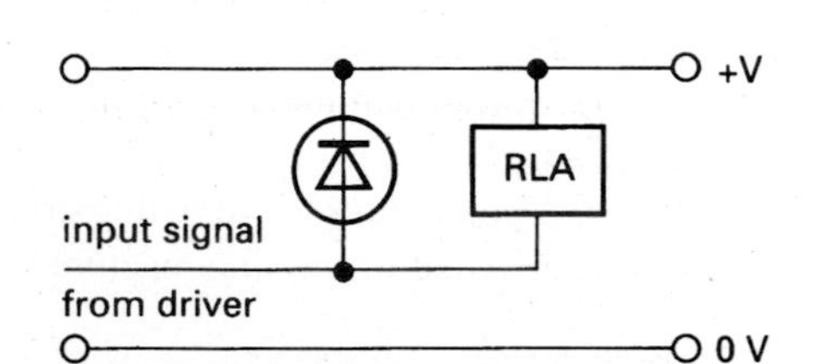

Figure 14.31 Relay with flywheel diode.

LEDs come in three main colours: red, yellow and green. Blue is available but it is expensive. There is a wide range of sizes and shapes – look at the manufacturers' catalogues for more details.

Seven-segment displays consist of seven LEDs as shown in Fig. 14.30. (There are actually eight since there is usually a decimal point.)

Because of the multiple switching required to show the numbers 0 to 9, a decoder-driver chip is usually used. The displays are available as either common cathode or common anode depending on whether the current is to be sunk or sourced. This involves joining all the cathodes or anodes together to reduce the number of connections needed. Since they are LEDs it is still necessary to use a series resistor for each LED.

It is possible to get the displays in different sizes and there are packages with two seven-segment LEDs connected together. Multiplexed displays, which involve switching each seven-segment on in turn, require more sophisticated signal processing.

There are other types of display, notably liquid crystal and fluorescent. Both of these require more complex driving techniques using special integrated circuits.

14.5.5 Relays

The relay is not really an output device but a switch. It may however switch on an output device and it needs driving in exactly the same way as other output transducers. Since it is an inductive device it needs a flywheel diode, Fig. 14.31, in exactly the same way as the solenoid (see section 14.5.2).

Relays can be rated according to the maximum current they can switch, their operating voltage, their coil resistance and their size. They also have different switching possibilities; double pole double throw (DPDT) and single pole double throw (SPDT) are the most common.

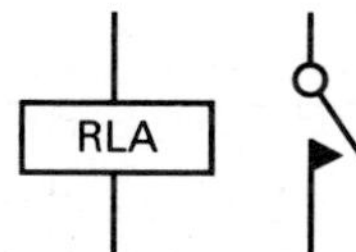

a Relay with SPST (single pole, single throw) contacts

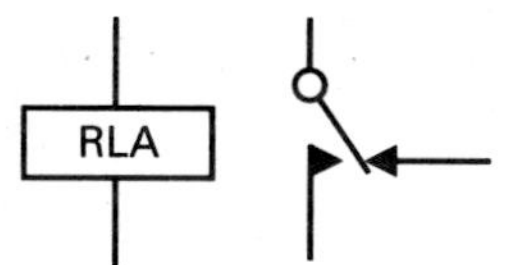

b Relay with SPDT contacts

Figure 14.32

Relay as a latch: thyristor as a latch

Before electronic latches existed a common way of providing latching was to use either a relay or a thyristor to latch the output and prevent it resetting if the signal that triggered the system was removed.

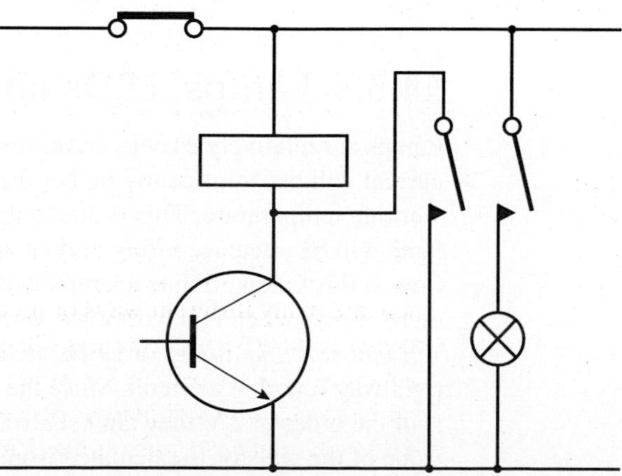

Figure 14.33 Relay as a latch

In Fig. 14.33, when the relay is activated both pairs of switches are activated. The first switch effectively shorts out the transistor so even if the transistor turns off the relay won't. The signal lamp (or bell or buzzer) will now be always on. A reset switch is usually provided that breaks the power supply to the relay

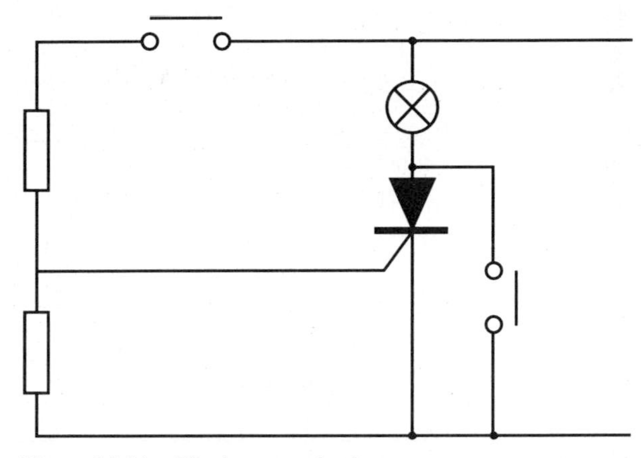

Figure 14.34 Thyristor as a latch

A thyristor (or silicon-controlled rectifier) is a three terminal device which passes current from anode to cathode, only if a gate current of about 20 mA flows (Fig. 14.34). Once the current has started, removing the gate current will not turn off the main current. Oncc again manual reset needs to be provided.

14.5.6 Meters and oscilloscopes

Initially, meters and oscilloscopes may not be considered to be output devices but further thought will show that they are converting the electrical signal into another form of energy so they do indeed fall into this category.

They are slightly different in that they do not just appear as the final stage of a system, but can be used between any of the stages to investigate the form of the signal. It is important to use high impedences so that the characteristics of the meter or oscilloscope do not significantly affect the form of the signal.

The choice between meter or oscilloscope is not always obvious. An oscilloscope is a voltage measuring device so it can be used, at any time, in place of a voltmeter. It has the ability to display a trace of the signal on its screen and many oscilloscopes can deal with more than one signal at the same time. It is therefore more useful in looking at signal shape and comparing two signals, particularly rapidly changing ones.

If the sole concern is to give a value to the size of the signal then a voltmeter would be the more appropriate device, but there may still be a choice to be made between a digital and an analogue meter.

14.6 Analogue and digital signal processing

There are many different ways of processing signals; Chapter 15 deals with processing analogue signals and Chapter 16 is concerned with digital signal processing. It is stressed that it would be poor design to take an analogue signal and send it directly into a digital processing stage although it might operate correctly. It would be far better to convert it to a digital signal first, as explained in section 16.4.2, before trying to process it.

14.7 Matching of sub-system

The traditional system diagram (INPUT, PROCESS, OUTPUT) can be expanded in two different ways. One way is to include a feedback loop as in Chapter 17. Another way is to take into account the complexities of matching the signals between sub-systems.

When we draw the diagram we assume the signal can cross from input to process and from process to output. Manufacturers of system kits have gone to great lengths to ensure that the system works no matter what inputs, processes and outputs are connected together.

Let us consider a slightly different system diagram (Fig. 14.35):

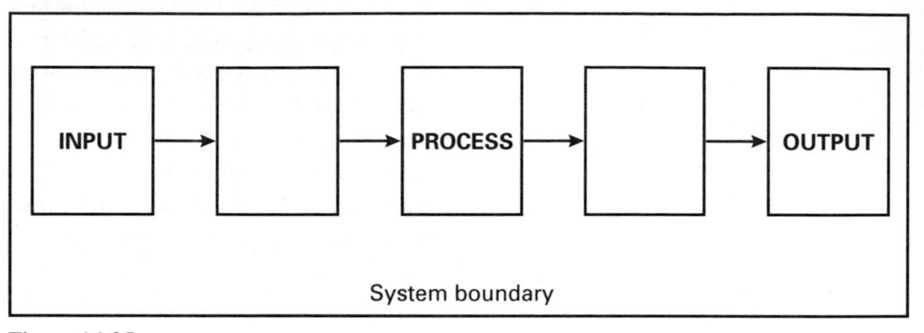

Figure 14.35

This shows the system boundary, and a diagram with our usual boxes together with two unlabelled boxes. What are these boxes? Let's call them 'signal modifiers'.

If we go back to first principles – the minimum system requirements would be an input and an output. We do not always have to process a signal. A quick example of this would be an automatic lamp – it comes on when it goes dark. The minimum requirements are a light sensor and a lamp (Fig. 14.36) (or relay connected to a lamp).

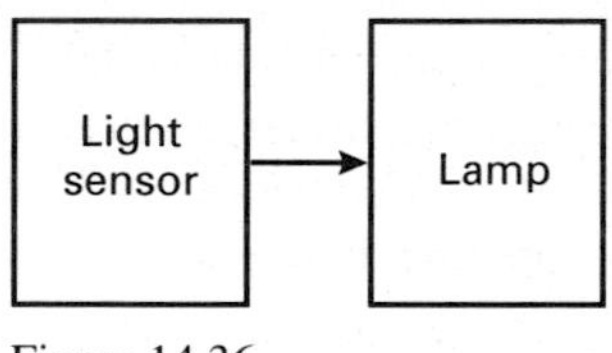

Figure 14.36

This could be achieved as shown in Fig. 14.37

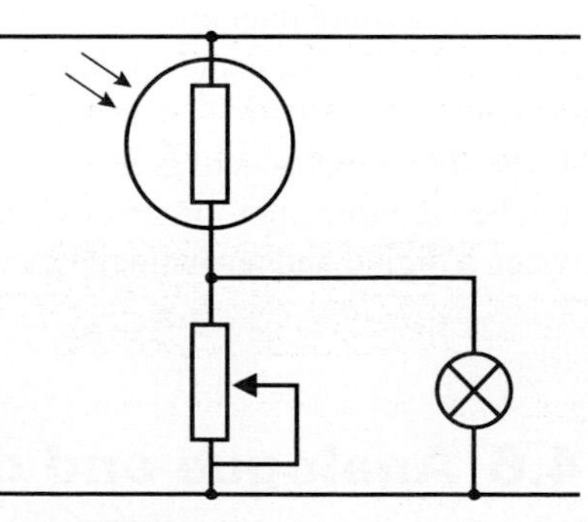

Figure 14.37

If you set this up you will find nothing happens when it goes dark! If you separate the two sub-systems and check the signal you will find they each work perfectly but they will not work when connected together. Of course, it is not true to say they do not work – they just do not work as you expected!

The problem is due to the impedance of the sub-systems. In this case, the load (lamp) has a low impedance. (You can think about impedance as being the same as resistance for the moment.) That means it will draw a large current from the input sub-system in order to operate. The input has a high output impedance – it does not have the ability to deliver large currents.

So we can say that the signal depends on both the current and the voltage. As more current is drawn the signal gets smaller and is therefore corrupted. (Think about the total power of the signal. Power in = Power out etc.) (The maximum power is transferred when the impedances are the same.)

Digital signals are less affected by this kind of corruption but it can happen with low impedance output devices. Analogue signals are more easily corrupted.

It is obvious that signals should be fed from low impedances into high impedances in order to reduce the current drawn and hence reduce corruption of the signal. If this is not possible by adjusting components within the sub-system then extra modules must be inserted which will provide for this matching *without corrupting the signal themselves*.

The most common matching sub-system you have used so far is a driver. This is a transistor based current amplifier system (or power amplifying sub-system since it boosts the current while keeping the voltage the same). It is often referred to as a PROCESS sub-system but, since it does not corrupt or change the form of the signal, except by inverting it, it is easier to think of it as a *signal modifier or signal conditioner*. It has a high impedance input (takes low current to operate) and a low impedance output (provides high output current).

It is not the only signal modifier!

Others include: debouncing a switch input, latching, buffers – both inverting and non-inverting – unity gain amplifiers – both inverting and non-inverting – the inverter and all drivers.

14.8 Component information

14.8.1 Resistor information

Types of resistor

Most resistors available today are of the carbon film type. Metal film resistors are a slightly more expensive alternative but are more stable under severe conditions. Wirewound resistors are used where high power dissipation is expected.

4-band colour code

Most resistors are marked with a colour code consisting of several bands, usually four, as shown in Fig. 14.38.

Figure 14.38 A four-band resistor.

The colour code is given in Table 14.1.

The three bands that are closer together give the resistor value whilst the single band gives the tolerance. The first two bands are converted into their number equivalents whilst the third band gives the number of zeros to be added. For example, if the first three bands are red, white, orange then the first two numbers are 2 (red) followed by 9 (white). The orange band tells us to add three zeros making 29 000 ohms or 29 kΩ.

Colour	*Number*
black	0
brown	1
red	2
orange	2
yellow	4
green	5
blue	6
violet	7
grey	8
white	9

Table 14.1

5-band colour code

Many of the higher tolerance metal film resistors use a five-band colour code. In this case the first three bands are converted directly to their number equivalents whilst the fourth band gives the number of zeros to be added. For example, if the first four bands are white, black, brown, orange then the first three numbers are 9 (white), 1 (brown) and 0 (black). The orange band tells us to add three zeros so the value is 910 000 Ω or 910 kΩ.

Tolerance

The final band on each type of resistor gives the tolerance. This indicates the variability of the measured value given by the colour code. All four-band resistors have a 5% tolerance which means that the measured value should be within ±5% of the value given by the colour code. The tolerance band showing a 5% tolerance is coloured gold. (Older resistors had a 10% tolerance shown by a silver band.) For example, a 1 kΩ resistor with a 5% tolerance can have a measured value between 950 and 1050 Ω. The five-band resistors have a tolerance of 2% shown by a red band or 1% shown by a brown band.

Numerical coding

Colour codes work well provided the colours are consistent and you are not colour blind. It is surprisingly easy to confuse red and orange if they are not together on the same resistor! There is a numerical code that avoids the use of the Ω sign as well as any colour. In this code (BS 1852) the capital letters R, K and M are used in place of the decimal point. At least two figures should always be shown.

Examples:	0.22 Ω	is marked R22
	1 Ω	is marked 1R0
	4.7 Ω	is marked 4R7
	39 Ω	is marked 39R
	100 Ω	is marked 100R
	1k Ω	is marked 1K0
	10k Ω	is marked 10K
	1M Ω	is marked 1M0
	2.2M Ω	is marked 2M2

The coding is often used for resistor values in circuit diagrams. The tolerance value is given by another letter at the end. F = 1%, G = 2%, J = 5% and K = 10%. Another type of numerical code in use involves just three figures. The first two figures are significant whilst the third figure is the multiplier (number of zeros). Thus 102 would be 1000 Ω .

Preferred values

Resistors are only available in certain values known as preferred values. It is not necessary to have resistors in all values due to the tolerances. The two sets of values normally obtainable are known as E12 and E24. The E12 set is based on 10% tolerance and contains twelve plus their multiples, whilst the E24 series is based on 5% tolerance and has twenty-four values plus multiples.

E12: 10 12 15 18 22 27 33 39 47 56 68 82

E24: 10 11 12 13 15 16 18 20 22 24 27 30
33 36 39 43 47 51 56 62 68 75 82 91

Power values

The power dissipation of resistors is calculated from the formulae shown in section 20.3. For most applications 0.25 W resistors are the normal choice.

Resistors in series and parallel

To calculate the total resistance of resistors in series use the formula:

$$R_{\text{total}} = R_1 + R_2 + R_3 \text{ etc.}$$

To calculate the total resistance of resistors in parallel use the formula:

$$\frac{1}{R_{\text{total}}} = \frac{1}{R_1} + \frac{1}{R_2} + \frac{1}{R_3} \text{ etc.}$$

14.8.2 Capacitor information

Different types of capacitor

Capacitors can be classified into two main types: polarised and non-polarised. Polarised capacitors can be either electrolytic or tantalum. Electrolytic capacitors tend to be of large value but are very leaky. Tantalum capacitors, on the other hand, are only available in a restricted range of values and only at low working voltages. The main types of non-polarised capacitors are ceramic, polyester and mica. It is rare to find a non-polarised capacitor with a value greater than $1\,\mu$F. Mica capacitors are used for high frequency work when accurate values are required, ceramic capacitors are used for decoupling and polyester capacitors for general use.

Colour code

Some polyester capacitors use a colour code to identify their values but they are rapidly being replaced with the value printed on the side. The colour code is identical to that used for resistors (see section 14.8.1) but it should be remembered that the value is in picofarad (pF).

Numerical coding

There are a large number of numerical codes shown on capacitors. The use of a three-number code is common where the first two numbers are significant and the third number is the multiplier. This is widely used on small ceramic capacitors and the value will be in picofarad. Thus 100 will be 10 pF (1,0 for 10, followed by 0 (no) zeros!). This can be confusing. It is also common for p, n or μ to be used in place of the decimal point as with resistors. (In some older capacitors μ was shown as m.)

Tolerance

The tolerance of some capacitors can be high: 20% for polarised capacitors, generally below 20% for non-polarised capacitors. Manufacturers' catalogues give more details.

Preferred values

The range of values is much more restricted than for resistors. Often it is limited to 10, 22, 33 and 47 with multiples. Again, more details can be found in manufacturers' catalogues.

Working voltage

This is the maximum voltage, a.c. peak-to-peak or d.c., that the capacitor can stand before the dielectric breaks down. It is important to ensure that the working voltage is never exceeded.

Capacitors in series and parallel

To calculate the total capacitance of capacitors in parallel, use the formula:

$$C_{total} = C_1 + C_2 + C_3 \text{ etc.}$$

To calculate the total capacitance of capacitors in series, use the formula:

$$\frac{1}{C_{total}} = \frac{1}{C_1} + \frac{1}{C_2} + \frac{1}{C_3} \text{ etc.}$$

Note the difference between these formulae and the resistor formulae for series and parallel.

Assembly techniques

There is no doubt that large amounts of time are spent within project work getting a prototype to function as expected. There is no suitable alternative to using a PCB technique for a final product. The UV method produces the most reliable results. For more complex circuits, the use of suitable PCB drawing software can reduce the time spent and may be able to simulate the final circuit before committing to the final PCB. It should, however, be remembered that many of these software packages use 'idealised components'. That is to say that they have zero tolerance.

Using a system kit that includes a breadboard is the fastest way to prototype. Use existing modules for standard building blocks and use the breadboard for 'missing' sub-systems or variations on a standard sub-system.

Stripboard is a popular half-way house but it is a difficult skill to learn and prone to errors. Its main use is by experts as a prototyping system for temporary use. The skills needed are not transferable and so this cannot be recommended for school and college use. Wire wrapping is a similar skill but does not require as much learning and so could be an acceptable alternative. As the move to surface-mount component continues it is becoming more difficult to fabricate prototype circuits. Surface mount components cannot be wire wrapped or put into a breadboard!

Bibliography

Brimicombe M 1985 *Electronic Systems*. Nelson

Hartley-Jones M 1989 *A Practical Introduction to Electronic Circuits*, Cambridge University Press

Horrowitz P and Hill W, 2nd edn. 1989 *The Art of Electronics*. Cambridge University Press

NEMEC 1994 *An Introduction to the Systems Approach Using Alpha Student Worksheets*. NEMEC Publications, University of Southampton (also available from Unilab, Novara House, Excelsior Rd, Ashby-de-la-Zouch, LE65 1NG)

The Alpha Resource. 1992. Unilab

Catalogues

Farnell Electronics Components Catalogue www.farnell.co.uk

RS Catalogue and RS Data Sheets rswww.com

Rapid Electronics Catalogue. Rapid Electronics Ltd, Severalls Lane, Colchester, Essex, CO4 5JS www.rapidelectronics.co.uk

Assignments

1 Explain the difference between an analogue sensor and a simple digital sensor giving an example of each type.

2 Explain with the aid of sketches what you understand by the following specifications for mechanical switches:
(a) SPDT slide switch;
(b) DPDT toggle switch;
(c) Momentary push-to-make switch.

3 A linear potentiometer is to be used as a reference voltage device in a comparator circuit as shown in Fig. 14.39.

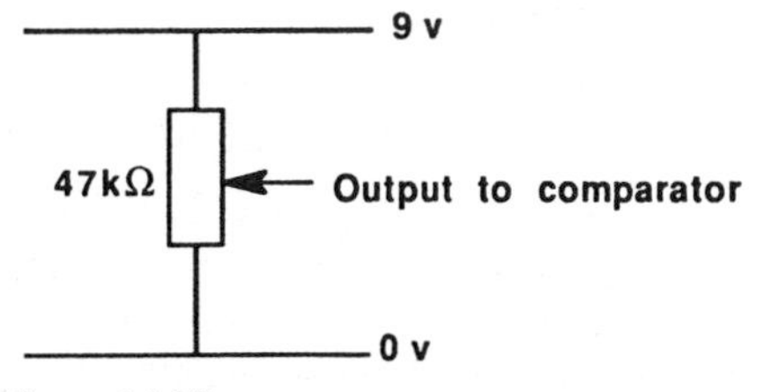

Figure 14.39

(a) Explain what happens to the output voltage as the potentiometer shaft is rotated.
(b) Draw a graph to show how the output voltage varies with the rotations of the potentiometer shaft.

Additional electronics assignments

Some examination questions on electronics relate to more than one chapter. They may be capable of being solved by digital or analogue technology and they may also require input signals as described in the present chapter. In solving the assignments that follow you may be expected to use Chapters 14, 15 and 16.

4 Develop a circuit which would display the speed of a motor vehicle, giving warning when predetermined speed limits have been exceeded. Special consideration should be given to each of the following:
(a) power supply
(b) speed sensor
(c) processing circuitry
(d) display device
(e) working environment
Explain, with reference to block and flow diagrams, the principles of operation at each stage.

5 A security alarm system for a large building, such as a school, will usually have several sensing devices connected to a central control unit which will trigger the alarm.
Using an annotated schematic diagram, show:
(a) the main components of such a system explaining their functions
(b) how the system can be tested
(c) how the location of any breach of security can be identified.

6 A temperature control unit is to be fitted to the sleeping quarters of an outdoor aviary. The designer has decided to investigate two possibilities prior to final selection.

method 1 – switch the heater on and off repeatedly to limit the temperature rise,

method 2 – to use a temperature control system to monitor the temperature inside the sleeping area and switch the heater on and off as appropriate.

The control system will be powered by a power supply unit (psu) whilst the heater will need to run off the mains (240 V/250 mA).

(a) Produce a system diagram for both possible methods of control.
(b) Consider and select which is the more appropriate type of system for the needs of the aviary birds and list the reasons for your choice.
(c) Design a circuit by considering individually each 'black box' of your system diagram, and describe its function.
(d) Produce a finalised circuit diagram; specific component values are not required.

7 An automatic clock tea-maker has a number of necessary and important control functions so that the overall system can operate safely, reliably and effectively. If we ignore the clock/timekeeping aspect of the device which will trigger the tea-maker at a chosen preset time, the other control functions are incorporated to ensure that the tea-making process can only start if certain safety conditions are met.

These are:

(i) the water heating chamber has water in it to cover the mains powered heating element.
(ii) the teapot is in the correct position to receive the boiling water which is transferred through a tube from the water heating chamber by pressure created by the steam from boiling the water.

(a) Draw a block diagram for the safe and reliable operation of the internal control function.
(b) For each individual block of your diagram, show an electronic/microelectronic circuit which will fulfil the required control function. (No specific values of components are required. Pin numbers for any integrated circuits (ICs) used will not be required.)
(c) Draw a complete circuit diagram with all the elements of the system matched to create the safe, reliable and effective product as described above.

8 A manufacturer wishes to incorporate a control system in a child's battery powered toy so that it will not fall off the edge of a raised surface or run into obstructions.

(a) Explain two different ways of sensing obstructions/edges.
(b) Draw a system diagram showing the main components of a toy control system.
(c) Explain the function of each component and the relationships between them.
(d) Explain in detail how a control problem created by sensing an obstruction and an edge at the same time might be overcome.

9 (a) The digital logic system of four 2-input NAND gates shown in Fig. 14.40 has two inputs, A and B, and one output S.

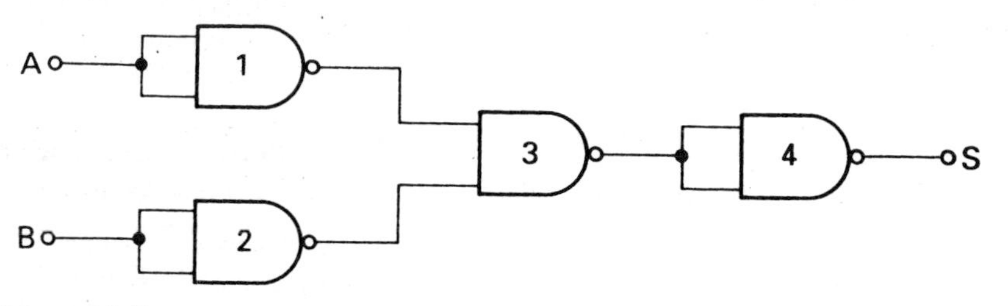

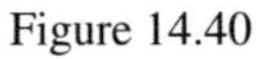
Figure 14.40

(i) Why is this digital logic system called combinational logic?
(ii) Produce a truth table which shows the logic state of the output for all combinations of the logic states of the two inputs. State the logic function of the system.
(iii) Draw diagrams similar to Fig. 14.40 which show how to produce
(a) an OR gate
(b) an Exclusive-OR gate from NAND gates.
(iv) State one advantage of using combinations of only NAND gates to produce other logic functions.

(b) Fig. 14.41 shows a block diagram of a simple digital instrument for detecting the 'highs' and 'lows' of the signals in a digital circuit. It is called a logic probe. The instrument is designed to operate from the 5V supply of the logic circuit. When the input probe is 'high' the red LED lights; when the input probe is 'low', the green LED lights.

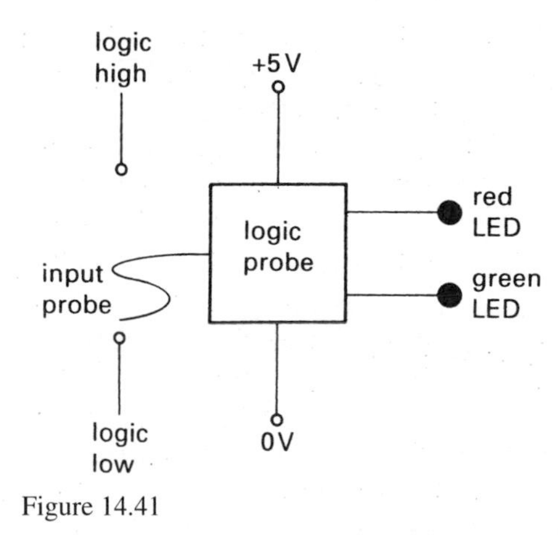

Figure 14.41

(i) One possible design of a circuit for this simple logic probe is shown in Fig. 14.42 and is based on two operational amplifiers, IC_1 and IC_2. The red LED is to light for input voltages greater than 2.4 V and the green LED is to light for input voltages less than 0.8 V. Calculate approximate values of the resistors R_1 and R_2 needed for the circuit to respond to the voltages stated above.
(ii) State whether each of the inputs marked P, Q, R and S is an inverting or noninverting terminal of the relevant operational amplifier.
(iii) Explain briefly the operation of the logic probe and state one of its limitations for testing logic circuits.

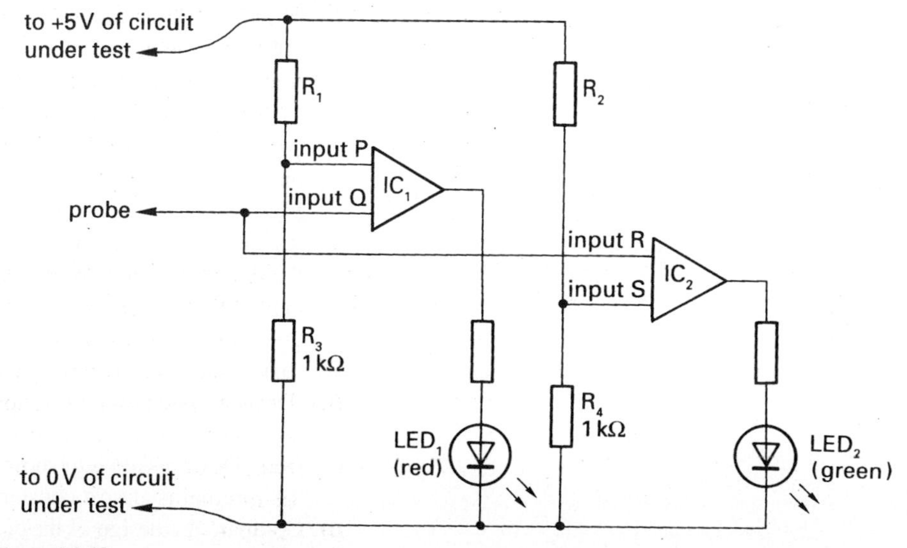

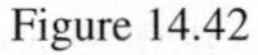
Figure 14.42

(c) A food-processing system comprises four containers that hold different materials. Containers A and B have liquid level sensors fitted to them and tanks C and D have temperature sensors fitted to them.

Assume that the liquid level sensors produce a logic high when the level is too high and a logic low when the level is acceptable. Also assume that the temperature sensors indicate logic high when the temperature is too low and logic low when the temperature is acceptable.

(i) Write down the Boolean equation for the logic function of this system which provides a logic high alarm signal when the level in tanks A or B (or both) is too high at the same time that the temperature in tank C or D (or both) is too low.

(ii) Design a logic circuit, using 2-input NAND gates only, which performs the logic function required.

10 (a) In each of the two cases shown in Fig. 14.43, explain briefly how the diode D protects the circuit or component from possible damage.

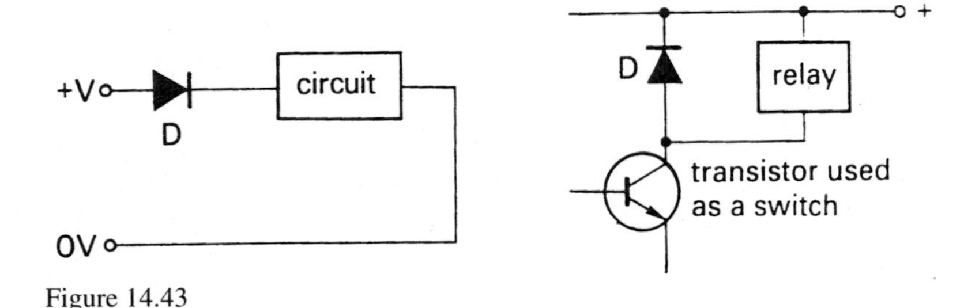

Figure 14.43

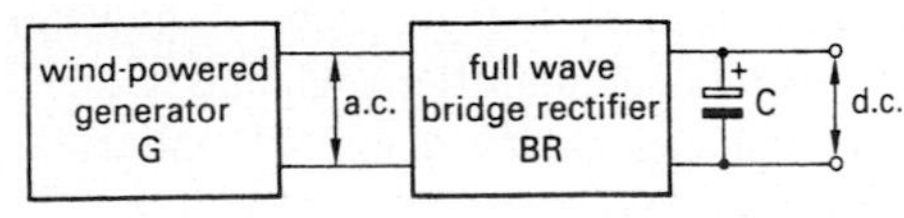

Figure 14.44

(b) Diodes were to be important components in the design of the system shown in Fig. 14.44. This system was to be used by a team of explorers for recharging a 6 V battery pack from a portable wind-powered electrical generator G. The generator produced a sinusoidal alternating current, the amplitude of which increased with wind speed. Since a direct current supply was needed for charging the battery pack, a full-wave bridged rectifier BR was used to convert the a.c. to d.c.

(i) Draw a diagram which shows the arrangement of diodes in the bridge rectifier and explain its effect on the waveform of the alternating current from the generator.

(ii) Draw a diagram to explain the purpose of component C.

(c) The d.c. voltage required to charge the battery pack was not to exceed 7.4 V. However, on measuring the d.c. voltage produced by the generator for the range of wind speeds expected on the expedition, the explorers found that the maximum rectified d.c. voltage produced by the strongest winds expected was 11 V. In order to limit the charging voltage to not more than 7.4 V, the explorers designed the voltage regulator circuit shown in Fig. 14.45. It is based on a Zener diode D_1 and npn silicon transistor Tr_1. Assume that the maximum charging current for the battery is 250 mA and the d.c. current gain of Tr_1 is 25.

(i) What breakdown voltage should the Zener diode have if the maximum charging voltage of 7.4 V is not exceeded?

(ii) Calculate the base current of the transistor when the battery is being charged at its maximum current. Comment on the relationship between this current and the current flowing through R_1.

(iii) Draw a sketch to show how the voltage drop across the Zener diode remains almost constant as the unstabilised input voltage varies.

(d) Compare the design of the constant voltage power supply shown in Fig. 14.45 with one that uses a purpose-designed integrated circuit voltage regulator.

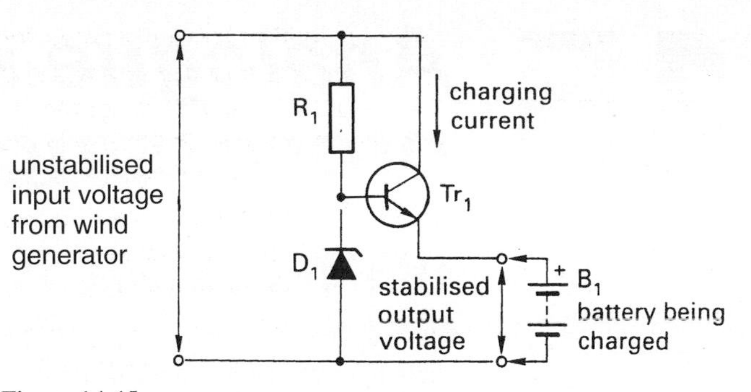

Figure 14.45

(e) Imagine you were one of the team of explorers and you had to advise the rest of the team on the relative merits of using a wind or solar-powered battery charger on your expedition. In terms of portability, cost, reliability, ease and convenience of use, and safety, evaluate in about 100 words the two types of charger given that you expect both sunshine and wind in about equal amounts.

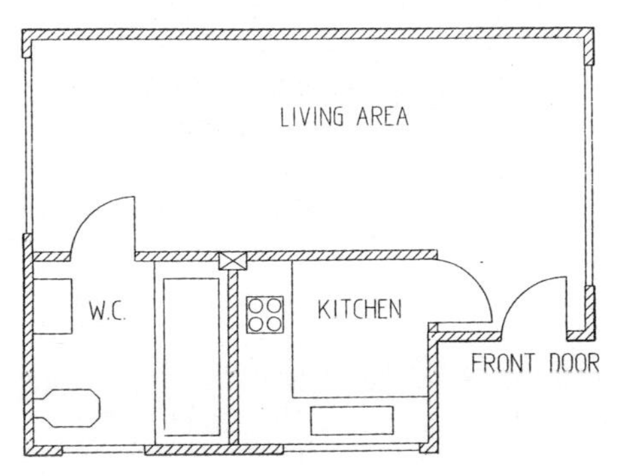

Figure 14.46

11 Fig. 14.46 shows a small ground floor studio flat which has suffered a series of burglaries. As a designer of security systems, your brief is to develop an electronic system to deter burglars. In order to fulfil the brief you should complete the following:

(a) draw up a range of design proposals in response to the brief;
(b) show by means of a systems diagram how you may start to develop your prototype system;
(c) state the component parts of each module, and explain their function;
(d) produce a finalised circuit diagram.

12 Automatic control systems for tuning on lights, opening doors, or activating alarms requires a means of detecting the presence of a person.

(a) Describe three different ways of detecting the presence of a person.
(b) Choose one situation where such a system may be used and state at least three factors which need to be taken into account in the design of a suitable means of detection.
(c) For each of the three means of detection explain how the factors in (b) influence their design and use.

15 Analogue signal processing

15.1 Transferring the analogue signal

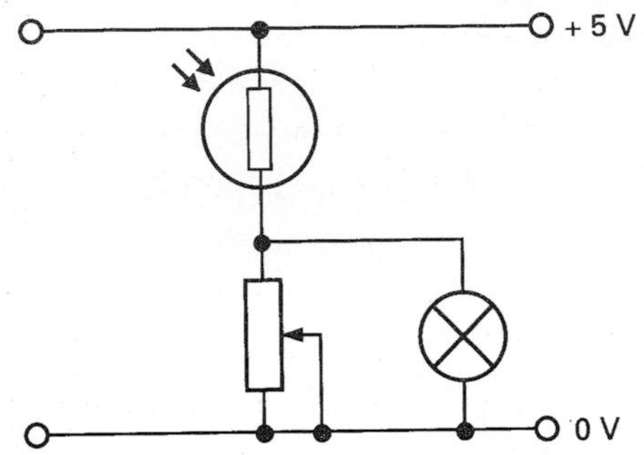

Figure 15.1

When an analogue signal is transferred from one subsystem to another there can be a challenge. Whilst it is quite difficult to lose an analogue signal totally, it is relatively easy to corrupt the information it carries. There are two possible ways this information can be corrupted. Firstly, it can be corrupted by adding *noise* to the signal. Electrical noise is caused whenever a signal passes through an electronic component but certain components are known to be very noisy. Carbon composition resistors are a particularly good example of components that introduce a lot of noise into a signal. If the next stage of a system is designed to amplify the signal, then any noise introduced before amplification will be further amplified and corrupt the signal still more. It is impossible to remove noise but a careful choice of components in a preamplification subsystem should help to preserve the integrity of the signal.

The second method of corrupting a signal is to get the impedance of the two subsystems wrong. (Impedance is examined in section 20.6.) If a high impedance signal feeds into a low impedance stage the information will certainly be corrupted and may even be totally lost. A simple example of this is a light sensor trying to drive a lamp directly, Fig. 15.1. The light sensor has a high impedance output but the lamp has a low impedance input and will not light.

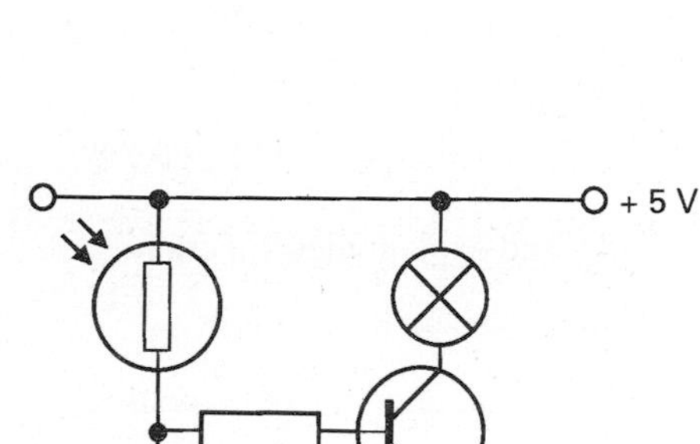

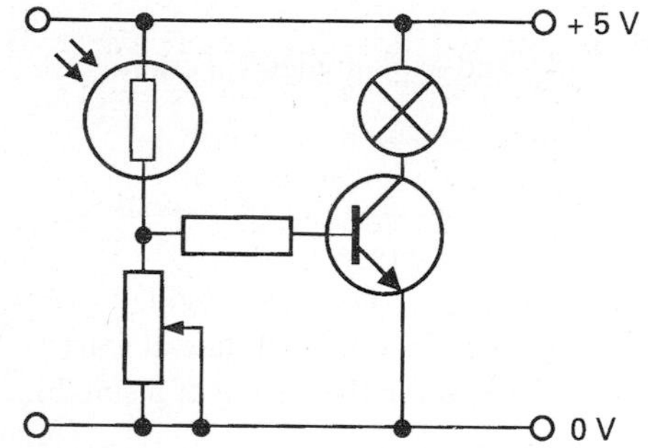

Figure 15.2

There are two possible solutions to this problem. The first is to put an *analogue buffer* between the stages whilst the second is to ensure that the next subsystem has a high input impedance. In the present case a current amplifier, based on a transistor, would be the most suitable arrangement since it has a high input impedance and a low output impedance, Fig. 15.2.

The most common analogue buffer (not to be confused with a digital buffer) is a special case of the noninverting amplifier shown in section 15.2.2. It should be remembered that all op-amp based configurations have a relatively high input impedance.

15.2 Amplifying the analogue signal

One of the most common processes applied to analogue signals is amplification. It should be emphasised that this is voltage amplification, since it is the magnitude of the p.d. that carries the information. There are two main types of signal amplifier – the *inverting amplifier* and the *noninverting amplifier*. This section will examine the differences and similarities between these two functions. The *op-amp* is commonly used for both these functions although it is possible to achieve the same effects with transistor-based subsystems. It is generally held that the transistor-based circuits are more expensive and offer no significant advantages over op-amp based circuits. Op-amps have removed many of the inherent instabilities of transistors by building a more complex circuit on one piece of silicon.

15.2.1 The inverting amplifier

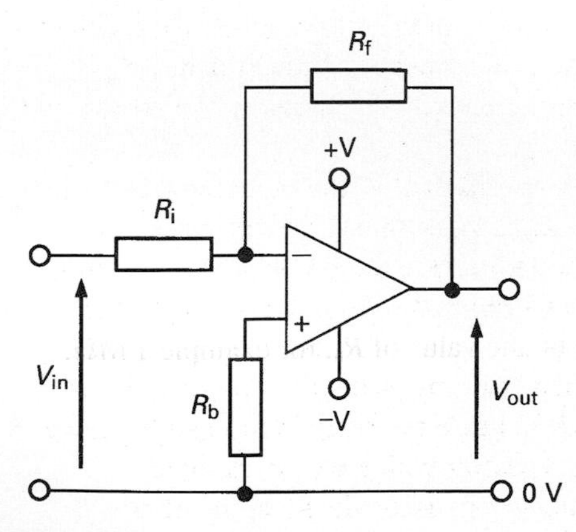

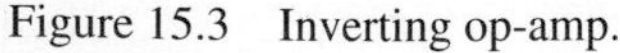

Figure 15.3 Inverting op-amp.

The standard circuit for this function is shown in Fig. 15.3.

The signal gain, A, is given by V_{out}/V_{in}. In this arrangement the signal gain is determined entirely by the resistors, not by the gain of the op-amp.

$$A = \frac{V_{out}}{V_{in}} = -\frac{R_f}{R_i}$$

The use of the minus sign indicates that the output is 180° out of phase with the input, that is, it is *inverted*. The limit to the amount of gain is the value of the power supply since it is never possible for V_{out} to exceed the value of the power supply. Indeed, the maximum or minimum value of V_{out} can be expected always to be within, and not equal to, the value of the power supply. On a ±12 V supply, it is unlikely that V_{out} will exceed ±11 V.

Note also that the minimum value of the signal gain can be less than one, so this circuit can attenuate signals as well as amplify them. Making R_f equal to R_i will produce an inverting analogue buffer. The output impedance of this arrangement is of the order of 1 kΩ whilst the input impedance is the value of R_i. This means that the input impedance is adjustable to meet the needs of the previous subsystem.

The inclusion of R_b improves the performance of this arrangement. It improves the symmetry by preventing any unwanted d.c. bias on the output. Its value is determined by the formula

$$\frac{1}{R_b} = \frac{1}{R_f} + \frac{1}{R_i}$$

The choice of the op-amp is important. Most op-amps are designed to work on a dual rail supply but there are some that will work well on a single rail supply. The venerable 741 is often the one shown in textbooks (as well as syllabuses and exam papers) but the 081 offers a better performance whilst the 071 is a low noise equivalent.

It may be necessary to include a decoupling capacitor in order to amplify a.c. signals. The impedance of the capacitor must be carefully calculated to allow through the range of frequencies it is proposed to amplify. The whole circuit is frequency dependent at higher frequencies. The exact frequency at which the gain will start to fall is determined by the choice of op-amp and the required gain. The data sheet for the op-amp will give a figure for the *gain-bandwidth product*. If the gain is 1000 and the gain-bandwidth product is 1 MHz, full gain can only be obtained up to 1 000 000/1000 = 1000 Hz. Above this value of frequency, the gain will reduce by a factor of ten as the frequency increases by a factor of ten.

15.2.2 The noninverting amplifier

The standard op-amp based circuit is shown in Fig. 15.4.

The signal gain, A, is again given by V_{out}/V_{in} in and this is governed entirely by the resistor values.

$$A = \frac{V_{out}}{V_{in}} = 1 + \frac{R_f}{R_i}$$

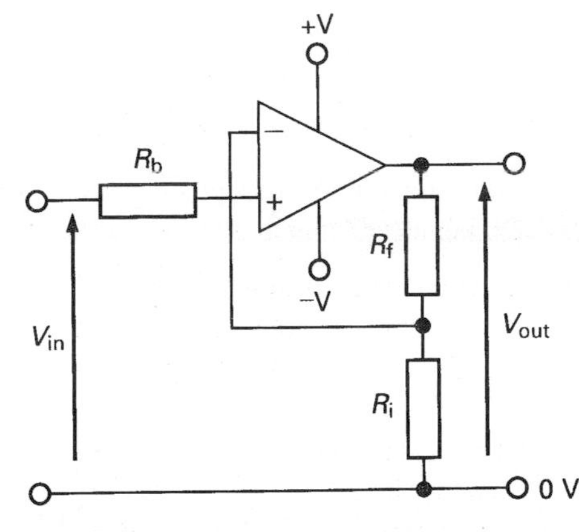

Figure 15.4 Noninverting op-amp.

In this case, the output signal is in phase with the input signal and it should be noted that this circuit cannot attenuate signals as the minimum value for V_{out}/V_{in} is one. As with the inverting amplifier, V_{out} cannot exceed, or even reach, the power supply rails so this can limit the gain.

The output impedance is again in the order of 1 kΩ whilst the input impedance is the input impedance of the op-amp. Since this can be as high as 10^{12} Ω it provides a very high input impedance. R_b has been added to increase the stability of the circuit, particularly when the decoupling capacitor has been added for a.c. signals. In this case, the input impedance of the system will be the value of R_b, for example 1 MΩ. This circuit is also frequency dependent in the same way as the inverting amplifier.

If R_f is made to be zero ohm (a wire) and R_i is made to be infinity (a gap) then we have a gain of one and a noninverting analogue buffer with a very high input impedance, Fig. 15.5. This is a very useful subsystem used when loading of an analogue input can be a problem.

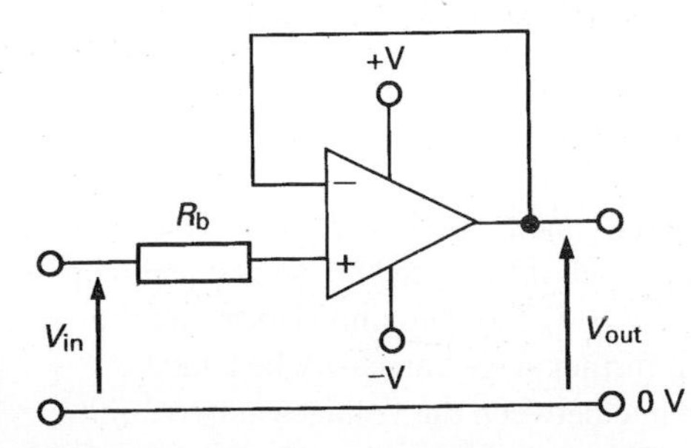

Figure 15.5 Noninverting analogue buffer.

15.3 Combining analogue signals

To *mix* two or more signals, it is necessary to use an active subsystem rather than just connect the two signal leads together. (If this happens the two signals just corrupt each other.) The signals may be mixed in audio systems, for example, as when the output from a tape deck is required to be mixed with the output from a microphone. The *summing amplifier* achieves this function. A *difference amplifier*, on the other hand, is particularly used to eliminate noise in microphone systems.

15.3.1 The summing amplifier

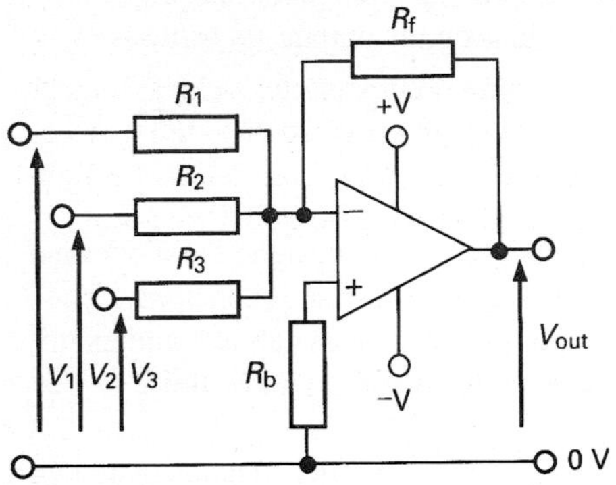

The general circuit is shown in Fig. 15.6.

The output signal, V_{out}, is given by

$$V_{out} = -R_f\left(\frac{V_1}{R_1} + \frac{V_2}{R_2} + \frac{V_3}{R_3}\right)$$

Figure 15.6 Summing amplifier.

Note that, as the circuit is based on the inverting amplifier, the output is inverted. If this is a problem an inverting buffer can be used to reinvert the output. Making the input resistors variable gives control over each of the signals whilst making R_f variable gives overall control of the size of the output signal.

The impedance of each input is determined by the magnitude of each input resistor. In general, the impedance of this stage should be ten times the output impedance of the previous stage if the signal is not to be corrupted. This can cause problems if high input resistor values are needed. One solution is to buffer the inputs before they reach the summing amplifier.

The balancing resistor, R_b, is calculated from the formula

$$\frac{1}{R_b} = \frac{1}{R_1} + \frac{1}{R_2} + \frac{1}{R_3}$$

If the input resistors are equal in value and the feedback resistor, R_f, is the same value then the general formula simplifies to

$$V_{out} = -(V_1 + V_2 + V_3)$$

that is, the output signal is the arithmetic sum of the inputs with the sign changed. To subtract an input signal it needs to be inverted to a negative value before it reaches the summing amplifier.

15.3.2 The difference amplifier

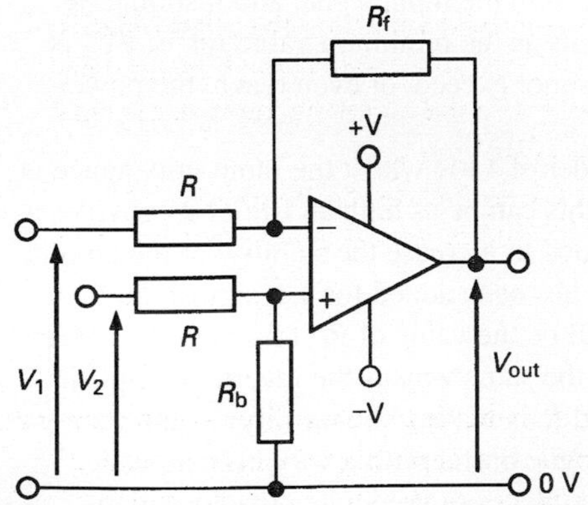

The op-amp is a difference amplifier but it has a very high gain so that even a small difference in signal between the two inputs results in saturation. This would corrupt the analogue information. The use of feedback, Fig. 15.7, reduces the gain and therefore preserves the analogue information.

The output signal is given by

$$V_{out} = \frac{R_f}{R}(V_2 - V_1)$$

Figure 15.7 Difference amplifier.

giving a gain of R_f/R which is usually reduced to one by making $R_f = R$. The impedances of the two inputs are not the same, and making the resistors equal helps to reduce this problem. If R_b is also made equal to R_f then the impedances are the same. If amplification is still required then a further stage can easily be added.

The circuit is best used to find the difference between the voltages at two points rather than act as a mathematical subtractor. This arrangement is particularly useful for eliminating noise on microphone leads. Microphone leads tend to be long and, as the signal is not very large, it can easily be corrupted. If the signal is fed into a difference amplifier the noise should be eliminated. This property of an op-amp is known as a high *common-mode rejection* ratio.

15.4 Changing the form of the analogue signal

In sections 15.2 and 15.3, ways were examined of changing the form of the analogue signal. It was shown how the signal could be amplified, attenuated and inverted as well as mixed with other signals or separated from other signals. However, the signal remains in an analogue form. There are occasions when it is necessary to convert the signal into a digital form in order to process it further. For example, a signal may be required to tell us when a temperature has exceeded a preset value or to tell us when one light is brighter than another light. This is best achieved by converting the signal into a digital form, and the device used for this purpose is a threshold detector or comparator.

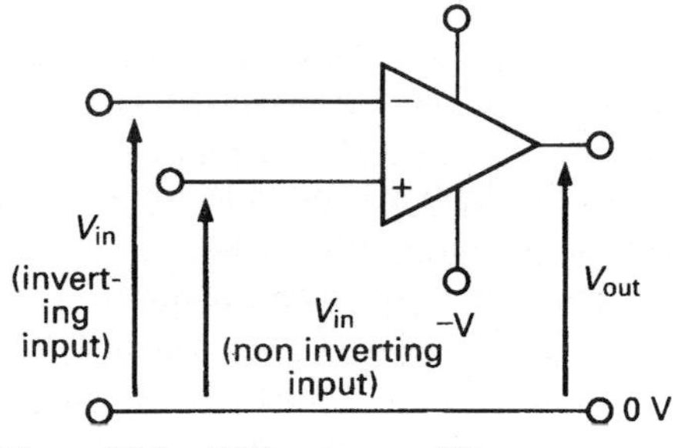

Figure 15.8 Difference amplifier.

15.4.1 The comparator and threshold detector

The comparator, Fig. 15.8, has two inputs and one output.

If the signal to the noninverting input is greater in magnitude than the signal to the inverting input, then the output will be *high*. If the signal to the inverting input is greater in magnitude than the signal to the noninverting input, then the output will be *low*. Since the gain of the comparator is very large, a minimum of 10^5, it takes only a very small difference in signal p.d.s for the comparator to be saturated. It is very difficult to make the signals exactly equal. If they are exactly equal, then the output will be between high and low. The actual value of the high and low states depends on the device being used and the value of the power supply. In trying to create a digital signal that is acceptable to a logic gate it would be sensible to use the same power supply which means +5 V for TTL type logic gates and +3 V to +15 V for CMOS type logic gates (for more detail on these types of logic gates see section 16.1).

Many of the devices used as comparators are op-amps which are designed to run on split rail supplies. This means that the output in the low state would be negative which would not be acceptable to the logic gate. Most op-amps will run on single rail power supplies but the outputs may not be on the correct range to be accepted by logic as 0 and 1. The 741 op-amp, for example, will give about +4.5 V in the high state when run on + 5 V but its low state will be about +1.7 V which is far too high to be reliably recognised as logic 0. The TL271 is one comparator which runs on +5 V and gives an output that will work satisfactorily with logic gates.

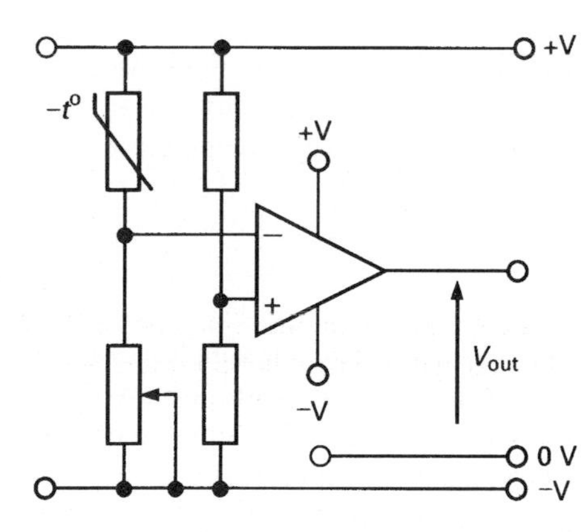

Figure 15.9 Temperature threshold detector.

If one of the inputs to a comparator is set at a fixed value and the other input is variable, it is known as a threshold detector. The fixed value sets the threshold and when the variable value exceeds this threshold then the output changes state. This circuit, Fig. 15.9, shows a temperature threshold detector.

If both inputs are variable then it is known as a comparator, Fig. 15.10. This circuit will give a high output if LDR1 detects a larger signal than LDR2.

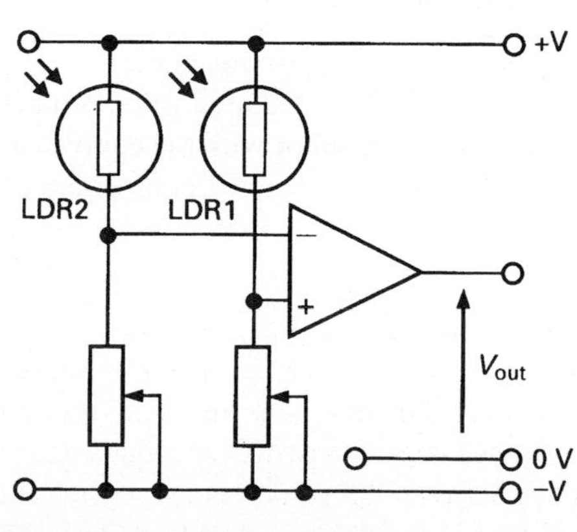

Figure 15.10 Light comparator.

15.4.2 The Schmitt trigger

There are some occasions when the very sharp switching action of the comparator or threshold detector is unwanted – if, for example, the input signal has noise added to it. This may mean that the output of the comparator or threshold detector will oscillate rather than stay in one state or the other. Fig. 15.11 shows such a signal.

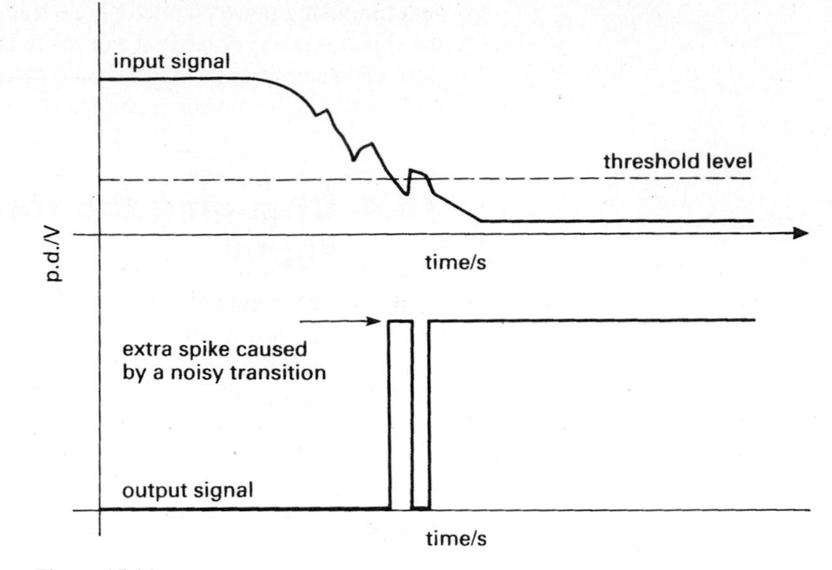

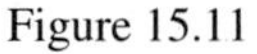

Figure 15.11

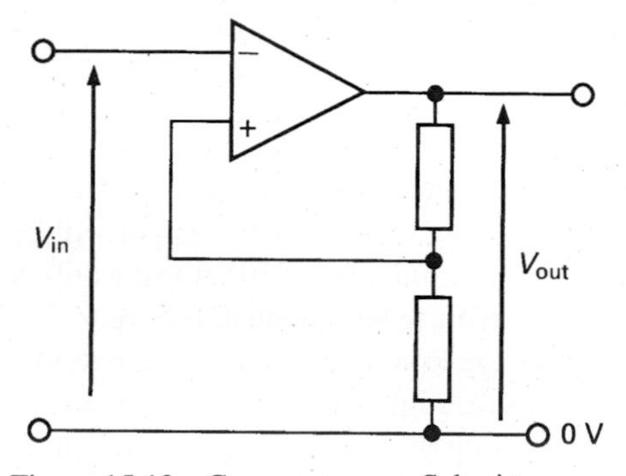

Figure 15.12 Comparator as a Schmitt trigger.

A similar condition occurs with light sensors if the lighting is provided by an alternating current. Fluorescent lights are particularly prone to this flicker.

The Schmitt trigger is used to overcome this problem. Positive feedback is applied to one terminal of the comparator to create a *dead band*. This means that whilst the comparator may switch to the high state at 2.1 V (say) it will not switch back to the low state until the input signal falls to 1.8 V. This gives a 0.3 V dead band. The size of the dead band is adjustable by changing the values of the resistors. The circuit in Fig. 15.12 shows such an arrangement.

This arrangement is particularly useful in giving sharp transitions and is used to clean up noisy signals as well as converting sine waves into square waves. Some logic gates have Schmitt triggers built into their input stages which can replace this arrangement at a much lower cost.

15.4.3 Filtering

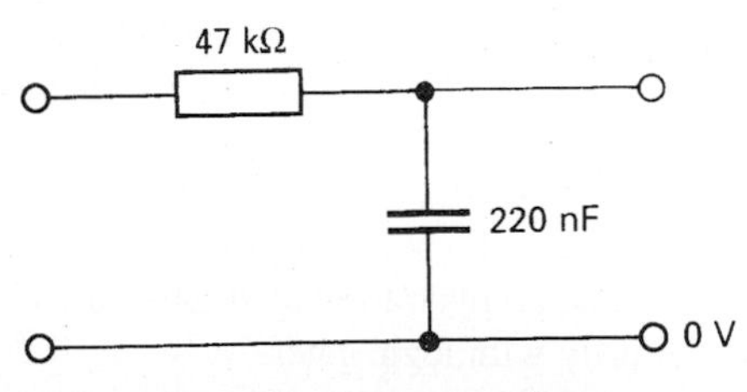

Figure 15.13 Simple low-pass filter.

Filtering involves the removal or selection of certain frequencies. The simplest example of filtering is the tone control on a radio or hi-fi system. When the control is moved the range of frequencies allowed through to the next stage is changed. The simplest circuit to achieve this effect is a frequency dependent potential divider, Fig. 15.13.

The capacitor has a reactance X_c (see section 20.6), given by

$$X_c = \frac{1}{2\pi fC}$$

and it can be seen that this depends on the frequency. If we approximate the reactance to the impedance (this is reasonable as long as the calculated value is much less than the value of the other resistor) and replace the capacitor with the equivalent resistor, it is possible to calculate the effect on different frequencies. Let us examine the effect of this arrangement at 30 Hz and 3 kHz.

at 30 Hz:

$$X_c = \frac{1}{2\pi \times 30 \times 220 \times 10^{-9}}\,\Omega$$

$$= \frac{10^9}{13\,200\pi}\,\Omega$$

$$= 24.1\ \text{k}\Omega$$

at 3 kHz:

$$X_c = \frac{1}{2\pi \times 3 \times 10^3 \times 220 \times 10^{-9}}\,\Omega$$

$$= \frac{10^6}{13\,200\pi}\,\Omega$$

$$= 241\ \text{k}\Omega$$

If we now replace the capacitor by this resistor we can investigate the effect of applying a 5 V alternating signal, see Fig. 15.14.

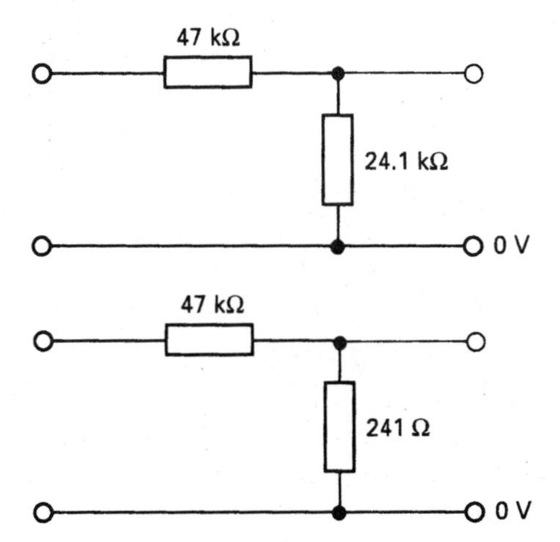

Figure 15.14 Potential divider equivalences at the two frequencies.

The p.d. across the resistor we have introduced will be:

at 30Hz:

$$V_{out} = V_{in}\left(\frac{R_2}{R_1 + R_2}\right)$$

$$= 5.0\left(\frac{24.1\ \text{k}\Omega}{47\ \text{k}\Omega + 24.1\ \text{k}\Omega}\right)V$$

$$= 5.0 \times 0.339\ \text{V}$$

$$= 1.695\ \text{V}$$

at 3 kHz:

$$\text{V}_{out} = 5.0\left(\frac{241\ \Omega}{47\ \text{k}\Omega + 241\ \Omega}\right)V$$

$$= 5.0\left(\frac{241\ \Omega}{47.241\ \text{k}\Omega}\right)V$$

$$= 5.0 \times 0.510 \times 10^{-2}\ \text{V}$$

$$= 0.0255\ \text{V}$$

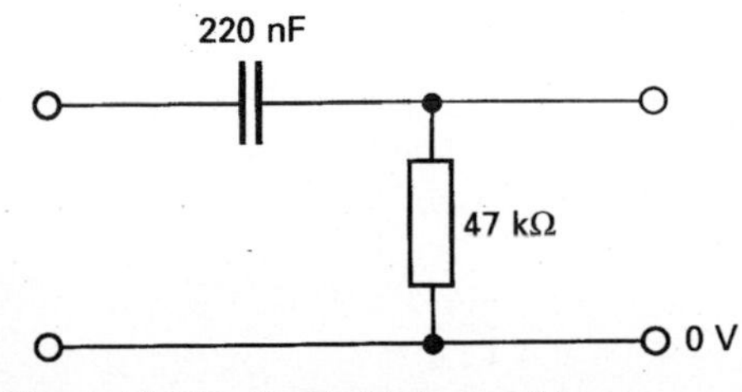

Figure 15.15 Simple high-pass filter.

It can therefore be seen that the output signal is greatly reduced at 3 kHz compared to the signal at 30 Hz.

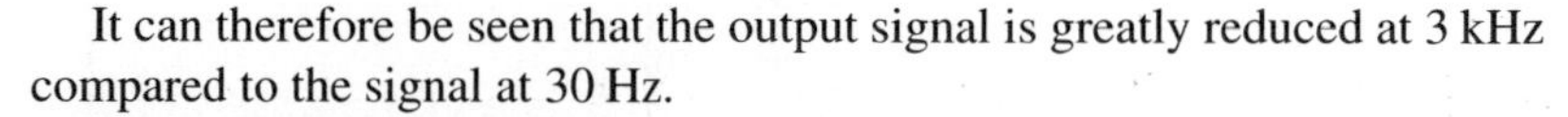

This arrangement is called a *low-pass* filter since it allows the lower frequencies to pass with less attenuation than the higher frequencies. It may also be referred to as a treble cut filter.

If the position of the two components is reversed we get a different function, Fig. 15.15.

An identical mathematical reasoning will show that this has the reverse effect in allowing the higher frequencies through whilst attenuating the lower frequencies. This is known as a *high-pass filter* or a bass cut filter.

For the simple RC filter, the key frequency is the half-power frequency. As explained in section 20.7 this is the frequency at which the output power has fallen by 3 dB. If we plot the frequency on a linear scale and the voltage gain in decibel on a logarithmic scale (a Bode plot) then, for a low-pass filter, we obtain the graph of Fig. 15.16.

The half-power frequency is shown by f_o and is given by the formula

$$f_o = \frac{1}{2\pi RC}$$

After the break frequency the voltage gain falls off at a rate of −20 dB/decade. An 'ideal' filter would show zero voltage gain up to the half-power frequency and infinite voltage gain above this frequency. This is represented in Fig. 15.17.

Cascading passive filters have the effect of increasing the slope of the graph but they also decrease the half-power frequency, and the filters must be separated by buffer amplifiers to remove the effects of loading, Fig. 15.18.

It is theoretically possible to include inductors in these circuits to improve the cut off, but at audio frequencies they introduce many more problems and are not to be recommended. Instead, active filters must be used and the active element is usually an op-amp. Active low-pass and high-pass filters are shown in Figs 15.19 and 15.20 respectively.

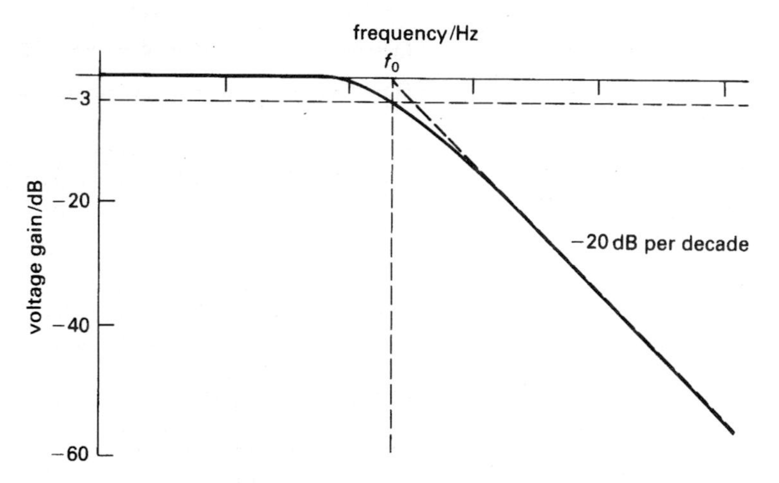

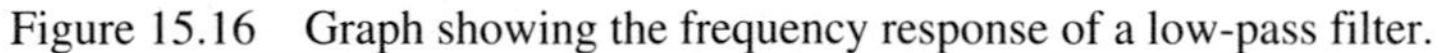

Figure 15.16 Graph showing the frequency response of a low-pass filter.

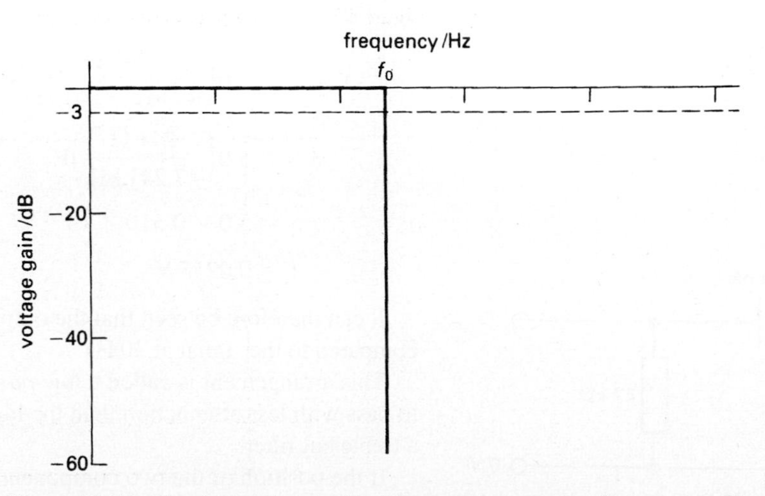

Figure 15.17 Graph showing the ideal frequency response of a low-pass filter.

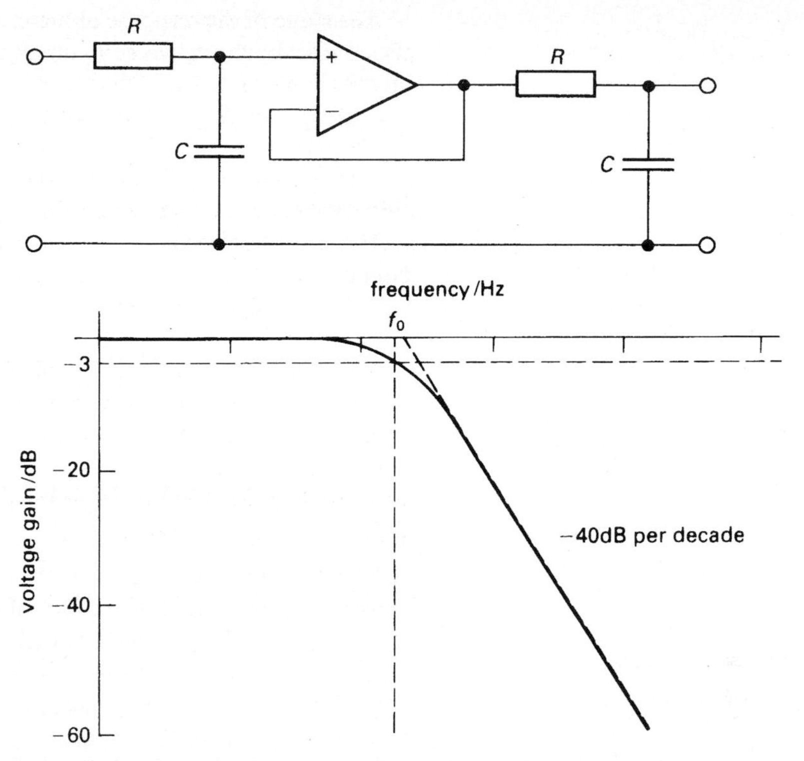

Figure 15.18 Second order low-pass filter and a graph to show its frequency

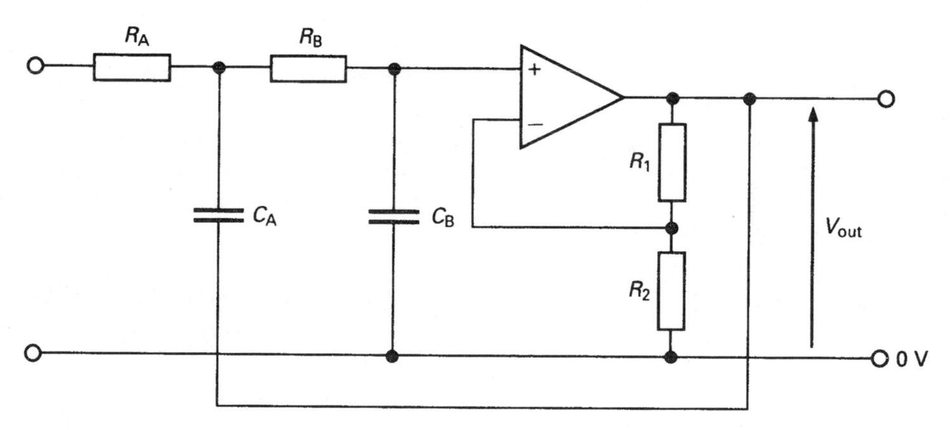

Figure 15.19 Two-pole active low-pass filter.

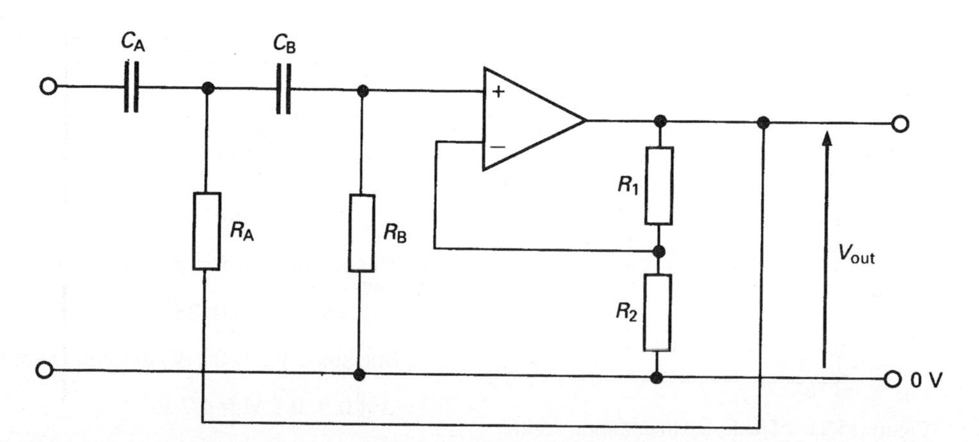

Figure 15.20 Two-pole active high-pass filter.

The shape of the response of these circuits is set by the gain of the op-amp system which is set by the values of R_1 and R_2. Note that there are four main types of filter circuit:

Low-pass The passband is from zero up to the highest allowed frequency.

High-pass The passband is from a lowest allowed frequency upwards.

Band-pass The passband is from a lowest frequency to a highest allowed frequency.

Notch Here a gap (or gaps) exist in a passband which stretches from zero to high frequency. Frequencies which lie in the gap are greatly attenuated.

The bibliography at the end of this chapter lists details of texts which contain details of more advanced active filters to achieve these functions.

15.5 Current amplification

When a signal emerges from a processing subsystem it has little energy. It rarely has sufficient energy to activate output devices, with the possible exception of a LED. The power of the signal needs boosting so there is sufficient energy to drive these output devices. The subsystem required to achieve this function is a power amplifier. However, the p.d. applied to the load is usually the p.d. applied to the processing subsystem. If the p.d. is constant, remembering that power = VI, it can be seen that it is the current that needs to be amplified. This current amplification is usually achieved with a transistor arrangement – often misleadingly referred to as a transistor switch. There are several different transistor arrangements possible depending on the need.

The first circuit (*common emitter*) is often used to switch on devices that need a higher current than the source can supply. It only needs a small input voltage change to saturate the transistor. The *emitter follower* is usually used when proportional current amplification is required since there is a linear relationship between the input and output currents. It should be noted that this circuit does not give voltage amplification and that the input impedance of the emitter follower circuit is greater than the input impedance of the common emitter circuit.

15.5.1 Bipolar transistor common emitter configuration

In the most common circuit, the device (a lamp in this case) is connected between the collector and the positive supply rail, Fig. 15.21.

Suppose the lamp requires 60 mA to be illuminated at normal brightness and also suppose that the current gain h_{FE} for this transistor is 150, then

$$I_b = \frac{I_c}{h_{FE}} = \frac{60}{150}\ \text{mA} = 0.4\ \text{mA}$$

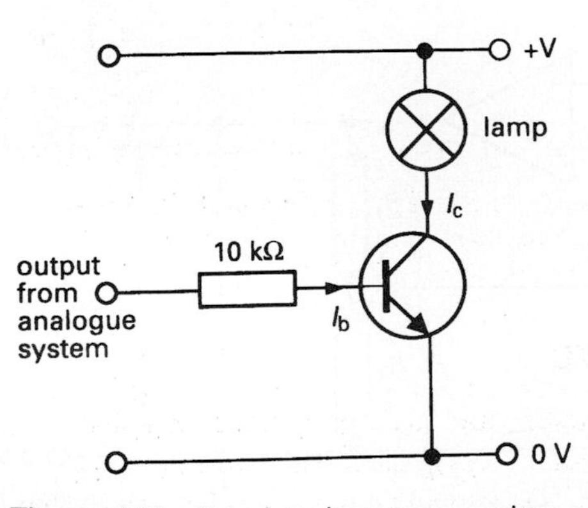

Figure 15.21 Transistor in common emitter configuration.

The potential difference across the 10 kΩ base resistor will be given by

$$0.4\ \text{mA} \times 10\,000\ \Omega = 4.0\ \text{V}$$

But, since $V_{be} = 0.7$ V, the voltage at the wiper of the potential divider must be

$$4.0 + 0.7\ \text{V} = 4.7\ \text{V}$$

in order for the lamp to be on at normal brightness.

15.5.2 Bipolar transistor emitter follower configuration

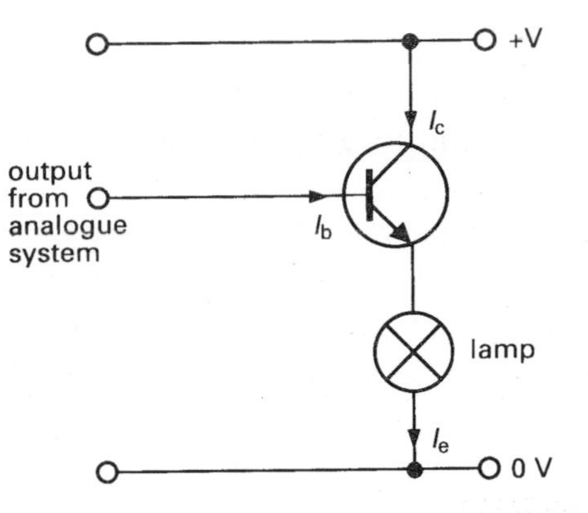

Figure 15.22 Transistor in emitter follower configuration.

An alternative approach is to use the transistor as an emitter follower, the circuit for which is shown in Fig. 15.22.

In this arrangement, the lamp is between the emitter and the zero volt line. It can be seen that the voltage across the lamp is always 0.7 V less than the voltage applied to the base.

If it is assumed that this is the same transistor and lamp as in the previous example then, for I_e to equal 60 mA, the base current must be 0.4 mA (making the assumption that $I_e = I_c$).

If the resistance of the lamp is 100 Ω then the p.d. across the lamp is (60 mA × 100 Ω) which is 6.0 V.

The p.d. at the base of the transistor must be

$$6.0\ \text{V} + 0.7\ \text{V} = 6.7\ \text{V}$$

in order for the lamp to be on at normal brightness.

The advantage of the emitter follower is that the p.d. across the lamp is always 0.7 V less than the p.d. applied to the base. This gives a large range of input p.d.s, 0.7 V to 6.7 V in this case, which will cause a change of current through the lamp. The common emitter arrangement gives a smaller range of input p.d.s before it saturates.

15.5.3 Darlington pair

The current gain for both configurations can be increased by using two transistors connected together. This arrangement is known as a *Darlington pair*. As well as increasing the current gain it also increases the input impedance. Current gain = $h_{fe1} \times h_{fe2}$. Note that V_{be} increases to 1.4 V.

Fig. 15.23 shows the arrangement for the common emitter configuration and Fig. 15.24 shows the emitter follower version.

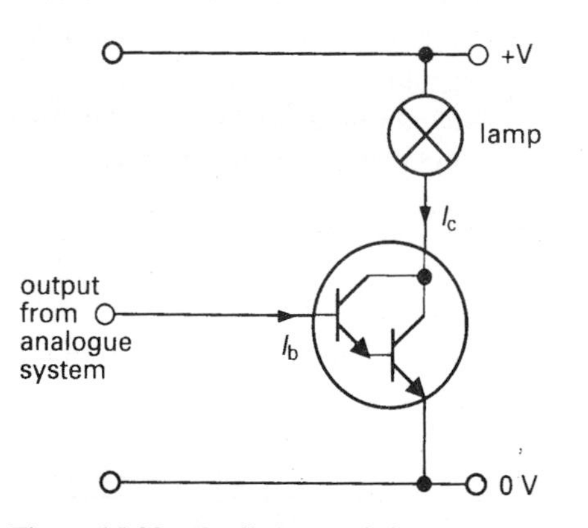

Figure 15.23 Darlington pair in common emitter configuration.

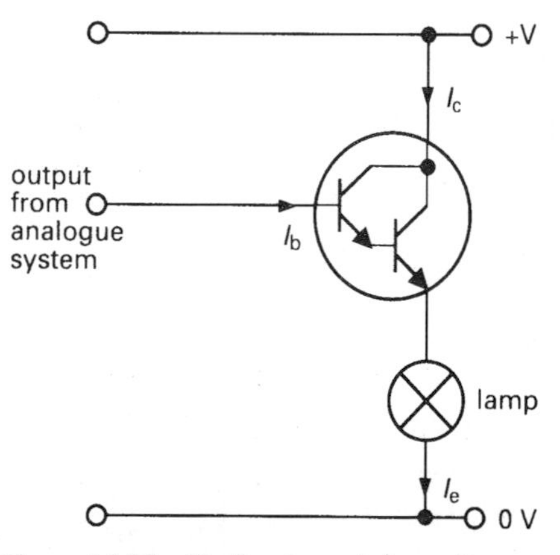

Figure 15.24 Darlington pair in emitter follower configuration.

15.5.4 Push-pull amplifiers

The *push-pull* amplifier is used when current control is required rather than just switching and when split rail power supplies are being used. This makes it particularly appropriate for the output from operational amplifiers as it amplifies the current when it goes both positive and negative.

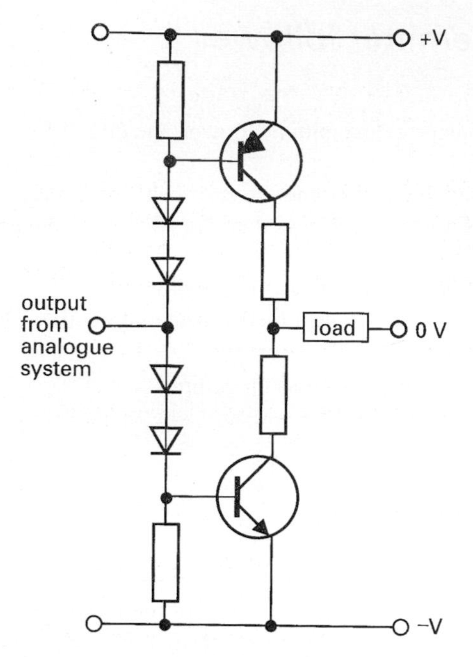

Figure 15.25 A push-pull amplifier.

Two transistors are used: they should be a complementary pair: one *n-p-n* and one *p-n-p* with similar characteristics. Manufacturers' catalogues and data sheets give details of appropriate devices. Darlington pairs are often used; TIP121 and TIP126, for example.

The transistors are both configured as emitter followers and the diodes are included to remove the dead band that would otherwise exist, due to the need to have 1.4 V to turn on a Darlington pair. If the transistors are not Darlington pairs, only one diode would be needed for each transistor. A suitable circuit is shown in Fig. 15.25.

15.5.5 The FET amplifier

FETs look similar to bipolar transistors in that they have three terminals. One terminal acts as a control for the current that passes between the other two. Bipolar transistors act as current-controlled current controllers, whilst FETs have two distinct functions: they can be either a voltage-controlled current controller or a voltage-controlled resistor.

The terminology for an *n*-channel junction gate FET (or *n*-channel JUGFET or JFET) is shown in Fig. 15.26.

The circuits that have already been examined which are made with bipolar transistors can also be made with FETs. In some respects, the circuits may be better but, in others, they may have much poorer performance. One particular problem is the variability of FET characteristics between devices. Fig. 15.27 shows the normal arrangement for an *n*-channel JUGFET in a circuit.

The drain is held above the source so that the drain current flows from drain to source. (The names appear to be the wrong way round: 'source' refers to the source of electrons.) The source current has the same value as the drain current and is known as the drain current. The voltage between the gate and the source will always be zero or negative.

The graph of Fig. 15.28 shows how the drain current depends on the voltage between the drain and source when the gate voltage is less than zero. A possible voltage amplifier circuit is shown in Fig. 15.29.

Note that the gains of FETs are much less than bipolar types (of the order of ten) and that the input impedance is very high. Note also that this circuit is a.c. coupled.

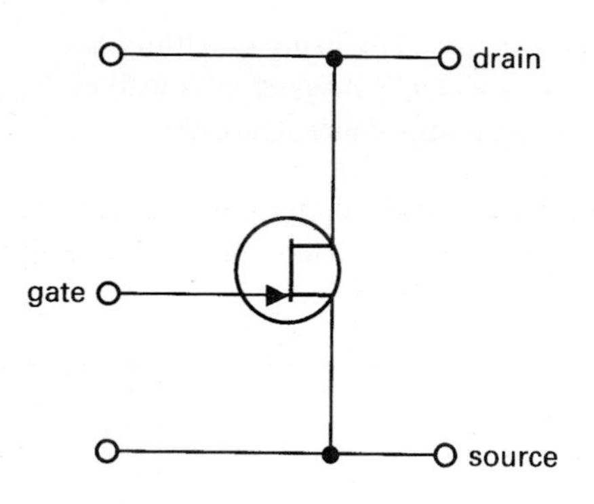

Figure 15.26 An *n*-channel JUGFET.

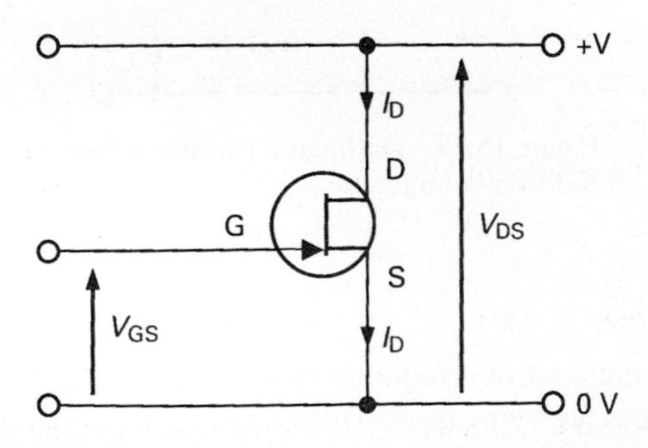

Figure 15.27 An *n*-channel JUGFET in a circuit.

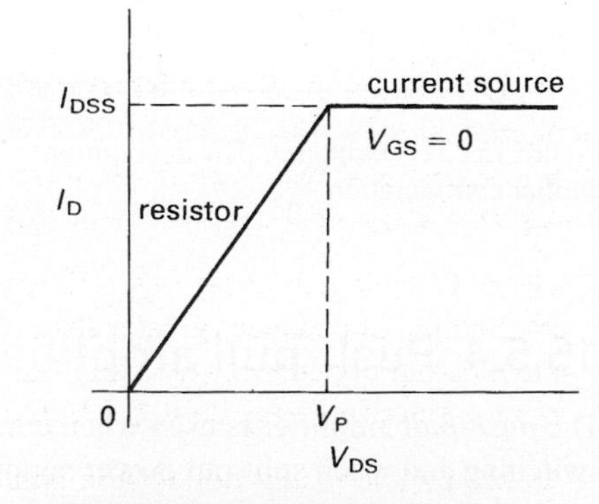

Figure 15.28 Graph showing the characteristics of an ideal FET.

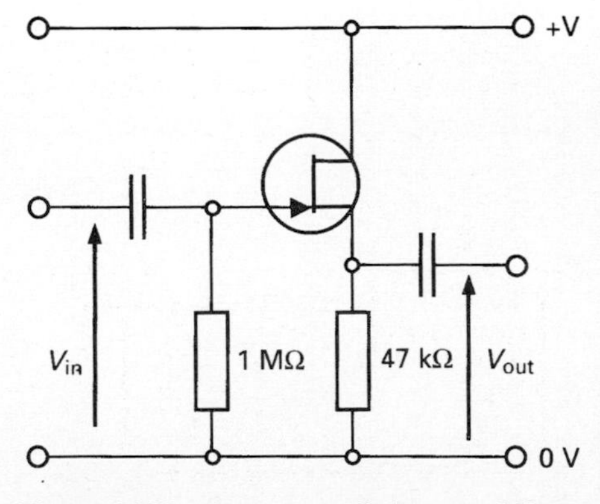

Figure 15.29 An a.c. coupled source follower.

MOSFET

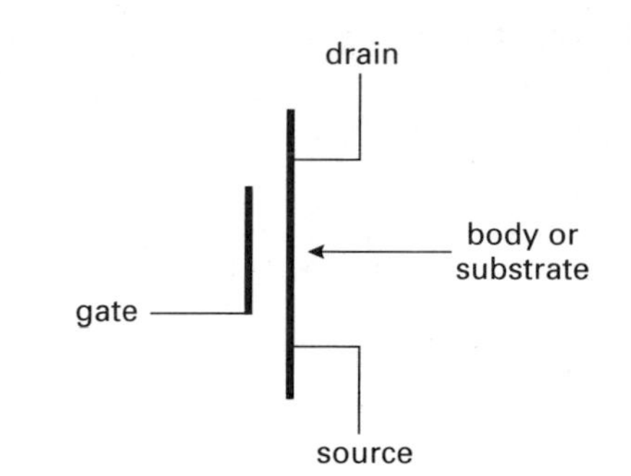

Figure 15.30

Another type of FET is a MOSFET. MOSFET stands for Metal Oxide Semiconductor FET but is sometimes known as an Insulated Gate FET or IGFET which more correctly describes its construction (Fig. 15.30).

The gate is insulated from the drain-source circuit by a layer of glass (silicon oxide). There is therefore no electrical connection between the gate and the drain-source circuit. The electric field from the gate-source controls the drain-source current.

If a bipolar transistor is a 'current controlled current controller' then a MOSFET is a 'voltage controlled current controller'.

The extra terminal on a MOSFET is the body or substrate. This is usually connected directly to the source.

There are several different types of MOSFET. We are looking at an *n*-channel, enhancement mode MOSFET.

MOSFET as a current amplifier (switch)

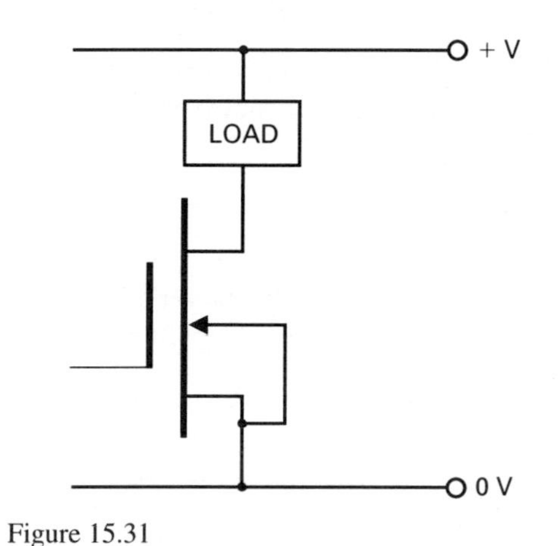

Figure 15.31

The MOSFET conducts when the voltage of the gate is brought towards the drain voltage (which can be thought of as the supply voltage) (Fig. 15.31).

MOSFETs have a threshold voltage, V_T or $V_{GS(th)}$, which has a value somewhere between 0.5 V and 5 V. This is not a fixed value as even the same batch of MOSFETs will show a spread of values. It is also not an absolute cut off point. Some sources give the definition of the threshold voltage to be when the drain current has dropped to 10 μA.

15.6 Component information

15.6.1 Operational amplifiers

Although it is possible to construct operational amplifiers from discrete components, it is not sensible to do so since op-amps are available as integrated circuits. Examples of these are:

709	bipolar and old
741	bipolar and old
748	741 without frequency compensation
759	bipolar with power output
351	JFET
TL061	BIFET
TL071	low noise BIFET
TL081	BIFET
3140	MOSFET
7611	CMOS

These op-amps are distinguished by the type of device used in the input circuits, bipolar transistors, JFETS, MOSFETS, etc., and also whether their output circuits were designed to drive a significant load directly.

A practical op-amp is a linear voltage amplifier with:

- a large voltage gain (typically greater than 10^5)
- a high input impedance (greater than 2 MΩ)
- a low output impedance (typically less than 200 Ω)
- a limited slew rate (typically 0.5 V/μs)

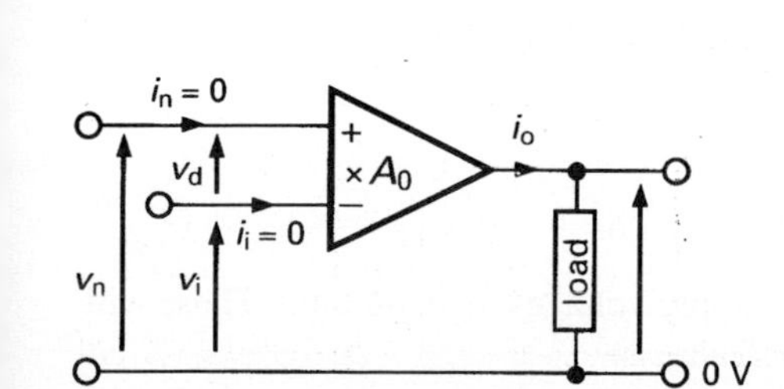

Figure 15.32 An ideal operational amplifier.

An ideal op-amp does not exist in practice but it is a useful concept in deriving the mathematical relationships between op-amp parameters. Fig. 15.32 shows some of these parameters for an ideal op-amp.

Terminology

The terminal marked + is known as the noninverting input.
The terminal marked − is known as the inverting input.
The gain is A_0 and is assumed to be infinity for an ideal op-amp.

$$v_d = v_n - v_i \text{ (the difference between the input p.d.s)}$$

$$v_o = v_d \times A_0 = v_d \times \text{infinity}$$

$$i_o = \frac{v_o}{R_l}$$

Therefore:
if $v_n > v_i$ then $v_o = +$infinity and
if $R_l <$ infinity then $i_o = +$infinity
if $v_n < v_i$ then $v_o = -$infinity and
if $R_l >$ infinity then $v_o = -$infinity
if $v_n = v_i$ then $v_o = 0$ and $i_o = 0$

Clearly this is impossible in practice. With a real op-amp i_n and i_o will be small (μA or pA) and the gain will be finite, although large, and frequency dependent. The output current, given by

$$i_o = \frac{v_o}{R_l}$$

will be available up to a maximum determined by the op-amp (20 mA for a 741). The minimum load resistance for a 741 is about 2 kΩ. The value of v_o will be limited by the value of the supply p.d.s.

Power supplies

Most op-amps require dual rail power supplies which consist of two separate supplies connected in series. Fig. 15.33 shows how this can be achieved with two 9 V batteries.

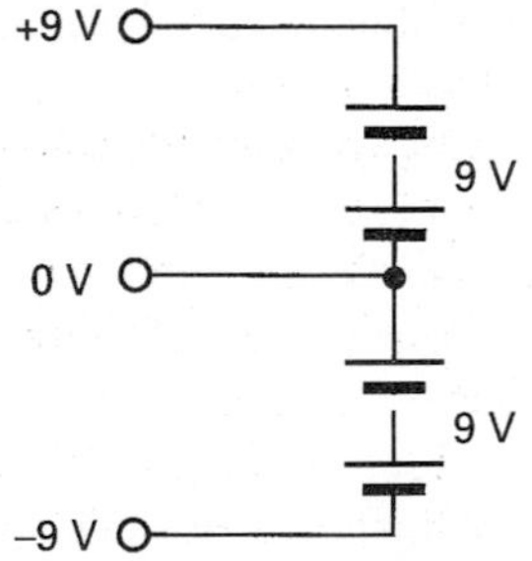

Figure 15.33 Dual rail power supply from 9 V batteries.

This would be a symmetrical supply but some op-amps will operate on a single rail supply or a nonsymmetrical dual rail supply. Details can be found in data sheets on individual op-amps.

The power supply provides the voltage limit for v_o. In all cases the output voltage, v_o, will never exceed or reach the value of the supply. Thus:

$$-v_s < v_o < +v_s$$

There are also practical limits to the value of the input voltage. These limits will depend on the op-amp chosen but, in general, v_i and v_n must not exceed the limits of the power supply to the ic and v_d must not exceed the total value of the power supply. Therefore, on a +15 V and −15 V supply, the value of v_d could not safely exceed 30 V.

This is not true for all op-amps so it will be necessary to study the data sheets carefully. For example, a 709 has maximum values as shown below.

$$-10\text{V} < v_n < +10\text{ V}$$
$$-10\text{V} < v_i < +10\text{ V}$$
$$-5\text{V} < v_d < +5\text{ V}$$

There are also maximum values for the supply voltages to an op-amp. These will be shown on the data sheet. For the 741 the value is:

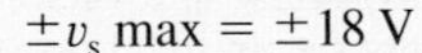

$$\pm v_s \text{ max} = \pm 18\text{ V}$$

15.6.2 Transistor characteristics

Let us consider the graphs of Fig. 15.34 showing the input, output, and transfer characteristics for an *n-p-n* bipolar transistor operated in the common emitter mode.

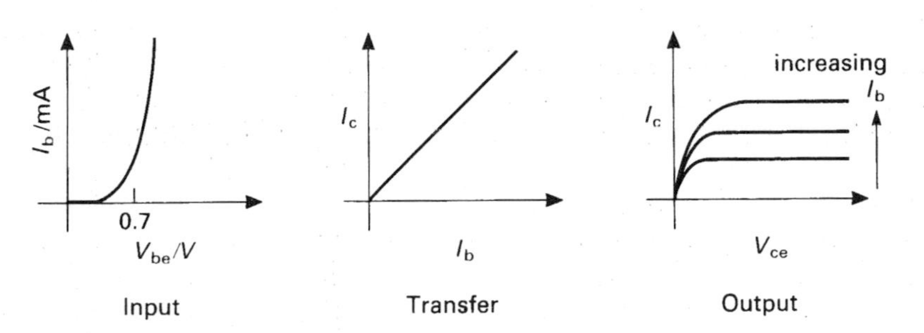

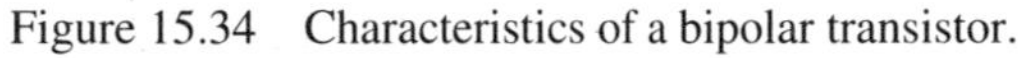

Figure 15.34 Characteristics of a bipolar transistor.

From these graphs it is possible to make the following statements:

1 V_{ce} is always greater than or equal to a value usually in the range 0.3–5 V. This implies that current always flows from the collector to the emitter. When V_{ce} is at this value it is said to be saturated.
2 The ratio I_c/I_b is constant as long as the transistor is not saturated. This value is known as the current gain, h_{FE}.
3 $I_b + I_c = I_e$ but if $I_b << I_c$ then I_c is approximately equal to I_e.
4 No base or collector current will flow until V_{be} exceeds a value of about 0.7 V.

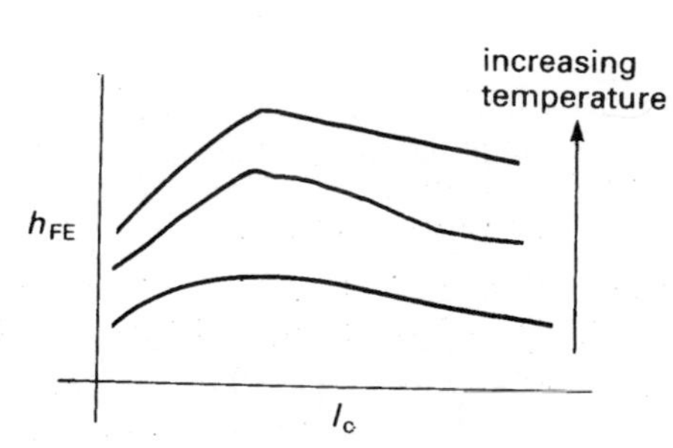

Figure 15.35 Graph to show the variability of h_{FE}.

These statements are a good first approximation.

The current gain, h_{FE}, is often taken to be a constant but this is not so. It is related to the magnitude of I_c and decreases substantially as I_c nears its maximum and minimum values. This is shown in Fig. 15.35.

h_{FE} is also a function of the temperature of the transistor.

It is impossible to make transistors of the same type with exactly the same value of h_{FE}. Therefore, if data sheets are consulted, there is often a range of values quoted, for example a ZTX 300 may have a current gain between 50 and 300.

It is an essential property of circuit design that the performance of a circuit does not depend on such a complex variable – it is designed out, usually using feedback techniques, so that the circuit will operate with a wide range of possible devices.

There are also leakage currents to be considered. Whilst these are small at room temperature they increase quite rapidly as the temperature increases. Hot transistors have a far from ideal behaviour and tend to self-destruct if the temperature rises too much.

Careful consideration of the data sheets is necessary to ensure that the maximum values of these parameters for a particular device are not exceeded. Finally, the value of V_{be} at 0.7 V before current will flow is an approximation since it is a function of I_c (and therefore I_b as well).

Bibliography

As for Chapters 20 and 14 together with the following.

Clayton G B 1979 *Operational Amplifiers* 2nd edn. Butterworths
Morris J C 1991 *Analogue Electronics*. Hodder and Stoughton
Thompson D S C 1989 *A Look at the Operational Amplifier.* NEMEC publications

Assignments

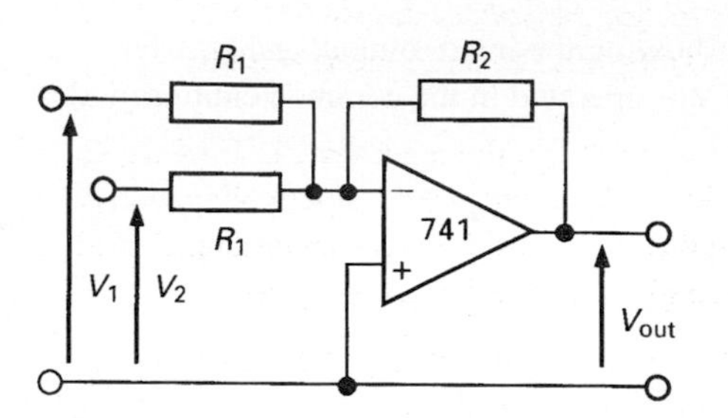

Figure 15.36

1 (a) Explain the principal characteristics of an operational amplifier.
(b) Compare the functions of an 'inverting' and 'non-inverting' operational amplifier.
(c) Briefly explain how a summing amplifier may be used as a digital to analogue converter.
(d) In Fig. 15.36, $R_1 = 10\ \text{k}\Omega$, $R_2 = 20\ \text{k}\Omega$, $V_1 = 3$ V and $V_2 = 2$ V. Calculate V_0.

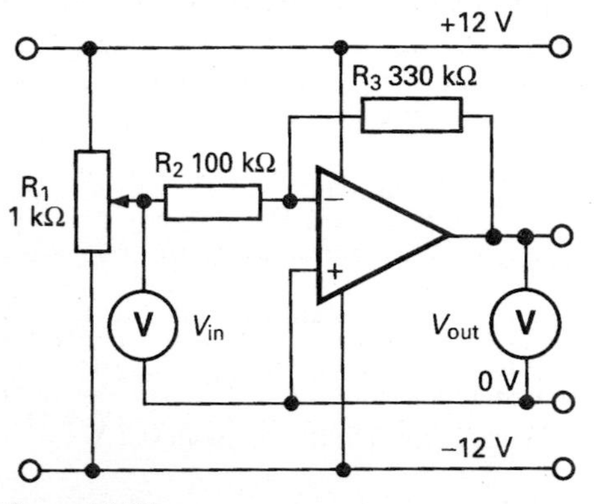

Figure 15.37

2 (a) Explain what is meant by the terms input impedance and output impedance when referred to operational amplifiers. Specify the ideal values of these impedances and explain why the designer of practical amplifiers tries to approach these values. How do the specifications for these impedances for real devices compare with the ideal values?
(b) Design a noninverting amplifier to have a d.c. gain of 30 using an operational amplifier. Assume that the largest value of resistor to be used has a resistance of 470 kΩ. Specify, following the appropriate calculations, the values of any other components required. State all the assumptions you have made in your design.
(c) An inverting amplifier with component values as shown is connected across a +12 V to −12 V supply (Fig. 15.37). If the operational amplifier is assumed to be ideal, sketch the output characteristics of the amplifier as the input voltage varies from −12 V to +12 V by varying the potentiometer R_1. What changes would you expect to find in the characteristic if a real operational amplifier was used? (Your characteristics should have labelled axes.)
(d) Using an operational amplifier in an a.c. mode, draw a circuit diagram for both simple low pass and high pass filters. (Component values are not required.) Explain with the aid of sketched graphs how the circuits function and describe briefly one typical application for each circuit.

3 For each amplifier circuit shown in Fig. 15.38 the input voltage V_i is +2 volts. What is the output voltage V_o? Assume that each amplifier has ideal characteristics.

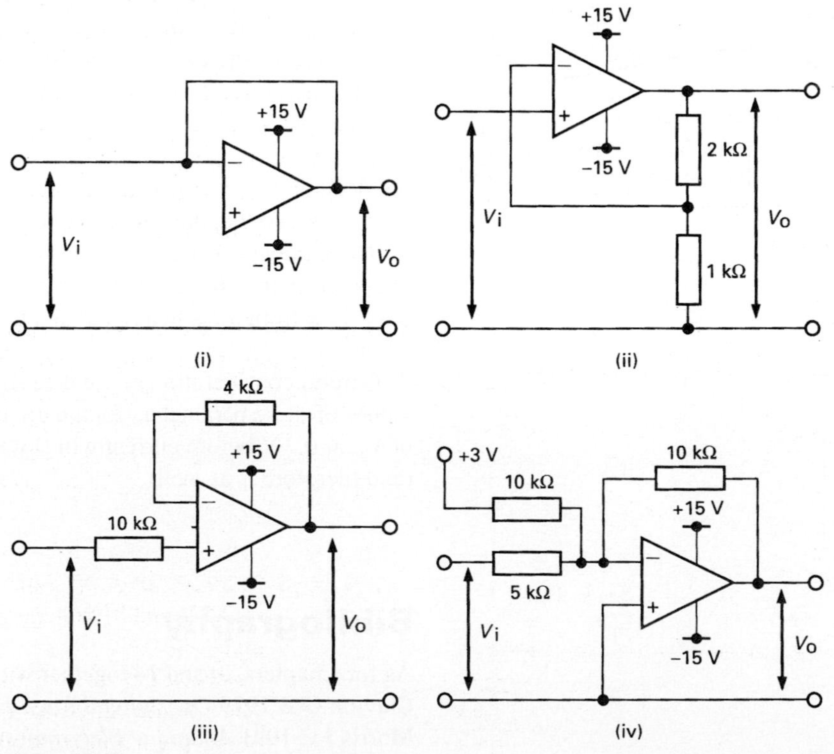

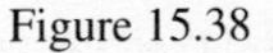
Figure 15.38

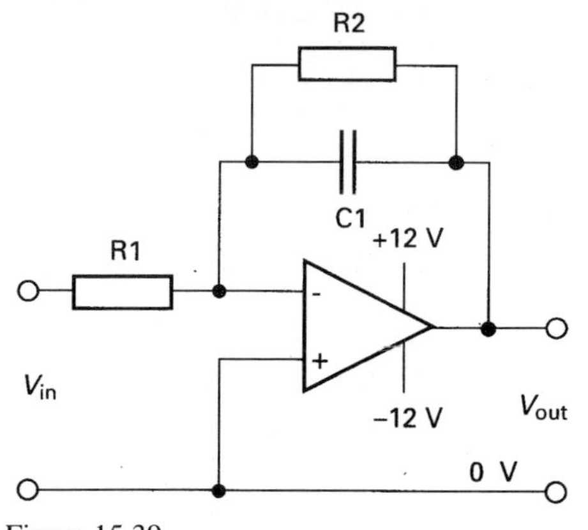

Figure 15.39

4 (a) In audio amplifiers it can be an advantage for the gain to vary with respect to the frequency over the frequency range. Give one reason why it can be an advantage for the gain of an audio amplifier to be reduced at high frequencies.

(b) The circuit shown in Fig. 15.39 was tested in the laboratory to investigate how the gain of the circuit varied with frequency. The input voltage was kept constant at 100 mV r.m.s. and the two resistors, $R1$ and $R2$, each had a resistance of 1 kΩ. The results are given in the following table:

Frequency (Hz)	0.1	0.3	1.0	3.0	10.0	30.0	56.0
Output Volts (mV r.m.s.)	100	100	98	92	60	24	15

Plot a graph of voltage gain in dB against $\log_{10}$ of frequency and explain why the shape is as produced.

(c) The point of the intersection of the horizontal line of constant gain, with the straight part of the sloping line extrapolated to cut this horizontal line, is known as the *break point*. At the break point the value of the reactance $C1$ numerically equals that of the resistance $R2$. Identify the break point on your graph and determine the value of the capacitor, $C1$, being used.

(d) (i) Consider the circuit in Fig. 15.40. Sketch the approximate relationship you would expect between the gain of the system and frequency of the input signal. (An accurate graph based on calculations is not expected.)

(ii) Determine the break point frequency of the circuit for the component values shown on the figure. At the break point the value of the reactance of $C1$ equals that of the resistance of $R2$.

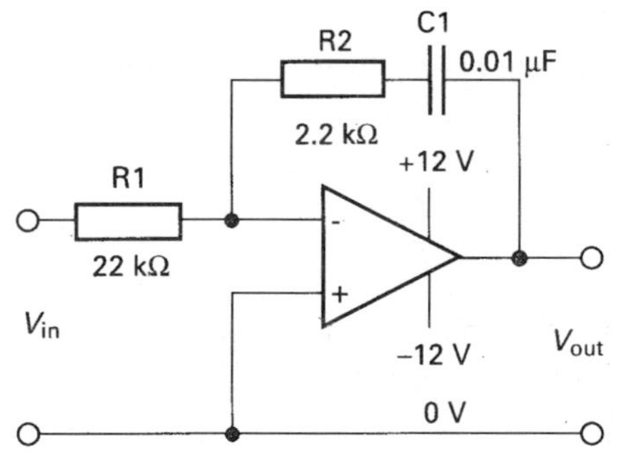

Figure 15.40

5 (a) Figure 15.41 shows a circuit which demonstrates the action of a comparator. Explain in detail the function of a comparator and how it operates, including in your answer the role of the zener diode and 741 operational amplifier. If the reading on V_1 was 5.2 volts, what would be the likely reading on V_0?

(b) A comparator type circuit is to be used to control the temperature of a tropical fish tank.

(i) Draw a black box diagram for a proposed control system.

(ii) Design a possible electronic circuit that will provide accurate temperature control.

(iii) Make sketches to show how various parts of the system can be kept waterproof but still create accuracy control.

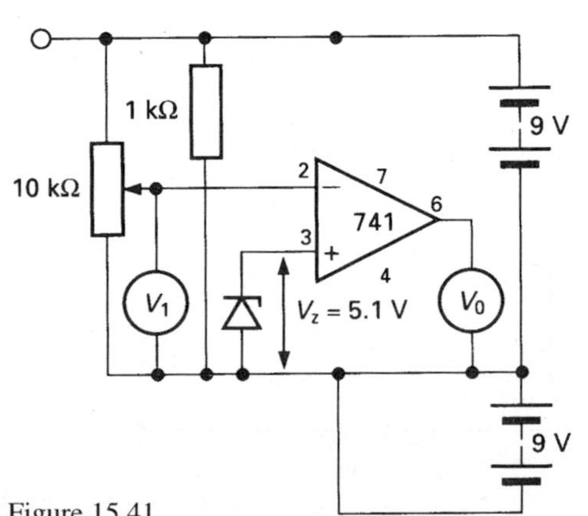

Figure 15.41

6 (a) A wildlife enthusiast has asked you to design an audio voltage amplifier for listening to the songs of wild birds by the use of headphones. She has asked that the amplifier meets the following four criteria: good sensitivity, wide bandwith, high gain and low noise. What is your understanding of each of these criteria as they apply to the performance of the amplifier?

(b) You decide that the amplifier is to be based on an operational amplifier as shown in Fig. 15.42.

(i) What type of amplifier is shown in Fig. 15.42?

(ii) If the output of the amplifier is limited to a maximum voltage swing of 0.5 V and its voltage gain is 60 dB, what is the magnitude of the input voltage?

(iii) Calculate the value of R_3.

(iv) The wildlife enthusiast has asked to be able to adjust the voltage gain of this pre-amplifier while she uses it. Describe one way of modifying the circuit of the pre-amplifier to provide this facility.

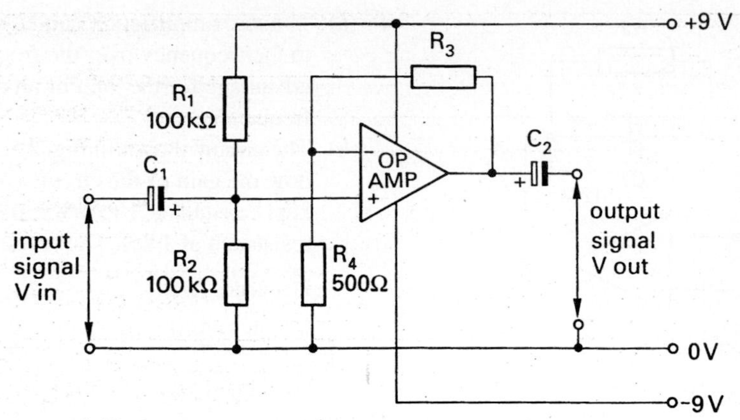

Figure 15.42

(v) In the absence of an input signal, a d.c. voltmeter is used to check the voltage on pin 3 (the noninverting input) with respect to 0 V. What reading would you expect?

(vi) Why are components C_1 and C_2 required in this design?

(c) On looking at the frequency response curves (Fig. 15.43) for the operational amplifier you are using, you see that there is a problem in achieving the voltage gain demanded of the circuit design for frequencies up to about 20 kHz.

(i) Explain how the curves illustrate this problem and calculate by how much you need to reduce the gain, in decibels, of the operational amplifier to achieve the bandwidth that the wildlife enthusiast had asked for.

(ii) What must be the new value of R_3 in Fig. 15.42 to provide this reduced voltage gain?

(iii) Explain why the frequency response curves for some operational amplifiers have the shape shown in Fig. 15.43.

(d) Draw a labelled diagram of a single-stage transistor amplifier which could provide the power amplification required for operating the headphones. Show clearly how you would connect this amplifier to the pre-amplifier circuit of Fig. 15.42.

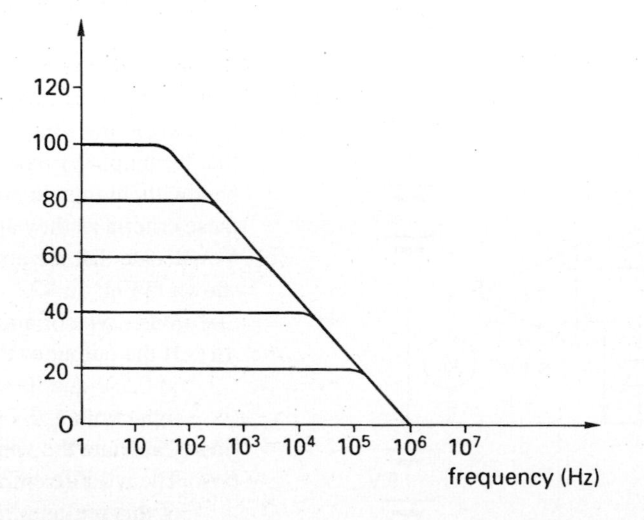

Figure 15.43

7 Fig. 15.44 shows a circuit diagram for an 'Excess temperature indicator'.

(a) Make a concise explanation of the input/process/output stages of the circuit. Indicate characteristics of components where necessary.

(b) The 741 operational amplifier can be used with or without feedback. Give an example of a 741 employing feedback and explain the operation of your circuit.

(c) When the circuit of Fig. 15.44 was constructed on a 'proto board' it was found not to operate correctly. Explain with the aid of a flow diagram how you would set about making the circuit function. Indicate clearly the function of any test instruments you would use.

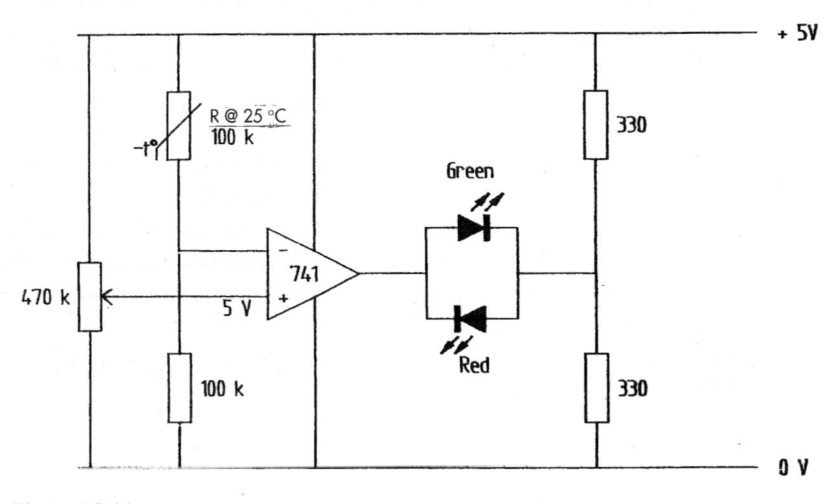

Figure 15.44

16 Digital signal processing

16.1 Combining digital signals – combinational logic

Digital signals are essentially simple. The signal is either: high or low; true or false; 1 or 0; logic 1 or logic 0. The information the signal carries is encoded into this form. Thus logic 1 may mean 'the light is on' if the signal is coming from a light-operated switch and logic 0 may mean 'the temperature is below a set value' if the signal comes from a temperature-operated switch.

Although the signals are simple, other aspects of digital signal processing are more complicated. To begin with there are different families of logic integrated circuits that can be used to process the signals. Also, the symbols used to show these logic gates are not always consistent. Many examination questions are concerned with the use of simple logic gates. In practice, simple logic gates are used for solving simple problems, but more complex problems may require different techniques rather than an extension of the logic gate solution. This section will examine these problems as well as the use of logic gates to combine digital signals.

16.1.1 Alternatives – logic families and symbols

Logic families

Traditionally, there have been two logic families from which to choose. The TTL (transistor transistor logic) family whose devices are numbered from 7400 and is often known as the 74 series, and the CMOS (complementary metal oxide semiconductors) family whose numbers start at 4000. There are advantages and disadvantages with each type and a table comparing some features of the two families is shown below:

	TTL	*CMOS*
operating voltage	5 ± 0.25 V	3–15 V
noise immunity	poor	excellent
input impedance	moderate	high
logic voltages		
– low	0–0.7 V	0–30% V_{cc}
– high	2.4–5 V	70–100% V_{cc}
output current		
– sourcing	1.6 mA	4 mA at 5 V
– sinking	16 mA	4 mA at 5 V
speed	fast	slow

However, new versions of these families, particularly TTL, have appeared in recent years. The 74 standard series is now becoming expensive as production is switched to the newer versions. The 74LS series is perhaps the best known alternative to the 74 standard but there are others. The use of the 74 standard series is no longer recommended because of its high cost and the limited range that remains available. For a more detailed description of the differences between different TTL series it will be necessary to consult a manufacturer's data sheet. However, some general guidance can be given if you are unsure about which logic family to use.

To begin with, unless you are very experienced, try to design your circuit within the same family of logic ICs. In many circumstances the chips from one family are not interchangeable with those of another. Note too that not all families contain all the same functions.

One of the most important decisions concerns the power supply. TTL requires a supply in the range 4.75 V to 5.25 V which means you will need a regulated 5 V supply. This is not an easy voltage range to achieve with batteries. CMOS, on the other hand, will function over a wide range of supply voltages.

CMOS has better noise immunity than TTL. Electrical noise can appear as spikes in any circuit. This can cause false signals to appear as the ICs respond to these changes. TTL ICs can be protected by the use of capacitors connected across the power supply and located very close to the IC but with CMOS ICs this precaution is not needed.

One of the enduring myths of electronics concerns the vulnerability of CMOS to static charge. ICs from the first series of 4000 devices were indeed liable to fail through the kind of voltages generated naturally by static. This led to warnings about never touching the pins on the IC, soldering with an earthed iron, tying an earthed band round your wrist when soldering and always placing the ICs on a metal plate or other conducting surface. Whilst these are sensible precautions, the need for them was largely removed when the improved 4000B series appeared. This included protection within the chip to guard against damage from static. There still are some specialised CMOS ICs that are not protected in this way but all the ICs in the 4000B series are safe from normal static damage.

Another point to be considered concerns the behaviour of inputs to gates when they are not connected to a signal. In TTL series ICs, the inputs always float high. That means that they appear to be connected to logic 1 even if they are not actually connected to anything. This can be a little disconcerting at first but it can reduce the number of connections needed. Unused CMOS inputs float about when they have no input signal connected to them and must be connected to either logic 1 or logic 0. If this is not done the IC will probably not perform its logic correctly. Finally it should be noted that this distinction between logic families is being eroded by the introduction of 74HC and 74HCT logic families which, although appearing as a TTL family by the numbering, are constructed from CMOS technology. It is likely that these will replace the existing TTL and CMOS series in the near future.

Symbols

There are two systems of symbols in common use. One is known as the American system and has distinctive shapes for each different logic gate whilst the other, which conforms to the British Standard BS 3939, has box shapes, with different symbols within the box to represent the different gates. Details of both sets are shown in the next section (see Fig. 16.1).

Some manufactured equipment is marked with the British symbols, some with the American. Most examination boards prefer the American symbols – but use the rest of the British Standard for the drawing of all other components. Some textbooks use the American symbols, others use the British system. This may seem chaotic but it is not important which symbols you use provided you are sure what they mean. There are signs that a common standard is emerging – albeit slowly. An international standard was established in the early 1980s and it is similar to the British system. Some American data books are now being drawn with the new symbols.

In this book the British symbols will be used, once the two systems have been described in the next section.

16.1.2 Logic gates and truth tables

Before examining the differences between logic gates it is useful to examine the function of a *logic gate*. *Combinational logic* is about bringing together two or more signals and producing one or more outputs according to a set of rules governing the behaviour of each *gate*. An alternative approach focuses on the meaning of the word *gate*. A gate is a controlling device that may be either open or shut. A signal may pass through a gate or it may not depending on the rules governing the behaviour of the gate.

Each definition is useful in certain circumstances and both are illustrated in this section. It should be made clear, however, that the function NOT is not a gate. NOT is a signal modifier, an inverter. Its purpose is to change the logical state of the incoming signal. It does not control the signal nor combine it with another signal, so it fails to meet the functional requirement of a logic gate. The confusion arises because the NOT function can be fabricated from logic gates.

Fig. 16.1 shows the name of each logic gate, its American and British symbols, a truth table and a Venn diagram showing its operation. For simplicity only two input logic gates have been included.

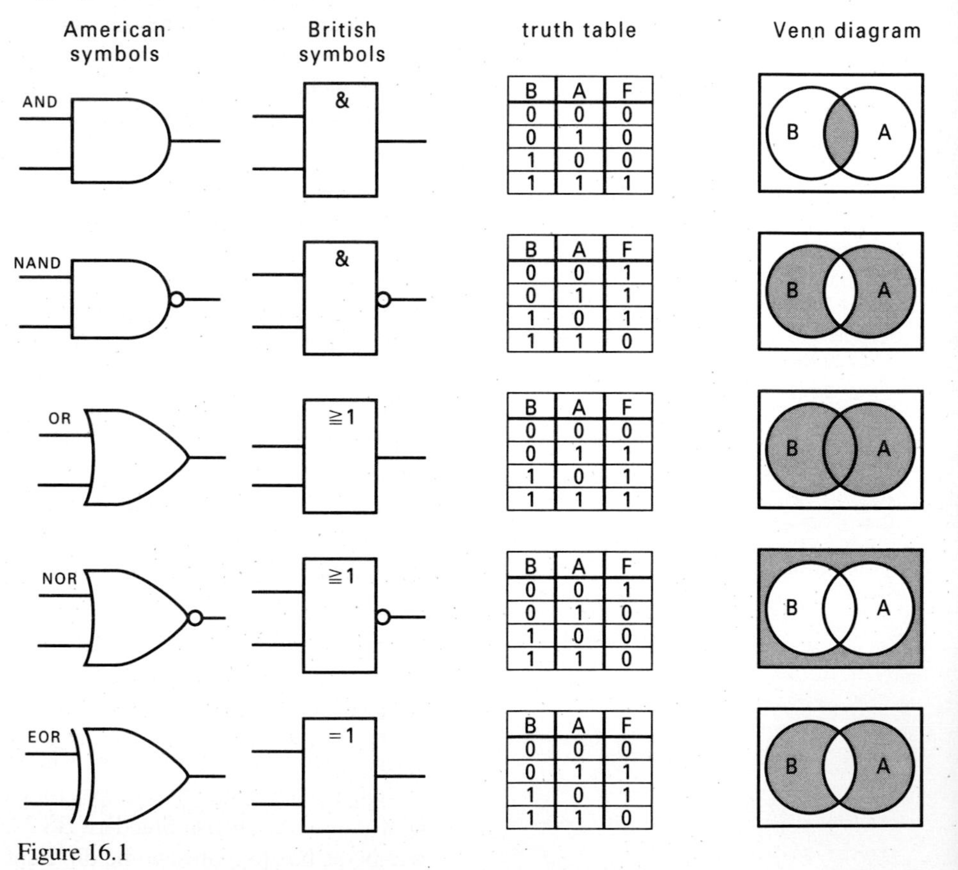

Figure 16.1

Notes:

1 On all of these diagrams the small circle is used to show negation.

2 Extension to three (or more) inputs is relatively easy if the relationship is understood. For example, a four input AND gate will only give a logic 1 output if inputs A AND B AND C AND D are all at logic 1.

Dynamic combining of digital signals

Another way of looking at the combination of digital signals is to consider the system as a dynamic operation. Each logic gate has two inputs, one is the *data* and the other is the *control*. For the sake of convenience, the Data consists of a square wave of equal mark-space ratio. The options for the *control line* are either 0 or 1 [high or low; true or false].

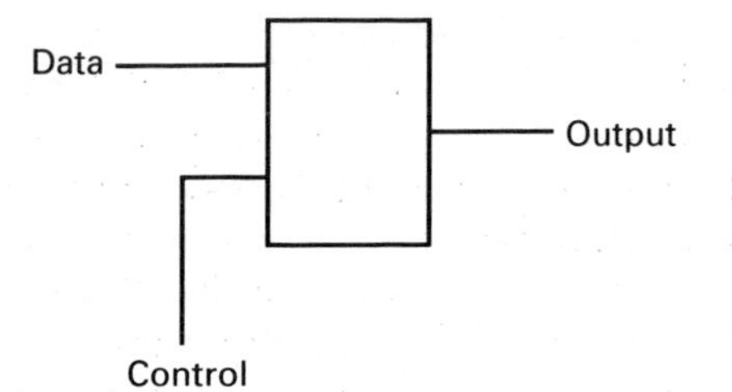

Figure 16.2 Dynamically combining signals with a two input logic gate

We can now compare the input data with the output data for the two possible states of the control line.

Type of gate	*Data*	*Control*	*Output*
AND		0	0
		1	
OR		0	
		1	1
NAND		0	1
		1	(inverted)
NOR		0	(inverted)
		1	0
EOR		0	
		1	(inverted)

Table 16.1 The results of combining signals with different logic gates

To summarise:

The AND gate opens and allows the signal through when the control line is high.

The OR gate opens and allows the signal through when the control line is low.

The NAND gate opens and allows an inverted signal through when the control line is high.

The NOR gate opens and allows an inverted signal to go through when the control line is low.

The EOR gate is the simplest programmable gate: if the control line is low it allows an inverted signal through, if the control line is high it allows the signal through in its original form. It can therefore be thought of as an inverter when the control line is high and a piece of conducting wire when the control line is low.

16.1.3 Boolean algebra

Boolean algebra provides a way of simplifying a logic problem to produce a solution with the minimum number of gates. Many simple logic problems may be solved without needing to use Boolean algebra. This process is known as solving by inspection. Let us look at an example of this process before moving on to Boolean algebra.

PROBLEM We wish to create a selection system which will allow one of two inputs to pass its data directly to the output when that input is selected. This is sometimes known as a data switch.

SOLUTION

1 Draw a block diagram of the system, Fig. 16.3.

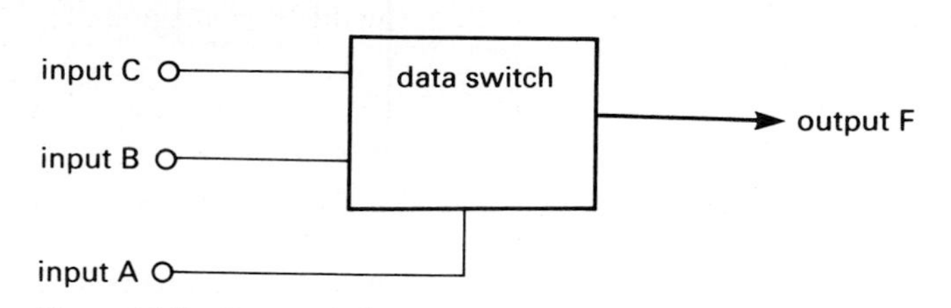

Figure 16.3 Data switch

2 Define the solution in words.

Input A is the select line.
If A is 0 then input C is selected.
If A is 1 then input B is selected.

The signal at the output, F, is the same as the signal at the selected input.

3 Construct the *truth table*.

Input C	Input B	Input A	Output F
0	0	0	0
1	0	0	1
0	1	0	0
1	1	0	1
0	0	1	0
1	0	1	0
0	1	1	1
1	1	1	1

It can be seen that for the top four lines of the table, when input $A = 0$, the output F follows input C whilst in the bottom four lines of the table, when input $A = 1$, the output F follows input B.

4 Write out a logic statement.
The output must be 1 when:

C is 1 AND B is 0 AND A is 0 OR C is 1 AND B is 1 AND A is 0
OR C is 0 AND B is 1 AND A is 1 OR C is 1 AND B is 1 AND A is 1

If this were now implemented in gates we would need (at least) eight two-input AND gates and three two-input OR gates.

5 Simplify this statement by inspection.
If the top line is considered first, B does not affect the outcome and can therefore be eliminated. The output will be 1 if C is 1 and A is 0. B can be either 1 or 0 and the output will still be 1. So the top line becomes:

the output must be 1 if C is 1 AND A is 0

Likewise with the second line, C is redundant in this case. The line simplifies to:

the output must be 1 if B is 1 AND A is 1

Putting both parts together gives the total logic statement as:

the output must be 1 if C is 1 AND A is 0 OR B is 1 AND A is 1

6 Express all terms in relation to logic 1.
The only term in this expression involving a 0 is A is 0. This can be easily removed by considering the inverse of A which is written as $\overline{A}$. If A is 0 then $\overline{A}$ is 1.
The full expression now becomes:

the output must be 1 if C is 1 AND $\overline{A}$ is 1 OR B is 1 AND A is 1

7 Draw the logic diagram, Fig. 16.4.

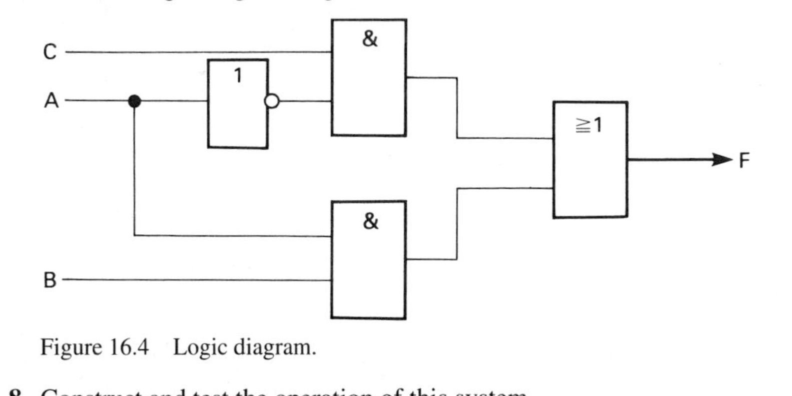

Figure 16.4 Logic diagram.

8 Construct and test the operation of this system.
Many suitable logic tutors are available for this stage or alternatively a breadboard can be used with individual integrated circuits.

If the problem is too complicated to be solved by inspection then it is possible to use Boolean algebra to arrive at a solution. Boolean algebra is a series of mathematical rules that enable us to reduce a complex problem to a simple form.

Postulates of Boolean algebra

1 A variable can assume only one of two possible values. These values are known as 0 and 1.
2 From the definitions of the AND, OR and NOT functions the following expressions are derived:

$.$ = AND, $+$ = OR, a bar above the term indicates inversion

$0 . 0 = 0$	$0 + 0 = 0$	
$0 . 1 = 0$	$0 + 1 = 1$	$\overline{1} = 0$
$1 . 0 = 0$	$1 + 0 = 1$	
$1 . 1 = 1$	$1 + 1 = 1$	$\overline{0} = 1$

Laws of Boolean algebra

Defining Laws

$A + 0 = A$
$A + 1 = 1$
$A . 0 = 0$
$A . 1 = A$

Identity law

$A + A = A$
$A . A = A$

Commutative law

$A + B = B + A$
$A . B = B . A$

Associative law

$$(A + B) + C = A + (B + C)$$
$$(A \,.\, B) \,.\, C = A \,.\, (B \,.\, C)$$

Distributive law

$$(A \,.\, B) + (A \,.\, C) = A \,.\, (B + C)$$
$$(A + B) \,.\, (A + C) = A + (B \,.\, C)$$

Complementary law

$$A + \overline{A} = 1$$
$$A \,.\, \overline{A} = 0$$

Involutional law

$$\overline{A} = A$$

Dualisation law (de Morgan's theorem)

$$(\overline{A \,.\, B}) = \overline{A} + \overline{B}$$
$$(\overline{A + B}) = \overline{A} \,.\, \overline{B}$$

Boolean algebra and logic gates

If two input gates are considered, the inputs being A and B, then the following expressions are obtained for the standard logic gates:

AND	$A \,.\, B$
OR	$A + B$
NAND	$\overline{A \,.\, B}$
NOR	$\overline{A + B}$

Prove for yourself, by examining the truth table, that the expression for the exclusive OR gate is as follows:

$$(\overline{A} \,.\, B) + (A \,.\, \overline{B})$$

To examine Boolean algebra in action we can look at the previous problem on page 662, which we solved by inspection:

$$(C \,.\, \overline{B} \,.\, \overline{A}) + (C \,.\, B \,.\, \overline{A}) + (\overline{C} \,.\, B \,.\, A) + (C \,.\, B \,.\, A)$$

using the distribution law gives:

$$C \,.\, \overline{A}(\overline{B} + B) + B \,.\, A(\overline{C} + C)$$

which, by the complementary law gives:

$$(B + \overline{B}) = 1$$
$$(C + \overline{C}) = 1$$

and, therefore,

$$C \,.\, \overline{A} + B \,.\, A$$

which is the same as the previous result.

16.1.4 Karnaugh mapping

It has been shown above that the most efficient realisation of a logic system can be obtained by simplification of the Boolean expression describing the system. This can become a complex procedure if there are more than two variables. The minimalisation of systems with three of four variables is often more easily accomplished using the Karnaugh map procedure.

A *Karnaugh map* is an ordered array of every possible combination of the variables in the expression. Rather in the manner of a truth table, the presence of a particular combination is shown by a 1.

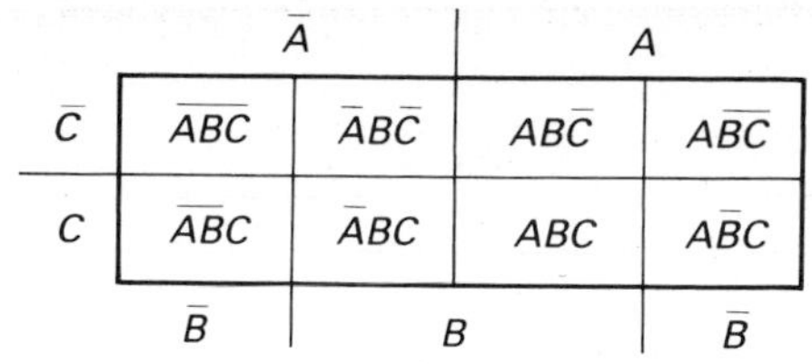

	$\bar{A}$		A	
$\bar{C}$	$\bar{A}\bar{B}\bar{C}$	$\bar{A}B\bar{C}$	$AB\bar{C}$	$A\bar{B}\bar{C}$
C	$\bar{A}\bar{B}C$	$\bar{A}BC$	ABC	$A\bar{B}C$
	$\bar{B}$	B		$\bar{B}$

Figure 16.5

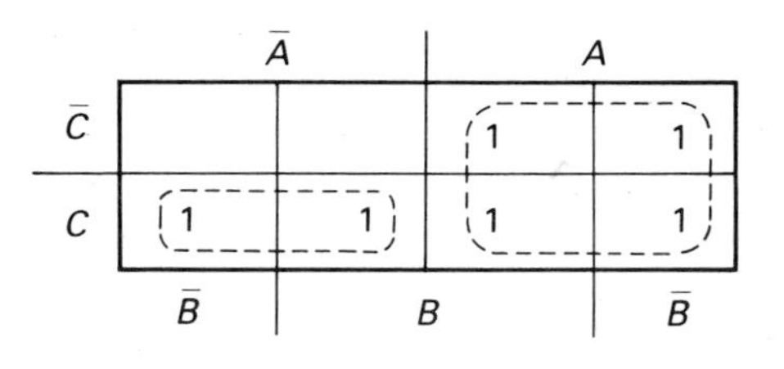

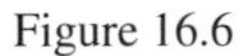
Figure 16.6

For a three variable system the array is arranged as shown in Fig. 16.5. Suppose we are confronted by the function

$$F = \bar{A}BC + \bar{A}\bar{B}C + AB\bar{C} + A\bar{B}\bar{C} + ABC + A\bar{B}C$$

This would lead to a Karnaugh map as shown in Fig. 16.6.

The two linked 1s indicate the presence of $\bar{A} . C$ and the block of four show the presence of A. Hence we have in its simplified form

$$F = A . \bar{C} + A$$

It should be noted that the Karnaugh map is continuous, as if it were drawn on the curved surface of a cylinder. Hence the block for B would consist of 1s in the columns on the left- and right-hand sides of the array. The interpretation of the arrangements of 1s is shown in Fig. 16.7.

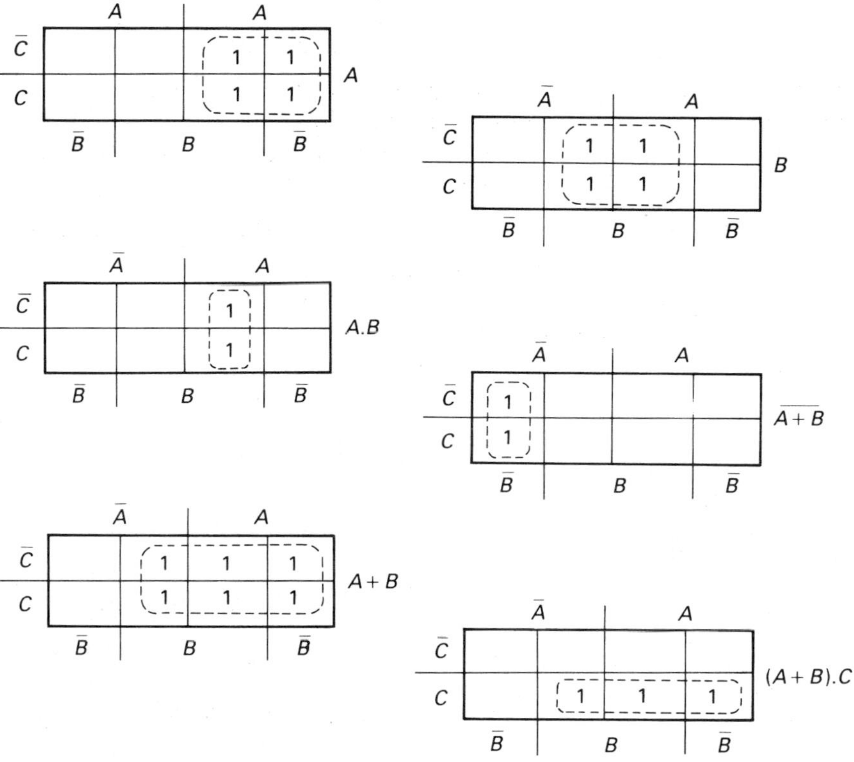

Figure 16.7

As an example of the use of the Karnaugh map consider the simplification of the following function:

$$F = (A . B + \bar{A} + B) + (A . C + \bar{A} + B)$$

We construct the Karnaugh map, Fig. 16.8, and mark on it all of the terms of the expression.

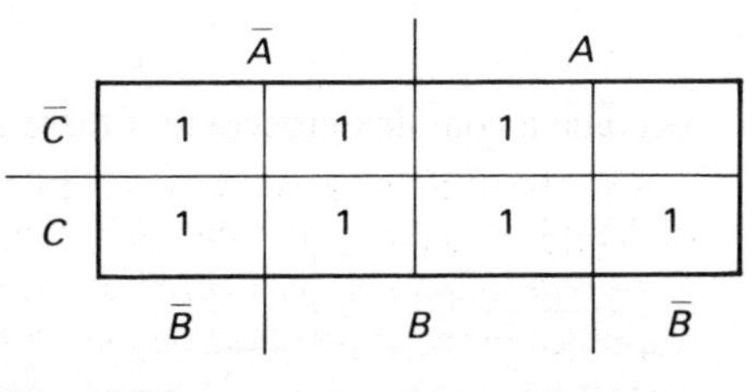

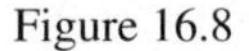
Figure 16.8

By inspection, the expression producing the same map in the simplest way is that of Fig. 16.9.

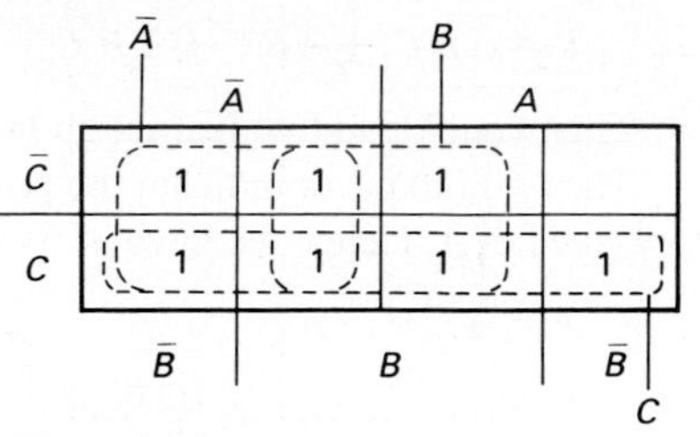

Figure 16.9

This gives:

$$F = \bar{A} + B + C$$

The Karnaugh map for four variables is only marginally more complicated. The general arrangement is shown in Fig.16.10. As before, the letters round the edge show the location of the variables' blocks.

	$\bar{A}$		A		
$\bar{C}$	$\bar{A}\bar{B}\bar{C}\bar{D}$	$\bar{A}B\bar{C}\bar{D}$	$AB\bar{C}\bar{D}$	$A\bar{B}\bar{C}\bar{D}$	$\bar{D}$
$\bar{C}$	$\bar{A}\bar{B}\bar{C}D$	$\bar{A}B\bar{C}D$	$AB\bar{C}D$	$A\bar{B}\bar{C}D$	D
C	$\bar{A}\bar{B}CD$	$\bar{A}BCD$	$ABCD$	$A\bar{B}CD$	D
C	$\bar{A}\bar{B}C\bar{D}$	$\bar{A}BC\bar{D}$	$ABC\bar{D}$	$A\bar{B}C\bar{D}$	$\bar{D}$
	$\bar{B}$	B	B	$\bar{B}$	

Figure 16.10

In this case it is the blocks for $\bar{B}$ and $\bar{D}$ which wrap around the horizontal and vertical directions respectively.

16.1.5 Design using only one type of gate

There are advantages in producing solutions to logic problems which only involve one type of gate. The solutions are usually made up of NAND gates or possibly NOR gates. The most obvious advantage is that it may reduce the number of ICs needed. This can be seen by examining a solution which is expressed in terms of AND, OR and NOT functions. As a minimum this will require three types of IC and it is unlikely that each IC will be fully utilised. This is because there will be at least four logic gates on each IC. If we consider the arrangement of logic gates shown in Fig. 16.11, we can see that there are AND, OR and NOT functions involved.

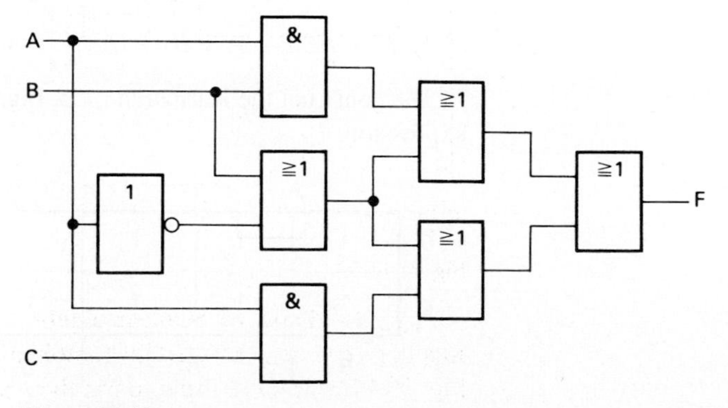

Figure 16.11

Each of these functions is now replaced with its NAND gate equivalent, Fig. 16.12. (See Fig. 16.14 for the full table of equivalents.)

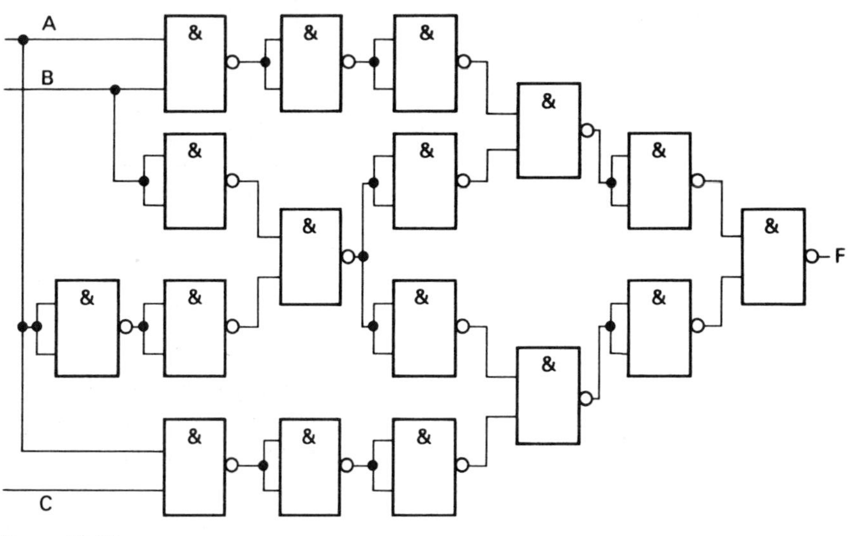

Figure 16.12

It can be seen that there are several places where two NOT functions occur adjacent to each other. This double negation is superfluous and both functions can be removed as shown in Fig. 16.13.

This results in a considerable saving in the number of ICs required.

It is also possible to get into the problem of propagation time delays. It requires a small, but finite, time for the signal at the inputs to a gate to affect the output of that gate. As the solutions become more complex and the number of AND, OR and NOT functions increases so the propagation time increases. It is likely that not all paths through the functions have the same number of gates so, at some stage, the probability of an incorrect signal being generated increases to an unacceptable level. Changing to one type of gate will reduce the complexity of the solution and therefore reduce the propagation time delays.

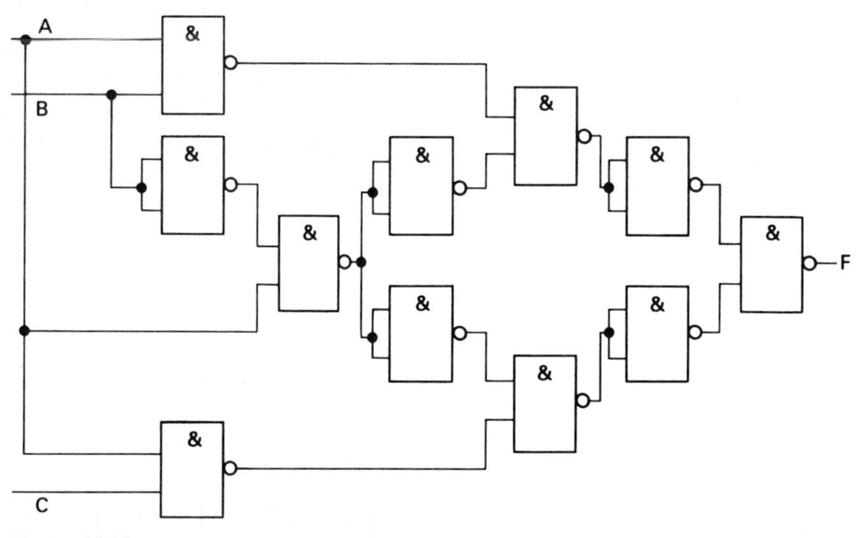

Figure 16.13

It can be shown by Boolean algebra, or by inspection, that each of the common functions can be generated by combinations of either NAND gates or NOR gates, Fig. 16.14. It is more usual to use the NAND equivalences but, for completeness, the NOR versions are shown as well as the NAND.

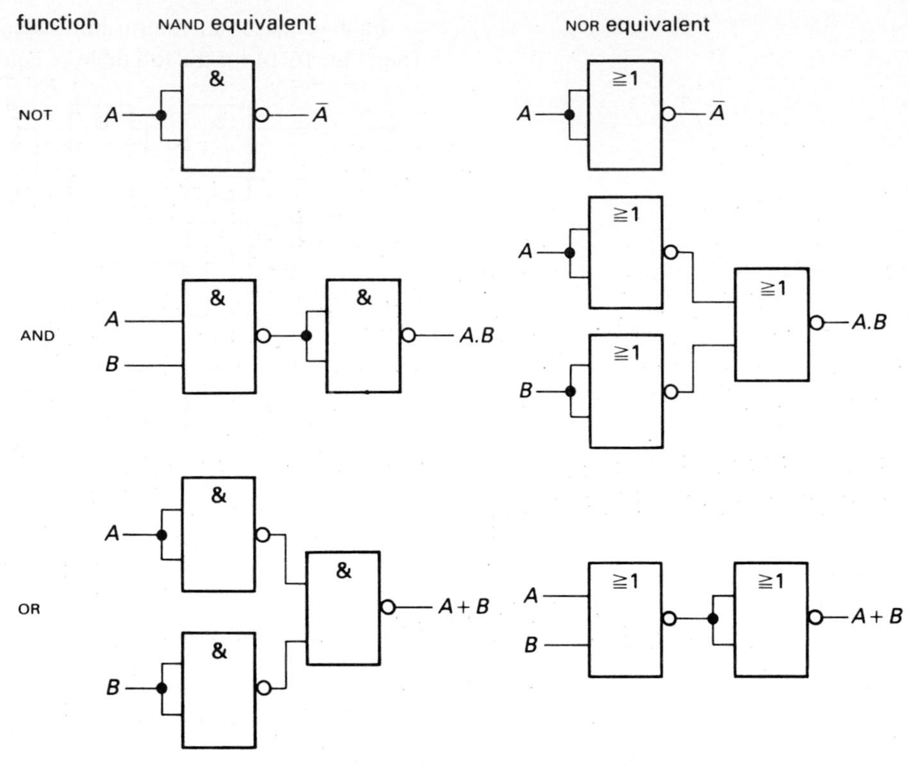

Figure 16.14

16.1.6 Alternatives to logic gates

Changes in technology are giving new and more flexible alternatives to logic gates. These alternatives can remove the need for Boolean algebra and Karnaugh mapping.

The first of these alternatives is the *programmable gate*, Fig. 16.15 on page 669. It is based on an 8-input analogue multiplexer such as the 4051 CMOS device. The cost of this device is about twice that of a standard logic gate. It has eight select lines, each of which can be set to either high or low by a switch (or a link). Inside the IC there are eight switches only one of which can be closed at any one time. The switch to be closed is selected by the pattern on the lines A_0 to A_2 according to the following table:

A_2	A_1	A_0	*Switch closed*
0	0	0	Y0
0	0	1	Yl
0	1	0	Y2
0	1	1	Y3
1	0	0	Y4
1	0	1	Y5
1	1	0	Y6
1	1	1	Y7

Closing a switch connects the output, Z, to one of the select lines. Z is connected to 0 volt through a resistor *R*. The value of *R* will depend on the type of gate employed in the next stage. The value is 180 Ω for standard TTL, 1 kΩ for LSTTL and 10 kΩ for CMOS.

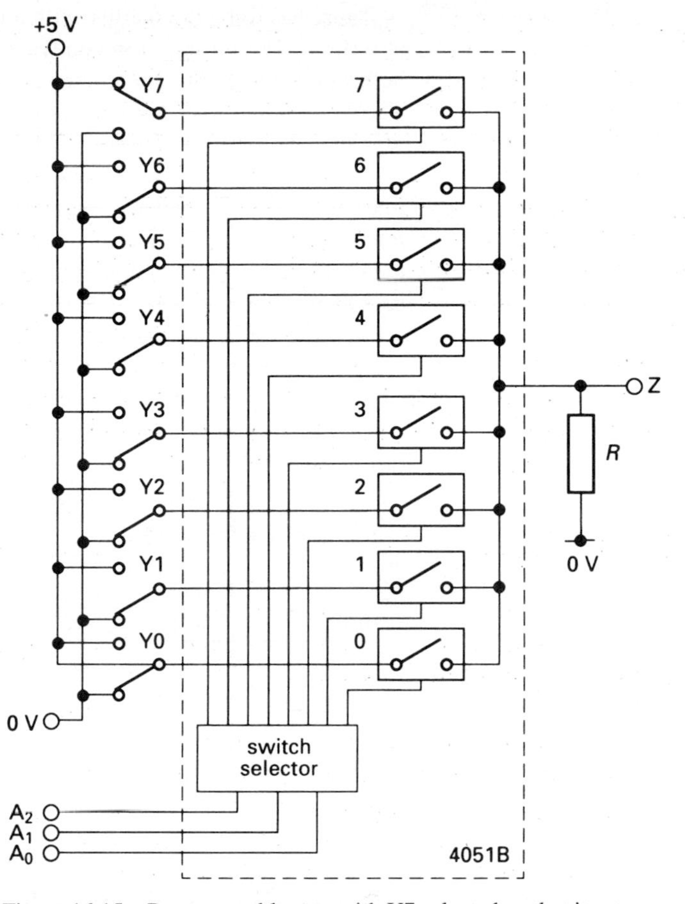

Figure 16.15 Programmable gate with Y7 selected so that it acts as a three-input AND gate.

If the lines A_0 to A_2 are considered as the inputs to this system then it can behave as any three input logic system. For example, to make it behave as a three input AND gate it is only necessary to select Y7 to be logic 1 since only when A_0 AND A_1 AND A_2 are at logic 1 will Z be at logic 1 (switch 7 closed). For all other conditions Z will be at logic 0. This is a particularly important device when nonstandard logic functions are required.

The second alternative is the use of memory. The truth table can be stored in memory directly – there is no need to work out the logic.

Let us consider a simple memory which can store 64 bits of information. This is arranged as 16 lines of four bits, Fig. 16.16.

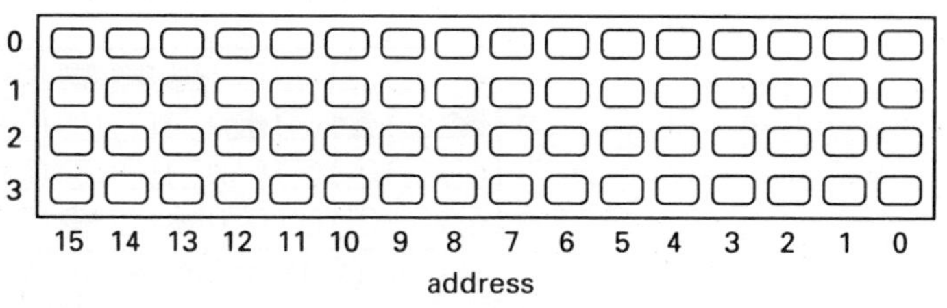

Figure 16.16 A 16 × 4-bit memory

The lines are numbered from 0 to 15, the bits from 0 to 3. Each 4-bit word is considered as an entity in itself and selecting any of the lines 0 to 15 therefore gives access to the whole of one 4-bit word. The numbers 0 to 15 are known as the address. In binary these are the numbers 0000 to 1111.

There has to be the facility with a memory IC, either to read the information or to write it. The information you want to store has to be written into the memory before it can be read. One line connected to the IC is known as the read/write line. The state of that line (high or low) determines whether the information is being read or written.

The advantages of this method are that up to four input lines and up to four output lines are available. (Other memory ICs allow a larger number of possibilities.) The four input lines are connected to the address lines whilst the outputs are connected to the data lines. When an address is selected the contents of that store appear at the data lines.

Suppose we need to save a sequence resembling a pelican crossing. The specification is that there are three lights, red (bit 0), amber (bit 1), green (bit 2) and a buzzer (bit 3). The specification is that when the red light is on the buzzer must go on and off alternately. At all other times the buzzer is off. The red is followed by flashing amber which is in turn followed by green. When the sequence is activated the green changes to steady amber before turning to red.

This can be shown in a table as shown below:

Address	*Red*	*Amber*	*Green*	*Buzzer*
0	0	0	1	0
1	0	1	0	0
2	0	1	0	0
3	0	1	0	0
4	1	0	0	1
5	1	0	0	0
6	1	0	0	1
7	1	0	0	0
8	1	0	0	1
9	1	0	0	0
10	1	0	0	1
11	0	1	0	0
12	0	0	0	0
13	0	1	0	0
14	0	0	0	0
15	0	1	0	0

The memory store is shown in Fig. 16.17.

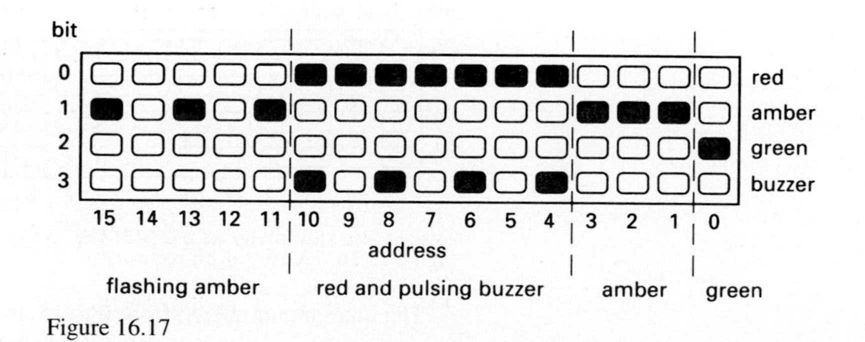

Figure 16.17

This assumes the output is from address 0 until the sequence is activated.

16.2 Sequential logic

With combinational logic the output depends only on the state of the inputs. With *sequential logic* the output depends on the state of the input but the order in which inputs (and possibly outputs) change state can also determine the output. A simple example is a *latch*: if the input signal changes state then the output changes state but if the input changes state again nothing will happen to the output state. If the input again changes state – the original change again – still nothing will change the output state. The state of the output no longer just depends on the state of the inputs but depends on the order or timing of these events.

The latch is a simple type of memory system. (A more complex memory system was considered in the previous section.) Here we are concerned with the action of the individual memory cells and the alternative ways of storing this information. Another obvious building block of sequential logic is the counter. This will tell us how many times a certain event has occurred. Finally in this section we are concerned with how the data stored in memory can be moved about within a larger system.

16.2.1 Types of memory – latches to flip-flops

There are many names for circuits/systems that remember. Memory, latch, bistable, bistable multivibrator or flip-flop are all possibilities. The functional description of a 'remember system' is that it has two states both of which are stable. It will change from one state to the other under the action of a certain set of input conditions. Different types of these systems respond to different types of input conditions and that is what we will be examining in this section.

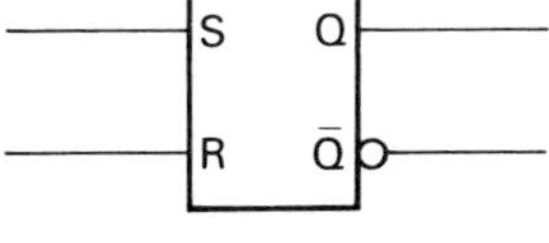

Figure 16.18 S-R flip-flop.

The S-R flip-flop

The name *flip-flop* has been chosen for the simple reason that it seems to describe the action of these systems clearly. Some may disagree... Fig. 16.18 shows the simplest type of flip-flop – the S stands for set and the R for reset.

It has two outputs, Q and $\overline{Q}$. They should never be in the same state. When Q is high the flip-flop is said to be *set* and when Q is low it is said to be *reset*. When S goes high the flip-flop will go into the set state. When R goes high then the flip-flop is reset. Unfortunately if both inputs are taken high at the same time then both outputs are forced into the low state. This is both illogical and undesirable. It may not matter in some circumstances but if it is possible to have both inputs high together then this particular type of flip-flop should be avoided. It is a useful starting point, however, to examine the operation of a circuit, Fig. 16.19, that might achieve this function.

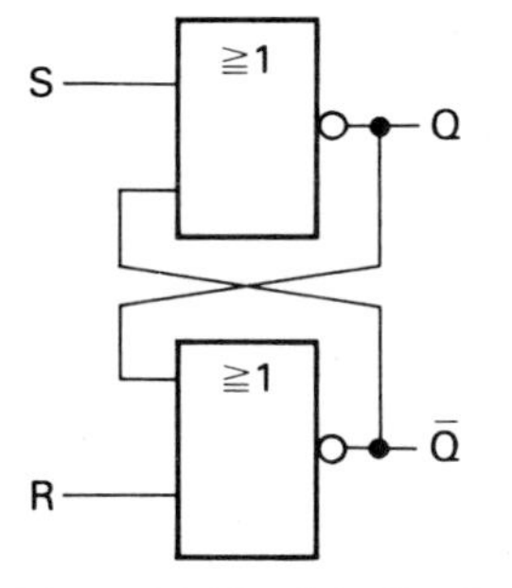

Figure 16.19 S-R flip-flop made from NOR gates.

This circuit is based on NOR gates. At switch on, with both inputs low, suppose that Q goes high. (Either Q or $\overline{Q}$ could go high on switch on.) If Q is high then $\overline{Q}$ is low. Taking input S high, on its own, will not affect the output but taking input R high, on its own, will cause Q to go low and $\overline{Q}$ to go high. If you are not sure of this action you must remember the truth table for a NOR gate; the output is high only if both inputs are set to low. As explained earlier, taking both inputs high together forces both outputs to go low.

An alternative implementation based on NAND gates is shown in Fig. 16.20. It suffers from exactly the same problems. The inverters make the circuit behave in exactly the same way as the NOR based system.

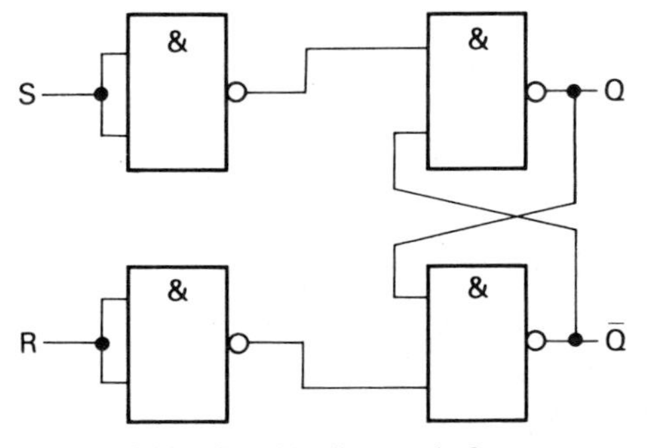

Figure 16.20 S-R flip-flop made from NAND gates.

The clocked S-R flip-flop

One way of making sure that undesirable states do not occur at the inputs is by preventing the information from being presented to the S-R flip-flop until it is convenient for us. The most obvious way of achieving this is to gate the inputs to the

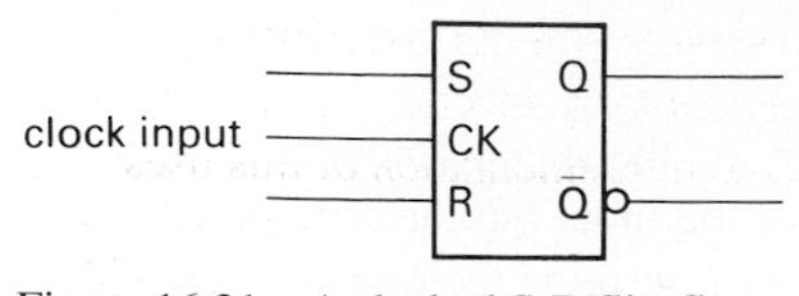

Figure 16.21 A clocked S-R flip-flop.

S-R flip-flop. The signals will now only go through to the S-R flip-flop if the *clock input* is high. The clock input controls the gates. A systems representation of this arrangement is shown in Fig. 16.21 and a circuit that will achieve this function is given in Fig. 16.22.

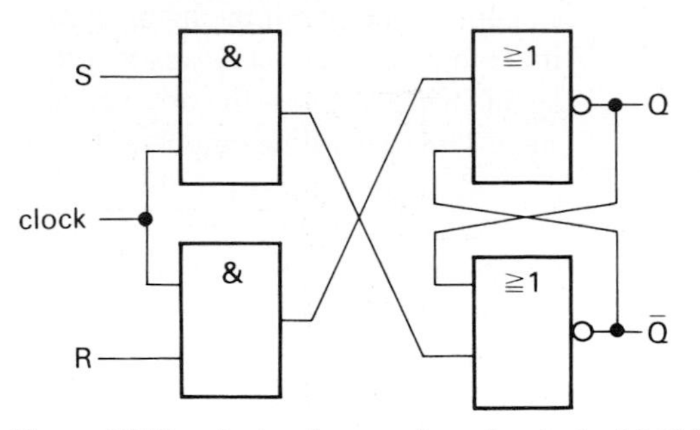

Figure 16.22 An implementation of a clocked S-R flip-flop.

The clock input may be pulsed continually. This arrangement gives a further advantage in that there is a finite time required before the outputs will have settled at the fixed state determined by the inputs. If changes at the inputs occur too quickly the outputs may not have settled before the next change is forced upon them. By clocking the flip-flop, it will only change when the gate is open and it cannot change again until the gate is open next time. This gives the necessary time for the outputs to settle. This propagation time is important in all logic systems but particularly where large numbers of gates are connected together as in sequential systems.

The D-type flip-flop

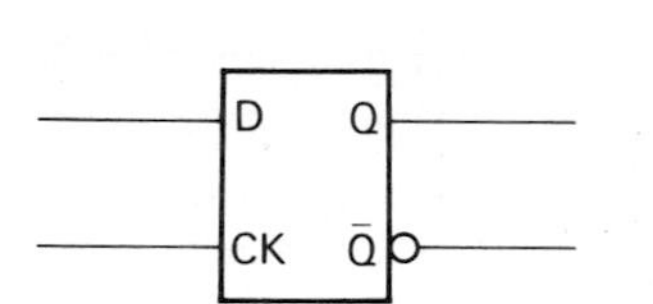

Figure 16.23 A D-type flip-flop.

As we have already seen, the S-R flip-flop can be confused if both inputs go high together. It is being told to set and reset at the same time. One simple way to prevent this occurring is to connect the S and R inputs together but with an inverter between them. R is therefore always NOT S. This arrangement, with a single input, is called a *D-type flip-flop* (D for data), Fig. 16.23. One possible implementation from logic gates is given in Fig. 16.24.

The signal on the data input is transferred to the Q output when the flip-flop is clocked.

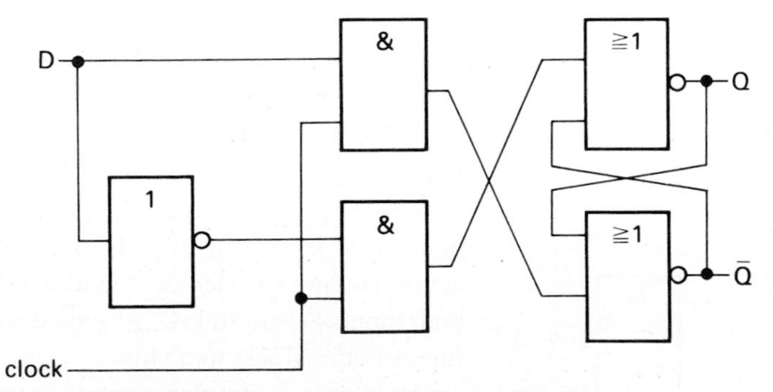

Figure 16.24 An implementation of a D-type flip-flop.

The master–slave and J-K flip-flop

One further problem that can occur with flip-flops is that the data on the input lines can change as it is being read. This will cause inaccurate data to be passed on. This problem can be overcome by replacing each single flip-flop by two. One is called a master and the other the slave. The master takes in the input signal when the clock signal is high and passes the information on to the slave when the clock input goes low. When the clock next goes high again the slave can be read since it is stable and the master can also take in any new information. One implementation of this uses two clocked S-R flip-flops with the clocks connected by an inverter so that they are always in opposite states, Fig. 16.25. Note that the addition of an inverter from S to R would create a D-type flip-flop.

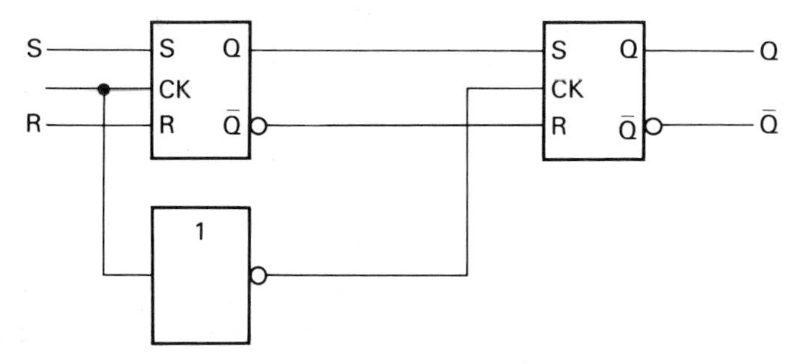

Figure 16.25 Master–slave flip-flop.

The circuit for a clocked S-R flip-flop made with NAND gates is shown in Fig. 16.26.

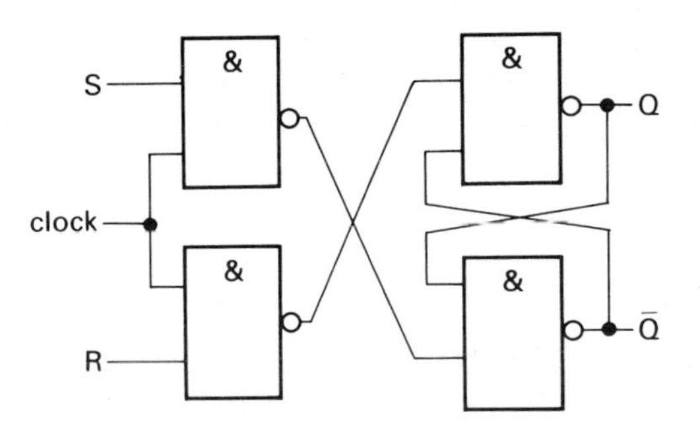

Figure 16.26 A clocked S-R flip-flop made from NAND gates.

This master–slave arrangement can easily be converted into the most useful flip-flop of all, the J-K master–slave. Fig. 16.27(a) shows the *J-K flip-flop* and Fig. 16.27(b), the *J-K master–slave flip-flop*.

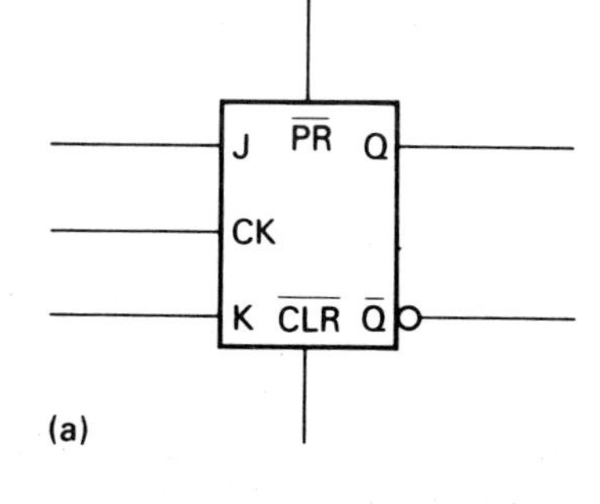

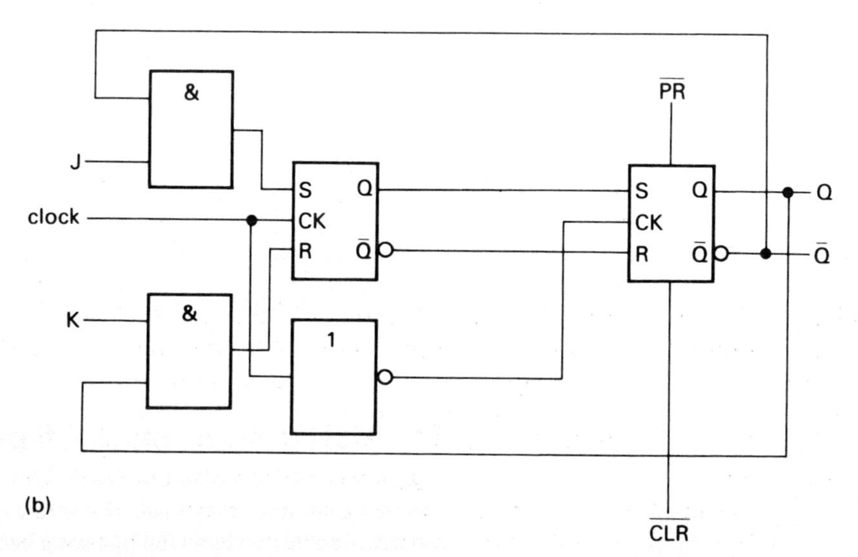

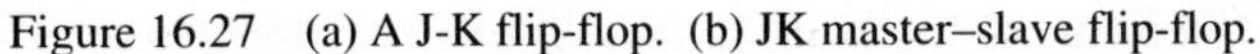

Figure 16.27 (a) A J-K flip-flop. (b) JK master–slave flip-flop.

The J and K inputs are allowed through to the S and R inputs only when the opposite output is high. This means that the J input can set the system only when it needs setting and the K input can only reset the system when it needs resetting. If both J and K are high, they switch alternately to S and R so that the state of the outputs changes on every clock pulse. This is known as *toggling*. When J and K are both low no changes occur when the system is clocked. S is the set or preset input and R the reset or clear input – both of these are usually active low, that is, a low signal causes them to set or reset the system.

It was stated that the J-K is the most useful member of the flip-flop family. The reason for this assertion is that it can be configured to perform each of the other functions we have so far examined. The arrangements for a clocked S-R flip-flop, a clocked D-type flip-flop, and a T (toggle) flip-flop are given in Figs 16.28, 16.29 and 16.30 respectively.

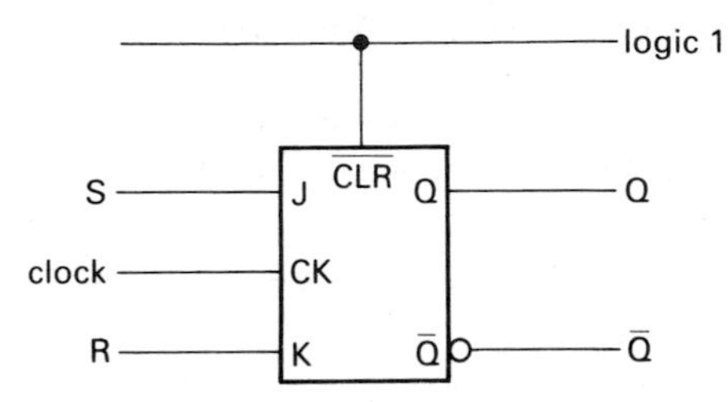

Figure 16.28 Clocked S-R flip-flop.

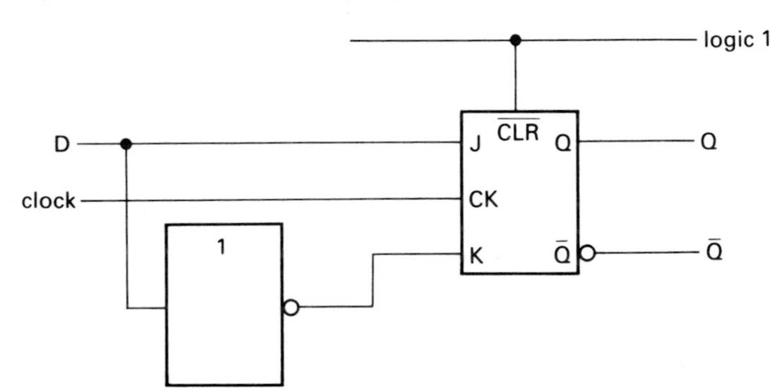

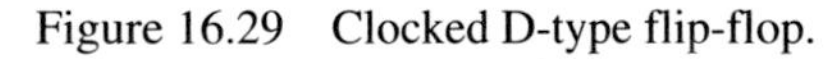

Figure 16.29 Clocked D-type flip-flop.

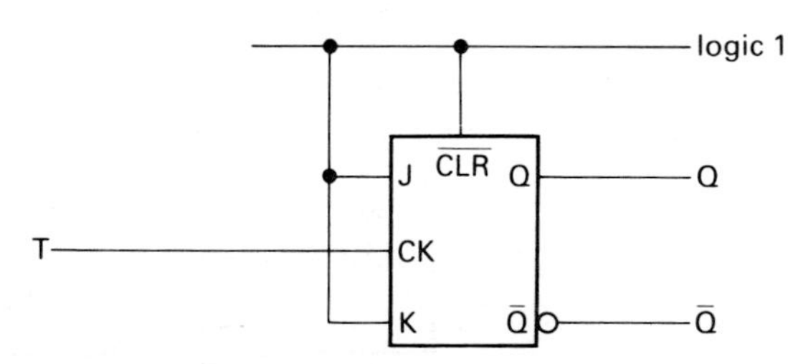

Figure 16.30 T (toggle) flip-flop.

16.2.2 Counting circuits

The basic building block of all counters is the binary counter. If we consider a 4-bit binary counter it will have one input line and four output lines with at least one control line to reset the counter to zero (0000), Fig. 16.31.

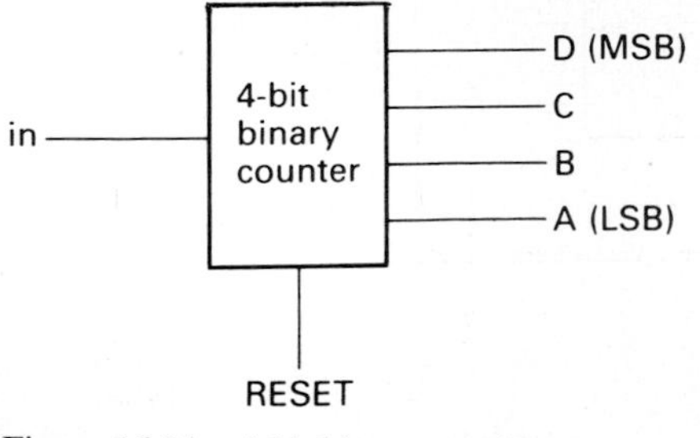

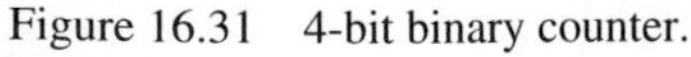

Figure 16.31 4-bit binary counter.

This can be achieved with four flip-flops connected in series. This will ensure that, as pulses are fed in to them, the number of pulses is given in binary form by the states of the flip-flop outputs taken in order. Any of the types of flip-flops considered in the previous section could be used but we shall confine ourselves to an arrangement based on J-K flip-flops, configured as toggle flip-flops. Adding more flip-flops would produce a larger counter – the principles can easily be examined with the 4-bit version shown in Fig. 16.32.

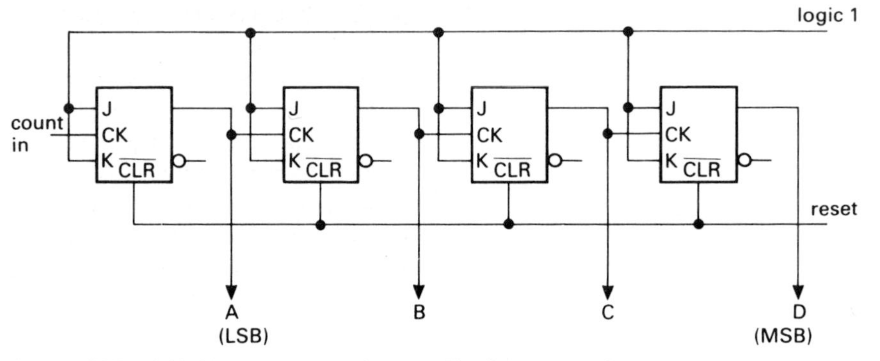

Figure 16.32 4-bit binary counter using J-K flip-flops as toggles.

This type of counter is known as the *ripple through counter* or, more correctly, the *asynchronous counter*. This is because there is a time delay between the pulse arriving and the counter settling to a fixed value, due to the propagation time of each flip-flop. When the last bit is about to toggle the change ripples through each preceding flip-flop so giving it its name. This can be illustrated by considering the timing diagram for this arrangement, Fig. 16.33. The dotted line shows how the effect of the propagation time is cumulative and can become significant, causing errors in the next subsystem.

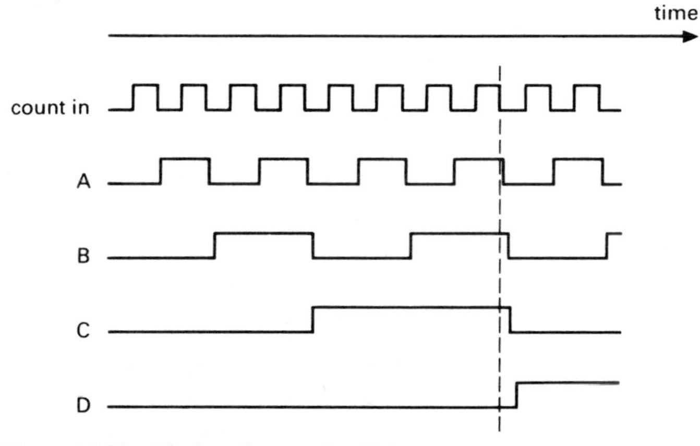

Figure 16.33 Timing diagram for 4-bit counter.

To obtain the correct operation of the counter it may be necessary to ensure that each bit is updated at the same time. This is known as a *synchronous counter*. A realisation of a 4-bit synchronous counter using J-K flip-flops is shown in Fig. 16.34.

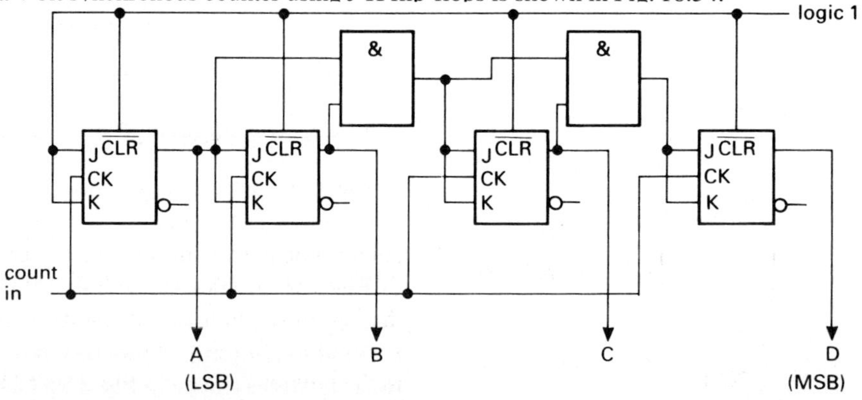

Figure 16.34 4-bit synchronous binary counter.

All the stages are pulsed together at their clock inputs to ensure that they all change simultaneously. The AND gates ensure that a stage only toggles when all preceding stages are at logic 1. For both counters the reset is obtained by taking the CLR line to logic 0.

At the present time both counters will go from 0000 to 1111, a total of 16 counts. By using AND gates and inverters, it can be arranged for the counter to reset at any required number. The most common example would be for a counter to reset after 9 making it a decade counter. In this particular case the D line and the B line, representing a value of 8 and a value of 2 respectively, are connected to the inputs of an AND gate, Fig. 16.35. This will give a logic 1 at the output when 10 (1010 in binary) appears at the outputs. Since this particular arrangement requires logic 0 to reset the signal would then be passed through an inverter before the CLR terminal. (Of course, a NAND gate could replace both functions.) It resets the counter so quickly that no visible display would appear on any indicators connected to the output lines.

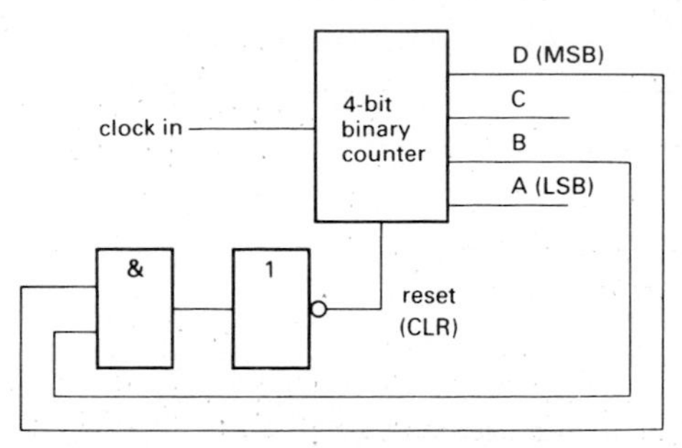

Figure 16.35 4-bit binary counter with reset making a decade counter.

16.2.3 Serial and parallel shifting of data

Information, in the form of a pattern of binary bits, can be transferred one bit at a time (serial) or several bits at the same time (parallel). Whilst this information is being transferred it will need to be placed in temporary stores known as registers. These registers, usually composed of arrangements of flip-flops, can be made for both serial and parallel transmission. They can also convert serial to parallel and vice versa.

The parallel input/output register (or data latch)

This register can be constructed from a suitable number of D-type flip-flops. When the clock is pulsed the data is stored until the next clock pulse. The operation of such a register is shown in Fig. 16.36.

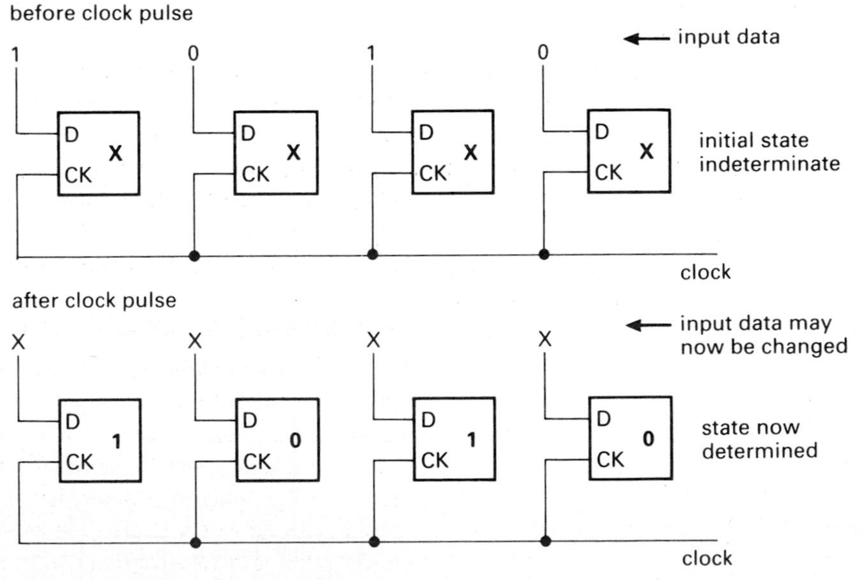

Figure 16.36 Parallel input register.

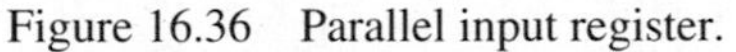

The serial input/output (shift) register

The shift register is usually an array of D-type flip-flops into which a bit pattern can be fed in serial form. It may then be output in either serial or parallel form. A 4-bit shift register is shown in Fig. 16.37 and the sequence of action of the 4-bit shift register is given in Fig. 16.38.

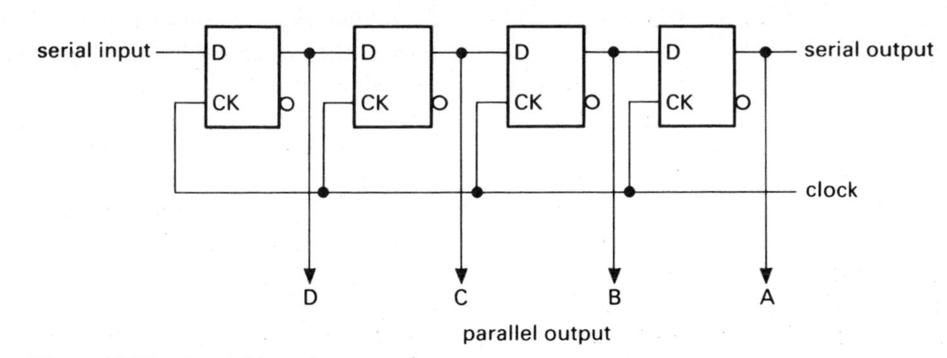

Figure 16.37 A serial input/output register.

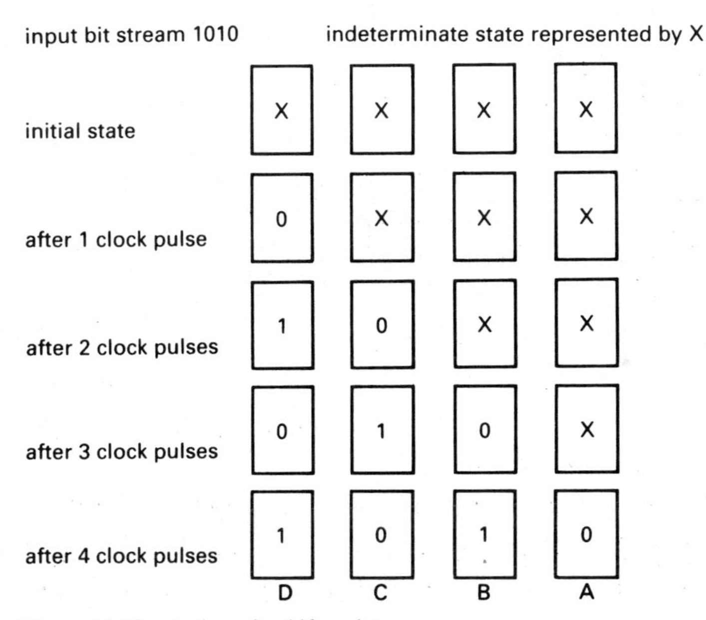

Figure 16.38 Action of a shift register.

Let us assume that the input bit stream is 1010 and that X represents an indeterminate state, that is, it can be either 1 or 0. After four clock pulses the data can then be taken from the parallel outputs connected to the individual flip-flops. Applying four further clock pulses will send the data out from the last flip-flop still in serial form.

The parallel or serial input/output register

The final circuit in this section, Fig. 16.39, shows a possible implementation of a register that will allow parallel or serial information in and send it out as either parallel or serial information.

It should also be noted that registers have other roles to play. In particular they permit the manipulation of data. If the bit pattern in a register is shifted one bit to the right this is equivalent to dividing its numerical value by two. Conversely, a shift one bit to the left is equivalent to a multiplication by two.

16.3 Driving digital output devices

16.3.1 Current amplification

Section 15.5 gives more detail about this topic. The current needs to be amplified in order to drive an output device from a digital process subsystem such as a logic gate. The available current from a gate is a few milliamp at best and this is insufficient to

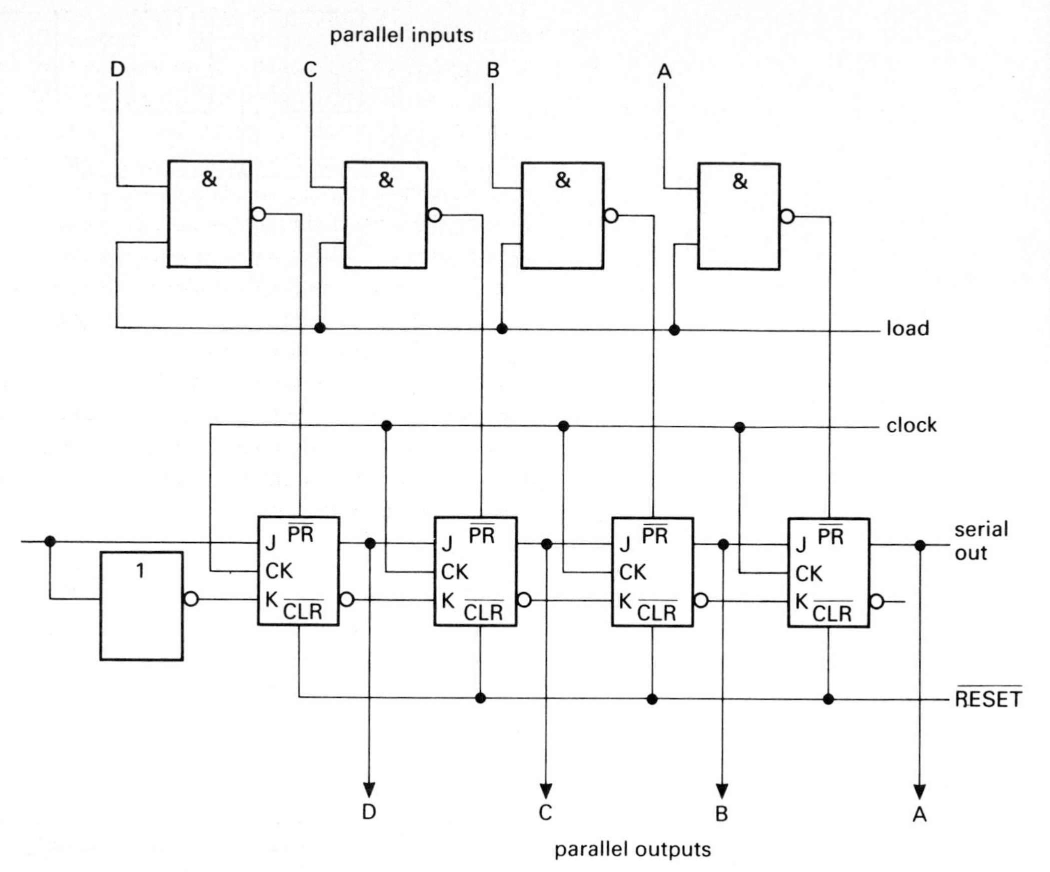

Figure 16.39 A parallel/serial input/output register.

drive most output devices. A single transistor, connected in the common emitter configuration, is the normal output arrangement to ensure sufficient current, Fig. 16.40. The values of resistors are calculated to ensure that the transistor is saturated.

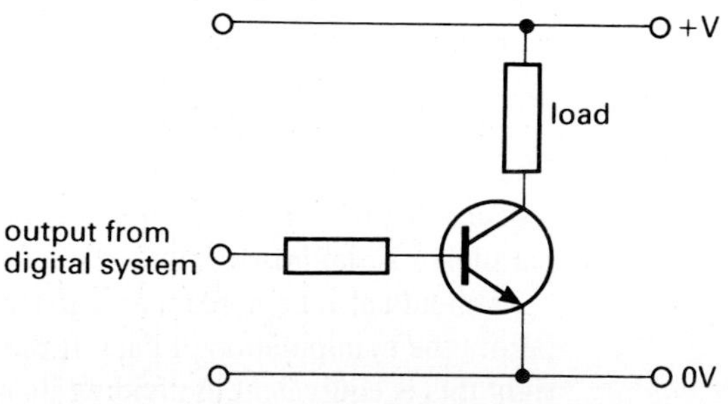

Figure 16.40 Driving loads from digital systems.

The transistor is selected with a maximum value of I_c greater than the current required by the output device. Note that this can only be termed current amplification if the supply p.d. to the transistor is the same value as that supplied to the logic gate (or other digital process subsystem). If the supply p.d. changes it will be power amplification. A Darlington driver may also be used. Remember to include a flywheel diode if the output is an inductive load (for example, solenoid or relay).

16.3.2 Decoders and drivers for optical displays

Although decoding and driving are two separate functions they are included here, in the same section, since they are often contained in the same integrated circuit. Driving is achieved by the same method as driving any other output device, namely a transistor. For optical devices it is usually necessary to include a resistor in series with the optical device. The data sheet or manufacturer's catalogue should give details. The other main point to note is that drivers can either sink or source the output current. This is particularly important with seven-segment displays which can be either common anode (for current sinking) or common cathode (for current sourcing). Many projects have floundered when a mismatch has occurred between the driver and the display.

Decoding is necessary because the signal obtained from the previous stage – usually a counter – is not in a form that will correctly illuminate the display. The most obvious example is when trying to take the output of a BCD (binary coded decimal) counter to illuminate a seven-segment display. Let us just consider the first few lines of the truth table:

Input from counter				*Signals required by display*						
D	C	B	A	a	b	c	d	e	f	g
0	0	0	0	1	1	1	1	1	1	0
0	0	0	1	0	1	1	0	0	0	0
0	0	1	0	1	1	0	1	1	0	1

These three lines show the decoding necessary to display the figures 0, 1 and 2 on the (common cathode) display. Whilst it would be possible to achieve this with gates, it is much more economical to employ a decoder that is already fabricated. As with any system based on gates there are several to choose from depending on whether CMOS or TTL is being used. It is essential to study the data sheets before making a choice.

16.3.3 Stepping motor driver ICs

Stepping motors (or stepper motors) have become popular in project work as their prices have come down, and as the prices of the specialised ICs to drive them have fallen considerably. The most common device is the SAA1027 IC, Fig. 16.41, since, with the addition of a few passive components, it has the capability to drive the most common four-phase stepping motors.

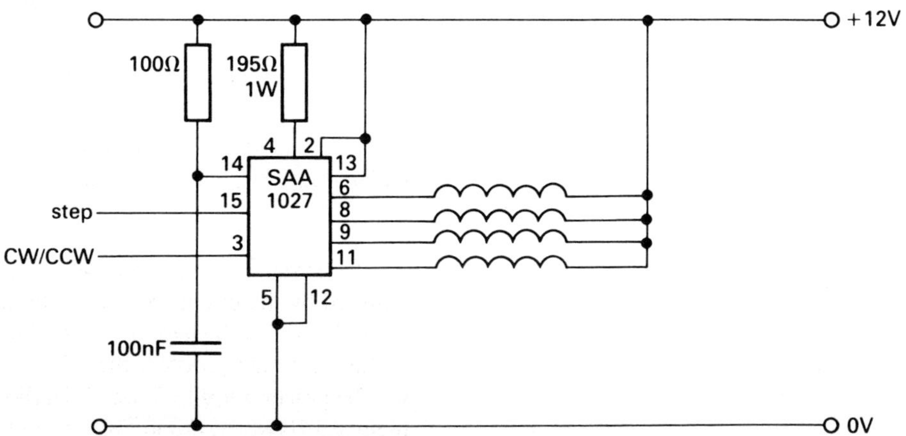

Figure 16.41 SAA1027 stepper motor driver.

Its additional advantage is that it only needs two lines to achieve this control; one line controls the direction of rotation whilst the other is pulsed to control the speed of rotation. These are both digital lines so making it ideal for connecting to a computer. Indeed, four such devices may be controlled by eight output lines from a computer. An examination of the functions contained inside this IC reveals several subsystems that have already been considered, Fig. 16.42.

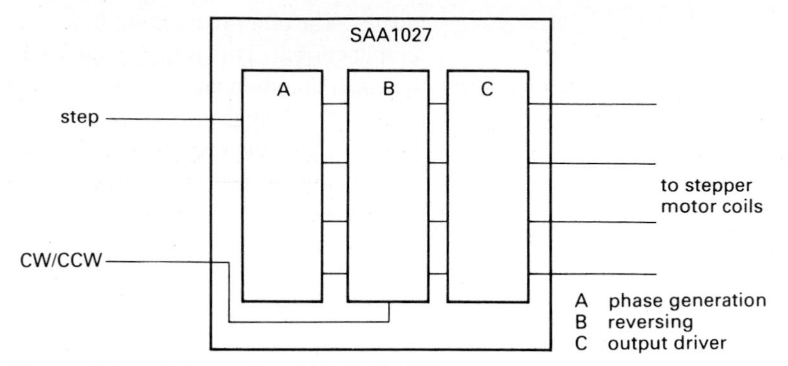

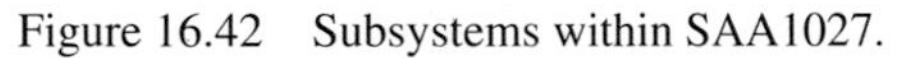
Figure 16.42 Subsystems within SAA1027.

The linked flip-flops, Fig. 16.43, create the required algorithm for the four lines (one for each set of coils in the motor).

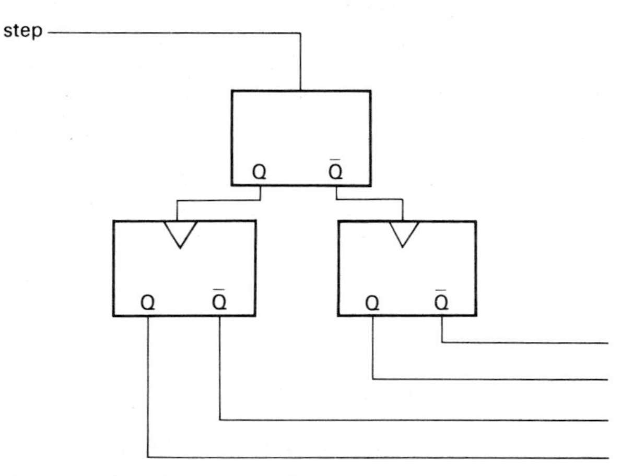

Figure 16.43 Phase generation.

The EOR gates can reverse the pattern to two of the lines so that the motor will run the opposite way, Fig. 16.44.

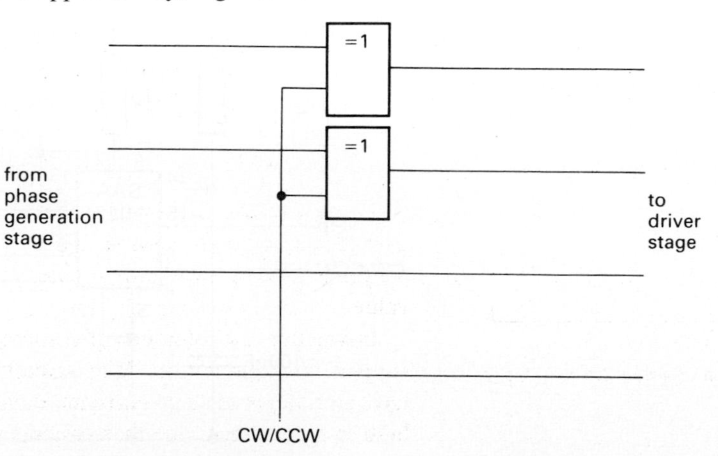

Figure 16.44 Reversing circuit.

The drive transistors have a maximum collector current of 350 mA which is sufficient for the common types of motor, Fig. 16.45.

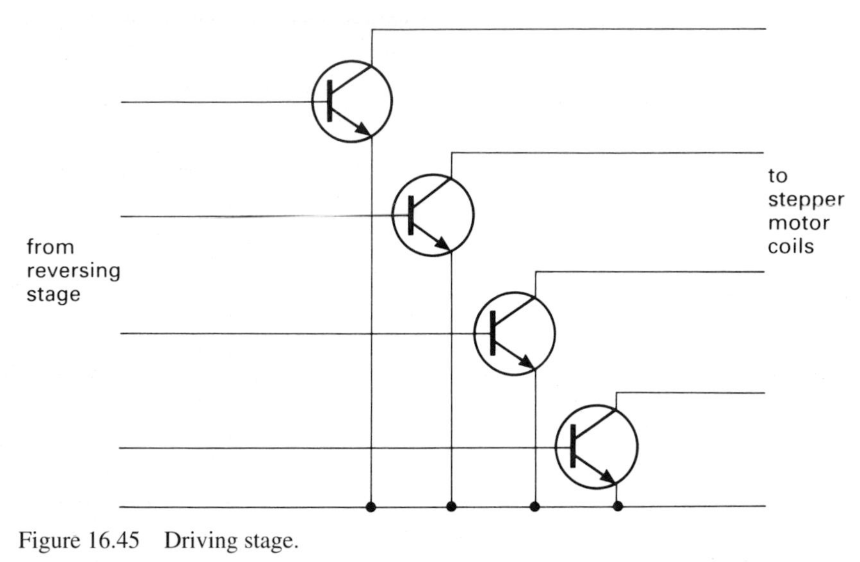

Figure 16.45 Driving stage.

16.4 Timing

16.4.1 The RC circuit

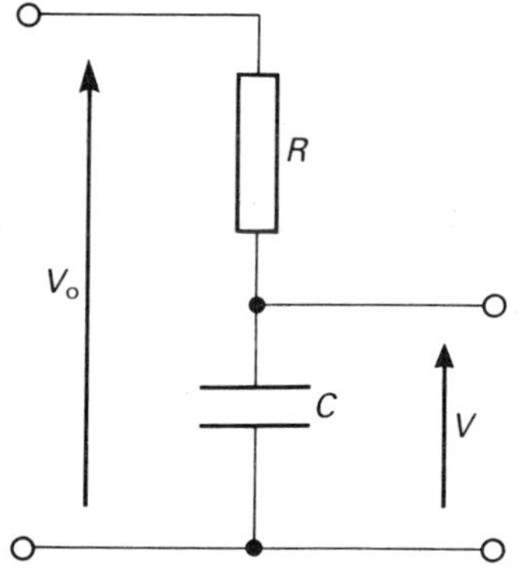

Figure 16.46 RC circuit.

Most timing methods depend upon the charging and/or discharging of a capacitor through a resistor. This circuit is a fundamental building block in electronics, not only because of its use in timing but also because of its application in filtering alternating signals. Consider the circuit of Fig. 16.46.

If a signal of constant magnitude, V_0, is applied to the input then the capacitor will charge up. The graph of this charging process will take the form shown in Fig. 16.47.

This trace is exponential and it can be shown that the p.d., V, after time t, is given by

$$V = V_0[1 - e^{(-t/CR)}]$$

where C is the capacitance in farad, and
R is the resistance in ohm.

CR is known as the time constant of the circuit. It can be seen that, when $t = CR$,

$$V = V_0[1 - e^{(-1)}]$$

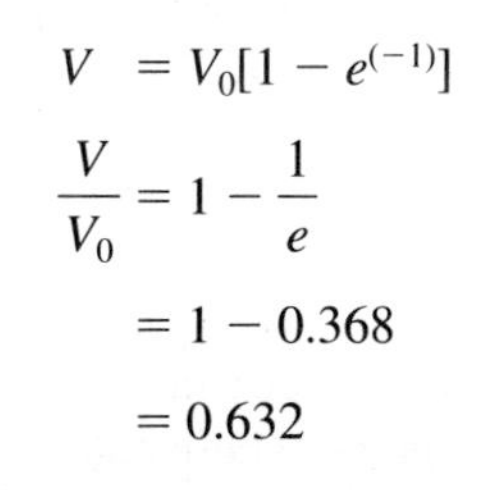

$$\frac{V}{V_0} = 1 - \frac{1}{e}$$

$$= 1 - 0.368$$

$$= 0.632$$

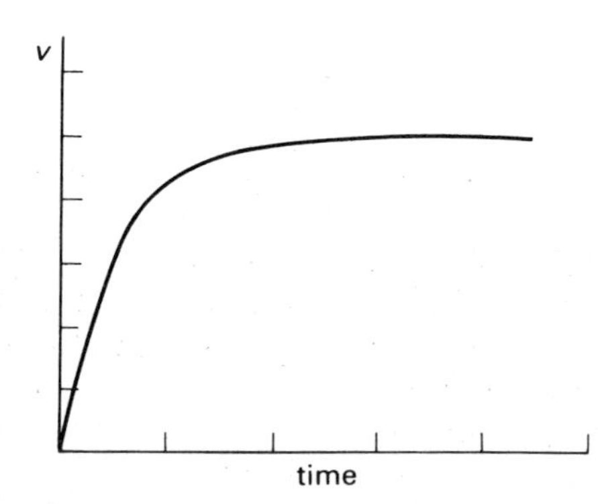

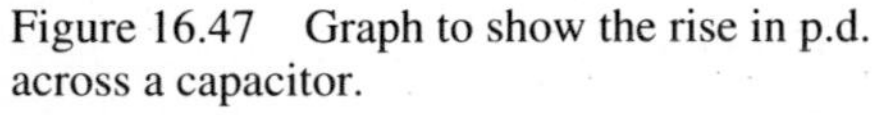
Figure 16.47 Graph to show the rise in p.d. across a capacitor.

In other words, the time constant is the time it takes for the p.d. to rise to 63.2 per cent of its maximum possible value. A similar argument shows that for discharging, the time constant is the time it takes for the p.d. to fall to 36.8 per cent of its original value.

In practice, the tolerance of components is a considerable handicap in designing simple and accurate time delay circuits as is the fact that large values of capacitance have significant leakage currents. In addition, the output impedance of this system is high so that connecting other systems which do not have very high input impedances can cause the actual value to be wildly different from the calculated value.

If adjustments to the time constant are required, it is usual to have a variable resistor in series with a fixed value capacitor. For timing periods longer than a few seconds it is usual to set up a counter together with a circuit to produce continuous pulses, based on the RC circuit. These circuits are examined in the next section.

16.4.2 Alternatives – single and continuous pulses

There are a multitude of circuits to produce both single and continuous pulses. In some textbooks you will find them under the heading of monostables and astables. They are all based on the RC circuit examined in the previous section. Some circuits for single pulse producers and continuous pulse producers are shown in Fig. 16.48 and Fig. 16.49 respectively.

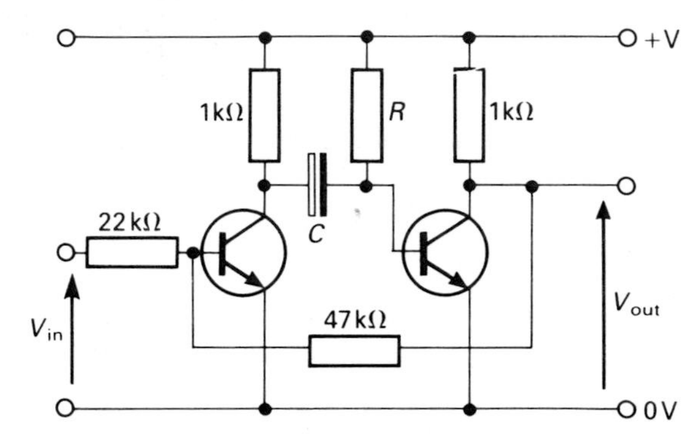

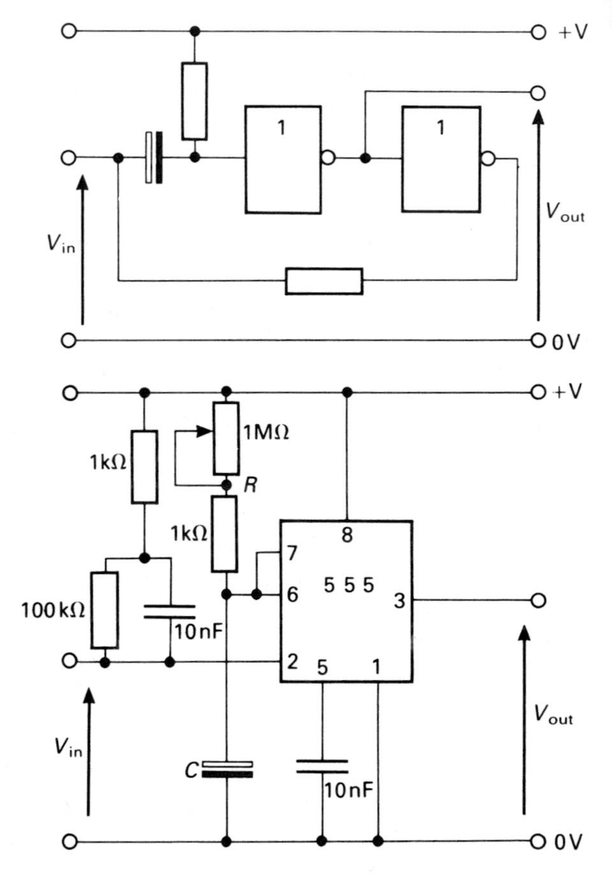

Figure 16.48 Single pulse producers.

When selecting a circuit from several alternatives you will need some criteria for making your choice. These considerations may include:

- How much does it cost?
- How many components are used?
- How much space will it take up?
- How long will it take to construct?
- Over what range is it adjustable?
- How well will it interface with the rest of the circuit?

This is not a full list but it should give you some ideas about the kind of criteria you should be establishing for yourself.

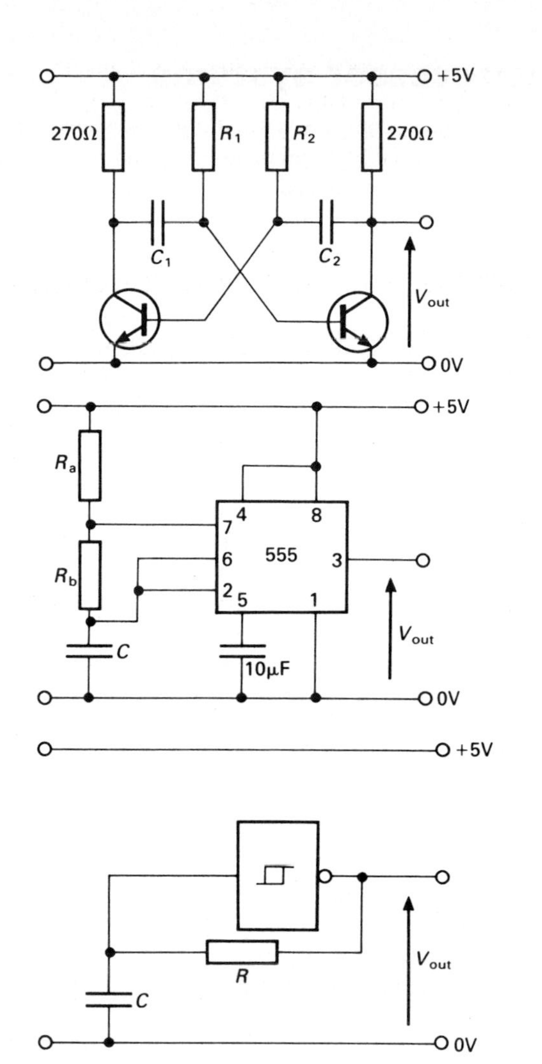

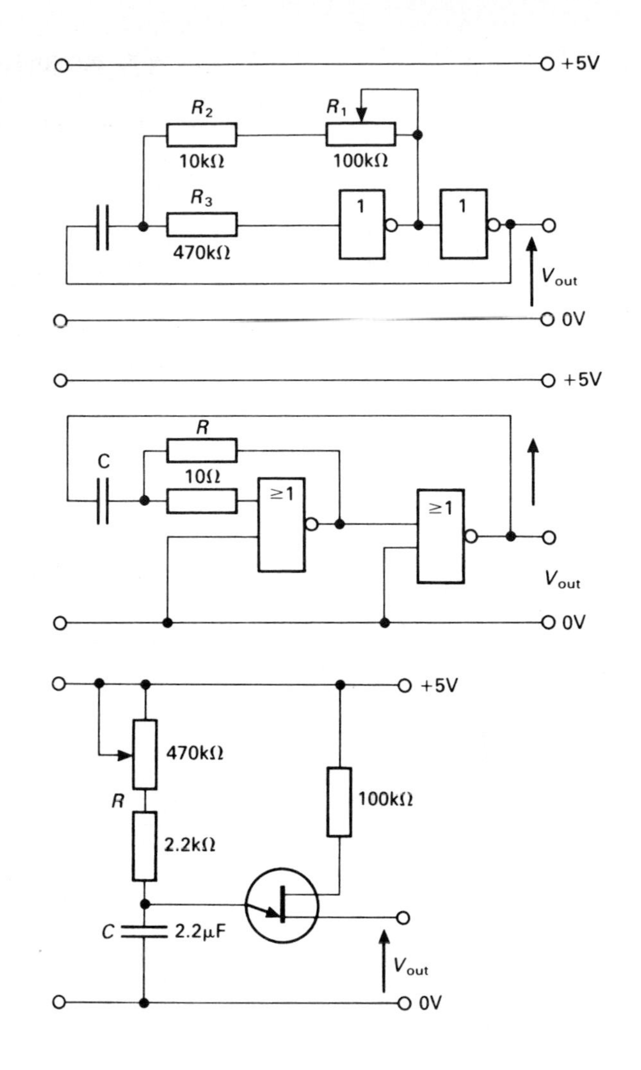

Figure 16.49 Continuous pulse producers.

16.4.3 Crystal oscillators

For particularly accurate timing pulses a *quartz crystal oscillator* can be used, Fig. 16.50. This will oscillate once it is energised at a frequency determined by the size and shape of the crystal. The most common frequency used is 32 768 Hz since, if this is fed into a 15-bit binary counter, the output from the most significant bit is 1 Hz.

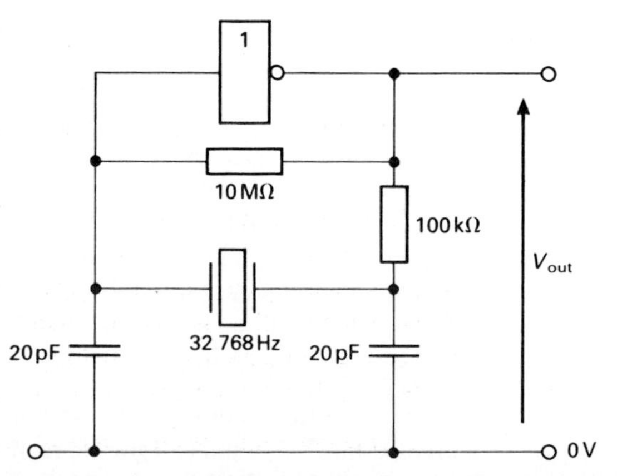

Figure 16.50 One implementation of a crystal oscillator.

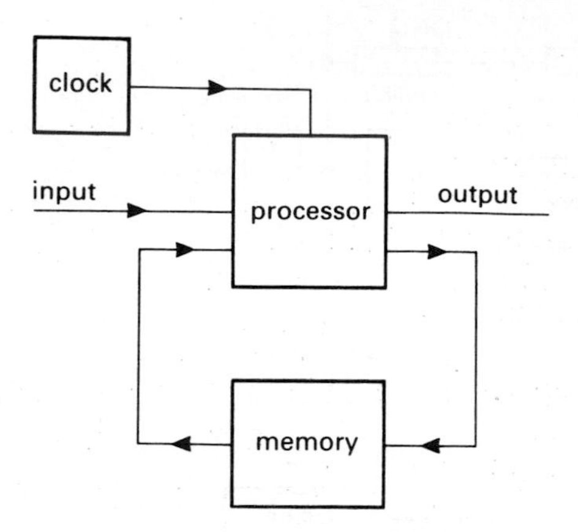

Figure 16.51 A simple microprocessor system.

16.5 Microprocessor systems

To many people, the ultimate in complexity for a digital system is the microcomputer. However, it is readily understandable from a systems perspective. All computer systems are based on a microprocessor. A block diagram for a simple microprocessor system is shown in Fig. 16.51.

The microprocessor itself is the processing part of the system. The inputs to the processing unit may come from tape, disk, CD-ROM, DVD, keyboard or modem. There is also an internal input, a clock running at a few megahertz, which keeps everything in step. The outputs may be to tape, disc, modem, VDU or printer.

The memory is where information may be stored until the computer is ready to use it, or it may be where the instructions are stored. The next three sections look, in more detail, at how the blocks are connected to each other, at how the system may be connected to the outside world and at how analogue signals can be converted to and from digital signals, for processing within the microprocessor system.

16.5.1 General architecture and control sequencing

Bus structure

Information inside a microprocessor system is transmitted in parallel. This means that a large number of connecting wires will be needed to transfer information between different parts of the system. The most economical way of doing this is to create one set of parallel connectors and join everything on to it. This arrangement is known as a bus. As it is going to carry data it is called the *data bus*.

Although this has reduced the number of connectors required it creates another problem. Suppose the microprocessor wants to send data to the memory; if it just puts the data out on the data bus how does it inform the memory that it has to store these data? The answer is to use a second bus, called the *address bus*, to select which block is to receive the information. The microprocessor is the only block in the system which can put out an address.

Since some blocks can transmit data as well as receive them, there is a need for a further line, known as the read/write line, to choose whether to read information from a particular block or to write it into the block. This is an example of a control function. There may be other control functions needed in other parts of the system. Together they make up the third bus, the *control bus*. A simple representation of the bus structure is shown in Fig. 16.52.

In a microprocessor system information is stored and processed digitally. The information is in the form of numerical codes stored as a combination of *bits*. Bits is derived from *binary digits*. If eight bits are used then the total number of different combinations is 256 (2 to the power of 8). We will assume that this system has an 8-bit data bus. These eight bits are usually referred to as one byte.

When information is transferred from one part of the system to another, it is transferred one byte at a time. In order to do this, all eight bits must be transferred simultaneously. This is achieved by having a data bus made up of eight parallel connectors.

If the address bus had only eight connectors the maximum number of locations to send the data would be only 256. Since each location in a memory needs addressing separately, this would be very restrictive. One solution is to use a 16-bit address bus which gives 65 536 (2 to the power of 16) different locations.

Types of memory

There are different types of memory inside a microprocessor system. These can be divided into two main categories; read-only memory (ROM) and random-access memory (RAM). ROM cannot be changed by the system so it contains information

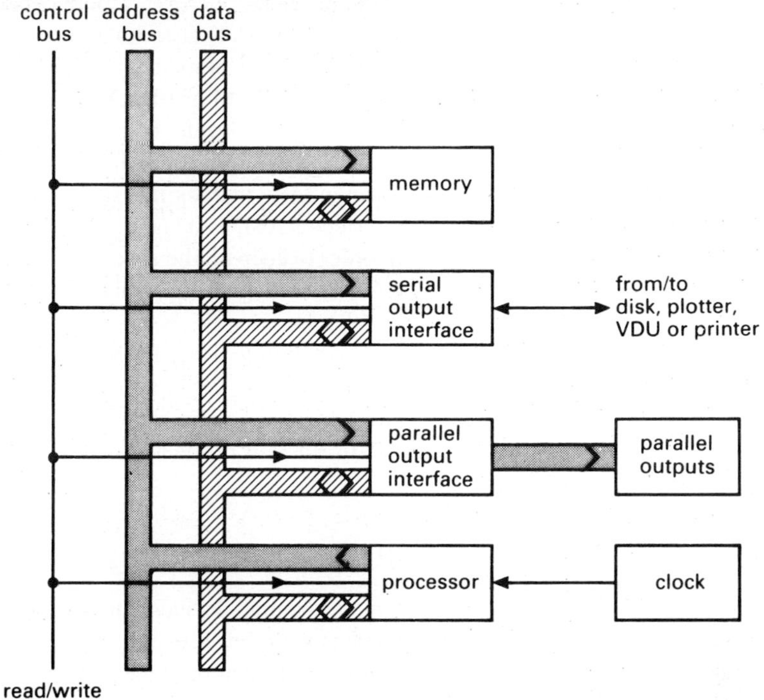

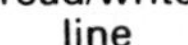

Figure 16.52 Microprocessor bus structure.

that is permanently required. At its simplest, this will be the program which allows the microprocessor to be booted up when it is turned on. The ROM is sustained by a small long-life battery inside a microprocessor system.

RAM is used as a temporary store. A microprocessor system will normally keep the operating system, application programs and data in current use within RAM. It is volatile, i.e. it loses all the information stored within it when the power is switched off. The information in RAM can be continually overwritten by the microprocessor. It is called Random Access Memory as the microprocessor can access any part of it when it wants – this contrasts with older storage systems, such as magnetic tape, where the data could only be accessed sequentially.

The input and output devices are connected to the address bus. The microprocessor treats these devices as being part of the memory. In other words, to input information the microprocessor selects the address of the input device required and reads the information onto the data bus. Likewise, to output information, the data is placed on the data bus and the address of the output device is selected by the microprocessor and placed on the address bus. For programming purposes the input and output devices can be thought of as 'memory locations'.

The RAM family has grown rapidly in recent years as the quest to get faster and more powerful computers has accelerated. The key performance indicator of RAM is 'access time' – the time to access and read the data once the request has been received. Times as low as 9 nanoseconds are currently achievable although it is felt that pure speed is not the only performance indicator. Co-ordination with the computer clock cycle can give improved performance at a lower access speed.

There are two different kinds of RAM within a microprocessor system: main RAM and video RAM. Main RAM can be either static RAM (SRAM) or dynamic RAM (DRAM). Dynamic RAM relies on a kind of capacitor to maintain its charge and keep the data stored. Reading a DRAM causes it to discharge and so there is a need for a power refresh after each read. Even if it is not read it requires the capacitor to be recharged about every 15 microseconds. However, DRAM is the cheapest form of RAM.

Static RAM (SRAM) is more expensive but it does not need to be power refreshed and so is faster to access.

Newer forms of memory chip contain both SRAM and DRAM on the same chip (enhanced DRAM or EDRAM).

The biggest development has been in SDRAM (synchronous DRAM) where the access is synchronised to the clock speed of the microprocessor. Access times are given in MHz making it easier to compare to the bus speed of the processor.

Video RAM, as its name suggests, stores the data for the display on the monitor. In order to keep up the speed of display most video RAM is dual-ported. As the RAM is writing to the display it is also reading in the next frame information through its other port.

In addition to using RAM, the computer can use its hard drive for extra memory. This is much slower to access than static or dynamic RAM. This is a fast developing area which will experience further technological development.

More details of the action of microprocessor systems can be found in Chapter 19.

16.5.2 Interfacing

Input and output devices can be connected to a microprocessor system through a *port*. Some systems have several different ports; others have very few. Some ports are for analogue information if the system has a built in analogue-to-digital converter (ADC); other ports may be for serial communication to modems or serial printers. We are concerned here with the *parallel digital port(s)* for inputting or outputting information.

Some of these parallel digital ports are bidirectional. That is to say, they can be either inputs or outputs or they can be configured so that any combination of the eight bits can be input or output. This can be highly confusing so, if it is possible to use one port for input and a separate one for output, it is to be recommended. If that is not possible, then arrange for the inputs to be together rather than split up by the outputs; for example, if four lines in and four lines out are required, then make the bottom four lines input and the top four lines output. The data direction register controls the bidirectional port and it will have to be told what combination of input and output you require. This is achieved by the control program.

An *interface* is something you connect to the port to allow access to the lines. It may also contain protection and signal conditioning. The simplest form of interface is wire. It will connect from the port to a set of sockets that will allow you access to the lines. The port may well also have some power supply lines and must have 0 V present. The power supplies are usually sufficient to power any protection or signal conditioning systems needed but should not be used to power output devices such as motors or solenoids.

One obvious form of signal conditioning is current amplification of output signals. Since these signals are digital a transistor arranged in the common emitter mode is most suitable. There is an IC made specifically for interfacing, containing eight Darlington pairs and flywheel diodes for inductive loads. This octal Darlington driver is the most cost-effective way of providing current amplification for an 8-bit port.

Even if current amplification is not required, it is good practice to buffer the output lines. This is because one of the easiest ways to damage a microprocessor system is to send an input into a line designated as an output. A buffer should prevent the input reaching the port. There are several possible buffers that could be used depending on circumstances. A non-inverting buffer, such as the 7407, can be used which gives the added advantage of an open-collector output so allowing it to be connected to other power supplies up to 30 V. Alternatively two inverting buffers, such as 74HCT04, can be connected in series to give the same degree of protection but still in a 5 V environment. A single transistor acts as an inverting buffer but is usually more expensive than IC-based solutions. Some possible solutions are shown in Fig. 16.53.

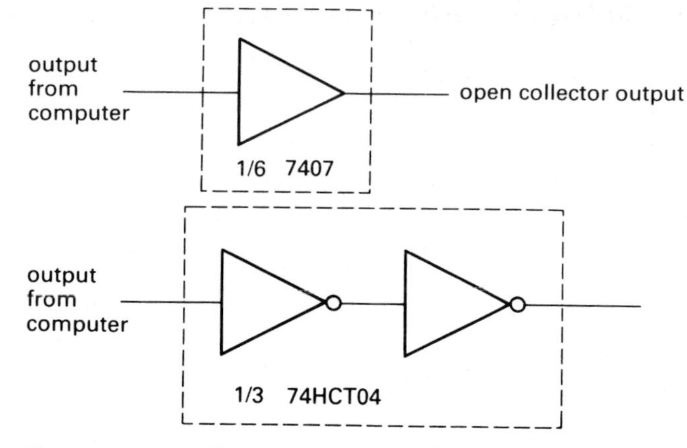

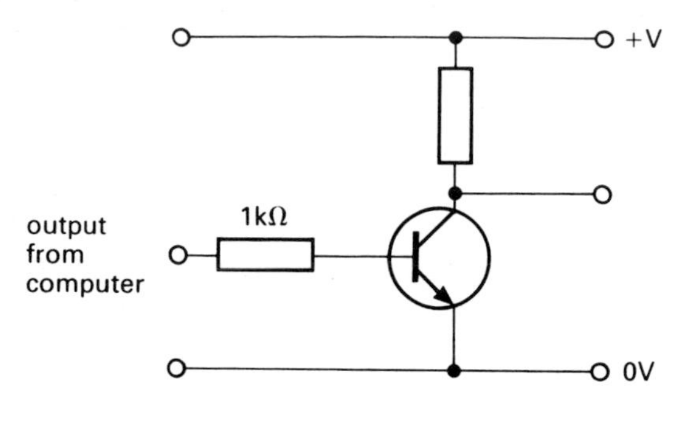

Figure 16.53 Buffering output lines.

Inputs need to be protected against excess p.d. and reverse polarised signals. The signal may also need conditioning. One way of protecting against both excess p.d. and reverse polarised signals is to make use of a zener diode and resistor combination, Fig. 16.54.

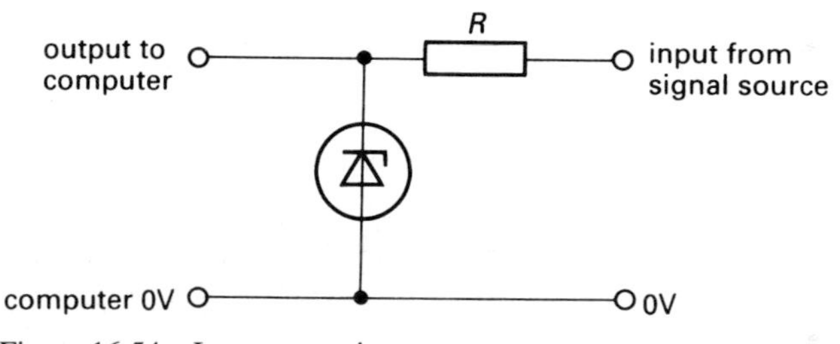

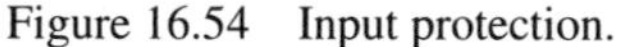
Figure 16.54 Input protection.

The zener diode is chosen to be just above the required maximum voltage and just below the value at which damage might occur. For most systems a 5.1 V zener diode is a suitable value. The resistor value and power rating can be calculated by assuming the maximum voltage applied to be 20 V, by looking up the maximum current that can be safely applied to the microprocessor input and knowing the power rating of the zener diode. In this way, the resistor will act as a fuse if the p.d. exceeds 20 V. For reverse p.d.s the zener diode will conduct preventing the reverse p.d. from exceeding – 0.6 V (a value the microprocessor seems capable of withstanding).

The signal conditioning required may be to improve the quality of any digital signals applied to input lines. The usual subsystem employed is the Schmitt trigger.

It is also common practice to opto-isolate the lines. This involves converting the electrical signal into an optical signal, passing it into a detector and then turning it back to an electrical signal. The two parts of the circuit can be electrically isolated. The symbol for an opto-isolator is shown in Fig. 16.55.

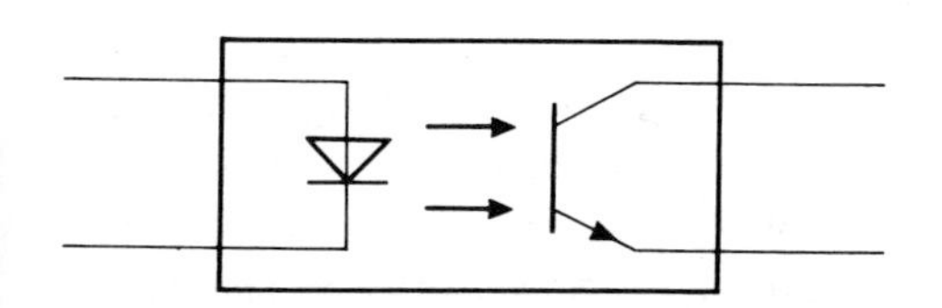

Figure 16.55 Opto-isolator.

It should be stressed that it is probably illegal for you to be using your microprocessor system to control mains electricity in an A-level course. The Health and Safety at Work Act implies that this is an unsafe practice unless a qualified electrician inspects and certifies the system before it is turned on. Opto-isolators cannot be considered as safe since the pin spacing is small enough to be shorted easily in the presence of moisture. Even condensation could cause enough moisture for this to occur.

16.5.3 DAC and ADC systems

The world around us is essentially an analogue world. Microprocessors only operate on digital signals. The success of microelectronics has been its ability to process and control analogue information using digital processing techniques. This large processing power has been made available due to the low cost of microelectronic systems. But why should we bother? Surely there are analogue processing techniques available? Yes – but... is the simple answer! Digital signal processing reduces noise and distortion and is much more versatile. Once the signal is in a digital form, it is possible to store it as well as perform all manner of mathematical operations upon it. The main problem is its lower speed of operation but speed is increasing as can be seen by the advances in compact disc and digital tape technology. In order to process analogue signals by digital techniques, the analogue signal must first be converted into a digital signal. After processing, the digital signal must then be turned back into an analogue signal. The two system building blocks that achieve these functions are *analogue-to-digital* and *digital-to-analogue* converters, often abbreviated to *ADC* and *DAC*.

It would be possible write a whole book on the techniques, problems and implementation of ADC and DAC but here we will first consider DAC, which is the easier process, and then look briefly at some of the possibilities using ADC.

Digital-to-analogue conversion (DAC)

There is a range of techniques available to convert a digital signal into an analogue form. The most popular technique used is termed *weighted addition*. Each bit is given its appropriate analogue value, by the value of the resistor, and the bits are then added by a summing amplifier. In an 8-bit system, the resistor connected to the *LSB* (least significant bit) must be 128 times the resistor value connected to the *MSB* (most significant bit), Fig. 16.55. The value of the feedback resistor determines whether the system also has gain as well as converting ability.

If the gain of the system is one then V_{out} is calculated from the formula:

$$V_{out} = -5(128H + 64G + 32F + 16E + 8D + 4C + 2B + A)$$

assuming that logic 1 is +5 V. The negative sign is easily removed by adding a unity gain inverting buffer.

It should be obvious that it would be very easy to saturate this system. If, for example, the input were 01110011, the output would be 115×5 V. It is usual, therefore, to have a gain much less than one. The circuit shown in Fig. 16.56 has a gain of 1/200 so V_{out} is given by the formula:

$$V_{out} = -\frac{5}{200}(128H + 64G + 32F + 16E + 8D + 4C + 2B + A)$$

As the output is negative this sytem would usually be followed by a unity gain inverting amplifier to give a positive output.

Analogue-to-digital conversion (ADC)

There is a wide range of techniques available for A-to-D conversion. The method chosen will depend on how quickly the conversion is required and the acceptable level of resolution. The other important criterion will be the cost.

The fastest conversion technique is termed *flash encoding* or *parallel encoding*. In this method the analogue signal is fed to one input of each of several comparators with the other input of each comparator connected to an equally spaced chain of reference voltages. The resulting signal is passed to a binary encoder to generate the appropriate digital output. The speed of conversion may be as fast as 50 ns. Fig. 16.57 shows the general principle for a 3-bit system.

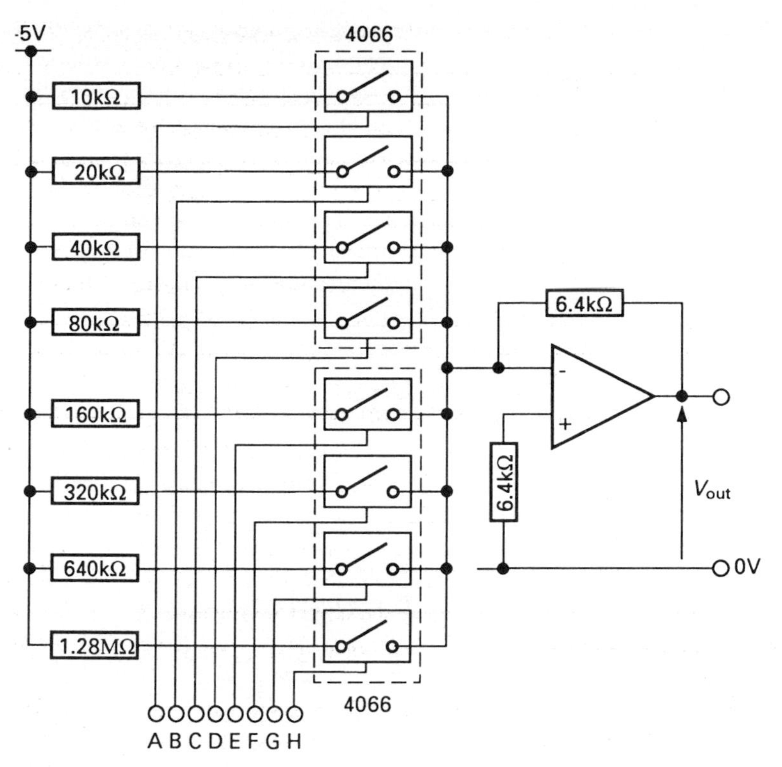

Figure 16.56 8-bit D-to-A converter.

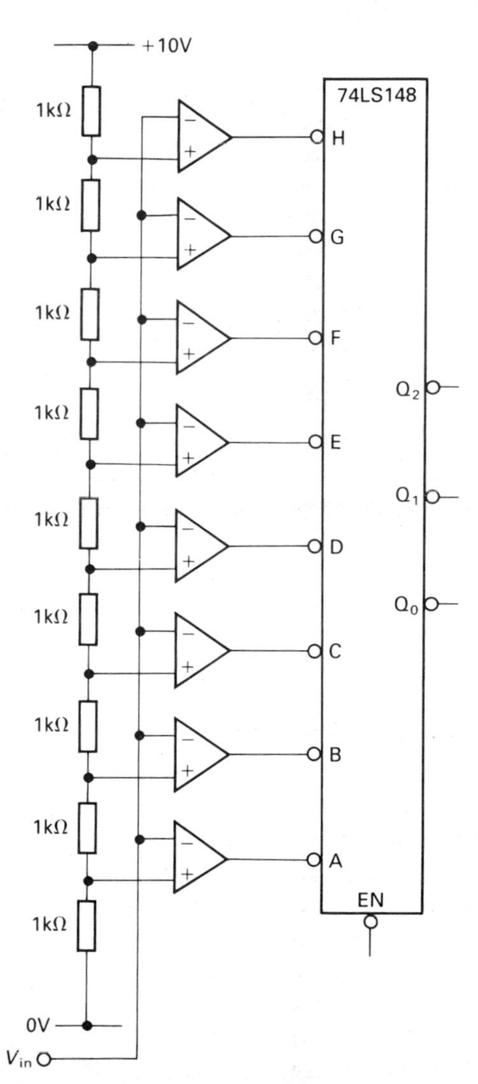

Figure 16.57 A three-bit parallel encoded A-to-D converter.

An 8-bit system will require 256 comparators. The resolution is relatively low in that it is only possible to say that the input is somewhere between the value of the reference signal for two comparators. In this case it will resolve to one part in eight (about 13 per cent). Flash encoders for 8-bit (and more) systems exist but are expensive. It is more usual to use a slower technique but with increased resolution at lower cost. There are several possible methods based on the use of a DAC. The simplest method is illustrated in Fig. 16.58.

The analogue signal is applied to one terminal of the comparator which controls the action of the counter. The clock pulses are fed to the counter which increments the digital display. The outputs of the counter are fed to the inputs of the DAC which feeds its output to the other input of the comparator. When the output of the DAC

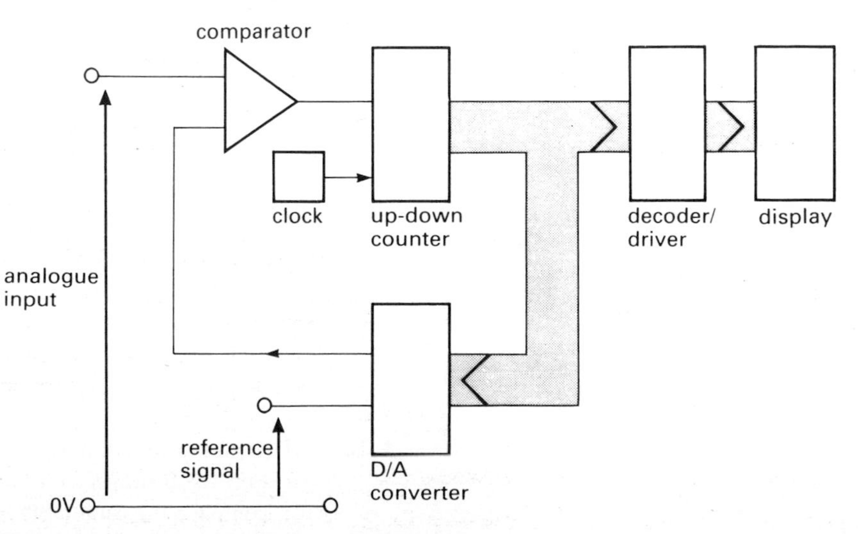

Figure 16.58 Simple tracking A-to-D converter.

exceeds the input analogue signal then the comparator changes state and the clock is stopped. Additional circuitry will be required to allow the control signal, *start conversion* and there is also a *busy* line which will indicate when the conversion is complete. In addition there needs to be a facility to *reset* ready for the next conversion.

The texts listed in the bibliography at the end of this chapter contain details of other, more advanced, techniques. It should be noted that most D-to-A and A-to-D conversion is achieved with specific ICs and precise details of their use can be found in the manufacturer's data sheets.

Bibliography

As for Chapter 20 together with the following:
Green D 1998 *Digital Electronics*

Assignments

1 (a) Copy out and complete the truth table for the logic network shown in Fig. 16.59.

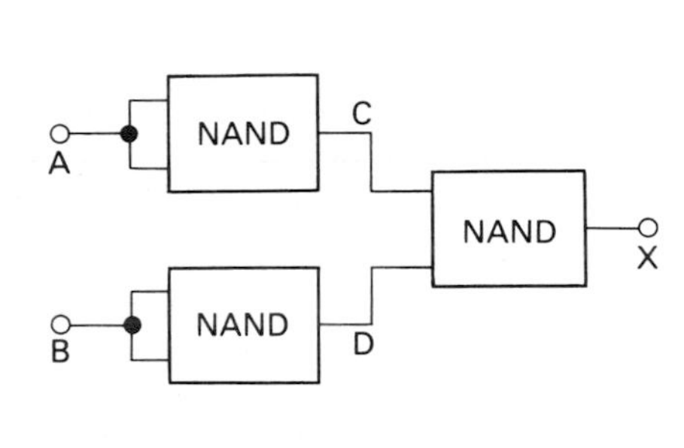

INPUTS		OUTPUTS		
A	B	C	D	X
0	0			
0	1			
1	0			
1	1			

Figure 16.59

(b) Name the single logic gate which is equivalent to this network of NAND gates.

2 (a) With the aid of *two* examples discuss the advantages and disadvantages of the transmission of information by *Analogue* and *Digital* means.

(b) Four sensors placed in different rooms of a house are used to control the central heating system. The output from each sensor is a voltage between 0 and 2.55 volts that is dependent on the temperature of the room.

Show with the aid of a block diagram how it is possible to sample the output of each sensor in turn and then convert the analogue signal to an 8-bit digital code.

(c) Describe a suitable heat sensor for use in part (b) and show how it could be used to produce the type of signal required.

3 (a) (i) With the help of a truth table, explain the function of a NAND gate.

(ii) Fig. 16.60 shows a circuit made up from NAND gates. Draw up and complete the truth table for all input combinations, and suggest the function of the circuit.

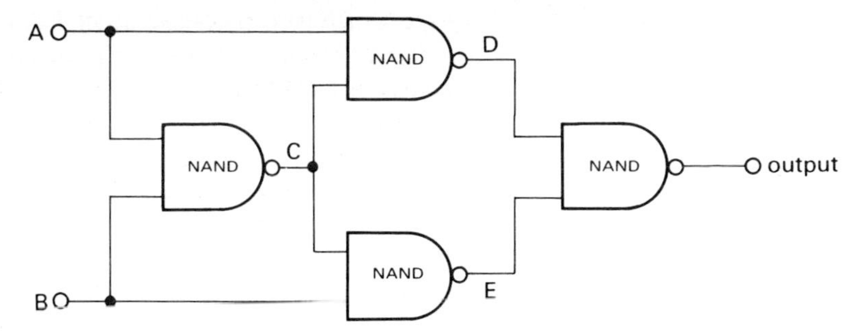

Figure 16.60

4 (a) The following extract is taken from a specification.
Explain briefly the terms in italics:
'... and an *up-down counter* Integrated Circuit which is *negative-edge clocked* has a binary output QA-AD which is *decoded* by the 7447 decoder/driver Integrated Circuit with an *active low* output driving a *common anode* seven segment display.'

(b) Explain, with the aid of a simple diagram, how a four-phase stepper motor might be connected to a microcomputer to obtain rotation in either direction. Explain how the rotation is obtained. The clockwise sequence state table (full step mode) is given in Fig. 16.61.

stepper motor switching sequence chart					
full step mode	half step mode	switched state of each of the four windings			
step no.	step no.	1	2	3	4
UNUSED	1	ON	OFF	ON	OFF
1	2	ON	OFF	OFF	OFF
UNUSED	3	ON	OFF	OFF	ON
2	4	OFF	OFF	OFF	ON
UNUSED	5	OFF	ON	OFF	ON
3	6	OFF	ON	OFF	OFF
UNUSED	7	OFF	ON	ON	OFF
4	8	OFF	OFF	ON	OFF
SEQUENCE 1234 . . . etc WILL CAUSE CLOCKWISE ROTATION					

Figure 16.61

(c) Using the 'clockwise half step mode' sequence in Fig. 16.61 and a binary count step through the sequence, design a complete combinational logic circuit for which AND, OR and NOT gates are available to switch all four windings, if appropriate, simultaneously and in the required sequence for clockwise rotation.

(d) (i) If the logic circuit you have designed is stepped by a 4-bit binary counter Integrated Circuit and a suitable astable clock pulse, how could the eight step sequence be made to rotate the motor continuously in one direction only when a button is depressed?

(ii) What problem might be encountered with this method of drive if the motor is required to step through an accurate angular displacement?

5 (a) Two commonly available IC logic families are the TTL 74 series and the CMOS 4000 series. Make a simple comparison, in tabular form, of at least six of the common characteristics of these two families.

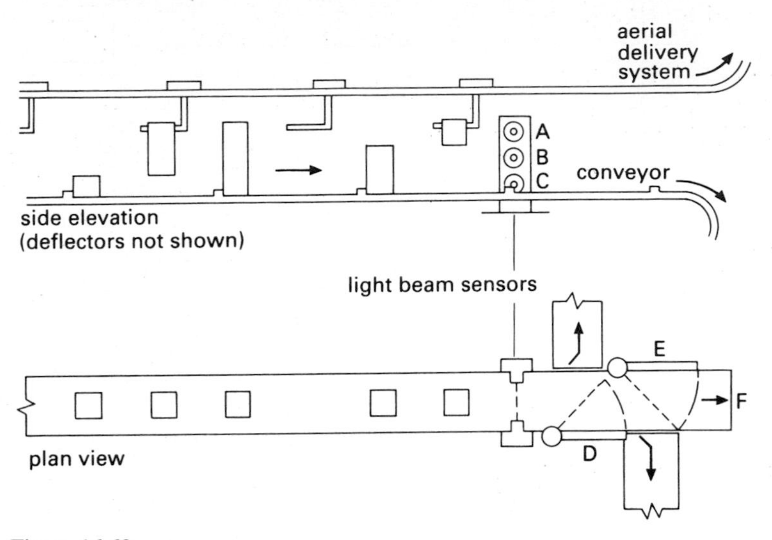

Figure 16.62

(b) Fig. 16.62 shows a conveyor system along which three different sizes of metal casings pass.
A logic circuit is required to sort the metal casings as follows:
All small casings are deflected by deflector D.
All medium casings are deflected by deflector E.
All large casings continue to collection point F.
The problem of the time delay between sensing and activating the deflectors is to be ignored.

(i) Write down and simplify three Boolean expressions which describe how the sensors can distinguish between each size of casing.

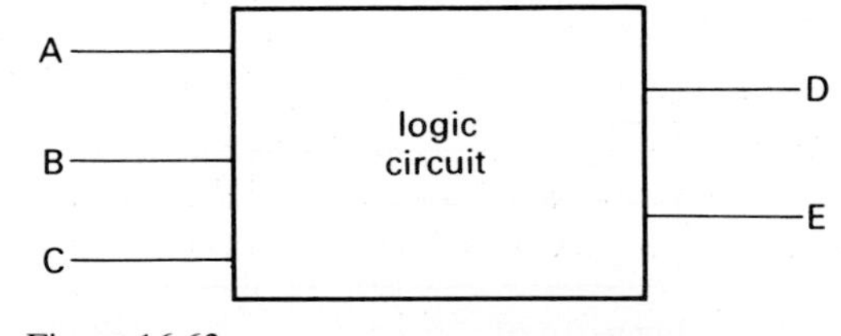

Figure 16.63

(ii) In the form of Fig. 16.63, draw a logic circuit which will activate the deflectors, as described in (b), using AND, OR, NOT and XOR gates only.

(iii) Draw and simplify a logic circuit to perform the same function as that in (ii), but using 2-input NAND gates only.

(iv) State briefly the advantages of designing circuits using only one type of gate.

(v) How appropriate are fixed logic circuit solutions if the basis for sorting the components on the conveyors is likely to be changed from time to time?

(c) (i) Draw a simple circuit for just one of the three light sensor circuits on the conveyor system in Fig. 16.62. The sensor should give an output which is compatible with any 5-volt logic circuit.
(ii) Suggest how you would refine the sensor circuit to avoid the problems of ambient light and possible analogue output to the logic circuit.

6 Part of an automated sequence for packing rice in sealed plastic bags is represented by the truth table shown in Fig. 16.64, and is pulsed by a simple binary count AB.

count		Q_0 hold bag open	Q_1 motor feeding rice to hopper	Q_2 hopper chute opened to fill bag	Q_3 seal plastic bag
B	A				
0	0	1	1	0	0
0	1	1	1	1	0
1	0	1	0	1	0
1	1	0	0	0	1

Figure 16.64

(a) (i) Derive a Boolean algebra expression in its simplest form for each of the output requirements $Q_0 - Q_3$.
(ii) Draw a complete logic circuit, incorporating all four output functions, and using any of the following gates, AND, OR, NOT, NAND, NOR, XOR.
(iii) Lay out a circuit diagram of the complete system using any quantity of the integrated circuit shown in Fig. 16.65. Assume that the 2-bit binary count is separately generated and clocked.

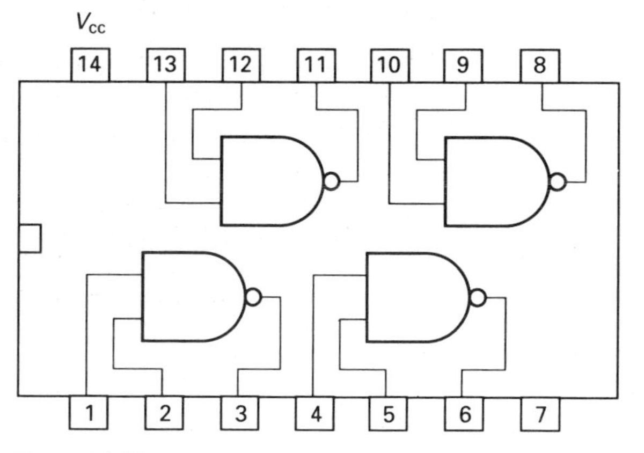

Figure 16.65

(iv) Explain why the packaging company may have difficulty maintaining a consistent weight of rice in each bag.
(v) The Auger feeding rice to the delivery hopper is powered by a 2 volt 1 A DC motor. Design a simple interface circuit capable of switching this motor from the logic circuit output Q_1.

7 (a) With the aid of a table, or otherwise, explain the operation of a J-K flip-flop in terms of its inputs and their effect on the resultant output.
(b) Design a circuit using a J-K flip-flop which will produce an output (logic 1) on receiving the first of many pulses and yet does not turn off (logic 0) until it is reset.

(c) Explain why BCD counters are often used in preference to 4-bit binary counters in instruments.

With the aid of a block diagram and table, show how a 4-bit binary counter using J-K flip-flops can by using other logic gates be changed to a BCD counter. Describe any limitations which exist in your circuit.

(d) A 4-bit binary counter can be connected to a BCD to decimal decoder chip (e.g. 7442) whose logic diagram and pin configuration are shown in Fig. 16.66. Explain the function of the BCD to decimal decoder and describe an application for its use.

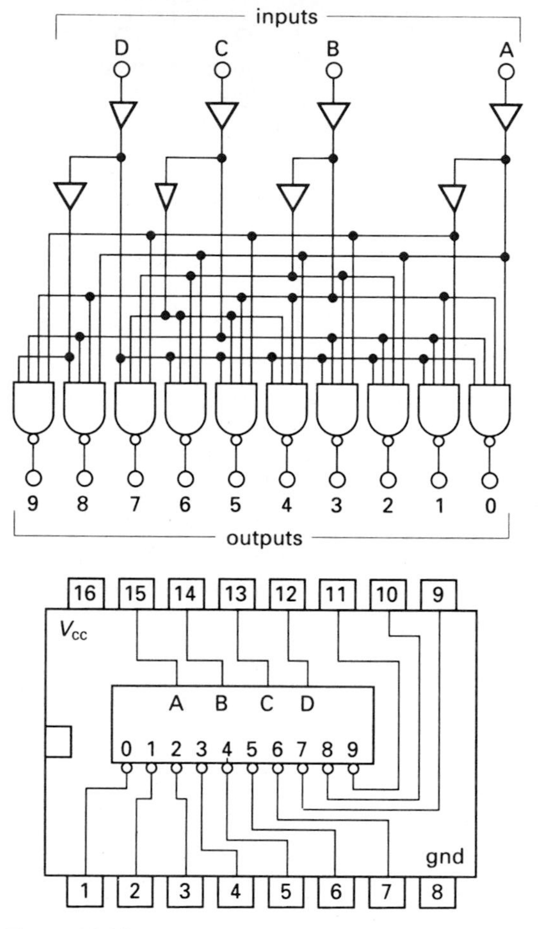

Figure 16.66

8 (a) Show how J-K flip-flops can be connected together with one or more logic gates and a clock to produce a binary counter which counts to base six and explain how the counter works.

(b) The pulse to the clock input could be fed from an approximate sine wave via a Schmitt trigger circuit. Explain why the Schmitt trigger circuit is used in such a situation and, with the aid of wave-form sketches, the meaning of the term hysteresis when applied to Schmitt triggers.

(c) Explain how a static RAM, with a 4-bit address and 4-bit locations (i.e. a total of 16 4-bit locations) can be used to produce the following lighting sequence using three lamps. You may assume there is a suitable interface between the RAM and each lamp. The lights are coloured red, amber and green.

The sequence is: red ON for six seconds, red and amber ON for three seconds, green ON for six seconds, amber ON for three seconds and back to red ON again.

You should state in your answer, with the help of a truth table or otherwise, the contents of the addresses to be used and how the content is written to, and read from, the RAM.

9 (a) Explain, with the use of a block diagram, the basic functioning of a simple microprocessor system.

(b) The use of the microprocessor for controlling equipment is of increasing importance. Choose one relatively simple piece of equipment and explain in detail how it is controlled by the microprocessor, giving details of the interface device.

10 (a) Explain the difference between an analogue sensor and a simple digital sensor, giving an example of the use of each type.

(b) A student wishes to build a computer-based weather station. Explain why an analogue-to-digital converter (ADC) will be needed for the temperature-measuring circuit.

(c) Figure 16.67 shows a pin diagram for a typical ADC. Explain the words in italics in the following extract from the Data Sheet for this ADC chip. '...this is a high speed ADC with 8-*bit resolution*. The digital outputs can be enabled by means of logic 1 signal to pin 5 (*output enable*) and they are *buffered* so as to be fully *microprocessor compatible*. The chip can be powered by a wide range of supply voltages (5 to 15 V) and has its own on-chip *voltage reference* output. *CMOS* circuitry is used for low power consumption.'

(d) Complete Fig. 16.68 to show how the temperature reading from the thermistor circuit can be read by the computer as an 8-bit binary number.

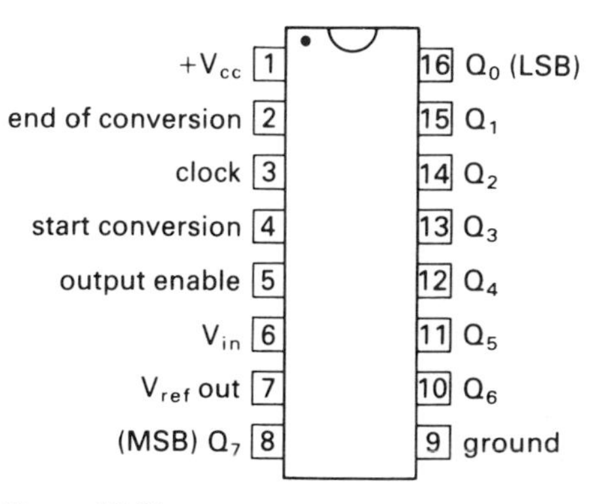

Figure 16.67

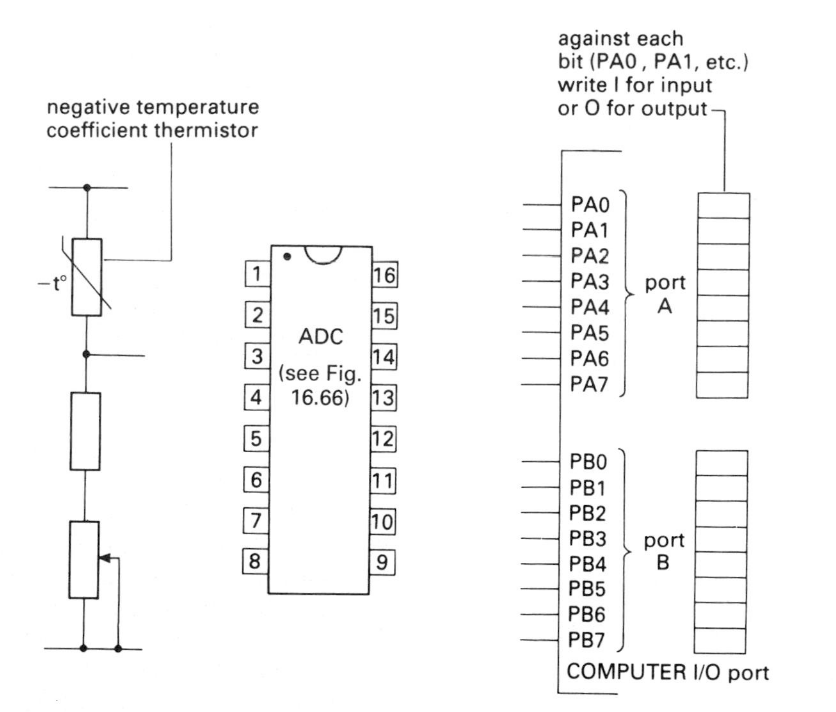

Figure 16.68

Include in your circuit a suitable clock circuit to generate a clock frequency of about 1 MHz. Show all component values.

The chip shown on Fig. 16.68 refers to the ADC on Fig. 16.69. A schematic diagram, rather than a p.c.b. layout, is required, i.e. it is acceptable for lines to cross over thus:

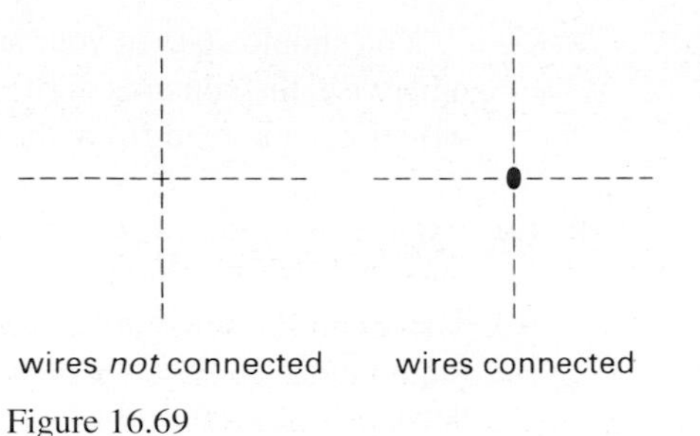

Figure 16.69

(e) If V_{ref} output is +2.55 V, what analogue voltage is being applied to V_{in} when the digital outputs are as follows?

	Q_7	Q_6	Q_5	Q_4	Q_3	Q_2	Q_1	Q_0
Logic level	0	1	1	0	1	0	0	1

11 Many desktop computers have the facility to interface with external devices by means of analogue and digital I/O ports.

(a) Explain the differences between analogue and digital input signals.
(b) Show by means of a block diagram how you would interface a stepper motor to a digital output port of a microcomputer.
(c) Draw up a circuit diagram for the stepper motor interface. Include the stepper motor and all necessary component parts (discrete component values are not required).
(d) The position of the motor shaft needs to be known when the system is turned on. Show by means of a block diagram how this may be done. Indicate the type of I/O port to be used.

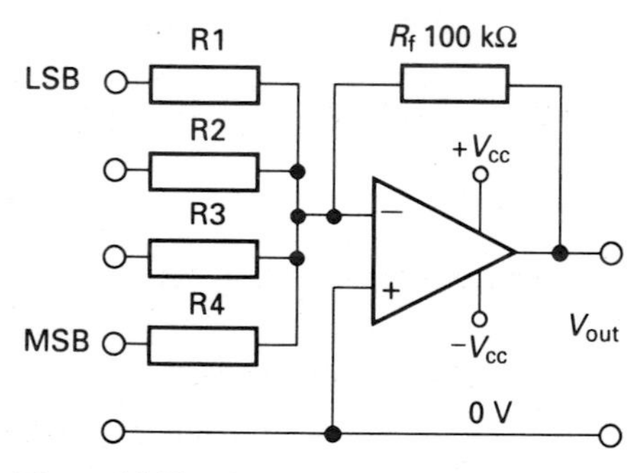

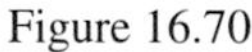

Figure 16.70

12 (a) Explain why digital to analogue and analogue to digital conversion has become so important in recent years.
(b) Using an operational amplifier in the configuration shown in Fig. 16.70 design a simple 4-bit digital to analogue converter.
The voltages of the 4-bit input signal are 5 volts for logic 1 and 0.0 volt for logic 0. The output from the converter should be 1 volt per unit input. Assume a feedback resistor (R_f) for the operational amplifier of 100 kΩ.
(c) Describe an instrumentation application which can make use of an analogue to digital conversion circuit. Draw a block diagram of the system and explain why the A to D converter is necessary. Discuss whether the same measurements could be made as effectively using only analogue signals.

13 A d.c. stepper motor is to be interfaced to a microprocessor controller, in order to drive part of an X–Y plotter.

(a) State the requirements of an interface to run a d.c. stepping motor with which you are familiar.
(b) Explain the nature of the signals required in order to make the motor rotate clockwise and anti-clockwise as appropriate.
(c) Explain how the rotational speed of a d.c. motor can be changed.
(d) Produce a block diagram to show how the microprocessor is interfaced with the stepper motor.

17 Control systems

17.1 Systems and cybernetics

The control of complex and simple systems, often spanning many disciplines, is of increasing importance to our society. So fundamental is the subject that it is now regarded as a distinct discipline, referred to as control engineering. The key concept for the successful operation of control systems is *feedback*.

Cybernetics is concerned with the acquisition, processing, communication and manipulation of information and its use in controlling the behaviour of natural and manufactured systems. The ability of the human race to evolve and determine its own destiny depends on an ability to control processes which govern its environment and its own activities in it. Cybernetics embodies many fundamental concepts and principles which are related to many of the processes which have determined human biological and social evolution. These basic principles are well illustrated in automatic control systems and most particularly in servomechanisms. Many of the terms and concepts specifically defined for this field of engineering are found with analogous meanings in fields such as medicine, biology and sociology. Within the strict confines of classical instrumentation and control, systems are implemented using electrical, electromechanical, hydraulic, pneumatic and electronic elements.

A general definition of a control system is an arrangement of physical components connected or related in such a manner as to command, direct or regulate itself or another system.

All control systems embody the same fundamental principles but there are several ways in which they can be classified. One method is to distinguish between manufactured and natural systems, as follows.

17.1.1 Manufactured and natural control systems

Manufactured control systems

A very simple example is a thermostatically controlled central heating system.

Natural systems including biological control systems

Probably the most sophisticated of all control systems are to be found in humans and other living creatures. A simple, but critical, example in the human body is temperature control where a number of mechanisms are used to maintain body temperature within very fine limits, despite a huge variation in ambient temperatures.

Control systems whose components are both manufactured and natural

An example of this type is the person controlling a motor car. The total system is one which embodies both the manufactured mechanisms within the car and the human control processes implemented by the driver.

We can also classify control systems according to their functions, as follows.

17.1.2 Control system functions

Regulating systems

This type includes systems whose control function is to maintain the state of the system at a particular level. One example is the thermostatically controlled heating system.

Linear systems

In this type of control system the output is constrained to follow the demand of the system inputs continuously. These are servosystems which may embody position, velocity or acceleration control. An example is the servosystem used to control the position of a ship's rudder.

Incremental control systems such as programmed control systems

This form of control system, which is particularly suited to digital technology, is one where the process under control follows a sequence of defined operations. In some systems the sequence of operations is established and set, and then run. This takes no account of the performance and effect of the system. An example of this type is an automatic washing machine. Other systems in this category would respond to information about the state of the control process and change their actions depending upon the information they received. An example of this category would be the control of a chemical process in an industrial plant where the parameters of the process being monitored determine aspects of the process sequence itself.

These are by no means the only ways of classifying control systems. Perhaps the most common classification is between systems which have *open-loop control* and those with *close-loop control*. These are examined in more detail in the next two sections.

17.2 Open-loop control

Fig. 17.1 below shows the organisation of a simple open-loop control system. Here the system controls some form of output and this output is affected by a single appropriate signal being input to the controlling element. A limitation of this system should be immediately evident; the input to the system is not influenced in any way by the actual output obtained. No corrective action is taken if the output is incorrect.

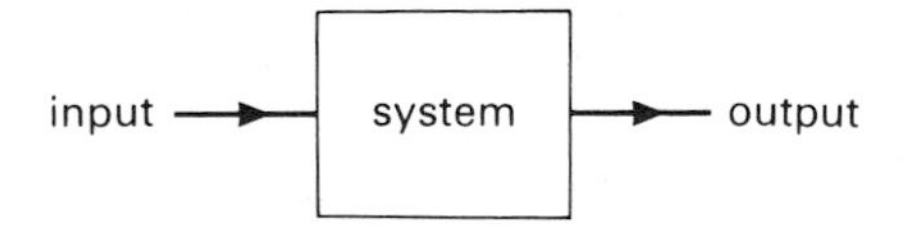

Figure 17.1 Open-loop control system.

The input is the stimulus or excitation applied to a control system from an external energy source usually in order to provide a specified response from the control system.

The output is the actual response obtained from the control system. It may or may not be equal to the specified response implied from the input.

A good example of an open-loop system is the ordinary domestic toaster. Here the required output is a piece of toast cooked to a particular requirement. The time required to make 'good toast' must be estimated by the user, who is not part of the toaster itself. Control over the quality of the toast (the output) is removed from the process once the input has been set. This is an example of an open-loop control system which is satisfactory, providing the parameters which determine the quality of the product are fixed. These parameters include the thickness, type and degree of staleness of the bread, ambient temperature and humidity and the number of times the toaster has been used immediately before this piece of toast is inserted. As these are hardly fixed parameters it is not surprising how difficult it is to get a piece of toast that is just right.

Electromechanical toasters are now being replaced by microchip toasters; the advantage being that the actuating mechanism is electronically determined rather than electromechanically determined and therefore is more reliable in the long term. However, the actual control process is in no way enhanced since the microchip provides only the timing of the process and there is no sensing of the output.

17.3 Closed-loop control and feedback

For an example of a closed-loop control system we can stay with the example of the automatic toaster from the previous section. If the customer for the toast has the time, patience and inclination, he or she is able to monitor the progress of the control action to be sure that the desired output is obtained. This means using the human capability to sense the progress of the toasting action and making a decision to turn off the toaster when the toast reaches the desired state. The system has now become a closed-loop system in which the output of the system has been monitored and the information obtained from this action has been used to control the overall system process. Besides the obvious difference in the control process itself, this closed-loop system is distinguished from the open-loop type by the fact that feedback has been introduced.

Feedback is that property of a closed-loop control system which permits the output (or some other control variable in the system) to be compared with the input of the system (or an input to some internally situated component or subsystem) so that an appropriate control action may be formed as some function of output and input.

There are two types of feedback – positive and negative. *Positive feedback* has the effect of moving the output of a system away from its existing or commanded state. Such a system is unstable and the output of the system moves until it is constrained by limits on it, which may be mechanical or energy dependent. A simple demonstration of positive feedback, and the resulting instability, is to place a microphone near the speaker in a public address system.

Negative feedback constrains changes in a system and is used to direct the output of a system towards the commanded state. The use of negative feedback therefore enables precise control of a system and is essential to the satisfactory action of machine tools, human systems and all automatic control systems.

As we have already seen, many systems rely on a human operator as the sensor and decider within the system process. Therefore, any self-regulating system needs to embody the same functions as are found in human operated systems and it is useful to examine the role of the human operator in considering the requirements for a fully automated, closed-loop control system.

Firstly, the human operator observes the output and compares it with the desired state, so generating information about the error or difference between the actual state and the desired state.

Secondly, the human operator determines an appropriate action in order to redress the imbalance in the system and reduce the error to zero, so that the actual output becomes that defined by the input requirement.

Therefore, a fully automated closed-loop control system needs:

- A reference signal (the input signal defining the required output).
- A feedback signal (defining the actual output).
- A comparing stage where the feedback signal is compared with the reference signal and the error or difference signal produced.
- An actuator or driving stage whose input, and therefore the system output, is controlled by the magnitude of the error signal.

These requirements can be represented by a simple systems diagram, Fig.17.2. This representation is universal and is to be found in the description of hydraulic, electrical, electromechanical, biological, mechanical, pneumatic, electronic and sociological systems. A more detailed diagram is given in Fig. 17.3.

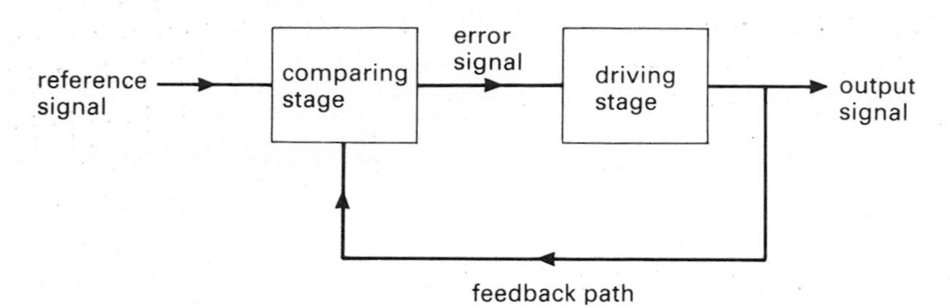

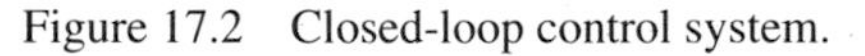

Figure 17.2 Closed-loop control system.

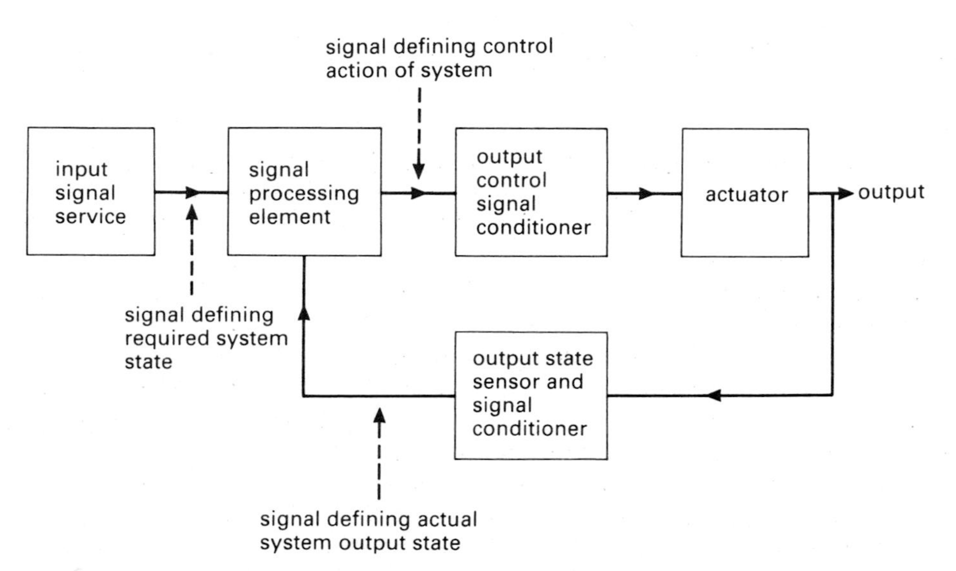

Figure 17.3 Detailed diagram of closed-loop control system.

Historically, the development of technology has moved through a number of distinct phases. The first industrial revolution enabled us to substitute machines for human manual skills and provided an improvement in power capability, speed and precision. These machines were used in industrial processes and the control was provided by the human operator. The human operator provided the feedback signal by sensing the process and also provided the decision-making part of the control loop (brain), by determining how the machine should operate throughout the whole process in which the operator and machine were involved. As the use of this technology developed, automatic mechanisms were found to replace the human operator where the processes were clear, repetitive and could be reliably operated by self-regulating mechanisms. This is clearly desirable where dull repetitive work is involved and particularly where the process is performed in a hostile environment.

The advent of new technology, involving the silicon chip as the signal processing element and decider, together with the new devices for sensing information, has enabled the human controller to be replaced entirely by electronic semiconductor systems in many control tasks. In all of the classes of systems already referred to in this chapter there are examples where the process of sensing and deciding may be done electronically. This produces automatic systems for a range of tasks from the simple to the complex.

17.4 Examples of control systems

There is a wide range of possible control systems to consider. It is instructive to examine positional and velocity control since they exemplify many of the points made in the previous sections and employ many of the electronic subsystems described in Chapters 14, 15 and 16.

17.4.1 Position control

Many problems in position control can be solved by simple *two-state systems*. These are not always ideal but the costs of better solutions may be prohibitive.

The railways provide two simple examples – points and signals. In both cases, a two-state system is needed. Points will be either open or closed: signals will indicate go or stop (in this simple case). The feedback from both systems can be provided by switches, informing the operator that the change he has initiated has in fact taken place. There are no intermediate positions so simple limit switches are all that are required. Many control problems can be solved by simple two-state solutions so these should always be considered first.

The stepper motor is often considered not to need feedback since its rotor rotates by a fixed amount when the correct signal is fed to it (see section 16.3.3). However if a stepper stalls for any reason it has immediately 'lost' its position. One technique employed to check the position of a stepper motor regularly is to use an optical switch or a mechanical limit switch at the extreme position of the stepper's movement. At switch on, and possibly at other times during the operation, the motor is sent into this extreme position to give it a datum point. Many computer printers employ this technique and run the print carriage to the extreme left-hand side when switched on. Disc drives use a similar technique.

If continuous feedback of position is required, then a servopotentiometer is used. This is attached to a moving part of the mechanism which is usually the final drive shaft of a motor-gearbox combination. This arrangement is then known as a *servomotor*.The principles of (analogue) position control are shown in Fig. 17.4.

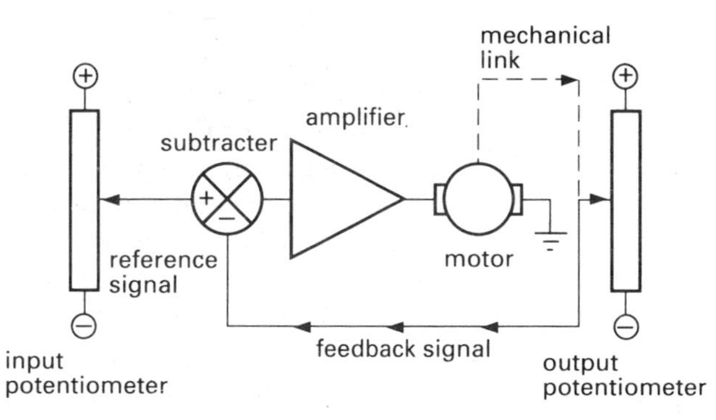

Figure 17.4 Typical position control system.

The reference signal carries the information concerning the position the motor is required to produce. The signal from the feedback potentiometer, which is mechanically connected to the motor, carries the information to tell the system actually where it is. The comparison stage, which acts as a subtracter, takes the difference between the two signals, the error signal, and amplifies it to produce the drive current. When the mechanism is where it should be, the error signal will be zero making the drive current zero. That is the ideal case. However, in practice, a number of problems are encountered, which we can list as follows.

- The motor has static friction which means that small signals will not cause it to move. When it does start, it will start with a jerk and accelerate to a significant speed. The range of signals for which no motion is caused is known as the *dead band*.
- The load, in a real system, will have inertia which will take time to accelerate. The response will therefore lag behind the stimulus.
- Because the load has inertia it will also acquire significant kinetic energy which must be removed before it can be brought to rest. This tends to cause overshoot – the response will lead the stimulus.
- The power supply must be very stable even if a large motor is used. If the power supply changes then the signals from the feedback potentiometer and the reference signal will change causing the whole system to be unstable.
- Any mechanical linkages will suffer from backlash.
- The potentiometers cannot be entirely wear resistant.

These problems lead to two conflicting requirements. The need for maximum sensitivity and precision suggests a high-gain amplifier so that the system will respond to very small error signals. On the other hand, a low-gain amplifier is required for maximum stability so that the load does not acquire too much kinetic energy. No solution can totally solve these conflicting requirements. There has to be a trade-off between the two desirable states.

In order to investigate the response of a particular system it is usual to examine its response to a step input. This is a sudden change in the reference signal. The possible responses are shown on the graph of Fig. 17.5.

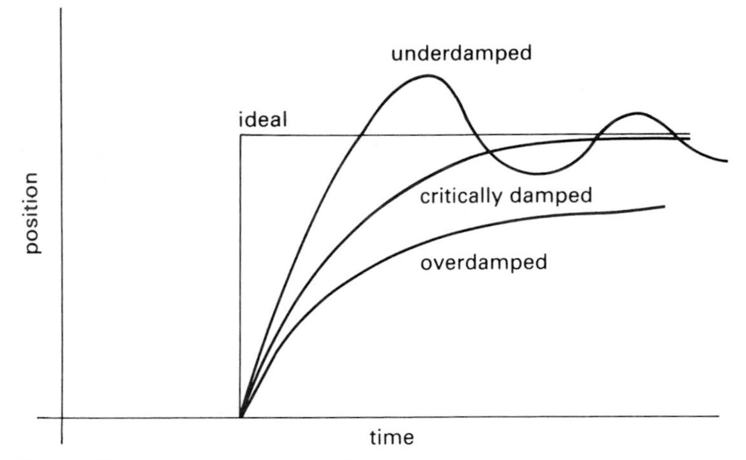

Figure 17.5 Responses to a step input.

Critical damping is the nearest it is possible to get to the ideal. *Overdamping* occurs when there is excessive friction and/or a low-gain amplifier. The response is sluggish and is unlikely ever to reach the required position.

Underdamping occurs when the load has a high moment of inertia, a high-gain amplifier is used, there is low friction in the system or any combination of these conditions.

Note that damping relates principally to the effects of dynamic friction whilst dead band is caused by static friction. There is no causal relationship between them but it is true that systems with a large amount of damping tend to have a large dead band and vice versa.

There are commercially available systems (or you can construct your own) to investigate analogue position control. You will also need a data capture system, probably connected to a microcomputer, if you are to see the effects of applying a step function. The subtracter is made by using a summing amplifier and reversing the connections to the servopotentiometer. The power amplifier will have to be selected to provide enough current for the particular motor being used. A significant load needs to be attached to the final drive shaft of the motor system to make sure it has sufficient inertia (Fig. 17.6).

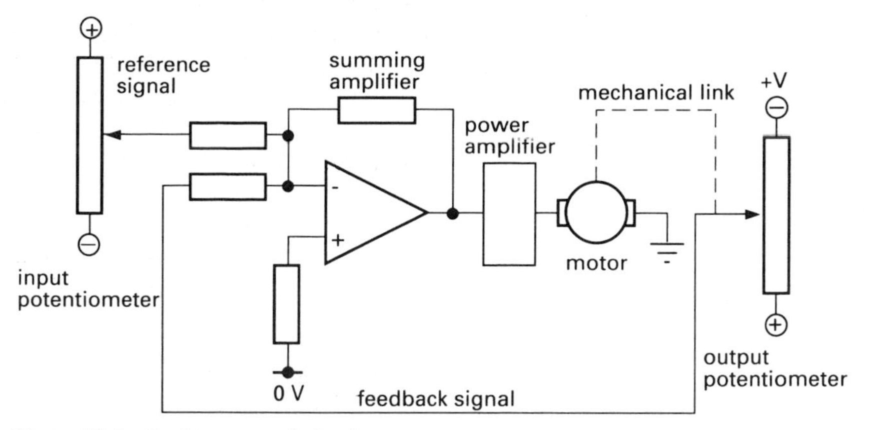

Figure 17.6 Position control circuit.

This is known as *simple proportional control* where the signal fed to the motor is directly proportional to the error signal. As long as the inertia of the load is reasonably large it should be found that the system is underdamped and that this effect increases as the gain of the power amplifier is increased. This overshoot is shown on the graph of Fig. 17.7.

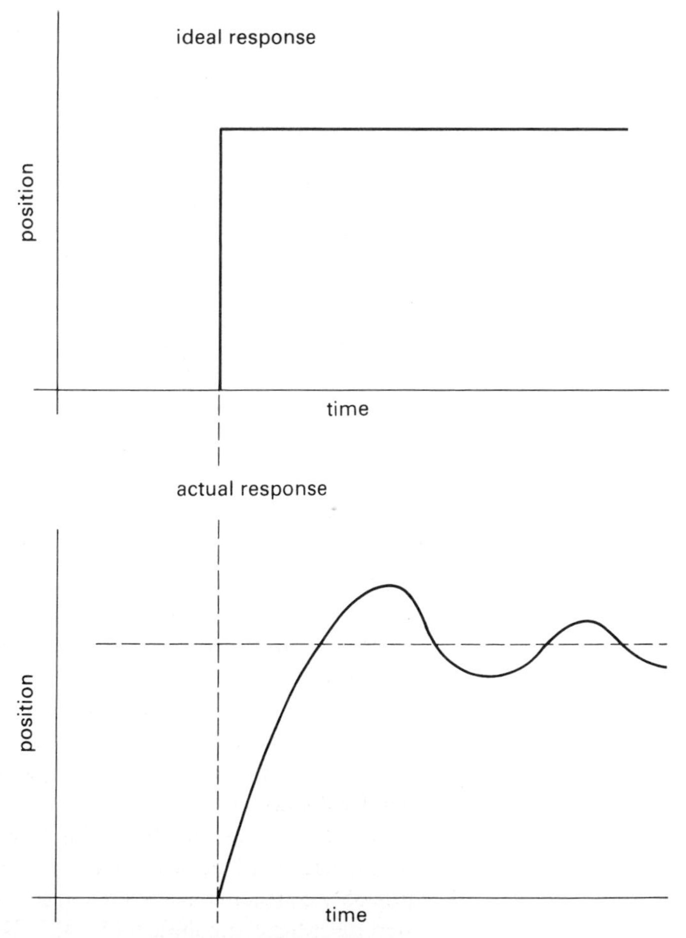

Figure 17.7 Graph showing overshoot.

It should be clear that some degree of braking is required if critical damping is to occur. The voltage to the motor must be reversed before it reaches the equilibrium position. The usual way of achieving this is to add an extra feedback signal relating to the velocity. If the ideal graph of velocity against time is examined, Fig. 17.8, it can be seen that it reaches a maximum before the equilibrium position is reached and falls to zero as the load reaches the equilibrium position.

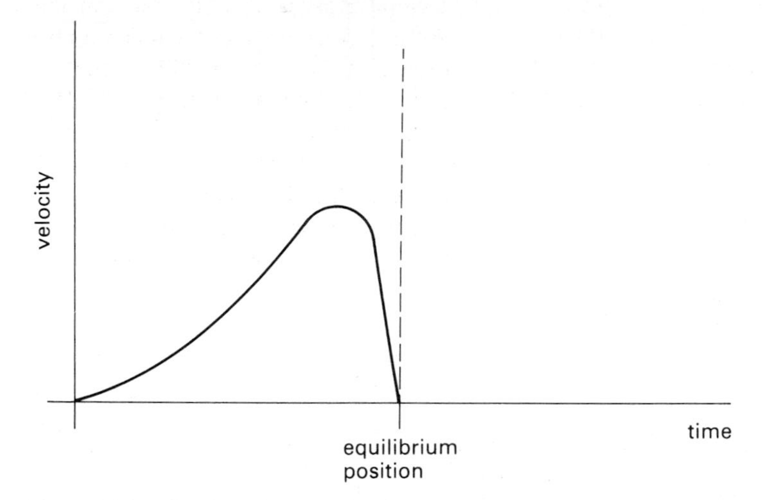

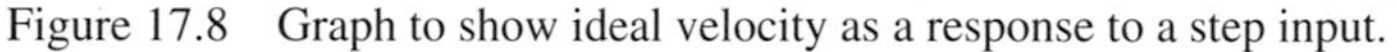

Figure 17.8 Graph to show ideal velocity as a response to a step input.

This is added to the error signal as negative velocity feedback and the magnitude of this feedback signal needs to be adjustable. Since velocity is the differential of position it can be achieved by an op-amp configured as a differentiator as in Fig. 17.9. This should improve the performance of the system making it more stable. Note however that the motor speed will be reduced whilst the dead band has been increased. It should be clear that excessive velocity feedback is undesirable as this will make the system heavily overdamped.

In order to improve the response still further it is possible to add a feedback term relating to the acceleration.

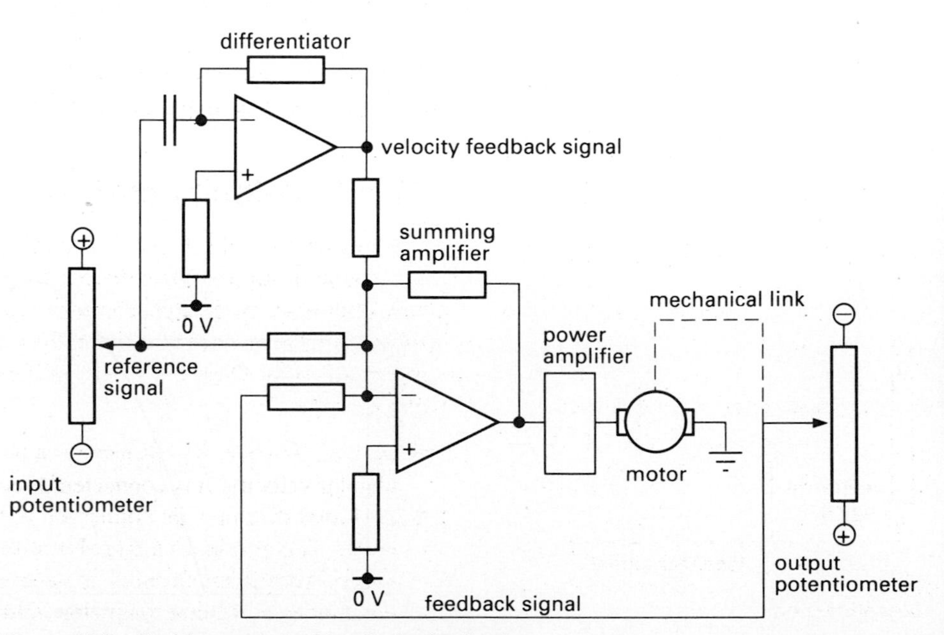

Figure 17.9 Position control with velocity feedback.

Position control with p.d.m. servos

A quite different technique is used for the control of pulse duration modulation (p.d.m.) servomotors. These motors originated in the hobby market and are much used in radio-controlled models. The motor, gearbox, servopotentiometer and electronics are all contained inside the same case and only three wires emerge from this case. The wires are +5 V, 0 V and signal which is a pulse input. The method of control involves supplying the system with a train of pulses with a periodic time of 20 ms. The duration of the pulse controls the position of the final drive shaft which can only move through 180°. A pulse duration of 0.5 ms gives a position of 0° whilst a pulse duration of 2.5 ms gives 180°. This is shown on the graph of Fig. 17.10.

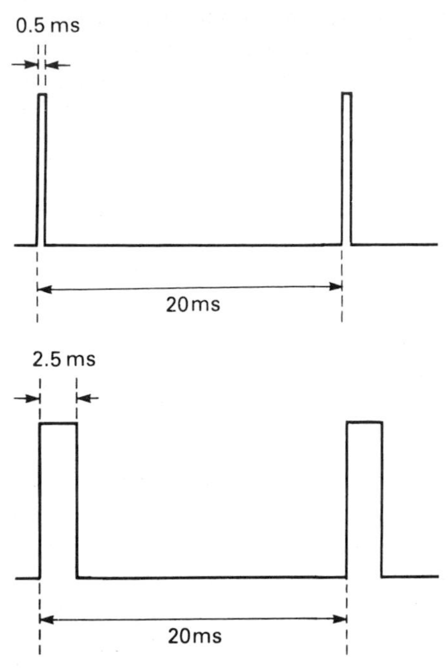

Figure 17.10 Traces showing the minimum and maximum signals for a p.d.m. servo.

Since only 2.5 ms of the total period have been used it can be seen that information for up to eight servos can be placed within this period. This is the usual method when the devices are radio-controlled. Alternatively, control may be achieved by setting up a pulse generator circuit with a period of 20 ms and using it to trigger a set of monostables – one for each servo. A possible arrangement is shown in Fig. 17.11.

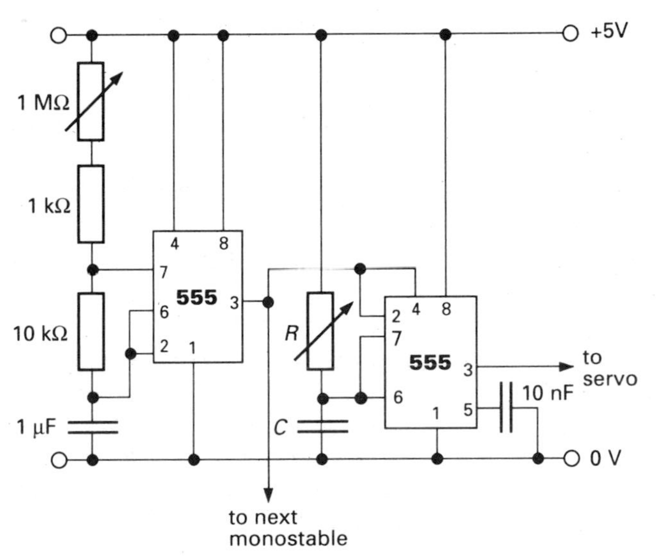

Figure 17.11 One method of driving p.d.m. servos.

Although these devices are not cheap they are economical if compared to the cost of buying the system in separate parts.

17.4.2 Velocity control

Velocity control, Fig. 17.12, is superficially similar to position control.

As with position control there is a compare stage consisting of a subtraction between the set speed signal and the feedback from the speed measurement subsystem. However, in this case, the error signal will not be zero otherwise the motor will stop. There are several different techniques possible for the measurement of motor speed or velocity.

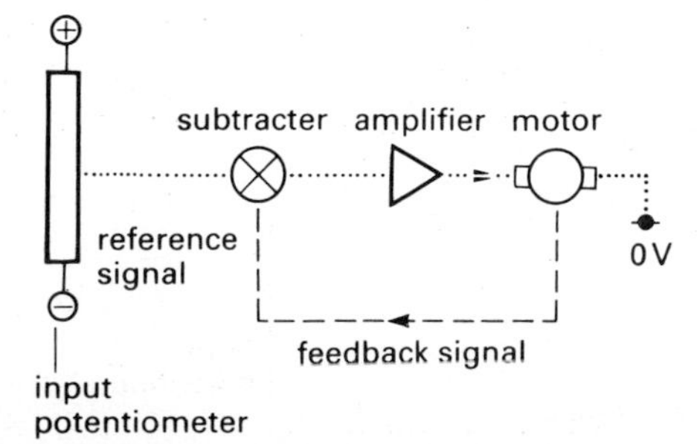

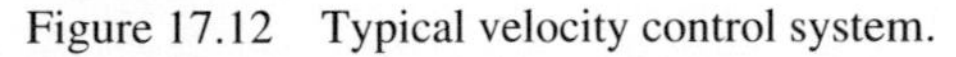

Figure 17.12 Typical velocity control system.

- A small dynamo may be used as a tachograph to produce a signal proportional to angular velocity. It is connected directly to the motor shaft.
- A slotted disc may be connected to the motor shaft and used to interrupt a light beam. This produces a signal whose frequency is proportional to angular speed. Additional circuitry is necessary to establish direction. Very similar techniques involving magnetic field sensing or reflective light sensing can also be employed.

- A computational technique may be used to calculate the back e.m.f. by including a resistor in series with the armature whose resistance is equal to that of the armature. Since the voltage drop across this resistor is equal to the internal resistive loss across the armature, the back e.m.f. can be calculated by subtracting twice the voltage across this resistor from the supply voltage. The back e.m.f. is directly proportional to the angular velocity.
- The motor current may be interrupted briefly and the back e.m.f. directly measured. This is a sophisticated technique requiring special integrated circuits for its operation. The range of possible speeds is limited since the ICs are designed for particular operations. The output is unidirectional and therefore contains no information about the direction of rotation of the motor.

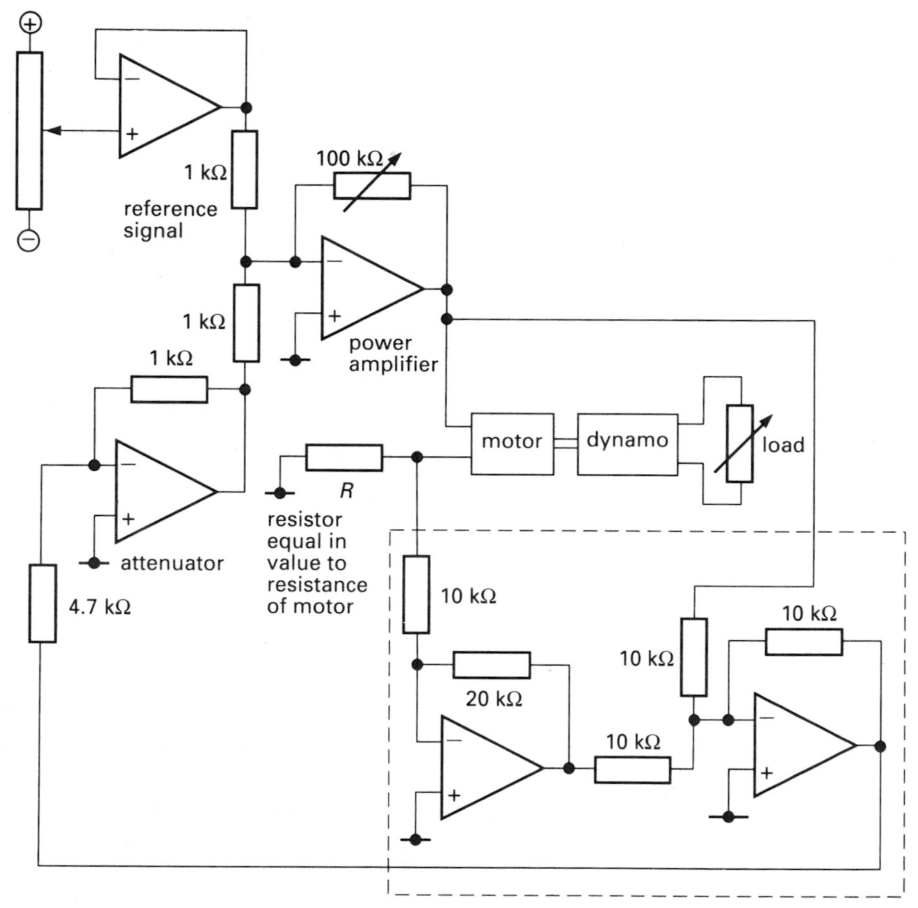

Figure 17.13 Velocity control with feedback.

If the direction of rotation is not important then the second method is attractive but it can be fairly bulky. The first method tends to be expensive but is commonly used in industry. The third method is cheap and provides directional information as well. A possible circuit for this method is shown in Fig. 17.13.

The power amplifier chosen must be capable of providing sufficient current for the particular motor used.

Bibliography

As for Chapter 14.

Assignments

1 (a) Apart from manual control, there are three commonly used methods of producing a force in a control system, namely electrical, pneumatic and hydraulic force.

Using a table similar to that shown overleaf, compare these three systems in terms of:

(i) relative magnitude of force;
(ii) controllability of movement;
(iii) speed of movement;
(iv) environmental considerations.

		Electrical	*Pneumatic*	*Hydraulic*
(i)	Force			
(ii)	Controllability			
(iii)	Speed			
(iv)	Environmental			

(b) Using a suitable control system as an example, explain the following terms: feedback, hunting, lag, damping.

(c) Fig. 17.14 shows a d.c. electric motor driving a worm gear which is linked to the steering mechanism of a remote-controlled model car.

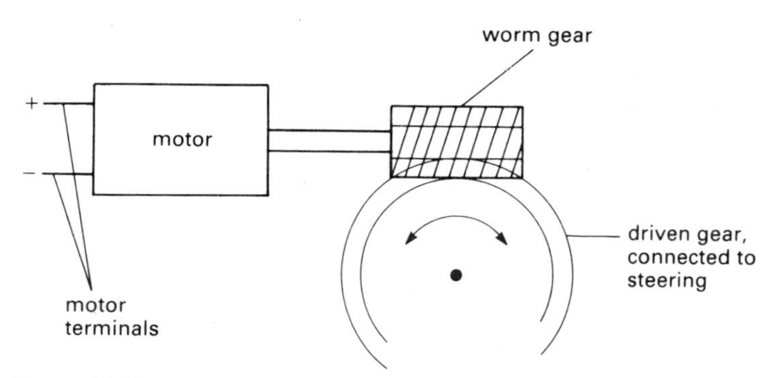

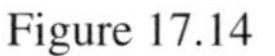
Figure 17.14

Complete Fig. 17.15 to show how the two relays, RLA and RLB, should be wired up so that the RLA switches the motor in one direction and RLB switches the motor in the opposite direction. When RLA and RLB are both off the motor stops.

(You may assume that RLA and RLB will never both be on at the same time.)

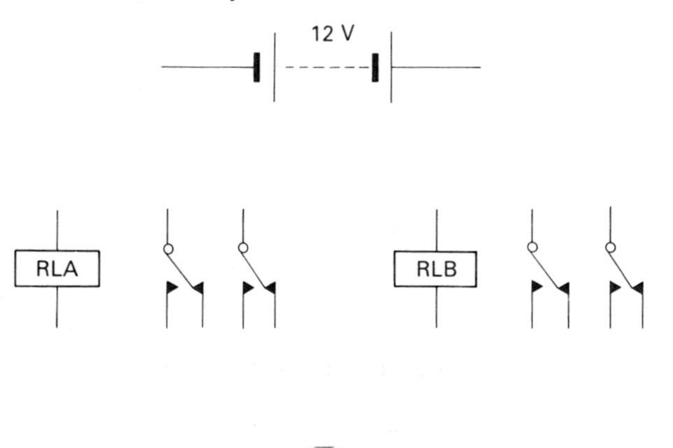

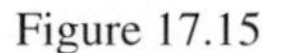
Figure 17.15

(d) When the steering mechanism referred to in part (c) reaches the limit of travel in either direction, a limit switch (normally closed) is needed to switch off the motor, in order to prevent the motor from burning out. Mark, on your circuit diagram, a cross where you would place each limit switch. The limit switches should be clearly labelled 'LIMIT SWITCH A' and 'LIMIT SWITCH B'.

(e) Fig. 17.16 shows a control circuit for the motor and relays in part (c). VR1 is the 'steering wheel' used by the operator. VR2 is the feedback potentiometer which is turned by the steering mechanism.

In theory, if the steering wheel, VR1, is turned, the comparator switches either RLA or RLB so that the motor turns until VRl equals VR2. The motor stops when VR1 = VR2.

What problem would occur in practice and what is the reason for this problem?

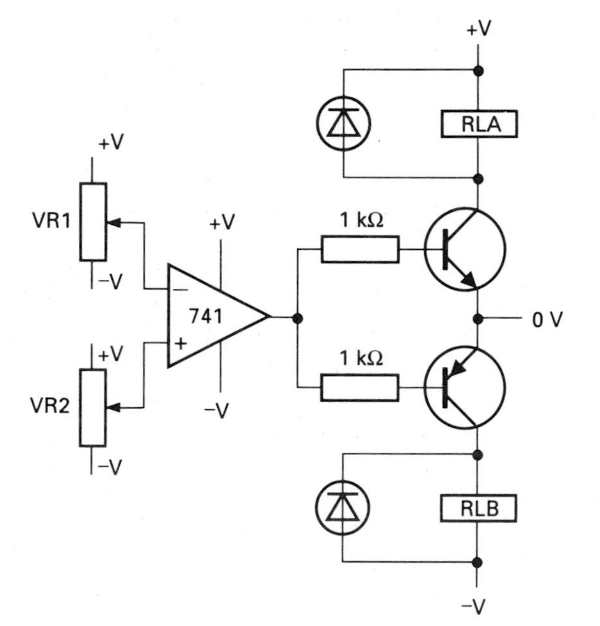

Figure 17.16

(f) What modification could be made to the circuit to overcome the problem in part (e)? Redraw part of the circuit to illustrate your answer.

2 (a) Describe the difference between a mechanised system and an automated system.

(b) With reference to an automated production process (or processes) which you have visited, explain briefly the following terms:

capital intensive	dedication of equipment
buffer store	safety precautions
quality control	occupations of workers.

Name the product or plant to which your answers refer.

(c) List *three* ways in which increased automation has benefited our society and *three* ways in which it has had a detrimental effect.

(d) With reference to two examples of your choice explain the differences between an *open-loop* control system and a *closed-loop* control system.

(e) Most electric toasters are controlled by a simple timing mechanism which causes the toast to 'pop-up' (and the heater to switch off) after a certain time has elapsed. The timer usually consists of a bi-metal strip which bends as it is

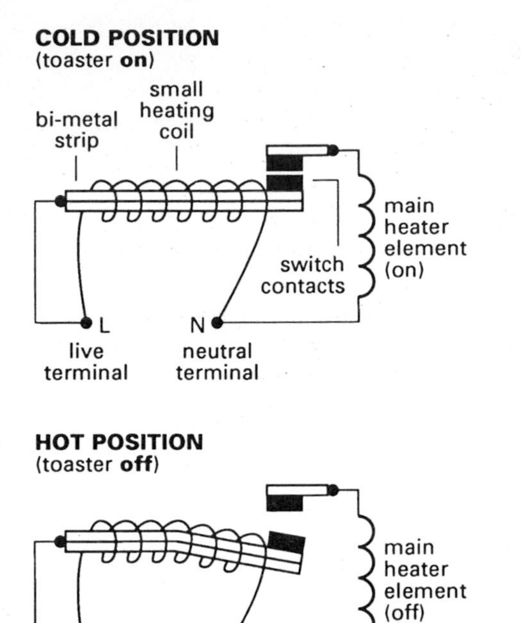

Figure 17.17

heated by a small electric coil, connected in parallel with the toaster heating element. When the bi-metal strip reaches a certain temperature, it breaks the circuit, thus switching the toaster off (see Fig. 17.17).

With reference to your answer to part (d) of this question, indicate the limitations of this method of control and give, in outline, two alternative control systems which would, in your opinion, give better control of toast colour. State clearly your reasons for suggesting these alternative systems.

3 Using examples of your choice supported by diagrams, explain the principles of control in the following:
(a) manual control;
(b) automatic control;
(c) a digital system;
(d) an analogue system.

18 Pneumatics

18.1 Pneumatic systems

What do we mean by pneumatics?

Pneumatic devices are activated by air. By compressing air, either electrically or manually, stored energy is available for the operation of pneumatic systems. A simple example is the car or bicycle tyre, in which a pump compresses the air into the tyre.

Common pneumatic devices are either *reciprocating* – that is, moving linearly – or *rotary*.

Reciprocating devices	*Rotary devices*
pneumatic road drill	pneumatic car wrench
pneumatic brakes on train or lorry	pneumatic drill
pneumatic paint sprayer	dentistry drills
automatic doors on a bus or train	pneumatic mixers
	pneumatic hoist

The use of pneumatics in industry became established in the 1950s, when automation was needed to reduce the unit costs of production to increase profit margins. Today the use of compressed air in industry is very common and pneumatics is essential to every low-cost automation production plant. Fig. 18.1 shows a typical application. The main advantages of pneumatics are:

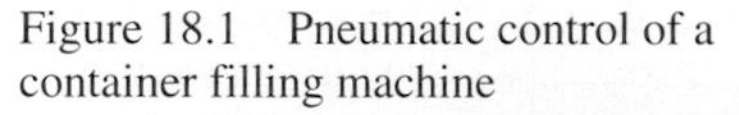

Figure 18.1 Pneumatic control of a container filling machine

Availability Compressed air is widely used in industry for a variety of applications from rock drilling to paint spraying and from operating a lathe to pumping liquids. Thus most factories and manufacturing businesses have a compressed air supply or portable compressor as a standard power source in addition to their electrical power. Air is an easily available substance which can be extracted from the atmosphere for compression and can also be restored to the atmosphere after it has been utilised. This contrasts with hydraulics which must have a return system and is also very messy if any leakage occurs.

Reliability Pneumatics equipment is extremely reliable over a long working life and is much more durable than electronic components.

Adaptability Existing machinery can be pneumatically automated with the minimum of alterations. The extra equipment is relatively simple in design, preparation and connection to the original system.

Operates in adverse conditions Pneumatics components are not affected by dust or corrosive atmospheres. Unlike electrical or hydraulic control circuits, pneumatics can be used in damp and inflammable conditions and, for this reason, are essential in the manufacture of chemicals and explosives.

Safety Compressed air has a very good safety record in comparison with electrical or hydraulic power but there may be the possibility of large loads being left dangerously unsupported if the air supply fails and this must be protected against mechanically, or with hydropneumatics.

Reciprocating motion Many automation processes involve repeated reciprocating motion which can be expensive to produce electrically. If a stroke movement greater than 50 mm is required an electromechanical system must be implemented which can become very cumbersome. Simple pneumatic components, on the other hand, can achieve a stroke of about 3 m cheaply and easily. However, it is necessary to choose devices with the correct stroke length as intermediate positioning is very difficult.

Variable speed and power Pneumatic circuitry can easily be adjusted to produce different speeds of operation. Short surges of power can also be provided by the source of stored compressed air which would be sufficient to overload an electrical power supply.

Economy Pneumatic equipment has both low set-up costs and low maintenance costs. The compression of air causes heat losses and is the main continuous expense in pneumatics. Compressed air must therefore be conserved where possible. Rotary pneumatic motors, used in hoists and mixers for example, are relatively inefficient (c.20%) in comparison to electric motors (c.90%). However, they are much more compact than electric motors, are not damaged by stalling or overloading and can achieve very high speeds of rotation, such as those apparent in a dentist's drill.

It is possible for control systems to be totally pneumatic. In industry, however, we find that it is often more expedient to combine pneumatic devices with other systems, especially electrical and electronic systems. The larger forces generated by the pneumatic devices can be positioned and controlled electrically, perhaps from remote positions, so making possible many automatic production processes. In a modern production plant, the materials are positioned, machined, assembled and welded with the aid of pneumatic automation lines and robots.

18.2 Compressed air supply

Fig. 18.2 shows a sketch of a typical *compressor* unit which is used to pressurise and store air. Air is drawn into the compressor through a *filter* and an electric motor compresses the air into a *receiver* which stores the air at a much increased pressure.

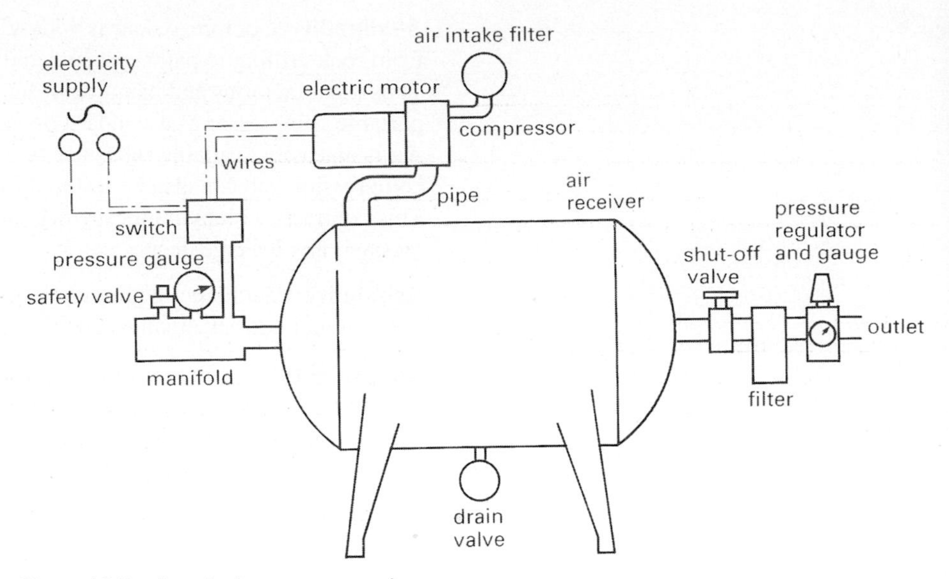

Figure 18.2 A typical compressor unit.

The compressor is controlled by maximum and minimum pressure settings on the receiver and the receiver pressure is shown on a gauge. The gauge shows the air pressure in bar; one bar is 0.1 N/mm^2 or 100 kPa, which is approximately atmospheric pressure. The supply from the receiver is passed through a filter to ensure that it is as clean and dry as possible, and then through the supply *regulator* which ensures that the system has a steady flow of air at a lower pressure than that in the receiver, Fig. 18.3. It is then *lubricated* by fine droplets of oil to enable the pneumatic components to work smoothly and efficiently. The supply air is directed to valves which act as switches and allow air to pass through the system, when required, to the pneumatic cylinders and other components which convert the energy to useful motion.

The supply to the school laboratory, Fig. 18.4, will be similar to that described above. A suitable compressor supplies the receiver at 8–10 bar or 800 kPa to 1 MPa. The system will operate at the pressure set on the supply regulator, which can be from 2–6 bar even if the receiver fluctuates around 8 bar. In addition to the supply pressure regulator, it is advisable to have a pressure regulator for each work station which can be set to the desired level.

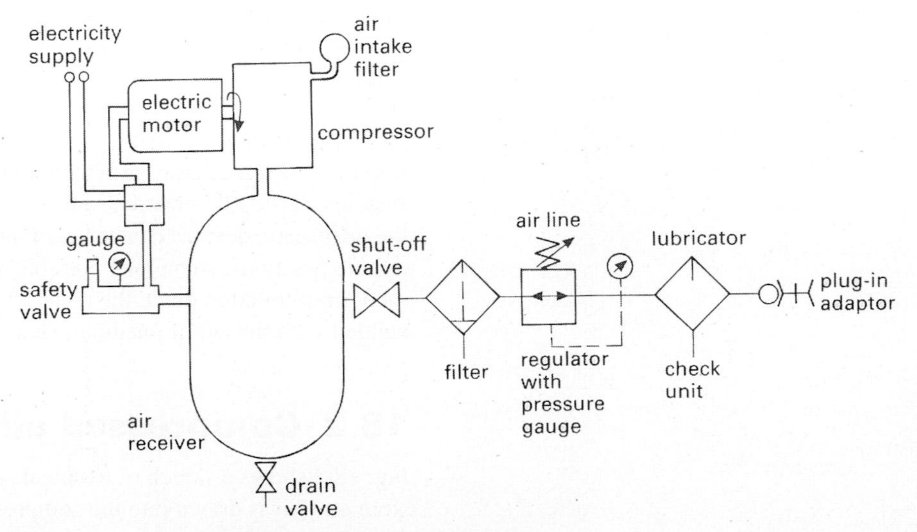

Figure 18.3 Schematic diagram of a compressor unit.

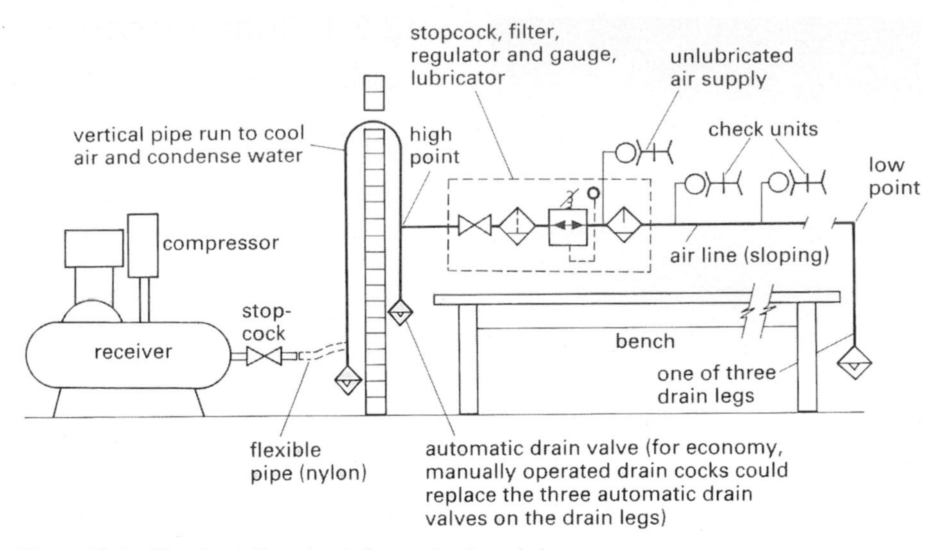

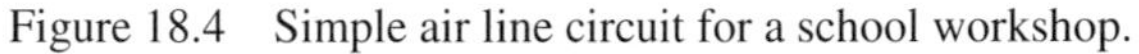

Figure 18.4 Simple air line circuit for a school workshop.

The main air supply that is directed to different work stations in an area is called the *air line*. Most factory installations operate at an air line pressure of 500–700 kPa (75–100 p.s.i.). The compressor and receiver must of course have enough capacity to operate all the equipment attached to an air line. The capacity of a compressor is usually quoted in m^3/s of air drawn into the unit before being compressed to the required level. In a large industrial installation, it is usual for all the operations to lead off from a circular or *ring main* air line to ensure a quick supply to all the connections, Fig. 18.5.

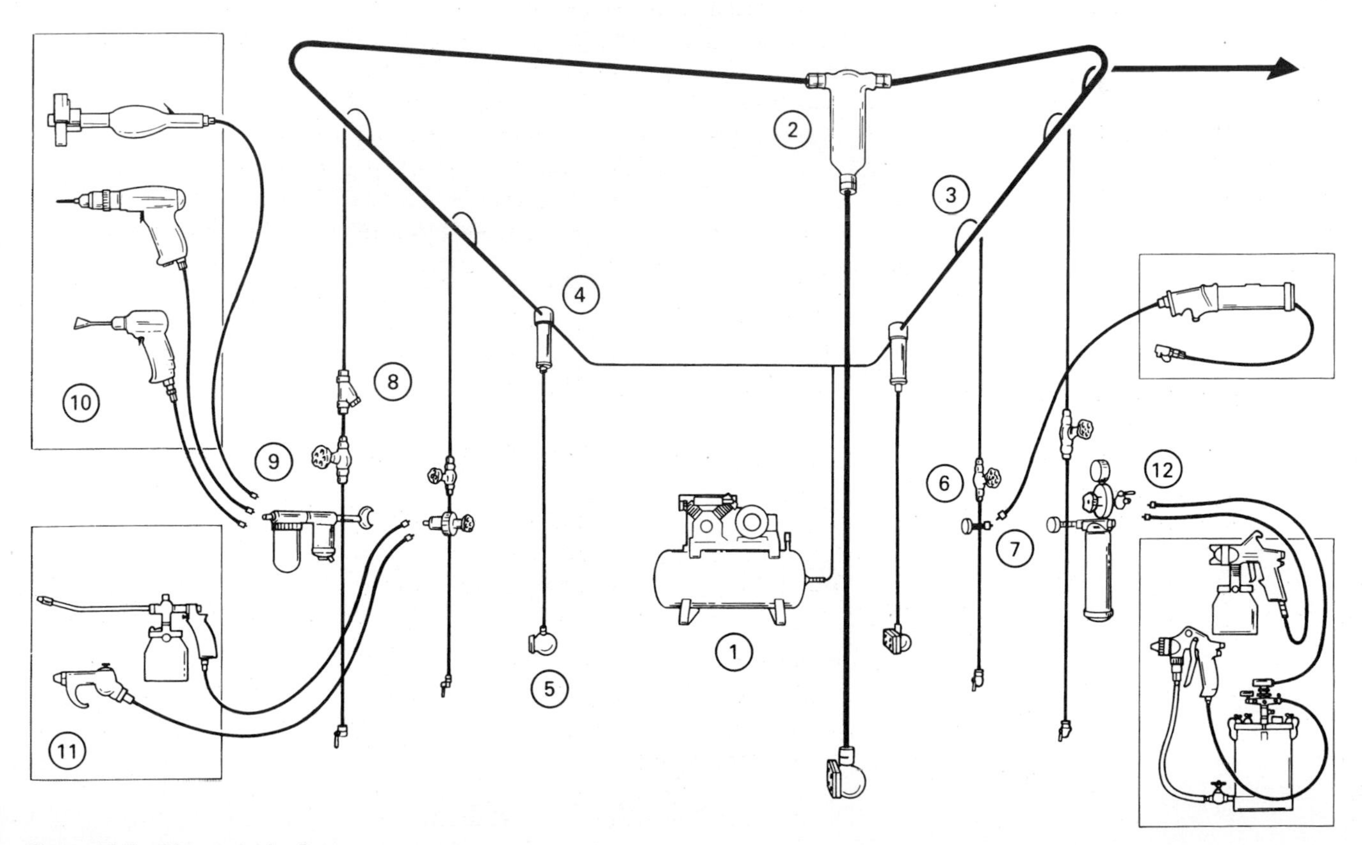

Figure 18.5 Ring main for factory or garage.

Figure 18.6 Single-stage reciprocating compressor

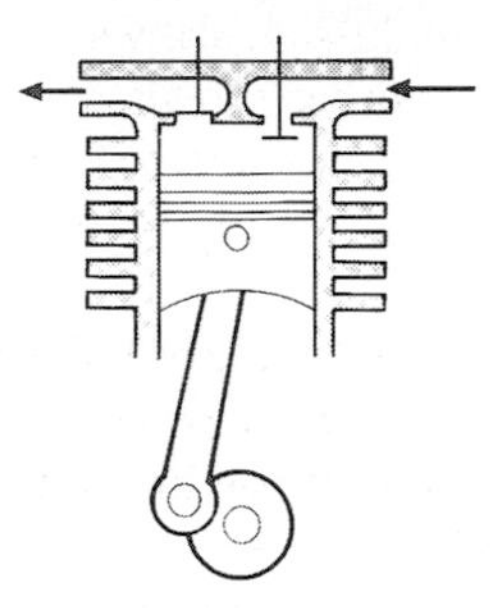

Figure 18.7 Diagram of single-stage reciprocating compressor

18.2.1 Compressors

A compressor draws in air from the atmosphere and compresses it into the enclosed space, so raising its pressure. Ideally this compression would be isothermal but in practice it is approximately adiabatic and there is a temperature rise (see section 12.8). The air that is drawn into the compressor should be as cool as possible so that the greatest mass of air can be drawn in on each stroke which will have as low a moisture content as possible. There are three main types of compressor that can be used for pneumatics:

- reciprocating,
- rotating vane,
- rotary screw.

Reciprocating compressors

In a reciprocating compressor a single-phase electric motor drives a piston which continually moves to and fro in a cylinder, compressing the air as it does so. A *single-stage reciprocating compressor*, Fig. 18.6, can have one or more cylinders of equal size, as shown in Fig. 18.7, and can produce an air pressure of up to 10 bar which is suitable for most applications. If an air pressure of up to 40 bar is required a *two-stage compressor*, Fig. 18.8, driven by a three-phase electric motor, can achieve this. A two-stage compressor has two cylinders of different sizes, as shown in Fig 18.9; the larger one is the low-pressure cylinder where the air is first compressed and then passed to the smaller cylinder where further compression takes place and high-pressure air is produced. This two-stage compression greatly increases the efficiency of the process.

Figure 18.8 Two-stage reciprocating compressor (with horizontal receiver)

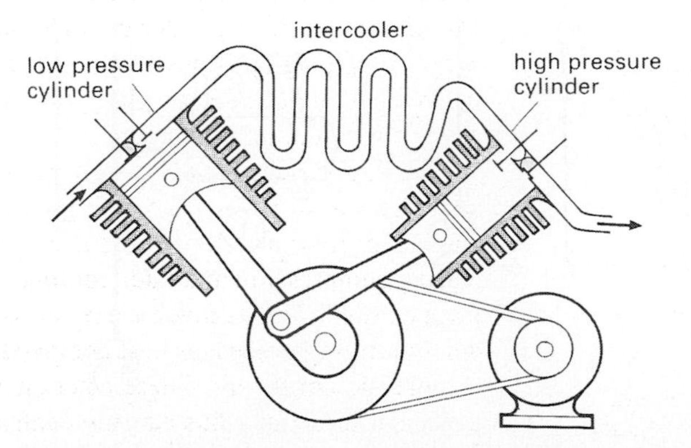

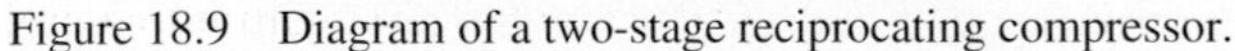

Figure 18.9 Diagram of a two-stage reciprocating compressor.

Compressors with three or more cylinders of unequal size compress the air in several stages and are called *multi-stage compressors*; they can achieve air pressures up to 1000 bar. Each time the air is compressed, heat is produced (see Chapter 12) and this must then be removed. Intercoolers are usually fitted between each stage of compression for this purpose. Turbine designs in particular are very suited to multi-stage compression techniques.

Other compressors

The rotating vane compressor has sliding vanes seated in an eccentrically mounted rotor. When the rotor is driven by a motor the centrifugal action causes the vanes to move and compress the air. High speeds can be produced and the volume of air that can be produced is greater than that of a similar sized reciprocating compressor. Large rotating vane compressors are available for general pneumatics and pneumatic road repair implements while smaller units are used to compress air for brazing hearths.

The rotating screw compressor has two rotating, concentric, multi-start screws. Air enters at one end and passes through the spaces between the meshed screws being forced into smaller and smaller spaces as it travels along, being compressed all the time to the outlet. This is a good compressor for large industrial uses but as it is expensive would not be cost-effective on smaller applications.

There are other compressors which are unsuitable for pneumatics but are useful in other applications. A centrifugal compressor with a fast rotating fan driven by an electric motor or internal combustion engine is typically used in dust extraction units where low pressures and large volumetric flow rates are required. Turbine compressors driven by diesel engines are used in jet engines as they can produce large volumes of compressed air for combustion. This feature also makes them useful in foundries, quarries and mine ventilation systems.

18.2.2 Conditioning the compressed air

It can be seen, Fig. 18.10, that the main stages in conditioning the compressed air are filtration, pressure regulation and lubrication. These are vital as pneumatic tools and equipment will jam and corrode if the air that is used to drive them is not prepared properly.

(a)

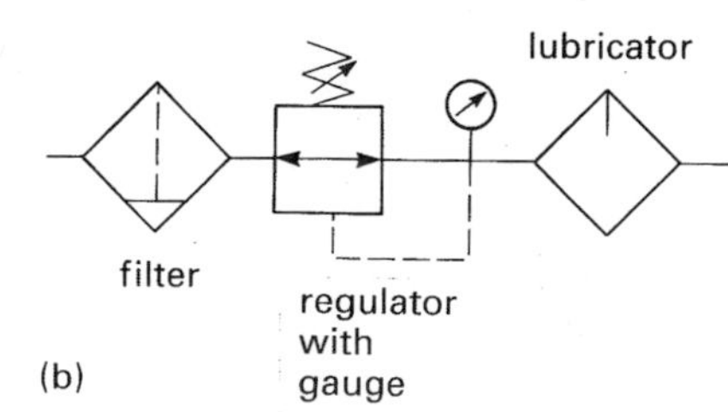

(b)

Figure 18.10
(a) Olympian system filter, regulator, gauge and lubricator unit.
(b) Circuit diagram symbol for a combined filter, regulator and lubricator unit.

Filtration

Water, solid impurities and old oil must all be removed completely or partially from the air stored in the receiver before it can be utilised. However, as this process is costly, care must be taken to prepare the air only to the level necessary for its final use.

Solids such as dirt, sand, carbon products, drops of weld or pipe scale can be filtered out by a number of means depending on the required level of filtration. Commonly, particles greater than 25 μm (0.000 025 m) in diameter should be removed although it is possible to filter down to about 3 μm size. The common materials used in filters are glass fibre, ceramics, nylon, felt, sintered metal or glass and a variety of strainer screens.

The liquids present in the compressed air are composed of water vapour, which sometimes absorbs fumes creating weak acid solutions, and old oil from the system and compressor which oxidises and breaks down into a thick, sticky 'treacle'. During compression and later cooling, some of the water vapour is condensed to form a liquid. This is much easier to remove than the vapour and therefore the dehumidifying process is best achieved by using mechanical filters after compression of the air. These filters will generally force the air to flow in a particular direction, thus causing centrifugal forces to separate the water and oil droplets from the airstream into a drain tap. Fig. 18.11 shows one example.

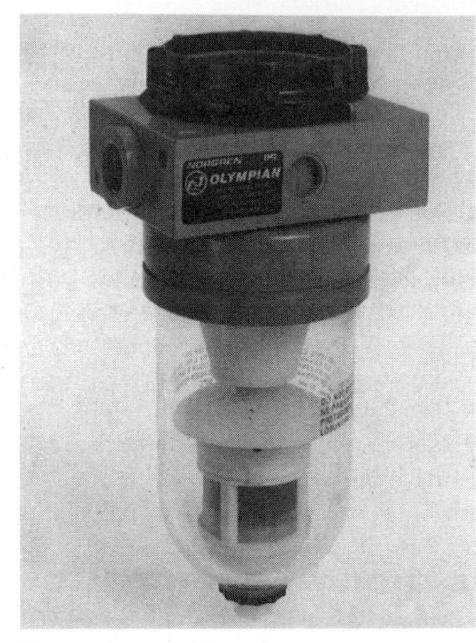

Figure 18.11 A filter.

The amount of liquid that can be condensed from the vapour present in a volume of air increases proportionally with pressure and inversely with temperature. Therefore, by deliberately increasing the pressure and decreasing the temperature, water can be removed from the air in the most effective and efficient way. In industrial systems, the compressed air is passed through an *aftercooler* to extract the heat of compression from the air, reduce the temperature and condense more vapour. During distribution of the air, a pressure drop in the lines may cause further cooling and more condensation. Consequently, drain valves should be incorporated into the air distribution lines. Finally, small filters must be placed in line near to the point of application to extract water before using the air and so protect the individual components and instruments. It is possible now to use filters which automatically empty any collected fluids when the line is not in use. This prevents the continual maintenance that would be necessary to prevent any liquids being reabsorbed into the system.

After all the liquids and solids have been filtered from the compressed air, it is still saturated with water vapour. If it is necessary to remove this water vapour from the air, an *air dryer* has to be incorporated into the preparation system after the liquids have been removed and when the air is at its lowest temperature. There are three principal types of air dryer:

- hydroscopic absorption by an agent,
- refrigeration dryer,
- chemical deliquescent dryer.

These filters are very effective but the removal of water vapour is a costly extra process and should only be done where necessary. Special multi-stage mechanical filters may also be necessary to remove oil mist from the airstream.

Pressure regulation

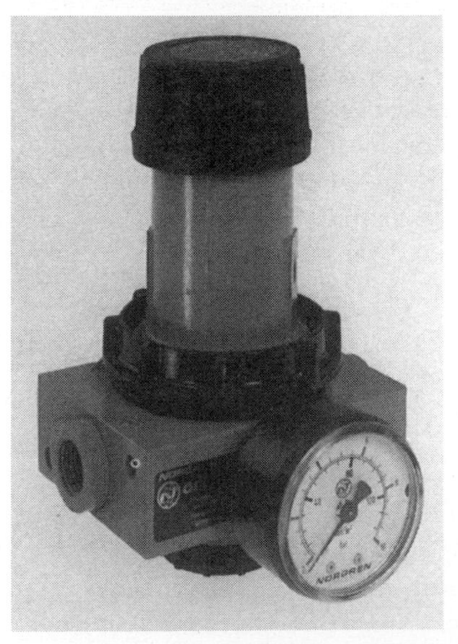

Figure 18.12 Pressure regulator with gauge.

Once the air has been cleaned, it must be regulated to operate at the optimum pressure for the system. All pneumatic equipment needs to operate at the correct pressure to maximise the efficiency of the system and the durability of the individual components. Pressure regulators, Fig. 18.12, are fitted at different points in the circuit in order to maintain a constant pressure in the line and also to control the forces generated by the cylinders, irrespective of changes in the line pressure caused by moving components. The commonest type of pressure regulator is a *nonrelieving* regulator which is placed in the air line. The primary air from the main supply enters at one side and, by moving an adjusting screw, the secondary or outflow of air can be controlled. A *relieving* regulator, however, has a vent which allows excess pressure to escape when the preset pressure is decreased and this is obviously preferable to the nonrelieving regulator which takes some time to adjust to variations in the required pressure. The secondary pressure is usually shown on a pressure gauge calibrated in bar. Regulators can be of differing sensitivities and the most appropriate model of regulator must be chosen for each system.

The adjusting screw in these regulators deflects either a diaphragm or a piston. The *piston regulator* gives good regulation characteristics and is more compact than the more sensitive *diaphragm regulator*. The pressure regulation can be *direct*, where the pressure is altered by an operator adjusting a screw, or *pilot-operated*, where the regulator is controlled by the airflow. More sophisticated pilot-operated regulators known as *precision controllers* are required for certain applications. These accurately sense the downstream pressure and compensate for any fluctuations to produce an airstream of an accurately preset pressure.

Lubrication

It is essential to lubricate the moving components of a pneumatic circuit and this is usually done by spraying a fine mist of oil into the clean, pressurised air.

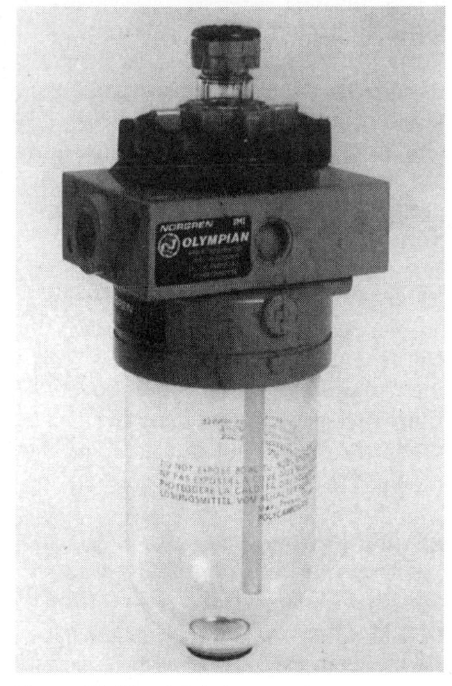

Figure 18.13 A lubricator unit.

An *oil fog* lubricator, Fig. 18.13, is the most common although it is not normally used for an air line greater than 10 metres. Suction, from the airstream passing through the lubricator, causes the oil to flow up a syphon tube where a needle valve controls the volume of oil which atomises into the air. Recent designs have emerged in which the degree of lubrication alters with the fluctuations of air flow giving *constant density lubrication* or, in the *microfog lubricator*, producing an oil mist composed of very fine drops of oil less than 2 μm in diameter.

Lubrication should not be used in fluid logic or low-pressure component circuits where there are no moving parts, as it serves no useful function in these circuits and will block the small-diameter airways.

18.3 Operation and application of pneumatic components

Valves and *cylinders* are the most important pneumatic components in a linear motion system. The valves control the cylinders which convert the stored energy to linear motion.

18.3.1 Cylinder operation and control

Single-acting cylinders

A diagram showing the cross-section of a single-acting cylinder is shown in Fig. 18.14. If compressed air is forced into the cylinder as shown, the piston is pushed out to the right and the piston rod can exert pressure on an external object. When the cylinder is operating pneumatically, the exhaust air from around the spring exits from the air outlet while the piston is moving out. When the air is turned off a spring returns the piston to its original position. During this in-stroke, air is drawn in from the air outlet and the exhaust air from the left-hand side of the cylinder now exits through the original air inlet. The single-acting cylinder is used for applications which need a small force and small-scale movement.

Fig.18.15 shows the British Standard symbol for a single-acting cylinder as it is drawn in circuit diagrams, and the symbols to be used for drawing the main air feed and supply lines.

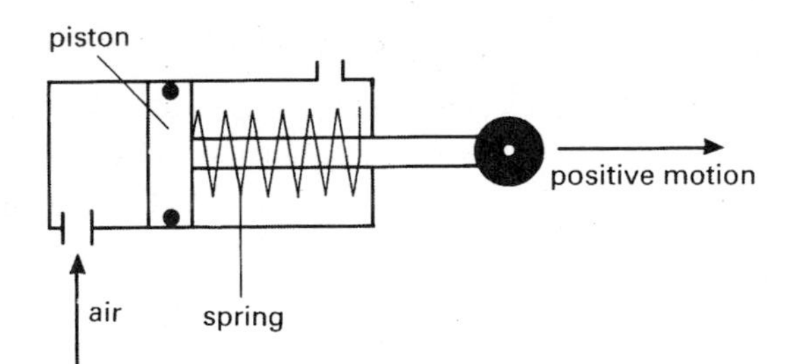

Figure 18.14 Cross-section of a single-acting cylinder..

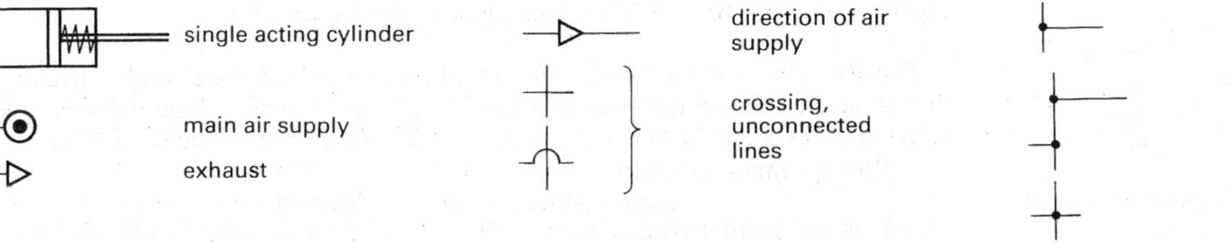

Figure 18.15 Pneumatic circuit symbols.

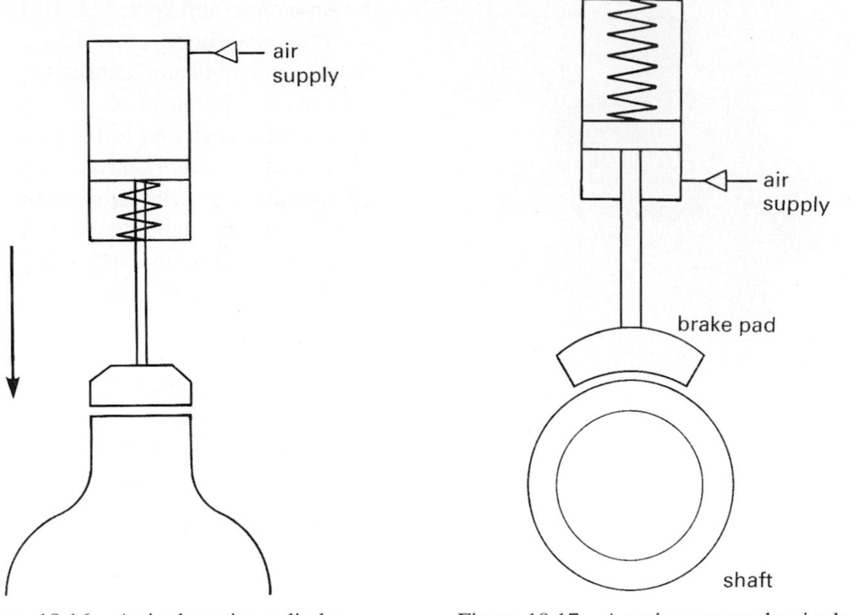

Figure 18.16 A single-acting cylinder pneumatically sealing a bottle.

Figure 18.17 A spring out-stroke single-acting cylinder used in a brake mechanism.

An application of a single-acting cylinder is shown in Fig. 18.16. The cylinder has a spring-action in-stroke and is used to press a lid on a bottle to seal it.

The single-acting cylinder can also have a spring out-stroke which is used to apply a pneumatic brake in Fig. 18.17. The piston is kept in its positive position by the spring until air is supplied to the cylinder which forces it to retract and release the brake.

Figure 18.18 A push-button operated, spring-returned 3-port value.

Control of single-acting cylinders

Fig. 18.18 shows an example of a *3-port valve* which is used to control the single-acting cylinder. The cylinder has two phases of operation (see Fig. 18.19):

- the OFF state when the air supply is prevented from reaching the cylinder but any air in the cylinder is free to exhaust so allowing the cylinder to return to its original position;
- the ON phase when the air is forced into the cylinder and moves the piston.

The BS symbol for a 3-port valve combines these two states and is usually drawn in the OFF state, shown in Fig. 18.20. The 3-port, 2-state valve is often referred to as a 3/2 *valve*. Its ports are always numbered in the same way:

Port 1 – main air supply connection
Port 2 – normal output connection or cylinder port
Port 3 – normal exhaust port

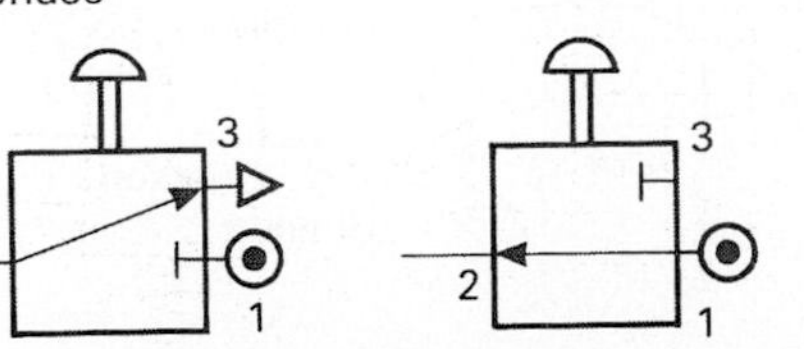

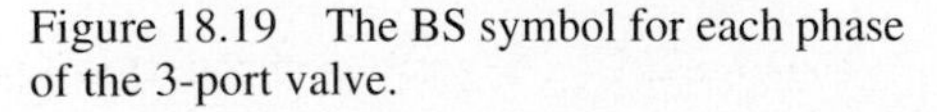

Figure 18.19 The BS symbol for each phase of the 3-port valve.

Figure 18.20 The complete symbol for a push-button 3-port valve.

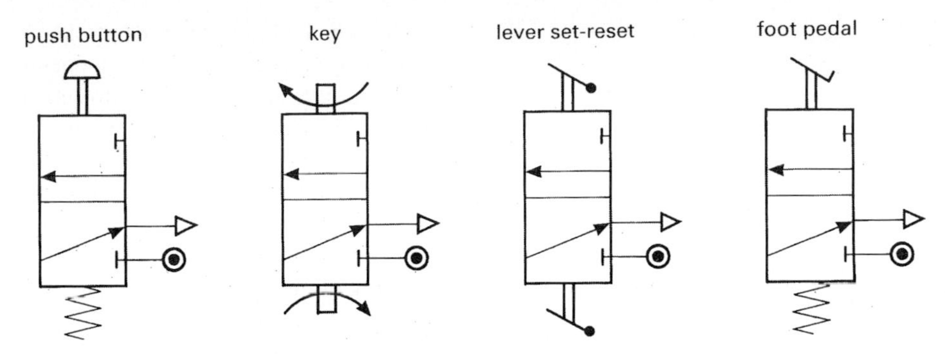

Figure 18.21 The four main types of manually operated 3-port valves.

Fig. 18.21 shows the symbols for the main types of manually operated 3-port valves. The *push-button valve* is useful for quick and easy starting of a sequence, or as a safety circuit breaker. A *key-operated valve* is used when a high degree of security is required. The lever valve stays in the position set until it is reset. Should it be necessary for the operator to keep both hands free, a *foot valve* can be used. It can be seen from the valve symbols that the push-button and foot-pedal valves have a spring-return mechanism when the actuator is released, whereas the lever and key valves have to be manually returned.

Fig. 18.22 shows other 3-port valves that can be operated by components and signals within the circuit. The *plunger valve* is similar to a push-button valve and can be operated by the piston rod of a cylinder. The *roller-trip valve* will activate when a cylinder trip pad on the piston passes over the roller in either direction. Because the *unidirectional roller-trip valve* will only operate in one direction it will normally be

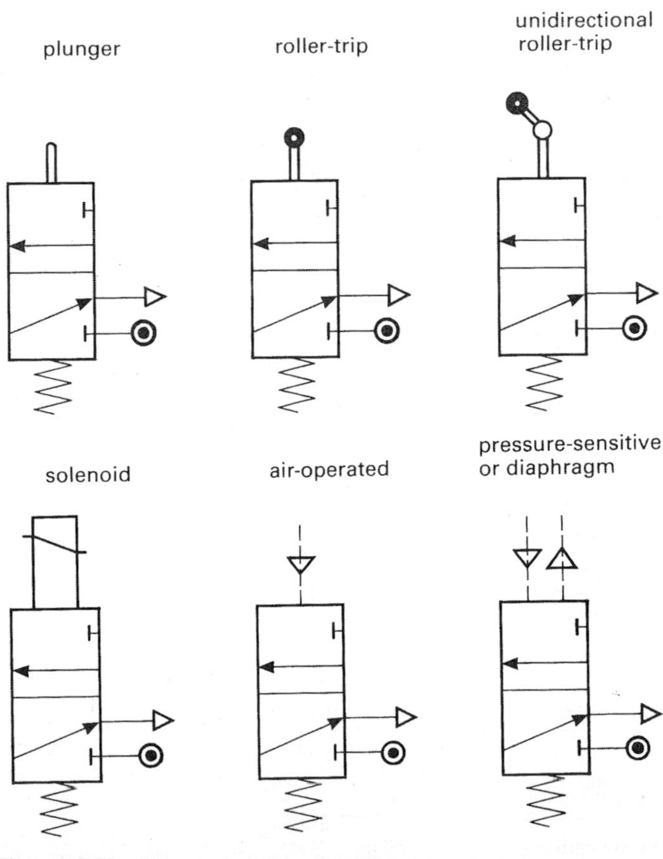

Figure 18.22 The symbols of activated 3-port valves.

used near the limit of the out-stroke, or in-stroke, of a piston. The *solenoid valve* converts an electrical pulse to a pneumatic signal, and enables a combined system to operate. When it is necessary to activate a component by an air signal from another part of the circuit, the *air-pressure operated* or *pilot valve* enables such automatic signalling to be achieved. The *diaphragm valve* will detect small changes in air pressure and switch states accordingly. All of these valves have a spring-return mechanism.

The valves can be either *spool* or *poppet* valves. The spool valve changes the routing of the air by means of an internal moving spool which seals off port 3 in the ON position and port 1 in the OFF position. The poppet valve has a hollow piston which allows air to pass up the centre to port 3 and seals off port 1 in its OFF position only. Fig. 18.23 shows the routing of air through a 3-port poppet valve.

Fig. 18.24 shows the circuit diagram for a single-acting cylinder operated by a push-button, 3-port valve in both the OFF and the ON positions. The out-stroke is termed the positive part of the stroke and the in-stroke the negative.

For the control of cylinders, 2-*port valves* are also available and two of these may be used in place of one 3-port valve.

In the circuit for a pneumatic lift, Fig. 18.25, valve **a** feeds the supply air into the system while valve **b** exhausts air from the unit. The advantage of this system is that there are intermediate rest positions where neither valve is being operated. However, a delay of more than ten minutes would give occasional slipping of the lift because of air leakage. This type of operation is often used for platform lifts, hoists and furnace door control. When used with double-acting cylinders the other sides of the cylinders are left to vent to the atmosphere.

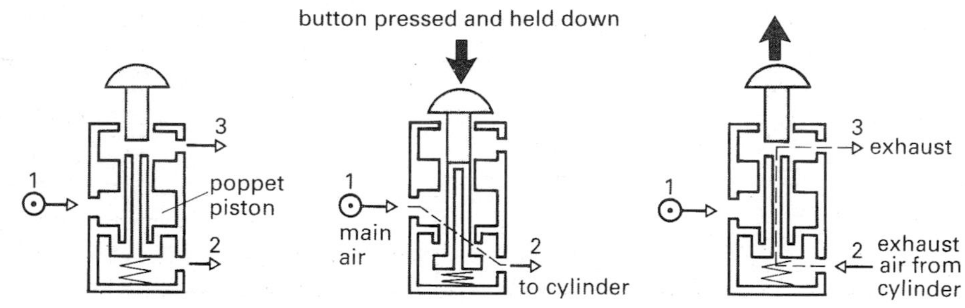

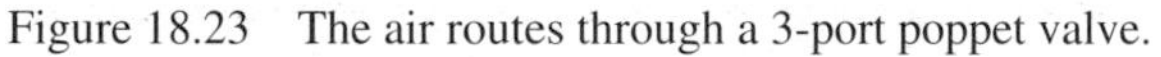
Figure 18.23 The air routes through a 3-port poppet valve.

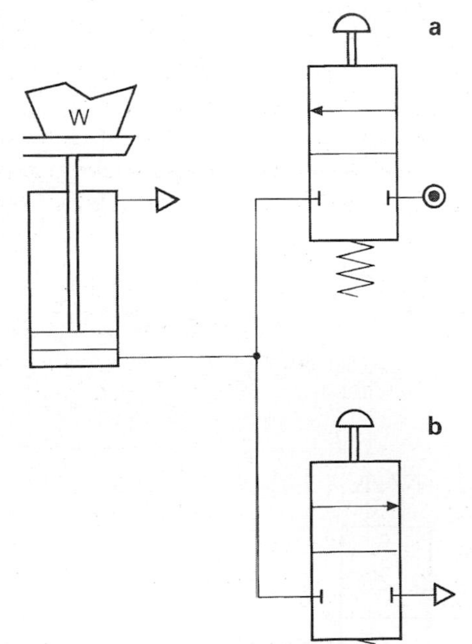

Figure 18.25 Two-port valve control of a pneumatic lift.

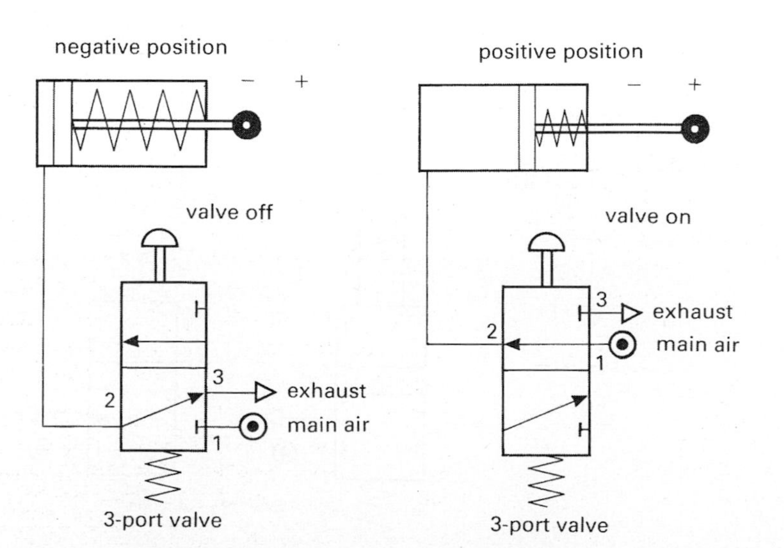

Figure 18.24 The control of a single-acting cylinder.

Double-acting cylinders

A *double-acting cylinder* is one which, instead of having a spring-return action for the piston, has an air port at each end. Thus, the movement in both directions can be controlled pneumatically by using two 3-port valves as shown in Fig. 18.26.

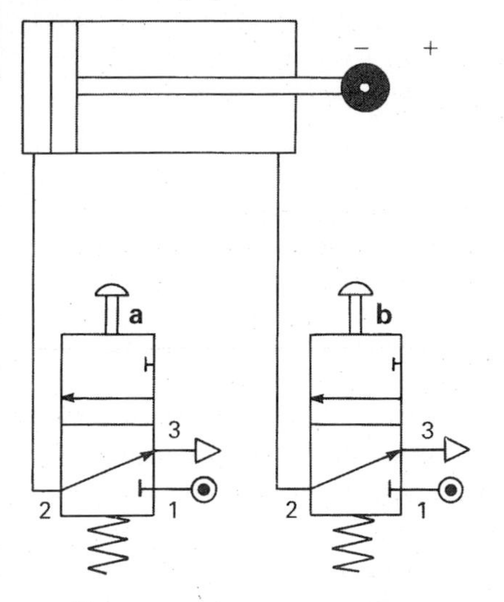

a and b are push-button 3-port valves

Figure 18.26 The control of a double-acting cylinder using 3-port valves.

The main air supply would need to be connected to both the valves **a** and **b** in the circuit shown in Fig. 18.26. When button **a** is pressed, air from the supply enters the left-hand side of the cylinder and the piston moves in the positive direction. When button **a** is released, the main air supply is cut off and the piston remains in a positive position, providing there is no load on the piston rod. When button **b** is pressed the main air supply enters the right-hand side of the cylinder, the piston moves in the negative direction, and the air is exhausted through port 3 of valve **a**.

There is a wide variety of cylinders, commonly 10 mm to 200 mm diameter and up to 300 mm stroke length although cylinders with a stroke as large as 6 metres can be obtained. Operating pressures of up to 10 bar are common but cylinders operating at 17.5 bar are now being used in industry. Most modern cylinders are designed to reduce the speed at the end of each stroke in order to lessen the impact shock on the system and the external contact. This is done by including an adjustable air cushion on the piston ring. Fig. 18.27 shows the operating principle of a *cushioned double-acting cylinder* and Fig. 18.28 shows the BS symbol for the same cylinder.

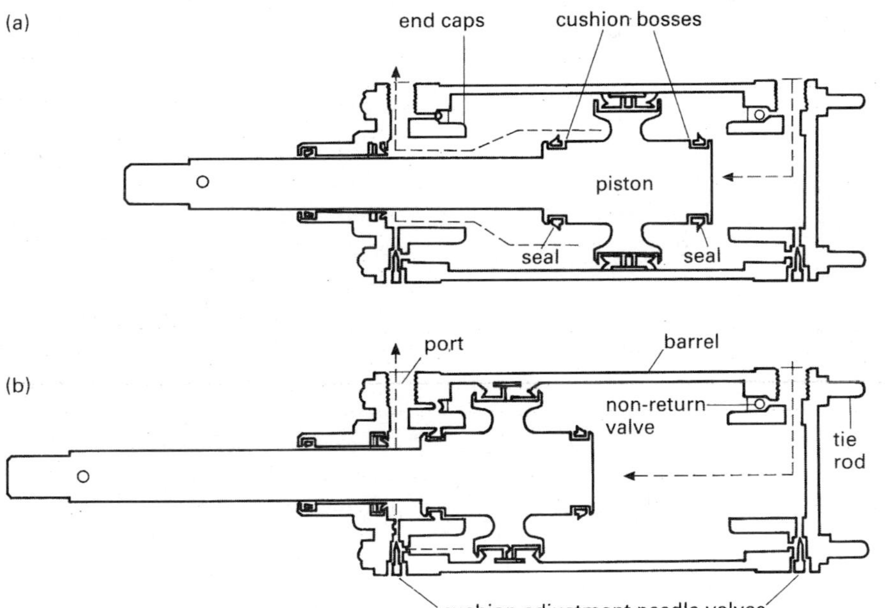

Figure 18.27 Cushioning effect of the double-acting cylinder.

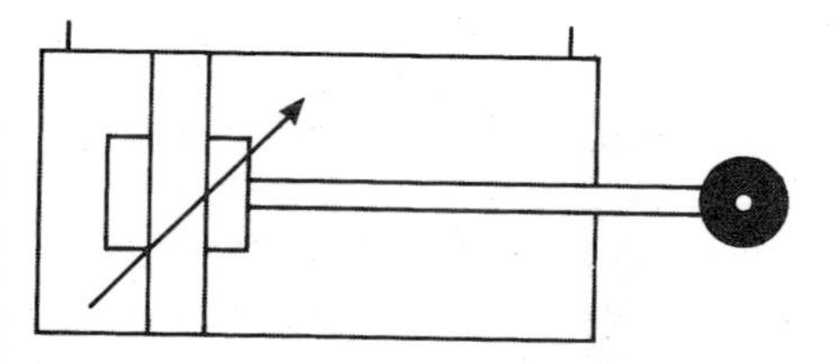

Figure 18.28 The BS symbol for a cushioned double-acting cylinder.

In situations where a large force is required and space is limited a *tandem cylinder* can be used. This has two double-acting cylinders linked together as shown in Fig. 18.29, so that double the force of one cylinder of the same cross-sectional area is achieved.

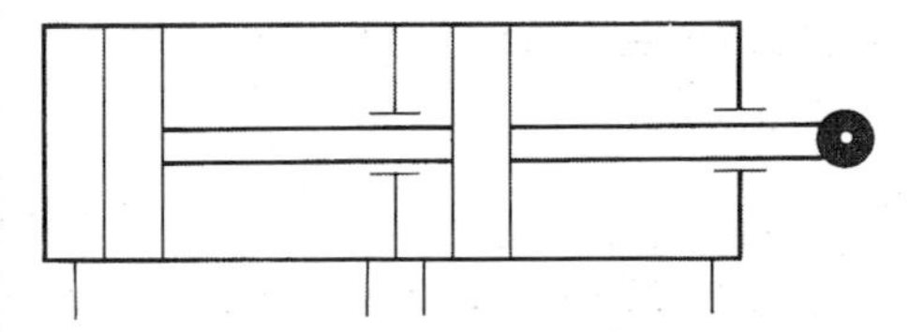

Figure 18.29 A tandem cylinder.

By connecting two cylinders of opposite sense, as shown in Fig. 18.30, accurate placement to certain positions can be obtained, regardless of the load that is applied. This is a *duplex cylinder* system.

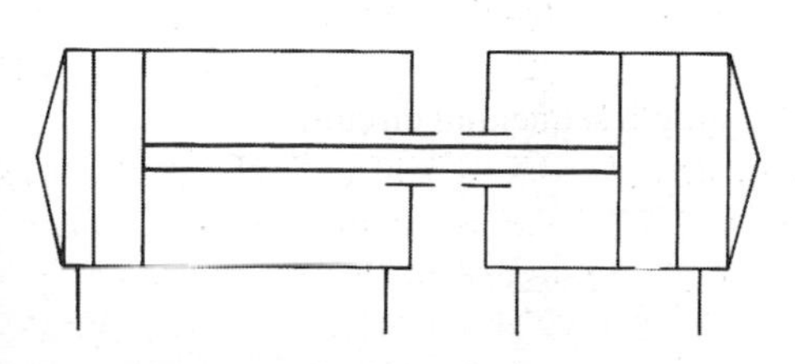

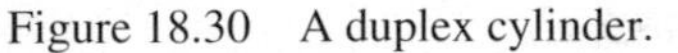

Figure 18.30 A duplex cylinder.

When components on an assembly line have to be moved on a circular path from one position to another, a lever connected to the piston rod of a conventional cylinder can rotate through angles of up to 120°. By connecting a rack and pinion between two pistons, a cylinder can provide a much larger rotary output. The total angle turned through is determined by the stroke length, the gear ratio, the cylinder

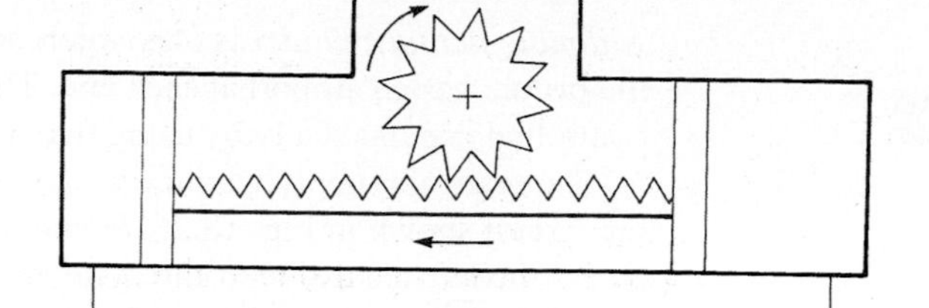

Figure 18.31 A torque cylinder.

diameter and the working pressure, and by adjusting the working pressure the output torque can be varied. This device is called a *torque cylinder*, Fig. 18.31.

An *adjustable cylinder* can be obtained in which, by sliding in the end cover of the cylinder, the stroke length can be adjusted to any smaller value. This can be useful in press operations where the full stroke length is not always required and it also conserves compressed air.

In many industrial applications requiring large forces the size of the conventional double-acting cylinder required would be excessive. An *impact cylinder* contains a reservoir which, once air is directed to the cylinder, is brought to full working pressure while the exhaust back pressure, decreases. When it is released, the piston will accelerate rapidly and, if sheet metal, for instance, is placed at the position of maximum kinetic energy of the piston rod boss, then the most effective means of punching holes can be obtained.

Control of a double-acting cylinder using a 5-port valve

A *5-port valve* can replace both the 3-port valves shown in Fig. 18.26 and will operate both strokes of a double-acting cylinder. It has one main air port, labelled 1, two cylinder ports (2 and 4), and two exhaust ports (3 and 5). The lever set-reset 5-port valve is shown in one of its two stable states in Fig. 18.32. The main air is routed from port 1 to port 2 while port 4 routes the exhaust air to port 5. When the valve is switched to its other state, the right-hand side of the diagram is shown connected and port 1 routes the air to port 4 while port 2 exhausts to port 3. A 5-port valve with two stable states can be referred to as a *5/2 valve*.

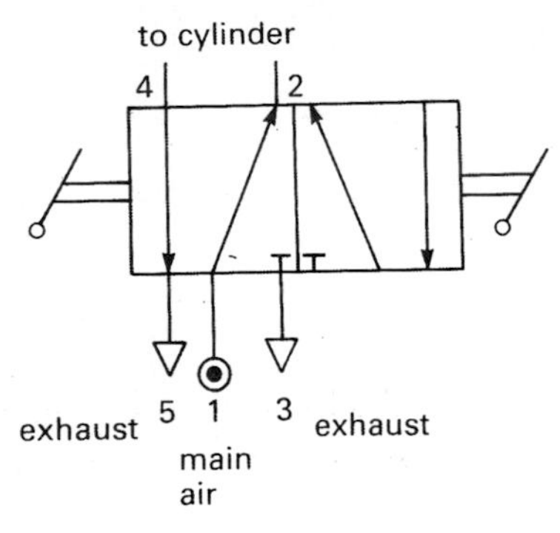

Figure 18.32 A 5-port lever set-reset valve.

Fig. 18.33 shows a 5-port lever valve controlling both phases of motion of a double-acting cylinder. The cylinder, main air and exhaust lines are only shown connected to one half of the 5-port valve symbol at a time, depending on which part is active in the circuit. The 5-port valve can also be manually operated in the different ways shown for the 3-port valves in Fig. 18.21 and 18.22.

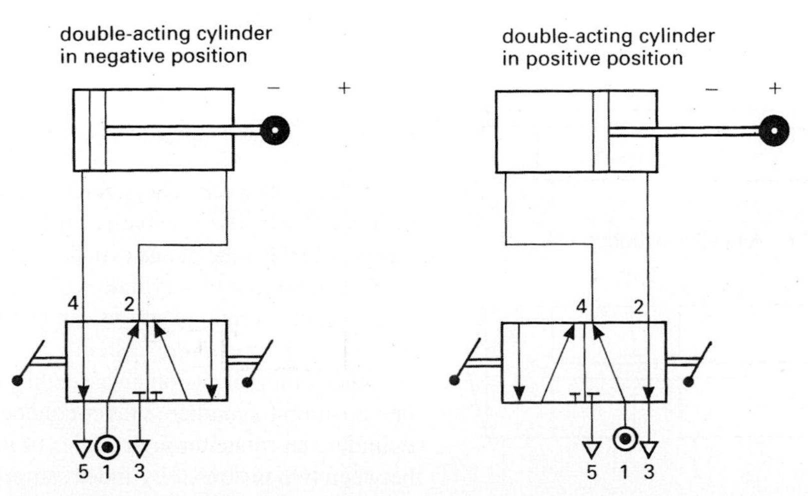

Figure 18.33 Five-port valve control of a double-acting cylinder.

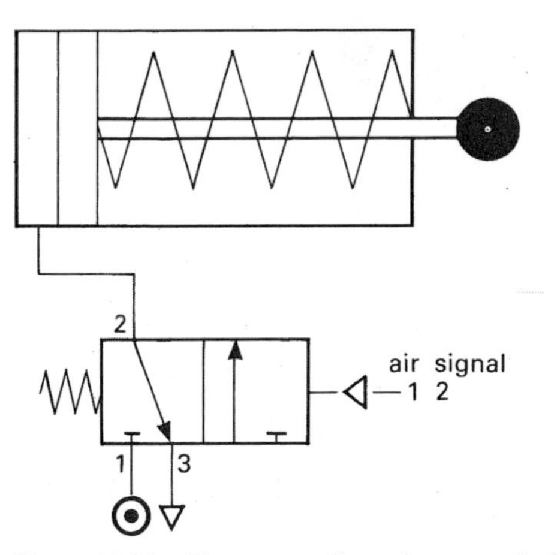

Figure 18.34 Three-port pilot valve control of a single-acting cylinder.

Control valves

In many instances, we need to operate a 3- or 5-port valve automatically by air signals, rather than by mechanical means. A *double air-pressure operated 5-port valve* or *pilot-controlled 5/2 valve* is used for this purpose, as is the air-signalled 3-*port valve*. Inside one of these valves is a spool which is moved into one of two positions by an air signal acting at either end of the valve. Fig. 18.34 shows how a 3-port control valve can actuate a single-acting cylinder after receiving an air signal.

Fig. 18.35(a), (b) and (c) show the BS/ISO symbol for the 5-port double air-pressure operated valve in each of its two possible states of operation and the alternative symbol for a 5-port valve. In both types of symbol each half of the valve has the ports connected as indicated by the nearest actuation signal, that is, ports 1 and 2 are connected next to the signal line 1 2.

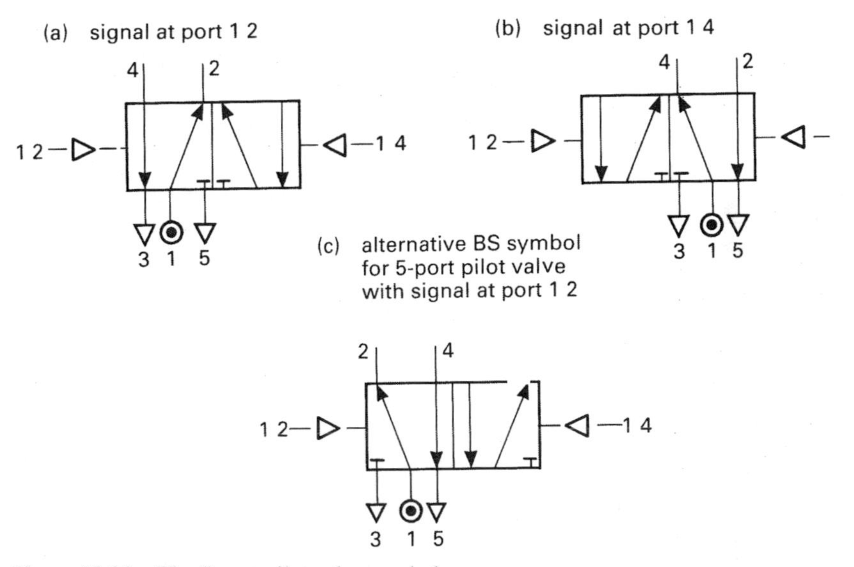

Figure 18.35 The 5-port pilot valve symbol.

It is also possible for any valves to have different methods of signalling the return although most examples so far have been spring return valves. A push-button air-return 3-port valve is illustrated in Fig. 18.36.

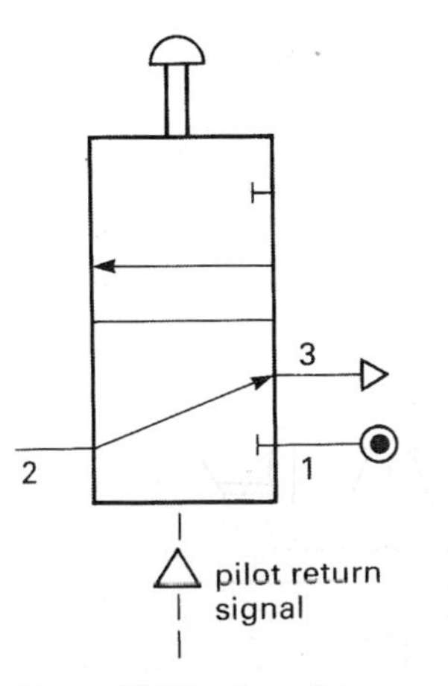

Figure 18.36 A push-button, pilot-return 3-port valve.

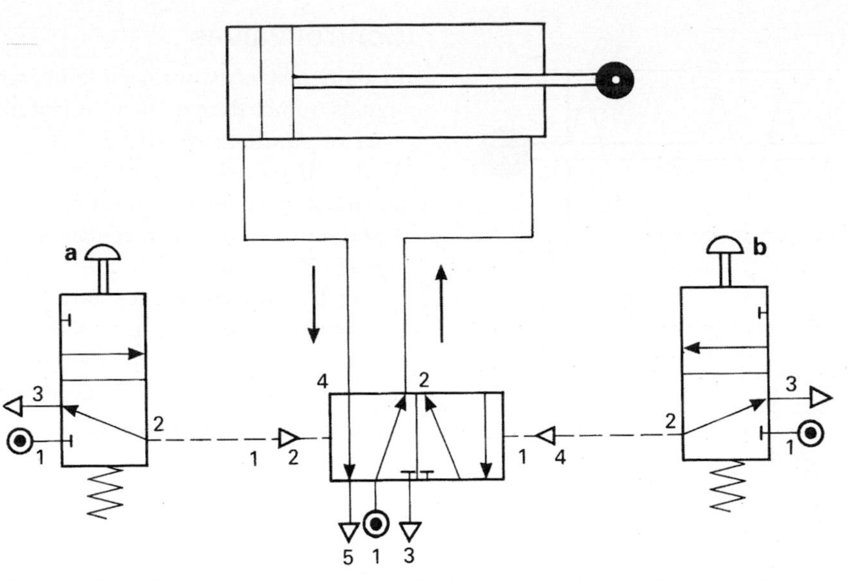

Figure 18.37 Five-port valve control of a double-acting cylinder following a pilot signal at port 1 2.

If we need to signal the control valve by two operators from remote positions we can use two spring-return 3-port valves. Fig. 18.37 shows the remote control circuit after valve **a** has been pressed. A pulse will change over the control valve to signal port 1 2 and so retract the piston until a pulse is received at port 1 4. Where there is positive supply pressure to a cylinder in both phases of its motion, as in the case of control valve operation of the cylinder, it is called a *pressure applied* system. This means that valve **a** does not have to be held down for the out-stroke signal (and thus the cylinder pressure on an external object) to be maintained. However, if valve **a** is held down it will be impossible for the control valve to receive a pulse at port 1 4 to change its output signal. Note that the control signal pipes are shown as broken lines and the air pipes connecting the control valve to the cylinder are shown as continuous lines.

18.3.2 Two-position control of cylinders

Control by either of two operators

If a cylinder is to be controlled by either of two operators, we can use a *shuttle valve*, shown in Fig. 18.38. The shuttle valve allows air to pass from a port 1 to port 2, but not straight through between the two number 1 ports and can connect two operators

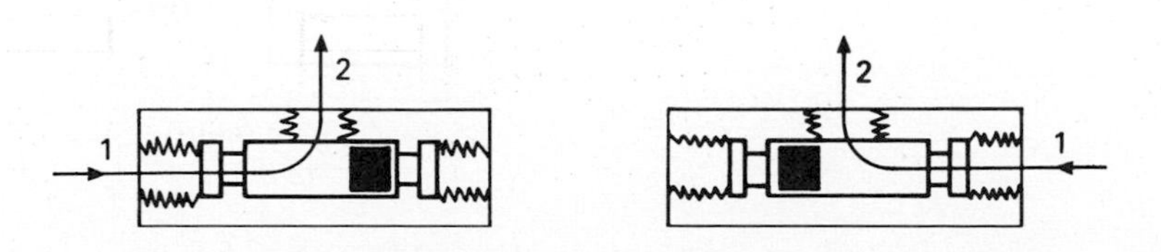

Figure 18.38 The shuttle valve.

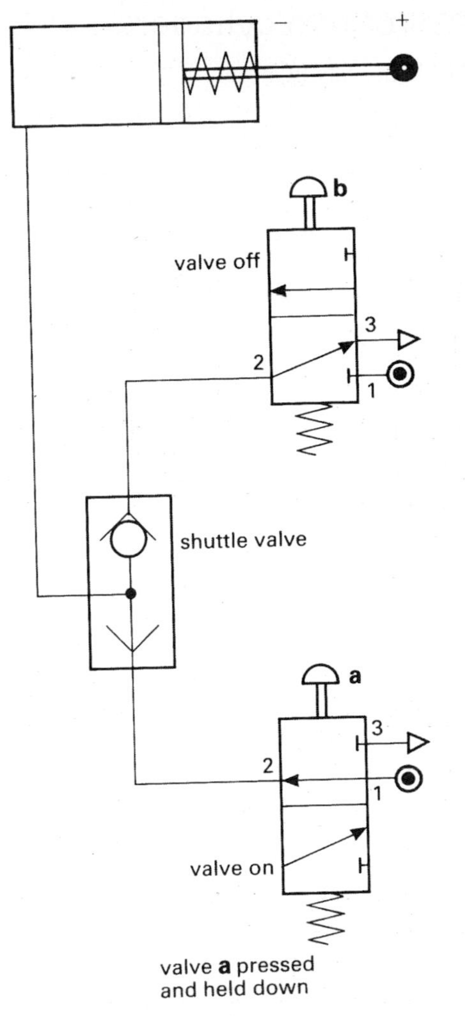

Figure 18.39 **a** OR **b** control of a single-acting cylinder.

to the cylinder as shown in Fig. 18.39. If a *T-connector*, shown in Fig. 18.40, were to be used in place of the shuttle valve it would not constrain the internal routing of the air. Air would simply take the path of least resistance so that, when either valve was pressed, the air supply would pass straight to the exhaust port of the other; consequently, the cylinder would never move.

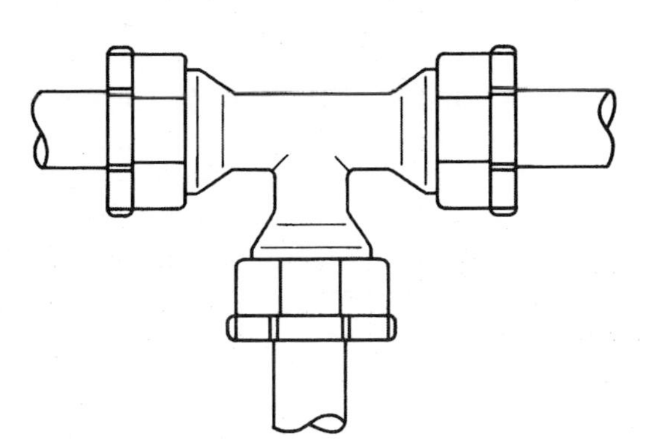

Figure 18.40 A T-connector.

Fig. 18.39 therefore shows a cylinder operated by either **a** OR **b** and it is clear that both the valves are in parallel in the circuit.

An alternative method of achieving the OR logic function circuit is shown in Fig. 18.41.

If valve **a** is pressed a signal is passed through valve **b** to the control valve. If **b** is pressed it signals the control valve directly. This circuit can be extended to four alternative positions which would achieve the logic function:

a OR **b** OR **c** OR **d**

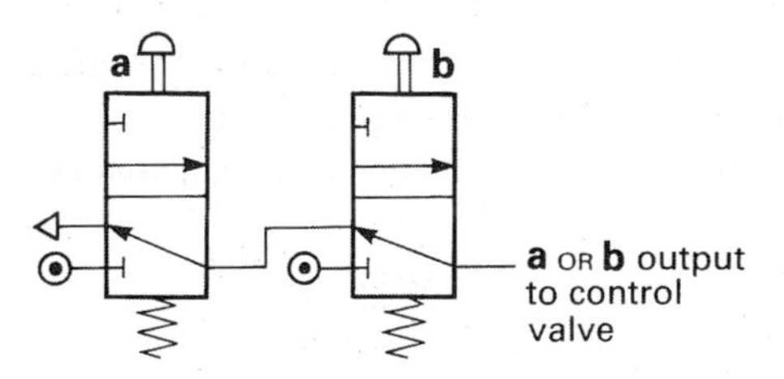

Figure 18.41 **a** OR **b**. an alternative method of two-position control.

Control by both of two operators

If we require both the operators **a** AND **b** to signal the operation of the cylinder, having checked that the line is clear, we could connect the valves in series as shown in Fig. 18.42. The main air supply is connected only to the first valve, which will then power the second valve when it is pressed. Clearly, valve **b** cannot function without valve **a** being activated.

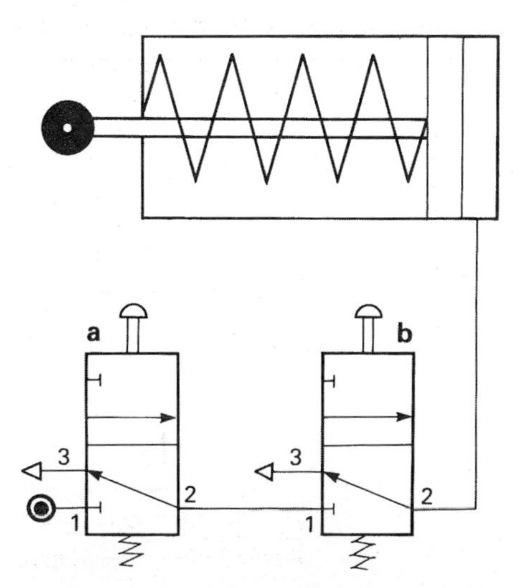

Figure 18.42 **a** OR **b**. an alternative method of two-position control.

18.3.3 Simultaneous control of two cylinders

Two or more cylinders may be operated simultaneously by means of T-connectors and equal lengths of piping. One 3-port valve could operate two single-acting cylinders as shown in Fig. 18.43, or one 5-port valve could operate two double-acting cylinders. *Simultaneous control* may be necessary when clamping a workpiece securely or lifting a horizontal platform, for example. Both cylinders would need to perform together and in the same sense (for example, out-stroke) in order to achieve either of the above tasks.

If they acted in opposite senses they could be used to open and load a furnace, as shown in Fig. 18.44. Clearly, the size, stroke length and timing of the cylinders would need careful consideration.

Unless the loading on two simultaneously acting cylinders is exactly the same and the piping is symmetrical, quite often one cylinder will complete its stroke before the other starts to move. The piston which is more lightly loaded starts to move first and its exhaust air then exerts extra back pressure on the other piston, thereby preventing its movement until the first has finished.

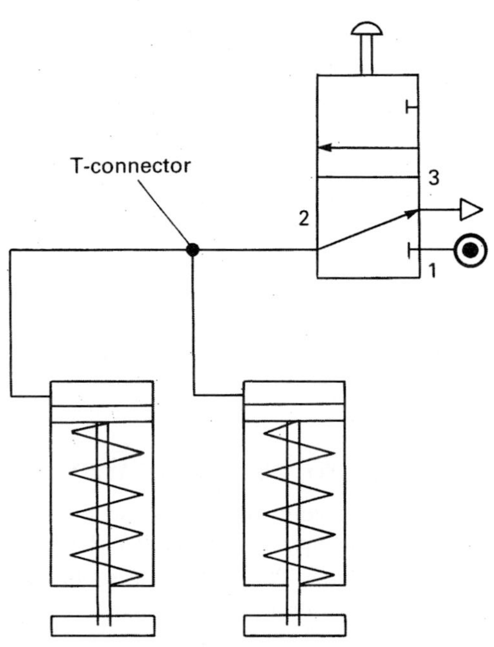

Figure 18.43 Simultaneous control of two cylinders.

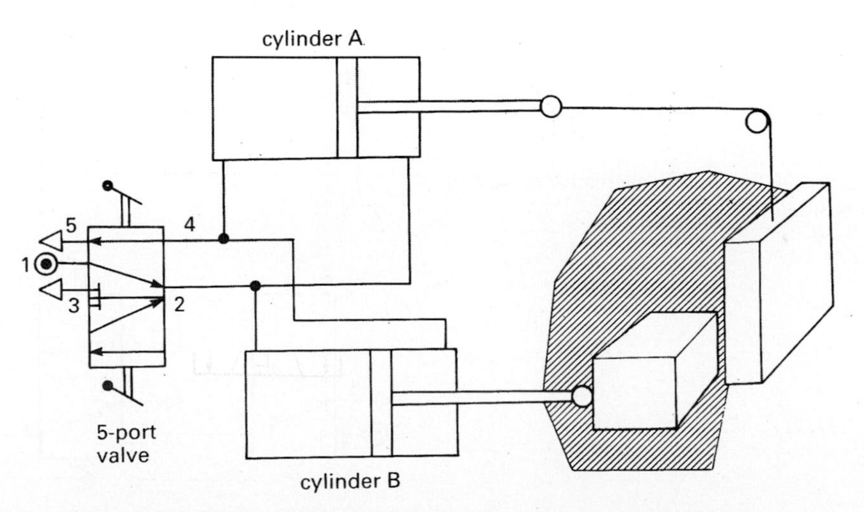

Figure 18.44 Two cylinders operated simultaneously to open and load a furnace.

Synchronisation of cylinders

The problem of synchronising cylinders may be solved by fitting *flow regulators*, as shown in Fig. 18.45. (See section 18.3.6 for details of flow regulators.) These can be used to adjust the rate of flow in each line until the cylinders are synchronised. However, they can be put out of phase again by changes of loading and friction in subsequent operations.

The best method of control is to fit individual main control valves to each cylinder so that the exhausts are not connected and the movement will essentially be in phase. The arrangement shown in Fig. 18.46 is used in many pneumatically operated machine circuits.

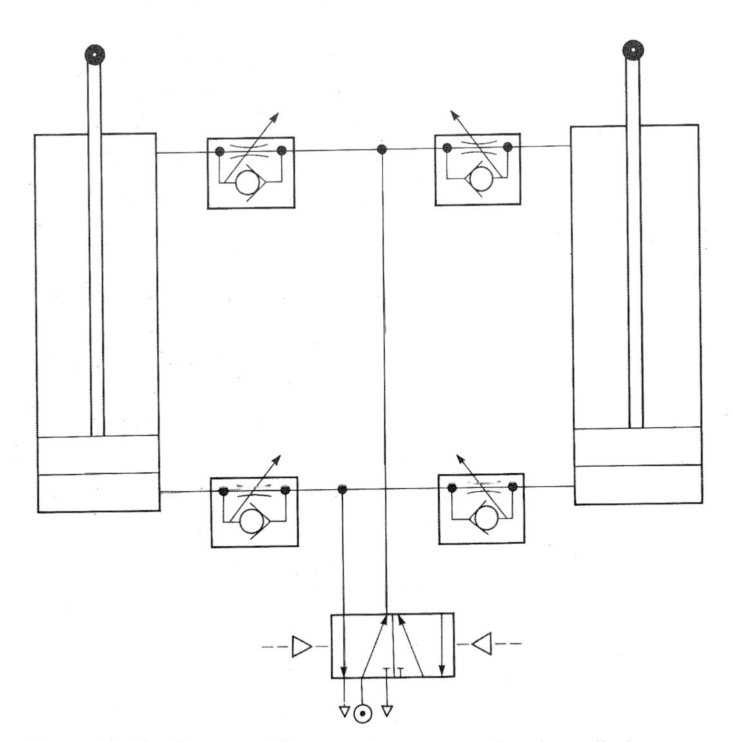

Figure 18.45 The use of flow regulators to synchronise cylinders.

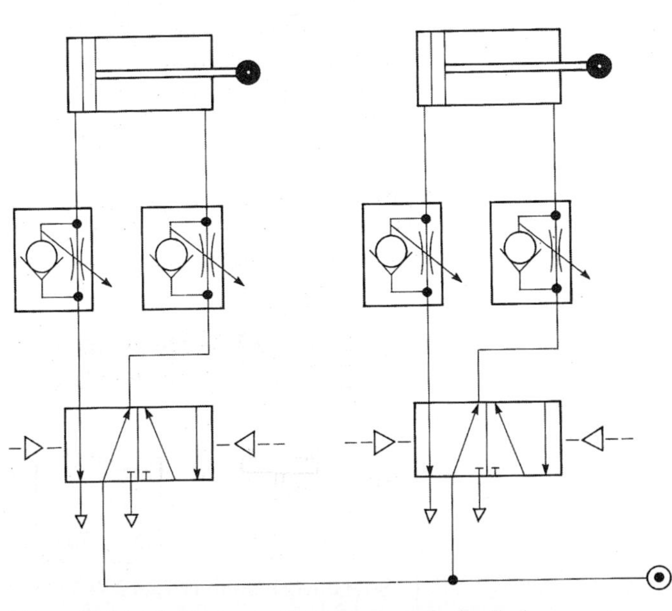

Figure 18.46 Accurate synchronous control of cylinders.

18.3.4 The force and speed generated by a cylinder

The out-stroke force of a cylinder

If we were designing a pneumatic circuit to push a component from a conveyor belt into a box we would need to ensure that the cylinder exerted the correct force, so that it neither left the component sitting on the conveyor nor projected it violently across the room. If we assume 100% efficiency in the cylinder and use the fact that

force = (pressure of supply air) × (area of piston)

that is:

$$F = p \times A$$

then, for a known air pressure and cylinder bore diameter (or piston diameter), we can calculate the force exerted on the piston.

EXAMPLE 18.1 If a cylinder has a piston diameter of 40 mm and air is supplied to it at a pressure of 600 kPa, what is the force generated on the piston?

SOLUTION The cross-sectional area of the piston,

$$A = \pi \times R^2 = \pi \times \left(\frac{40}{2}\right)^2$$

The air pressure is 600 kPa or 600 kN/m^2, but the dimensions are in millimetre. To convert the pressure to N/mm^2, divide by 10^6, so

$$p = \frac{600\,000}{10^6}\ \text{N/mm}^2 = 0.6\ \text{N/mm}^2$$

Now using $F = p \times A$,

$$F = 0.6 \times \pi \times 20^2\ \text{N} = 754\ \text{N}$$

So the thrust of the piston is 754 N.

If the other side of the piston is open to the atmosphere, the force that the cylinder could exert on an object is 754 N. In practice the force exerted will depend on the efficiency of the cylinder, which is determined by the frictional losses in the system. The efficiency is approximately 95% at pressures of 5–6 bar and no load, so that 95% of the previously calculated thrust, 716.3 N, could be generated by this cylinder.

The work done in a cylinder

The work done W in one stroke, is given by the force F multiplied by the piston stroke length L.

$$W = F \times L$$

EXAMPLE 18.2 If the cylinder in Example 18.1 had a stroke length of 100 mm and an assumed efficiency of 100%, what would be the work done on the out-stroke?

SOLUTION The force generated F was 754 N. The stroke length L is 0.1 m. Therefore

$$W = 754 \times 0.1\ \text{Nm} = 75.4\ \text{Nm or } 75.4\ \text{J}$$

So the work done is 75.4 J.

The in-stroke force of a cylinder

In a double-acting cylinder, where air is supplied to perform the in-stroke and exert a pull on an external object, the piston area that the air in the cylinder makes contact with is smaller because the piston rod occupies some of the space. This will reduce the force available.

If the internal cylinder radius is R and the piston radius is r the area of contact is now $\pi R^2 - \pi r^2$

or $$A = \pi(R^2 - r^2)$$

so $$F_{in} = p\pi(R^2 - r^2)$$

For the cylinder in Example 18.1, with a piston rod diameter of 10 mm and neglecting frictional losses,

$$F_{in} = 0.6\pi(20^2 - 5^2)\text{ N} = 706.9\text{ N}$$

Hence the in-stroke force of the cylinder is 706.9 N.

Calculation of cylinder diameter

If a cylinder with an air supply pressure of 600 kPa is required to exert 250 N on its out-stroke, the diameter of such a cylinder can be calculated as follows:

$$F = p \times A$$

Therefore

$$250 = 0.6 \times A$$

where A is the cross-sectional area of the piston in square millimetre so

$$A = \frac{250}{0.6} = 416.7\text{ mm}^2$$

$$\pi R^2 = 416.7$$

Hence $R^2 = 132.63$ and therefore $R = 11.52$ mm

The required diameter of the cylinder is therefore 23.04 mm.

However, since the efficiency of the cylinder, η, is less than 100%, a larger cylinder is necessary. Allowing for this,

$$250 = \eta \times 0.6 \times A$$

If the efficiency is 95%, then $\eta = 0.95$ and the diameter required is 23.63 mm. In practice, the next available larger size would be used.

Force exerted by a cylinder with exhaust pressure

We have looked at a mathematical model of the cylinder force out-stroke using:

force = (pressure of supply air) × (area of piston)

which ignores the fact that there may be an opposing force. For instance, the supply pressure may be p_1 but the exhaust pressure may be p_2 as shown in Fig. 18.47.

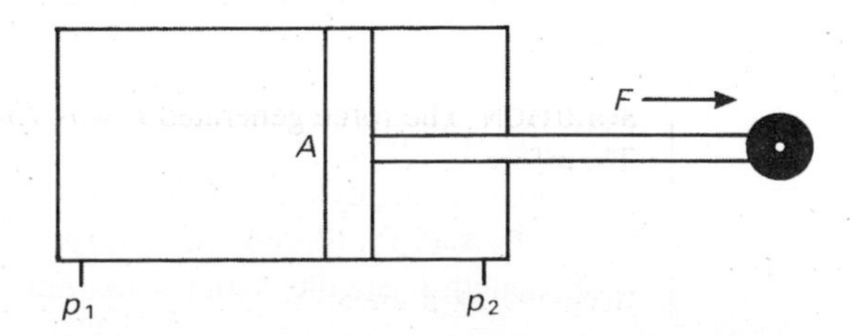

Figure 18.47 The force exerted by a double-acting cylinder.

The effective pressure is equal to $(p_1 - p_2)$, so now

$$\text{force} = A(p_1 - p_2) = \pi R^2(p_1 - p_2)$$

If the cylinder bore is 25 mm, the supply pressure 600 kPa, and the exhaust pressure 400 kPa, then the initial force is given by

$$F = \pi \times 12.5^2(0.6 - 0.4) = 98.2\ \text{N}$$

This may not be high enough initially but because the gas is exhausting continuously p_2 will fall to zero eventually. The force exerted will therefore gradually rise to a maximum of

$$\pi \times 12.5^2 \times 0.6\ \text{N or } 294.5\ \text{N}$$

If it is necessary for the cylinder to move a heavy weight quickly, then the required force should be assumed to be 150% of that at no load. The extra force will be used to overcome air and friction losses and the exhaust back pressure.

Piston speed

The cylinder speed is determined by the size of the control valve and connecting piping, the magnitude and type of load and the supply pressure. Generally:

$$V_H = \frac{\text{v.p.a.}}{\text{c.p.a.}} \times 100$$

where V_H is the average speed for a heavy load, v.p.a. is the valve port area, c.p.a. is the cylinder piston area.

V_0 is the average speed at zero load:

$$V_0 = \frac{\text{v.p.a.}}{\text{c.p.a.}} \times 150$$

V_H is normally about 300 mm/s and V_0 about 450 mm/s. In practice, the speed will vary. Initially, the exhaust back pressure will prevent the cylinder from moving as will a heavy load. As the input and output pressure differential grows, the cylinder will gain speed. The final speed can be twice the average speed. In calculating the kinetic energy of a cylinder, the maximum speed should be used.

18.3.5 Air consumption of cylinders

The compressor displacement of a piston is the amount of air displaced per minute and not the total volume of air actually delivered to the piston by the compressor. The compressor displacement for each piston D is given by:

$$D = L \times A \times N$$

where L is the length of the piston stroke,
A is the piston head area,
N is the number of piston strokes per minute.

The cost of operating a pneumatic system is determined by the amount of compressed air that is used. Consequently it is necessary to calculate the air required to operate the components.

Single-acting cylinder

The single-acting cylinder is only operated by air on the out-stroke. The volume of air V necessary for this out-stroke is given by the product of the piston area A, the stroke length L and the compression ratio of the air, c.r.

$$V = L \times A \times \text{c.r.}$$

The compression ratio is given by

$$\text{c.r} = \frac{\text{gauge pressure } P + \text{atmospheric pressure}}{\text{atmospheric pressure}}$$

When all the pressures are in bar, atmospheric pressure = 1 approximately. Hence:

$$\text{c.r.} = \frac{P+1}{1} = P + 1$$

Therefore the volume of air used in one cycle = $L \times A \times (P + 1)$.

If the diameter of the piston is d_1, where both d_1 and L are quoted in centimetres, and the piston completes n cycles in a minute, then

$$V = \frac{L \times n \times \pi \times d_1^2 \times (P+1)}{4} \text{ cm}^3\text{/min}$$

If the air consumption in litre/min is required, we can use the fact that 1000 cm^3 = 1 litre. Hence:

$$V = \frac{L \times n \times \pi \times d_1^2 \times (P+1) \times 10^{-3}}{4} \text{ l/min}$$

Double-acting cylinder

The above equation for air consumption would give the amount consumed only on the out-stroke of a double-acting cylinder. To obtain the value for the whole cycle, we would use the above model for the first half of the cycle and then use the reduced piston area to calculate the air consumption on the return stroke. If the piston rod diameter is d_2 cm then the in-stroke area, in square centimetre, is given by:

$$A_2 = \frac{\pi(d_1^2 - d_2^2)}{4}$$

Hence the volume of air consumed when operating at n cycles per minute for a double-acting cylinder is given by:

$$V = \left[L \times n \times \frac{\pi d_1^2}{4} \times (P+1)\right] + \left[L \times n \times \frac{\pi(d_1^2 - d_2^2)}{4} \times (P+1)\right]$$

$$= L \times n \times \frac{\pi}{4} \times (P+1)(2d_1^2 - d_2^2) \text{ cm}^3\text{/min}$$

18.3.6 Piston speed control

In many cases, it is necessary to control the speed of the out-stroke or in-stroke of a piston. Imagine the problems it would cause if the doors on buses and trains always closed at top speed.

Flow regulators

A *unidirectional flow control valve*, also known as a *flow regulator*, is used to control the flow of air in one direction but give free flow of air in the other direction. The regulator and its BS symbol are shown in Figs 18.48 and 18.49.

The flow regulator is usually clearly marked showing ports 1 and 2 and an arrow to show the free-flow direction. The diagram in Fig. 18.50 shows how the air inside the regulator has two alternative routes. If the air flows from port 2 to port 1 it will flow along the lower path, forcing the valve away from its seating to give unimpeded flow. However, when the air flows from port 1 to port 2 the pressure will force the valve shut, sealing this route and forcing the air through the restricted upper path. An adjusting screw varies the size of the aperture and therefore the rate of airflow along this path.

Figure 18.48 The flow regulator, a unidirectional flow control valve.

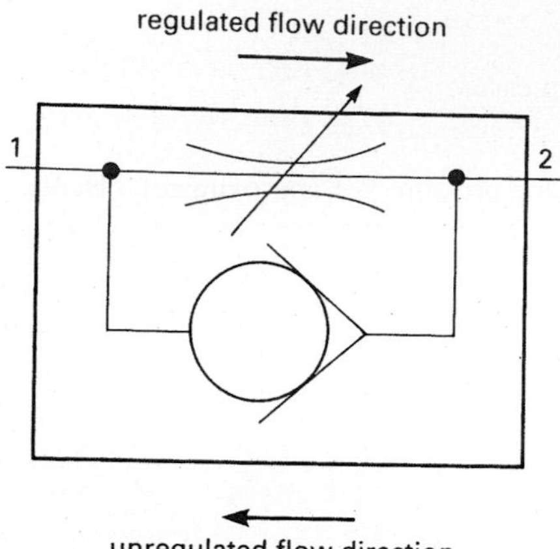

Figure 18.49 The BS symbol for a unidirectional flow regulator.

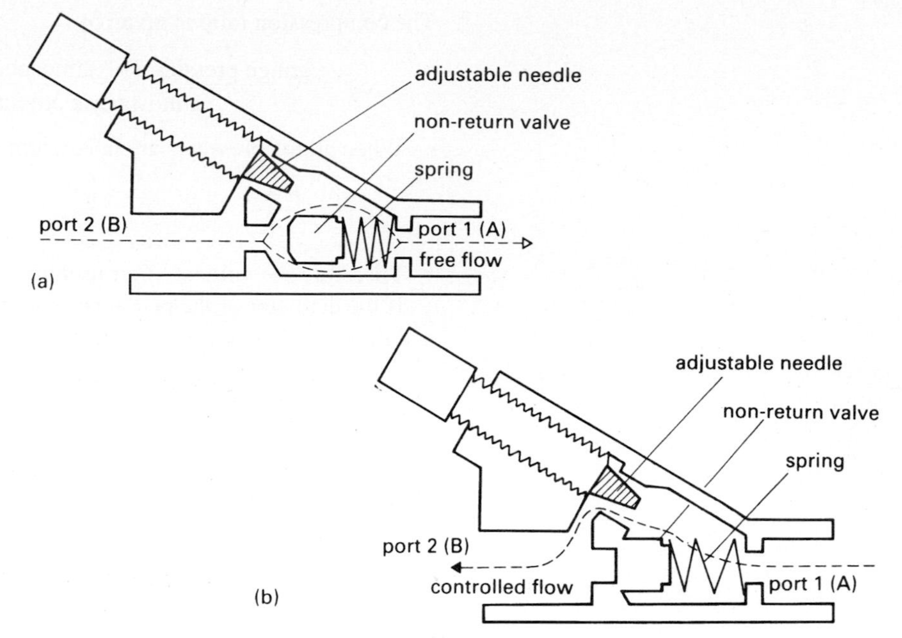

Figure 18.50 Cross-section of a flow regulator.

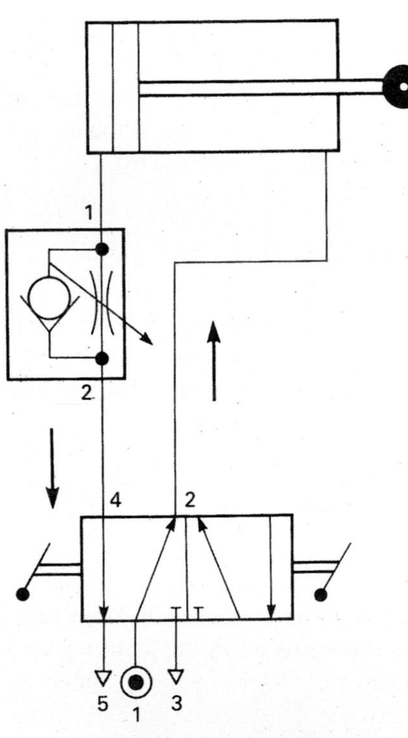

Figure 18.51 In-stroke speed control of a double-acting cylinder.

Should a double-acting cylinder, controlled by a lever 5-port valve, be required to have a slower in-stroke than normal we would connect the flow regulator in the position shown in Fig. 18.51, with port 1 connected as close as possible to the cylinder. If the adjusting screw is tightened when the piston is on the in-stroke a back pressure is created which will slow the cylinder down. When the valve is switched to the opposite position air will pass freely through the regulator into the cylinder, causing a rapid out-stroke.

Should we wish to control the piston speed in both directions for this same circuit, the best solution is given in Fig. 18.52. Here we have inserted another flow regulator to restrict the exhaust air when the piston is moving out, and so now we have controlled the speed of the out-stroke as well as the in-stroke.

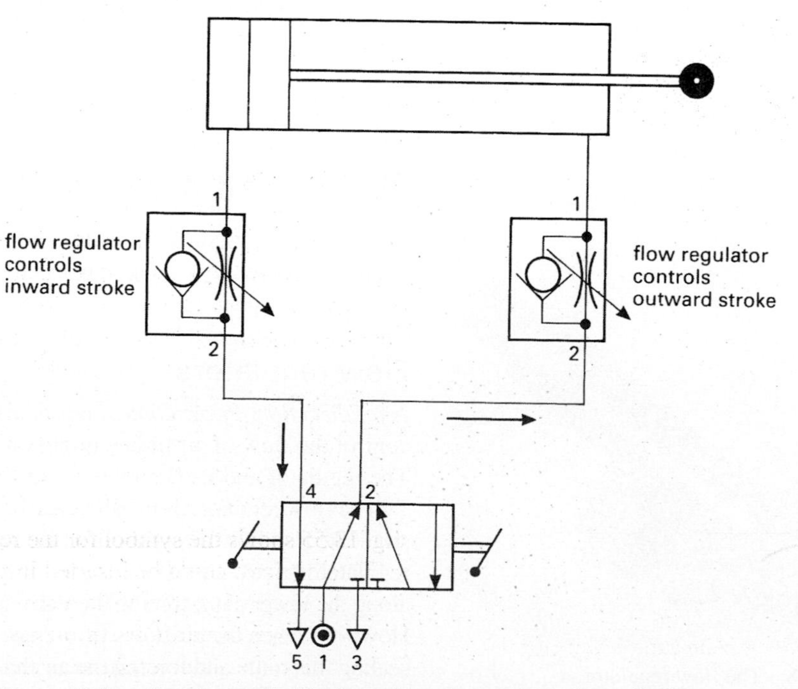

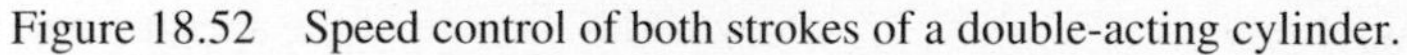

Figure 18.52 Speed control of both strokes of a double-acting cylinder.

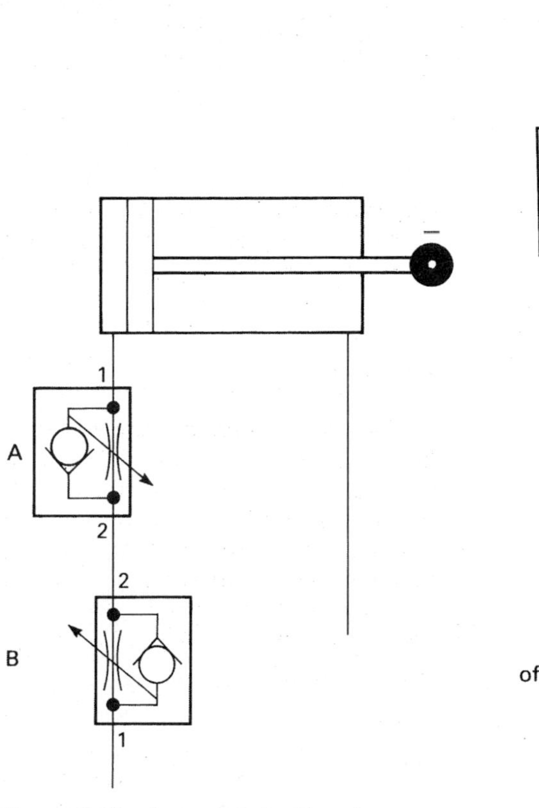

Figure 18.53 A circuit that will produce unsatisfactory speed control of both strokes of a double-acting cylinder.

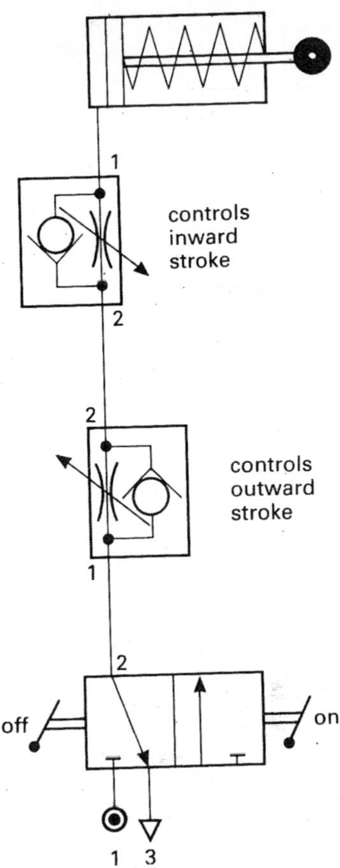

Figure 18.54 Speed control of both strokes of a single-acting cylinder.

If we had put two flow regulators 'back to back' as shown in Fig. 18.53, it might be thought that we could achieve the same speed control of the piston.

However, this does not prove a satisfactory control method because flow regulator B is controlling the air being supplied to the piston instead of the exhaust gas. When the piston is moving against a load, it will perform in an unsatisfactory manner, moving jerkily and with the speed becoming very unstable at the end of its stroke. Consequently, it is always advisable to control the exhaust air from a cylinder whenever possible as it gives smooth, stable control of the piston speed.

Should we wish to control the in-stroke and out-stroke of a single-acting spring-return cylinder, there is no exhaust gas to control on the out-stroke and it is therefore necessary to use two 'back to back' regulators as illustrated in Fig. 18.54. The spring prevents the jerky action of the piston that is apparent in a double-acting cylinder and it is an effective method to control out-stroke and in-stroke of a single-acting cylinder.

Flow restrictors

A simpler device for achieving airflow control in both directions is a *bidirectional restrictor*; this has just one route for the air, which is restricted by means of an adjusting screw. The same flow rate is therefore achieved in either direction. Fig. 18.55 shows the symbol for the restrictor.

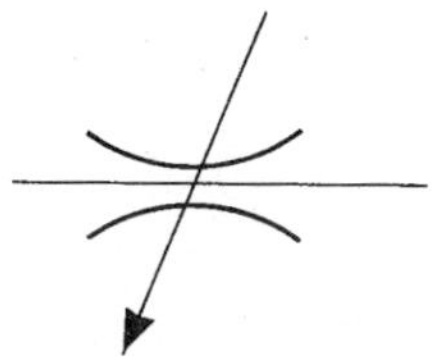

Figure 18.55 A bidirectional restrictor.

The restrictor could be inserted in place of both the flow regulators in Fig. 18.53 if the same speed control were required in both directions. This would obviously be more economical and efficient. In practice, it is usually necessary to provide independent speed control in each direction, so the two original regulators would still be required.

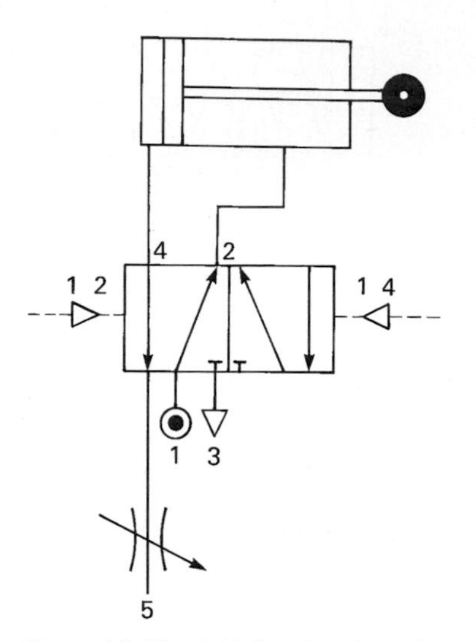

Figure 18.56 A bidirectional restrictor used to control the in-stroke of a double-acting cylinder.

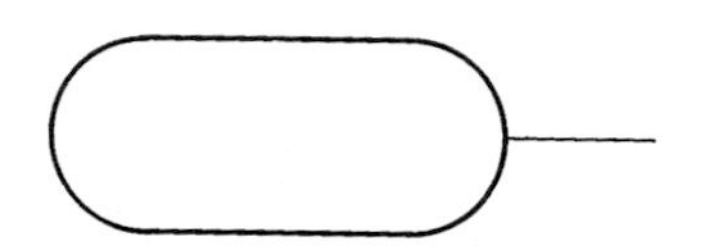
Figure 18.57 A pneumatic reservoir.

A flow restrictor can, in some cases, be used to control the air exhausted from the ports of a control valve, but not all makes of valve will allow this. An illustration of this is given in Fig. 18.56, where port 5 of a control valve is connected to a restrictor. The flow of exhaust air from the left-hand side of the piston can now be restricted and so the in-stroke speed can be controlled. This would achieve the same effect as Fig. 18.51.

18.3.7 Time delay circuits

A *time delay circuit*, in which a piston rod is held positive for a few seconds before automatically returning, is useful in clamping or pressing applications in which pressure must be applied for a predetermined time. The delay of one part of a pneumatic sequence is accomplished by using an empty cylindrical vessel called a *reservoir*. It has the same effect as a long line of tubing; the line volume is greatly increased and it takes a few seconds to bring it back to the working pressure of the system. An illustration of the BS symbol for a reservoir is shown in Fig. 18.57.

The size of the reservoir determines the time delay that can be achieved, consequently a larger reservoir will give a longer time delay. A flow regulator may also be used to restrict the flow of air into the reservoir, which results in much greater time delays being achieved. A 250 cm^3 reservoir connected in this way can provide a time delay of from 4 seconds to 32 seconds.

Fig. 18.58 shows how a reservoir may be used to hold the piston in its out-stroke for a pressing operation. When valve **a** is depressed the piston extends until it actuates valve **b** at the end of its out-stroke. The air supply is then fed through valve **b** and the flow regulator to the reservoir which, when charged, will signal the return of the piston. The piston will not return until the pressure in the reservoir is high enough to signal the control valve. Any excess air in the reservoir will exhaust through valve **b** once the piston has been signalled to retract, so that a constant delay is achieved each time.

To delay the out-stroke of a single-acting cylinder a 3-port pilot-operated control valve can be used, as illustrated in Fig. 18.59.

An automatic door system could be designed so that, when the doors opened, there was a predetermined time delay before they automatically closed again. A circuit such as that shown in Fig. 18.60, using only one operating valve, could be used. When the pressure pad is activated by a person treading on it, a signal activates port 1 4 of the double pressure 5-port valve, which then opens the door. Once the left-hand side of the cylinder is brought up to pressure and the piston has moved out, the air supply is routed through the correctly adjusted regulator and reservoir. After

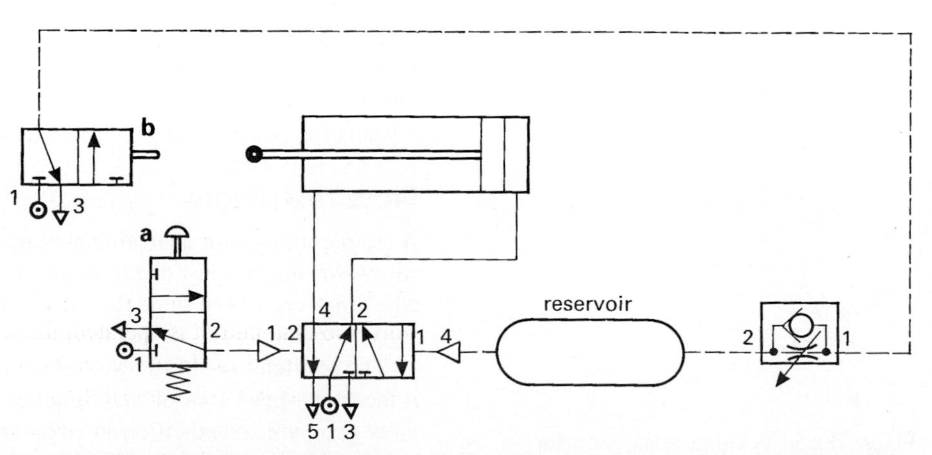

Figure 18.58 Control of a double-acting cylinder with a time delay on the out-stroke to allow a pressure operation to take place.

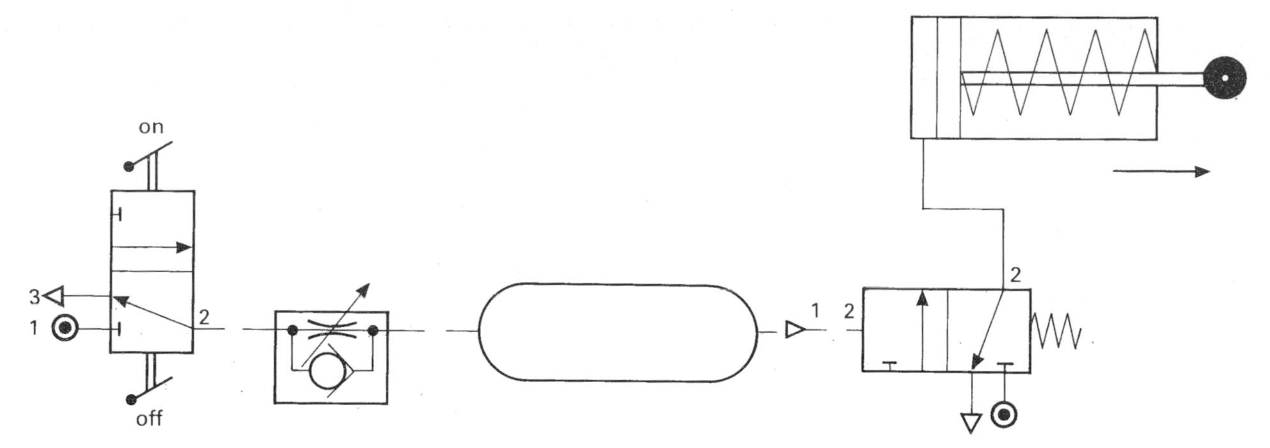

Figure 18.59 Time delay of the out-stroke of a single-acting cylinder.

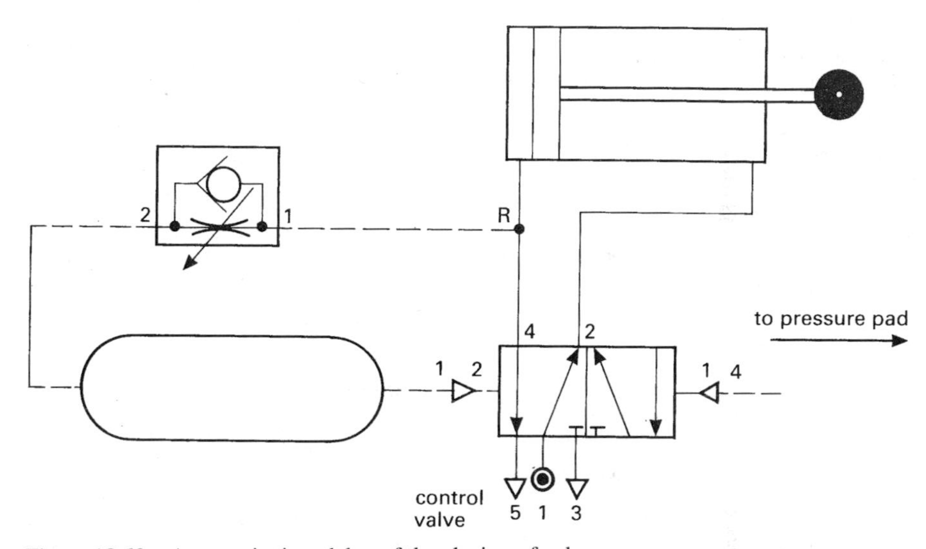

Figure 18.60 Automatic time delay of the closing of a door.

the required time delay the reservoir signals port 1 2 of the control valve and the door is signalled to close. The flow regulator should be placed after R to close the doors slowly. In practice, electrical transducers may be used to activate the control valve to open the doors, and a safety system would be required so that if a person hesitated in the doorway they would not be trapped by the closing door.

A reservoir is analogous to a water tank in a household water system or a capacitor in an electronic circuit. It builds up pressure in the same way that a capacitor stores electric charge.

Pulse circuits

A *pulse circuit* avoids the problem of conflicting signals which occur on a 5-port control valve if a pilot valve is held on longer than necessary. Such circuits are important in sequential control.

Valve **a**, a 3-port spool valve, in Fig. 18.61 has the main air supply connected to port 3 so that, until it is operated, the reservoir is kept in a charged state. When this valve is pressed, air from the reservoir will signal the control valve to operate the out-stroke of the cylinder and then will slowly bleed away to atmosphere via the flow regulator. Thus after some time the signal on port 1 4 of the control valve is removed and even if the original valve is kept depressed, the control valve can accept another signal at port 1 2.

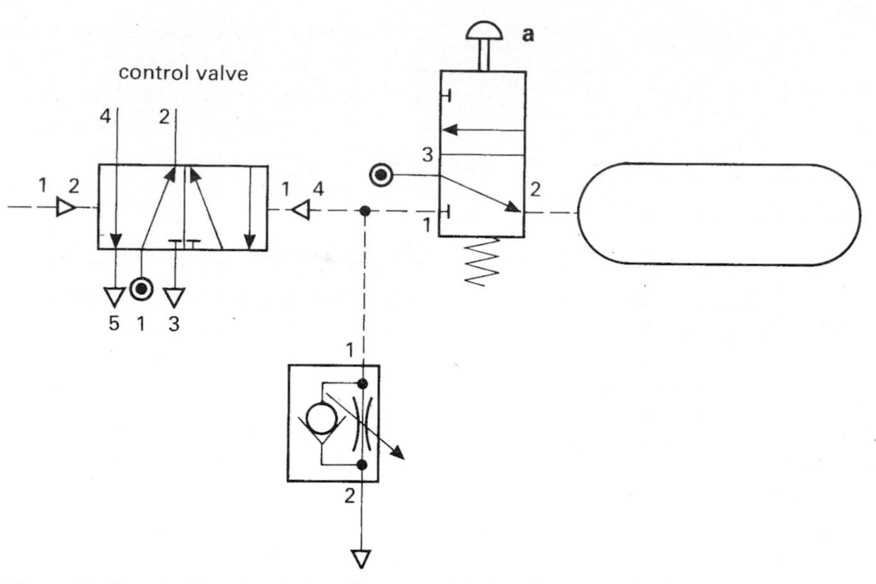

Figure 18.61 A pulse circuit signalling port 1 4 of a pilot control valve.

18.3.8 Air-bleed and pressure-sensing circuits

Pressure-sensitive valves

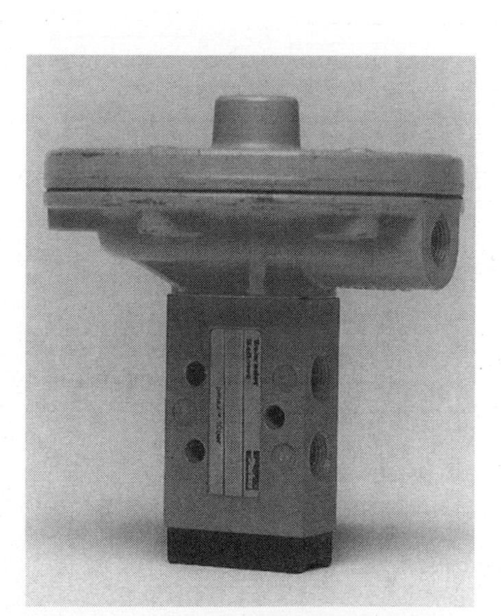

Figure 18.62 A diaphragm-operated pressure-sensitive valve.

The *diaphragm* valve or *pressure-sensitive* valve, illustrated in Fig. 18.62, was introduced in section 18.3.1 and the symbol illustrated in Fig. 18.22. It can be obtained as either a 3-port or a 5-port valve. The function of the diaphragm valve is to amplify a low-pressure signal by means of a diaphragm in order to provide the high pressure necessary to trigger the small area of spool; it can be triggered by a signal as low as 0.5 bar. It can also be operated by a vacuum port below the diaphragm, which is why there are always two pilot lines shown on the circuit diagram. The pilot line labelled 1 2 is the normal positive pressure-sensitive line.

It is also possible to obtain pressure-sensitive valves for which the pressure setting is adjustable, so that they can be operated by a variation of signal pressure or by a set difference in pressure. These adjustable pressure-sensitive valves are useful either for controlling time delays accurately or for ensuring that the next phase of an operation, such as pressing or welding, is only activated when the system is at the correct pressure.

Air-bleed circuits

The diaphragm valve can be used effectively in conjunction with an *air bleed*. This is a device which allows low-pressure air to escape to atmosphere. When it is blocked off or occluded the line becomes pressurised and signals the diaphragm valve to operate the circuit.

The type of circuit shown in Fig. 18.63 is used to signal the arrival of cars on a garage forecourt. A restrictor feeds low-pressure air to an air bleed. Only when this tube is blocked will the necessary pressure build up to activate the pressure-sensitive valve to channel the air supply from port 1 to port 2 and make the single-acting cylinder move on the positive out-stroke via the control valve to sound the bell. The spring then retracts the piston when the car has passed over the pressure pad.

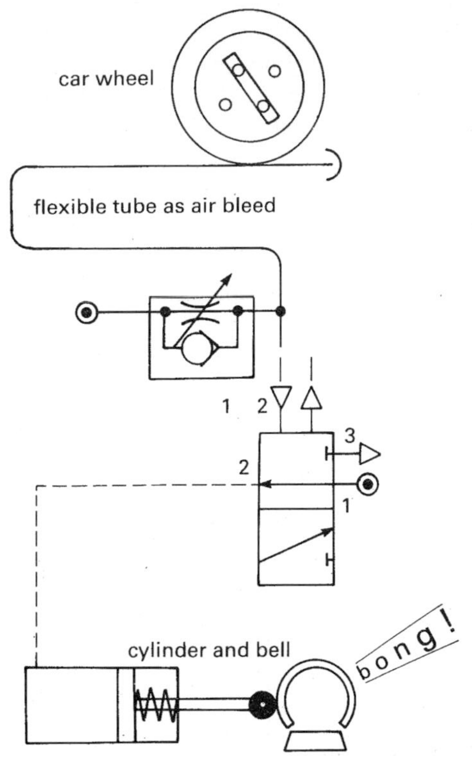

Figure 18.63 An air-bleed signalling system.

Air-bleed circuits can be used in machine safety circuits to ensure that the system operates only when a safety guard is in place, or in small panel pressing operations to indicate when the metal has been pressed to the extremities of the die.

A *pneumatic counter* can be used with an air-bleed circuit, Fig. 18.64, and will record a signal whenever an air bleed is blocked and released. It is safe to use in chemical and explosive plants, unlike its electrical equivalent.

Pressure-sensing circuits

The diaphragm valve can be used to sense the decay of the exhaust in a cylinder as the piston reaches the end of its stroke. In order to do this, the main air supply must be connected to port 3 of the valve and air is allowed to exhaust from port 1. The diaphragm valve can only be connected in this way if it is a spool valve, recommended for this use, and not a poppet valve. When connected in this way it can be considered to be acting as a NOT logic gate, because an output is obtained when there is no incoming signal and is cancelled when a pilot signal is received.

When the push button on the 3-port valve in Fig. 18.65 is pressed and released, a signal is sent to port 1 2 of the double pressure-operated 5-port valve. This will send the piston positive, and the flow regulator will control the exhaust air and hence the piston speed. The flow regulator also maintains a pressure at X until the piston nears the end of its out-stroke. When this pressure has fallen to below 0.5 bar the diaphragm valve switches to open the route between its port 2 and port 3. Now it can send a signal to port 1 4 of the pilot control 5/2-valve to return the piston.

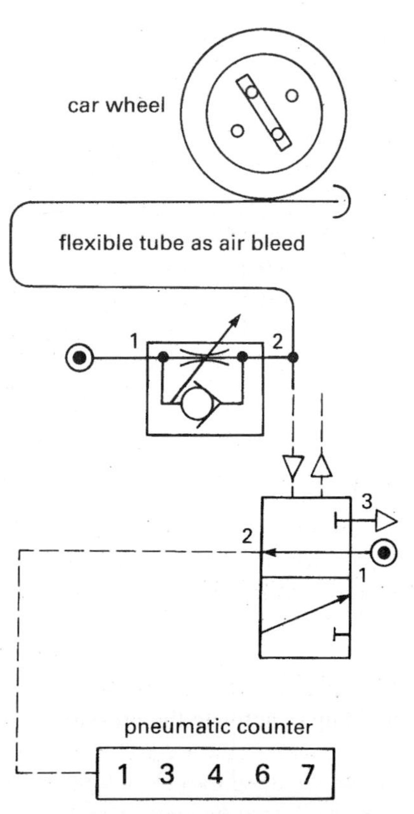

Figure 18.64 An pneumatic counter used in conjunction with an air bleed.

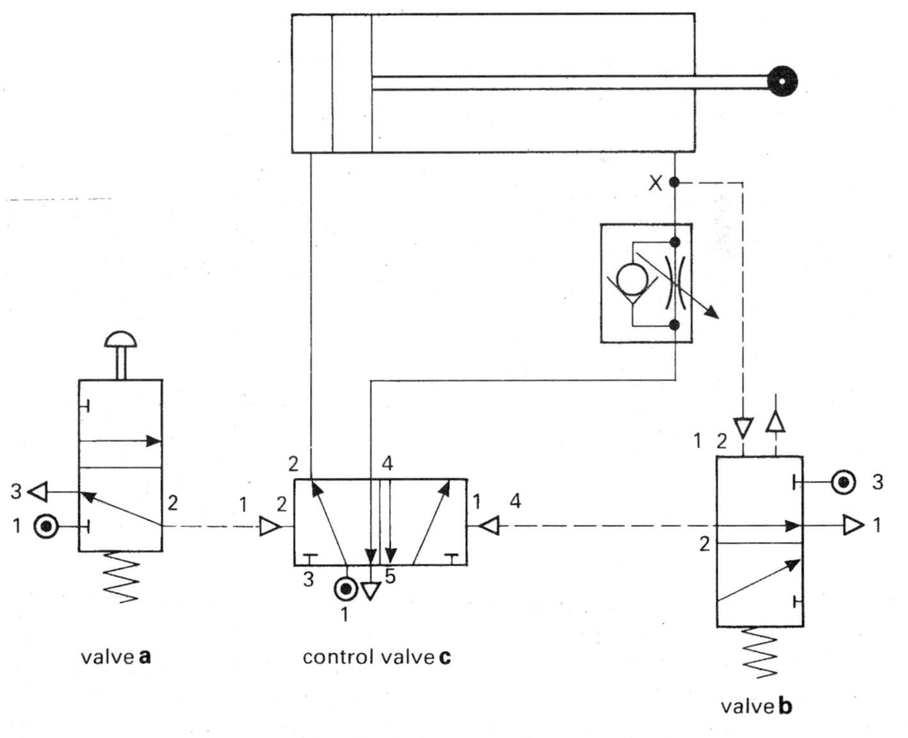

Figure 18.65 A pressure-sensitive circuit for automatic return of a piston.

The actuation of the valve by a decay of pressure can also be achieved by using the vacuum port; in this way a reverse signal is obtained by applying a signal to the opposite side of the diaphragm when the main air supply and exhaust ports are connected normally.

Solenoid valves

Figure 18.66 A solenoid-operated 3-port valve.

For activation of a pneumatic circuit by an electrical signal, 3-port and 5-port solenoid valves (Fig. 18.66), first introduced in section 18.3.1, are needed. They can be activated in less than 1/40 second by a signal from a variety of electrical devices such as microswitches, reed switches, thermistors, photoelectric cells and variable resistor circuits. The movement of a spring-loaded armature is used to operate the valve, as shown in Fig. 18.67. When no current flows the armature prevents the air supply from passing to port 2, but when the electrical signal is given, the armature moves up towards the centre of the coil. This allows a free path for the supply air to port 2, thus converting the original electrical signal into a pneumatic signal. When the electrical signal is removed, the armature springs back to its original position and shuts off the air signal.

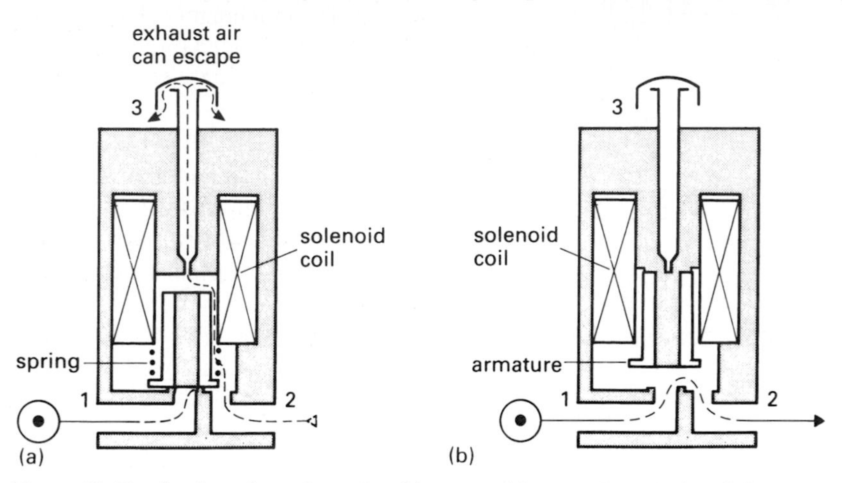

Figure 18.67 Sections through a solenoid-operated 3-port valve (a) closed (b) open.

Solenoid valves are useful because electrical signals can be transmitted over large distances faster than pneumatic signals, electrical components are cheaper and more efficient, and the circuits use less energy. They can be obtained for different working voltages, from 6 volt to 50 volt d.c. or a.c., and for a range of frequencies. The voltage should always be shown on the circuit. For safety reasons, 12-volt d.c. solenoids are usually used in educational applications.

Fig. 18.68 shows how a solenoid valve is used to control a single-acting cylinder.

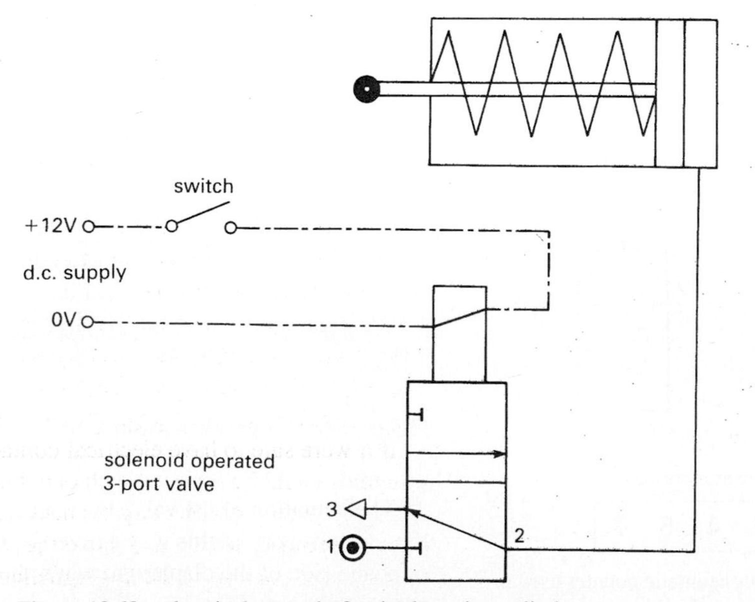

Figure 18.68 electrical control of a single-acting cylinder.

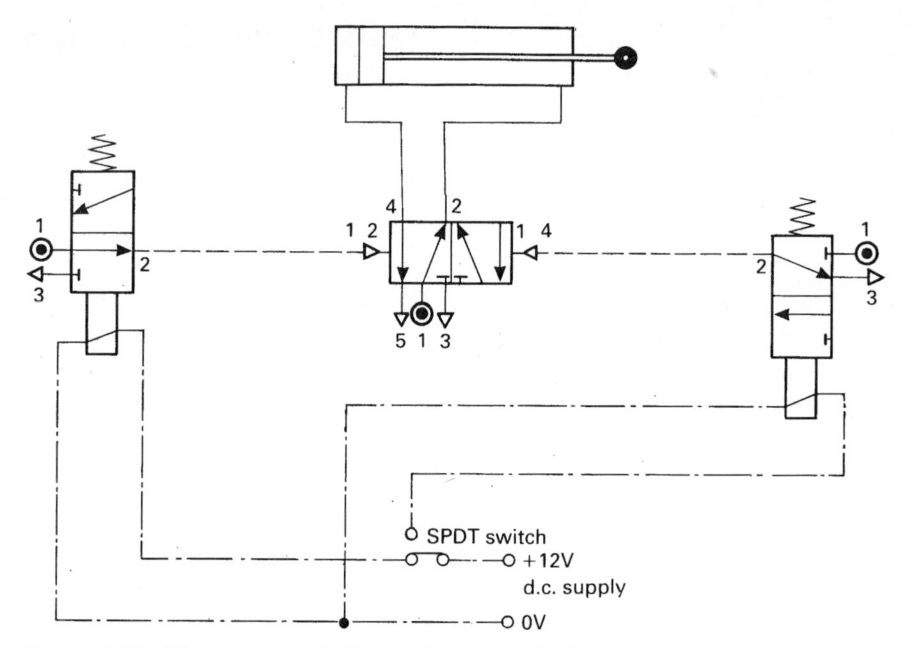

Figure 18.69 Electrical control of a double-acting cylinder.

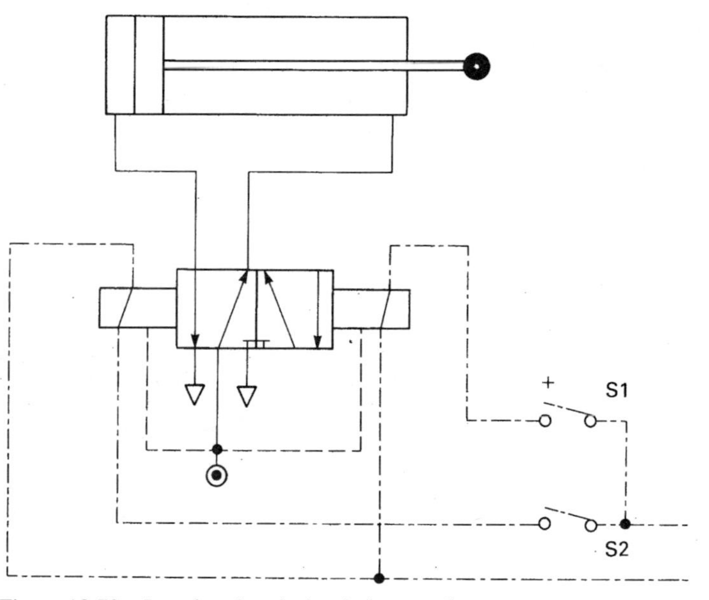

Figure 18.70 Impulse electrical switch control.

A double-acting cylinder could have a manual control on one direction of motion and a solenoid on the other, or each direction could be controlled by a 3-port solenoid valve with its own switch.

If it is necessary to control a simple reciprocating operation in a paint factory where dangerous chemicals necessitate remote control, the electrical signals to the two solenoids could perhaps be combined in a single-pole double-throw switch. Fig. 18.69 shows how this switch operates each action of the piston alternately.

If it were safe to have electrical contacts close to the operation, the two 3-port solenoids could be replaced with one double-solenoid operated valve or electric relay, Fig. 18.70. Momentary connection of electrical switch S1 causes the right-hand solenoid to actuate and move the cylinder in a positive direction. On opening S1 the valve remains in the last signalled position until a signal is received from switch S2, when it will cause the in-stroke of the piston.

18.4 Control circuits

In many industrial applications, such as polishing machines and the movement of components from conveyors, automatic circuits that will operate continuously are required. In others, a sequence of operations is required that will automatically perform the elements of the sequence in the correct order. The next sections look at how to use the pneumatic components to achieve the various methods of automatic control.

18.4.1 Automatic control

Semi-automatic control

The manually operated pilot control of a double-acting cylinder that was looked at in Fig. 18.37 is shown again in Fig. 18.71 with valve **a** positioned at the end of the piston stroke. A boss on the piston rod now activates the pilot valve to signal the 5-port pilot-controlled valve. The in-stroke is now signalled automatically, although the out-stroke is still manual.

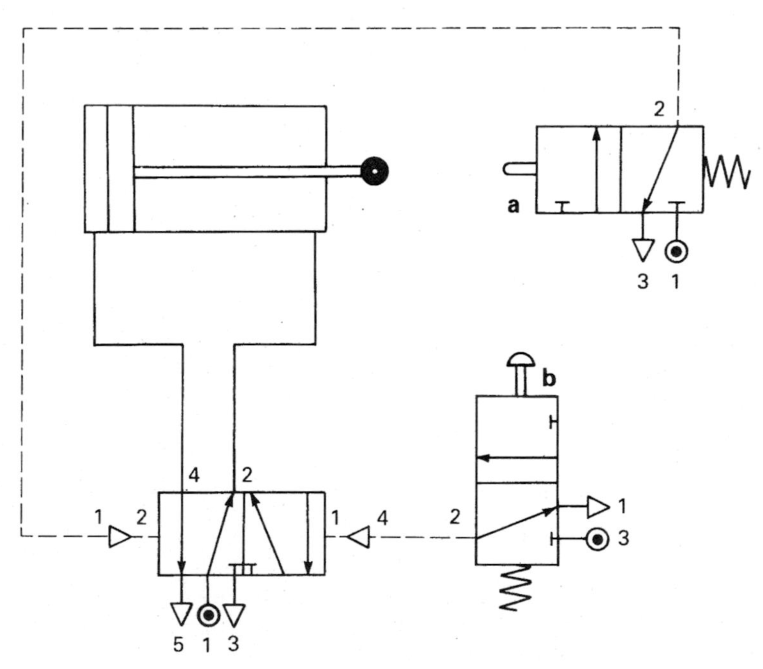

Figure 18.71 Semi-automatic control of a double-acting cylinder.

This is now a semi-automatic circuit. In a semi-automatic operation, either the in-stroke or the out-stroke of the piston is automatic. Such a semi-automatic circuit could equally well be electrically activated instead of using the mechanically operated pneumatic valves used previously. Valve **b** becomes an electrical switch to a solenoid relay; valve **a** is replaced by a microswitch which, when activated, signals the other port of the solenoid relay to return the cylinder. This new circuit is illustrated in Fig. 18.72.

The pressure-sensing circuit in Fig. 18.65 is another form of semi-automatic control as the piston returns automatically after the pressure on the rod side of the piston falls to the required level.

Interlocking

Sequential circuits can be designed so that each operation is dependent on the completion of the previous one. This facility is called *interlocking* and should be designed into circuits where possible. Interlocking serves to:

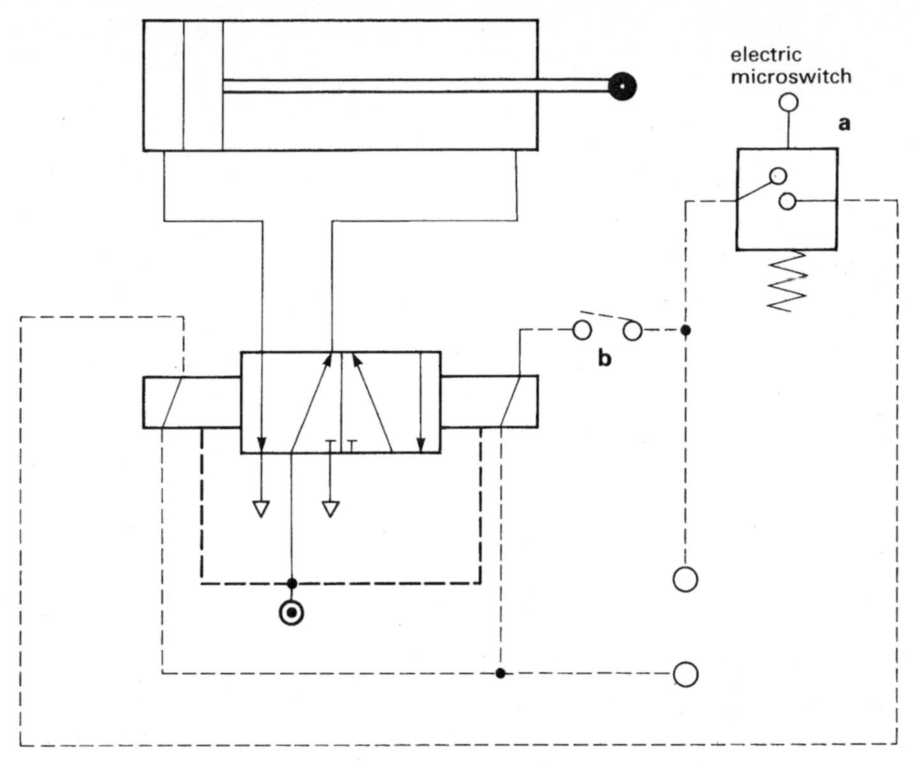

Figure 18.72 Semi-automatic control of a double-acting cylinder.

- minimise human intervention and possible error,
- lessen the risk of injury to operators,
- avoid damage to machinery caused by interrupting a cycle.

A manually initiated interlock cycle for the sequence where cylinder A performs an out-stroke, then cylinder B an out-stroke, A an in-stroke and finally B an in-stroke is represented in Fig. 18.73. This sequence can be represented as:

A+, B+, A−, B−

When a signal is received at port 1 2 of control valve **a**, main air is directed to port 2 of this valve, which in turn will pressurise the cylinder A, send the piston positive and then pressurise port 1 of valve **c**. Therefore, it is only after cylinder A has gone

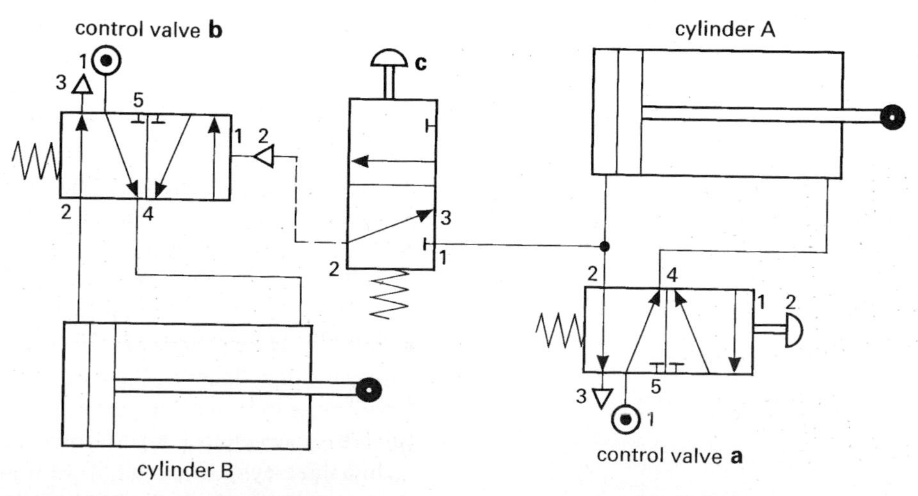

Figure 18.73 An interlock circuit.

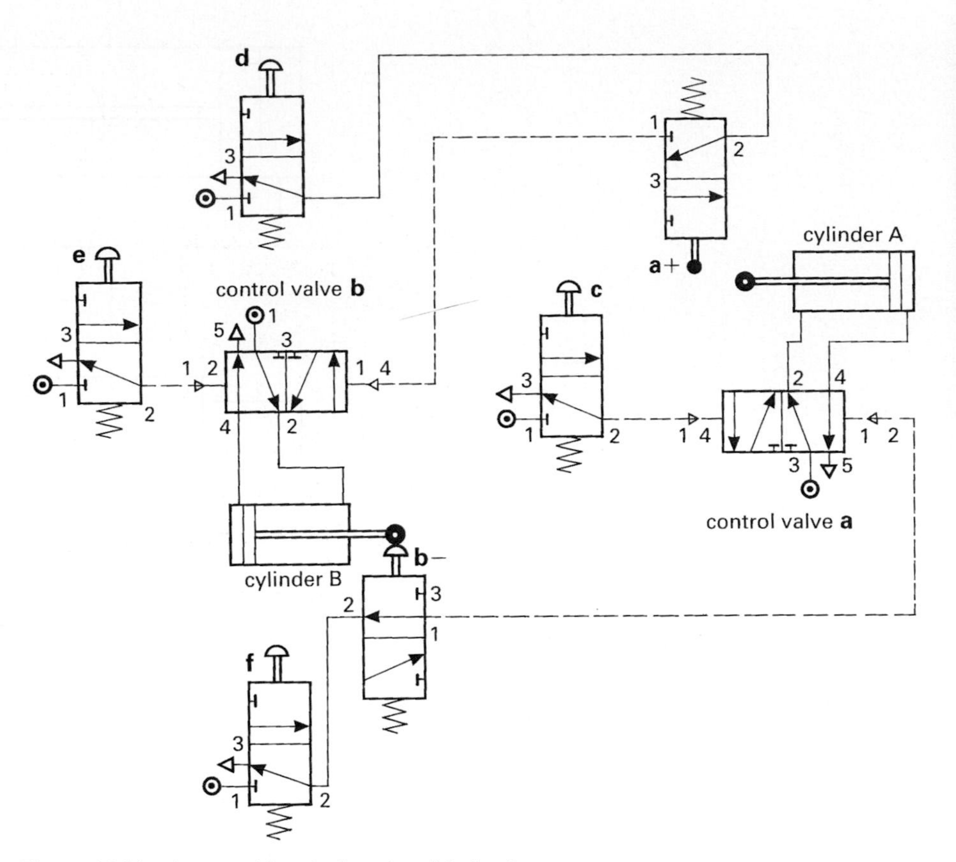

Figure 18.74 A manual interlock sequential circuit.

positive that valve **c** can be activated to send a pilot signal to control valve **b** and send cylinder B positive. When the signal on port 1 2 of control valve **a** ceases the spring return on this valve ensures that cylinder A is signalled to perform an in-stroke, the air supply to valve **c** is discontinued and so control valve **b** switches back to signal port 1 4 and cylinder B's in-stroke.

Fig. 18.74 shows a manual interlock circuit for the sequence:

$$A+, B+, B-, A-$$

If valve **c** is depressed then control valve **a** will signal cylinder A to go positive and contact valve **a**+. Only when this movement has been accomplished can valve **d** signal control valve **b** to move cylinder B positive. In the other half of the sequence you will see that A− cannot be signalled before B− as the roller-trip valve **b**− must be in contact with the retracted cylinder B before **f** can signal control valve **a** to retract cylinder A.

Fig. 18.75 shows the semi-automatic circuit equivalent to the manual circuit shown in Fig. 18.74 for the sequence A+, B+, B−, A−. The circuit is still interlocked so that each part of the sequence cannot start until the one before has been completed, but now valve **c** is the only manual actuation and it signals either A+ or B– in its two separate modes. The rest of the sequence follows automatically from the previous movement. When control valve **c** is switched to 1 2, control valve **a** actuates the positive movement of cylinder A. Only when cylinder A has moved to its full out-stroke position and actuated **a**+ can it signal control valve **b** to actuate the out-stroke of cylinder B. If control valve **c** is switched over to 1 4 then control valve **b** will cause cylinder B to perform an in-stroke and only when contact is made with valve **b**− can cylinder A perform its in-stroke.

In a three-cylinder circuit, if we wished a third cylinder, C, not to make its out-stroke until two other cylinders had completed their positive movement, then two

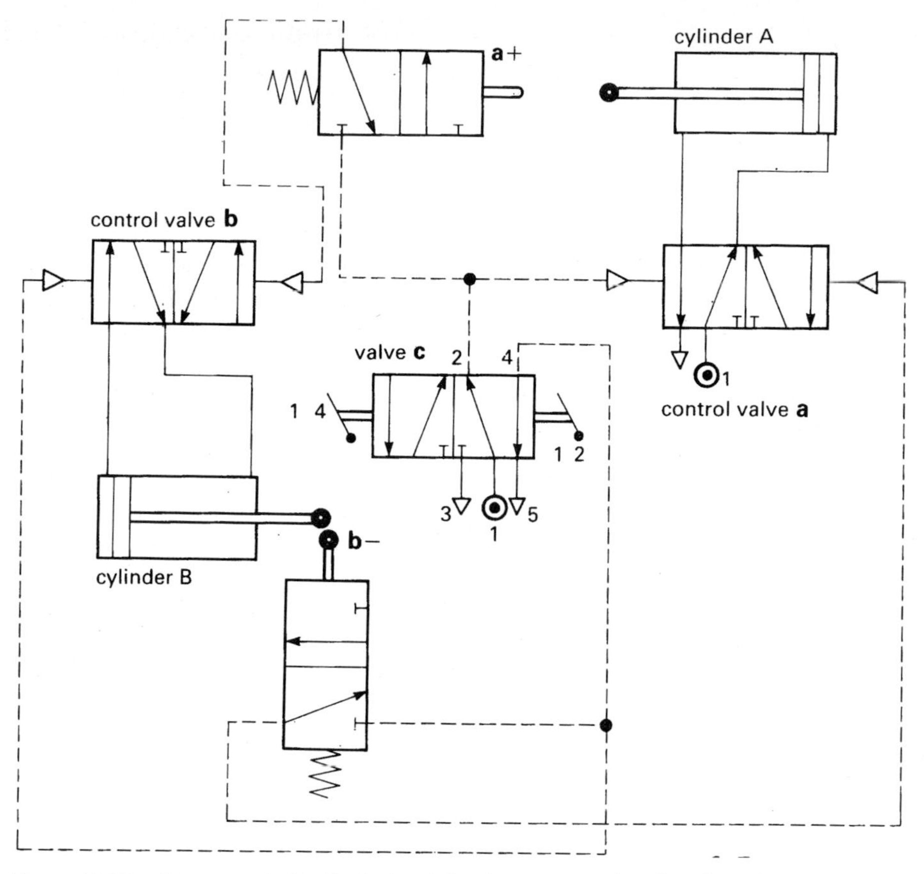

Figure 18.75 An automatic interlock circuit for the sequence A+, B+, B−, A−.

roller-trip valves could be placed at the outer limit positions of cylinders A and B and connected to provide a three-way AND function with the operator's signal which will actuate the 5/2 valve controlling cylinder C's out-stroke. This is illustrated in Fig. 18.76.

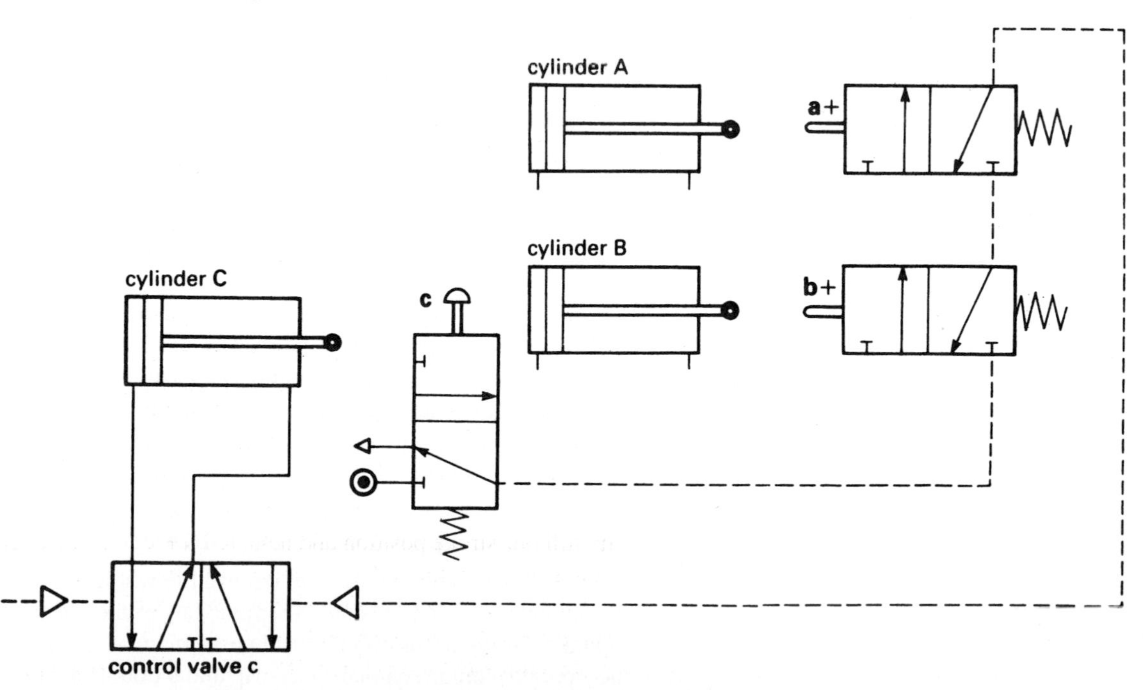

Figure 18.76 A three-cylinder interlock sequence.

Automatic oscillating cylinder

In the same way that valve **a** in Fig. 18.71 was positioned so that the cylinder would trip the valve and automatically return, we could change valve **b** to a mechanical trip 3/2-valve and automatically signal the out-stroke. For an automatic circuit the valve that is tripped by cylinder A when it is positive is termed **a**+ and the valve tripped by cylinder A when it is negative is termed **a**−.

The circuit in Fig. 18.77 is now fully automatic and will operate continuously without any manual intervention. However, it must be possible to stop and start the cycle if a problem occurs. Valve **c** is the interrupt valve. If this is in the ON position and valve **a**− is tripped, the cycle will start. If valve **c** is switched OFF at any time the cylinder will automatically come to rest after **a**− has been activated and the cylinder is in the negative position. An interrupt valve is normally placed in a fail-safe position in order to minimise the danger. If it were necessary to stop the cylinder in the out-stroke position valve **c** could be placed at point X on the circuit or it could be used to route the supply air to **a**+.

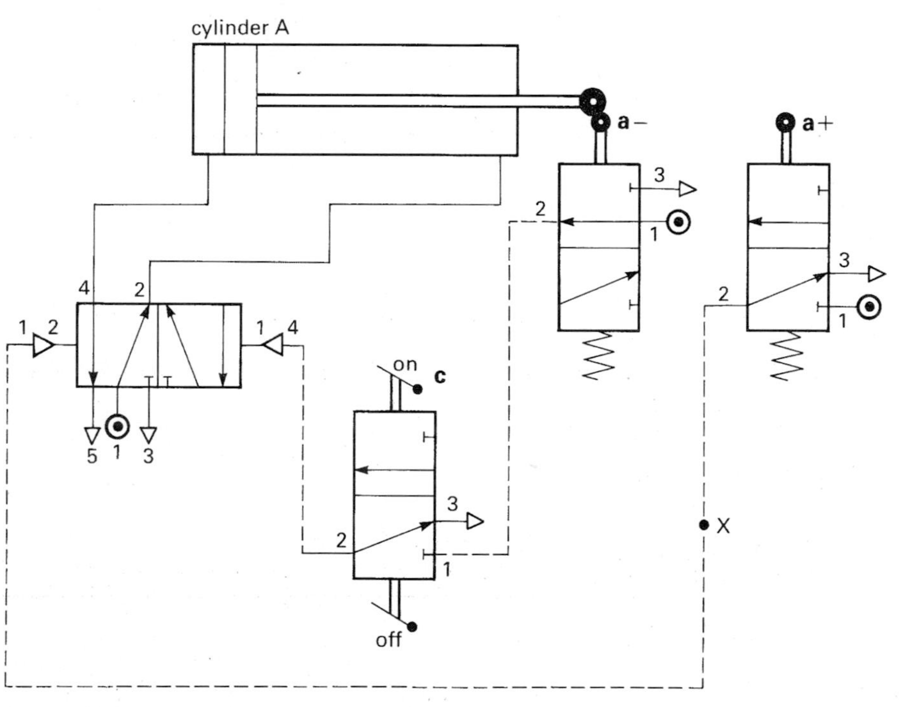

Figure 18.77 Automatic reciprocating cylinder.

Self-cancelling signals

In the above circuit it can be seen that the pilot signalling valves signal an operation that will remove their own signal, that is, the valve **a**+ signals for the piston to retract and therefore ceases to maintain the signal on **a**+. Had this not been the case, the subsequent signal from valve **a**− could not have activated the control valve.

Self-cancelling signals are essential in an interlocking circuit where conflicting signals on a pilot control valve will prevent the sequence being completed.

Part-stroke speed control of automatic circuits

Flow regulators will enable us to control the whole out-stroke or in-stroke of a piston, but very often it is only the final part of the out-stroke that must be controlled to make the system as efficient and cost-effective as possible. To do this we would incorporate a unidirectional roller-trip and a double air-pressure operated 3-port valve, as in Fig. 18.78.

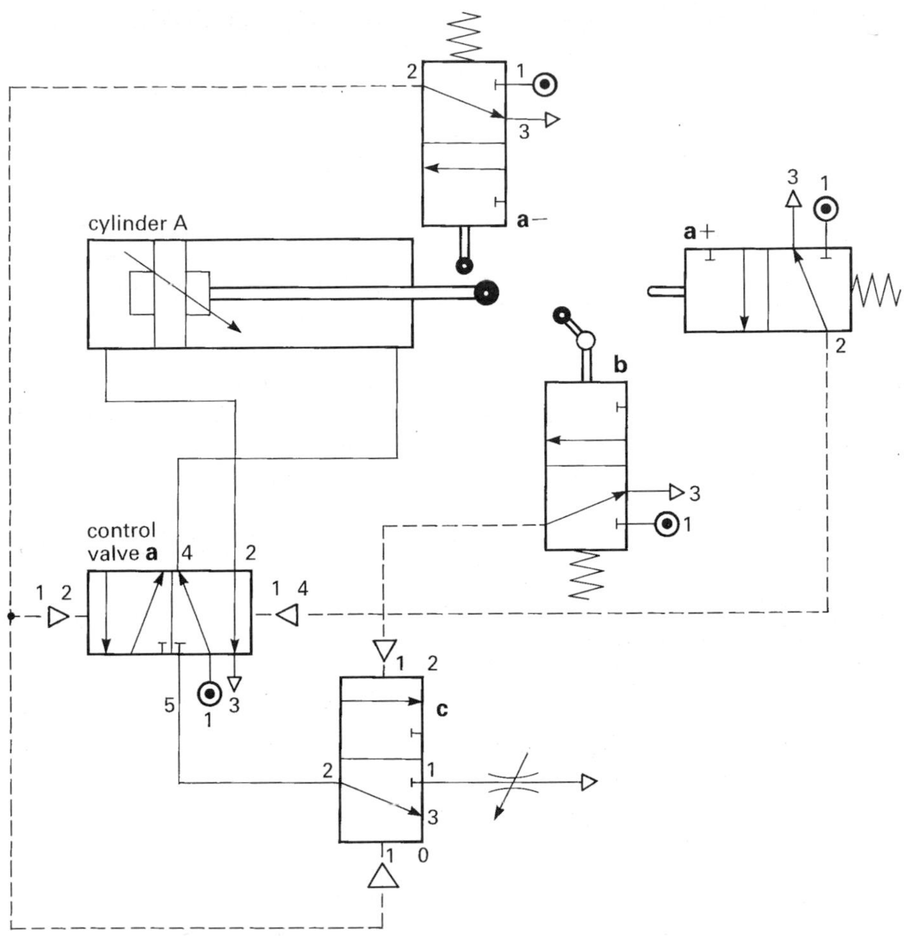

Figure 18.78 A part-stroke speed control unit.

Valve **a**− signals the out-stroke of the cylinder, which will then contact valve **b** at the position that it is required to slow down; if this is a unidirectional roller-trip it will only operate in one direction. We have positioned it to send a signal to valve **c** on the out-stroke only. This will then cause the exhaust port 5 from control valve **a** to be restricted and hence back pressure on the right-hand side of the piston will cause it to slow down. The piston next strikes **a**+ and moves inward at full speed, as exhaust port 3 of control valve **a** is unrestricted. When valve **a**− is tripped a signal is sent to control valve **a** to start the out-stroke again, but also to valve **c** to remove the exhaust restriction until valve **b** is next tripped.

Automatic time delay circuits

An automatic time delay circuit, such as that shown in Fig. 18.79, is used when a reciprocating motion is required with known times between starting the out-stroke and in-stroke. At the end of each piston stroke the reservoir that signalled the movement discharges through the free flow route of its regulator to an exhaust port of the control valve. The flow regulators require careful adjustment to accomplish this control but enable a time delay to be achieved at either end of the stroke.

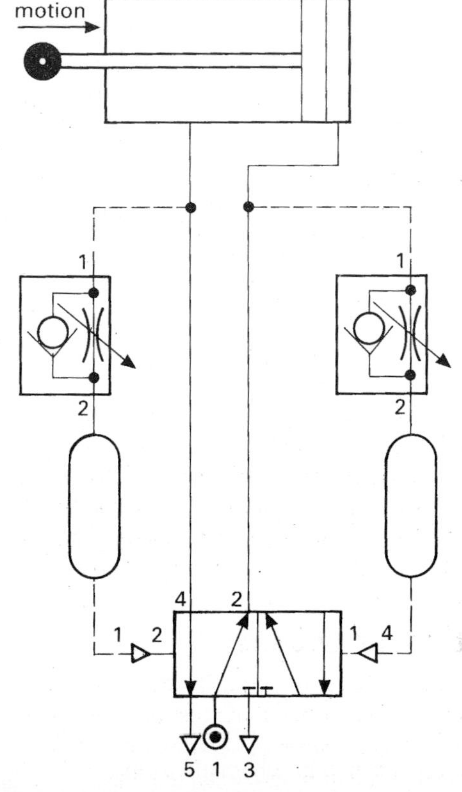

Figure 18.79 An automatic time delay circuit.

Automatic pressure-sensing circuits

Pressure-sensitive valves, connected as a NOT logic function, can sense when the piston of a cylinder is fully extended or retracted and its exhaust pressure falls below 0.5 bar, and can then signal the next phase of the motion of an automatic circuit. This

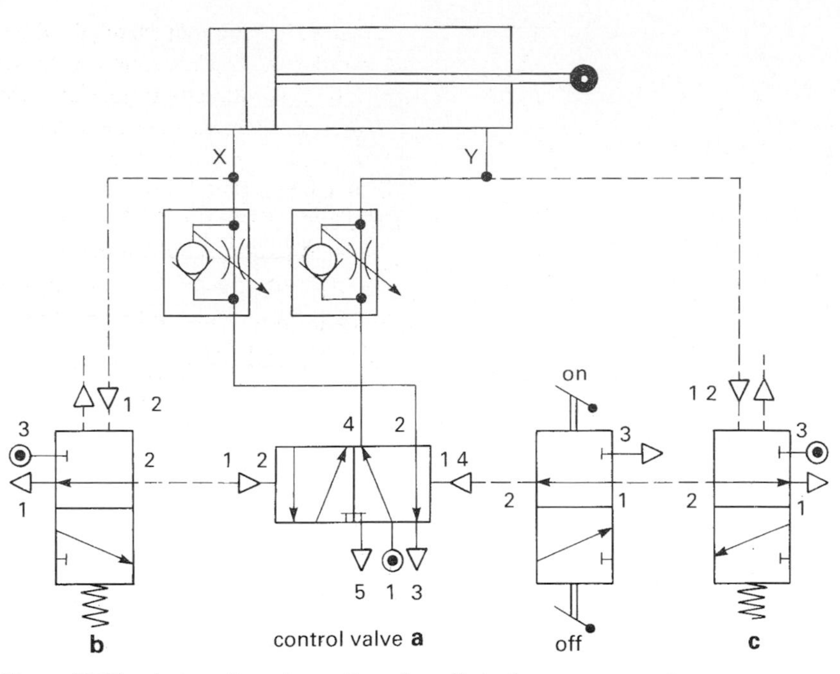

Figure 18.80 Automatic reciprocation of a cylinder by pressure sensing.

is shown in Fig. 18.80. When the sequence is started, a signal is received at port 1 4 of control valve **a**. This will cause motion of the piston in the direction shown and high pressure at both X and Y. Consequently, both valves **b** and **c** have route 1 2 open and no signal is transmitted by either of them. When the piston has moved to full in-stroke position the signal at Y is still high but the pressure at X falls as the exhaust air is expelled. When the pressure at X falls below a preset level (or about half a bar for a diaphragm valve), the signal on port 1 2 of pressure-sensitive valve **b** ceases and route 2 3 is now open. This gives a signal to port 1 2 of control valve **a**, thus signalling the out-stroke of the cylinder. The second half of the sequence then operates in the same way as the first half.

Electrical control of a pressure-sensing circuit can be achieved by replacing the roller-trip and plunger 3-port valves by microswitches and solenoid valves.

18.4.2 Sequential control

Many completely pneumatic automatic machines are designed to perform actions such as positioning objects, operating clamps and pressing in a predetermined order.

Consider the design of a robot arm which will swing through an arc, clamp an object, swing back to its previous position and release the object on to a conveyor belt. Two cylinder movements are required, one to swing through the arc and the other to operate the clamp. If we assign alphabetic symbols to each movement of the cylinders and list the order in which they must take place, A+ will represent the extension of the first cylinder to swing the arm round; B+ will represent the clamping of the object; A− the swing back to the conveyor; and B− the release of the object. The whole cycle is represented by:

$$A+, B+, A-, B-$$

This sequence of operations can also be represented in schematic form, Fig. 18.81, showing clearly the cyclic nature of an automatic circuit.

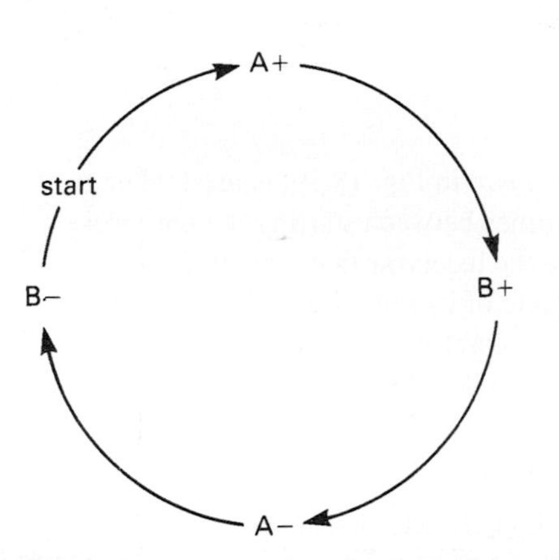

Figure 18.81 A schematic representation of the sequence of cylinder movements.

You will notice that each half of the sequence contains the same letters in the same order: A, B. Consequently the sequencing circuit is of the easiest type, where one action will directly trigger the next of the sequence.

When we start the operation represented by the circuit in Fig. 18.82 with the interrupt valve we signal the out-stroke of cylinder A. Valve **a**+ is then tripped, which actuates port 1 4 of control valve **b**, causing cylinder B to perform an out-stroke and contact **b**+. Valve **b**+ causes cylinder A to retract and valve **a**− causes cylinder B to retract, which in turn starts the whole sequence again.

In order to achieve the clamping function originally required in our robot arm, we could replace cylinder B by two simultaneously acting cylinders mounted opposite each other to grip the workpiece directly, in which case control valve **b** would be

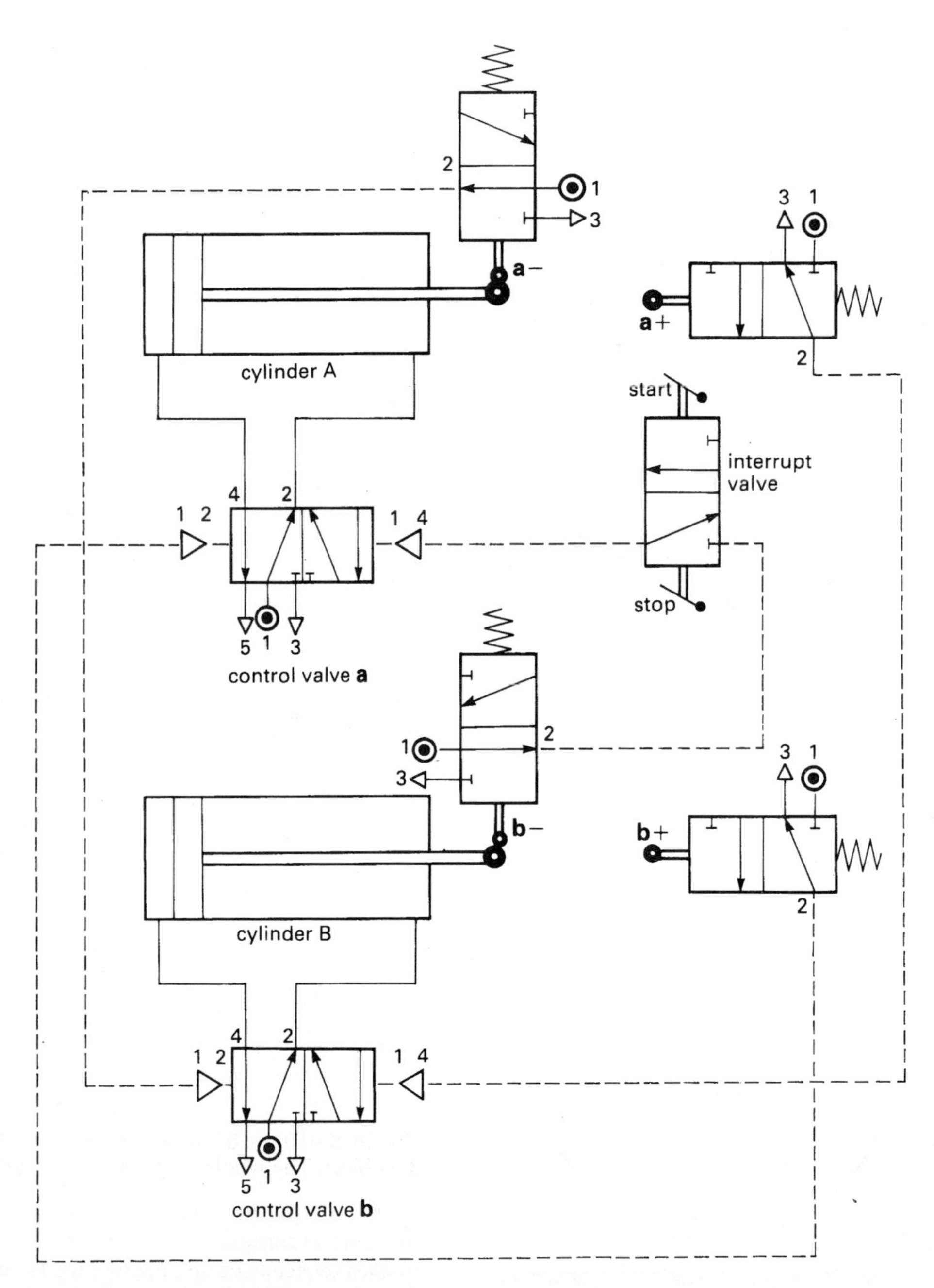

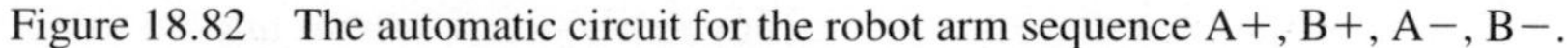

Figure 18.82 The automatic circuit for the robot arm sequence A+, B+, A−, B−.

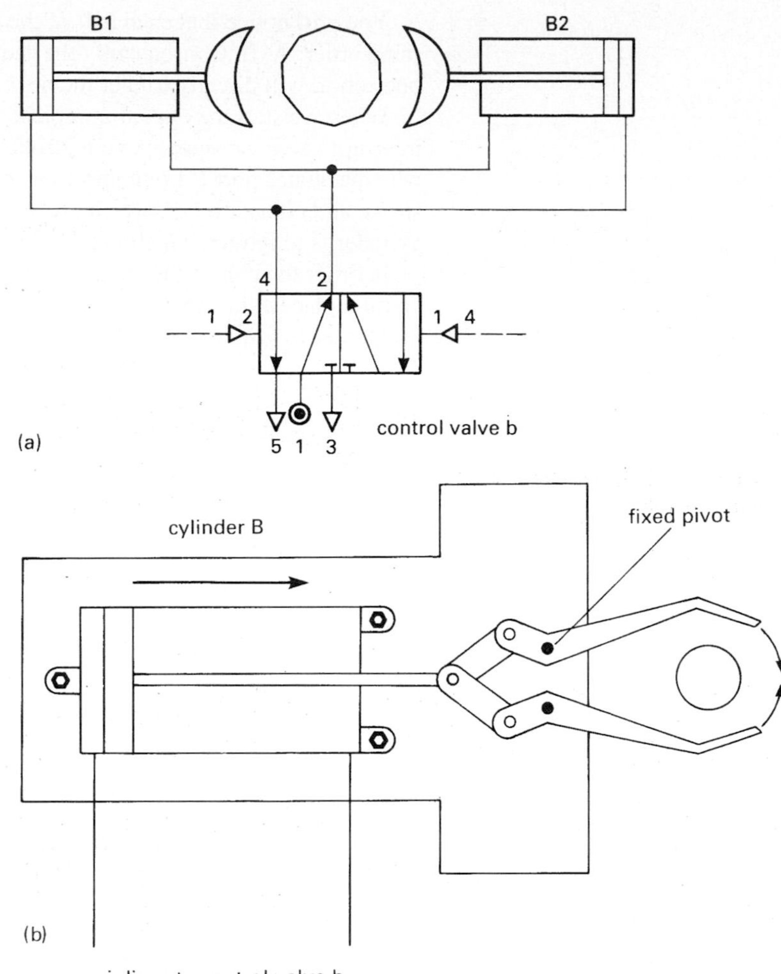

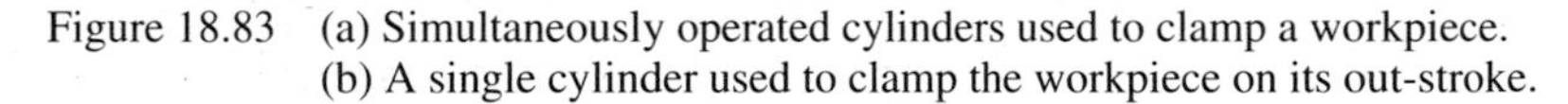

Figure 18.83 (a) Simultaneously operated cylinders used to clamp a workpiece. (b) A single cylinder used to clamp the workpiece on its out-stroke.

connected to operate both cylinders B1 and B2 as shown in Fig. 18.83(a). Alternatively, the linkage shown in Fig. 18.83(b) will clamp the object on the out-stroke of one cylinder B.

Another method of achieving the same effect would be to have cylinder B connected to two pivoted jaws which would clamp on the in-stroke of cylinder B, as shown in Fig. 18.84. The in-stroke force of cylinder B would need to be calculated carefully, bearing in mind the mechanical advantage of the jaws, in order to ensure correct clamping of the workpiece.

The sequence for this new method of operation would be:

$$A+, B-, A-, B+$$

As each half of the sequence is still an A operation followed by a B operation the only change is to trigger the valves in a different sequence, as shown in Fig. 18.85.

Time delay sequential circuits

A time delay may be necessary in the above circuit, perhaps to allow a component to be moved into place before the start of each sequence. This delay can be achieved by inserting a reservoir and flow regulator at position R in Fig. 18.85. Port 2 of the interrupt valve leads to port 1 of the flow regulator; air then passes through the flow regulator to the reservoir and on to port 1 4 of control valve **a**.

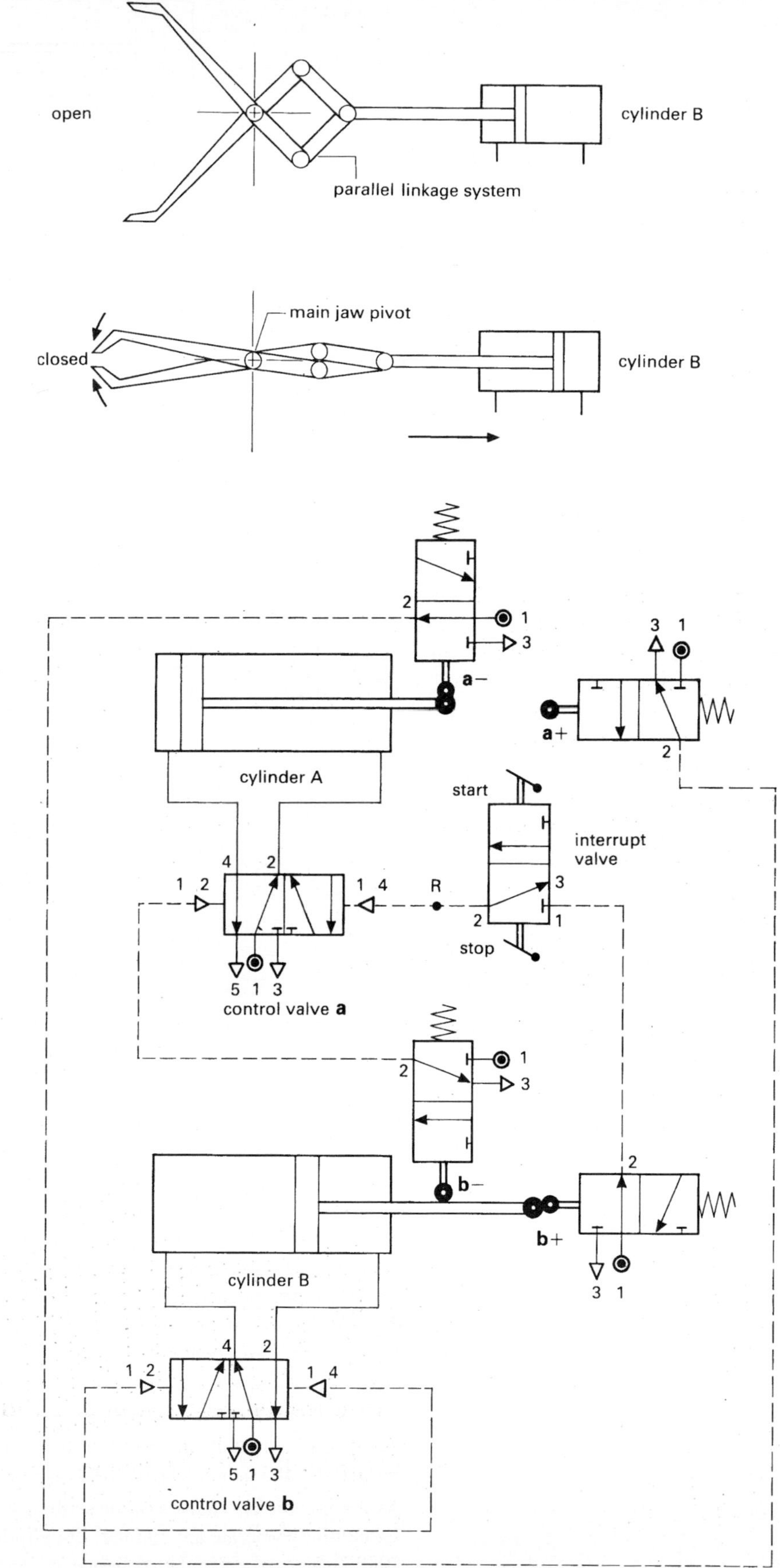

Figure 18.84 A single cylinder used to clamp the workpiece on its in-stroke.

Figure 18.85 Automatic circuit for the sequence A+, B−, A−, B+.

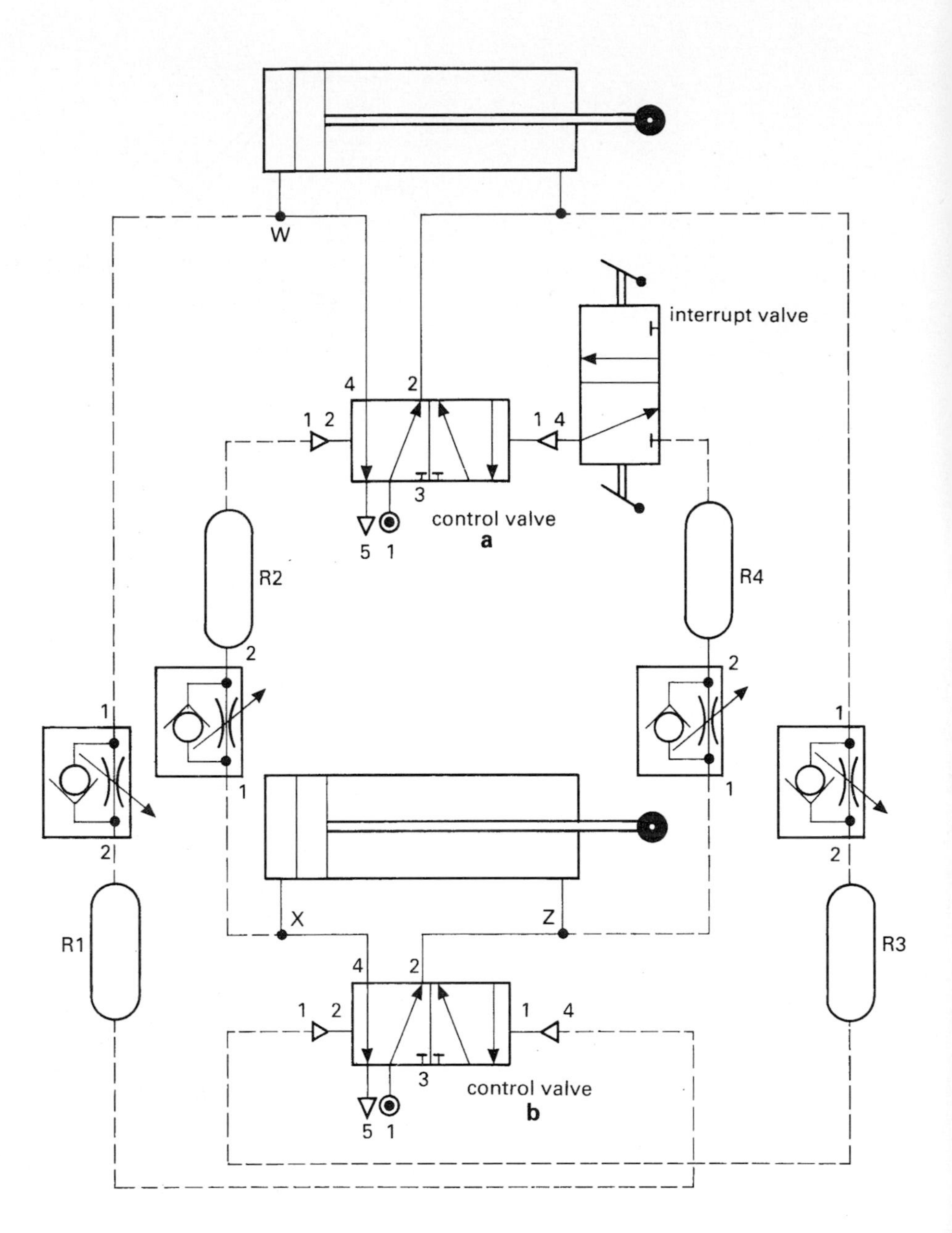

Figure 18.86 Automatic time delay circuit for the sequence A+, B−, A−, B+.

In the same way that Fig. 18.79 illustrated how the automatic reciprocation of one cylinder can be achieved by a time delay method, the sequential control of two cylinders can also be achieved by a time delay method as shown in Fig. 18.86.

The roller-trips have now been replaced by reservoirs and flow regulators so that each operation has a time delay before activating the next part of the sequence. When the interrupt valve is switched to start the sequence, cylinder A will perform an out-stroke and the pressure maintained at W will then charge reservoir R1, which, after the required delay, signals control valve **b** to perform an out-stroke of cylinder B. This principle is repeated for each stage of the operation. A time delay circuit such as this obviously needs careful adjustment to operate correctly and is not generally reliable enough to use in a complex sequential control circuit.

Multi-cylinder sequential circuits

As long as the two halves of the sequence have the same order of letters, we can sequentially operate any number of cylinders in this way. A three-cylinder operation that was to operate in the sequence

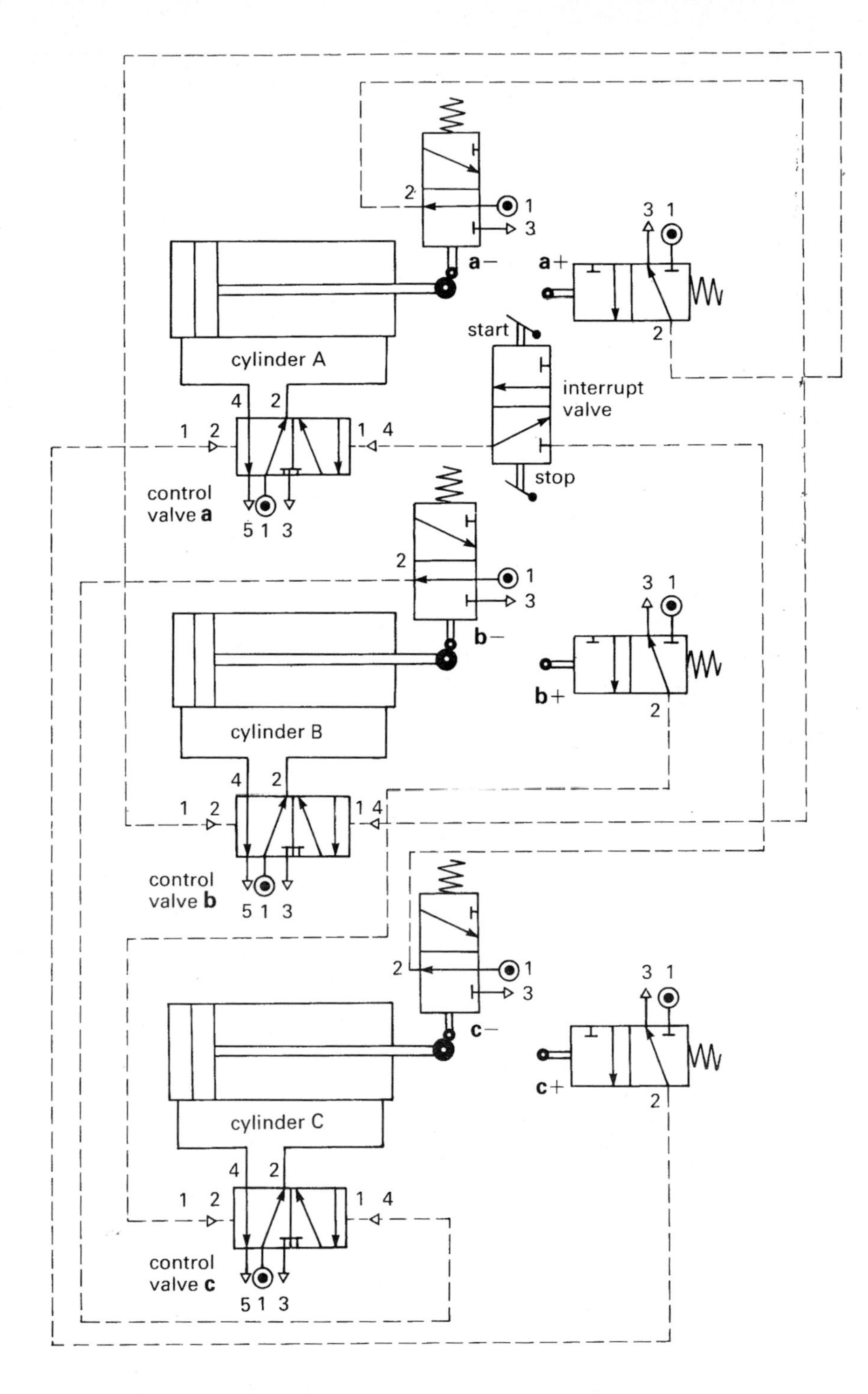

Figure 18.87 Automatic circuit for the sequence A+, B−, C+, A−, B+, C−.

A+, B−, C+, A−, B+, C−

again conforms to a simple sequential control circuit and we would connect the components each to trigger the next as shown in Fig. 18.87.

Reverse sequence pattern

A+, B+, C+, C−, B−, A− is an example of a reverse sequence pattern in which the cylinder return motions take place in reverse order. This needs a more complex circuit design to enable the sequence to be completed.

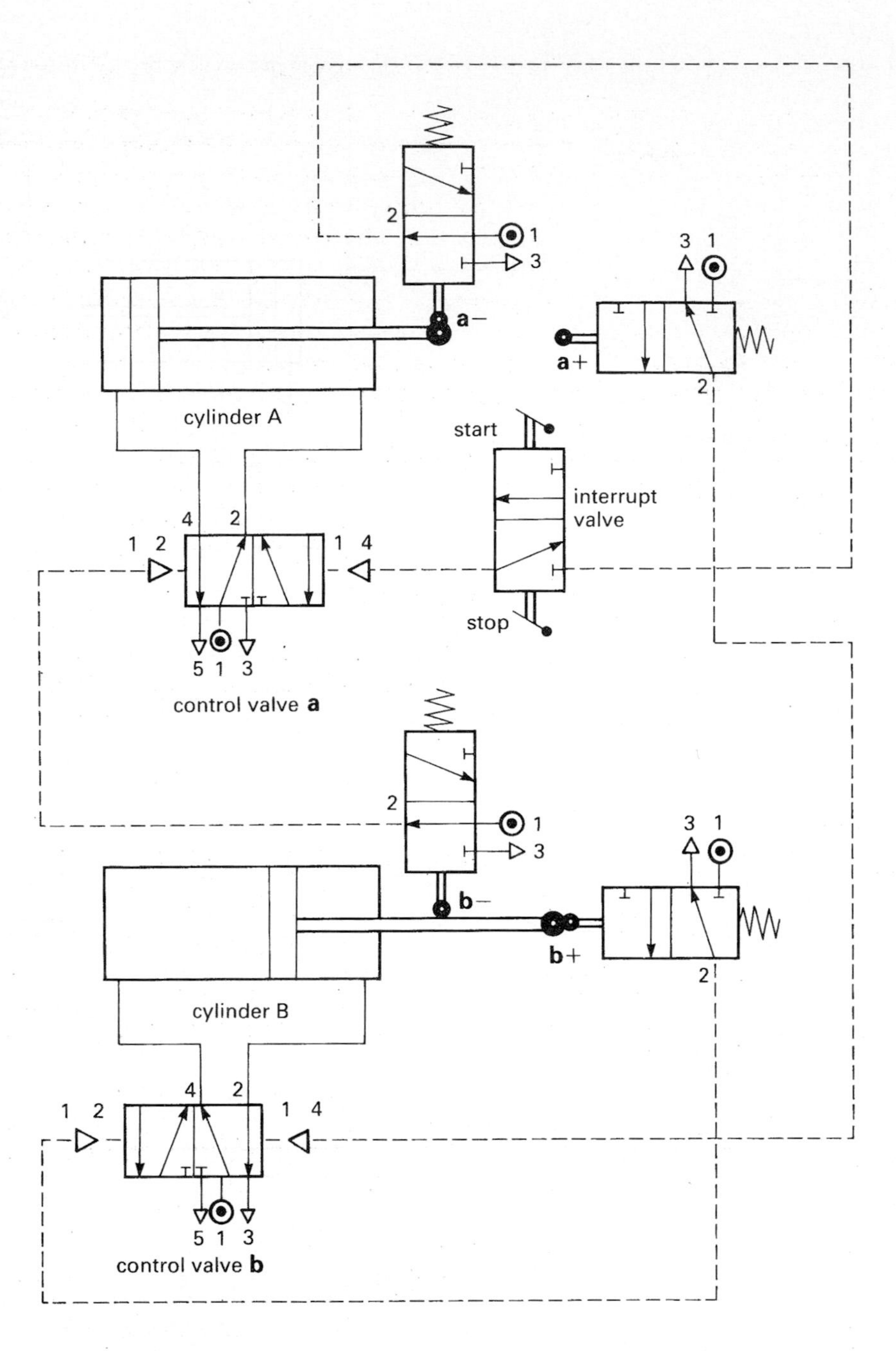

Figure 18.88 An inoperable sequential circuit for A+, B+, B−, A−. Control valve **b** will jam.

Inoperable sequential circuit

If a process needs to move a component, stamp it, remove the stamp and move the component back, a circuit would be designed for the sequence

A+, B+, B−, A−

This has the operations in the second half of the sequence operating in reverse order. Connecting the simple trigger circuit as before, but with the cylinder operations in this new order, illustrates that this sequence jams the air signals and is inoperable as a simple sequential circuit.

The circuit in Fig. 18.88 will operate in the following way: when the start lever is activated cylinder A will extend and valve **a**+ will signal port 1 4 of valve **b**, which is left live until cylinder A is retracted. Cylinder B now extends to trigger valve **b**+, which brings a signal to port 1 2 of control valve **b**. This obviously is unable to have any effect while the signal at port 1 4 is maintained, and so the circuit is inoperable.

Pulse signals

A simple method of overcoming the problem of this maintained signal is to use a one-way trip device for valve **a**+ which will operate and release during the extension of cylinder A thus providing a pulse signal to control valve **b**.

A pulse signal provided by a reservoir, Fig. 18.89, is another way of ensuring that a signal is not maintained beyond the initial actuation and could be used in the above circuit in order to enable the completion of the sequence. On actuating valve **a**+, connected as in Fig. 18.89, the main air supply to the reservoir is sealed off and the reservoir discharges to port 1 4 of control valve **b**, the remaining reservoir charge then bleeds away enabling valve **b**+ to signal control valve **b** to move cylinder **B** negative and the sequence:

A+, B−, B+, A−

to be completed.

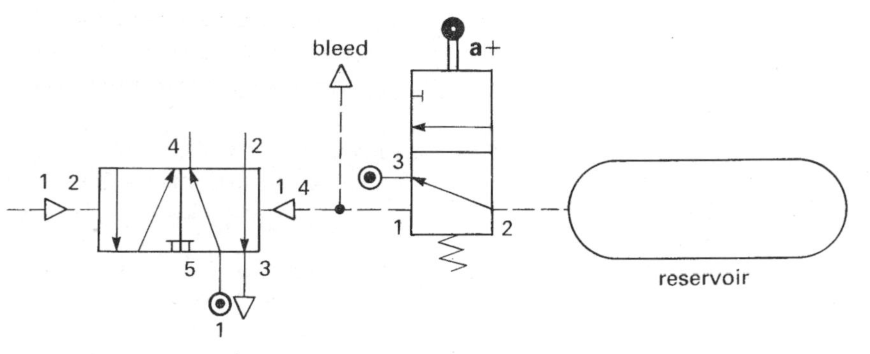

Figure 18.89 A circuit that provides a limited period signal to the control valve.

These *pulse* or *limited period* signals do not have the reliability or general stability of the earlier circuits in which the signals were positively maintained, and so in many cases we need to consider a more complex circuit design.

18.4.3 Cascade system of sequential control

The *cascade system* provides a means of cancelling the maintained signals on control valves that prevent the completion of sequences such as A+, B−, B+, A− in Fig. 18.89. It is achieved by the addition of more pilot control valves and rerouting of some signals, but otherwise does not change the mechanism or location of the original cylinders and control and trip valves.

Two-cylinder, two-group cascade circuits

Consider the circuit that earlier proved problematic for the clamping and moving operation,

A+, B+, B−, A−

The sequence must be divided into groups in such a way that any letter, regardless of sign, appears only once in any group. This gives A+, B+/ B−, A−. Now by assigning a number to each group we have

A+, B+ /	B−, A−
group 1	group 2

In order to signal cylinder B to retract, the signal maintained at control valve **b** from trip valve **a**+ must be cancelled after cylinder B has extended. This changeover must

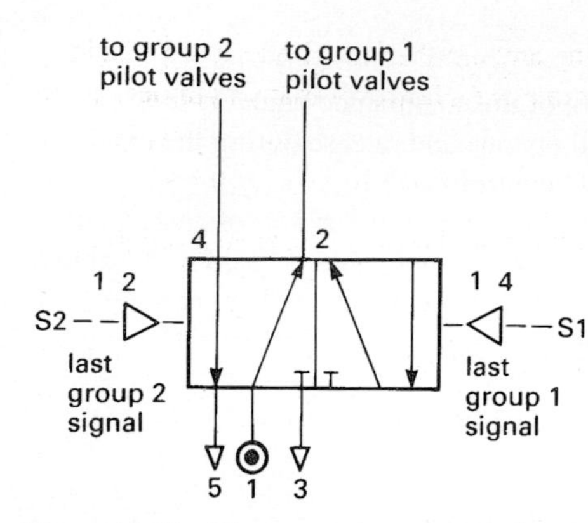

Figure 18.90 The group changeover valve.

therefore occur at the division between group 1 and group 2. If we now control the air supplied to each group so that group 1 is switched off at this point and group 2 is switched on, we can actuate the next signal for cylinder B to retract. After the retraction of A has been activated we can switch the air supply back from group 2 to group 1.

In a two-group cascade system this switching of the air supply can be achieved by using a 5-port double air-signalled control valve. Fig. 18.90 shows a *group changeover valve*, where the signal S1 will switch the air supply over from group 1 to group 2 and signal S2 will switch it back again.

It is also clear that several air lines in each group will have to be connected to ports 2 and 4 of this changeover valve. This can be achieved in a small circuit by the use of T-connectors or 4-way connectors, or in a larger circuit by the use of a *busbar* or *manifold*. This is usually a cylindrical or square section of metal or nylon with several connectors to enable all the connecting lines to be at the same pressure.

Starting valve **c**, in the circuit shown in Fig. 18.91, with the air initially live to group 1, moves cylinder A positive, which trips valve **a**+. This then actuates control valve **b** to move cylinder B positive, which in turn trips **b**+. Trip valve **b**+ signals the group changeover valve to switch the air from group 1 to group 2. This means that port 4 of the group changeover valve is now transmitting an air signal to valves **a**− and **b**− so that they can operate when signalled. It is also signalling control valve **b** to actuate the in-stroke of cylinder B which is the first motion in the group 2 sequence. We can now operate the sequence:

A+, B+, B−, A−

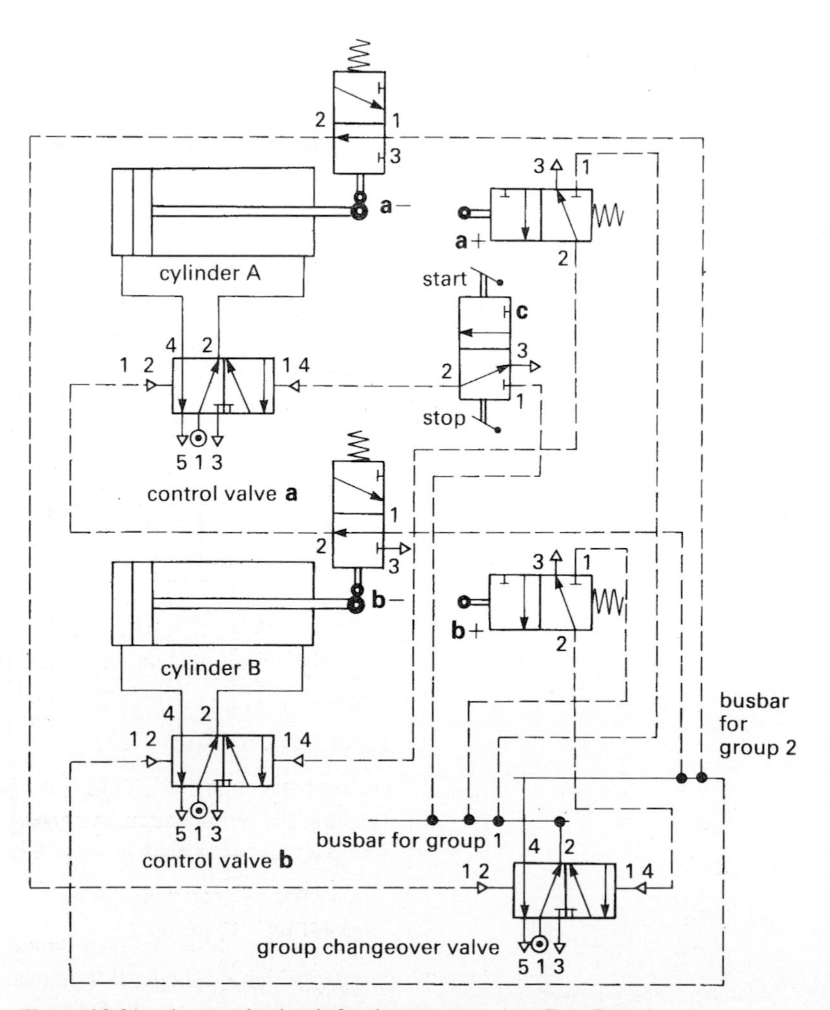

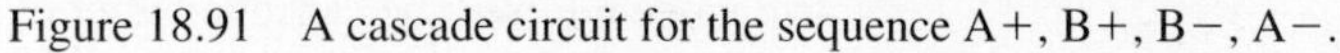
Figure 18.91 A cascade circuit for the sequence A+, B+, B−, A−.

For any other two-group system containing any number of cylinders, we would use a 5-port group changeover valve in the same way, however many cylinders were in use. The sequence:

A−, B+/B−, A+

would obviously be similar to Fig. 18.91, with the signal lines connected slightly differently.

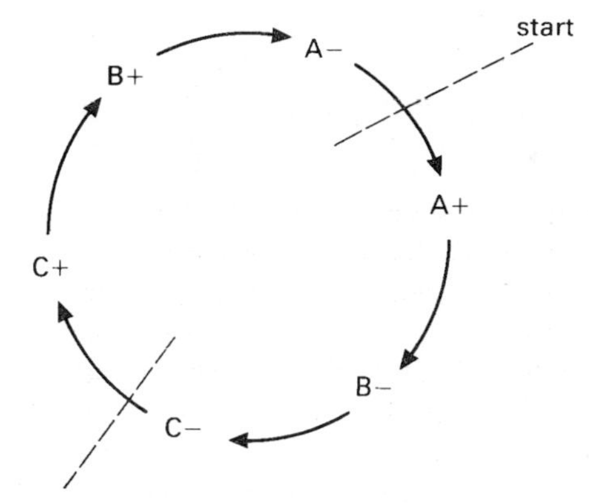

Figure 18.92 The cyclic representation of the sequence A−, A+, B−, C−, C+, B+.

Three-cylinder, two-group cascade circuits

Now consider the sequence:

A−, A+, B−, C−, C+, B+

At first sight it may appear that we need three air groups:

A−	/	A+, B−, C−	/	C+, B+
1		2		3

However, the last two operations C+ and B+ could be grouped with the first, A−. This becomes much clearer if we draw the cycle as in Fig. 18.92.

It is essential to break the sequence into as few groups as possible. This cycle plainly reduces to two air groups and can be connected as in Fig. 18.93. An interrupt valve could then be placed in a suitable position in the circuit.

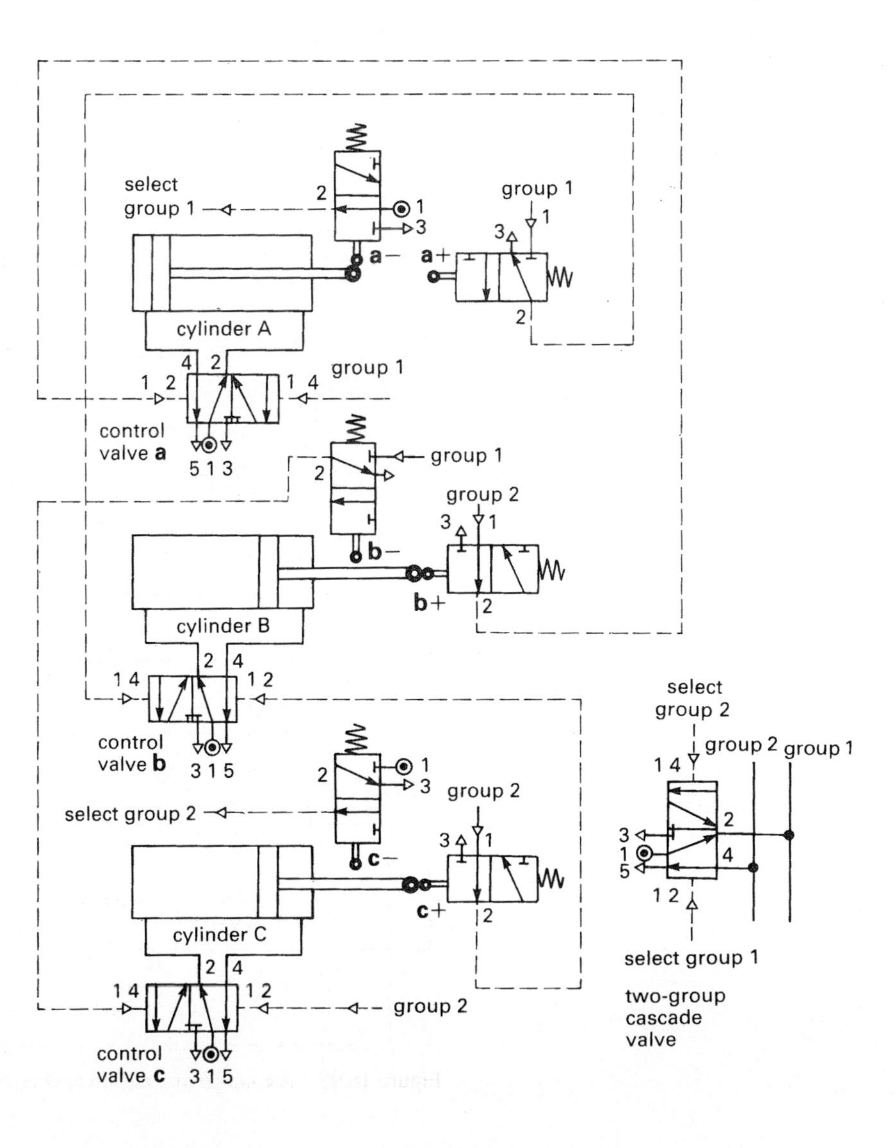

Figure 18.93 Automatic circuit for the sequence A+, B−, C−, C−, B+, A−.

Multi-group cascade circuits

There are two methods of designing group changeover valve systems. A cascade circuit with more than two air groups can use as many 5-port double air-signalled group changeover valves as there are groups. Fig. 18.94 shows this changeover system for a three-group cascade circuit while Fig. 18.95 shows a four-group cascade circuit.

Alternatively for a four-group cascade circuit we can use the system shown in Fig. 18.96, which has just three changeover valves.

These circuits still require a 5-port double air-signalled control valve supplied with main air and two pilot trip valves for each cylinder in the circuit. They will also require an ON/OFF valve, which is placed prior to the first movement in the cycle so that when stopped, all the elements in the system will be in their initial positions. If a

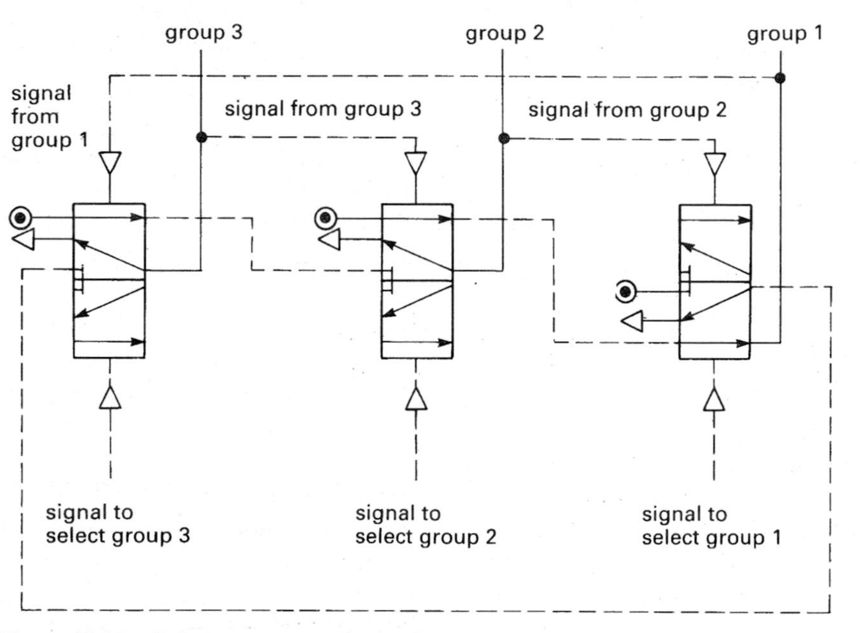

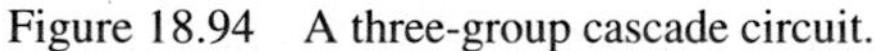

Figure 18.94 A three-group cascade circuit.

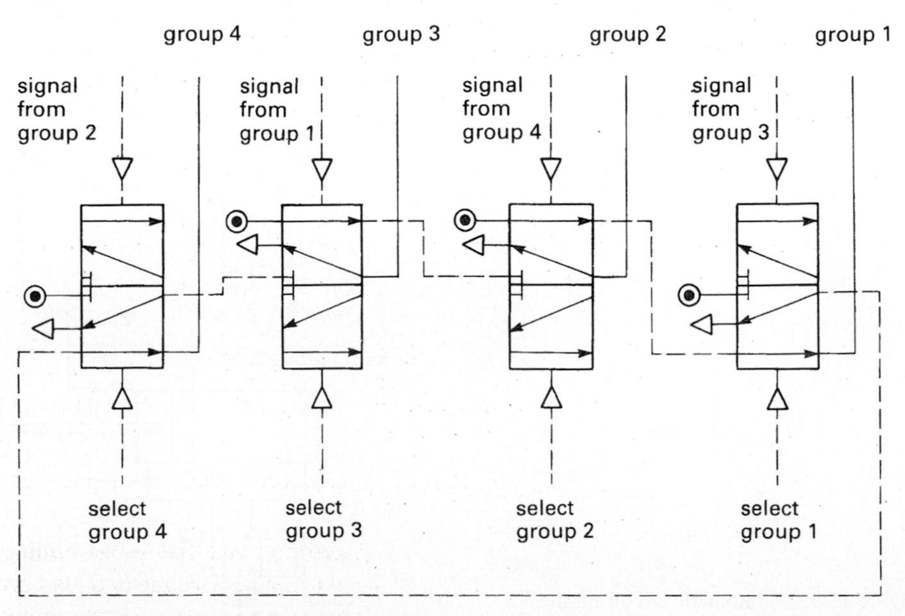

Figure 18.95 A four-group cascade circuit.

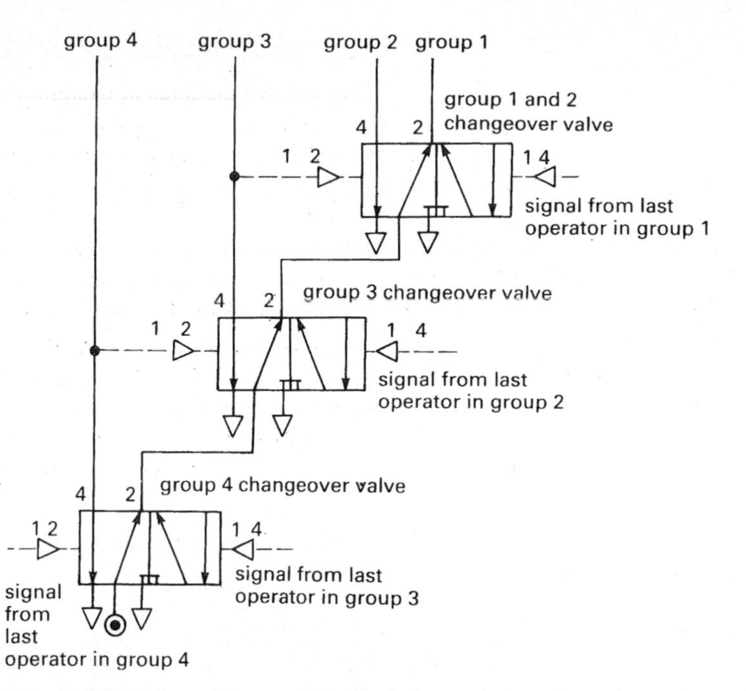

Figure 18.96 A four-group cascade circuit that uses these three changeover valves.

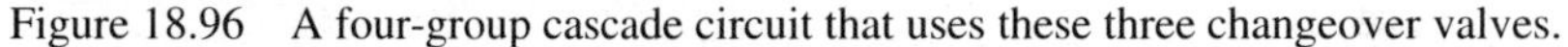

single-acting cylinder is to be included in the circuit, the control valve need only be a 3-port air-signalled valve. Additionally, an emergency stop or interrrupt, fail-safe valve would sometimes be included in the circuit. The circuit is then connected up so that each operation signals the next operation or changeover in the sequence.

Design of a multi-group cascade circuit

Suppose the flowing sequence is required:

A+, B+, B−, C+, C−, D+, A−, D−

The steps are as follows:

1 Split up the sequence into as few groups as possible and number the air groups, as in Fig. 18.97.

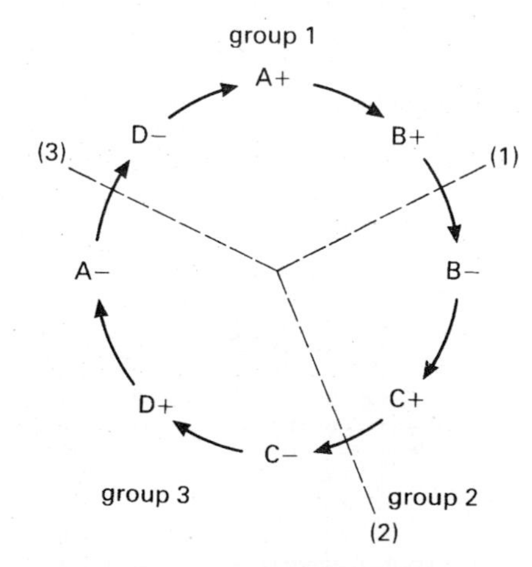

Figure 18.97 The cyclic representation of the sequence A+, B+, B−, C+, C−, D+, A−, D−.

The lines 1 and 2 must be dividing points if the same letter cannot appear twice in one group. This method also gives line 3. The sequence is now clearly a three-group cascade circuit, numbered as shown.

2 Decide on the components in the circuit. It will need four cylinders, four double air-signalled control valves, eight 3-port trip valves, a minimum of two group changeover valves and an ON/OFF lever valve.

3 Draw the components in the circuit, with busbars connected to each group changeover valve. Group 1 air should be connected to port 1 of trip valves **d**−, **a**+, **b**+; group 2 air to port 1 of **b**− and **c**+; and group 3 air to **c**−, **d**+ and **a**−, as in Fig. 18.98.

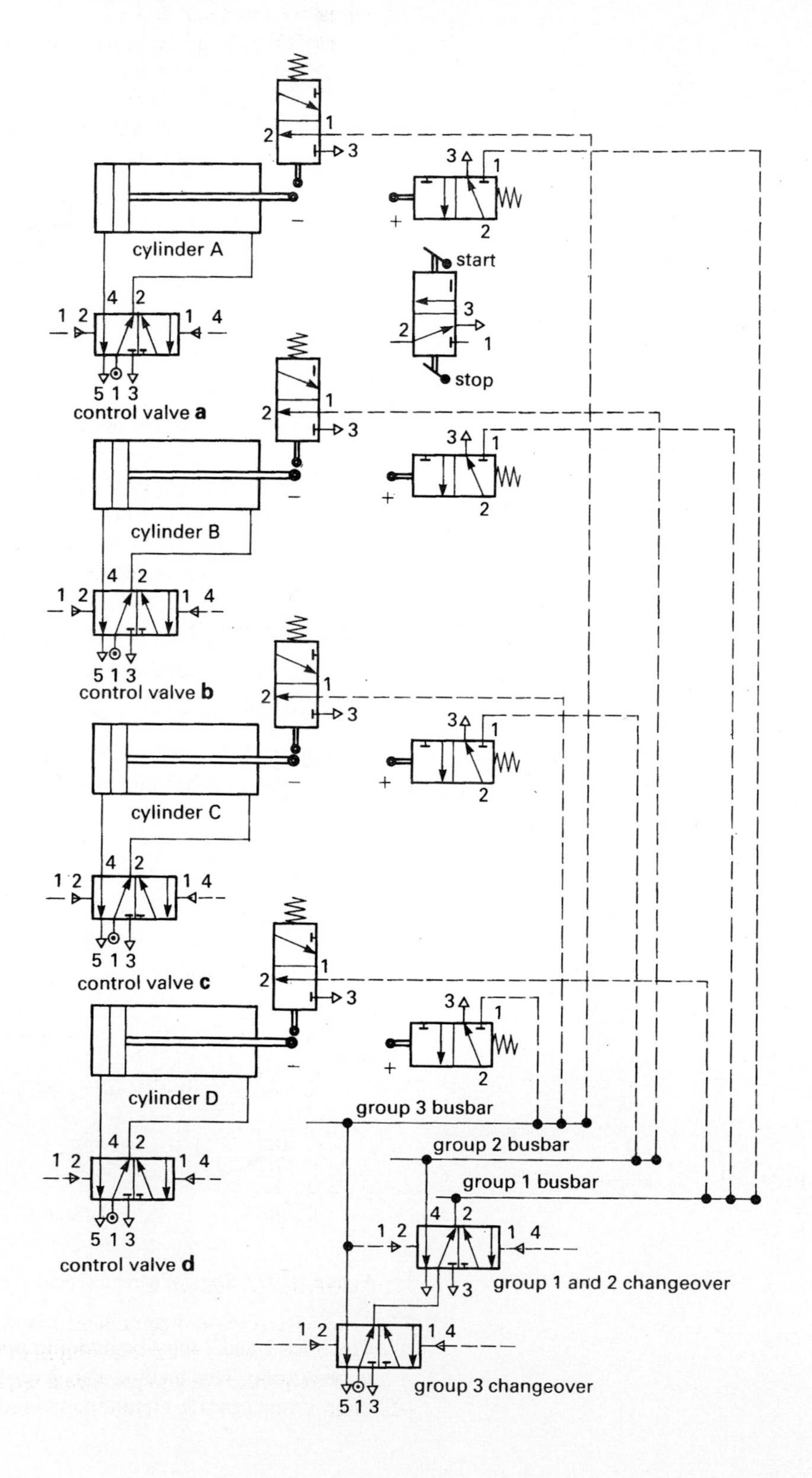

Figure 18.98 Automatic circuit for the sequence A+, B+, B−, C+, C−, D+, A−, D−.

4 Connect the circuit sequentially. Starting at the ON/OFF valve, connect its output to signal the first movement of the sequence, A+, through control valve **a**; valve **a**+ should then signal B+; **b**+ signals groups 1 and 2 changeover; the output from this changeover valve should then signal for the next operation B−; valve **b**− then signals C+; **c**+ signals the changeover to group 3 which then signals C−; **c**– signals D+; **d**+ signals A−; **a**− signals changeover back to group 1 which then signals D−, and **d**− returns to the starting point of the signal to start again at A+.

The final circuit is shown in Fig. 18.99.

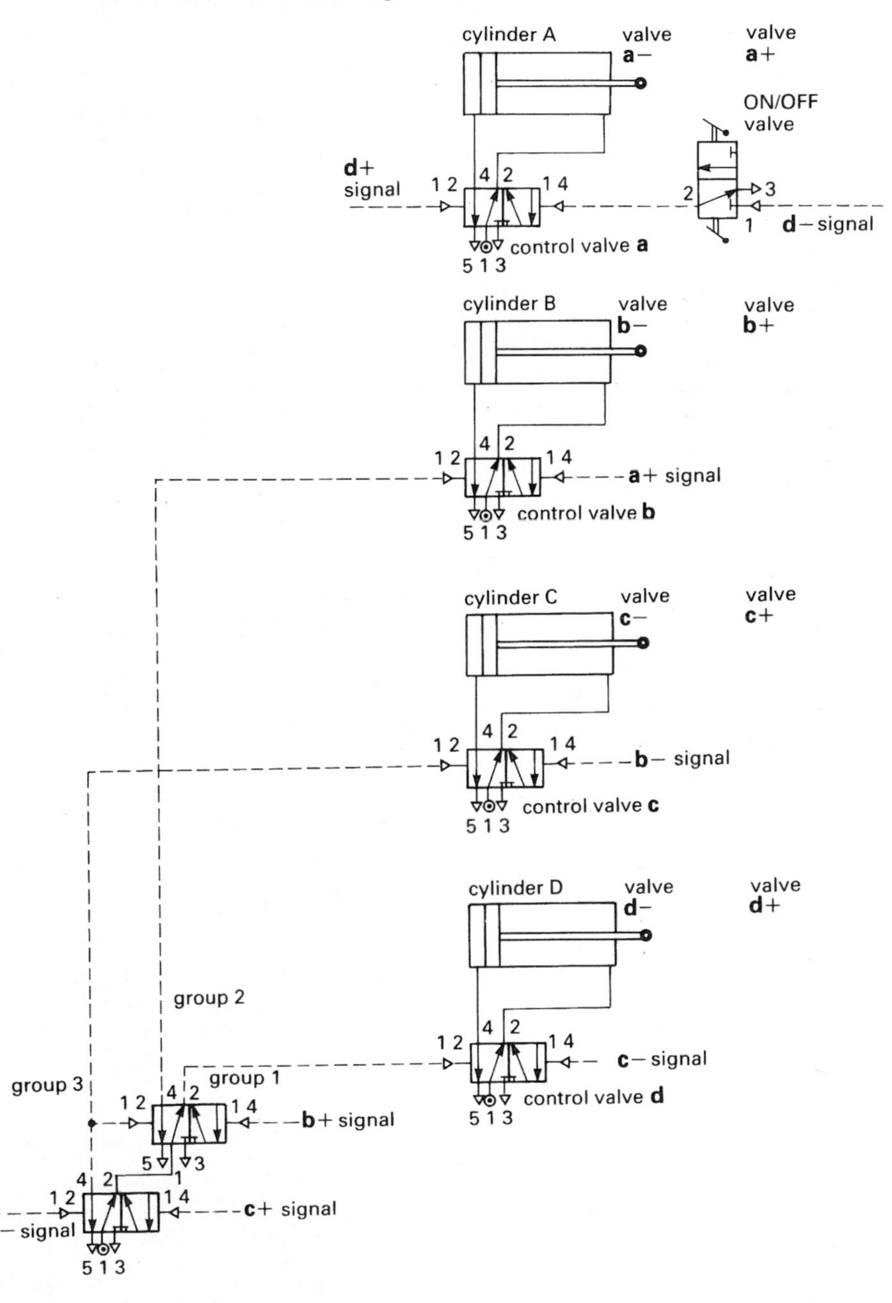

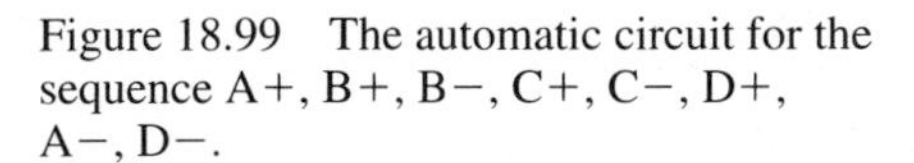
Figure 18.99 The automatic circuit for the sequence A+, B+, B−, C+, C−, D+, A−, D−.

Sequence diagrams

In addition to a circuit diagram and a layout drawing of components a *sequence diagram* may also be needed to simplify the explanation of the operation of a circuit. It may also be necessary to produce a sequence diagram to help with the design of the circuit.

Consider the circuit for the sequence A+, B+, A−, B− shown in Fig. 18.100. Valves **a**+ and **a**− are positioned at the out-stroke and in-stroke positions of cylinder A; **a**+ signals B to perform an out-stroke and **a**− signals B to perform an in-stroke. This, and the movements caused by **b**+ and **b**−, can be represented on a sequence diagram which will show the time that the cylinders are in the rest position by horizontal lines and their movements by diagonal lines. A positive gradient diagonal joining the − to the + symbol represents an out-stroke and vice versa. At the end of this line any pilot valve that is automatically contacted and which then signals the next operation is shown.

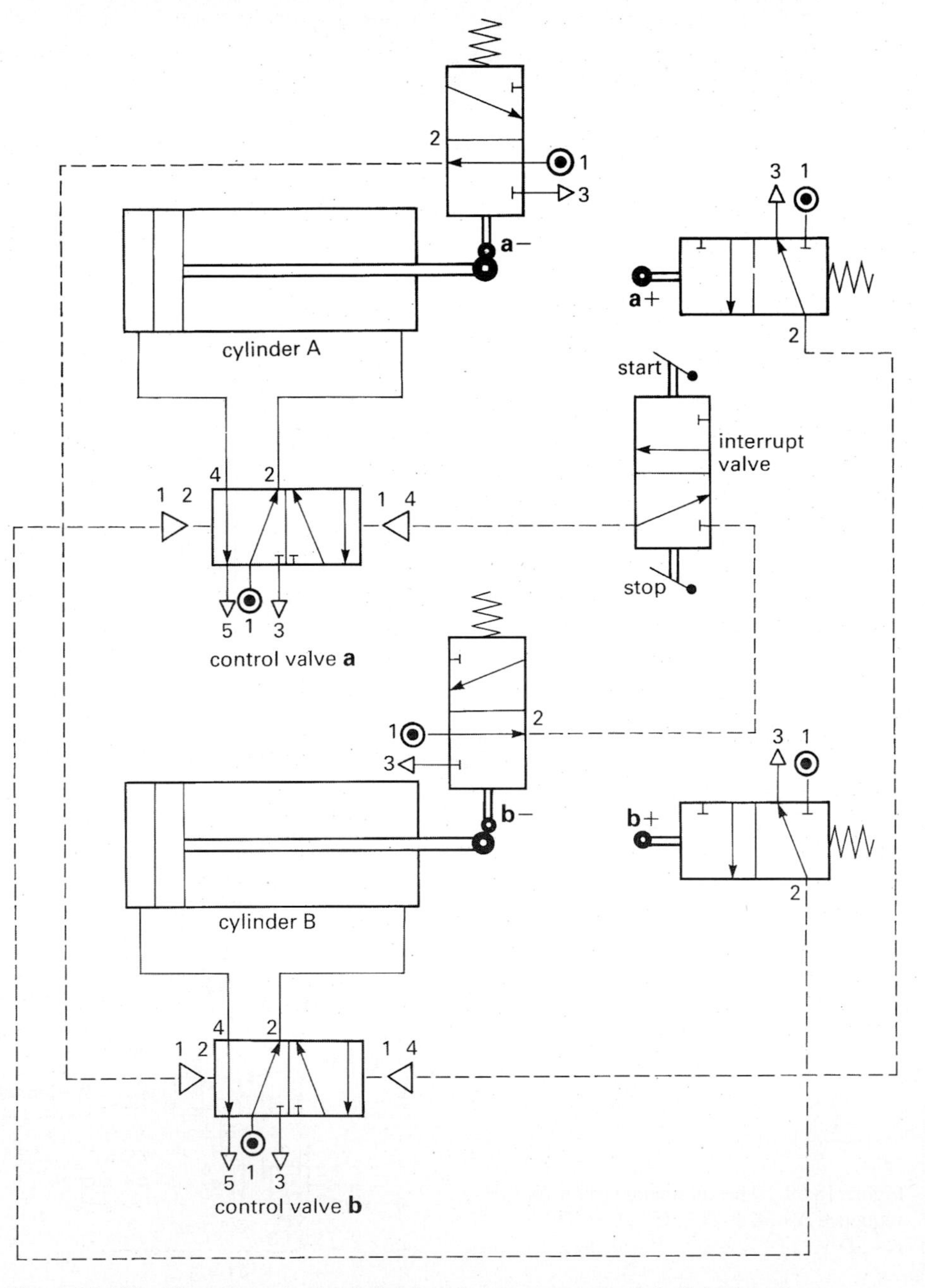

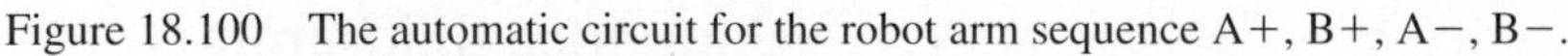
Figure 18.100 The automatic circuit for the robot arm sequence A+, B+, A−, B−.

Fig. 18.101 shows the sequence diagram for the circuit shown in Fig. 18.100. At the top of the diagram, cylinder A is shown to move positive and contact **a**+ which then causes B to perform an out-stroke. Therefore at the end of the diagonal

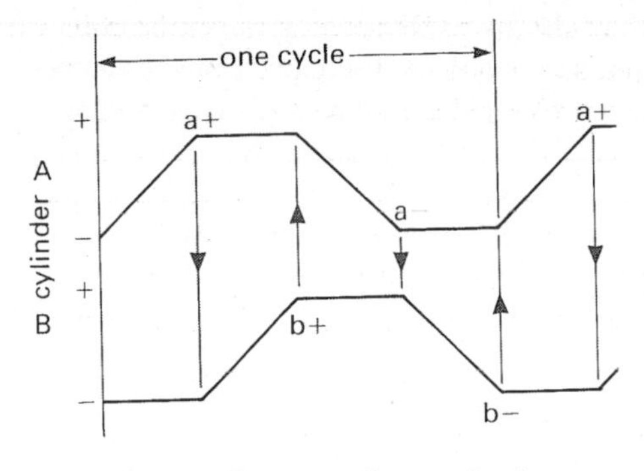

Figure 18.101 Sequence diagram for A+, B+, A−, B− circuit.

indicating the A+ movement the symbol **a**+ is shown and this point is connected by a vertical line to the B− position and joins it diagonally to the B+ position. Here the pilot signal **b**+ is connected vertically to the A+ position and is drawn diagonally to indicate the in-stroke of A. This is then connected vertically to the B+ position and the last diagonal line in the first cycle indicates the retraction of B. The cycle is then shown to start again as it is an automatic sequence.

The circuit for the sequence A+, B−, C+, A–, B+, C− is shown in Fig.18.102 and its sequence diagram is given in Fig. 18.103. Here we can see clearly the sequence of cylinder movements and actuators. In general, the cylinder operation lines are equally inclined to the horizontal but, if an idea of piston speed were also required, the slope could be changed to indicate this.

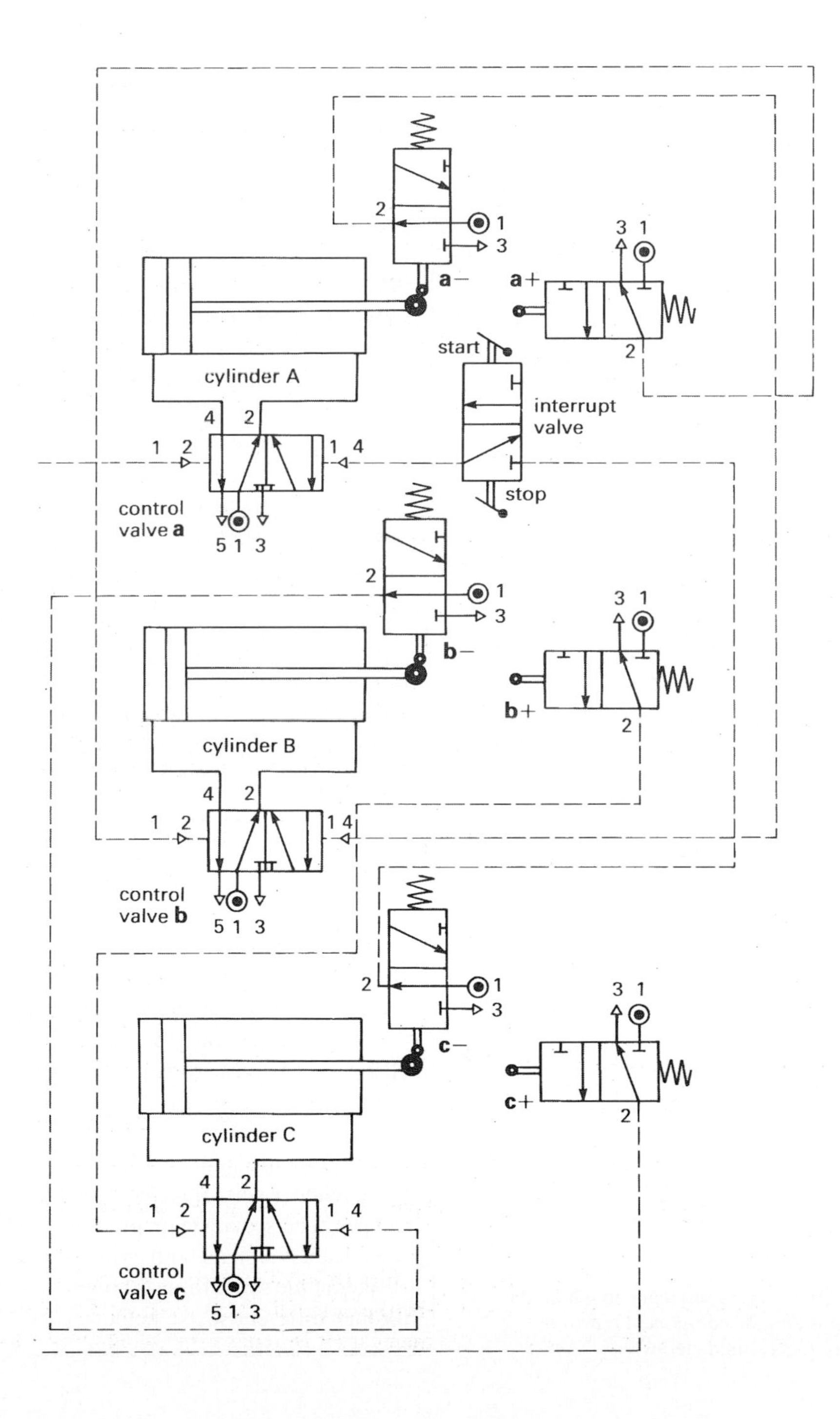

Figure 18.102 Automatic circuit for the sequence A+, B−, C+, A−, B+, C−.

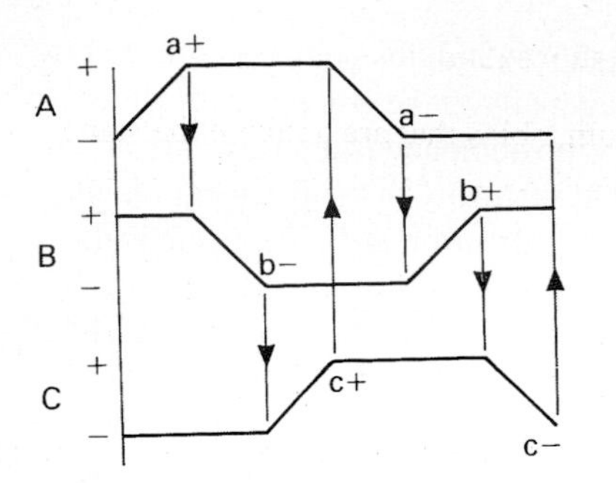

Figure 18.103 One cycle of the sequence A+, B−, C+, A−, B+, C−.

A chart showing the signal sequence may also be required to establish the order of signals and movements in the design of a more complicated sequence such as in the example below.

EXAMPLE 18.3 Design a pneumatic circuit to operate a small plastic forming press. The sequence and method of actuation of the press is described below. The press is to be loaded and unloaded manually.

Operation	*Actuation method or time*
guard to close	two handed AND function circuit arrangement
press to downstroke slowly	automatically when the guard is closed
press to retract	three-input air-bleed AND function
component to be ejected	automatically after the retraction of press ejected
guard to open	automatically when the component has been ejected
in-stroke of component ejection cylinder	automatically when the guard is open

SOLUTION There are clearly three cylinders required in this circuit. If we call the cylinder operating the guard A, the press cylinder B and the component ejection cylinder C, we can represent their movements appropriately. Two push-button valves are required to start the sequence and we shall call these valves **d** and **e**. The three-way air bleed can be represented by **b**+ as this shows that the press cylinder is fully extended and the pressing operation complete before the press is signalled to retract.

A chart showing the operation in the sequence, their methods of actuation and the signals that are transmitted on completion of each movement are given below.

Actuator	*Operation*	*Signal*
manual	load	
2-hand AND, d.e	close guard, A+	a+
a+	press down, B+	3-way-air bleed, b+
b+	press up, B−	b−
b−	eject component, C+	c+
c+	guard open, A−	a−
a−	eject cylinder return, C− unload	signal operator to unload

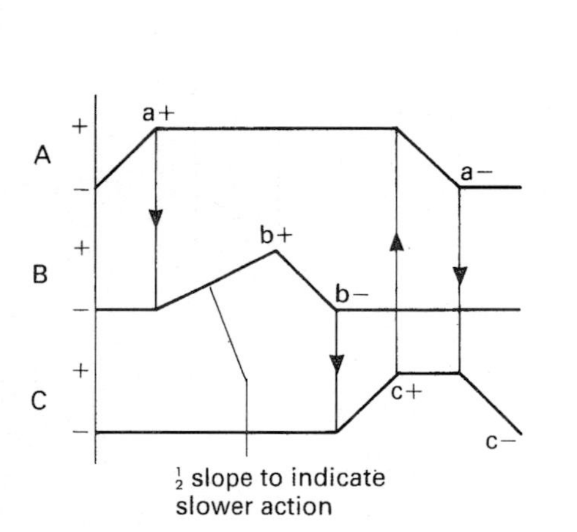

Figure 18.104 Sequence diagram for the pneumatic press.

This chart makes it possible to draw the sequence diagram for this cycle, shown in Fig. 18.104. The slope of the line indicates that the movement B+ is different from the others and represents the slow downstroke of the press.

The sequence diagram will also help to determine where air-group changeover valves are required in the circuit as it is clear that cylinder B must retract again as soon as it has reached its out-stroke position and therefore the **b**+ signal must be removed before we signal the **b**−.

Suppose, in Example 18.3, as soon as the component had been ejected, the eject cylinder was signalled to return at the same time as the guard was signalled to open. The sequence diagram would then be as in Fig. 18.105. It can clearly be seen that **c**+ signals both cylinder A and cylinder C which now start their in-strokes at the same time although they may be moving at different speeds or through different stroke lengths.

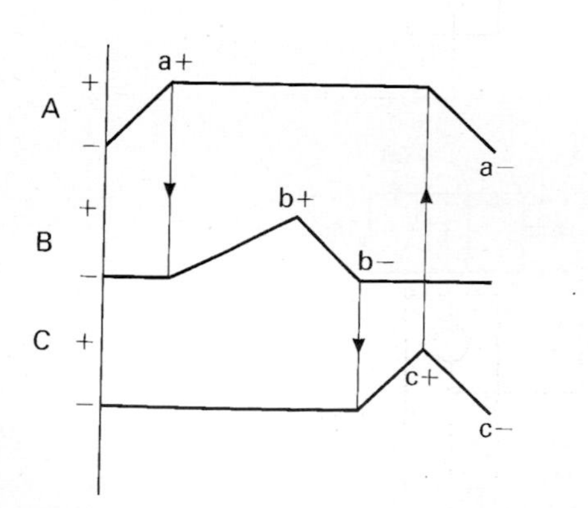

Figure 18.105 Press sequence in which the guard opens and the component ejection cylinder retracts simultaneously.

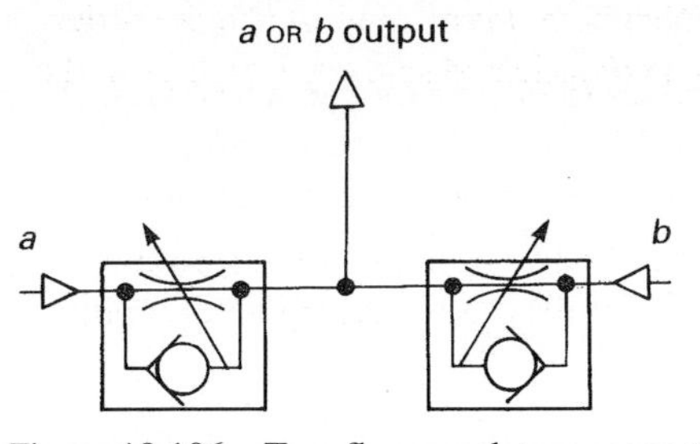

Figure 18.106 Two flow regulators connected to give an OR output.

18.5 Logic devices and circuits

For pneumatic logic circuits the symbols and truth tables that are used are the same as those used in electronic logic circuits. Full details of the logic symbols and truth tables can be found in Fig. 16.1 in Chapter 16.

18.5.1 OR logic circuits

Shuttle-valve circuits

We have already seen that the OR function for two manually operated signals can be achieved by using two 3-port valves, as shown in Fig. 18.41 or by using a shuttle valve, as shown in Fig. 18.39. Two flow regulators connected as shown in Fig. 18.106 would also have the same effect as the shuttle valve. The regulators would prevent the pilot signal from one valve passing directly to the exhaust of the other valve.

For multiple signals we can use shuttle valves in parallel. Fig. 18.107 and 18.108 show two possible circuits for a six-function logic OR circuit. The parallel arrangement of shuttle valves shown in Fig. 18.108 is a better method for circuits with many possible signals, as *a* and *b* traverse fewer valves and therefore the pressure drop associated with these signals is less.

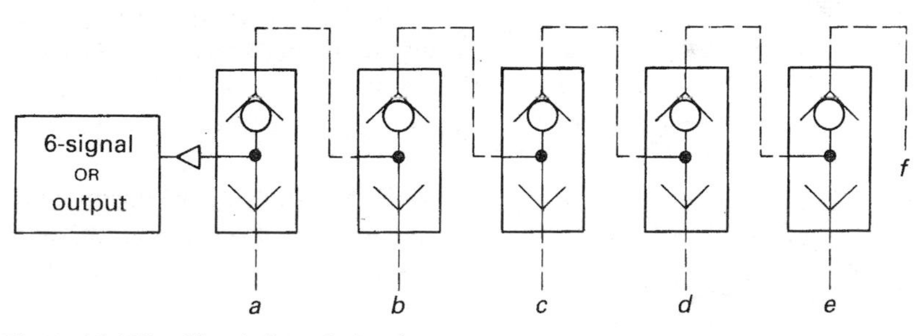

Figure 18.107 Circuit for a 6-signal OR output.

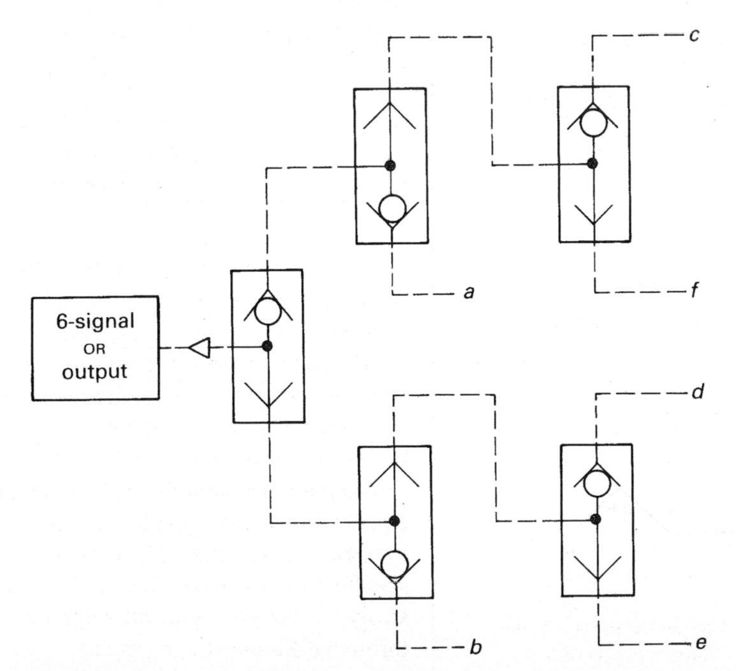

Figure 18.108 A more efficient 6-signal OR output circuit.

Alternative OR circuits

If we wish to use the method of connecting valves to give an OR operation without using shuttle valves, we could connect each of them to the next, as shown in Fig. 18.109, up to a maximum of four before we should relay to a new group of four. The system for six alternative pilot signals is shown in Fig. 18.109.

This method of relaying to a new group of four valves can be used to create any number of alternative inputs.

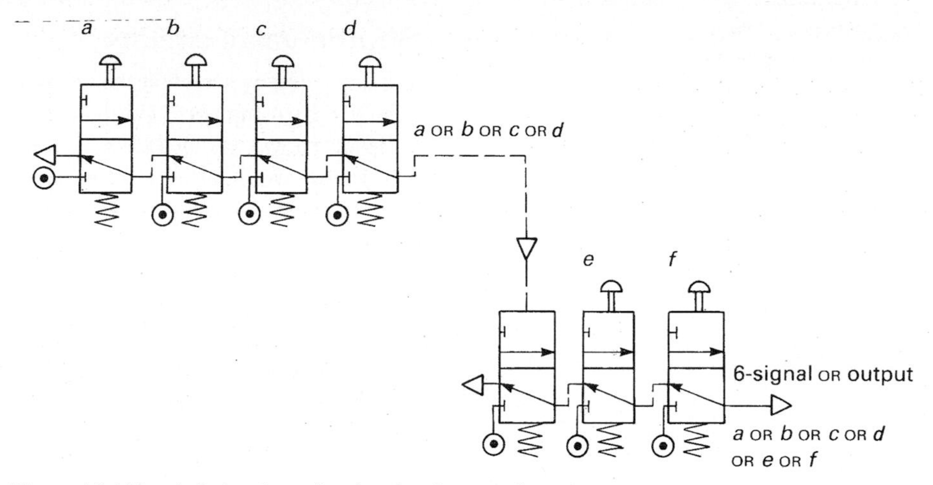

Figure 18.109 A 6-signal OR circuit using 3-port valves.

Pilot valve OR function

Should an OR function be required for one manually operated and one air signal, a 3/2 manual valve could be connected as in Fig. 18.110.

If an OR function were required for two air signals there are two alternative methods. The first is to replace the manually operated valve in Fig. 18.110 with a single-pilot operated 3/2-valve connected with signal *a* as the pilot actuator and signal *b* to port 3, as in Fig. 18.111. The output would again be *a* OR *b*.

It is possible to use a double air pilot valve to provide an OR function for two air signals by connecting it as shown in Fig. 18.112. The input *a* is connected to port 1 of our valve as well as the signal port 1 2 in a self-signalling arrangement. Input *b* is similarly connected to port 3 and port 1 0 and the output is taken from port 2. If signal *a* is on it will activate the pilot line to connect its port 1 to the output. Similarly, *b*'s signal is routed to switch the valve to enable an output to be obtained.

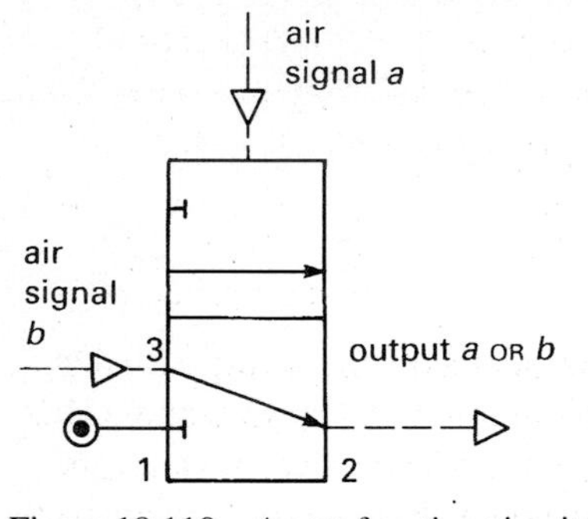

Figure 18.110 An OR function circuit for one pilot and one manual signal.

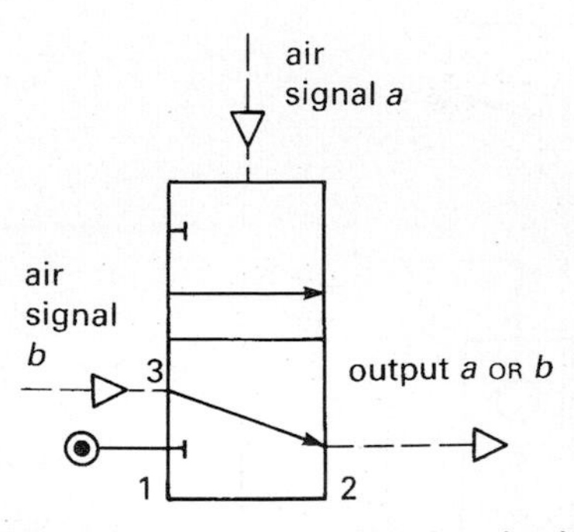

Figure 18.111 An OR function circuit for two pilot signals.

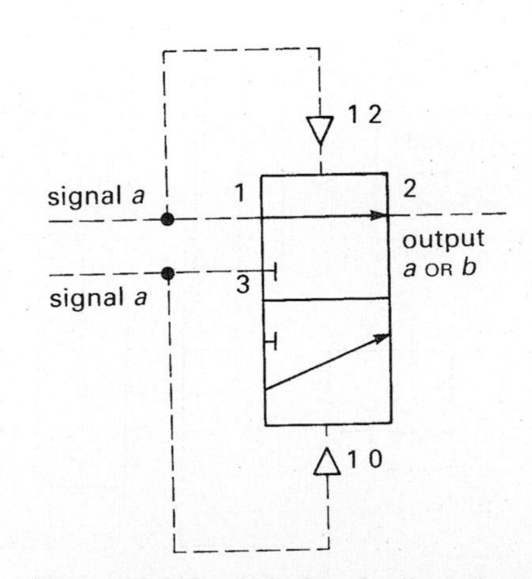

Figure 18.112 A double air pilot valve connected to provide an OR output.

18.5.2 AND logic circuits

Simple AND circuit using 3/2-valves

A simple AND function is achieved by connecting the output of one pilot valve to the main air port of the second valve, as was illustrated in Fig. 18.42. It is obviously possible to extend this system to any number of inputs by interposing extra valves connected in the same way as valve **b** between valve **b** and the output. An alternative method is to connect the output of both valves to a T-connector as in Fig. 18.113 so that one will exhaust through the other unless both valves are actuated.

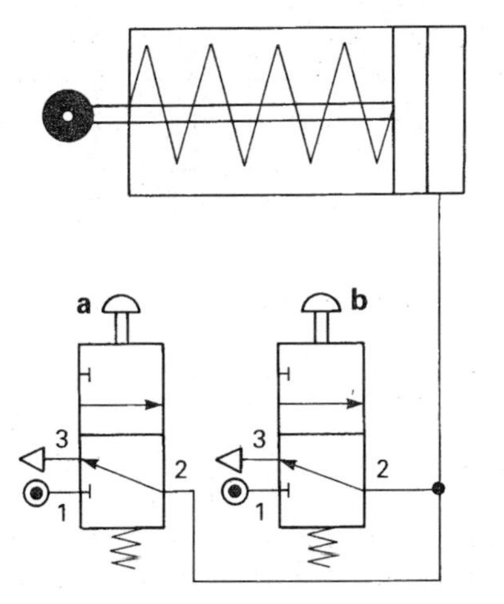

Figure 18.113 An alternative AND circuit.

These circuits are intended for use when the operator's hands are to be kept away from the machinery, as they will not function without both hands on the control buttons. In practice, operators will often try to jam one control down mechanically so that they will have one hand free. Consequently more sophisticated two-handed circuits had to be devised to overcome this.

Two-handed safety circuit

The circuit shown in Fig. 18.114 for the direct control of a double-acting cylinder needs the application of both valves to actuate the extension of the cylinder as **b** has main air connected to port 3. The valves must continue to be pressed down for the whole out-stroke and it is then necessary to release both valves in order to return the cylinder. This ensures that one control cannot be jammed down permanently.

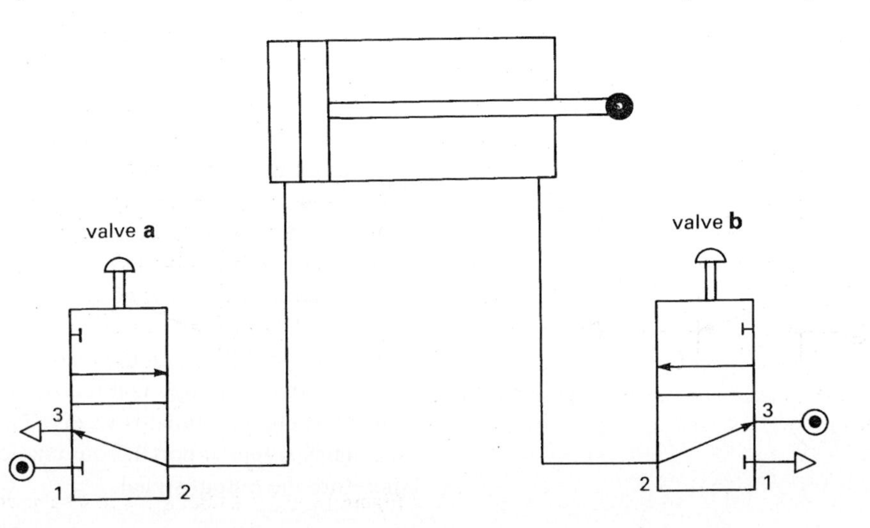

Figure 18.114 A two-handed safety circuit.

Simultaneous AND circuit

In order to ensure that the operator has both hands safe and away from the pneumatic cylinders it may be necessary for the two controls to be activated simultaneously. Fig. 18.115 shows a circuit that makes this possible.

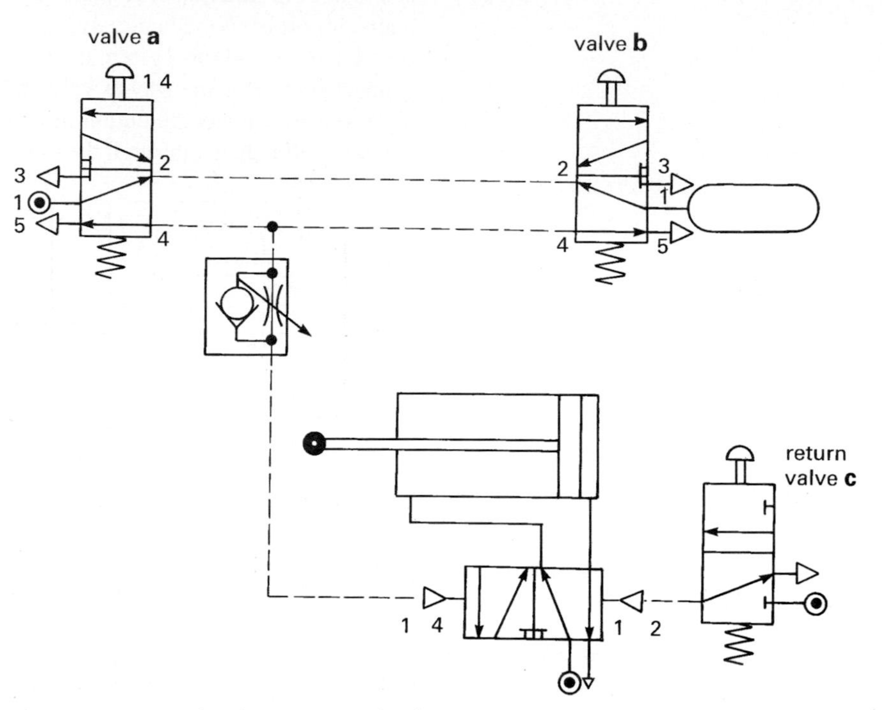

Figure 18.115 A simultaneous AND circuit.

When valves **a** and **b** are both OFF, then air from **a** will charge the reservoir at **b**. When both valves are pressed together, main air from **a** and the reservoir charge from **b** will enable the circuit to be signalled. If valve **a** is pressed before valve **b**, its main air will exhaust back through port 5 of valve **b** and the reservoir is free to discharge through port 3 of valve **a**. If valve **b** is activated before valve **a**, then air flows from the reservoir at **b** and discharges through the exhaust port 5 of valve **a**. Therefore, there is a limited time span in which to operate both pilot valves and send a positive signal to the control valve.

18.5.3 NOT logic circuits

In the pressure-sensing circuit shown in Fig. 18.116 (originally Fig. 18.65), the diaphragm valve was connected to give a NOT circuit function. No output is obtained when a pilot signal is received, and a signal is output when there is no pilot signal by connecting main air to port 3 and leaving port 1 open to exhaust.

An alternative method of achieving the same function is to use the NOT valve shown in Fig. 18.117. When no pilot signal is received at port 1 0 there is a continuous output at port 2. This is because port 1 2, which activates the main air from port 1 to port 2, receives mains air from its internal connection to port 1 which is only disconnected when a signal is received at the other actuator port. The actuator symbol for port 1 0 is obviously larger than that for port 1 2 and represents the magnification of a low-pressure signal large enough to exceed the signal at port 1 2. When this happens, port 2 is connected to port 3, the main air line at port 1 no longer is directed through the valve, and so no signal is output when an actuator signal is received at 1 0. When the pilot signal at port 1 0 ceases, the mains signal at port 1 2 automatically reinstates the internal 1 2 routing and therefore the output signal.

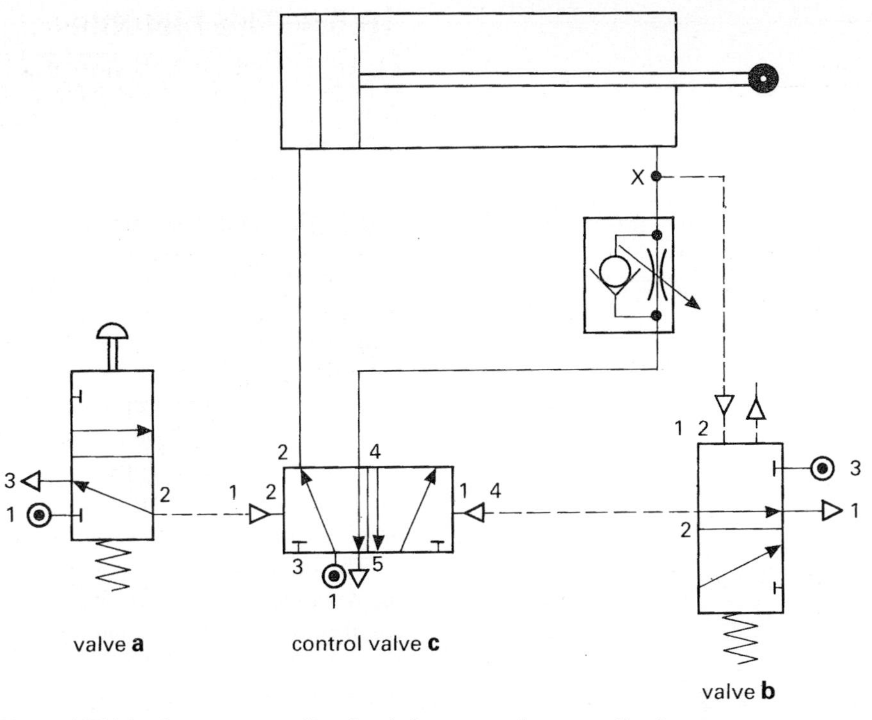

Figure 18.116 A pressure-sensing circuit for automatic return of a piston.

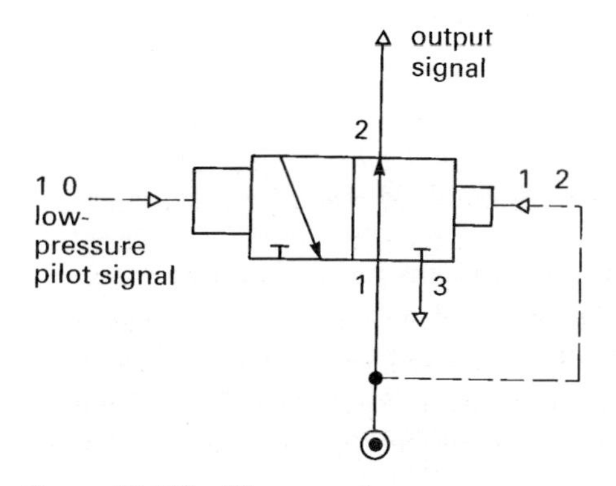

Figure 18.117 The NOT value.

Fig. 18.118 shows the use of a NOT valve in a pressure-decay sensing circuit. This is the equivalent circuit to Fig. 18.116 which ensures that there is not a cylinder return signal to the control valve as long as the cylinder exhaust pressure is high. As the NOT valve operates on a pressure differential between the main air signal and the amplified exhaust signal, it can be seen that by varying the supply pressure to the valve the return pressure can be varied. Thus the length of time that the cylinder moves on out-stroke can be adjusted.

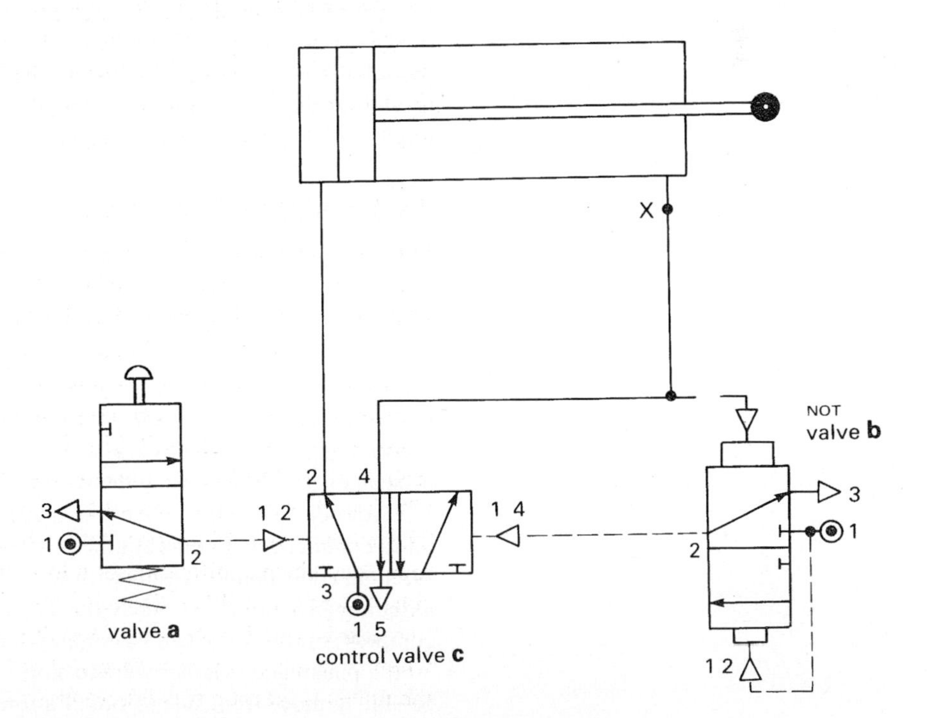

Figure 18.118 The NOT valve used in a pressure-decay sensing circuit.

18.5.4 The bistable

The double pressure-operated 5-port valve can be considered to be a bistable device as it has two stable outputs which are triggered by momentary pilot signals which can be considered to be set and reset inputs.

18.5.5 Fluid logic devices

The air bleed system introduced in section 18.3.8 was shown to be a sensing device which passes a signal when an object makes contact with it. This idea has been refined and developed to produce *air-jet devices* which are now used in many control circuits and which operate on a continuous, oil-free, low-pressure supply of less than 1 bar. The development of *fluidic control devices* (fluid logic control devices) and *blow nozzles* has widened the use of low-pressure pneumatics because complete systems can now be designed with touch-free sensors. As the operating pressure is so low there is very little wastage of compressed air but circuits require high amplication of the low-pressure signals.

The simplest fluid devices are based on a blow nozzle. This creates an interruptible jet which will sense when an object has passed through a flow of air. It can therefore be used to count components travelling past on a conveyor belt, or as the pulse unit for a revolution counter. It has a similar function to the electronic photocell but has the advantage that it is not adversely affected by dust and can also sense transparent materials. Therefore it can be used to sense the edge of plastic sheeting and hence control the rolling of the sheet. The blow nozzle can also be incorporated into a *proximity sensor* which can sense when an object is within a certain range of the sensor. This can also be used to count passing components without risk of damage to them. The proximity sensor also has special application to the control of containers being filled with fluids such as paint, because the level of the fluid can be sensed without coming into contact with the sensor.

Fluidic control devices have proved very successful in controlling air-jet systems. These logic devices can provide the correct output with no internal moving parts, merely the appropriate internal routing of the air jet. This means that the components have far less wear and are more reliable. A *turbulence amplifier* will provide a NOR logic function for slow, long-range sensing. A *wall attachment device* is a bistable fluid logic device and when the output signal is amplified significantly it can be used in place of a double pressure-operated 5-port valve. It utilises the boundary layer effect of the fluid flow to provide the correct output. Similar devices can provide OR, AND and NAND functions. Integrated circuits of fluidic elements can be constructed to provide all the fluid control elements and connections.

18.6 Hydraulics

If pneumatics is a pervasive technology in automated industrial systems then hydraulics is pervasive in the world around us. Automobile braking systems and power steering are two obvious uses and the ubiquitous digger and tipper truck can be seen on streets all around the world.

Hydraulics, like pneumatics, uses a fluid to operate cylinders. This fluid is usually oil in commerical systems but the principles can be demonstrated and investigated safely in educational systems with water as the fluid. The main property of a gas and a liquid which affects its behaviour in pneumatic or hydraulic systems is its compressibility. A gas can be compressed – a liquid cannot.

If a pneumatic cylinder were to stop in the middle of its stroke, the piston would be able to move from side to side if a force were applied. If a hydraulic cylinder stops in the middle of its stroke it is firmly locked in that position. Hydraulic systems

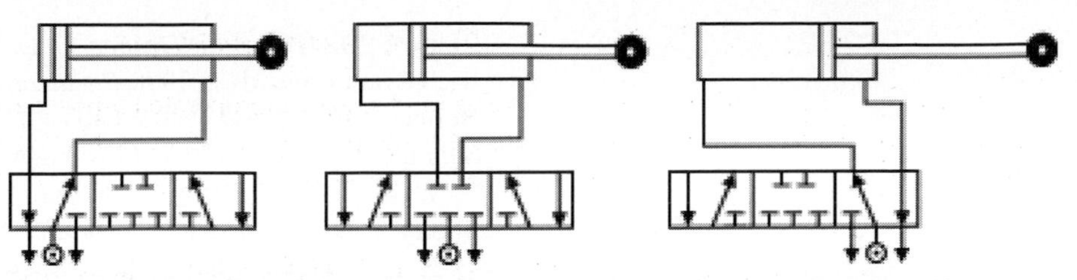

Figure 18.119 Five port control of a hydraulic cylinder with a centre lock off position.

are therefore designed to stop the cylinder in any position. You could liken pneumatics to a digital system – the cylinder is always either at one end or the other or on its way there. Hydraulics is an analogue system – systems can be locked in any position. To do this we need an extra position on our valves in which the system is said to be locked off.

Fig. 18.119 shows the three positions of this five-port valve. The middle position shows the locked off state. The springs at either end show that when the actuator is released, in either end position, the springs return the valve to the locked off state. it is instructive to compare this diagram with Fig. 19.33, showing the pneumatic two-state equivalent. Note that in hydraulics diagrams the arrows are shown filled in to distinguish them from the open arrows of pneumatics.

Hydraulic power has an advantage over pneumatic power in that it can produce high forces and accurate positioning. Where a load must be inverted or where a constant feed speed under sudden or intermittent loading is required, hydraulic and mechanical elements must be used. The higher compressibility of air makes pneumatics unsuitable for these applications. Hydraulic power is most useful at pressures of 60 bar to 150 bar. Hydraulic motors compare favourably to electrical motors; they can provide high torque at low speed and have an efficiency of about 80% to 85%. The oil used to power a hydraulic system must have an enclosed feed and return system, unlike that for air, and consequently it is about three times as costly as pneumatics. Care must be taken to seal the components, as any leaks of the hydraulic oil can be messy and dangerous.

For these reasons, if a steady output under varying load is required but not excessive forces, a combination of hydraulics and pneumatics, *hydropneumatics*, can prove expedient in industry. In this system, compressed air is used to pressure a cylinder of oil which operates the output. This is obviously cheaper, easier and cleaner to manage than hydraulics if high pressures are not required.

Hydraulic and pneumatic circuits have very similar control systems and pneumatic control techniques can be transferred to their hydraulic counterparts, remembering that the return flow of the hydraulic fluid must be included. The properties and behaviour of fluids and gases are considered in more detail in Chapters 12 and 13. The bibliography at the end of this chapter includes books and publications that detail hydraulic circuit design.

Bibliography

Festo Didactic *Introduction to Pneumatics* and *Fundamentals of Pneumatic Control Engineering*. Festo Didactic Ltd Brentford

Kemperman *Pneumatic Mechanisation*. Kemperman

Martonair 1985 *A Course in Applied Pneumatics*. Martonair Ltd St Margarets Road, Twickenham, Middlesex

Mecman 1986 *Manual of Pneumatics and Circuitry*. Mecman Ltd Mecman House, Sutton Park Avenue, Reading RG6 1AZ

Parr A 1998 *Hydraulic Pneumatics*, Butterworth-Heinemann

Patient P, Pickup R and Powell N 1986 *Schools Council Modular Courses in Technology: Pneumatics*. Oliver and Boyd

Rexroth *The Hydraulic Trainer*. G. L. Rexroth Ltd, St Neots, Cambridge
Rich S and Edwards A 1990 *Pneumatics*. Stanley Thomas
Rohner P 1979 *Fluid Power Logic Circuit Design*. Macmillan
Schrader Bellows *Technical Course Notes*: *Basic Pneumatics*. Schrader Bellows, Walkmill Lane, Bridgetown, Cannock WS11 3LR
Sperry Vickers *Industrial Hydraulics Manual*, *Mobile Hydraulics Manual* and *Practical Hydraulics Booklet*. Sperry Vickers Ltd, Southampton
Weaving C 1984 *Introducing Pneumatics*. Teaching Media Resource Service, Russell House, 14 Dunstable Street, Ampthill, Bedford

Assignments

1 (a) Explain the advantages of using pneumatics in industrial manufacturing applications.
(b) A pneumatic road drill is an example of a reciprocating device. What is the other main category of pneumatic device and how do the two categories differ? Give two examples of each.

2 (a) Explain the main stages in the production of compressed air for use in pneumatics.
(b) How can an air pressure of 500 bar best be achieved?

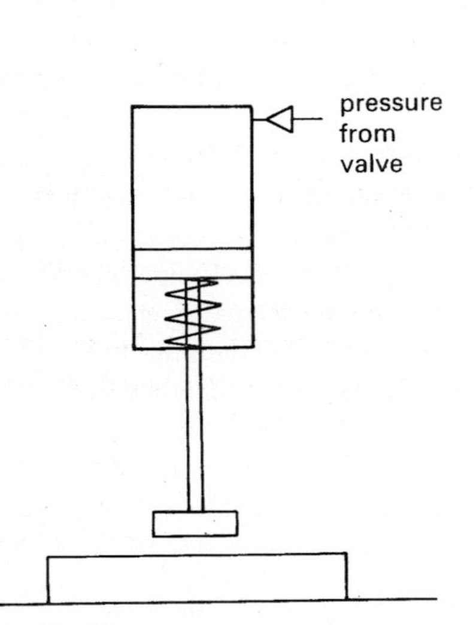

Figure 18.120

3 (a) Specify as completely as possible the type of cylinder being used in the clamping operation shown in Fig. 18.120. What type of valve would you use to operate the clamp manually?
(b) Explain the movement and effect of both the main air supply and the exhaust air when the clamp is activated, held down for a few seconds and then released.
(c) If it was required to keep both hands free to process the work once clamped, suggest two alternative valves that would allow this to be done and explain any disadvantages of each system.

4 (a) If a double-acting cylinder is controlled in each direction by a 3-port valve and both valves are pressed at the same time, what is the result?
(b) Explain the cushioning effect of a double-acting cylinder.
(c) What cylinder would be required if an exact stroke length was needed that was between two standard sizes?

5 (a) Explain the function of a pneumatic shuttle valve and describe how it operates. Give one example of its use.
(b) Explain the action of a 5-port lever set-reset valve.
(c) How could a 5-port pilot valve be used in remote control in a chemical plant and how would this valve differ from the one described in part (b)?

6 (a) Find the working pressures of typical car and bicycle tyres. Give the values in SI units.
(b) Does the compressed air stroke or the spring return of the single-acting cylinder produce the greater force?
(c) (i) Calculate the out-stroke force produced by a 20 mm diameter cylinder when it operates at 90% efficiency and a supply pressure of 6 bar.
(ii) If the back pressure is initially 3 bar what will the initial force be?

7 (a) Explain how a single valve could operate the simple shaft brake shown in Fig. 18.121.
(b) If the required force on the brake pad was 360 N, what would be the minimum diameter of cylinder? Why would it need to be greater than this in practice?
(c) If a two-handed operation of the braking mechanism were necessary, how could this be achieved?

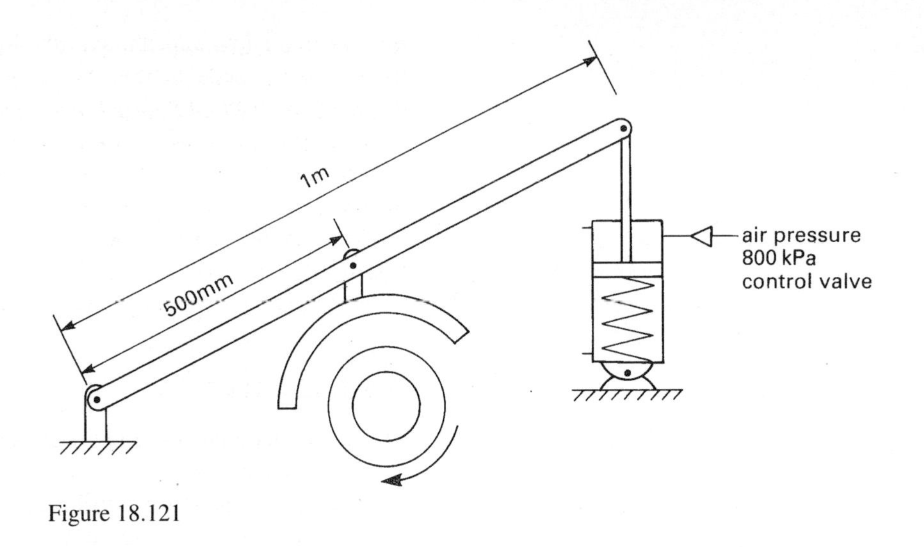

Figure 18.121

8 Consider the circuit shown in Fig. 18.122. When air is routed from port 1 to port 2 in the 5-port valve, cylinder A is used to open the furnace door and cylinder B is used to load the component. They are activated simultaneously. When the valve is switched over cylinder B retracts and cylinder A closes the door again.

(a) Draw the completed circuit that will accomplish the above task.

(b) If the furnace door weighs 100 N, and the supply air pressure is 600 kPa what is the minimum possible diameter for cylinder A?

(c) In practice a time delay would be necessary to ensure that the door was open before loading commenced. How could this be achieved?

(d) The out-stroke of cylinder B would also need to be performed more slowly so that the component would not be damaged. Explain in detail how this could be done without affecting the in-stroke of cylinder B.

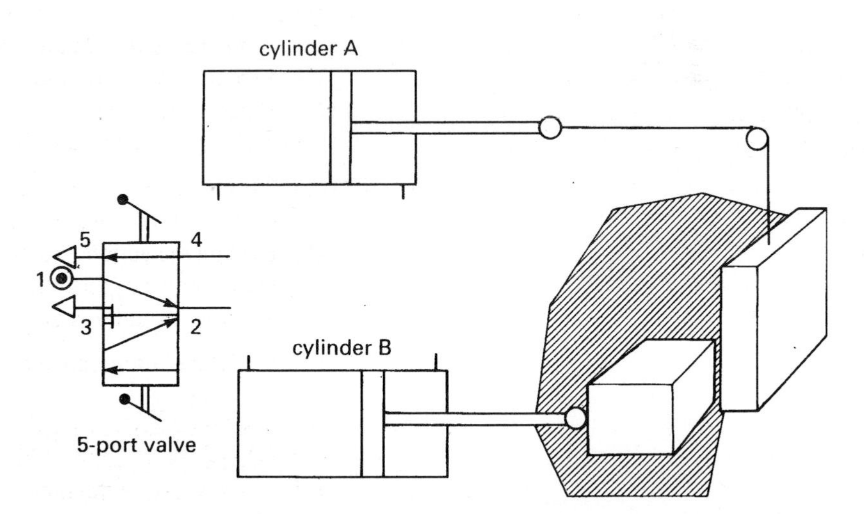

Figure 18.122

9 (a) What is an air bleed and how is it used?

(b) Explain the action of a diaphragm-operated pressure-sensitive valve.

(c) State one application where both the air bleed and the diaphragm valve could be used together.

(d) When is it useful to have a pressure-sensitive valve that is a spool valve rather than a poppet valve?

10 (a) Explain the meaning of the following terms:
(i) mechanisation
(ii) automation
(b) Select a suitable product and describe how mechanisation and automation could have been introduced into the manufacturing process.
(c) Discuss the social advantages and disadvantages of introducing automated production.

11 (a) Fig. 18.123 shows an automatic return circuit. Explain in detail the function and operation of each component during one complete cycle of operation. Give an example of the possible use of this circuit.
(b) If a fully automatic circuit is required in which the out-stroke is also to be signalled by sensing the pressure decay, how could we achieve this? Describe where a set-reset safety valve, which would stop the cylinder at its in-stroke position, would be placed.

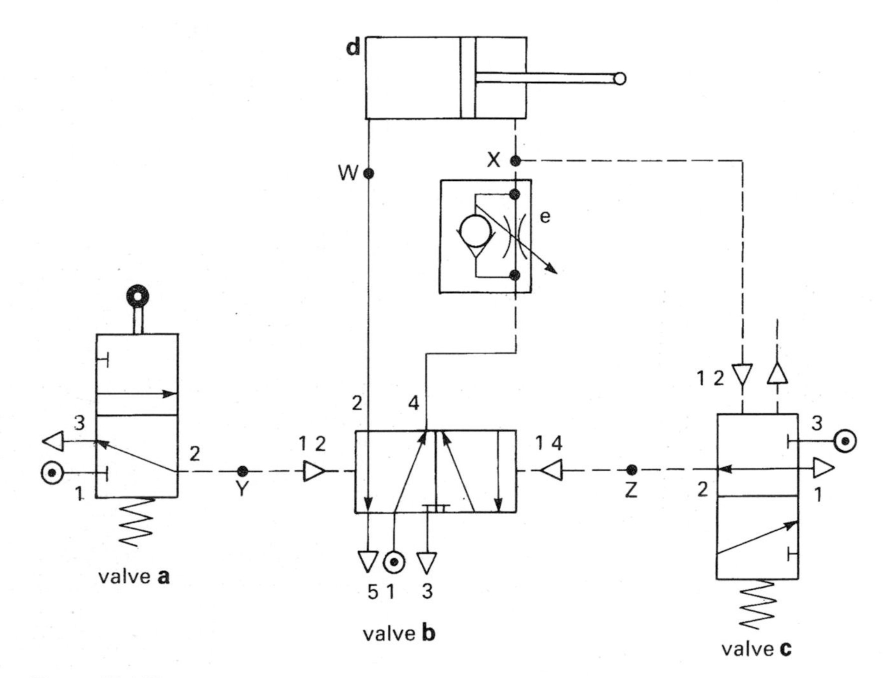

Figure 18.123

12 (a) What advantages are there in linking pneumatic and electronic circuits and what components enable us to do this?
(b) In what situations would it be dangerous to include electrical components in the control of an industrial process?

13 (a) Establish whether the following circuit can be achieved using a simple sequential circuit explaining your reasons clearly.

$$A+, C-, B+, A-, C+, B-$$

A+ signifies the out-stroke of cylinder A and A− the in-stroke.
(b) If a sequence is wrongly assumed to be a simple sequential circuit what will be the outcome?
(c) Describe briefly how the problem encountered in part (b) can be overcome.

14 In a sequential-control hot pressing operation, two dies are activated by pneumatic cylinders A and B, so that A is pressed on to the plastic sheet, then B is pressed down, A is released and finally B is released.

(a) If this is an automatic sequence after activating a start switch, draw a schematic representation of the cylinder movements using A+, A− etc. to represent the out-stroke and in-stroke of each cylinder. Is more than one air group required in this circuit?
(b) Is Fig. 18.124 a correct representation of this circuit? Explain in detail how this circuit operates and how we could alter it to make it fully automatic.
(c) Where could a 5-port lever set-reset valve be placed to interrupt the cycle when both cylinders are retracted?

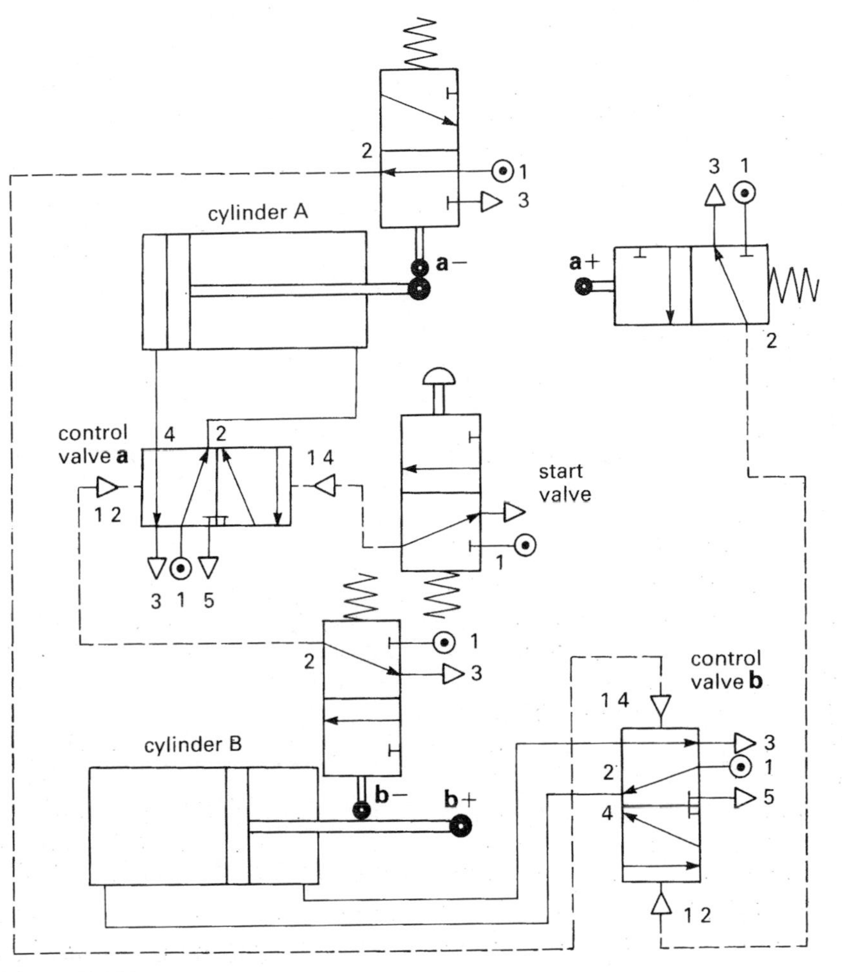

Figure 18.124

15 Fig. 18.125 shows the circuit for a pneumatically operated car park barrier. Initially, valve A, a 3-port roller-trip valve is operated by the pressure of a passing car.

(a) Explain the sequence of operations in the circuit and any problems that could arise when it is in use. How could these problems be overcome?
(b) Explain the purpose of:
   (i) the reservoir, D, and flow regulator, C.
   (ii) the flow regulator, E.
(c) By approximating each side of the barrier to a triangle, find the minimum internal diameter of the cylinder, G that could raise the barrier. The supply pressure is 600 kPa and the barrier is 50 mm thick and made of timber of density 700 kg/m^3. Suggest a suitable bore diameter of cylinder G for the operation of the car park barrier.

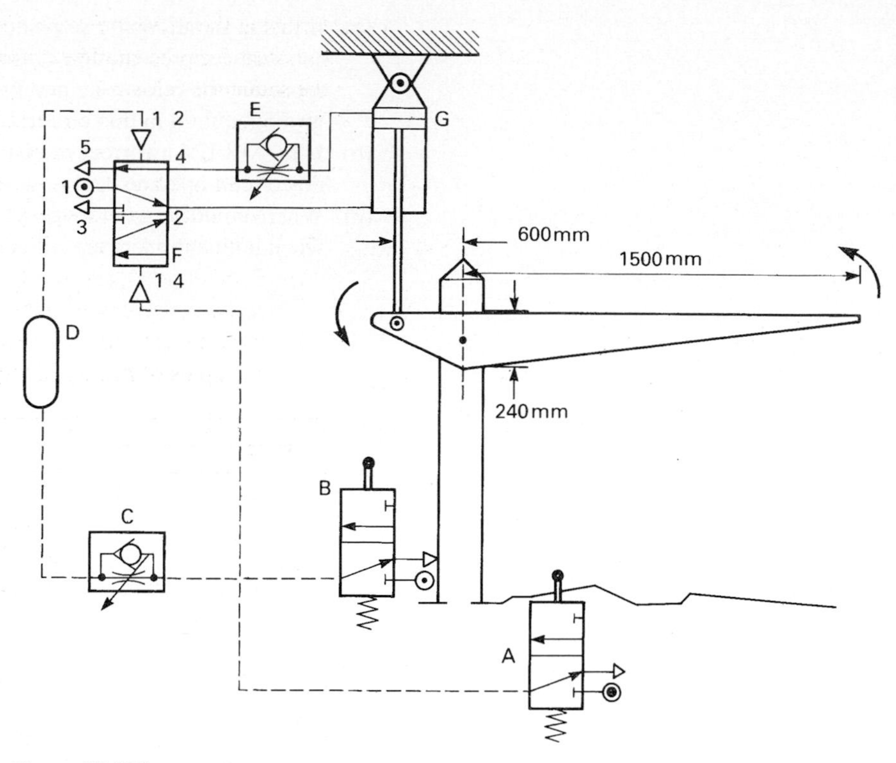

Figure 18.125

16 A pneumatic control circuit is required for an ultrasonic welding machine. The sequence of operations is as follows.

1. An ultrasonic horn is to be mounted on a crosshead between two vertical, rodless cylinders moving in parallel. A small plastic keyring that is to be ultrasonically welded is positioned manually in a jig directly beneath the crosshead.
2. With the crosshead positioned at the top, motion is started by simultaneously pressing two push button valves. These valves must be held down until the ultrasonic horn contacts the workpiece, and the release of either, before contact is made, must result in the crosshead reversing back to its start position.
3. The ultrasonic horn should descend slowly until it contacts the workpiece. At this point an electrical timer should be actuated by a microswitch which starts the ultrasonic welder and after a preset time has elapsed, a second timer should be signalled. This should maintain the pressure for a preset dwell time after which the crosshead should quickly retract to its start position. Once the timing process has started it must run to completion, regardless of whether the operator has released one, both or none of the start buttons. The crosshead should retract after the weld and dwell timing is finished even if the operator continues to hold the start buttons down.

Plan, design and draw a circuit which will perform the ultrasonic welding operation to the specification given above. List the components needed for your circuit and show on a planning chart the actuation and output signals for each operation in the sequence.

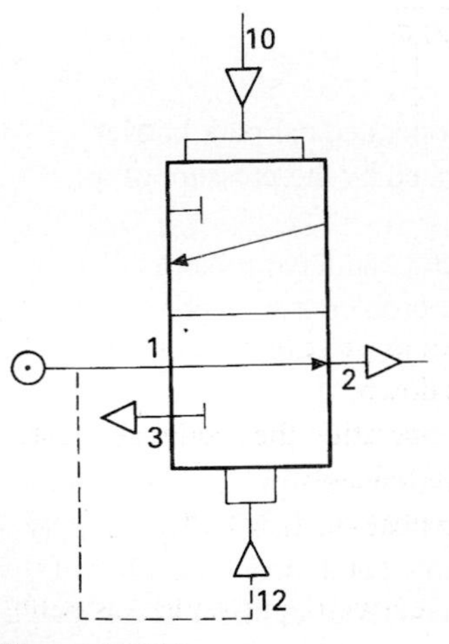

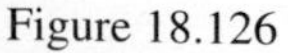
Figure 18.126

17 (a) Describe, individually, the main stages in the preparation of compressed air suitable for use in an industrial manufacturing process.
Suitable symbols may be used to help in your explanation.
(b) What is the logic function of the valve shown in Fig. 18.126? Illustrate and describe its operation.

(c) Utilising the valve shown in Fig. 18.126, design a circuit which uses pressure decay to reopen a guard door should it fail to close completely in the sequence below.
   (i) Guard door closes (cylinder A out-strokes).
   (ii) A 'press' moves down (cylinder B out-strokes).
   (iii) After a delay, the press retracts with the guard (B and A in-stroke almost simultaneously).

(d) The circuit shown in Fig. 18.127 demonstrates an 'interlocking' technique.
   (i) Explain how the circuit operates, giving a detailed stage-by-stage account starting with the operation of valve X.
   (ii) Why are circuits designed to interlock in this or any other way? Give examples of *two* applications.

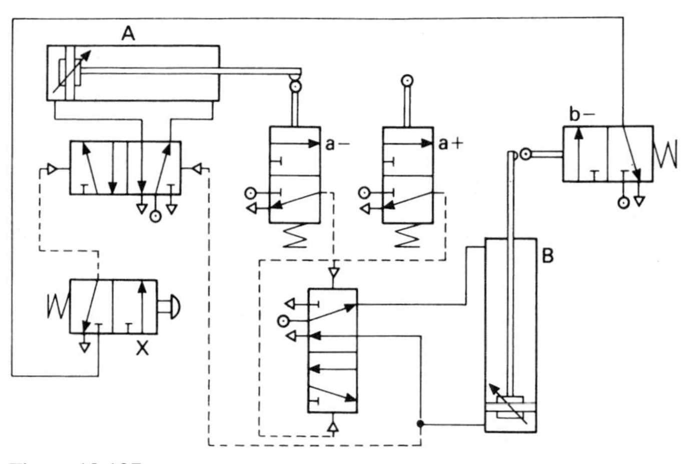

Figure 18.127

18 (a) Sketch a labelled, cross-sectional view of a pneumatic OR valve. Explain briefly how this valve works and give *one* example of its use.

(b) Fig. 18.128 shows a pneumatic press for bending metal brackets. The bending process is successfully completed when air bleed A_2 is blocked by the metal bracket.

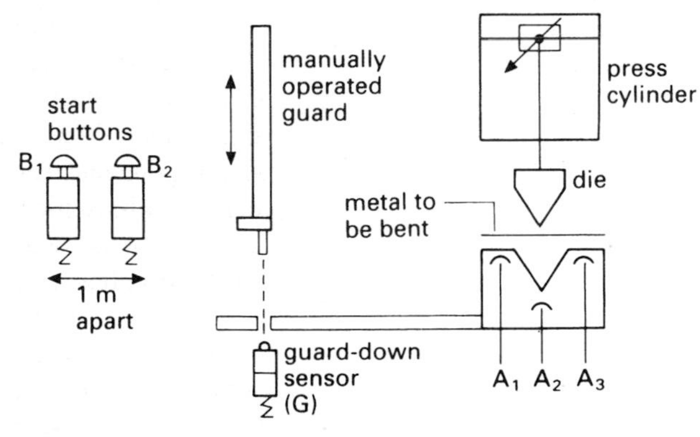

Figure 18.128

In order to operate safely the following conditions must be met.

1 The press must operate only if the guard is down.
2 If the guard is lifted at any time during the operation the sequence must be interrupted and the press must retract immediately.
3 The sequence must start only if both start buttons are pressed simultaneously.
4 The sequence must start only if the metal is correctly positioned covering air bleeds A_1 and A_3.

Using Fig. 18.129, design a pneumatic circuit which will satisfy all of the above criteria.

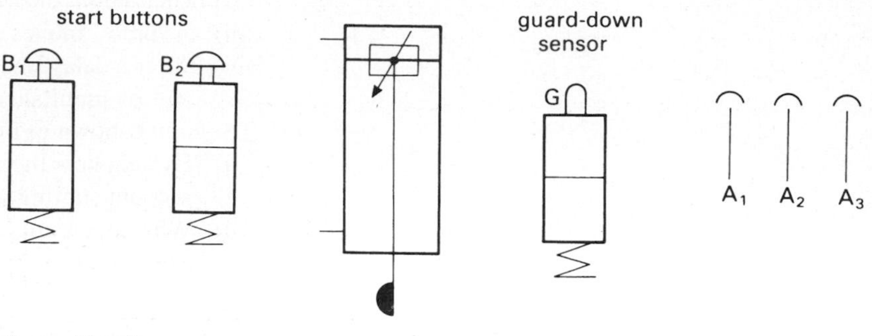

Figure 18.129

(c) The circuit in (b) can also be expressed as a logic diagram. Complete the diagram below using logic symbols to explain the correct operation of the press.
Assume that pressing a valve or blocking an air bleed creates a logic 1 signal.

Inputs		*Outputs*
start	B_1	press cylinder
buttons	B_2	down (logic 1)
guard	G	press cylinder up (logic 1)
air	A_1	
bleeds	A_2	
	A_3	

19 (a) (i) Describe, with the aid of a clearly labelled sketch, the operation of a simple 2-stage reciprocating compressor.
(ii) Explain why a reservoir is required for this type of compressor and describe, with the aid of a simple sketch, one method of controlling the reservoir pressure.

(b) In Fig. 18.130 the double-acting cylinder is simply controlled by pilot signals to the main direction control 5/2-valve. However, three additional 3/2-valves (A, B and C) have been added to produce a manual safety feature.
(i) By investigating the action of the additional valves, explain their action and hence that of the safety circuit.
(ii) Suggest a suitable application for this circuit modification.

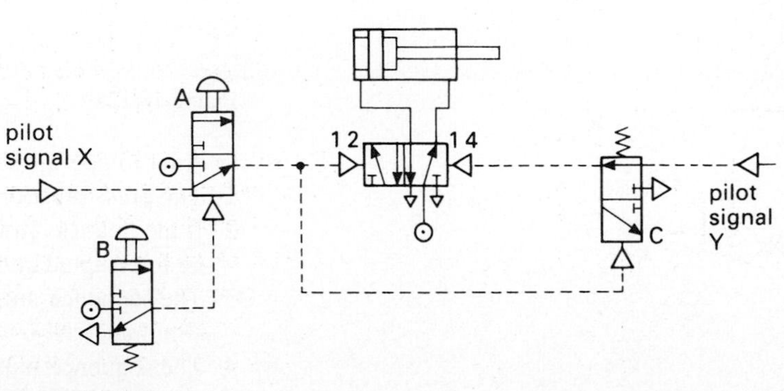

Figure 18.130

(c) Fig. 18.131 shows the arrangement of a machine for the automatic assembly of roller-skate wheels.

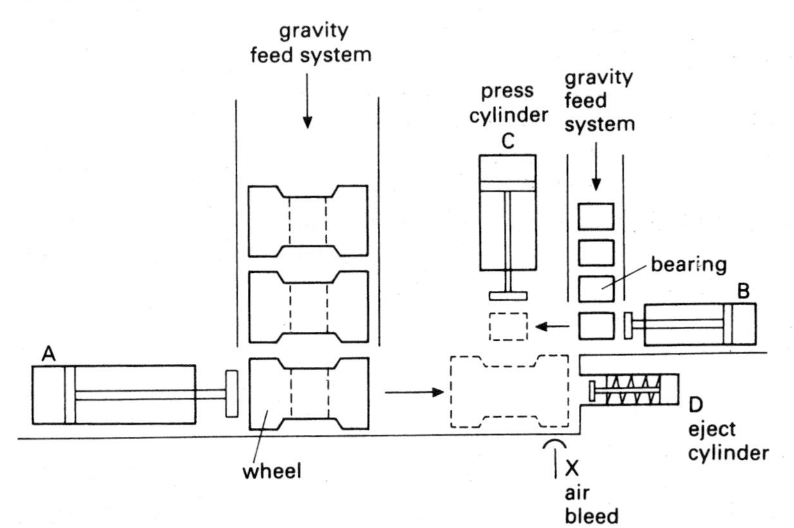

Figure 18.131

The sequence of operations is as follows:
Cylinder A out-strokes to deliver the next wheel to its correct position for pressing. When air bleed X senses the correct position of the wheel, cylinder B out-strokes to position a bearing above the wheel and then retracts to collect the next bearing from a gravity-fed system. Press cylinder C then out-strokes to press-fit the bearing into the wheel. Cylinder C then in-strokes, followed by cylinder A, then single-acting cylinder D out-strokes momentarily to eject the assembled wheel.
The sequence then repeats automatically.

(i) Write down the required operational sequence of the cylinders and, utilising a cascade air group control system, complete a planning chart in the form shown in Fig. 18.132. An air-bleed circuit is to be used to sense the out-stroke position of cylinder A, but you may use any suitable feedback method for other parts of the sequence. Speed control should be used where appropriate.

Operation		*Feedback*	*Next Operation*
start	A+	a+ (air bleed)	B+
bearing positioned	B+	b+	

Figure 18.132

(ii) Draw a neat diagram of your complete circuit and group cascade memory circuit.

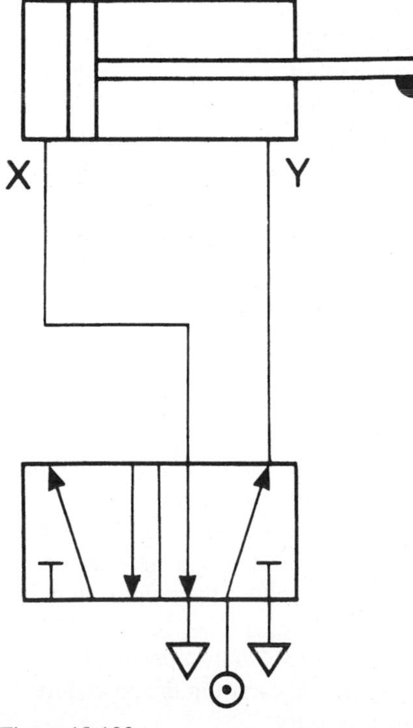

Figure 18.133

20 (a) In order to control the outstroke speed of the pneumatic cylinder shown in Fig. 18.133, a flow restrictor is required. State, with reasons, which position, X or Y, is the most suitable for placing the restrictor.

(b) Draw a circuit diagram incorporating a pilot-operated spring-return valve and a reservoir to show how a brief pilot signal could be made to trigger a fixed-volume air blast for removal of swarf from a drilling machine.

(c) In an industrial weaving process synthetic yarn is supplied wound on aluminium tubes. At the end of a production run there is always a small amount of yarn left on each tube. This yarn can be profitably recycled if it can be removed by a cheap enough method. The machine illustrated in Fig. 18.134 is used to remove scrap yarn by the following process:

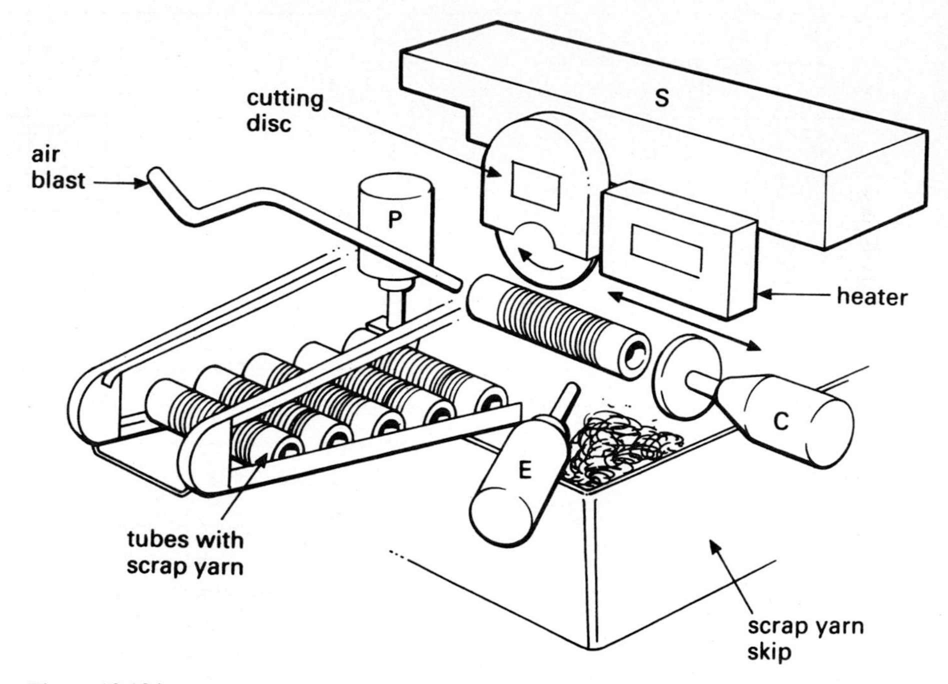

Figure 18.134

Cylinder P instrokes, then outstrokes, allowing one tube to enter machine. Cylinder C clamps tube in position. Rodless cylinder S moves a cutting wheel from right to left and back again to cut through the yarn to within 0.3mm of the tube surface. (A heater element positioned behind the cutting wheel melts through the remaining 0.3mm of yarn.)
An air blast then blows the scrap yarn into a skip.
The tube is unclamped, then ejected by cylinder E.

(i) Plan out the sequence using a planning chart or other suitable method.
Use the cascade method of control and include any air group changes in your planning chart.

(ii) Using the components shown in Fig. 18.135, draw a circuit diagram which would execute this sequence continuously so long as valve X is in the ON state.

21 The production area of a microchip manufacturer has to be ultra clean, so personnel entering must pass through an air-lock. (Layout as shown in Fig. 18.136.) The system must never allow doors A and B to be opened at the same time. Door A is opened by pressing valve X and, after personnel have entered the air-lock through A, it is closed by pressing valve Y. Both doors must remain closed for 10 seconds after door A closes, then door B must open automatically. After passing through door B, the workers press valve Z to close the door behind them. (The air-lock is designed for travel in one direction only, i.e. from A to B.) The closing force of the doors is small, so they will stop if an obstruction is encountered.

(a) Starting with the components shown in Fig. 18.136, draw a pneumatic circuit diagram for the control system required to carry out the procedure. Add any components you consider to be necessary.

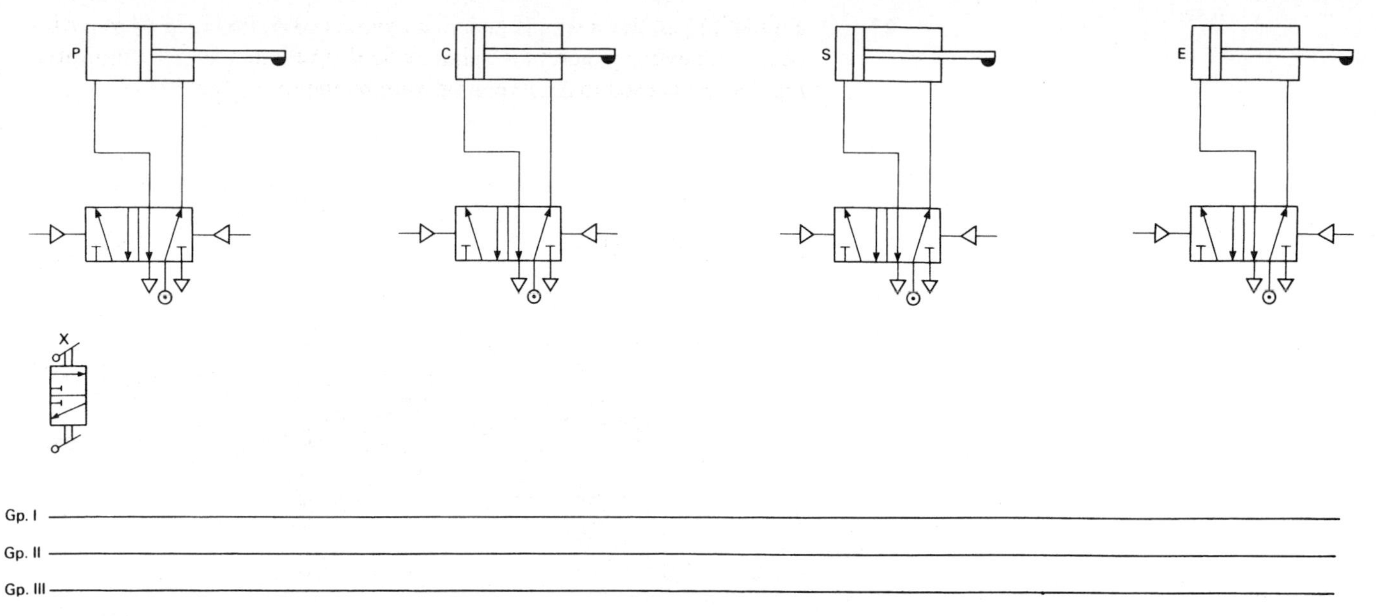

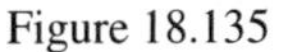

Figure 18.135

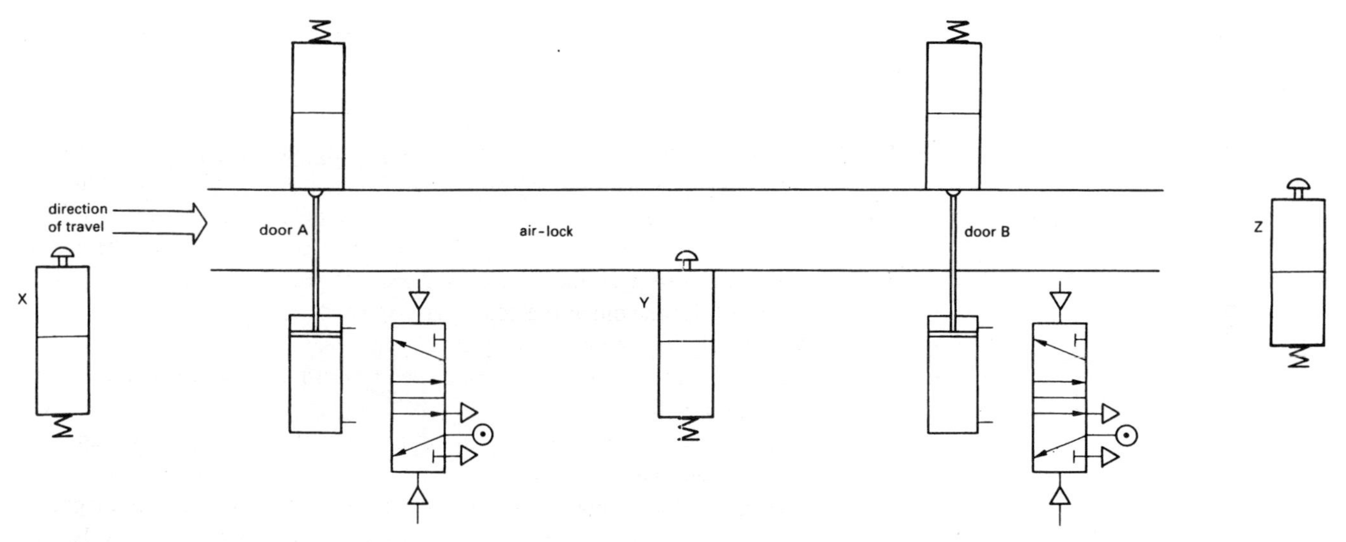

Figure 18.136

(b) Suggest three different ways of detecting people entering the air-lock and describe their relative merits.

(c) After four people have entered door A, one at a time, a red light warns that this is the maximum and stays on until they have all passed right through and the last person pressed valve Z, which closes door B and resets the green light.

Design an electronic system, using JK flip-flops, to operate the warning lights as described. Indicate where the entry sensor and valve Z are connected to the JK flip-flops.

(**Note**: Part (c) is covered in Chapter 19.)

22 (a) Fig. 18.137 shows a simple pneumatic circuit operating on a 4 bar pressure supply. The graph shows the variation of pressure on gauge P during one instroke/outstroke cycle. Explain the shape of the graph.

(b) Describe with the aid of a simple labelled sketch how a low flow of hydraulic fluid at 100 bar pressure can be produced *without* the use of a hydraulic pump. Assume that pneumatic air supply of 4 bar pressure is available.

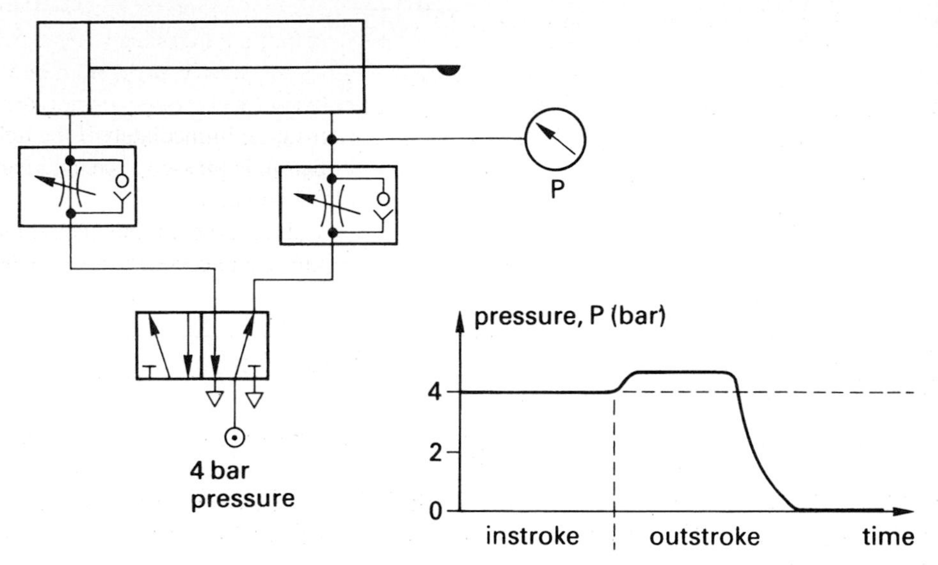

Figure 18.137

(c) Fig. 18.138 shows a hydraulic press used to form car body panels out of sheet steel. The press is to be operated by a combination of hydraulic, pneumatic and electrical systems.

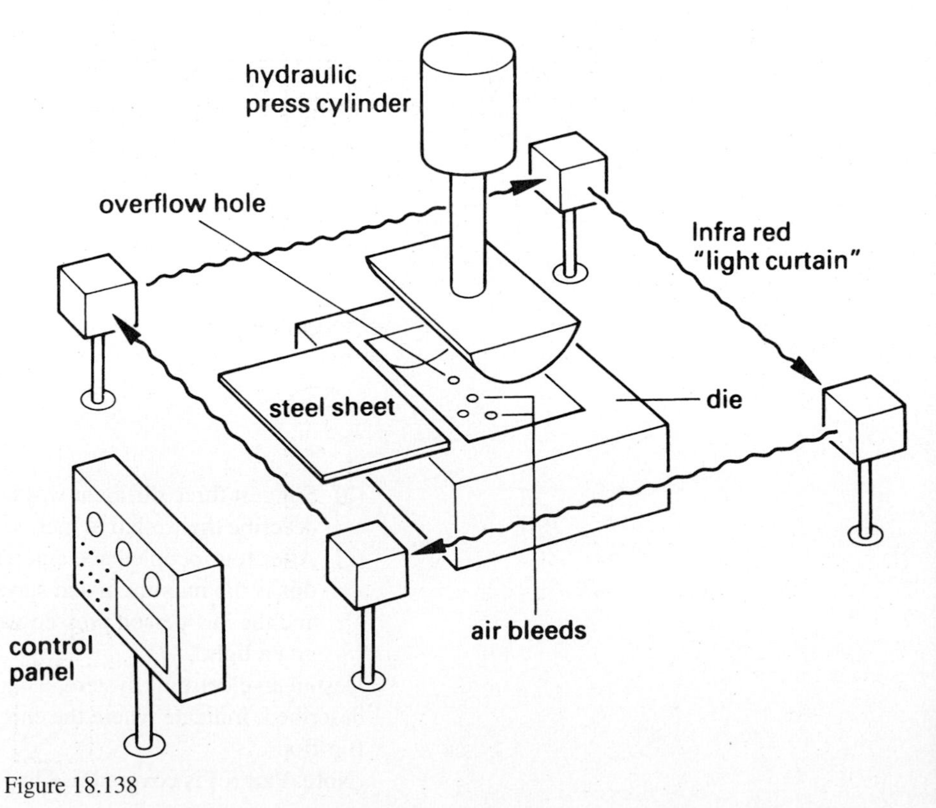

Figure 18.138

The press cylinder is hydraulic and is powered by a 100 bar hydraulic pump unit.

The press must not outstroke until a pneumatic key-operated 3/2 valve is **on** and two push-button 5/2 valves are pressed **simultaneously**. The press must *not* be operable if one of the 5/2 valves is jammed permanently on.

The metal in the press is sensed by three air bleeds in the bottom of the die. The metal is correctly formed when *all three* air bleeds are blocked. When the metal is correctly formed there should be a short time-delay before the press instrokes.

Around the press is an infra-red 'light curtain'. If anyone breaks the beam by entering the press area, a 24 V output is produced. The press must instroke immediately if the light curtain is broken or if the emergency stop button is pressed. The emergency stop button is a pneumatic push-button 3/2 valve.

Using the outline diagram in Fig. 18.139, design a circuit which will safely and correctly operate the press according to the above specification.

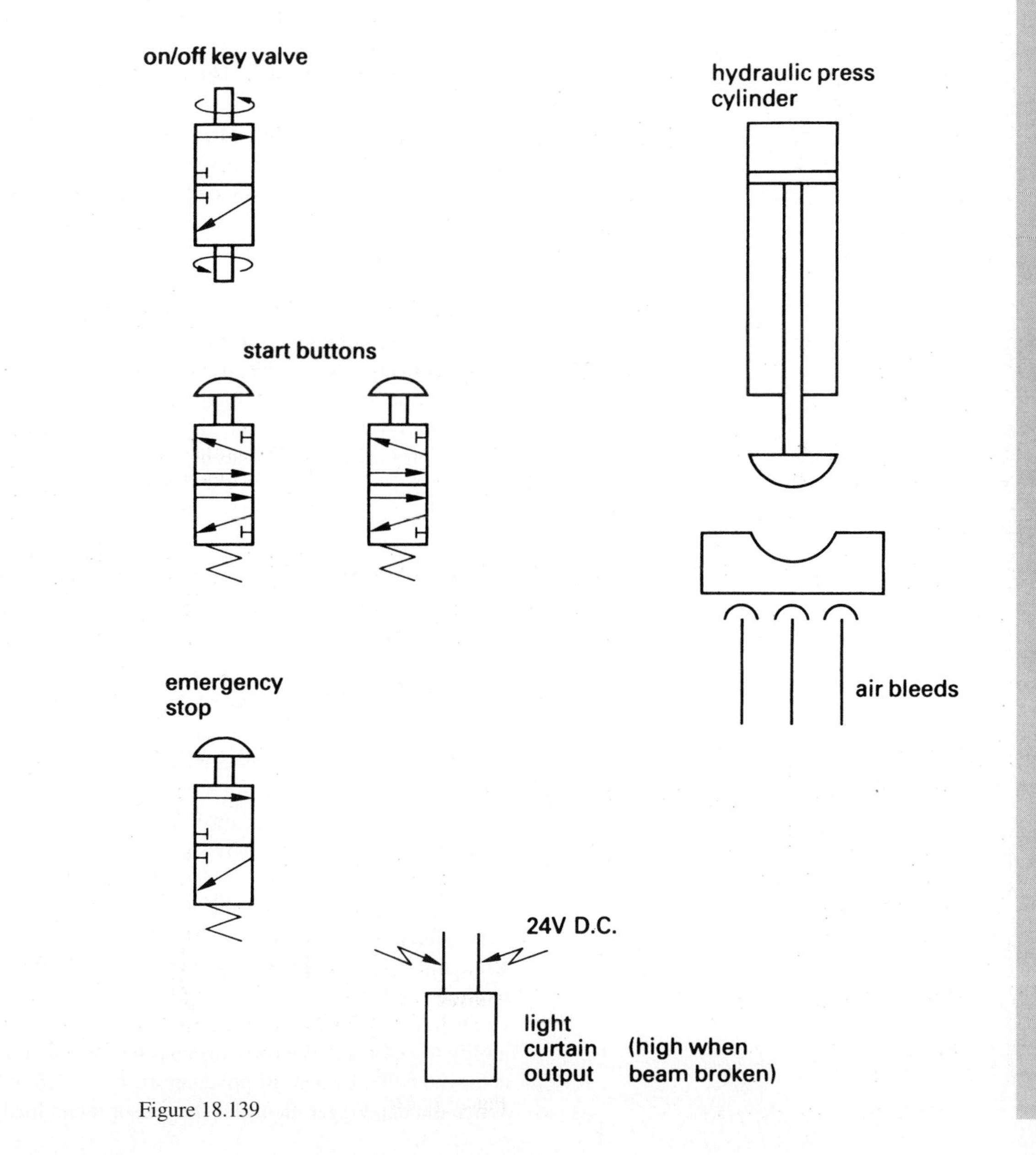

Figure 18.139

19 Programmable control systems

19.1 Overview of programmable control systems

Section 16.5 dealt with a general overview of the parts of a microprocessor system. Much has changed since the first edition of this text was written in the 1980s. Control is now predominately achieved using PCs with commercially made interfaces (or buffer boxes) connected to the serial port of the computer. There is still a choice of manufacturer for these systems and developments in this area can be expected to continue. Single board computers (SBCs) are still available but have evolved into Programmable Interface Controllers (PICs) which are becoming the standard way of fabricating stand-alone programmable control systems.

Choosing a suitable language for control is another area in which there has been a great deal of rationalisation. These languages are studied in more detail in Section 19.4.

There may be occasions when you are faced with the possibility of achieving a particular function in either hardware or software. A trivial example is latching an input line, i.e. storing the value of the input in order to use it again later in the program. It is possible to achieve this in either software or hardware. You should use the criteria for choice you have established in Chapters 14, 15 and 16 and not keep the two environments, software and hardware, separate. There is always a temptation to use the computer for everything whether it is appropriate or not.

We should not ignore the fact that many programmable control systems are used for datalogging. Some systems are capable of both datalogging and control, although it is still true that a system designed for datalogging does not make a particularly good controller and vice versa. A system designed to both log and control to the same degree will probably not be one of the 'best buy' dataloggers or 'best buy' controllers.

19.1.1 Features of datalogging and control systems

Datalogging

Datalogging used to be carried out by connecting sensors directly to the computer A/D port. As these ports have now disappeared it is necessary to connect a box to the computer to provide the interface. Any analogue values have to be converted into digital values in order to be read by the computer. These sensor values are stored sequentially in the memory of a computer or in the memory of a separate datalogging system. The readings are taken at regular time intervals. Usually up to eight input lines can be saved to memory. The values are saved at pre-determined time intervals. A table of time against stored value can be recalled from the programmable system.

The main advantage of a datalogger is to be able to record changes that happen very quickly when it would be physically difficult to take the readings manually, or when the changes happen very slowly, when it would be very boring to take readings.

Different dataloggers have different capabilities. Newer dataloggers have the ability to automatically recognise the sensor when it is plugged in. This can be achieved either by a multi-pin connection system or by using a reference resistor. When the datalogger identifies the sensor it can load any calibration files and make the name of the sensor and its unit(s) appear in the correct place (in graphs, tables etc.). This is usually termed *autorecognition*.

Some dataloggers have small LCD screens built in. This allows the user to view the data as either text or graph before deciding whether to transfer the data to a computer for more permanent storage. This is useful with those dataloggers that can be used remotely from the computer. The computer can program some dataloggers and the instructions can be saved. The datalogger is then disconnected from the computer and taken to a remote site where it can then run the stored program. The data is collected and then the logger is taken back to the computer so the data can be downloaded. It is usual that remote dataloggers can run on batteries rather than mains. It is even better if remote dataloggers can run on either mains or batteries, so they can use batteries out in the field but mains in the laboratory or workshop.

For very fast recording some dataloggers offer a single channel recording with the whole of the memory space dedicated to saving the data at an interval of a few microseconds. It is quite usual to see perhaps a four channel datalogger with the ability to store say 2000 data points on one channel or 500 points per channel when all four channels are in use. The maximum number of points that can be stored depends on whether the data is being saved in the datalogger (in its remote mode for example) or being saved in the computer. In general, the computer has the ability to store more data.

The data is then displayed as either a table or a graph. Most software systems allow the data to be seen in both modes. Some of the more sophisticated (and expensive) software packages allow for analysis of the data. Also, many systems have the ability to export the data they have collected to a spreadsheet such as Excel.

Control

There are different ways in which a programmable control system may be used. Some systems can only be used only when connected to a computer. Some systems can be used remotely, but battery life becomes a problem as the current drain of most output devices is significant. Some must be programmed by a computer, others have a built-in programming capability which tends to be more difficult to program than computer program systems.

Modern programmable control systems are capable of autorecognition of both inputs and outputs. As we have already indicated, some have simple logging facilities.

One particularly nice feature is the ability to multitask: that is, running several different control programs independently of each other. This obviously requires a degree of planning because there is a limited number of inputs and outputs that can be used by each program.

The remainder of this chapter concentrates on the control capabilities of programmable control systems.

19.2 Input and output ports

At the present time almost all interfaces use the bi-directional serial port facility. The serial port connection is usually transparent to the user.

There may still be some older machines in use that use parallel ports but these are becoming rare. In parallel port interfaces it was necessary to set up the ports to let the machine know whether the lines were input or output. It is not necessary to do this with a serial port connection.

A typical interface will have four or eight digital input lines, the same number of digital output lines, and possibly analogue input and output facilities.

Multi-pin connection systems on input and output ports are also becoming more common as autorecognition of sensors and actuators becomes the standard. This feature of the modern software makes programming a much easier task than it used to be. This also allows a single output port be used as a simple digital port or alternatively gives analogue control as well over the output device. An example of this would be a motor, which is plugged into a single output line. From this one line the motor can be run forward, backwards or turned off. When it is running forward or backwards the power can be controlled so its speed can vary between a maximum value and zero.

Early interfaces and buffer boxes, with eight output lines only used to be able to run four motors as each motor had to have two lines in order to be run in reverse.

PICs can be purchased in a wide range of input and output configurations. One chooses a PIC to match the control requirements. Many PICs have other onboard facilities, such as A to D converters, and careful study is required to avoid buying more than you need. At the same time you will be implementing many electronic systems within the IC which would have had to be separate systems previously.

PIC programming is done in assembly language (see below). There are several systems that use high level languages on a PC to make the programming easier.

19.2.1 Interfacing with electrical, mechanical and pneumatic devices

Although we are now dealing with a programmable control system rather than the electronic systems referred to in Chapters 14, 15 and 16 the same interfacing principles apply. Section 14.5 gave details of the kinds of output transducers that can be connected to an electronic system and section 16.3 contained the necessary information to drive these output devices. Most microprocessor-based systems are 5V systems so, unless the output transducers are 5V rated, it will be necessary to either use a relay to switch the transducer or use an output driver which allows a change in power supply value. This means that the signal level at the output of the microprocessor system will still be 5V maximum, and this must be sufficient to activate an output driver and device which is operating on (say) a 12V supply. Fig. 19.1 shows two possible ways of achieving this effect.

It is becoming popular to use a pneumatic cylinder at the output of electronic systems to deliver more displacement than an electronic transducer, such as a solenoid. Although it is possible to use a pneumatic system for the whole of a system, it is recognised that the increased costs may make the system uneconomic. A range of small pneumatic valves which operate on a 5V supply are now available, and these are being increasingly incorporated into project work.

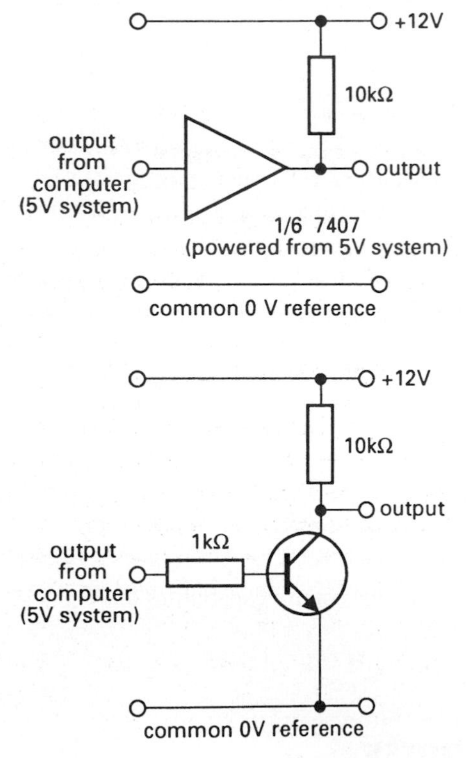

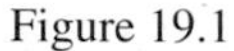
Figure 19.1

19.3 General programming principles

Due to the wide variety of available systems and languages, it is not feasible in a text of this kind to give comprehensive programming examples. Instead the following paragraphs give some general advice about the principles to be taken into account when writing control programs.

1. Use a 'top-down' approach

For the whole of the work on electronics we urged that a systems approach be adopted. This was explained in Section 14.2. There is no reason why this approach should not be continued into programming. Code can be created in modules, each of which will perform a particular function. The modules are equivalent to electronic subsystems. The program will then just list the modules that are needed in the order in which they are to be implemented. This is the same as just listing the electronic subsystems needed to produce an electronic system. Each module will have to be defined in a separate part of the program. This is equivalent to unpacking an electronic sub-system to see which components are used in the circuit. Icon-based programming illustrates this best.

2. Use names that have meaning

In early control work, modules of code were given nonsense names or even just letters. This makes it very difficult when looking through code at a later date to remember just what function 'FRED' or 'B$' was supposed to achieve. Wherever possible use names that have a precise function associated with them, even if the name becomes long. 'Reverse', 'Forward' or 'Detect_end_of_travel' are acceptable functional names. Notice the use of '_' to join words together, since many languages use a space as the terminator of a name. Where names are not possible, as in assembly language for example, it is possible to include lines of text to explain the purpose of a section of code. Although this slows down the writing of code it greatly increases the speed of sorting out problems afterwards. Always include this text at the time of writing the code: don't put it on one side and intend to include it when you finish. You never will!

Unfortunately, the use of long procedural names can slow execution speeds in many languages so, as usual, it will have to be a compromise. If short names are used try to document the program by including lines of text explaining what you are doing. These lines of text can often be included in the initialisation module so they only run once when the program is run (see next paragraph). They might otherwise slow up the program to an unacceptable amount.

3. Create an initialisation module

One of the first modules you will find it necessary to create is one that initialises the system. This will clear any programs still in the memory, set the input lines and output lines to a preset value, and select which lines (or ports) are being used for input and which for output. Labelling can also be accomplished in the initialisation module (see 5 below). Once this has been created it can be saved and used time and time again. The procedure MAIN in the text-based program shown below to solve the 'bears' problem is an example of this.

4. Decide whether a module is input, output or process

There will be input modules, output modules and process modules just as there are in hardware subsystems. An output module could be 'forward', 'reverse' or 'stop'. When an output module is called it results in a signal being transmitted from the software system to the external microelectronic system. No information is returned to the software system for further processing. When the module has been executed control returns to the software system.

Process modules, on the other hand, require at least one input signal from the software system and will return a signal to the software system. Examples of process modules include binary to decimal conversion, decimal to b.c.d. conversion and all kinds of masking of bits.

Input modules take information from the external microelectronic system and pass this information on to the software system. Modules that read the signals at ports or look at the state of individual bits are input modules.

5. Name lines/bits

It is helpful to name all lines/bits with sensible names if you wish to make the program more understandable. Commands like 'Make bit 5, high' will not help you a great deal in understanding the program since you need to know both what bit 5 is connected to, and what effect a high signal at bit 5 will produce. Small modules can be written naming each bit and allowing the use of words like 'on' and 'off'. 'Make bit 5, high' may now become 'make buzzer on'. There is an unfortunate tendency to use the words 'on' and 'off' in place of high and low. This is particularly noticeable in introductory software. It makes the unfortunate assumption that high signals always turn things on and low signals always turn things off. You should not need convincing that this is not always the case! It is also useful to consider using 'true' and 'false' when looking at the state of input lines/bits. True and false are logical operators and can save programming time and the need for extra modules. Not all languages allow the use of true and false.

6. Consider the use of flow charts or structure diagrams

Many texts recommend the use of flow charts before starting to write a program. In contrast, many professional programmers will tell you that they only use flow charts to show themselves what they have done after they have written the code. Many programmers *do* use diagrams and pencil and paper before actually sitting down to write any code. This process of specifying what task the code should do is vital and has its parallel in solving problems with electronic systems. You should never attempt to write code or put electronic subsystems together until you have written down a clear specification and a possible solution.

7. Include error trapping

It is good practice to include techniques that stop errors from disrupting the program. For example, if the program requires the use of the 'Y' or 'N' keys on a computer make sure that pressing any other key will not cause anything to happen. These techniques help to create a more 'user friendly' program.

8. Program the 'abort' state

If the program is aborted at any stage it is wise to ensure that the output is then set to some value which will prevent anything happening. If this is not done the artefact is left with the final output set and will continue to operate from this output pattern. It can be funny to watch a buggy continue over the edge of a table or up a wall, but it should be prevented with good programming.

19.4 Simple control programming

Programming a computer for control is no different from programming a computer to do other tasks. We need to know a little more about how computers work before looking at control in particular.

There are three stages involved in the use of a computer. Firstly, information or data must be input to the computer. The computer then processes the data in accordance with the instructions given to it in a program. Finally, the new data that results from this processing is produced as output by the machine.

A program is a set of instructions, which must be written in a language the computer can interpret. These instructions are arranged in order and the computer must work through the program in a sequential manner.

Both the data and the program are represented in a way that the computer can interpret. This can be by digits, letters, punctuation marks or other symbols, which we refer to collectively as characters. The next section looks at characters the computer can recognise and can convert into the binary form required by the processor.

19.4.1 Communicating with the computer

High-level and low-level languages

The increasing speed of the processor in computers has changed the emphasis in programming away from machine code and towards high level languages. Improved GUIs (graphical user interfaces) do not now carry such a speed penalty as they used to do. The consequence is that most programming has been made a lot easier.

Languages such as BASIC are known as high-level languages, a term that relates to the complexity of action associated with a single command. If very high speed is essential then it may be necessary to write the program in a low-level language such as assembly language, machine code or a compiled high-level language such as C++ or Visual Basic. The following information starts with low-level languages and moves towards high-level languages.

Machine code

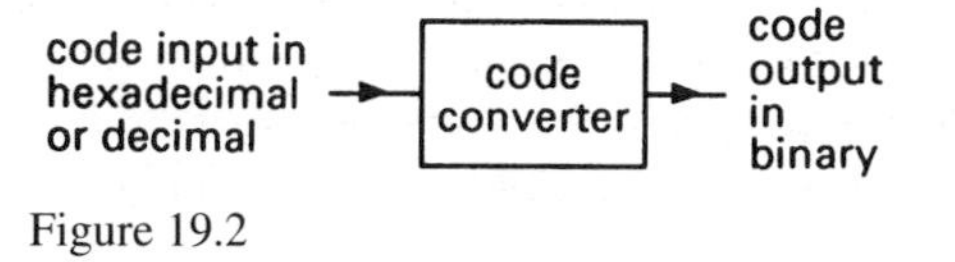

Figure 19.2

Any number system is nothing more than a code. For each distinct quantity there is an assigned symbol. There are several different codes a microcomputer can convert into the binary form required by the processor (Fig. 19.2).

Binary

Since the microcomputer is a digital processing system the base form of instruction that the system could interpret would be a set of 1s and 0s. If it were an 8 bit machine there would have to be at least eight 1s and/or 0s for each instruction. An instruction might just look like this:

10001010 or 10011110

Switches could be used to set the state of the bits and l.e.d.s used to indicate the state of the lines. The chances of making a mistake whilst setting a bit are considerable and it would be an almost impossible task to 'unpack' the program to find a fault. Programming in binary is not to be recommended.

All other codes will be converted, by the computer, into binary code. Since this will obviously require a certain time to do the conversion, it makes all other codes slower than binary.

While this was just about possible when 8 bit machines were common, it has to be recognised that this is not a sensible option for 32 bit machines.

Decimal

We are most familiar with the decimal system of coding information. It is relatively simple to convert the binary numbers given above into decimal. The decimal equivalent of the binary number 10001010 is 138 [calculated as $(1 \times 128) + (1 \times 8) + (1 \times 2)$] while 10011110 is 158 [$(1 \times 128) + (1 \times 16) + (1 \times 8) + (1 \times 4) + (1 \times 2)$]. It can be a tedious process to convert numbers from decimal to binary and vice versa.

Hexadecimal

It is possible to program in code called hexadecimal, which converts the 8 bit binary number into two digits that are in base 16. This is achieved by breaking each 8 bit binary number down into two halves. Let us use the two examples already given:

10001010 and 10011110

10001010 can be subdivided to give 1000 and 1010. 1000 in binary gives 8 in decimal and 1010 gives 10 in decimal. Hexadecimal numbers from 0 to 9 are the same as decimal numbers. The hexadecimal equivalents for the decimal numbers 10 to 15 are shown below.

Decimal	*Hexadecimal*
10	A
11	B
12	C
13	D
14	E
15	F

Table 19.1

Therefore, 10001010 in binary gives 8A in hexadecimal whilst 10011110 (binary) gives 9E (hex).

It is relatively easy to convert a 4 bit binary number into a hexadecimal digit.

Each arrangement of two hexadecimal digits is a code representing an 8 bit binary number. Some of these codes will be instructions whilst others will have no meaning. Each different type of microprocessor has a different set of codes so a table of instruction codes will have to be consulted to find out which codes have meaning. It is much easier to program and debug than the simple binary code. It still suffers from the fact that it is code and as such needs translating before you can make sense of it. If you use any code a lot it becomes easier, just as it is easier to remember telephone numbers which you use on a regular basis. Modern PCs work on 32 bit architecture so we are looking at eight hexadecimal digits representing one instruction or data. Computers with 64 bit architecture are said to be almost here.

Assembly language

The next type of language takes us nearer to the English language but we are moving still further away from the binary code recognised by the processor. This is assembly language and uses mnemonics to describe the instructions to the processor. So the instruction to 'load the accumulator' has the mnemonic LDA whilst 'jump to subroutine' becomes JSR. An instruction word will start with the operation to be performed by the processor followed by the required data. The first byte of the instruction word is known as the 'operation code' or 'op-code' it is this part of the instruction that tells the processor which operation to perform. The remaining byte or bytes are referred to as operands.

Once the program has been written in these mnemonics it has to be 'assembled' into its binary equivalents. This can be done by hand or by a special program called an assembler.

High-level languages

High-level languages, such as BASIC, take us still closer to English. The computer requires a sophisticated program called a compiler or interpreter, which can convert these words into binary code. (An assembler is an early form of a compiler.)

Interpreted languages, such as BASIC, can issue commands directly whilst compiled languages, such as C, cannot.

19.4.2 The choice of programming language

There is a wide range of possible programming languages that can be used for control purposes. There is a lot of discussion about the 'best' language for control. This section asks you to think about possible criteria for a control language by looking at some of the more popular languages.

Despite what you might hear or read there is not a perfect computer language. Each language has advantages and disadvantages. Sometimes these depend on the microcomputer as much as on the language itself. For example, the speed of execution of an instruction is both machine and language dependent. As the speed of the microprocessor increases the speed of execution will also increase. There has also been a move to a 32 bit instruction set which makes the transfer of information faster.

There are three main types of language in use at the present time: Flow chart-based, language-based and icon-based. Before we look at these in turn it is a good idea to consider the operations a programmable control system will need to do.

- Direct Actions. Turn outputs on or off. Set power levels for outputs capable of analogue control. Wait for a specified time. Set a timer to a certain value. Stop. Set variables such as counters to particular values. Increase or decrease variables. Log data.
- Delayed actions. Do (something) until or while (something else) is happening.
- Repeat. Repeat a certain number of times or repeat forever.
- Conditional responses. If (condition) then (action) else (alternative action).
- Call another procedure. This includes, if you must, the ubiquitous GOSUB.
- Apply logical operators (AND & OR) within most of the above.

Flow chart-based

Flow charts were originally used as a means of designing the structure of a program. There are now several programs that allow a control program to be written as a flow chart on the computer by selecting the flow chart symbols and inserting them into the chart. The code is then compiled behind the scenes and then run.

The main symbols used are shown in Fig. 19.3:

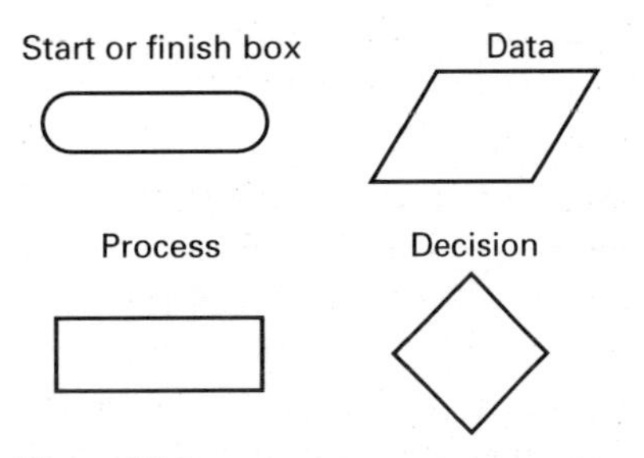

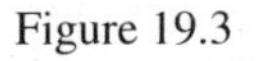

Figure 19.3

As with other programming languages it is possible to use a procedural style. This allows sections of code that will be used a lot to be written as a separate procedure. Every time this code is required simply call the procedure. Here we have a flow chart designed to flash a lamp.

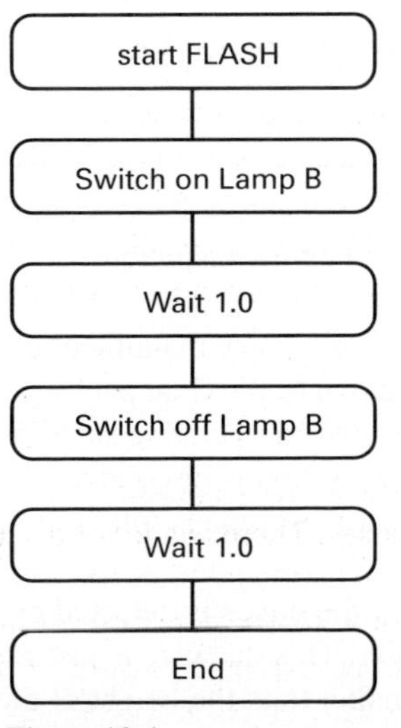

Figure 19.4

To make it flash ten times a new procedure can be used which calls Flash.

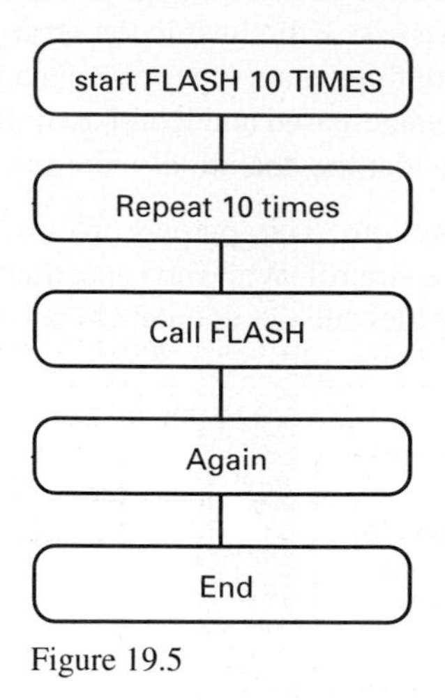

Figure 19.5

If we now make this conditional on some other factor, such as a switch being pressed, we incorporate the procedure FLASH 10 TIMES within the procedure SWITCH.

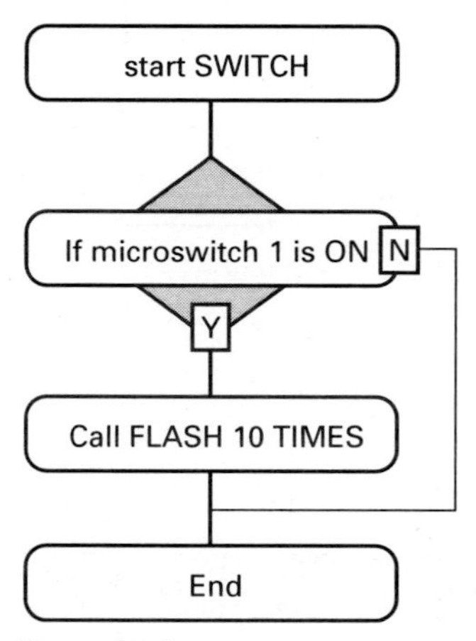

Figure 19.6

Finally we can improve the switch procedure by enclosing the existing code within a 'Repeat forever' loop (Fig. 19.7).

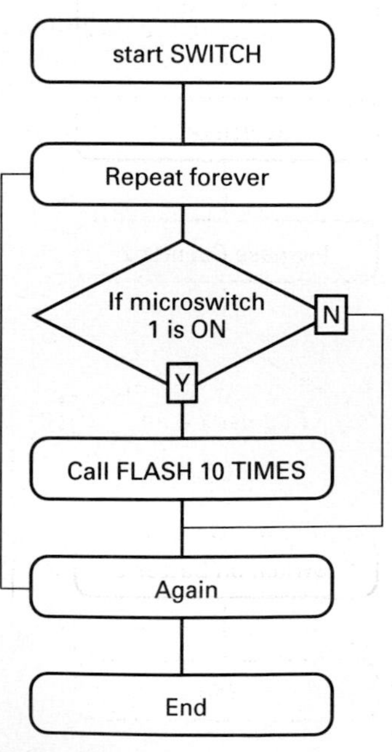

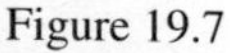
Figure 19.7

You should be able to see how procedures can be nested within one another to save writing the same code again and again. We also used several different actions and a simple conditional response.

It is good practice to keep procedures short. Any procedure of more than ten lines should be treated with suspicion.

Problem: A toy factory manufactures large and small teddy bears. These are boxed and then go down to the final packing area on the same conveyor belt. It is not economic to have an operative standing there all the time to separate the bears so a simple control system is devised to separate the bears and to indicate when the storage box for each type is full. Twenty small bears fit in a storage box but only ten large bears.

The size of the boxes is detected by two light sensors arranged a certain distance apart vertically. This distance is just greater than the height of the box for small bears, but smaller than the height of the box for large bears. If both light sensors are simultaneously at a low value then it is a large bear; if just sensor 1 is low then it is a small bear. A ram, controlled by a pneumatic solenoid connected to a pneumatic valve, pushes large bears off the belt. Small bears go to the end of the belt and fall into a different box.

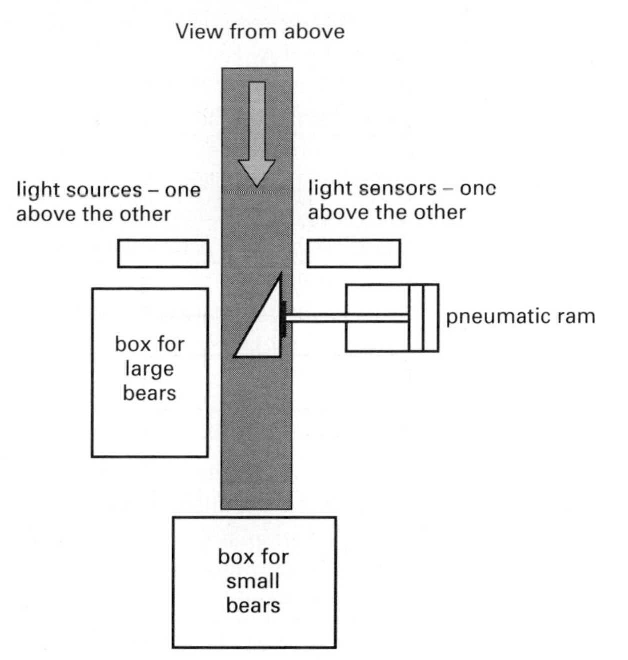

Figure 19.8

FLOW CHART SOLUTION

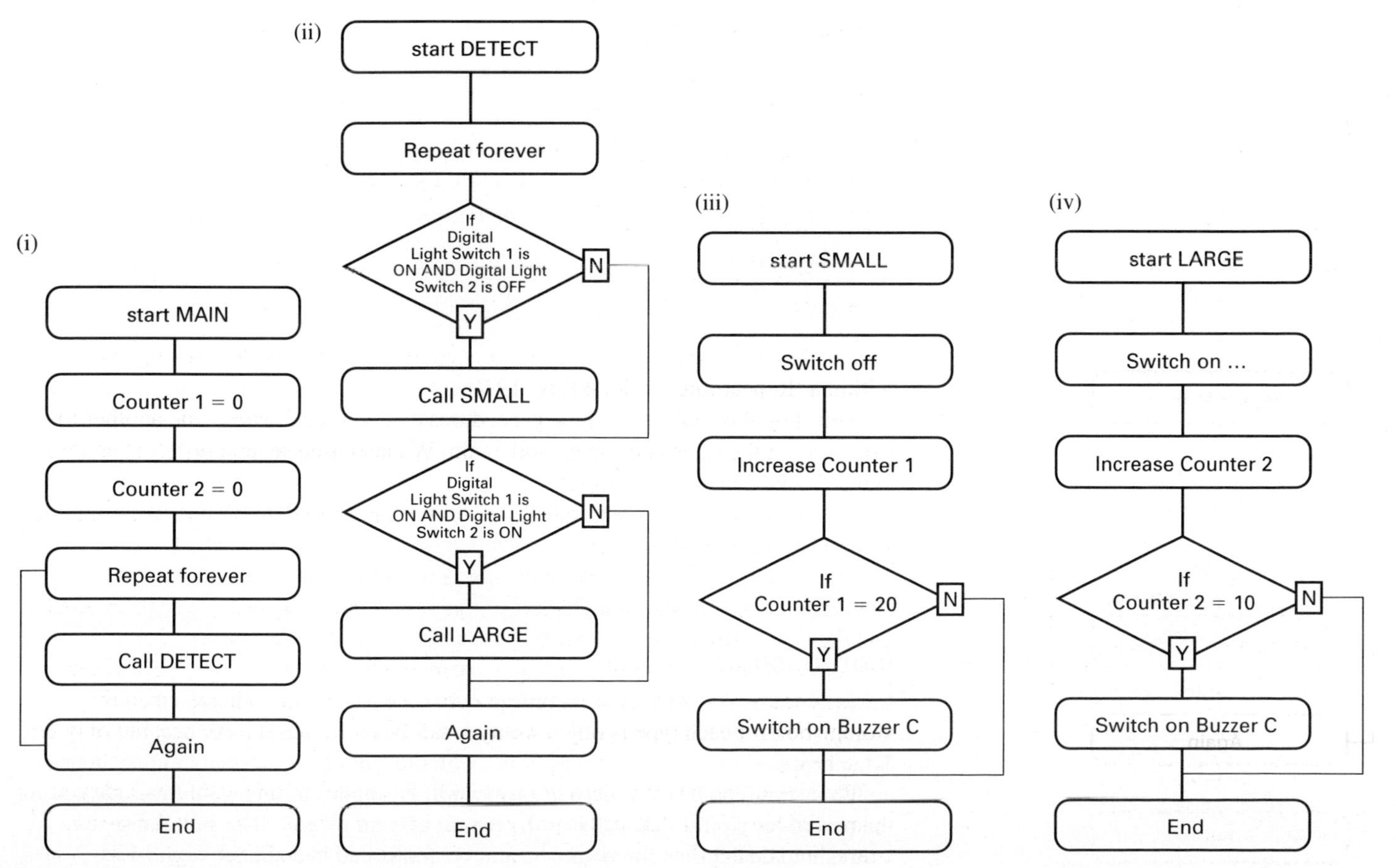

Figure 19.9

It is, of course, imperfect. WAIT statements would need to be added to allow for the time the bear takes to pass the sensors. Can you work out where these should be? Why will you need more than one WAIT statement? The buzzer needs to be reset and counters reset when someone comes to remove the full box and put a new empty box there. How could this be done? We have also used digital light switches. Not all systems have them so an analogue light sensor would have to be used instead. Experimentation should give suitable analogue values to switch but ambient lighting can be a problem.

One of the problems with flow chart-based software is the size of the solution. A complex problem can give a solution that runs to several pages. The use of a TRACE facility (where the part of the program being executed is highlighted on the screen) helps considerably as does SINGLE STEP mode (the next step does not execute until a key on the keyboard is pressed). These facilities are also available in language or text-based programs.

Text-based solution

If we now repeat the previous section but use a text-based language we should be able to see the similarities and differences. There are many alternative words in use but the structure is essentially the same.

The FLASH procedure would be as follows (assumes an output line has been designated 'lamp')

```
Switch on lamp
Wait 1
Switch off lamp
Wait 1
```

FLASH 10 TIMES then calls FLASH.

```
Repeat 10
  FLASH
End
```

Notice how the indenting is added to make the procedure easier to read.

If we now incorporate the procedure 'FLASH 10 TIMES' within the procedure SWITCH and enclose the existing code within a repeat ... Forever loop.

```
Repeat forever
Wait Until InputOn? 1
FLASH_10_TIMES
End
```

You should be able to see how procedures can be nested within one another to save writing the same code again and again. We also used several different actions and a wait until command instead of the simple conditional response.

Fig. 19.10 shows a possible text-based solution to the problem outlined on page 790.

There are some wait statements included in this version. The actual times would be checked by investigation. We have not tackled the reset problem.

Icon-based solution

In many ways icon-based programming is similar to working with electronic systems. The 'boxes' or icons are assembled on screen and then connected together. The properties of each icon are usually available from the right-hand mouse button and this allows adjustment of their properties. Examples of this would include setting the period for a pulse generator or the length of a time delay. The switching point of a threshold detector or comparator could be controlled in a similar way. Most programs now use picture icons. For clarity in these examples we will use text icons.

Procedure - MAIN

Reset
Label Input 0 *sensor_1*
Label Input 1 *sensor_2*
Label Output 3 *buzzer*
Label Output 1 *pneumatic_solenoid*
Make counter 1=0
Make counter 2=0
DETECT

Procedure - DETECT

Repeat forever
Wait until InputOn? *sensor_1*
If InputOn? *sensor_1* And **InputOff?** *sensor_2* [SMALL]
If InputOn? *sensor_1* And **InputOff?** *sensor_2* [LARGE]
End

Procedure - SMALL

Make *counter 1* 1 **more**
SwitchOff *pneumatic_solenoid*
Wait 3
If *counter 1* = 20 [SwitchOn buzzer]

Procedure - LARGE

Make *counter 2* 1 **more**
SwitchOff *pneumatic_solenoid*
Wait 5
If *counter 2* = 10 [SwitchOn buzzer]

Figure 19.10

It is interesting to note that some things that are simple to achieve with flow charts or text-based programming are quite difficult with icons, and vice versa. We will follow the same exercises as in the flow chart and text-based sections.

Flash a lamp

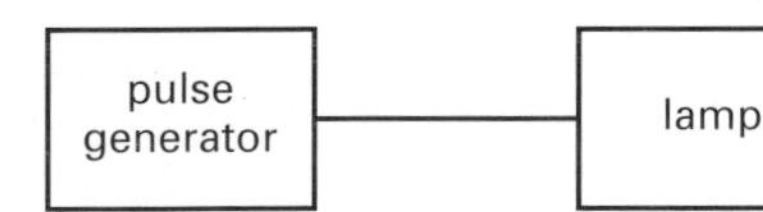

Figure 19.11

Flash a lamp ten times. This may be configurable from the properties of the time delay, in which case no extra icons are required (Fig. 19.11). If not, a more complex solution is possible (Fig. 19.12). Again it depends how the counter is configured – there may be an easier way to get the '10' signal.

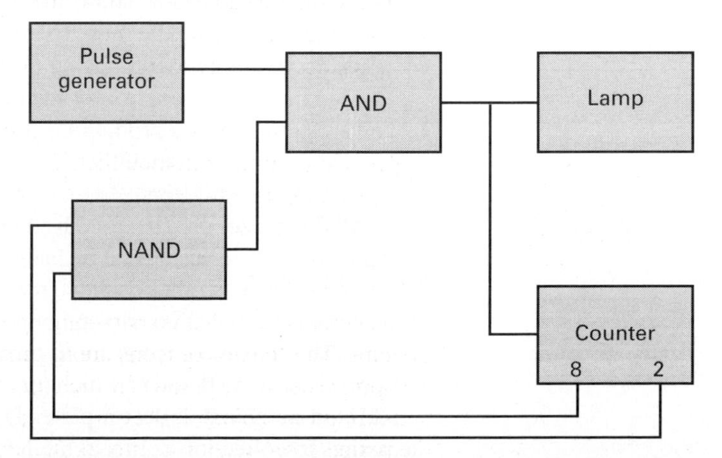

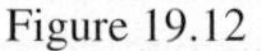

Figure 19.12

For switching on a flashing light we will use the simpler example above.

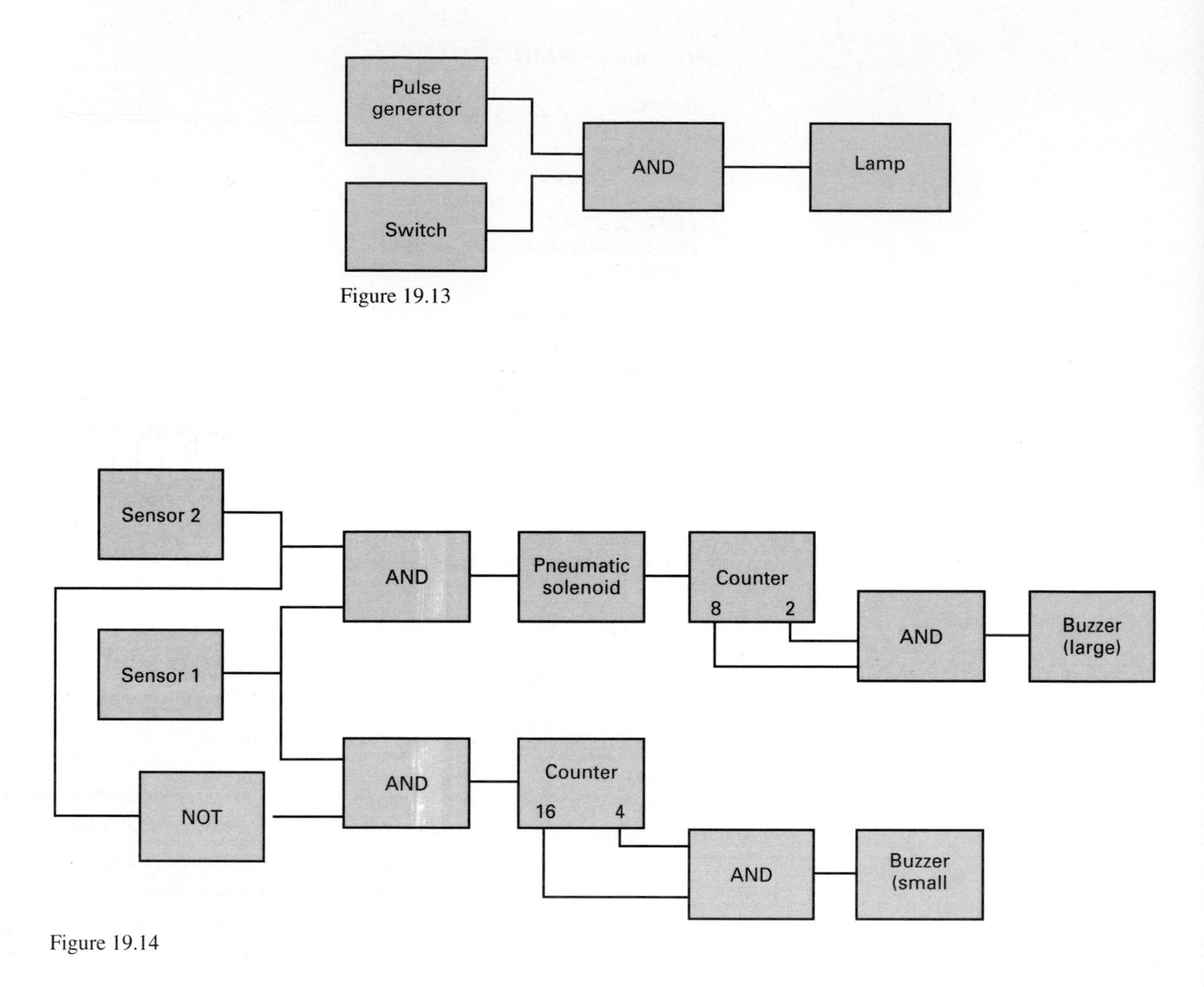

Figure 19.13

Figure 19.14

Assignments

1 (a) With the aid of flow charts describe the control of a Pelican Crossing. Reference should be made to the traffic and pedestrian control lights.

(b) With reference to interfacing discuss the merits of micro-electronic, electrical and mechanical control for the Pelican Crossing.

2 Towards the end of a production process, three different sizes of metal casings pass, in their natural metallic finish, along the conveyor belt shown in Fig. 19.15.

All three sizes are also produced with a paint finish and pass through a stove enamelling oven, suspended by hangers, on an aerial delivery system. This aerial delivery system is directly above the conveyor belt and their motions are synchronised so that no two components arrive simultaneously.

All of the metal casings, both painted and unpainted, pass the bank of light-beam sensors A, B and C which are used to sort them as required using solenoid-activated deflector plates D and E. These deflector plates can divert casings to collection points to the left or right of the conveyor system. Undeflected casings pass unhindered to a third collection point F at the end of the conveyor system.

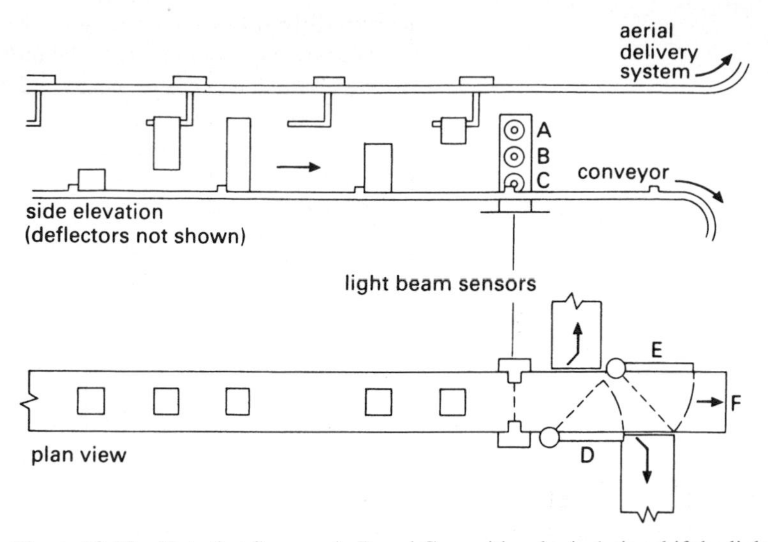

Figure 19.15 Note that Sensors A, B and C provide a logic 1 signal if the light beams are interrupted: Solenoid operated deflectors D and E have their own interface circuits and are moved to the dotted positions shown by a logic signal 1 from the control circuit.

(a) Referring to the conveyor system described above, the following sampling procedure is to be carried out by microcomputer:

- 10% of all *large* castings must be deflected by deflector D for testing and inspection
- 1% of all *medium* casings must be deflected by deflector E for inspection.
- All other casings pass unhindered to collection point, F.

The speed of the system is such that there is at least a two second gap between casings. The problem of the time delay between sensing and activating the deflectors is to be ignored. The deflectors need to be activated for one second after the casing passes the sensors, in order to ensure its removal from the system.

(i) Using a flow chart, or similar method, plan a computer program to control the system as described above. Provide adequate information for your program to be understood (e.g. port addresses, input/output connections). Note: circuit details are not required.

(ii) Write a simple program to implement the flow chart in part (i).

(b) (i) Describe how the sensors would need to be modified if the difference in surface finish or colour were to be detected.

(ii) Give a brief outline of the circuitry which would be required if the computer were to sort the metal casings on the basis of their surface finish or colour, rather than their size.

3 (a) Using flow charts and sketches as required, illustrate the operation of either a washing machine or a central heating system, showing clearly that you understand the terms: automatic control, open loop control, closed loop control and lag.

(b) Identify, and make a case for, the automation of an operation in your home or school which is currently manually controlled.

(c) The increasing automation of industry is seen variously as desirable, disastrous or inevitable. Present more objectively and in tabular form, the possible benefits and disadvantages of increased mechanisation/automation in the workplace.

(d) What effect, if any, has this change had upon the labour skills and specialisations subsequently required?

4 Fig. 19.16 shows the basic functions of an automatic washing machine designed for microprocessor control. The **outputs** of the controller are as follows: pump, wash motor, spin motor, heater, inlet water valve, soap inlet valve.

The inputs to the controller are as follows: start button, temperature sensor, high water level sensor, low water level sensor.

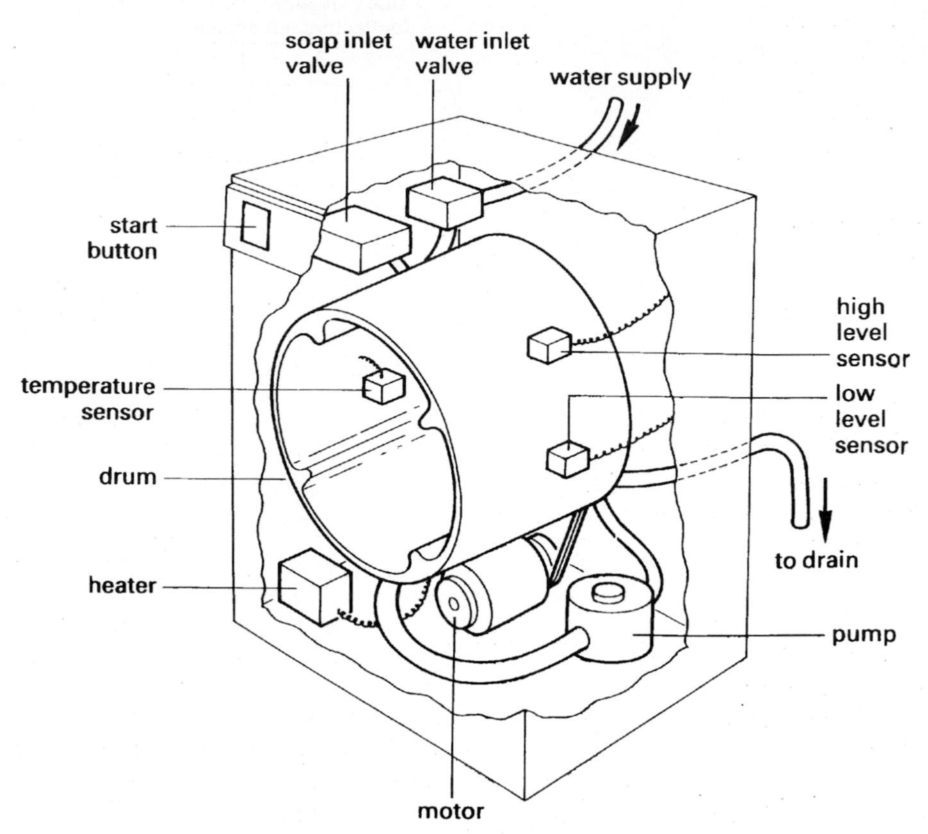

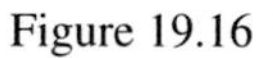
Figure 19.16

(i) Draw a flow chart to describe the control program needed to operate the machine as follows:

- Start button pressed
- Machine fills with water and soap until high level sensor is reached.
- Heater heats water to correct temperature.
- Wash-motor turns drum alternately clockwise and anticlockwise for five seconds in each direction. This process is repeated ten times.
- Spin motor spins drum and pump removes water and suds until low level sensor is reached.
- Drum refills with water up to high level sensor.
- Wash motor rotates drum in alternate directions for five seconds in each direction for a total of five cycles.
- Spin motor spins drum and pump removes water until low level sensor is reached then continues for a further two minutes.

(ii) Given the following information about the connections between the washing machine and the computer, write a program to control the process described in your flow chart:

Port A	*Bit O*	*Pump*	*logic 1 = ON*
	Bit 1	Wash motor	logic 1 = ON clockwise
	Bit 2	Wash motor	logic 1 = ON anticlockwise
	Bit 3	Spin motor	logic 1 = ON
	Bit 4	Water inlet valve	logic 1 = OPEN
	Bit 5	Soap inlet valve	logic 1 = OPEN
	Bit 6	Heater	logic 1 = ON
Port B	*Bit 0*	*Start button*	*Pressed = logic 1*
	Bit 1	Temperature sensor	Hot = logic 1
	Bit 2	High level sensor	Wet = logic 1
	Bit 3	Low level sensor	Wet = logic 1

Table 19.2

Your program should be logically written with explanatory notes. The structure of the program should correspond to the structure of the flow chart.

20 Electrical principles

20.1 Electricity in daily life

We depend on electricity in almost every aspect of modern life. We use it for lighting and heating our homes and schools, and for powering the tools and equipment needed in workshops and factories.

Electricity itself is not a form of energy but it is often thought of as a convenient 'fuel' and its use is measured in the units of energy. For many practical purposes the SI energy unit, the joule, is very small and electricity bills are calculated in terms of kilowatt-hours (see section 11.2.1). 1 kW h = 3.6 MJ. Our dependence on electricity can be judged by the annual UK electricity sales – about 250×10^9 kW h in 1990.

The electrical properties of materials and some of the relationships between them were introduced in Chapter 5 (section 5.1.2). The present chapter describes some electrical principles, some electrical devices, the circuits in which they are used and the power supplies that are needed to energise them.

The storage of electricity is a major limitation. A fully charged 12 V car battery has a typical capacity of 72 ampere-hours – equivalent to a current of 3 A flowing for 24 hours. This current corresponds to a power of $12 \times 3 = 36$ W or, over 24 hours, energy amounting to $36 \times 24 = 864$ W h or 0.864 kW h. This is a small proportion of a normal household daily requirement and, for lighting or heating a building by electricity and running electric power tools, we need a mains supply. Nowadays many hand tools are powered by rechargeable batteries but these too are dependent on mains electricity.

In addition to heating and power, electricity has many applications in the field of electronics (Chapter 14) and, in recent years, these have led to rapid developments in radio, television, terrestrial and satellite communications, and computers. In many cases only small amounts of power are needed and the devices are often battery powered.

Circuit diagrams can be drawn in two ways and these are shown in Fig. 20.1 for a simple electric circuit containing a resistor R and a cell (or battery) with positive (+) and negative (−) terminals. The resistor could be replaced by a device such as a bell or buzzer. If the circuit is closed as shown an electrical current will flow and energy will be dissipated in the resistor in the form of heat.

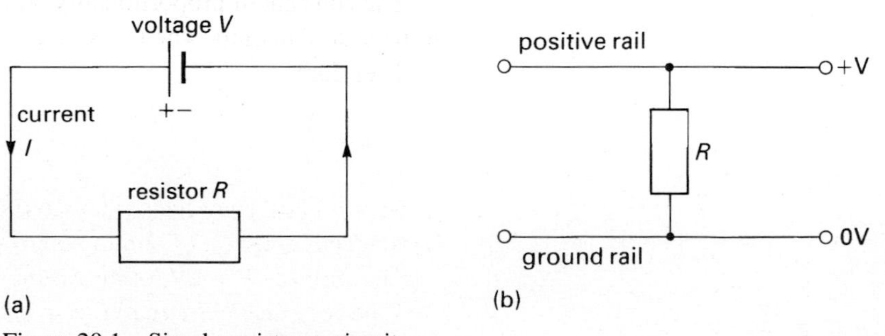

Figure 20.1 Simple resistance circuit.

Although the circuit can be drawn as a complete loop including the battery as in Fig. 20.1(a) it is also possible to represent it in electronics diagrams by positive and ground rails labelled + and 0 V as in Fig. 20.1(b); it is then assumed that the circuit is energised by a battery or mains supply (see section 20.9) as this is not shown.

20.2 Current, potential difference and Ohm's law

An electric current is a flow of charged particles. In metals the charge is carried largely by electrons, in semiconductors the charge is carried by both electrons and holes.

Electric current is measured in ampere (A). This is a measure of how fast charge is flowing. It does not measure how fast the charge carriers themselves are travelling!

Whenever a current flows from one point to another in a circuit, it does so because the voltage at the two points is different. If the voltages were the same then no current would flow.

Although we talk about voltages at points we would be more correct to talk about *potential differences* (p.d.s). The voltage at a point is measured between the point and a reference level usually given the value zero volt and may be called 'earth' or 'ground'. Thus, when we say the voltage at point X is 3 V what we mean is that the p.d. between X and the reference level is 3 V. It is quite acceptable to talk about the voltage at a point as long as it is understood that it is measured with respect to the reference level. It is essential that the same reference level is kept throughout a circuit or confusion will result. The maximum voltage available from a battery is called the electromotive force (e.m.f.).

Ohm's law is one of the most misused laws of physics! It states:

Provided that the temperature and other physical conditions remain constant then the current flowing through a material is proportional to the potential difference applied across the material.

Mathematically:

$$V \propto I$$

where V is the p.d. in volt and I is the current in ampere, or

$$V = kI \text{ where } k \text{ is a constant}$$

and

$$k = \frac{V}{I}$$

The constant of proportionality, k, is defined to be the resistance of the material, R, measured in ohm, when V is measured in volts and I in amperes.

Therefore,

$$R = \frac{V}{I}$$

Herein lies the confusion: a material can only be said to obey Ohm's law if its resistance is constant (assuming constant temperature and other physical conditions).

The formula $R = V/I$, often called the Ohm's law formula, can be applied to find the resistance of a material in a particular configuration even if the material does not obey Ohm's law. We may use the formula without necessarily applying Ohm's law.

In order to discover whether a material obeys Ohm's law it is necessary to plot pairs of values for V and I on a graph. If the material obeys Ohm's law the graph will be a straight line passing through the origin. Any graph other than this shows that the material does not obey Ohm's law. The resistance of a nonohmic conductor can be calculated at any point by taking the gradient of the graph, see Fig. 20.2.

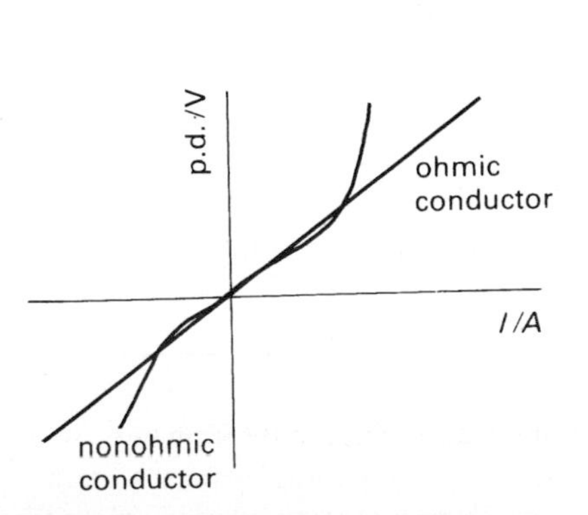

Figure 20.2 Graph showing ohmic and nonohmic conductors.

The Ohm's law formula is often needed as part of the design process. A good example is calculating the value of a series resistor for a light emitting diode (LED). The circuit is similar to that shown in Fig. 20.26 but with the following LED characteristics:

Typical current = 10 mA
Maximum voltage = 2.2 V

The voltage to be applied across the resistor/LED combination is 9 V. What is the required value of the series resistor?

If the voltage across the LED must not exceed 2.2 V then the voltage across the resistor must be $9 - 2.2\ \text{V} = 6.8\ \text{V}$.

The current through the LED must be the same as the current through the LED as they are in series = 10 mA.

Using the Ohm's law formula

$$R = \frac{V}{I}$$

gives

$$R = \frac{6.8\ \text{V}}{2.2\ \text{mA}} = 3.09\ \text{k}\Omega$$

As this rarely comes out to be a preferred value it is usual to take the next highest value (3.3 kΩ) and look at the brightness of the LED. If it still appears to be quite dim the next lower value (3.0 kΩ) would be taken. An explanation of preferred values can be found in section 14.7.1.

20.3 Electrical power

The rate of dissipation of energy is known as the *power*, P, measured in watt (W), and is defined as

$$P = VI$$

When the power is being dissipated in a resistor, it is in the form of heat. Combining $P = VI$ with the previous relationship $R = V/I$, leads to the results

$$P = I^2R$$

and

$$P = V^2/R$$

These formulae are used to calculate the power rating of components and electrical equipment.

EXAMPLE 20.1 What power rating of resistor is needed if a 150 Ω resistor is required to drop 20 V across it?

SOLUTION Using the last result,

$$P = V^2/R = 20^2/150\ \text{W} = 2.67\ \text{W}$$

In practice, it is likely that a 4 W resistor would be used as it is the next readily available value above the calculated value.

EXAMPLE 20.2 Calculate the current and resistance in each of the following electrical applications:

(a) an electric kettle rated at 1.2 kW on a 240 V domestic supply.
(b) an electric train operating at 4 MW on a supply voltage of 25 000 V.

SOLUTION (a) From the relationship $I = P/V$ the current is:

$$I = 1.2 \times \frac{1000}{240} = 5\ \text{A}$$

From the equation $V = IR$, the corresponding resistance is:

$$R = \frac{V}{I} = \frac{240}{5} = 48\ \Omega$$

(b) With the new values,

$$I = 4 \times \frac{1\,000\,000}{25\,000} = 160\ \text{A}$$

$$R = \frac{25\,000}{1640} = 156.25\ \Omega$$

20.4 Potential division and the Wheatstone bridge

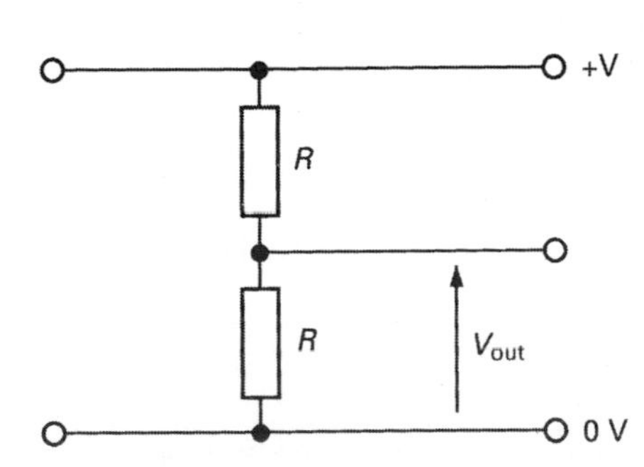

Figure 20.3 Potential divider with two equal value resistors.

We often need to produce a potential difference less than that of the supply. The simplest form of potential divider consists of two equal value resistors connected between the power supply rails as in Fig. 20.3.

Since the same current must pass through both resistors, and their resistances are equal, it follows that the potential difference across each resistor must be the same and therefore equal to half the supply voltage. In practice, remember that each resistor may have a slightly different value due to the tolerance of resistors and also that connecting a meter to the junction of the resistors to measure this p.d. may change the value since some of the current will have gone through the meter rather than the bottom resistor.

Any number of resistors may be connected in series in this way to produce a potential divider, Fig. 20.4.

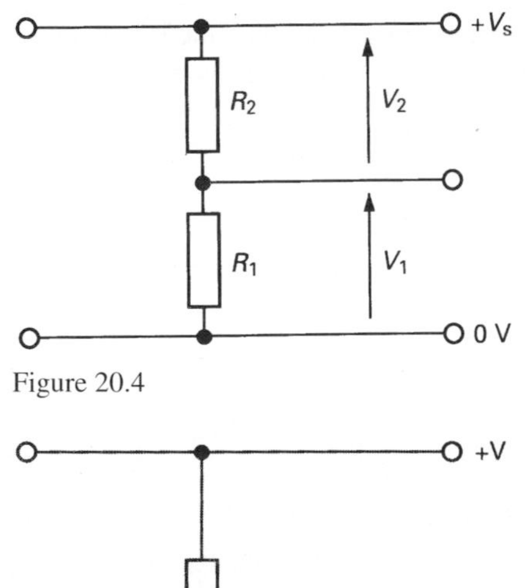

Figure 20.4

In general if we have two resistors R_1 and R_2 connected between ground and $+V_s$, as shown in Fig. 20.2, then a p.d. of V_1 will appear across R_1 and a p.d. of V_2 will appear across R_2.

We know that $V_s = V_1 + V_2$ and that the same current, I, flows through both resistors. Applying the result $I = V/R$ three times, we have

$$I = V_1/(R_1 + R_2) \qquad I = V_1/R_1 \qquad I = V_2/R_2$$

Therefore, by equating the first two expressions for I,

$$V_1/R_1 = V_s/(R_1 + R_2)$$

and

$$V_1 = V_sR_1/(R_1 + R_2)$$

In words: the p.d. across a resistor in a potential divider is equal in value to the product of the supply p.d. and the ratio of that resistance to the total resistance of the potential divider.

If a continuously variable output p.d. is required the separate resistors can be replaced by a potentiometer as in Fig. 20.5.

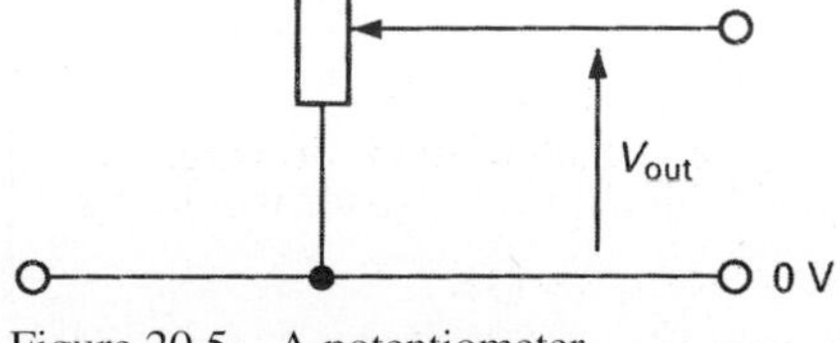

Figure 20.5 A potentiometer.

This is capable of giving any output signal between $+V_s$ and 0 V.

Wheatstone bridge

The *Wheatstone bridge* consists of two potential dividers both connected to the same power supply, Fig. 20.6. At least one of the resistors must be variable so let us assume one is a strain gauge (R_4).

The output signal is taken from the midpoints of each potential divider. V_{out} will be zero if $R_1 = R_2$ and $R_3 = R_4$. It can be shown that if the resistance of the strain gauge changes by δR then the output, V_{out} is approximately $V_s\ \delta R/4R$.

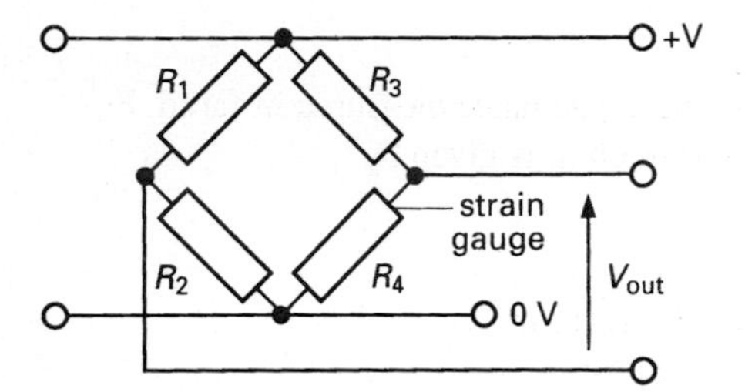

Figure 20.6 Wheatstone bridge.

20.5 Kirchoff's laws

These laws are useful in solving some circuit problems – particularly those relating to resistor networks.

Law 1 The algebraic sum of currents at a junction is zero. In symbols,

$$\Sigma I = 0$$

Consider a junction in a circuit with currents as shown in Fig. 20.7.

If currents flowing into the junction are regarded as positive and those flowing out of the junction are regarded as negative, then:

$$I_1 + (-I_2) + (-I_3) = 0$$

or

$$I_1 = I_2 + I_3$$

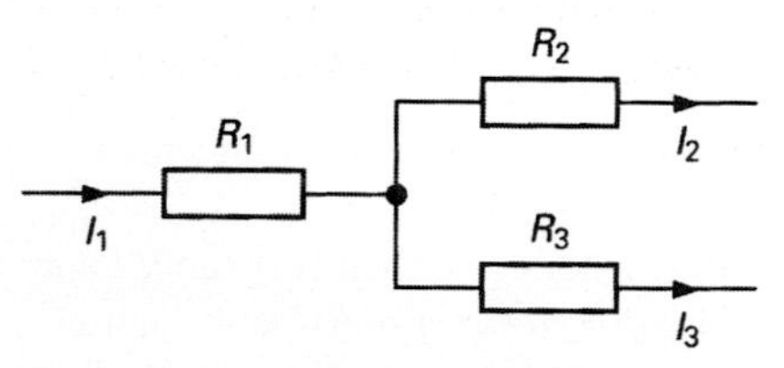

Figure 20.7 Currents at a junction.

Law 2 In any closed loop, the algebraic sum of the e.m.f.s is equal to the algebraic sum of the products of current and resistance. Thus:

$$\Sigma E = \Sigma IR$$

EXAMPLE 20.3 Consider the recharging of a model car battery pack from a 12 V supply, Fig. 20.8. The battery pack is of nominal 7.2 V but this will rise to about 9.6 V on charge. The charging current is not to exceed 0.12 A at full charge but this may be safely applied indefinitely. What value series resistor will be needed to ensure this continuous current will not be exceeded? It is to be assumed that both the lead acid battery and the rechargeable battery pack have very small input impedances.

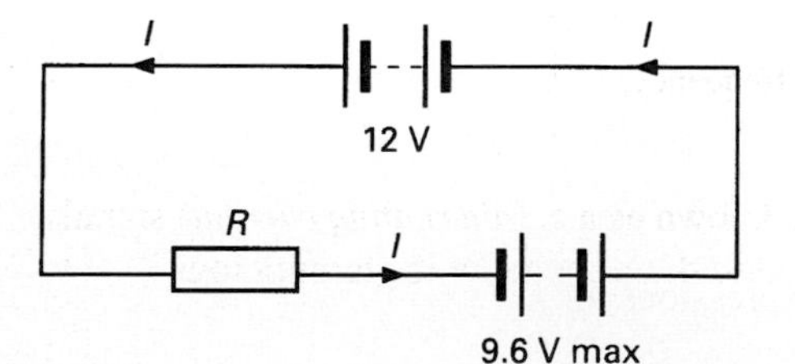

Figure 20.8 Charging a battery pack.

On full charge $\Sigma E = 12.0 + (-9.6)$ V and $\Sigma IR = 0.12R$ V
Since

$$\Sigma E = \Sigma IR$$

$$12.0 + (-9.6) = 0.12\,R$$

$$2.4 = 0.12\,R$$

and

$$R = 2.4/0.12\ \Omega = 20\ \Omega$$

If the battery pack were totally discharged then the initial charging current would be:

$$I = E/R = 12.0/20\ \text{A} = 0.6\ \text{A}$$

Application of the power formula shows that a high power resistor would be needed. Can you work out why?

20.6 Impedance

In introductory texts, the term *impedance* is used interchangeably with resistance. This is a reasonable approximation in many cases, but it is not totally accurate. Circuits containing capacitors and/or inductors are more complicated than simple resistive circuits. These components exhibit 'reactance' (as well as resistance). Reactance is frequency-dependent so, for a given circuit containing reactive elements, the amount of corruption of the signal depends upon the frequency of that signal.

The reactance of a capacitor, X_c, measured in ohm, is given by

$$X_c = \frac{1}{2\pi fC}$$

where f is the frequency in hertz, Hz, and C the capacitance measured in farad, F.

The reactance of an inductor, X_j, measured in ohm, is given by

$$X_j = 2\pi fL$$

where f is the frequency in Hz and L is the inductance in henry, H.

The impedance, Z, measured in ohm, is given by

$$Z = \sqrt{(R^2 + X^2)}$$

Therefore, for a resistor, the impedance is equal to the resistance since it has no reactance. For capacitors and inductors the resistance and reactance must both be taken into account.

20.7 A.C. signals

Many signals do not remain at a constant level. For example, when a microphone is detecting a steady musical note, the signal it produces repeats a set pattern continually as an analogue of the changes in air pressure, Fig. 20.9.

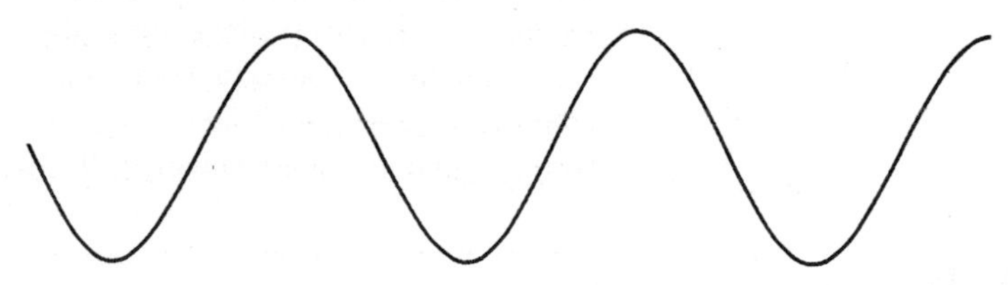

Figure 20.9 Signal due to sound of constant frequency.

Signals which alternate in this way are known as a.c. (*alternating current*) signals. We will consider a simple sine wave signal to define some of the terminology associated with a.c. signals, Fig. 20.10.

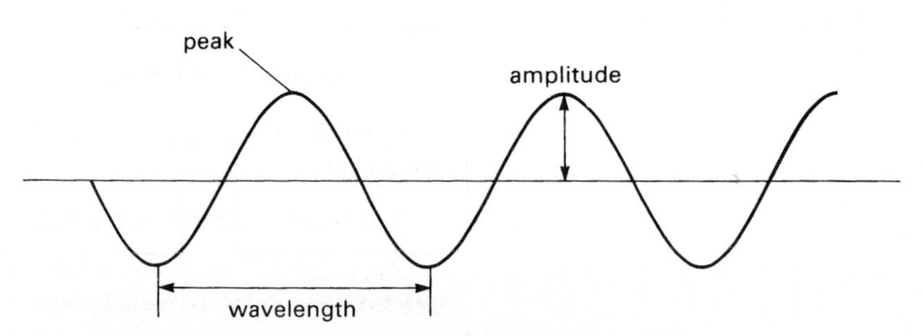

Figure 20.10 Terminology associated with a.c. signals.

two waves in phase

two waves 180° out of phase

Figure 20.11

The mean position is the line drawn through the trace so that the wave is symmetrical about this line. The amplitude is the maximum displacement from the mean position. The peak is the top of a wave.

The *wavelength* is the length of one wave. Since the trace is continuous it is difficult to know where to measure from or to! If any point on the wave is chosen as the starting point then, keeping parallel to the mean position, mark the next point along the wave where the signal is of the same value and is at the same stage of its motion. This distance is the wavelength. The easiest place to measure it is as the distance between two successive crests or troughs.

Points on a wave which are of the same value and at the same stage of motion are said to be *in phase*, Fig. 20.11. All other points on a single wave must therefore be *out of phase*. It is common to compare two waves and measure the *phase difference* between them.

In the first diagram the two waves are in phase; in the second diagram the two waves are totally out of phase, usually expressed as 180° out of phase.

Frequency is the number of waves per second. The *period* of a wave is the time for one complete wave be generated.

Square waves (or rectangular waves) will be encountered in electronics as well as sine waves (and sometimes triangular waves as well). All the terminology introduced so far applies to square waves. A further term is needed – the *mark-space ratio*. This is the ratio of the time the square wave is in the high state compared to the time it is in the low state. Square waves with a 1:1 and 1:4 mark-space ratios are shown in Fig. 20.12.

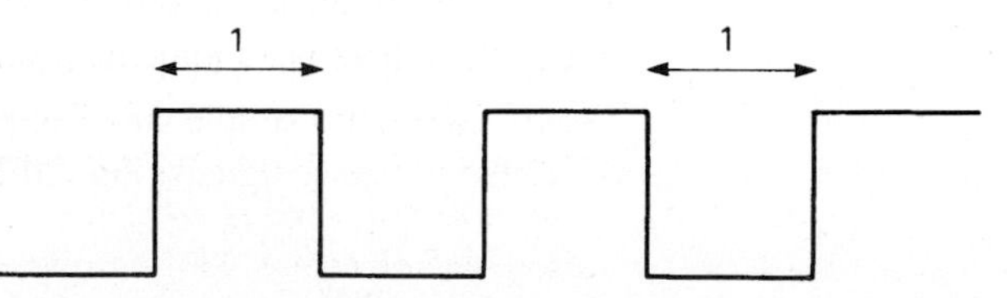

square wave with 1:1 mark-space ratio

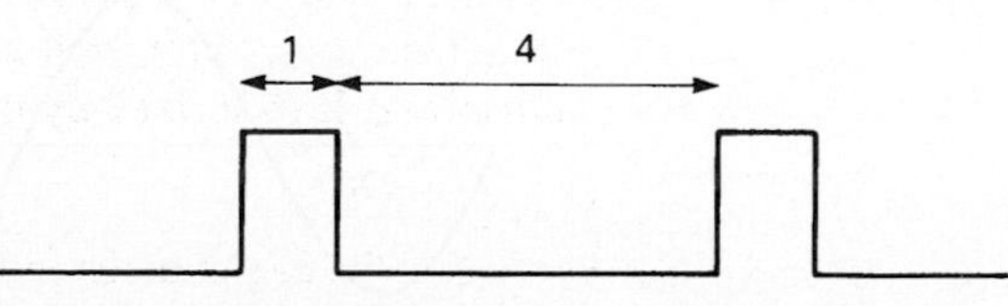

square wave with 1:4 mark-space ratio

Figure 20.12

When a voltmeter is used to measure the strength of an a.c. signal it measures the r.m.s. (*root mean square*) value of the signal rather than its peak value. The peak value and the r.m.s. value are related by the formula:

$$V_{peak} = V_{rms} \times \sqrt{2}$$

The decibel

The bel was originally defined as a measure of cable quality for the transmission of signals by wire telegraphy (by the Bell Corporation of the USA). Losses in cables do not increase linearly with the length of a cable and as a result of this the bel (and decibel) are logarithmic functions. The bel is too large a unit to use with small electronic circuits so the decibel is used instead.

The power gain of a circuit, in decibel, is defined as:

$$\text{power gain in dB} = 10 \log_{10} P_{out}/P_{in}$$

Power gain in dB	P_{out}/P_{in}
−30	1/1000
−20	1/100
−10	1/10
−6	1/4
−3	1/2
0	1
+3	2
+6	4
+10	10
+20	100
+30	1000

Table 20.1

Table 20.1 shows the relationship between the power gain and the ratio P_{out}/P_{in}.

It can be seen from this table that if the power gain is doubled it results in an increase of 3 dB whilst if the power gain is halved it results in a decrease of 3 dB. Note also that the scale is logarithmic and if, for instance, P_{out}/P_{in} is increased from 10 to 1 000 000 then the power gain increases from +10 dB to +60 dB.

It is often necessary to consider voltage gain rather than power gain in electronic circuits. It is generally accepted that the voltage gain can be expressed by the following formula:

$$\text{voltage gain in dB} = 20 \log_{10} V_{out}/V_{in}$$

This is an approximation since it assumes that the input and output impedances are equal. It is only true that the voltage gain in decibel equals the power gain in decibel if the input and output impedances of the system have the same value.

The half-power frequency

The *half-power frequency* of a system is that frequency at which the power gain of the system has fallen by −3 dB. This can be referred to by a number of different names some of which are less than accurate, for example, 'the half-power point', 'the −3dB point', 'the breakpoint' and 'the break frequency'.

The concepts of power gain, voltage gain and half-power frequency are particularly useful in Chapter 15 when considering the action of op-amps and filters.

20.8 Capacitance

A *capacitor* is a device which stores charge and simply consists of two metal plates separated by an insulator known as the 'dielectric'. If a capacitor is connected to a battery then there is a momentary current in the circuit as electrons are transferred from one plate to the positive terminal of the battery whilst, at the same time, electrons are transferred from the negative terminal of the battery to the other plate of the capacitor, see Fig. 20.13.

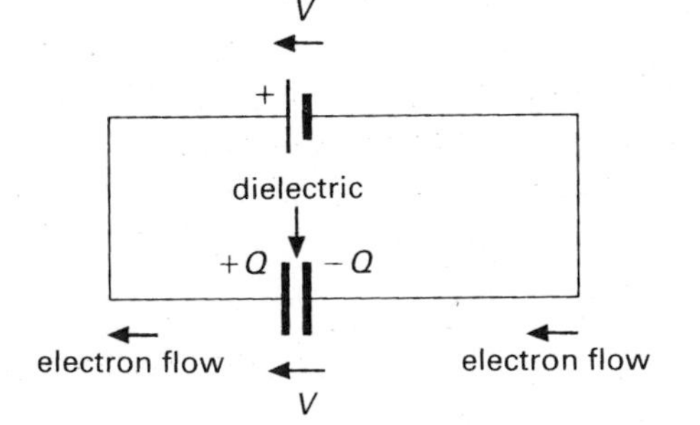

Figure 20.13

Note that electrons do not flow across the insulator! After a very short time the potential across the plates of the capacitor is equal and opposite to the potential of the battery. No more charge can then flow and the capacitor is said to be charged.

The measure of how well a capacitor can store charge is known as the *capacitance*. It is defined by the equation

$$C = Q/V$$

where C is the capacitance in farad, Q is the magnitude of the charge on either plate in coulomb and V is the p.d. between the plates in volt. The capacitance of any particular capacitor depends upon the size and separation of the plates and the type of material used as the dielectric.

20.9 Power supplies

There are two different areas covered by this title. One is the use of a suitable power supply for a project and the other is investigating the characteristics of power supplies.

Firstly let us consider the choice of a suitable power supply for project work. The choice is between batteries and mains operated power supplies. Some years ago it was very popular to build power supplies for use in technology and electronics courses. This can no longer be permitted due to the safety regulations now in force. It is however possible to buy completed power supplies relatively cheaply but most of these supplies are unregulated.

A *regulated supply* is one which will maintain the same output voltage regardless of the amount of current being drawn, up to a given maximum value. An *unregulated supply* gives an output that falls progressively as the amount of current drawn increases. This would be particularly inappropriate for many ICs which require a supply p.d. in a very precise range. As long as the current being drawn is small this will not be a problem, but many output devices require significant currents which would pull down the value of the supply p.d. For that reason it is suggested that regulated supplies are used whenever possible.

It is possible to convert the unregulated supply to a regulated one by the use of a regulator integrated circuit. The 7805 gives a +5 V output if up to 1 A is drawn from it provided it is given an input power supply of 8 to 20 V. The 7905 gives −5 V for the same conditions. Both circuits are shown in Fig. 20.14.

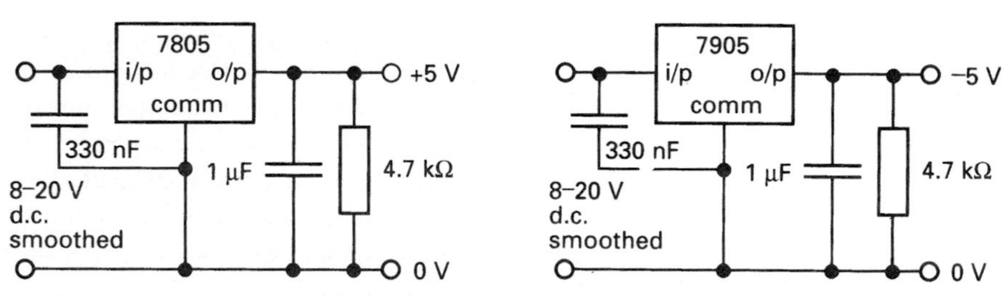

Figure 20.14 Regulator circuits.

It is possible to buy regulators with many different output voltages and maximum currents. The manufacturers' catalogues and data sheets give more information. A battery is not regulated but it would be unlikely to be used if large currents were required. The principles involved in the construction of a power supply can be investigated if a low voltage a.c. supply is available. The parameters of this a.c. supply will determine the parameters of the final output supply. In general, the subsystems required for a power supply are shown in Fig. 20.15.

The transformer is the part that needs to be commercially manufactured to meet the safety specifications. Its purpose is to change the value of the mains supply from about 240 V down to 20 V or less. There are two main options for transformers; they can be made with one coil in the secondary or two, see Fig. 20.16.

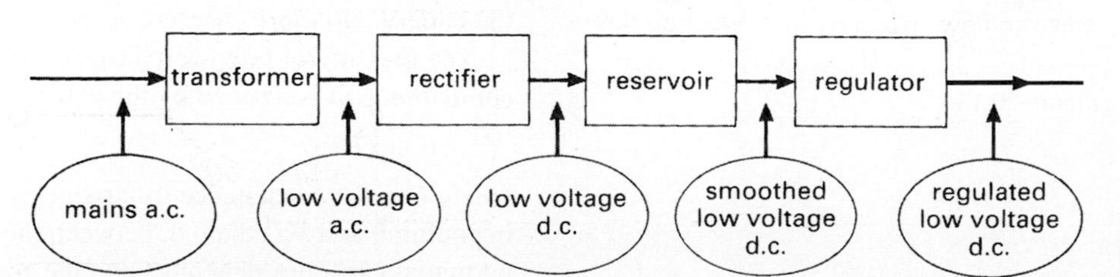

Figure 20.15 Functional diagram of a power supply.

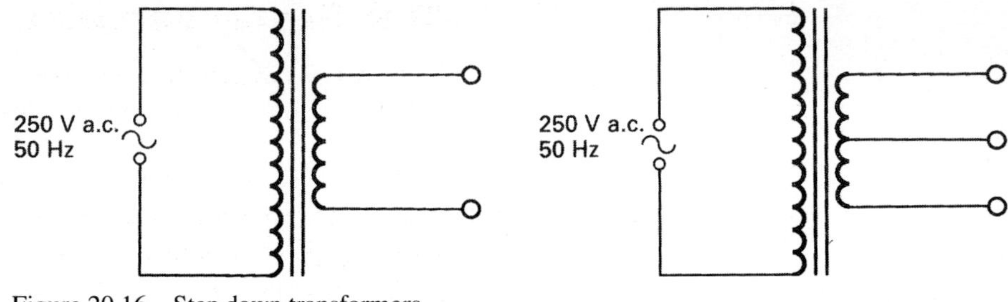

Figure 20.16 Step down transformers.

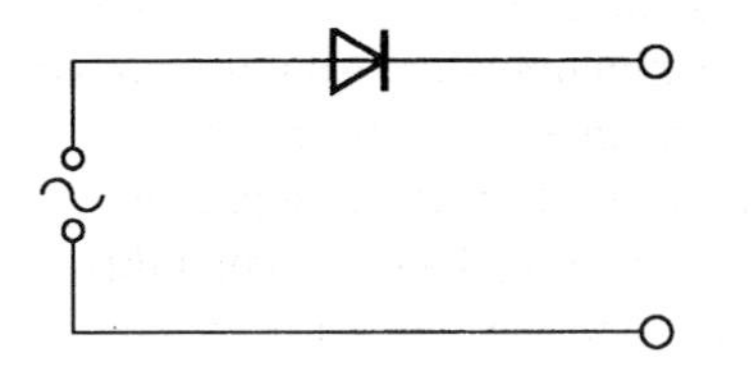

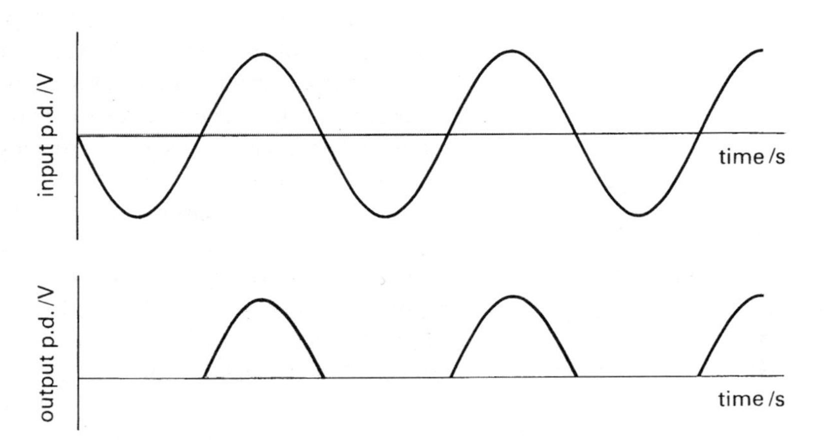

Figure 20.17 A half-wave rectifier. Graph to show the relationship between the input and output p.d.s. for a half-wave rectification circuit.

One coil tends to be cheaper but two coils give increased flexibility. They can be connected either in parallel to give double the output current or in series to give double the output voltage. In addition this gives a connection in the centre of the two coils, known as the centre tapping, which allows for a power supply with dual rails to be constructed.

Rectification is the changing of the alternating signal into one that flows in one direction only. The simplest way of achieving this is to use a semiconductor diode, Fig. 20.17.

This will totally remove the negative half of the signal and is known as *half-wave rectification*. Unfortunately the output is now zero for half a cycle. To incorporate the other half of the cycle, for *full-wave rectification*, requires an extra three diodes in the diode bridge arrangement, Fig. 20.18. Note the alternative way of drawing this circuit, which should make its operation easier to understand.

If the centre-tapped transformer is being used it will be necessary to use two diodes, one for each coil, taking the centre as the ground or reference voltage, Fig. 20.19.

The smoothing of this signal is usually achieved by a large, electrolytic capacitor – known as the reservoir capacitor, Fig. 20.20. Whilst the signal is rising the

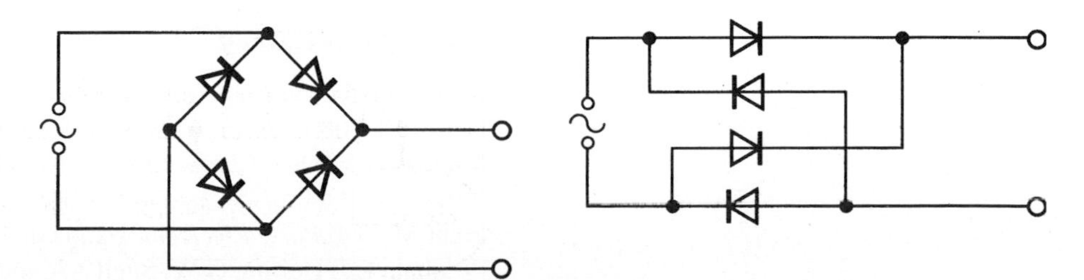

Figure 20.18 Full-wave rectifier – alternative diagrams.

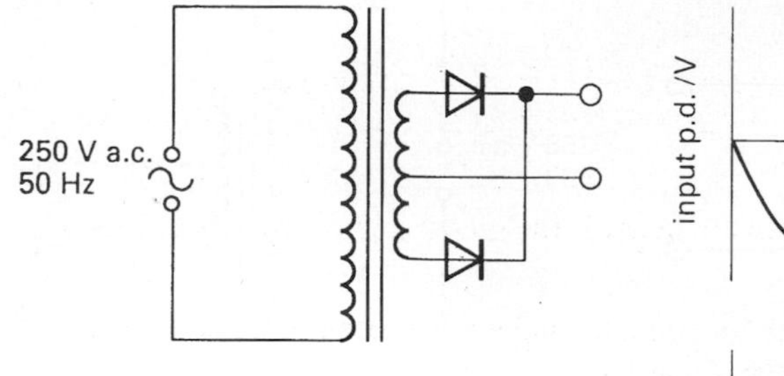

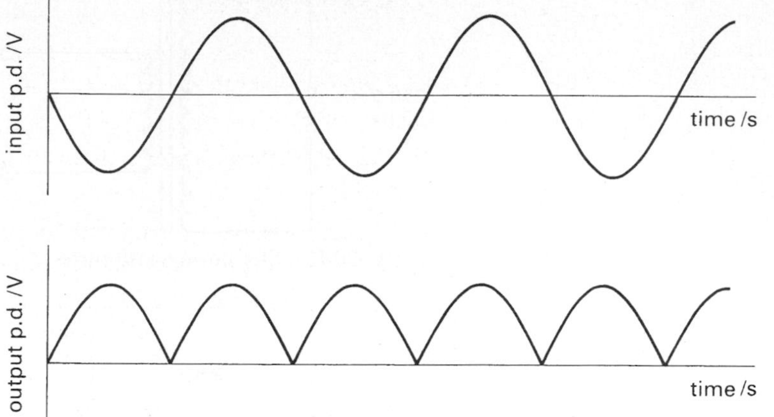

Figure 20.19 Full-wave rectification with a centre-tapped transformer. Graph to show the relationship between the input and output p.d.s. for a full-wave rectification circuit.

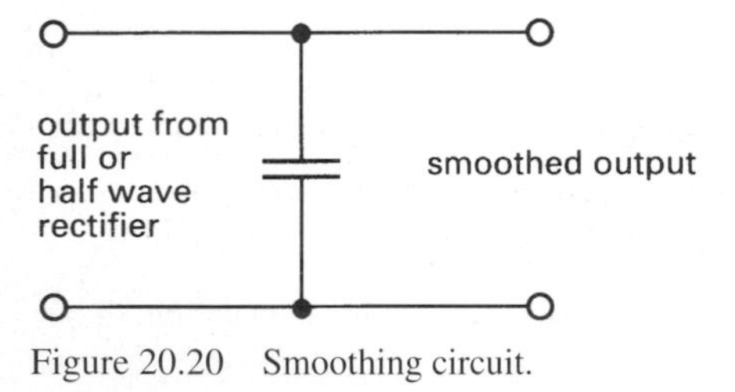

Figure 20.20 Smoothing circuit.

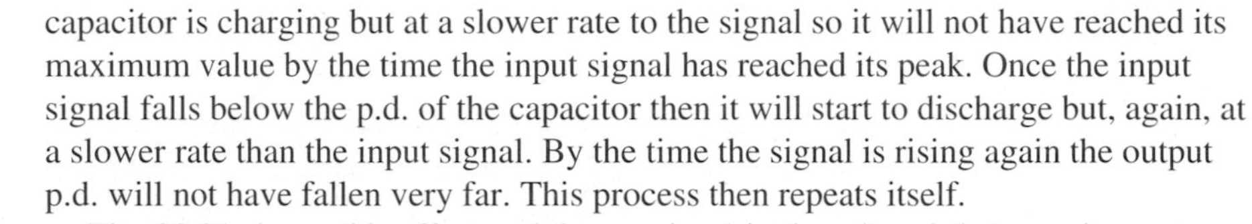

capacitor is charging but at a slower rate to the signal so it will not have reached its maximum value by the time the input signal has reached its peak. Once the input signal falls below the p.d. of the capacitor then it will start to discharge but, again, at a slower rate than the input signal. By the time the signal is rising again the output p.d. will not have fallen very far. This process then repeats itself.

Fig. 20.21 shows this effect and the ensuing 'ripple voltage' that remains on top of the steady signal.

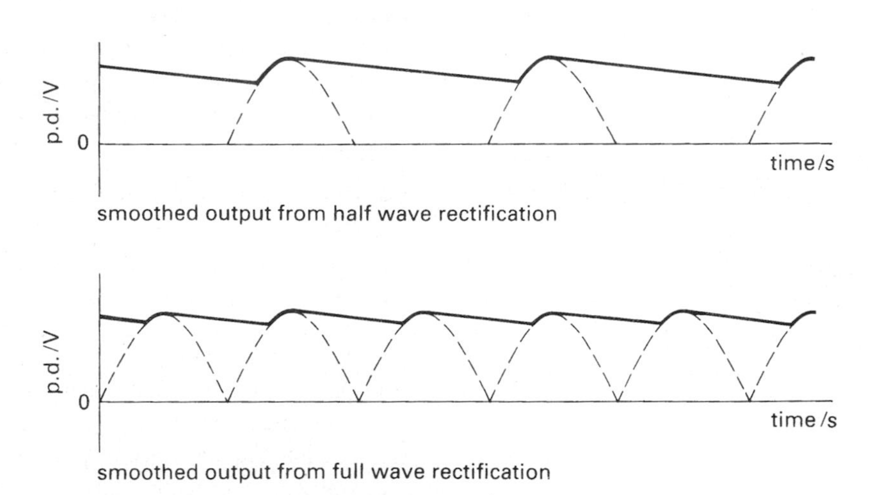

Figure 20.21

Bibliography

Briethaupt J 2000 *Understanding Physics for Advanced Level*. Nelson Thornes
Duncan T 2000 *Advanced Physics*. John Murray
Muncaster R 1993 *A Level Physics*. Nelson Thornes
Nelkon M 1994 *Advanced Physics*. Heinemann
Plant M 1990 *Basic Electronics* 2nd edn. Hodder and Stoughton
Wenham E J, Dorling G W, Snell J A and Taylor B 1986 *Physics: Concepts and Models*. Longman

Assignments

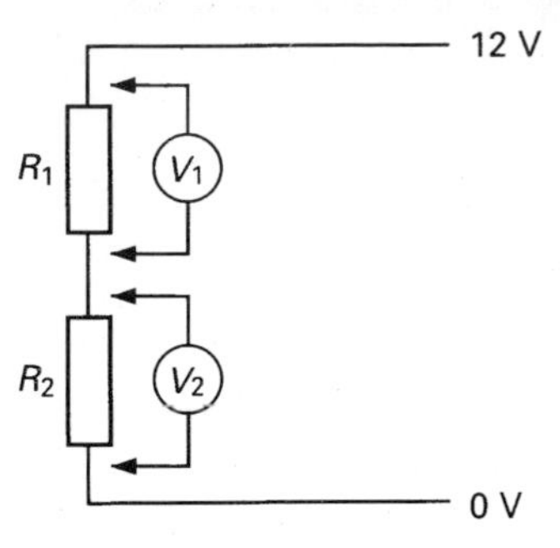

Figure 20.22

1 Fig. 20.22 shows a voltage divider as used in an electronic circuit. Calculate the values of V_1 and V_2 if $R_1 = 2\ \text{k}\Omega$ and $R_2 = 4\ \text{k}\Omega$ and determine the current flow through R_1 and R_2.

2 Fig. 20.23 shows a circuit diagram. What is the potential difference between A and B?

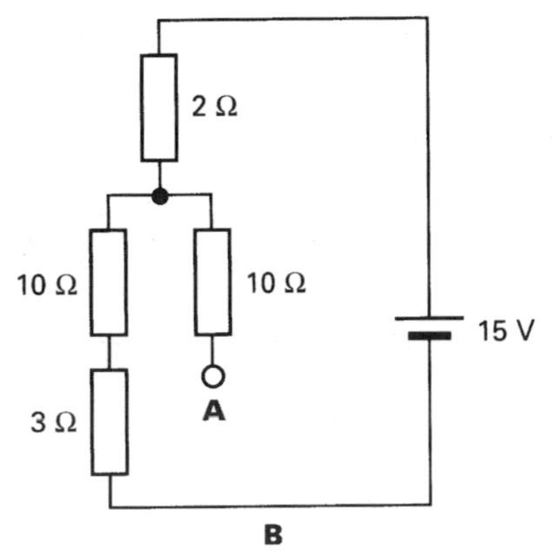

Figure 20.23

3 Fig. 20.24 shows a circuit with an ammeter and a voltmeter connected in a circuit. Determine the readings on the two meters.

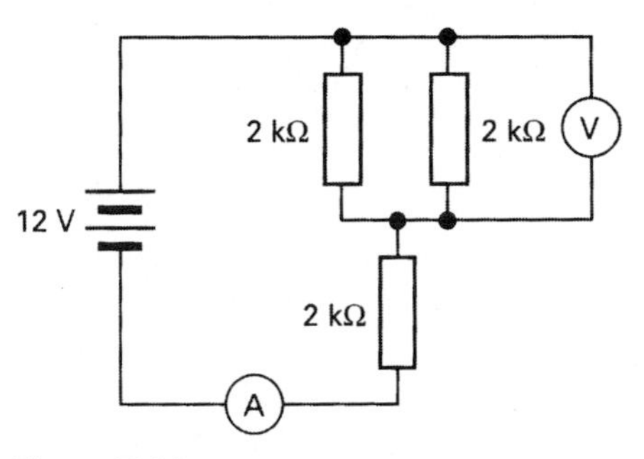

Figure 20.24

4 The circuit shown in Fig. 20.25 is of a simple Zener stabilised power supply. Having studied the circuit and component data you should answer the questions that follow. Initially, ignore the effect of the ripple voltage across the capacitor, C_1.

Component data
Diodes D_1, D_2 $I_{max} = 1.0$ A, $V_d = 0.7$ V
Zener diode D_3 $V_2 = 6.0$ V for $I_2 = 5.0 - 15$ mA
Resistor R_1 56 Ω, 1 W
Capacitor C_1 Greater than 5000 μF
Transformer T_1 Primary 240 V, 50 Hz
Secondaries 2 × 9 V 0.11 A
Regulation 0%

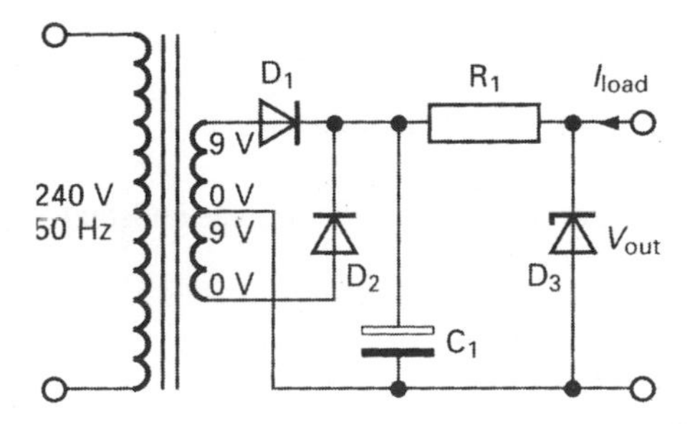

Figure 20.25

(a) Calculate the peak voltage across the capacitor, C_1.
(b) Calculate the maximum reverse voltage across the diode D_1.
(c) Calculate the power dissipated in the resistor R_1 and diode D_3 when the load current from the supply is 50 mA.
(d) If the load is disconnected, what is the new value of the power dissipated in the Zener diode D_3?
(e) Describe the benefits of using a Zener diode in a power supply when real components are used.
(f) Assume that the value of capacitance is reduced so that the peak to peak voltage ripple across C_1 is 1.0 V at a load current of 110 mA. Sketch with care the waveform of the voltage you would expect to find across C_1 on full load conditions.
What would the waveform look like if the diode D_1 failed so that it became an open circuit?

5 A light emitting diode (LED) is to be used as an indicator to show that the power is switched on for a microelectronic circuit. A resistor is placed in series with the LED, as shown in Fig. 20.26, to protect it from voltage and current overload. The LED requires 1.7 V and 10 mA to work effectively.
(a) Calculate the value of the resistor R.
(b) If two LEDs were used in parallel in a project, sketch the circuit diagram for this arrangement and explain what the effect on brightness of the two LEDs is, when compared with the single LED system.

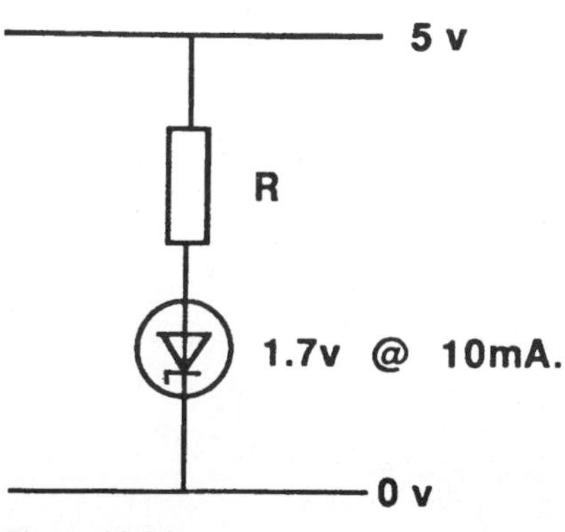

Figure 20.26

Appendix 1 Symbols

The following table is divided into upper case, lower case and Greek symbols. The names of the Greek letters are added to indicate their pronunciation. Many symbols have two or more meanings. The table does not include the abbreviations for units which are given in Appendix 2.

Symbol	*Quantity*
A	area, piston area
B	inductance, breadth of rectangular notch or weir
C	electrical capacitance
C_D, C_L	drag and lift coefficients
$\mathrm{C_d}$	coefficient of discharge
D	aerodynamics or hydrodynamic drag
E	modulus of elasticity (Young modulus), energy, effort, electromotive force
F	force, friction force, shearing force
Fr	Froude number
G	modulus of rigidity (shear modulus)
H	horizontal component of force, enthalpy, head of liquid, magnetic field
I	electric current, moment of inertia, second moment of area
J	polar second moment of area
L	electrical inductance, span of beam, piston stroke, aerodynamic lift, load, length as a primary quantity in dimensional analysis
M	bending moment, moment of resistance, molecular mass, mass as a primary quantity in dimensional analysis
Ma	Mach number
N	normal reaction, number of working strokes per second
P	pressure, mean effective pressure, power, force, load
Q	electric charge, heat (energy in transition), heat transfer rate, volumetric flow rate
R	electrical resistance, resultant force, radius of curvature, gas constant, thermal resistance
Re	Reynolds number
R_{mol}	universal gas constant
S	stiffness of spring
T	torque, absolute temperature, periodic time, time as a primary quantity in dimensional analysis
U	internal energy, strain energy, overall heat transfer coefficient
V	electrical potential, electromotive force, vertical component of force, volume
W	weight, load on beam, work
Z	section modulus, impedance
a	linear acceleration
b	breadth of rectangle
c	velocity of light, specific heat, fluid velocity in steady flow equation
c_p, c_v	specific heats at constant pressure and constant volume
d	diameter, depth of rectangle
f	frequency, friction coefficient in pipe flow
g	acceleration due to gravity (9.81 m/s^2)

h	height, specific enthalpy, coefficient of surface heat transfer (film coefficient)
k	thermal conductivity
l	length
m	mass
n	polytropic index, refractive index
p	pressure, pence
q	reference pressure
r	radius, compression ratio
s	distance, displacement
t	time interval, temperature change, wall thickness, track of vehicle
u	linear velocity, specific internal energy
v	linear velocity, specific volume
y	distance from neutral axis, deflection
z	height above datum
α alpha	angle, angular acceleration, coefficient of linear expansion
β beta	angle
γ gamma	angle, shear strain, ratio of specific heats (c_p/c_v)
δ delta	deflection, small increment of a quantity
ϵ epsilon	linear strain, emissivity
η eta	efficiency
θ theta	angle, angular displacement, angle of twist
λ lambda	angle of friction, wavelength
μ mu	permeability, coefficient of friction, dynamic viscosity
ν nu	kinematic viscosity, Poisson's ratio
π pi	ratio of circumference to diameter of circle (3.142)
ρ rho	density, electrical resistivity
σ sigma	direct stress, Stefan-Boltzmann constant, electrical conductivity
τ tau	shear stress, viscous shear stress
ϕ phi	phase angle, a function of …
ω omega	angular speed, simple harmonic motion constant

Appendix 2 Units

Whenever you give the size of a physical quantity, you must state the units in which it is measured. To say that a vehicle has 'a speed of 50' is meaningless unless you specify the units – such as kilometres per hour, metres per second or knots. In this book the units of each physical quantity are explained when it first appears. This appendix summarises the details and explains how systems of units are built up.

Although we use dozens of different physical quantities, their units are all related to a small number of fundamental or *base* units. For instance, if we decide to measure distances in kilometres and time in hours, the units of speed will be kilometres per hour. The full names of many units are so long that we use an agreed system of symbols or abbreviations for them all.

This book uses the Système International d'Unités (known as SI) as set out in BS 5555.[1] The system is founded on seven base units, together with two supplementary units (for plane angle and solid angle). The units and their symbols are shown in Table A2.1.

Quantity	*Unit*	*Abbreviation*
length	metre	m
mass[1]	kilogram	kg
time[2]	second	s
electric current	ampere	A
thermodynamic temperature[3]	kelvin	K
amount of substance	mole	mol
luminous intensity	candela	cd
plane angle[4]	radian	rad
solid angle[4]	steradian	sr

Notes 1 Large masses are often given in tonnes (or metric tons). 1 tonne = 1000 kg.
2 Time may also be given in minutes (min) and hours (h)
3 For changes in temperature 1 K = 1 °C (degree Celsius)
4 Plane and solid angles may be regarded as base units or derived units. Since the radian is defined as a length divided by a length, and the steradian as an area divided by the square of a length, both maybe considered as dimensionless.

Table A2.1

The units of all other quantities can be expressed in terms of those given in Table A2.1. Where the units involve ratios there are two ways of expressing them. Velocity, for example, is measured in metres per second and the abbreviation may be written as m/s or $m\ s^{-1}$. This book follows BS 5555 in using the fraction form (m/s) but the alternative ($m\ s^{-1}$) is recommended by the Association for Science Education[2] and, as the assignments show, the various examining boards do not all use the same convention.

In the same way, the units of acceleration may be given as m/s^2 or $m\ s^{-2}$ and you should be prepared for both forms.

Some quantities have a more complex combination of the base units and, in these cases, special names and symbols have been adopted. Force, for instance, can be regarded as mass × acceleration and it is therefore measured in $kg\ m/s^2$ or $kg\ m\ s^{-2}$. This unit has been given the name *newton* and the symbol N.

Units which are products, such as the newton metre, should be shown with a gap (N m).

The derived units, together with the special names and symbols, are given in Table A2.2

Table A2.2

Quantity	*Derivation*	*Derived units*	*Name*	*Abbreviation*
Geometrical				
area	$(\text{length})^2$	m^2		
volume	$(\text{length})^3$	m^3		
first moment of area	area × length	m^3		
second moment of area	area × length^2	m^4		
second modulus	second moment/length	m^3		
Motion				
linear velocity	distance/time	m/s^2		
linear acceleration	velocity/time	m/s^2		
angular velocity[1]	angle/time	rad/s		
angular acceleration	$\dfrac{\text{angular velocity}}{\text{time}}$	rad/s^2		
frequency	1/time	s^{-1}	hertz	Hz
Dynamics				
force	mass × acceleration	$kg\ m/s^2$	newton	N
work, energy	force × distance	N m	joule	J
power	work/time	J/s	watt	W
density	mass/volume	kg/m^3		
moment of inertia	mass × $(\text{length})^2$	$kg\ m^2$		
linear momentum	mass × velicity	kg m/s		
angular momentum[2]	moment of inertia × angular velocity	$kg\ m^2/s$		
Structures				
moment of a force[2]	force × distance	N m		
stress, pressure	force/area	N/m^2	pascal	Pa
strain[4]	length/length	–		
modulus of elasticity	stress/strain	N/m^2	pascal	Pa
distributed load	force length	N/m		
Thermodynamics				
heat, enthalpy, internal energy, flow work	energy	J	joule	J
specific heat, gas constant	$\dfrac{\text{energy}}{\text{mass} \times \text{temperature}}$	J/(kg K)		
specific enthalpy, specific internal energy	energy/mass	J/kg		

Notes
1. Rotational speeds are often given in revolutions per minute (rev/min). These must be multiplied by $2\pi/60$ to obtain SI values in rad/s.
2. Although angular momentum implies radians the symbol rad is usually omitted from its unit.
3. Moment, like work and energy, is the product of force and distance. However, its unit is always given as N m, not J. It must be shown as N m, and not m N, to avoid confusion with the millinewton (see Table A2.3).
4. Strain is a ratio of two lengths and is therefore a 'pure' number. It is dimensionless and its value will be the same in all systems of units

Quantity	*Derivation*	*Derived units*	*Name*	*Abbreviation*
Thermodynamics (continued)				
specific volume	volume/mass	m^3/kg		
thermal conductivity	$\frac{(\text{energy/time}) \times \text{length}}{\text{area} \times \text{temperature}}$	W/(m K)		
U-value, coefficient of heat transfer	$\frac{\text{energy/time}}{\text{area} \times \text{temperature}}$	W/(m^2 K)		
specific volume	volume/mass	m^3/kg		
Fluids				
velocity gradient	velocity/distance	s^{-1}		
dynamic viscosity[5]	$\frac{\text{stress}}{\text{velocity gradient}}$	N s/m^2 or kg/(m s)	poise (= 0.1 n s/m^2)	P
kinematic viscosity[6]	$\frac{\text{dynamic viscosity}}{\text{density}}$	m^2/s	stokes (= 10^{-4} m^2/s)	St
Electricity				
electric charge, quantity of electricity	current × time	As	coulomb	C
electric potential, e.m.f.	energy/charge	J/C	volt	V
capacitance	charge/potential	C/V	farad	F
resistance	potential/current	V/A	ohm	Ω
conductance	1/resistance	Ω^{-1}	siemens	S
magnetic flux	potential × time	Vs	weber	Wb
inductance	flux/current	Wb/A	henry	H

Notes 5 The poise is a c.g.s. unit which is still used.
6 The stokes is a c.g.s. unit which is still used.

Table A2.2 (continued).

Multiple units

Practical values of quantities often lead to very large or very small numbers. It is convenient to express them using factors that are powers of 10 and, in the SI system, it is recommended that these powers should normally be multiples of 3. This convention is often referred to as engineer's notation. An alternative is to use prefixes with the unit symbols and Table A2.3 lists those that are used in this book. Note that the combination of 'kilo' and 'ohm' for large values of resistance is shortened to 'kilohm'.

The chief exception to the 'multiples of 3' rule is the prefix centi- (10^{-2}), but you may also come across the deci- (10^{-1}), deca- (10) and hecto- (10^2).

Both the factor and prefix methods should be learned and it is important to be able to convert quickly from one to the other. In stress analysis, for example, we meet values such as:

$$\text{stress} = 238\,100\,000 \text{ N/m}^2 = 238.1 \times 10^6 \text{ N/m}^2 = 238.1 \text{ MN/m}^2 = 238.1 \text{ MPa}$$

$$\text{modulus} = 89\,100\,000\,000 \text{ N/m}^2 = 89.1 \times 10^9 \text{ N/m}^2 = 89.1 \text{ GN/m}^2 = 89.1 \text{ GPa}$$

When units are raised to powers, as in the case of area or volume, the power applies also to the prefix. Thus 13.7 mm^2 = 13.7 × 10^{-6} m^2, since the prefix milli- represents 10^{-3} and this has to be squared.

In tackling numerical problems, the prefixes should first be replaced by the appropriate powers of 10. Scientific calculators will accept numbers in this form. At the end of the working the prefix form can be adopted for the answer.

Factor	*Prefix*	*Abbreviation*
10^9	giga	G
10^6	mega	M
10^3	kilo	k
10^{-2}	centi	c
10^{-3}	milli	m
10^{-6}	micro	μ (mu)
10^{-9}	nano	n
10^{-12}	pico	p

Table A2.3

EXAMPLE The central deflection δ of a centrally loaded uniform beam is given by $\delta = WL^3/48EI$. Calculate δ when $W = 144$ kN, $L = 4$ m, $E = 200$ GPa and $I = 8000\ \text{cm}^4$.

SOLUTION

The work may set out as follows:

$$\delta = \frac{144\ \text{kN} \times (4\ \text{m})^3}{48 \times 200\ \text{GN/m}^2 \times 8000\ \text{cm}^4}$$

$$= \frac{(144 \times 10^3\ \text{N}) \times (4\ \text{m})^3}{48 \times (200 \times 10^9\ \text{N/m}^2) \times (8000 \times 10^{-8}\ \text{m}^4)}$$

$$= \frac{144 \times 10^3 \times 4^3}{48 \times (200 \times 10^9 \times 8000 \times 10^{-8})} \quad \frac{\text{N} \times \text{m}^3}{(\text{N/m}^2) \times \text{m}^4}$$

$$= 12 \times 10^{-3}\ \text{m}$$

$$= 12\ \text{mm}$$

This layout may seem elaborate but it provides a check on the units of the answer and it may be shortened as confidence in the method grows. Note that symbols such as Pa (pascal) must be replaced by their equivalents (N/m^2 in this case) so that cancelling can take place.

Other SI-related units

The earlier c.g.s. system of units was based on the centimetre (cm), gram (g) and second (s). The conversion factors between c.g.s. and SI units are therefore multiples of 10. In the c.g.s. system the force unit was the *dyne* (10^{-5} N) and the work or energy unit was the *erg* (10^{-7} J).

The c.g.s. units for viscosity have been retained in the SI system (see Table A2.2).

Length, area and volume

For small lengths the micrometre ($10^{-6}\ \text{m} = 1\ \mu\text{m}$) is used. This is sometimes known as the *micron*, although this terminology is out of date, and should be avoided. The wavelengths of electromagnetic radiations can be very much smaller and these are expressed in terms of 10^{-10} m, or 0.1 nanometre, a unit that has been given the name *angstrom*, denoted by Å.

Land areas are measured in hectares (ha), 1 ha being $10\,000\ \text{m}^2 = 10^4\ \text{m}^2$.

Volumes are frequently denoted in litres (l). The litre is 1000 cubic centimetres so that $1\ \text{l} = 1000\ \text{cm}^3 = 0.001\ \text{m}^3$.

The volume of liquids is often given in gallons and it is important to distinguish between the UK and USA definitions. For most practical purposes, we can take 1 UK gallon = 4.5461 litres and 1 US gallon = 3.7851 litres.

Stress and pressure

Both of these quantities are defined as force/area and, in terms of base units, they are therefore measured in N/m^2 for which the name pascal (Pa) has been adopted.

Practical values of stress run into millions of pascal and are therefore expressed in MN/m^2 or MPa. However, we can reduce the area by 10^6 rather than increase the force, giving the units as N/mm^2. Note that the various versions $10^6\ \text{N/m}^2$, MN/m^2, N/mm^2 and MPa are all equivalent and numerical values will be the same which ever is used. You should also be prepared to meet any of the equivalent symbols of $10^9\ \text{N/m}^2$, GN/m^2, kN/mm^2 and GPa for modulus of elasticity. See sections 8.2.1 and 8.2.2.

Several special units are used for pressure, particularly in meteorology and fluid mechanics. These include the *bar* (b), the standard atmosphere (atm), and the *torr* (the pressure due to a column of mercury 1 mm high). The equivalent values in SI units are:

$$1 \text{ bar} = 100 \text{ kN/m}^2$$
$$1 \text{ atm} = 101.3 \text{ kN/m}^2$$
$$1 \text{ torr} = 133.3 \text{ N/m}^2$$

Gravitational units

The acceleration due to gravity varies slightly from place to place and between different heights above sea level but a value of 9.806 65 m/s^2 has been adopted internationally. In mechanics, the rounded value of 9.81 m/s^2 is generally used so that the force of gravity on a 1 kg mass (1 kgf) is taken to be 9.81 N. In examinations you may be given a slightly different value, such as 9.8 m/s^2.

Conversion of units

Examination papers in the UK now use SI units but you may encounter the former British system (foot-pound-sector) inbooks and journals, and you may have to use apparatus calibrated in these units. The internationally agreed conversion factors are:

$$1 \text{ ft} = 0.3048 \text{ m} \quad \text{or} \quad 1 \text{ in} = 25.4 \text{ mm}$$

and

$$1 \text{ lb} = 0.453\,592\,37 \text{ kg}.$$

It follows from the last result and the agreed value for g that the weight of 1 lb equals 4.448 N, to four significant figures.

Conversion factors for other quantities can be derived from these basic results[3]. One that is often required is the factor for converting pressures and stresses from the former British system to SI.
It is 1 lb/in^2 (or p.s.i.) = 6.895 kN/m^2.

References

1 BS 5555: 1981 (ISO 1000 1981) *SI Units and Recommendations for the Use of their Multiples and Certain Other units*. British Standards Institution.
2 ASE 1981 *SI Units, Signs, Symbols and Abbreviations*. Association for Science Education.
3 Biggs A J 1969 *Direct Reading Two-Way Metric Conversion Tables*. Pitman.

Answers to numerical assignments

Chapter 2

19. Minimum project time, 32 days; critical activities: B, C, L, J

Chapter 5

7. (b) 100 GPa (or GN/m^2)
11. (ii) 672 kM/mm^2 (or GM/m^2) (iii) 255 M/mm^2 (or MN/m^2)

Chapter 7

18. Batch size = 1000, cost per part = £5.42 (IM) and £1.70 (RM); choose RM
    Batch size = 50 000, cost per part = £0.52 (IM) and £0.72 (RM); choose IM

Chapter 8

Note The forces in the bars of frames are in kN, positive (+) being tensile and negative (−) compressive.

1. (a) 1.44 kN at 29.4° to vertical (b) 0.410 and 22.3° (c) 1.29 kN
3. 5.83 kN at 239° from the direction AB and intersecting the side AD (projected) at a point 1.60 m from A.
5. AB +6.67, BC −10, CD −10, DE −17.0, EF −21.54, BD +17.95, BE −6.67, AE +3.89
6. R_A = 5 kN, R_B = 6 kN, AB −7.07, BC −5, CD −6, DE −8.49, EF +6, FG +8, GH +8, HA +5, BH +5, HC −4.24, CG +5, CF −2,83, FD +6
7. R_A = 3 kN, R_B = 4 kN, AB −1.73, BC −4.04, CD −2.31, DE +4.62, EF +4.62, FG +3.46, GA +3.46 GB −3.46, BF +1.15, FC −1.15, CE −4.62
8. (a) −120, (b) −25, (c) +135 (d) +60
9. (a) R_A = 141.4 kN upwards to the left at an angle of 79.8° to the horizontal, R_E = 160.8 kN vertically upwards (b) AB −160.7, BD −185.7, DE −185.7, EC +92.9, CA +105.4, BC +160.7, CD +185.7
10. AC $F\sqrt{(x^2 + y^2)}/y$ compressive, BC Fx/y tensile
11. (a) 1.26 kN (b) AB +0.80, BC +1.13, CD −0.80, DE +1.26, EA 0, AD +0.92, BD −0.8 (c) 1.22 kN downwards to the right at 32.4° to the horizontal
12. 6.98 kN compressive in each leg, 7.25 kN tensile
13. (a) centre (b) (i) middle bar in bottom chord +1156, two middle bars in top chord −1281 (each) (c) 6500, 7880, 8686
14. (a) left-hand reaction = 11.56 kN upwards to the left at 52.7° to the horizontal, right-hand reaction = 12.70 kN vertically upwards (b) −32.46 (upper), 26.50 (lower) (c) 5.22 mm (d) 35 kN m
15. (a) four diagonals required, answers depend on which are chosen

Chapter 9

2. (a) 156.1 MN/m^2, 4.53 mm
   (b) 99.47 MN/m^2, 2.40 mm
   (c) 120.17 MN/m^2 in shorter rods, 78.77 MN/m^2 in longer rods
3. 5.64 mm
4. 48 MN/m^2, 40 MN/m^2, 16.98 MN/m^2
5. 108.4 MN/m^2 compressive, 85.74 kN, 13.3 MN/m^2 and 10.53 kN

6. (a) (all in kN) (i) $R_A = 46$, $R_B = 54$ (ii) $R_B = 55$, $R_D = 80$,
   (iii) $R_A = 50$, $R_B = 130$
   (b) SF (kN), left-up positive; BM (kN m), sagging positive
   (i) SF: A to B 46, B to C 16, C to D −34, D to E −54; BM: straight lines from 0 at A to 92 at B, to 156 at C, to 54 at D, to 0 at E
   (ii) SF: A to B −20; B to C 35, C to D −45, D to E 35; BM: straight lines from 0 at A to −40 at B, to 100 at C, to −35 at D, to 0 at E
   (iii) SF: A to B 50, straight line from 50 at B to −90 at D, D to E 40; BM: straight line from 0 at A to 100 at B, parabola from 100 at B to −40 at D passing through maximum of 162.5 at 4.5 m from A; point of contraflexure 8.53 m from A
7. SF (kN): straight lines from 0 at left-hand end to −40 at the left hand support; from 80 at left-hand support to −80 at right-hand support, from 40 at right-hand support to 0 at right-hand end; BM (kN m): parabolas from 0 at left-hand end to −40 at left-hand support; from −40 at left-hand support to −40 at right-hand support passing through a maximum of 120 at midpoint from −40 at right-hand support to 0 at right-hand end; points of contraflexure at 2.536 and 9.464 m from left-hand end; with end supports, maximum BM is 360 kN m, three times that of the present case
8. (a) 457 × 191 mm (b) 533 × 120 mm (c) 11.87 MN/m^2
9. (a) Centroid/neutral axis in mm from bottom edge, second moment in mm^4, section modulus in mm^3
   (a) 55, 5,887 × 10^6, 107 × 10^3 and 107 × 10^3
   (b) 30, 553.7 × 10^3, 18.46 × 10^3 and 18.46 × 10^3
   (c) 15, 208.3 × 10^3, 13.89 × 10^3 and 5.952 × 10^3
   (d) 42.7, 9.268 × 10^6, 217.3 × 10^3 and 119.8 × 10^3
   (e) 48, 2.088 × 10^6, 43.5 × 10^3 and 36.6 × 10^3
   (f) 11.2, 71.96 × 10^3, 6.44 × 10^3 and 3.82 × 10^3
   (b) (i) 12.84 kN m, 13.68 kN m (ii) 130.0 MN/m^2, 112 mm
   (iii) 126.0 MN/m^2 tensile (maximum compressive stress is 54.0 MN/m^2), 109 mm (iv) 4.35 kN actual maximum stresses are 10 MN/m^2 tensile and 18.14 MN/m^2 compressive (v) 55.2 MN/m^2 tensile and 65.6 MN/m^2 compressive, 93 mm (vi) 306 m
10. (a) 51.7 kN (b) 88 kN (c) 38.8 kN (d) 31.5 kN
11. (a) 56 (b) 122 (c) 34 (d) 50 (e) 62
12. 23.3 mm
13. 23.34 kN
14. SF (N), left-up positive; straight lines from 953 at A to 511 at B, from −273 at B to −862 at C, from 147 at C to 0 at D; maximum SF, 953 N immediately to the right of A
15. 1.97 m
16. (a) (i) 81.7 mm^2 (ii) 85.7 MN/m^2 (iii) 45 mm
17. 20 kN left-down between B and C
18. 116.8 mm, 90.7 mm

Chapter 10

3. (a) (i) 1190 rev/min (ii) 1.575 kN m assuming 100% efficiency
   (b) 0.833 or 83.3%
4. 10.6 m
5. (c) (i) 3.25 : 1 (ii) 2 min
6. 0.25 or 25%
7. (b) (i) 15 mm/s (ii) 9 W (iii) 20 W
8. 3.86
11. (a) (i) 9 mm (ii) 450 N

13. (a) (ii) 30.6 s
14. (a) 0.417 m/s^2 (b) 24 s (c) 30 km/h (d) 9.718 m/s or 35.0 km\h
15. 0.375 rad/s^2

Chapter 11

1. 210 MJ, 3.125 MW
2. 22.5 MW
5. (a) 32.6 m/s (b) 2.37 m/s^2
6. (i) 33.93 kN, 8.31 kN (ii) 34.43 kN m (iii) 6.75 m
7. (i) 924.6 W (ii) 682.2 kJ (iii) 1476
9. (b) (i) 7500π J or 23.56 kJ (ii) 6.54 W
10. (b) 1503 MJ
11. (b) (i) 50.4 m/s (ii) 17.02 m/s (c) 15.51 m/s
12. (a) (i) 750 N (ii) 900 N (iii) 0.115 (iv) 6.54°
13. (a) 35 mm (b) 37.5 J
14. 5.625 km/h
15. 20.14 kW
16. (a) 44.44 m/s^2 (b) 66.67 m/s (c) 656.7 kN (d) 185.2 kN (e) 44.5 MW
17. (b) 150 kW (c) 2.16 MJ, 2.727 h

Chapter 12

1. 0.982 g 1.002 kg/min
2. 380.2 kJ
3. 2.35 *P*
4. 293 kW
5. (a) 6.888 MW (b) 2.808 MW (c) 0.319 (d) 478 °C
6. (a) 408.7 °C, 12.33 kJ (b) 17 °C, 10.16 kJ, 10.16 kJ rejected
8. 2.304 MW
9. (a) 0.401 (b) 439 W
10. (a) (i) 93.8 W (ii) 851 MJ (iii) £5.67 (b) 75 mm
11. (i) 60.33 kW (ii) 21.40 kW (iii) 18.22 kW (iv) 0.851 (v) 0.303
12. (a) 4/7 = 0.571 (b) 37.5 MW

Chapter 13

1. Long side 33.1 kN, short side 22.1 kN, base 88.3 kN
2. 68.7 kN
3. (a) 96 mm, stable (b) 160 mm, unstable
4. 2238 N/m^2, 1.09%
5. 9 m/s, 65.6 kM/m^2 (lower at the first section)
6. (a) 61.5 M/m^2 (b) 6150 M/m^2, 3.99
7. 3.67 m
8. (a) 0.003 44 m^3/s and 0.019 44 m^3/s (b) 107 mm
9. (a) 3.84 m/s (b) 325.5 W
11. (a) 2845 N/m^2 (b) 3922 N/m^2 (c) 0.725 (d) 0.0451 (e) 15.92 kN (f) 1.639 MW
13. 53.2 kPa
15. (a) 3433 Mg (b) 20.47 m
16. 6.087 kg/s (b) 0.538 m/s (c) 18.5 mm
17. (a) 1.965 MPa^2 (b) 62.68 m/s, 138.3 kg/s (c) 16.18 kN, 6.470 kN m

Chapter 14

9. (b) (i) $R_1 = 923\ \Omega$; $R_2 = 190\ \Omega$; (ii) P inverting; Q non-inverting; R inverting; S non-inverting
10. (c) (i) 8.0 V (ii) 10 mA

Chapter 15

1. (d) −10 V
3. (i) +2 V (ii) +6 V (iii) −2.4 V (iv) 7 V
6. (b) (ii) ±0.5 mV (iii) 500 kΩ (v) 4.5 V (c) (ii) 50 kΩ

Chapter 16

10. (e) 1.05 V

Chapter 18

6. (c) (i) 196.6 N (ii) 84.8 N
7. (b) 16.9 mm
8. (b) 14.6 mm
15. (c) 9.58 mm (theoretical minimum)

Chapter 20

1. $V_1 = 4$ V, $V_2 = 8$ V, I = 2 mA
2. 13 V
3. A = 4 mA, V = 4 V
4. (a) 9 V (b) 8.3 V (c) 0.14 W (d) 246 mW
5. (a) 330 Ω

Index

A

B

C

D

E

F

G

H

I

J

K

L

N

O

P

S

T

U

V

W

WITHDRAWN

X

Y

Z

THE LIBRARY
GUILDFORD COLLEGE
of Further and Higher Education
Author NORMAN Eddie
Title Advanced Design & Technology
Class 620.0042 NOR
Accession 97606